Methods in Enzymology

Volume 126

BIOMEMBRANES

Part N

Transport in Bacteria, Mitochondria, and Chloroplasts: Protonmotive Force

METHODS IN ENZYMOLOGY

EDITORS-IN-CHIEF

Sidney P. Colowick   Nathan O. Kaplan

Methods in Enzymology

Volume 126

Biomembranes

Part N

Transport in Bacteria, Mitochondria, and Chloroplasts: Protonmotive Force

EDITED BY

Sidney Fleischer
Becca Fleischer

DEPARTMENT OF MOLECULAR BIOLOGY
VANDERBILT UNIVERSITY
NASHVILLE, TENNESSEE

1986

ACADEMIC PRESS, INC.
Harcourt Brace Jovanovich, Publishers
Orlando San Diego New York Austin
London Montreal Sydney Tokyo Toronto

ACADEMIC PRESS, INC.
Orlando, Florida 32887

United Kingdom Edition published by
ACADEMIC PRESS INC. (LONDON) LTD.
24–28 Oval Road, London NW1 7DX

LIBRARY OF CONGRESS CATALOG CARD NUMBER: 54-9110

ISBN 0–12–182026–2

PRINTED IN THE UNITED STATES OF AMERICA

86 87 88 89 9 8 7 6 5 4 3 2 1

Table of Contents

CONTRIBUTORS TO VOLUME 126 xi

PREFACE xvii

VOLUMES IN SERIES xix

1. Protonmotive Force and Secondary Transport: Historical Perspectives and Unifying Principles — YASUO KAGAWA — 1

Section I. Electron Transfer

A. Cytochrome Oxidase

2. Measurement of the H^+-Pumping Activity of Reconstituted Cytochrome Oxidase — ROBERT P. CASEY — 13

3. Structure of Beef Heart Cytochrome-*c* Oxidase Obtained by Combining Studies of Two-Dimensional Crystals with Biochemical Experiments — RODERICK A. CAPALDI AND YU-ZHONG ZHANG — 22

4. Isozymes of Cytochrome-*c* Oxidase: Characterization and Isolation from Different Tissues — BERNHARD KADENBACH, ANNEMARIE STROH, MARGIT UNGIBAUER, LUCIA KUHN-NENTWIG, URSULA BÜGE, AND JOCHEN JARAUSCH — 32

5. Techniques for the Study of Bovine Cytochrome-*c* Oxidase Monomer–Dimer Association — KATARZYNA A. NAŁĘCZ, REINHARD BOLLI, AND ANGELO AZZI — 45

6. Affinity Chromatography Purification of Cytochrome-*c* Oxidase from Bovine Heart Mitochondria and Other Sources — CLEMENS BROGER, KURT BILL, AND ANGELO AZZI — 64

7. Resolution of Cytochrome-*c* Oxidase — ANGELO AZZI, CLEMENS BROGER, KURT BILL, AND MICHAEL J. CORBLEY — 72

8. Functional Reconstitution of Proton-Pumping Cytochrome-*c* Oxidase in Phospholipid Vesicles — MICHELE MÜLLER, MARCUS THELEN, PAUL O'SHEA, AND ANGELO AZZI — 78

9. Purification and Reconstitution of the Cytochrome *d* Terminal Oxidase Complex from *Escherichia coli* — MICHAEL J. MILLER AND ROBERT B. GENNIS — 87

10. Purification and Properties of Two Terminal Oxidase Complexes of *Escherichia coli* Aerobic Respiratory Chain — KIYOSHI KITA, KIYOSHI KONISHI, AND YASUHIRO ANRAKU — 94

11. Purification and Reconstitution of the Cytochrome *o*-Type Oxidase from *Escherichia coli* — KAZUNOBU MATSUSHITA, LEKHA PATEL, AND H. RONALD KABACK — 113

12. Functional Reconstitution of Bacterial Cytochrome Oxidases in Planar Lipid Bilayers — TOSHIRO HAMAMOTO AND MAURICIO MONTAL — 123

13. Purification of the aa_3-Type Cytochrome-*c* Oxidase from *Rhodopseudomonas sphaeroides* — ANGELO AZZI AND ROBERT B. GENNIS — 138

14. Cytochrome Oxidase from Thermophilic Bacterium PS3 — NOBUHITO SONE — 145

15. Cytochrome *c* Oxidase from *Paracoccus denitrificans* — BERND LUDWIG — 153

16. Purification, Enzymatic Properties, and Reconstitution of Cytochrome-*c* Oxidase from *Bacillus subtilis* — WIM DE VRIJ, BERT POOLMAN, WIL N. KONINGS, AND ANGELO AZZI — 159

B. Cytochrome bc_1 Complex

17. Preparation of Complex III from Yeast Mitochondria and Related Methodology — LIVIU CLEJAN AND DIANA S. BEATTIE — 173

18. Purification of Cytochrome bc_1 Complexes from Phylogenically Diverse Species by a Single Method — PER O. LJUNGDAHL, JEFFREY D. PENNOYER, AND BERNARD L. TRUMPOWER — 181

19. Preparation of Membrane Crystals of Ubiquinol–Cytochrome-*c* Reductase from *Neurospora* Mitochondria and Structure Analysis by Electron Microscopy — HANNS WEISS, SVEN HOVMÖLLER, AND KEVIN LEONARD — 191

20. Reconstitution of Ubiquinol–Cytochrome-*c* Reductase from *Neurospora* Mitochondria with regard to Subunits I and II — PETRA LINKE AND HANNS WEISS — 201

21. Resolution and Reconstitution of the Iron–Sulfur Protein of the Cytochrome bc_1 Segment of the Mitochondrial Respiratory Chain — CAROL A. EDWARDS AND BERNARD L. TRUMPOWER — 211

22. Isolation of the Eleven Protein Subunits of the bc_1 Complex from Beef Heart — H. SCHÄGGER, TH. A. LINK, W. D. ENGEL, AND G. VON JAGOW — 224

23. Preparation of Hinge Protein and Its Requirement for Interaction of Cytochrome *c* with Cytochrome c_1 — TSOO E. KING AND CHONG H. KIM — 238

24. Use of Specific Inhibitors on the Mitochondrial bc_1 Complex — G. VON JAGOW AND TH. A. LINK — 253

25. Preparations of Electrogenic, Proton-Transporting Cytochrome Complexes of the b_6f-Type (Chloroplasts and Cyanobacteria) and bc_1-Type (*Rhodopseudomonas sphaeroides*) — GÜNTER HAUSKA — 271

26. Reconstitution of H^+ Translocation and Photophosphorylation with Photosystem I Reaction Centers, PMS, and CF_1CF_0 — GÜNTER HAUSKA AND NATHAN NELSON — 285

27. Construction of the Photosynthetic Reaction Center–Mitochondrial Ubiquinol–Cytochrome-*c* Oxidoreductase Hybrid System — CHRISTOPHER C. MOSER, KATHLEEN M. GIANGIACOMO, KATSUMI MATSUURA, SIMON DEVRIES, AND P. LESLIE DUTTON — 293

28. Isolation of Ubiquinol Oxidase from *Paracoccus denitrificans* — EDWARD A. BERRY AND BERNARD L. TRUMPOWER — 305

C. Other Electron Transfer

29. Isolation of a Three-Subunit Cytochrome bc_1 Complex from *Paracoccus denitrificans* — XIAOHANG YANG AND BERNARD L. TRUMPOWER — 316

30. A Simple, One-Step Purification for Cytochrome *b* from the bc_1 Complexes of Bacteria — WILLIAM E. PAYNE AND BERNARD L. TRUMPOWER — 325

31. Cooperative Proton-Transfer Reactions in the Respiratory Chain: Redox Bohr Effects — S. PAPA, F. GUERRIERI, AND G. IZZO — 331

32. Preparation of Two-Dimensional Crystals of Complex I and Image Analysis — EGBERT J. BOEKEMA, MARIN G. VAN HEEL, AND ERNST F. J. VAN BRUGGEN — 344

33. Purification and Reconstitution of Bovine Heart Mitochondrial Transhydrogenase — LICIA N. Y. WU, JULIA A. ALBERTA, AND RONALD R. FISHER — 353

34. Isolation of the Iron–Sulfur-Containing Polypeptides of NADH : Ubiquinone Oxidoreductase — C. IAN RAGAN AND YOUSSEF HATEFI — 360

35. Monoclonal Affinity Purification of D-Lactate Dehydrogenase from *Escherichia coli* — EUGENIO SANTOS AND H. RONALD KABACK — 370

36. Fumarate Reductase of *Escherichia coli* — BERNARD D. LEMIRE AND JOEL H. WEINER — 377

37. Reconstitution of a Functional Electron-Transfer Chain from Purified Formate Dehydrogenase and Fumarate Reductase Complexes — GOTTFRIED UNDEN AND ACHIM KRÖGER — 387

38. Molecular Properties, Genetics, and Biosynthesis of *Bacillus subtilis* Succinate Dehydrogenase Complex — LARS HEDERSTEDT — 399

Section II. Reversible ATP Synthase (F_0F_1-ATPase)

A. Preparation and Reconstitution

39. Preparation of a Highly Coupled H^+-Transporting ATP Synthase from Pig Heart Mitochondria — DANIÈLE C. GAUTHERON, FRANÇOIS PENIN, GILBERT DELÉAGE, AND CATHERINE GODINOT — 417

40. Resolution and Reconstitution of F_0F_1-ATPase in Beef Heart Submitochondrial Particles — L. ERNSTER, T. HUNDAL, AND G. SANDRI — 428

41. Electron Microscopy of Single Molecules and Crystals of F_1-ATPases — CHRISTOPHER W. AKEY, STANLEY D. DUNN, VITALY SPITSBERG, AND STUART J. EDELSTEIN — 434

42. Isolation and Reconstitution of Membrane-Bound Pyrophosphatase from Beef Heart Mitochondria — I. S. KULAEV, S. E. MANSUROVA, AND YU. A. SHAKHOV — 447

43. Use of Monoclonal Antibodies to Purify Oligomycin Sensitivity-Conferring Protein and to Study Its Interactions with F_0 and F_1 — PHILIPPE ARCHINARD, FRANÇOIS PENIN, CATHERINE GODINOT, AND DANIÈLE C. GAUTHERON — 455

44. Purification and Properties of the ATPase Inhibitor from Bovine Heart Mitochondria — MAYNARD E. PULLMAN — 460

45. Purification of the Proton-Translocating ATPase from Rat Liver Mitochondria Using the Detergent 3-[(3-Cholamidopropyl)dimethylammonio]-1-propane Sulfonate — MAUREEN W. MCENERY AND PETER L. PEDERSEN — 470

46. Rapid Purification of F_1-ATPase from Rat Liver Mitochondria Using a Modified Chloroform Extraction Procedure Coupled to High-Performance Liquid Chromatography — NOREEN WILLIAMS AND PETER L. PEDERSEN — 477

47. Purification of α and β Subunits and Subunit Pairs from Rat Liver Mitochondrial F_1-ATPase — NOREEN WILLIAMS AND PETER L. PEDERSEN — 484

48. Isolation and Hydrodynamic Characterization of the Uncoupling Protein from Brown Adipose Tissue — MARTIN KLINGENBERG AND CHI-SHUI LIN — 490

49. Nucleotide Binding Assay for Uncoupling Protein from Brown Fat Mitochondria — MARTIN KLINGENBERG, MARIA HERLT, AND EDITH WINKLER — 498

50. Isolation and Characterization of an Inactivated Complex of F_1F_0-ATPase and Its Inhibitory Factors from Yeast — KUNIO TAGAWA, TADAO HASHIMOTO, AND YUKUO YOSHIDA — 504

51. Isolation and Reconstitution of CF_0-F_1 from Chloroplasts — URI PICK — 512

52. Extraction, Purification, and Reconstruction of the Chloroplast *N,N'*-Dicyclohexylcarbodiimide-Binding Proteolipid — ANGELO AZZI, KRISTINE SIGRIST-NELSON, AND NATHAN NELSON — 520

53. Selective Extraction and Reconstitution of F_1 Subunits from *Rhodospirillum rubrum* Chromatophores — ZIPPORA GROMET-ELHANAN AND DANIEL KHANANSHVILI — 528

54. Preparation and Reconstitution of the Protein-Pumping Membrane-Bound Inorganic Pyrophosphatase from *Rhodospirillum rubrum* — MARGARETA BALTSCHEFFSKY AND PÅL NYRÉN — 538

55. Purification of F_1F_0 H^+-ATPase from *Escherichia coli* — ROBERT H. FILLINGAME AND DAVID L. FOSTER — 545

56. Use of λ-*unc* Transducing Phages in Genetic Analysis of H^+-ATPase Mutants of *Escherichia coli* — ROBERT H. FILLINGAME AND MARY E. MOSHER — 558

57. Proton-Conducting Portion (F_0) from *Escherichia coli* ATP Synthase: Preparation, Dissociation into Subunits, and Reconstitution of an Active Complex — ERWIN SCHNEIDER AND KARLHEINZ ALTENDORF — 569

58. Preparation and Reconstitution of F_1F_0 and F_0 from *Escherichia coli* — PETER FRIEDL AND HANS ULRICH SCHAIRER — 579

59. Use of Isolated Subunits of F_1 from *Escherichia coli* for Genetic and Biochemical Studies — MASAMITSU FUTAI AND HIROSHI KANAZAWA — 588

60. Analysis of *Escherichia coli* Mutants of the H^+-Transporting ATPase: Determination of Altered Site of the Structural Genes — HIROSHI KANAZAWA, TAKATO NOUMI, AND MASAMITSU FUTAI — 595

61. Proton Permeability of Membrane Sector (F_0) of H^+-Transporting ATP Synthase (F_0F_1) from a Thermophilic Bacterium — NOBUHITO SONE AND YASUO KAGAWA — 604

B. Kinetics, Modification, and Other Characterization

62. Rate Constants and Equilibrium Constants for the Elementary Steps of ATP Hydrolysis by Beef Heart Mitochondrial ATPase — HARVEY S. PENEFSKY — 608

63. Refinements in Oxygen-18 Methodology for the Study of Phosphorylation Mechanisms — KERSTIN E. STEMPEL AND PAUL D. BOYER 618

64. ATP Formation in Mitochondria, Submitochondrial Particles, and F_0F_1 Liposomes Driven by Electric Pulses — YASUO KAGAWA AND TOSHIRO HAMAMOTO 640

65. Synthesis of Enzyme-Bound Adenosine Triphosphate by Chloroplast Coupling Factor CF_1 — RICHARD I. FELDMAN 643

66. Divalent Azido-ATP Analog for Photoaffinity Cross-Linking of F_1 Subunits — HANS-JOCHEN SCHÄFER 649

67. Synthesis and Use of an Azido-Labeled Form of the ATPase Inhibitor Peptide of Rat Liver Mitochondria — KLAUS SCHWERZMANN AND PETER L. PEDERSEN 660

68. Benzophenone-ATP: A Photoaffinity Label for the Active Site of ATPase — NOREEN WILLIAMS, SHARON H. ACKERMAN, AND PETER S. COLEMAN 667

69. Use of ADP Analogs for Functional and Structural Analysis of F_1-ATPase — GÜNTER SCHÄFER, UWE LÜCKEN, AND MATHIAS LÜBBEN 682

70. Modifiers of F_1-ATPases and Associated Peptides — MICHEL SATRE, JOËL LUNARDI, ANNE-CHRISTINE DIANOUX, ALAIN DUPUIS, JEAN PAUL ISSARTEL, GÉRARD KLEIN, RICHARD POUGEOIS, AND PIERRE V. VIGNAIS 712

71. A Nonlinear Approach for the Analysis of Different Models of Protein–Ligand Interaction: Nucleotide Binding to F_1-ATPase — FRENS PETERS AND UWE LÜCKEN 733

72. Identification of Essential Residues in the F_1-ATPases by Chemical Modification — WILLIAM S. ALLISON, DAVID A. BULLOUGH, AND WILLIAM W. ANDREWS 741

73. Monoclonal Antibodies to F_1-ATPase Subunits as Probes of Structure, Conformation, and Functions of Isolated or Membrane-Bound F_1 — CATHERINE GODINOT, MAHNAZ MORADI-AMELI, AND DANIÈLE C. GAUTHERON 761

74. Determination of the Stoichiometry and Arrangement of α and β Subunits in F_1-ATPase Using Monoclonal Antibodies in Immunoelectron Microscopy — KARIN EHRIG, HEINRICH LÜNSDORF, PETER FRIEDL, AND HANS ULRICH SCHAIRER 770

AUTHOR INDEX 777

SUBJECT INDEX 803

Contributors to Volume 126

Article numbers are in parentheses following the names of contributors.
Affiliations listed are current.

SHARON H. ACKERMAN (68), *Laboratory of Biochemistry, Department of Biology, New York University, Washington Square Campus, New York, New York 10003*

CHRISTOPHER W. AKEY (41), *Department of Cell Biology, Stanford University School of Medicine, Stanford, California 94305*

JULIA A. ALBERTA (33), *Program of Cell and Developmental Biology, Harvard Medical School, Boston, Massachusetts 02115*

WILLIAM S. ALLISON (72), *Department of Chemistry, University of California, San Diego, La Jolla, California 92093*

KARLHEINZ ALTENDORF (57), *Universität Osnabrück, Fachbereich Biologie/Chemie, 4500 Osnabrück, Federal Republic of Germany*

WILLIAM W. ANDREWS (72), *Synbiotics, 11011 Via Frontera, San Diego, California 92127*

YASUHIRO ANRAKU (10), *Department of Biology, Faculty of Science, University of Tokyo, Hongo, Tokyo 113, Japan*

PHILIPPE ARCHINARD (43), *Laboratoire de Biologie et Technologie des Membranes du CNRS, Université Claude Bernard de Lyon, 69622 Villeurbanne Cédex, France*

ANGELO AZZI (5, 6, 7, 8, 13, 16, 52), *Medizinisch-chemisches Institut, Universität Bern, CH-3012 Bern, Switzerland*

MARGARETA BALTSCHEFFSKY (54), *Department of Biochemistry, Arrhenius Laboratory, University of Stockholm, S-106 91 Stockholm, Sweden*

DIANA S. BEATTIE (17), *Department of Biochemistry, West Virginia University School of Medicine, Morgantown, West Virginia 26506*

EDWARD A. BERRY (28), *Department of Physiology and Biophysics, University of Illinois at Urbana–Champaign, Urbana, Illinois 61801*

KURT BILL (6, 7), *Ciba-Giegy AG, CH-4002 Basel, Switzerland*

EGBERT J. BOEKEMA (32), *Fritz-Haber-Institut der Max-Planck-Gesellschaft, D-1000 Berlin 33, Federal Republic of Germany*

REINHARD BOLLI (5), *Medizinisch-chemisches Institut, Universität Bern, CH-3012 Bern, Switzerland*

PAUL D. BOYER (63), *Department of Chemistry and Biochemistry, and Molecular Biology Institute, University of California, Los Angeles, California 90024*

CLEMENS BROGER (6, 7), *Medizinisch-chemisches Institut, Universität Bern, CH-3012 Bern, Switzerland*

URSULA BÜGE (4), *Biochemie, Fachbereich Chemie, Philipps-Universität, D-3550 Marburg, Federal Republic of Germany*

DAVID A. BULLOUGH (72), *Department of Chemistry, University of California, San Diego, La Jolla, California 92093*

RODERICK A. CAPALDI (3), *Institute of Molecular Biology, University of Oregon, Eugene, Oregon 97403*

ROBERT P. CASEY[1] (2), *Department of Medical Chemistry, University of Helsinki, SF-00170 Helsinki, Finland*

LIVIU CLEJAN (17), *Department of Biochemistry, Mount Sinai School of Medicine, New York, New York 10029*

PETER S. COLEMAN (68), *Laboratory of Biochemistry, Department of Biology, New York University, Washington Square Campus, New York, New York 10003*

MICHAEL J. CORBLEY (7), *Department of Biological Chemistry, Harvard University*

[1] Deceased.

Medical School, Boston, Massachusetts 02115

GILBERT DELÉAGE (39), *Laboratoire de Biologie et Technologie des Membranes du CNRS, Université Claude Bernard de Lyon, 69622 Villeurbanne Cédex, France*

SIMON DEVRIES (27), *Laboratory of Biochemistry, BCP Jansen Institute, University of Amsterdam, 1018 TV Amsterdam, The Netherlands*

WIM DE VRIJ (16), *Laboratorium voor Microbiologie, Groningen Biologisch Centrum, NL-9751 NN Haren, The Netherlands*

ANNE-CHRISTINE DIANOUX (70), *Laboratoire de Biochimie, Département de Recherche Fondamentale, Centre d'Etudes Nucléaires, 85 X, 38041 Grenoble Cedex, France*

STANLEY D. DUNN (41), *Department of Biochemistry, Health Sciences Center, The University of Western Ontario, London, Ontario, Canada N6A 5C1*

ALAIN DUPUIS (70), *Laboratoire de Biochimie, Département de Recherche Fondamentale, Centre d'Etudes Nucléaires, 85 X, 38041 Grenoble Cedex, France*

P. LESLIE DUTTON (27), *Department of Biochemistry and Biophysics, University of Pennsylvania, Philadelphia, Pennsylvania 19104*

STUART J. EDELSTEIN (41), *Section of Biochemistry, Molecular and Cell Biology, Cornell University, Ithaca, New York 14853*

CAROL A. EDWARDS (21), *New York Blood Center, Inc., Blood Derivatives Program, New York, New York 10021*

KARIN EHRIG (74), *Gesellschaft für Biotechnologische Forschung, D-3300 Braunschweig, Federal Republic of Germany*

W. D. ENGEL (22), *D-8121 Pähl, Federal Republic of Germany*

L. ERNSTER (40), *Department of Biochemistry, Arrhenius Laboratory, University of Stockholm, S-106 91 Stockholm, Sweden*

RICHARD I. FELDMAN (65), *Department of Biochemistry, University of California, Berkeley, California 94720*

ROBERT H. FILLINGAME (55, 56), *Physiological Chemistry, University of Wisconsin Medical School, Madison, Wisconsin 53706*

RONALD R. FISHER (33), *Department of Chemistry, University of South Carolina, Columbia, South Carolina 29208*

DAVID L. FOSTER (55), *Lawrence Berkeley Laboratory, University of California, Berkeley, California 94720*

PETER FRIEDL (58, 74), *Gesellschaft für Biotechnologische Forschung, Department of Cytogenetics, D-3300 Braunschweig, Federal Republic of Germany*

MASAMITSU FUTAI (59, 60), *Institute of Scientific and Industrial Research, Osaka University, Ibaraki, Osaka 567, Japan*

DANIÈLE C. GAUTHERON (39, 43, 73), *Laboratoire de Biologie et Technologie des Membranes du CNRS, Université Claude Bernard Lyon, 69622 Villeurbanne Cédex, France*

ROBERT B. GENNIS (9, 13), *Department of Chemistry, University of Illinois at Urbana–Champaign, Urbana, Illinois 61801*

KATHLEEN M. GIANGIACOMO (27), *Department of Biochemistry and Biophysics, University of Pennsylvania, Philadelphia, Pennsylvania 19104*

CATHERINE GODINOT (39, 43, 73), *Laboratoire de Biologie et Technologie des Membranes du CNRS, Université Claude Bernard Lyon, 69622 Villeurbanne Cédex, France*

ZIPPORA GROMET-ELHANAN (53), *Biochemistry Department, Weizmann Institute of Science, Rehovot 76100, Israel*

F. GUERRIERI (31), *Institute of Medical Biochemistry, University of Bari, 70124 Bari, Italy*

TOSHIRO HAMAMOTO (12, 64), *Department of Biochemistry, Jichi Medical School, Tochigi-ken, Japan 329-04*

TADAO HASHIMOTO (50), *Department of Physiological Chemistry, Medical School, Osaka University, Osaka 530, Japan*

YOUSSEF HATEFI (34), *Division of Biochem-*

istry, Department of Basic and Clinical Research, Scripps Clinic and Research Foundation, La Jolla, California 92037

GÜNTER HAUSKA (25, 26), *Institut für Botanik, Universität Regensburg, 8400 Regensburg, Federal Republic of Germany*

LARS HEDERSTEDT (38), *Department of Bacteriology, Karolinska Institutet, S-104 01 Stockholm, Sweden*

MARIA HERLT (49), *Institut für Physikalische Biochemie, Universität München, 8000 München 2, Federal Republic of Germany*

SVEN HOVMÖLLER (19), *Strukturkemi, Arrheniuslaboratoriet, Stockholms Universitet, S-106 91 Stockholm, Sweden*

T. HUNDAL (40), *Department of Biochemistry, Arrhenius Laboratory, University of Stockholm, S-106 91 Stockholm, Sweden*

JEAN PAUL ISSARTEL (70), *Laboratoire de Biochimie, Département de Recherche Fondamentale, Centre d'Etudes Nucléaires, 85 X, 38041 Grenoble Cedex, France*

G. IZZO (31), *Institute of Medical Biochemistry, University of Bari, 70124 Bari, Italy*

JOCHEN JARAUSCH (4), *Biochemie, Fachbereich Chemie, Philipps-Universität, D-3550 Marburg, Federal Republic of Germany*

H. RONALD KABACK (11, 35), *Department of Biochemistry, Roche Institute of Molecular Biology, Nutley, New Jersey 07110*

BERNHARD KADENBACH (4), *Biochemie, Fachbereich Chemie, Philipps-Universität, D-3550 Marburg, Federal Republic of Germany*

YASUO KAGAWA (1, 61, 64), *Department of Biochemistry, Jichi Medical School, Tochigi-ken, Japan 329-04*

HIROSHI KANAZAWA (59, 60), *Oncogene Division, National Cancer Center Research Institute, Chou-ku, Tsukiji, Tokyo 104, Japan*

DANIEL KHANANSHVILI (53), *Biochemistry Department, Weizmann Institute of Science, Rehovot 76100, Israel*

CHONG H. KIM (23), *Department of Chemistry and Laboratory of Bioenergetics, State University of New York at Albany, Albany, New York 12222*

TSOO E. KING (23), *Department of Chemistry and Laboratory of Bioenergetics, State University of New York at Albany, Albany, New York 12222*

KIYOSHI KITA (10), *Department of Parasitology, Juntendo University, School of Medicine, Hongo, Tokyo 113, Japan*

GÉRARD KLEIN (70), *Laboratoire de Biochimie, Département de Recherche Fondamentale, Centre d'Etudes Nucléaires, 85 X, 38041 Grenoble Cedex, France*

MARTIN KLINGENBERG (48, 49), *Institut für Physikalische Biochemie, Universität München, 8000 München 2, Federal Republic of Germany*

WIL N. KONINGS (16), *Laboratorium voor Microbiologie, Groningen Biologisch Centrum, NL-9751 NN Haren, The Netherlands*

KIYOSHI KONISHI (10), *Department of Biochemistry, Faculty of Medicine, Toyama Medical and Pharmaceutical University, Sugitani, Toyama 930-01, Japan*

ACHIM KRÖGER (37), *Institut für Mikrobiologie, Fachbereich Biologie, J. W. Goethe-Universität, 6000 Frankfurt am Main, Federal Republic of Germany*

LUCIA KUHN-NENTWIG (4), *Biochemie, Fachbereich Chemie, Philipps-Universität, D-3550 Marburg, Federal Republic of Germany*

I. S. KULAEV (42), *Institute of Biochemistry and Physiology of Microorganisms, Academy of Sciences of USSR, Pustchino on the Oka, Moscow Region, USSR*

BERNARD D. LEMIRE (36), *Department of Biochemistry, University of Alberta, Edmonton, Alberta, Canada T6G 2H7*

KEVIN LEONARD (19), *European Molecular Biology Laboratory, 6990 Heidelberg, Federal Republic of Germany*

CHI-SHUI LIN (48), *Shanghai Institute of Biochemistry, Academia Sinica, Shanghai 200031, People's Republic of China*

TH. A. LINK (22, 24), *Institut für Physikalische Biochemie, Universität München, 8000 München 2, Federal Republic of Germany*

PETRA LINKE (20), *Institut für Biochemie, Universität Düsseldorf, 4000 Düsseldorf 1, Federal Republic of Germany*

PER O. LJUNGDAHL (18), *Department of Biochemistry, Dartmouth Medical School, Hanover, New Hampshire 03756*

MATHIAS LÜBBEN (69), *Institut für Biochemie, Medizinische Universität zu Lübeck, 2400 Lübeck, Federal Republic of Germany*

UWE LÜCKEN (69, 71), *Institute of Molecular Biology, University of Oregon, Eugene, Oregon 97403*

BERND LUDWIG (15), *Institut für Biochemie, Medizinische Universität Lübeck, D-2400 Lübeck, Federal Republic of Germany*

JOËL LUNARDI (70), *Laboratoire de Biochimie, Département de Recherche Fondamentale, Centre d'Etudes Nucléaires, 85 X, 38041 Grenoble Cedex, France*

HEINRICH LÜNSDORF (74), *Gesellschaft für Biotechnologische Forschung, D-3300 Braunschweig, Federal Republic of Germany*

S. E. MANSUROVA (42), *A. N. Belozersky Laboratory of Molecular Biology and Bioorganic Chemistry, Moscow State University, Moscow 119899, USSR*

KAZUNOBU MATSUSHITA (11), *Yamaguchi University, Faculty of Agriculture, Yamaguchi 753, Japan*

KATSUMI MATSUURA (27), *Department of Biology, Faculty of Science, Tokyo Metropolitan University, Fukazawa 2-1-1, Setagaya, Tokyo 158, Japan*

MAUREEN W. MCENERY (45), *Laboratory of Cellular Metabolism, NHLBI, National Institutes of Health, Bethesda, Maryland 20892*

MICHAEL J. MILLER (9), *Department of Physiological Chemistry, University of Wisconsin, Madison, Wisconsin 53706*

MAURICIO MONTAL (12), *Department of Neurosciences, Roche Institute of Molecular Biology, Nutley, New Jersey 07110*

MAHNAZ MORADI-AMELI (73), *Laboratoire de Biologie et Technologie des Membranes du CNRS, Université Claude Bernard Lyon, 69622 Villeurbanne Cédex, France*

CHRISTOPHER C. MOSER (27), *Department of Biochemistry and Biophysics, University of Pennsylvania, Philadelphia, Pennsylvania 19104*

MARY E. MOSHER (56), *Botany, University of Minnesota, St. Paul, Minnesota 55108*

MICHELE MÜLLER (8), *Medizinisch-chemisches Institut, Universität Bern, CH-3012 Bern, Switzerland*

KATARZYNA A. NAŁĘCZ (5), *Nencki Institute of Experimental Biology, Polish Academy of Sciences, PL-02093 Warsaw, Poland*

NATHAN NELSON (26, 52), *Department of Biochemistry, Roche Institute of Molecular Biology, Nutley, New Jersey 07110*

TAKATO NOUMI (60), *Institute of Scientific and Industrial Research, Osaka University, Ibaraki, Osaka 567, Japan*

PÅL NYRÉN (54), *Department of Biochemistry, Arrhenius Laboratory, University of Stockholm, S-106 91 Stockholm, Sweden*

PAUL O'SHEA (8), *The Open University, Milton Keynes MK7 6AA, England*

S. PAPA (31), *Institute of Medical Biochemistry and Chemistry, University of Bari, 70124 Bari, Italy*

LEKHA PATEL (11), *Department of Biochemistry, Roche Institute of Molecular Biology, Nutley, New Jersey 07110*

WILLIAM E. PAYNE (30), *Department of Biochemistry, Dartmouth Medical School, Hanover, New Hampshire 03756*

PETER L. PEDERSEN (45, 46, 47, 67), *Laboratory for Molecular and Cellular Bioenergetics, Department of Biological Chemistry, The Johns Hopkins University School of Medicine, Baltimore, Maryland 21205*

HARVEY S. PENEFSKY (62), *Department of*

Biochemistry, Public Health Research Institute, New York, New York 10016

FRANÇOIS PENIN (39, 43), *Laboratoire de Biologie et Technologie des Membranes du CNRS, Université Claude Bernard de Lyon, 69622 Villeurbanne Cédex, France*

JEFFREY D. PENNOYER (18), *Department of Biochemistry, Dartmouth Medical School, Hanover, New Hampshire 03756*

FRENS PETERS (71), *Medizinische Hochschule Hannover, Biophysikalische Messgeräteabteilung, D-3000 Hannover, Federal Republic of Germany*

URI PICK (51), *Department of Biochemistry, Weizmann Institute of Science, Rehovot 76100, Israel*

BERT POOLMAN (16), *Laboratorium voor Microbiologie, Groningen Biologisch Centrum, NL-9751 NN Haren, The Netherlands*

RICHARD POUGEOIS (70), *Laboratoire d'Hormonologie, Centre Hospitalier Régional et Universitaire, 38043 Grenoble Cedex, France*

MAYNARD E. PULLMAN (44), *Department of Biochemistry, The Public Health Research Institute of The City of New York, Inc., New York, New York 10016*

C. IAN RAGAN (34), *Department of Biochemistry, The University of Southampton, School of Biochemical and Physiological Sciences, Southampton SO9 3TU, England*

G. SANDRI (40), *Dipartimento di Biochimica, Biofisica e Chimica delle Macromolecole, Università di Trieste, Trieste, Italy*

EUGENIO SANTOS (35), *Laboratory of Molecular Microbiology, NIAIE, National Institutes of Health, Bethesda, Maryland 20205*

MICHEL SATRE (70), *Laboratoire de Biochimie, Département de Recherche Fondamentale, Centre d'Etudes Nucléaires, 85 X, 38041 Grenoble Cedex, France*

GÜNTER SCHÄFER (69), *Institut für Biochemie, Medizinische Universität zu Lübeck, 2400 Lübeck, Federal Republic of Germany*

HANS-JOCHEN SCHÄFER (66), *Institut für Biochemie, Johannes-Gutenberg-Universität, D-6500 Mainz, Federal Republic of Germany*

H. SCHÄGGER (22), *Institut für Physikalische Biochemie, Universität München, 8000 München 2, Federal Republic of Germany*

HANS ULRICH SCHAIRER (58, 74), *Gesellschaft für Biotechnologische Forschung, Department of Cytogenetics, D-3300 Braunschweig, Federal Republic of Germany*

ERWIN SCHNEIDER (57), *Universität Osnabrück, Fachbereich Biologie/Chemie, 4500 Osnabrück, Federal Republic of Germany*

KLAUS SCHWERZMANN (67), *Department of Anatomy, University of Berne, 3000 Berne 9, Switzerland*

YU. A. SHAKHOV (42), *Institute of Preventive Cardiology of the National Cardiology Research Center, Medical Academy of Science USSR, Moscow 101837, USSR*

KRISTINE SIGRIST-NELSON (52), *Institut für Biochemie, Universität Bern, CH-3012 Bern, Switzerland*

NOBUHITO SONE (14, 61), *Department of Biochemistry, Jichi Medical School, Tochigi-ken, Japan 329-04*

VITALY SPITSBERG (41), *Section of Biochemistry, Molecular and Cell Biology, Cornell University, Ithaca, New York 14853*

KERSTIN E. STEMPEL (63), *Department of Chemistry and Biochemistry, University of California, Los Angeles, California 90024*

ANNEMARIE STROH (4), *Biochemie, Fachbereich Chemie, Philipps-Universität, D-3550 Marburg, Federal Republic of Germany*

KUNIO TAGAWA (50), *Department of Physiological Chemistry, Medical School, Osaka University, Osaka 530, Japan*

MARCUS THELEN (8), *Theodor Kocher Institut, Universität Bern, CH-3012 Bern 9, Switzerland*

BERNARD L. TRUMPOWER (18, 21, 28, 29, 30), *Department of Biochemistry, Dartmouth Medical School, Hanover, New Hampshire 03756*

GOTTFRIED UNDEN (37), *Institut für Mikrobiologie, Fachbereich Biologie, J. W. Goethe-Universität, 6000 Frankfurt am Main, Federal Republic of Germany*

MARGIT UNGIBAUER (4), *Biochemie, Fachbereich Chemie, Philipps-Universität, D-3550 Marburg, Federal Republic of Germany*

ERNST F. J. VAN BRUGGEN (32), *Biochemisch Laboratorium, Rijksuniversiteit Groningen, 9747 AG Groningen, The Netherlands*

MARIN G. VAN HEEL (32), *Fritz-Haber-Institut der Max-Planck-Gesellschaft, D-1000 Berlin 33, Federal Republic of Germany*

PIERRE V. VIGNAIS (70), *Laboratoire de Biochimie, Département de Recherche Fondamentale, Centre d'Etudes Nucléaires, 85 X, 38041 Grenoble Cedex, France*

G. VON JAGOW (22, 24), *Institut für Physikalische Biochemie, Universität München, 8000 München 2, Federal Republic of Germany*

JOEL H. WEINER (36), *Department of Biochemistry, University of Alberta, Edmonton, Alberta, Canada, T6G 2H7*

HANNS WEISS (19, 20), *Institut für Biochemie, Universität Düsseldorf, 4000 Düsseldorf 1, Federal Republic of Germany*

NOREEN WILLIAMS (46, 47, 68), *McCollum-Pratt Institute, and Department of Biology, The Johns Hopkins University, Baltimore, Maryland 21218*

EDITH WINKLER (49), *Institut für Physikalische Biochemie, Universität München, 8000 München 2, Federal Republic of Germany*

LICIA N. Y. WU (33), *Department of Chemistry, University of South Carolina, Columbia, South Carolina 29208*

XIAOHANG YANG (29), *Department of Biochemistry, Dartmouth Medical School, Hanover, New Hampshire 03756*

YUKUO YOSHIDA (50), *Department of Physiological Chemistry, Medical School, Osaka University, Osaka 530, Japan*

YU-ZHONG ZHANG (3), *Institute of Molecular Biology, University of Oregon, Eugene, Oregon 97403*

Preface

Volumes 125 and 126 of *Methods in Enzymology* initiate the transport volumes of the Biomembranes series. Biological transport represents a continuation of methodology for the study of membrane function, Volumes 96–98 having dealt with membrane biogenesis, assembly, targeting, and recycling.

This is a particularly good time to cover the topic of biological membrane transport because a strong conceptual basis for its understanding now exists. Membrane transport has been divided into five topics. Topic 1 is covered in Volumes 125 and 126. The remaining four topics will be covered in subsequent volumes of the Biomembranes series.

1. Transport in Bacteria, Mitochondria, and Chloroplasts
2. ATP-Driven Pumps and Related Transport
3. General Methodology of Cellular and Subcellular Transport
4. Cellular and Subcellular Transport: Eukaryotic (Nonepithelial) Cells
5. Cellular and Subcellular Transport: Epithelial Cells

We are fortunate to have the advice and good counsel of our Advisory Board. Additional valuable input to these volumes was obtained from many individuals. Special thanks go to Giovanna Ames, Angelo Azzi, Ernesto Carafoli, Hans Heldt, Lars Ernster, Peter Pedersen, Youssef Hatefi, Dieter Oesterhelt, Saul Roseman, and Thomas Wilson. The enthusiasm and cooperation of the participants have enriched and made these volumes possible. The friendly cooperation of the staff of Academic Press is gratefully acknowledged.

These volumes are dedicated to Professor Sidney Colowick, a dear friend and colleague, who died in 1985. We shall miss his wise counsel, encouragement, and friendship.

SIDNEY FLEISCHER
BECCA FLEISCHER

METHODS IN ENZYMOLOGY

EDITED BY

Sidney P. Colowick and Nathan O. Kaplan

VANDERBILT UNIVERSITY
SCHOOL OF MEDICINE
NASHVILLE, TENNESSEE

DEPARTMENT OF CHEMISTRY
UNIVERSITY OF CALIFORNIA
AT SAN DIEGO
LA JOLLA, CALIFORNIA

I. Preparation and Assay of Enzymes
II. Preparation and Assay of Enzymes
III. Preparation and Assay of Substrates
IV. Special Techniques for the Enzymologist
V. Preparation and Assay of Enzymes
VI. Preparation and Assay of Enzymes (*Continued*)
Preparation and Assay of Substrates
Special Techniques
VII. Cumulative Subject Index

METHODS IN ENZYMOLOGY

EDITORS-IN-CHIEF

Sidney P. Colowick and Nathan O. Kaplan

VOLUME VIII. Complex Carbohydrates
Edited by ELIZABETH F. NEUFELD AND VICTOR GINSBURG

VOLUME IX. Carbohydrate Metabolism
Edited by WILLIS A. WOOD

VOLUME X. Oxidation and Phosphorylation
Edited by RONALD W. ESTABROOK AND MAYNARD E. PULLMAN

VOLUME XI. Enzyme Structure
Edited by C. H. W. HIRS

VOLUME XII. Nucleic Acids (Parts A and B)
Edited by LAWRENCE GROSSMAN AND KIVIE MOLDAVE

VOLUME XIII. Citric Acid Cycle
Edited by J. M. LOWENSTEIN

VOLUME XIV. Lipids
Edited by J. M. LOWENSTEIN

VOLUME XV. Steroids and Terpenoids
Edited by RAYMOND B. CLAYTON

VOLUME XVI. Fast Reactions
Edited by KENNETH KUSTIN

VOLUME XVII. Metabolism of Amino Acids and Amines (Parts A and B)
Edited by HERBERT TABOR AND CELIA WHITE TABOR

VOLUME XVIII. Vitamins and Coenzymes (Parts A, B, and C)
Edited by DONALD B. MCCORMICK AND LEMUEL D. WRIGHT

VOLUME XIX. Proteolytic Enzymes
Edited by GERTRUDE E. PERLMANN AND LASZLO LORAND

VOLUME XX. Nucleic Acids and Protein Synthesis (Part C)
Edited by KIVIE MOLDAVE AND LAWRENCE GROSSMAN

VOLUME XXI. Nucleic Acids (Part D)
Edited by LAWRENCE GROSSMAN AND KIVIE MOLDAVE

VOLUME XXII. Enzyme Purification and Related Techniques
Edited by WILLIAM B. JAKOBY

VOLUME XXIII. Photosynthesis (Part A)
Edited by ANTHONY SAN PIETRO

VOLUME XXIV. Photosynthesis and Nitrogen Fixation (Part B)
Edited by ANTHONY SAN PIETRO

VOLUME XXV. Enzyme Structure (Part B)
Edited by C. H. W. HIRS AND SERGE N. TIMASHEFF

VOLUME XXVI. Enzyme Structure (Part C)
Edited by C. H. W. HIRS AND SERGE N. TIMASHEFF

VOLUME XXVII. Enzyme Structure (Part D)
Edited by C. H. W. HIRS AND SERGE N. TIMASHEFF

VOLUME XXVIII. Complex Carbohydrates (Part B)
Edited by VICTOR GINSBURG

VOLUME XXIX. Nucleic Acids and Protein Synthesis (Part E)
Edited by LAWRENCE GROSSMAN AND KIVIE MOLDAVE

VOLUME XXX. Nucleic Acids and Protein Synthesis (Part F)
Edited by KIVIE MOLDAVE AND LAWRENCE GROSSMAN

VOLUME XXXI. Biomembranes (Part A)
Edited by SIDNEY FLEISCHER AND LESTER PACKER

VOLUME XXXII. Biomembranes (Part B)
Edited by SIDNEY FLEISCHER AND LESTER PACKER

VOLUME XXXIII. Cumulative Subject Index Volumes I–XXX
Edited by MARTHA G. DENNIS AND EDWARD A. DENNIS

VOLUME XXXIV. Affinity Techniques (Enzyme Purification: Part B)
Edited by WILLIAM B. JAKOBY AND MEIR WILCHEK

VOLUME XXXV. Lipids (Part B)
Edited by JOHN M. LOWENSTEIN

VOLUME XXXVI. Hormone Action (Part A: Steroid Hormones)
Edited by BERT W. O'MALLEY AND JOEL G. HARDMAN

VOLUME XXXVII. Hormone Action (Part B: Peptide Hormones)
Edited by BERT W. O'MALLEY AND JOEL G. HARDMAN

VOLUME XXXVIII. Hormone Action (Part C: Cyclic Nucleotides)
Edited by JOEL G. HARDMAN AND BERT W. O'MALLEY

VOLUME XXXIX. Hormone Action (Part D: Isolated Cells, Tissues, and Organ Systems)
Edited by JOEL G. HARDMAN AND BERT W. O'MALLEY

VOLUME XL. Hormone Action (Part E: Nuclear Structure and Function)
Edited by BERT W. O'MALLEY AND JOEL G. HARDMAN

VOLUME XLI. Carbohydrate Metabolism (Part B)
Edited by W. A. WOOD

VOLUME XLII. Carbohydrate Metabolism (Part C)
Edited by W. A. WOOD

VOLUME XLIII. Antibiotics
Edited by JOHN H. HASH

VOLUME XLIV. Immobilized Enzymes
Edited by KLAUS MOSBACH

VOLUME XLV. Proteolytic Enzymes (Part B)
Edited by LASZLO LORAND

VOLUME XLVI. Affinity Labeling
Edited by WILLIAM B. JAKOBY AND MEIR WILCHEK

VOLUME XLVII. Enzyme Structure (Part E)
Edited by C. H. W. HIRS AND SERGE N. TIMASHEFF

VOLUME XLVIII. Enzyme Structure (Part F)
Edited by C. H. W. HIRS AND SERGE N. TIMASHEFF

VOLUME XLIX. Enzyme Structure (Part G)
Edited by C. H. W. HIRS AND SERGE N. TIMASHEFF

VOLUME L. Complex Carbohydrates (Part C)
Edited by VICTOR GINSBURG

VOLUME LI. Purine and Pyrimidine Nucleotide Metabolism
Edited by PATRICIA A. HOFFEE AND MARY ELLEN JONES

VOLUME LII. Biomembranes (Part C: Biological Oxidations)
Edited by SIDNEY FLEISCHER AND LESTER PACKER

VOLUME LIII. Biomembranes (Part D: Biological Oxidations)
Edited by SIDNEY FLEISCHER AND LESTER PACKER

VOLUME LIV. Biomembranes (Part E: Biological Oxidations)
Edited by SIDNEY FLEISCHER AND LESTER PACKER

VOLUME LV. Biomembranes (Part F: Bioenergetics)
Edited by SIDNEY FLEISCHER AND LESTER PACKER

VOLUME LVI. Biomembranes (Part G: Bioenergetics)
Edited by SIDNEY FLEISCHER AND LESTER PACKER

VOLUME LVII. Bioluminescence and Chemiluminescence
Edited by MARLENE A. DELUCA

VOLUME LVIII. Cell Culture
Edited by WILLIAM B. JAKOBY AND IRA PASTAN

VOLUME LIX. Nucleic Acids and Protein Synthesis (Part G)
Edited by KIVIE MOLDAVE AND LAWRENCE GROSSMAN

VOLUME LX. Nucleic Acids and Protein Synthesis (Part H)
Edited by KIVIE MOLDAVE AND LAWRENCE GROSSMAN

VOLUME 61. Enzyme Structure (Part H)
Edited by C. H. W. HIRS AND SERGE N. TIMASHEFF

VOLUME 62. Vitamins and Coenzymes (Part D)
Edited by DONALD B. MCCORMICK AND LEMUEL D. WRIGHT

VOLUME 63. Enzyme Kinetics and Mechanism (Part A: Initial Rate and Inhibitor Methods)
Edited by DANIEL L. PURICH

VOLUME 64. Enzyme Kinetics and Mechanism (Part B: Isotopic Probes and Complex Enzyme Systems)
Edited by DANIEL L. PURICH

VOLUME 65. Nucleic Acids (Part I)
Edited by LAWRENCE GROSSMAN AND KIVIE MOLDAVE

VOLUME 66. Vitamins and Coenzymes (Part E)
Edited by DONALD B. MCCORMICK AND LEMUEL D. WRIGHT

VOLUME 67. Vitamins and Coenzymes (Part F)
Edited by DONALD B. MCCORMICK AND LEMUEL D. WRIGHT

VOLUME 68. Recombinant DNA
Edited by RAY WU

VOLUME 69. Photosynthesis and Nitrogen Fixation (Part C)
Edited by ANTHONY SAN PIETRO

VOLUME 70. Immunochemical Techniques (Part A)
Edited by HELEN VAN VUNAKIS AND JOHN J. LANGONE

VOLUME 71. Lipids (Part C)
Edited by JOHN M. LOWENSTEIN

VOLUME 72. Lipids (Part D)
Edited by JOHN M. LOWENSTEIN

VOLUME 73. Immunochemical Techniques (Part B)
Edited by JOHN J. LANGONE AND HELEN VAN VUNAKIS

VOLUME 74. Immunochemical Techniques (Part C)
Edited by JOHN J. LANGONE AND HELEN VAN VUNAKIS

VOLUME 75. Cumulative Subject Index Volumes XXXI, XXXII, and XXXIV–LX
Edited by EDWARD A. DENNIS AND MARTHA G. DENNIS

VOLUME 76. Hemoglobins
Edited by ERALDO ANTONINI, LUIGI ROSSI-BERNARDI, AND EMILIA CHIANCONE

VOLUME 77. Detoxication and Drug Metabolism
Edited by WILLIAM B. JAKOBY

VOLUME 78. Interferons (Part A)
Edited by SIDNEY PESTKA

VOLUME 79. Interferons (Part B)
Edited by SIDNEY PESTKA

VOLUME 80. Proteolytic Enzymes (Part C)
Edited by LASZLO LORAND

VOLUME 81. Biomembranes (Part H: Visual Pigments and Purple Membranes, I)
Edited by LESTER PACKER

VOLUME 82. Structural and Contractile Proteins (Part A: Extracellular Matrix)
Edited by LEON W. CUNNINGHAM AND DIXIE W. FREDERIKSEN

VOLUME 83. Complex Carbohydrates (Part D)
Edited by VICTOR GINSBURG

VOLUME 84. Immunochemical Techniques (Part D: Selected Immunoassays)
Edited by JOHN J. LANGONE AND HELEN VAN VUNAKIS

VOLUME 85. Structural and Contractile Proteins (Part B: The Contractile Apparatus and the Cytoskeleton)
Edited by DIXIE W. FREDERIKSEN AND LEON W. CUNNINGHAM

VOLUME 86. Prostaglandins and Arachidonate Metabolites
Edited by WILLIAM E. M. LANDS AND WILLIAM L. SMITH

VOLUME 87. Enzyme Kinetics and Mechanism (Part C: Intermediates, Stereochemistry, and Rate Studies)
Edited by DANIEL L. PURICH

VOLUME 88. Biomembranes (Part I: Visual Pigments and Purple Membranes, II)
Edited by LESTER PACKER

VOLUME 89. Carbohydrate Metabolism (Part D)
Edited by WILLIS A. WOOD

VOLUME 90. Carbohydrate Metabolism (Part E)
Edited by Willis A. Wood

VOLUME 91. Enzyme Structure (Part I)
Edited by C. H. W. HIRS AND SERGE N. TIMASHEFF

VOLUME 92. Immunochemical Techniques (Part E: Monoclonal Antibodies and General Immunoassay Methods)
Edited by JOHN J. LANGONE AND HELEN VAN VUNAKIS

VOLUME 93. Immunochemical Techniques (Part F: Conventional Antibodies, Fc Receptors, and Cytotoxicity)
Edited by JOHN J. LANGONE AND HELEN VAN VUNAKIS

VOLUME 94. Polyamines
Edited by HERBERT TABOR AND CELIA WHITE TABOR

VOLUME 95. Cumulative Subject Index Volumes 61–74 and 76–80
Edited by EDWARD A. DENNIS AND MARTHA G. DENNIS

VOLUME 96. Biomembranes [Part J: Membrane Biogenesis: Assembly and Targeting (General Methods; Eukaryotes)]
Edited by SIDNEY FLEISCHER AND BECCA FLEISCHER

VOLUME 97. Biomembranes [Part K: Membrane Biogenesis: Assembly and Targeting (Prokaryotes, Mitochondria, and Chloroplasts)]
Edited by SIDNEY FLEISCHER AND BECCA FLEISCHER

VOLUME 98. Biomembranes [Part L: Membrane Biogenesis (Processing and Recycling)]
Edited by SIDNEY FLEISCHER AND BECCA FLEISCHER

VOLUME 99. Hormone Action (Part F: Protein Kinases)
Edited by JACKIE D. CORBIN AND JOEL G. HARDMAN

VOLUME 100. Recombinant DNA (Part B)
Edited by RAY WU, LAWRENCE GROSSMAN, AND KIVIE MOLDAVE

VOLUME 101. Recombinant DNA (Part C)
Edited by RAY WU, LAWRENCE GROSSMAN, AND KIVIE MOLDAVE

VOLUME 102. Hormone Action (Part G: Calmodulin and Calcium-Binding Proteins)
Edited by ANTHONY R. MEANS AND BERT W. O'MALLEY

VOLUME 103. Hormone Action (Part H: Neuroendocrine Peptides)
Edited by P. MICHAEL CONN

VOLUME 104. Enzyme Purification and Related Techniques (Part C)
Edited by WILLIAM B. JAKOBY

VOLUME 105. Oxygen Radicals in Biological Systems
Edited by LESTER PACKER

VOLUME 106. Posttranslational Modifications (Part A)
Edited by FINN WOLD AND KIVIE MOLDAVE

VOLUME 107. Posttranslational Modifications (Part B)
Edited by FINN WOLD AND KIVIE MOLDAVE

VOLUME 108. Immunochemical Techniques (Part G: Separation and Characterization of Lymphoid Cells)
Edited by GIOVANNI DI SABATO, JOHN J. LANGONE, AND HELEN VAN VUNAKIS

VOLUME 109. Hormone Action (Part I: Peptide Hormones)
Edited by LUTZ BIRNBAUMER AND BERT W. O'MALLEY

VOLUME 110. Steroids and Isoprenoids (Part A)
Edited by JOHN H. LAW AND HANS C. RILLING

VOLUME 111. Steroids and Isoprenoids (Part B)
Edited by JOHN H. LAW AND HANS C. RILLING

VOLUME 112. Drug and Enzyme Targeting (Part A)
Edited by KENNETH J. WIDDER AND RALPH GREEN

VOLUME 113. Glutamate, Glutamine, Glutathione, and Related Compounds
Edited by ALTON MEISTER

VOLUME 114. Diffraction Methods for Biological Macromolecules (Part A)
Edited by HAROLD W. WYCKOFF, C. H. W. HIRS, AND SERGE N. TIMASHEFF

VOLUME 115. Diffraction Methods for Biological Macromolecules (Part B)
Edited by HAROLD W. WYCKOFF, C. H. W. HIRS, AND SERGE N. TIMASHEFF

VOLUME 116. Immunochemical Techniques (Part H: Effectors and Mediators of Lymphoid Cell Functions)
Edited by GIOVANNI DI SABATO, JOHN J. LANGONE, AND HELEN VAN VUNAKIS

VOLUME 117. Enzyme Structure (Part J)
Edited by C. H. W. HIRS AND SERGE N. TIMASHEFF

VOLUME 118. Plant Molecular Biology
Edited by ARTHUR WEISSBACH AND HERBERT WEISSBACH

VOLUME 119. Interferons (Part C)
Edited by SIDNEY PESTKA

VOLUME 120. Cumulative Subject Index Volumes 81–94, 96–101 (in preparation)

VOLUME 121. Immunochemical Techniques (Part I: Hybridoma Technology and Monoclonal Antibodies)
Edited by JOHN J. LANGONE AND HELEN VAN VUNAKIS

VOLUME 122. Vitamins and Coenzymes (Part G)
Edited by FRANK CHYTIL AND DONALD B. MCCORMICK

VOLUME 123. Vitamins and Coenzymes (Part H)
Edited by FRANK CHYTIL AND DONALD B. MCCORMICK

VOLUME 124. Hormone Action (Part J: Neuroendocrine Peptides)
Edited by P. MICHAEL CONN

VOLUME 125. Biomembranes (Part M: Transport in Bacteria, Mitochondria, and Chloroplasts: General Approaches and Transport Systems)
Edited by SIDNEY FLEISCHER AND BECCA FLEISCHER

VOLUME 126. Biomembranes (Part N: Transport in Bacteria, Mitochondria, and Chloroplasts: Protonmotive Force)
Edited by SIDNEY FLEISCHER AND BECCA FLEISCHER

VOLUME 127. Biomembranes (Part O: Protons and Water: Structure and Translocation) (in preparation)
Edited by LESTER PACKER

VOLUME 128. Plasma Lipoproteins (Part A: Preparation, Structure, and Molecular Biology) (in preparation)
Edited by JERE P. SEGREST AND JOHN ALBERS

VOLUME 129. Plasma Lipoproteins (Part B: Characterization, Cell Biology, and Metabolism) (in preparation)
Edited by JOHN ALBERS AND JERE P. SEGREST

VOLUME 130. Enzyme Structure (Part K) (in preparation)
Edited by C. H. W. HIRS AND SERGE N. TIMASHEFF

VOLUME 131. Enzyme Structure (Part L) (in preparation)
Edited by C. H. W. HIRS AND SERGE N. TIMASHEFF

VOLUME 132. Immunochemical Techniques (Part J: Phagocytosis and Cell-Mediated Cytotoxicity) (in preparation)
Edited by GIOVANNI DI SABATO AND JOHANNES EVERSE

[1] Protonmotive Force and Secondary Transport: Historical Perspectives and Unifying Principles

By YASUO KAGAWA

Concepts of Translocators

Physiological Concept and Single-Channel Recording

In the 1940s, Hodgkin and Huxley[1] proposed the concept of a pump and channel to explain the passage of electric current through an excitable membrane. A pump means a translocator that drives ions against their electrochemical potential gradient using energy derived from ATP or other sources (a group of translocators for primary active transport). A channel is a pore in a membrane through which a certain solute is transported down its electrochemical potential gradient (a group of translocators for passive transport). Neher and Stevens measured the functions of ion channels directly by a single-channel recording technique[2] and confirmed Hodgkin and Huxley's prediction of a channel with controlling gate and ion-specific filter.[1] A carrier binds a specific solute,[3] and thus shows saturation kinetics and inhibitor sensitivity, and some movement of the binding site explains exchange diffusion. Some carriers exert secondary active transport, i.e., transport of a solute against its electrochemical potential gradient at the expense of energy of ion flux down the electrochemical potential gradient of the ion which is maintained by a pump.[3,4]

Genetic Concept and Primary Structure of Translocators

In the 1940s, Monod and his colleagues initiated studies on the lactose operon.[5] Lactose permease, i.e., the *y* gene product, was shown to be essential for the translocation of lactose.[5] The inductions of lactose permease and β-galactosidase were strongly inhibited by an uncoupler of oxidative phosphorylation. Later, this permease was shown to be a proton-driven lactose carrier.[6] Genetic analysis of transport systems coded

[1] A. L. Hodgkin and A. F. Huxley, *J. Physiol.* (*London*) **117,** 500 (1952).
[2] E. Neher and C. F. Stevens, *Annu. Rev. Biophys. Bioeng.* **6,** 345 (1977).
[3] R. K. Crane, *Rev. Physiol. Biochem. Pharmacol.* **78,** 99 (1977).
[4] P. Overath and J. K. Wright, *Trends Biochem. Sci.* **8,** 405 (1983).
[5] J. Monod, J. P. Changeux, and F. Jacob, *J. Mol. Biol.* **6,** 306 (1963).
[6] H. R. Kaback, *Science* **186,** 882 (1974).

on different loci of chromosomes confirmed the presence of many different types of passive transport (nonspecific and specific facilitated diffusion) and active transport (e.g., Na^+- and H^+-driven secondary transport, binding protein-mediated systems, and a phosphoenolpyruvate-dependent system). DNA sequencing of the structural genes for translocators revealed the molecular properties of the translocators, such as their subunits, molecular weights, and primary structure. Thus, lactose permease[7] and Na^+ channel of both acetylcholine receptor[8] and electric organ[9] were completely sequenced. Recently, primary structures of subunits of pumps such as Na^+, K^+-ATPase,[10,11] Ca^{2+}-ATPase,[12] and H^+-ATPase[13–19] (bacteria,[13,14] chloroplasts,[15–17] and yeast[18] and human[19] mitochondria) have been determined. Genetic studies also revealed the complicated metabolic pathways in the membrane and the control system of translocators; the so-called *unc* mutant of *Escherichia coli* was shown to have a defective protonmotive ATPase of oxidative phosphorylation.[20]

Chemiosmotic Theory of Energy Transduction in Membranes

At the end of the 1950s, Mitchell became concerned with the mechanism by which the energy provided by metabolism caused the active

[7] D. E. Büchel, B. Gronenborn, and B. Müller-Hill, *Nature (London)* **283,** 541 (1980).

[8] M. Noda, H. Takahashi, T. Takabe, M. Toyosato, S. Kiyotani, T. Hirose, M. Asai, H. Takashima, S. Inayama, T. Miyata, and S. Numa, *Nature (London)* **301,** 251 (1983).

[9] M. Noda, S. Shimizu, T. Tanabe, T. Takai, T. Kayano, T. Ikeda, H. Takahashi, H. Nakayama, Y. Kanaoka, N. Minamio, K. Kangawa, H. Matsuo, M. A. Raftery, T. Hirose, S. Inayama, H. Hayashida, T. Miyata, and S. Numa, *Nature (London)* **312**, 121 (1984).

[10] K. Kawakami, S. Noguchi, M. Noda, H. Takahashi, T. Ohta, M. Kawamura, H. Nojima, K. Nagano, T. Hirose, S. Inayama, H. Hayashida, T. Miyata, and S. Numa, *Nature (London)* **316,** 733 (1985).

[11] G. E. Shull, A. Schwartz, and J. B. Lingrel, *Nature (London)* **316,** 691 (1985).

[12] D. H. MacLennan, C. J. Brandl, B. Korczak, and N. M. Green, *Nature (London)* **316,** 696 (1985).

[13] H. Kanazawa and M. Futai, *Ann. N.Y. Acad. Sci.* **402,** 45 (1982).

[14] J. E. Walker, A. Eberle, N. J. Gay, M. J. Runswick, and M. Saraste, *Biochem. Soc. Trans.* **10,** 203 (1982).

[15] G. Zurawski, W. Bottomley, and P. R. Whitfeld, *Proc. Natl. Acad. Sci. U.S.A.* **79,** 6260 (1982).

[16] E. T. Krebbers, I. M. Larrinua, L. McIntosh, and L. Bogorad, *Nuc. Acid Res.* **10,** 4985 (1982).

[17] K. Shinozaki and M. Sugiura, *Nuc. Acid Res.* **10,** 4923 (1982).

[18] J. Saltzgaber-Muller, S. P. Kunapuli, and M. G. Douglas, *J. Biol. Chem.* **258,** 11465 (1983).

[19] S. Ohta and Y. Kagawa, *J. Biochem. (Tokyo)* **99,** 135 (1986).

[20] F. Gibson, *Annu. Rev. Biochem.* **48,** 103 (1979).

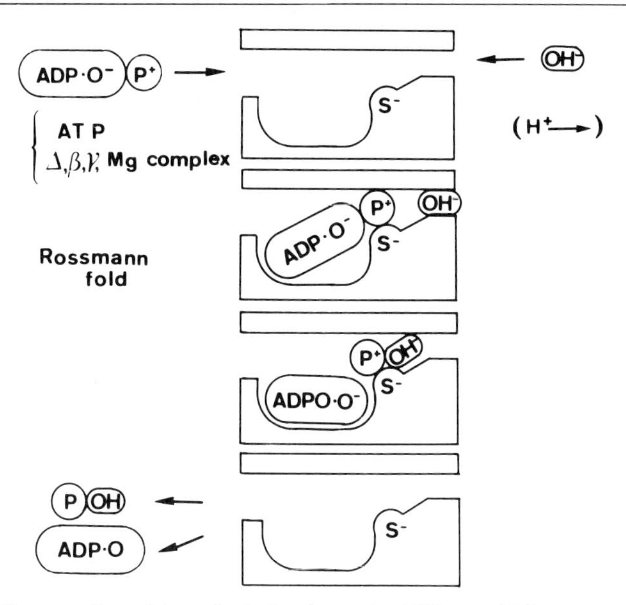

FIG. 1. Cross section of hypothetical anisotropic ATPase, which converts phosphate bond energy into the work of proton (or OH^-) translocation. This scheme was originally proposed by Mitchell in 1962,[22] and during 23 years of research, it has been modified by the findings that (1) the enzyme is composed of F_0 (H^+ channel) and F_1 (catalytic portion)[33]; (2) the H^+/ATP stoichiometry is 3; (3) the true substrate is a Δ,β,γ, bidentate ATP–Mg complex[46]; (4) F_1-bound ATP is formed from F_1, ADP, Mg^{2+}, and P_i without a protonmotive force (no direct interaction of the translocated ion with phosphate)[48]; (5) the ATP-binding site is not a cysteinyl residue,[13-19] but the Rossman fold of the β subunit[61]; (6) the reaction is in-line nucleophilic substitution (S_N2), as shown by experiments with [^{16}O,^{17}O,^{18}O]thiophosphate; and (7) multiple ATP-binding sites show positive cooperativity in V_{max} and strong negative cooperativity in K_m. For details, see recent reviews.[46-49]

transport of solutes across membranes.[21] He regarded the biomembrane not simply as an osmotic barrier and an osmotic link between the media on either side of it, but as a chemical link allowing the exchange of one covalently linked group with another. Figure 1 depicts his idea on the hypothetical H^+- (or OH^-) translocating ATPase (i.e., H^+ pump), which converts chemical energy liberated by ATP hydrolysis into electrochemical energy for H^+ translocation.[22] ATP is synthesized via the reverse reaction of this enzyme, when an electrochemical potential gradient of H^+ is maintained by the respiratory chain. Any enzyme molecule has its own anisotropic structure (e.g., the dipole axis of cytochrome *c* during oxidoreduction[23]), and the vectorial nature of ATPase and oxidoreductases

[21] P. Mitchell, *Nature (London)* **191,** 144 (1961).
[22] P. Mitchell, *J. Gen. Microbiol.* **29,** 25 (1962).
[23] W. H. Koppenol and E. Margoliash, *J. Biol. Chem.* **257,** 4426 (1982).

becomes apparent when these pumps are organized into anisotropic structures to translocate protons unidirectionally. In fact, outward transport of protons is observed when ATP is hydrolyzed or a substrate is oxidized in mitochondria or the bacterial plasma membrane. Secondary transport of a solute may also be driven by H^+ flux.[24] Mitchell first defined the simple terminology now generally used for secondary transport: Symport means translocation of solutes such as H^+ and sugar by a translocator (symporter) in the same direction, while antiport means a translocation of solutes such as Na^+ and H^+ by a translocator (antiporter) in the reverse direction. A translocator for a single solute is called a uniporter. The porters need not be intrinsically anisotropic and are simply required to couple the flow of pump-driven ions to the flows of specific solutes by a secondary mechanism via ion circuits across the membrane.[24]

Biochemical Isolation of Lipid-Dependent Translocators

In 1957, Skou described an ATPase stimulated by Na^+, K^+, and Mg^{2+},[25] which is the pump predicted by Hodgkin and Huxley.[1] The activities of both the pump and ATPase were inhibited by ouabain. Likewise, Ca^{2+}-stimulated ATPase was found in the sarcoplasmic reticulum.[26] Crystallization of the galactose-binding protein[27] and the partial solubilization of the phosphoenolpyruvate-glucose-phosphotransferase system[28] were successful examples of the application of classical enzymological methods to soluble proteins. However, as shown by Danielli and Davson,[29] and later in more detail by Singer and Nicolson (fluid mosaic model),[30] the basic structure of biomembranes is a lipid bilayer, and many membrane proteins were shown to be hydrophobic and inactive in the absence of phospholipids.[31–33] The classical studies by Fleischer *et al.*[32] on restoration of electron transfer activity in mitochondria depleted of lipid conclusively demonstrated a role for phospholipid in electron transport. In this regard, the earlier study of Kakiuchi (1927) is worth noting.[31] Intrinsic membrane proteins, including translocators, were thus purified by the use

[24] P. Mitchell, *Nature (London)* **206,** 1148 (1979).
[25] J. C. Skou, *Biochim. Biophys. Acta* **23,** 394 (1957).
[26] S. Ebashi and F. Lipmann, *J. Cell Biol.* **14,** 389 (1962).
[27] Y. Anraku, *J. Biol. Chem.* **243,** 3116 (1968).
[28] W. Kundig and S. Roseman, *J. Biol. Chem.* **246,** 1393 (1971).
[29] J. F. Danielli and H. J. Davson, *J. Cell. Comp. Physiol.* **5,** 495 (1935).
[30] S. J. Singer and G. L. Nicolson, *Science* **175,** 720 (1972).
[31] S. Kakiuchi, *J. Biochem.* **7,** 263 (1927).
[32] S. Fleischer, G. Brierley, H. Klouwen, and D. B. Slawtterback, *J. Biol. Chem.* **237,** 3264 (1962).
[33] Y. Kagawa and E. Racker, *J. Biol. Chem.* **241,** 2461 (1966).

of detergents, and their activities were measured in the presence of phospholipids.[33] After solubilization of membranes with detergents, translocators, such as human acetylcholine receptor, could be purified by the use of monoclonal antibodies,[34] affinity chromatography, and many new methods described in this volume. Soluble proteins of solute transport and soluble coupling factors (F_1, F_2, etc.)[35] of oxidative phosphorylation were thus shown to have essential membrane portion called Enzyme II,[36] and F_0,[33] respectively.

The isolation of a protein factor by itself does not prove its role in transport.[35] For example, even after the isolation of lipid-dependent proton-translocating ATPase (F_0F_1),[33] the role of coupling factors was assumed to be in forming a high-energy intermediate ($ADP\text{-}F_1\text{-}F_0\text{-}F_4 \sim F_2 + P_i \rightarrow ADP\text{-}F_1\text{-}F_0\text{-}F_4\text{-}F_2 \sim P \rightarrow ATP$).[35] This idea was based on the chemical hypothesis of oxidative phosphorylation which postulated that the energy of oxidoreduction is used to synthesize a high-energy intermediate $X \sim Y$, which is used for ATP synthesis via $X \sim P$.[37]

Reconstitution of Active Proteoliposomes

For proof that an isolated translocator is responsible for translocation, it must be reconstituted into a lipid bilayer and its activity must be measured.[38] Proton-translocating ATPase was reconstituted into liposomes, and its activity was measured as predicted by Mitchell.[21] The reverse reaction, ATP synthesis driven by proton flux through reconstituted ATPase liposomes, was also demonstrated by applying either an ion gradient[39] or an electric pulse.[40] Reconstitution of proteoliposomes has been used for many types of translocators, for instance, the nonspecific channel of mitochondrial outer membranes (organic solvent method),[41] a glucose carrier for facilitated diffusion (freeze-thawing method),[42] the galactose/H^+ symporter (dilution method with octylglucoside),[43] and Ca^{2+}-ATPase (dialysis method).[44]

[34] M. Yoshida-Momoi and V. A. Lennon, *J. Biol. Chem.* **257,** 12757 (1982).
[35] E. Racker, "Mechanisms in Bioenergetics." Academic Press, New York, 1965.
[36] W. Kundig and S. Roseman, *J. Biol. Chem.* **246,** 1407 (1971).
[37] E. C. Slater, *Q. Rev. Biophys.* **7,** 401 (1975).
[38] Y. Kagawa, *Biochim. Biophys. Acta* **265,** 297 (1972).
[39] N. Sone, M. Yoshida, H. Hirata, and Y. Kagawa, *J. Biol. Chem.* **252,** 2956 (1977).
[40] M. Rögner, K. Ohno, T. Hamamoto, N. Sone, and Y. Kagawa, *Biochem. Biophys. Res. Commun.* **91,** 362 (1979).
[41] L. S. Zalman, H. Nikaido, and Y. Kagawa, *J. Biol. Chem.* **255,** 1771 (1980).
[42] M. Kasahara and P. C. Hinkle, *J. Biol. Chem.* **252,** 7384 (1977).
[43] P. J. F. Henderson, Y. Kagawa, and H. Hirata, *Biochim. Biophys. Acta* **732,** 204 (1983).
[44] C. T. Wong, A. Saito, and S. Fleischer, *J. Biol. Chem.* **254,** 9209 (1979).

A combination of macroliposomes and a single-channel recording method is also useful for analysis of translocators.[45] Black membranes can be used only when the channel allows passage of sufficient ions per second to be detectable as a current.

Protonmotive Force

The molecular mechanism by which ATP is synthesized by a protonmotive force has been discussed since Mitchell proposed the scheme shown in Fig. 1 (see recent reviews[46–49]). A protonmotive force is defined as[50,51]

$$\Delta p = \Delta\psi - Z\Delta\text{pH} = (\Delta\bar{\mu}_{\text{H}^+})/F$$

where $\Delta\psi$ and ΔpH are the differences in electrical potential and pH between the outside and inside of the membrane and $Z = 2.3\ RT/F$, where R (J mol^{-1} K^{-1}), T (°K), F (°C mol^{-1}), and $\Delta\bar{\mu}_{\text{H}^+}$ are the gas constant, absolute temperature, Faraday constant, and electrochemical potential difference of H^+ across the membrane, respectively.

Reconstitution of F_0F_1 liposomes confirmed that the artificially imposed bulk phase Δp is used to synthesize ATP,[39] and the direct contact of respiratory enzymes or other factors with F_0F_1 to transfer localized protons[52] is not essential. However, any molecule reacting with protons should have localized protons at a given moment, and it is still uncertain whether the bulk phase Δp or localized proton is transferred to F_0F_1 in physiological conditions. Application of an electric pulse to chloroplasts[53] or F_0F_1 liposomes[40,54] (this volume [66]) should produce a localized electric potential in the vesicles, but still results in ATP synthesis. The problem of the localized proton including ATP synthesis in alkalophilic bacteria, the effect of uncouplers, etc., is discussed in recent reviews.[55] It is interesting that even the simplest reaction, such as electron transfer from

[45] D. W. Tonk, R. L. Huganir, P. Greengard, and W. W. Webb, *Proc. Natl. Acad. Sci. U.S.A.* **79,** 7749 (1982).

[46] Y. Kagawa, *in* "New Comprehensive Biochemistry, Bioenergetics" (L. Ernster, ed.), p. 149. Elsevier, Amsterdam, 1984.

[47] L. M. Amzel and P. L. Pedersen, *Annu. Rev. Biochem.* **52,** 801 (1983).

[48] R. L. Cross, *Annu. Rev. Biochem.* **50,** 681 (1981).

[49] C. Tanford, *Annu. Rev. Biochem.* **52,** 379 (1983).

[50] P. Mitchell and J. Moyle, *Eur. J. Biochem.* **7,** 471 (1968).

[51] A. G. Lowe and M. N. Jones, *Trends Biochem. Sci.* **9,** 11 (1984).

[52] R. J. P. Williams, *J. Theor. Biol.* **1,** 1 (1961).

[53] H. T. Witt, *Biochim. Biophys. Acta* **505,** 355 (1979).

[54] Y. Kagawa, *Biochim. Biophys. Acta* **505,** 45 (1978).

[55] S. J. Ferguson and M. C. Sorgato, *Annu. Rev. Biochem.* **51,** 185 (1982).

cytochrome *c* to oxidase, proceeds via a strictly stereospecific localized area.[23] The stoichiometry of the F_0F_1 reaction (H^+/ATP ratio) is still controversial because of leakage of protons through the membrane and release of protons caused by differences of pK_a values of ATP, ADP, and P_i. The H^+/ATP ratio of three has been reported.[55]

Unifying Principles of Translocators

1. Protein structure: Translocators have a transmembrane structure and are oriented unidirectionally in a biomembrane. They are hydrophobic proteins and only their hydrophobic side chains are exposed at the inner side of the membrane. Within the membrane their structure appears to be a primary α-helix of 22–26 amino acid residues (e.g., F_0 subunits[13,14] and lactose permease[56]); the c subunit of F_0 has two α-helices, and lactose permease has 12.[56] In contrast to soluble proteins, translocators, such as bacteriorhodopsin, F_0,[13,14] Na^+-channel,[9] and acetylcholine receptor,[8] have charged groups on their inside surface.

2. Indirect action of driving ions: Figure 1 suggests that the translocated hydroxy ion interacts directly with phosphate. However, F_1-bound ATP is synthesized without a Δp from P_i, ADP, and F_1.[48] Although the formation of an anhydride bond between ADP and P_i requires a large free-energy change in aqueous solution, the energy-requiring step of ATP synthesis by F_0F_1 is the release of ATP from F_1.[48] Direct measurement of enzyme-bound nucleotide with ^{31}P NMR confirmed equilibration of ATP and ADP on the enzyme (equalization of internal thermodynamics).[57] It was also confirmed that proton translocation by electron transport components, such as cytochrome oxidase, is caused by the pump action of the enzyme.[58] In the case of secondary transport of sugars driven by the symport of either H^+ or Na^+, direct interaction of these ions with the sugar molecule is impossible.[3,4] Thus, the protonmotive force may cause a conformational change in ATPase, oxidoreductase, or secondary transporter, and thus indirectly release the substrates (or solute) which are bound to the translocators.

3. Ion flux and conformation change via acid–base residues: There are many hypotheses on proton translocation through these translocators, such as that it is like H^+ conductance through ice.[59] However, any hypothesis must include the essential role of the acidic and basic residues of the translocator. Chemical modification and genetic mutation of specific

[56] D. L. Foster, M. Boublik, and H. R. Kaback, *J. Biol. Chem.* **258,** 31 (1983).
[57] B. D. Nageswara Rao, F. K. Kayne, and M. Cohn, *J. Biol. Chem.* **254,** 2689 (1979).
[58] M. Wikström, K. Krab, and M. Saraste, *Annu. Rev. Biochem.* **150,** 623 (1981).
[59] J. F. Nagle, M. Mille, and H. J. Morowitz, *J. Chem. Phys.* **72,** 3959 (1980).

carboxyl and arginyl groups of translocators often result in loss of proton translocation[46]; for instance, reconstituted F_0 liposomes are inhibited by dicyclohexylcarbodiimide or phenylglyoxal. The proton pump activities of cytochrome oxidase[58] and other electron carriers are also inhibited by dicyclohexylcarbodiimide.

Conformational change of translocators during solute transport may result in at least two states of the binding site: one facing inside with high affinity and the other facing outside with low affinity.[49] There are other possible mechanisms, such as a kinetic constant model[4] or rotation of a binding site (not a whole molecule). The central problem is how protons (or other ions) change the conformation. A good example of a proton-induced affinity change is that of hemoglobin, which has two conformations that are mainly governed by the salt bridges of acid–base residues between the two kinds of subunits.[60] A Rossmann fold (β-α-β-α-β structure) is a nucleotide binding site (or a similar structure) that is also found in the β subunit of F_1.[13,61] Changes in the Rossmann fold and the roles of acid–base residues have been discussed.[46,62] An acid–base cluster hypothesis has been proposed to explain the $3H^+$/ATP stoichiometry and many other experimental findings.[62]

Acid–base cluster may be an essential transmembrane structure forming an aqueous pore of acetylcholine receptor,[63] mitochondrial porin, and many other translocators. In fact, site-directed mutagenesis of acetylcholine receptor and electrical properties of the mutated products revealed the essential role of these clusters.[64]

Methods for Unsolved Problems in Molecular Mechanisms

Voltage-sensitive Na^+ channels predicted by Hodgkin and Huxley are now purified, reconstituted, and sequenced.[9] Lactose permease,[7] F_0F_1,[13–19] electron carriers, and many other translocators have also been analyzed in the same manner. Physicochemical methods revealed some molecular mechanisms, as written in the legend of Fig. 1.[46,47] But in order to estab-

[60] G. E. Schuly and R. H. Schirmer, *in* "Principles of Protein Structure." Springer-Verlag, Berlin, 1979.

[61] M. Hollemans, M. J. Runswick, I. M. Fearnley, and J. E. Walker, *J. Biol. Chem.* **258,** 9307 (1983).

[62] Y. Kagawa, *J. Biochem.* **95,** 295 (1984).

[63] H. R. Guy, *Biophys. J.* **45,** 249 (1984).

[64] M. Mishina, T. Tobimatsu, K. Imoto, K. Tanaka, Y. Fujita, K. Fukuda, M. Kurasaki, H. Takahashi, Y. Morimoto, T. Hirose, S. Inayama, T. Takahashi, M. Kuno, and S. Numa, *Nature (London)* **313,** 364 (1985).

lish unifying principles at a molecular level, the following information[64-68] is essential: (1) oligomer reconstitution by biosynthetic assembly methods (acetylcholine receptor,[67] etc.) or with thermophilic subunits[46] combined with the genetic method[64] to correlate the amino acid sequence to the higher structure with functions; (2) three-dimensional crystals of translocators and their crystallographic analysis (X-ray crystallography,[66] etc.); (3) movements of relevant residues of the translocators and the solute molecules (NMR, EPR, with induced mutation, etc.); and (4) direct measurement of functions of translocators (electrical devices for biosensors,[68] etc.).

Mitchell's recent review on the correlation of chemical and osmotic forces in biochemistry describes "the nanometer force," which will drive the solutes through membrane proteins.[69]

[65] R. L. Barchi, *Trends Biochem. Sci.* **9,** 358 (1984).
[66] J. Deisenhofer, H. Michel, and R. Huber, *Trends Biochem. Sci.* **10,** 243 (1985).
[67] J. P. Merlie, *Cell* **36,** 573 (1984).
[68] M. Gronow, *Trends Biochem. Sci.* **9,** 336 (1984).
[69] P. Mitchell, *J. Biochem.* (*Tokyo*) **97,** 1 (1985).

Section I

Electron Transfer

A. Cytochrome Oxidase
Articles 2 through 16

B. Cytochrome bc_1 Complex
Articles 17 through 28

C. Other Electron Transfer
Articles 29 through 38

[2] Measurement of the H^+ Pumping Activity of Reconstituted Cytochrome Oxidase

By ROBERT P. CASEY*

Introduction

Despite the fact that an H^+-translocating function for cytochrome oxidase[1] does not contradict any of the essential postulates of the chemiosmotic hypothesis,[2] the feasibility of such an activity and its empirical demonstration have been the subject of intense controversy in recent years. Although many crucial experiments indicating this function have been performed using mitochondria or submitochondrial particles,[1,3,4] extremely important supportive evidence has been obtained using purified mitochondrial cytochrome oxidase reconstituted into vesicles.[5,6] The chief empirical advantages of the latter system are due largely to the relative simplicity of its composition with respect to the intact systems. This latter feature has also led to reconstituted vesicles emerging as the system of choice for the measurement of H^+ translocation by bacterial cytochrome oxidases.[7–9]

In this chapter, the types of experiments employing reconstituted vesicles in the study of the H^+ pumping activity of cytochrome oxidase will be summarized, with particular attention to some important technical points.

Experimental Materials

Cytochrome oxidase is prepared from bovine heart mitochondria.

Asolectin: Commercially available soybean phospholipid extracts are subjected to successive acetone washes and ether extractions to remove free fatty acids and neutral hydrophobic components.[10] It is essential that

* Deceased August 2, 1985.

[1] M. Wikström, *Nature (London)* **266,** 271 (1977).

[2] P. Mitchell, "Chemiosmotic Coupling in Oxidative and Photosynthetic Phosphorylation." Glynn Research, Bodmin, England, 1966.

[3] M. Wikström and H. T. Saari, *Biochim. Biophys. Acta* **462,** 347 (1978).

[4] M. Wikström and K. Krab, *FEBS Lett.* **91,** 8 (1978).

[5] K. Krab and M. Wikström, *Biochim. Biophys. Acta* **504,** 200 (1978).

[6] R. P. Casey, *Biochim. Biophys. Acta* **768,** 319 (1984).

[7] M. Solioz, E. Carafoli, and B. Ludwig, *J. Biol. Chem.* **257,** 1579 (1982).

[8] R. B. Gennis, R. P. Casey, A. Azzi, and B. Ludwig, *Eur. J. Biochem.* **125,** 189 (1982).

[9] T. Yoshida and J. A. Fee, *J. Biol. Chem.* **259,** 1031 (1984).

[10] Y. Kagawa and E. Racker, *J. Biol. Chem.* **246,** 5477 (1971).

throughout this procedure a suitable antioxidant (e.g., 1 m*M* dithiothreitol) is present to avoid extensive lipid peroxidation.

Potassium cholate: Commercially available analytical grade cholic acid is further purified, e.g., by acid precipitation from solution in acetone, and dissolved at a concentration of 24.5 m*M* (10 mg/ml) with addition of KOH to maintain the pH at 7.

Ferrocytochrome *c*: A solution of high-grade ferricytochrome *c* from horse heart is reduced with dithionite and passed down a short Sephadex column to remove unreacted dithionite and reaction products. The concentration of ferrocytochrome *c* is determined from its absorbance at 550 nm minus 540 nm using an extinction coefficient of 21 mM^{-1} cm^{-1} (reduced minus oxidized[11]).

Other reagents are at the highest quality available.

Reconstitution of Cytochrome Oxidase in Phospholipid Vesicles

Cytochrome oxidase may be reconstituted in vesicles by a wide variety of techniques,[6] though the great majority of experiments on H^+ translocation have employed vesicles prepared by the detergent dialysis method. Although the details of this procedure may vary somewhat, the following description is given by way of illustration.

To 100 mg of asolectin is added 2.5 ml of 24.5 m*M* potassium cholate, 0.1 *M* HEPES/K^+,[11a] pH 7.4, and the suspension is sonicated to clarity under N_2 in an iced-water bath. Cytochrome oxidase is added to give a final concentration of 4–8 μM cytochrome aa_3 and the suspension transferred to a dialysis bag. The suspension is then dialyzed at 4° for 4 hr vs 100 volumes of 0.1 *M* HEPES/K^+, pH 7.4 then a further 4 hours vs 200 volumes of 10 m*M* HEPES/K^+, 39.6 m*M* KCl, 50.4 m*M* sucrose, pH 7.4, then a further 20 hr vs 200 volumes of 1 m*M* HEPES/K^+, 43.6 m*M* KCl, 55.4 m*M* sucrose, pH 7.4. By using these media, the K^+ activity and total osmolarity are kept constant throughout the preparation while the extravesicular HEPES is diluted 100-fold.

Although cholate is usually the detergent of choice, octylglucoside[8] and deoxycholate[12] also function well in this type of procedure. Similarly, pure phospholipid preparations may substitute for asolectin.[13]

[11] V. Massey, *Biochim. Biophys. Acta* **34,** 255 (1959).

[11a] Abbreviations: CCCP, carbonyl cyanide *m*-chlorophenylhydrazone; FCCP, carbonyl cyanide *p*-trifluoromethoxyphenylhydrazone; HEPES, *N*-2-hydroxyethylpiperazineethanesulfonic acid; TMPD, tetramethyl-*p*-phenylenediamine; TPP^+, tetraphenylphosphonium ion.

[12] N. Sone, M. Yoshida, H. Hirata, and Y. Kagawa, *J. Biochem.* **81,** 519 (1977).

[13] R. P. Casey and A. Azzi, *FEBS Lett.* **154,** 237 (1983).

A convenient check for the successful incorporation of cytochrome oxidase into intact vesicles is the acceleration of its electron-transport activity in the presence of $\Delta\mu_{H^+}$-collapsing ionophores such as a combination of a protonophore and valinomycin. The requirements of the K^+/valinomycin couple, in addition to the protonophore, for maximal accelerations of enzyme turnover is not fully understood, although there are some indications that K^+/valinomycin facilitates the diffusion of the negatively charged form of the proton carrier across the hydrophobic core of the phospholipid bilayer.[14] For vesicles to be useful in proton pumping measurements this acceleration (termed the respiratory control ratio) must be at least 4-fold.

Measurement of Cytochrome Oxidase H^+ Pumping Activity as Extravesicular pH Changes

Experimental Sample

Although the types of measurement of H^+ pumping may differ considerably in their technical features, the essential details of the experimental sample are usually approximately the same. Here, oxidase vesicles are suspended in a medium containing ~0.1 *M* monovalent salt or its equivalent, including at least 5 m*M* potassium. The latter is to prevent the buildup of $\Delta\psi$ during oxidase turnover; valinomycin is included to catalyze transmembrane K^+ equilibration, optimally at a concentration of from 0.8 to 1 nmol per μmol phospholipid[15] (assuming a phospholipid molecular weight of 700). Similarly, to avoid the buildup of a large transmembrane ΔpH, high internal buffering power is ensured by entrapping 0.1 *M* HEPES/K^+ or its equivalent.

pH Measuring System

Owing to the extremely low intravesicular volume, only proton translocation linked to the first few oxidase turnovers may be detected before a restrictive $\Delta\mu_{H^+}$ builds up, forcing the extruded protons to flow back rapidly into the vesicles. Thus, the extravesicular acidifications obtained (~0.02 pH units[16]) are considerably smaller than those often reported for pulse measurements using intact mitochondria where a much higher number of enzyme turnovers may be induced. If glass electrodes are em-

[14] T. A. O'Brien, D. Nieva-Gomez, and R. B. Gennis, *J. Biol. Chem.* **253,** 1749 (1978).

[15] R. P. Casey, P. S. O'Shea, J. B. Chappell, and A. Azzi, *Biochim. Biophys. Acta* **765,** 30 (1984).

[16] R. P. Casey, J. B. Chappell, and A. Azzi, *Biochem. J.* **182,** 149 (1979).

ployed as the pH monitoring system in this type of measurement, this requires the electrode and the associated electronics to be extremely sensitive, stable, and rapidly responding.

An alternative approach is to measure H^+ activity changes spectrophotometrically using pH-sensitive indicator dyes which have a pK_a close to 7, e.g., Phenol Red[16] and Bromocresol Purple.[5] A clear advantage of this approach is that it presents the possibility of carrying out H^+ translocation measurements in the millisecond time range.

Reductant Pulse Measurements

The most extensive measurements of H^+ translocation by cytochrome oxidase vesicles have been those where a relatively small number of enzyme turnovers is induced, either by the addition of reduced cytochrome *c* to an aerobic sample,[5,16,17] or of O_2 to an anaerobic sample in the presence of excess reduced substrate.[4,18] A typical reductant-pulse measurement using Phenol Red as the H^+ activity monitoring system is described below (see Fig. 1).

Cytochrome oxidase vesicles (0.1 ml) (0.3–0.5 nmol cytochrome aa_3) and valinomycin (5 nmol) are added to 1.4 ml of 75 m*M* choline chloride, 25 m*M* KCl, 50 μM Phenol Red, pH 7.4, at 15°, stirred continuously in a cuvette. Changes in pH are monitored by following the absorbance difference at 556.5 − 504.5 nm; these wavelengths have been carefully chosen to give minimal interference from cytochrome *c* absorbance. Sufficient ferrocytochrome *c* for 1 or 2 turnovers (here, 5 μl of 0.52 m*M*) is added, resulting in a transient acidification of the extravesicular medium. To determine the precise amount of H^+ extruded, the pulse of acidification should be extrapolated to the point of complete cytochrome *c* oxidation. The change in $A_{556.5-504.5}$ may be calibrated using a standard pulse of acid to indicate the *extravesicular* buffering power. Here, the extravesicular acidification induced by acid addition immediately begins to decay, indicating equilibration of the acid with the internal buffer, to a final level which indicates the *total* buffering power. Thus, in order to obtain an accurate value for the external buffering power, this acidification must be extrapolated to zero time. In the experiment of Fig. 1, the cytochrome *c*-induced acidification corresponds to 0.8 H^+ per electron before extrapolation.

This measurement is then repeated in the presence of 3.5 μM CCCP. Addition of 5 μl of ferrocytochrome *c* to the sample will now cause a rapid

[17] J. T. Coin and P. C. Hinkle, *in* "Membrane Bioenergetics" (C. P. Lee, L. Ernster, and G. Schatz, eds.), p. 405. Addison-Wesley, Reading, Massachusetts, 1979.

[18] T. Penttilä, *Eur. J. Biochem* **133,** 355 (1983).

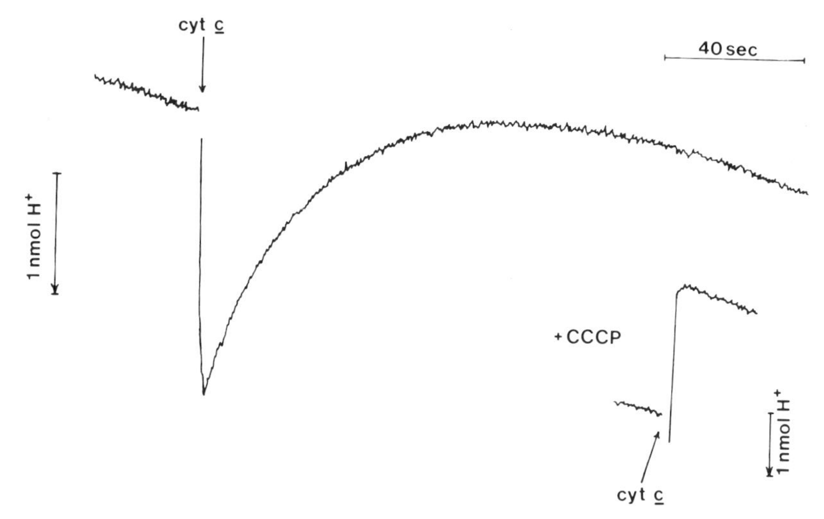

FIG. 1. Ferrocytochrome *c*-induced H^+ extrusion from cytochrome oxidase vesicles. 0.1 ml of cytochrome oxidase vesicles (0.33 nmol enzyme) and 5 nmol valinomycin were added to 1.4 ml of 75 m*M* choline chloride, 25 m*M* KCl, 50 μM Phenol Red, pH 7.4, at 25°. For the lower trace, 3.5 μM CCCP was also present. At the points indicated 2.6 nmol of ferrocytochrome *c* were added, and the resulting changes in H^+ activity were followed as the changes in absorbance of Phenol Red at 556.5 − 504.5 nm. The vertical discontinuities represent small step absorbance changes resulting from cytochrome *c* addition (see text). The calibration bars to the left and right of the traces show the absorbance changes caused by the addition of 1 nmol of H^+ in the absence or presence, respectively, of CCCP.

alkalinization; the acidification phase is no longer detectable owing to the extremely high membrane proton permeability in the presence of CCCP. The alkalinization should correspond to the consumption of *exactly* 1 proton per electron transferred. This strictly stoichiometric proton uptake is dictated by the chemistry of water formation through O_2 reduction, i.e., in this case, 4 ferrocytochrome $c + O_2 + 4H^+ \rightarrow 4$ ferricytochrome c + $2H_2O$.

Any deviation from this value indicates a net uptake or release of protons, and the stoichiometry of H^+ extruded (i.e., in the absence of CCCP) must be corrected for this artifactual contribution. This alkalinization may be calibrated by addition of a standard pulse of acid to indicate the *total* buffering power of the sample.

Even at the optimal wavelengths used there is a small step change in $A_{556.5-504.5}$ caused by the addition of cytochrome *c* to initiate enzyme turnover. To determine the size of this step artifact, 5 μl of cytochrome *c* is added to 1.5 ml of 75 m*M* choline chloride, 25 m*M* KCl, 5 m*M* HEPES, pH 7.4; the change in the $A_{556.5-504.5}$ gives the artifactual absorbance change (A) due to ferrocytochrome *c* addition. Then 5 μl of water is added to 1.5 ml of 75 m*M* choline chloride, 25 m*M* KCl, and 50 μM Phenol Red, pH 7.4; the change in absorbance at 556.5 − 504.5 nm gives the artifactual absorbance change (B) due to dilution of the Phenol Red. Addition of 5 μl of ferrocytochrome *c* should now give a small step change in $A_{556.5-504.5}$ of size (A + B). If there is any deviation from this, then the pH of the cytochrome *c* solution must be adjusted until the step change has the predicted magnitude.

Cytochrome *c* is itself strongly charged, and thus the possibility arises that the protons appearing extravesicularly on addition of cytochrome *c* may be released from the vesicle surface owing to a differential interaction with the oxidized and reduced cytochrome[19] (though this would have to be uncoupler sensitive; see above). This possibility may be tested by adding ferrocytochrome *c* to vesicles in the presence of 2 m*M* NaN_3, which strongly inhibits electron flow through the oxidase. Subsequent addition of 30 μM potassium ferricyanide causes the cytochrome *c* to become rapidly oxidized, bypassing the oxidase; this procedure normally leads to no measurable changes in the external pH.[7,13,16] In principle, cyanide should substitute for azide in this experiment, though one must take into account the fact that cyanide does not inhibit the first turnover of cytochrome oxidase in its oxidized, "resting" state.[20]

A possible pitfall in the reductant pulse measurement of H^+ translocation arises from the oxidation of cytochrome *c* by peroxidation products which are inevitably present in the vesicle phospholipid,[21] a reaction which under some conditions is proton consuming (see, e.g., Fig. 2). Although this interfering reaction is usually much slower than oxidation of cytochrome *c* by the oxidase, it may lead to considerable errors in the measurement of H^+ extrusion under conditions where the enzyme is turning over slowly, e.g., at low ionic strength. The consumption of cytochrome *c* without consequent H^+ extrusion and the concomitant uptake of protons by this process will cause an additive underestimation of the true H^+/electron ratio.

O_2 Pulse Measurements

An alternative approach to the limited turnover measurement of H^+ translocation by cytochrome oxidase vesicles is to add a small pulse of O_2 to an anaerobic vesicle suspension. The essential features of the experi-

[19] P. Mitchell and J. Moyle, *FEBS Lett.* **151,** 167 (1983).

[20] K. J. H. van Buuren, P. F. Zuurendonk, B. F. van Gelder, and A. O. Muijsers, *Biochim. Biophys. Acta* **256,** 243 (1972).

[21] M. de Cuyper and M. Joniau, *Eur. J. Biochem.* **104,** 397 (1980).

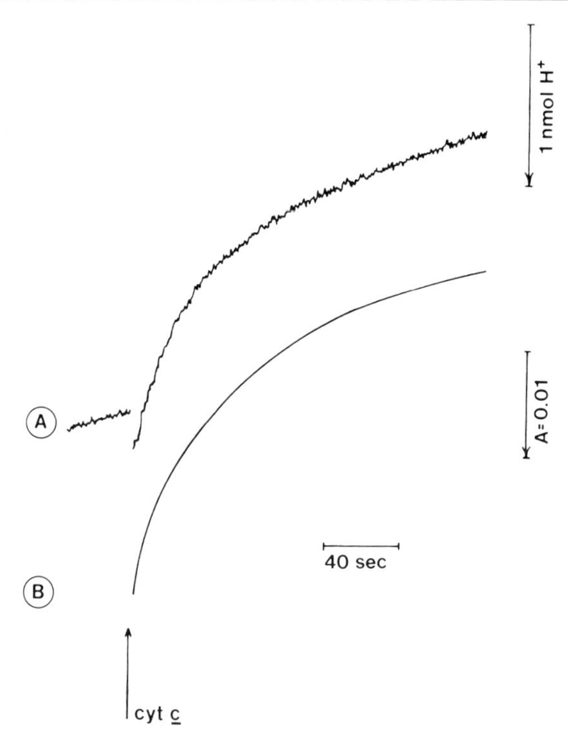

FIG. 2. The proton-consuming oxidation of cytochrome *c* by phospholipid vesicles. 0.1 ml of phospholipid vesicles, prepared by the same method as for cytochrome oxidase vesicles (see text), but in the absence of enzyme, and 5 nmol of valinomycin were added to 1.4 ml of 75 m*M* choline chloride, 25 m*M* KCl, 50 μM Phenol Red, pH 7.4, at 25° in the presence (B) or absence (A) of 10 m*M* HEPES. At the point indicated, 2.5 nmol of ferrocytochrome *c* were added. Trace A shows the resulting alkalinization as the change in absorbance of Phenol Red at 556.5 − 504.5 nm. Trace B shows the concomitant oxidation of cytochrome *c* as the change in absorbance at 550 − 540 nm. (NB: An upward change represents a decrease in $A_{550-540}$.) Under these conditions the proton consumption accompanying cytochrome *c* oxidation corresponded to ~0.5 H^+/electron.

mental sample are similar to those for a reductant pulse experiment (see above), the main differences being that an excess of ferrocytochrome *c* (or ascorbate plus cytochrome *c*) is present to keep the system reduced and that the sample is made anaerobic before O_2 addition. In assessing the data from this type of experiment, it is essential to correct the pH changes resulting from cytochrome *c* oxidation for the *scalar* release of protons (0.5 per electron at pH 7) from ascorbate as the cytochrome *c* becomes re-reduced. Under the conditions described, however, this net proton release is much slower than the pH changes linked directly to oxidase turnover, and thus is clearly separated kinetically. Similarly, the rapid release of one proton per electron when naphthoquinol is used as the electron donor to cytochrome *c*[22] must also be taken into account. It is advisable not to include TMPD to catalyze electronic equilibration between ascorbate and cytochrome *c*; this substance being membrane permeant may reduce oxidase molecules oriented inwardly in the vesicle membranes[23] and thus cause a transient proton *uptake* consequent on O_2 addition.

An advantage of the O_2 pulse measurement is that any artifactual proton release resulting from differential binding of oxidized or reduced cytochrome *c* to the vesicles[19] is automatically excluded, provided a reductant for cytochrome *c* is present, as the redox state of the cytochrome *c* is the same before and after enzyme turnover.

Comparative Rate Measurements

Here instead of comparing the extents of H^+ extrusion and electron transfer, the rate of H^+ pumping is determined relative to that of O_2 consumption,[24] the latter being measured with a rapidly responding oxygen electrode. The composition of the sample is essentially as for the other measurements in this section except that a low concentration of cytochrome *c* (~2 μM) is present before initiation of enzyme turnover by the addition of a pulse of ascorbate and TMPD to give final concentrations of ~1 m*M* and 100 μM, respectively.

Measurement of the Number of Charges Translocated per Electron Transferred

A crucial test of H^+ translocation by cytochrome oxidase which is not affected by any hypothetical proton-release artifacts is the determination of the charge-translocation stoichiometry. Any proton ejection due to genuine transport will entail a loss of intravesicular positive charges in excess of the 1 per electron due to the consumption of 1 *internal* H^+ per electron in water formation.[6,25]

Measurement of charge translocation may be achieved using experimental approaches such as reductant pulse,[17] O_2 pulse,[18] or rate[24] measurements. In addition, here an ion-sensitive electrode is employed to

[22] P. C. Hinkle, this series, Vol. 55, p. 748.

[23] R. P. Casey, B. H. Ariano, and A. Azzi, *Eur. J. Biochem.* **122,** 313 (1982).

[24] E. Sigel and E. Carafoli, *Eur. J. Biochem.* **89,** 119 (1978).

[25] P. Mitchell and J. Moyle, *in* "Biochemistry of Mitochondria" (E. C. Slater, Z. Kanjuga, and L. Wojtczak, eds.), p. 53. Academic Press, London, 1967.

monitor the uptake of extravesicular permeant cations, e.g., K^+ (in the presence of valinomycin) or TPP^+. The essential features of the sample are as described above except that the measuring cations are present at concentrations within the range of sensitivity of the respective electrode, i.e., ~0.5–2 m*M* for K^+ and 1–50 μ*M* for TPP^+.[15,26] By way of illustration, the determination of the charge-translocation stoichiometry using O_2 pulses[18] will be considered.

Two milliliters of 200 m*M* sucrose, 100 m*M* HEPES/Na^+, 0.5 m*M* KCl, 7 m*M* Tris/ascorbate, 4.2 μ*M* cytochrome *c*, 1.2 μ*M* valinomycin, pH 7.0, containing cytochrome oxidase vesicles (2–2.5 nmol of enzyme) at 25° are made anaerobic by bubbling with argon. Changes in K^+ activity are monitored using a K^+-sensitive, liquid-membrane electrode in conjunction with a reference electrode filled with saturated choline chloride; the response is calibrated using standard pulses of KCl. Enzyme turnover is initiated by the addition of 10 μl of aerated suspension medium (corresponding to 2.55 nmol of O_2). There follows a transient decrease in extravesicular K^+ activity; the size of this decrease relative to the quantity of added electron acceptor indicates the extent of total positive charge translocation; the uptake of one positive charge per electron is accounted for by internal charge consumption in water formation (see above), and thus any K^+ uptake in excess of this is indicative of H^+ translocation from the vesicle interior. Published reports[18,27] have shown the total charge translocation to be close to 2 per electron transferred. In the presence of 0.7 μ*M* FCCP, any changes in K^+ activity resulting from O_2 addition should be abolished.

If a TPP^+-sensitive electrode is used to measure charge translocation, then the relatively low effective TPP^+ concentration range must be taken into account. Owing to the extremely low intravesicular volume, a limiting gradient of TPP^+ activity may be established before the transmembrane $\Delta\psi$ resulting from oxidase turnover has been fully dissipated. This may consequently lead to underestimations in the true charge per electron stoichiometry.

[26] N. Kamo, M. Muratsugu, R. Hongoh, and Y. Kobatake, *J. Membr. Biol.* **49,** 105 (1979).

[27] T. Penttilä and M. Wikström, *in* "Vectorial Reactions in Electron and Ion Transport in Mitochondria and Bacteria" (F. Palmieri, E. Quagliariello, E. C. Slater, and N. Siliprandi, eds.), p. 71. Elsevier, Amsterdam, 1981.

[3] Structure of Beef Heart Cytochrome-c Oxidase Obtained by Combining Studies of Two-Dimensional Crystals with Biochemical Experiments

By RODERICK A. CAPALDI and YU-ZHONG ZHANG

Cytochrome-*c* oxidase (EC 1.9.3.1) is the terminal enzyme in the mitochondrial electron transfer chain, catalyzing the four electron reduction of molecular oxygen and coupling this reaction to the generation of a proton gradient across the mitochondrial inner membrane.[1-3] The enzyme has been isolated from several prokaryote and eukaryote sources, but is best characterized structurally from beef heart. Cytochrome-*c* oxidase is readily isolated in gram quantities from this source. Moreover, as purified under certain conditions, the beef heart enzyme is arranged in two-dimensional crystals, and this allows for detailed structural analysis.

Two-dimensional crystals of beef heart cytochrome-*c* oxidase have been made by three different procedures, one involving purifying the enzyme in Triton X-114 followed by Triton X-100,[4,5] the second using deoxycholate alone as the detergent,[6,7] and the third using cholate followed by deoxycholate.[8]

Preparation of Two-Dimensional Crystals by Triton Detergent Treatment of Mitochondria

The use of Triton detergents to isolate cytochrome-*c* oxidase originated with Crane and colleagues,[9,10] and the method to prepare two-

[1] M. Wikström, K. Krab, and M. Saraste, "Cytochrome Oxidase: A Synthesis." Academic Press, New York, 1981.

[2] R. A. Capaldi, F. Malatesta, and V. M. Darley-Usmar, *Biochim. Biophys. Acta* **726,** 135 (1983).

[3] A. Azzi, *Biochim. Biophys. Acta* **594,** 231 (1980).

[4] G. Vanderkooi, A. E. Senior, R. A. Capaldi, and H. Hayashi, *Biochim. Biophys. Acta* **274,** 38 (1972).

[5] R. Henderson, R.A. Capaldi, and J. S. Leigh, *J. Mol. Biol.* **112,** 631 (1977).

[6] S. Seki, H. Hayashi, and T. Oda, *Arch. Biochem. Biophys.* **138,** 110 (1970).

[7] S. D. Fuller, R. A. Capaldi, and R. Henderson, *J. Mol. Biol.* **134,** 305 (1979).

[8] S. D. Fuller, R. A. Capaldi, and R. Henderson, *Biochemistry* **21,** 2525 (1982).

[9] F. F. Sun, K. S. Prezbindowski, F. L. Crane, and E. E. Jacobs, *Biochim. Biophys. Acta* **153,** 804 (1968).

[10] E. E. Jacobs, E. C. Andrews, W. Cunningham, and F. L. Crane, *Biochem. Biophys. Res. Commun.* **25,** 87 (1966).

dimensional crystals follows closely the isolation procedure described by Sun *et al.*[9] In this method, beef heart mitochondria are first treated with Triton X-114 at a concentration of 0.6 mg/mg protein in a buffer of 0.25 *M* sucrose, 10 m*M* Tris–HCl, pH 7.8. Solid KCl is added to 0.2 *M* and the solution allowed to incubate on ice. The pellet, recovered by centrifugation at 78,000 *g* for 20 min, is resolubilized in Triton X-100 (1 mg/mg protein) in the same buffer. Solid KCl is added to 1 *M* and the suspension incubated at 4° for 20 min. A green pellet is again collected by centrifugation at 105,000 *g*, and this is examined in the electron microscope. In our hands the pellet, after this first Triton X-100 step, shows two-dimensional crystals in one in five preparations. If no crystalline structures are seen or if the arrays that are present are small, the Triton X-100 step is repeated. This second treatment improves crystallization, presumably by removing more phospholipids from the membranous preparation. Our success rate with two Triton X-100 steps has been to obtain good crystalline arrays 35–45% of preparation.

Preparation of Two-Dimensional Crystals with Deoxycholate as Detergent[7]

For this procedure, mitochondria are suspended in 0.66 *M* sucrose, 50 m*M* Tris–HCl, 1 m*M* histidine (pH 8.0), and sodium deoxycholate (10% w/v solution) is added slowly with stirring to a final concentration of 0.3 mg/mg protein. Solid KCl is then added to a final concentration of 72 g/liter and the suspension incubated for 12–24 hr at 4°. The suspension is centrifuged at 105,000 *g* for 30 min and the middle dark green layer of the three-layered pellet retained. This is resuspended to 20 mg/ml in the sucrose Tris–histidine buffer, and sodium deoxycholate (10% w/v) is added slowly with stirring on ice to between 0.3 and 1.0 mg/mg protein. This suspension is incubated for 2 hr and then centrifuged at 105,000 *g* for 30 min to collect the crystalline, membranous pellet. The amount of deoxycholate to be added in the last step can be estimated from the appearance of the suspension. Crystalline preparations have been obtained most often by adding just enough deoxycholate to convert the suspension from very turbid and brown in appearance to lightly turbid and greenish in color. Alternatively several aliquots can be used, adding 0.3, 0.5, 0.7, 0.9, and 1.0 mg/ml of deoxycholate to different samples and then examining the pellets obtained for crystals in the electron microscope. As with the Triton crystals, the protein-to-lipid ratio appears to be a key determinant in whether two-dimensional crystals are formed by the deoxycholate procedure.

Crystals from Purified Enzyme and Lipid[8]

Crystals can be obtained from isolated cytochrome-*c* oxidase and purified phospholipids by insertion of protein into preformed lipid vesicles. To prepare vesicles, L-α-phosphatidylcholine (0.5 ml, 100 mg/ml) in chloroform–methanol (2 : 1) is mixed with 5 μl of 10 m*M* butylated hydroxytoluene in ethanol (used to prevent lipid peroxidation), and then the solution is dried with nitrogen. Ethanol (1 ml) is added and the mixture dried first with nitrogen and then under vacuum for 1 hr. The lipid is then suspended in 10 ml of 50 m*M* Tris-acetate, 0.5 m*M* EDTA, and 5 μ*M* butylated hydroxytoluene (pH 7.0) by sonication with a Branston bath sonicator for ~1 hr, by which time the solution is optically clear. Purified cytochrome-*c* oxidase, made according to Steffens and Buse,[11] works best in the crystallization, probably because of the low lipid composition of the preparation. The enzyme (200 μl, 30 mg/ml) is added to 2.4 ml of the sonicated vesicles (5 mg/ml), and this mixture is stirred on ice for 12 hr and then dialyzed against 0.66 *M* sucrose, 50 m*M* Tris–HCl, and 5 μ*M* butylated hydroxytoluene for 7 hr to remove residual cholate.

The reconstituted vesicles are treated with 1 mg of deoxycholate per milligram protein, solid KCl is added to a final concentration of 1 *M*, and the solution is stirred on ice. After 48 hr, the reconstituted membranes are placed on top of a sucrose gradient (1.5–2.0 *M*/5 ml of total volume) in 10 m*M* Tris–HCl and 2 m*M* sodium deoxycholate (pH 8.0) and spun in an SW 50.1 rotor for 20 hr at 40,000 *g*. The crystals are collected in a tight band near 1.8 *M* sucrose.

Gross Structure of Cytochrome-c Oxidase as Revealed from Studies of Two-Dimensional Crystals

Figure 1 summarizes the packing arrangements of the different crystal forms. The enzyme is dimeric and inserted into a lipid bilayer in the Triton crystals[5] and monomeric and in detergent-rich sheets in the deoxycholate crystals.[7] The enzyme molecule is revealed in both crystal forms as characteristically Y-shaped and divided into three major domains as shown in Fig. 2. The C domain is around 55 Å tall (from the surface of the bilayer) and appears to be contoured into three subdomains.[5,7,12,13] The M_1 domain is larger than the M_2 domain, having a surface area in projection of about 900 Å compared to 650 Å for M_2. Low-angle X-ray studies show that the

[11] G. Steffens and G. Buse, *Hoppe-Seyler's Z. Physiol. Chem.* **357,** 1125 (1976).

[12] J. F. Deatherage, R. Henderson, and R. A. Capaldi, *J. Mol. Biol.* **158,** 501 (1982).

[13] J. F. Deatherage, R. Henderson, and R. A. Capaldi, *J. Mol. Biol.* **158,** 487 (1982).

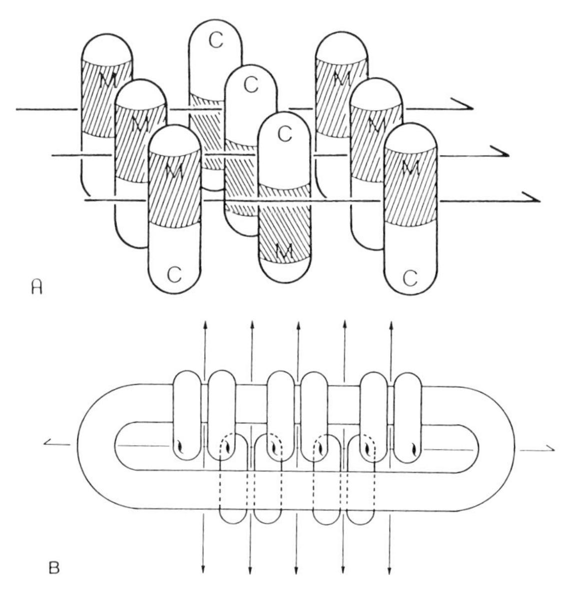

FIG. 1. Packing arrangement of two-dimensional crystals of cytochrome-c oxidase. (A) Deoxycholate crystals; (B) Triton crystals.

bilayer-intercalated part of cytochrome-c oxidase is organized as α-helices running approximately perpendicular to the plane of the membrane. The M_1 and M_2 domains are large enough to fit 9–13 and 7–9 closely packed α-helices, respectively. Neither of these domains extend far from the surface of the bilayer.

In the mitochondrial inner membrane, cytochrome-c oxidase is oriented with the C domain in the intracristal space and the M domains protruding a short way into the matrix space. Thus, most of the mass of cytochrome-c oxidase is on the side of the membrane at which substrate cytochrome c binds.

Topography of the Polypeptide Components of Cytochrome-c Oxidase Revealed by Biochemical Approaches

The resolution of the cytochrome-c oxidase structure obtained so far by electron microscopy and image analysis of the two-dimensional crystals is 20 Å, far lower than needed to see the heme and copper prosthetic groups or to follow the folding of polypeptide chains. However, the arrangement of these components within the cytochrome-c oxidase com-

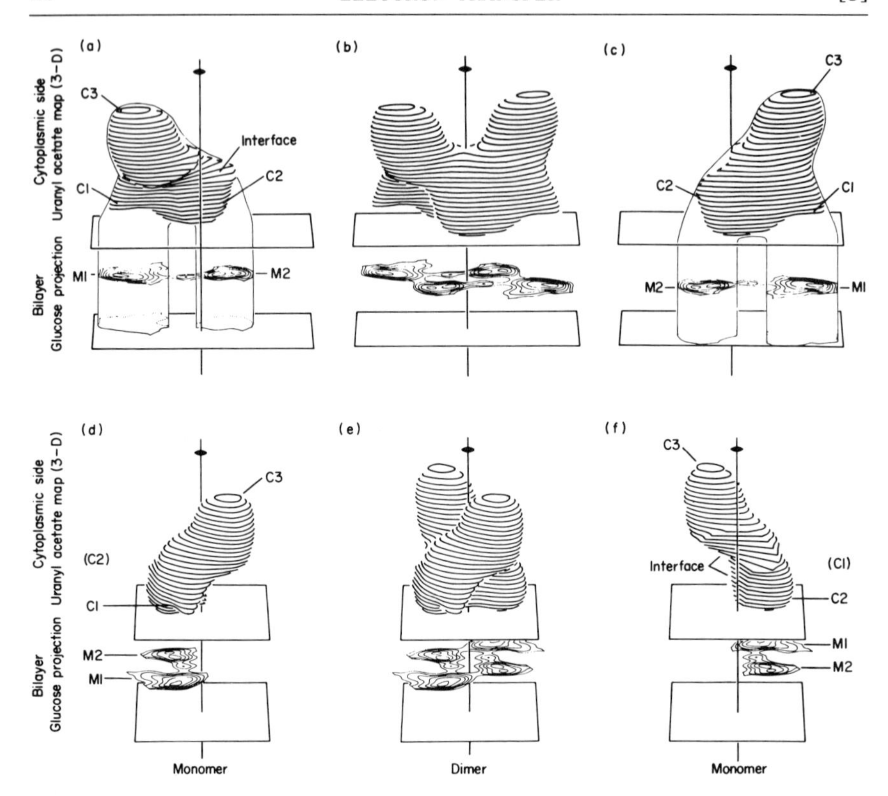

FIG. 2. Perspective drawings to show the Y-shaped structure of the cytochrome-*c* oxidase monomer and interaction between monomers in the Triton crystals. The upper and lower surfaces of the bilayer are represented by the planes perpendicular to the 2-fold axis. Reproduced from Deatherage *et al.*[12] with permission.

plex has been resolved in part by biochemical approaches such as protease digestion,[14–17] chemical labeling,[17–23] and cross-linking experi-

[14] Y.-Z. Zhang, G. Georgevich, and R. A. Capaldi, *Biochemistry* **23,** 5616 (1984).

[15] F. Malatesta, V. M. Darley-Usmar, C. deJong, L. Prochaska, R. Bisson, R. A. Capaldi, G. C. M. Steffens, and G. Buse, *Biochemistry* **22,** 4405 (1983).

[16] S. H. P. Chan and R. P. Tracy, *Eur. J. Biochem.* **89,** 595 (1978).

[17] F. Malatesta, G. Georgevich, and R. A. Capaldi, *in* "Structure and Function of Membrane Proteins" (F. Palmieri *et al.*, eds.), p. 223. Elsevier, New York, 1983.

[18] G. D. Eytan and R. Broza, *J. Biol. Chem.* **253,** 3196 (1978).

[19] G. D. Eytan, R. C. Carroll, G. Schatz, and E. Racker, *J. Biol. Chem.* **250,** 8598 (1975).

[20] B. Ludwig, N. W. Downer, and R. A. Capaldi, *Biochemistry* **18,** 1401 (1979).

[21] L. Prochaska, R. Bisson, and R. A. Capaldi, *Biochemistry* **19,** 3174 (1980).

[22] R. Bisson, C. Montecucco, H. Gutweniger, and A. Azzi, *J. Biol. Chem.* **254,** 9962 (1979).

[23] J. Jarausch and B. Kadenbach, *Eur. J. Biochem.* **146,** 219 (1985).

POLYPEPTIDES CONTAINED IN BEEF HEART CYTOCHROME-*c* OXIDASE

N terminus	Nomenclatures: Capaldi[28]	Nomenclatures: Kadenbach[26]	Nomenclatures: Buse[27]	MW	Yeast enzyme[48]	Site of synthesis[a]
MFI	I	I	I	56993	I	M
MAY	II	II	II	26049	II	M
MTH	III	III	III	29918	III	M
AHG	IV	IV	IV	17153	V	C
SHG	V	Va	V	12434	VI	C
ASG	a	Vb	VIa	10670	IV	C
ASA	b	VIa	VIb	9419	—	C
Ac-AED	c	VIb	VII	10068	—	C
STA	VI	VIc	VIc	8480	—	C
FEN	VIIs	VIIa	VIIIc	6244	[b]	C
SHY	VIIs	VIIc	VIIIa	5541	VII	C
ITA	VIIs	VIII	VIIIb	4962	[b]	C
—	—	VIIb	—	5900	(?)	C

[a] M, Mitochondrial; C, cytoplasmically synthesized.

[b] The yeast enzyme has three polypeptides in the molecular weight range 7000 and less, only one of which has been sequenced.

ments.[24,25] Each of these methods has its intrinsic difficulties, and the overall problem of defining the interactions among polypeptides is complicated by the number of components involved. Beef heart cytochrome-*c* oxidase can contain as many as 13 different polypeptides, depending on the method of isolation.[26-28] These are listed in the table by N terminal sequence, site of biosynthesis, molecular weight according to amino acid sequences, and by the various nomenclatures used for the different components. Finally, those polypeptides that have counterparts in yeast cytochrome-*c* oxidase are indicated.

A major point of uncertainty is whether all of the polypeptides listed in the table should be considered as subunits of cytochrome-*c* oxidase. This relates to a general difficulty in defining membrane proteins. If the cytochrome-*c* oxidase complex is defined as that group of polypeptides which copurifies in stoichiometric amounts with hemes a and a_3 and the copper atoms, then the enzyme contains 12 or 13 different subunits.[26,27]

[24] M. M. Briggs and R. A. Capaldi, *Biochemistry* **16,** 73 (1977).

[25] M. M. Briggs and R. A. Capaldi, *Biochem. Biophys. Res. Commun.* **80,** 553 (1978).

[26] B. Kadenbach, J. Jarausch, R. Hartmann, and P. Merle, *Anal. Biochem.* **129,** 517 (1983).

[27] G. Buse, G. C. M. Steffens, L. Meinecke, R. Biewald, and M. Erdweg, *Eur. Bioenerg. Conf. Rep.* **2,** 163 (1982).

[28] N. W. Downer, N. C. Robinson, and R. A. Capaldi, *Biochemistry* **15,** 2930 (1976).

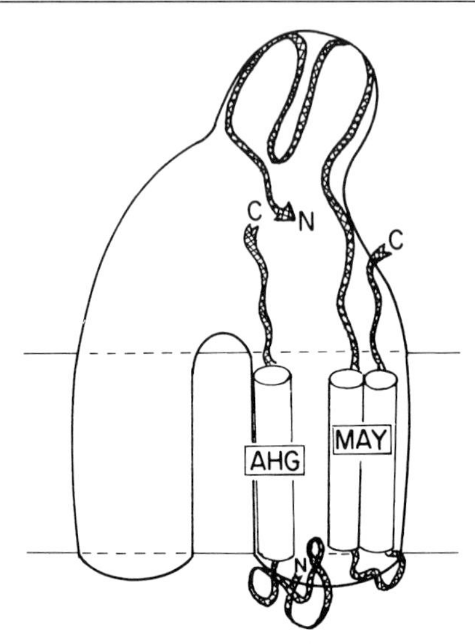

FIG. 3. Schematic of the folding of polypeptides beginning MAY (subunit II) and AHG (subunit IV) in the cytochrome-*c* oxidase complex.

However, at least two polypeptides (those with N-termini AED and ASA) can be removed with little or no alteration of electron transfer or proton pumping activity.[17,28] Therefore, if cytochrome-*c* oxidase is defined functionally, i.e., as the minimum unit competent in electron transfer from cytochrome *c* to molecular oxygen and able to couple this reaction to generation of a proton gradient across the membrane, then less subunits must be included in the complex.

By analogy with bacterial cytochrome-*c* oxidase, a functional enzyme could contain as few as three subunits, the three large mitochondrially synthesized polypeptides.[29] Of course, this disregards the probable role of the smaller nuclear-coded polypeptides in assembly of the complex, in regulating cytochrome-*c* oxidase activity, and possibly in energy-linked protein translocation across the mitochondrial inner membrane.

Figure 3 summarizes the folding of the polypeptides beginning MAY and AGH, respectively, in the plane or the membrane. These arrangements are based on labeling experiments with both water-soluble[14,20] and bilayer-intercalated protein modifying reagents,[17,21] on protease digestion[15] of intact mitochondria and submitochondrial particles, and on studies in reconstituted membranes of cytochrome-*c* oxidase and phospholipid.[14]

[29] B. Ludwig, *Biochim. Biophys. Acta* **594,** 177 (1980).

Most of the component polypeptides of cytochrome-*c* oxidase appear to span the membrane,[2,17] and there are, in all, around 20 long (19–23 residue) stretches of sequence containing predominantly hydrophobic amino acids in the different subunits (10 in the polypeptide that begins MFI, 2 in MAY, 6–7 in MTH, 1 each in AHG, FEN, SHY, and ITA).[2] These hydrophobic sequences probably identify transmembrane helices.

Locus of Prosthetic Groups in the Cytochrome-c Oxidase Complex

Cytochrome-*c* oxidase contains four prosthetic groups per monomer, two hemes (heme *a* and a_3), and two copper atoms. The available evidence suggests that heme *a* is liganded by two His[30] and Cu_a by two Cys and two His residues.[31] Heme a_3 and Cu_{a3} together form the oxygen-binding site and appear to be closely positioned in the same pocket. Isotopic substitution studies have established that the distal ligand to heme a_3 is an His while Cu_{a3} is liganded by two His residues.[32] These two closely placed centers also appear to be bridged by a common ligand, the candidates for which are a peroxy[33] or a sulfur[34] group. Several approaches have been used to determine which subunits of the enzyme provide the ligands for these metals. There is a general agreement that all four prosthetic groups are in polypeptides MFI and MAY. This is based mainly on the fact that bacterial cytochrome-*c* oxidases have prosthetic groups spectrally similar to the mammalian enzyme, but only two (or sometimes three) subunits, which have extensive sequence homology to the two largest polypeptides in beef and yeast cytochrome-*c* oxidase.[29] The most definitive location is for Cu_a. The locus of this metal center in polypeptide MAY is indicated by the analysis of Cys groups in the protein.[35] Only two Cys residues in the cytochrome-*c* oxidase are conserved in different species, as would be expected of ligands to a functionally critical prosthetic group.[35] These are in the C-terminal part of polypeptide MAY in a sequence with significant homology to the blue copper proteins.[36] A Cu_a site can be constructed using Cys-196, Cys-200, and His-

[30] G. T. Babcock, J. van Steelant, G. Palmer, L. E. Vickery, and I. Saleem, *Dev. Biochem.* **5,** 105 (1979).

[31] T. H. Stevens, C. T. Martin, H. Wang, G. W. Brudvig, C. P. Scholes, and S. I. Chan, *J. Biol. Chem.* **257,** 12106 (1982).

[32] S. I. Chan, G. W. Brudvig, C. T. Martin, and T. H. Stevens, *in* "Electron Transport and Oxygen Utilization" (C. Ho, ed.), p. 171. Elsevier, Amsterdam, 1982.

[33] W. E. Blumberg and J. Peisack, *Dev. Biochem.* **5,** 153 (1979).

[34] L. Powers, B. Chance, Y. Ching, and P. Angiolillo, *Biophys. J.* **34,** 465 (1981).

[35] V. M. Darley-Usmar, R. A. Capaldi, and M. T. Wilson, *Biochem. Biophys. Res. Commun.* **103,** 1223 (1981).

[36] G. Buse, *in* "Biological Chemistry of Organelle Formation" (T. H. Bucher *et al.*, eds.), Vol. 31, p. 59. Springer-Verlag, Berlin, 1983.

204 as ligands,[2] with the polypeptide chain folded into a loop similar to that of azurin and plastocyanin.[37,38] The fourth ligand would then be one of the other two conserved His residues, most likely His-161.

The locus of hemes *a* and a_3 and Cu_{a3} remains unclear. However, if four of the five conserved His, Cys, and Met residues in polypeptide MAY are involved in liganding Cu_a, there are insufficient liganding residues for other prosthetic groups in this polypeptide, implying that they are associated with polypeptide MFI instead.

The High-Affinity Site for Cytochrome c on the Cytochrome-c Oxidase Complex

The cytochrome-*c* oxidase monomer with its associated tightly bound phospholipids binds two molecules of cytochrome *c*, one with high affinity (10^{-8} *M*), the second with lower affinity (10^{-6} *M*).[39] The high-affinity site for cytochrome *c* has been localized by cross-linking[25,40–43] and chemical modification experiments.[44,45] Cytochrome *c* modified at Lys-13 with an arylazido group cross-links into the polypeptide beginning MAY close to His-161.[40,46] Chemical modification of cytochrome-*c* oxidase with the water-soluble carbodiimide 1-ethyl-3-[3(trimethylamino)propyl]carbodiimide (ETC) inhibits electron transfer by altering the K_m for cytochrome *c*,[44] as a result of modifying selected carboxyls in this same subunit. These carboxyls, Asp-112, Glu-114, and Glu-198, are shielded from reaction by ETC by the presence of cytochrome *c* in the high-affinity site.[45] Thus, the functional interaction of cytochrome *c* through the front face (heme cleft) of this electron donor to cytochrome-*c* oxidase is with polypeptide MAY. Cytochrome *c* can also be cross-linked through its back face to polypeptide MTH.[41–43] Given that the cytochrome *c* molecule is ~30 Å in diameter, the high-affinity site for this substrate must be a cleft of close to this size on the cytochrome-*c* oxidase molecule. The only such cleft seen in

[37] E. T. Adman and L. H. Jensen, *Isr. J. Chem.* **21,** 8 (1981).

[38] P. M. Colman, H. C. Freeman, J. M. Guss, M. Murata, V. A. Norris, J. A. M. Ramshaw, and M. P. Veukatuppa, *Nature* (*London*) **272,** 319 (1978).

[39] S. Ferguson-Miller, D. L. Brautigan, and E. Margoliash, *J. Biol. Chem.* **251,** 1104 (1976).

[40] R. Bisson, B. Jacobs, and R. A. Capaldi, *Biochemistry* **19,** 417 (1980).

[41] W. Birchmeier, C. E. Kohler, and G. Schatz, *Proc. Natl. Acad. Sci. U.S.A.* **73,** 4334 (1976).

[42] R. N. Moreland and M. E. Docktor, *Biochem. Biophys. Res. Commun.* **99,** 339 (1981).

[43] S. D. Fuller, V. M. Darley-Usmar, and R. A. Capaldi, *Biochemistry* **20,** 7046 (1982).

[44] F. Millett, V. M. Darley-Usmar, and R. A. Capaldi, *Biochemistry* **21,** 3857 (1982).

[45] F. Millett, C. deJong, L. Paulson, and R. A. Capaldi, *Biochemistry* **22,** 546 (1983).

[46] R. Bisson, G. C. M. Steffens, R. A. Capaldi, and G. Buse, *FEBS Lett.* **144,** 359 (1982).

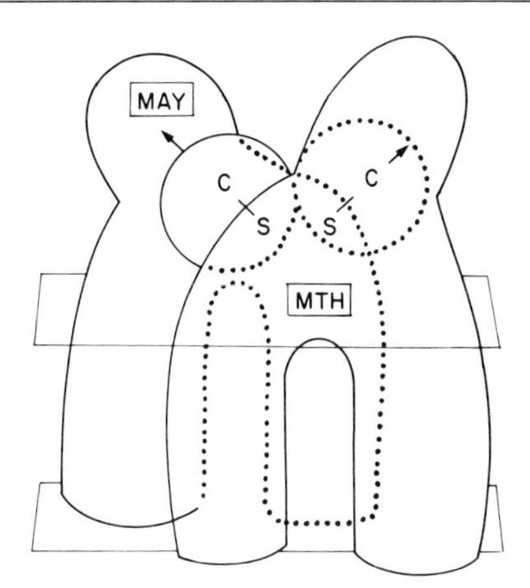

FIG. 4. Schematic depiction of the high-affinity binding site for cytochrome *c* in the cytochrome-*c* oxidase dimer based on cross-linking studies.

the low-resolution structure is between monomers in the cytochrome-*c* oxidase dimer[12,13] (Fig. 4).

Conclusions and Prospects

This brief review shows a picture of mammalian cytochrome-*c* oxidase built up by combining the data of biochemical studies with a low-resolution structure provided by electron microscopy and image reconstruction. It is now possible to combine the two approaches directly. Recently, electron microscopy has been done using ice rather than phosphotungstate or uranylacetate as the contrast stain for water against protein.[47] This allows higher resolution imaging of the protein than was possible previously, and should allow direct visualization of a cytochrome *c*–cytochrome-*c* oxidase complex. The locus of other polypeptides in the cytochrome-*c* oxidase complex can in principle be visualized by using monoclonal antibodies as an electron-dense stain, and it has recently proved possible to remove polypeptides, including that beginning MTH[17] (subunit III in all nomenclatures), and make two-dimensional crystals for analysis. These approaches are being followed in this laboratory to confirm and extend our present picture of beef heart cytochrome-*c* oxidase.

[47] R. A. Milligan, A. Brisson, and P. N. T. Unwin, *Ultramicroscopy* **13,** 1 (1984).

[48] S. D. Power, M. A. Lochrie, T. E. Patterson, and R. O. Poyton, *J. Biol. Chem.* **259,** 6571 (1984).

[4] Isozymes of Cytochrome-*c* Oxidase: Characterization and Isolation from Different Tissues

By BERNHARD KADENBACH, ANNEMARIE STROH, MARGIT UNGIBAUER, LUCIA KUHN-NENTWIG, URSULA BÜGE, and JOCHEN JARAUSCH

Cytochrome-*c* oxidase (EC 1.9.3.1), the terminal enzyme of the respiratory chain of most aerobic organisms (complex IV), is a membrane-bound protein complex composed of a variable number of subunits, depending on the evolutionary stage of the organism. Two to three different subunits have been identified in bacteria,[1,2] 9 in yeast,[3] and 13 in mammalian tissues.[4,5] In eukaryotes subunits I–III are encoded on mitochondrial, and subunits IV–VIII on nuclear DNA. The mitochondrially encoded polypeptides represent the catalytic subunits, since they contain the two heme *a* groups, the two copper ions, and the proton channel.[6,7] The nuclear encoded subunits are suggested to have a regulatory function.[8,9]

Tissue-specific isozymes of mammalian cytochrome-*c* oxidase have been detected, based on differences in (a) apparent molecular weight after SDS–gel electrophoresis,[10] (b) N-terminal amino acid sequence,[11,12] (c) immunological reactivity of nuclear encoded subunits,[13,13a] (d) kinetic properties,[14,40] and (e) reactivity of carboxylic groups in the presence and absence of cytochrome *c*.[15] The mitochondrially encoded subunits of cy-

[1] B. Ludwig, *Biochim. Biophys. Acta* **594,** 177 (1980).

[2] R. K. Poole, *Biochim. Biophys. Acta* **726,** 205 (1983).

[3] S. D. Power, M. A. Lochrie, K. A. Sevarino, T. E. Patterson, and R. O. Poyton, *J. Biol. Chem.* **259,** 6564 (1984).

[4] B. Kadenbach, J. Jarausch, R. Hartmann, and P. Merle, *Anal. Biochem.* **129,** 517 (1983).

[5] L. Kuhn-Nentwig and B. Kadenbach, *FEBS Lett.* **172,** 189 (1984).

[6] M. Wikström, K. Krab, and M. Saraste, "Cytochrome Oxidase: A Synthesis." Academic Press, New York, 1981.

[7] R. A. Capaldi, F. Malatesta, and V. M. Darley-Usmar, *Biochim. Biophys. Acta* **726,** 135 (1983).

[8] B. Kadenbach and P. Merle, *FEBS Lett.* **135,** 1 (1981).

[9] B. Kadenbach, *Angew. Chem.* **95,** 273 (1983); *Angew. Chem. Int. Ed. Engl.* **22,** 275 (1983).

[10] P. Merle and B. Kadenbach, *Hoppe-Seyler's Z. Physiol. Chem.* **361,** 1257 (1980).

[11] B. Kadenbach, R. Hartmann, R. Glanville, and G. Buse, *FEBS Lett.* **138,** 236 (1982).

[12] B. Kadenbach, M. Ungibauer, J. Jarausch, U. Büge, and L. Kuhn-Nentwig, *Trends Biochem. Sci.* **8,** 398 (1983).

[13] J. Jarausch and B. Kadenbach, *Hoppe-Seyler's Z. Physiol. Chem.* **363,** 1133 (1982).

[13a] L. Kuhn-Nentwig and B. Kadenbach, *Eur. J. Biochem.* **149,** 147 (1985).

[14] P. Merle and B. Kadenbach, *Eur. J. Biochem.* **125,** 239 (1982).

[15] B. Kadenbach and A. Stroh, *FEBS Lett.* **173,** 374 (1984).

tochrome-*c* oxidase from different tissues of the same organism are suggested to be identical.[16,40]

The following procedure for the isolation of cytochrome-*c* oxidase from mitochondria of different vertebrate tissues is based on a previously described method,[17] involving extraction of matrix proteins and most other membrane proteins with nonionic detergents,[18-20] chromatography on DEAE-cellulose in the presence of Triton X-100,[19] and ammonium sulfate fractionation in the presence of sodium cholate.[21-26] In our laboratory numerous cytochrome-*c* oxidase preparations from different mammalian and avian tissues and species always yielded enzymes containing 13 different subunits. It should be pointed out that with most previously described procedures similar enzyme preparations are obtained, containing also 13 different subunits. The discrepancy in the claimed number of subunits in publications from other laboratories (mainly seven) and those from our group is mainly due to incomplete separation by the SDS–gel electrophoretic system applied in other laboratories. Only by detergent treatment of the enzyme at alkaline pH[27,28] or by incubation with proteases[28-30] some subunits are separated from the detergent-solubilized complex, leading to preparations with less than 13 subunits. In recent publications the purification of mammalian cytochrome-*c* oxidase by affinity chromatography on Sepharose, containing covalently bound cytochrome *c*, has been described.[31-33] In our hands neither the purity nor the enzyme

[16] S. S. Potter, J. E. Newbold, C. A. Hutchinson, and M. H. Edgell, *Proc. Natl. Acad. Sci. U.S.A.* **72,** 4496 (1975).

[17] P. Merle and B. Kadenbach, *Eur. J. Biochem.* **105,** 499 (1980).

[18] E. E. Jacobs, E. C. Andrews, W. Cunningham, and F. L. Crane, *Biochem. Biophys. Res. Commun.* **25,** 87 (1966).

[19] E. E. Jacobs, F. H. Kirkpatrick, E. C. Andrews, W. Cunningham, and F. L. Crane, *Biochem. Biophys. Res. Commun.* **25,** 96 (1966).

[20] F. F. Sun, K. S. Prezbindowski, F. L. Crane, and E. E. Jacobs, *Biochim. Biophys. Acta* **153,** 804 (1968).

[21] T. Yonetani, *J. Biol. Chem.* **236,** 1680 (1961).

[22] T. Yonetani, this series, Vol. 10 [59].

[23] D. E. Griffiths and D. C. Wharton, *J. Biol. Chem.* **236,** 1850 (1961).

[24] D. C. Wharton and A. Tzagoloff, this series, Vol. 10 [45].

[25] L. R. Fowler, S. H. Richardson, and Y. Hatefi, *Biochim. Biophys. Acta* **96,** 103 (1962).

[26] C. R. Hartzell, H. Beinert, B. F. van Gelder, and T. E. King, this series, Vol. 53 [10].

[27] M. Saraste, T. Penttilä, and M. Wikström, *Eur. J. Biochem.* **115,** 261 (1981).

[28] T. Pentillä, *Eur. J. Biochem.* **133,** 355 (1983).

[29] R. C. Carroll and E. Racker, *J. Biol. Chem.* **252,** 6981 (1977).

[30] B. Ludwig, N. W. Downer, and R. A. Capaldi, *Biochemistry* **18,** 1401 (1979).

[31] H. Weiss and J. Kolb, *Eur. J. Biochem.* **99,** 139 (1979).

[32] K. Bill, C. Broger, and A. Azzi, *Biochim. Biophys. Acta* **679,** 28 (1982).

[33] D. A. Thompson and S. Ferguson-Miller, *Biochemistry* **22,** 3178 (1983).

yield, as obtained by the procedure described below, could be reached with the affinity chromatography method.

The following methods for the identification of cytochrome-*c* oxidase isozymes always include separation of the 13 subunits by SDS–gel electrophoresis,[4] representing the most difficult step. Unfortunately no other method is available yet for the complete separation of all subunits in one step.

Isolation of Cytochrome-c Oxidase from Different Tissues

Step 1: Isolation of Mitochondria

Mitochondria are isolated from liver, kidney, and brain,[34] and heart and diaphragm[35] by published procedures. Large-scale preparations of mitochondria from liver and kidney (pig and bovine) include a short homogenization of diced tissue in a Waring blender for 10 sec, instead of with a Potter–Elvehjem-type homogenizer.

Step 2: Extraction of Mitochondria with Triton X-114

All the following steps are carried out at 0°. About 1 g of mitochondrial protein is adjusted to a concentration of 25 mg/ml with isolation medium (0.25 *M* sucrose, 10 m*M* Tris–Cl, pH 7.4, 2 m*M* EDTA). After addition of 1 *M* potassium phosphate, pH 7.2, to a final concentration of 200 m*M*, a 20% (v/v) solution of Triton X-114 is added under stirring up to 0.5–1 mg/mg protein. The optimal amount is attained when the solution becomes translucent. Heart and muscle mitochondria usually need more Triton X-114 than liver or kidney mitochondria. The mixture is stirred for 30 min, centrifuged for 30 min at 250,000 *g* (av), and the supernatant is carefully poured off and discarded. An occasionally occurring lipid layer on top of the supernatant may be suctioned off and the wall may be cleaned with tissue. The green-brownish pellet is homogenized in a Potter–Elvehjem-type homogenizer in 200 m*M* potassium phosphate, pH 7.2, and centrifuged for 30 min at 250,000 *g*. A second wash with the same buffer containing in addition 0.1% Triton X-100 may be included for a more complete removal of cytochrome bc_1.

Step 3: Solubilization and Chromatography on DEAE–Cellulose

The sediment is dissolved in 20–40 ml 5% Triton X-100 (v/v), 200 m*M* potassium phosphate, pH 7.2, by homogenization using a Potter–Elveh-

[34] D. Johnson and H. Lardy, this series, Vol. 10 [15].
[35] A. L. Smith, this series, Vol. 10 [13].

TABLE I
PURIFICATION OF CYTOCHROME-*c* OXIDASE FROM RAT LIVER[a]

Fraction	Total protein (mg)	Total heme *a* (nmol)	Concentration heme *a* (nmol/mg protein)	Yield (%)
1. Mitochondria	600	126	0.21	100
2. Triton X-114 extraction	112	93	0.83	74
3. Triton X-100 supernatant	63	88	1.42	70
4. DEAE-Sephacel, green eluate	20	52	2.6	41
5. 38% ammonium sulfate precipitate	2.2	20	9.1	16

[a] Protein was determined by the biuret method (mitochondria) or by the Lowry method[38] (fraction 2–5) after precipitation with 90% acetone and once washing the sediment with distilled water. The heme *a* content was calculated from the difference spectra (dithionite-reduced minus air-oxidized) using a $\Delta\varepsilon_{605-630\ nm} = 12\ mM^{-1}\ cm^{-1}$ [G. von Jagow and M. Klingenberg, *FEBS Lett.* **24,** 278 (1972)].

jem-type homogenizer and centrifuged for 30 min at 250,000 *g*. The green supernatant is diluted with three volumes, double-distilled water and layered on a DEAE-Sephacel (Pharmacia) column (1.5 × 20 cm), equilibrated with 50 m*M* potassium phosphate, pH 7.2, 0.05% Triton X-100, and washed with two column volumes of the same buffer. Cytochrome-*c* oxidase is eluted as a green band with 200 m*M* potassium phosphate, pH 7.2, 0.05% Triton X-100.

Step 4: Fractionation with Ammonium Sulfate

The green fractions are combined and supplemented with solid sodium cholate to 1% final concentration (w/v). A saturated (0°) solution of neutral ammonium sulfate is slowly added under stirring until a slight, but persistent turbidity is formed (25–27% saturation) and the solution is kept overnight at 0°. After centrifugation for 15 min at 12,000 *g*, the supernatant is brought to 38% saturation by addition of saturated ammonium sulfate solution. The precipitated cytochrome-*c* oxidase is centrifuged after 30 min for 15 min at 12,000 *g*, dissolved in a small volume of 0.25 *M* sucrose, 10 m*M* Tris–Cl, pH 7.4, 2 m*M* EDTA (>20 mg/ml), and stored at −76°.

Yield and purification of a typical cytochrome-*c* oxidase preparation is presented in Table I for the rat liver enzyme.

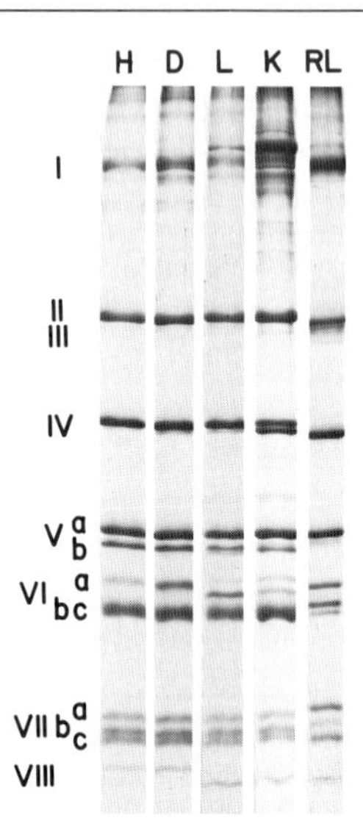

FIG. 1. Polypeptide pattern of isolated cytochrome-*c* oxidases from pig heart (H), diaphragm (D), liver (L), and kidney (K), and from rat liver (RL). SDS–gel electrophoresis was done as described.[4] The pig enzymes were precipitated with 80% acetone in order to remove the detergent, leading to aggregation of subunit III, not visible on the gel. In this gel subunits VIb and VIc of the pig enzymes are not separated. Subunit IV of kidney and VIa of liver and kidney show double bands, possibly due to different conformational states with different SDS/protein ratios. Enzyme preparations from liver and kidney usually show some high-molecular-weight impurities in contrast to those from muscle tissues.

Properties of Cytochrome-*c* Oxidase Isozymes from Different Tissues

Polypeptide Composition

With an efficient SDS–polyacrylamide gel electrophoretic system[4] isolated cytochrome-*c* oxidase from different vertebrate tissues can be resolved into 13 different polypeptides which are suggested to occur in stoichiometric (1 : 1) amounts.[17,36] This has been shown in our laboratory for enzyme preparations from liver, heart, kidney, skeletal muscle, and/or brain from rat, beechmarten, rabbit, deer, pig, bovine, human, and chicken. The polypeptide patterns of the isolated enzymes from pig heart, diaphragm, liver, and kidney and of rat liver are shown in Fig. 1. Tissue-specific differences in apparent molecular weights are found for subunit

[36] G. Buse, G. C. M. Steffens, L. Meinecke, R. Biewald, and M. Erdweg, *Eur. Bioenerg. Conf. Rep.* **2**, 163 (1982).

VIa and VIII between heart (diaphragm) and liver (kidney). With some isozymes, no complete separation of all subunits is obtained (e.g., subunits VIIb and VIIc from human heart). In this case subunit-specific antisera against the enzyme of another species (e.g., rat[5]) have been successfully used to demonstrate the presence of two different polypeptides in one and the same band.[37]

Spectral Properties

The visible and ultraviolet spectra of isolated isozymes from different pig tissues were found to be almost identical. However, some minor differences cannot be excluded. The isolated isozymes from fetal pig kidney and heart, obtained from the same organism and prepared in parallel, showed a variation of the maxima of the reduced and oxidized α- and γ-band of about 1–2 nm (unpublished results).

The ratio of absorbance A_{280}/A_{420} and the heme *a* content of isolated enzymes from four pig tissues are presented in Table II. No marked differences can be seen. The theoretical heme *a* content, based on a molecular weight of 204,000, would be 9.8 mol heme *a*/mg protein. The molecular weight of 204,000 is calculated from the known amino acid sequences of 12 bovine heart polypeptides[36] and an apparent molecular weight of 6000 for subunit VIIb. It should be pointed out, however, that an exact heme *a* content is difficult to determine due to inaccurate protein determination by the Lowry method.[38]

Kinetic Properties

In previous studies kinetic differences have been described between bovine heart and liver cytochrome-*c* oxidase.[14] The liver enzyme was found to have a higher activity as measured polarographically in the presence of deoxycholate and phospholipids. When the activity of isolated cytochrome-*c* oxidases from pig heart, diaphragm, liver, and kidney are measured photometrically in the presence of laurylmaltoside and without added phospholipids, no differences are found, as shown in Table II. Cytochrome-*c* oxidases from bovine liver and heart, after reconstitution in liposomes, however, do show differences in K_m and V_{max} values, if measured photometrically.[40] Some minor differences are also found when the cytochrome-*c* oxidase activity of total mitochondria is determined

[37] L. Kuhn-Nentwig and B. Kadenbach, submitted (1985).

[38] O. H. Lowry, N. J. Rosebrough, A. L. Farr, and R. J. Randall, *J. Biol. Chem.* **193,** 265 (1951).

[39] D. L. Brautigan, S. Ferguson-Miller, and E. Margoliash, this series, Vol. 53 [18].

[40] U. Büge, Dissertation, Fachbereich Chemie, Philipps-Universität, Marburg (1985).

TABLE II
PROPERTIES OF CYTOCHROME-*c* OXIDASES FROM DIFFERENT PIG TISSUES

	Heart	Diaphragm	Liver	Kidney
Spectral purity				
A_{280}/A_{420}	2.6	2.7	2.6	3.0
Heme *a* content[a] (nmol/mg protein)	9.5	8.6	8.7	7.8
Enzymatic activity[b]				
Photometric assay[39]				
Mitochondria, fresh isolated[c]	910	990	1360	1120
Isolated enzyme[c]	550	520	540	500
Polarographic assay[39]				
Isolated enzyme[d]	170	160	180	180
Isolated enzyme[e]	340	400	380	360

[a] Heme *a* was calculated from the difference spectra (dithionite-reduced minus air-oxidized) using a $\Delta\varepsilon_{605-630\ nm} = 12\ mM^{-1}\ cm^{-1}$ [G. von Jagow and M. Klingenberg, *FEBS Lett.* **24,** 278 (1972)]. Protein was determined by the Lowry method.[38]

[b] The activity was calculated as turnover number (mol cytochrome *c*/sec × mol cytochrome aa_3).

[c] Activity was measured at 25° in 80 m*M* potassium phosphate, pH 6.5, 2 m*M* EDTA, 0.25 m*M* laurylmaltoside, 40 μ*M* reduced cytochrome *c*, and 2 n*M* cytochrome-*c* oxidase.

[d] Activity was measured at 25° in 25 m*M* Tris–acetate, pH 7.6, 7 m*M* Tris-ascorbate, 0.014 m*M*, EDTA, 0.7 m*M* TMPD, 1 m*M* laurylmaltoside, 20 n*M* cytochrome-*c* oxidase, and was calculated for infinite cytochrome *c* concentration.

[e] Activity was measured at 25° in 25 m*M* Tris-acetate, 30 m*M* potassium phosphate, pH 7.6, 7 m*M* Tris-ascorbate, 0.014 m*M* EDTA, 0.7 m*M* TMPD, 1 m*M* laurylmaltoside, 40 μ*M* cytochrome *c*, and 20 n*M* cytochrome-*c* oxidase.

after solubilization in laurylmaltoside (Table II). No differences are observed, however, when the kinetic parameters are determined polarographically in the presence of laurylmaltoside and tetramethyl-*p*-phenylenediamine (TMPD)[41] at various cytochrome *c* concentrations, as shown in the Eadie–Hofstee plot in Fig. 2. The measurement of differences in the activity of the isozymes, due to different nuclear-coded "regulatory subunits,"[8,9] obviously requires native conditions, i.e., the absence of strong detergents, the presence of phospholipids, and the more "physiological"[39] photometric assay which, in contrast to the polarographic assay, does not involve artificial electron carriers.

The strong dependence of cytochrome-*c* oxidase activity on the assay conditions (Table II) is further demonstrated in Fig. 3, where the influence

[41] S. Ferguson-Miller, D. L. Brautigan, and E. Margoliash, *J. Biol. Chem.* **251,** 1104 (1976).

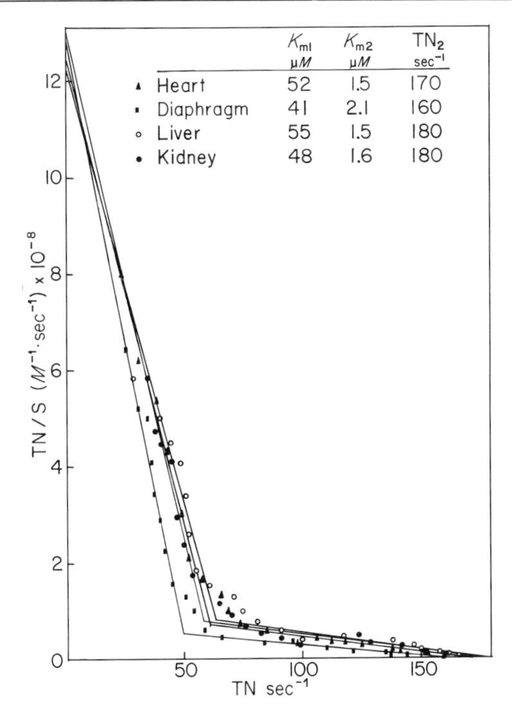

FIG. 2. Comparison of the kinetic properties of cytochrome-*c* oxidases from pig heart (▲), diaphragm (■), liver (○), and kidney (●). The activities, expressed as turnover number (TN, mol cytochrome *c*/sec × mol cytochrome aa_3), were measured polarographically[39] at 25° in 25 m*M* Tris-acetate, pH 7.6, 7 m*M* Tris-ascorbate, 0.014 m*M* EDTA, 1 m*M* laurylmaltoside, 0.7 m*M* TMPD, 0.02–40 μM cytochrome *c*, and 20 n*M* cytochrome-*c* oxidase. K_{m1} and K_{m2} represent the Michaelis constants for the high- and low-affinity binding sites for cytochrome *c*,[41] respectively, and TN_2 the activity at infinite cytochrome *c* concentrations.

of various anions on the oxygen uptake of pig heart cytochrome-*c* oxidase is presented. From this figure it is evident that in order to compare cytochrome-*c* oxidase activities, a thorough analysis of the assay conditions is required.

Reactivity of Carboxylic Groups of Isozyme Subunits in the Absence and Presence of Cytochrome c

The reaction of cytochrome-*c* oxidase with its positively charged substrate cytochrome *c* is mainly due to electrostatic interactions.[42] The binding domain for the oxidase at the cytochrome *c* surface has been mapped

[42] W. H. Koppenol and E. Margoliash, *J. Biol Chem.* **257,** 4426 (1982).

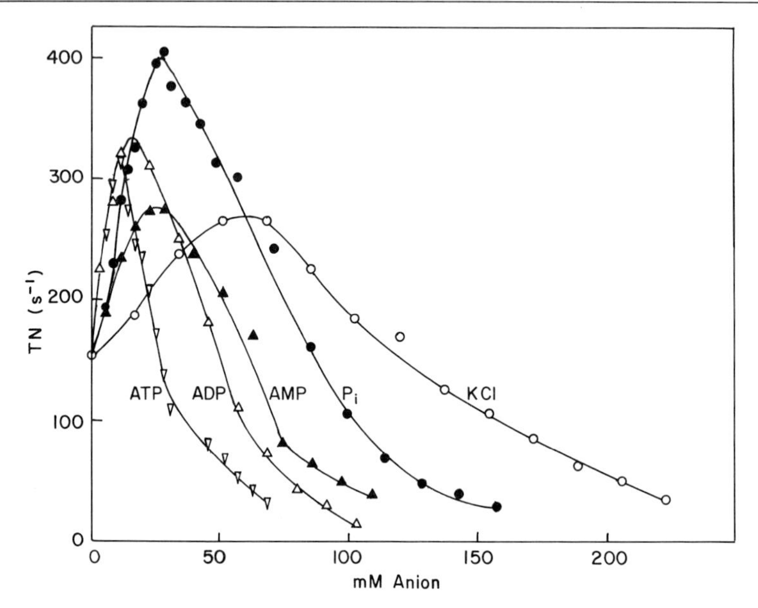

FIG. 3. Effect of different concentrations of various anions on the activity of isolated pig heart cytochrome-*c* oxidase. The activity was measured polarographically[39] at 25° in 25 m*M* Tris-acetate, pH 7.6, with or without the given concentration of the indicated anion, 7 m*M* Tris-ascorbate, 0.7 m*M* TMPD, 0.014 m*M* EDTA, 1 m*M* laurylmaltoside, 40 μM cytochrome *c*, and 20 n*M* cytochrome-*c* oxidase.

by measuring the reaction of labeled acetic anhydride with specific lysines in the absence and presence of cytochrome-*c* oxidase.[43] Consequently the binding domain for cytochrome *c* at the cytochrome-*c* oxidase complex can be determined by measuring the reaction of [^{14}C]glycine ethyl ester with carboxylic groups of subunits in the absence and presence of cytochrome *c*, after their activation with 1-ethyl-3-(3-dimethylaminopropyl)carbodiimide (EDC). In previous investigations on the labeling of cytochrome-*c* oxidase with [^{14}C]glycine ethyl ester and EDC, an identification of subunits, labeled in addition to subunit II, was not obtained.[44–46] With the efficient SDS–gel electrophoretic separation system[4] a tissue-specific and cytochrome *c*-protected labeling of several nuclear-encoded subunits was found in addition to labeling of subunit II.[15] As shown in a typical experiment in Fig. 4, in the liver and kidney enzyme subunits VIa,

[43] R. Rieder and H. R. Bosshard, *J. Biol. Chem.* **255,** 4732 (1980).

[44] F. Millett, V. Darley-Usmar, and R. A. Capaldi, *Biochemistry* **21,** 3857 (1982).

[45] R. Bisson, and C. Montecucco, *FEBS Lett.* **150,** 49 (1982).

[46] F. Millett, K. de Jong, L. Paulson, and R. A. Capaldi, *Biochemistry* **22,** 546 (1983).

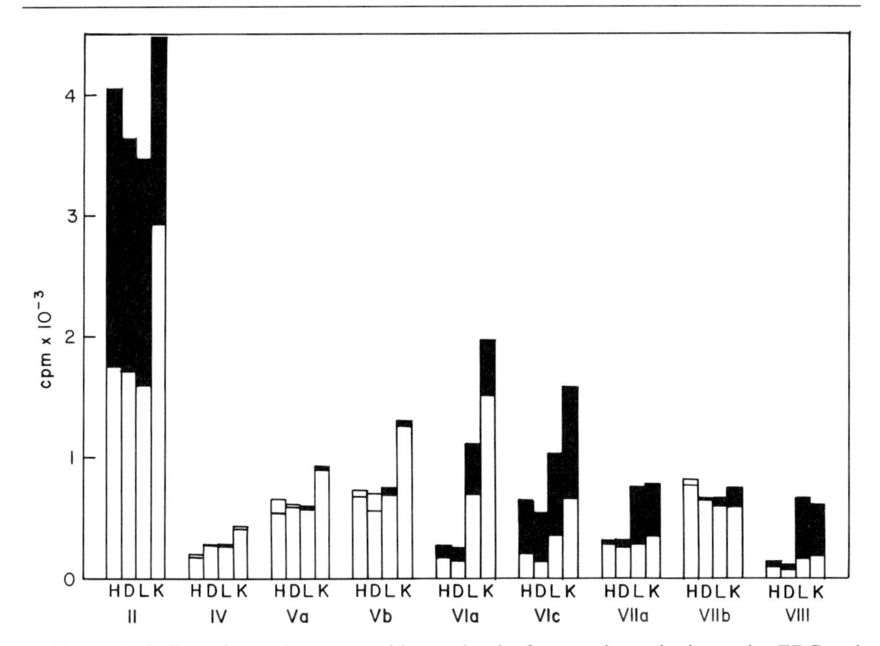

FIG. 4. Labeling of cytochrome-c oxidase subunits from various pig tissues by EDC and [^{14}C]glycine ethyl ester in the absence and presence of cytochrome c. H, Heart; D, diaphragm; L, liver; and K, kidney. Labeling conditions: 0.3 nmol of the indicated isolated cytochrome-c oxidase were incubated for 20 hr at 0° in a buffer containing 10 mM sodium phosphate, pH 7.0, 5 mM laurylmaltoside, 2 mM EDC, 0.5 mM [^{14}C]glycine ethyl ester (51.5 mCi/mmol, New England Nuclear, Dreieich, Federal Republic of Germany), and 0.3 nmol of cytochrome c, if indicated. The reaction was stopped by addition of ammonium acetate (0.1 M final concentration) and the enzyme was precipitated with acetone (80% final concentration). After SDS–gel electrophoresis,[4] the gel was dried and radioactive bands, visualized by fluorography,[47] were cut out and counted in a scintillation counter. In the figure each column shows the radioactivity bound to the indicated enzyme and subunit in the absence and presence of cytochrome c. The cytochrome c-protected labeling is represented by the black part of the columns.

VIIa, and VIII are labeled to a much higher degree than in the heart or diaphragm enzyme, and this labeling is largely suppressed in the presence of cytochrome c. Thus, cytochrome-c oxidase isozymes appear to differ in their binding domains for cytochrome c.

Immunological Characterization of Isozymes

The immunological relationship between enzymes of different species and tissues was investigated with a rabbit antiserum against isolated rat liver cytochrome-c oxidase by applying an immunoblotting method

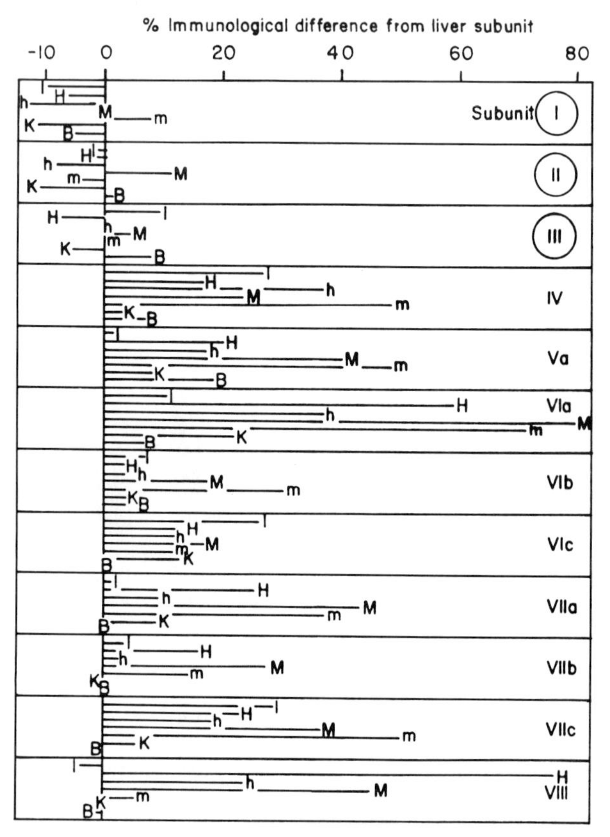

FIG. 5. Immunological relationship between corresponding subunits of cytochrome-*c* oxidase from different rat tissues. Monospecific antisera to the indicated subunits of the adult rat liver enzyme were titrated by a quantitative nitrocellulose ELISA with mitochondria from fetal liver (l), heart (h), and skeletal muscle (m), and from adult heart (H), skeletal muscle (M), kidney (K), and brain (B). The immunological reactivities were related to those of adult rat liver mitochondria, with 0% difference.[13a]

(Western blot). Comparison of cytochrome-*c* oxidases from different rat tissues showed a strongly reduced reactivity of subunit VIa of heart and skeletal muscle as compared to liver and kidney.[13]

Monospecific antisera against 12 of the 13 rat liver cytochrome-*c* oxidase subunits were applied in a quantitative nitrocellulose-ELISA to compare the immunological relationship of corresponding subunits from various rat tissues.[13a] As shown in Fig. 5, differences were observed for all nuclear-coded subunits between isozymes from various adult tissues and between fetal and adult enzymes of the same tissue (liver, heart, skeletal muscle). The mitochondrial coded catalytic subunits appear to be identical in all tissues.

Isolation of Subunits

Subunits of cytochrome-*c* oxidase isozymes can be isolated from Coomassie Brilliant Blue stained gel bands by extraction either with 1% SDS or with 66% acetic acid.[47] Slab gels (16 × 23 × 0.15 cm) are loaded with 1–2 mg enzyme. SDS–urea–gel electrophoresis, and staining and destaining of gels are performed as described.[4] In order to reduce carbamylation of amino groups by isocyanate,[48] the urea is deionized by use of a mixed-bed ion exchanger.

Extraction with 1% SDS.[49] Individual bands from 5–10 gels are cut out, dried under vacuum, and homogenized in 15–30 ml 1% SDS solution with a loose-fitting 30 ml Potter–Elvehjem-type homogenizer. The suspension is brought to pH 8.0 by addition of triethylamine and is shaken overnight. After centrifugation for 15 min at 15,000 *g*, the sediment is washed with distilled water and centrifuged again. The combined supernatants are lyophilized. Alternatively the protein is extracted from sliced gel bands by electroelution for 24 hr[50] with a buffer containing 0.1% SDS, 50 m*M* ethylmorpholine acetate, pH 9.0. The extract is dried under vacuum. SDS is removed by ion-pair extraction.[51] The dried residue is dissolved in a mixture of anhydrous acetone/triethylamine/acetic acid/water (85 : 5 : 5 : 5, v/v) equivalent to the volume of the original SDS solution. After standing overnight at 0°, the precipitated protein is centrifuged, washed once with the above mixture, once with acetone, and dried under vacuum. The yield of subunit protein varied between 5 and 20%.

Extraction with 66% Acetic Acid.[52] Stained gel bands are homogenized with a minimum volume of 66% acetic acid in a mortar, diluted with 5 volumes 66% acetic acid, and shaken for 2–3 days at 0°. After centrifugation for 20 min at 15,000 *g* the blue supernatant is removed and the sediment again extracted with 66% acetic acid. The combined supernatants are concentrated under vacuum to a final volume of about 1 ml. The extract is separated from Coomassie Brillant Blue and other impurities by gel chromatography on a BioGel P4 column (100–200 mesh, 1.8 × 21 cm),

[47] W. M. Bonner and R. A. Laskey, *Eur. J. Biochem.* **46,** 83 (1974).

[48] G. R. Stark and D. G. Smyth, *J. Biol. Chem.* **238,** 214 (1963).

[49] P. Merle, J. Jarausch, M. Trapp, R. Scherka, and B. Kadenbach, *Biochim. Biophys. Acta* **669,** 222 (1981).

[50] A. S. Bhown, J.E. Mole, F. Hunter, and J. C. Bennett, *Anal. Biochem.* **103,** 184 (1980).

[51] L. E. Henderson, S. Oroszlan, and W. Konigsberg, *Anal. Biochem.* **93,** 153 (1979).

[52] C. Barnabeu, F. P. Conde, and D. Vazquez, *Anal. Biochem.* **84,** 97 (1978).

TABLE III
DIFFERENT N-TERMINAL AMINO ACID SEQUENCES OF BOVINE AND PIG HEART AND LIVER CYTOCHROME-*c* OXIDASE SUBUNITS

Subunit	Source	N-terminal sequence
VIa	Bovine heart	Ala-Ser-Ala-Ala-Lys-Gly-Asp-His
	Bovine liver	Ser-Ser-Gly-Ala-His-Gly-Glu-Glu
VIIa	Bovine heart	Phe-Glu-Asn-Arg-Val-Ala-
	Bovine liver	Phe-Glu-Asn-Lys-Val-Ala-
	Pig heart	Phe- ? -Asn-Arg-Val-
	Pig liver	Phe-Glu-Asn-His-Val-
VIIb	Pig heart	Ser-Gly-Tyr-Ser-Val-Val-
	Pig liver	Ser-Gly-Tyr-Gly-
VIIc	Pig heart	Ser-His-Tyr-
	Pig liver	Ser-His-Ser-
VIII	Bovine heart	Ile-Thr-Ala-Lys-Pro-Ala-
	Bovine liver	Ile-His-Ser-Lys-Pro-Pro-
	Pig heart	Ile-Tyr-Ala- ? -Pro-Ala-Pro-Val
	Pig liver	Ile-His-Ser-Lys-Pro-Pro-Pro-Pro

equilibrated with 66% acetic acid. The protein fractions, detected at 280 nm, are combined, concentrated under vacuum, and stored frozen. Aliquots are taken for protein determination by the Lowry method[38] and for purity analysis by SDS–gel electrophoresis.[4] The yield of individual subunits varied between 20 and 40%. It should be pointed out that some subunits (in particular VIIb) tend to aggregate irreversibly in 66% acetic acid, i.e., remain insoluble in sample buffer, containing 8% SDS, or in 50% pyridine, used for manual Edman degradation. Therefore these solutions should not be stored for longer time.

N-Terminal Amino Acid Sequence Analysis

The complete amino acid sequence of bovine heart cytochrome-*c* oxidase has been determined for subunit II and for nine nuclear-coded subunits by Buse and co-workers.[36] The sequences of the mitochondrially encoded subunits I–II have also been deduced from the complete DNA sequence of the mitochondrial genome.[53] The subunits of bovine liver cytochrome-*c* oxidase were extracted from gels with 1% SDS (see above) and the N-terminal amino acid sequences of subunits VIa[11] and IV, Va, VIIa, and VIII were determined with an automatic sequencer by Buse (G. Buse and B. Kadenbach, unpublished results). Table III shows the N-terminal amino acid sequences of those subunits which are different in liver and heart enzymes from bovine and pig.[12] The subunits of the pig enzymes were extracted from gels by electroelution and sequenced by manual Edman degradation.[54] In 5 of the 10 nuclear-coded subunits, amino acid exchanges between the heart and liver enzyme were found, thus providing the molecular basis of tissue-specific isozymes of cytochrome-*c* in vertebrates.

[53] S. Anderson, M. H. L. de Bruijn, A. R. Coulson, I. C. Eperon, F. Sanger, and I. G. Young, *J. Mol. Biol.* **156,** 683 (1982).

[54] J. Y. Chang, D. Brauer, and B. Wittmann-Liebold, *FEBS Lett.* **93,** 205 (1978).

[5] Techniques for the Study of Bovine Cytochrome-c Oxidase Monomer–Dimer Association

By KATARZYNA A. NAŁĘCZ, REINHARD BOLLI, and ANGELO AZZI

Cytochrome-*c* oxidases isolated from different sources contain a variable number of subunits.[1,2] Enzymes, composed of at least 7 polypeptides, were isolated from bovine, rat, chicken, and camel heart as well as from the heart of several elasmobranch fish.[3–6] Recently, a method involving a high-resolution sodium dodecyl sulfate electrophoresis was developed[7,8] which allows the separation of 13 different polypeptides in cyto-

[1] A. Azzi, *Biochim. Biophys. Acta* **594,** 231 (1980).

[2] A. Azzi, K. Bill, R. Bolli, R. P. Casey, K. A. Nalecz, and P. O'Shea, *in* "Structure and Properties of Cell Membranes" (G. Benga, ed.), Vol. 2. CRC Press, Boca Raton, Florida, in press (1985).

[3] B. Ludwig, N. W. Downer, and R. A. Capaldi, *Biochemistry* **18,** 1401 (1979).

[4] M. T. Wilson, W. Lalla-Maharajh, V. Darley-Usmar, J. Bonaventura, C. Bonaventura, and M. Brunoir, *J. Biol. Chem.* **255,** 2722 (1980).

[5] V. M. Darley-Usmar, N. Alizai, A. I. Al-Ayash, G. D. Jones, A. Sharpe, and M. Wilson, *Comp. Biochem. Physiol.* **68B,** 445 (1981).

[6] G. Georgevich, V. M. Darley-Usmar, F. Malatesta, and R. A. Capaldi, *Biochemistry* **22,** 1317 (1983).

[7] P. Merle and B. Kadenbach, *Eur. J. Biochem.* **105,** 499 (1980).

[8] B. Kadenbach, J. Jaraush, R. Hartmann, and P. Merle, *Anal. Biochem.* **129,** 517 (1983).

chrome-*c* oxidases from different mammalian sources (beef and pig heart, and beef and rat liver). How many of the polypeptides in cytochrome-*c* oxidase preparations are true subunits (i.e., are necessary for complete functioning and regulation) is still unresolved.

From the N-terminal amino acid sequences of the separated polypeptides (Kadenbach *et al.*[8]) a 1 : 1 stoichiometry was suggested. The molecular weight of cytochrome-*c* oxidase was calculated to be 197,933, based on the molecular weight of the 13 polypeptides, taken either from the DNA sequences or from the amino acid sequences (see Ref. 8). A minimal molecular weight of about 200,000 can also be calculated from the usually reported heme *a*/protein ratio, in the vicinity of 10 $nmol \cdot mg^{-1}$ (for discussion of this calculation, see Wikström *et al.*[9]).

Ultracentrifugation studies on oxidases from a number of higher eukaryotic organisms have revealed that in nonionic detergents, such as Tween 80 or Triton X-100, the enzyme can be either monomeric or dimeric, or in both forms, depending on the organism. Naturally occurring monomers of cytochrome-*c* oxidase were found in enzymes isolated from hammerhead shark, rat, and camel,[4,5] and both molecular forms were found in chicken and dogfish enzyme.[4,5] The majority of the results on the molecular form of cytochrome-*c* oxidase were obtained from experiments either on the enzyme from elasmobranch fish or on that from bovine heart. In the first case the enzyme is mainly monomeric, while the aggregation state of bovine heart oxidase depends on experimental conditions, such as pH, salt, or detergent concentration and type.

At neutral pH and in the presence of nonionic detergents, the bovine heart oxidase is mainly in the form of dimers,[4,5,10–13] although at least two more independent species were detected by analytical ultracentrifugation of the enzyme in deoxycholate.[11] This influence of detergents on the aggregation state of cytochrome-*c* oxidase was confirmed by electron diffraction analyses of bidimensional crystals of oxidase. In crystals prepared in deoxycholate the enzyme exists as a monomer in detergent-rich sheets, and in crystals obtained in Triton X-100 the oxidase is a dimer inserted across the lipid bilayer (for references, see Capaldi *et al.*[14]). At

[9] M. Wikström, K. Krab, and M. Saraste, "Cytochrome Oxidase. A Synopsis," p. 38. Academic Press, London, 1981.

[10] B. Love, S. H. P. Chan, and E. Stotz, *J. Biol. Chem.* **245,** 6664 (1970).

[11] N. C. Robinson and R. A. Capaldi, *Biochemistry* **16,** 375 (1977).

[12] M. Saraste, T. Pentillä, and M. Wikström, *Eur. J. Biochem.* **115,** 261 (1981).

[13] P. Rosevear, T. VanAken, J. Baxter, and S. Ferguson-Miller, *Biochemistry* **19,** 4108 (1980).

[14] R. A. Capaldi, F. Malatesta, and V. M. Darley-Usmar, *Biochim. Biophys. Acta* **726,** 135 (1983).

alkaline pH (pH 9.5 and 10.5) in Emasol the enzyme has been reported to dissociate into monomers.[10] A similar effect of high pH was reported by Georgevich *et al.*,[6] after incubation of the oxidase at pH 8.5 in the presence of 5% Triton X-100.

Several factors seem to regulate interconversion of the bovine enzyme from the monomeric to the dimeric form. It was shown that an increase in ionic strength facilitated dimerization of the enzyme in dodecyl-β-D-maltoside.[15] Lowering pH and detergent concentration reversed the monomerization of oxidase, while monomers were obtained at high pH and high Triton X-100 concentration.[16]

The composition of the protein–lipid–detergent complex can influence the molecular form of the enzyme as well. It was proposed[17] that the monomerization of the enzyme could be correlated with lower lipid content.

Since the enzyme preparation obtained by chymotrypsin digestion[16] lacks subunit III and is monomeric (but cf. Ref. 12), a role for subunit III in determining the state of aggregation has been inferred.

Gel filtration and centrifugation studies (equilibrium centrifugation, sedimentation velocity measurements, and density gradient centrifugation) were applied most often to the study of the molecular form of cytochrome-*c* oxidase. Each of those methods has its own advantages and limitations. In all cases calculation of the molecular weight of the protein in regard to the amount of bound detergent and bound phopholipids is necessary.

Gel Filtration

Gel exclusion chromatography is a well-established technique used to separate the forms of an enzyme that differ in molecular weight. However, with asymmetric membrane proteins such as cytochrome-*c* oxidase, which is analyzed as a detergent–protein complex, molecular weights cannot be calculated simply on the basis of their elution position with reference to standard, water-soluble proteins. By interaction of the gel matrix with the enzyme or detergent, retardation[18] or even exclusion[19] may occur, producing significant errors in the estimated molecular

[15] K. A. Nalecz, R. Bolli, and A. Azzi, *Biochem. Biophys. Res. Commun.* **114,** 822 (1983).

[16] F. Malatesta, G. Georgevich, and R. A. Capaldi, *in* "Structure and Function of Membrane Proteins" (F. Palmieri, ed.), pp. 223–235. Elsevier, Amsterdam, 1983.

[17] R. Bolli, K. A. Nalecz, and A. Azzi, *Arch. Biochem. Biophys.* **240,** 102 (1985).

[18] Y. Nozaki, N. M. Schechter, J. A. Reynolds, and C. Tanford, *Biochemistry* **15,** 3884 (1976).

[19] M. Le Maire, E. Rivas, and J. V. Moller, *Anal. Biochem.* **106,** 12 (1980).

weights. Other methods are recommended in this case, such as hydrodynamic or light-scattering measurements to confirm the conclusions drawn from gel filtration experiments. Furthermore, the amount of bound detergent and lipid has to be established. According to our studies, purified bovine heart oxidase, when dispersed in dodecyl-β-D-maltoside, reveals an equilibrium between the monomeric and dimeric form, which depends on ionic strength and protein concentration.

Ultrogel AcA 34, under the conditions described below, can be used to separate the two forms as well as to estimate the amounts of bound detergent ([^{3}H]dodecyl-β-D-maltoside).

Reagents

Ultrogel AcA 34 (LKB, Sweden) preswollen and well equilibrated in the buffer of appropriate KCl concentration

Running buffer Tris–HCl (10 m*M*), pH 7.4, 0.1% dodecyl-β-D-maltoside (Calbiochem, or synthesized according to Rosevear *et al.*[13]) and potassium chloride as described

Bovine heart cytochrome-*c* oxidase (stock solution). Bovine heart cytochrome-*c* oxidase is purified under conditions which result in a low lipid-containing enzyme, according to Yu *et al.*,[20] and stored at −80° at concentrations of 0.3–1.0 m*M* in 50 m*M* potassium phosphate buffer, pH 7.4. The heme aa_3 concentration is measured spectrophotometrically, using $\varepsilon = 24\ \mathrm{cm}^{-1}\ \mathrm{m}M^{-1}$ at 605–630 nm, in the dithionite reduced minus oxidized spectrum. The stock enzyme contains usually about 8–9 nmol heme *a* per milligram of protein and a lipid content of about 14 mol/mol of heme aa_3, determined by phosphorus estimation.[17,21]

[^{3}H]Dodecyl-β-D-maltoside. The product, which is obtained from Amersham after catalytic exchange (TR7) of cold dodecyl-β-D-maltoside with Tritium, is heavily contaminated with radioactive degradation products. It must be purified by thin layer chromatography (TLC) in ethyl acetate/methanol (4 : 1), and after extraction with methanol, it is passed through a short Sephadex G-100 column equilibrated with 10 m*M* Tris–HCl, pH 7.4, to separate the detergent micelles from small molecular contaminants. The final stock solution of [^{3}H]dodecyl-β-D-maltoside, used for binding experiments, is then diluted in buffers containing cold dodecyl-β-D-maltoside. The final specific radioactivity of dodecyl-β-D-maltoside in the buffer is 5 Ci/mol.

[20] C. Yu, L. Yu, and T. E. King, *J. Biol. Chem.* **250,** 1383 (1975).

[21] P. S. Chen, T. Y. Toribara, and H. Warner, *Anal. Biochem.* **28,** 1756 (1956).

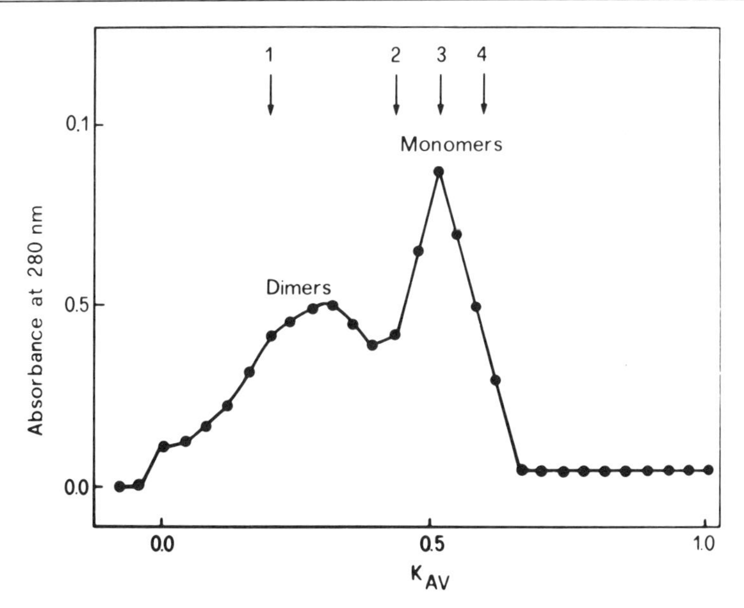

FIG. 1. Gel filtration of bovine cytochrome-*c* oxidase on Ultrogel AcA 34. Cytochrome-*c* oxidase (1.5 μM) was preincubated in 10 m*M* Tris–HCl, pH 7.4, 0.5% dodecyl-β-D-maltoside for 2 hr at 4° and next loaded onto Ultrogel AcA 34 column (1 × 45 cm) and eluted with the same buffer containing 0.1% dodecyl-β-D-maltoside. The following proteins used for calibration are indicated by arrows: (1) ferritin, (2) catalase, (3) immunoglobulin G, and (4) aldolase.

Scintillation fluid. PPO (5 g) (Merck) and POPOP (0.2 g) are dissolved in 1 liter of toluene; then 270 ml of this solution is mixed with 230 ml of Triton X-100 and 10.2 ml of acetic acid (Merck).

Separation of Monomers and Dimers

Cytochrome-*c* oxidase is diluted to 1.5 μM concentration of heme aa_3 in buffer without potassium chloride, but with a dodecyl-β-D-maltoside concentration of 0.5%. After 2 hr of incubation at 4° in the dark (to prevent loss of heme), 100–150 μl are loaded on an Ultrogel AcA 34 column (1 × 45 cm), equilibrated with buffer without KCl, and 0.1% dodecyl-β-D-maltoside. The column is run in the dark at 4° at a speed of 5 ml/hr. The absorbance at 280 nm is recorded by a Uvicord unit (LKB, Sweden; full range 0.01 absorbance units). The elution profile (Fig. 1) shows two main peaks, corresponding to monomers (K_{av} = 0.50–0.55) and dimers (K_{av} = 0.3). K_{av} represents the fraction of the stationary gel

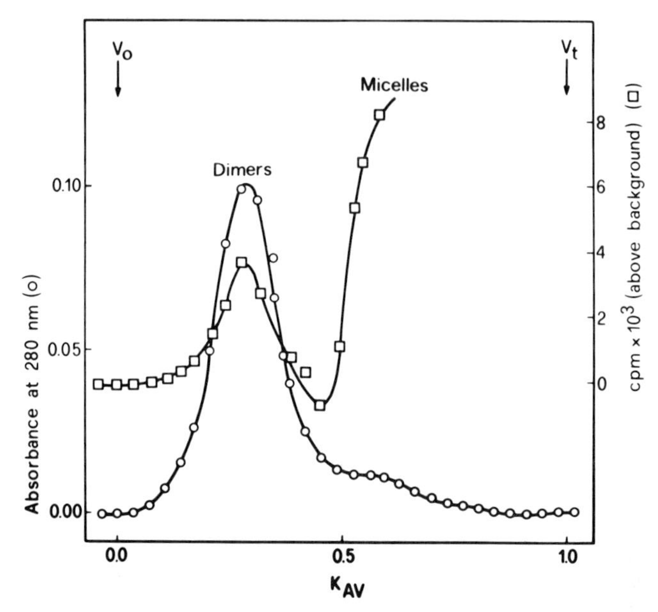

FIG. 2. Binding of [^{3}H]dodecyl-β-D-maltoside to dimers of bovine cytochrome-*c* oxidase. Cytochrome-*c* oxidase (10 μM) was preincubated in 50 mM KCl, 10 mM Tris–HCl, pH 7.4, 0.5% [^{3}H]dodecyl-β-D-maltoside (5 Ci/mol) for 2 hr at 4° and next loaded onto Ultrogel AcA 34 column (1 × 45 cm) equilibrated with 50 mM KCl, 10 mM Tris–HCl, pH 7.4, and 0.1% [^{3}H]dodecyl-β-D-maltoside (5 Ci/mol), and eluted with the same buffer. Radioactivity and absorbance at 280 nm were measured in each fraction. The void volume was determined with blue dextran and the total volume with ferricyanide.

volume which is available for diffusion of a given solute species and is defined as the ratio $(V_e - V_0)/(V_t - V_0)$, where V_e is the elution volume, V_0 is the void volume, and V_t is the total volume. To estimate K_{av}, a separate run with blue dextran and ferricyanide is performed to mark the void volume and the total volume, respectively. The fractions containing the higher- or lower-molecular-weight species are pooled, and if needed, concentrated by ultrafiltration in Amicon B15 boxes and used for sedimentation analysis.

Comments. Monomer/dimer equilibrium is dependent on the ionic strength and the concentration of the enzyme. Incubation and gel filtration in the presence of KCl lead to a shift of equilibrium in favor of the dimeric form; at 50 mM KCl the enzyme is mainly dimeric (Fig. 2). If the protein concentration is increased above 5 μM, not only the ratio of the amounts of dimers to monomers is increased, but also higher aggregates are formed, which elute with the void volume. For the isolation of higher amounts of monomers it is possible to increase the size of the column; one has to take care, however, that the elution time is not longer than 24 hr, that the enzyme is kept strictly in the dark, and that the elution procedure should be run at 4°. Otherwise degradation and loss of heme are observed.

Determination of the Amount of Bound Detergent

Procedure for Determining the Amount of Detergent Bound to the Dimeric Form of Cytochrome-c Oxidase

The same Ultrogel AcA 34 column (1 × 45 cm) used for the separation of monomers and dimers is equilibrated with running buffer containing 50 m*M* KCl and 0.1% [^{3}H]dodecyl-β-D-maltoside, with a specific activity of 5 Ci/mol (10 μCi/mg).

Cytochrome-*c* oxidase (7 nmol of heme aa_3) is taken up in 200 μl of running buffer, containing 0.5% [^{3}H]dodecyl-β-D-maltoside with the same specific activity as the equilibration buffer and incubated for 2 hr in the dark at 4° before loading on the column. The column is run with 5 ml/hr, and fractions of ~0.8 ml are collected. After measurement of the absorbance of each fraction at 280 nm, 0.5 ml is used for the spectra (estimation of the heme *a* content), 0.1 ml for determining the protein concentration (Lowry *et al.*[22]), and 50 μl are dissolved in 5 ml of scintillation fluid (in triplicate) to measure the radioactivity. The elution profile (Fig. 2) shows a good separation of the dimeric enzyme (K_{av} = 0.3) and of the dodecyl-β-D-maltoside micelles, a necessary prerequisite for an accurate binding estimation. The background radioactivity is taken from the fractions in the void volume. The amount of the bound detergent is then calculated from the radioactivity in the protein peak (exceeding the background) and the known specific radioactivity in the buffer. The central fractions of the protein peak show a constant value of 11 mol of heme per mol of protein and a constant detergent binding of 320 mol of dodecyl-β-D-maltoside per mol of cytochrome-*c* oxidase (Table I).

Comments. The same procedure may be applied to estimate the amount of bound Triton X-100, using either tritiated detergent (NEN) or the ratio of the absorptions at 277 and 422 nm for quantification.[11,12] Since in all cases an error of 10–15% has to be considered, several experiments should be performed to get an accurate result. The calculated detergent bound per protein should be a constant value, at least in the central

[22] O. H. Lowry, N. J. Rosebrough, A. L. Farr, and R. J. Randall, *J. Biol. Chem.* **193,** 265 (1951).

TABLE I
BINDING OF DETERGENT AND LIPID CONTENT OF BOVINE CYTOCHROME-*c* OXIDASE

Detergent	Bound detergent (mol/mol oxidase)[a]	Bound phospholipid (mol/mol oxidase)[b]	Reference
Triton X-100	180	15.7	11
Triton X-100	150 (dimer)	ND[c]	6
	190 (monomer)	ND	6
Triton X-100	240	ND	12
Tween 80	ND	14.6	11
Deoxycholate	80	8.8	11
Dodecyl-β-D-maltoside	320 (dimer)	14 (dimer)	17
	362 (monomer)	10 (monomer)	17

[a] Oxidase (1 mol) corresponds to 1 mol of heme aa_3.
[b] An average molecular weight of 750 for phospholipids is considered.
[c] ND, No data.

fractions of the protein peak. The fractions at the borders of the protein peak show apparent lower detergent binding and therefore should not be taken into account.

Determination of the Amount of Detergent Bound to the Monomeric Enzyme

Since the monomeric form of cytochrome-*c* oxidase is not well separated from the dodecyl-β-D-maltoside micelles on Ultrogel AcA 34, the detergent binding cannot be measured directly. In this case, the monomers are collected after gel filtration on the Ultrogel AcA 34 column equilibrated with radioactive dodecyl-β-D-maltoside and rechromatographed on a Sephadex G-100 column (1 × 20 cm) equilibrated with the same buffer. On this column, the enzyme elutes in the void volume, while the detergent micelles are retarded. The calculation of the bound detergent is the same as described for the dimeric enzyme. A slightly increased binding of 362 mol of dodecyl-β-D-maltoside per mol of heme aa_3 is observed (Table I).

Comments. Phospholipid analysis showed that the monomeric bovine cytochrome-*c* oxidase, when dispersed in dodecyl-β-D-maltoside, contains less lipid (10 mol of phosphorus per mol of heme aa_3) in comparison to the dimeric form (14 mol of phosphorus per mol of heme aa_3; Table I). The displacement of lipids by detergent during the monomerization process might be a reason for this enhanced binding of detergent. Furthermore, the monomer seems to be more hydrophobic (dissociation by lowering the ionic strength) and more asymmetric than the dimer, which also favors an increase in detergent binding.

Centrifugation Techniques

The molecular weight of a protein in a detergent solution can be determined unambiguously by measurement of either sedimentation equilibrium or sedimentation velocity, taking into consideration the multicomponent nature of the system (containing protein, detergent, and lipids). Data concerning the amount of bound detergent and the residual phospholipid are therefore necessary in order to evaluate a precise molecular weight. The phospholipid content in the preparation of cytochrome-*c* oxidase can be calculated on the basis of its phosphate content, assuming an average molecular weight of phospholipid as 750 per atom of phosphorus. A convenient method of phosphorus estimation is that of Chen *et al.*,[21] after the extraction of phospholipids with chloroform/methanol (2 : 1).[17]

The amount of bound detergent is usually determined by gel filtration, in the presence of the radioactive detergent, as described above. In the case of Triton X-100 the amount of bound detergent can be calculated from the sum of the absorbance due to the detergent and protein, using $\varepsilon_{277} = 1465\ \mathrm{cm}^{-1}\ \mathrm{m}M^{-1}$ as the extinction coefficient for this detergent.[23]

In order to calculate the molecular weight of the protein–lipid–detergent complex from centrifugation studies, the partial specific volume of this complex should be known.

Estimation of Partial Specific Volumes

When a substrate is dissolved in a solvent at constant temperature and pressure, the change in volume of this solution after addition of 1 g of the given substance is defined as its partial specific volume, v. For many proteins, partial specific volumes were calculated from their amino acid content. Cohn and Edsall,[24] using Traube's rule of additive increments in volume, corresponding to different chemical functional groups, calculated specific volumes of all the amino acid residues. Their values (see p. 372 of ref. 24 or this series[25]) can be used in the following equation:

$$v = \Sigma N_i(W_i v_i)/\Sigma N_i W_i \tag{1}$$

[23] S. Makino, J. A. Reynolds, and C. Tandord, *J. Biol. Chem.* **248,** 4926 (1973).

[24] E. J. Cohn and J. T. Edsall, "Proteins, Amino Acids and Peptides as Ions and Dipolar Ions," Reinhold, New York, 1943.

[25] J. C. Lee and S. N. Timasheff, this series, Vol. 61, p. 49.

where v_i represents the specific volumes of amino acids, N_i is the number of residues of amino acid of type i, and W_i is the residue weight (i.e., molecular weight minus 18).

This formula has been applied to calculate the partial specific volume of cytochrome-*c* oxidase, and the value 0.743 ml/g was obtained from the amino acid composition of the enzyme and used for further calculations.[9,11] If only seven subunits (and the amino acid composition presented by Capaldi[26]) are considered, the partial specific volume of the enzyme would be 0.730 ml/g.

Another possibility of establishing partial specific volume is based on density measurements. A detailed description of this method, applied for lipid-associated proteins, is given by Steele *et al.*[27] This method is based on measurements of the densities of a series of solutions containing varying concentrations of that component. The apparent partial specific volume is then calculated from the following relationships:

$$v = 1/\rho_0 \cdot (1 - d\rho/dc) \qquad (2)$$

where $d\rho/dc$ is the deviation of the density of the solution with the concentration of the solute and ρ_0 is the density of the solvent. Density measurements can be performed by a column technique or by a mechanical oscillator densitometer. With use of a column method for the density measurements of cytochrome-*c* oxidase dispersed in Emasol,[10] a partial specific volume of 0.720–0.730 ml/g was found. The density measurements for dodecyl-β-D-maltoside and those for the enzyme, using a mechanical oscillator densitometer (Praezisions-Dichtemesseinrichtung DMA-50, Paar) at varying concentrations of the solute, are shown in Fig. 3. The calculated values of partial specific volumes are given in Table II. It is important to note that the partial specific volume of the oxidase as well as that of the detergent are considerably dependent on the temperature (therefore thermostatization is necessary) and the ionic strength of the buffer. The density of cytochrome-*c* oxidase measured at 20° in the presence of 50 m*M* KCl, 10 m*M* Tris–HCl, pH 7.4, and 0.5% dodecyl-β-D-maltoside reveals a nonlinear dependence on the concentration. This phenomenon is difficult to interpret, but has been reported as well for another membrane protein, namely, the bc_1 complex.[28]

The partial specific volume of the multicomponent complex (v^*) is calculated from the sum of its individual components,[27]

[26] R. A. Capaldi, *in* "Membrane Proteins in Energy Transduction" (R. A. Capaldi, ed.), p. 201. Dekker, New York, 1979.

[27] J. C. H. Steele, C. Tanford, and J. A. Reynolds, this series, Vol. 48, p. 11.

[28] S. De Vries, "The Pathway of Electrons in QH_2: Cytochrome-*c* Oxidoreductase," Ph.D. thesis, University of Amsterdam, 1983.

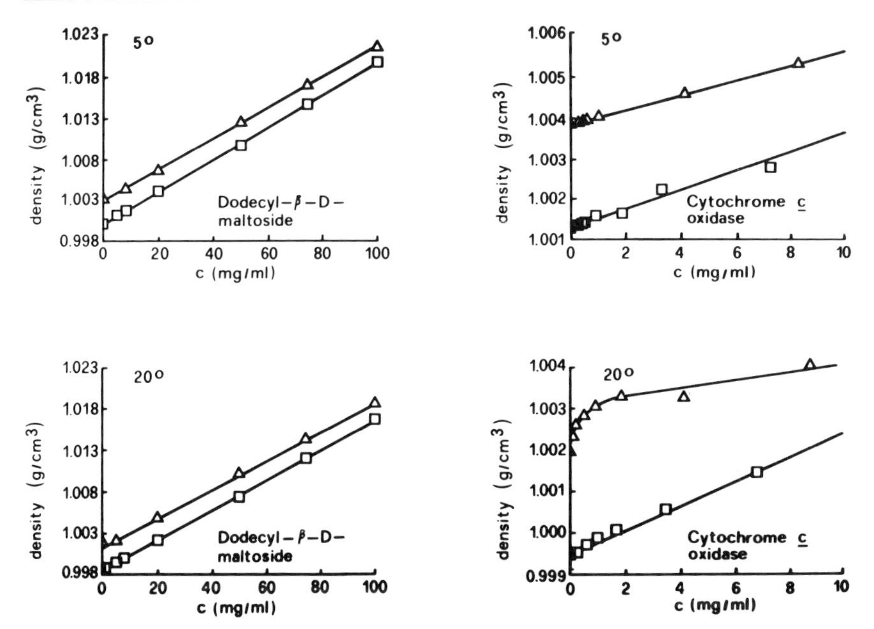

FIG. 3. Density of a solution as a function of solute's concentration. Density measurements were performed with an oscillator densitometer (Praezisions-Dichtemesseinrichtung DMA-50, Paar) at 5° and 20°, as indicated. In the case of dodecyl-β-D-maltoside, the medium contained 10 m*M* Tris–HCl, pH 7.4, without (□) or with (△) 50 m*M* KCl. Densities of cytochrome-*c* oxidase were performed in 10 m*M* Tris–HCl, pH 7.4, 0.5% dodecyl-β-D-maltoside in the absence (□) or presence (△) of 50 m*M* KCl. Protein content was estimated by the method of Lowry *et al.*[22]

TABLE II
PARTIAL SPECIFIC VOLUMES OF CYTOCHROME-*c* OXIDASE AND DODECYL-β-D-MALTOSIDE MICELLES

Compound	Temperature (°C)	Partial specific volume (ml/g)	
		50 m*M* KCl	No KCl
Dodecyl-β-D-maltoside	5	0.813	0.808
Cytochrome-*c* oxidase	5	0.822	0.768
Detergent–cytochrome-*c* oxidase complex	5	0.823	0.806
Dodecyl-β-D-maltoside	20	0.825	0.821
Cytochrome-*c* oxidase	20	0.763	0.713
Detergent–cytochrome-*c* oxidase complex	20	0.816	0.802

$$v^* = \Sigma x_{ijk} v_{ijk} \tag{3}$$

where x and v are the weight fractions and partial specific volumes of the individual components (protein, detergent, phospholipid), respectively. The values of partial specific volumes of many ionic and nonionic detergents, as well as for the most common lipids, are given in the chapter by Steele *et al.*[27]; see this series.

It is interesting to note that cytochrome-*c* oxidase dispersed in dodecyl-β-D-maltoside at 5° has a partial specific volume for the complex very similar to that of the protein alone. It is difficult to derive a model, based on the above observation, for the packing of the detergent molecules bound to the oxidase, but a minimum conclusion may be that the Stokes radius of the complex should be similar to that of the protein. This conclusion is confirmed by the fact that molecular weights calculated from sedimentation experiments are similar to those obtained from gel filtration (also carried out at 5°) without subtracting the bound detergent. At 20°, however, the partial specific volumes of the protein and the complex are different.

These differences may be due to the fact that a good agreement between the values calculated from the amino acid composition and those experimentally obtained by densitometric measurements can be obtained only in the case of diluted buffers, and in the absence of a detergent.

The method for measuring partial specific volume from equilibrium centrifugation in H_2O/D_2O mixtures[29,30a] has been recently applied to cytochrome-*c* oxidase.[30b]

Sedimentation Equilibrium Measurements

Principle. If the solution of a substance is rotated at low speed, the concentration near the meniscus of the cell decreases, but solute, because of its diffusion, does not leave this region completely. After a certain time an equilibrium between sedimentation and diffusion is reached. By means of optical methods the concentration gradient in the cell may then be scanned and recorded.

Calculations. The equilibrium equation for a protein–lipid–detergent complex takes the following form:

$$M^*(1 - v^*\rho_s) = 2RT/\omega^2 \cdot (d \ln c/dr^2) \tag{4}$$

[29] S. J. Edelstein and H. K. Schachman, *J. Biol. Chem.* **242,** 306 (1967).

[30a] J. A. Reynolds and C. Tanford, *Proc. Natl. Acad. Sci. U.S.A.* **73,** 4467 (1976).

[30b] M. D. Suarez, A. Revzin, R. Narlock, E. S. Kempner, D. A. Thompson, and S. Ferguson-Miller, *J. Biol. Chem.* **259,** 13791 (1985).

where R is the gas constant, T is the absolute temperature, ω is the angular velocity of the rotor, c is the concentration of the redistributed solute (in this case the oxidase–lipid–detergent complex), r is the distance from the axis of rotation, ρ_s is the density of the solvent, and v^* is the partial specific volume of the complex. The molecular weight of the protein moiety is then calculated taking into account all the bound components, expressed as the weight fraction ($\Sigma\delta_i$):

$$M = M^*(1 + \Sigma\delta_i) \tag{5}$$

Application to Cytochrome-c Oxidase. Sedimentation equilibrium centrifugation was applied to the Triton X-100 and deoxycholate complexes of the enzyme.[6,11,12] Usually the samples are centrifuged for 24 hr at 10° or 22°, with velocities in the range of 6400–9800 rpm in a Model E Beckman-Spinco analytical ultracentrifuge equipped with a photoelectric scanner. The initial protein concentration for such experiments has to be in the range of 0.17–0.50 mg/ml, which is 10 times more than the protein concentrations used for gel filtration studies. Sedimentation equilibrium is reached when the plots of ln c vs r^2 after different centrifugation times are superimposable.

Precautions. Equation (4) is valid for an ideal two-component system and can be applied to a protein in monodispersed form, as in the case of the cytochrome-*c* oxidase–Triton X-100 complex. Under these conditions the enzyme forms only dimers of mw 345,000–385,000,[11,12] after correcting for the bound detergent. However, when other components are present, the inclusion of additional terms in Eq. (4) is mandatory and the calculations become more complicated. Such limitation of this method appears particularly relevant when more than one protein species is present and, as a consequence of this, a nonlinear function of ln c vs r^2 is observed. Cytochrome-*c* oxidase treated at high pH, in high Triton X-100 concentration,[6] or analyzed in deoxycholate[11] offers an example of such a problem, with the consequence that only an approximate estimation of the molecular weight can be obtained (~129,000–200,000 for a monomer).

The determination of the partial specific volume and of the extent of bound detergent can be avoided by running the equilibrium centrifugation in media of different densities (using H_2O with D_2O in different proportions[29,30a]) and with detergent solution having a density identical to that of the medium.

Sedimentation Velocity Measurements

Principle. When a solution containing only one solute is submitted to high enough centrifugal force, the solute will start to sediment and a boundary (only slightly influenced by diffusion) will appear between solvent and solution. If the solution contains two solutes with different sedimentation coefficients, two boundaries will be formed, one between solvent and the slower moving component and one between the latter and the solution containing both the components. For a system of several solutes, each new component will give rise to a new boundary, and from the movement of the boundary, the corresponding sedimentation coefficient can be determined.

Calculations. Formation and migration of these boundaries are followed by an optical detection system. Most often centrifuges are equipped with schlieren optics. The sedimentation is monitored by taking photographs at defined time intervals. The sedimentation coefficient(s) is evaluated from the plot of the logarithm of the distance of the boundary from the center of rotation vs time, according to the equation:

$$s = 1/\omega^2 \cdot \ln(x_2/x_1)/(t_2 - t_1) \quad (6)$$

where x_1 and x_2 are the distances of the boundaries from the rotor axis when measured at times t_1 and t_2, respectively, and ω is the angular velocity of the rotor.

The molecular weight of the complex (M^*) can be calculated from the following function:

$$M^* = (s_{20,w} \cdot RT)/[D_{20,w}(1 - v^*\rho_s)] \quad (7)$$

where R is the universal gas constant, T the absolute temperature, v^* the partial specific volume of the complex, ρ_s the density of the solvent, and $s_{20,w}$ represents the sedimentation coefficient (normalized to 20° and water conditions).

The diffusion coefficient (D) should be established in a parallel ultracentrifugation run at low speed (about 4000 rpm). A plot of the height/area ratio of the peak vs time can be used to evaluate the diffusion coefficient. An apparent diffusion coefficient can then be calculated from the equation

$$4\pi Dt = A^2/\mathrm{H}^2 \quad (8)$$

where D is the diffusion coefficient ($\mathrm{cm}^2\ \mathrm{sec}^{-1}$), A is the area under the boundary (cm^2), H is the maximum ordinate (cm), and t is time (sec). The diffusion constant can be replaced by the following equation containing the Stokes radius (R_S) as a parameter[31]:

$$D = kT/(6\pi\eta R_S) \quad (9)$$

where k is the Boltzmann constant and η is the solvent viscosity.

[31] C. Tanford, Y. Nozaki, J. A. Reynolds, and S. Makino, *Biochemistry* **13,** 2369 (1974).

The Stokes radius can be measured by gel filtration experiments. Since it is known, however, that in the case of asymmetric proteins values of R_S can be underestimated,[18] the laser light-scattering method gives more reliable results[17,32] in measuring the particle's size.

The molecular weight (M) of the protein is calculated after subtracting all the bound components (phospholipids, detergent) from the molecular weight of the complex (M^*).

Application for Cytochrome-c Oxidase. Several laboratories have used sedimentation velocity to measure the molecular weight of cytochrome-*c* oxidase. The same type of centrifuge employed for the equilibrium centrifugation experiments (Beckman-Model E analytical ultracentrifuge) can be used, but much higher velocities of 56,000[17] or 60,000[10] rpm should be applied in order to estimate *s* values. The photographs are usually taken after the boundary begins to form at 4 min[17] or 8 min[10] intervals. To estimate the diffusion constant, much lower speeds, 6000 or 12,000 rpm,[10] should be applied, and the photographs should be taken at 2-hr intervals. Diffusion coefficient values of 4.8×10^{-10} and 5.1×10^{-10} $cm^2 sec^{-1}$ have been found for sedimentation in Emasol,[10] and one of 3.06×10^{-10} $cm^2 sec^{-1}$ was calculated using an $R_S = 70$ Å[17] in dodecyl-β-D-maltoside.

In order to determine the positions of the boundaries when using schlieren optics, high enzyme concentrations must be applied (5–10 mg protein/ml have been used[4,5,10,17]). The values of sedimentation coefficients estimated in different detergents are shown in Table III. Since the diffusion coefficients, the Stokes radii, and the amount of bound detergent were not always measured, it is difficult to draw firm conclusions about the molecular weight of the protein moiety. It seems, however, that monomeric species may exist in Triton X-100 and dodecyl-β-D-maltoside with a sedimentation coefficient of 8.2–9.6 S, while the dimers have a sedimentation coefficient of 11.8–18.5 S.

Precautions and Comment. The advantage of this technique, in comparison with equilibrium centrifugation, is its use even in a polydispersed system in which the relative amounts of different forms (polymers, dimers, monomers) can be estimated quantitatively (see Refs. 4, 5, and 17). Sedimentation velocity measurements require amounts of protein higher than the equilibrium technique, and centrifugation should be run twice, once with higher speed for estimation of sedimentation coefficient and a second time with low speed for establishing the diffusion coefficient. High protein concentration is connected of necessity with high detergent concentration, which can change per se the molecular form of the enzyme.

[32] N. A. Mazer, P. Schürtenberger, M. C. Carey, R. Preisig, K. Weigand, and W. Känzig, *Biochemistry* **23,** 1994 (1984).

TABLE III
SEDIMENTATION COEFFICIENT AND MOLECULAR WEIGHT OF CYTOCHROME-*c* OXIDASE SPECIES

Detergent	Cytochrome-*c* oxidase species	Assumed aggregation state	$s_{20,w}$	Molecular weight	Reference
Tween 80 (Emasol)	Bovine	Dimer	10.2–14.2	184,000–199,000	10
	Bovine (high pH treated)	Monomer	6.0–6.4	100,000–106,000	10
	Bovine	Dimer	11	ND[a]	4
	Elasmobranch fish	Monomer	5–7	ND	4
	Bovine, dogfish, chicken	Dimer	11	ND	5
	Bovine, rat	Polymer	17–20	ND	5
	Shark, dogfish, rat, chicken, camel	Monomer	5–7	ND	5
	Bovine (high pH treated), shark	Monomer	6.5–7.5	ND	6
Triton X-100	Bovine	Dimer	11.8	345,000[b]	12
	Bovine (subunit III depleted)	Monomer	8.2	208,000[c]	12
	Bovine	Dimer	10–13	326,000[c]	6
	Bovine (high pH treated)	Monomer	6.5–7.5	129,000[c]	6
	Shark	Monomer	6.5–7.5	158,000[c]	6
Lysophosphatidylcholine	Bovine (high pH treated)	Monomer	6.5–7.5	ND	6
Dodecyl-β-D-maltoside	Bovine (50 m*M* KCl)	Dimer	15.5–18.5	340,000–406,000[d]	17
	Bovine (no KCl)	Monomer	9.6	170,000[d]	17

[a] ND, No data.
[b] Calculated after subtraction of the bound detergent, assuming $R_S = 72$ Å.
[c] Estimated by sedimentation equilibrium analysis.
[d] Calculated according to Eqs. (7) and (9), with $R_S = 70$ Å for dimers and $R_S = 61$ Å for monomers, and after subtracting of the bound detergent.

Centrifugation in Sucrose Density Gradient

Principle. Sucrose gradient centrifugation is a simple and fast method to separate macromolecules according to their sedimentation velocity (not buoyant densities); in addition, the sedimentation coefficients of a molecule may be easily approximated by this method. The migration of a molecule through the sucrose gradient, however, is only occasionally linear with time, whatever gradient is applied (linear, convex, or concave). The sedimentation velocity is dependent on the density and viscosity of the sucrose medium in the gradient, and the changes of those parameters are normally not equivalent to the change in the centrifugal force during sedimentation. Calibration of sucrose gradient with molecules of known sedimentation coefficients can be applied only to estimate roughly the s value of a particle, especially because the sedimentation coefficient is also dependent on the particle's density. McEven[33] calculated the sedimentation coefficient as a function of the sucrose concentration in linear gradients for different particle densities and temperatures. The tables of McEven are then used to calculate the sedimentation coefficients of the monomeric and dimeric cytochrome-c oxidase detergent complexes.

Calculations. The rate of sedimentation, dr/dt, is expressed by the following function:

$$dr/dt = s(r)\omega^2 r \tag{10}$$

where r is the distance from the rotor axis, t is time, s is the sedimentation coefficient, and ω is the angular velocity.

For any sucrose concentration the sedimentation coefficient can be determined according to the following equation:

$$s(z,T) = s_{20,\mathrm{w}} \cdot [\eta_{20,\mathrm{w}}/\eta(z,T)] \cdot [\rho_\mathrm{p} - \rho(z,T)]/(\rho_\mathrm{p} - \rho_{20,\mathrm{w}}) \tag{11}$$

where z is solute concentration (weight fraction), T is temperature, η is solution viscosity, and ρ solution density. The subscripts p or w refer to the sedimenting particle or to water, respectively.

When the distribution of sucrose is linear, then Eq. (10), being a function of distance from the rotor axis, can be expressed as the function of sucrose concentration. The tables of McEven[33] allow calculation of the values of sedimentation coefficients from the sucrose concentration at the position of a band containing the sedimenting protein. For this calculation it is necessary to know the theoretical concentration of sucrose at 0 distance (rotor axis), and the particle density (reciprocal of partial specific volume of the complex). The tables give, for any sucrose concentration

[33] C. R. McEven, *Anal. Biochem.* **20,** 114 (1967).

(z_i), integrated functions, $I(z_i)$, of viscosity and density for different temperatures and particle densities. The sedimentation coefficient can then be calculated from the following relationship:

$$s_{20,w}\int_{t_1}^{t_2}\omega^2 dt = I(z_1) - I(z_2) \tag{12}$$

Application for Cytochrome-c Oxidase

Gradient Formation. Sucrose gradient solutions are made in 10 m*M* Tris–HCl, pH 7.4, 50 m*M* KCl, and 0.1% dodecyl-β-D-maltoside. Linear sucrose gradients (5–20%) are formed using a conventional mixing chamber.[34] Exponential gradients may be used when the sedimenting protein species is not allowed to sediment more than halfway through the gradient. Up to this point the sucrose density in the gradients can be approximated by a linear function of the radius. Exponential sucrose gradients are more stable and are made by pumping a 20% sucrose solution through a fixed volume of an initially 5% sucrose solution, which is continuously mixed. The sucrose concentration at certain positions in the gradient are either measured by estimating the density in a densitometer (DMA 50, Paar) or calculated according to the following equation[35]:

$$c_V = c_R \cdot [1 - \exp(-V/M)] + c_m \exp(-V/M) \tag{13}$$

where c_V is the sucrose concentration after the volume (V) eluted from the mixing chamber, c_R is the sucrose concentration in the reservoir (20%), and M and c_m are the volume and concentration (5%) in the mixing chamber, respectively. In this case, M is the volume of the gradient formed.

A total of 150 μl of a minimal concentration of 0.25 μ*M* cytochrome-*c* oxidase (if necessary, concentrated on Amicon B15) in gradient buffer without sucrose is layered on top of each gradient.

Centrifugation. The Beckman SW 60 rotor is used. The gradients are centrifuged at 55,000 rpm (400,000 *g*) either for 5.25 hr (linear gradient) or for 4.25 hr (exponential gradient) at 4°.

Analysis of the Gradient. The centrifugation tubes containing the sucrose gradients are punctured from the bottom. The absorbance of the effluent is then monitored at 280 nm by a UV detection unit (Uvicord I, LKB) connected with a logarithmic recorder (Servogor, Goerz). The recorder plot gives the position of the protein according to the eluted volume of the gradient. From the rotor and tube sizes the position of the enzyme in the gradient can now be calculated and from this the sedimen-

[34] E. H. McConkey, this series, Vol. XII, p. 620.

[35] H. Noll, "Technics in Proteinsynthesis," Vol. 2, p. 116. Academic Press, London, 1969.

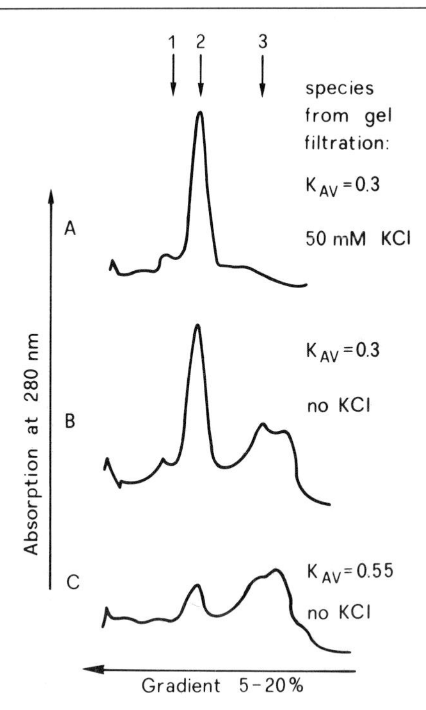

FIG. 4. Sucrose gradient centrifugation of cytochrome-c oxidase. The enzyme was sedimented through 5–20% sucrose gradient in the presence of 50 mM KCl, 10 mM Tris–HCl, pH 7.4, 0.1% dodecyl-β-D-maltoside. The centrifugation was run for 4.25 hr at 55,000 rpm in an SW 60 rotor on a Beckman L8-80 ultracentrifuge at 5°. The cytochrome-c oxidase samples were obtained from the gel filtration. (A) Peak (K_{av} = 0.3) obtained at 50 mM KCl; (B) peak of K_{av} = 0.3; and (C) peak of K_{av} = 0.55 obtained in the absence of KCl. Observed cytochrome-c oxidase species: (1) not identified species; (2) dimers; (3) monomers.

tation coefficient may be estimated according to the tables of McEven[33] (see above).

Results and Comments. Figure 4 shows the sucrose density sedimentation analysis of monomeric and dimeric forms of cytochrome-c oxidase, obtained after gel filtration on Ultrogel AcA 34 (see above). Dimers obtained in the presence of potassium chloride (50 mM) are monodisperse also in the sedimentation analysis (Fig. 4A). From the main species 2, a sedimentation coefficient of 17 S is calculated, which corresponds, according to Eqs. (7) and (9), to a molecular weight of the protein moiety (after subtraction of the bound dodecyl-β-D-maltoside) of 400,000 (Table III), as expected from two sets of the 13 subunits described for this enzyme.[7] Dimers (K_{av} = 0.3) eluted from Ultrogel AcA 34 in the absence of KCl (Fig. 4B) are not devoid of monomers (species 3) and monomers

(K_{av} = 0.55; Fig. 4C) also contain dimers. From the estimated sedimentation coefficient of 9.6 S for the monomer–detergent complex (species 3), a net molecular weight for the protein is calculated as 170,000 (Table III). This value differs from the theoretically expected one (200,000). However, it is neither possible to estimate directly the Stokes radius of the monomeric complex by light-scattering measurements nor to evaluate the diffusion coefficient by sedimentation analysis. Therefore the calculated molecular weight, which is based on an assumed Stokes radius of 61 Å,[17] is less accurate and an error of 20% has to be admitted in this case.

[6] Affinity Chromatography Purification of Cytochrome-c Oxidase from Bovine Heart Mitochondria and Other Sources

By CLEMENS BROGER, KURT BILL, and ANGELO AZZI

In the mitochondrial respiratory chain cytochrome-*c* oxidase catalyzes the electron transfer from ferrocytochrome *c* to molecular oxygen. The molecular details of the interaction of the enzyme with its electron donor, cytochrome *c*, have been studied by using various labeling techniques. It has been established that the "front side" of cytochrome *c*, which contains the exposed heme edge, is in contact with the oxidase,[1,2] and that subunit II of the enzyme complex is interacting with cytochrome *c*[3] through a domain on its surface which has also been mapped.[4] It has been recognized for a long time that the specific interaction between cytochrome-*c* oxidase and its electron donor could be useful in developing an affinity chromatography purification technique for cytochrome-*c* oxidase based on cytochrome *c* immobilized on a gel matrix.

Essentially two types of affinity gels have been used which differ in the way cytochrome *c* is linked to the gel: The first was developed before details about the interaction of the oxidase with its electron donor were known. In this method cytochrome *c* is attached to the Sepharose through lysine residues which are mainly located on the front side of the protein and which are now known to be important for the interaction of this protein with the oxidase. In the development of the second method the

[1] S. Ferguson-Miller, D. L. Brautigan, and E. Margoliash, *J. Biol. Chem.* **253,** 149 (1978).
[2] R. Rieder and H. R. Bosshard, *J. Biol. Chem.* **253,** 6045 (1978).
[3] R. Bisson, A. Azzi, H. Gutweniger, R. Colonna, C. Montecucco, and A. Zanotti, *J. Biol. Chem.* **253,** 1874 (1978).
[4] R. Bisson, G. C. M. Steffens, R. A. Capaldi, and G. Buse, *FEBS Lett.* **144,** 359 (1982).

new information about the interaction between the two proteins has been taken into consideration by attaching cytochrome *c* through its rear side.

The application of the two gels for the purification of enzymes which interact specifically with cytochrome *c* is straightforward. In general, the solubilized proteins are loaded at low salt concentration to a column containing the gel and, after washing the column, they are eluted by increasing the ionic strength of the buffer. Various detergents, mainly of the nonionic type, have been used in the solubilization step and during the chromatography procedure. If the solubilization step is optimized and the protein of interest can be released from the membrane in a rather selective way, it is possible in many cases to purify it to homogeneity in one affinity chromatography step. The cytochrome *c* gels described below can be used for several purifications. They are washed after every use with 1 *M* NaCl and detergent and subsequently stored at 4°.

Horse Heart Cytochrome c Linked to CNBr-Activated Sepharose 4B

Materials

For the preparation of 4 ml of gel:

- Freeze-dried CNBr-activated Sepharose 4B (1 g) (Pharmacia Fine Chemicals AB, Uppsala, Sweden)
- 200 ml of 1 m*M* HCl
- 10 ml of 0.1 *M* $NaHCO_3$, pH 8.3
- 20 ml of 0.1 *M* $NaHCO_3$, pH 8.0, containing 0.5 *M* NaCl
- Horse heart cytochrome *c* (4 mg) (Sigma Chemical Company, St. Louis, MO 63178; type VI)
- 50 ml of 1 *M* ethanolamine/HCl, pH 8
- 400 ml of 1 *M* NaCl
- 200 ml of 0.1 *M* sodium acetate, pH 4.0
- 200 ml of 0.1 *M* sodium borate, pH 8.0

Preparation[5]

CNBr-activated Sepharose 4B (1 g) is swollen and washed for 15 min on a sintered glass filter with about 200 ml of 1 m*M* HCl, yielding about 4 ml of swollen gel. Alternatively, Sepharose 4B can be activated with cyanogen bromide, as published in this series.[6] Part of the active groups are hydrolyzed by shaking the gel at room temperature for 4 hr in about 10 ml of 0.1 *M* $NaHCO_3$, pH 8.3. The gel is then incubated at room tempera-

[5] H. Weiss, B. Juchs, and B. Ziganke, this series, Vol. 53, p. 98.

[6] I. Parikh, S. March, and P. Cuatrecasas, this series, Vol. 34, p. 77.

TABLE I
PURIFICATION OF CYTOCHROME-*c* OXIDASE FROM VARIOUS SOURCES BY AFFINITY CHROMATOGRAPHY ON HORSE HEART CYTOCHROME *c*-CNBr-ACTIVATED SEPHAROSE 4B

Source	Comment	Detergent	References
Bovine heart	Further purification of isolated enzyme	Deoxycholate	*a*
	One-step purification	Laurylmaltoside	*b*
Neurospora crassa	Main purification step	Triton X-100	*c*
Rat liver	Main purification step	Tween 80, Triton X-100	*d*
	Final purification step	Laurylmaltoside	*b*

[a] T. Ozawa, M. Okumura, and K. Yagi, *Biochem. Biophys. Res. Commun.* **65,** 1102 (1975).
[b] D. A. Thompson and S. Ferguson-Miller, *Biochemistry* **22,** 3178 (1983).
[c] H. Weiss and H. J. Kolb, *Eur. J. Biochem.* **99,** 139 (1979).
[d] R. J. Rascati and P. Parsons, *J. Biol. Chem.* **254,** 1586 (1979).

ture for 2 hr in 5 ml of 0.1 *M* $NaHCO_3$, pH 8, containing 0.5 *M* NaCl and 4 mg of horse heart cytochrome *c* (1 mg per milliliter of swollen gel). The nonbound cytochrome *c* is removed by washing the gel with 10 ml of coupling buffer, and the remaining active groups are reacted at 4° for 1 hr with 50 ml of 1 *M* ethanolamine–HCl, pH 8. The gel is then washed at 4° in turn with 1 *M* NaCl, 0.1 *M* sodium acetate, pH 4, 1 *M* NaCl, and 0.1 *M* sodium borate, pH 8, and subsequently with the buffer which will be used for the affinity chromatography procedure, e.g., 50 m*M* Tris–HCl, pH 7.4.

Applications

These are summarized in Table I. A detailed procedure for the purification of cytochrome-*c* oxidase from *Neurospora crassa* has been published in an earlier volume in this series,[5] and the procedures for the purification of the enzyme from other sources can be found in the original literature as indicated in Table I.

Comments

The above-described procedure is not applicable to all starting materials and with all detergents. For example, it was reported that the bovine heart and yeast enzymes cannot be purified in Triton X-100 by using this method.[7] It was assumed that Triton X-100 might not be able to disperse

[7] H. Weiss and H. J. Kolb, *Eur. J. Biochem.* **99,** 139 (1979).

the enzyme completely. On the other hand, it was shown that bovine heart cytochrome-*c* oxidase dispersed in Triton X-100 can be purified by using the affinity gel described below. It has to be stressed that cytochrome *c* may be attached to the CNBr-activated gel through some of its lysine residues which are known to be crucial for the interaction with the oxidase. This drawback can apparently be overcome, at least partially, by inactivating some of the CNBr-activated groups of the gel, as mentioned above, in order to avoid multipoint attachment of cytochrome *c*. Indeed the gel is useful in some applications.

Yeast Cytochrome c Linked to Activated Thiol-Sepharose 4B

Materials

For preparation of 30 ml of gel:

- Freeze-dried activated thiol-Sepharose 4B (8 g) (Pharmacia Fine Chemicals AB, Uppsala, Sweden)
- 75 ml of 50 m*M* Tris–HCl, pH 7.2
- Cytochrome *c* (90 mg) from *Saccharomyces cerevisiae* (Sigma Chemical Company, St. Louis, MO 63178; type VIII)
- 50 ml of 50 m*M* sodium acetate, pH 4.5, containing 1.5 m*M* 2-mercaptoethanol
- 1 liter of 50 m*M* Tris–HCl, pH 7.2, containing 1 *M* NaCl, 1% Triton X-100, and 1 m*M* potassium hexacyanoferrate(III)

Preparation[8]

Activated thiol-Sepharose 4B (8 g) is swollen and washed at room temperature for 15 min with 50 m*M* Tris–HCl, pH 7.2, yielding about 30 ml of swollen gel. Alternatively, the gel can be prepared from Sepharose 4B according to a method published in a previous volume in this series.[9] The gel is suspended in 50 m*M* Tris–HCl, pH 7.2, and 3 mg of *S. cerevisiae* cytochrome *c* per milliliter of swollen gel are added in the same buffer. The final volume should be about 50 ml. The suspension is shaken gently overnight at 4°, after which time practically all the cytochrome *c* is bound to the gel. The almost colorless buffer is removed and the gel is incubated for 30 min in 50 ml of 50 m*M* sodium acetate, pH 4.5, containing 1.5 m*M* 2-mercaptoethanol, in order to inactivate residual active groups. Under these conditions no covalently bound cytochrome *c* is removed

[8] A. Azzi, K. Bill, and C. Broger, *Proc. Natl. Acad. Sci. U.S.A.* **79,** 2447 (1982).

[9] K. Brocklehurst, J. Carlsson, M. P. J. Kierstan, and E. M. Crook, this series, Vol. 34, p. 531.

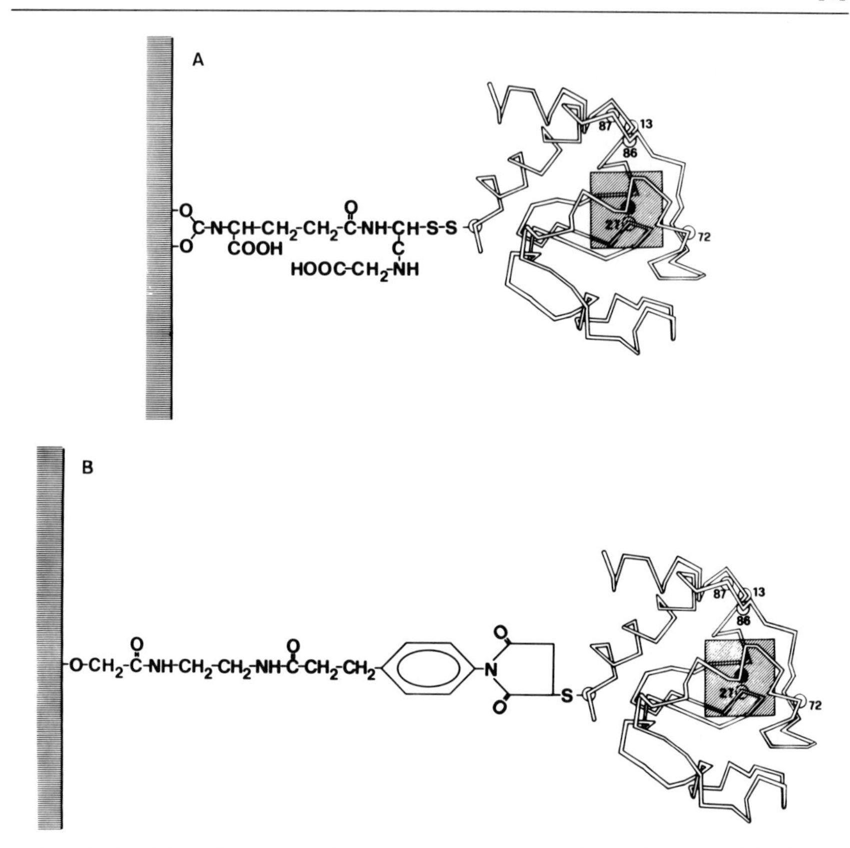

FIG. 1. (A) Schematic drawing of yeast cytochrome *c* linked to activated thiol-Sepharose 4B. The hatched square in cytochrome *c* represents the heme plane and the numbers indicate the lysine residues, which are important in the interaction of this protein with cytochrome-*c* oxidase. Notice that cytochrome *c* is attached through a cysteine which is located on the opposite side of the molecule. (B) Schematic drawing of yeast cytochrome *c* linked to Affi-Gel 102 through the heterobifunctional reagent SMPB. The spacer between the gel and cytochrome *c* is longer as compared to the gel above and not sensitive to reducing agents.

from the gel. The gel is then washed with 1 liter of 50 m*M* Tris–HCl, pH 7.2, containing 1 *M* NaCl, 1% Triton X-100 (the detergent to be used for the purification procedure described below), and 1 m*M* potassium hexacyanoferrate(III) in order to remove not covalently bound cytochrome *c* and to reoxidize the gel fully. The amount of cytochrome *c* remaining bound to the gel is about 125 nmol/ml. A schematic drawing of the gel is shown in Fig. 1A.

As an example for the use of the affinity chromatography technique, the purification of cytochrome-*c* oxidase from bovine heart mitochondria is presented in detail.

Materials

The following are used for preparation of 30 ml of affinity gel:

Bovine heart mitochondria isolated according to a standard method, such as described in Ref. 10, about 1 g of protein
100 ml of 10 m*M* Tris–HCl, pH 7.2, containing 150 m*M* KCl
1 liter of 10 m*M* Tris–HCl, pH 7.2
500 ml of 50 m*M* Tris–HCl, pH 7.2
Triton X-100 stock solution, 25% in water
200 ml of 50 m*M* Tris–HCl, pH 7.2, containing 1% Triton X-100
200 ml of 50 m*M* Tris–HCl, pH 7.2, containing 0.1% Triton X-100
100 ml of 50 m*M* Tris–HCl, pH 7.2, containing 0.1% Triton and 50 m*M* NaCl

The entire procedure is carried out at 4°. The mitochondria are suspended in 50 ml of 150 m*M* KCl buffered with 10 m*M* Tris–HCl, pH 7.2, and depleted of their outer membrane and cytochrome *c* by dilution into 10 volumes of 10 m*M* Tris–HCl, pH 7.2, and subsequent centrifugation at 27,000 *g* for 20 min. The pellet of mitoplasts is suspended, diluted, and centrifuged again, and then taken up in 50 m*M* Tris–HCl, pH 7.2, at a protein concentration of 2 mg/ml. Triton X-100 is added to a final concentration of 1%. After incubation for about 30 min the suspension is centrifuged at 27,000 *g* for 1 hr. The supernatant has a protein concentration of about 1.1 mg/ml. About 400 ml of this solution are applied at a flow rate of 35 ml/hr to the affinity column (2 × 10 cm), equilibrated in 50 m*M* Tris–HCl, pH 7.2, containing 1% Triton X-100. Only cytochrome-*c* oxidase and reductase bind to the gel; the other proteins are eluted without retardation. After loading, the column is washed with 50 m*M* Tris–HCl, pH 7.2, containing 0.1% Triton X-100 until the eluate is free of protein and hemes. Elution of the cytochrome-*c* oxidase is carried out by adding 50 m*M* NaCl to the washing buffer or by applying a gradient of NaCl from 0 to 150 m*M* in the same buffer. The enzyme is eluted at a salt concentration of 50 m*M* or less, whereas cytochrome-*c* reductase remains bound to the column and may be eluted at higher salt concentration. The eluted enzyme can be concentrated by using an Amicon filter PM30. Ratios of 10 nmol heme *a* per milligram protein are reached with this preparation. The enzyme contains about 30–40 mol of phospholipid per mol of heme aa_3. It

[10] A. L. Smith, this series, Vol. 10, p. 81.

TABLE II
PURIFICATION OF CYTOCHROME-*c* OXIDASE FROM VARIOUS SOURCES BY AFFINITY CHROMATOGRAPHY ON YEAST CYTOCHROME *c*-ACTIVATED THIOL-SEPHAROSE 4B

Source	Comment	Detergent	References
Bovine heart	One-step purification	Triton X-100	*a*
Bacillus subtilis	One-step purification	Laurylmaltoside	*b*
Rhodopseudomonas sphaeroides	Main purification step	Triton X-100	*c*
Paracoccus denitrificans	Final purification step	Triton X-100	*d*

[a] A. Azzi, K. Bill, and C. Broger, *Proc. Natl. Acad. Sci. U.S.A.* **79,** 2447 (1982).
[b] W. de Vrij, A. Azzi, and W. N. Konings, *Eur. J. Biochem.* **131,** 97 (1983).
[c] R. B. Gennis, R. P. Casey, A. Azzi, and B. Ludwig, *Eur. J. Biochem.* **125,** 189 (1982).
[d] M. Solioz, E. Carafoli, and B. Ludwig, *J. Biol. Chem.* **257,** 1579 (1982).

consists of 12–13 polypeptides, as in the traditional preparations, as analyzed by polyacrylamide gel electrophoresis according to Kadenbach *et al.*[11] The spectral characteristics of the enzyme are identical to the best conventional preparations.

Further Applications

Table II summarizes further applications of the method for the purification of cytochrome-*c* oxidases from various sources. The general procedures are similar to the one described in detail above. Some technical details can be found in the original literature as indicated in Table II.

Comments

It is important to block activated thiol groups after coupling cytochrome *c* to the gel in order to prevent cysteine-containing proteins to bind covalently to the resin.

It was observed that under some circumstances a gel with lower substitution of cytochrome *c* has a relatively higher capacity for binding of cytochrome-*c* oxidase (Bernd Ludwig, personal communication). This might be due to less steric hindrance.

The method is very useful also for microscale purifications and can be used for the preparation of other proteins which interact specifically with cytochrome *c,* as cytochrome-*c* reductase[8] or photosynthetic reaction

[11] B. Kadenbach, J. Jarausch, R. Hartmann, and P. Merle, *Anal. Biochem.* **129,** 517 (1983).

centers, which in fact have been copurified with cytochrome-*c* oxidase from *Rhodopseudomonas sphaeroides*.[12]

The preparation of the affinity gel just described is extremely simple and its applicability is wide; it cannot be used, however, under reducing conditions because the S–S bridge between cytochrome *c* and the column would be broken. Recently, a new type of gel has been proposed[13] which not only is stable under reducing conditions, but also has a higher capacity for cytochrome-*c* oxidase, probably as a consequence of less steric hindrance due to a longer spacer. The preparation of this gel is also simple and uses commercially available chemicals. It is described below.

Yeast Cytochrome c Linked to Affi-Gel 102-SMPB

Materials

For preparation of 15 ml of gel:
- 15 ml of Affi-Gel 102 (Bio-Rad, Richmond, CA 94804)
- 500 ml of 25 m*M* sodium phosphate, pH 7.4
- 75 ml of anhydrous dioxane
- 20 mg of succinimidyl-4-(*p*-maleimidophenyl)butyrate (SMPB) (Pierce Chemical Company, Rockford, IL 61105)
- 12 mg of cytochrome *c* from *S. cerevisiae* (Sigma Chemical Company, St. Louis, MO 63178; type VIII)
- 20 ml of swollen Sephadex G-25 (Pharmacia Fine Chemicals AB, Uppsala, Sweden)
- 200 ml of 25 m*M* sodium phosphate, pH 7.4, containing 1 *M* NaCl

Preparation[13]

Swollen Affi-Gel 102 (15 ml) is washed on a sintered glass funnel with 25 m*M* sodium phosphate, pH 7.4. After being drained from buffer, the gel is washed three times with about 10 ml of anhydrous dioxane and finally suspended in 5 ml of dioxane. SMPB (20 mg) is dissolved in 2 ml of dioxane and added to the gel. The suspension is stirred under nitrogen for 3 hr in the dark. The gel is washed with dioxane to remove unbound SMPB and subsequently with several small portions of 25 m*M* sodium phosphate, pH 7.4. Finally, it is suspended in 10 ml of the buffer. Yeast cytochrome *c* (12 mg) is dissolved in the same buffer, reduced by the addition of sodium dithionite, and passed through a Sephadex G-25 column (2 × 20 cm) equilibrated in the same buffer. About 900 nmol of the

[12] R. B. Gennis, R. P. Casey, A. Azzi, and B. Ludwig, *Eur. J. Biochem.* **125,** 189 (1982).
[13] K. Bill and A. Azzi, *Biochem. Biophys. Res. Commun.* **120,** 124 (1984).

reduced cytochrome *c* are added to the gel, and the suspension is stirred for 12 hr at 4° in the dark. The gel is washed free of nonbound cytochrome *c* with 25 m*M* sodium phosphate, pH 7.4, containing 1 *M* sodium chloride. More than 80% of the added cytochrome *c* remains bound to the resin. A schematic drawing of the gel is shown in Fig. 1B.

Comments

This gel has originally been used for measuring the binding affinities of cytochrome-*c* oxidase and reductase for oxidized and reduced cytochrome *c*. It will result in the useful purification of cytochrome-*c* oxidases from various sources not only because of its high capacity, but also because it can be used under reducing conditions where the binding affinity for cytochrome-*c* oxidase is higher.

[7] Resolution of Cytochrome-*c* Oxidase

By ANGELO AZZI, CLEMENS BROGER, KURT BILL, and MICHAEL J. CORBLEY

Cytochrome-*c* oxidase from bovine heart mitochondria is a multisubunit complex consisting of 13 polypeptides[1] and four prosthetic groups, two hemes and two coppers.[2] There have been mainly two approaches to study the role of the individual subunits in the function of the enzyme and the location of the prosthetic groups. Bacterial cytochrome-*c* oxidases[3] have been found to consist of a smaller number of subunits than mammalian enzymes, and in some cases it has been shown that they possess immuno-cross-reactivity with the largest subunits of the mitochondrial enzymes. In a second approach, bovine heart cytochrome-*c* oxidase has been split into subcomplexes and their polypeptide content and prosthetic groups have been analyzed. Alternatively, cytochrome-*c* oxidase has been depleted specifically of certain subunits and the function of the depleted complex has been studied. The purification of bacterial cytochrome-*c* oxidases by using affinity chromatography is described in an-

[1] B. Kadenbach, J. Jarausch, R. Hartmann, and P. Merle, *Anal. Biochem.* **129,** 517 (1983).
[2] M. Wikström, K. Krab, and M. Saraste, *in* "Cytochrome Oxidase, A Synthesis," p. 55. Academic Press, 1981.
[3] B. Ludwig, *Biochim. Biophys. Acta* **594,** 177 (1980).

other chapter in this volume.[4] The present chapter is concerned with the second approach.

Splitting of Bovine Heart Cytochrome-c Oxidase into Subcomplexes

Materials

The following are used with 3.5 mg of oxidase:

- Cytochrome-*c* oxidase with a low lipid content, isolated according to a standard procedure, as described in Ref. 5, 3.5 mg of protein
- 500 ml of 30 m*M* HEPES/Li^+, pH 7.4, containing 0.05% Triton X-100 and 0.0005% lithium dodecyl sulfate
- Stock solution of lithium dodecyl sulfate (25%)
- Swollen Sephadex G-150 superfine (100 ml) (Pharmacia Fine Chemicals AB, Uppsala, Sweden)
- 200 ml 30 m*M* HEPES/Li^+, pH 7.4, containing 0.05% lithium dodecyl sulfate and 0.05% Triton X-100

Method

The following is used for the resolution of bovine heart cytochrome-*c* oxidase into subcomplexes[6]: The entire procedure is carried out at 4°. The frozen sample of 3.5 mg of cytochrome-*c* oxidase is thawed and dissolved at a concentration of 5 mg/ml in 30 m*M* HEPES/Li^+, pH 7.4, containing 0.05% Triton X-100 and 0.0005% lithium dodecyl sulfate. Lithium dodecyl sulfate (0.0005%) is added to increase the enzymatic activity of the native oxidase, and the lithium salt is used to avoid precipitation of the detergent in the cold. The oxidase is dialyzed for 4 hr against 500 volumes of the same buffer in order to remove the cholate and the sucrose present in the storage buffer. Subsequently lithium dodecyl sulfate is added from a stock solution to a final concentration of 0.2%, and the sample is kept on ice for 10 min. A sample of about 700 μl is then immediately applied to a Sephadex G-150 superfine column (2 × 30 cm) and eluted first with 5 ml of the buffer in which the oxidase is dissolved and then with 30 m*M* HEPES/Li^+, pH 7.4, 0.05% lithium dodecyl sulfate, and 0.05% Triton X-100. The flow rate is about 2.5 ml/hr.

The fractions are analyzed for heme *a* and protein content and the polypeptide composition is determined by polyacrylamide gel electrophoresis. The subcomplexes are eluted from the column in the following

[4] C. Broger, K. Bill, and A. Azzi, this volume [6].

[5] C. Yu, L. Yu, and T. E. King, *J. Biol. Chem.* **250,** 1383 (1975).

[6] M. J. Corbley and A. Azzi, *Eur. J. Biochem.* **139,** 535 (1984).

order: (1) subunits I,III; (2) subunits I,II,III; (3) subunits II,IV,V,VI,VII; (4) subunits IV,V,VI,VII; and (5) free heme.

Comments

The cytochrome-*c* oxidase subcomplexes, as prepared by the method given above, are enzymatically not active in a polarographic assay. However some of them contain heme. Heme is present only in the complexes that contain the major three subunits of cytochrome-*c* oxidase. Subunits III can be shown not to contain heme, since the heme/protein ratio of cytochrome-*c* oxidase is increased by depletion of subunit III, as shown below.

The spectrum of free heme is different from the spectrum of the partially denatured subcomplexes, indicating that the heme is not possibly associated to a denatured system. The heme-containing subcomplexes can be reduced and oxidized, and they exhibit also CO binding capacity.

Depletion of Bovine Heart Cytochrome-c Oxidase of Subunit III

Several methods are available to deplete cytochrome-*c* oxidase of its subunit III. The different published methods are listed and compared in the table. In all procedures, in addition to subunit III, two or three of the smaller subunits are removed.

As an example, one of these techniques is presented in more detail[7] below. The procedure is based on the specific formation of a disulfide bridge bewteen cysteine-102 located in the rear of the yeast cytochrome *c* molecules and a cysteine of subunit III of cytochrome-*c* oxidase.[8] Yeast cytochrome *c* is linked to CNBr-activated Sepharose 4B through lysine residues which are located mainly on the front side of the protein, leaving the domain of cysteine-102 accessible for the interaction with subunit III of the oxidase. After formation of the disulfide bridge between the two cysteines, the subunit III depleted oxidase can be eluted from the gel under nonreducing conditions (see Fig. 1).

Materials

The following are needed for subunit III depletion of 2.5 mg of oxidase:

Cytochrome-*c* oxidase from bovine heart, with a low lipid content, isolated according to a standard procedure, as described in Ref. 5, and frozen in small portions, 2.5 mg of protein

[7] K. Bill and A. Azzi, *Biochem. Biophys. Res. Commun.* **106,** 1203 (1982).

[8] W. Birchmeier, C. E. Kohler, and G. Schatz, *Proc. Natl. Acad. Sci. U.S.A.* **73,** 4334 (1976).

METHODS TO DEPLETE CYTOCHROME-*c* OXIDASE OF SUBUNIT III

Source of oxidase	Detergent/method	Subunits removed other than III[a]	Reference
Bovine heart	Triton X-100, specific binding of subunit III to gel	VIa,b VIIa	*b*
Bovine heart	Triton X-100, high Tris concentration, high pH, DEAE chromatography	VIa,b VIIa	*c*
Bovine heart	Triton X-100, chymotrypsin	VIa,b VIIa	*d*
Rat liver	Laurylmaltoside, affinity chromatography on cytochrome *c* gel[e]	ND	*f*

[a] B. Kadenbach, J. Jarausch, R. Hartmann, and P. Merle, *Anal. Biochem.* **129,** 517 (1983).

[b] K. Bill and A. Azzi, *Biochem. Biophys. Res. Commun.* **106,** 1203 (1982).

[c] T. Penttilä, *Eur. J. Biochem.* **133,** 355 (1983).

[d] F. Malatesta, G. Georgevich, and R. A. Capaldi, *in* "Structure and Function of Membrane Proteins" (F. Palmieri *et al.*, eds.). Elsevier, Amsterdam, 1983.

[e] Detachment of subunit III in this preparation may be caused by proteolytic digestion induced by endogenous proteases.

[f] D. A. Thompson and S. Ferguson-Miller, *Biochemistry* **22,** 3178 (1983).

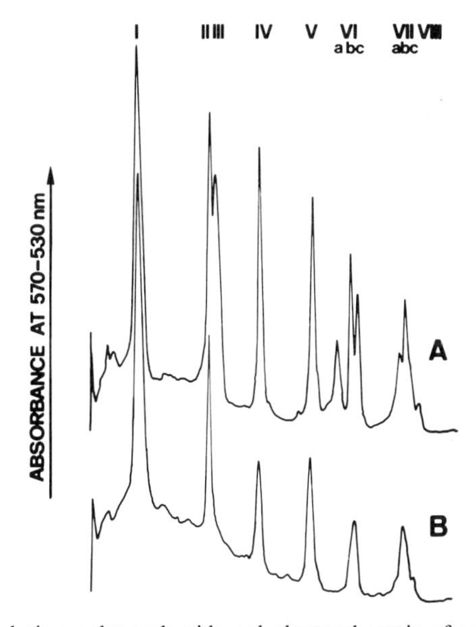

FIG. 1. High-resolution polyacrylamide gel electrophoresis of cytochrome-*c* oxidase. Scans of the gel after staining with Coomassie Brilliant Blue. (A) Oxidase isolated by a standard procedure. (B) Oxidase depleted of subunit III and lower molecular weight subunits by the method described here.

Swollen Sephadex G-100 (30 ml) (Pharmacia Fine Chemicals AB, Uppsala, Sweden)

100 ml of 10 m*M* sodium phosphate, pH 7.2, containing 2% sodium cholate

Yeast cytochrome-*c*–Sepharose 4B gel (2 ml) prepared according to the method of Weiss, described in another article in this volume,[4] but by using yeast instead of horse heart cytochrome *c*

25 ml of 10 m*M* sodium phosphate, pH 7.2, containing 0.3 *M* 2-mercaptoethanol

400 ml of 10 m*M* sodium phosphate, pH 7.2

25 ml of 10 m*M* sodium phosphate, pH 8.0, containing 1.5 m*M* 2,2′-dipyridyl disulfide

100 ml of 10 m*M* sodium phosphate, pH 7.2, containing 150 m*M* NaCl and 2% sodium cholate

60 ml of 10 m*M* sodium phosphate, pH 7.2, containing 200 m*M* NaCl and 1.5% Triton X-100

1.5 ml of 10 m*M* sodium phosphate, pH 7.2, containing 150 m*M* NaCl, 2% sodium cholate, and 0.3 *M* 2-mercaptoethanol

Cytochrome-*c* oxidase (2.5 mg) is thawed and incubated for 2 hr at 4° in 100 μl of 10 m*M* sodium phosphate, pH 7.2, containing 2% sodium cholate and subsequently passed through a Sephadex G-100 column (1 × 34 cm) equilibrated in the same buffer. Cytochrome *c* gel (2 ml) is reduced with 25 ml of 0.3 *M* 2-mercaptoethanol in 20 m*M* sodium phosphate buffer. After incubation in this buffer for 20 min, the gel is washed free of mercaptoethanol by using 10 m*M* sodium phosphate buffer. The thiol group of cytochrome *c* is activated by incubating the gel for 30 min at room temperature in 25 ml of 10 m*M* sodium phosphate, pH 8, containing 1.5 m*M* 2,2′-dipyridyl disulfide. The excess dipyridyl disulfide is removed by washing with 10 m*M* phosphate buffer.

About 11 nmol of the oxidase, as eluted from the Sephadex G-100 column, are diluted with the same buffer to a volume of 3 ml and titrated with NaOH to pH 8.5. After incubation for 5 min at 37°, the activated cytochrome *c* gel is added and the suspension is stirred gently at 37° for 20 min. The gel is then washed with 50 ml of phosphate buffer, containing 150 m*M* NaCl and 2% sodium cholate, and finally with 25 ml of the buffer alone. More than 60% of the oxidase remains covalently bound to the gel.

The gel is drained of buffer and incubated overnight at 4° with constant gentle stirring in 1.5 ml of 10 m*M* sodium phosphate buffer, containing 200 m*M* NaCl and 1.5% Triton X-100. Over this time cytochrome-*c* oxidase lacking subunit III is released from the gel, and 80% of the originally bound oxidase is recovered in the buffer.

The gel is washed with 50 ml of 10 m*M* sodium phosphate, containing 200 m*M* NaCl and 1.5% Triton X-100, then with 25 ml of the phosphate buffer alone, and finally with the phosphate buffer containing 150 m*M* NaCl and 2% sodium cholate.

To recover subunit III, the gel is then incubated for 1 hr at 4° with stirring in 1.5 ml of the last washing buffer supplemented with 0.3 *M* 2-mercaptoethanol. Subunit III is eluted slightly contaminated with other subunits.

Triton X-100 can be exchanged practically quantitatively by cholate by passing the subunit III-depleted enzyme through a Sephadex G-100 column equilibrated with 10 m*M* sodium phosphate, pH 7.2, containing 2% sodium cholate. Alternatively, the enzyme can be bound to DEAE–BioGel (Bio-Rad), washed with the desired detergent, and eluted with a buffer containing 300 m*M* KCl.

The visible spectrum of the subunit III-depleted oxidase is similar but not identical to the one of the native oxidase. The heme/protein ratio is increased to 15 nmol of heme *a* per milligram of protein. The electron transfer activity of the enzyme is slightly affected by the subunit III depletion procedure. No heme is found in the subunit III-containing fraction eluted from the cytochrome *c* gel after reduction with 2-mercaptoethanol.

Comments

The subunit III depletion procedure can also be performed in Tris–HCl buffers using the same molarity and pH as indicated for the phosphate buffers.

The cytochrome *c* gel can be regenerated and used several times, although best yields and purity are obtained with freshly prepared gels.

This method avoids the use of proteases and too high concentrations of detergents and is carried out mostly at physiological pH, thus ensuring that the enzyme remains as much as possible in its native state.

[8] Functional Reconstitution of Proton-Pumping Cytochrome-*c* Oxidase in Phospholipid Vesicles

By MICHELE MÜLLER, MARCUS THELEN, PAUL O'SHEA, and ANGELO AZZI

Introduction

One of the most important methodological approaches utilized to study the structure and function relationships of membrane-bound proteins has been their reconstitution in closed phospholipid vesicles of known composition and size.

Although the cholate dialysis technique, introduced by Racker and coworkers,[1] is the one still preferentially used to reconstitute membrane proteins, the synthesis and employment of new detergents has forstered new research in this field and the development of alternative reconstitution techniques, such as detergent dialysis, detergent dilution, sonication, spontaneous incorporation in the presence or absence of detergents, fusion, and hydrophobic adsorption.[2-9]

Since the method of choice is that of detergent dialysis, the selection of alternative reconstitution procedures is related to the type of detergent (whether it can be dialyzed) and the type of protein (whether it is stable enough to remain native after relatively long dialysis times). With the use of the dialysis technique, under controlled conditions, it is possible to produce vesicles of defined size[10,11] having the protein component vectorially oriented.

[1] Y. Kagawa and E. Racker, *J. Biol. Chem.* **246,** 5477 (1971).

[2] E. Racker, B. Violand, S. O'Neal, M. Alfonzo, and J. Telford, *Arch. Biochem. Biophys.* **198,** 470 (1979).

[3] E. Racker, this series, Vol. 55, p. 699.

[4] F. Szoka and D. Papahadjopoulos, *Ann. Biophys. Bioenerg.* **9,** 467 (1980).

[5] H. Hauser, *Trends Pharmacol. Sci.* **3,** 274 (1982).

[6] J. Brunner, P. Skabral, and H. Hauser, *Biochim. Biophys. Acta* **455,** 322 (1976).

[7] L. T. Mimms, G. Zampighi, Y. Nozaki, C. Tanford, and J. A. Reynolds, *Biochemistry* **20,** 833 (1980).

[8] A. J. Furth, *Anal. Biochem.* **109,** 207 (1980).

[9] P. S. J. Cheetham, *Anal. Biochem.* **92,** 447 (1979).

[10] O. Zumbuehl and H. G. Weder, *Biochim. Biophys. Acta* **640,** 252 (1981).

[11] R. A. Schwendener, M. Asanger, and H. G. Weder, *Biochem. Biophys. Res. Commun.* **100,** 1055 (1981).

This chapter describes some of the techniques used for the reconstitution of bovine heart cytochrome-*c* oxidase.

Hydrophobic Adsorption

This technique has not been widely used to reconstitute cytochrome-*c* oxidase due to the fact that the vesicles obtained with it appear in general not to be competent in catalyzing vectorial proton translocation. Nevertheless, there have been reports of successful reconstitutions with a high respiratory control ratio (the ratio between maximal uncoupler-stimulated electron transfer rate and that in the absence of the uncoupler, RCR) in the case of rat liver[12,13] and rat brown fat[14] cytochrome-*c* oxidase. The principle of this type of reconstitution consists of the removal of Triton X-100, used as detergent for the isolation of the enzyme, in the presence of added phospholipid by use of adsorbing polystyrene material such as Bio-Beads SM-2[15] or Amberlite XAD-2.[9] A batch or a column procedure can be employed. The same principle can also be used for the removal of other types of nonionic detergents, such as those belonging to the group of the alkylpolyoxyethylenes.[16]

Treatment of Amberlite XAD-2

The resin contains in general a large amount of impurities, originated from large-scale industrial preparation, which should be removed before use. The beads are washed with occasional gentle stirring using pure methanol in a glass beaker. After 5–10 min the turbid supernatant is decanted or filtered. The operation is repeated with fresh methanol until no more impurities can be extracted. The beads are then washed extensively with double-distilled water until all of them are rapidly sedimenting. Floating beads are discarded. Use and storage at room temperature should be limited to about 1 month. After this time a short treatment with methanol and water is necessary.

Preparation of Lipids

Pure phospholipids or soybean phospholipids (asolectin; Associated Concentrates, NY) are dried from organic solvent solutions with an N_2

[12] M. Mattenberger, G. Hugentobler, and P. Gazzotti, *Experientia* **36,** 728 (1980).

[13] P. Gazzotti, *in* "Enzymes, Receptors and Carriers of Biological Membranes" (A. Azzi, U. Brodbeck, and P. Zahler, eds.), p. 106. Springer-Verlag, Berlin, 1984.

[14] P. Gazzotti, Laboratory for Biochemistry, ETH Zürich, personal communication.

[15] P. W. Holloway, *Anal. Biochem.* **53,** 304 (1973).

[16] M. Veno, C. Tanford, and J. A. Reynolds, *Biochemistry* **23,** 3070 (1984).

stream. Traces of the organic solvent are removed by vacuum evaporation. The phospholipids are hydrated in 100 m*M* choline-Cl, 30 m*M* KCl, 10 m*M* HEPES, pH 7.4, containing a nonionic detergent (typically 1%) belonging to the series of Triton X or of the alkylpolyoxyethylene derivatives and sonicated under N_2 until the suspension is clarified (45 min at 4° with a Branson B15 sonifier at 50% duty cycle). Crude asolectin may be purified according to Kagawa and Racker.[1] An additional step can be introduced consisting of stirring the crude asolectin solution with Al_2O_3 in order to bind peroxy compounds, which have been found to be a contaminant of commercially available crude phospholipid preparations. This step is followed by filtration to remove the Al_2O_3 together with the contaminating peroxides.[17]

Reconstitution

Purified protein is added to the phospholipid detergent mixture to the desired final ratio and incubated for 5–10 min at 4°. Reconstitutions carried out at 4° and with a lipid-to-protein ratio of 30 (w/w) give typically monolamellar vesicles with a size of about 65 nm, larger than those obtained by cholate dialysis (25–40 nm). Reconstitutions carried out at room temperature under the same conditions result in larger, multilamellar vesicles (1–2 μm).

To the protein–phospholipid–detergent mixture the adsorbing material is added after preequilibration with the same buffer in which the lipid/detergent mixture (~0.5 g/mg detergent) is dispersed.[18] After 3 hr of incubation with gentle stirring the supernatant is decanted and fresh beads are added. After 2–4 changes, the vesicles are formed. Removal of detergent can be followed by adding traces of radioactive detergent. Radioactivity of the vesicles is then counted after each bead change. In the case of Triton X-type detergents the absorbance at 276–300 nm can be used as well [0.01% Triton X-100 (Fluka AG, Buchs, Swizerland) has an absorbance of $A_{277-300} \simeq 0.53$]. The presence of detergent can be detected empirically by shaking the tube. The formation of foam indicates the presence of detergent. Another alternative is to pipette a drop (20–40 μl) of vesicles on a piece of parafilm and to compare its diameter with that of a drop of the buffer without detergent. A larger diameter indicates the presence of detergent.

Comments

An advantage of this reconstitution relative to the cholate dialysis technique is that it can be used for a wide range of starting volumes (from

[17] P. S. O'Shea, G. Petrone, R. P. Casey, and A. Azzi, *Biochem. J.* **219,** 719 (1984).
[18] M. Müller, D. Cheneval, and E. Carafoli, *Eur. J. Biochem.* **140,** 447 (1984).

0.5 ml to liters) and that it minimizes wasting of expensive buffers. On the other hand, adsorption of protein to the beads may occur. This phenomenon can be diminished by preincubating the beads with the buffer and/or with low amounts of phospholipids. Particular attention in establishing the reconstitution conditions should be given to the temperature and to the lipid/protein ratio. For example, a reconstitution carried out at 4° at low lipid/protein ratio (1–2) results in large multilamellar vesicles, as it occurs when the reconstitution is carried out at 20° with a high lipid/protein ratio (30).

Detergent Dialysis

The best technique to obtain a functional reconstitution of cytochrome oxidase is cholate dialysis.[19–24] With this procedure, cytochrome-*c* oxidase vesicles were obtained which were used for studying the lipid requirement for activity,[25] the rotational mobility of tightly associated cardiolipin molecules,[26] temperature-induced conformational changes,[27,28] electron spin resonance studies,[29] subunit topology,[30,31] and rotational mobility of the enzyme.[32]

Experimental Procedure

Preparation of Lipids. This step is essentially identical to the one described before, with the difference that cholate is used as detergent to a final concentration of 1–2%.

"Normal" and "Slow" Cholate Dialysis[33]

Asolectin (40 mg/ml) is sonicated with a Branson B15 sonifier (50% duty cycle) for 10 min in 1.5% K^+-cholate and 185 m*M* K^+-HEPES under

[19] K. Krab and M. K. F. Wikström, *Biochim. Biophys. Acta* **504,** 200 (1978).
[20] R. P. Casey, J. B. Chappell, and A. Azzi, *Biochem. J.* **182,** 149 (1979).
[21] E. Sigel and E. Carafoli, *Eur. J. Biochem.* **111,** 299 (1980).
[22] R. P. Casey, M. Thelen, and A. Azzi, *J. Biol. Chem.* **255,** 3994 (1980).
[23] R. P. Casey, P. S. O'Shea, J. B. Chappell, and A. Azzi, *Biochim. Biophys. Acta* **765,** 30 (1984).
[24] M. Thelen, P. S. O'Shea, G. Petrone, and A. Azzi, *J. Biol. Chem.* **260,** 3626 (1985).
[25] T. O. Madden, M. J. Hope, and R. P. Cullis, *Biochemistry* **22,** 1970 (1983).
[26] M. B. Cable and G. L. Powell, *Biochemistry* **19,** 5679 (1980).
[27] S. Kawato, A. Ikegami, S. Yoshida, and Y. Orii, *Biochemistry* **19,** 1598 (1980).
[28] S. Kawato, S. Yoshida, Y. Orii, A. Ikegami, and K. Kinoshita, Jr., *Biochim. Biophys. Acta* **634,** 85 (1981).
[29] R. P. Casey, C. Broger, and A. Azzi, *Biochim. Biophys. Acta* **638,** 86 (1981).
[30] R. Bisson, G. C. M. Steffens, and G. Buse, *J. Biol. Chem.* **257,** 6716 (1982).
[31] R. Bisson, C. Montecucco, H. Gutweniger, and A. Azzi, *J. Biol. Chem.* **254,** 9962 (1979).
[32] S. Kawato, E. Sigel, E. Carafoli, and R. J. Cherry, *J. Biol. Chem.* **256,** 7518 (1981).
[33] M. Thelen, P. S. O'Shea, and A. Azzi, *Biochem. J.* **227,** 163 (1985).

N_2 at 4°, pH 7.3. The phospholipid/detergent mixture is then centrifuged at 30,000 *g* for 10 min. To the supernatant cytochrome-*c* oxidase is added to a final concentration of 7.5 μM heme aa_3. The enzyme is preincubated in 1.5% K^+-cholate for 10 min on ice before the addition. The suspension is dialyzed according to procedure A (slow) or B (normal) of Table I. The vesicles taken from the dialysis tubing are then centrifuged as before to remove any large lipid particles which may have formed during the dialysis process (see Fig. 1, page 84).

Reconstitution of Cytochrome-c Oxidase Vesicles Containing Fluorescein-Phosphatidylethanolamine[24,34]

These vesicles are prepared with the following modifications of the above-described procedure. Asolectin and fluorescein-phosphatidylethanolamine (40:1, w/w) are mixed in $CHCl_3$, dried under N_2, and sonicated in 125 m*M* KCl, 25 m*M* sucrose, and 1.8 m*M* $CaCl_2$, pH 7.3. The final enzyme concentration is 7.5 μM in terms of heme aa_3 content.

The dialysis is carried out as described in Table I, procedure C.

Characterization of the Reconstituted Vesicles

Respiratory Control Ratio (RCR). This can be defined as the rate of electron transfer in the presence of uncoupler and valinomycin divided by the rate in their absence.

The determination of the respiratory control ratio may be carried out spectroscopically or polarographically by an oxygen electrode. The main difference between these two assays is that in the spectroscopic assay cytochrome *c* becomes oxidized and the reaction is started by the addition of the vesicles containing the oxidase. In the polarographic assay cytochrome *c* remains practically reduced by ascorbate, and the reaction is started by the addition of cytochrome *c* in order to correct the oxygen consumption for autoxidation. It is important to keep the concentration of *N,N,N',N'*-tetramethyl-*p*-phenylenediamine low enough to avoid reduction of the oxidase and high enough not to be limiting in the reduction of cytochrome *c* by ascorbate.

Spectroscopic Assay. After removal from the dialysis bag, an aliquot of the vesicle suspension is diluted 50 times with a medium having the composition of that used in the last dialysis. The diluted preparations (3 μl) are added to 1.8 ml of 10 μM ferrocytochrome *c*, 20 m*M* HEPES, and 50 m*M* KCl, pH 7.3, thermostatted at 20°. The rate of oxidation of cytochrome *c* in the absence (controlled) and presence (uncoupled) of 2.2

[34] M. Thelen, G. Petrone, P. S. O'Shea, and A. Azzi, *Biochim. Biophys. Acta* **766,** 161 (1984).

TABLE I
PROCEDURES FOR THE PREPARATION OF COVs BY DIALYSIS[a]

Procedure	Time (hr)	Volume of buffer per volume of COVs suspension	Medium composition
Procedure A			
1	4	20	0.68% K^+-cholate, 200 m*M* K^+-HEPES
2	4	40	0.3% K^+-cholate, 200 m*M* K^+-HEPES
3	14	80	200 m*M* K^+ HEPES
4	5	80	200 m*M* K^+-HEPES
5	5	100	200 m*M* K^+-HEPES
6	16	800	0.25 m*M* K^+-HEPES, 50 m*M* KCl, 200 m*M* sucrose
Procedure B			
1	4	100	200 m*M* K^+-HEPES
2	4	100	200 m*M* K^+-HEPES
3	14	100	200 m*M* K^+-HEPES
4	3	800	0.25 m*M* K^+-HEPES, 50 m*M* KCl, 200 m*M* sucrose
Procedure C			
1	4	200	125 m*M* KCl, 25 m*M* sucrose, 1.8 m*M* $CaCl_2$, 6.67 g/liter bovine serum albumin
2	4	200	125 m*M* KCl, 25 m*M* sucrose, 1.8 m*M* $CaCl_2$, 0.87 g/liter bovine serum albumin
3	14	400	125 m*M* KCl, 25 m*M* sucrose, 1.8 m*M* $CaCl_2$, 100 μM HEPES

[a] The suspension of phospholipid, cholate, and cytochrome-*c* oxidase was placed in a dialysis tubing and sequentially transferred to the dialysis media (where they were kept for the time indicated) according to procedure A, B, or C. All dialysis media were adjusted to pH 7.3 at 4° before use. COVs, Cytochrome-*c* oxidase vesicles.

μM carbonyl cyanide *m*-chlorophenylhydrazone and 1.1 μM valinomycin is recorded at 550–540 nm with an Aminco DW2a spectrophotometer. The maximum rate (equivalent to the uncoupled rate) for the native preparation is typically 280–290 e^-/sec. The ratio of the uncoupled rate over the controlled rate (RCR) is normally greater than 8.5. Vesicles containing fluorescein-phosphatidyl-ethanolamine obtained according to procedure C of Table I show an RCR of 4–5 (90–120 e^-/sec).

Polarographic Assay.[35] Vesicles containing 20–30 pmol of enzyme are diluted into 0.8 ml of the same medium for the spectroscopic assay, but

[35] R. W. Estabroock, this series, Vol. 10, p. 41.

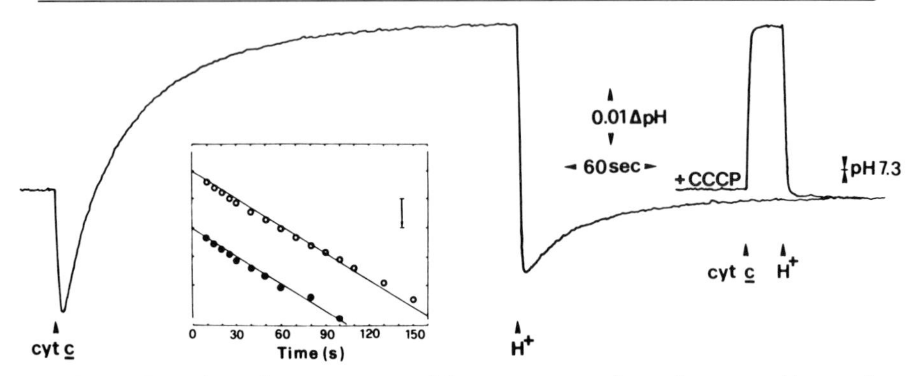

FIG. 1. Potentiometric measurement of the proton pump of cytochrome-*c* oxidase vesicles as a criterion for reconstitution.

The pH measurements were carried out with a rapid-responding pH electrode attached to a Radiometer PHM 64. The experimental conditions were as follows: 0.58 nmol of cytochrome-*c* oxidase reconstituted by cholate dialysis, as in procedure A of Table I, was diluted in the last dialysis medium (Table I, Procedure A6) to a final volume of 1.5 ml; 2.1 nmol of valinomycin per milligram of phospholipid, and 2.67 nmol of CCCP/liter were added; T = 20°; pH as indicated. The trace marked "+CCCP" was obtained in the presence of 1.33 μmol of CCCP/liter. At the arrows 28.6 nmol of ferrocytochrome *c* or 30 nmol of HCl (10 m*M* calibrated solution) were added. The inset is the semilogarithmic analysis of the pH equilibration kinetics. The pH decay after the ferrocytochrome *c* addition (open symbols) and the acid addition (closed symbols) in the absence of the uncoupling concentration of CCCP were plotted semilogarithmically against time. The calibration bar represents the half-decay of the ΔpH.

with 5 m*M* ascorbate and 10 μM *N,N,N′,N′*-tetramethyl-*p*-phenylenediamine. The reaction is started by the addition of cytochrome *c* in order to correct the oxygen consumption derived from the ascorbate/*N,N,N′,N′*-tetramethyl-*p*-phenylenediamine autoxidation. The stimulation of the reaction is performed as described before with carbonyl cyanide *m*-chlorophenylhydrazone and valinomycin.

Measurement of the Orientation of Cytochrome-c Oxidase in the Vesicular Preparation. An aliquot (100 μl) of the vesicles is suspended in 1.4 ml of a medium equal to the one used in the last dialysis and supplemented with 100 μl 200 m*M* HEPES, pH 7.3. Cytochrome-*c* oxidase oriented with the ferrocytochrome *c* binding site facing the extravesicular compartment is estimated by taking a baseline spectrum (580–640 n*M*) in the presence of 2.2 μM CCCP, 1.1 μM valinomycin, 1.1 μM ferricytochrome *c*, and 2 m*M* KCN. The sample is then reduced by the addition of 1 μl of 2 *M* K-ascorbate, pH 7.3, and another spectrum is recorded after 5 min. The amount of cytochrome-*c* oxidase reduced represents the outward-facing enzyme. Finally, a few grains of sodium dithionite are added to reduce all

TABLE II
ORIENTATION AND RESPIRATORY CONTROL RATIO (RCR) OF CYTOCHROME-c OXIDASE IN PHOSPHOLIPID VESICLES

Type of phospholipid and procedure	RCR[a]	Ascorbate plus cytochrome *c*-induced oxidase reduction (%)	$Na_2S_2O_4$-induced oxidase reduction after cytochrome *c*-ascorbate reduction (%)
Asolectin, dialysis A[b]	8.5	81	19
Asolectin, dialysis B	7	90	10
PC/PE/CL[c] (2/2/1), dialysis B	11	92	8
PC/CL,[c] dialysis B	6.5	81	19
Asolectin/FPE[c] (40/1), dialysis C	4.5	85	15

[a] Spectroscopically assayed at 10 μM cytochrome *c*.
[b] A, B, and C refer to the dialysis type referred to in Table I.
[c] PC, Phosphatidylcholine; PE, phosphatidylethanolamine; CL, cardiolipin; FPE, fluorescein-phosphatidylethanolamine.

the cytochrome-*c* oxidase present. After correction for the cytochrome *c* contribution to the absorbance spectrum, the ratio of the two reduced spectra provides a measure of the respective orientations (Table II).

Electron Microscopy

The samples are frozen in a propane jet[36] without addition of glycerin, with the cooling rate high enough (10^4 °C/sec) to vitrify the sample. Freeze-fracturing is carried out in a Balzer's BAF 300 freeze-etching apparatus at 123°K and a pressure of 10^{-5} Pa. Specimens are replicated with Pt/C and were examined in a Philips EM 301 electron microscopy at 100 kV. Micrographs were taken on Agfa Scientia 23 D56 (Fig. 2).

Comments

The result of the cholate dialysis depends on the starting lipid/detergent mixture, which should be sonicated to clarity. It depends also on efficient stirring to avoid the formation of cholate gradients in the dialysis bag.

Table II summarizes some of the factors affecting the reconstitution of cytochrome-*c* oxidase, taking, as reconstitution parameters, the respira-

[36] M. Müller, N. Meister, and H. Moor, *Mikroskopie (Vienna)* **36,** 129 (1980).

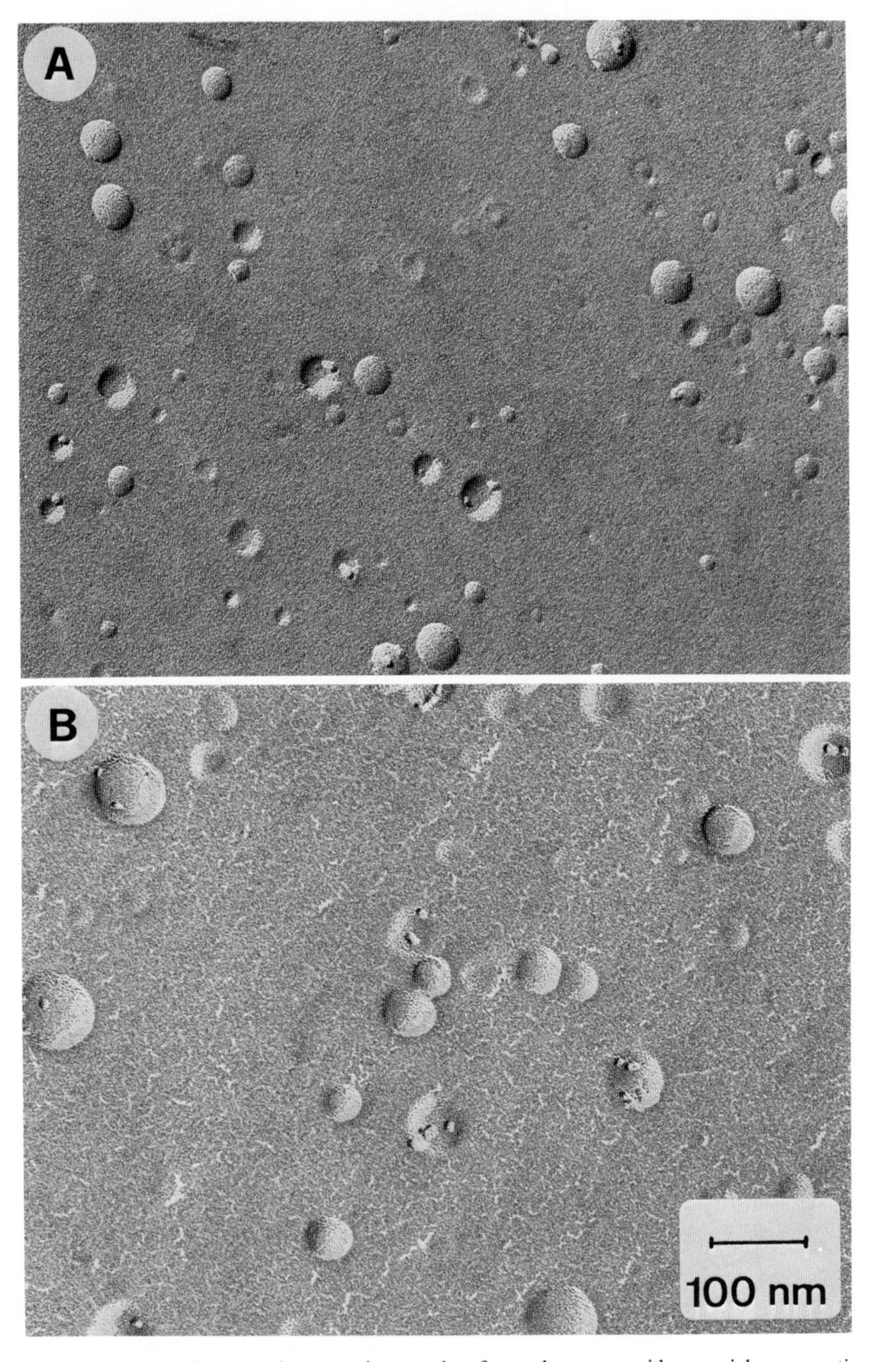

FIG. 2. Freeze-fracture electron micrographs of cytochrome-*c* oxidase vesicles reconstituted by the cholate dialysis procedure. (A) Vesicles reconstituted according to procedure B of Table I; 75% of the vesicles had a diameter ranging from 15 to 37 nm. (B) Vesicles reconstituted according to procedure C of Table I. These vesicles containing fluorescein-phosphatidylethanolamine and dialyzed in the presence of 1.8 m*M* $CaCl_2$ showed a diameter of 20–50 nm for 80% of their population.

tory control ratio and the vectorial insertion of the enzyme. In this case, purified natural lipids gave the best coupling degree, indicating that crude or partially purified asolectin still contains some contaminant which may catalyze proton transfer across the membrane. Also the composition of purified lipids may affect the respiratory control ratio.[12]

The asymmetric binding of cytochrome *c* to the oxidase offers the possibility, through differential reduction of the enzyme, to determine the vesicles' orientation. The vesicles' population, having inverted orientation, may be separated from the main pool of right-oriented vesicles by DEAE–Sephacel column chromatography.[37,38]

Cryofixation and freeze-fracturing of reconstituted vesicles is the most reliable technique to control, by electron microscopy, the incorporation of membrane proteins, although a quantitative analysis of the incorporation appears to be difficult at the present time.

Acknowledgments

We are indebted to Dr. Martin Müller and his collaborators (Laboratory for Electronmicroscopy I, Swiss Federal Institute of Technology) for the use of the electron microscopy facilities.

[37] T. D. Madden, M. J. Hope, and P. R. Cullis, *Biochemistry* **23,** 1413 (1984).
[38] T. D. Madden and P. R. Cullis, *J. Biol. Chem.* **259,** 7655 (1984).

[9] Purification and Reconstitution of the Cytochrome *d* Terminal Oxidase Complex from *Escherichia coli*

By MICHAEL J. MILLER and ROBERT B. GENNIS

Introduction

Escherichia coli is capable of producing two different terminal oxidases, the cytochrome *o* and cytochrome *d* complexes. The cytochrome *d* complex is produced when oxygen is limited and the cytochrome *o* complex predominates when high concentrations of oxygen are present.[1] The cytochrome *d* complex contains cytochromes *b*-558 and a_1 as well as

[1] P. D. Bragg, *in* "Diversity of Bacterial Respiratory Systems" (C. V. Knowles, ed.), Vol. 1, P. 115. CRC Press, West Palm Beach, Florida, 1979.

cytochrome *d*.[2,3] The prosthetic group of both cytochromes a_1 and *b*-558 is probably heme *b* (iron-protoheme IX) whereas the cytochrome *d* prosthetic group is a chlorin.[3] This chapter describes the purification of the cytochrome *d* complex and the incorporation of the complex into artificial proteoliposomes.

Spectrophotometric Assay for Cytochrome *d*

The amount of cytochrome *d* in a preparation can be determined from a reduced-minus-oxidized spectrum. Sodium dithionite is used as a reductant and ferricyanide or air serves as oxidant. The cytochrome is rapidly autoxidizable. It should be noted that the autoxidized form of the cytochrome probably is an oxygenated intermediate.[3] If the enzyme is reduced anaerobically by dithionite and then reoxidized by ferricyanide, another spectral form is observed which is presumably the oxidized form not complexed with oxygen. The extinction coefficient of cytochrome *d* is 7500 M^{-1} cm^{-1} for the difference between the absorbance at 628 and 605 nm (R. M. Lorence and R. B. Gennis, unpublished). There are 2 mol of cytochrome *d* in the oxidase complex. The total amount of protoheme IX in the complex can be determined[2] from the difference between the absorption at 562 and 583 nm using an extinction coefficient of 10,800 M^{-1} cm^{-1}. This includes the heme *b* in both cytochrome a_1 and in cytochrome *b*-558 for a total of about two to three *b* hemes per active complex. This extinction coefficient only applies to the purified complex. Membranes also contain cytochromes *b*-556 and *o*, both of which contribute to the absorption in the portion of the spectrum.

Ubiquinol-1 Oxidase Activity

Oxygen utilization is measured at 37° with a Clark-type oxygen electrode (Yellow Springs Instrument Co., Yellow Springs, OH). The assay buffer consists of 0.025 *M* Tris–HCl, 0.025% Tween 20, 4 m*M* dithioerythritol, and 250 μM ubiquinol-1, pH 7.0. Ubiquinol-1 is usually added in a small volume of DMSO or ethanol ($<0.3\%$ v/v). High concentrations of monovalent salts (e.g., 0.5 *M* KCl) or divalent cations (e.g., 50 m*M* $MgCl_2$) inhibit activity by $>90\%$. Triton X-100 (0.05%) can be used in place of Tween 20, but the specific activity obtained is reduced by about 50%. High concentrations of Triton X-100 (0.5%) are inhibitory, as are octylglucoside (100 m*M*) and cholate (1%, measured at pH 8.0).

[2] M. J. Miller and R. B. Gennis, *J. Biol. Chem.* **258,** 9159 (1983).

[3] J. G. Koland, M. J. Miller, and R. B. Gennis, *Biochemistry* **23,** 1051 (1984).

Protein Assay

Protein is assayed by the method of Lowry *et al.*,[4] including SDS to a final concentration of 1%. Some of the samples contain Tris, which interferes with the Lowry reaction, and this is compensated for by including Tris with the BSA standards. When BSA is used as a standard, the Lowry assay overestimates the amount of protein in the samples. This can be corrected by multiplying the amount of protein determined by Lowry by 0.88. The concentration of the BSA standard should be determined spectrophotometrically from the absorbance at 280 nm.[2]

Strain and Growth Conditions

The purification protocol reported here requires a strain of *E. coli* that overproduces the cytochrome *d* complex, such as strain MR43L/F2 (Shipp[5]). The growth conditions are critical, since the purity of the final product is dependent on starting with cells which have a high level of cytochrome *d* (about 1 nmol heme *d* per milligram membrane protein). The bacteria are grown on Cohen and Rickenburg salts[6] containing 0.15% casamino acids (technical grade, Difco) and 0.5% sodium DL-lactate, pH 7.0, using low aeration of the culture. Batches (200 liters each) can be grown in a New Brunswick FM250 fermentor with a sparge rate of 1 ft^3 air/min and an agitation rate of 150–200 rpm. The cells are harvested 24 hr into stationary phase and stored at −80° until use. The cell yield is typically about 3–4 g/liter.

Purification of the Cytochrome *d* Complex[2]

All of the steps in this procedure should be carried out at 4°. Approximately 60 g of frozen cells are suspended, using a blender, in 300 ml of a buffer consisting of 100 m*M* Tris/HCl, 10 m*M* EDTA, 10 m*M* benzamidine, 1 m*M* phenylmethylsulfonyl fluoride (PMSF), pH 8.5. The last three are added after the pH has been adjusted. The high EDTA concentration is to help inhibit endogenous proteases. The PMSF is added from an isopropanol solution. The final concentration of isopropanol should be 0.3% (v/v) or less. A short spray of Antifoam A (Dow-Corning) prevents foaming during blending. The suspension is passed through a French pressure cell (SLM-Aminco, Urbana, IL) three times at 18,000 psi at a

[4] O. H. Lowry, N. J. Rosebrough, A. L. Farr, and R. J. Randall, *J. Biol. Chem.* **193,** 265 (1951).

[5] W. S. Shipp, *Arch. Biochem. Biphys.* **150,** 459 (1972).

[6] G. N. Cohen and H. W. Rickenburg, *Ann. Inst. Pasteur (Paris)* **20,** 693 (1956).

flow rate of 5 ml/min. This step is important, since low yields result from incomplete breakage of the cells. The specific heme content of the purified cytochrome is lower if sonication is used to disrupt the cells. Unbroken cells are pelleted by centrifugation for 20 min at 10,000 *g* in a Sorvall GSA rotor. (All centrifugal forces are specified for the bottoms of the tubes.) The supernatant is then centrifuged at 200,000 *g* in a Beckman-type 60 Ti rotor for 1 hr, and the resulting pellet is solubilized in 50 ml of 75 m*M* potassium phosphate, 150 m*M* KCl, 5 m*M* EDTA, and 60 m*M* *N*-dodecyl-*N,N*-dimethyl-3-ammonio-1-propane sulfonate (Zwittergent 3-12 from Calbiochem-Behring or SB-12 from Serva), pH 6.4. A Ten Broeck tissue homogenizer is used to disperse the pellet. The solution is centrifuged at 200,000 *g* for 1 hr, and the supernatant loaded on a 30 × 4.8 cm DEAE–Sepharose CL-6B column which is eluted at 200 ml/hr with a 1600-ml gradient running from 150 to 250 m*M* KCl in buffer containing 75 m*M* potassium phosphate, 5 m*M* EDTA, 6 m*M* Zwittergent 3-12, pH 6.3. Fractions of 15 ml are collected.

All fractions with a ratio of A_{412}/A_{280} greater than 0.6 are pooled and loaded on a 30 × 2.0-cm DNA-grade hydroxyapatite column (Bio-Rad). This column serves both to further purify the cytochrome and to remove the Zwittergent 3-12 which sometimes denatures the cytochrome after long periods of time. It is important to use fresh hydroxyapatite. The loaded column is first washed with 75 ml of 150 m*M* potassium phosphate, 25 m*M* sodium cholate, pH 8.2. The cytochrome is eluted with 450 m*M* potassium phosphate, 25 m*M* sodium cholate, pH 8.2, at a flow rate of 25 ml/hr. Fractions of 2 ml are collected and those fractions with an A_{412}/A_{280} ratio greater than 0.8 are pooled. The ratio may appear to be less than this due to light scattering by small crystals of hydroxyapatite that wash through the column. The pooled fractions are concentrated to 4 ml by ultrafiltration with an Amicon XM50 membrane and dialyzed overnight against 1 liter of 10 m*M* sodium phosphate, 16 m*M* sodium cholate, 1 m*M* EDTA, pH 8.2, to yield the final product. The table shows the effectiveness of the purification protocol.

Characterization of the Cytochrome *d* Complex

The pure cytochrome contains 19 nmol heme *b*/mg protein and two subunits by SDS–PAGE analysis using the Laemmli system.[7] For SDS–PAGE analysis, the samples should not be boiled before loading, since this results in the formation of aggregates that will not enter the separation gel. The apparent molecular weights of the two subunits are 50,000 and

[7] U. K. Laemmli, *Nature (London)* **227,** 680 (1970).

PURIFICATION OF CYTOCHROME d COMPLEX

Step	Total heme b (nmol)	Total protein (mg)	Specific heme content[a] (nmol heme b/ mg protein)	Yield (%)
Whole cells (60 g)	9350	7150	1.3	100
Membranes	6280	1440	4.4	67
Zwittergent extract	5210	1030	5.1	56
DEAE-Sepharose CL-6B pool	1620	143	11.3	17
Hydroxyapatite pool	617	31.8	19.4	7

[a] Computed using $\varepsilon_{562-580} = 10{,}800\ M^{-1}\ cm^{-1}$ for heme b in the complex. This value is obtained by extraction and measurement of protoheme IX.[2]

28,000 on a 12.5% polyacrylamide gel. The upper band is diffuse, but inclusion of 8 M urea in the gel produces a tight band. The apparent molecular weight of the smaller subunit is dependent on the percentage of acrylamide in the gel. If a silver stain is used, an additional component at the dye front is also evident, which has been identified as lipopolysaccharide.[8] The only metal present is iron (34 nmol/mg protein).[2] The ubiquinol-1 oxidase activity of the pure cytochrome in the presence of Tween 20 is about 39 μmol O_2/nmol heme d/min. A reduced-minus-oxidized spectrum and a CO difference spectrum are shown in Fig. 1.

Reconstitution of Artificial Proteoliposomes

Formation of artificial proteoliposomes containing the cytochrome d complex is accomplished by the detergent dialysis method originally described by Racker and co-workers.[9] The vesicles are made from a mixture of three parts (by weight) of *E. coli* phosphatidylethanolamine to one part egg phosphatidylglycerol, which is similar to the ratio found in the *E. coli* membrane. Ubiquinone-8 (1 mol%), the major quinone species in the *E. coli* membrane,[1] can also be included. A chloroform/methanol solution containing these lipids is dried under nitrogen flow and put under vacuum for several hours in order to remove all the organic solvents. Enough buffer is then added to give a lipid concentration of 5 mg/ml. The buffer consists of 50 mM potassium phosphate, 1 mM EDTA, pH 7.5, with 50 mM sodium cholate. The low concentration of lipid and high concentra-

[8] R. G. Kranz and R. B. Gennis, *J. Biol. Chem.* **259,** 7998 (1984).

[9] E. Racker, this series, Vol. 55, p. 699.

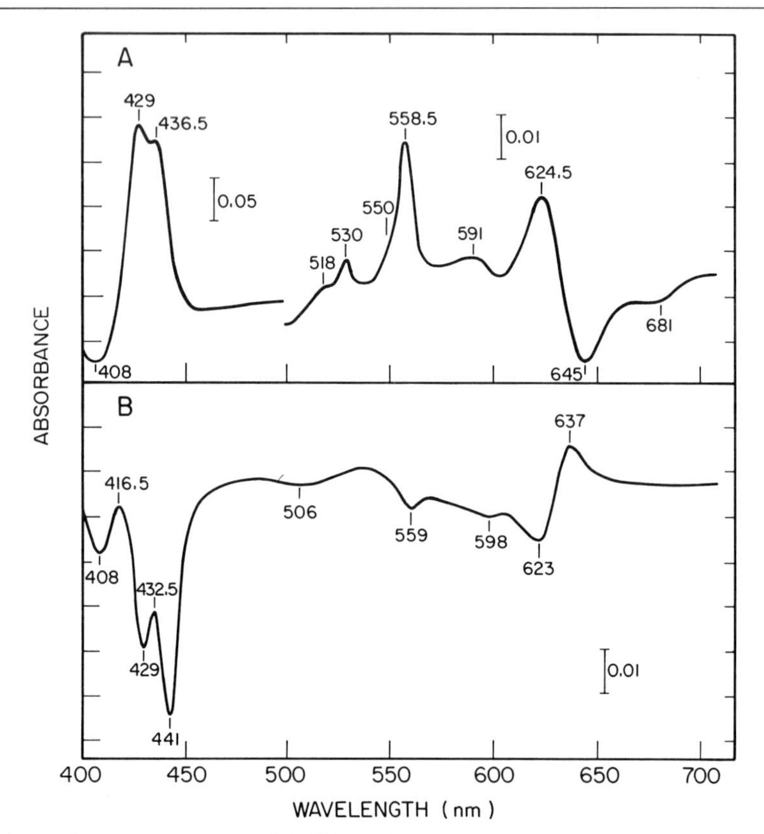

FIG. 1. Low-temperature (77°K) difference spectra of the purified cytochrome *d* complex. Samples contain 0.14 mg/ml protein in 10 m*M* Tris, 16 m*M* sodium cholate, 20% glycerol (v/v), pH 7.8. Higher concentrations of Tris are deleterious to the cytochrome. (A) Dithionite-reduced minus air-oxidized. (B) Dithionite-reduced plus CO minus dithionite-reduced.

tion of cholate are needed to prevent the *E. coli* phospholipids from precipitating in the cold. The solution is sonicated at 4° with a probe-type sonicator until clear. Cytochrome *d* in cholate is then added to the solution. The amount of cytochrome added to the lipids can be varied (~15 μg/mg lipid). Higher protein-to-lipid ratios produce higher ubiquinol-1 oxidase specific activity. The solution is dialyzed against three or four changes of 200 volumes of buffer (without cholate) for 6–8 hr each. Spectra taken after dialysis show that all of the cytochrome is recovered. Sucrose density gradient ultracentrifugation with a 5–20% linear sucrose gradient can be used to show that about one-third of the protein is aggregated and not incorporated into the vesicles. The cytochrome that has been incorporated in the membrane is oriented so that about 75% has the quinol oxidase site on the outside (M. J. Miller and R. B. Gennis, unpublished). The total ubiquinol-1 oxidase activity of the reconstituted enzyme (both orientations) is measured using the standard assay conditions except that Tween 20 is omitted. After dialysis, the specific activity is about half that observed prior to reconstitution, but drops substantially following sucrose density centrifugation.

Cytochrome *d* incorporated in the proteoliposomes will generate a proton gradient across the membrane with ubiquinol-1 as a substrate. The quenching of the fluorescence of carbocyanine dyes [e.g., 3,3′-bis(propylthio)dicarbocyanine] can be used to measure the electrical potential gradient formed concomitant with quinol oxidase activity.[10] Ubiquinol-1 by itself quenches carbocyanine fluorescence, so low concentrations, 10–25 μM, should be used. Samples containing 20 μg/ml phospholipid (0.1 μg protein) and 1 μM dye produce a good response. The response is eliminated by uncouplers, but enhanced by nigericin (60 mV in the absence of nigericin and 140 mV in the presence of 0.3 μM nigericin). This proteoliposome system may also be reconstituted with other enzymes, e.g., pyruvate oxidase, in which case it is necessary to incorporate ubiquinone-8 in the liposomes.

Reconstitution of the Pyruvate:Oxygen Oxiodoreductase Chain[11]

Pyruvate oxidase[10,11] is a flavoprotein dehydrogenase found in *E. coli* which, although it couples to the aerobic respiratory chain, can be isolated in a water-soluble form. When the flavin prosthetic group is reduced, as by the addition of the substrate pyruvate, the enzyme acquires an enhanced affinity for lipids and membranes.[11] This enzyme will function as a ubiquinone-8 reductase *in vitro* and, presumably, *in vivo* as well.[10] Pyruvate oxidase is purified using a published procedure.[12] The pyruvate:oxygen oxidoreductase chain is reconstituted[10] by adding pyruvate oxidase to preformed proteoliposomes containing the cytochrome *d* complex and ubiquinone-8 (1 mol%) in a buffer consisting of 200 mM sodium pyruvate, 100 mM sodium phosphate, 20 mM $MgCl_2$, 100 μM thiamine pyrophosphate, pH 6.3. Activity is measured with the oxygen electrode. A wide range of protein concentrations can be used in this assay. Reasonable activity is obtained with 1 μg/ml cytochrome *d* and 20 μg/ml pyruvate oxidase. It should be noted that the phospholipid com-

[10] J. G. Koland, M. J. Miller, and R. B. Gennis, *Biochemistry* **23,** 445 (1984).

[11] H. L. Schrock and R. B. Gennis, *J. Biol. Chem.* **252,** 5990 (1977).

[12] T. A. O'Brien, H. L. Schrock, P. Russell, R. Blake, II, and R. B. Gennis, *Biochim. Biophys. Acta* **452,** 13 (1976).

position of the proteoliposomes is critical to successful reconstitution of this electron transport chain. Pyruvate oxidase will not effectively reduce ubiquinone-8 incorporated in vesicles prepared with soybean phospholipids (asolectin), for example. The phosphatidylglycerol appears to be particularly important, though no systematic experimentation has been performed to address the question.

The observed activity is absolutely dependent on the presence of pyruvate, flavoprotein, ubiquinol-8, and the cytochrome *d* complex. The cytochrome *d* complex can turn over in this system at about 100 electrons/sec/heme *d*. Although this is only about 4% of the activity observed with ubiquinol-1 as substrate, it is faster than the observed velocity in whole cells.[13] The reaction also results in developing a transmembrane potential of about 180 mV, negative inside, as indicated by carbocyanine dye quenching.[10] It should also be possible to use the protonmotive force generated by this system to power other enzymes, such as transport proteins or ATPases, which could be incorporated in the same proteoliposomes.

[13] C. W. Rice and W. P. Hempfling, *J. Bacteriol.* **134,** 115 (1978).

[10] Purification and Properties of Two Terminal Oxidase Complexes of *Escherichia coli* Aerobic Respiratory Chain

By KIYOSHI KITA, KIYOSHI KONISHI, and YASHUHIRO ANRAKU

Introduction

In the respiratory chain of aerobically grown *Escherichia coli* there are two terminal oxidases, cytochrome *o* and cytochrome *d*.[1] In the early exponential phase of aerobic batch culture, cells contain only the terminal oxidase cytochrome *o*, with cytochrome *b*-562. During the late exponential or early stationary phase of aerobic batch culture, cytochrome *d* is also synthesized together with cytochrome *b*-558. Organisms in which this type of phenomenon has been observed include *Proteus vulgaris*,[2] *Pseudomonas putida*,[3] and many other bacteria.[1]

[1] B. A. Haddock and C. W. Jones, *Bacteriol. Rev.* **41,** 47 (1977).
[2] L. N. Castor and B. Chance, *J. Biol. Chem.* **234,** 1587 (1959).
[3] W. J. Sweet and J. A. Peterson, *J. Bacteriol.* **133,** 217 (1978).

We purified both oxidase complexes from the inner cytoplasmic membrane of *E. coli*.[4,5] The unity and diversity of structure and function of terminal oxidases of *E. coli* in the aerobic energy transduction are comprehensively described based on the comparative biochemical studies of the purified oxidases.

Purification of Cytochrome *b*-562–*o* Complex[4]

Growth of Cells and Preparation of Inner Membrane Vesicles

Escherichia coli K12 strain KL251/ORF4 (CGSC strain no. 4282) is used. The content of cytochrome *o* in this strain is found to be twice that in the wild strain MR43L.[6] Cells are grown aerobically at 37° in a Magnaferm fermentor (New Brunswick Scientific Co.) in 10 liters of medium containing, per liter: 3.4 g of KH_2PO_4, 0.5 g of sodium citrate, 7.3 g of Tris, 4 g of sodium lactate, 4 g of casamino acid (Difco), and 0.1 g of $MgSO_4 \cdot 7H_2O$, pH 7.4. The cells are harvested in the early exponential phase of growth.

The cells (about 30 g wet weight) are washed once with 30 m*M* Tris–HCl, pH 8.0, and suspended at a density of 2×10^{10} cells per milliliter in about 500 ml of 30 m*M* Tris–HCl, pH 8.0, containing 20% (w/v) sucrose. The suspension is incubated for 5 min on ice, and then lysozyme (final concentration of 90 μg/ml) and EDTA (final concentration of 10 m*M*; disodium salt) are added simultaneously. Incubation on ice is continued for 30 min and spheroplasts are collected by centrifugation at 10,000 *g* for 30 min.

Inner membrane vesicles are prepared by the method of Yamato *et al.*[7] with a slight modification: The spheroplasts (about 30 g wet weight) obtained as above are suspended in about 70 ml of 20% (w/v) sucrose–3 m*M* EDTA, pH 8.0, disrupted by a French press (Ohtake Co., Tokyo) at 1000 kg/cm^2, diluted with an equal volume of 3 m*M* EDTA, pH 8.0, and centrifuged at 190,000 *g* for 1.5 hr. Unless otherwise noted, all procedures should be performed at below 4°. The pellet is suspended by vigorous homogenization with a Dounce Teflon and glass homogenizer in 50 ml of 10% (w/v) sucrose–3 m*M* EDTA, pH 8.0 (E-S-10), and centrifuged in swinging buckets at 2000 *g* for 15 min to remove intact cells. Then 150 ml of E-S-10 is added to the supernatant and the mixture is centrifuged at

[4] K. Kita, K. Konishi, and Y. Anraku, *J. Biol. Chem.* **259,** 3368 (1984).
[5] K. Kita, K. Konishi, and Y. Anraku, *J. Biol. Chem.* **259,** 3375 (1984).
[6] K. Kita, M. Kasahara, and Y. Anraku, *J. Biol. Chem.* **257,** 7933 (1982).
[7] K. Kita, I. Yamato, and Y. Anraku, *J. Biol. Chem.* **253,** 8910 (1978).

190,000 *g* for 1 hr. This procedure for washing the crude membrane fractions is repeated again and the washed membrane fraction is suspended in 20 ml of 3 m*M* EDTA, pH 8.0, by homogenization. A sample of 2.5 ml of the crude membrane fraction is layered on 15 ml of 44% (w/w) sucrose–3 m*M* EDTA, pH 8.0, and centrifuged at 69,000 *g* for about 12 hr at 4° using an RP50-2 angle rotor (Hitachi). The upper layer (inner membrane vesicles) and the pellet (crude outer membrane vesicles) are collected separately with Pasteur pipettes. The upper layer fraction is diluted with 2 volumes of 3 m*M* EDTA, pH 8.0, and collected by centrifugation at 190,000 *g* for 1.5 hr. The inner membrane vesicles thus obtained are suspended in E-S-10 (20 mg of protein/ml) and stored at −75° until use. See the original method[7] for further details and distributions of marker enzymes in the inner and outer membrane vesicles.

Purification Procedure

Step 1: Triton X-100 Extraction and DEAE–Sepharose. The inner membrane vesicles (100–150 mg) are solubilized in 5% Triton X-100, 10 m*M* $MgCl_2$, 10 m*M* Tris–HCl buffer (pH 8.0), and the protein concentration of the resulting mixture is adjusted to 5 mg/ml. The mixture is centrifuged at 270,000 *g* for 60 min and the resultant supernatant is applied to a column of DEAE-Sepharose (2 × 15 cm). The column is washed with 1% Triton X-100, 1 m*M* EDTA, and 10 m*M* Tris–HCl (pH 8.0); then cytochromes are eluted with 300 ml of the same buffer containing a linear gradient of 0–0.3 *M* NaCl (from fraction 28) at a flow rate of 25 ml/hr (Fig. 1). Most of the cytochrome *b*-556 is eluted in flow-through fractions, and the cytochrome *b*-562–*o* complex is eluted with 0.13 *M* NaCl as a single peak. Fractions 57–62 (2.5 ml each) are collected and dialyzed against 5 liters of 1 m*M* EDTA, 10 m*M* Tris–HCl (pH 8.0) for 5 hr and applied to a small column of DEAE-Sepharose (1 × 5 cm). The cytochrome *b*-562–*o* complex is eluted in the same conditions as for the first DEAE-Sepharose column.

Step 2: Sephacryl S-200. The fractions containing the cytochrome *b*-562–*o* complex are collected and brought to 50% saturation of ammonium sulfate by adding saturated ammonium sulfate solution. The precipitate is suspended in 2 ml of 0.05% Sarkosyl, 0.6 *M* NaCl, 10 m*M* Tris–HCl (pH 8.0) and applied to a column of Sephacryl S-200 (2 × 140 cm). Elution is carried out with the Sarkosyl/NaCl buffer mentioned above at a flow rate of 5 ml/hr.

Step 3: BioGel HT. The peak fraction (5 ml) containing cytochrome *b*-562–*o* complex eluted from Sephacryl S-200 is dialyzed against 1 liter of 0.05% Sarkosyl, 10 m*M* potassium phosphate (pH 7.4) for 12 hr and then

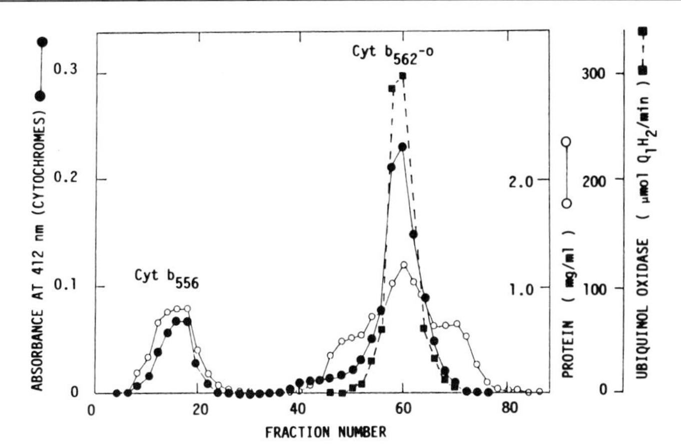

FIG. 1. Chromatography of cytochrome *b*-562–*o* complex on DEAE-Sepharose. Inner membranes were obtained from *E. coli* harvested in the early exponential phase of aerobic growth and solubilized with 5% (w/v) Triton X-100, 10 m*M* $MgCl_2$, and 100 m*M* Tris–HCl (pH 7.5). The supernatant was applied to a column (2 × 15 cm), equilibrated with 1% (w/v) Triton X-100, 1 m*M* EDTA, and 10 m*M* Tris–HCl (pH 8.0). Cytochrome *b*-562–*o* complex was eluted with 300 ml of the same buffer containing a linear gradient of 0–0.3 *M* NaCl at a flow rate of 25 ml/hr. (●——●), Absorbance at 412 nm; (■--■), ubiquinol oxidase; (○——○), protein.

applied to a column of BioGel HT (1.5 × 5 cm). The column is washed with 0.05% Sarkosyl solution containing a 100-ml linear gradient of 10–500 m*M* potassium phosphate (pH 7.4). The cytochrome *b*-562–*o* complex is then eluted with 0.05% Sarkosyl, 0.7 *M* potassium phosphate (pH 7.4) at a flow rate of 15 ml/hr. Fractions of cytochrome *b*-562–*o* complex with a heme content of more than 19.5 nmol/mg of protein are collected and used for experiments. Table I summarizes the purification.

Purification of Cytochrome *b*-558–*d* Complex[5]

Growth of Cells

Escherichia coli K12 strain MR43L (F^-, *gal, recA, thi, lac*) is used. Cells are grown in the medium described above at 37° in a Magnaferm fermentor (New Brunswick Scientific Co.) in a 10-liter vessel. For oxygen-limited growth, the pressure and flow rate of air are 3 lb/in. and 3 liters/min, respectively, and the cultures are stirred at 300 rpm. The cells are harvested in the late exponential phase or early stationary phase of growth.

TABLE I
PURIFICATION OF CYTOCHROME *b*-562–*o* COMPLEX

Step	Protein (mg)	Cytochrome *o* (nmol)[a]	Specific content (nmol/mg)	Recovery (%)	Ubiquinol oxidase (μmol QH_2/min)	Specific[c] activity
Inner membranes	126	110	0.873	100	3540	32.2
Triton X-100 extract	113	99.4	0.880	90.4	3280	33.0
DEAE–Sepharose	5.94	47.5	8.0	43.2	1560	32.8
Sephacryl S-200	3.22	30.6	9.50	27.8	1010	33.0
BioGel HT	1.91	18.6	9.74[b]	16.9	605	32.5

[a] A molar extinction coefficient of 145 cm^{-1} mM^{-1} was used.
[b] The heme content of this preparation was 19.5 nmol/mg of protein, determined by the pyridine hemochromogen method.
[c] Measured as μmol QH_2/min/nmol cytochrome *o*.

Purification Procedure

Step 1: Inner Membranes. Inner membrane vesicles are prepared from the freshly grown cells as described in "Purification of Cytochrome *b*-562–*o* Complex, Growth of Cells." Inner membranes containing more than 0.8 nmol cytochrome *d*/mg of protein are used as an enzyme source.

Step 2: Triton X-100 Extraction and DEAE–Sephacel Chromatography. Inner membrane vesicles (2 mg/ml) are solubilized in 5% Triton X-100, 10 m*M* $MgCl_2$, 10 m*M* Tris–HCl (pH 7.4), and the mixture is centrifuged at 100,000 *g* for 60 min. The resultant supernatant is diluted 2-fold with 10 m*M* Tris--HCl (pH 7.4) and applied to a column of DEAE-Sephacel (2 × 15 cm). The column is washed with 1% Triton X-100, 1 m*M* EDTA, and 10 m*M* Tris–HCl (pH 7.4); then cytochromes are eluted with 300 ml of the same buffer containing a linear gradient of 0–0.3 *M* NaCl at a flow rate of 25 ml/hr. As shown in Fig. 2, the cytochrome *b*-558–*d* complex is eluted with 0.12 *M* NaCl (fractions 30–35). Under these conditions, a small amount of cytochrome *b*-562–*o* complex is eluted in fractions 36–39 and cytochrome *b*-556 appears in fractions 40–50 as a broad shoulder. Fractions 32–34 (5 ml each) containing the cytochrome *b*-558–*d* complex are combined, diluted 2-fold with 10 m*M* Tris–HCl (pH 7.4), and rechromatographed under the same conditions as in the first chromatography.

Step 3: Exchange of Detergent and Sephacryl S-300. The fractions containing the cytochrome *b*-558–*d* complex eluted from the second

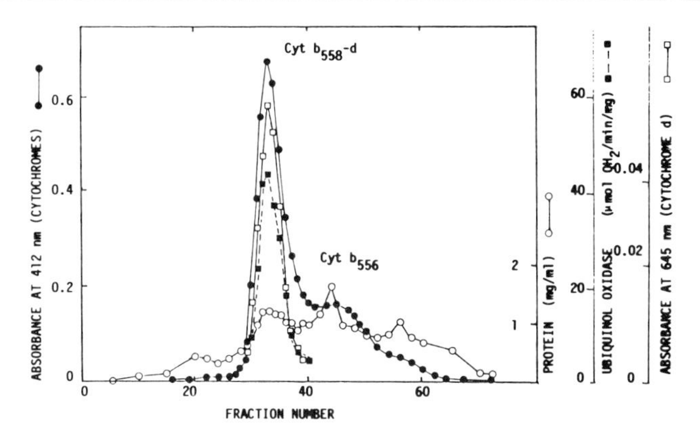

FIG. 2. Chromatography of cytochrome *b*-558–*d* complex on DEAE-Sephacel. The supernatant from the inner membranes of *E. coli* harvested in the early stationary phase of aerobic growth and solubilized with 5% (w/v) Triton X-100, 10 m*M* $MgCl_2$ was applied to a column (2 × 15 cm) equilibrated with 1% (w/v) Triton X-100, 1 m*M* EDTA, and 10 m*M* Tris–HCl (pH 7.4). Cytochrome *b*-558–*d* complex was eluted with 300 ml of the same buffer containing a linear gradient of 0–0.3 *M* NaCl at a flow rate of 25 ml/hr. (●——●), Absorbance at 412 nm (total cytochromes); (■--■), ubiquinol oxidase; (□——□), absorbance at 645 nm (cytochrome d); (○——○), protein.

DEAE–Sephacel column are collected, diluted with an equal volume of 10 m*M* Tris–HCl (pH 7.4), and applied to a small DEAE-Sephacel column (1 × 5 cm). The column is washed with 0.05% (w/v) Sarkosyl, 10 m*M* Tris–HCl (pH 7.4) until Triton X-100 has been washed out. Then cytochrome *b*-558–*d* is eluted with 0.05% (w/v) Sarkosyl, 0.6 *M* NaCl, and 10 m*M* Tris–HCl (pH 8.0). The fractions containing cytochrome *b*-558–*d* are concentrated by ultrafiltration (YM-10; Amicon) and applied to column of Sephacryl S-300 (2 × 150 cm). Elution is carried out with 0.05% Sarkosyl, 0.6 *M* NaCl, 10 m*M* Tris–HCl (pH 8.0) at a flow rate of 5 ml/hr. Fractions 50–53 (2.5 ml each) containing the cytochrome *b*-558–*d* complex are collected. The preparation is rechromatographed on Sephacryl S-300 under the same condition as before.

The protoheme content of the final preparation, determined by the pyridine hemochromogen method, is 12.3 nmol protoheme/mg of protein in the peak fraction. Table II summarizes the purification.

Polypeptide Composition and Purity[4,5]

The purified cytochrome *b*-562–*o* complex is composed of two polypeptides with molecular weights of 33,000 and 55,000, determined by gel electrophoresis in the presence of sodium dodecyl sulfate. These proteins

TABLE II
PURIFICATION OF CYTOCHROME *b*-558–*d* COMPLEX

Step	Protein (mg)	Cytochrome *d* (nmol)	Specific content (nmol/mg)	Recovery (%)	Ubiquinol[a] oxidase (μmol QH_2/min)
Triton X-100 extract	294	268	0.91	100	2080
DEAE–Sephacel	17.7	92.2	5.21	34.4	612
Rechromatography	6.58	57.1	8.68	21.3	310
Sephacryl S-300	1.48	12.7	8.58	4.74	72.4
Rechromatography	0.38	3.54	9.32	1.32	21.2

[a] Ubiquinol oxidase was assayed in the presence of 4 m*M* asolectin and 200 μM ubiquinol.

are present in equimolar proportions judging from the intensity of Coomassie Brilliant Blue staining of the bands, and 1 mol of complex contains 1 mol of cytochrome *b*-562 and 1 mol of cytochrome *o*. (See Table III and "Spectral Properties.")

Cytochrome *b*-562–*o* complex containing 19.5 nmol heme/mg of protein contains iron and copper, respectively, at 22.3 and 16.8 nmol/mg of

TABLE III
CHEMICAL COMPOSITION AND SPECTRAL PROPERTIES OF PURIFIED OXIDASES

Oxidase	Chemical composition (nmol/mg protein): Cytochrome *b*-562–*o* complex	Chemical composition (nmol/mg protein): Cytochrome *b*-558–*d* complex
Heme *b*	19.5	12.3
Heme *d*	—[a]	9.54
Iron	22.3	26.6
Copper	16.8	1.26
	ε(cm^{-1} mM^{-1})	
	18.7 (560–580)[b]	14.8 (560–580)[b]
		18.8 (628–649)[b]
	145 (416–430)[c]	12.6 (622–642)[c]

[a] Not detected.
[b] Reduced-minus-oxidized difference spectrum.
[c] (Reduced plus CO)-minus-reduced difference spectrum.

protein. Assuming that heme and copper are not detached during the purification, the cytochrome *b*-562–*o* complex contains 2 hemes *b*, 2 coppers, and no nonheme iron.

The purified cytochrome *b*-558–*d* complex is found to consist of two main polypeptides with molecular weights of 51,000 and 26,000, as shown by gel electrophoresis in the presence of sodium dodecyl sulfate, and the molar ratio of these two polypeptides is estimated as 1 : 1 by Coomassie Brilliant Blue staining.

The purified cytochrome *b*-558–*d* complex with 12.3 nmol of protoheme/mg of protein contains cytochrome *d* and iron, respectively, at 9.54 and 26.6 nmol/mg of protein, indicating that 1 mol of oxidase complex contains 1 mol of cytochrome *b*-558, 1 mol of cytochrome *d*, and no nonheme iron. This preparation contains copper at less than 1.3 nmol/mg of protein.

The oxidase complexes thus obtained have no detectable dehydrogenase activity with D-lactate, L-lactate, succinate, or NADH as substrate, and no detectable ubiquinone, menaquinone, or phospholipid.

Spectral Properties[4,5]

Cytochrome b-562–o Complex

In the absolute spectra, the oxidized form has peaks at 410, 527, and 565 nm, and in that of reduced form, obtained with sodium dithionite, the peaks are shifted to 427, 531, and 560 nm with increase in their absorption intensities.

The difference spectrum at room temperature (Fig. 3A) has a single absorption peak at 560 nm, and the second-order finite difference spectrum has two troughs (Fig. 3B). These observations indicate that there are two components in the cytochrome *b*-562–*o* complex, and that the absorption peaks of these components are at 555 and 562 nm at 77°K (Fig. 3C). These components are present in equal amounts, since the heights of their peaks in the second-order finite difference and low-temperature difference spectra are approximately the same (Fig. 3B,C).

The CO difference spectrum of the cytochrome *b*-562–*o* complex at room temperature (Fig. 4A) has the typical absorbance characteristics of cytochrome *o* with a peak at 416 nm and trough at 430 nm. In this spectrum, the trough in the α region is at 554 nm, and the trough at lower wavelength in the second-order finite difference spectrum is decreased by treatment with CO (Fig. 4B). From these results and the observation by

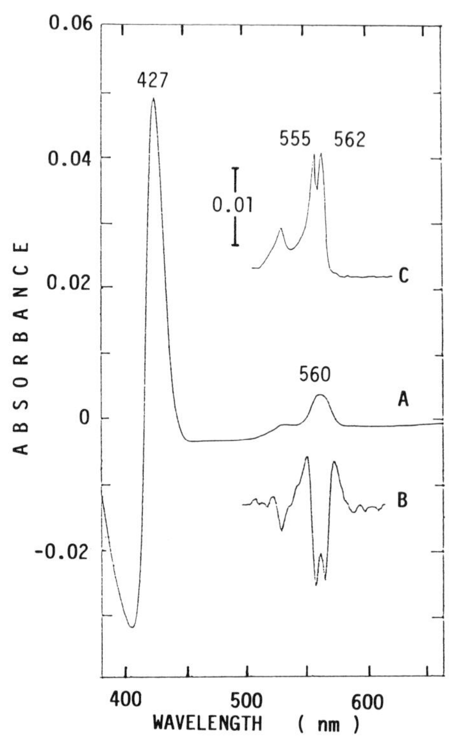

FIG. 3. Reduced-minus-oxidized difference spectra of cytochrome *b*-562–*o* complex. (A) Spectra were recorded with a Hitachi 320 spectrophotometer at 20°, with a light pass of 10 mm. The sample and reference cuvettes contained 63.8 μg of purified cytochrome *b*-562–*o* complex in 2.5 ml of 50 m*M* Tris–HCl, pH 8.0. Reduced form in sample cuvette was obtained with sodium dithionite. (B) Second-order finite difference spectrum of the difference spectrum shown in (A). $\Delta\lambda = 3$ nm. (C) Low-temperature difference spectrum of cytochrome *b*-562–*o* complex (50 μg of protein/ml) at 77°K in liquid nitrogen with a light path of 3 mm.

Bragg,[8] the component having a peak at 555 nm at 77°K is determined to be cytochrome *o*.

A solution of purified cytochrome *b*-562–*o* (25.5 μg of protein/ml), containing 19.5 nmol heme/mg of protein, shows an absorbance of 0.00465 unit in the difference spectrum of the sodium dithionite reduced-minus-oxidized form at the wavelength pair 560–580 nm. From these values, the molar extinction coefficient of the purified cytochrome *b*-562–*o* complex is calculated to be 18.7 cm^{-1} mM^{-1} (Table III). In the same

[8] P. D. Bragg, *in* "Diversity of Bacterial Respiratory Systems" (C. W. Knowles, ed.), Vol. 1, p. 115. CRC Press, Boca Raton, Florida, 1980.

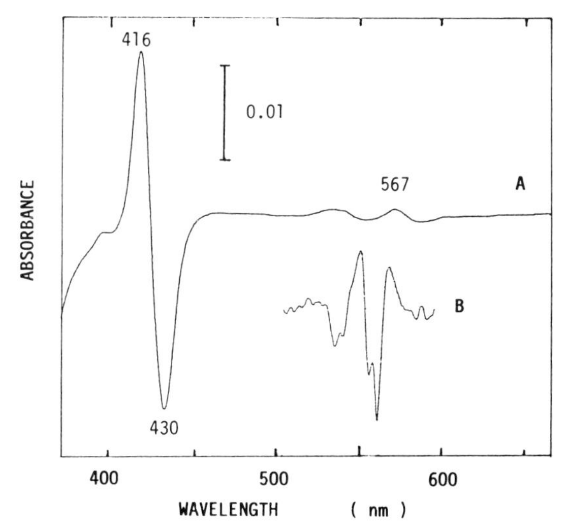

FIG. 4. Effect of CO on spectra of cytochrome *b*-562–*o* complex. (A) CO-reduced-minus-reduced difference spectrum of cytochrome *b*-562–*o* complex. Conditions were as for Fig. 3. Half the cytochrome *b*-562–*o* complex reduced with dithionite was placed in the reference cuvette. The other half was treated with CO gas for 1 min and placed in the sample cuvette. (B) Second-order finite difference spectrum of the absolute spectrum of reduced cytochrome *b*-562–*o* complex in the presence of CO. The CO-reduced absolute spectrum was recorded with the same sample cuvette as for (A) and buffer in the reference cuvette. Then the second-order finite difference spectrum was obtained. $\Delta\lambda = 3$ nm.

way, the molar extinction coefficient of cytochrome *o* is calculated to be 145 cm^{-1} mM^{-1} (416–430 nm) from the CO difference spectrum of the purified complex (Fig. 4).

Cytochrome b-558–d Complex

The absolute spectrum of the oxidized form has peaks at 645 and 412 nm, and that of the reduced form obtained with sodium dithionite has peaks at 629, 560, and 429 nm.

The difference spectrum at room temperature (Fig. 5A) exhibits typical absorbance characteristics of cytochrome *b* at 560 nm and cytochrome *d* at 628 nm. At 77°K, these peaks shift to 558 and 624 nm (Fig. 5B,C), and the single peak seen at 558 nm indicates that the purified preparation contains no cytochrome *b*-556, *b*-562, or *o*.

The CO-reduced-minus-reduced difference spectrum at room temperature (Fig. 6) has peaks at 420 and 642 nm and troughs at 430, 442, 560, and 622 nm. This indicates that cytochrome *b*-558 and cytochrome *d* in the purified oxidase combine with CO. The trough at 560 nm is not found

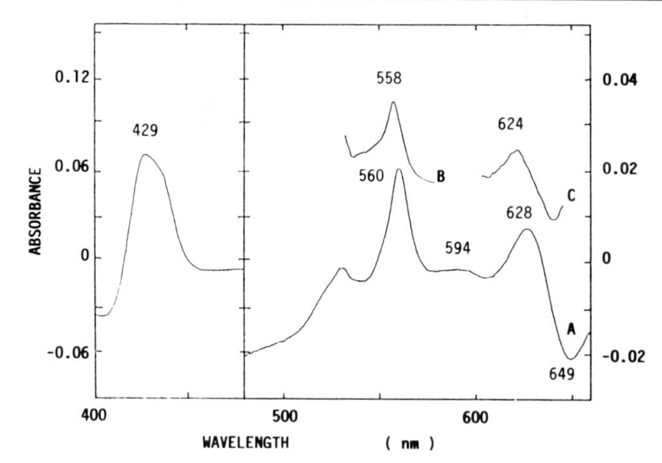

FIG. 5. Reduced-minus-oxidized difference spectra of the purified cytochrome *b*-558–*d* complex. (A) Spectrum recorded with a Hitachi 356 dual-wavelength spectrophotometer at 20°. Conditions were as described in the legend to Fig. 3. Both cuvettes contained 0.3 mg of purified cytochrome *b*-558–*d* complex. (B, C) Low-temperature difference spectra of the cytochrome *b*-558–*d* complex at 77°K in liquid nitrogen with a light path of 3 mm.

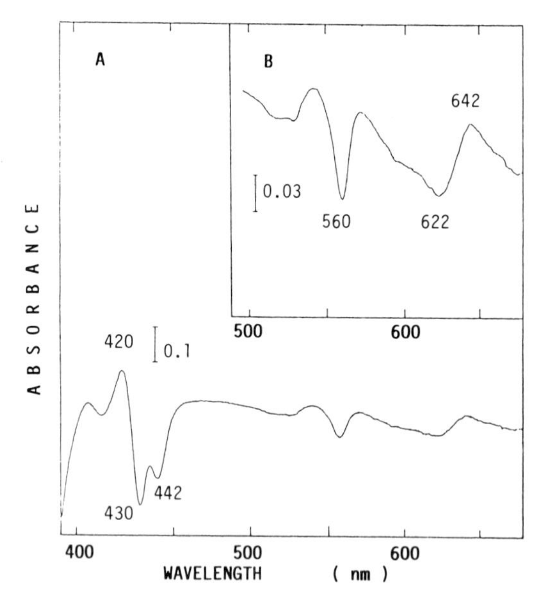

FIG. 6. CO-reduced-minus-reduced difference spectrum of the purified cytochrome *b*-558–*d* complex. (A) Conditions were as described in the legend to Fig. 5. Half the solution of cytochrome *b*-558–*d* complex reduced with dithionite was placed in the reference cuvette. The other half was treated with CO gas for 1 min and placed in the sample cuvette. (B) Spectrum between 490 and 670 nm.

in the spectra of inner membranes or of purified oxidase in the presence of excess soybean phospholipids.

A solution of purified cytochrome *b*-558–*d* (121 μg of protein/ml), which contains 12.3 nmol of protoheme/mg of protein, shows absorbances of 0.022 and 0.028 unit in the difference spectrum of the sodium dithionite reduced-minus-oxidized form at the wavelength pairs 560–580 nm and 628–649 nm, respectively. From these values, the molar extinction coefficients of purified cytochrome *b*-558 and cytochrome *d* are calculated to be 14.8 cm^{-1} mM^{-1} at the wavelength pair 560–580 nm, and 18.8 cm^{-1} mM^{-1} at the wavelength pair 628–649 nm, respectively (Table III).

The molar extinction coefficient of CO–cytochrome *d* complex is calculated in the same way, from the CO difference spectrum of purified oxidase, as 12.6 cm^{-1} mM^{-1} (622–642 nm).

The difference spectrum of the purified cytochrome *b*-558–*d* complex shows a broad peak at 594 nm which seems to be that of cytochrome a_1. The ratio of cytochrome a_1 to cytochrome *d* in the purified oxidase is one-tenth of that in the inner membranes, and the amount of cytochrome a_1 in the purified enzyme is estimated to be less than 3% using a molar extinction coefficient of 16.4 cm^{-1} mM^{-1}.[9]

Oxidation–Reduction Potential[4,5]

The oxidation–reduction potential is determined by potentiometric titration as described by Takamiya and Dutton.[10] Cytochrome *b*-562 and cytochrome *o* have apparently the same E'_m value of 125 mV (pH 7.4), and this is pH dependent (−60 mV/pH) in medium of pH 6.0–7.4.[10a] The reason why cytochrome *b*-562 and cytochrome *o* appear to have the same E'_m value is that there is actually only a small difference in their E'_m values. Recently, this was confirmed by the finding of van Wielink *et al.*[11]

The E'_m values of cytochrome *b*-558 and cytochrome *d* are determined to be +10 mV and +240 mV, respectively. With membrane preparations,

[9] Dr. T. Yamanaka, personal communication.

[10] K. Takamiya and P. L. Dutton, *Biochim. Biophys. Acta* **546,** 1 (1979).

[10a] The abbreviations used are: E'_m, oxidation-reduction potential at neutral pH; TMPD, *N,N,N',N'*-tetramethyl-*p*-phenylenediamine dihydrochloride; MES, 2-(*N*-morpholino)-ethanesulfonic acid monohydrate; HEPES, 4-(2-hydroxyethyl)-1-piperazineethane-sulfonic acid; PMS, phenazine methosulfate; diS-C_3-(5), 3,3'-dipropylthiodicarbocyanine iodide; FCCP, carbonyl cyanide *p*-trifluoromethoxyphenylhydrazone; SF6847, 3,5-di-*tert*-butyl-4-hydroxybenzylidenemalononitrile; HQNO, 2-*n*-heptyl-4-hydroxyquinoline *N*-oxide.

[11] J. E. van Wielink, L. F. Oltmann, F. J. Leeuweric, J. A. De Hollander, and A. H. Stouthamer, *Biochim. Biophys. Acta* **681,** 177 (1982).

the E'_m values of cytochrome *d* were estimated to be +260 mV by Pudek and Bragg,[12] and +280 mV by Reid and Ingledew.[13] The E'_m value of purified cytochrome *d* determined in this work is consistent with these values. The E'_m value of cytochrome *b*-558 in the membranes of +160 mV estimated by Pudek and Bragg[12] is higher than that obtained with purified cytochrome *b*-558 in this work and reported by Koland *et al.*[14] The E'_m value of cytochrome *b*-558 may decrease during purification.

Ubiquinol Oxidase Activity[4,5]

Assay Method

Before the assay, Sarkosyl is removed by treating the enzyme preparation with Bio-Beads SM-2 as described by Kasahara and Hinkle.[15] A mixture (100 μl) of oxidase complex (10 μg), phospholipid (1 m*M*), and 60 m*M* Tris–HCl (pH. 7.5) is incubated at 4° for 5 min. Ubiquinol oxidase activity is measured by recording the absorbance change of ubiquinol-1 at 278 nm in the initial 5 min in a quartz cuvette (2.5 ml) containing 150 μ*M* ubiquinol-1 and 60 m*M* Tris–HCl (pH 7.5) at 25° (molar extinction coefficient, 15,000). The reaction is started by adding oxidase complex. Ubiquinol-1 is prepared as described previously.[16]

Ubiquinol oxidase activity is also measured by recording the rate of oxygen consumption with an oxygen electrode. Assay condition is the same as described above, but more than 10-fold of purified oxidase complex is required for this assay.

Electron Donors and Kinetics[4,5]

Since the final preparations of two terminal oxidase complexes contain no phospholipid, preincubation with phospholipid is essential for studying the properties of the oxidase activities. Total phospholipids from the cytoplasmic membrane of *E. coli* and asolectin are most effective, and an acidic phospholipid, such as phosphatidylinositol or cardiolipin, alone also stimulates the enzyme activities. This activation is found to be accompanied by a change in the K_m value, and the low affinities of the purified oxidases for ubiquinol-1 increased on addition of phospholipids. The cytochrome *b*-562–*o* complex requires 0.5 m*M* asolectin and the

[12] M. Pudek and P. D. Bragg, *Arch. Biochem. Biophys.* **174,** 546 (1976).
[13] G. A. Reid and W. J. Ingledew, *Biochem. J.* **182,** 465 (1979).
[14] J. G. Koland, M. J. Miller, and R. B. Gennis, *Biochemistry* **23,** 1051 (1984).
[15] M. Kasahara and P. C. Hinkle, *J. Biol. Chem.* **252,** 7384 (1977).
[16] J. S. Rieske, this series, Vol. 10, p. 239.

TABLE IV
KINETIC CHARACTERISTICS OF OXIDASE ACTIVITY WITH VARIOUS ELECTRON DONORS[a]

	Cytochrome *b*–562–*o* complex		Cytochrome *b*-558–*d* complex	
Substrate	K_m (μM)	V_{max}[b]	K_m (μM)	V_{max}[b]
Ubiquinol-1	48	15.0	230	7.5
Menadiol	38.4	0.688	1.67[c]	3.85
TMPD	9.5[c]	0.28	18.2[c]	2.84
O_2	2.9	—[d]	0.38	—[d]

[a] Oxidase activities were measured with an oxygen electrode.
[b] μmol O_2/min/nmol cytochrome *o* or cytochrome *d*.
[c] Millimolar.
[d] Not determined.

cytochrome *b*-558–*d* complex requires 3 m*M* asolectin for their maximal activation.

Ubiquinol-6 and ubiquinol-10, which have a long isoprenoid side chain, are oxidized by the oxidase complexes as well as ubiquinol-1 and menadiol. The artificial electron donor TMPD is a good substrate for the oxidases, and the rate of oxygen consumption is the same in the presence and absence of ascorbate.

The kinetic characteristics of the oxidation reaction with these substrates are summarized in Table IV.

Inhibitors

The ubiquinol oxidase activities of the purified cytochrome *b*-562–*o* complex and cytochrome *b*-558–*d* complex are inhibited by the respiratory inhibitors HQNO, piericidin A, and hydroxylamine. The activity of the cytochrome *b*-558–*d* complex is less sensitive than that of the cytochrome *b*-562–*o* complex to these inhibitors (Fig. 7). On the other hand, the oxidase activity of the cytochrome *b*-562–*o* complex is less sensitive to hydrogen peroxide than that of the cytochrome *b*-558–*d* complex. The concentrations required for 50% inhibition of activity of the cytochrome *b*-562–*o* complex are 2 μM for HQNO and piericidin A and 2.2 m*M* for hydroxylamine, while those for 50% inhibition of the cytochrome *b*-558–*d* complex are 7 μM for HQNO and 15 μM for piericidin A. The oxidase activity of the cytochrome *b*-558–*d* complex is fairly resistant to hydroxylamine and is inhibited only 20% by 5 m*M* hydroxylamine.

Great differences are observed in the inhibitions of the two oxidase complexes by cyanide and by azide (Fig. 7E,F). The oxidase activity of

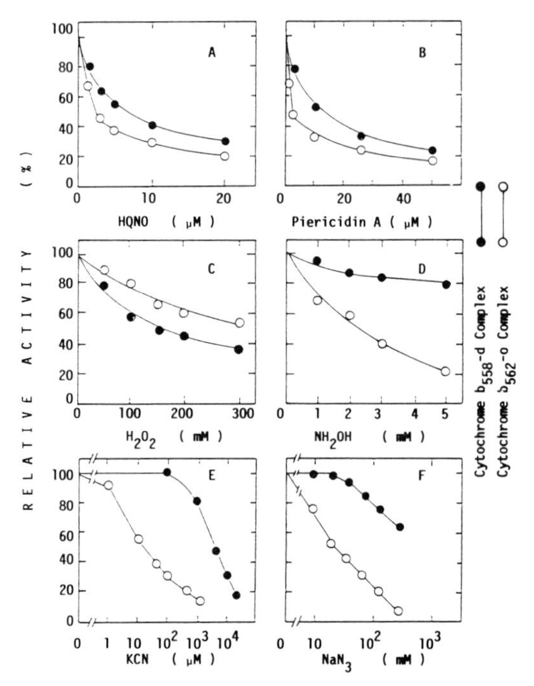

FIG. 7. Effects of inhibitors on ubiquinol-1 oxidase activities of purified cytochrome *b*-562–*o* complex and cytochrome *b*-558–*d* complex. Control activity corresponded to a value of 250 μmol of ubiquinol-1 oxidized/min/mg of protein for cytochrome *b*-562–*o* complex or 125 μmol of ubiquinol-1 oxidized/min/mg of protein for cytochrome *b*-558–*d* complex, and was taken as 100%. HQNO (A) and piericidin A (B) were dissolved in ethanol. Addition of 0.2% ethanol to the assay mixture did not affect the activity. (○——○), Cytochrome *b*-562–*o* complex; (●——●), cytochrome *b*-558–*d* complex.

the cytochrome *b*-558–*d* complex is extremely resistant to cyanide, and 100-fold higher concentration of cyanide is required for 50% inhibition than the cytochrome *b*-562–*o* complex.

These results explain why in earlier experiments *E. coli* cells containing cytochrome *d* were resistant to cyanide[17] and why the synthesis of cytochrome *d* was induced in the presence of cyanide.[18]

The ubiquinol oxidase activity of purified cytochrome *b*-562–*o* complex is strongly inhibited by zinc ion. The concentration for 50% inhibition is 1 μ*M*. On the other hand, the ubiquinol oxidase activcity of purified cytochrome *b*-558–*d* complex is resistant to zinc ion, and the concentration for 50% inhibition of its activity is 60 μ*M*.

[17] M. R. Pudek and P. D. Bragg, *Arch. Biochem. Biophys.* **164,** 682 (1974).
[18] J. R. Aschroff and B. A. Haddock, *Biochem. J.* **148,** 349 (1975).

Effect of Oxygen Concentration on Oxidase Activity[5]

Assay Method

The kinetics of oxygen consumption by suspension of purified oxidases as functions of the dissolved oxygen concentration is determined with a Clark-type oxygen electrode (Sumitomo Denko Type PO-100) which is sensitive to low oxygen concentration and can follow the reaction rapidly, as described elsewhere.[19]

A mixture (0.1 ml) of purified cytochrome *b*-562–*o* complex (10 μg) or cytochrome *b*-558–*d* complex (20 μg), phospholipids (1 m*M*), and 100 m*M* Tris–HCl (pH 7.0) is incubated at 4° for 5 min, transferred in 1.5 ml of reaction mixture, and ubiquinol-1 (0.2 m*M*) is added to start reaction. All experiments are carried out anaerobically at room temperature, and oxygen is introduced into the mixture by addition of the same Tris–HCl buffer equilibrated with air. The time course of change in dissolved oxygen concentration is followed and the K_m values are calculated by the method of Dixon and Webb[20] from the slope of a plot of $(2.303/t)$ log $s/(s - y)$ against (y/t), where s is the oxygen concentration at a given time and y is the amount of oxygen change in time t from a given time.

Oxygen Affinity of the Oxidases

The K_m value of the purified cytochrome *b*-558–*d* complex (0.38 μM) is lower than that of the purified cytochrome *b*-562–*o* complex (2.9 μM) when ubiquinol-1 is used as substrate.

These results indicate that the cytochrome *b*-558–*d* complex can utilize oxygen even at low oxygen concentration and that the apparent affinity for oxygen of cells and membranes at various growth phases and in various conditions depends on the relative content of cytochrome *b*-558–*d* complex.

Formation of a Membrane Potential by Reconstituted Liposomes

Reconstitution Procedure

The method of freeze-thaw/sonication is used.[15] A mixture (0.5 ml) of cytochrome *b*-562–*o* complex (8 μg) or cytochrome *b*-558–*d* complex (15 μg), acetone-washed soybean phospholipid (asolectin 15 mg), 20 m*M*

[19] B. Hagihara, *Anal. Biochem.*, in press (1986).

[20] M. Dixon and E. C. Webb, *in* "Enzymes" (M. Dixon and E. C. Webb, eds.), 3rd Ed., p. 55. Longman, London, 1979.

HEPES/choline (pH 7.5), and 25 m*M* potassium sulfate is flushed with nitrogen gas, frozen at − 70°, warmed to room temperature for 15 min, and sonicated for 15–20 sec in a bath-type sonicator at 20°. Then 1 *M* $MgCl_2$ is added to a final concentration of 2 m*M*.

Assay Method and Calibration of Membrane Potential

The membrane potential is measured at 25° as fluorescence quenching of diS-C_3-(5)[10a] with excitation and emission wavelengths of 620 and 670 nm, respectively.[21] The assay mixture (2 ml) contains 50 μl of reconstituted liposomes (0.8 μg of protein), 0.15 μM diS-C_3-(5), 20 m*M* HEPES/choline (pH 7.4), and 2 m*M* $MgSO_4$.

Fluorescence quenching is measured by addition of 20 μM ubiquinol-1, 5 μM ubiquinol-6, or 5 m*M* ascorbate/50 μM PMS to standard assay mixture (see below), and the membrane potential generated is expressed as $\Delta F/F$. Ubiquinol-1 oxidase activity is not inhibited by diS-C_3-(5) at concentrations of about 0.15 μM but is inhibited at a concentration of 1 μM or more. The membrane potential is calibrated by the valinomycin-dependent potassium diffusion potential. For removal of extravesicular potassium, reconstituted liposomes are passed by centrifugation[22] through a column which has been filled with 1 ml of Sephadex G-50 and equilibrated with 20 m*M* HEPES/choline (pH 7.4) and 2 m*M* $MgSO_4$. They are then added to standard assay mixture (2 ml) containing 0.15 μM diS-C_3-(5), 20 m*M* HEPES/choline (pH 7.4), 2 m*M* $MgSO_4$, 20 m*M* NaSCN, and K_2SO_4 at various concentrations. The membrane potential generated by ubiquinol-1 oxidation is the same value in the presence and absence of NaSCN. The diffusion potential of potassium is generated by addition of 0.1 μM valinomycin to standard assay mixture. In the absence of NaSCN, a much higher concentration (5–10 μM) of valinomycin is required to obtain an adequate speed of fluorescence quenching by potassium diffusion.

Formation of Membrane Potential by Reconstituted Liposomes

Formations of membrane potential by reconstituted cytochrome *b*-562–*o* complex and reconstituted cytochrome *b*-558–*d* complex in liposomes are observed with the fluorescent dye diS-C_3-(5) on addition of ubiquinol-1. This is the first indication that there are coupling sites in the terminal oxidases, which contain cytochrome *o* and cytochrome *d* (Fig. 8). The membrane potential formed is dissipated by the addition of the protonophore uncoupler SF6847 and the inhibitors of the oxidase system.

[21] P. J. Sims, A. S. Waggoner, C. H. Wang, and F. Hoffman, *Biochemistry* **13,** 3315 (1974).
[22] M. Kasahara and H. S. Penefsky, *J. Biol. Chem.* **253,** 4180 (1978).

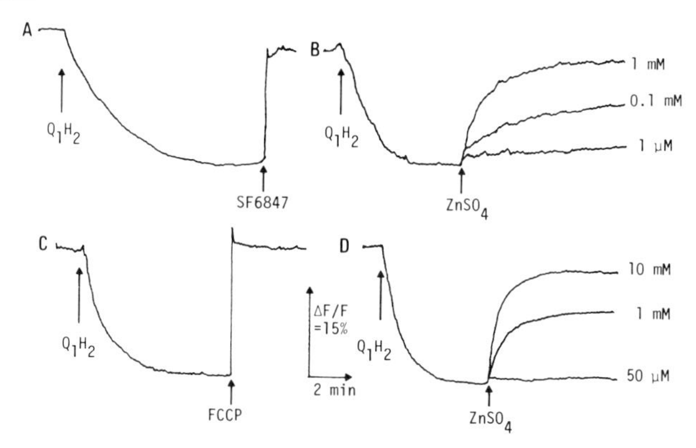

FIG. 8. Generation of membrane potential by reconstituted liposomes made with purified oxidases and effects of protonophore uncouplers and oxidase inhibitors. Purified cytochrome *b*-562–*o* complex (A, B) or purified cytochrome *b*-558–*d* complex (C, D) were reconstituted by freeze-thaw/sonication, and membrane potential was measured with diS-C_3-(5), as described under "Assay Method and Calibration of Membrane Potential." The assay mixture (2 ml) contained 50 μl of reconstituted liposomes. Ubiquinol-1 was added at the time indicated by arrow. After a few minutes, 0.1 μ*M* SF6847 in (A), 0.16 μ*M* FCCP in (C), and $ZnSO_4$ at the concentration indicated in (B) and (D) were added.

Freeze-thawing and subsequent sonication are essential for generation of a membrane potential.

The magnitude of the membrane potential formed is determined by measuring the potassium diffusion potential generated by addition of valinomycin (Fig. 9). A linear response of fluorescence quenching is observed between −90 and −200 mV. The membrane potential formed with ubiquinol-1 as an electron donor is determined to be −145 mV in the case of cytochrome *b*-562–*o* complex. The same value is obtained when the reconstituted cytochrome *b*-558–*d* complex is examined.

Summary

Two terminal oxidase complexes, cytochrome *b*-562–*o* complex and cytochrome *b*-558–*d* complex, are isolated in highly purified forms which show ubiquinol oxidase activities.

From the result of steady-state kinetics of cytochromes in the membrane[23] and E'_m values of purified cytochromes,[4,5,7] we propose a branched arrangement of the late exponential phase of aerobic growth, as shown in Fig. 10. Cytochrome *b*-556 is reduced by several dehydrogenases and the gene for this cytochrome (*cybA*) is located in the *sdh* gene

[23] K. Kita and Y. Anraku, *Biochem. Int.* **2**, 105 (1981).

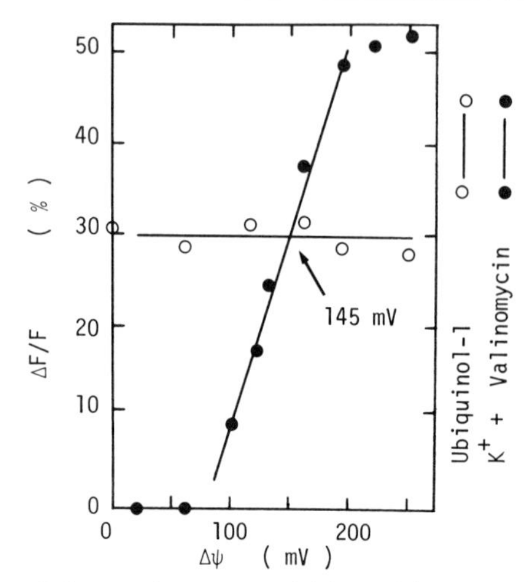

FIG. 9. Calibration of the membrane potential formed in reconstituted liposomes made with cytochrome *b*-562–*o* complex. The membrane potential generated by the potassium diffusion potential (●) was measured as described (see legend to Fig. 8). The membrane potential generated by ubiquinol-1 oxidation (○) was measured in the same ionic conditions with variable potassium concentrations. Relative quenching is plotted on the abscissa and corresponds to diffusion potential of potassium ion calculated by a Nernst equation.

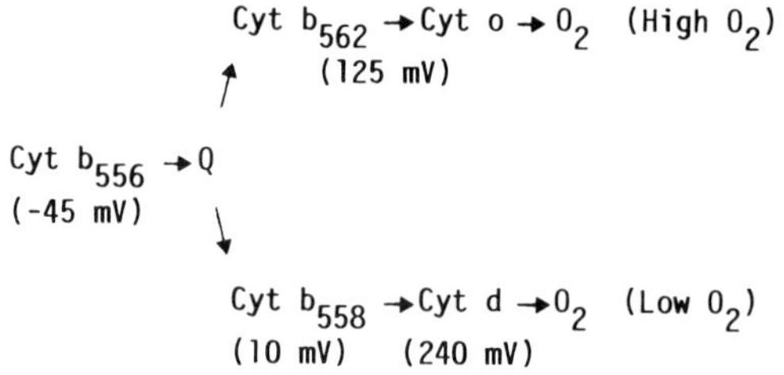

Cyt b_{562} → Cyt o → O_2 (High O_2)
(125 mV)
↑
Cyt b_{556} → Q
(-45 mV)
↓
Cyt b_{558} → Cyt d → O_2 (Low O_2)
(10 mV) (240 mV)

FIG. 10. Arrangement of cytochromes in the respiratory chain of cells of *E. coli* in the late exponential phase of aerobic growth. The E'_m values in parentheses were reported.[4,5,7] cyt, Cytochrome; Q, ubiquinone-8.

cluster.[24,25] Recently, we found another low-potential *b*-type cytochrome, cytochrome *b*-561 (E'_m = 20 mV),[26] which is also reduced by dehydrogenases. The position of this new cytochrome in the aerobic respiratory chain is under investigation. Two terminal oxidase complexes branch at the site of ubiquinone-8, and the K_m value for oxygen of the purified cytochrome *b*-558–*d* complex is about 8-fold lower than that of the purified cytochrome *b*-562–*o* complex when ubiquinol-1 is used as substrate.

[24] H. Murakami, K. Kita, H. Oya, and Y. Anraku, *Mol. Gen. Genet.* **196,** 1 (1984).
[25] H. Murakami, K. Kita, H. Oya, and Y. Anraku, *FEMS Microbiol. Lett.* **30,** 307 (1985).
[26] H. Murakami, K. Kita, and Y. Anraku, *Mol. Gen. Genet.* **198,** 1 (1984).

This result is consistent with the idea that the cytochrome *b*-558–*d* complex is synthesized as an alternative oxidase for more efficient utilization of oxygen at low oxygen concentration. Thus, *E. coli* cells can maintain efficient oxidative energy conservation over a wide range of oxygen pressures by simply changing the contents of the two terminal oxidases, each of which functions as a coupling site.

Acknowledgments

This work was supported in part by Grant-in-Aid for Scientific Research 544088 from the Ministry of Education, Science, and Culture of Japan, Special Coordination Funds for Promotion of Science and Technology from the Science and Technology Agency of Japan, and a grant from the Yamada Science Foundation, Japan.

[11] Purification and Reconstitution of the Cytochrome *o*-Type Oxidase from *Escherichia coli*

By KAZUNOBU MATSUSHITA, LEKHA PATEL, and H. RONALD KABACK

The cytochrome *o*-type oxidase from *Escherichia coli* catalyzes the following overall reaction:

$$Q_8H_2 + 2H^+_{in} + 1/2O_2 \rightarrow Q_8 + 2H^+_{out} + H_2O \qquad (1)$$

where Q_8H_2 and Q_8 are ubiquinol-8 and ubiquinone-8, respectively.

In aerobically growing *E. coli,* as in mitochondria, a transmembrane electrochemical gradient of hydrogen ion ($\Delta\bar{\mu}_{H^+}$) is generated primarily by substrate oxidation via a membrane-bound respiratory chain with oxygen as terminal electron acceptor. The respiratory chain of *E. coli* contains two terminal oxidases, cytochromes *o* and *d*, that are induced under conditions of high and low aeration, respectively, during growth. Both oxidases function as $Q_8H_2:O_2$ oxidoreductases, and some of the free energy generated from the electron-transfer reactions is used to generate $\Delta\bar{\mu}_{H^+}$.[1-5]

In this article, purification of the cytochrome *o*-type oxidase from the membrane of *E. coli* and its functional reconstitution into proteoliposomes are described.[2,5] In addition, a procedure for the simultaneous reconstitution of the *o*-type oxidase and *lac* permease into proteolipo-

[1] K. Kita, M. Kasahara, and Y. Anraku, *J. Biol. Chem.* **257,** 7933 (1982).

[2] K. Matsushita, L. Patel, R. B. Gennis, and H. R. Kaback, *Proc. Natl. Acad. Sci. U.S.A.* **80,** 4889 (1983).

[3] J. G. Koland, M. J. Miller, and R. B. Gennis, *Biochemistry* **23,** 445 (1984).

[4] K. Kita, K. Konishi, and Y. Anraku, *J. Biol. Chem.* **259,** 3375 (1984).

[5] K. Matsushita, L. Patel, and H. R. Kaback, *Biochemistry* **23,** 4703 (1984).

somes is described.[2] An alternative procedure for purification of cytochrome *o* has also been published.[6]

Assay Methods[2,5]

Enzyme Activity

Oxidation of ubiquinol-1 (Q_1H_2) is measured at room temperature by following absorbance at 275 nm in a spectrophotometer. Reaction mixtures contain 80 μM Q_1H_2, 0.1% Tween 20 and/or 0.25 mg/ml of *E. coli* phospholipids, 50 m*M* potassium phosphate (K-phosphate; pH 7.5), and enzyme in a total volume of 1.0 ml. Activity is calculated by using a millimolar extinction coefficient of 12.25 for Q_1H_2. Q_1H_2 is prepared from Q_1 as described by Rieske,[7] and *E. coli* phospholipid is prepared as described in this series.[8]

Alternatively, oxygen uptake is measured polarographically at 25°. Reaction mixtures contain 5 m*M* dithiothreitol (DTT), 80 μM Q_1, 0.1% Tween 20 and/or 0.25 mg/ml of phospholipid, 50 m*M* K-phosphate (pH 7.5), and enzyme in a total volume of 1.0 ml. Activity is calculated by assuming that the initial oxygen concentration is 258 μM at 25°.

Determinations of Membrane Potential ($\Delta\Psi$) and pH Gradient (ΔpH)

$\Delta\Psi$ (interior negative) is determined by measuring either the steady-state distribution of [^{3}H]tetraphenylphosphonium (TPP^+) using flow dialysis[9] or fluorescence quenching of 3,3′-diisopropylthiodicarbocyanine [diS-C_3-(5)],[10] and both methods yield comparable results.[5] For the latter determinations, the reaction mixtures contain 1 μM diS-C_3-(5), 50 m*M* K-phosphate (pH 7.5), and proteoliposomes in a total volume of 2.0 ml. Data are quantitated by comparison to valinomycin-induced potassium diffusion potentials ($K^+_{in} \rightarrow K^+_{out}$).

ΔpH (interior alkaline) is determined from the steady-state distribution of [^{3}H]acetate using flow dialysis.[9]

Lactose Transport

Uptake of [1-^{14}C]lactose by proteoliposomes is measured by filtration as described.[2,8]

[6] K. Kita, K. Konishi, and Y. Anraku, *J. Biol. Chem.* **259,** 3368 (1984).
[7] J. S. Rieske, this series, Vol. 10, p. 239.
[8] P. V. Viitanen, M. J. Newman, D. L. Foster, T. H. Wilson, and H. R. Kaback, this series, Vol. 125 [32].
[9] S. Ramos, S. Schuldiner, and H. R. Kaback, this series, Vol. 55, p. 680.
[10] A. S. Waggoner, this series, Vol. 55, p. 689.

SUMMARY OF TYPICAL PURIFICATION

Step	Protein (mg)	Q_1H_2 (units)	Oxidase[b] (units/mg)	Cytochrome[c] (nmol/mg)
Initial membrane[a]	1024	1880	1.84	0.13
Washed membrane	115	448	3.89	—
Octylglucoside extract	27	502	18.6	1.29
DEAE-Sepharose	4.6	381	82.0 (128)[d]	7.50

[a] Membranes were prepared from 40 g of wet cells.
[b] Enzyme assay was done in the presence of 0.1% Tween 20.
[c] Cytochrome *o* content was calculated from CO-difference spectra by using a millimolar extinction coefficient of 170 at 415–430 nm.
[d] Enzyme activity in the presence of 0.25 mg/ml of *E. coli* phospholipid and 0.1% Tween 20.

Purification Procedures[2,5]

The cytochrome *d*-deficient mutant of *E. coli* isolated by Green and Gennis[11] is useful for the purification of cytochrome *o*, since membranes from the mutant contain only cytochrome *o* as a terminal oxidase. As described for *lac* permease,[8] sequential extraction of *E. coli* membranes with relatively high concentrations of urea and cholate provides significant *in situ* purification of the *o*-type oxidase. Furthermore, by centrifugation after urea treatment, inner and outer membranes are separated readily by differential resuspension of the layered precipitate. The oxidase is then solubilized with octyl-β-D-glucopyranoside (octylglucoside) in the presence of exogenous phospholipids and subjected to further purification (see the table).

Growth of Cells

Escherichia coli GR19N (cyd^-) is grown aerobically at 37° into late logarithmic phase in the following medium: 0.5% sodium DL-lactate, 0.15% casamino acids, 0.2% ammonium sulfate, 1 m*M* $MgSO_4$, 0.1 m*M* $CaCl_2 \cdot 2H_2O$, 0.001% $FeSO_4 \cdot 7H_2O$, and 0.1 *M* K-phosphate (pH 7.0).

Preparation of Membranes

Cells are harvested by centrifugation and washed once in 50 m*M* K-phosphate (pH 7.5). Washed cells are resuspended in 50 m*M* K-phosphate (pH 7.5) containing 5 m*M* $MgSO_4$ (4 ml/g wet weight of cells), and pancre-

[11] G. N. Green and R. B. Gennis, *J. Bacteriol.* **154,** 1269 (1983).

atic DNase (type I; Sigma) is added to a concentration of 10 μg/ml. The suspension is passed through a French pressure cell at 20,000 psi, and unbroken cells are removed by centrifugation at 10,000 *g* for 10 min. The supernatant is then centrifuged at 120,000 *g* for 2 hr to sediment the membrane fraction which is washed once in 50 m*M* K-phosphate (pH 7.5).

Purification in Situ and Solubilization

Washed membranes are suspended to a final concentration of about 10 mg of protein/ml in 50 m*M* K-phosphate (pH 7.5) and mixed with an equal volume of 10 *M* urea at room temperature (the urea solution is prepared freshly). The suspension is incubated on ice for 20 min and centrifuged at 150,000 *g* for 2 hr, and the supernatant is carefully decanted and discarded. The pellet contains two clearly demarcated portions, a lower transparent layer and an upper red layer. The loose, red layer is resuspended manually by agitation in 50 m*M* K-phosphate (pH 7.5) and centrifuged at 120,000 *g* for 1 hr. The pellet is resuspended in 50 m*M* K-phosphate (pH 7.5), adjusted to a concentration of about 4 mg of protein/ml, and 20% sodium cholate (pH 7.8) is added to a final concentration of 6%. The mixture is incubated on ice for 20 min and centrifuged at 120,000 *g* for 1 hr, and the supernatant is discarded.

The red pellet which is almost devoid of a lower transparent layer is resuspended in 50 m*M* K-phosphate (pH 7.5) and washed once by centrifugation at 120,000 *g* for 1 hr. The pellet is dispersed in 50 m*M* K-phosphate (pH 7.5), and *E. coli* phospholipid [50 mg/ml in 50 m*M* K-phosphate (pH 7.5)] and 12.5% octylglucoside [in 50 m*M* K-phosphate (pH 7.5)] are added to final concentrations of 3.8 mg/ml and 1.25%, respectively (the final protein concentration is 1.5–2.0 mg/ml). The suspension is stirred for 10 min and centrifuged at 120,000 *g* for 1 hr. The supernatant (octylglucoside extract) is carefully aspirated and used in subsequent steps.

DEAE–Sepharose Column Chromatography

The octylglucoside extract is applied to a DEAE–Sepharose CL-6B column (Pharmacia; about 1 ml bed volume/mg of protein applied) that had been equilibrated with 50 m*M* K-phosphate (pH 7.5) and washed with 1 bed volume of 50 m*M* K-phosphate (pH 7.5) containing 1% octylglucoside. After application of the sample, the column is washed with 1 bed volume of 50 m*M* K-phosphate (pH 7.5) containing 1% octylglucoside, followed by 2 bed volumes of 100 m*M* K-phosphate (pH 7.5) containing 1% octylglucoside. Cytochrome *o* is then eluted with a linear gradient of 100–175m*M* K-phosphate (pH 7.5) containing 1% octylglucoside (each

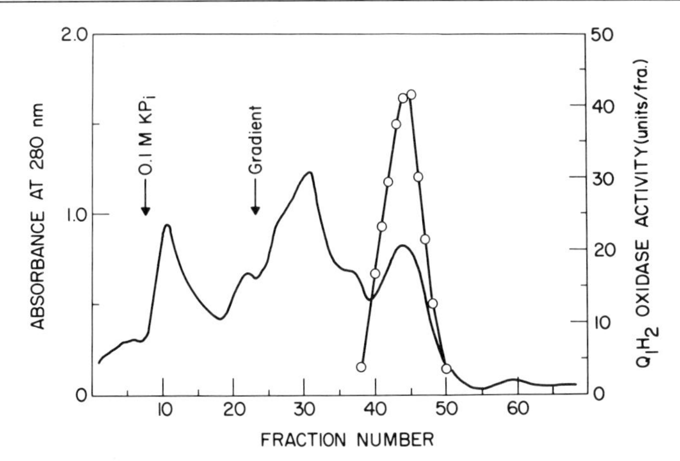

FIG. 1. Typical elution profile of cytochrome *o*-type oxidase from DEAE-Sepharose. In this case, 35 ml of octylglucoside extract (15 mg protein) was applied to a column (1.5 × 10 cm, 17.7 ml), which was washed with 30 ml of 50 m*M* K-phosphate (pH 7.5) containing 1% octylglucoside, then with 30 ml of 100 m*M* K-phosphate (pH 7.5) containing 1% octylglucoside. Oxidase was eluted with a linear gradient consisting of 50 ml each of 100 m*M* K-phosphate (pH 7.5) and 175 m*M* K-phosphate (pH 7.5) containing 1% octylglucoside. Each fraction contained 1.8 ml. (——), Absorbance at 280 nm; (○——○), enzyme activity.

reservoir contains 2 bed volumes of the appropriate buffer) at a flow rate of about 20 ml/hr. The enzyme elutes as a single symmetrical peak coincident with protein at about the middle of the gradient (Fig. 1).

Sephacryl S-200 Column Chromatography

Although the oxidase is substantially homogeneous after DEAE-Sepharose chromatography, the material sometimes contains minor impurities. In this case, the enzyme may be subjected to further purification by ammonium sulfate fractionation and gel filtration through Sephacryl S-200. To pooled DEAE-Sepharose fractions containing Q_1H_2 oxidase activity, solid ammonium sulfate is added to a concentration of 0.361 g/ml, and the mixture is stirred for 10 min on ice. After centrifugation at 10,000 *g* for 10 min, a small pellet and an orange supernatant are obtained. The supernatant is decanted, and 0.201 g/ml of solid ammonium sulfate is added. After 10 min incubation on ice with stirring followed by 10 min centrifugation at 10,000 *g*, a red precipitate floating on the surface is observed. The precipitate is collected and dissolved in 50 m*M* K-phosphate (pH 7.5) containing 1% octylglucoside (1–2 ml, final volume). The concentrated sample is applied to a Sephacryl S-200 column (Pharmacia; 1.5 × 83 cm) that had been preequilibrated with 50 m*M* K-phosphate (pH

7.5) containing 1% octylglucoside. The column is eluted with the same buffer at a flow rate of about 5 ml/hr, and Q_1H_2 oxidase activity and protein chromatograph as single coincident peaks. Fractions containing activity are pooled, frozen in liquid nitrogen, and stored at −180° without significant loss of activity over at least 6 months.

Reconstitution[2,5]

Reconstitution of Proteoliposomes with Purified Cytochrome o

Proteoliposomes containing various ratios of purified oxidase and *E. coli* phospholipids are prepared by octylglucoside dilution or dialysis, and a preparation with a protein : lipid ratio of 1 : 125 (w/w) is described.

A solution of purified oxidase (0.5 mg of protein) containing 1% octylglucoside is mixed with sonicated *E. coli* phospholipid (62.5 mg of lipid), and octylglucoside is added to a final concentration of 1.25% in a final volume of 7 ml (the octylglucoside : phospholipid ratio is about 1.4, w/w). The mixture is incubated on ice for 10–20 min and then diluted into 30 volumes of 50 m*M* K-phosphate (pH 7.5) that had been equilibrated to room temperature. After stirring with a magnetic bar for 10 min, proteoliposomes are collected by centrifugation at 120,000 *g* for 2 hr. The supernatant is discarded, and the proteoliposomes are resuspended in 50 m*M* K-phosphate (pH 7.5) to a protein concentration of 200–250 μg/ml and frozen rapidly in liquid nitrogen. Just prior to use, the samples are thawed at room temperature and sonicated in a plastic tube for 10–20 sec with a bath-type sonicator (80W, 80 Hz, generator model G80-80-1, tank model T80-80-IRS from Laboratory Supplies, Hicksville, NY).

Reconstitution of Proteoliposomes with Cytochrome o and lac Permease[2]

In order to obtain proteoliposomes in which turnover of cytochrome *o* drives active transport of β-galactosides, octylglucoside dilution is performed in the presence of purified oxidase, purified *lac* permease,[8] and DTT.

Sonicated *E. coli* phospholipids [0.56 ml of a solution containing 28 mg of phospholipid in 50 m*M* K-phosphate (pH 7.5)] are mixed with octylglucoside [0.16 ml of 12.5% octylglucoside dissolved in 50 m*M* K-phosphate (pH 7.5)], purified cytochrome *o* [1.0 ml of a solution containing 170 μg of protein and 1% octylglucoside in 50 m*M* K-phosphate (pH 7.5)], *lac* permease [2.0 ml of a solution containing 68 μg of permease and 1.5% octylglucoside in 10 m*M* K-phosphate (pH 6.0)] purified as described,[8] and

50 m*M* K-phosphate (pH 7.5) (0.68 ml containing 1 m*M* DTT) in a total volume of 4.4 ml. The sample is mixed by hand, incubated on ice for 20 min, and then diluted rapidly into 132 ml of 50 m*M* K-phosphate (pH 7.5) containing 1 m*M* DTT at room temperature. After stirring with a magnetic bar for 10 min, proteoliposomes are collected by centrifugation at 120,000 *g* for 2 hr. The supernatant is discarded, and the proteoliposomes are resuspended in 50 m*M* K-phosphate (pH 7.5) containing 1 m*M* DTT to a final volume of 0.75 ml (300 μg of protein/ml). The suspension is then rapidly frozen in liquid nitrogen, thawed at room temperature, and sonicated with a bath-type sonicator as described above. Proteoliposomes prepared in this manner contain a 1 : 1 molar ratio of oxidase to *lac* permease.

Characteristics of the Enzyme[5]

Molecular Properties

The purified *o*-type oxidase contains two *b*-type cytochromes (*b*-558 and *b*-563), one of which is a cytochrome *o* that reacts with carbon monoxide. The oxidase also contains two copper atoms in addition to the two *b*-type hemes.[6] Purified oxidase exhibits four polypeptide species on sodium dodecyl sulfate–polyacrylamide electrophoresis followed by silver staining, and the apparent molecular weights (M_r) of subunits I, II, III, and IV are 55,000, 35,000, 22,000 and 17,000, respectively (Fig. 2). The electrophoretic mobility of subunit I is highly dependent on the concentration of acrylamide in the gel, suggesting that it is hydrophobic, and the corrected M_r is 66,000. The hydrophobicity of subunit I is also supported by amino acid analysis and by photolabeling experiments with 3-(trifluoromethyl)-3-(*m*-[^{125}I]iodophenyl)diazirine, a hydrophobic probe that reacts from within the bilayer.[12] Subunits III and IV are not clearly visualized with Coomassie Brilliant Blue, and the stoichiometry of these subunits with respect to I and II is uncertain (Fig. 2). Thus, with the exception of subunit I, which contains heme, and subunit II, which is hydrolyzed by chymotrypsin with loss of oxidase activity, it cannot be concluded definitively that the other polypeptides are subunits of the oxidase, as opposed to contaminants that copurify with the enzyme. It is noteworthy, however, that each of the polypeptide species remains after further purification of the oxidase by ammonium sulfate fractionation and gel filtration chromatography and that the composition of the oxidase is unaffected by

[12] K. Matsushita and H. R. Kaback, unpublished data.

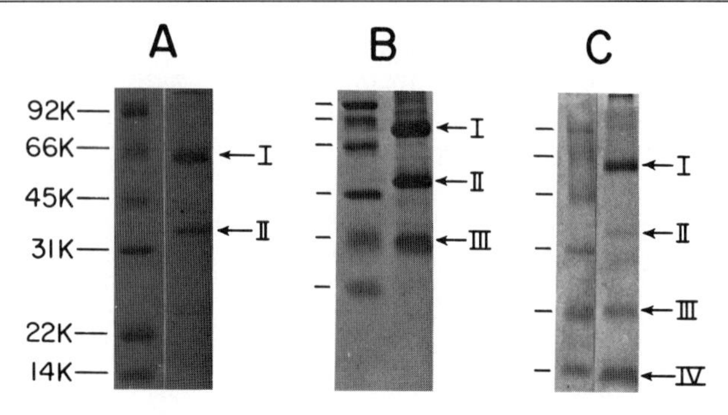

FIG. 2. SDS–PAGE of cytochrome *o*-type oxidase. (A) Oxidase (4.5 μg protein, right column) and marker proteins (1 μg each, left column) in 12.5% acrylamide gel stained with Coomassie Brilliant Blue. (B) Oxidase (7.5 μg protein, right column) and marker proteins (1 μg each, left column) in 12.5% acrylamide gel containing 10% glycerol and 3.6 *M* urea (high-resolution gel) stained with Coomassie Brilliant Blue. (C) Oxidase (2.2 μg protein, right column) and marker protein (0.3 μg each, left column) in high-resolution gel stained with silver.

Marker proteins include phosphorylase *b* (92.5K), bovine serum albumin (66.2K), ovalbumin (45K), carbonate dehydratase (31K), trypsin inhibitor (21.5K), and lysozyme (14.4K).

purification in the presence of a protease inhibitor. Furthermore, on the basis of immunoelectrophoretic studies,[13] it has been concluded that the oxidase complex contains four subunits.

Kinetic Properties

Cytochrome *o*-type oxidase catalyzes the oxidation of Q_1H_2 (K_m = 10 μM), *N,N,N′,N′*-tetramethyl-*p*-phenylenediamine (TMPD; K_m = 4.2 m*M*), reduced phenazine methosulfate and other artificial electron donors with redox potentials between +50 and +260 mV, but exhibits no activity toward ascorbate, ferrocyanide, or horse and yeast ferrocytochrome *c*. The oxidase also appears to utilize Q_8H_2, the physiological ubiquinol in *E. coli,* since proteoliposomes containing cytochrome *o*, Q_8, and D-lactate dehydrogenase or pyruvate oxidase catalyze oxidation of D-lactate or pyruvate, respectively.[14] The optimum pH of the oxidase is about 7.5. Cyanide (K_i = 23 μM) and the quinone analogs 2-heptyl-4-hydroxyquinoline *N*-oxide (K_i = 0.8 μM) and 5-*n*-undecyl-6-hydroxy-4,7-dioxobenzothiazole (K_i = 0.3 μM) are potent inhibitors of the oxidase with either Q_1H_2 or TMPD as electron donors. The activity of purified oxidase is highly dependent upon the presence of detergent and/or phospholipid, particularly after gel filtration. *Escherichia coli* phospholipid stimulates

[13] R. G. Kranz and R. B. Gennis, *J. Biol. Chem.* **258,** 10614 (1983).

[14] K. Matsushita and H. R. Kaback, *Biochemistry,* in press (1986).

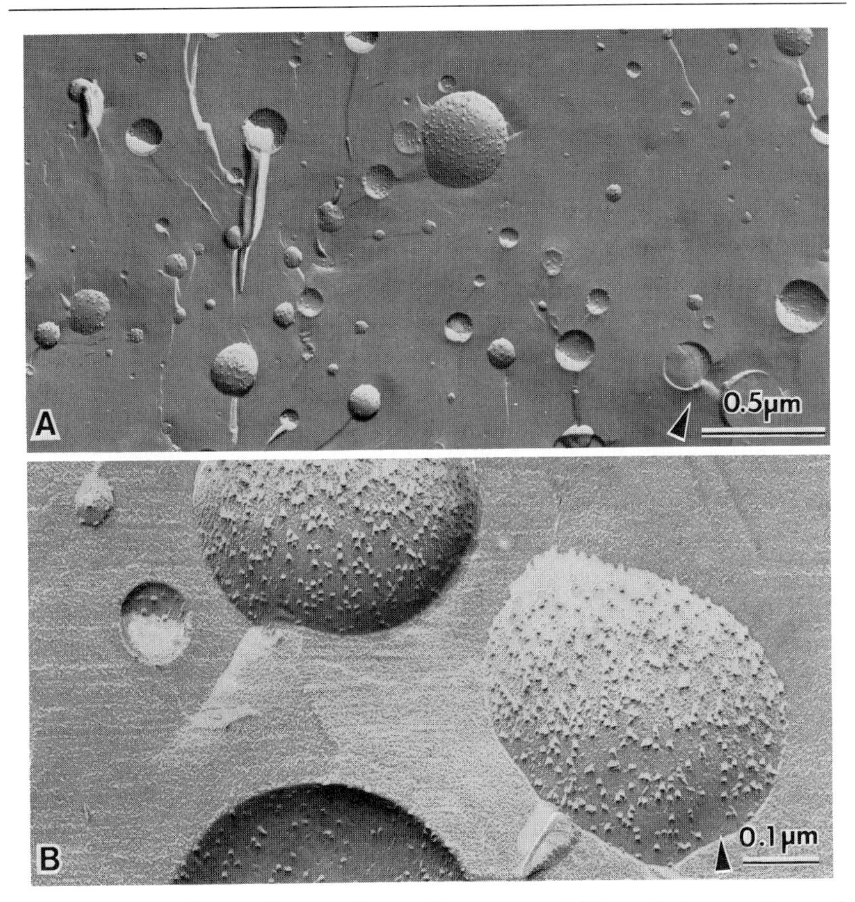

FIG. 3. Freeze-fracture images of proteoliposomes containing cytochrome *o*-type oxidase. Proteoliposomes with an oxidase-to-phospholipid ratio of 1 : 10 (w/w) were frozen ultrarapidly in the absence of fixatives or cryoprotectants. Replicas were formed by fracturing at −160°, followed by unidirectional deposition (arrows) of platinum. The study was performed by Dr. M. J. Costello in the Department of Anatomy, Duke University Medical Center.

Q_1H_2 oxidase activity about 20-fold at a concentration of 0.25 mg/ml. Maximum activity is observed in the presence of phospholipid and 0.1% Tween 20.

Characterization of Proteoliposomes Containing Cytochrome *o*[5]

Relatively low magnification electron microscopy of platinum replicas of freeze-fractured proteoliposomes containing purified cytochrome *o* demonstrates that the great majority of these vesicles are unilamellar (Fig.

3A). Higher magnification reveals that both the convex and concave fracture surfaces exhibit a relatively uniform distribution of particles that are 85–90 Å in diameter (Fig. 3B). Furthermore, all of the particles visualized are observed within fracture faces, indicating that most, if not all, of the oxidase is associated with the proteoliposomes.

Proteoliposomes containing cytochrome *o* generate a $\Delta\bar{\mu}_{H^+}$ (interior negative and alkaline) of over -100 mV during oxidase turnover. A variety of experiments, including reconstitution of the oxidase into planar bilayers formed at the tip of patch pipets,[15] are consistent with the notion that oxidase turnover generates an electrical potential (interior negative) due to vectorial electron flow from the outer to the inner surface of the membrane. The pH gradient (interior alkaline), on the other hand, appears to result from scalar (i.e., nonvectorial) reactions that consume and release protons at the inner and/or outer surfaces of the membrane, respectively; that is, cytochrome *o* oxidase from *E. coli* does not appear to catalyze vectorial proton translocation.

Characteristics of Proteoliposomes Containing Cytochrome *o* and *lac* Permease[2]

Proteoliposomes reconstituted with cytochrome *o* and *lac* permease generate a $\Delta\bar{\mu}_{H^+}$ (interior negative and alkaline) comparable to that observed in proteoliposomes reconstituted with cytochrome *o* only. In addition, these proteoliposomes accumulate lactose against a concentration gradient during oxidase turnover, and the phenomenon is abolished by addition of valinomycin and nigericin or by cyanide. Importantly, the magnitude of the lactose concentration gradient is commensurate with the magnitude of the $\Delta\bar{\mu}_{H^+}$ that is established, providing strong evidence that the steady-state level of lactose accumulation is in equilibrium with the bulk phase $\Delta\bar{\mu}_{H^+}$. Thus, proteoliposomes containing cytochrome *o* and the *lac* permease catalyze the active transport of lactose in a fashion that closely resembles intact cells and rightside-out membrane vesicles.

[15] T. Hamamoto, N. Carrasco, K. Matsushita, H. R. Kaback, and M. Montal, *Proc. Natl. Acad. Sci. U.S.A.* **82,** 2570 (1985).

[12] Functional Reconstitution of Bacterial Cytochrome Oxidases in Planar Lipid Bilayers

By TOSHIRO HAMAMOTO and MAURICIO MONTAL

The elucidation of the transduction mechanism of redox energy into transmembrane potential catalyzed by cytochrome oxidase has been a challenge since its postulation by Mitchell.[1] The electrogenic activity of cytochrome oxidase in mitochondria, bacterial membrane vesicles, and reconstituted vesicles has been inferred from bulk pH measurements,[2] and from the distribution of lipophilic ions across membranes.[3–5] These indirect measurements of transmembrane potential and pH have led to the notion that the electron-transfer reaction, catalyzed by cytochrome oxidases, is associated with the generation of a transmembrane electrical potential.

The purpose of this chapter is to describe a method to obtain electrical measurements of the electrogenic nature of the redox reaction catalyzed by cytochrome oxidases. It involves the reassembly of the purified enzyme into a well-defined planar lipid bilayer that exhibits functional activity.[6–8] The transmembrane events associated with the electron-transfer reaction can be investigated with this system, namely, (1) the electrogenic nature of the redox reaction can be recorded directly; (2) transmembrane potential and pH can be precisely controlled; and (3) the effects of applied electric fields on the electron-transfer reaction proceeding in the reconstituted cytochrome oxidase are readily measurable.

The purified bacterial oxidases that were selected for this study have a well-defined subunit composition and their electrogenic activity has been demonstrated in reconstituted vesicles. They are a cytochrome-*c* oxidase

[1] P. Mitchell, "Chemiosmotic Coupling in Oxidative and Photosynthetic Phosphorylation." Glynn Research, Bodmin, Conwall, England, 1966.

[2] M. Wikström, K. Krab, and M. Saraste, *Annu. Rev. Biochem.* **50,** 623 (1981).

[3] H. Rottenberg, this series, Vol. 55, p. 547.

[4] P. C. Hinkle, *Fed. Proc., Fed. Am. Soc. Exp. Biol.* **32,** 1988 (1973).

[5] L. A. Drachev, A. A. Jasaitis, A. D. Kaulen, A. A. Kondrachin, L. V. Chu, A. Y. Semenov, I. I. Severina, and V. P. Skulachev, *J. Biol. Chem.* **251,** 7072 (1976); V. P. Skulachev, this series, Vol. 55, p. 586.

[6] B. A. Suarez-Isla, K. Wan, J. Lindstrom, and M. Montal, *Biochemistry* **22,** 2319 (1983).

[7] T. Hamamoto and M. Montal, *Biophys. J.* **45,** Abstr. 38a (1984).

[8] T. Hamamoto, N. Carrasco, K. Matsushita, H. R. Kaback, and M. Montal, *Proc. Natl. Acad. Sci. U.S.A.* **82,** 2570 (1985).

from the thermophilic bacterium PS3[9,10] and an *o*-type cytochrome oxidase from *Escherichia coli*.[11,12] The PS3 oxidase is composed of three polypeptide subunits with $M_r \sim$ 56,000, 38,000, and 22,000, respectively.[9,10] The *E. coli* oxidase consists of four polypeptide subunits with $M_r \sim$ 66,000, 35,000, 22,000, and 17,000, respectively.[11,12]

Materials and Apparatus

Materials

Ascorbate (sodium salt), phenazine methosulfate (PMS), dithiothreitol (DTT), and *N*-trishydroxymethylmethylglycine (tricine from Sigma), 4-morpholinopropanesulfonic acid (MOPS) from Behringer, and ubiquinone (Q_1) generously provided by Hoffman-La Roche, Inc.

Solutions

Buffer A: 10 m*M* KMOPS, 50 m*M* KCl, 1 m*M* $CaCl_2$, pH 7.0
Buffer B: 10 m*M* KMOPS, 10 m*M* KCl, 20 m*M* K_2SO_4, 2 m*M* $CaCl_2$, pH 7.0
0.5 *M* Na ascorbate, pH 7.0
0.1 *M* PMS
0.1 *M* KCN, pH 7.0, prepared immediately before use
4 m*M* cytochrome *c* of *Saccharomyces cerevisiae* (Sigma); cytochrome *c* was reduced with sodium hydrosulfite and separated from reductant by chromatography on a Sephadex G-15 column

Preparation and Reconstitution of Bacterial Cytochrome Oxidases

Cytochrome-c Oxidase from the Thermophilic Bacterium PS3. Cytochrome oxidase was solubilized from the bacterial membrane, purified, and reconstituted into proteoliposomes, as described.[9,10] The proteoliposomes were prepared with acetone-washed α-tocopherol-treated soybean phospholipids (Associated Concentrates, Woodside, NY, or L-α-lecithin from soybeans, Sigma). The protein-to-lipid ratio was 1 : 5 (w/w) (4 mg/ml protein and 20 mg/ml phospholipid) unless otherwise indicated. The proteoliposomes were stored at −70°.

o-Type Cytochrome Oxidase from E. coli. Cytochrome oxidase was solubilized from the membrane *E. coli* GR19N (*cytd*$^-$), purified, and re-

[9] N. Sone, this volume [14].
[10] N. Sone and Y. Yanagita, *J. Biol. Chem.* **259,** 1405 (1984).
[11] K. Matsushita, L. Patel, and H. R. Kaback, this volume [11].
[12] K. Matsushita, L. Patel, and H. R. Kaback, *Biochemistry* **23,** 4703 (1984).

constituted into proteoliposomes with *E. coli* phospholipids, as described.[11,12] Proteoliposomes were washed once in 10 m*M* sodium tricine (pH 7.5), frozen in liquid nitrogen, and stored at −70°.

Apparatus

Fabrication of the Patch Pipet. A detailed description of pipet fabrication is provided in Refs. 13 and 14. Standard hematocrit capillaries made of flint glass (BLU-TIP, plain; i.d. 1.1–1.2 mm, Lancer, St. Louis, MO) are used. The capillaries are immersed in 1 N nitric acid overnight, rinsed thoroughly with deionized water, and stored in deionized water. Immediately before pulling, the capillaries are rinsed with methanol and dried by a nitrogen gas flush. The pulling procedure is performed in a vertical pipet puller (David Kopf 700C, Tujunga, CA) and consists of two steps. First, the capillary is extended (8 mm) to produce an intermediate thin section. This section is then recentered with respect to the heating element by an upward displacement of ~5 mm. The final pull produces 2 pipets with an opening (tip diameter) of ~1.0 μm with a steep taper. The tip size can be controlled by adjusting the heater current in the second pull. No further treatment such as fire polishing or coating is necessary. The tip size is adjusted to yield 20–40 MΩ of open pipet resistance when filled and immersed in the buffers described in "Solutions." Pipets should be used within a few (≤4) hours after fabrication.

Electrical Recordings (*Fig. 1*). A commercially available extracellular patch clamp system[13,14] (List L-M EPC-5, List Electronics, Darmstadt, Federal Republic of Germany, and Medical Systems Corp., NY) is used. This system provides a voltage clamp mode and a current clamp mode. The voltage clamp mode is used to measure the formation of the membrane and the membrane resistance (R_m). This is achieved by imposing a constant voltage across the pipet and measuring the current flowing between the two electrodes. The current clamp mode is used to measure the open circuit membrane potential (ΔV) generated by the reconstituted oxidase. Constant voltage is applied from a DC source (Omnical 2001, WPI Instruments, New Haven, CT). The signal output from the clamp is recorded on FM tape (Racal 4 DS, Hythe, Southampton, England) and displayed on an oscilloscope and a strip chart recorder (Gould 2200S or Hewlett-Packard 7100) (see Fig. 1). The two compartments separated by

[13] O. P. Hamill, A. Marty, E. Neher, B. Sakmann, and F. J. Sigworth, *Pfluegers Arch.* **391,** 85 (1981).

[14] B. Sakmann and E. Neher, eds., "Single-Channel Recording," p. 503. Plenum, New York, 1983.

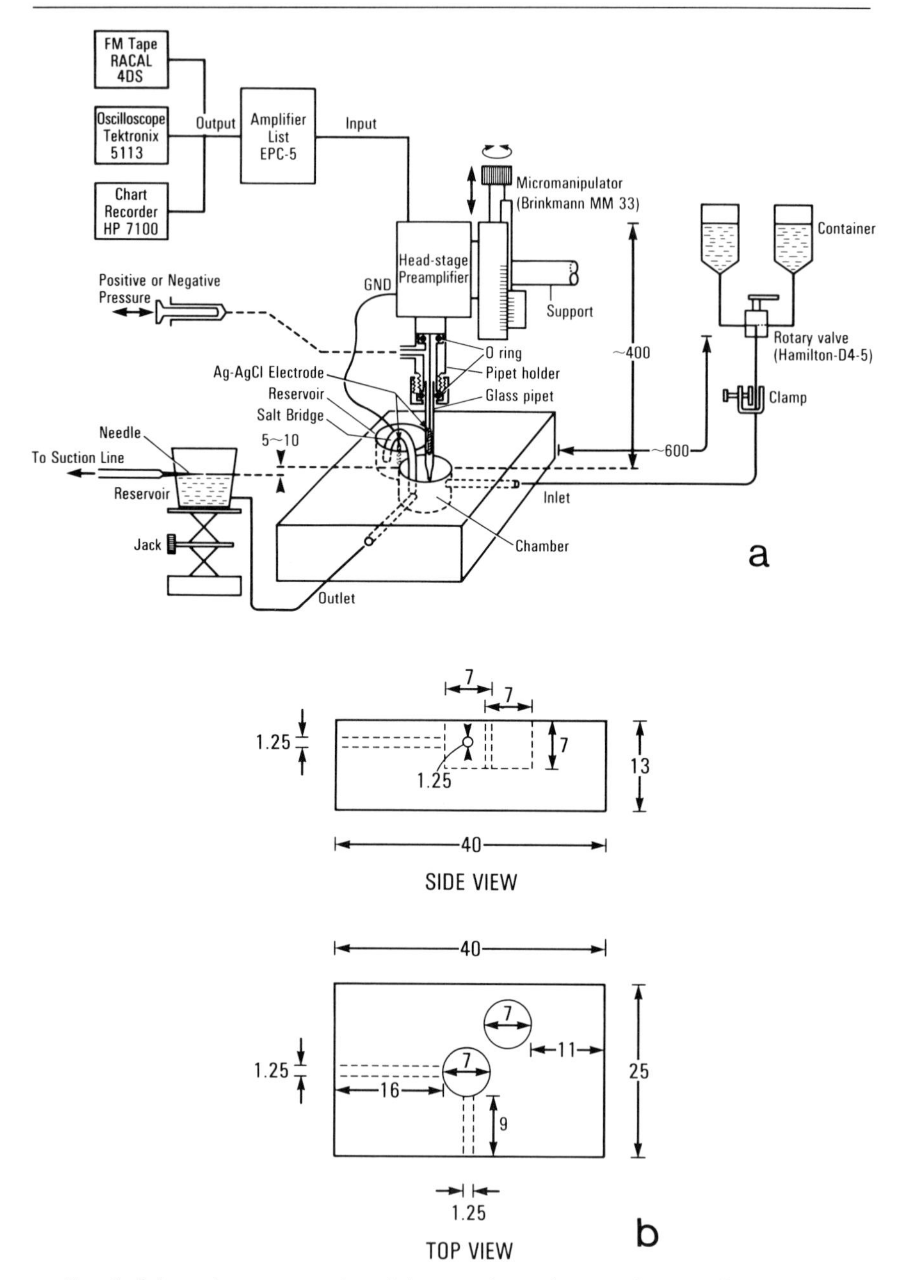

FIG. 1. Schematic representation of the experimental system for recording open-circuit membrane potentials of the reconstituted bilayers. (a) Arrangement of components (see text). (b) Diagram of the chamber used for bilayer formation (see text).

the membrane are connected to the amplifier by means of two Ag/AgCl electrodes. The active electrode, identified as the pipet electrode, is connected to a BNC connector pin of the shielded pipet holder, a component of the EPC-5 patch clamp system. The second electrode, the reference or bath electrode, is immersed in the chamber used by bilayer formation and is connected to the high-quality ground connection on the head stage of the patch clamp system (see Fig. 1). The pipet potential is determined with reference to this ground. The Ag/AgCl pellet electrodes are commercially available [pellet electrodes (E255 and E206) from In Vivo Metric Systems, Healdsburg, CA].

The Chamber (*Fig. 1*). The pipet, previously filled with buffer, is placed in the pipet holder. The holder is connected to the head stage which is designed to be mounted directly on a micromanipulator. The second Ag/AgCl electrode, which is connected to the ground terminal of the head stage, is placed in the reservoir that contains the same buffer. Chamber and reservoir are connected via a salt bridge (2% agar in 1 *M* KCl). Positive or negative pressures are applied to the pipet with a syringe, as indicated.

The chamber is fabricated in white Teflon (Fig. 1B). Two wells are drilled in a Teflon block. One of them is the actual chamber for bilayer formation. The second is the reservoir where the reference electrode is placed. Chamber and reservoir are connected via the salt bridge. Inlet and outlet indicate the paths used for perfusion of the membranes. The dimensions are given in millimeters.

Perfusion System (*Fig. 1*). Pertubation of the reconstituted membrane by substrates or inhibitors is performed via the continuous perfusion of reactants in and out of the bilayer chamber. Disposable syringes (50 ml) are used as containers for the buffer containing substrates and inhibitors. These containers are located ~0.4 m above the level of the chamber and the flow rate is manually controlled with a clamp. Substrates or inhibitors are selected by the manipulation of a rotary valve (Hup valve, D4-5 Hamilton, Reno, NV) (see Fig. 1). The inlet line (polyethylene tubing, i.d. 0.86 mm) connects the containers with the chamber. The outlet of the chamber (Tygon tubing, i.d. 2.4 mm) is led to a reservoir. To achieve the selected flow rate, the height of the reservoir is adjusted with a manual jack. Routinely, the water level of the outlet reservoir is 5–10 mm below that of the membrane chamber. The experimental setup is contained within a Faraday cage to shield the membrane from electrical artifacts and is mounted on a vibration-isolation platform. The dimensions shown in Fig. 1 are given in millimeters. All the experiments are conducted at room temperature (22 ± 2°).

Principle of Bilayer Formation from Lipid Monolayers at the Tip of Patch Pipets

Figure 2 shows an idealized representation of the process of bilayer formation from monolayers at the air–water interface at the tip of patch pipets. This method is a refinement of the well-established technique of lipid bilayer formation from two monolayers at the air–water interface across an aperture in a Teflon partition that separates two aqueous compartments.[21–23]

Step 1. Lipid protein monolayers are derived from the reconstituted vesicles.[6,15–17] A patch pipet is introduced into the solution under positive pressure in order to avoid its occlusion.

Step 2. The pipet is removed from the solution. The lipid head groups of the monolayers are attached to the polar glass pipet while the hydrocarbon tails contact air.

Step 3. The pipet is reimmersed into the solution. This leads to the apposition of the hydrocarbon tails of the attached monolayer to those of the original monolayer, thereby forming a bilayer in the tip of the patch pipet.[6,18–20]

Routine Procedure to Reconstitute Cytochrome Oxidase into Lipid Bilayers Formed at the Tip of Patch Pipets

Reconstitution of the Enzyme in Lipid Vesicles and Assay of Functional Activity

The proteoliposomes formed as described above are thawed immediately before use and sonicated for 30 sec (*E. coli*) or 90 sec (PS3) in a water bath sonicator (G112SP1, Laboratory Supplies, Hicksville, NY). The integrity of cytochrome oxidase function in the reconstituted vesicles is assessed by measurements of the proton electrochemical gradient generated by substrate oxidation of both transmembrane pH (ΔpH) and potential ($\Delta\Psi$) with appropriate probes.[9,10–12]

[15] H. Schindler, *Biochim. Biophys. Acta* **555,** 316 (1979).

[16] H. Schindler, *FEBS Lett.* **122,** 77 (1980).

[17] N. Nelson, R. Anholt, J. Lindstrom, and M. Montal, *Proc. Natl. Acad. Sci. U.S.A.* **77,** 3057 (1980).

[18] R. Coronado and R. Latorre, *Biophys. J.* **43,** 231 (1983).

[19] T. Schuerholz and H. Schindler, *FEBS Lett.* **152,** 187 (1983).

[20] V. Wilmsen, C. Methfessel, W. Hanke, and G. Boheim, *in* "Physical Chemistry of Transmembrane Ion Motions" (G. Spach, ed.), p. 479. Elsevier, Amsterdam, 1983.

[21] M. Montal and P. Mueller, *Proc. Natl. Acad. Sci. U.S.A.* **69,** 3561 (1972).

[22] M. Montal, *in* "Perspectives in Membrane Biology" (S. Estrada and C. Gitler, eds.), p. 591. Academic Press, New York, 1974.

[23] M. Montal, A. Darszon, and H. Schindler, *Q. Rev. Biophys.* **14,** 1 (1981).

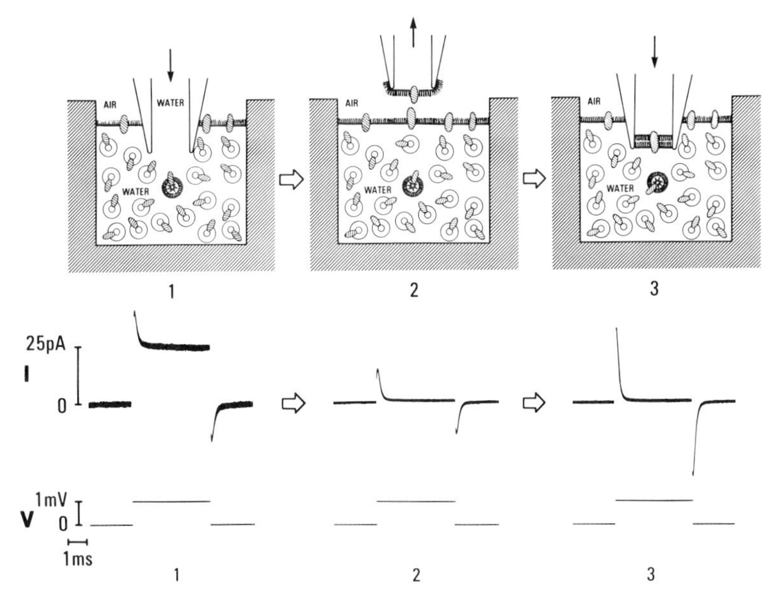

FIG. 2. Principle of bilayer formation from lipid monolayers at the tip of a patch pipet.[6,18-21] The upper panel is an idealized representation of the process of bilayer formation. The lower panel illustrates the actual electrical measurements corresponding to the individual steps in the process of bilayer formation. Step 1: Cytochrome oxidase proteoliposomes are introduced into the Teflon chamber. Monolayers spontaneously form from the proteoliposomes. First, a patch pipet is immersed into the solution; then, step 2, the pipet is removed from the solution. As a result, a patch of monolayer is attached to the glass pipet with the lipid head groups interacting with the glass while the hydrocarbon tails contact air. Next, step 3, the tip of the pipet is reimmersed into the chamber. The hydrocarbon tails of the attached monolayer contact those of the original monolayer, thereby forming a bilayer.

The electrical correlates of these steps in membrane formation are shown in the lower panels. Step 1: Application of a voltage step of 1 mV between the pipet electrode and the ground electrode results in an initial capacitative transient followed by a steady level of current. From the magnitude of the current (I = 25 pA) and the applied voltage, it follows from Ohm's law ($V = IR$) that the resistance of the open pipet is equal to 40 MΩ. Step 2: Upon removal of the pipet from the suspension, the two electrodes are out of contact and the current generated by the applied voltage step is not measurable. Step 3: Reimmersion of the pipet results in the formation of a bilayer reflected by the appearance of a measurable steady current. In this case, the membrane resistance was 20 GΩ. The capacitative transient increases due primarily to the capacity of the glass pipet.

Transformation of Reconstituted Vesicles into Monolayers at the Air–Water Interface

A necessary condition for the assembly of planar lipid bilayers from vesicle-derived monolayers is that the surface pressure of the monolayer be ≥25 dyne/cm.[15,16] Surface pressure is extremely sensitive to the composition of the aqueous medium, as described in detail by Schindler.[15,16] The procedure routinely used to derive monolayers from the reconstituted vesicles with surface pressures sufficient to support the assembly of bilayers is the following: A 2-μl aliquot of the reconstituted vesicles [3–4 mg protein/ml at a protein-to-lipid ratio of 1 : 5 (w/w)] is added to the Teflon chamber containing 200 μl of buffer and stirred. Under these conditions, the reconstituted vesicles generate monolayers at the air–water interface.[7,8,17,24]

Assay of Membrane Formation

A few minutes after the addition of the reconstituted vesicles into the chamber, a patch pipet is introduced into the suspension, and the electrical resistance of the open pipet is measured. This is achieved by applying test pulses (e.g., 1 mV, 5 msec pulses) to the circuit. When the pipet tip enters the solution, the current trace shows a deflection (Fig. 2, lower panel, step 1). From the amplitude of the current (I) and the known applied voltage, the resistance of the open pipet is calculated ($R = V/I$). With the pipets fabricated as described, the typical open pipet resistance is 20 M$\Omega \leq R_m \leq$ 40 MΩ. Thereafter, the pipet is removed from the solution and immediately reimmersed into the same solution. The formation of the bilayer is reflected by the immediate appearance of a resistance in the GΩ range (Fig. 2, lower panel, step 3). The capacitance is typically in the range of 10–50 pF and is generated by the capacity of the glass pipet. Occasionally, positive or negative pressure is applied in order to increase the bilayer resistance.[13,14] This procedure indeed leads to the formation of bimolecular membranes as established by measurements of single-channel currents from channel-forming antibiotics, such as gramicidin A and alamethicin,[18] and purified *Torpedo* acetylcholine receptors,[6] entities which form ion-conducting channels only in membranes that are bimolecular in thickness.

Membrane Stability

The bilayers are stable for several hours. The dielectric breakdown voltage varies with lipid composition and lipid-to-protein ratio. Practically, all bilayers are stable at applied voltages up to 150 mV and break at applied voltages $\geq$300 mV. At zero applied voltage, the bilayers are stable to differences in pH across the membrane of up to 2 pH units. To assay the electrogenic activity of cytochrome oxidase, only membranes that exhibit high stability to an applied voltage of 100 mV should be used.

[24] P. Labarca, J. Lindstrom, and M. Montal, *J. Gen. Physiol.* **83,** 473 (1984).

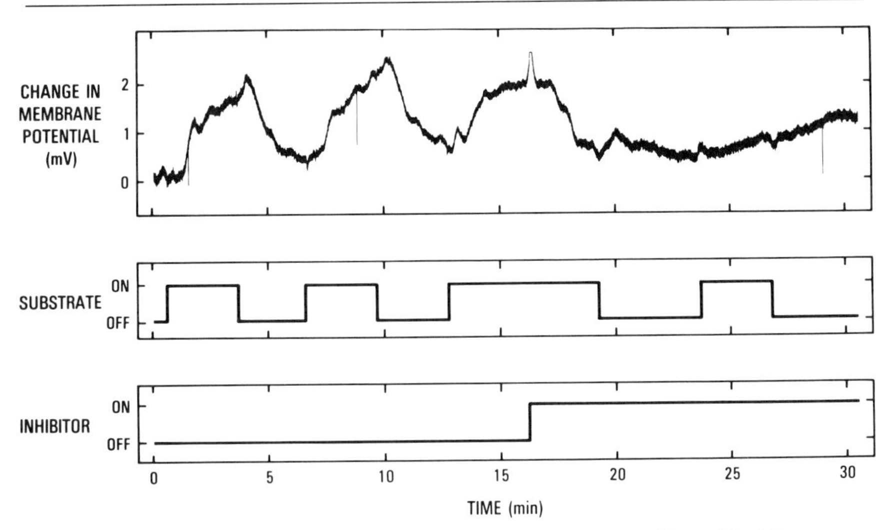

FIG. 3. Membrane potential generation by PS3 cytochrome-*c* oxidase. The bilayer was continuously perfused with buffer (10 m*M* sodium tricine, 200 m*M* KCl, 10 m*M* $CaCl_2$, pH 7.2) with or without substrate (2.5 m*M* sodium ascorbate and 50 μM PMS) or inhibitor (1 m*M* KCN), as indicated. The medium inside the pipet was the same as the perfusion buffer. The membrane resistance was 10 GΩ. The record illustrates the time course of change in membrane potential upon perfusion of the bilayer with the substrate and inhibitor. The transient spikes are electrical artifacts introduced by changing one medium for another.

Assay of the Electrogenic Activity of Cytochrome Oxidase

Membrane Potential Generation by the Thermophilic Bacterium PS3 Oxidase

Figure 3 illustrates an actual recording of the time course of change in membrane potential upon perfusion of a reconstituted lipid bilayer with substrate, in this case, ascorbate-PMS, and/or inhibitor (1 m*M* KCN) according to the pulse sequence shown in the lower panel of the figure. Perfusion with buffer containing substrate generates a membrane potential of ~2 mV, positive in the substrate side, i.e., the outside of the pipet. As the substrate is washed out of the chamber (substrate pulse turned off), the membrane potential returns to the initial baseline. This response can be evoked virtually indefinitely by repeating the substrate pulse sequence illustrated in the figure. Perfusion with buffer containing the enzyme inhibitor KCN collapses the membrane potential generated by the substrate and prevents further generation of membrane potentials by additional introduction of buffer containing substrate. The amplitudes of the membrane potential generated by substrate oxidation vary from 0.5 to 4

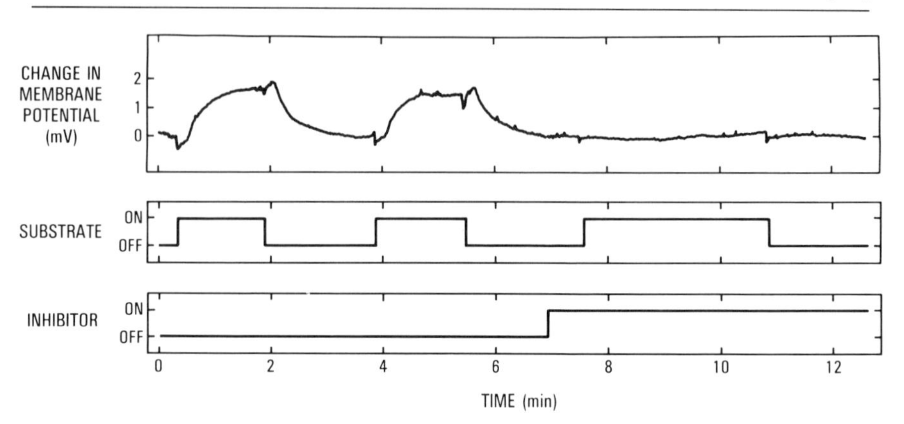

FIG. 4. Membrane potential generation by *E. coli* cytochrome-*o* oxidase. The bilayer was perfused continuously with buffer A supplemented with 5 mM sodium ascorbate with or without substrate (25 μM PMS) and/or inhibitor (1 mM KCN), as indicated. The pipet was filled with buffer A. The membrane resistance was 8 GΩ. Other conditions were as for Fig. 3.

mV, depending on the bilayer resistance and the nature of the substrate. However, the polarity of the generated potential is always substrate side positive. Similar results are obtained when reduced cytochrome *c* is the substrate. The amplitude of the generated membrane potential is largest when 20 μM reduced cytochrome *c* or 10 μM PMS are used.

Membrane Potential Generation by E. coli Oxidase

Figure 4 illustrates the time course of membrane potential generation upon perfusion of a reconstituted bilayer with substrate (in this case, ascorbate PMS) and/or inhibitor (1 mM KCN) in a similar pulse sequence as illustrated in Fig. 3. Similar results are obtained when ubiquinone-1 supplemented with dithiothreitol is the substrate. The amplitude of the signal is largest with 50 μM PMS and 5 mM ascorbate. However, 25 μM PMS is used routinely in order to prevent oxygen depletion by the autoxidation of PMS. The amplitude of the ΔV generated using either reduced PMS or reduced quinone as electron donors ranges from 0.6 to 3.8 mV, with an average of 1.2 mV $\pm$ 0.7 SD. The polarity of the ΔV generated is always positive on the substrate side, just as in the case of the PS3 oxidase.

Specificity of the Assay: Substrates and Inhibitors

As indicated above, the membrane potential generated by the redox reaction catalyzed by cytochrome oxidase is measured with established substrates of the enzymes according to their specificity. For example,

OPEN-CIRCUIT MEMBRANE POTENTIAL (ΔV) AS FUNCTION OF PROTEIN/LIPID RATIO IN THE PROTEOLIPOSOMES[a]

Protein/lipid (w/w)	ΔV (mV)	Number of experiments
1:2–3	1.9 ± 1.3	6
1:4–5	1.0 ± 0.6	30
1:20	≤ 0.2	10

[a] PS3 oxidase proteoliposomes reconstituted to yield different protein-to-lipid ratios, as indicated, were used to form the bilayers. All other conditions were as described in Fig. 3.

reduced cytochrome *c* is an electron donor for the PS3 oxidase but not for *E. coli* oxidase; in contrast, reduced ubiquinone is an effective electron donor only for the *E. coli* oxidase. Both enzymes can accept electrons from the artificial electron donor PMS in the presence of ascorbate. Both oxidases are selectively inhibited cyanide. Thus, the membrane potential generated by the reconstituted cytochrome oxidase displays the substrate specificity and inhibitor sensitivity characteristics of the native enzymes.

Effect of Lipid/Protein Ratio on the Amplitude of the Signal

The amplitude of the generated membrane potential increases as the protein-to-lipid ratio of the reconstituted proteoliposomes (and presumably of the monolayers derived from these vesicles) increases. This effect is illustrated in the table. It is clear that as the density of the enzyme in the bilayer is increased, the magnitude of the generated membrane potential increases. There is, however, a limit to this favorable effect, which is reached approximately at a protein-to-lipid ratio of 1 : 3 (w/w). This ratio has to be titrated for any given enzyme. In general, the higher the lipid content of the lipid bilayer, the more stable the reconstituted membrane is bound to be. In contrast, the higher the protein content in the reconstituted membrane, the more prone to breakage it will become. The optimum titer, therefore, is that which leads to the largest signal amplitude without compromising the stability of the bilayer.

Effect of Membrane Resistance on the Amplitude of the Signal

Figure 5a displays the time course of membrane potential generation by the thermophilic bacterial PS3 oxidase upon perfusion of reduced cytochrome *c*. The upper and middle panels show the signals generated from reconstituted bilayers with $R_m = 50$ GΩ and $R_m = 10$ GΩ, respectively.

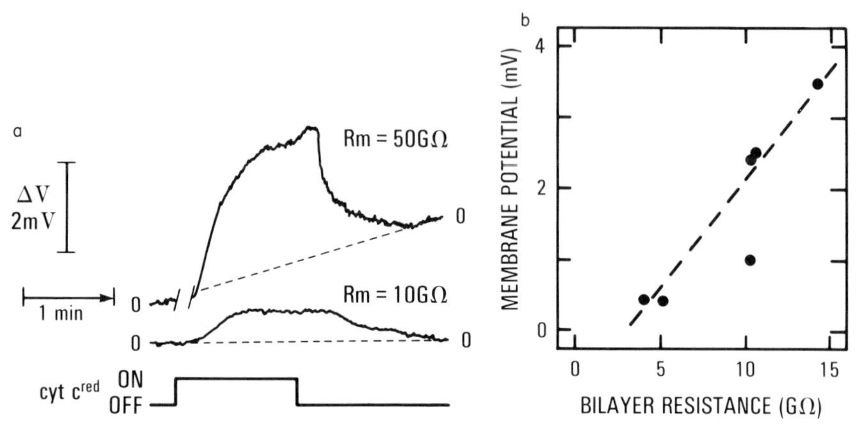

FIG. 5. Membrane potential generation as function of membrane resistance. (a) Time course of change in membrane potential upon perfusion of two reconstituted bilayers of different membrane resistance. In these experiments, the PS3 cytochrome-*c* oxidase was used. Both bilayers were continuously perfused with buffer B supplemented with 5 m*M* sodium ascorbate. The pipets were filled with buffer B. The substrate in both records was 10 μM reduced cytochrome-*c*. The upper record is from a bilayer with R_m = 50 GΩ while the lower record is from a bilayer with R_m = 10 GΩ. The two membranes were derived from the same preparation of proteoliposomes at the protein-to-lipid ratio of 1 : 5. Notice the drift of the baseline in the recording obtained from the 50-GΩ membrane. (b) The amplitude of the generated membrane potential is plotted as a function of the membrane resistance. The reconstituted PS3 cytochrome-*c* oxidase was used for these experiments. All other conditions were the same as for Fig. 3.

The two sets of experiments were performed under identical experimental conditions, including the same protein-to-lipid ratio of the reconstituted proteoliposomes. Figure 5b displays the amplitude of the oxidase-generated membrane potential as a function of the resistance of the reconstituted bilayer. Clearly, the amplitude of the signal increases with membrane resistance. This result implies that a higher signal-to-noise ratio may be obtained by improvements in the passive resistance of the membrane. This notion can be explored by varying the specific phospholipids used in the reconstituted membrane that may promote the interaction between the lipid and the glass and thereby increase the membrane resistance. This consideration arises from the presumption that a nonspecific leak current flows through the junction of the phospholipid bilayer with the glass. A common finding in open-circuit membrane potential measurements, such as those described here, is a considerable drift in the baseline. This point is purposely emphasized in the upper panel of Fig. 5a. This baseline drift is more apparent the higher the membrane resistance.

Effect of the Glass Pipet Composition on the Activity of the Reconstituted System

The structure of the interface between the phospholipid bilayer and the glass pipet and the nature of the interaction occurring at this interface are at present not well understood. From this consideration, it follows that the properties of the reconstituted bilayer may be determined to a large extent by the chemical nature of the glass used in the patch pipet. All the results thus far described were obtained with flint glass (BLU-TIP). In addition, the use of hard borosilicate glass (Kimax-51, Kimble, Vineland, NJ) and Kovar (Corning #7052, Corning, NY[25]) was explored. Bilayers were formed in the tip of patch pipets made from these capillary glasses, and the activity of the thermophilic PS3 oxidase was determined. In general, the reconstituted membranes exhibited higher membrane resistance (20 GΩ $\leq R_m \leq$ 100 GΩ), especially when the pipet tips were fire polished[13,14] after the pulling step. However, the efficiency of the reconstitution process was low (≤10%). The reconstitution efficiency is defined as the fraction of the total number of membranes formed that exhibit functional activity. Low efficiency reflects the formation of many bilayers devoid of enzyme and composed exclusively of phospholipids. This point is discussed in more detail later. The fact that the nature of the glass used in the patch pipet affects the reconstitution efficiency suggests that glass–protein interactions are important in determining the eventual composition of the reconstituted bilayer. Presumably, strong glass–protein interactions lead to the exclusion of enzyme from the lipid bilayer.

Control Experiments

The reliability of the reconstituted bilayer is ascertained by forming lipid bilayers in the patch pipet from liposomes prepared under identical conditions to the proteoliposomes, but in the absence of the purified enzyme. Those membranes routinely show higher membrane resistances (10 GΩ $\leq R_m \leq$ 25 GΩ) and are more stable. Perfusion with buffer containing substrates causes no change in membrane potential (≤0.2 mV). However, the experimenter should be aware of artifacts introduced by spurious changes in the flow rate while the perfusion is proceeding or by pH misadjustments. The use of agar bridges eliminates the artifactual signals which arise from the interaction of a strong reductant such as ascorbate with the Ag/AgCl electrodes.

[25] J. L. Rae and R. A. Levis, *Biophys. J.* **45,** 144 (1984).

Efficiency and Reliability of the Reconstituted System

As previously defined, efficiency is operationally used to describe the fraction of the total membranes formed, under a given condition, that exhibits functional activity. Under the conditions described, both the thermophilic PS3 oxidase and the *E. coli* oxidase exhibit an efficiency better than 60%, i.e., $\geq$60% of the reconstituted bilayers generate a cyanide-sensitive membrane potential upon perfusion with substrates. Efficiency is stringently dependent on the quality of the enzyme preparation: The higher the specific activity, the higher the reconstitution efficiency. Preparations with reduced activity after storage show a concomitant decrease in efficiency. Since the technique of forming a bilayer samples discrete areas of monolayer with the patch pipet, nonrandom distribution of protein in the monolayer will affect the reconstitution efficiency. Therefore, it is desirable to have a homogeneous distribution of oxidase molecules in the reconstituted vesicles. This is achieved by repeating the freeze-thaw–sonication steps (up to four times for the PS3 oxidase).

Comparison of Predicted and Measured Membrane Potentials Generated by the Cytochrome Oxidase Reaction

The amplitude of the membrane potential, with an average of 2 mV, is considerably lower than that anticipated[1,2] ($\sim$100 mV) and therefore requires qualification. Considering a few parameters of the system under study, one can estimate the membrane potential generated by the oxidase using the following assumptions (see Ref. 8): (1) The area of the bilayer in the tip of the patch pipet is $\sim$1 μm^2; (2) the protein-to-lipid ratio in the reconstituted bilayer is the same as in the original reconstituted vesicles (1 : 750 mol/mol for PS3); (3) the turnover number, T, of the oxidase in the bilayer is essentially equivalent to that measured in the reconstituted vesicles (200 electrons/sec for PS3); and (4) the area occupied by one phospholipid molecule is analogous to that occupied in a condensed monolayer (60 Å^2). Then the current, i, generated by the oxidase reaction can be estimated according to Eq. (1),

$$i = q/t = (\text{Number of oxidase molecules})(N)(T)(1.6 \times 10^{-19}\ \text{coul/molecule}) \quad (1)$$

where q is the charge transferred across the membrane per unit time, t and N is the stoichiometry of charge translocation. The estimated current is $\sim 5 \times 10^5 \times N$ charges/sec or 0.1 NpA transferred across the bilayer. If we now consider a typical bilayer resistance of 10 GΩ, it follows from Ohm's law that a membrane potential of 1 mV would be generated by the oxidase reaction. If the resistance of the native membrane were 1000 GΩ

(1TΩ), then the same current would produce a potential difference of 100 mV, a value more in concert with the estimated intrinsic limit of the oxidase (150 mV).

Advantages and Limitations

Planar bilayers offer the unique advantage of being accessible for the direct measurement of membrane potential as well as for the direct application of an electric field across the cytochrome oxidase and, consequently, for the direct recording of its effect on the electron-transfer reaction proceeding in the reconstituted enzyme. For example, the amplitude of the ΔV generated by the reconstituted *E. coli* oxidase with PMS was inhibited by negative applied voltages and approached zero at -150 mV.[8] In addition, the pH on both sides of the membrane can be readily manipulated, allowing one to examine the effects of transmembrane pH gradients on the electron-transfer reaction. This provides an opportunity to test the specific predictions implicit in the two reaction mechanisms which are currently under experimental evaluation: vectorial electron or proton translocation.[1,2]

A limitation of the reconstituted system is the small amplitude of the generated membrane potential. As discussed before, the signal appears to result from a balance between the oxidase-generated electric current and the nonspecific current leakage at the junction between the reconstituted bilayer and the glass pipet. On one hand, there appears to be a limit above which the density of oxidase molecules in the bilayer cannot be increased, because at higher protein-to-lipid ratios, bilayers do not form or are unstable, especially at applied voltages larger than ± 100 mV. The nonspecific current leakage may be improved by surveying the most appropriate combination of phospholipid species and glass compositions. A possible complication of the system concerns the orientation of the cytochrome oxidase in the reconstituted bilayer. The procedure described to assemble the membrane presumably leads to the formation of a symmetric membrane with a random orientation of oxidase molecules. To selectively activate one population of enzymes, a nonpermeant substrate for the cytochrome oxidase is used, such as for the PS3 oxidase. This is not the case for the *E. coli* oxidase, and one is limited to use substrates, such as PMS or Q_1, which are permeant across the membrane and therefore can activate the two enzyme populations. It is conceivable that the charge flow generated by the turnover of molecules oriented in both directions across the bilayer balance each other electrogenically. The fact that a signal is generated under the conditions described is attributed to a limited permeability of the substrates across the membrane.

Acknowledgments

We are indebted to N. Sone for providing samples of the purified cytochrome-*c* oxidase from the thermophilic bacterium PS3, to K. Matsushita, N. Carrasco, and H. R. Kaback for their collaboration with the experiments on the *E. coli* oxidase, and to R. Greenblatt for his participation in the design of the perfusion system. This investigation was supported by a research grant from the National Institutes of Health (EY-02084) and in part by the Department of the Army Medical Research (17-82-C-2221). Toshiro Hamamoto is a fellow of the Japanese Society for the Promotion of Science (JSPS).

[13] Purification of the aa_3-Type Cytochrome-*c* Oxidase from *Rhodopseudomonas sphaeroides*

By ANGELO AZZI and ROBERT B. GENNIS

Introduction

Rhodopseudomonas sphaeroides is a purple nonsulfur bacterium which, when grown anaerobically in the light, develops a photosynthetic apparatus. In the dark and in the presence of oxygen, the photosynthetic apparatus is suppressed and the bacterium utilizes a respiratory chain.[1] This chain is similar to that of mitochondria[2] and to that of *Paracoccus denitrificans*,[3] an organism phytogenetically related to *R. sphaeroides*, and proposed to be very similar to the prokaryote postulated to be the ancestor from which mitochondria were derived.

The above properties make it interesting to isolate and characterize the respiratory enzymes of *R. sphaeroides*.

Preparation of the Bacterial Membranes and Purification of the Oxidase

Bacterial Growth

Rhodopseudomonas sphaeroides, strain 2.4.1, was obtained from the laboratory of S. Kaplan (Microbiology Department, University of Illinois, Urbana, IL 61801). It may be grown in the dark in a 20-liter fermentor to a density of 1.5×10^9 cells/ml at 30°, on medium A of Sistrom.[4] To ensure aerobic growth, vigorously stirred cultures must be sparged with air. The cell paste (obtained after low-speed centrifugation of the growth medium)

[1] F. R. Whale and O. T. G. Jones, *Biochim. Biophys. Acta* **223,** 146 (1970).

[2] P. L. Dutton and D. F. Wilson, *Biochim. Biophys. Acta* **346,** 165 (1970).

[3] P. John and F. R. Whatley, *Nature (London)* **254,** 495 (1975).

[4] W. R. Sistrom, *J. Gen. Microbiol.* **22,** 778 (1960).

and the various membrane preparations, to be discussed later, can be stored frozen for several months at −70° without apparent loss of enzymatic activity.

Preparation of Cytoplasmic Membranes

The success of the purification of the oxidase critically depends on the purity of the cytoplasmic membrane preparation. Frozen cells (e.g., 30 g) are suspended in two volumes of buffer containing 50 m*M* Tris–HCl, 1 m*M* EDTA, pH 7.4. DNase and the protease inhibitor phenylmethylsulfonyl fluoride (PMSF) (final concentration 5 m*M*, dissolved in methanol just prior to use) are added immediately before cell disruption. The cells are broken by two passages through a Ribi press at 137.8 MPa (20,000 lb/in.^{2}). Centrifugation for 10 min at 10,000 *g* is employed to remove whole cells and debris. The membranes are then collected in a pellet by centrifugation for 2 hr at 150,000 *g*, and, following resuspension in a buffer containing 0.1 *M* Tris–HCl, 10 m*M* EDTA, pH 8.8, the above centrifugation step is repeated. Both the cytoplasmic membrane and the outer membrane are present in this preparation. After resuspension in 1–2 volumes of Tris–EDTA buffer, pH 7.4, the membrane pellet is layered on a sucrose step gradient consisting of equal volumes of 40% sucrose and 20% sucrose (w/v). Following centrifugation overnight at 150,000 *g*, the cytoplasmic membranes are collected at the 20%/40% sucrose interface. The membrane suspension is diluted in several volumes of Tris–EDTA buffer, pH 7.4, and the membranes sedimented by centrifugation. A typical cytoplasmic membrane preparation contains 0.5–1 nmol heme *a*/mg of protein.

Preparation of the Yeast–Cytochrome c Affinity Column

Activated thiol-Sepharose 4B (Sepharose-glutathione-2-pyridyldisulfide) is either purchased (Pharmacia) or prepared according to published procedures.[5] Best results are obtained by coupling 10 ml of the commercial resin with 30 mg of yeast cytochrome *c*. The resulting resin is washed with 1 *M* KCl and equilibrated and resuspended in the Tris–EDTA buffer, pH 7.4, containing 0.1% Triton X-100.

Preparation of the R. sphaeroides Cytochrome-c Oxidase[6]

A suspension of cytoplasmic membranes containing ~5 nmol heme *a* (~2 ml) is supplemented with Triton X-100 (20% stock solution) to give a

[5] K. Bill, R. P. Casey, C. Broger, and A. Azzi, *FEBS Lett.* **120,** 248 (1960).
[6] R. B. Gennis, R. P. Casey, A. Azzi, and B. Ludwig, *Eur. J. Biochem.* **125,** 189 (1982).

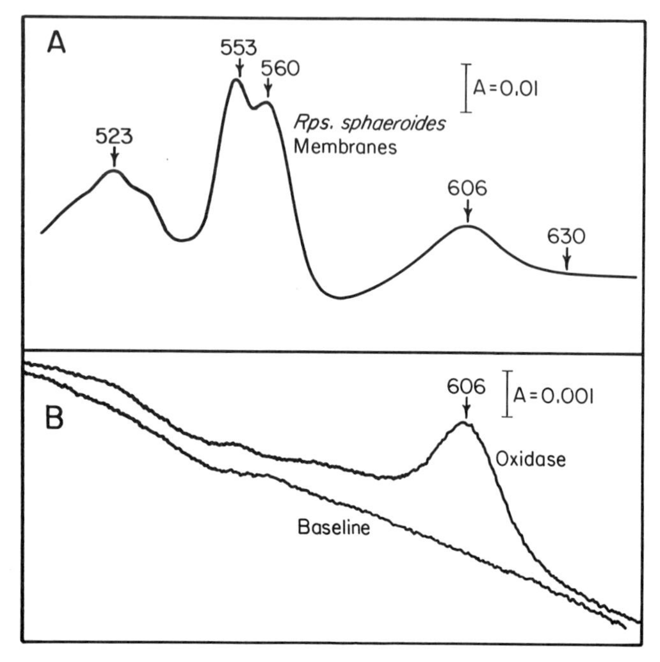

FIG. 1. Reduced-minus-oxidized spectra of (A) a suspension of cytoplasmic membranes of *R. sphaeroides,* and (B) purified cytochrome-*c* oxidase from *R. sphaeroides*. Spectra were recorded at 25° as described in the text. The membrane concentration was 1 mg/ml protein. The oxidase was eluted from the DEAE-BioGel A column in the presence of 0.1% Triton X-100. (From Gennis *et al.*,[6] with permission.)

final concentration of 2% and kept for 10 min on ice. The clarified suspension is centrifuged for 1 hr at 150,000 *g*. The supernatant, which contains virtually all of the cytochrome-*c* oxidase, can be loaded directly onto a small column (1 × 3 cm) filled with the cytochrome *c*-Sepharose 4B.

The *R. sphaeroides* oxidase binds tightly to freshly prepared yeast-cytochrome *c*-Sepharose 4B. The capacity is ~2 nmol of heme *a*/ml of packed resin. Under the conditions described, approximately two-thirds of the protein present in the Triton X-100 extract is retained by the cytochrome *c* resin.

This column is run at room temperature, with a flow rate of ~1 ml/5 min. During the washing of the column, carried out with a total of 9 ml of 50 m*M* Tris–HCl, 1 m*M* EDTA buffer, pH 7.4, containing 0.1% Triton X-100, no heme *a* should be eluted. Elution of the oxidase at this step may indicate column overloading or a resin preparation which is too old. Step elution with KCl results in nearly complete recovery of the cytochrome-*c* oxidase activity between 50 mM and 100 mM KCl, causing an increase in specific heme *a* content to between 1.5 and 3 nmol/mg protein. Elution occurs more specifically with a solution of the Tris–EDTA buffer containing 1 mg/ml horse heart cytochrome *c*. Between 10 and 15 ml of this buffer are usually required to elute all the activity. A further purification of the oxidase and removal of the horse heart cytochrome *c* is obtained by passing the suspension rapidly through a Pasteur pipette containing 2 ml of DEAE-BioGel A equilibrated in the same Tris–EDTA buffer plus 0.1% Triton X-100 at 4°. Following a wash with 4 ml of the Tris–EDTA buffer with 0.1% Triton X-100, the oxidase is eluted using a linear gradient (20 ml total volume of 25–120 mM NaCl) with the same buffer. Near the end of the gradient, from 50 to 80% of the loaded oxidase is eluted in about 7-ml volume. The oxidase can be easily concentrated and, if required, Triton X-100 can be replaced by octylglucoside by using the following procedure. The pooled, active fractions are diluted 3-fold with cold 0.1% Triton X-100 and loaded onto a 0.5-ml column containing DEAE-BioGel A. Elution in a small volume can be achieved by eluting with 0.1 M NaCl in the standard Tris–EDTA–Triton X-100 buffer. Alternatively, the column can first be washed with the Tris–EDTA buffer containing 30 mM octylglucoside (5–10 ml) and then eluted using the same buffer containing 0.1 M NaCl.

The entire preparation can be carried out easily in 1 day. The yield, based on heme *a*, is between 25 and 80%.

Characterization of the Membrane-Bound and Solubilized Cytochrome-c Oxidase

Spectral and Activity Analyses

A reduced-minus-oxidized spectrum of a suspension of *R. sphaeroides* cytoplasmic membranes is shown in Fig. 1, with the cytochromes *b, c,* and *a* clearly visible.

Cytochrome concentrations are determined from dithionite-reduced-minus-oxidized difference spectra recorded at 25° using an Amicon DW-2A spectrophotometer. The following molar absorption coefficients and wavelength pairs are utilized.[7] Heme $a + a_3$: 605–630 nm, $\varepsilon = 24.0$ mM^{-1} cm^{-1}; heme *b:* 560–575 nm, $\varepsilon = 23.4$ mM^{-1} cm^{-1}; heme *c:* 553–542 nm, $\varepsilon = 18.7$ mM^{-1} cm^{-1}.

Reduced-minus-oxidized difference spectra of the *R. sphaeroides* oxidase (Fig. 1B) show the presence of heme *a* with, in some cases, a small

[7] P. Rosevear, T. VanAken, J. Baxter, and S. Ferguson-Miller, *Biochemistry* **19,** 4108 (1980).

amount of contaminating horse heart cytochrome *c*. The preparation (80% pure) contains 14 nmol heme *a*/mg protein, consistent with the oxidase containing two heme *a* groups (heme *a* + heme a_3) and one each of the three polypeptides. Because of the small scale of these preparations, metal analyses and phospholipid determination cannot be easily carried out.

The Membrane-Bound Enzyme

The membrane exhibits substantial ascorbate-TMPD (*N,N,N',N'*-tetramethylphenylenediamine) oxidase activity which cannot be further stimulated by the addition of 100 μM horse heart or yeast cytochrome *c*. Assays of this activity may be performed by utilizing an oxygen electrode (Clark-type oxygen electrode, Yellow Springs Instruments). The reaction is carried out at 25° in a glass, water-jacketed chamber in a volume of 2 ml of buffer containing 50 m*M* Tris–HCl, 1 m*M* EDTA, 100 μM TMPD, 10 m*M* ascorbate, and 0.1% Triton X-100, pH 7.4. The absence of cytochrome c_2 in the membrane preparation can be documented by potentiometric titration.[6]

The Extracted Enzyme

The 2% Triton extract of the membranes exhibits very little ascorbate-TMPD oxidase activity, in contrast to the intact membranes, but activity can be completely restored by addition of cytochrome *c*. This substantial cytochrome-*c* oxidase activity is fully sensitive to the addition of 10 μM cyanide. The activity of cytochrome-*c* oxidase can also be measured spectrophotometrically at 25° using either yeast or horse heart reduced cytochrome *c*. Cytochrome *c* is reduced by dithionite or ascorbate and then passed through a Sephadex G-25 column prior to storage at −20°. The assay buffer contains 50 m*M* Tris–HCl, 1 m*M* EDTA, and 0.1% Triton X-100, pH 7.4, and ~30 μM cytochrome *c*. Activity of cytochrome-*c* oxidase is calculated using an extinction coefficient of 19 mM^{-1} cm^{-1} for the wavelength pair 550–540 nm.

Unlike the mitochondrial enzyme, the *R. sphaeroides* oxidase shows the same apparent K_m (about 1 μM cytochrome *c* in 50 m*M* Tris–HCl) for cytochrome *c* when assayed using either the oxygen electrode with ascorbate-TMPD as the reductant or using the spectroscopic assay. The oxidase activity is highly sensitive to ionic strength in both types of activity. At concentrations of KCl above 200 m*M*, the cytochrome-*c* oxidase activity is virtually eliminated when the concentration of cytochrome *c* is 30 μM. In contrast, the ascorbate-TMPD oxidase activity exhibited by the intact membrane is not sensitive to ionic strength. The origin of this activity in the membrane is not known, but may result from the association of the oxidase with another membrane-bound component which is rapidly reduced by TMPD.

The Purified Enzyme

The purified *R. sphaeroides* oxidase, in contrast to the mitochondrial enzyme, is highly active in the presence of Triton X-100 up to a concentration of 2%. When assayed in the presence of 0.1% Triton X-100 with reduced horse heart cytochrome *c*, the maximal velocity of the purified oxidase is typically 300 sec^{-1}/mol of oxidase. The activity of the purified oxidase is sensitive to both cyanide, azide, and to salts as in the solubilized membranes. The concentrations required for 50% inactivation are 1.3 μM and 100 μM for cyanide and azide, respectively.

Structure Analysis

The oxidase, as it elutes from the cytochrome *c* affinity resin, with 10-fold purification and recoveries of over 75% of cytochrome-*c* oxidase activity, is ~30% pure as judged from the dodecyl sulfate/polyacrylamide gels. The subunit composition of the partially purified enzyme can be easily analyzed by two-dimensional dodecyl sulfate/polyacrylamide gel electrophoresis. Prior to electrophoresis, the protein solutions should be concentrated in the presence of dodecyl sulfate using an Aminco B-15 unit. Two-dimensional polyacrylamide gel electrophoresis can then be carried out using the Laemmli system,[8] with 0.1% Triton X-100 in place of dodecyl sulfate in the first dimension. Electrophoresis in the first dimension is carried out using a 4% polyacrylamide slab gel at 4°. A 4-mm vertical strip from this gel is soaked for a few minutes in the dodecyl sulfate-containing sample buffer and then placed on top of a 12% polyacrylamide slab gel and sealed in place with agarose in the same buffer. Parallel strips can be stained for protein (Coomassie Brilliant Blue), heme,[9] and TMPD oxidase activity. The second dimension of the electrophoresis is run using the Laemmli system in the presence of dodecyl sulfate.

In the first dimension, three bands are visible, the central one containing heme and TMPD oxidase activity. The central band analyzed in the second dimension shows the presence of three major polypeptides. The band having the lowest mobility in the first dimension also contains three polypeptides which have the same mobilities in dodecyl sulfate/polyacrylamide gel electrophoresis as the three polypeptides of the *R. sphaeroides*

[8] U. K. Laemmli, *Nature (London)* **227,** 680 (1970).

[9] P. E. Thomas, D. Ryan, and W. Levin, *Anal. Biochem.* **75,** 168 (1976).

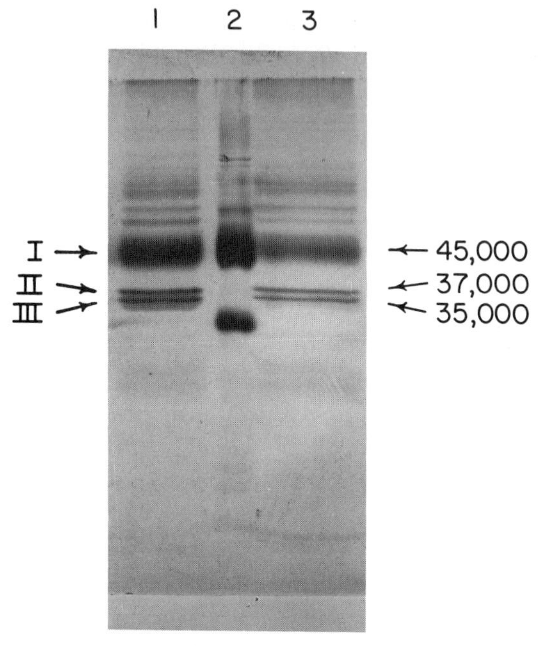

FIG. 2. Dodecyl sulfate/polyacrylamide gel electrophoresis. A slab gel containing 12% polyacrylamide was used. Two preparations of the *R. sphaeroides* oxidase are shown (lanes 1, 3) and a preparation of the *P. denitrificans* oxidase (lane 2). Details are given in the text. (From Gennis *et al.*,[6] with permission.)

reaction center.[10] This is apparently the major impurity in the oxidase preparation. Since the reaction center also interacts directly with a *c*-type cytochrome, it is not surprising that this is the major contaminant eluted from the cytochrome *c* affinity resin.

The subsequent purification of the oxidase by chromatography on DEAE-BioGel A gives a product that in dodecyl sulfate/polyacrylamide gel electrophoresis has three major polypeptides with M_r values 45,000, 37,000, and 35,000 (Fig. 2) and 15% impurities as indicated by integration of the densitometric scan following staining by Coomassie Brilliant Blue.

Immunological Analysis

An immunological study of the cross-reactivity of the *R. sphaeroides* oxidase subunits with other oxidases can be carried out with the technique of immunoblotting.[11]

[10] M. Y. Okamura, L. A. Steiner, and G. Feher, *Biochemistry* **13,** 1394 (1974).
[11] W. Burnette, *Anal. Biochem.* **112,** 195 (1981).

The 45,000 M_r band is immunologically related to subunit I of the oxidase isolated from *P. denitrificans*. The 35,000 and 37,000 M_r bands appear to be equally cross-reactive with antibodies raised against subunit II of the oxidases from either *P. denitrificans* or yeast. Thus, it is possible that these two polypeptides in the *R. sphaeroides* oxidase may, in fact, be derived from a single subunit.

Conclusions

The cytochrome-*c* oxidase from *R. sphaeroides* can easily and quickly be prepared in small amounts. The enzyme is immunologically related to that found both in yeast mitochondria and *P. denitrificans*. This oxidase appears to be structurally similar to other bacterial aa_3-type oxidases,[12] though does not function as a proton pump.[6]

[12] B. Ludwig, *Biochim. Biophys. Acta* **594,** 117 (1980).

[14] Cytochrome Oxidase from Thermophilic Bacterium PS3

By NOBUHITO SONE

The thermophilic bacterium PS3, isolated from a hot spring in Japan, is a gram-positive spore former similar to *Bacillus stearothermophilus*. It produces *a*-, *b*-, and *c*-type cytochromes. Under highly aerobic conditions, the terminal oxidase is an aa_3-type cytochrome oxidase, but under air-restricted conditions cytochrome *o* seems to replace cytochrome aa_3.[1] From cells grown under highly aerated conditions, aa_3-type cytochrome oxidase can be prepared by following the α-band spectrum of the reduced form.[2,3] The enzyme is composed of a set of three subunits containing one heme *c*, two heme *a* (cytochrome aa_3), and two Cu atoms as chromophores.[3] The enzyme shows active H^+ pumping with a reduced cytochrome *c* pulse when reconstituted into liposomes.[4,5]

[1] N. Sone, Y. Kagawa, and Y. Orii, *J. Biochem.* (*Tokyo*) **93,** 1329 (1983).
[2] N. Sone, T. Ohyama, and Y. Kagawa, *FEBS Lett.* **109,** 39 (1979).
[3] N. Sone and Y. Yanagita, *Biochim. Biophys. Acta* **682,** 216 (1982).
[4] N. Sone and P. Hinkle, *J. Biol. Chem.* **257,** 12600 (1982).
[5] N. Sone and Y. Yanagita, *J. Biol. Chem.* **259,** 1405 (1984).

Assay Methods

Absorption Spectrum. The α-band of the reduced form is recorded 2–5 min after the addition of $Na_2S_2O_4$ in the medium containing 50 m*M* Tris–HCl buffer (pH 8.0) and 0.5% Triton X-100 with a conventional recording spectrophotometer. An E_{mM}(604–630 nm) value of 16.6 m*M* cm^{-1} is used to calculate heme *a* content.

Cytochrome Oxidase Activity. Ascorbate oxidation via a substrate of cytochrome oxidase such as cytochrome *c* from various sources and TMPD (*N,N,N',N'*-tetramethyl-*p*-phenylenediamine) is measured at 40° with an oxygen electrode (YS1 4001) in the reaction medium containing 20 m*M* sodium phosphate buffer (pH 6.5) and 2–10 m*M* sodium ascorbate. By reducing the buffer concentration and adding 20 m*M* K_2SO_4, the reaction was also followed with a pH meter.[6]

$$\text{Ascorbate(H)}^- + \tfrac{1}{2}O_2 + H^+ = \text{Dehydroascorbate} + H_2O$$

H^+ Pump Activity. Cytochrome oxidase vesicles (0.1 ml) are suspended in 2 ml of the reaction medium containing 25 m*M* K_2SO_4, 2.5 m*M* $MgSO_4$, 0.25 m*M* MOPS- (morpholinoethanesulfonic acid) KOH buffer (pH 6.6–6.8), and 0.02 μg/ml valinomycin at 32°. After 5–10 min incubation, a 5–10 μl aliquot containing 4–8 nmol of reduced cytochrome *c* from *Candida krusei* (Kyowa Hakko Co., available from Sigma Chemicals) is added to initiate the reaction. The pH change is measured with a Beckman combination pH electrode No. 39505. The electrode signal is amplified with a Beckman pH meter (model 4500) and feeds into a strip chart recorder (Rikendenshi, SP-G6V). The response time (90%) is less than 3 sec.

Ferrocytochrome *c* is prepared by reduction of cytochrome *c* with $Na_2S_2O_4$ and successive removal of $Na_2S_2O_4$ by the centrifuge-column method,[7] using BioGel P-6 (Bio-Rad) equilibrated with the reaction medium.

Purification of PS3 Cytochrome Oxidase

Culture of Bacteria. This is the same as described previously,[8] except that log phase cells (about 600 g wet weight per 250 liters of the culture medium) grown at 66–69° are used.

[6] P. Nicholls and N. Sone, *Biochim. Biophys. Acta* **767,** 240–247 (1984).

[7] H. S. Penefsky, this series, Vol. 56 [47].

[8] Y. Kagawa and M. Yoshida, this series, Vol. 55 [84].

Reagents

DEAE-cellulose (DE-52, Whatman)
DEAE-toyopearl (DEAE derivative of vinyl polymer gel manufactured by Toyo Soda Co., Tokyo, also available from Merck under the name of Fractgel TSK DEAE-650)
Hydroxyapatite, prepared according to the modified method of Tiselius[9]
Tris–H_2SO_4 buffer, 1 *M*, pH 8.0
Tris–HCl buffer, 1 *M*, pH 8.0
Tricine–NaOH buffer, 5 m*M*, pH 8.0
Sodium phosphate buffer, 0.5 *M*, pH 7.0
NaCl, 2 *M*
LiCl, 2.5 *M*
Sodium cholate, 10%
Ammonium sulfate, reagent grade
Sodium dithionite, reagent grade
Sodium deoxycholate, Difco

Procedure[3]

Step 1. Bile Acid Treatment. Washed membranes[8] from 500 g of wet cells are suspended in a solution containing 2% sodium cholate, 1% sodium deoxycholate, 0.2 *M* Na_2SO_4, 30 m*M* Tris–H_2SO_4 buffer (pH 8.0), and 0.5 m*M* dithiothreitol in a final volume of 1000 ml. The mixture is sonicated in an ice bath for 10 min, three times, and then centrifuged at 140,000 *g* for 40 min. This treatment is repeated once more.

Step 2. LiCl Treatment. The pellets obtained in Step 1 are resuspended in 2.5 *M* LiCl, and the mixture is sonicated in an ice bath for 10 min and then centrifuged at 22,000 *g* for 20 min. The pellets are resuspended in 50 m*M* Tris–HCl buffer (pH 8.0) and centrifuged again.

Step 3. Triton X-100 Extraction. The resulting residues of Step 2 are extracted with 200–250 ml of a mixture of 6% Triton X-100, 0.05 *M* NaCl, and 50 m*M* Tris–HCl buffer (pH 8.0). Almost all membrane proteins including cytochromes are solubilized at this step.

Step 4. DEAE-Cellulose Column Chromatography. The amber-colored Triton X-100 extract from Step 3 is diluted with water (about 3 vol) until its conductance becomes 0.7 mmho (cm^{-1}). It is advisable to add Triton X-100 (about 0.5% of the total volume). The extract is then applied to a DEAE-cellulose (Whatman DE-52) column (8 × 6 cm) equilibrated

[9] K. Raymond, J. Marjorie, J. Wilkins, and L. Cole, *J. Am. Chem. Soc.* **81,** 6490 (1959).

with 50 m*M* Tris–HCl buffer (pH 8.0) containing 0.5% Triton X-100. Most of the cytochrome oxidase (cytochrome caa_3) is not absorbed.

Step 5. DEAE-toyopearl (DEAE-derivative of vinyl polymer gel) is used instead of Whatman DE-52 because it gives a superior separation. The unabsorbed fraction from Step 4 is diluted with water until its conductance becomes 0.35 mmho (cm^{-1}), and then it is applied to a DEAE-toyopearl column (3 × 8 cm). The column has been equilibrated with 25 m*M* Tris–HCl buffer (pH 8.0) containing 0.25% Triton X-100. Cytochrome oxidase is adsorbed, while most *b*-type cytochromes are not. The column is then washed with 300 ml of 50 m*M* Tris–HCl buffer (pH 8.0) containing 1.5% Triton X-100, followed by 50 m*M* Tris–HCl buffer (pH 8.0) containing 2 m*M* NaCl and Triton X-100. Then the concentration of NaCl is raised to 6 m*M* which causes a greenish-brown band to slowly move down the column, leaving a red band containing *b*- and *c*-type cytochromes on the top of the column.

Step 6. Hydroxyapatite Column Chromatography. The greenish-brown eluate from Step 5 is applied on a hydroxyapatite column (3 × 2.5 cm) equilibrated with 0.5% Triton X-100 and 1 m*M* phosphate buffer (pH 7.0), washed with 5 m*M* phosphate containing 0.5% Triton X-100, and eluted by raising the phosphate concentration to 50 m*M*.

Step 7. Ammonium Sulfate Precipitation. The eluate (about 100 ml) is mixed with sodium cholate (at a final concentration of 1.5%) and precipitated by adding 230 mg/ml of solid ammonium sulfate while stirring at 2°–5°. The precipitate is collected, suspended in 5 m*M* Tricine-NaOH (pH 8.0), and kept at 80° until use.

Comments on the Purification

1. Results of a typical purification of the PS3 cytochrome oxidase are summarized in the table.

2. At Step 4, sometimes a large part of the PS3 cytochrome oxidase is adsorbed on the column. It is advisable to add Triton X-100 to the diluted extract of Step 4 before applying to a DEAE-cellulose column in order to prevent the adsorption. But if this has occurred, the column is treated as in Step 5, and the resulting eluate contaminated with *b*- and *c*-type cytochromes is passed through a DEAE-cellulose column (3.5 × 6 cm) equilibrated with 50 m*M* Tris–HCl (pH 8.0) containing 0.5% Triton X-100 at a conductance of 0.4 mmho (cm^{-1}) to remove (absorb) the contaminants. The greenish-brown eluate can be used as the eluate of Step 4.

3. At Step 5, in order to raise the yield, 10 m*M* NaCl containing Tris–HCl and Triton X-100 can be applied to the column after the 6 m*M* NaCl elution. The resulting eluate containing cytochrome oxidase and some

PURIFICATION OF PS3 CYTOCHROME OXIDASE

Step	Protein (mg)	Heme *a* (nmol)	Heme *a*/protein (nmol/mg)	Yield (%)
Membranes	11,500	4,140	0.36	100
1. Bile acid treatment	4,020	3,294	0.82	80.0
2. LiCl treatment	2,700	2,971	1.10	71.8
3. Triton X-100 extract	2,185	2,647	1.21	63.9
4. DEAE-cellulose eluate	707	2,142	3.03	51.7
5. DEAE-toyopearl eluate	81.6	750	9.19	18.1
6. Hydroxyapatite eluate	27.5	410	14.9	9.9
7. $(NH_4)_2SO_4$ precipitate	23.8	349	16.3	8.4

b-type and *c*-type cytochromes is also treated as described in Comment 2 to remove the contaminating *b*- and *c*-type cytochromes. The greenish-brown eluate can be used for Step 6.

4. Procedures described in Comments 2 and 3 raise the final yield from around 10% to 15–20%.

Properties of PS3 Cytochrome Oxidase

Spectra and Prosthetic Groups. A *c*-type cytochrome is copurified with cytochrome aa_3. Figure 1 shows the spectra of oxidized and reduced

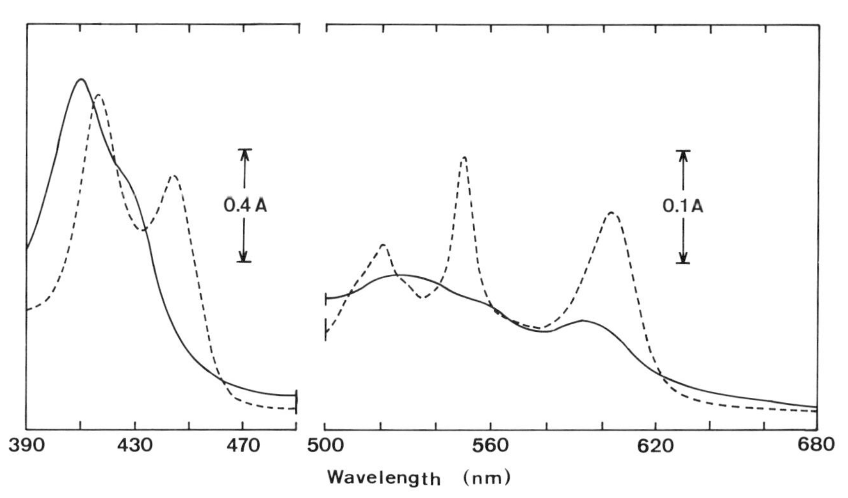

FIG. 1. Absorption spectra of PS3 cytochrome oxidase. (——), Oxidized (as purified); (---), reduced with $Na_2S_2O_4$ (peaks at 416, 444, 549.5, and 604 nm).

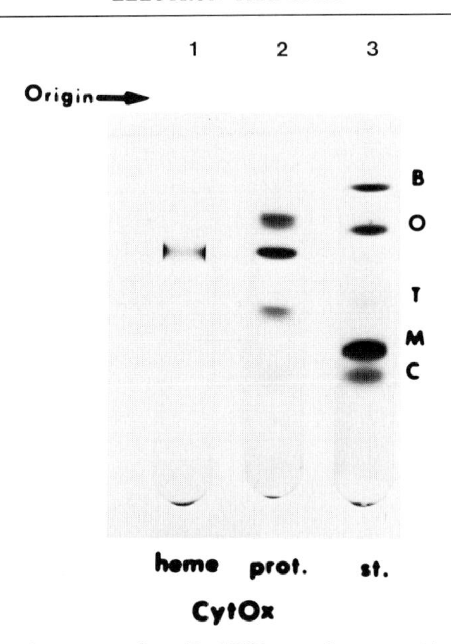

FIG. 2. Electrophoretic pattern of purified PS3 cytochrome oxidase. Lane 1 was stained with *o*-toluidine for heme and lane 2 was stained with Coomassie Brilliant Blue R-250 for the protein of PS3 cytochrome oxidase. Lane 3 was stained with Coomassie Brilliant Blue for standard proteins. B, Bovine serum albumin; O, ovalbumin; T, trypsin; M, myoglobin; C, equine cytochrome *c*. The polyacrylamide gel (7.5%) contained 0.1% sodium dodecyl sulfate and 8 *M* urea.

forms of the purified enzyme, which are very similar to those of the 1 : 1 cytochrome *c*–mammalian cytochrome oxidase complex.

The CO-difference spectrum is also similar to that of mammalian cytochrome oxidase (CO-cytochrome a_3^{2+} minus cytochrome a_3^{2+}). Contents of heme *a*, heme *c*, and copper have been estimated to be 15.9, 8.3, and 17.6 (atomic absorption) or 16.2 (bathocuproine) nmol/mg protein, respectively.

Catalytic Activity. PS3 cytochrome oxidase oxidizes cytochrome *c*-551 from this bacterium and cytochrome *c*-552 from *Thermus thermophilus* with K_m values around 4 μM, and it oxidizes yeast cytochrome *c* from *Saccharomyces cerevisiae* and *Candida krusei* with K_m values around 7 μM. The V_{max} values are almost the same turnover number of 80–120, expressed as cytochrome *c* oxidized/aa_3 sec^{-1}) at 32°.[6] Equine cytochrome *c* is a poor substrate with a higher K_m and a lower V_{max}.[5] *N,N,N′,N′*-Tetramethyl-*p*-phenylenediamine, phenazine methosulfate, and ruthenium hexaamine are also oxidized without adding one cytochrome *c*, indicating that the intrinsic cytochrome *c* of PS3 oxidase plays the same role as cytochrome *c* bound to the high-affinity site of mitochondrial cytochrome oxidase.[5,6] Phospholipids are necessary for the activity. Cyanide, azide, hydroxylamine, and hydrazine are inhibitory to the PS3 enzyme as they are to the mitochondrial enzyme, but slightly higher concentrations of the respective inhibitors are necessary for inhibitions.[5]

Stability. The oxidase activity is stable up to 64°,[2] but the H^+ pumping activity was lost at 60°.[10] Oxidase activity is also stable against denaturing reagents such as 7 *M* urea, 5 *M* LiCl, and 40% ethanol.[2] The purified enzyme can be kept at −80° without significant loss of either the oxidase or H^+ pumping activity.

Subunit Composition. PS3 cytochrome oxidase is composed of three kinds of subunits (I, 56 kDa; II, 38 kDa, III, 22 kDa), as shown in Fig. 2. Since the ratio is 1 : 1 : 1, one enzyme unit may be composed of one molecule each of the three subunits and may contain two heme *a* (aa_3), one heme *c*, and two copper atoms.

Very hydrophobic subunits I and III show abnormal Ferguson plots and form aggregates when denaturation with sodium dodecyl sulfate is carried out at above 70°. The intrinsic *c*-type cytochrome is found with subunit II (Fig. 2, left lane).[11]

Reconstitution of Vesicles Capable of H^+ Pumping

Proteoliposomes suitable for demonstration of H^+ pumping can be prepared by the freeze-thaw method[12] from purified PS3 cytochrome oxidase and partially purified soybean phospholipids.[4,5]

Preparation of Phospholipids.[13] Soybean *P*-lipids (Asolectin) can be purchased from Associated Concentrates (Woodside, NY). Asolectin (25 g) is suspended in acetone (400 ml) containing butyrated hydroxytoluene (500 mg) and stirred for 3 hr at 25° in the dark. The acetone-soluble fraction is discarded by decantation. After reextraction with acetone (200 ml), the insoluble fraction is dissolved in ethyl ether (100 ml) containing α-tocopherol (200 mg) and centrifuged at 8000 *g* for 10 min. The soluble fraction is concentrated to dryness under reduced pressure and stored at −80° until use.

Procedure for Reconstitution.[4,5] Phospholipids prepared as described above (40 mg) are suspended in 1 ml of 2.5 m*M* MOPS-KOH (pH 6.5) containing 0.2 m*M* EDTA and 4 m*M* K_2SO_4 and sonicated for 3 min at 5–25° with a probe-type sonicator (Branson model 200 at an amplitude of 2)

[10] T. Ogura, N. Sone, K. Tagawa, and T. Kitagawa, *Biochemistry* **23,** 2826 (1984).

[11] Y. Yanagita, N. Sone, and Y. Kagawa, *Biochem. Biophys. Res. Commun.* **113,** 575 (1983).

[12] M. Kasahara and P. C. Hinkle, *J. Biol. Chem.* **252,** 7384 (1977).

[13] N. Sone, M. Yoshida, H. Hirata, and Y. Kagawa, *J. Biochem.* (*Tokyo*) **81,** 519 (1977).

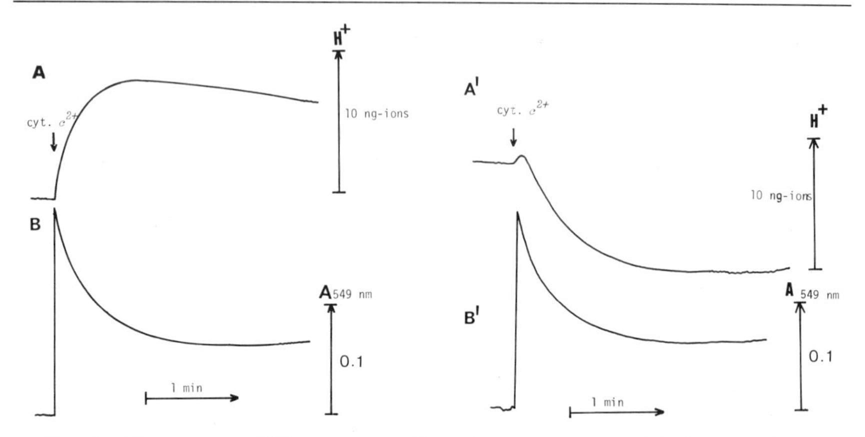

FIG. 3. Time course of H^+ transport (A) and cytochrome *c* oxidation (B) when PS3 cytochrome oxidase vesicles were pulsed with ferrocytochrome *c*.[5] The reaction was started by the addition of 8.2 nmol of *Candida krusei* ferrocytochrome *c* (cyt. *c*). In the case of (A′) and (B′), 0.6 μg FCCP was also added. PS3 cytochrome oxidase vesicles (75 μl) were suspended in the reaction mixture containing 25 m*M* K_2SO_4, 2.5 m*M* $MgSO_4$, 0.25 m*M* K-MOPS (pH 6.6) and 0.15 μg of valinomycin. (Redrawn from Fig. 1 of Sone and Yanagita[5].)

with a microtip. PS3 cytochrome oxidase (1–3 nmol as cytochrome aa_3) is then added and the mixture is sonicated for 5 sec (10 pulses of 0.5 sec). The mixture in the test tube is then quickly frozen at −80°, thawed at room temperature, and sonicated for 5–10 sec. The freeze-thaw sonication cycle is repeated once more.

Comments on Reconstitution and Measurement of H^+ Pump Activity

1. The freeze-thaw method described above was also successful for reconstituting beef heart cytochrome oxidase and *Thermus thermophilus* cytochrome oxidase into vesicles.[14,15]

2. The cholate-deoxycholate dialysis method, which was successfully used for incorporation of PS3 H^+-ATP synthase (TF_0 F_1), was not good for reconstitution of PS3 cytochrome oxidase.

3. Two or three cycles of the freeze-thaw sonication gave the best results, i.e., the highest H^+ pumping activity and a slow leakage.

4. PS3 cytochrome oxidase vesicles thus prepared pump H^+ upon addition of reduced cytochrome *c* under aerobic conditions. The time courses of H^+ pump and cytochrome *c* oxidation are slow due to a rela-

[14] N. Sone and P. Nicholls, *Biochemistry* **23,** 6550–6554 (1984).

[15] N. Sone, Y. Yanagita, K. Hon-nami, Y. Fukumori, and T. Yamanaka, *FEBS Lett.* **155,** 150 (1983).

tively small amount of the oxidase molecule inlayed and a high K_m value for cytochrome *c,* and thus can be simultaneously followed with a conventional spectrophotometer and a pH meter, as shown in Fig. 3. A high vectorial H^+/e ratio exceeding 1, as high as 1.4, has been obtained as a ratio of H^+ ejected per cytochromes oxidized.

[15] Cytochrome *c* Oxidase from *Paracoccus denitrificans*

By BERND LUDWIG

Introduction

In recent years, interest in bacterial cytochrome-*c* oxidases has increased since the available evidence (see Refs. 1,2) suggests that these enzymes could be viewed as simpler models for the structurally far more complex oxidases from mitochondria. Whereas most bacterial- (heme aa_3) type oxidases appear to closely match the eukaryotic enzyme in terms of redox ligands, spectral and kinetic properties, and, in some cases, even their ability to act as redox-linked proton pumps (Refs. 3–5a; see also this volume [14]), the most prominent distinguishing feature of bacterial oxidases is the fact that they are composed of only between one and three different subunits, in contrast to the enzymes from mitochondria, where 12–13 different subunits have been reported (see Refs. 6,6a).

Among the best-characterized bacterial oxidases is the one from *Paracoccus denitrificans,* a bacterium long known for its close relationship to mitochondria.[7,8] For this oxidase, composed of two subunits of molecular weight 45,000 and 28,000 (Fig. 1), the expected structural homology has been substantiated by partial protein sequence homologies of its two subunits to the two largest subunits of the mitochondrial oxidase.[9] This sug-

[1] B. Ludwig, *Biochim. Biophys. Acta* **594,** 177 (1980).

[2] R. K. Poole, *Biochim. Biophys. Acta* **726,** 205 (1983).

[3] M. Solioz, E. Carafoli, and B. Ludwig, *J. Biol. Chem.* **257,** 1579 (1982).

[4] N. Sone and P. C. Hinkle, *J. Biol. Chem.* **257,** 12600 (1982).

[5] N. Sone, Y. Yanagita, K. Hon-Nami, Y. Fukumori, and T. Yamanaka, *FEBS Lett.* **155,** 150 (1983).

[5a] M. Wikström, K. Krab, and M. Saraste, *Annu. Rev. Biochem.* **50,** 623 (1981).

[6] G. Buse, G. C. M. Steffens, and L. Meinecke, *in* "Structure and Function of Membrane Proteins" (E. Quagliariello and F. Palmieri, eds.), p. 131. Elsevier, Amsterdam, 1983.

[6a] B. Kadenbach and P. Merle, *FEBS Lett.* **135,** 1 (1981).

[7] P. John and F. R. Whatley, *Nature (London)* **254,** 495 (1975).

[8] P. John and F. R. Whatley, *Biochim. Biophys. Acta* **463,** 129 (1977).

[9] G. C. M. Steffens, G. Buse, W. Oppliger, and B. Ludwig, *Biochem. Biophys. Res. Commun.* **116,** 335 (1983).

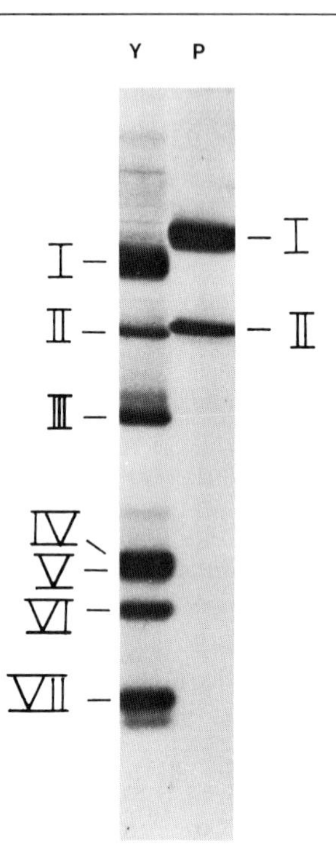

FIG. 1. SDS–gel electrophoresis of *Paracoccus* and yeast cytochrome *c* oxidase. Yeast (Y) (20 μg) and *Paracoccus* (P) (8 μg) oxidase were electrophoresed on a 12.5% polyacrylamide gel and stained with Coomassie Brilliant Blue.

gests that also in mitochondria the two largest subunits comprise the basic unit catalyzing electron transport and proton translocation, since the *Paracoccus* oxidase has been shown to pump protons in a reconstituted system,[3,10] although with a reduced rate when compared to mitochondria[5a] or to whole cells of *Paracoccus*.[11]

Such a simple, well-defined bacterial oxidase should be particularly useful in attempts to obtain three-dimensional crystals of oxidase and in attempts to prove structure/function relationships by site-directed mutagenesis.

The following protocol provides a comprehensive and updated description of important steps in the growth of the organism, the preparation of membranes, and the purification of reasonable amounts of the enzyme suitable for structural and functional studies.

[10] I. Püttner, M. Solioz, E. Carafoli, and B. Ludwig, *Eur. J. Biochem.* **134,** 33 (1983).

[11] H. W. van Verseveld, K. Krab, and A. H. Stouthamer, *Biochim. Biophys. Acta* **635,** 525 (1981).

TABLE I
SUCCINATE GROWTH MEDIUM

Succinate growth medium[a]		Salt stock solution[b]	
Chemical	Concentration	Salt	Concentration
K-phosphate,	50 m*M*	$CaCl_2$	0.1 *M*
NH_4Cl	75 m*M*	$FeCl_3$	90 m*M*
Na_2SO_4	11.5 m*M*	$MnCl_2$	50 m*M*
$MgCl_2$	1.25 m*M*	$ZnCl_2$	25 m*M*
Citric acid	1 m*M*	$CoCl_2$	10 m*M*
Biotin	4 μg/liter	$CuCl_2$	5 m*M*
Salt stock	1 ml/liter	H_3BO_3	5 m*M*
Succinic acid	10 g/liter	Na_2MoO_4	10 m*M*
Adjust pH to 6.2 with KOH before autoclaving		Dissolve in half-concentrated HCl	

[a] Modified from H. G. Lawford, *Can. J. Biochem.* **56,** 13 (1978).
[b] Modified from P. A. Light and P. B. Garland, *Biochem. J.* **124,** 123 (1971).

Growth of *Paracoccus denitrificans*

Paracoccus denitrificans (ATCC 13 543) stock cultures are stored at −70° in the succinate growth medium (Table I) supplemented with 15% glycerol, or for short periods of time on 1.5% agar plates at 4°.

Liquid cultures are grown at 30° in small volumes on gyratory shakers with good aeration. If larger volumes are to be cultivated in fermenters with forced aeration, a silicone-based antifoam agent is added, but only scarcely as needed during the run. In large-volume fermenters, the oxygen concentration in the medium is monitored and kept at above 50% saturation, and the pH is maintained at neutral by addition of phosphoric acid. Cultures are harvested at mid-log phase, corresponding to 300 Klett units (red filter); cells from the stationary phase show a decreased specific heme *a* content. Large-volume fermenters are cooled down to 4°, and the medium is applied to a continuous-flow centrifuge. The cell paste is shock-frozen at −40° in 1-kg bars and stored at −20°. Yields range between 10 and 15 g wet cell weight/liter of growth medium.

Cell Breakage and Isolation of Crude Membranes

To 10 kg of frozen cell paste, sodium phosphate buffer (10 m*M*, pH 7.6) of room temperature is added to 25 liters total volume; the cells are allowed to thaw under constant stirring, and 500 mg of DNase I are added. The following steps are performed at 4°. At a rate of 5 liters/hr, the

TABLE II
PURIFICATION SCHEME[a]

Material/step	Protein (mg)	Heme *a* (nmol)	Heme/protein (nmol/mg)	Yield heme *a* (%)
Crude membranes	81,180	17,466	0.22	(100)
Preextracted membranes	62,540	16,855	0.27	96.5
Supernatant of solubilization	32,760	15,467	0.47	88.6
AS fractionation[b]	11,084	12,860	1.2	73.6
DEAE-cellulose column	2,287	11,148	4.9	63.8
Gel filtration column	632	6,037	9.6	34.6
Affinity column	142	4,195	29.5	24.0

[a] For details, refer to "Enzyme Purification."
[b] AS, Ammonium sulfate.

suspension is applied to a continuously operating mechanical grinder (Dynomill, Basel) using glass beads of 0.1 mm diameter for breakage.

Efficient cooling is achieved by a dry ice/ethanol bath to maintain a temperature below 10° in the eluate of broken cells. In a first centrifugation at 7000 *g* for 30 min, cell debris is removed; in a subsequent centrifugation at 50,000 *g* for 4 hr, a crude membrane fraction is obtained as a fluffy, brownish layer whereas the dense, gray sediment below is discarded. The membrane fraction is washed once by resuspension in the same buffer and subsequent centrifugation as before; it is stored frozen at 20–40 mg protein/ml. The yield is around 250 g membrane protein with a specific heme *a* content of 0.15–0.30 nmol/mg protein.

Enzyme Purification

The following protocol is based on the originally published procedure[12] with some modifications. (All steps are performed at 4°; yields of each step are given in Table II.)

Preextraction of Crude Membranes

Protein loosely attached to membranes is removed by adding 0.1 mg of Na-deoxycholate/mg of protein, 0.4 *M* KCl, 1 m*M* EDTA, and 0.1 *M* K-phosphate, pH 7.6, at a protein concentration of 20 mg/ml; membranes are resedimented as above. The supernatant contains only little spectrally detectable heme *a* and is discarded.

[12] B. Ludwig and G. Schatz, *Proc. Natl. Acad. Sci. U.S.A.* **77,** 196 (1980).

Membrane Solubilization

The preextracted membrane pellet is resuspended at 15 mg/ml in K-phosphate/EDTA as above plus 1 *M* KCl, and Triton X-100 is added from a 25% stock solution to a final concentration of 1.5% (w/v) with stirring. A subsequent centrifugation step (as above) removes detergent-insoluble material, which is discarded.

Ammonium Sulfate Fractionation

The clear supernatant is immediately diluted with an equal volume of cold, distilled water and made 0.7% in sodium cholate. Ammonium sulfate solution saturated at 4° is added with stirring to a concentration of 30%, and the precipitate removed by centrifugation for 30 min at 8000 *g*. To the supernatant, ammonium sulfate is added to 42% saturation, and the solution is spun as before. The protein is recovered in the oily, detergent-rich floating layer; the bulk phase and any sediment is removed as far as possible by aspiration and the protein/detergent layer immediately taken up in 4 liters of 0.5% Triton X-100, K-phosphate, 10 m*M*, pH 7.6.

Ion-Exchange Chromatography

Before loading the material on the DEAE-cellulose column (Whatman DE-52, 1.5-liter volume), the conductivity of the solution is measured and adjusted, if required, to a value below 2.5 mmho by addition of 0.5% Triton X-100 solution. The column is preequilibrated with 0.5% Triton X-100, K-phosphate, 10 m*M*, pH 7.6. After binding to the exchange resin, oxidase is eluted by a continuous gradient of 0–500 m*M* NaCl in 0.2% Triton X-100, K-phosphate, 10 m*M*, pH 7.6; total volume of the gradient is 4 liters, elution speed 500 ml/hr. Fractions of highest heme/protein ratios are pooled, made 0.4% in Na-cholate, and ammonium sulfate fractionated at 30%, 33%, 42%, 45%, and 47% saturation. (Here, sedimenting pellets should be obtained upon centrifugation.) Again, fractions of highest purity are combined and taken up in a minimal volume (below 100 ml) of gel filtration buffer (see below).

Gel Filtration Chromatography

A 2-liter column of Ultrogel AcA 34 (LKB) is equilibrated with 0.2% Triton X-100, 0.2 *M* KCl, 1 m*M* EDTA, 50 m*M* K-phosphate, pH 7.6. After applying the material from the previous ammonium sulfate fractionation, the column is developed with the same buffer at a speed of 50 ml/hr. Fractions of highest specific heme content are pooled, diluted with an equal volume of column buffer, made 0.4% in cholate, and ammonium sulfate fractionated to 33%, 42%, and 45% saturation. Suitable fractions are combined and taken up in 0.2% Triton X-100, 1 m*M* EDTA, Tris–HCl, 10 m*M*, pH 7.3.

Affinity Chromatography

Yeast cytochrome *c* linked to thiol-Sepharose 4B (Pharmacia) via a disulfide bridge[13] is used to specifically bind cytochrome *c* oxidase. (If this material is not available, this last step may be replaced by repeating the ion-exchange chromatography and gel filtration step until a sufficiently pure enzyme preparation is obtained.)

One-fifth of the material obtained in the previous step is diluted with the equilibration buffer (see below) until its conductivity is below 1.2 mmho. The affinity column (packed volume 500 ml; binding capacity for *Paracoccus* oxidase ~40 mg) is equilibrated with 0.2% Triton X-100, 1 m*M* EDTA, 10 m*M* Tris–HCl, pH 7.3, loaded with the partially purified oxidase preparation, washed with 500 ml of equilibration buffer, and eluted by a gradient of 0–300 m*M* NaCl in the same buffer (total volume of the gradient 1.5 liter; flow rate 150 ml/hr).

The remaining four-fifths of the material are processed accordingly in subsequent runs of the column. Fractions of highest purity are pooled, made 0.4% in Na-cholate, and precipitated with ammonium sulfate at 33%, 36%, 39%, 42%, and 45% saturation. Each fraction is taken up in a minimal volume of 0.1% cholate, 10 m*M* Tris–HCl, pH 7.3, heme/protein ratios are determined, and the material is shock-frozen in small aliquots at −70°.

Related Techniques

Protein estimation in the presence of sodium dodecyl sulfate, spectral heme determinations, metal quantitations by atomic absorption spectroscopy, enzyme activity measurements, reconstitution, immune precipitation, and gel electrophoresis were done as described earlier[12]; proton pumping measurements of reconstituted *Paracoccus* oxidase have been published in Refs. 3,10, kinetic data in Refs. 14,15, EPR studies in Ref. 16, sedimentation analysis in Ref. 17, cytochrome *c* binding in Ref. 18, and a partial amino acid sequence analysis in Ref. 9.

[13] K. Bill, R. P. Casey, C. Broger, and A. Azzi, *FEBS Lett.* **120,** 248 (1980).

[14] B. Ludwig and Q. H. Gibson, *J. Biol. Chem.* **256,** 10092 (1981).

[15] J. Reichardt and Q. H. Gibson, *J. Biol. Chem.* **258,** 1504 (1983).

[16] A. Seelig, B. Ludwig, J. Seelig, and G. Schatz, *Biochim. Biophys. Acta* **636,** 162 (1981).

[17] B. Ludwig, M. Grabo, I. Gregor, A. Lustig, M. Regenass, and J. P. Rosenbusch, *J. Biol. Chem.* **257,** 5576 (1982).

[18] J. Reichardt and B. Ludwig, *in* "Vectorial Reactions in Electron and Ion Transport in Mitochondria and Bacteria" (F. Palmieri *et al.*, eds.), p. 3. Elsevier, Amsterdam, 1981.

Acknowledgments

I wish to thank G. Schatz, Basel, for his support and encouragement during this project, and B. Paetow, W. Oppliger, H. Bracher, and S. Smit for excellent technical assistance. I am also indebted to Sandoz AG, Basel, for generous help in large-scale fermentation. This investigation was supported by Grant 3.606.80 from Schweizer Nationalfond and in part by Grant Lu 318/1-1 from Deutsche Forschungsgemeinschaft.

[16] Purification, Enzymatic Properties, and Reconstitution of Cytochrome-*c* Oxidase from *Bacillus subtilis*

By WIM DE VRIJ, BERT POOLMAN, WIL N. KONINGS, and ANGELO AZZI

Introduction

In prokaryotes several classes of cytochrome oxidases can be found which function as terminal electron-transfer components in aerobic metabolism.[1] The bacterial cytochrome oxidase that resembles most closely the mitochondrial oxidase is the cytochrome-*c* oxidase (aa_3 type). In the past 5 years several cytochrome aa_3-type oxidases from different bacteria have been isolated,[2–6] including cytochrome-*c* oxidase of *Bacillus subtilis*.[6] The respiratory chain of the gram-positive, obligate aerobe was one of the first extensively studied electron-transfer systems in bacteria and the first organism in which a 603-nm absorbance band, indicative for cytochrome-*c* oxidases, was revealed.[7] An interesting feature of this organism is that the respiratory chain, located in the cytoplasmic membrane, resembles closely the respiratory chain of mitochondria and contains as functional electron carriers the cytochromes b, c_1, c, and the oxidase c.[8,9] Methods have been developed for the isolation of very pure

[1] R. K. Poole, *Biochim. Biophys. Acta* **726,** 205 (1983).
[2] B. Ludwig and G. Schatz, *Proc. Natl. Acad. Sci. U.S.A.* **77,** 196 (1980).
[3] J. A. Fee, G. C. Miles, K. R. Findling, R. Lorence, and T. Yoshida, *Proc. Natl. Acad. Sci. U.S.A.* **77,** 147 (1980).
[4] N. Sone and Y. Yanagita, *Biochim. Biophys. Acta* **682,** 216 (1982).
[5] R. B. Gennis, R. P. Casey, A. Azzi, and B. Ludwig, *Eur. J. Biochem.* **125,** 189 (1982).
[6] W. de Vrij, A. Azzi, and W. N. Konings, *Eur. J. Biochem.* **131,** 97 (1983).
[7] D. Keilin, "The History of Cell Respiration and Cytochrome." Cambridge Univ. Press, London, 1966.
[8] K. Miki and K. Okunuki, *J. Biochem.* (*Tokyo*) **66,** 845 (1969).
[9] K. Miki, I. Sekuzu, and K. Okunuki, *Annu. Rep. Sci. Works, Fac. Sci. Osaka Univ.* **15,** 33 (1967).

and well-characterized cytoplasmic membrane vesicles from *B. subtilis*.[10,11] The membrane vesicles have been proved to be useful model systems for the study of energy-transducing processes in the cytoplasmic membrane,[12,13] and several aspects of energy transduction by the respiratory chain of this organism have been described.[14–16] These membrane vesicles also offer excellent starting material for the isolation of membrane-bound proteins and in particular the components of the respiratory chain.

In this chapter we describe the methods used for the isolation and purification of cytochrome-*c* oxidase of *B. subtilis* W23. The purified enzyme has been biochemically and biophysically characterized and reincorporated into liposomes. The incorporation procedure and the physiological properties of the incorporated enzyme will also be described.[17]

Isolation Procedures of Cytochrome-c Oxidase

Growth of B. subtilis

Bacillus subtilis W23 is grown to the mid-log phase at 37° in 10-liter batches under vigorous aeration in a medium containing 0.8% (w/v) tryptone (Difco), 0.5% (w/v) NaCl, 0.2% (w/v) KCl, 50 mg/liter $CaCl_2 \cdot 2H_2O$, 50 mg/liter $FeSO_4 \cdot 7H_2O$, and 150 μl/liter micronutrient solution.[6] The micronutrient solution contains in percentage (w/v): $MnCl_2$ (2.2), $ZnSO_4 \cdot 7H_2O$ (0.05), H_3BO_3 (0.5), $CuSO_4 \cdot 5H_2O$ (0.016), $Na_2MoO_4 \cdot 2H_2O$ (0.025), $Co(NO_3)_2 \cdot 6H_2O$ (0.46), and concentrated H_2SO_4 (0.5%, v/v).

Isolation of Membrane Vesicles

Cells from a 10-liter culture are harvested at an absorbance at 660 nm of 0.8–1.0 (cell density, about 0.2 g of protein per liter) and suspended in 20 ml of a 50 m*M* K_2PO_4 solution.[10] The concentrated cell suspension is added to 2 liters of 50 m*M* K_2HPO_4 containing 250 mg/liter egg lysozyme (E. Merck AG, Darmstadt, FRG), 15 mg/liter DNase (Miles Laboratories Ltd., Slough, UK), and 15 mg/liter RNase (Miles Laboratories Ltd.).

[10] W. N. Konings, A. Bisschop, M. Veenhuis, and C. A. Vermeulen, *J. Bacteriol.* **116,** 1456 (1973).

[11] A. Bisschop and W. N. Konings, *Eur. J. Biochem.* **67,** 357 (1976).

[12] H. R. Kaback, this series, Vol. 22, p. 99.

[13] W. N. Konings, *Adv. Microb. Physiol.* **15,** 175 (1976).

[14] W. N. Konings and E. Freese, *J. Biol. Chem.* **247,** 2408 (1972).

[15] A. Bisschop, Ph.D. thesis, University of Groningen, 1976.

[16] J. Bergsma, Ph.D. thesis, University of Groningen, 1983.

[17] D. H. Thompson and S. Ferguson-Miller, *Biochemistry* **22,** 3178 (1983).

Subsequently, 10 m*M* $MgSO_4$ is added and the solution is incubated for 30 min at 37°. Following this incubation, 15 m*M* of sodium-EDTA, pH 7.0, is added which results in clearance of the solution within 1 min. The $MgSO_4$ concentration is increased to 40 m*M* and the mixture is incubated for another 15 min and subsequently centrifuged (45 min, 23,000 *g*, 4°). The pellet containing the membrane vesicles is resuspended in 40 ml of 50 m*M* Tris–HCl, pH 7.4, and 1 m*M* EDTA (for the affinity chromatography purification of cytochrome-*c* oxidase) or in 50 m*M* Tris–HCl, pH 7.0 (for the ammonium sulfate precipitation procedure). Membrane vesicles are collected by centrifugation (30 min, 48,000 *g*, 4°) and suspended in buffer to a concentration of 20 mg of membrane proteins per milliliter. Aliquots of 0.5 ml are rapidly frozen in liquid nitrogen and stored at −80° or in liquid nitrogen.

The yield of cytoplasmic membranes obtained by this procedure is 10–15 mg membrane protein per liter of cells. The membrane vesicles contain 0.4–0.5 nmol heme *a* per milligram membrane protein, 0.7–0.8 nmol heme *b* per milligram membrane protein and varying small amounts of hemes *c* and c_1.

Purification of Cytochrome-c Oxidase by Affinity Chromatography

Preparation of Yeast Cytochrome c-Thiol-Sepharose 4B Gel. Activated thiol-Sepharose 4B (Pharmacia, Uppsala, Sweden) is swollen (2 g in 20 ml of distilled water) for 30 min and washed on a sintered glass filter (G3) with, successively, 1 liter of distilled water and 0.5 liter of 50 m*M* Tris–HCl, pH 7.4, and 1 m*M* EDTA. To avoid mechanical damage of the gel beads, solutions are added without stirring of the gel. After washing the thiol-Sepharose 4B, the gel is suspended in 10 ml 50 m*M* Tris–HCl, pH 7.4, and 1 m*M* EDTA. *Saccharomyces cerevisiae* cytochrome *c* (type VIII) (30 mg) is added to this suspension, and the mixture is incubated with stirring overnight at 4°. Subsequently, the gel is washed with 250 ml 50 m*M* Tris–HCl, pH 7.4, containing 1% (v/v) Triton X-100 and 1 *M* NaCl to remove noncovalently bound cytochrome *c*. The gel is washed with another 0.5 liter 50 m*M* Tris–HCl buffer, pH 7.4, 1 m*M* EDTA, and then poured into a column (1 × 10 cm). Finally, the column is equilibrated with 50 m*M* Tris–HCl, pH 7.4, 1 m*M* EDTA, and 0.1% (w/v) dodecyl-β-D-maltoside.

Solubilization of Cytochrome-c Oxidase. Membrane vesicles of *B. subtilis* (10 mg membrane protein/ml) are incubated with 1% (w/v) dodecyl-β-D-maltoside[18] in 50 m*M* Tris–HCl, pH 7.4, and 1 m*M* EDTA for 2 hr

[18] P. Rosevear, T. Van Aken, J. Baxter, and S. Ferguson-Miller, *Biochemistry* **19**, 4108 (1980).

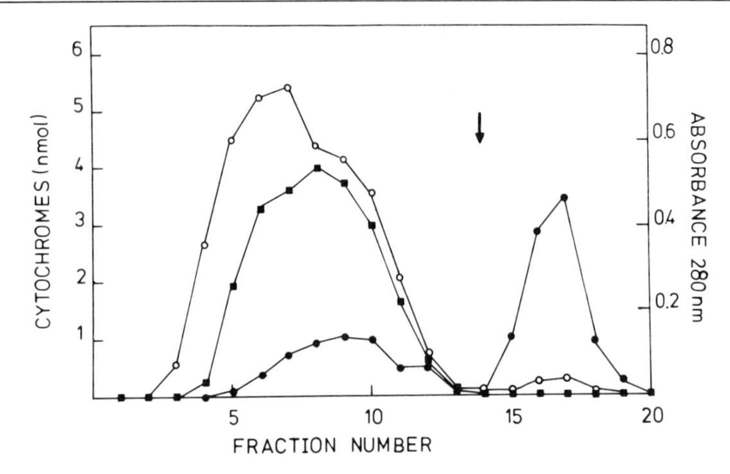

FIG. 1. Elution pattern of cytochromes and protein after affinity chromatography. Elution was obtained with 50 m*M* Tris–HCl, pH 7.4, containing 1 m*M* EDTA, 0.1% (w/v) dodecyl-β-D-maltoside, and 0 or 100 m*M* KCl (indicated by the arrow). (○) Protein; (●) heme *a*; (■) heme *b*.

at 4° with gentle stirring. The mixture is centrifuged for 30 min, 48,000 *g* at 4°. The solubilized material contains 70–80% of the membrane proteins, including the different types of cytochromes. To prevent loss in enzymatic activity of cytochrome-*c* oxidase, membranes are solubilized just before the affinity chromatography.

Affinity Chromatography. A diluted dodecyl-β-D-maltoside membrane extract (final detergent concentration 0.2%, w/v) is applied to the affinity column at a flow rate of 15 ml/hr. Subsequently, the column is washed with 50 m*M* Tris–HCl, pH 7.4, 1 m*M* EDTA, and 0.1% (w/v) dodecyl-β-D-maltoside until no more protein is detectable in the collected fractions. In the initial eluate mainly heme *c* absorbance is found. Hereafter, other proteins, including all *b*-type cytochromes and some heme *a*, elute from the column. The eluation is continued at a flow rate of 40 ml/hr with 50 m*M* Tris–HCl, pH 7.4, 1 m*M* EDTA, 0.1% (w/v) dodecyl-β-D-maltoside, and 100 m*M* KCl to remove heme *a* from the column. Raising the ionic strength with 100 m*M* KCl also results in the release of some yeast cytochrome *c* from the thiol-Sepharose 4B matrix.

In Fig. 1 a typical experiment is given in which 25 mg of a membrane extract was loaded on the column. At a flow rate of 15 ml/hr, 4.8-ml fractions were collected and analyzed for their protein and cytochrome contents using an Aminco DW2a UV-visible spectrophotometer. In the experiment, most proteins eluted from the column with about 60 ml of

50 mM Tris–HCl, pH 7.4, 1 mM EDTA, and 0.1% (w/v) dodecyl-β-D-maltoside.

Since the binding capacity of the column is about 6 pmol of heme *a*/nmol *S. cerevisiae* cytochrome *c,* 12–15 nmol heme *a* can maximally be bound. The binding capacity of the column decreases gradually in time after intensive use, but satisfactory results are still acquired after at least 3 months.

With the cytochrome *c* affinity column, a 25- to 30-fold purification of heme *a* can be achieved, resulting in heme *a*-to-protein ratios of 10–13 (nmol/mg protein). The recovery of *B. subtilis* cytochrome-*c* oxidase in the fractions with the highest heme *a*-to-protein ratios is about 80% of the amount retained by the gel.

Ion-Exchange Chromatography. The main impurity in the cytochrome-*c* oxidase fractions after the cytochrome *c* affinity chromatography is *S. cerevisiae* cytochrome *c*. Therefore, fractions containing the highest heme *a*-to-protein ratios are combined, diluted twice with 50 mM Tris–HCl, pH 7.4, and 0.1% (w/v) dodecyl-β-D-maltoside and loaded on a 2-ml Amberlite CG-50 column, which is equilibrated with 50 mM Tris–HCl, pH 7.4, 50 mM KCl, and 0.1% dodecyl-β-D-maltoside. Due to the high isoelectric point (pK = 10.5) cytochrome *c* binds to the cation exchanger, while cytochrome-*c* oxidase is eluted. After the cation-exchange column, cytochrome-*c* oxidase is spectroscopically pure and heme *a*-to-protein ratios of 15–17 nmol/mg protein are reached. Depending on the cytochrome *c* content of the cytoplasmic membranes, this corresponds to a 35- to 40-fold purification.

In order to concentrate the enzyme and to remove final contaminations, the fractions containing cytochrome-*c* oxidase are pooled and loaded on a 1-ml DEAE-Cellex D column, which is equilibrated with 50 mM Tris–HCl, pH 7.4, 50 mM KCl, and 0.1% (w/v) dodecyl-β-D-maltoside. Heme *a* is eluted from the anion-exchange column with ~10 ml of 50 mM Tris–HCl, pH 7.4, containing 1 mM EDTA, 0.1% (w/v) dodecyl-β-D-maltoside, and 200 mM KCl at a flow rate of 20 ml/hr. Under these conditions cytochrome-*c* oxidase runs as a sharp green band on the column and more than 50% of the enzyme that is applied to the column can be collected at a concentration of 20–30 nmol heme *a* per milliliter. An alternative to the DEAE-Cellex D column is the use of Amicon B15 ultrafilters. However, the latter method has the disadvantage that more than 50% of the enzyme is lost due to binding to the filter. In addition, detergents like dodecyl-β-D-maltoside with a micellar weight of 76 kDa[19]

[19] A. Azzi, K. Bill, and C. Broger, *Proc. Natl. Acad. Sci. U.S.A.* **79,** 2447 (1982).

are also concentrated by ultrafiltration, which leads to a loss of enzymatic activity.

Regeneration of the Cytochrome c Affinity Column. To regenerate the cytochrome *c* affinity column, the gel is washed with 5 column volumes of 50 m*M* Tris–HCl, pH 7.4, containing 1 m*M* EDTA, 1 *M* NaCl, and 1% Triton X-100, followed by 250 ml of 50 m*M* Tris–HCl, pH 7.4, and 1 m*M* EDTA at a flow rate of 40 ml/hr.

Comments. Starting from the cytoplasmic membranes, the described procedure for the purification of cytochrome-*c* oxidase takes about 10–12 hr. To avoid inactivation of the enzyme during the isolation procedure all actions have to be carried out at 4°.

The binding of heme *a* to the cytochrome *c* affinity column depends strongly on the detergent used. In contrast to what is observed for the beef heart enzyme,[19] *B. subtilis* cytochrome-*c* oxidase does not bind to the column with nonionic detergents like Triton X-100 (0.1% v/v) and Tween 80 (0.1%, v/v). On the other hand, when *n*-octyl-β-D-glucopyranoside (0.7%, w/v) is used, the binding of cytochrome-*c* oxidase to the cytochrome *c* matrix can only be disrupted with KCl concentrations higher than 500 m*M*. Binding of heme *b* occurs with octylglucoside, but also with low concentrations of dodecyl-β-D-maltoside (0.02%, w/v).

Purification of Cytochrome-c Oxidase by Ammonium Sulfate Precipitation Procedure[20]

Solubilization of Cytochrome-c Oxidase. Cytochrome-*c* oxidase is solubilized by incubating 40 ml of membrane vesicles [10 mg membrane protein/mg 2% (w/v) octyl-β-D-glucoside in 50 m*M* Tris–HCl, pH 7.0] for 1 hr at 4° with gentle stirring. The mixture is centrifuged (30 min, 48,000 *g*, 4°), and the supernatant is used for further purification.

Ammonium Sulfate Precipitation. The solubilized material is diluted once in 50 m*M* Tris–HCl, pH 7.0, followed by a stepwise ammonium sulfate precipitation. The solubilized material is adjusted to 0.8 ammonium sulfate saturation, and the protein is allowed to precipitate by incubation for 15 min on ice under continuous stirring. The precipitated protein is removed by centrifugation for 20 min, 48,000 *g* at 4°. The supernatant is adjusted to 0.85 ammonium sulfate saturation. Again the protein is allowed to precipitate and the precipitated protein is removed by centrifugation. The resulting greenish supernatant, containing cytochrome-*c* oxidase (81 ml), is dialyzed against 10 liters of 50 m*M* Tris–HCl, pH 7.0, at 4° for 2 hr in order to remove the ammonium sulfate. To prevent precipitation of the hydrophobic proteins Triton X-100 is added to

[20] W. de Vrij, B. Poolman, A. Azzi, and W. N. Konings, unpublished.

a final concentration of 0.1% (v/v). Subsequently the supernatant is concentrated overnight by dialysis against 0.5 liter polyethylene glycol 6000 30% (w/v) in 50 m*M* Tris–HCl, pH 7.0, at 4° (final volume 7 ml).

Anion Exchange Chromatography. The concentrated material (7 ml) is applied on a high-performance liquid chromatography (HPLC) DEAE-Si 100 polyol derivative column (Serva, size 3 μm, 7.1 × 250 mm, operating pressure 350 psi), which is equilibrated with 50 m*M* Tris–HCl, pH 7.0, and 1% (w/v) octyl-β-D-glucoside. After loading the concentrated material, elution is performed with 30 ml 50 m*M* Tris–HCl, pH 7.0, and 1% (w/v) octyl-β-D-glucoside in order to exchange Triton X-100 for octyl-β-D-glucoside. Subsequently, a linear ionic strength gradient (total volume 50 ml) is initiated from 50 m*M* Tris–HCl, pH 7.0 to 50 m*M* Tris–HCl, pH 7.0 and 500 m*M* KCl, containing both 1% (w/v) octyl-β-D-glucoside (flow rate 1 ml/min). Cytochrome-*c* oxidase elutes from the column in a narrow band as observed by measuring the absorbance at 280 nm and the enzyme activity. The cytochrome-*c* oxidase fractions are pooled (6 ml).

Gel Filtration. Further purification of cytochrome-*c* oxidase can be achieved by using an HPLC Protein Pak 300-SW column (Waters Associates, 7.5 × 300 mm, operating pressure 350 psi), which is equilibrated with 50 m*M* Tris–HCl, pH 7.0 and 1% (w/v) octyl-β-D-glucoside. Before loading the oxidase sample on this column, the enzyme is concentrated by ultrafiltration with an Amicon B15 filter. Elution during gel filtration is performed with 50 m*M* Tris–HCl buffer, pH 7.0, containing 1% (w/v) octyl-β-D-glucoside (flow rate 1 ml/min).

All HPLC experiments are performed at room temperature.

The purification procedure (Table I) results in a spectral pure cytochrome-*c* oxidase with nmol heme *a* per milligram protein ratios varying from 14 to 16. Only this cytochrome-*c* oxidase preparation is used in the reconstitution experiments described.

Biochemical and Biophysical Properties

Structural Properties

Electrophoretic analysis[21] of the purified denatured cytochrome-*c* oxidase shows three polypeptides with apparent molecular weights of 57,000, 37,000, and 21,000. Native gel electrophoresis[22] of purified cytochrome-*c* oxidase shows one single band staining for *N,N,N',N'*-tetramethyl-*p*-phenylenediamine oxidase activity as well as for heme.[6]

[21] U. K. Laemmli, *Nature (London)* **277,** 680 (1970).

[22] T. Pentillä, M. Saraste, and M. Wikström, *FEBS Lett.* **101,** 295 (1979).

TABLE I
PURIFICATION OF CYTOCHROME-*c* OXIDASE BY AMMONIUM SULFATE PRECIPITATION[a]

Step	Volume (ml)	Protein (mg)	Heme *a* (nmol)	Heme *a* (nmol/mg protein)	Yield (%)
Cytoplasmic membranes	40	402	169.8	0.42	100
Solubilized material	39	290	160.4	0.55	94.4
Ammonium sulfate supernatant (80% saturated)	81	52	118.8	2.28	69.9
PEG-6000 concentrate	7	42.2	97.4	2.31	57.4
Pool DEAE-fractions	6	9.0	58.2	6.47	34.2
Pool Protein-Pak 300 SW fractions	4	1.48	20.9	14.1	12.3

[a] The original material consisted of a 20-liter culture of *B. subtilis* W23, which gave an absorbance at 660 nm of 0.8.

Gel filtration of the purified oxidase using an Ultragel Aca 34 column in the presence of dodecyl-β-D-maltoside results in a single band with an apparent molecular weight of 290,000–315,000. Since hydrophobic membrane-bound proteins in solubilized state are associated with detergent,[23] the amount of bound detergent must be experimentally established and subtracted from the apparent molecular mass.[24]

Crossed immunoelectrophoresis (CIE) is an excellent technique to determine some additional characteristics of (purified) membrane proteins.[25,26] The CIE pattern of the purified cytochrome-*c* oxidase against anti- (*B. subtilis* W23 membrane vesicle) immunoglobulins shows one precipitation line after staining with Coomassie Brilliant Blue. The immunoprecipitate also stains for *N,N,N′,N′*-tetramethyl-*p*-phenylenediamine oxidase activity and heme. Quantitative information about the localization of cytochrome-*c* oxidase in the cytoplasmic membrane can be obtained by using an absorption technique.[27] The accessibility of an enzyme, known to be located at the inner surface of the cytoplasmic membrane, succinate dehydrogenase,[25] is compared with the accessibility of

[23] D. Lichtenberg, R. J. Robson, and E. A. Dennis, *Biochim. Biophys. Acta* **737,** 285 (1983).
[24] M. Suarez, A. R. Revzin, M. Swaisgood, D. A. Thompson, and S. Ferguson-Miller, *Am. Soc. Biol. Chem. Biophys. Soc. Abstr.* 1782 (1983).
[25] J. Bergsma, R. Strijker, J. Y. E. Alkema, H. G. Seijen, and W. N. Konings, *Eur. J. Biochem.* **120,** 599 (1981).
[26] M. G. L. Elferink, K. J. Hellingwerf, P. A. M. Michels, H. G. Seijen, and W. N. Konings, *FEBS Lett.* **107,** 300 (1979).
[27] P. Owen and H. R. Kaback, *Proc. Natl. Acad. Sci. U.S.A.* **75,** 3148 (1978).

cytochrome-*c* oxidase.[6] The fractions accessible from the outer surface of the vesicles were calculated to be 18% for succinate dehydrogenase and 65% for cytochrome-*c* oxidase. From other studies it is already known that at least 80% of the cytoplasmic membranes have the rightside-out orientation.[10,28] These results strongly suggest that cytochrome-*c* oxidase possesses antigenic sites at both sides of the membrane, indicative for a transmembranal localization.

Spectral Properties

The cytochrome content of *B. subtilis* W23 cells varies with the growth conditions and growth stage. The spectral characteristics of the cytochrome present in the cytoplasmic membranes are shown in Fig. 2A. Besides the heme *a* component, several heme *b* and *c* components can be found. The purified enzyme shows spectral characteristics of an aa_3-type oxidase (Fig. 2B), with absorption maxima at 414 and 598 nm in the oxidized form and at 443 and 601 nm in the reduced form. The presence of cytochrome a_3 can be demonstrated by the absolute and difference spectra of the CO-saturated enzyme.[6] The complex of CO and cytochrome-*c* oxidase (in the reduced form) shows peaks at 431 and 598 nm.

Enzymatic Properties

Ionic Strength Effects. Steady-state kinetic analysis of the ferrocytochrome *c* oxidation by cytochrome-*c* oxidase shows that the enzyme exhibits positive cooperativity at low ionic strength with an apparent Hill coefficient of 2.4 (Fig. 3). Increasing the KCl concentration from 0 to 25 m*M* does not affect the kinetic parameters V_{max} and K_m but causes loss of the cooperative behavior. In addition, it is observed that up to 75 m*M* KCl, the maximum turnover number of both the membrane solubilized and the purified enzyme is unaffected by ionic strength. The Michaelis constant, on the other hand, increases with increasing ionic strength (from about 25 m*M* KCl), pointing to an electrostatically governed binding reaction between cytochrome *c* and cytochrome-*c* oxidase, as has also been reported for the bovine heart[29] and the *Thermus thermophilus*[30] enzyme.

The occurrence of the biphasic pattern in the Hill plot may be interpreted in terms of two populations of cytochrome-*c* oxidase with one and two cytochrome *c* binding sites, respectively (Fig. 3). Alternatively, it

[28] W. N. Konings, *Arch. Biochem. Biophys.* **167,** 570 (1975).

[29] J. Wilms, E. C. I. Veerman, B. W. König, H. L. Dekker, and B. F. van Gelder, *Biochim. Biophys. Acta* **635,** 13 (1981).

[30] T. Yoshida and J. A. Fee, *J. Biol. Chem.* **259,** 1031 (1984).

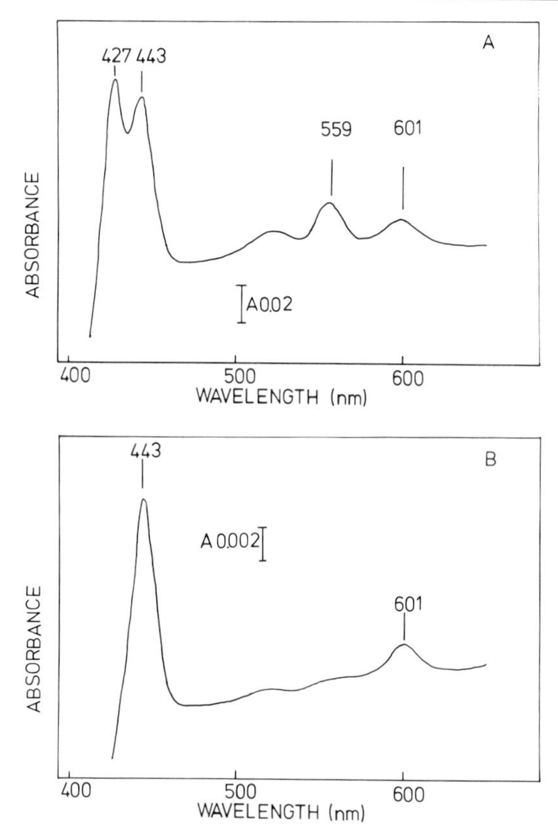

FIG. 2. Absorption spectra of solubilized membranes and purified cytochrome-*c* oxidase of *B. subtilis* W23. (A) Dithionite-reduced-minus-oxidized spectrum of a dodecyl-β-D-maltoside cytoplasmic membrane extract, containing 0.8 μM cytochrome-*c* oxidase. (B) Dithionite-reduced-minus-oxidized spectrum of 0.12 μM purified cytochrome-*c* oxidase. Spectra were recorded at room temperature using an Aminco DW-2A spectrophotometer at a scan speed of 2 nm/sec and bandwidth of 3 nm.

may be suggested that under this condition two cytochrome *c* binding sites exist on the oxidase exhibiting positive cooperativity.[31]

It is suggested that in the association/dissociation reaction of cytochrome-*c* oxidase and cytochrome *c*, charges of opposite signs are important.[32] In fact, when the ascorbate–cytochrome *c*–cytochrome-*c* oxidase activity is assayed, turnover numbers of ~150 sec^{-1} can be measured,

[31] S. Ferguson-Miller, D. L. Brautigan, and E. Margoliash, *J. Biol. Chem.* **251,** 1104 (1976).
[32] J. Wilms, J. M. L. L. van Rijn, and B. F. van Gelder, *Biochim. Biophys. Acta* **593,** 17 (1980).

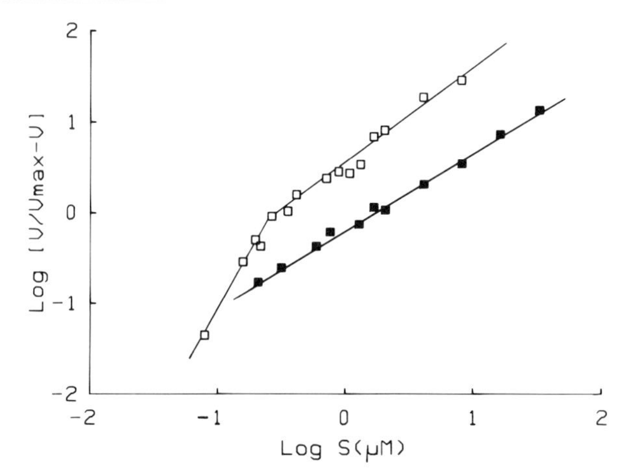

FIG. 3. Effect of ionic strength on the kinetics of oxidation of ferrocytochrome *c* by purified cytochrome-*c* oxidase. The data are reported as Hill plots. The activity measurements were carried out spectroscopically with *S. cerevisiae* cytochrome *c* (type VIII) concentrations ranging from 0.08 to 33.5 μM. The cuvette contained 2.0 ml of 10 mM Tris–HCl, pH 7.4, containing 0.1% (w/v) dodecyl-β-D-maltoside, 4.9 pmol purified heme aa_3 oxidase with (■) and without (□) 50 mM KCl.

indicating that in the spectroscopic assay the binding and/or dissociation of cytochrome *c* and cytochrome-*c* oxidase is the rate-limiting step. In the polarographic assay the highest turnover rates are obtained at low ionic strength and relatively high pH,[6] thus favoring the tight binding of cytochrome *c* to the oxidase[29,33] and permitting rapid reduction of the cytochrome *c*–cytochrome-*c* oxidase complex by ascorbate.

Electron Donors. The effects of ionic strength on the steady-state kinetics of cytochrome-*c* oxidase are the same with both *S. cerevisiae* (type VIII) and horse heart (type VI) cytochrome *c*, except that with the latter enzyme the turnover rates are about 5 times lower. Efficient artificial electron-donating systems for cytochrome-*c* oxidase are found in ascorbate-*N*,*N*,*N′*,*N′*-tetramethyl-*p*-phenylenediamine (TMPD) and ascorbate-hexaamine ruthenium. Under these conditions the oxidase activity is insensitive to ionic strength.

Inhibitors. Typical inhibitors of cytochrome-*c* oxidase such as cyanide and azide block the purified *B. subtilis* enzyme for 50% at 1 and 80 μm, respectively. Tenfold higher concentrations of both inhibitors are required for the same level of inhibition of cytochrome-*c* oxidase in membrane vesicles.

[33] S. P. G. Brocks and P. Nicholls, *Biochim. Biophys. Acta* **680,** 33 (1982).

Role of Detergents and Phospholipids. Cytochrome-*c* oxidase isolated with the affinity column in the presence of 0.1% (w/v) dodecyl-β-D-maltoside has a maximum turnover number of 8–12 sec^{-1} in the spectroscopic assay. From the detergents tested, i.e., octyl-β-D-glucoside, cholate, lauryldimethylammonium *N*-oxide, and Tween 80, only the latter gives rise to higher activities (50–60%) than dodecyl-β-D-maltoside. In general, the highest activities are found when the detergents are used in concentrations just above the critical micellar concentration (CMC). Further stimulation of the cytochrome-*c* oxidase turnover rate is observed in the presence of negatively charged phospholipids such as cardiolipin and phosphatidylserine.

Reconstruction

Procedure. To determine the energy-transducting capacity of cytochrome-*c* oxidase, it is essential to reconstitute the purified enzyme into artificial membranes. The efficiency of energy transduction is strongly dependent on the incorporation of the enzyme in the artificial membrane. Several factors influence the incorporation, e.g., the phospholipid composition of the membrane and the protein-to-lipid ratio used for reconstitution. Reconstitution with the affinity column purified enzyme has not been successful until now, and thus only the reconstitution will be described using the enzyme prepared by the ammonium sulfate precipitate, after DEAE and gel filtration chromatography.

Cytochrome-*c* oxidase of *B. subtilis* W23, isolated as described earlier, is used at concentrations ranging from 3 to 5 nmol heme *a*/ml 50 m*M* Tris–HCl, pH 7.0, containing 250–300 KCl and 1% (w/v) octyl-β-D-glucoside. To reconstitute cytochrome-*c* oxidase, varying amounts of different phospholipids in chloroform/methanol are dried to a film and subsequently suspended in 0.5 ml of 10 m*M* HEPES-KOH, pH 7.0, 25 m*M* KCl, 30 m*M* choline chloride, and 0.7% (w/v) octyl-β-D-glucoside (final lipid concentrations 10–20 mg/ml). This mixture is sonicated to clarity at 0° under nitrogen. To this varying amounts of purified enzyme are added and the mixture (total volume 1 ml) is dialyzed against 1 liter of 10 m*M* HEPES-KOH, pH 7.0, containing 25 m*M* KCl and 30 m*M* choline chloride, for 4 hr at 4° with rapid stirring. The buffer is changed after 4 and 8 hr of dialysis.

Orientation of Cytochrome-c Oxidase. In order to determine the orientation of cytochrome-*c* oxidase in the artificial membrane (i.e., location of cytochrome *c* binding site), the hemes of externally facing oxidase molecules are reduced by ascorbate and cytochrome *c*, followed by the addition of the membrane permeant *N,N,N′,N′*-tetramethyl-*p*-phenylenediamine to the system to reduce the hemes of the inwardly facing

TABLE II
EFFECT OF PHOSPHOLIPID COMPOSITION ON RECONSTITUTION AND ORIENTATION OF CYTOCHROME-*c* OXIDASE

Lipids[a]	Nanomole heme *a*/mg lipid	Orientation[b]	RCI[c]
PC : CL (1 : 1)[d]	0.4	40	2.55
PC	0.4	60	1.3
CL	0.4	35	1.75
PE : PG (1 : 1)[d]	0.3	65	2.1
PE : PG : CL (2 : 5.5 : 2.5)[d]	0.2	75	2.75

[a] PC, phosphatidylcholine; CL, cardiolipin; PE, phosphatidylethanolamine; PG, phosphatidylglycerol.
[b] Orientation: Defined as percentage cytochrome-*c* oxidase molecules with the cytochrome *c* binding site accessible to cytochrome *c*.
[c] RCI: Respiratory Control Index.
[d] Mole ratio.

molecules. To prevent reoxidation of the reduced hemes, cyanide has to be added before the reductants.[34]

The phospholipid composition of the artificial membrane influences the incorporation of the cytochrome-*c* oxidase (Table II). The highest net incorporation (i.e., cytochrome *c* binding site facing outward) and respiratory control index for oxidase preparations is achieved when membranes were used, which had a phospholipid composition comparable to the *B. subtilis* cytoplasmic membrane (i.e., phosphatidylethanolamine/phosphatidylglycerol/cardiolipin in a mole ratio of 2 : 5.5 : 2.5). Relatively high respiration control indices are measured in mixtures of neutral and negatively charged phospholipids.

Another phenomenon which should be mentioned is the increase of cytochrome-*c* oxidase molecules with cytochrome *c* binding sites facing outward with increasing protein-to-lipid ratios, used during liposome formation. This is connected with an increase of respirating control index.

Electrical Potential Measurements. To quantitate the electrical potential (inside negative) generated across the cytochrome-*c* oxidase containing liposomal membrane after addition of the electron donor system ascorbate-cytochrome *c*, the uptake of the lipophilic cation tetraphenylphosphonium (TPP^+) can be followed with a tetraphenylphosphonium-sensitive electrode[35] (Fig. 4). Cytochrome-*c* oxidase (0.4 nmol heme *a*/mg

[34] R. P. Casey, B. H. Ariano, and A. Azzi, *Eur. J. Biochem.* **122,** 313 (1982).
[35] J. S. Lolkema, K. J. Hellingwerf, and W. N. Konings, *Biochim. Biophys. Acta* **681,** 85 (1982).

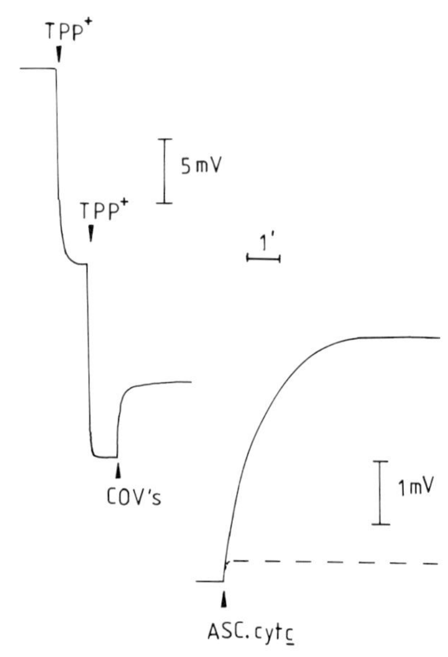

FIG. 4. Registration of tetraphenylphosphonium uptake in cytochrome-*c* oxidase containing liposomes after addition of the electron donor system ascorbate-cytochrome *c*. The incubation medium (2 ml) was composed of air-saturated 10 m*M* HEPES-KOH, pH 7.0, containing 25 m*M* KCl, 30 m*M* choline chloride, and 4 μM tetraphenylphosphonium. At the time indicated by the arrow, 200 μl cytochrome-*c* oxidase vesicles (COVs), prepared as described, were added to the incubation medium. After stabilization of the signal the sensitivity was scaled up, and, at the time indicated, 10 μl 1 *M* K-ascorbate, pH 7.0, and yeast cytochrome *c* were added to a final concentration of 5 m*M* and 10 μM, respectively. The dashed line represents the signal when COVs were preincubated with CCCP (carbonyl cyanide *m*-chlorophenylhydrazone) (final concentration 4 μM) or valinomycin (final concentration 0.8 μM). During the experiment the temperature was kept at 20°. The electrical potential was calculated assuming an internal volume of 5 μl/mg phospholipid.

lipid) is reconstituted with membranes composed of phosphatidylcholine and cardiolipin (mole ratio 1 : 1), essentially as described previously). During dialysis, however, 9 m*M* $CaCl_2$ is included in the dialysis buffer to initiate membrane fusion.[36] This procedure result is well-coupled relatively large unilamellar liposomes (respiratory control index varying from 2.4 to 2.7).

The magnitude of the membrane potential is calculated with the Nernst equation. For $\lambda\psi$ measurements a correction for tetraphenylphosphonium binding must be applied.[35] Correction for probe binding to the artificial membrane is based on the assumption that all binding to the

[36] J. Wilschut, M. Holsappel, and R. Jansen, *Biochim. Biophys. Acta* **690,** 297 (1982).

membrane occurs symmetrically and is dependent on both external and internal probe concentration. Information concerning the reliability of the method used can be obtained from K^+-diffusion potential measurements.[37] An excellent correlation can be observed between imposed and calculated potentials in the liposome preparations. A slight deviation of the measured potential relative to the calculated one is only observed with relatively high imposed membrane potentials (i.e., >120 mV).

In cytochrome-*c* oxidase vesicles membrane potentials can be determined varying from 30 to 40 m*M* (inside negative). A slight increase of the potential can be observed with increasing pH (varying from pH 5 to pH 8). Addition of nigericin, which catalyzes an electroneutral exchange of protons and potassium and should thus dissipate the proton gradient, leads to an increase of the membrane potential.

[37] S. Schuldiner and H. R. Kaback, *Biochemistry* **14,** 5451 (1976).

[17] Preparation of Complex III from Yeast Mitochondria and Related Methodology

By LIVIU CLEJAN and DIANA S. BEATTIE

Introduction

Complex III, also called the cytochrome bc_1 complex, is a multiprotein enzyme complex that catalyzes a ubiquinol–cytochrome-*c* oxidoreductase activity with a concomitant unidirectional and electrogenic movement of protons across the inner mitochondrial membrane. The isolation and purification of complex III from beef heart mitochondria was pioneered by Hatefi *et al.*,[1] Rieske,[2] and Yu *et al.*[3] These procedures are based on the extraction of the complex from the inner membrane in the presence of bile salts and subsequent fractionation with ammonium sulfate. Later, these procedures were adapted for the purification of the complex from other sources such as rat liver,[4] yeast,[5–8] and *Neurospora*

[1] Y. Hatefi, A. G. Haavik, and D. E. Griffiths, *J. Biol. Chem.* **237,** 1681 (1962).
[2] J. S. Rieske, this series, Vol. 10, p. 239.
[3] C.-A. Yu, L. Yu, and T. E. King, *J. Biol. Chem.* **249,** 4905 (1974).
[4] P. Gellerfors, T. Johansson, and B. D. Nelson, *Eur. J. Biochem.* **115,** 275 (1981).
[5] M. B. Katan, L. Pool, and G. S. P. Groot, *Eur. J. Biochem.* **65,** 95 (1976).
[6] J. N. Siedow, S. Power, F. F. de LaRosa, and G. Palmer, *J. Biol. Chem.* **253,** 2392 (1978).
[7] D. S. Beattie, C. A. Battie, and R. A. Weiss, *J. Supramol. Struct.* **14,** 139 (1980).
[8] A. Sidhu and D. S. Beattie, *J. Biol. Chem.* **257,** 7879 (1982).

crassa[9] as well as the analogous cytochrome b_6f complex from chloroplasts[10–12] and photosynthetic bacteria.[13,14]

In the first reported purification of complex III from yeast mitochondria,[5] antimycin was added during the isolation procedure to prevent the cleavage of cytochromes *b* and c_1. Consequently, the complex obtained was without enzymatic activity, but was suitable for chemical characterization. An enzymatically active complex III was isolated from yeast mitochondria by Siedow *et al.*,[6] who introduced potassium cholate as a solubilizing agent. More recently, new approaches for the isolation and purification of complex III from various species have been published. These procedures include the use of nonionic detergents such as Triton X-100 for the extraction of the complex from the membrane and chromatography on hydroxyapatite columns,[15] immobilized cytochrome *c*-Sepharose 4B columns,[16] or cytochrome *c*–thiol-activated Sepharose columns[17] for fractionation of the complex. Unfortunately, the complexes obtained using these methods often have lost varying amounts of the different subunits of the complex.

In this chapter are presented the procedures for purification of complex III from yeast mitochondria, for biochemical characterization of the soluble complex, and for the study of the bioenergetic properties of the complex reconstituted into proteoliposomes.

Large-Scale Preparation of Complex III

Reagents

KPEP buffer
- 0.9% KCl
- 50 m*M* K_2HPO_4
- 1 m*M* Na_2 EDTA
- 0.5 m*M* PMSF (phenylmethylsulfonyl fluoride) added prior to use, as a 100 m*M* stock solution in DMSO (dimethyl sulfoxide)

[9] H. Weiss and H. J. Kolb, *Eur. J. Biochem.* **99,** 139 (1979).
[10] N. Nelson and J. Neumann, *J. Biol. Chem.* **247,** 1917 (1972).
[11] P. M. Wood and D. S. Bendall, *Eur. J. Biochem.* **61,** 337 (1976).
[12] E. Hurt and G. Hauska, *Eur. J. Biochem.* **117,** 591 (1981).
[13] M. Krinner, G. Hauska, E. Hurt, and W. Lockau, *Biochim. Biophys. Acta* **681,** 110 (1982).
[14] N. Gabellini, J. R. Bowyer, E. Hurt, B. A. Melandri, and G. Hauska, *Eur. J. Biochem.* **126,** 105 (1982).
[15] G. von Jagow, H. Schagger, W. D. Engel, P. Riccio, H. J. Kolb, and M. Klingenberg, this series, Vol. 53, p. 92.
[16] H. Weiss and H. J. Kolb, *Eur. J. Biochem.* **99,** 139 (1979).
[17] K. Bill, C. Broger, and A. Azzi, *Biochim. Biophys. Acta* **679,** 28 (1982).

SKEP buffer
- 0.25 *M* sucrose
- 0.01 *M* K_2HPO_4
- 1 m*M* Na_2 EDTA
- 0.5 m*M* PMSF (added fresh as stated above)

KEP buffer
- 0.1 *M* K_2HPO_4, pH 7.4
- 1 m*M* Na_2 EDTA
- 0.5 m*M* PMSF (added fresh)

TASTP buffer
- 40 m*M* Tris-acetate, pH 7.0
- 5% sucrose
- 0.05% Triton X-100
- 0.2 m*M* PMSF (added fresh)

KTD buffer
- 0.1 *M* K_2HPO_4, pH 7.4
- 0.1% Triton X-100
- 0.1% K-deoxycholate

K-cholate, 20%

Ammonium sulfate, analytical reagent grade, finely ground

KOH, 4 *N*

Choice of Yeast Strain

Subtle differences in the various strains and/or batches of commercial yeast can affect the reproducibility and the quality of the complex III obtained. The primary criterion for the choice of a specific strain of yeast is the specific content of cytochrome *b* and c_1. Currently, in our laboratory Red Star yeast has been used successfully for the purification of complex III. The fresh pressed (but not dried) yeast cells must be kept at 6–8° and processed within 6 days of purchase. Cells kept for longer periods of time become progressively richer in lipids with a detrimental effect on the yield and purity of the mitochondria obtained. Before breakage, the pressed cells are crumbled manually into ~5 mm pieces, suspended in KPE buffer without PMSF in a ratio of 450 ml of buffer per pound of yeast, and incubated at 20° overnight with constant aeration of the solution. This procedure typically increases 3–4 times the yield of mitochondria per milligram of cell protein.

Cell Breakage

The yield of broken cells and the quality of the mitochondria obtained are the maximum when the cells are broken with the Dyno-Mill Type

KD-L disintegrator as compared with those obtained with either the Manton-Gaulin homogenizer or with a Waring blender and liquid nitrogen. Currently, in our laboratory, 4–6 lb of yeast can be processed to mitochondria in 1 day using the Dyno-Mill. The suspension of yeast cells after overnight aeration is cooled in an ice bath and 0.5 m*M* PMSF (in DMSO) is added. Either a continuous pumping of the suspension into the chamber and cell breakage can be performed as described by the manufacturer or a discontinuous method can be used, depending on the cooling capacity of the system. It is essential to avoid overheating of the cell suspension during breakage. The suspension is maintained at a temperature of 13° or less as it leaves the chamber. For the discontinuous method, ~500 ml of the cell suspension is pumped into the chamber, which contains 500 ml of glass beads (0.45–0.50 mm in diameter) and has been previously cooled to −10°. The cells are broken 4 times for 60 sec at a stirring shaft speed of 2000 rpm and pumped out of the chamber. All subsequent steps are performed at 4–6°.

Preparation of Mitochondria and SMP

The broken-cell suspension is diluted 1 : 1 with KPEP buffer and centrifuged at 2600 *g* for 10 min. The supernatant is then brought back to the original volume and recentrifuged twice at 2600 *g* for 10 min. The subsequent supernatant is centrifuged at 16,000 *g* for 25 min. The mitochondrial pellets are washed twice with KPEP buffer (homogenized in glass–Teflon homogenizer). The final pellets are resuspended in KPEP buffer to a final protein concentration of 40–50 mg/ml and stored in plastic bottles at −20°. Usually, 3–3.5 g mitochondrial protein are obtained from 1 lb of pressed yeast.

The mitochondria are pooled, rehomogenized, and diluted with KPEP buffer to a concentration of 14–16 mg/ml prior to sonication. Sonication is performed on 50-ml batches at 4° with a Q Horn Flat tip, in a "Rossette" cup, for 3 × 45 sec at maximum output with 45-sec pauses. The sonicated mitochondria are centrifuged twice at 3000 *g* for 10 min, and then at 106,000 *g* for 30 min. The pellets (submitochondrial particles) are resuspended in SKEP buffer to one-half the original volume, using a motor-driven glass–Teflon homogenizer. The suspension is recentrifuged and resuspended to a final protein concentration of 27 mg/ml. Starting material suitable for purification should contain at least 0.275 nmol cytochrome *b*/mg protein.

Potassium Cholate Solubilization and Ammonium Sulfate Fractionation

Potassium cholate (20%) is added to a final ratio of 0.6 mg/mg protein, and the final volume (V_i) is recorded. The mixture is stirred at 4° for 10 min; then solid ammonium sulfate (114 g/liter) is slowly added with constant stirring at 4°. The pH is adjusted to 7.2–7.3 with 4 *N* KOH.

After stirring overnight at 4°, the ammonium sulfate is brought to 35% saturation by slowly adding 91 g/liter of V_i. The mixture is stirred for 30 min and centrifuged at 18,500 *g* for 40 min. All subsequent ammonium sulfate additions are done with stirring at 4° for 15 min and centrifugations at 18,500 *g* for 40 min. The supernatant is brought to 45% ammonium sulfate saturation (63 g/liter of actual volume), stirred, and centrifuged. The resulting greenish-brownish pellet (consisting mainly of cytochrome-*c* oxidase) is discarded. The supernatant is filtered through two layers of cheesecloth to remove the layer of floating lipid, the ammonium sulfate is brought to 50% saturation (32 g/liter of actual volume), and the pH is adjusted to 8.0. The mixture is stirred and centrifuged. (Starting at this step, the purification of the complex can be easily monitored by spectral analysis.) The resulting pellet is discarded and the ammonium sulfate brought to 52.5% saturation (33 g/liter of actual volume). The stirring and centrifugation steps are repeated as described above, and the supernatant brought to 55% saturation (33 g/liter of actual volume). After stirring and centrifugation, as described above, the supernatant is discarded and the pink precipitate (sometimes appearing as a slanted floating pellet) is resuspended with a homogenizer (glass–glass) in KEP buffer to 15% of V_i.

In the following steps, the bc_1 complex is partially separated from succinate dehydrogenase and, hence, assay of the latter activity will indicate the degree of purification. A saturated solution of ammonium sulfate is added to the suspension to 40% saturation and the pH is maintained above 7.0. The mixture is stirred and centrifuged for 30 min; ammonium sulfate is added to the supernatant to 45% saturation. The stirring and centrifugation steps are repeated as described above. The resulting pellet is rich in SDH and is discarded. The cherry red bc_1 complex is precipitated between 45 and 55% ammonium sulfate, depending on the preparation of mitochondria.

For further separation of the complex from high-molecular-weight contaminants and to remove ammonium sulfate, the pellet is dissolved in TASTP buffer and chromatographed twice on an Ultrogel AcA 34 column (1.2 × 50 cm), preequilibrated with the same buffer containing 0.1% sodium azide.[8] The fractions containing complex III, determined by the absorbance at 420 nm, are concentrated using Diaflo XM300 ultrafilters and then stored in small aliquots at −70°.

As an alternative to the Ultrogel chromatography, desalting can be achieved by homogenization of the bc_1 complex in 0.1 *M* K_2HPO_4, pH 7.4, 1 m*M* EDTA, and centrifugation in 65 ml of the same buffer at 106,000 *g* for 3 hr. The resulting red-pink pellet is solubilized in a small volume of KTD buffer and stored as described above.

Usually, 90 mg of bc_1 complex can be obtained from 50 g of mitochondrial protein.

Enzymatic Assays

The ubiquinol–cytochrome-*c* oxidoreductase is assayed in a 1 ml cuvette containing 500 μM cytochrome *c*, 0.2 mM EDTA, 5 mM sodium azide, 1 g percent bovine serum albumin (Sigma Fraction V), in 25 mM sodium phosphate, pH 7.6. (The sodium azide can be omitted if there is a complete absence of cytochrome-*c* oxidase in the complex III preparation.) After adding the appropriate quantity of complex III (0.5–2 μl) to the reaction mixture, a baseline rate is recorded at 550 nm for 1 min. The reaction is started with the addition of 150 μM of the quinol analog in methanol. The sensitivity to antimycin is checked by adding 1 μM of the inhibitor. The specific activity is calculated based on cytochrome c = 18.5 mM^{-1} cm^{-1}.

Succinate dehydrogenase is assayed using succinate as the substrate, dichlorophenolindophenol as the electron acceptor, and phenazine methosulfate as an intermediate electron carrier.[18]

Difference spectra are recorded on a dual beam/dual wavelength spectrophotometer Perkin-Elmer Model 557.[7] Complex III is diluted 1 : 50 (v : v) in 0.1 M potassium phosphate, pH 7.5. It is essential that the complex in the reference cuvette is first oxidized with ferricyanide.

A modified Lowry reagent containing 0.5% SDS is used for protein determination of the Triton-solubilized complex,[19] with bovine serum albumin as a standard.

Properties of the Soluble Complex

The polypeptide composition of the complex is determined by SDS–gel electrophoresis of the purified complex, dissociated in 5% SDS.[8] The apparent molecular weights of the subunits calculated from their mobility on a 15% acrylamide gel using the standards obtained from Pharmacia are I = 49,000; II = 39,000; III = 34,000; IV = 29,000; V = 22,400; VI = 13,400; VII = 11,100; VIII = 10,000; and IX = 9000.[20]

The p*I* of the subunits obtained from isoelectric focusing[8] are I = 6.53; II = 6.37; III = 8.30; IV = 6.45; V = 6.57; VI = 6.20; and VII = 7.15.

The enzymatic activity of the complex with DBH_2 as substrate averaged 15–17 nmol of cytochrome *c* reduced per minute per milligram protein in several purifications.

[18] I. C. Kim and D. S. Beattie, *Eur. J. Biochem.* **36,** 509 (1973).
[19] J. R. Dalley and P. A. Grieve, *Anal. Biochem.* **64,** 136 (1975).
[20] D. S. Beattie, L. Clejan, and C. G. Bosch, *J. Biol. Chem.* **259,** 10526 (1984).

The contents of cytochromes b and c_1 are generally 7.0 nmol mg^{-1} protein and 4.2 nmol mg^{-1} protein, respectively, using the absorption of the α-band (cytochrome b = 25.6 mM^{-1} cm^{-1} and cytochrome c_1 = 20.9 mM^{-1} cm^{-1}).

Based on the specific heme content and the cytochrome c reductase activity, a 52- to 56-fold purification of complex III from mitochondria is achieved.

Bioenergetic Properties of Complex

Reconstitution of Complex III into Liposomes and Preparation of Proteoliposomes for H^+ Ejection and Respiratory Control Measurements

A modified cholate-dialysis method is used as follows: 160 mg phosphatidylcholine (soybean, type IIS, Sigma) are dissolved in 2.5 ml of chloroform, in a 15-ml conical tube. The solvent is evaporated to dryness under a stream of O_2-free N_2 by continuously rotating the tube in a slanted position. A homogenous film formed on the wall of the tube is left, under vacuum, in a desiccator for at least 3 hr. The phospholipid film is subsequently dispersed by vortexing in 8 ml of 100 mM KCl, 3 mM K^+ HEPES, pH 7.2, containing 40 mg K-cholate and 0.01 mg hydroxybutylated toluene (methanolic solution, 100 mg/ml) as an antioxidant. Sonication under N_2 is performed using a Heat Systems, Inc. Ultrasonics microtip at an output control setting of 1–2, in 30-sec bursts, with intermittent cooling in ice, until the suspension becomes translucent. Complex III, containing 3.5–4.5 nmol heme b, is added to the suspension. Dialysis of 4-ml aliquots in Spectropor dialysis tubing (diameter 7.6 mm; molecular weight cutoff: 2000) at 4–6° is performed for 16 hr against 800 ml of 100 mM KCl, 3 mM K^+ HEPES, pH 7.2, with 4 changes of buffer.[21]

The rates of electron flow and respiratory control in the reconstituted complex III are determined by measuring ferricyanide reductase at 420–460 nm in a dual-wavelength spectrophotometer.[22] The reaction mixture contains 30 μg of cytochrome c, 7.5 mg liposomes, and 200 μM potassium ferricyanide in 3 ml of 100 mM KCl, 3 mM K^+ HEPES, pH 7.2. The reaction is started by the addition of 100–195 μM reduced Q analog. The respiratory control rate is expressed as the ratio of the rate of reduction of ferricyanide in the presence of 2.6 μM CCCP to that in its absence. The enzymatic specificity of the reaction is tested by addition of 0.66 μM antimycin.

[21] Y. Kagawa and E. Racker, *J. Biol. Chem.* **246,** 5477 (1971).

[22] D. S. Beattie and A. Villalobo, *J. Biol. Chem.* **247,** 14745 (1982).

Proton movements elicited by the reconstituted complex are measured with a combination of a small diameter Beckman glass pH electrode and a Sargent-Welch recorder Model DSRG-2, in a 3-ml chamber equipped with a magnetic stirrer, at 25°.[22] The assay medium contains 100 m*M* KCl, 1 m*M* K^+ HEPES, pH 7.2, 30 mg proteoliposomes (150 μg complex III), 20 μg cytochrome *c*, 1 μg/ml^{-1} valinomycin, and 1.07 m*M* DBH_2. The extent of acidification is determined by the addition of a pulse of 20–50 nmol ferricyanide and calculated from a known quantity of HCl added after each experiment. The ratio $H^+/2e^-$ is calculated by dividing the total nanograms of H^+ ejected by the equivalent number of nanomoles ferricyanide reduced by $2e^-$. The scalar H^+ ejected is obtained in the presence of CCCP (4μ*M*). An $H^+/2e^-$ ratio of 3.75 is usually obtained.[22]

The Binding of DCCD to Cytochrome b in Complex III

Complex III (2–3 nmol cytochrome *b*) is incubated with 50–100 nmol [^{14}C]DCCD/nmol heme *b* in 1 ml of 100 m*M* KCl, 3 m*M* K^+ HEPES, pH 7.2, for 1–3 hr at 12°.[20,23] The mixture is subsequently centrifuged through 8 ml of 10% sucrose at 106,000 *g* for 3 hr. The resulting pellet is dissociated overnight, at room temperature, in 100 μl of 5% SDS, 2 m*M* EDTA, 50 m*M* Tris–HCl, pH 6.8, 10% glycerol, 5% 2-mercaptoethanol. Aliquots are quantitatively analyzed on 15% acrylamide-SDS gels, either by slicing the gel and counting the radioactivity,[23] or by fluorography-autoradiogram.[20] A preferential binding of DCCD to cytochrome *b* is observed under these conditions.

The Effect of DCCD on H^+ Ejection and Electron Flow

An aliquot of soluble complex III (7–8 nmol cytochrome *b*) is incubated with 1400–1600 nmol DCCD (1 m*M* freshly prepared methanolic solution) in 1 ml of 100 m*M* KCl, 3 m*M* K^+ HEPES, pH 7.2, for 2 hr at 12–15° prior to incorporation into liposomes, as described above.[24] Alternatively, liposomes containing complex III are incubated for 3 hr with DCCD at 20° with 100 nmol of DCCD/nmol of cytochrome *b*. Removal of the unbound DCCD was achieved by centrifugation of the DCCD-treated liposomes through a 1-ml column of Sephadex G-50 previously equilibrated with the dialyzing buffer. The column is centrifuged for 2 min at 800 rpm in a swing-out bucket Sorvall GLC-1, Type HL-4 rotor.[24] At a ratio of 200 nmol DCCD/nmol heme *b*, the electrogenic H^+ ejection decreases by 60–80%, while the electron flow is decreased by a maximum of 26%.[24]

[23] D. S. Beattie and L. Clejan, *FEBS Lett.* **149,** 245 (1982).
[24] L. Clejan and D. S. Beattie, *J. Biol. Chem.* **258,** 14271 (1983).

[18] Purification of Cytochrome bc_1 Complexes from Phylogenically Diverse Species by a Single Method

By PER O. LJUNGDAHL, JEFFREY D. PENNOYER, and BERNARD L. TRUMPOWER

Introduction

Cytochrome bc_1 complexes have been purified from a variety of sources over the past 20 years. The majority of published methods rely on detergent extraction of isolated membranes and subsequent salt precipitations. These methods require a series of tedious manipulations that are difficult to reproduce, possibly due to differing lipid content of the starting material, and often yield preparations with low enzymatic activity.

Recently, cytochrome bc_1 complexes have also been purified from Triton X-100 solubilized membranes by a variety of chromatographic procedures. Though easily reproduced, these preparations exhibit reduced enzymatic activity.

The nonionic alkyl glycoside detergent, dodecyl maltoside, has proved to be an effective dispersing agent of a multitude of energy-transducing membranes.[1] Utilizing this detergent, we have developed a general method which uses anion-exchange chromatography to isolate bc_1 complexes from yeast and beef heart mitochondria, and the photosynthetic bacterium *Rhodopseudomonas sphaeroides*. With slight modifications, the general method is also suitable for purification of the bc_1 complex from *Paracoccus denitrificans* (Yang and Trumpower, this volume [29]). This method is easily reproduced and yields preparations which are pure and highly active. It is likely that with minor modifications this chromatographic method using dodecyl maltoside will be applicable to species other than those described here.

Materials and Analytical Methods

Serum albumin (essentially fatty acid free), horse heart cytochrome *c*, and DEAE-Sepharose CL-6B are obtained from Sigma. DEAE-BioGel A and sodium dodecyl sulfate are obtained from Bio-Rad. Dodecyl maltoside is obtained from Boehringer, Mannheim. This detergent was initially purchased from Calbiochem; however, we switched sources when subse-

[1] P. Rosevear, T. van Aken, J. Baxter, and S. Ferguson-Miller, *Biochemistry* **19**, 4108 (1980).

quent batches contained a yellow-tinted impurity, possibly lauryl alcohol, which adversely affected enzymatic activity. All other reagents were of the highest grade available.

Protein determinations were made according to the method of Lowry as modified by Markwell *et al.*[2] Cytochrome concentrations of the bc_1 complexes are determined spectrophotometrically using oxidation–reduction difference spectra as described by Vanneste.[3] Ubiquinol–cytochrome *c* oxidoreductase and cytochrome *c* oxidase activities are assayed as previously described (Berry and Trumpower, this volume [28]).

SDS–polyacrylamide gel electrophoresis is carried out according to Laemmli,[4] except that SDS is omitted from the gel and lower electrode buffer. Linear gradient gels (13–17%) are made from a stock acrylamide solution of 30% T, 2.7% C. Individual samples were denatured as described in the figure legend.

Yeast submitochondrial particles are prepared from commercially available Red Star yeast according to Siedow *et al.*,[5] with the following modifications. The concentration of phenylmethylsulfonyl fluoride is increased from 0.5 to 1 m*M* in all buffers. In addition, 0.5 m*M* diisopropyl fluorophosphate and 4 m*M* potassium fluoride are added to the suspended cells just prior to cell disruption.

Beef heart mitochondria are prepared according to Smith.[6] Both yeast and beef heart membranes are stored at −70° at a protein concentration of 60 mg/ml.

Rhodopseudomonas sphaeroides wild-type strain NCIB 8253 are grown phototrophically in the medium of Cohen-Bazire *et al.*[7] at 30° in sealed 1-liter Rous bottles to mid-logarithmic growth. Light is provided by 75-W flood lamps at 1800 lux. French pressure cell extracts are prepared from washed cells, centrifuged at 10,000 *g* for 10 min, and the supernatant material is subjected directly to rate-zone sedimentation on sucrose density gradients, as described by Niederman *et al.*,[8] on linear 5–35% (w/w) sucrose gradients over a 60% (w/w) cushion and centrifuged at 96,000 *g* for 4 hr. The chromatophore fraction is removed with a Pasteur pipette, diluted 1 : 1 with 1 m*M* Tris–HCl, pH 7.5, and sedimented at 368,000 *g* for

[2] M. K. Markwell, S. M. Haas, L. L. Bieber, and N. E. Tobert, *Anal. Biochem.* **87,** 206 (1978).

[3] W. H. Vanneste, *Biochim. Biophys. Acta* **113,** 175 (1966).

[4] U. K. Laemmli, *Nature (London)* **227,** 680 (1970).

[5] J. N. Siedow, S. Power, F. F. De La Rose, and G. Palmer, *J. Biol. Chem.* **253,** 2392 (1978).

[6] A. L. Smith, this series, Vol. 10, p. 81.

[7] G. Cohen-Bazire, W. R. Sistrom, and R. Y. Stanier, *J. Cell. Comp. Physiol.* **49,** 25 (1957).

[8] R. A. Niederman, D. E. Mallon, and J. J. Langen, *Biochim. Biophys. Acta* **440,** 429 (1976).

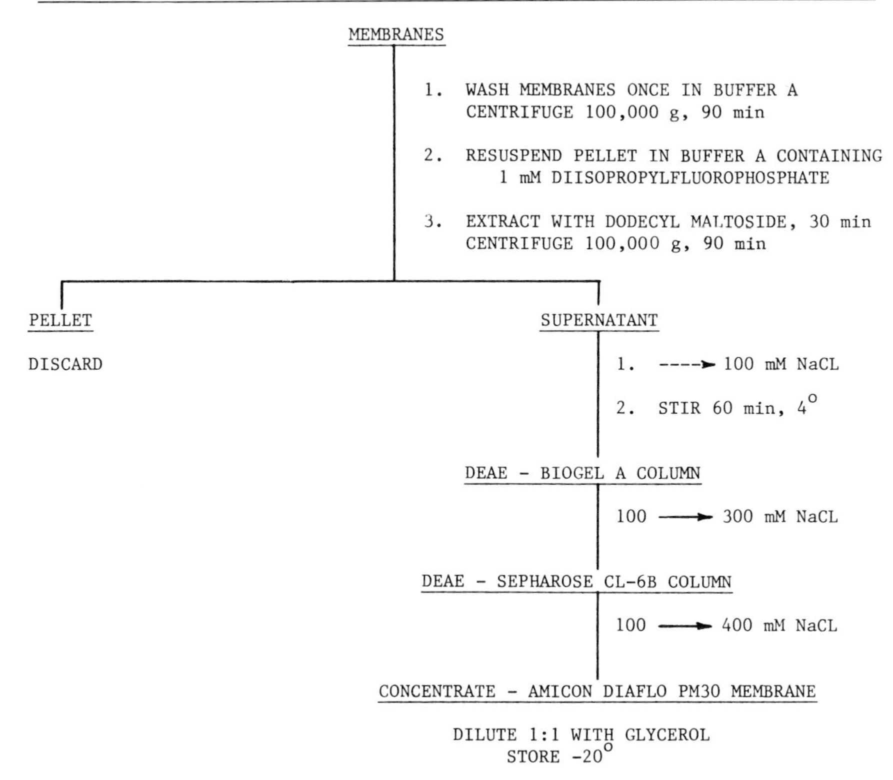

FIG. 1. General flow scheme for the isolation of ubiquinol–cytochrome-*c* oxidoreductase.

60 min. The resulting pellet is suspended in 50 m*M* Tris–HCl, pH 8.0, 1 m*M* diisopropylfluorophosphate.

General Method

The isolation procedure is summarized in a flow scheme (Fig. 1). All steps are performed at 4°. Starting membrane preparations are suspended at 30 mg/ml protein in Buffer A (50 m*M* Tris–HCl, pH 8, at 4°, 1 m*M* $MgSO_4$, 1 m*M* phenylmethylsulfonyl fluoride), and the pH of the suspension is adjusted to 8. The suspension is centrifuged at 100,000 *g* for 90 min, and the resulting pellet resuspended to a protein concentration of 10 mg/ml in Buffer A containing 1 m*M* diisopropylfluorophosphate. Suspended membranes are solubilized with dodecyl maltoside, using an amount of detergent determined to be optimal for each species, as described below. In all cases the optimal solubilization conditions are established with regard to detergent/protein ratios vs amount and stability of the solubilized activity. In the cases described here the optimal ratio is between 0.5 and 1.0 g dodecyl maltoside per gram of membrane protein.

Appropriate volumes of a 100 mg/ml dodecyl maltoside stock solution are added to suspended membranes; the mixture is stirred for 30 min and centrifuged at 100,000 *g* for 90 min. To the resulting supernatant a 4 *M* solution of NaCl is added to obtain a final concentration of 100 m*M* NaCl, and this mixture is allowed to stir for 60 min.

The above extract is applied to a DEAE-BioGel A column equilibrated with Buffer B (50 m*M* Tris–HCl, pH 8, at 4°, 1 m*M* $MgSO_4$, 0.1 mg/ml dodecylmaltoside) containing 100 m*M* NaCl. The column with adsorbed sample is washed with three column volumes of equilibration buffer. The column is then eluted with four column volumes of a linear 100–300 m*M* NaCl gradient in Buffer B. Fractions are analyzed for absorption at 415 nm, and for ubiquinol–cytochrome *c* oxidoreductase and cytochrome *c* oxidase activities. Fractions containing maximal ubiquinol–cytochrome *c* oxidoreductase activity and minimal cytochrome *c* oxidase activity, typically eluting with 225–275 m*M* NaCl, are pooled.

These pooled fractions are directly applied to a DEAE-Sepharose CL-6B column equilibrated as above. The column is washed with a half column volume of equilibration buffer, and adsorbed protein is eluted with a linear 100–400 m*M* NaCl gradient in Buffer B. Pure ubiquinol–cytochrome *c* oxidoreductase (bc_1 complex) elutes between 350 and 375 m*M* NaCl. These fractions are free of succinate dehydrogenase and cytochrome *c* oxidase activities.

Both DEAE columns are 15 cm in length. The width is varied with the amount of protein applied. The collected fraction volumes are one-tenth the column volume. The elution profiles are nearly identical for the species to which the method has been applied. Representative column profiles are shown in Fig. 2.

The fractions containing pure bc_1 complex are concentrated on an Amicon Diaflo PM30 membrane and diluted 1 : 1 with glycerol. Glycerol-diluted preparations are stable for at least 1 year at −20°. In fact, glycerol dilution typically increases ubiquinol–cytochrome *c* oxidoreductase activity in about 2 weeks. This activation is possibly due to partitioning of excess detergent away from the complex. These conditions do not promote detectable intercomplex aggregation, as preparations remain completely disperse. This method of storage also eliminates the detrimental effects of freezing and thawing.

Optical spectra of the purified complexes are shown in Fig. 3. A comparison of the subunit compositions of the bacterial and mitochondrial complexes is shown in the SDS–polyacrylamide gel in Fig. 4.

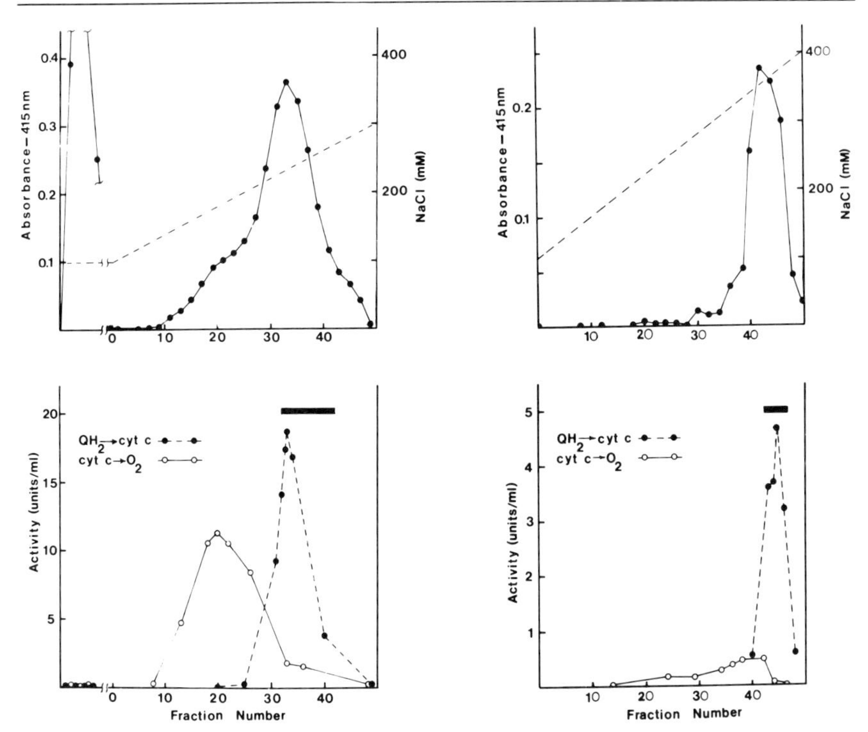

FIG. 2. DEAE column chromatography profiles obtained during the purification of the yeast cytochrome bc_1 complex. The upper and lower left panels show the BioGel A profile, and the right two panels show the Sepharose CL-6B profile. The upper panels depict absorbance at 415 nm (heme) and the NaCl gradients (dashed lines). The lower panels show enzymatic activity profiles as indicated; the solid bars indicate the ubiquinol–cytochrome c oxidoreductase containing fractions that were pooled.

Comments on Specific Species

Yeast

Submitochondrial membranes are initially prewashed by the method of Jacobs and Sanadi[9] in the presence of 1 mM phenylmethylsulfonyl fluoride. Membranes are suspended at 30 mg/ml protein in each of the successive KCl wash steps. The pellet resulting from the final 0.15 M KCl wash is resuspended in buffer A. This prewash decreases the level of nonintergral membrane protein. Submitochondrial membranes are extracted with 0.8 g of dodecyl maltoside per gram of protein, and the cytochrome

[9] E. E. Jacobs and D. R. Sanadi, *J. Biol. Chem.* **235,** 531 (1960).

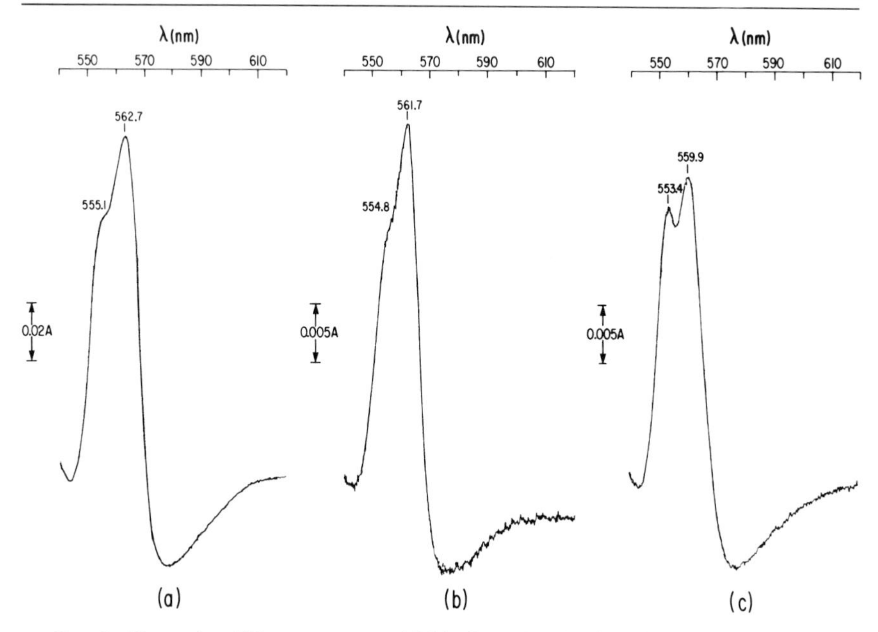

FIG. 3. Absorption difference spectra (dithionite reduced minus ferricyanide oxidized) of purified cytochrome bc_1 complexes from (a) beef heart suspended at 1 mg/ml protein; (b) yeast suspended at 0.17 mg/ml protein; and (c) *Rhodopseudomonas sphaeroides* suspended at 0.09 mgml protein.

bc_1 complex is purified according to the general procedure described above. See Fig. 2 for elution profiles and Table I for purification data.

Beef Heart

Mitochondria are extracted with 1.0 g of dodecyl maltoside per gram of protein, and the cytochrome bc_1 complex is purified according to the general procedure as described above. See Table II for purification data.

Rhodopseudomonas sphaeroides

Chromatophore membranes are difficult to solubilize in comparison to the other membranes utilized in this study, and ubiquinol–cytochrome *c* oxidoreductase activity is more sensitive to high detergent concentrations. Membranes are extracted with 0.66 g of dodecyl maltoside per gram of protein, as indicated above. Upon centrifugation, the resulting loose upper pellet, unique to chromatophores, is removed along with the supernatant. This fraction, containing only 46% of the total ubiquinol–cytochrome *c* oxidoreductase activity, is brought to 100 m*M* in NaCl, stirred

TABLE I

PURIFICATION OF UBIQUINOL–CYTOCHROME-*c* OXIDOREDUCTASE FROM YEAST

Fraction	Protein (mg)	QCR activity (units)	Yield (%)	Purification (-fold)	Cytochrome c_1 (nmol)	Cytochrome *b* (nmol)	T*n* (sec^{-1})
SMP	3000	2016	—	—	—	266.9	252
SMP + DM	2979	3456	100	1	—	377.8	305
DM extract	920	2824	82	2.7	—	255.4	372
DEAE-BioGel A	138	3266	95	20.0	147.8	302.4	368
DEAE-Sepharose 6B	22	809	23	31.7	87.4	174.7	154
Glycerol diluted	22	1164	34	45.6	88.7	177.4	219

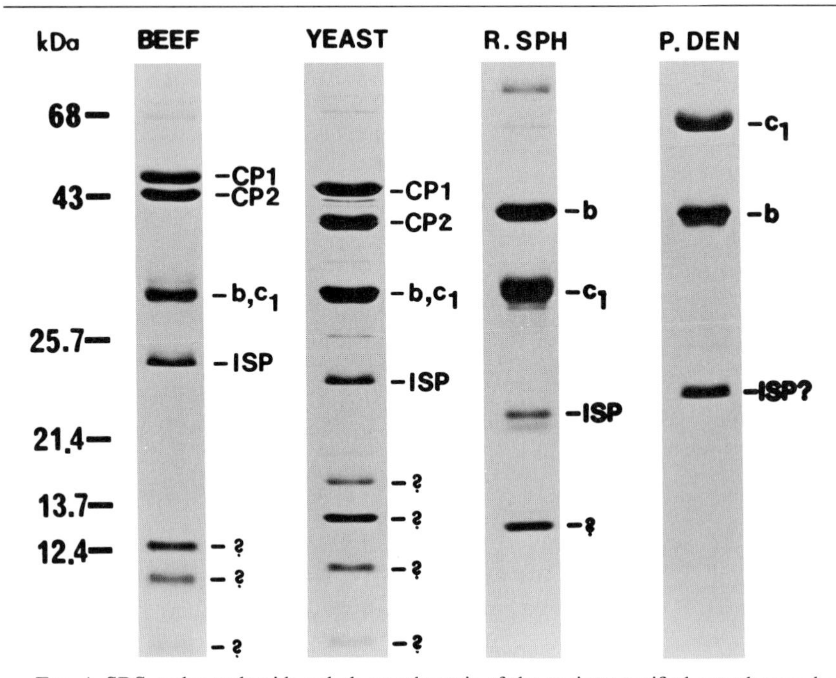

FIG. 4. SDS–polyacrylamide gel electrophoresis of the various purified cytochrome bc_1 complexes. For comparison, the *Paracoccus denitrificans* complex has been shown (Yang and Trumpower, this volume [29]). Samples were first reduced with 2% 2-mercaptoethanol (30 m*M* dithiothreitol was used for *Paracoccus denitrificans*) and then denatured with 2% SDS at 50° for 10 min (25° for *Rhodopseudomonas sphaeroides*). Glycerol (50%) was then added to a final concentration of 20%, and 0.2% Bromphenol Blue to a final concentration of 0.004%. The 1.5 mm thick (14 cm width by 12 cm length) gel was electrophoresed at 20 mA constant current through the 4% T stacking gel, 30 mA constant current through the running gel, and stained with Coomassie Brilliant Blue R-250 (in 50% methanol, 10% acetic acid) overnight. Destaining was performed in 25% methanol, 10% acetic acid. The protein standards utilized included bovine serum albumin (68 kDa), ovalbumin (43 kDa), chymotrypsinogen (25.7 kDa), trypsin inhibitor (21.4 kDa), ribonuclease A (13.7 kDa), and horse heart cytochrome *c* (12.4 kDa).

for 60 min, and centrifuged as before, resulting in a hard pellet. The supernatant contains about 30% of the total units.

Simultaneously, the hard portion of the pellet from the initial solubilization is resuspended to 10 mg/ml protein, reextracted with 0.5 g of dodecyl maltoside per gram of protein, and centrifuged as above. The resulting supernatant is incubated with 100 m*M* NaCl, as above. This second supernatant contains about 20% of the total units.

The cytochrome bc_1 complex in the combined supernatants, amounting to ~50% of the total ubiquinol–cytochrome *c* reductase activity, is

TABLE II
PURIFICATION OF UBIQUINOL–CYTOCHROME-*c* OXIDOREDUCTASE FROM BEEF HEART

Fraction	Protein (mg)	QCR activity (units)	Yield (%)	Purification (-fold)	Cytochrome c_1 (nmol)	Cytochrome *b* (nmol)	T*n* (sec^{-1})
SMP + DM	2051	55303	100	1	—	1004	1843
DM extract	1536	49178	89	1.2	—	916	1790
DEAE-BioGel A	152	24800	45	6.0	315	709	1312
DEAE-Sepharose 6B	96	13150	24	5.1	240	525	913
Glycerol diluted	92	16867	31	6.8	244	465	1152

TABLE III
PURIFICATION OF UBIQUINOL–CYTOCHROME *c* OXIDOREDUCTASE FROM *R. sphaeroides*

Fraction	Protein (mg)	QCR activity (units)	Yield (%)	Purification (-fold)	Cytochrome c_1 (nmol)	Cytochrome *b* (nmol)	T*n* (sec^{-1})
Chromatophores	105	315	—	—	—	—	—
Chromatophores + DM	104	1512	100	1	—	154	164
DM extract	40	763	51	1.3	—	106	176
DEAE-BioGel A	5	190	13	2.5	18.6	25.3	170
DEAE-Sepharose 6B	1	101	7	6.9	7.1	10.4	239

purified by anion-exchange chromatography with dodecyl maltoside, as described in the general method (see Table III for purification results).

The ubiquinol–cytochrome *c* oxidoreductase complex isolated from the above extract is spectrally pure, with no absorption from the carotenoids or bacteriochlorophyll (data not shown). This indicates no contamination of the complex with light-harvesting antennae (B800-850 and B875) or reaction center, and is corroborated by the electrophoresis profile, as there are no contaminating light-harvesting or reaction center polypeptides.

The activity of our isolated complex is the best yet reported for *Rhodopseudomonas sphaeroides* (for comparison, see Refs. 10 and 11) and is the first report of ubiquinol–cytochrome *c* oxidoreductase complex isolation from the wild-type strain of this bacteria. This is significant, as most of the mutants previously used are not well characterized and could contain pleotrophic mutations that might affect the integrity of this enzyme complex.

Acknowledgment

We would like to thank Robert A. Niederman, Dennis Farrelly, and David Purvis for providing phototrophically grown *Rhodopseudomonas sphaeroides*.

[10] L. Yu and C. A. Yu, *Biochem. Biophys. Res. Commun.* **108,** 1285 (1982).

[11] N. Gabellini, J. R. Bowyer, E. Hurt, A. Melandri, and G. Hauska, *Eur. J. Biochem.* **126,** 105 (1982).

[19] Preparation of Membrane Crystals of Ubiquinol–Cytochrome-*c* Reductase from *Neurospora* Mitochondria and Structure Analysis by Electron Microscopy

By HANNS WEISS, SVEN HOVMÖLLER, and KEVIN LEONARD

Structural information is a prerequisite for the understanding of molecular mechanisms of proteins. Until now, only a few membrane proteins have been obtained as three-dimensional crystals suitable for X-ray crystallographic studies (for a review, see Ref. 1). From a small number of

[1] H. Michel, *TIBS* 56 (1983).

membrane proteins, two-dimensional crystals have been prepared and used for three-dimensional structure analysis by electron microscopy (for a review, see Ref. 2). In this chapter, we report on a systematic approach to making such two-dimensional crystals from cytochrome reductase (ubiquinol–cytochrome-*c* reductase, EC 1.10.2.2) and a subcomplex of this enzyme, both isolated as homogeneous protein–detergent complexes from the mitochondrial membranes of *Neurospora crassa*.[3-7] The factors that influence the formation of these membrane crystals are reported in detail because they may be general and may facilitate the two-dimensional crystallization of other membrane proteins. From tilted electron microscopic views of the crystals at low resolutions, three-dimensional structures have been calculated. By combination of the structural properties obtained by electron microscopy with properties of isolated subunits obtained from biochemical studies, the topography of most of the subunits within the enzyme has become known.[6]

Preparation of Membrane Crystals

Single-Layer Membrane Crystals of Cytochrome Reductase. All steps were carried out at 4°. Cytochrome reductase in Triton X-100 solution was isolated from *Neurospora* mitochondria as described in the next chapter.[8,9] Only those preparations from the gel filtration step which eluted with a symmetrical and narrow peak with protein concentration of 5–8 mg/ml were used for crystallization. Appropriate amounts (see below) of soybean phosphatidylcholine (Roth) and bovine brain phosphatidylserine (Sigma) were diluted with chloroform and dried under a stream of nitrogen. The residue was twice dissolved in ether and dried. Routinely, 0.4% phosphatidylcholine and 0.1% phosphatidylserine were sonicated in 0.5% Triton X-100, 50 m*M* Tris-acetate, pH 7, 1 m*M* EDTA, and 5 μM butylated hydroxytoluene (Sigma) for 5 min until a homogeneous solution was obtained. The solution was centrifuged at 100,000 *g* for 30 min to remove any undispersed phospholipid and metal pieces from the sonicator tip. Approximately 90% of the phospholipid remained in solution as estimated using phosphatidyl [^{14}C]choline. Equal volumes of 5 mg/ml enzyme solu-

[2] N. Unwin and R. Henderson, *Sci. Am.* **250,** 56 (1984).
[3] P. Wingfield, T. Arad, K. Leonard, and H. Weiss, *Nature (London)* **280,** 696 (1979).
[4] S. Hovmöller, K. Leonard, and H. Weiss, *FEBS Lett.* **123,** 118 (1981).
[5] K. Leonard, P. Wingfield, T. Arad, and H. Weiss, *J. Mol. Biol.* **149,** 259 (1981).
[6] B. Karlsson, S. Hovmöller, H. Weiss, and K. Leonard, *J. Mol. Biol.* **165,** 287 (1983).
[7] S. Hovmöller, M. Slaughter, J. Berriman, B. Karlsson, H. Weiss, and K. Leonard, *J. Mol. Biol.* **165,** 401 (1983).
[8] P. Linke and H. Weiss, this volume [20].
[9] H. Weiss and H. J. Kolb, *Eur. J. Biochem.* **99,** 139 (1979).

tion and phospholipid–Triton solution were mixed. Triton was removed by gently stirring 0.25 g Bio-Beads SM-2 (Bio-Rad) per milliliter in the mixture for 2–3 hr.[10] The membrane suspension obtained was separated from the beads by filtration through loosely packed glass wool, 10% (w/v) sucrose was added, and the suspension was stored in portions at liquid nitrogen temperature. Prior to use, a 0.1–0.2 portion was thawed and gel filtrated through a 0.5 × 7 cm Sepharose 2B (Pharmacia) column in 50 m*M* Tris-acetate, pH 7. Fractions of 2 drops were collected. Large membranes were separated from smaller ones and from excess phospholipid and sucrose.

Multilayer Membrane Crystals of Cytochrome Reductase. Enzyme and phospholipid solution were prepared as above. For detergent removal equal aliquots of both solutions were combined and dialyzed for 2 to 3 days at room temperature against a solution containing 50 m*M* MES (morpholinoethane sulfonate) buffer, pH 5.5, 50 m*M* NaCl, 1 m*M* EDTA, and 1 m*M* ascorbate.

Membrane Crystals of the Cytochrome bc_1 Subcomplex. The cytochrome bc_1 subcomplex was isolated from cytochrome reductase by splitting off the subunits I, II, and V (core proteins and iron-sulfur subunits), as described in the next chapter.[8] The subcomplex was concentrated by ultrafiltration on a Diaflo PM30 filter to a protein concentration of 3–4 mg/ml. An equal amount of a solution of 0.4% phosphatidylcholine and 0.1% phosphatidylserine in 0.5% Triton X-100, 50 m*M* Tris-acetate, pH 7, 50 m*M* NaCl, 1 m*M* EDTA, 1 m*M* ascorbate, and 5 μM butylated hydroxytoluene, prepared as above, was added. Triton X-100 was removed by dialysis for 2–3 days at room temperature against 50 m*M* MES-buffer, pH 5.5, 50 m*M* NaCl, 1 m*M* EDTA, and 1 m*M* ascorbate.

Factors That Influence the Formation of Membrane Crystals

A large number of crystallization experiments on the above two proteins were carried out to understand the factors that influence the formation of the crystals. The following parameters did not appear to affect crystallization significantly: protein concentration (1–5 mg/ml); detergent concentration (0.5–2%); type of alkyl(phenyl)polyoxyethylene detergent for protein and phospholipid solubilization (Triton X-100, a *tert*-$C_8\emptyset E_{9.6}$[11] detergent, or Cemulsol LA 90, a $C_{12}E_{8-10}$ detergent could be used); type of phospholipid (100% phosphatidylcholine from soybean to a mixture of 50% phosphatidylcholine plus 50% phosphatidylethanolamine from bo-

[10] P. S. Holloway, *Anal. Biochem.* **53,** 304 (1973).

[11] A. Helenius, P. R. M. Caslin, E. Fries, and C. Tanford, this series, Vol. 56, p. 734.

vine heart were used with similar success); the ionic strength (0–50 m*M* NaCl); or the relative amounts of phospholipid to protein (in the range of 2 : 1 to 1 : 2); it appears that the material undergoes a phase separation such that, in addition to the membrane crystals that are lipid bilayers saturated with protein, there will be protein aggregates, if protein is in excess or protein-free lipid bilayers, if lipid is in excess.

Temperature, pH, and the method of detergent removal were found to be the most important factors affecting the quality of the crystals obtained. The procedure of detergent removal, by stirring for 2 hr with polystyrene beads,[10] resulted in considerable variation in their quality, and the preparations always contained a mixture of sheets, tubes, and small vesicles. Detergent removal by dialysis is a more controlled method for crystallization. Using dialysis, it became clear that pH, temperature, and time for completion of crystallization are important factors. The following observations were made for cytochrome reductase. Above pH 6.5, very small vesicles were obtained; at pH 6.0, more than 90% of the crystals were in the form of tubes; and at pH 5.5, the preparation contained almost entirely large sheets. The crystal quality depended upon the temperature at which crystals were grown: at 4°, the crystalline order was poor and the crystalline size small; at 25°, the crystalline order was good and membrane crystals up to 20 μM in diameter were formed.

Electron Micrographs of Membrane Crystals

Two different forms of membrane crystals prepared from cytochrome reductase by the detergent-adsorption method at pH 7 and 4° are shown in Fig. 1. Single-layer sheets, vesicles, or tubes of various size were obtained. The tubes collapse during fixation and staining and the electron micrographs often show moiré patterns arising from the overlap of the lattices on opposite sides of the flattened tubes. The diffraction pattern of these monolayer cytochrome reductase crystals extends to the fifth order, which corresponds to a resolution of about 2.5 nm. The pattern shows pgg symmetry which is consistent with the two-sided plane group p 22_12_1. In this symmetry the alternate dimeric enzyme molecules are packed up and down across the bilayer.

Multilayer membrane crystals of cytochrome reductase formed by the dialysis method at pH 5.5 and room temperature are shown in Fig. 2. These membrane crystals stack in register on top of each other to give three-dimensional crystals several layers thick. Optical diffraction patterns extending to the tenth order, what corresponds to 1.8 nm resolution, were obtained for these crystals.[7]

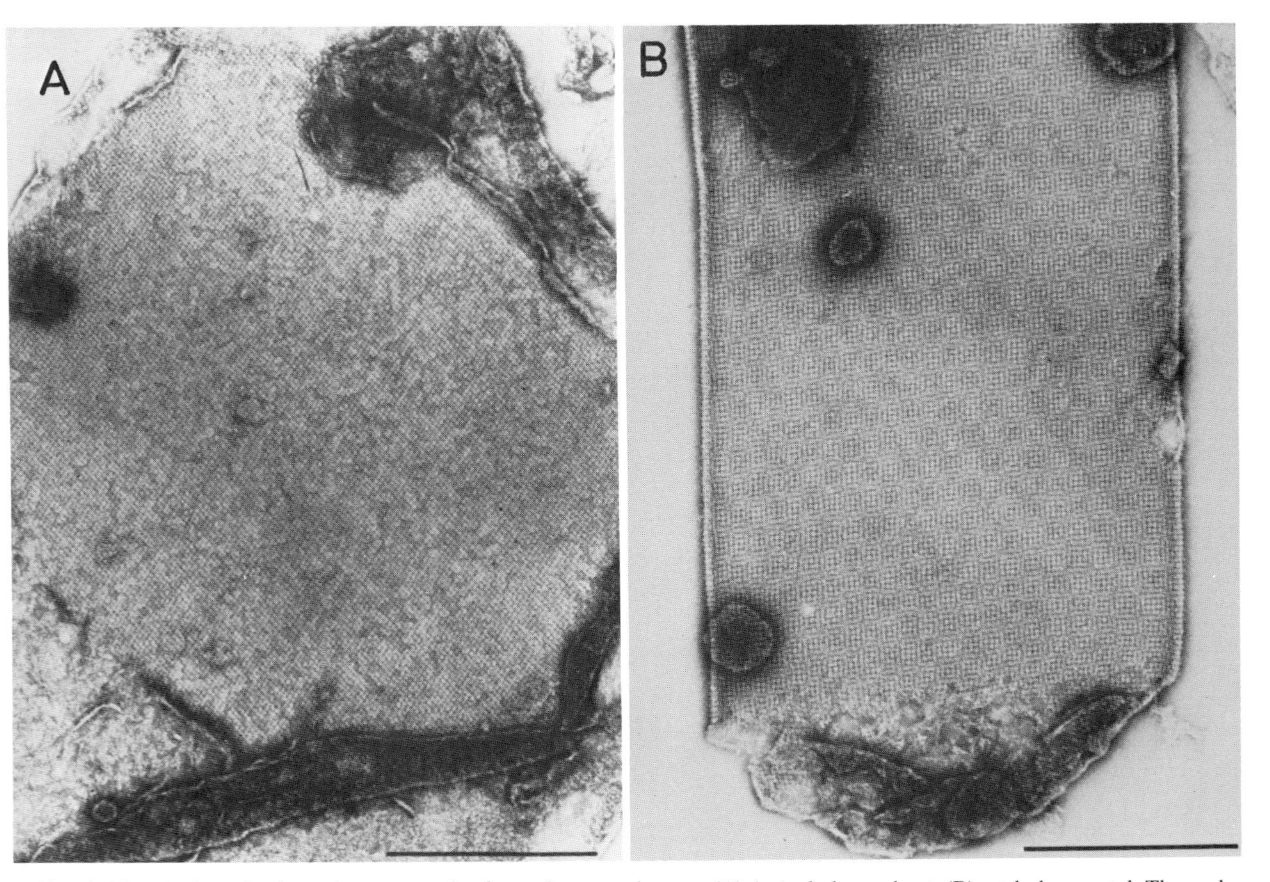

FIG. 1. Negatively stained membrane crystals of cytochrome reductase. (A) A single-layer sheet; (B) a tubular crystal. The scale bar represents 1 μm.

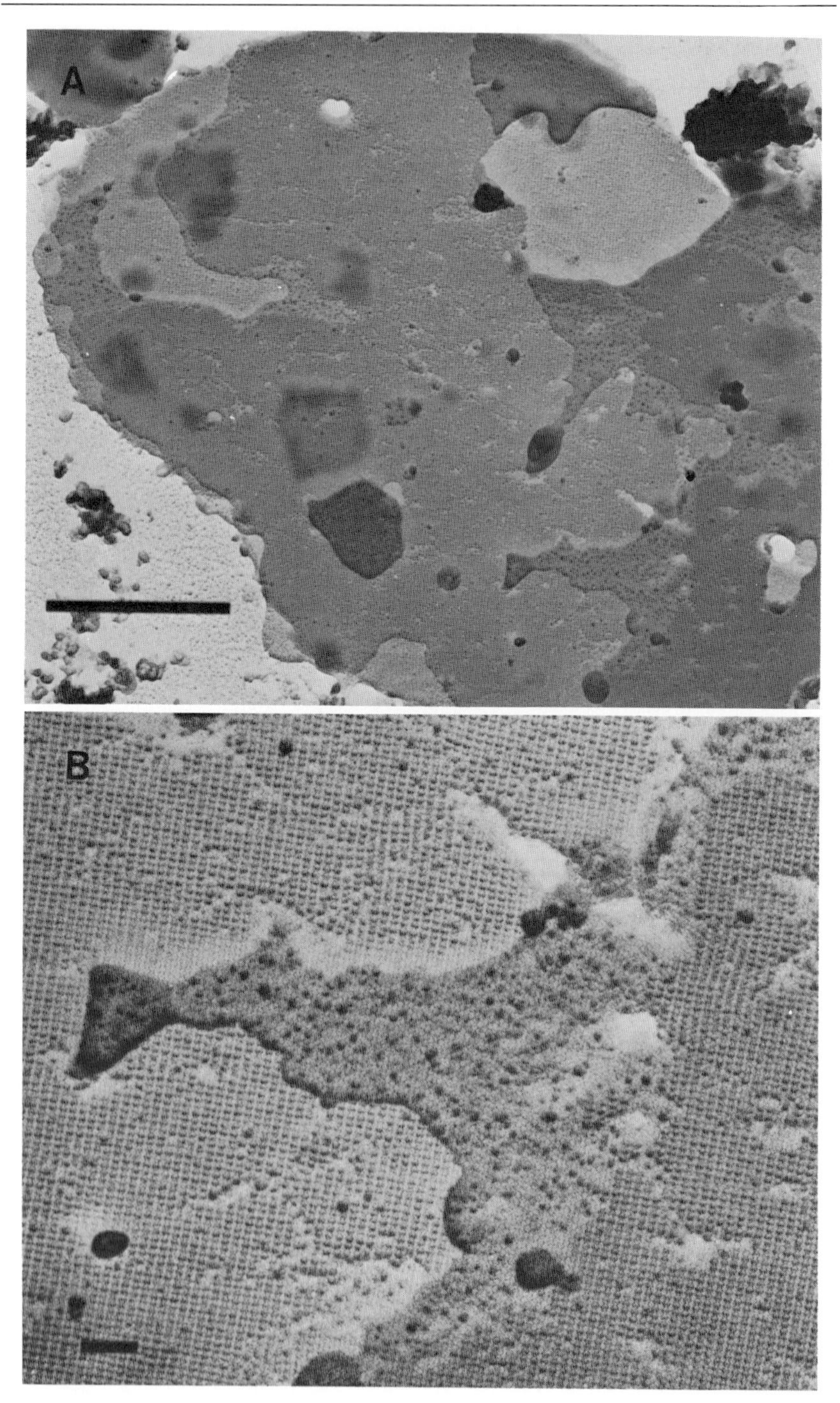
A
B

Membrane crystals of the cytochrome bc_1 subcomplex were obtained only by the dialysis method at pH below 6 and temperatures above 15°. Tubes, vesicles, and sheets typically were formed (Fig. 3). In these membrane crystals the dimeric protein also points up and down across the bilayer.

Three-Dimensional Structure Analysis

From sets of tilted electron microscopic views the three-dimensional structures were calculated as described in Ref. 6. The structure of the whole enzyme shows that the monomeric units are related by a 2-fold axis perpendicular to the plane of the membrane. They are elongated and extend ~15 nm across the membrane. The protein is unequally distributed with about 30% of the total mass located in the bilayer, 50% in a large peripheral section which extends 7 nm from one side of the bilayer, and 20% in a small peripheral section which extends 3 nm from the opposite side of the bilayer. The two monomeric units are in contact essentially in the membranous section (Fig. 4).

The three-dimensional structure of the cytochrome bc_1 subcomplex shows defined protein lobes correlated by a 2-fold axis perpendicular to the membrane plane. These lobes correspond in size and shape to the smaller peripheral sections of the structure for the whole enzyme. Below the lobes, the protein density merges into one less well-defined region, which is about 4 nm in thickness and corresponds roughly to the membranous section of the enzyme. The structure of the subunit complex shows no part large enough to fit to the large peripheral section of the enzyme (Fig. 4).

Topography of Subunits

As described in detail elsewhere,[4,6,12–14] cytochrome reductase from *Neurospora* in solution of alkyl(phenyl)polyoxyethylene detergents can

[12] Y. Li, K. Leonard, and H. Weiss, *Eur. J. Biochem.* **116,** 199 (1981).
[13] Y. Li, S. De Vries, K. Leonard, and H. Weiss, *FEBS Lett.* **135,** 277 (1981).
[14] S. Perkins and H. Weiss, *J. Mol. Biol.* **168,** 847 (1983).

FIG. 2. Multilayer membrane crystal of cytochrome reductase. The presentation was first negatively stained, dried, and then shadowed with platinum. The scale bar in (A) represents 1 μm; (B) an enlargement of (A), the scale bar represents 100 nm. The large peripheral section of the enzyme can be seen as rows running vertically. Between the rows are molecules facing the opposite way across the membrane, with the smaller peripheral section facing up. They are too small to be seen by metal shadowing.

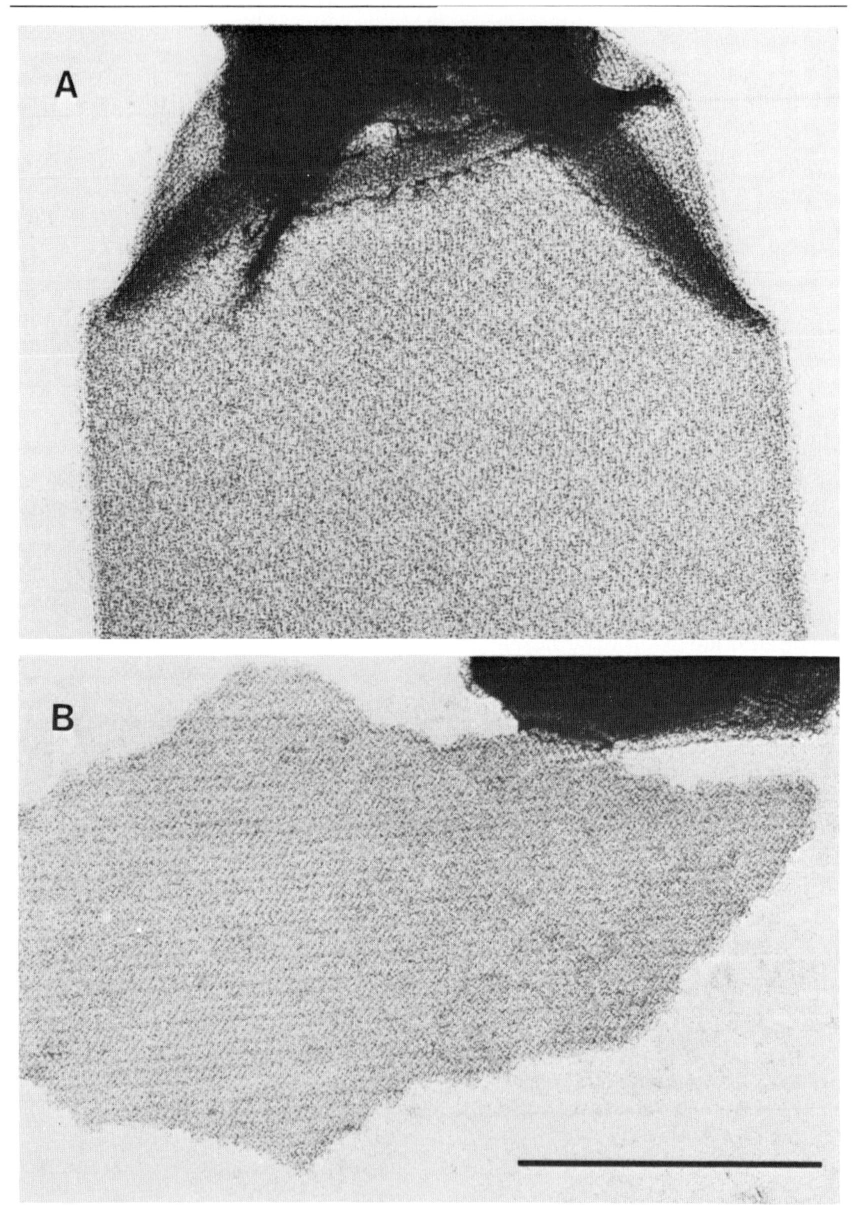

FIG. 3. Negatively stained membrane crystals of the cytochrome bc_1 subunit complex. (A) Tubular crystal; (B) single-layer sheet. The scale bar represents 0.5 μm.

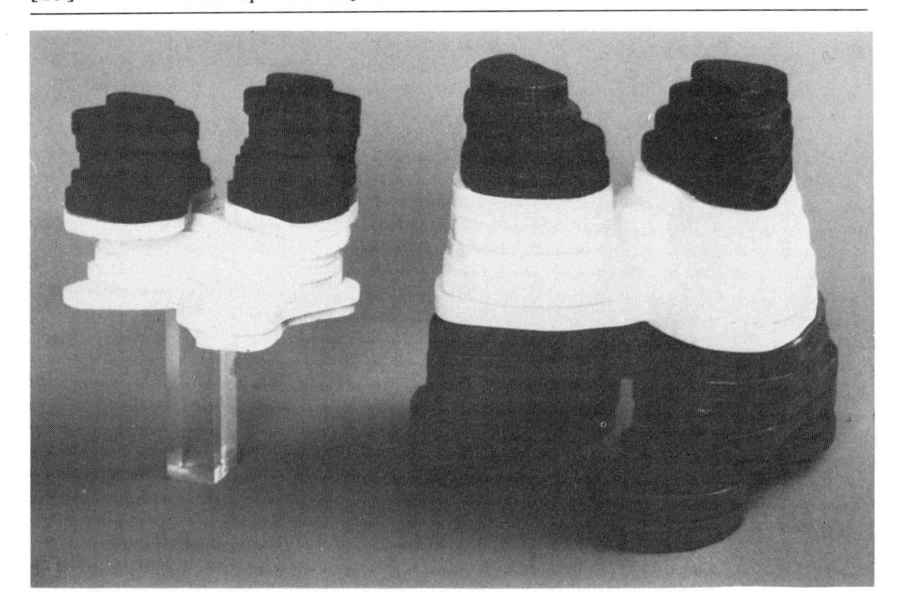

FIG. 4. Three-dimensional models of the dimeric cytochrome bc_1 subunit complex (left) and the dimeric cytochrome reductase (right). The darker parts extend into the aqueous phase, and the lighter parts lie within the bilayer; the view is parallel to the plane of the membrane.

be cleaved in stages and the parts obtained (subunit complexes, subunits, and subunit domains) can be characterized as hydrophilic, amphiphilic, or hydrophobic according to their solubility in aqueous buffers or detergent solution. This procedure assisted in the assignment of these parts to one of the two aqueous sections or the membranous section of the enzyme.

The subunits I and II (core proteins) of the enzyme are isolated as a hydrophilic complex which is water soluble without detergent. This subunit complex is therefore assumed to extend completely from the bilayer into the aqueous phase. The assignment of this complex within either of the two peripheral sections of cytochrome reductase can be made by comparing the structures of the whole enzyme to that of the cytochrome bc_1 subunit complex which lacks the subunits I, II, and V. A part corresponding to the larger peripheral section of cytochrome reductase is missing from the structure of the subunit complex, and the overall size of the complex fits well with that part of cytochrome reductase taken by the small peripheral section and the membrane section. The large peripheral section of cytochrome reductase therefore must be assumed to be contributed by the subunits I and II. These two subunits account well for the protein mass of this section.

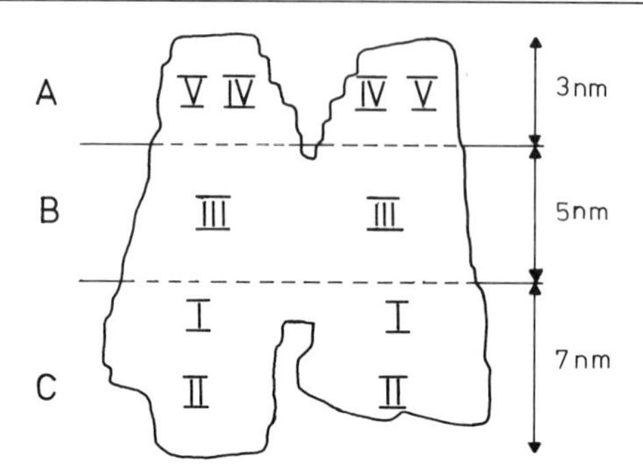

FIG. 5. Schematic drawing of the location of cytochrome reductase in the mitochondrial inner membrane and the topography of the subunits in the enzyme. The subunits I and II are also called "core proteins," and subunits II, IV, and V are cytochrome *b*, cytochrome c_1, and iron-sulfur subunit. The subunits VI–IX have not been included in the scheme because we have no information about their location. A, B, and C refer to intermembrane, membrane, and matrix space.

The subunits IV and V (cytochrome c_1 and iron-sulfur subunits) are amphiphilic; the isolated subunits are soluble only when a minor stretch of each is bound to a detergent micelle. After this stretch has been clipped off by proteolysis, the major part of each subunit, which carries the redox center, is water soluble without detergent.[12,13] The amphiphilic character of the two subunits also follows from their amino acid sequences. The sequence of cytochrome c_1 (from bovine heart,[15]) shows only one hydrophobic protein stretch long enough to span the membrane located near the C terminus and the two heme binding cysteines near the N terminus.

The sequence of the iron-sulfur subunit (from the *Neurospora* enzyme)[16] is characterized by the presence of one protein stretch 20 hydrophobic residues long near the N terminus, and the only 4 crysteines, which probably bind the 2Fe-2S center, close together near the C terminus. The major domain of the cytochrome c_1 and iron-sulfur subunit must therefore be located each in a peripheral section of the structure. Since the only peripheral part common to the structure of the whole enzyme and the cytochrome bc_1 subcomplex is the small peripheral section, this section must be the site of cytochrome c_1.

The question of where to locate the iron-sulfur subunit is more difficult to answer, since the resolution limit of the reconstruction is of the same order as the expected diameter of the subunit. The comparison of the

[15] S. Wakabayashi, H. Matsubara, C. H. Kim, K. Kawai, and T. E. King, *Biochem. Biophys. Res. Commun.* **97,** 1548 (1980).

[16] U. Harnisch, H. Weiss, and W. Sebald, submitted to *EMBO J.* (1985).

small peripheral section of cytochrome reductase with the (only) peripheral section of the cytochrome bc_1 complex shows that the distance between the center of mass of the latter is 10% less than that of the former. This difference suggests that protein is missing from the external faces of this section as compared with cytochrome reductase, thus shifting inward the mass centers of protein. The iron-sulfur subunit protein could thus be located on the outside of the small peripheral section of cytochrome reductase.[6,13]

The subunit III (cytochrome *b*) is hydrophobic; it is soluble only in detergent solution. The amino acid sequence, obtained from the sequencing of mitochondrial DNA (for *Neurospora*),[17,18] shows 8 protein stretches more than 20 hydrophobic residues long distributed over the whole sequence. Cytochrome *b* is therefore assumed to span the bilayer several times. The subunit must be located in the membrane section of the enzyme.

The orientation of the structure of cytochrome reductase in the mitochondrial inner membrane results from the topography of the cytochrome c_1 subunit. This subunit interacts with cytochrome *c* at the outer surface of the inner membrane. The heme-carrying part of cytochrome c_1 was located in the small peripheral section. Therefore this section is assumed to extend into the intermembrane space of mitochondria and the large peripheral section into the matrix space (Fig. 5).

[17] M. H. Citterich, G. Morelli, and G. Macino, *EMBO J.* **2,** 1235 (1983).
[18] J. M. Burke, C. Breitenberger, J. E. Heckmann, B. Dujon, and V. L. Rajbhandary, *J. Biol. Chem.* **259,** 504 (1984).

[20] Reconstitution of Ubiquinol–Cytochrome-*c* Reductase from *Neurospora* Mitochondria with regard to Subunits I and II

By PETRA LINKE and HANNS WEISS

Ubiquinol–cytochrome-*c* reductase (cytochrome reductase, EC 1.10.2.2) isolated from *Neurospora crassa* mitochondria has a molecular weight of ~550,000, is in a dimeric state, and consists of nine different subunits.[1-3] The subunits are the so-called core proteins, which do not

[1] H. Weiss and H. J. Kolb, *Eur. J. Biochem.* **99,** 139 (1979).
[2] B. Karlsson, S. Hovmöller, H. Weiss, and K. Leonard, *J. Mol. Biol.* **165,** 287 (1983).
[3] J. S. Rieske, *Biochim. Biophys. Acta* **456,** 195 (1976).

have redox centers (subunit I and II, M_r 50,000 and 45,000), the Rieske iron-sulfur protein (subunit V, M_r 22,000), and four subunits, probably without prosthetic groups (subunits VI–IX, M_r 14,000 to 8,000). In Triton solution the *Neurospora* cytochrome reductase is fragile and easily dissociates at high salt concentration into three parts. These parts are the cytochrome bc_1 subcomplex containing the subunits III, IV and VI–IX, the core complex of the subunits I and II, and the single iron-sulfur subunit. From these parts the enzyme can be reassociated in stages. In this chapter we report on the methods used for isolation, dissociation, and reassociation of the enzyme. We compare quinol : quinone and quinol : ferricytochrome-*c* oxidoreductase activities of the partially reassembled enzyme to the activities of the whole enzyme.

Isolation of Cytochrome Reductase

Isolation of cytochrome reductase as protein–Triton complex is achieved by the following steps: preparation of mitochondrial membranes, solubilization of the membrane proteins by Triton X-100, affinity chromatography on cytochrome *c* coupled to Sepharose, and gel filtration. All steps are carried out at 4°.

Preparation of Mitochondrial Membranes

Hyphae (500 g wet weight, stored at −70° before use) are disrupted by means of a grindmill in a solution containing 15% sucrose, 50 m*M* Tris-acetate, pH 7, 0.1 m*M* phenylmethylsulfonyl fluoride (PMSF) (5 liters), and mitochondria are isolated by centrifugation as described.[1] They are sonicated in 0.2 *M* Na-phosphate, pH 7, 0.1 m*M* PMSF (500 ml), and mitochondrial membranes are separated from the matrix and intermembrane proteins by ultracentrifugation.[1] Phosphate buffer is removed from the membranes by short sonication in 5% sucrose, 40 m*M* Tris-acetate, pH 7, 0.1 m*M* PMSF (200 ml), followed by ultracentrifugation. The membranes are suspended at a protein concentration of 40–50 mg/ml in the above sucrose–Tris buffer (50 ml). They can be stored at −70°.

Affinity Chromatography on Cytochrome c-Sepharose

Coupling of cytochrome *c* (from horse heart, Sigma, type III) to CNBr-activated Sepharose 4B (Pharmacia) is carried out as described.[1,4] The cytochrome *c*-Sepharose column (1.6 × 20 cm, from 12 g CNBr-activated Sepharose 4B and 60 mg cytochrome *c*) is equilibrated with 5%

[4] R. Axen, J. Porath, and S. Ernbach, *Nature (London)* **214,** 1302 (1967).

sucrose, 20 mM Tris-acetate, pH 7.0, 0.05% Triton X-100, and 0.1 mM PMSF. The suspension of the mitochondrial membranes is mixed with an equal volume of 10% Triton X-100 and centrifuged for 10 min at 20,000 g to remove unsolubilized material. The supernatant (about 100 ml, 8 μM with regard to cytochrome c_1) is pumped (100 ml/hr) through the column. Cytochrome reductase (60–80 ml, 5–8 μM cytochrome c_1) is eluted by a gradient (300 ml, 30–50 ml/hr) from 20 to 200 mM Tris-acetate, pH 7, in 5% sucrose, 2 mM ascorbate, and 0.1 mM PMSF.

Gel Filtration

The cytochrome reductase solution is concentrated to 0.1 mM cytochrome c_1 by ultrafiltration using Diaflo XM300 filters (Amicon) and passed through an Ultrogel AcA 34 (LKB) column (1.6 × 100 cm) in 40 mM Tris-acetate, pH 7, 0.05% Triton X-100, and 0.1 mM PMSF. The enzyme elutes as a symmetric peak with maximal protein concentration of 5–7 mg/ml well separated from the aggregated material and excess Triton. The enzyme can be stored at −70°.

Cleavage of Cytochrome Reductase

The *Neurospora* cytochrome reductase in Triton solution dissociates at pH 6 and 2 M NaCl into a subcomplex that contains the cytochromes *b* and c_1 and the four small subunits, a subcomplex that contains the two core proteins and the single iron-sulfur subunit (Fig. 1). The cytochrome bc_1 subcomplex is separated from the other two parts by gel filtration because of the different Stokes radii (Table I). The core complex, which is a pure protein, is separated from the detergent-bound iron-sulfur subunit by sucrose gradient centrifugation because of different buoyant densities.[2,5]

Isolation of the Cytochrome bc_1 Subcomplex

Cytochrome reductase is concentrated by ultrafiltration to about 20 mg/ml protein and successively brought to 2% Triton X-100 and 2 M NaCl. Immediately thereafter the solution (3–5 ml) is gel filtrated through an Ultrogel AcA 34 column (1.6 × 100 cm) in 50 mM MES (morpholinoethane sulfonate) buffer, pH 6.0, 0.2 M NaCl, 0.05% Triton X-100. The cytochrome bc_1 subcomplex elutes at K_{av} of 0.25, well separated with a peak at K_{av} of 0.36, which contains a mixture of the core complex, the single iron-sulfur subunit, and the excess Triton. Alternatively, the gel

[5] Y. Li, S. De Vries, K. Leonard, and H. Weiss, *FEBS Lett.* **135,** 277 (1981).

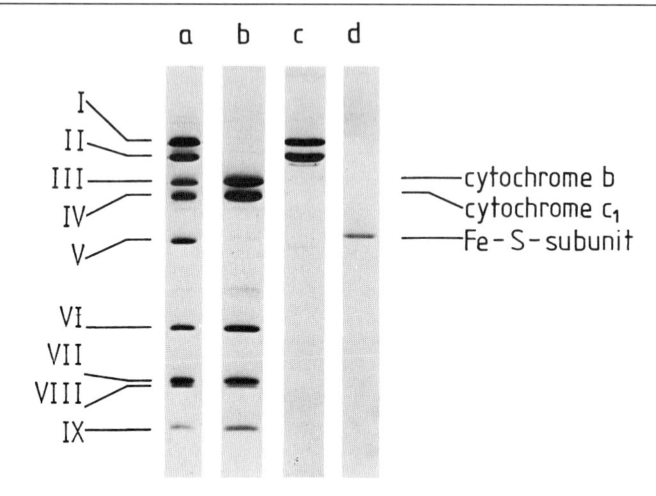

FIG. 1. SDS–gel electrophoresis of (a) cytochrome reductase; (b) the cytochrome bc_1 subcomplex; (c) the core complex, and (d) the iron–sulfur subunit. The roman numbers refer to the number of the subunits.

filtration step can be performed using a preparative (21.5 × 600 mm) high-performance liquid chromatography (HPLC) TSK 4000 GW (LKB) column in the above buffer at a flow rate of 1–2 ml/min.

Isolation of the Core Complex and the Iron-Sulfur Subunit

The fractions from the gel filtration step which contain the core complex and the iron-sulfur subunit are concentrated 5-fold by ultrafiltration on Diaflo PM30 filters (Amicon) laid in aliquots of 1.5 ml on 12-ml sucrose gradients from 5 to 20% sucrose in 50 m*M* Tris-acetate, pH 7, and centrifuged at 280,000 *g* for 24 hr. The core complex migrates halfway through the gradient at a protein concentration of 1–2 mg/ml, while the iron-sulfur subunit at a concentration of 5–10 μM and the excess Triton remain close to the top of the gradient.

The concentration of the whole enzyme routinely is determined by means of an absorption spectrum of the dithionite-reduced form using the extinction coefficient 27 mM^{-1} cm^{-1} at 560 nm for cytochrome *b* and a molecular weight of 550,000 for four cytochromes *b*. The concentration of the core complex is measured at 280 nm using the extinction coefficient 1 mg^{-1} cm^{-2}, and of the iron-sulfur subunit by means of labile sulfide[6] using the molecular weight of 22,000[7] for two S^{2-}.

[6] H. Beinert, *Anal. Biochem.* **131,** 373 (1982).

[7] U. Harnisch, H. Weiss, and W. Sebald, *Eur. J. Biochem.* **149,** 95 (1985).

TABLE I

STRUCTURAL PROPERTIES OF CYTOCHROME REDUCTASE AND THE CLEAVAGE PRODUCTS

Preparation	Subunit composition	Triton X-100 binding (g/mmol)	Hydrodynamic properties[a]		Molecular weight of protein determined by r_s and $s_{20,w}$ or		
			r_s (nm)	$s_{20,w}$ (S)	Sedimentation equilibrium	Neutron scattering	Subunit composition
Cytochrome reductase	I–IX	110–130	8.5	19	550,000	560,000	280,000–330,000
Cytochrome bc_1 subcomplex	III, IV, VI–IX	100–120	6.2	12.5	280,000	270,000	120,000–160,000
Core complex	I and II	0	5.4	8.1	170,000	190,000	140,000–190,000
Iron-sulfur subunit	V	60–90	5.0				22,000[b]

[a] For protein–detergent complexes when detergent is bound, for protein alone in the case of the core complex.

[b] From the primary structure.[7]

Structural Properties of Cytochrome Reductase and the Cleavage Products

The subunit composition of the enzyme and the three cleavage products as revealed by SDS–gel electrophoresis is shown in Fig. 1. Detergent binding, hydrodynamic properties, and calculated molecular weights of proteins are listed in Table I. The three-dimensional cross structures of the enzyme and the cytochrome bc_1 subcomplex analyzed by electron microscopy of membrane crystals are described in the accompanying chapter.[8] The low-resolution structures in solution of the whole enzyme, the cytochrome bc_1 subcomplex, and the core complex analyzed by small-angle neutron scattering are described elsewhere.[9]

Reconstitution of Cytochrome Reductase

Reassociation of cytochrome reductase from the cleavage parts was not possible in reverse of the dissociation, i.e., by decreasing the salt concentration in the presence of the detergent. Most likely, the detergent belts around the cytochrome bc_1 subcomplex[9] and the iron-sulfur subunit prevent the assemblage of the proteins. Reassociation occurs, however, when at low salt concentration the detergent micelles were replaced by phospholipid membranes.

Incorporation of Cytochrome Reductase and Ubiquinone-10 into Phospholipid Membranes

A mixture of 10 mg soybean phosphatidylcholine (Sigma, P5638) in 100 μl chloroform and 0.25 mg ubiquinone-10 (Serva) in 50 μl ethanol is dried under a stream of nitrogen and sonicated in 1 ml 50 m*M* Tris-acetate, pH 7, and 0.25% Triton X-100 five times for 10 sec at 4°. Equal volumes of the phospholipid and ubiquinone-10 solution and a solution of 4–6 mg/ml cytochrome reductase in 50 m*M* Tris-acetate, pH 7, 0.05% Triton X-100 are combined and 0.2 g/ml Bio-Beads SM-2 (BioRad), washed as described,[10] are added. The mixture is incubated under gentle stirring at 20° for 1–2 hr. The beads are allowed to sediment; the membrane suspension is used for enzymatic analyses.

Reassociation of Cytochrome Reductase from the Cleavage Parts

For reassemblage of the whole enzyme the following four solutions are combined: (1) 50 μl of 2 mg/ml cytochrome bc_1 subcomplex; (2) 200 μl of

[8] H. Weiss, S. Hovmöller, and K. Leonard, this volume [19].
[9] S. J. Perkins and H. Weiss, *J. Mol. Biol.* **168,** 847 (1983).
[10] P. S. Holloway, *Anal. Biochem.* **53,** 304 (1973).

0.2 mg/ml iron-sulfur protein; (3) 50 μl of 2 mg/ml core complex; and (4) 50 μl of the phospholipid and ubiquinone-10 solution. The mixture contains about 0.7 mg/ml protein, 1.4 mg/ml phospholipid, and 0.04 mg/ml ubiquinone-10. The molar ratio of cytochrome c_1 : iron-sulfur subunit : core complex is about 2 : 2 : 1. The Triton is removed by gently stirring 0.2 g/ml Bio-Beads in the solution for 1–2 hr at 20°. For partial reassociation either the iron-sulfur subunit or the core complex or both are substituted by their buffers.

Restoration of maximal activity is achieved by incubation at room temperature within 1 hr whereas incubation at 4° for 4 hr yields only one-third to one-half of the maximal activity. Varying the amount of phospholipid within 0.1–0.5 mg per nmol cytochrome c_1 does not affect the restored activity. The type of the phospholipids has little effect; 40% of the phosphatidylcholine can be substituted by phosphatidylserine; 20% cardiolipin increases the activity by 30%. Varying the pH between 7.0 and 7.5 has little effect, but at pH 6.5 only one-half the activity is restored.

Electron-Transfer Activities

Two electron-transfer pathways through cytochrome reductase are tested: from duroquinol or ubiquinol-10 to cytochrome *c* which is cyclic with regard to one electron and inhibited by myxothiazol and antimycin (for a review, see Ref. 11), and from duroquinol to ubiquinone-10 which is insensitive to myxothiazol but inhibited by antimycin.

Quinol : Ferricytochrome-c Reductase Activity

Quinol : ferricytochrome-*c* reductase activity is assayed in a dual-wavelength photometer using the wavelength pair 550 nm and 580 nm and the extinction coefficient 20 mM^{-1} cm^{-1} for cytochrome *c*. The temperature was 30°. The test solution contains 50 m*M* Tris-acetate, pH 7.0, 50 m*M* NaCl, 20 μM KCN, and 6 μM horse heart cytochrome *c* (Sigma, type III). The rate of nonenzymatic cytochrome *c* reduction is recorded for 1 min after addition of 5–100 μM quinol (see below) as ethanolic solution. The enzymatic reaction is started by enzyme addition to final cytochrome c_1 concentrations of 0.01–0.1 μM. The turnover number is determined by extrapolating the rates of the enzymatic reaction corrected for the nonenzymatic rates to infinite quinol concentration. For assaying the enzyme in detergent solution 0.1% Triton X-100 was present and ubiquinol-10 was used as reductant. The enzyme in phospholipid membranes is tested with duroquinol in detergent-free solution. As shown in Table II, a turnover

[11] P. R. Rich, *Biochim. Biophys. Acta* **768**, 53 (1984).

TABLE II
QUINOL : CYTOCHROME c AND QUINOL : QUINONE REDUCTASE ACTIVITIES OF CYTOCHROME REDUCTASE AND THE PARTLY REASSOCIATED ENZYME

Preparation	Turnover number per cytochrome c_1: in phospholipid, Duroquinol : cytochrome c (sec^{-1})	in Triton: Ubiquinol : cytochrome c (sec^{-1})	in Triton: Duroquinol : ubiquinone (sec^{-1})
Cytochrome reductase	58	6.7	5.8
Cytochrome bc_1 subcomplex	0.2	0.06	0.04
Cytochrome bc_1 subcomplex and core complex	0.5	0.09	1.05
Cytochrome bc_1 subcomplex and iron-sulfur subunit	0.5	0.08	0.10
Cytochrome bc_1 subcomplex and core complex and iron-sulfur subunit	18	2.08	2.10

number of about 7 sec^{-1} is obtained with the enzyme in Triton solution, but of 60 sec^{-1} with the enzyme in phospholipid membranes. The activity obtained with duroquinol depends on the amount of ubiquinone-10 present in the membrane (Fig. 2).

No quinol : cytochrome-c reductase activity is obtained with the cytochrome bc_1 subcomplex alone or with the subcomplex associated with either the core complex or the iron-sulfur subunit. Approximately one-third of the original activity is obtained with the cytochrome bc_1 subcomplex associated with the core complex and the iron-sulfur subunit. For maximal activity a 2- to 5-fold molar excess of the iron-sulfur subunit was necessary, this value varying from preparation to preparation. With regard to the core complex the molar ratio necessary for maximal activity was between 0.5 to 1.

Quinol : Quinone Oxidoreductase Activity

The duroquinol : ubiquinone-10 reductase activity is measured at 30° in microcuvettes of 2 mm path length at the wavelength pair 285 nm and 305 nm, which is isosbestic for duroquinol-duroquinone. The molar extinction coefficient for ubiquinol-10 of 5 mM^{-1} cm^{-1} is used.[12] The assay mixture

[12] A. Boveris, R. Oshino, M. Erecinska, and B. Chance, *Biochim. Biophys. Acta* **245,** 1 (1971).

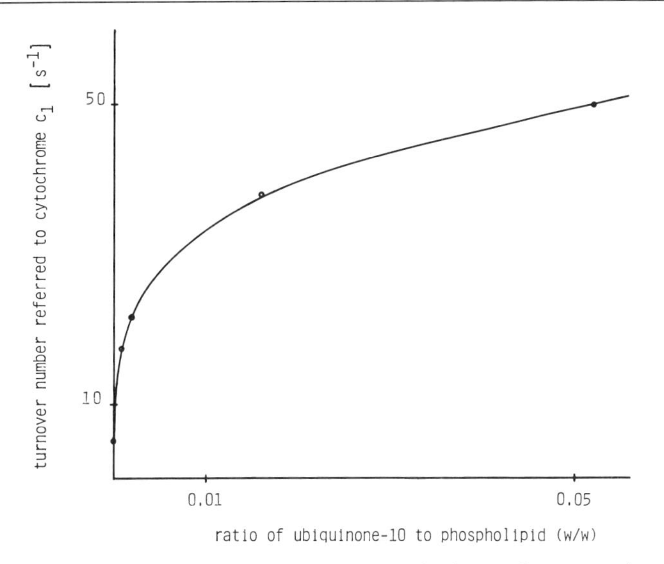

FIG. 2. Dependency on ubiquinone-10 of the duroquinol : cytochrome-c reductase activity.

contains 50 mM Tris-acetate, pH 7.0, 0.1% Triton X-100, 40 μM ubiquinone-10. The solution is preincubated until the turbidity caused by ubiquinone-10 addition has disappeared. The reaction is started by adding the enzyme preincubated with a 10-fold molar excess of myxothiazol[13] to a final cytochrome c_1 concentration of 0.05–0.5 μM and immediately thereafter 50–200 μM duroquinol. The rate of enzymatic ubiquinone-10 reduction is corrected for the rate of nonenzymatic reduction obtained when the enzyme was preincubated with myxothiazol and antimycin. The enzyme rate is extrapolated to infinite duroquinol concentration.

For the intact enzyme a maximal turnover number of 6 sec^{-1} is obtained (Table I). This value is comparable to the ubiquinol–cytochrome-c reductase activity in Triton solution. We were not yet able to determine exact values for duroquinol : ubiquinone reductase activity of the enzyme in phospholipid membranes because the small amount of ubiquinone-10 that can be incorporated into the membrane is used up too fast.

With the cytochrome bc_1 subcomplex alone the quinol : quinone reductase activity was less than 1% as compared to the whole enzyme, but with the cytochrome bc_1 subcomplex associated with the core complex it was about 20%. We do not yet understand why the quinol : quinone re-

[13] G. Thierbach and H. Reichenbach, *Biochim. Biophys. Acta* **638,** 282 (1981).

ductase activity was further increased to about 35% when the iron-sulfur protein was also present (Table II). It should be stressed that the quinol : quinone reductase activity is insensitive to myxothiazol but inhibited by antimycin, whereas the quinol : cytochrome *c* reductase activity is inhibited by both drugs.

Conclusion

According to the Q-cycle mechanism of ubiquinol : cytochrome-*c* reductase[14] (for a review, see Ref. 11), the quinol oxidation occurs in a two-step reaction. One electron is delivered to cytochrome *c* via the iron-sulfur center and cytochrome c_1, the other is transferred across the membrane via the low-potential and high-potential hemes of cytochrome *b* to ubiquinone, yielding a ubisemiquinone. After a second turnover the semiquinone is reduced to ubiquinol.[15] The ubiquinol oxidase site which binds the myxothiazol[16] is exposed to the positively charged membrane side, whereas the ubiquinone reductase site which binds the antimycin is located at the side of the negative protonic potential.

With our reconstitution studies in combination with our structural studies[8] on cytochrome reductase of *Neurospora* we try to give further details on the ubiquinol reductase site. From the observation that the duroquinol : ubiquinone oxidoreductase activity is restored by the cytochrome bc_1 subcomplex associated with the core complex we conclude that the ubiquinol reductase side is located at the contact plane between cytochrome *b* and the core complex. This plane lies just outside the bilayer at the (protonic negative) matrix side of the enzyme.[8,9] We propose that one function of the core complex is the stabilization of the bound ubisemiquinone and the provision of a binding site for duroquinol to deliver two electrons to the bound ubiquinone.

Acknowledgments

We thank Brigitte Kurze for technical assistance and Dr. W. Trowitzsch, Braunschweig, for his gift of myxothiazol. This work was supported by the Deutsche Forschungsgemeinschaft and the Fond der Chemischen Industrie.

[14] P. Mitchell, *J. Theor. Biol.* **62,** 327 (1976).

[15] A. R. Crofts, S. W. Meinhardt, K. R. Jones, and M. Snozzi, *Biochim. Biophys. Acta* **723,** 202 (1983).

[16] H. Schägger, T. A. Link, and G. von Jagow, this volume [22].

[21] Resolution and Reconstitution of the Iron–Sulfur Protein of the Cytochrome bc_1 Segment of the Mitochondrial Respiratory Chain

By CAROL A. EDWARDS and BERNARD L. TRUMPOWER

Introduction

The iron–sulfur protein of the cytochrome bc_1 segment of the mitochondrial respiratory chain, discovered in complex III by Hatefi[1] and isolated therefrom by Rieske and co-workers,[2,3] has been purified in a reconstitutively active form from bovine heart mitochondria by two methods.[4–6] The purified protein is a homogeneous polypeptide on polyacrylamide gel electrophoresis in sodium dodecyl sulfate and migrates with an apparent molecular weight of 24,500.[4–6] This iron–sulfur cluster has a characteristic EPR spectrum with a central resonance absorbance at g_y = 1.90.[3] An iron–sulfur protein having these properties is also present in yeast,[7,8] plant,[9] and *Neurospora crassa*[10] mitochondria, in chloroplasts,[11–13] and in photosynthetic and thermophilic bacteria.[14–23]

[1] Y. Hatefi, A. G. Haavik, and D. E. Griffiths, *J. Biol. Chem.* **237,** 1681 (1962).
[2] J. S. Rieske, D. H. MacLennan, and P. Coleman, *Biochem. Biophys. Res. Commun.* **15,** 338 (1964).
[3] J. S. Rieske, R. E. Hansen, and W. S. Zaugg, *J. Biol. Chem.* **239,** 3017 (1964).
[4] B. L. Trumpower, and C. A. Edwards, *FEBS Lett.* **100,** 13 (1979).
[5] B. L. Trumpower and C. A. Edwards, *J. Biol. Chem.* **254,** 8697 (1979).
[6] W. D. Engel, C. Michalski, and G. von Jagow, *Eur. J. Biochem.* **132,** 395 (1983).
[7] T. Ohnishi, T. G. Cartledge, and D. Lloyd, *FEBS Lett.* **52,** 90 (1975).
[8] J. N. Siedow, S. Power, de la Rosa, F. F. and G. Palmer, *J. Biol. Chem.* **253,** 2392 (1978).
[9] R. C. Prince, W. D. Bonner, and P. A. Bershak, *Fed. Proc., Fed. Am. Soc. Exp. Biol. Abstr.* 1667 (1981).
[10] Y. Li, S. de Vries, K. Leonard, and H. Weiss, *FEBS Lett.* **135,** 277 (1981).
[11] R. Malkin and P. J. Aparicio, *Biochem. Biophys. Res. Commun.* **63,** 1157 (1975).
[12] R. Malkin and H. P. Posner, *Biochim. Biophys. Acta* **501,** 552 (1978).
[13] E. Hurt, G. Hauska, and R. Malkin, *FEBS Lett.* **134,** 1 (1981).
[14] P. L. Dutton and J. S. Leigh, *Biochim. Biophys. Acta* **314,** 178 (1973).
[15] D. W. Reed and G. Palmer, *Biophys. J.* **13,** 63a (1973) (Abstr.).
[16] M. C. W. Evans, A. V. Lord, and S. G. Reeves, *Biochem. J.* **138,** 177 (1974).
[17] R. C. Prince, J. G. Lindsay, and P. L. Dutton, *FEBS Lett.* **51,** 108 (1975).
[18] D. B. Knaff and R. Malkin, *Biochim. Biophys. Acta* **430,** 244 (1976).
[19] J. R. Bowyer, P. L. Dutton, R. C. Prince, and A. R. Crofts, *Biochim. Biophys. Acta* **592,** 445 (1980).
[20] R. C. Prince, J. S. Leigh, and P. L. Dutton, *Biochem. Soc. Trans.* **2,** 950 (1974).
[21] R. Malkin and A. J. Bearden, *Biochim. Biophys. Acta* **505,** 147 (1978).
[22] T. Yoshida, R. M. Lorence, M. G. Choc, G. E. Tarr, K. L. Findling, and J. A. Fee, *J. Biol. Chem.* **259,** 112 (1984).

The iron–sulfur protein purified from bovine heart mitochondria will rebind to the cytochrome bc_1 complex from which it has been extracted and restore cytochrome *c* reductase activity to the complex. The two methods for purification of the reconstitutively active protein utilize this activity as an assay during purification, in which restoration of cytochrome *c* reductase activity is proportional to the amount of iron–sulfur protein added back to the membranous complex which has been depleted of iron–sulfur protein.

Resolution of the iron–sulfur protein from the membranous complex in both methods involves extraction with detergent and a chaotropic agent, although different in each case, in the presence of reducing agent to stabilize the complex. The iron–sulfur protein isolated with cholate and resuspended without added detergent appears to be aggregated in a protein micelle and must be incubated at higher temperature to disperse it and rebind to the membranous complex. The protein isolated with Triton appears to be monodispersed and can restore activity to the depleted complex without incubation, but is brought off the membrane again more easily. These two methods of purification are described below.

Purification of Reconstitutively Active Iron–Sulfur Protein of the Cytochrome bc_1 Complex and Iron–Sulfur Protein-Depleted Complex from Isolated Succinate–Cytochrome *c* Reductase Complex

Assay of Iron–Sulfur Protein Activity

Iron–sulfur protein activity can be measured in a succinate–cytochrome *c* reductase assay, in which case succinate dehydrogenase must also be reconstituted to the depleted complex,[5,24,25] or in a ubiquinol–cytochrome *c* reductase assay. In the latter assay DBH, a synthetic analog of ubiquinone-2, is used as substrate,[26] thus circumventing the need for reconstitutively active succinate dehydrogenase.

Succinate–cytochrome *c* reductase activity is measured in a mixture containing 50 μM cytochrome *c*, 40 mM sodium phosphate, 20 mM sodium succinate, 0.5 mM EDTA, and 0.25 mM potassium cyanide, pH 7.4.

[23] J. A. Fee, K. L. Findling, T. Yoshida, R. Hille, G. E. Tarr, D. O. Hearshen, W. R. Dunham, E. P. Day, T. A. Kent, and E. Munck, *J. Biol. Chem.* **259,** 124 (1984).

[24] B. L. Trumpower, C. A. Edwards, and T. Ohnishi, *J. Biol. Chem.* **255,** 7487 (1980).

[25] B. L. Trumpower, C. A. Edwards, A. Katki, and T. Ohnishi, *in* "Membrane Bioenergetics" (C. P. Lee, G. Schatz, and L. Ernster, eds.), p. 217. Addison-Wesley, Reading, Massachusetts, 1979).

[26] W. P. Wan, R. H. Williams, K. Folkers, K. H. Leung, and E. Racker, *Biochem. Biophys. Res. Commun.* **63,** 11 (1975).

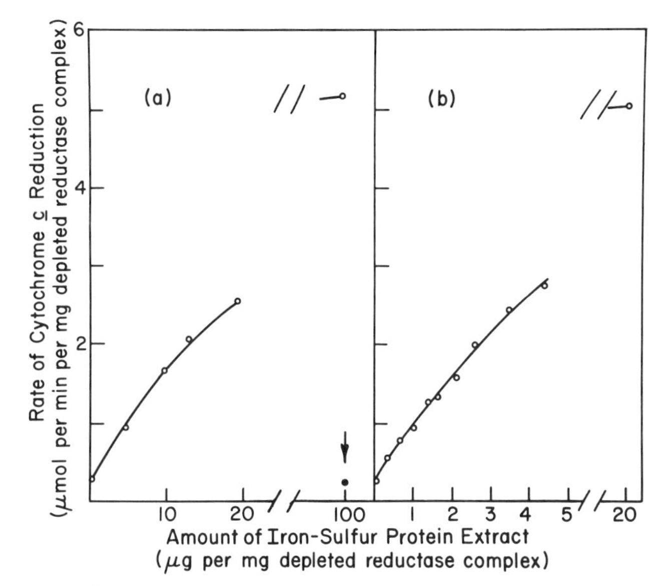

FIG. 1. Assay of iron–sulfur protein activity by reconstitution of succinate–cytochrome *c* reductase activity to depleted reductase complex.[4] The activities in (a) were reconstituted with partially purified iron-sulfur protein, after fractionation with ammonium sulfate (see Fig. 2), while those in (b) were obtained with purified iron–sulfur protein, after mercurial resin plus hydroxyapatite chromatography. The arrow shows the activity of the reconstituted complex in the presence of antimycin.

Ubiquinol–cytochrome *c* reductase activity is measured in the same buffer, using 60 μM reduced DBH as substrate, and 20 mM malonate is substituted for succinate to inhibit any succinate–cytochrome *c* reductase activity which might occur due to transfer of succinate into the assay mixture. In the latter assay the nonenzymatic rate of cytochrome *c* reduction is obtained by adding reduced DBH, allowing the reaction to proceed for 5 sec, and then adding reductase complex. The zero order rates of cytochrome *c* reduction are calculated from the initial absorbance at 550 nm, using $\Delta E_{\text{red-ox}} = 18.5\ \text{m}M^{-1}\ \text{cm}^{-1}$.

Use of the succinate–cytochrome *c* reductase assay to measure iron-sulfur protein specific activity at two steps during its purification is shown in Fig. 1. Specific activities of the iron–sulfur protein are calculated from the rates of cytochrome *c* reduction under conditions where the amount of activity restored to the depleted reductase complex is first order with respect to iron–sulfur protein. In the example shown the specific activity of the partially purified protein (Fig. 1a) is 140 units/mg and that of the purified protein (Fig. 1b) is 700 units/mg. A unit is defined as 1 μmol of cytochrome *c* reduced per minute. Identical specific activities are obtained with the ubiquinol–cytochrome *c* reductase assay.

Preparation of Isolated Succinate–Cytochrome c Reductase Complex

Succinate–cytochrome-*c* reductase complex is isolated from bovine heart mitochondria, previously washed with phosphate buffer, and stored at −70°.[5] The mitochondria are thawed, centrifuged for 60 min at 37,500 *g* to remove proteins solubilized by freezing, and suspended at 30 mg/ml in 0.1 *M* sodium phosphate, pH 7.4. While stirring at 4°, a solution of 20% sodium cholate is added to obtain 0.35 mg of cholate/mg of protein. Solid ammonium sulfate is then added to obtain 25% saturation (144 g/liter), while a pH of 7.4 is maintained by addition of ammonium hydroxide. The mixture is stirred for 60 min, after which the volume is measured and ammonium sulfate added to obtain 35% saturation (61 g/liter). After the mixture is stirred for 20 min, it is centrifuged 60 min at 37,500 *g* and the supernatant collected. Solid ammonium sulfate is added to obtain 50% saturation (94 g/liter), and the mixture is centrifuged for 30 min at 37,500 *g*.

The resulting dark red pellet is dispersed in 0.1 *M* sodium phosphate, 0.5% cholate, 0.5 m*M* EDTA, pH 7.4, by gentle homogenization in a loose-fitting Teflon–glass homogenizer and adjusted to 30 mg/ml. The transparent dark red mixture is stored 16–18 hr at 4°, during which time it becomes turbid due to aggregation of cytochrome *c* oxidase, which is then removed by centrifugation for 90 min at 78,500 *g*.

The bright red, detergent-dispersed reductase is used to prepare reductase complex depleted of iron–sulfur protein (see below), and to purify reconstitutively active iron–sulfur protein. To obtain active reductase complex for purification of iron–sulfur protein, cholate is removed by dialyzing 16–18 hr against 20 volumes of 0.25 *M* sucrose, 10 m*M* sodium phosphate, 0.5 m*M* EDTA, pH 7.4, with 3 changes of buffer, followed by centrifugation for 60 min at 37,500 *g*. The reductase complex is suspended to 15–20 mg/ml in 0.9 *M* sucrose, 10 m*M* TES, pH 7.4, and stored at −70°.

The reductase complex contains 1.4–1.6 nmol of cytochrome c_1 and 2.6–3.0 nmol of cytochrome *b*/mg of protein. It has succinate–cytochrome *c* reductase activity of 5–7 units/mg, succinate–ubiquinone reductase activity of 9–11 units/mg, and ubiquinol–cytochrome *c* reductase activity of 22–25 units/mg.

Purification of Reconstitutively Active Iron–Sulfur Protein

Isolated reductase complex, previously stored at −70°, is centrifuged 60 min at 37,500 *g* and suspended to 10 mg/ml in 0.1 *M* sodium phosphate,

0.5 m*M* EDTA, pH 7.4. While stirring at 4°, the suspension is adjusted to pH 9.5 by addition of 1 *N* NaOH. The mixture is incubated at 37° for 60 min, returned to 4°, and centrifuged for 60 min at 78,500 *g*. The pellet is washed by suspension in 0.1 *M* sodium phosphate, 0.5 m*M* EDTA, pH 7.4, and centrifuged, after which it is suspended in phosphate–EDTA buffer at 15 mg/ml and stored at −70°.

The alkaline-washed reductase is thawed and dispersed by homogenization. An ethanolic solution of antimycin (10 mg/ml) is then added to obtain 2 μg of antimycin/mg of protein. The mixture is stirred 60 min at 4°, after which 11.1 ml of 10% sodium cholate/100 ml of mixture plus 2 mg/ml of sodium hydrosulfite are added, while the pH is maintained at 8.0 with ammonium hydroxide. While continuing to maintain pH 8.0 with ammonium hydroxide, 42.9 ml of 5 *M* guanidine hydrochloride/100 ml of mixture is added.

The resulting mixture, containing 9.4 mg of protein/ml, 0.7% cholate, 1.4 mg/ml sodium hydrosulfite, and 1.5 *M* guanidine hydrochloride, is stirred for 30 min at 4°. It is then diluted with one-half volume of 0.1 *M* sodium phosphate, 0.5 m*M* EDTA, pH 7.4, and centrifuged 30 min at 78,500 *g*. The orange supernatant is dialyzed for 4 hr against a 20-fold volume of 20 m*M* sodium phosphate, 0.5 m*M* dithiothreitol, 0.1 m*M* EDTA, pH 7.4. The dialysis bags are inverted after 1, 2, and 3 hr to facilitate mixing, after which the dialysate is centrifuged 90 min at 78,500 *g*.

Ammonium sulfate is added to the supernatant to obtain 48% saturation (290 g/liter). The precipitate is recovered by centrifugation for 20 min at 37,500 *g* and dissolved in 0.1 *M* sodium phosphate, 0.5% cholate, 0.5 m*M* EDTA, pH 7.4, to a volume equal to one-quarter that of the original alkaline-washed reductase complex. The suspended pellet, at a protein concentration of ~1.5 mg/ml, is transparent and yellowish orange in color.

Saturated ammonium sulfate (3.9 *M*) is then added to obtain 15% saturation. After centrifuging 10 min at 37,500 *g*, the supernatant is recovered and additional saturated ammonium sulfate is added to 35% saturation. The 15–35% ammonium sulfate pellet is dissolved in 20 m*M* TES, 150 m*M* sodium chloride, pH 7.4, and either stored at −70° or applied directly to the two-stage chromatography column.

Final purification of iron-sulfur protein is by a two-stage chromatography procedure. Typically, 10 mg of 15–35% ammonium sulfate pellet is applied to a 0.9 × 4.5 cm column of hydroxyapatite, over which a 0.9 × 2.5 cm bed of mercurial resin is laid. The column is equilibrated with 20 m*M* TES, 150 m*M* sodium chloride, pH 7.4, and the sample applied in this same buffer. The column is eluted with 5–6 ml of TES/sodium chloride buffer followed by 30 ml of a 25–200 m*M* linear gradient of potassium phosphate, pH 7.0. Fractions are collected and monitored for protein by absorbance at 280 nm and for iron-sulfur protein activity by the reconstitution assay.

Purification of Iron–Sulfur Protein-Depleted Reductase Complex

After aggregated cytochrome oxidase is separated from the reductase complex by centrifugation (see above), the detergent-dispersed reductase is adjusted to 12–15 mg/ml by addition of 0.1 *M* sodium phosphate, 0.5% cholate, 0.5 m*M* EDTA, pH 7.4, after which 92 ml of glycerol/100 ml of mixture is slowly added. While stirring at 4°, 2 mg/ml of sodium hydrosulfite is added, followed by 186 ml of 2 *M* guanidine hydrochloride/100 ml of mixture to obtain a concentration of 1.30 *M* guanidine. Subsequent to the addition of hydrosulfite and during the addition of guanidine, the mixture is maintained at pH 8.0 by addition of ammonium hydroxide. The mixture is stirred for 60 min at 4°, during which time an additional 1 mg/ml of sodium hydrosulfite is added in approximately equal portions at 10-min intervals while maintaining pH 7.6.

The mixture is centrifuged 90 min at 78,500 *g*, and the orange–red supernatant is discarded. The bright red pellet and any loosely packed residue are suspended in 0.1 *M* sodium phosphate, 0.5% cholate, 0.5 m*M* EDTA, pH 7.4, to one-half the original volume of dispersed reductase complex, and centrifuged 60 min at 165,000 *g*. The pellet is suspended in 2 *M* urea to one-half the original volume of reductase complex, stirred for 15 min at 4°, then diluted with an equal volume of 40 m*M* sodium phosphate, 20% glycerol, pH 7.4, and centrifuged as above. The depleted reductase complex is suspended in 40 m*M* sodium phosphate, 20% glycerol, pH 7.4, to 15–20 mg/ml, and stored in aliquots at −70°.

Comments on Purification Procedures

The schemes in Fig. 2 summarize the procedures to purify reconstitutively active iron–sulfur protein from succinate–cytochrome *c* reductase complex and to prepare reductase complex depleted of iron–sulfur protein. The amount of reconstitutively active protein extractable from reductase complex depends on the extraction conditions.[5,25] Greater amounts of activity are extracted by guanidine plus cholate if the complex is first washed at pH 9.5, treated with antimycin, and reduced with dithionite during the extraction.

Latent activity of the iron–sulfur protein-depleted complex is preserved during extraction of the complex with guanidine and cholate by

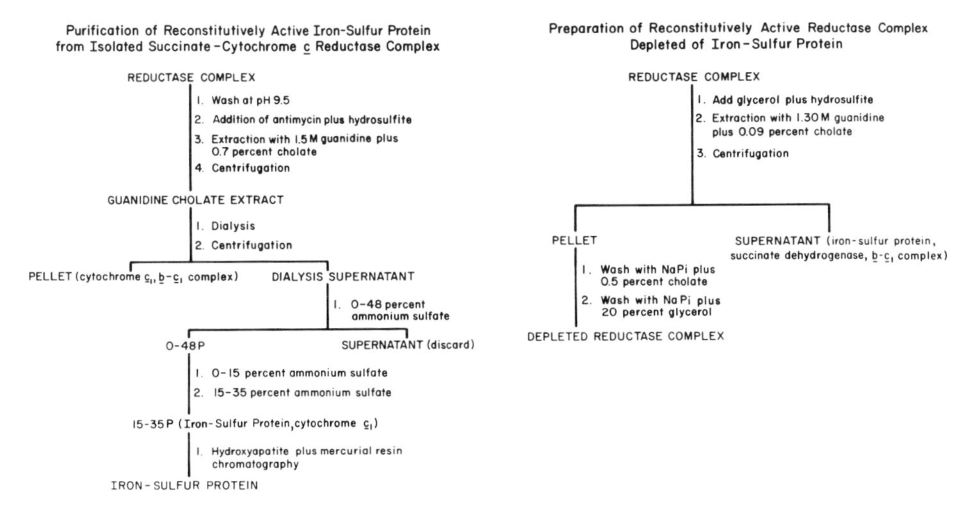

FIG. 2. Flow scheme summarizing the methods for purification of reconstitutively active iron–sulfur protein from isolated succinate–cytochrome *c* reductase complex and for preparation of reconstitutively active reductase complex depleted of iron–sulfur proteins.[5]

including 25% glycerol and 2 mg/ml dithionite, while maintaining the pH at 8.0. Antimycin is omitted during the extraction to prepare the depleted complex, since the inhibitor remains bound to the depleted complex and inhibits activity after reconstitution with iron–sulfur protein.

Electrophoresis profiles at different steps of the purification are shown in Fig. 3. After extraction from the reductase complex and concentration with ammonium sulfate, the partially purified 0-48P fraction (Fig. 3b) appears to be virtually free of the higher molecular weight polypeptides present in the reductase complex (Fig. 3a), including the 44,300 and 48,200 core proteins,[27–29] two larger peptides which may be the α and β subunits of the ATPase, and the 30,600 heme-containing peptide of cytochrome c_1.[27,30] Most of the c_1 which is extracted precipitates during dialysis[31] and can be recovered and used for purification of this cytochrome.[27] An unidentified inhibitor, present in the 0-48P fraction, is removed by fractionation with ammonium sulfate.[5,25]

[27] B. L. Trumpower and A. Katki, *Biochemistry* **14,** 3636 (1975).
[28] H. I. Silman, J. S. Rieske, S. H. Lipton, and H. Baum, *J. Biol. Chem.* **242,** 4867 (1967).
[29] P. Gellerfors, M. Lunden, and B. D. Nelson, *Eur. J. Biochem.* **67,** 463 (1976).
[30] C. A. Yu, L. Yu, and T. E. King, *J. Biol. Chem.* **247,** 1012 (1972).
[31] H. Nishibayashi-Yamashita, C. Cunningham, and E. Racker, *J. Biol. Chem.* **247,** 698 (1972).

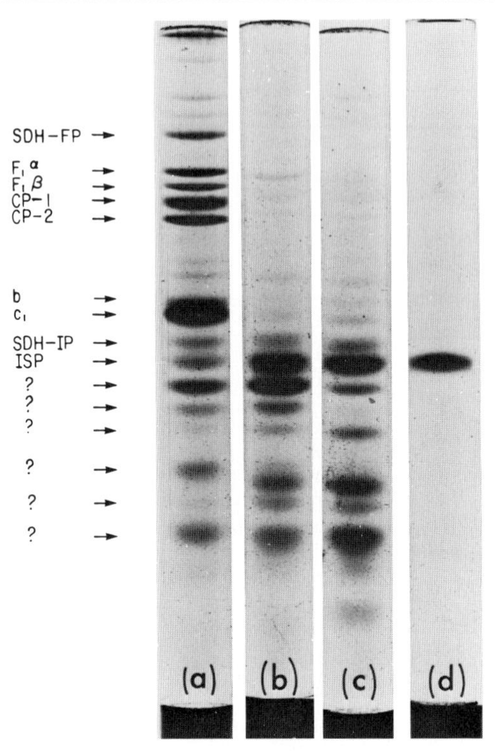

FIG. 3. Polyacrylamide gel electrophoresis in sodium dodecyl sulfate of iron–sulfur protein at different stages of purification. The electrophoresis gel in (a) is of alkaline-washed reductase complex, gel (b) is of the 0-48P fraction, and (c) is of the 15–35P fraction. Gel (d) is of the pure protein after two-stage chromatography. The arrows and numbers indicate the migration positions of the FP subunit of succinate dehydrogenase, contaminating subunits from F_1-ATPase, the 48,200 and 44,300 core proteins of the bc_1 complex, apocytochrome *b*, the 30,600 heme-containing polypeptide of cytochrome c_1, the 27,000 IP subunit of succinate dehydrogenase, and the 24,500 iron–sulfur protein polypeptide.

Final purification of the iron–sulfur protein is by two-stage chromatography in which the protein is passed through a mercurial resin and then adsorbed to hydroxyapatite.[5,25] Prior to chromatography the sample contains two low-molecular-weight polypeptide impurities (Fig. 3c) and small amounts of c_1, which binds to the mercurial resin through sulfhydryl groups.[27]

Resolution of Iron–Sulfur Protein and bc_1 Subcomplex from Complex III

Alternatively, the iron–sulfur protein can be dissociated from complex III by treatment with Triton plus urea and separated by density gradient

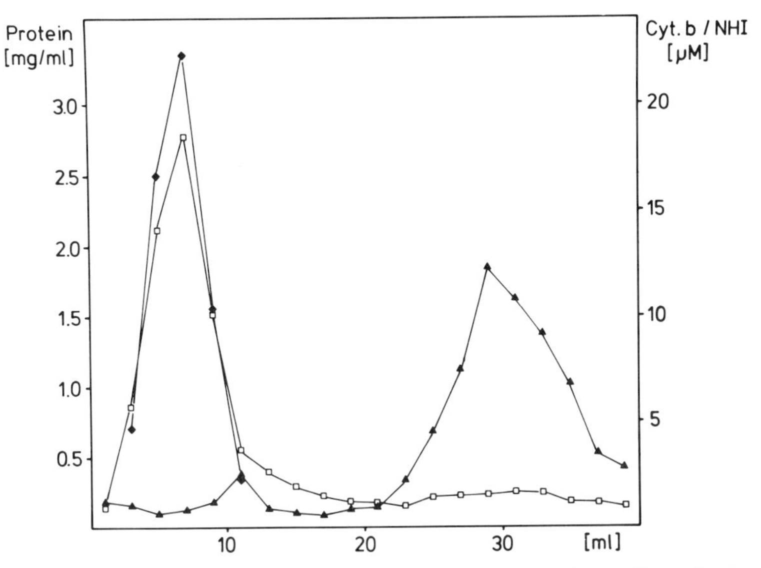

FIG. 4. Distribution pattern of complex III in the sucrose density gradient after incubation with urea and dithionite. Fractions were collected from the bottom upward. Cytochrome *b* (◆), a marker for the bc_1 subcomplex, is found in the lower third of the gradient with most of the protein (□). Nonheme iron (▲), a marker for the iron–sulfur protein, is found in the upper third of the gradient.

centrifugation, yielding reconstitutively active iron–sulfur protein and a reconstitutively active bc_1 subcomplex devoid of iron–sulfur protein.[6] Complex III, prepared by hydroxyapatite and gel filtration chromatography,[32] is precipitated by overnight dialysis against 100 m*M* NaCl, 20 m*M* MOPS, pH 7.2, to remove glycerol and detergent, and collected by centrifugation. The precipitate is solubilized in 100 m*M* NaCl, 20 m*M* MOPS, 2 *M* urea, pH 7.2, by addition of 20% (v/v) Triton X-100 with homogenization, after which 5 m*M* dithionite is added to reduce the redox components. A final concentration of ~0.5% Triton is adequate to solubilize the complex at 25 mg/ml.

Iron–sulfur protein dissociates from the bc_1 subcomplex and is separated therefrom by density gradient centrifugation, as shown in Fig. 4. Typically, 2.5 ml of Triton-solubilized complex III is loaded onto a 40-ml linear gradient containing 10–30% sucrose in 250 m*M* NaCl, 20 m*M* MOPS, 5 m*M* dithionite, 0.1% Triton X-100, 2 *M* urea, and 0.2 m*M* phenylmethylsulfonyl fluoride, pH 7.2.

[32] W. D. Engel, H. Schagger, and G. von Jagow, *Biochim. Biophys. Acta* **592,** 211 (1980).

Fractions from the top of the gradient contain iron–sulfur protein, slightly contaminated by other complex III subunits, as shown by electrophoresis of the gradient fractions in Fig. 5. The faster sedimenting bc_1 subcomplex is recovered from the bottom of the gradient.

Iron–sulfur protein-containing fractions, identified by nonheme iron (Fig. 5), are pooled and desalted on a Sephadex G-25 column equilibrated with 50 m*M* Tris, 50 m*M* NaCl, 2 m*M* dithioerythritol, 1 m*M* sodium azide, 0.1% Triton X-100, pH 9. The protein can be further purified by passage through a QAE-Sephadex A-25 column equilibrated with this same buffer. The active iron–sulfur protein elutes in the flow through fractions.

The bc_1 subcomplex fractions from the gradient are pooled and desalted on a Sephadex G-25 column equilibrated with 100 m*M* NaCl, 20 m*M* MOPS, 1 m*M* sodium azide, and 0.05% Triton X-100, pH 7.2. After adding glycerol to a final concentration of 50% (v/v), the complex is stored at −20°.

Reconstitution of Iron–Sulfur Protein to Depleted Reductase Complex and Restoration of Cytochrome c Reductase Activity to Complex III by Reversibly Dissociated Iron–Sulfur Protein

Reconstitution of Iron–Sulfur Protein to Depleted Reductase Complex

Succinate–cytochrome *c* reductase complex was first isolated by Hatefi and co-workers (for reviews, see Refs. 33 and 34). Prior to extraction of the iron–sulfur protein of the bc_1 complex, the reductase complex has succinate–ubiquinone reductase activity of 9–11 units/mg, ubiquinol–cytochrome *c* reductase activity of 22–25 units/mg, and succinate–cytochrome *c* reductase activity of 5–7 units/mg.[33,34] Extraction of the isolated reductase complex with guanidine plus cholate removes succinate dehydrogenase in addition to the iron–sulfur protein of the bc_1 complex; consequently, all of these activities are lost.

Extraction with guanidine plus cholate causes a loss of the $g_y = 1.90$ EPR signal and the 24,500 iron–sulfur protein polypeptide (Fig. 6a). Thus, the loss of cytochrome *c* reductase activities results from a bona fide resolution of iron–sulfur protein rather than *in situ* destruction of the iron-sulfur cluster. After reconstitution of the depleted complex with iron–sulfur protein, the 24,500 polypeptide remains bound to the reconsituted complex, as shown in Fig. 6b.

[33] Y. Hatefi, *in* "The Enzymes of Biological Membranes" (A. Martonosi, ed.), Vol. 4, p. 3. Plenum, New York, 1976.

[34] B. L. Trumpower and A. Katki, *in* "Membrane Proteins in Energy Transduction" (R. A. Capaldi, ed.), Vol. 2, p. 89. Dekker, New York, 1979.

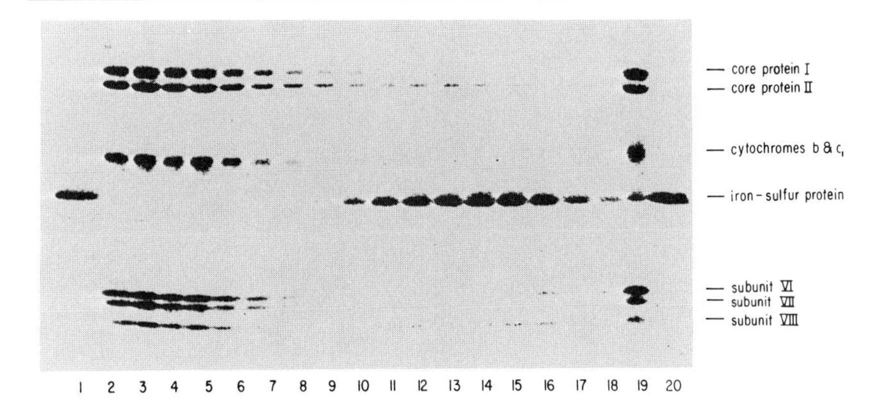

FIG. 5. Polyacrylamide gel electrophoresis in sodium dodecyl sulfate of the isolated iron–sulfur protein (lanes 1 and 20), and various fractions from the sucrose density gradient (lanes 2–18) shown in Fig. 4. Lane 19 shows the sample of complex III applied to the gradient.

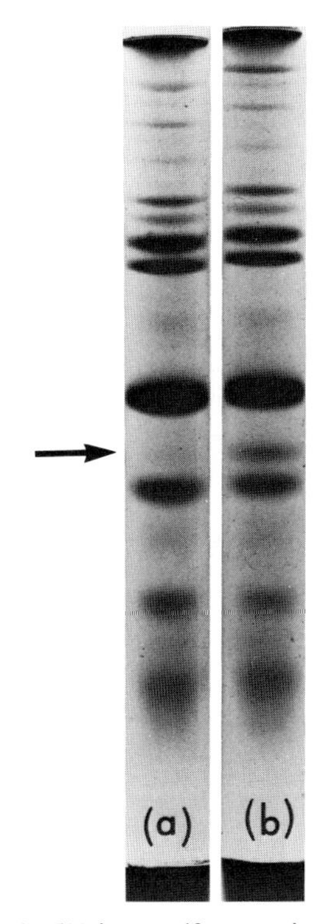

FIG. 6. Reappearance of the 24,500 iron–sulfur protein polypeptide after reconstitution of depleted reductase complex with purified iron–sulfur protein. The electrophoresis gel in (a) is of the depleted complex, and that in (b) is of depleted reductase complex after reconstitution with iron–sulfur protein. The arrow points to the migration position of the iron–sulfur protein.

Depleted reductase complex is reconstituted with iron–sulfur protein by mixing at 4°, in sequence, 20 μl of buffer containing 200 m*M* sodium phosphate, 200 m*M* sodium succinate, 10 m*M* EDTA, pH 7.4, 0.6 nmol of depleted reductase complex, 300 nmol of soybean phospholipid,[31] and 6 nmol of ubiquinone-10. Variable amounts of iron–sulfur protein are added, followed by water, to bring the volume to 200 μl. The samples are then incubated 60 min at 35°. The incubation is stopped by diluting the samples to 2 ml with 20 m*M* sodium phosphate, 20 m*M* sodium succinate, 1 m*M* EDTA, pH 7.4, previously chilled to 4°, after which the samples are retained at 4°. Aliquots containing 1–5 μg of reconstituted complex are withdrawn for measurement of electron-transfer activities.

The results in Fig. 7 show the reconstitution of succinate–cytochrome *c* reductase (solid lines) and ubiquinol–cytochrome *c* reductase (dashed lines) activities to depleted reductase complex by the iron–sulfur protein. In the depleted complex these activities are 0.3 and 0.6 unit/mg, respectively. If the iron–sulfur protein is added back to the depleted complex in the presence of succinate dehydrogenase, it restores both succinate and ubiquinol–cytochrome *c* reductase activities (Fig. 7a). Both of these reconstituted activities are fully inhibited by antimycin. If the dehydrogenase is omitted, the iron–sulfur protein reconstitutes ubiquinol–cytochrome *c* reductase, but not succinate–cytochrome *c* reductase activity (Fig. 7b). Restoration of the cytochrome *c* reductase activities is associated with reappearance of the $g_y = 1.90$ EPR signal and the 24,500 iron–sulfur protein polypeptide.[24]

The amount of cytochrome *c* reductase activity reconstituted depends on time and temperature during incubation of the depleted complex with iron–sulfur protein. At 35° restoration of activity is more than 90% complete after 60 min, whereas at 4° less than half of this activity is restored after 8 hr.[24] These results suggest that the iron–sulfur protein is a slowly equilibrating mixture of monomeric and oligomeric forms in which the reconstitutively active form is present in low concentrations. The amount of activity reconstituted also depends on the concentrations of depleted reductase complex and iron–sulfur protein in the incubation mixture, as expected for a second-order reaction between these two components.[35]

If the reconstituted reductase complex is to be used to examine oxidation–reduction reactions of the redox components in the complex, the amounts of reconstituted complex required are greater than described above and the reconstitution is scaled up appropriately.[35] In this case the succinate, which is included in the reconstitution incubation buffer, is

[35] C. A. Edwards, J. R. Bowyer, and B. L. Trumpower, *J. Biol. Chem.* **257,** 3705 (1982).

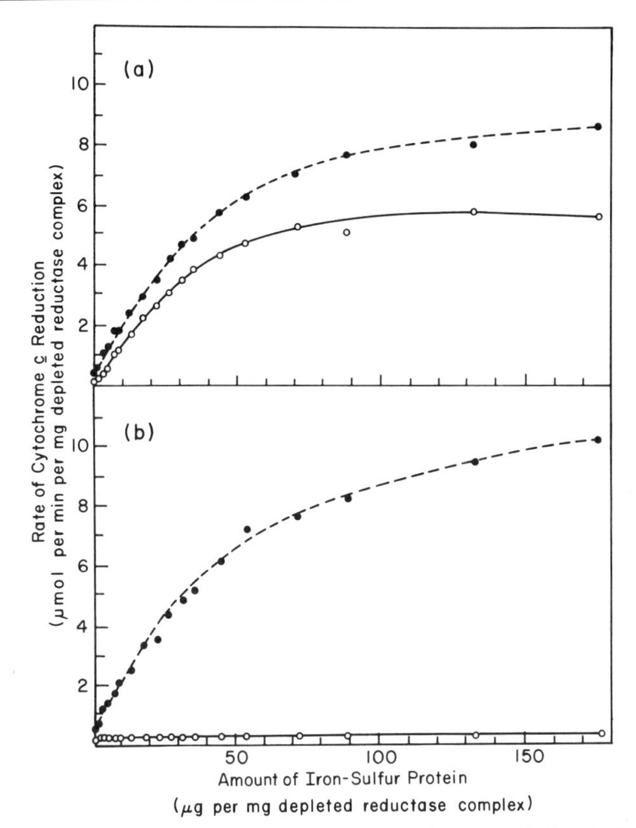

FIG. 7. Reconstitution of succinate–cytochrome *c* reductase and ubiquinol–cytochrome *c* reductase activities to depleted reductase complex by purified iron–sulfur protein. The curves in (a) show the succinate–cytochrome *c* reductase (○) and ubiquinol–cytochrome *c* reductase (●) activities of depleted reductase complex reconstituted with purified iron–sulfur protein in the presence of succinate dehydrogenase and phospholipid plus ubiquinone-10. The curves in (b) show the activities under the same conditions as in (a), except that succinate dehydrogenase was omitted from the reconstitution.

removed after reconstitution by centrifugal gel filtration columns so that redox reactions of the reconstituted complex can be initiated with the cytochromes in the oxidized state.[35]

Restoration of Cytochrome c Reductase Activity to Complex III by Reversibly Dissociated Iron–Sulfur Protein

In this method iron–sulfur protein and bc_1 subcomplex are mixed for only a short (seconds) preincubation, after which aliquots are measured for activity in the ubiquinol–cytochrome *c* reductase assay. The iron–sulfur protein and bc_1 subcomplex are apparently loosely associated such that they dissociate when transferred into the cytochrome-*c* reductase assay, and the rates of cytochrome *c* reduction drop off during the assay.[6]

To 2 μl of bc_1 (42 μ*M*) subcomplex in 20 m*M* MOPS, 100 m*M* NaCl, 0.1% Triton X-100, pH 7.2, is added 2–60 μl of iron–sulfur protein (9 μ*M*) in the same buffer. The volume is adjusted to 62 μl by addition of buffer, and after 15 sec aliquots are withdrawn for measurement of cytochrome-*c* reductase activity.

Acknowledgments

We thank Professor Gebhard von Jagow for providing Figs. 4 and 5 and for helpful discussions of the method for reversibly dissociating iron–sulfur protein and cytochrome bc_1 subcomplex.

[22] Isolation of the Eleven Protein Subunits of the bc_1 Complex from Beef Heart

By H. Schägger, Th. A. Link, W. D. Engel, and G. von Jagow

The middle segment of the respiratory chain of mammalian mitochondria, the segment of energy coupling site 2, is a multiprotein complex which is usually referred to as the bc_1 complex and which is composed of many more proteins than, for instance, the bc_1 complexes of chromatophores or the b_6f complex of chloroplasts.[1] The mammalian bc_1 complex consists of 11 individual proteins, only 3 of which carry a redox center. These are cytochrome *b*, a 43.7 kDa protein carrying two heme centers which are about 20 Å apart from each other[2]; cytochrome c_1, a 27.9 kDa protein[3]; and a ferredoxin-type iron-sulfur protein with an unusually posi-

[1] G. Hauska, E. Hurt, N. Gabellini, and W. Lockau, *Biochim. Biophys. Acta* **726,** 97 (1983).
[2] T. Ohnishi and G. von Jagow, *Biophys. J.* **47,** 241a (1985).
[3] S. Wakabayashi, H. Matsubara, C. H. Kim, K. Kawai, and T. E. King, *Biochem. Biophys. Res. Commun.* **97,** 1548 (1980).

tive redox center, designated, according to its discoverer, the Rieske iron-sulfur protein. Thus, about two-thirds of the protein mass consists of subunits lacking a redox center. The functional purpose of these subunits remains an enigma for the present.

In order to establish their structure and function, an isolation procedure for all 11 subunits has been developed. It uses a bc_1 complex isolated in Triton X-100[4,5] as starting material, since this preparation is in a monodisperse state of aggregation. A successful cleavage of the subunits cannot be obtained using a bc_1 complex prepared in cholate, since cholate complexes form huge aggregates.

The basic principles of the preparative approach used here are (1) a successive cleavage of the multisubunit complex into smaller subcomplexes and free proteins by urea and guanidine, and (2) the different binding behavior of the various subcomplexes and subunits to hydroxyapatite. For a detailed scheme see Fig. 1. The proteins are obtained in a Triton-dispersed state, with the exception of the two core proteins and the 9.5 kDa protein, which are present in an unfolded state after treatment with SDS.

The cytochrome *b* obtained still carries both heme centers. Although displaying CO sensitivity and loss of the splitting of the two heme signals in the absorption spectrum, both *b* centers still have two different potentials in the redox titration.[6,7] Cytochrome c_1 seems to be unperturbed in c_1 subcomplex, but displays CO reactivity when isolated. The isolated iron-sulfur protein is reconstitutively active. The properties of the remaining subunits are subject to further studies.

Preparation of the bc_1 Complex in Triton X-100

All procedures are performed at 4°. Mitochondria prepared according to Smith[8] are concentrated by a 15-min centrifugation step at 27,000 *g* to reduce the amount of sucrose. The hydroxyapatite used during all procedures is prepared according to Tiselius *et al.*[9] This material has a better flow rate than that commercially available.

[4] W. D. Engel, H. Schägger, and G. von Jagow, *Biochim. Biophys. Acta* **592,** 211 (1980).

[5] W. D. Engel, H. Schägger, and G. von Jagow, *Hoppe-Seyler's Z. Physiol. Chem.* **364,** 1753 (1983).

[6] G. von Jagow, H. Schägger, W. D. Engel, W. Machleidt, I. Machleidt, and H. J. Kolb, *FEBS Lett.* **91,** 121 (1978).

[7] G. von Jagow, W. D. Engel, H. Schägger, W. Machleidt, and I. Machleidt, *in* "Vectorial Reactions in Electron and Ion Transport in Mitochondria and Bacteria" (F. Palmieri, E. Quagliariello, N. Siliprandi, and E. C. Slater, eds.), p. 149. Elsevier, Amsterdam, 1981.

[8] A. L. Smith, this series, Vol. 10, p. 81.

[9] A. Tiselius, S. Hjerten, and O. Levin, *Arch. Biochem. Biophys.* **65,** 132 (1956).

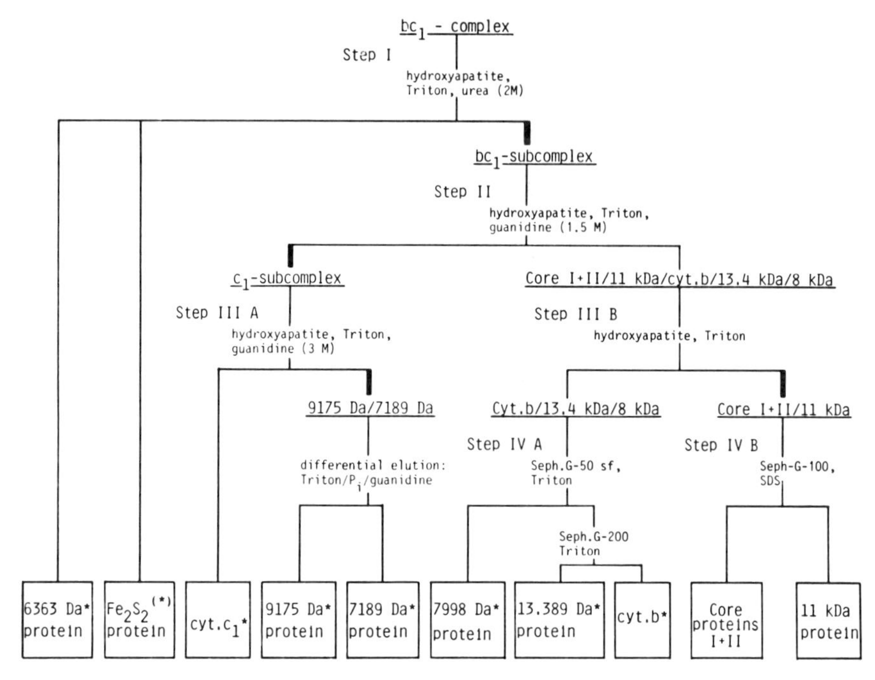

FIG. 1. Path of preparation of the 11 protein subunits of the bc_1 complex from beef heart mitochondria. Heavy lines indicate fractions bound to the hydroxyapatite column. The amino acid sequences of proteins marked by an asterisk have been elucidated.[3,16–21] The amino acid sequence of the iron-sulfur protein from *N. crassa* has been derived from the DNA sequence.[22] (*Note added in proof:* The amino acid sequence of the former 11 kDa protein has been determined (U. Borchart *et al., FEBS Lett.*, submitted). The molecular mass is 9507 Da.)

Negative Extraction of Mitochondria

Mitochondria are suspended in 20 m*M* MOPS, pH 7.2, to a protein concentration of 35 mg/ml. To 100 ml of the suspension 11.6 ml of a 20% Triton X-100 solution and 20 ml of 4 *M* NaCl are added, giving final concentrations of 26 mg protein/ml, 1.75% Triton X-100, and 600 m*M* NaCl. The suspension is centrifuged for 45 min at 100,000 *g*.

Solubilization of the bc_1 Complex

The sediment is resuspended in 300 m*M* sucrose, 20 m*M* MOPS, pH 7.2, to a protein concentration of 35 mg/ml and diluted by the same volume of a buffer containing 4% Triton, 1.2 *M* NaCl, 20 m*M* MOPS, 300 m*M* sucrose, 2 m*M* NaN_3, pH 7.2. After stirring for 5 min the solution is centrifuged for 45 min at 100,000 *g*.

Hydroxyapatite Batch Procedure

The reddish brown supernatant is bound to an equal volume of hydroxyapatite, equilibrated with 0.5% Triton, 250 m*M* NaCl, 80 m*M* Na-phosphate, 2 m*M* NaN_3, pH 7.2. The bc_1 complex bound to hydroxyapatite is sedimented by slow centrifugation. After suctioning off of the supernatant, the sediment is resuspended in 5 times the volume of hydroxyapatite of a buffer containing 0.05% Triton, 250 m*M* NaCl, 110 m*M* Na-phosphate, 2 m*M* NaN_3, pH 7.2, and sedimented again by slow centrifugation. The sediment is taken up in the same buffer, packed into a column, and rinsed with half a column volume of the same buffer. Then a buffer containing 0.25% Triton, 0.2 *M* K-phosphate, 2 m*M* NaN_3, pH 7.2, is applied to elute the bc_1 complex. The fractions of crude bc_1 complex are concentrated by pressure filtration on an Amicon YM100 membrane to a protein concentration of 15 mg/ml.

Gel Filtration

In a final step the bc_1 complex is purified by gel filtration on a Sepharose Cl-6B column, equilibrated with 0.05% Triton, 100 m*M* NaCl, 20 m*M* MOPS, 2 m*M* NaN_3, pH 7.2.

Storage

The bc_1 complex is stored at $-20°$ in 50% glycerol (w/w) and can be used for functional studies up to a month. For the preparation of the protein subunits, however, a freshly prepared bc_1 complex is used because storage leads to partial degradation of some subunits by proteases.

Preparation of the Protein Subunits of bc_1 Complex

Step I. Cleavage of bc_1 Complex into Three Fractions: The 6.4 kDa Protein, the Rieske Iron-Sulfur Protein, and bc_1 Subcomplex

Freshly prepared bc_1 complex (100 mg) in 100 m*M* NaCl, 100 m*M* MOPS, 0.05% Triton X-100, is applied onto a 100-ml hydroxyapatite column after addition of Na-phosphate buffer, pH 7.2, to a final concentration of 35 m*M*. The hydroxyapatite column has to be equilibrated in advance with buffer 1 (for all buffers used in the isolation of the subunits see Table I). After washing the hydroxyapatite-bound bc_1 complex with 50 ml of buffer 1, the 6.4 kDa protein and the Rieske iron-sulfur protein are split off by application of 30 ml of buffer 2. After consecutive application of 50 ml of buffer 3 and 50 ml of buffer 1, the 6.4 kDa protein elutes

TABLE I
BUFFERS FOR PREPARATION OF THE SUBUNITS OF THE bc_1 COMPLEX[a]

	1	2	3	4	5	6	7	8	9	10	11	12	13
Triton X-100 (%)	0.05	1.0	0.05	0.05	0.5	0.5	0.5	1	0.5	0.5	0.5	0.1	0.05
Na-chloride (*M*)	0.05	0.4	0.4	0.2	0.2	0.25	—	—	0.2	—	—	0.2	0.1
Na-phosphate (*M*)	0.035	0.025	0.05	—	0.025	0.085	0.35	0.075	0.01	0.15	—	—	—
K-phosphate (*M*)	—	—	—	—	—	—	—	—	—	—	0.25	—	—
MOPS (*M*)	—	—	—	0.01	—	—	—	—	—	—	—	0.01	0.01
Guanidine (*M*)	—	—	—	—	1.5	—	—	6.0	—	—	3.0	—	—
Na-dithionite (*M*)	—	0.002	0.002	—	—	—	—	—	—	—	—	—	—
PMSF (*M*)	—	0.0002	0.0002	0.0002	—	—	—	—	0.0002	—	—	0.0002	0.0002
Urea (*M*)	—	2.0	2.0	—	—	—	—	—	—	—	—	—	—

[a] All buffers are adjusted to pH 7.2 and contain 1 m*M* Na-azide.

TABLE II
YIELD OF bc_1 COMPLEX AND ITS 11 PROTEIN SUBUNITS[a]

Subunit	Molecular mass (kDa)	Yield (mg)	Yield (%)
bc_1 complex	248	100	100
bc_1 subcomplex	216	80	90
c_1 subcomplex	44	15	85
Core I + II	49 + 47	23	60
Cytochrome *b*	43.7	7	40
Cytochrome c_1	27.9	7	60
Fe_2S_2 protein	25	6	60
13389 Da protein	13.4	5.2	95
9507 Da protein	9.5	2.8	75
9175 Da protein	9.2	1.7	45
7998 Da protein	8.0	1.5	45
7189 Da protein	7.2	1.8	60
6363 Da protein	6.4	1.7	65

[a] For the calculation of percentage yield a 1 : 1 stoichiometry for all protein subunits was assumed.

before the iron-sulfur protein. The eluted protein fractions are passed through a Sephadex G-25 column preequilibrated with buffer 4 and are then stored at −20°. For all final yields see Table II.

Step II. Cleavage of the bc_1 Subcomplex into a Fraction Comprising 6 Proteins and a Cytochrome c_1 Subcomplex Composed of 3 Proteins

The bc_1 subcomplex still bound to hydroxyapatite is cleaved further by application of one column volume of buffer 5, leading to the elution of 6 proteins. The cytochrome c_1 subcomplex remains bound on the column and is washed with 100 ml of buffer 6. After the column has been brought to room temperature, the cytochrome c_1 subcomplex is eluted with buffer 7. It is then dialyzed against 10 m*M* MOPS, pH 7.2, at 4° and processed further according to step IIIA. The eluate containing the 6 proteins is processed according to step IIIB.

Step IIIA. Cleavage of the Cytochrome c_1 Subcomplex into the Heme-Carrying 27.9 kDa Cytochrome c_1, the 9.2 kDa Hinge Protein, and the 7.2 kDa Protein

The dialyzed cytochrome c_1 subcomplex (400 nmol heme) is concentrated to 5 ml by pressure filtration on an Amicon PM10 membrane. The sample is then diluted with an equal volume of buffer 8 and applied to a

10-ml hydroxyapatite column equilibrated with buffer 6. Ten milliliters each of buffer 9, buffer 10, and buffer 11 are applied consecutively. Thereby about 80% of the heme-carrying cytochrome c_1 passes through the hydroxyapatite column unbound. Buffer 10 causes the 9175 Da protein to elute, to a small extent still contaminated by the other two subunits. Buffer 11 finally elutes the 7189 Da protein, also slightly contaminated. For chemical purposes the 9175 Da protein and the 7189 Da protein have to be purified by the following procedure: gel filtration on Sephadex G-25, preequilibrated with buffer 12; precipitation by an equal volume of acetone and leaving overnight at $-20°$; resolubilization in 4% SDS and 1% mercaptoethanol, followed by gel filtration on Sephadex G-50 sf (superfine) in 1% SDS at room temperature.

Step IIIB. Separation of the Fraction Containing the Two Core Proteins and the 9.5 kDa Protein from the Fraction Containing Cytochrome b, the 13.4 kDa Protein, and the 8 kDa Protein

The fraction obtained in step II, containing 6 of the 11 proteins of the bc_1 complex, is separated from guanidine by passing through a 250-ml Sephadex G-25 column preequilibrated with buffer 12. The eluate is then applied to a hydroxyapatite column of 30 ml volume, preequilibrated with buffer 9, and subsequently washed by 30 ml of buffer 12. More than 60% of cytochrome *b* and almost 100% of the 13.4 kDa and the 8 kDa protein pass the column unbound, whereas the core proteins, the 9.5 kDa protein, and the rest of cytochrome *b* remain bound on the hydroxyapatite column.

Step IVA. Separation of Cytochrome b, the 13.4 kDa Protein, and the 8 kDa Protein from Each Other

The fraction containing cytochrome *b*, the 13.4 kDa protein, and the 8 kDa protein is concentrated by a factor of 10 by pressure filtration on an Amicon YM5 membrane. PMSF is present in the buffer to prevent protease degradation of the 8 kDa protein. The Triton concentration should not exceed 5%. It is then applied onto a Sephadex G-50 sf column of 600 ml volume, preequilibrated with buffer 13. The gel filtration leads to the separation of the 8 kDa protein. The cytochrome *b*/13.4 kDa fractions are pooled and concentrated by pressure filtration on an Amicon YM5 membrane so that the Triton concentration measured at 280 nm does not exceed 5%. Gel filtration on a Sephadex G-200 column (600 ml) preequilibrated with buffer 12 leads to a complete separation of cytochrome *b* from the 13.4 kDa protein.

Step IVB. Separation of the 9.5 kDa Protein from the Two Core Proteins

The fraction containing the 9.5 kDa protein and the two core proteins still bound on the hydroxyapatite column is eluted by buffer 11. After removing the guanidine by gel filtration on a 100-ml Sephadex G-25 column preequilibrated with buffer 12, the proteins are precipitated by an equal volume of acetone. The precipitate is washed with water and dissolved in 4% SDS, 1% mercaptoethanol, to a final protein concentration of 10 mg/ml. Then the 9.5 kDa protein is separated from the two core proteins by gel filtration at room temperature in 1% SDS on a Sephadex G-100 column of 300 ml. Core protein I is only partially separated from core protein II under these conditions.

SDS–Polyacrylamide Gel Electrophoresis

The SDS–PAGE method described was developed because none of the systems described in the literature[10–13] allowed the resolution of all the protein subunits of the bc_1 complex. The SDS–PAGE systems according to Swank and Munkres[10] and Merle and Kadenbach[13] both do not separate the 9175 Da protein from the 9.5 kDa protein. A modified staining procedure was necessary because the 8 kDa protein does not stain in the methanolic staining solution.

The resolving power of the SDS–PAGE described increases with falling molecular mass of the proteins. Very effective resolution in the range of 3–25 kDa is combined with an overall separation range of 2–100 kDa. Thus, only one gel is necessary for the resolution of all protein subunits of the bc_1 complex. In a semilogarithmic plot a linear correlation between molecular mass and migration distance is observed over the whole separation range. The gel system can also be used for the preparative separation of small proteins and protein fragments for amino acid sequencing and for the separation of protein subunits covalently labeled with inhibitors of proton and electron transport. Urea may lead to blocking of amino groups and is avoided. Glycine as used in the Laemmli system[14] is replaced by Tricine. Thus, difficulties in preparing samples for amino acid sequencing are overcome.

In the SDS–PAGE of the protein subunits of the bc_1 complex (Fig. 2), up to band VI the numerical order of the bands is the same as described by

[10] R. T. Swank and K. D. Munkres, *Anal. Biochem.* **39,** 462 (1971).

[11] M. Douglas, D. Finkelstein, and R. A. Bukow, this series, Vol. 56, p. 58.

[12] F. Cabral and G. Schatz, this series, Vol. 56, p. 602.

[13] P. Merle and B. Kadenbach, *Eur. J. Biochem.* **105,** 499 (1980).

[14] U. K. Laemmli and M. Faure, *J. Mol. Biol.* **80,** 575 (1973).

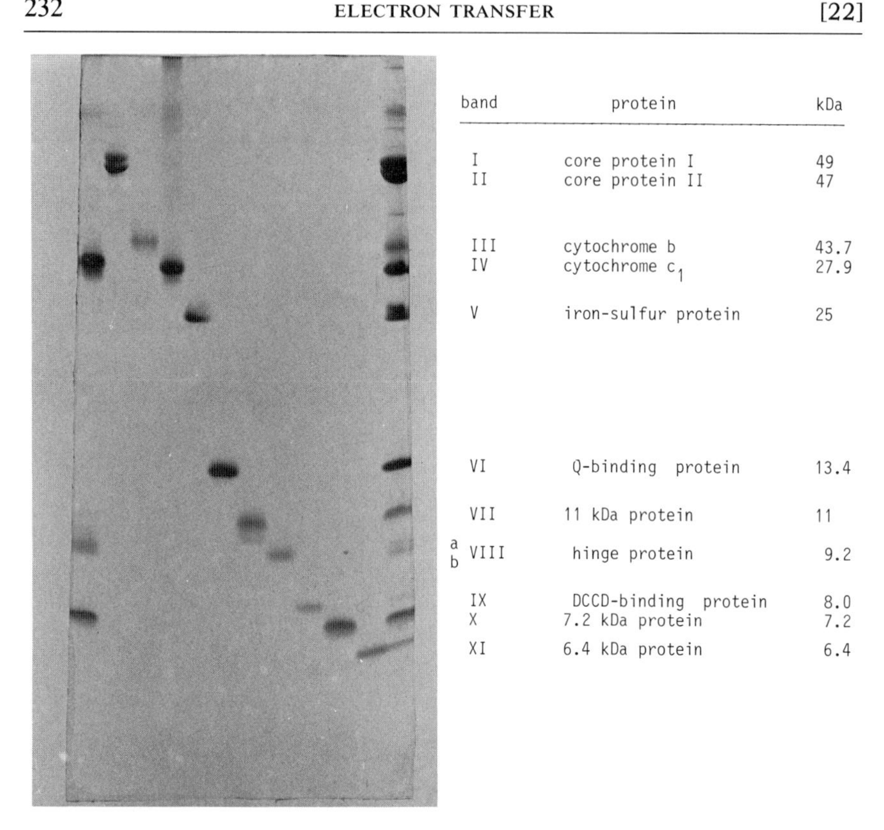

band	protein	kDa
I	core protein I	49
II	core protein II	47
III	cytochrome b	43.7
IV	cytochrome c_1	27.9
V	iron-sulfur protein	25
VI	Q-binding protein	13.4
VII	11 kDa protein	11
VIII a, b	hinge protein	9.2
IX	DCCD-binding protein	8.0
X	7.2 kDa protein	7.2
XI	6.4 kDa protein	6.4

FIG. 2. SDS–PAGE of the subunits of bc_1 complex from beef heart mitochondria. Outer left lane, c_1 subcomplex; outer right lane, bc_1 complex isolated in Triton X-100; middle lanes, subunits isolated by the procedure described. [*Note added in proof:* 11 kDa protein = 9.5 kDa protein (see Fig. 1).]

Marres and Slater[15] using the Swank and Munkres gel system. In the SDS–PAGE described here, band 7 of Marres and Slater is split into bands VII, VIIIa, and VIIIb. Band VII is the 9.5 kDa protein. The hinge protein is represented by two bands (VIIIa and VIIIb) appearing very closely to each other. The ratio of the intensities of bands VIIIa and VIIIb varies depending on the time of incubation with mercaptoethanol/SDS. Therefore the two bands seem to reflect different states of unfolding due to an incomplete splitting of intramolecular disulfide linkages. The amino acid analyses and the N-terminal sequences of the band VIIIa and VIIIb proteins, isolated by preparative SDS–PAGE, proved to be identical.

[15] C. A. M. Marres and E. C. Slater, *Biochim. Biophys. Acta* **462,** 531 (1977).

TABLE III
STOCK SOLUTIONS FOR SDS–PAGE

Buffer	Tris (*M*)	Tricine (*M*)	HCl (*M*)	SDS (%)
Anode buffer[a]	0.2	—	—[a]	—
Cathode buffer	0.1	0.1	—	0.1
Gel buffer	3.0	—	1.0	0.3
Overlay solution	1.0	—	0.33	0.1
Acrylamide–bisacrylamide mixture	48% Acrylamide (w/v)–1.5% bisacrylamide (w/v)			

[a] Adjusted to pH 8.9 with HCl.

Gel Preparation

The stock solutions prepared for gel electrophoresis are given in Table III. They are kept at room temperature with the exception of the acrylamide–bisacrylamide mixture which is stored at 4°. The separating gel (Table IV) is polymerized within 15 min by addition of 100 μl of 10% ammonium persulfate and 10 μl of TEMED per 30 ml. To ensure the formation of a smooth surface between separating gel and sample gel, the separating gel is covered with the overlay solution before polymerization. Some minutes after termination of the polymerization the overlay solution is decanted and replaced by the sample gel mixture (Table IV), which is polymerized by addition of 100 μl of ammonium persulfate (10%) and 10 μl of TEMED per 12 ml. The lower end of the inserted comb should be 2 cm away from the separating gel.

Sample Preparation

The protein samples are incubated in 5% SDS, 15% glycerol, 50 m*M* Tris, 2% mercaptoethanol, 0.003% Bromphenol Blue, pH 6.8, adjusted with HCl.

Electrophoresis Conditions

Slab gels are used in a vertical apparatus. Usually long slab gels with separating gel dimensions of 24 × 15.5 × 0.075 cm or 24 × 15.5 × 0.15 cm are used for analytical or preparative purposes. The electrophoresis is performed at room temperature; the gels warm up slightly under the following running conditions:[15a]

[15a] If the equipment for long gels is not available, shorter slab gels may be used. Independent of the thickness of the gels, the following running conditions for a 13-cm long separating gel are recommended: 1–2 hr at 30 V, 17 hr at 90 V constant.

TABLE IV
COMPOSITION OF SEPARATING AND SAMPLE GELS

Components	16% Separating gel		4% Sample gel
	0.15 cm	0.075 cm	
Acrylamide–bisacrylamide	20 ml	10 ml	1 ml
Gel buffer	20 ml	10 ml	3 ml
Glycerol	8 g	4 g	—
Add water to a final volume of	60 ml	30 ml	12 ml

0.075 cm gels (analytical):
 1–2 hr at 30 V (~7.5 mA), until the sample has completely entered the sample gel
 17 hr at 35 mA constant (the voltage rises from 110 V to 350 V)
0.15 cm gels (preparative):
 1–2 hr at 30 V (15 mA)
 24 hr at 50 mA constant (the voltage rises from 80 V to 250 V)

Staining

First the protein bands are fixed in a solution consisting of 50% methanol and 10% acetic acid. Gels [0.075 (0.15) cm] are fixed for 1 (3) hr before they are stained for at least 2 (4) hr in a solution containing 10% acetic acid and 0.1% Coomassie Brilliant Blue G-250. Staining of the 8 kDa protein could be achieved only by omitting methanol from the staining solution.

Discussion

The 11 protein subunits, the isolation of which has been described here, are obviously structural components of the bc_1 complex, since they are present in equimolar ratio both in the complex isolated in Triton X-100 and in the complex isolated in cholate. The amino acid sequences of 8 out of the 11 proteins from beef heart have been elucidated,[16–20] including

[16] H. Schägger, U. Borchart, H. Aquila, Th. A. Link, and G. von Jagow, *FEBS Lett.* **190,** 89 (1985).

[17] H. Schägger, G. von Jagow, U. Borchart, and W. Machleidt, *Hoppe-Seyler's Z. Physiol. Chem.* **364,** 307 (1983).

cytochromes b[21] and c_1.[3] Additionally, the sequence of the iron-sulfur protein from *Neurospora crassa* has been determined.[22] Therefore the sequences of all proteins carrying redox centers are available.

Cytochrome b and the iron-sulfur protein are part of the ubiquinol oxidation site (Q_o center).[23] One or more small proteins, probably ubiquinone-binding or semiquinone-stabilizing proteins, may be involved in the formation of this reaction site. The 13.4 kDa subunit designated as a quinone-binding protein (QP-C) of the bc_1 complex by Wakabayashi *et al.*,[19] may serve this function at the Q_i center.

Although not carrying redox centers, some small proteins may be involved in proton translocation. The most likely candidate for being involved in proton translocation is the 8 kDa DCCD-binding protein which is primarily labeled by DCCD.[24] The hydropathy pattern according to the algorithm of Kyte and Doolittle[25] did not reveal a hydrophobic stretch, which would be expected for a DCCD-binding protein. However, the existence of a sided α helix and an amphipathic β strand became obvious when an algorithm devised for the detection of amphipathic structures was used (Fig. 3).[20] The sided β strand extends over 22 residues. One side is strongly hydrophilic, comprising 5 charged residues. This side is probably directed toward the aqueous bulk phase, whereas the other side of the strand seems to face the lipid region. The glutamic acid residue (position 53), situated in a relative apolar region, may be the preferential site of reaction to DCCD.

The c_1 subcomplex seems to be a distinct structural and functional component of the bc_1 complex. Its three protein subunits are present in equimolar ratio and stick together very tightly; they are not separated even by 1.5 *M* guanidine.[17] The 9.2 kDa protein subunit of the c_1 subcomplex is the so-called hinge protein, which is supposed to be responsible for the interaction between cytochromes c and c_1.[26] It contains a hydrophilic

[18] S. Wakabayashi, H. Takeda, H. Matsubara, C. H. Kim, and T. E. King, *J. Biochem.* **91,** 2077 (1982).

[19] S. Wakabayashi, T. Takao, Y. Shimonishi, S. Kuramitsu, H. Matsubara, T.-y. Wang, Z.-p. Zhang, and T. E. King, *J. Biol. Chem.* **260,** 337 (1985).

[20] U. Borchart, W. Machleidt, H. Schägger, Th. A. Link, and G. von Jagow, *FEBS Lett.* **191,** 125 (1985).

[21] S. Anderson, M. H. L. de Bruijn, A. R. Coulson, I. C. Eperon, F. Sanger, and I. G. Young, *J. Mol. Biol.* **156,** 683 (1982).

[22] U. Harnisch, H. Weiss, and W. Sebald, *Eur. J. Biochem.* **149,** 95 (1984).

[23] G. von Jagow and T. Ohnishi, *FEBS Lett.* **185,** 311 (1985).

[24] G. Lenaz, M. Degli Eposti, J. Saus, M. Crimi, H. Schägger, U. Borchart, and G. von Jagow, Special FEBS Meeting, April 22–26, 1985, Algarve, Portugal.

[25] J. Kyte and R. F. Doolittle, *J. Mol. Biol.* **157,** 105 (1982).

[26] C. H. Kim and T. E. King, *Biochem. Biophys. Res. Commun.* **101,** 607 (1981).

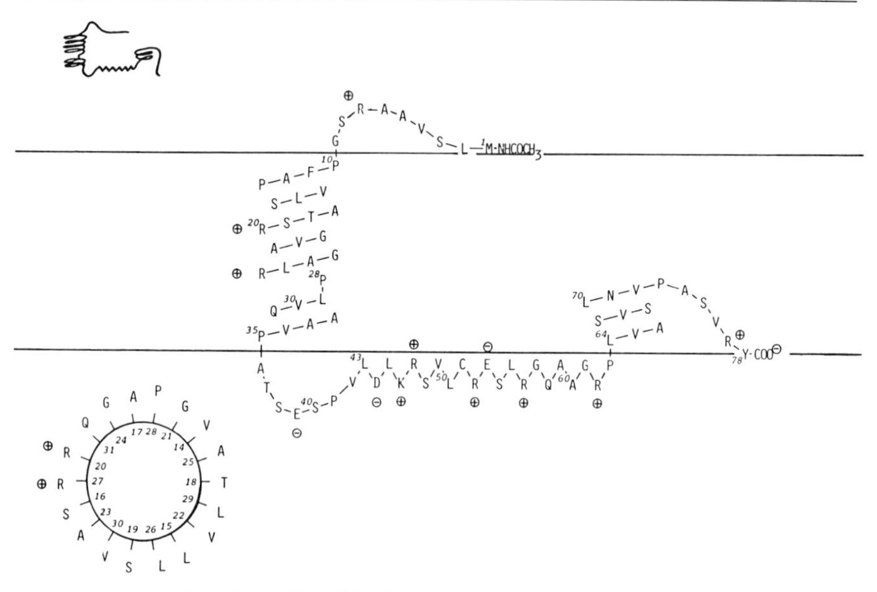

FIG. 3. Predicted folding pattern of the 8 kDa subunit.

segment encompassing 8 consecutive glutamic acid residues and a total of 30.8% of acidic residues. The 7.2 kDa protein also belonging to the c_1 subcomplex is a small membrane protein containing 33.9% aliphatic hydrophobic residues[17] and appears to have a membrane spanning segment.

The Rieske iron-sulfur protein and the 6.4 kDa protein can be cleaved off together from the bc_1 complex by 2 *M* urea. Since their separation from each other was not possible in the past, the reconstitution experiments with isolated iron-sulfur protein and an iron-sulfur protein-depleted bc_1 complex described by Engel *et al.*[27] were in fact carried out with a mixture consisting of the iron-sulfur protein and the 6.4 kDa protein. Therefore it remains an open question whether the 6.4 kDa protein is really involved in the reconstitution of electron transport. No definite function can be attributed so far to the two core proteins and the 9.5 kDa protein. In studies with bc_1 complex from *N. crassa,* dissociation of the two core proteins is found to abolish steady-state electron flow, which can be restored on reconstitution of the two core proteins.[28] This process has not yet been successfully achieved using mammalian bc_1 complex.

[27] W. D. Engel, C. Michalski, and G. von Jagow, *Eur. J. Biochem.* **132,** 395 (1983).
[28] P. Linke and H. Weiss, *Abstr. Eur. Bioenerg. Conf., 3rd* **3A,** 143 (1984).

Besides the analysis of the properties and function of the individual subunits, further information can be obtained from the analysis of the primary structures. The question was posed whether the small subunits traverse the membrane or whether they are attached to the central part of the complex on either side of the membrane. Membrane spanning segments with high polarity are not detected by the algorithm of Kyte and Doolittle.[25] Therefore we devised a program of our own to search for amphipathic α helices and β strands.[29] The folding patterns thus obtained (similar to that of the 8 kDa protein) indicate that each of the six small subunits contains membrane anchoring domains, even the most hydrophilic ones, the 13.4 kDa and the 9.2 kDa subunits.

Outlook

The isolation of the subunits of the mitochondrial bc_1 complex provides good starting material for two different approaches to an understanding of this highly sophisticated proton-translocating electron-transfer complex: (1) Elucidation of the amino acid sequences. The amino acid sequences of 8 out of the 11 subunits have already been established, and those of the remaining 3 components are under study and will permit structural predictions, which may provide further insight into protein–protein interactions. Crystallization of individual proteins or subcomplexes and subsequent X-ray crystallographic studies may open up a new dimension of structural resolution. (2) The functional role of the individual proteins or possible subcomplexes will be elucidated by means of reconstitution studies, resulting in an improved picture of the mechanism of electron and proton transfer. Immunological studies with antibodies specific for certain subunits may render valuable information.

[29] Th. A. Link and G. von Jagow, *J. Mol. Biol.*, submitted.

[23] Preparation of Hinge Protein and Its Requirement for Interaction of Cytochrome *c* with Cytochrome c_1

By TSOO E. KING and CHONG H. KIM

Method of Preparation of the Hinge Protein

Principle

The hinge protein is prepared by sequential resolution of the respiratory chain.[1,2] Succinate–cytochrome-*c* reductase is isolated from the submitochondrial particles from which two-band cytochrome c_1 is obtained. The hinge protein is prepared from two-band c_1.

Method

Two-band cytochrome c_1 prepared from the improved method described later in this chapter in "Determination of the Hinge Protein" at 6–8 mg/ml with a purity of 23–25 nmol of c_1/mg of protein is mixed with potassium cholate and urea to final concentrations of 1% and 2 *M*, respectively, in 50 m*M* Tris/succinate buffer, pH 7.4. The mixture is brought to 30% ammonium sulfate saturation with neutralized saturated ammonium sulfate solution to remove impurities, if any.

The clear pink supernatant is adjusted to pH 5.0–5.5 with 0.5 *N* HCl. It is then centrifuged at 20,000 rpm for 15 min in a Beckman Model 21. The supernatant is discarded. It is usually colorless, but occasionally a yellowish color appears when less pure two-band c_1 is used. The precipitate is dissolved in 50 m*M* Tris/succinate buffer, pH 7.4, containing 1% potassium cholate and 2 *M* urea to a protein concentration of 6–8 mg/ml and adjusted to pH 7.4 with 1 *N* NaOH. The solution is incubated at room temperature for 20 min, returned to an ice-water bucket, and adjusted to pH 5.0–5.5. The mixture is centrifuged at 20,000 rpm for 15 min. The colorless supernatant thus obtained is frozen at −20° for 3 hr or more. After thawing, it is centrifuged at 20,000 rpm for 15 min to remove any small amounts of denatured cytochrome c_1 precipitate that may be present. The colorless supernatant is adjusted to pH 7.4 with 1 *N* NaOH, and practically all cholate and urea is removed by continuous concentra-

[1] C. H. Kim and T. E. King, *J. Biol. Chem.* **258,** 13543 (1983).
[2] C. H. Kim and T. E. King, *Biochem. Biophys. Res. Commun.* **101,** 607 (1981).

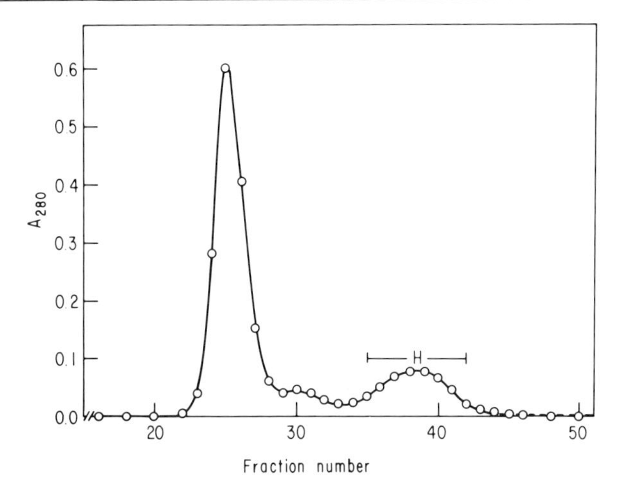

FIG. 1. Sephadex chromatography for purification of the hinge protein. The hinge protein preparation obtained according to the penultimate step of Fig. 2 is applied to a Sephadex G-75 column (1.6 × 35 cm). The elution is performed with 50 m*M* Tris/chloride buffer, pH 7.4. The flow rate is 7.2 ml/hr, and each fraction is 1.2 ml. Absorbance at 280 nm is recorded in a cell within a 1-cm light path. H designates a pooled fraction of the purified hinge protein. The first peak is not detectable by SDS–PAGE until more than 200 μg of sample are on the gel column, showing a faint high-molecular-weight band.

tion and dilution with 50 m*M* Tris/chloride buffer, pH 7.4, by ultrafiltration using a YM-5 Diaflo membrane.

The hinge protein thus obtained shows practically a single band by sodium dodecyl sulfate–polyacrylamide gel electrophoresis (SDS–PAGE) at the position of molecular weight about 11,000 by either the method of Weber and Osborn[3] or Swank and Munkres.[4] The yield of the hinge protein at this stage is 40–50%. In contrast, samples of two-band cytochrome c_1 obtained by the original methods[5] give less complete cleavage in above steps.

The hinge protein preparation, concentrated by ultrafiltration using YM-5 Diaflo membrane to about 10 mg/ml, is applied to a Sephadex G-75 column (1.6 × 35 cm) which has been equilibrated with 50 m*M* Tris/chloride buffer, pH 7.4, for further purification. The elution profile is shown in Fig. 1, and the fraction under H is the hinge protein. The purified sample thus obtained amounts to about 20–25% in yield. The purification procedure is summarized in Fig. 2.

[3] K. Weber and M. Osborn, *J. Biol. Chem.* **244,** 4406 (1969).

[4] R. T. Swank and K. D. Munkres, *Anal. Biochem.* **39,** 462 (1971).

[5] T. E. King, this series, Vol. 53, p. 181.

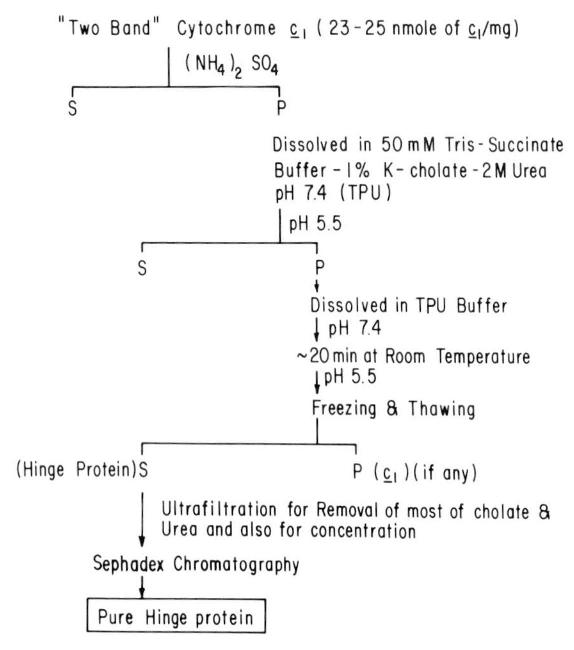

FIG. 2. Summary of isolation and purification of the hinge protein. P, precipitate; S, supernatant; TPU, 50 m*M* Tris/succinate buffer containing 1% potassium cholate and 2 *M* urea.

Comments on the Method

The hinge protein was first recognized as the colorless band of two-band cytochrome c_1[2,6] in the course of an endeavor to obtain a highly purified one-band cytochrome c_1. At the beginning, it was isolated in a crude form and named 15,000 protein, because in crude extracts it occupied a band at a position with a molecular weight of ~15,000 on SDS–PAGE performed according to the Weber and Osborn method.[3] However, the multiplicity of proteins with molecular weight of about 15,000 from mitochondria determined by this gel method and consequently the difficulty of their separation led us to change tactics and attempt isolation, not directly from mitochondria, but from highly purified two-band cytochrome c_1.[1,7]

Samples of the hinge protein obtained by either of two original methods[5] for the preparation of two-band cytochrome c_1, i.e., using either

[6] Unpublished results from this laboratory.

[7] T. E. King, *Adv. Enzymol.* **54,** 267 (1983).

2-mercaptoethanol or cholate-ammonium sulfate, do not give the purity and yield of the hinge protein as high as those from the improved method (see "Determination of the Hinge Protein"). These facts may indicate that the original methods for two-band c_1 preparations yield samples of its components which are much more tightly bound to each other apparently due to the very high degree of polymerization. By adjustment to pH 5.5 from a system of 50 m*M* Tris/succinate buffer, pH 7.4, containing 2 *M* urea and 1% cholate, the dissociation is much easier than other methods. The method as described seems to be the best and simplest procedure at present among the other possible methods tried during the course of study, such as high-performance liquid chromatography separation after the incubation with deoxycholate and dissociation by other agents.[7] The method reported herein is reliable and reproducible to obtain the pure, active, apparently natural form of the hinge protein so far tested.

The first peak of the elution profile as shown in Fig. 1 has yet to be identified. It should be mentioned that this peak is not clearly visible by SDS–PAGE with Coomassie Brillant Blue staining until 200 μg of protein are applied to the electrophoretic column. The molecular weight corresponding to the position of this peak determined by the Weber and Osborn method[3] is found to be about 60,000.

Other Methods of Preparation

Only one different method has been reported for the isolation of the hinge protein, M_r 11,000. It[8] is from the carboxymethylated and citraconylated cytochrome c_1 complex which has been found to consist of three polypeptides with M_r of 29,000, 11,000, and 9,000. The protein of M_r 11,000 proves to be identical with the hinge protein by amino acid sequence, although functionally it is not active due to the predesigned chemical derivatization.

Properties of the Hinge Protein: General Properties

The hinge protein does not show any spectral maxima, and the absorbance around 280 nm is relatively low either at pH 5.0 or 7.4 (Fig. 3). The absorption coefficients at these two pH values are not the same. The shoulders around the 258 and 265 nm regions are likely due to the disulfide bond and two phenylalanine residues in the proteins. The low absorbance around 280 nm is in agreement with the amino acid composition (Table I) showing no tyrosine and tryptophan residues in this protein.[8] This charac-

[8] S. Wakabayashi, H. Takeda, H. Matsubara, C. H. Kim, and T. E. King, *J. Biochem. (Tokyo)* **91,** 2077 (1982).

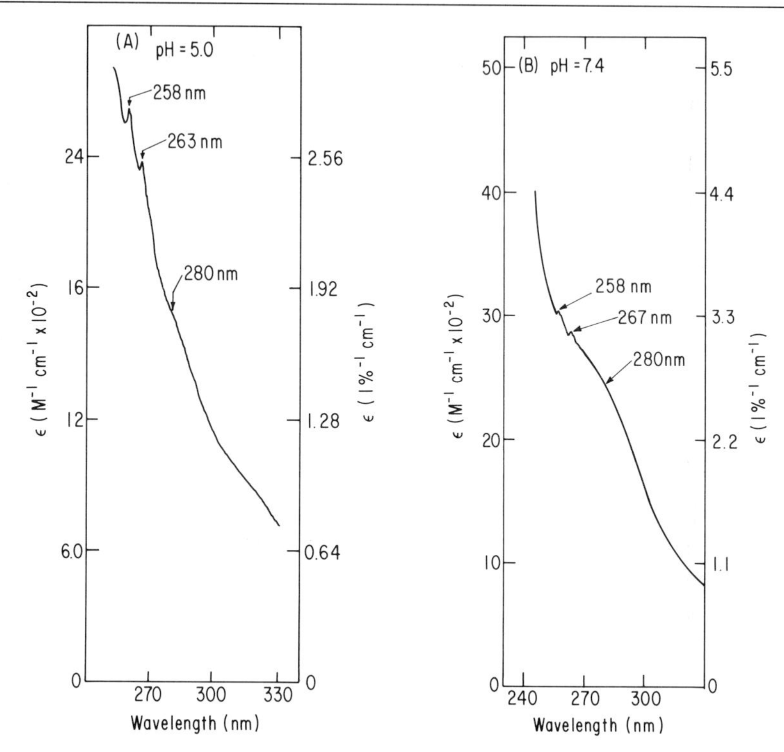

FIG. 3. Molar extinction coefficient and $A_{1\%}$ of the hinge protein (A) at pH 5.0 and (B) at pH 7.4 in 50 m*M* Tris/succinate buffer.

teristic of low absorption at 280 nm is important, e.g., the hinge protein can be easily missed in the course of isolation from mitochondrial proteins, especially in dilute eluant. Another remarkable characteristic of the hinge protein is the poor color value after staining with Coomassie Brilliant Blue. This fact may have led to neglecting the contamination of the hinge protein with the putative one-band cytochrome c_1,[9–11] which shows a high glutamic acid content in comparison with the sequence reported. Likewise, the diffuse characteristics of the hinge protein make it difficult to determine its exact position on SDS–PAGE columns. This is especially so when smaller quantities are applied to the SDS–PAGE column. Consequently, it has been considered close to the M_r 15,000 band position

[9] B. W. König, L. T. M. Schilder, M. J. Tervoort, and B. F. van Gelder, *Biochim. Biophys. Acta* **621,** 283 (1980).

[10] B. W. König, J. Wilms, and B. F. van Gelder, *Biochim. Biophys. Acta* **639,** 9 (1981).

[11] S. Wakabayashi, H. Matsubara, C. H. Kim, and T. E. King, *J. Biol. Chem.* **257,** 9335 (1982).

determined by SDS–PAGE of the bc_1 complex[12] and was considered a protein of M_r 15,000.[2]

The hinge protein is rather insensitive to temperature, urea, and low pH. However, when the preparation is kept in concentrated form (>10 mg/ml at pH 7.4 and 0°) for a few days, it tends to become turbid and evidently forms self-associated polymers. When such a preparation is tested by SDS–PAGE, the old, even still clear sample shows an additional band position at a molecular weight equivalent to a higher polymeric (usually the tetrameric) form, although the sample is routinely preincubated with 2% SDS. This band is completely abolished, however, when the sample is first adjusted to pH 5.0 before SDS–gel electrophoresis; then only one band at the M_r 11,000 position is observed.

Molecular Weight

A careful analysis of SDS–PAGE reveals the molecular weight of the hinge protein to be 11,000. From Sephadex G-75 gel filtration chromatography, the molecular weight of the hinge protein is found, however, to be 23,000 with a Stokes radius of 22.4 Å in 50 m*M* Tris/chloride buffer, pH 7.4, or in 50 m*M* Tris/succinate buffer, pH 5.0. In a medium of 20 m*M* cacodylate containing 1% Emasol 1130, the hinge protein shows the same M_r of 23,000, indicating no interaction with Emasol. The fact that the molecular weight of the hinge protein is found to be 23,000 by gel chromatography could be speciously interpreted as a dimeric form of this protein in the absence of detergent. However, in gel chromatography, in addition to molecular mass, the size, shape, viscosity, and distribution coefficient, among others, must be considered (e.g., Refs. 13, 14, and references cited therein). Moreover, sedimentation equilibrium in the same buffer gives an M_r of 9800, and the hinge protein is eluted at the same elution volume in either the presence or absence of 1% Emasol 1130 at pH 7.4 or 5.0. It is not likely that cacodylate and Tris act differently as far as the influence of the detergent on complex formation is concerned, so that Emasol or any detergent is not needed to keep the hinge protein monomeric. The hinge protein is quite different from the standard nonmembranous proteins employed for calibration of the gel filtration column for molecular weight and Stokes radius determinations, in addition to its unusual characteristics,

[12] T. E. King, C. A. Yu, L. Yu, and Y. L. Chiang, *in* "Electron Transfer Chains and Oxidative Phosphorylation" (E. Quagliariello, S. Papa, F. Palmieri, E. C. Slater, and N. Siliprandi, eds.), p. 105. North-Holland Publ., Amsterdam, 1975.

[13] C. Tanford, *in* "Physical Chemistry of Macromolecules," p. 317. Wiley, New York, 1961.

[14] G. K. Ackers, *in* "The Proteins" (H. Neurath and R. L. Hill, eds.), 3rd Ed., Vol. 1, p. 1. Academic Press, New York, 1975.

such as high glutamic acid content. These considerations might explain the disparity of observed molecular weights determined by the gel filtration method, on the one hand, and that by the sequence or hydrodynamic method on the other.

From sedimentation equilibrium experiments, the molecular weight of the hinge protein is estimated to be 9800 when the sample is run at 23,911 rpm for 22 hr. Molecular weight is calculated from the asymptotic slope at r^2 values of 0.342. This method of calculation is justified, although a small amount of "contaminant" is revealed in the system. The reason is that the molecular weight of the contaminant is so much bigger that it contributes very little to the concentration gradient near the meniscus.[1] The high-molecular-weight contaminant is due evidently to the existence of a polymer(s) of the hinge protein.

TABLE I
AMINO ACID COMPOSITIONS OF THE HINGE PROTEIN AND ONE-BAND CYTOCHROME c_1 FROM SEQUENCE DETERMINATION

Amino acid	Hinge protein[8]	Cytochrome c_1[11]
Asp	7	19[a]
Thr	4	7
Ser	4	16
Glu	21	21[b]
Pro	2	22
Gly	1	17
Ala	3	18
Half-Cys	5	5
Val	5	15
Met	0	10
Ile	0	4
Leu	10	25
Tyr	0	15
Phe	2	8
Lys	5	12
His	3	9
Arg	6	15
Try	0	3
Total	78	241

[a] 13 aspartic acid and 6 asparagine residue.
[b] 16 glutamic acid and 5 glutamine residue.

Isoelectric Point

Isoelectric focusing of the hinge protein is performed in a sucrose density gradient using a pH gradient range of 3–5 or 3–6. Only one opaque clean band is observed after 6 hr at 500 V, and p*I* is found to be 3.9.

Amino Acid Composition, Sulfhydryl and Disulfide Groups, and Primary Structure

The complete amino acid composition of the pure hinge protein is shown in Table I. This protein does not contain tyrosine, tryptophan, isoleucine, or methionine. Five sulfhydryl groups are found after carboxylmethylation. Direct determination of sulfhydryl groups gives only one in the free form, and the other four sulfhydryl groups are evidently in disulfide linkages. The result is the same even after the protein is incubated in 8 *M* urea at 30° for 30 min.

The primary structure[8] of the hinge protein is shown in Fig. 4. Its unique characteristics is that 8 glutamic acid residues are linked consecutively from positions 5–12. To our knowledge, this is the only naturally occurring protein with such a structure that has been reported. This protein contains 78 amino acid residues, and 21 of them, or 27%, are glutamic acid. Position 36 is occupied by either arginine or lysine due either to microheterogeneity or to the existence of isozymes. It does not contain a spectroscopically visible prosthetic group, or possibly the prosthetic group is lost during isolation.

Demonstration of the Requirement of the Hinge Protein for the Interaction of Cytochrome c_1 with Cytochrome c (Its Indispensability for the c_1–c Complex Formation): Basic Concept of the Determination of the Hinge Protein

The elution patterns of gel filtration chromatography of the mixture of the purified one-band cytochromes c_1 and c in the presence and absence of

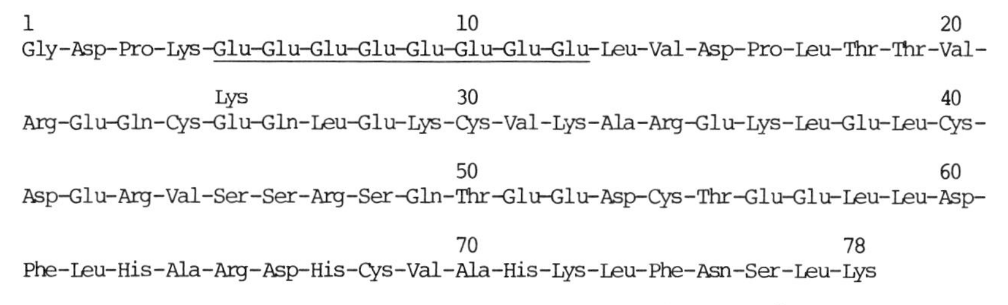

FIG. 4. The amino acid sequence of the hinge protein.

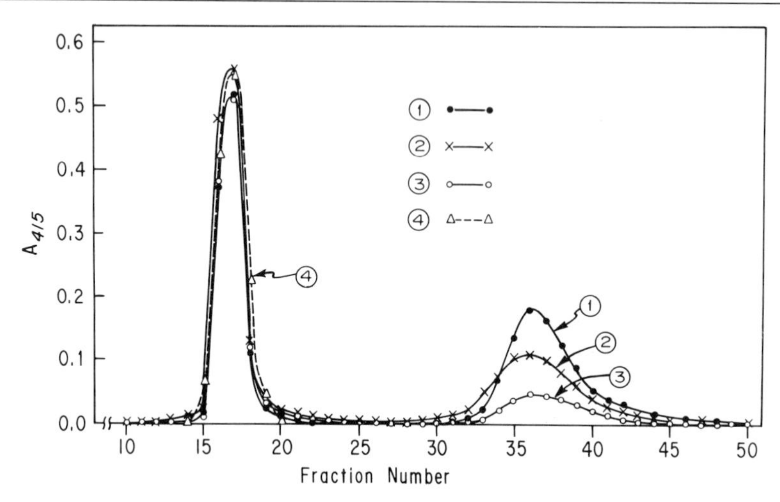

FIG. 5. The effect of the hinge protein on the c_1–c complex formation demonstrated by gel filtration chromatography. The sample (0.4 ml) is applied to a Sephadex G-75 column (1.6 × 35 cm). The elution is performed with 10 m*M* potassium phosphate buffer, pH 7.4. The flow rate is 8.4 ml/hr and each fraction is 1.4 ml. Absorbance at 415 nm is recorded in a cell with a 1-cm light path. c_1 concentration is 25 μM for all systems. The ratio of cytochrome c_1 to the hinge protein is about 1 : 1.2. Curve 1 represents an elution profile of a mixture of c_1 and c (1 : 2.6) in the absence of the hinge protein, and curve 2 is the same as curve 1, but in the presence of the hinge protein. Curve 3 represents an elution profile of a mixture of c_1 and c (1 : 1) in the absence of the hinge protein. Curve 4 is the same as curve 3, but in the presence of the hinge protein; notice only one peak was formed.

the hinge protein are presented in Fig. 5. It is clear that two separate peaks for cytochromes c_1 and c in the absence of the hinge protein are shown in curves 1 and 3. The $c_1 : c$ ratio in curve 1 is 1 : 2.6, while that of curve 3 is 1 : 1. The cytochrome c peak in curve 1 is larger that that of curve 3 accordingly, because of no complex formation in the absence of the hinge protein. On the other hand, only one peak is shown from elution in the system of cytochromes c_1 and c in the ratio of 1 : 1 in the presence of the hinge protein (curve 4). And, as shown in curve 2, the absorbance of cytochrome c is lower than that in curve 1 when the hinge protein is added to the system containing a $c_1 : c$ ratio of 1 : 2.6 because a part of c is used in the formation of the complex.

One characteristic of the c_1–c complex is the large increase of the Soret–Cotton effect over the summation of molecular ellipticities of these two cytochromes in the Soret region.[1,15] The indispensability of the hinge

[15] Y. L. Chiang, L. S. Kaminski, and T. E. King, *J. Biol. Chem.* **251,** 29 (1976).

protein in the c_1–c complex formation is further substantiated by titration study using the increase of the Soret–Cotton effect as a criterion.[1] The results are shown in Figs. 6 and 7. One-band c_1 does not form the complex with cytochrome c, in contrast to the two-band c_1.[1] However, the addition of the hinge protein to the mixture of one-band c_1 and cytochrome c causes the large increase of the Cotton effect in the Soret region. On the other hand, no effect of the hinge protein is observed on the circular

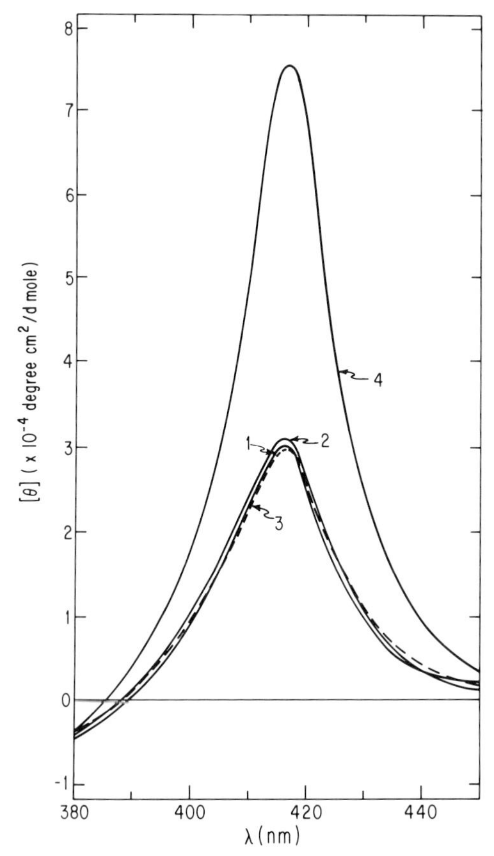

FIG. 6. The effect of the hinge protein on the CD spectra in the Soret region of the cytochrome c_1–c complex formation. Concentration of cytochrome c and c_1 is 4.2 μM in 10 mM potassium phosphate buffer, pH 7.4. Curve 1 is the spectrum of c_1 and c in separate cuvettes, and curve 2 is the spectrum of mixed c_1 and c in the absence of the hinge protein. Curve 3 is the spectrum of the system containing c_1 to which the hinge protein (6.6 μM) is added and c is a separate cuvette. Curve 4 is the spectrum of the system containing c_1–c and the hinge protein.

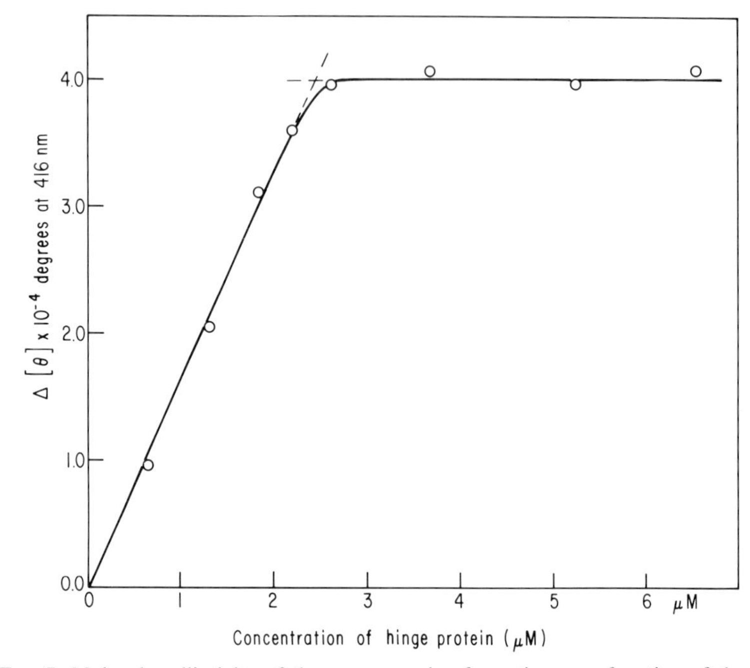

FIG. 7. Molecular ellipticity of the c_1–c complex formation as a function of the hinge protein. The plot is computed from the system containing 2.1 μM cytochrome c_1 and 2.1 μM cytochrome c in 5.2 ml. The molarity of the hinge protein is calculated on the basis of M_r = 9175 as the molecular weight. The ordinate is the molar ellipticity of the Soret region. Notice the extrapolated point showing the ratio of the hinge protein to c_1 to c to be 1 : 1 : 1.

dichroism (CD) spectra of either c_1 or c individually in the Soret region, and the degree of the increase of the molecular ellipticity is proportional to the amount of the hinge protein added (Fig. 7). Maximal molar ellipticity is reached when the equimolar amount of the purified hinge protein is added to the one-band cytochrome c_1 and cytochrome c system; further additions of the hinge protein do not increase the Cotton effect, showing a sharp inflection point and followed by a plateau, as shown in Fig. 7. At the inflection point, it is revealed that the ratio of c_1 to c to the hinge protein in the complex is 1 : 1 : 1. Polyglutamic acid and bovine serum albumin, individually or in combination, at even five times the c_1 amount do not take the place of the hinge protein for the complex formation.

The cytochrome c_1–c–hinge protein complex is dissociated to its components at ionic strength of the medium higher than 20 mM.[1,15]

Determination of the Hinge Protein

Principle

The principle (see above) of the determination of the hinge protein is based on the formation of the cytochrome c_1–c–hinge protein complex in 1 : 1 : 1 ratio. The Cotton effect in the Soret region of the complex is much larger than that of the summation of the individual components separately as shown in Fig. 7. The increase of molecular ellipticity is linear to the amount of the hinge protein in the system.

Reagents

Cytochrome *c*. Cytochrome *c* of best commercially available samples such as Sigma type III or type IV may be used after a simple purification of Sephadex gel chromatography to remove polymerized or other denatured forms. A solution of about 20 mg/ml of 50 m*M* phosphate buffer, pH 7.4, is mixed with potassium ferricyanide in about 50% of the molar amount of the cytochrome. The solution is then passed through a column of Sephadex G-75 "super-fine" particle size (Pharmacia) at a flow rate of about 0.2 ml/min. The concentration of cytochrome *c* must be accurately determined because it will serve as the basis of calculation of the cytochrome c_1–c complex. The extinction coefficient (mM^{-1} cm^{-1}) of the absorption spectrum at 550 nm is 29.5 for the reduced form and 9.0 for the oxidized.[16]

Cytochrome c_1. The method described here is essentially the procedure as originally reported (cf. Refs. 2, 6, 7) with the modifications and improvements accumulated in this laboratory during the past few years. Cytochrome c_1 may be prepared from the cytochrome bc_1 complex or directly from succinate–cytochrome-*c* reductase. The yield from the latter is better than that from the bc_1 complex. Succinate–cytochrome-*c* reductase (300 ml), which is prepared according to a previous method,[17,18] is adjusted to a protein concentration of 10–15 mg/ml in 50 m*M* potassium phosphate buffer, pH 7.4, containing 1.5% potassium cholate, 0.5% potassium deoxycholate, 8% of ammonium sulfate saturation, and 15% 2-mercaptoethanol, and incubated for 1 hr with slow stirring at 4°. The mixture is centrifuged at 20,000 rpm in a Beckman Model 21 for 20

[16] E. Margoliash and O. F. Walasek, this series, Vol. 10, p. 339.

[17] S. Takemori and T. E. King, *J. Biol. Chem.* **239,** 3546 (1964).

[18] C. A. Yu, L. Yu, and T. E. King, *J. Biol. Chem.* **247,** 1012 (1972).

min. The supernatant is quickly passed through the Sephadex G-25 ("medium" size particle) column, 5 × 60 cm, which has been equilibrated with 50 m*M* potassium phosphate buffer, pH 7.4. The slightly turbid cytochrome c_1 fraction from the column at the void volume is further clarified by centrifugation for 20 min as before. The clear supernatant obtained is immediately adjusted to 1% potassium cholate and adsorbed, in a beaker, to 80 ml of DEAE-cellulose equilibrated with 50 m*M* potassium phosphate buffer, pH 7.4, containing 1% potassium cholate and 0.1% 2-mercaptoethanol. The c_1-adsorbed DEAE-cellulose is repeatedly washed in the beaker by decantation until the wash shows no absorbance at 280 nm. Finally, the suspension is filtered through a sintered glass funnel but not suctioned dry. The washed c_1-adsorbed DEAE-cellulose is packed on the top of a layer of DEAE-cellulose (2.6 × 25 cm) equilibrated with the same buffer as a cushion. The whole column is then washed again with 2-bed volumes of the buffer. The elution is performed with a linear gradient of 0.0–0.5 *M* NaCl in the 50-m*M* potassium phosphate buffer, pH 7.4, containing 1% potassium cholate and 0.1% β-mercaptoethanol.

Cytochrome c_1 eluted from the DEAE-cellulose column is concentrated by an Amicon ultrafiltration cell, using PM30 Diaflo ultrafiltration membrane, to a protein concentration of 14–16 mg/ml. The concentrated c_1 fraction was applied on an Ultrogel AcA 44 column (2.6 × 44 cm), equilibrated with 50 m*M* potassium phosphate buffer, pH 7.4, containing 1% cholate and 0.1% 2-mercaptoethanol. The elution profile is shown in Fig. 8.

The purification profile with the purity as well as yield of one-band and two-band cytochrome c_1 is shown in Table II.

If the one-band c_1 is contaminated with the hinge protein, the latter may be removed by one or more cycles of DEAE-cellulose and Ultrogel AcA 44 column chromatography. One of the best ways to ascertain whether there is any contamination is by amino acid analysis. Pure one-band cytochrome c_1 has the amino acid composition shown in Table I. Any deviation should not be greater than, say, 10%. An unusually high glutamic acid content invariably indicates the contamination of the hinge protein. Cytochrome c_1 is determined[18] by using $A_{\text{red}}^{552.5} - A_{\text{red}}^{540} = 17.5\ \text{m}M^{-1}\ \text{cm}^{-1}$; ascorbate and ferricyanide may be used as the reduced and oxidized reagent, respectively. By adjusting the cytochrome c_1 concentration to be equal molar to cytochrome c and comparing the $[\theta]$ from the sample of c_1 with the calibration curve (e.g., Fig. 7), the amount of the hinge protein may be easily computed (see "Calculation").

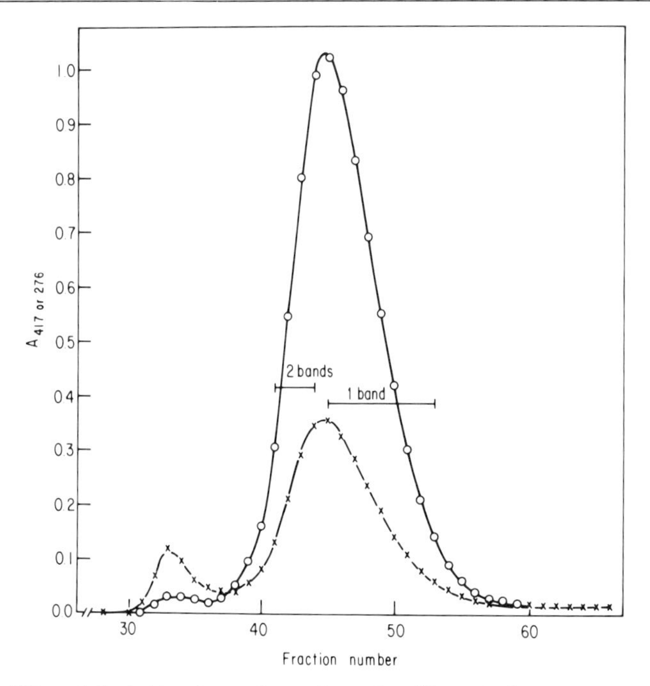

FIG. 8. Ultrogel AcA 44 column chromatography. The cytochrome c_1 preparation obtained from DEAE-cellulose chromatography was concentrated to 15 mg/ml and then applied to an Ultrogel AcA 44 column (2.6 × 44 cm). The elution was performed with 50 mM potassium phosphate buffer, pH 7.4, containing 1% cholate and 0.1% 2-mercaptoethanol. The flow rate was 6 ml/hr and each fraction was collected for 20 min. (x–x) represents absorbance at 276 nm; (O–O) indicates absorbance at 417 nm. All absorbance was recorded in a cell with 0.2-cm light path. The first peak showed "two-band" cytochrome c_1 contaminated with high-molecular-weight polymers as revealed by SDS–PAGE. The fractions labeled "2 band" and "1 band" are cytochrome c_1 preparations.

The Hinge Protein. It is not necessary to be absolutely pure. This is because the inflection point is always at $c_1 : c :$ hinge protein = 1 : 1 : 1 (see "Method of Preparation of the Hinge Protein").

Method

The buffer, potassium phosphate, pH 7.4, is at a final concentration of 10 mM. The calibration curve, as in Fig. 7, is made by titration of the hinge protein in about 0.2, 0.4, 0.6, 0.8, 1.0, 1.2, and 1.5 times the cytochrome c_1 or c concentration. The concentrations of these two cytochromes should be adjusted to the same values. The titration is conducted by measuring the CD of the complex in the Soret region. These values are

TABLE II
PURITY AND YIELD OF ONE-BAND CYTOCHROME c_1 FROM SUCCINATE–CYTOCHROME-c REDUCTASE BY THE IMPROVED METHOD

Purification step	A_{417}/A_{276}	Purity (nmol/mg)	Cytochorme c_1 (μmol)	Yield (%)
Succinate–cytochrome-c reductase		2.0 ~ 2.2	7.4	100
Sephadex G-25 column	1.1		4.8	65
DEAE-cellulose batch-wise washing	2.5		3.4	45
Ultrogel AcA 44 chromatography				
Two-band c_1	2.4 ~ 2.7	23 ~ 25	1.1	15
One-band c_1	2.9 ~ 3.0	31 ~ 32	1.9	25

subtracted from the CD reading of cytochrome c_1 plus cytochrome c, but without the hinge protein. This is done because the hinge protein does not contribute or affect the Cotton effect of the cytochrome c_1 or c in the Soret region. The CD spectrum from a scan of from 400 to 450 nm is performed to determine the maximum, which is usually at 415–416 nm.

Calculation

Let $[\theta_0]$ be the molecular ellipticity of the calibration curve in degree cm^2/decimol and $[\theta]$ be the observed molecular ellipticity; then $[\theta]/[\theta_0] \times 100$ is the percentage of purity of the hinge protein in the sample.

The molecular ellipticity is calculated from Eq. 1.

$$[\theta] = \frac{(\alpha_c - \alpha_m)\ 9175}{10\ LC} \tag{1}$$

Here, $[\theta]$ is in degree cm^2/decimol; α_c is the observed rotation in degree of the complex, i.e., cytochromes c_1, c, and the sample of hinge protein; α_m is the observed rotation in degree of the system without the hinge protein; 9175 is the molecular weight of the hinge protein; L is the optical path of the cuvette in centimeters; and C is the concentration in grams per cubic centimeters.

For example, a sample of the hinge protein used containing 0.125 mg in 5.2 ml gives α_c of 0.31×10^{-3} degree and α_m of 0.15×10^{-3} degree in a cuvette of 2-cm optical path.

Then

$$[\theta] = \frac{(0.31 - 0.15) \times 10^{-3} \times 9175}{(10 \times 2 \times 0.125 \times 10^{-3})/5.2} = 3053 \text{ degree cm}^2/\text{decimol}$$

From the calibration curve we find $[\theta_0]$ at this molar concentration, i.e., $(0.125 \times 10^{-3})/52$, would be 3570 degree cm²/decimol, if the hinge protein is 100% pure.

$$\therefore \qquad \frac{3053 \times 100}{3570} = 85\% \text{ pure}$$

Usually the sample is determined at two levels of concentration. The average of these two determinations is the result.

If the hinge protein is contaminated with cytochrome c_1, the method of determination would be the same as that without contamination. But the concentration of c_1 must be adjusted equal to that of c.

Acknowledgments

The original work was supported by grants from NIH (NIGMS-16767 and HLB-12576) and the American Cancer Society (Helen Willets Dutcher Memorial Grant BC-349). Discussion with Michael Seaman about the manuscript is gratefully acknowledged.

[24] Use of Specific Inhibitors on the Mitochondrial bc_1 Complex

By G. von Jagow and Th. A. Link

The chapter is intended to serve as a guide that will make it possible to choose the appropriate inhibitor when part of or the whole electron-transfer chain of the bc_1 complex is to be blocked, independent of whether single-turnover or steady-state experiments are to be performed.[1-10]

[1] P. Mitchell, *FEBS Lett.* **56,** 1 (1975).
[2] P. Mitchell, *FEBS Lett.* **59,** 137 (1975).
[3] P. Mitchell, *J. Theor. Biol.* **62,** 327 (1976).
[4] B. L. Trumpower, *J. Bioenerg. Biomembr.* **13,** 1 (1981).
[5] E. C. Slater, J. A. Berden, S. de Vries, and Q. S. Zhu, *in* "Vectorial Reactions in Electron and Ion Transport in Mitochondria and Bacteria" (F. Palmieri, E. Quagliariello, N. Siliprandi, and E. C. Slater, eds.), p. 163. Elsevier, Amsterdam, 1981.
[6] G. Hauska, E. Hurt, N. Gabellini, and W. Lockau, *Biochim. Biophys. Acta* **726,** 97 (1983).
[7] P. R. Rich, *Biochim. Biophys. Acta* **768,** 53 (1984).

Therefore special attention has been devoted to a classification of the various inhibitors into definite groups, in accordance with their point of action. The structural data (Tables I–IV[11–32]) and the physicochemical data (Table V[33–52]) of the different molecules have been compiled. Care has been taken to discuss in detail the structural segments indispensable

[8] G. von Jagow and Th. A. Link, *in* "Biomedical and Clinical Aspects of Coenzyme Q" (K. Folkers and Y. Yamamura, eds.), p. 87. Elsevier, Amsterdam, 1984.

[9] E. A. Berry and B. L. Trumpower, *in* "Coenzyme Q" (G. Lenaz, ed.), p. 365. Wiley, New York, 1985.

[10] P. R. Rich and D. S. Bendall, *in* "Vectorial Reactions in Electron and Ion Transport in Mitochondria and Bacteria" (F. Palmieri, E. Quagliariello, N. Siliprandi, and E. C. Slater, eds.), p. 187. Elsevier, Amsterdam, 1981.

[11] K. Gerth, H. Irschik, H. Reichenbach, and W. Trowitzsch, *J. Antibiot.* **33,** 1474 (1980).

[12] W. Trowitzsch, G. Reifenstahl, V. Wray, and K. Gerth, *J. Antibiot.* **33,** 1480 (1980).

[13] W. Trowitzsch, G. Höfle, and W. S. Sheldrick, *Tetrahedron Lett.* 3829 (1981).

[14] T. Anke, F. Oberwinkler, W. Steglich, and G. Schramm, *J. Antibiot.* **30,** 806 (1977).

[15] G. Schramm, W. Steglich, T. Anke, and F. Oberwinkler, *Chem. Ber.* **111,** 2779 (1978).

[16] T. Anke, G. Schramm, B. Schwalge, B. Steffan, and W. Steglich, *Liebigs Ann. Chem.* 1616 (1984).

[17] P. Sedmera, F. Nerud, V. Musilek, and M. Vondracek, *J. Antibiot.* **34,** 1069 (1981).

[18] M. Vondracek, J. Capkova, and K. Culik, *Chem. Abstr.* **93,** 204 (1980).

[19] T. Anke, H. Besl, U. Mocek, and W. Steglich, *J. Antibiot.* **36,** 661 (1983).

[20] T. Anke, J. Hecht, G. Schramm, and W. Steglich, *J. Antibiot.* **32,** 1112 (1979).

[21] B. Kunze, T. Kemmer, G. Höfle, and H. Reichenbach, *J. Antibiot.* **37,** 454 (1984).

[22] G. Höfle, B. Kunze, C. Zorzin, and H. Reichenbach, *Liebigs Ann. Chem.* 1882 (1984).

[23] M. D. Friedmann, P. L. Stotter, T. H. Porter, and K. Folkers, *J. Med. Chem.* **16,** 1314 (1973).

[24] H. Roberts, W. M. Choo, S. C. Smith, S. Marzuki, A. W. Linnane, T. H. Porter, and K. Folkers, *Arch. Biochem. Biophys.* **191,** 306 (1978).

[25] W. B. Wendel, *Fed. Proc., Fed. Am. Soc. Exp. Biol.* **5,** 406 (1946).

[26] E. G. Ball, C. B. Anfinsen, and O. Cooper, *J. Biol. Chem.* **168,** 257 (1949).

[27] J. P. Wan, T. H. Porter, and K. Folkers, *Proc. Natl. Acad. Sci. U.S.A.* **71,** 952 (1974).

[28] B. R. Dunshee, C. Leben, G. W. Keitt, and F. M. Strong, *J. Am. Chem. Soc.* **71,** 2436 (1949).

[29] E. E. van Tamelen, J. P. Dickie, M. E. Loomans, R. S. Dewey, and F. M. Strong, *J. Am. Chem. Soc.* **83,** 1639 (1961).

[30] K. Ando, S. Suzuki, T. Saeki, G. Tamura, and K. Arima, *J. Antibiot.* **22,** 189 (1969).

[31] K. Ando, I. Matsuura, Y. Nawata, H. Endo, H. Sasaki, T. Okytomi, T. Saeki, and G. Tamura, *J. Antibiot.* **31,** 533 (1978).

[32] J. W. Cornforth and A. T. James, *Biochem. J.* **63,** 124 (1956).

[33] G. Thierbach and H. Reichenbach, *Biochim. Biophys. Acta* **638,** 282 (1981).

[34] G. von Jagow and W. D. Engel, *FEBS Lett.* **136,** 19 (1981).

[35] G. von Jagow and W. F. Becker, *Bull. Mol. Biol. Med.* **7,** 1 (1982).

[36] W. F. Becker, G. von Jagow, T. Anke, and W. Steglich, *FEBS Lett.* **132,** 329 (1981).

[37] G. Thierbach, B. Kunze, H. Reichenbach, and G. Höfle, *Biochim. Biophys. Acta* **765,** 227 (1984).

[38] B. L. Trumpower and J. G. Haggerty, *J. Bioenerg. Biomembr.* **12,** 151 (1980).

[39] B. L. Trumpower, *J. Bioenerg. Biomembr.* **13,** 1 (1981).

for full efficacy of the inhibitors. This seems important in view of the fact that new structural data expected in the near future will no doubt soon permit a better understanding of the structure/function relationship.

Introduction

The inhibitors are discussed using the following model of the bc_1 complex.[1-9] The functional unit consists of at least 11 proteins, only 3 of which carry redox centers. These are cytochrome *b*, a monomeric two-heme peptide with a molecular weight of 43,700, cytochrome c_1 with a molecular weight of 27,874, and the ferredoxin-type iron–sulfur protein (Fe_2S_2). The reaction centers appear to be formed by several subunits each.

The reaction sequence is assumed to be as follows (Fig. 1): At the Q_o center, ubiquinol is oxidized to ubiquinone in two steps, one electron being transferred to the iron–sulfur center and then via cytochrome c_1 to cytochrome *c*, while the second electron is transferred to heme b_l (low potential, *b*-566) and from there to heme b_h (high potential, *b*-562). The oxidation of ubiquinol is accompanied by a release of two protons to the c-side.

At the Q_i center, ubiquinone is re-reduced in a two-step mechanism by the heme b_h center, so that half of the electrons set free during ubiquinol oxidation return to their point of origin.

During the flow of two electrons from ubiquinol to cytochrome c_1, four protons are released to the outer space while two protons are taken up at the inside.

The encircled Q/QH_2 in Fig. 1 represents the quinone pool which is present in 10-fold molar excess over the iron-sulfur protein or cytochrome

[40] J. R. Bowyer, C. A. Edwards, T. Ohnishi, and B. L. Trumpower, *J. Biol. Chem.* **257,** 8321 (1982).
[41] C. A. Edwards, P. Graf, B. Godde, and B. L. Trumpower, *Biophys. J.* **37,** 250a (1982).
[42] K. Matsuura, J. R. Bowyer, T. Ohnishi, and P. L. Dutton, *J. Biol. Chem.* **258,** 1571 (1983).
[43] Q. S. Zhu, J. A. Berden, S. de Vries, K. Folkers, T. H. Porter, and E. C. Slater, *Biochim. Biophys. Acta* **682,** 160 (1982).
[44] Q. S. Zhu, H. N. van der Wal, R. van Grondelle, and J. A. Berden, *Biochim. Biophys. Acta* **765,** 48 (1984).
[45] J. A. Berden and E. C. Slater, *Biochim. Biophys. Acta* **216,** 237 (1970).
[46] J. A. Berden and E. C. Slater, *Biochim. Biophys. Acta* **256,** 199 (1972).
[47] E. C. Slater, *Biochim. Biophys. Acta* **301,** 129 (1973).
[48] F. M. Strong, J. P. Dickie, M. E. Loomans, E. E. van Tamelen, and R. S. Dewey, *J. Am. Chem. Soc.* **82,** 1513 (1960).
[49] K. K. Moser and P. Walter, *FEBS Lett.* **50,** 279 (1975).
[50] B. D. Nelson, P. Walter, and L. Ernster, *Biochim. Biophys. Acta* **460,** 157 (1977).
[51] J. R. Brandon, J. R. Brocklehurst, and C. P. Lee, *Biochemistry* **11,** 1150 (1972).
[52] G. van Ark and J. Berden, *Biochim. Biophys. Acta* **459,** 119 (1977).

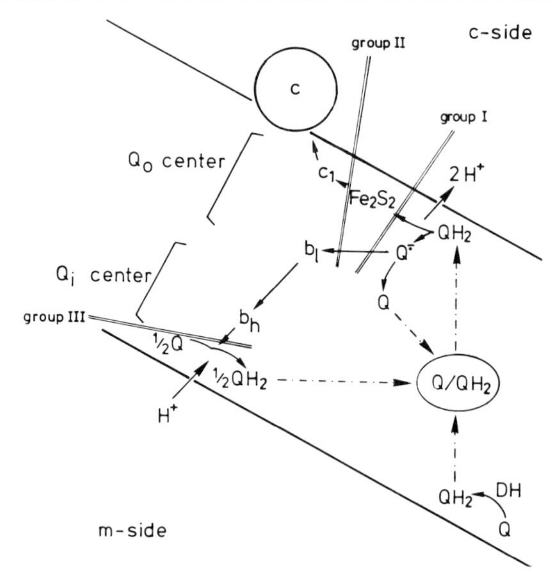

FIG. 1. Circuit scheme of the branched respiratory chain and points of action of the three groups of inhibitors of the bc_1 complex.

c_1. The dashed arrows of the circuit scheme indicate that the net oxidized ubiquinone formed during the reaction process at the c-side of the mitochondrial membrane is re-reduced at the m-side by various dehydrogenases.

At present a multitude of inhibitors is available which can be classified roughly into three groups, according to three different points of action, as follows:

Group I: Various β-methoxyacrylates, binding at the Q_o site and blocking two reactions at the same time, namely, electron transfer from QH_2 to the iron–sulfur center and electron transfer onto the heme b_l center

Group II: Hydroxyquinone analogs, binding at the Q_o site and blocking electron transfer from the iron–sulfur center to cytochrome c_1, as well as electron transfer onto the heme b_l center

Group III: Antimycin, funiculosin, and certain quinone analogs, binding at the Q_i site and blocking electron transfer from the heme b_h center to ubiquinone

As a fourth group the chromone inhibitors (stigmatellins) have recently been discovered. They also completely block the Q_o center, but show binding properties different to both group I and group II inhibitors.

Inhibitors of groups I and III bind to their corresponding *b* domains or heme *b* centers directly, as can be judged from the red shifts they induce in the spectra of the hemes. Smaller subunits may be involved in the formation of the reaction sites and thus in the binding of the inhibitors as well as in the binding of the substrate ubiquinone. Inhibitors of group II appear to bind to the Rieske iron–sulfur protein which contributes to the Q_o reaction center. It has been suggested that the mechanism of inhibition of group II inhibitors may be a form of competition between natural and inhibitory quinol for the reaction site.[10]

Evidence for the branching of electron flow and the existence of two ubiquinone reaction sites has been derived from many experiments. For the sake of clarity, a key experiment, the "oxidant-induced reduction of cytochrome *b*"—unique to the bc_1 complex—will briefly be discussed.[10a,b] When cytochrome c_1 and the iron–sulfur center are reduced (e.g., by ascorbate) while the heme b_h center is blocked by a group III inhibitor, reduction of cytochrome *b* by an excess of ubiquinol is not possible. This is due to the fact that oxidation of ubiquinol at the Q_o site is not possible when the iron–sulfur center is in the reduced state. When cytochrome c_1 and the iron–sulfur center are then oxidized by a pulse of ferricyanide, the reaction sequence takes place and cytochrome *b* is rapidly reduced. Therefore inhibitors of group I and group II block the oxidant-induced reduction of cytochrome *b* because they inhibit electron transfer at the Q_o site.

Group I

Inhibitors of UQH_2-Oxidation (Iron–Sulfur Center/Heme b_l Reduction): E-β-Methoxyacrylate (MOA) Inhibitors

These inhibitors all contain an *E*-β-methoxyacrylate (MOA) group, which resembles part of the structure of ubiquinone (Table I). They block reduction of the iron–sulfur center and of cytochrome *b* via the Q_o site in single- and multiple-turnover experiments, but permit reduction of cytochrome *b* in a group III-inhibitor-sensitive pathway by reversed electron flow through the b_h center.[53,54]

Binding of a group I inhibitor is competitive with the binding of another group I or of a group II inhibitor, but independent of binding of a group III inhibitor.

All of these group I inhibitors induce a red shift in the spectra of the reduced b_l. These red-shift spectra are obtained in a double-beam spec-

[53] G. von Jagow, P. O. Ljungdahl, P. Graf, T. Ohnishi, and B. L. Trumpower, *J. Biol. Chem.* **259,** 6318 (1984).

[54] P. L. Dutton and D. F. Wilson, *Biochim. Biophys. Acta* **346,** 165 (1974).

TABLE I
GROUP I: INHIBITORS OF UQH_2-OXIDATION (IRON–SULFUR CENTER/HEME b_l REDUCTION)[a]

Inhibitor	Structural formula	Molecular formula
Myxothiazol[11,12]	O, NH2, S, N, N, S, OCH3, OCH3	$C_{25}H_{33}N_3O_3S_2$
Strobilurin A[14-16] (mucidin[17,18])	O, OCH3, OCH3	$C_{16}H_{18}O_3$
Strobilurin B[19]	Cl, OCH3	$C_{17}H_{19}ClO_4$
Strobilurin C[19]	O	$C_{21}H_{26}O_4$
Oudemansin A[20]	O, OCH3, H3CO, H, H3C, H, OCH3	$C_{17}H_{22}O_4$
Oudemansin B[19]	H3CO, Cl	$C_{18}H_{23}ClO_5$

[a] References 11–20 quote the first publication on the respective inhibitor and publications elucidating the structural formula.

trophotometer by reducing both reference and measuring samples using dithionite, and then adding the inhibitor in saturating concentration to the measuring sample only. The red-shift spectra are pronounced in experiments performed with the bc_1 complex isolated in Triton X-100. They are less pronounced when the complex is isolated in cholate, where the shift spectra may be superimposed by band changes and broadening of the heme b_l signal. The same pertains to spectra obtained with submitochondrial particles.

Group I consists of myxothiazol and various homologs of strobilurin (including mucidin) and oudemansin. Myxothiazol is the most tightly binding antibiotic,[36] followed in binding affinity by the strobilurins and finally the oudemansins.

Myxothiazol

The antibiotic myxothiazol is produced by the myxobacterium *Myxococcus fulvus*. It is a complex molecule which includes two thiazole rings. The essential segment for binding is the amide of MOA, which is linked in the β position to the rest of the molecule (Table I). Modification of the MOA segment largely decreases binding; e.g., the K_d of demethylmyxothiazol (having a β-ketopropionate group instead of the MOA) is 500 times higher than the K_d of the parent compound.

Myxothiazol is fairly stable when stored in organic solvents at 0°. The concentration of this solution may be determined spectroscopically at 313 nm ($\varepsilon_{mM^{-1}\,cm^{-1}} = 10.5$) (Table V).[11]

The red shift of the α-absorbance band is 2 nm, resulting in a shifted spectrum with a maximum of 568 nm and a minimum of 560 nm.[34] By observing the red shift, the K_d has been estimated to be below 1×10^{-10} M. As expected from the low K_d, binding takes place in a linear concentration dependence and is saturated at a titer of one molecule of myxothiazole per heme b_l.

The blockage of electron flow to the iron–sulfur center was demonstrated by EPR spectroscopy.[55] The extent of reduction of the two b centers through the b_h site depends on the redox poise of the system under study and on the actual potential of the heme centers in the preparation being used. Thus, in mitochondria and submitochondrial particles half of the heme b is reduced.[34] This reduction has to be attributed mainly to the b_h center, as judged from the difference of the two heme b centers in electrochemical potential of 100 mV. In isolated succinate–cyto-

[55] S. de Vries, P. J. Albracht, J. A. Berden, C. A. Marres, and E. C. Slater, *Biochim. Biophys. Acta* **723,** 91 (1983).

chrome-*c* reductase, both *b*-type centers are reduced under these conditions.[53]

Strobilurins: Strobilurin A (Mucidin), B, and C

The strobilurins A and B, two antifungals, have been isolated from the mycelium of the basidiomycete *Strobilurus tenacellus;* strobilurin C has been isolated from cultures of *Xerula longipes* R.Mre., a genus closely related to *Oudemansiella*.

The strobilurins contain the methyl ester of the MOA group, to which an α-methyl-Ω-phenylbutadienyl unit is linked in the α position. The three strobilurins differ from each other in their respective substitutions at positions 2 and 3 of the phenyl ring (Table I).

The stereochemistry of strobilurin A has been established recently by chemical synthesis. The structure had to be corrected from the formerly proposed Δ-9,10-*E* to a Δ-9,10-*Z* structure.[16]

Strobilurin A is a colorless oil, whereas strobilurin B forms colorless crystals with a melting point of 95°. The UV spectrum of strobilurin A in ethanol shows maxima at 229 nm ($\varepsilon_{mM^{-1}\,cm^{-1}}$ = 17.8), and 293 nm ($\varepsilon_{mM^{-1}\,cm^{-1}}$ = 21.9), whereas the maxima of strobilurin B are at 225 nm ($\varepsilon_{mM^{-1}\,cm^{-1}}$ = 53.7) and 287 nm ($\varepsilon_{mM^{-1}\,cm^{-1}}$ = 30.9) (Table V).[15]

The antibiotic mucidin has been shown to be identical with strobilurin A.[17,56] It is suggested that the name strobilurin A be used for this compound, since the structure of strobilurin A has been established by chemical synthesis.[16] The binding of the strobilurins follows classical saturation equilibria with a K_d value below 1×10^{-7} *M*; strobilurin B binds more tightly than strobilurin A.

The red shifts of the α-absorbance band of heme b_1 induced by strobilurin are similar to the red shifts exerted by myxothiazol, with the same maxima and minima of the reduced shift spectra at 568 and 560 nm.

The inhibitory effects of the strobilurins are identical to the effects of myxothiazol; however, since the binding to the Q_o center is not as tight, the inhibition of the respective electron-transferring steps is usually not quite as complete.

Oudemansins: Oudemansin A and B

Oudemansin A, a crystalline, optically active antibiotic, has been isolated from mycelium cultures of *Oudemansiella mucida;* oudemansin B has been isolated from cultures of *Xerula melanotricha*.

[56] G. von Jagow, G. W. Gribble, and B. L. Trumpower, *Biochemistry*, submitted.

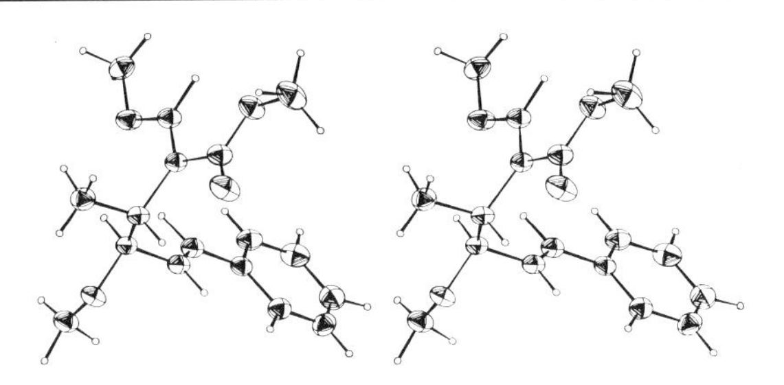

FIG. 2. ORTEP stereo plot of oudemansin A.

The oudemansins differ from the respective strobilurins by an addition of methanol to the Δ-9,10 double bond of the latter (Table I). The conformation of oudemansin A has been determined by X-ray crystallography.[20] The ORTEP stereo plot (Fig. 2) shows that the molecule is arranged in two planes which form an angle of 78°; one plane is formed by the terminal styryl moiety, the other by the planar MOA system.

Oudemansin A in methanol has an absorbance maximum at 245 nm ($\varepsilon_{mM^{-1}\,cm^{-1}} = 30.2$), while oudemansin B has two maxima at 219 nm ($\varepsilon_{mM^{-1}\,cm^{-1}} = 21.9$) and 248 nm ($\varepsilon_{mM^{-1}\,cm^{-1}} = 17.8$). The red-shift spectrum of reduced heme b_1 has a maximum at 568 nm and a minimum at 557 nm.[36]

Oudemansin A has the lowest binding affinity of all MOA inhibitors, $K_d = 5 \times 10^{-7}$ *M*. All other binding and inhibitory characteristics are similar to those of the strobilurins.

Chromone Inhibitors: Stigmatellin A and B

Like the group I inhibitors, the stigmatellins bind to the Q_o center. However, since their binding appears to be different from that of group I inhibitors, they are regarded as a separate group with properties of both group I and group II inhibitors.

The stigmatellins have been isolated from the myxobacterium *Stigmatella aurantiaca*. They contain a 5,7-dimethoxy-8-hydroxychromone system with a hydrophobic alkenyl chain in position 2. The two stigmatellins differ from each other in the stereochemistry of the terminal double bond of the side chain (Table II).

Stigmatellin A forms colorless microcrystals; it is stable at neutral and alkaline pH, but slowly decomposes below pH 5. The UV spectrum of stigmatellin A dissolved in methanol has absorbance maxima at 267 nm ($\varepsilon_{mM^{-1}\,cm^{-1}} = 65.5$) and 335 nm ($\varepsilon_{mM^{-1}\,cm^{-1}} = 5.2$) (Table V).

TABLE II
CHROMONE INHIBITORS[a]

Inhibitor	Structural formula	Molecular formula
Stigmatellin A[21,22]	H3CO O CH3 H3CO OCH3 H3CO O OH CH3 CH3	$C_{30}H_{42}O_7$
Stigmatellin B[22]		$C_{30}H_{42}O_7$

[a] References 21 and 22 quote the first publication on stigmatellin and the publication elucidating the structural formula.

Modification of the 8-hydroxy group, the 5-methoxy group, or the 4-keto group leads to partial or complete loss of inhibitory activity.[22] On the other hand, reduction of the chromone to a chromanone system does not alter the K_i significantly, nor may this be achieved by an alteration of the side chain as long as the overall hydrophobicity is not decreased, indicating that the partition coefficient of the inhibitor between the aqueous phase and the membrane/micelle is of decisive importance.

The chromone inhibitors induce a red-shift spectrum of the heme b_l,[37] which is about twice as large as that induced by myxothiazol. The maximum lies at 568 nm, the minimum at 558 nm. The red shift is superimposed by a broadening of the absorbance band which somewhat distorts the symmetry of the shift spectrum.[57] The K_d has been estimated to be below 1×10^{-10} *M*, an affinity comparable to that of myxothiazol. Therefore binding occurs in a linear concentration dependence, and the binding site is saturated at one molecule stigmatellin per heme b_l.

In a single-turnover experiment in submitochondrial particles, inhibition of electron flow leads to about 60% reduction of cytochrome *b*, as in the case of myxothiazol. Steady-state electron flow is blocked completely.

Stigmatellin binds to the heme b_l domain of cytochrome *b* as well as to the iron–sulfur protein. Concomitant with the red shift of the heme b_l

[57] G. von Jagow, unpublished.

TABLE III
GROUP II: INHIBITORS OF REOXIDATION OF THE IRON–SULFUR CENTER AND OF REDUCTION OF b_l[a]

Inhibitor	Structural formula	Molecular formula
UHDBT[23,24] (undecylhydroxydioxobenzothiazole)	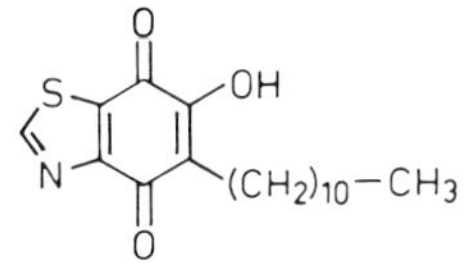 O, S, N, OH, $(CH_2)_{10}-CH_3$, O	$C_{18}H_{25}NO_3S$
PHDBT (pentadecylhydroxydioxobenzothiazole)	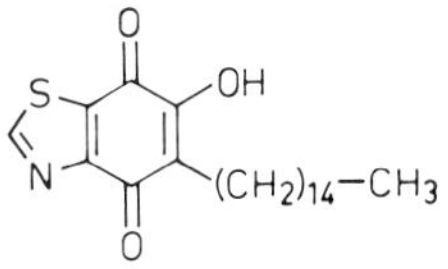 O, S, N, OH, $(CH_2)_{14}-CH_3$, O	$C_{22}H_{33}NO_3S$
UHNQ[25,26] (undecylhydroxynaphthoquinone)	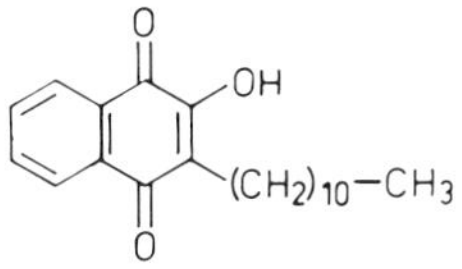 O, OH, $(CH_2)_{10}-CH_3$, O	$C_{21}H_{28}O_3$
HMHQQ[27] (heptadecylmercaptohydroxyquinoline quinone)	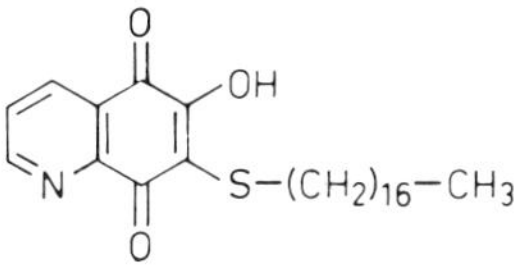 O, OH, N, $S-(CH_2)_{16}-CH_3$, O	$C_{26}H_{39}NO_3S$

[a] References 23–27 quote the first publication on the respective inhibitor and publications elucidating the structural formula.

spectrum, stigmatellin gives rise to an alteration of the EPR line shape of the Fe_2S_2 cluster. The midpoint redox potential of the iron–sulfur protein is thereby shifted from +290 to +540 mV [G. von Jagow and T. Ohnishi, *FEBS Lett.* **185,** 311 (1985)].

Group II

Inhibitors of Reoxidation of the Iron–Sulfur Center and of Reduction of b_l: Hydroxyquinones

This group of inhibitors has as a common structural element a 6-hydroxyquinone fragment, with different substitutions at positions 2 and 3 (Table III); a heterocyclic ring system fused to the benzoquinone ring seems to enhance inhibitory activity. This structure again mimics the ubiquinone structure. Hydroxyquinone systems change color, depending on the protonation–deprotonation equilibrium of the phenolic group.[58] Group II inhibitors do not exert a red shift in the spectra of either heme b_l or heme b_h, indicating a mechanism of binding different to that of group I inhibitors. Binding seems to take place at the iron–sulfur center, as the titer of inhibition is proportional to the content of iron–sulfur protein[40] and the K_d is strongly dependent on the state of reduction of the iron–sulfur cluster.[40,43] Oxidation of this center decreases binding by at least two orders of magnitude; therefore binding of these inhibitors changes the observed midpoint potential of the iron-sulfur cluster.[54]

As expected, group II inhibitors compete with each other, but bind independently of group III inhibitors. Although inhibitors of group I interact in a way different from that of group II inhibitors, group I inhibitors have been observed to displace group II inhibitors from their binding site.[53]

Group II inhibitors block steady-state electron transfer from succinate or NADH to oxygen. They completely block electron transfer between the iron–sulfur center and cytochrome c_1, thus preventing reoxidation of the iron–sulfur center after transfer of one electron from ubiquinol to Fe_2S_2.[59] This can be explained by binding of the inhibitor to the iron-sulfur protein and displacing the natural ubiquinol with accompanying raising of the potential of the iron–sulfur center.[7]

In the presence of group II inhibitors, reduction of cytochrome b is still possible by reversed electron flow through the Q_i site; this is only blocked by addition of a group III inhibitor. However, it has been shown that UHDBT[40] and HMHQQ[43] also bind to the Q_i site with lower affinity.

Alkylhydroxybenzothiazoles: UHDBT and PHDBT

These compounds contain a thiazole ring fused at positions 2 and 3 to the hydroxyquinone (Table III). They differ only in the length of their alkyl side chains, which is responsible for the hydrophobicity of the respective molecule.

The phenolic group is weakly acidic, pK_a 6.5.[38] Deprotonation leads to a change in color from yellow to rose-violet. The absorbance maxima of the fully deprotonated UHDBT dissolved in methanol + 0.1 mM acetic

[58] R. A. Morton, *in* "Biochemistry of Quinones" (R. A. Morton, ed.), p. 23. Academic Press, New York, 1965.

[59] J. R. Bowyer, P. L. Dutton, R. C. Prince, and A. R. Crofts, *Biochim. Biophys. Acta* **592,** 445 (1980).

acid are at 241 nm ($\varepsilon_{mM^{-1}\,cm^{-1}} = 10.6$), 287 nm ($\varepsilon_{mM^{-1}\,cm^{-1}} = 12.2$), and 445 nm ($\varepsilon_{mM^{-1}\,cm^{-1}} = 0.77$) (Table V).

Deprotonation also shifts the redox potential of the system; the half-reduction potential is ($436 - 68 \times$ pH) mV; this amounts to an E_m of -40 mV at pH 7.0.[38]

Deprotonation lowers the efficacy of the inhibitor drastically; therefore experiments should preferably be carried out below pH 7.2. Below pH 6.8, the K_d is less than 1×10^{-8} *M*.[38]

The binding to the reduced iron–sulfur protein seems to occur in almost stoichiometric manner; the half-reduction potential of the iron-sulfur cluster is thereby raised from +280 to +350 mV.[40]

Alkylhydroxynaphthoquinones: UHNQ

3-Alkyl-2-hydroxynaphthoquinones were among the first respiratory inhibitors detected during the search for antimalarials.[25] Their inhibitory activity is dependent on the hydrophobicity of the molecule and therefore on the length of the alkyl side chain.[26,42] UHNQ has the highest efficacy of the derivatives available, but requires a concentration about two times higher than that of UHDBT for half-inhibition. All effects, i.e., electron-transfer inhibition, midpoint potential shift of the iron–sulfur cluster, and the dependence of inhibition on the redox state of the iron–sulfur center and the quinone, are similar to those exerted by UHDBT.[42]

Hydroxyquinoline Quinones: HMHQQ

7-(*n*-Heptadecylmercapto)-6-hydroxy-5,8-quinoline quinone contains a pyridine ring fused at positions 2 and 3 to the hydroxyquinone system (Table III). The phenolic group is more acidic than that of the hydroxydioxobenzothiazoles; $pK_a = 5.5$.[43] The half-reduction potential of HMHQQ at pH 7.2 is -132 mV; therefore HMHQQ cannot be reduced by NADH dehydrogenase or succinate dehydrogenase, and the inhibitor is in an oxidized and deprotonated state during biological experiments.[43]

The absorbance maxima in ethanol/2 *M* Tris–HCl buffer, pH 8 (9 : 1, v/v), are at 260 nm ($\varepsilon_{mM^{-1}\,cm^{-1}} = 15.5$) and at 345 nm ($\varepsilon_{mM^{-1}\,cm^{-1}} = 7.5$) (Table V). The fluorescence has an excitation maximum at 275 nm and an emission maximum at 385 nm.

The binding is difficult to determine due to a high level of unspecific binding; K_d has been expressed as [$3 + 240 \times$ protein conc./(mg/ml)] n*M*,[43] when the iron–sulfur center is in the reduced state. This low K_d again indicates a stoichiometric binding of the inhibitor.

In contradiction to earlier reports,[43] Zhu *et al.* have recently found that, unexpectedly, HMHQQ did not significantly affect the equilibration between the iron–sulfur center and cytochrome c_1.[44]

Other Hydroxyquinones

Other hydroxyquinone derivatives, such as 5-*n*-decyl-2,3-dimethoxy-6-hydroxybenzoquinone, have been reported to inhibit electron transfer.[24] In all cases the hydroxy group is essential for inhibitory activity.[24] However, since the K_d values of these compounds are relatively high, they have not been used widely in biological experiments.

Group III

Inhibitors of Ubiquinone Reduction (Oxidation of the b_h Center)

The inhibitors of group III do not show structural analogies as the inhibitors of the two other groups do (Table IV). Between the inhibitors of group III, there even seem to be some differences in binding properties, as judged from the differences in their red-shift spectra. On the other hand, they block the same reaction, namely, the flow of electrons from the heme b_h center to oxidized ubiquinone and the respective reversed electron flow from UQH_2 to heme b_h, so that it seems to be justified to classify them within one group.

Group III inhibitors compete with each other, but bind independently of group I and II inhibitors; this is due to the fact that their point of action is at the Q_i site, which is most probably close to the middle of the membrane, compared to the Q_o site, which is located closer to the cytosolic surface of the mitochondrial inner membrane,[60] and where the group I and II inhibitors react.

In combined action with one of the inhibitors of group I or II, the group III inhibitors inhibit all electron transfer to or from either of the heme *b* centers[34]; this is the so-called double kill of cytochrome *b*.[5,61]

As described in the Introduction, group III inhibitors may serve to demonstrate the "oxidant-induced reduction of cytochrome *b*." Group III inhibitors generally induce superoxide generation when electrons are fed into the complex via the Q_o site and when cytochrome oxidase is active.[62] The $O_2^{\cdot-}$ generation is based on an autoxidation of the $Q_o^{\cdot-}$ radical. The presence of group I inhibitors, destruction of the iron–sulfur cluster

[60] T. Ohnishi and G. von Jagow, *Biophys. J.* **47,** 241a (1985).

[61] D. H. Deul and M. B. Thorn, *Biochim. Biophys. Acta* **59,** 426 (1962).

[62] M. Ksenzenko, A. Konstantinov, G. B. Khomutov, A. N. Tikhonov, and E. Ruuge, *FEBS Lett.* **155,** 19 (1983).

TABLE IV
GROUP III: INHIBITORS OF UBIQUINONE REDUCTION (OXIDATION OF THE b_h CENTER)[a]

Inhibitor	Structural formula	Molecular formula
Antimycin[28,29]	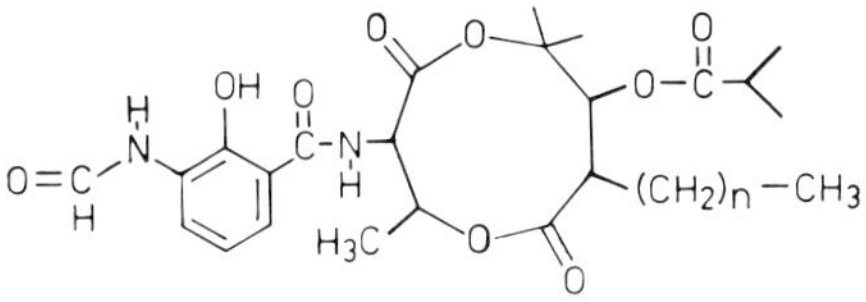	$C_{28}H_{40}N_2O_9$[b]
Funiculosin[30,31]	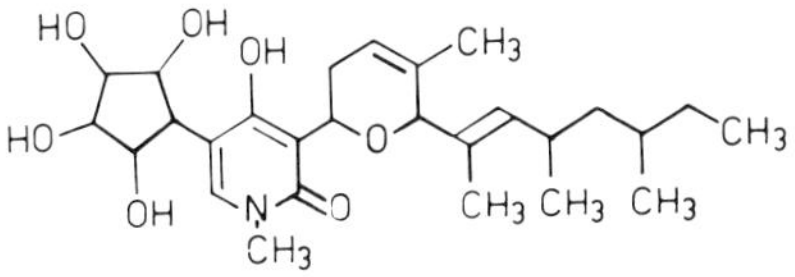	$C_{27}H_{41}NO_7$
HQNO[32] (heptylhydroxyquinoline-*N*-oxide)	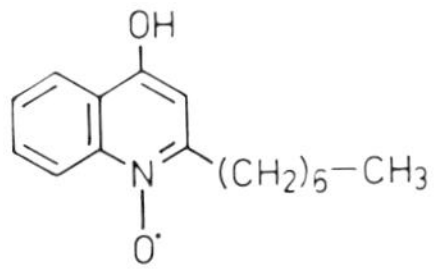	$C_{16}H_{21}NO_2$
NQNO (nonylhydroxyquinoline-*N*-oxide)	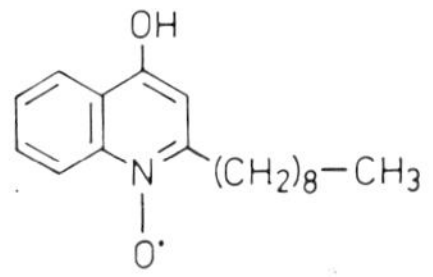	$C_{18}H_{25}NO_2$

[a] References 28–32 quote the first publication on the respective inhibitor and publications elucidating the structural formula.
[b] Natural antimycin A is a mixture of various homologs with different lengths of the alkyl side chain (n = 3–6); the main component (n = 5) is shown here.

by BAL, or KCN suppress the superoxide generation. In contrast to inhibition by myxothiazol, both *b* centers are reduced in the presence of antimycin when electrons are fed into the respiratory chain of submitochondrial particles on the route of the various dehydrogenases, e.g., succinate dehydrogenase or NADH dehydrogenase, as well as when electrons are delivered to an isolated bc_1 complex by ubiquinol-9 or a homolog thereof.

Antimycin

The antibiotic antimycin is a natural product of various species of *Streptomyces*. Its chemical structure is complex (Table IV); it consists of

3-formamidosalicylic acid linked via an amide bond to an alkyl- and acyl-substituted dilactone ring. Commercially available antimycin is a mixture of homologs with varying lengths of the alkyl side chain.

In ethanol containing 1 mM HCl, antimycin reveals an absorption spectrum with a maximum at 320 nm ($\varepsilon_{mM^{-1}\,cm^{-1}} = 4.8$). Additionally, the molecule exhibits a fluorescence spectrum with an excitation maximum at 350 nm and an emission maximum at 420 nm (also in ethanol) (Table V). The fluorescence is enhanced on binding to albumin, while it is completely quenched on binding to submitochondrial particles.

The binding of the inhibitor can be studied either by fluorescence quenching techniques or by measurement of the red shift exerted on the spectrum of the heme b_h center.

On binding of antimycin, the α-absorbance band of reduced heme b_h is shifted from 562 to 564 nm. Besides this, antimycin is the only inhibitor known so far which also induces a red shift in the γ-band of the heme; this γ-red shift is 5 times as large as the α-shift.[63] Binding of antimycin usually occurs with a linear concentration dependence; saturation is achieved at a titer of 1 molecule of antimycin per complex III. The affinity of antimycin in submitochondrial particles ($K_d = 3.2 \times 10^{-11}$ M) is the highest measured so far of all of the inhibitors of the bc_1 complex. In succinate-reduced submitochondrial particles, binding may occur in a positive cooperative manner.[46]

The binding of antimycin induces a conformational change of the bc_1 complex, indicated, e.g., by a change in affinity of the "iron–sulfur protein-depleted complex" for the iron-sulfur protein[63] or by the markedly increased stability to treatment by detergent.[63a] It has been shown by EPR studies that binding of antimycin abolishes the binding of the $Q_i^{\overline{\cdot}}$ radical at the heme b_h center.[64]

From fluorescence quenching data it has been estimated that the benzene ring is situated about 17 Å away from the heme b_h group.[46]

Funiculosin

Funiculosin is an antibiotic produced in *Penicillium funiculosum Thom.*[30] Its chemical structure consists of a *N*-methyl-substituted 4-hydroxy-2-pyridone ring with a hydrophobic side chain in position 1 and a tetrahydroxycyclopentane ring in position 3.[31] In ethanol made slightly alkaline with NaOH, funiculosin shows an absorbance spectrum with a maximum at 290 nm ($\varepsilon_{mM^{-1}\,cm^{-1}} = 5.5$). The K_d of funiculosin is one order

[63] G. von Jagow, unpublished.

[63a] J. S. Rieske, H. Baum, C. D. Stone, and S. H. Lipton, *J. Biol. Chem.* **242,** 4854 (1967).

[64] T. Ohnishi and B. L. Trumpower, *J. Biol. Chem.* **255,** 3278 (1980).

TABLE V

PHYSICOCHEMICAL DATA OF THE INHIBITORS OF THE bc_1 COMPLEX[a]

Inhibitor	Ref.	Molecular weight	Absorbance λ_{max}(nm)	Absorbance ε($mM^{-1}cm^{-1}$)	Fluorescence maxima $\lambda_{exc}/\lambda_{em}$(nm)	K_d (M)
Group I						
Myxothiazol	33, 34	487.7	313	10.5[11]		$<1 \times 10^{-10}$ [35]
Strobilurin A	36	258.3	229	17.8[15]		$<1 \times 10^{-7}$ [36]
(Mucidin)			293	21.9		
Strobilurin B	15	322.8	225	53.7[15]		$<1 \times 10^{-7}$ [36]
			287	30.9		
Oudemansin A	36	290.4	245	30.2[20]		5×10^{-7} [36]
Oudemansin B	36	354.8	219	21.9		—
			248	17.8		
Chromone inhibitors						
Stigmatellin A	37	514.7	267	65.5[21]		$<1 \times 10^{-10}$ [37]
			335	5.2		
Group II						
UHDBT	38–40	335.5	241	10.6[38]		$<1 \times 10^{-8}$ [38,b]
			287	12.2		
PHDBT	38, 41	391.6	Same as for UHDBT			$<1 \times 10^{-9}$ [41]
UHNQ	42	328.5	—	—		See text
HMHQQ	43	445.7	260	15.5[43]	275/385[43]	See text and Ref. 43
			345	7.5		
Group III						
Antimycin	45–47	548[c]	320	4.8	350/420[47]	3.2×10^{-11} [46]
Funiculosin	49, 50	491.6	290	5.5		2×10^{-10} [49]
HQNO	51, 52	259.3	253	24	355/480[52]	6.4×10^{-8} [52]
			346	9.5		
NQNO		287.4	Same as for HQNO[32]			—

[a] The references (33–52) quote the basic publications on the respective inhibitors and those giving the absorbance and fluorescence maxima, as well as K_d values.

[b] pH < 6.8

[c] Natural antimycin A is a mixture of various homologs with differing lengths of the side chain; the main component is shown here.

of magnitude higher than that of antimycin, but two orders of magnitude lower than the K_d of the hydroxyquinoline *N*-oxides.[49]

Inhibition of steady-state electron flow by funiculosin in submitochondrial particles as well as in isolated bc_1 complex leads to reduction of both *b* centers.[50]

Hydroxyquinoline N-Oxides: HQNO and NQNO

These quinone analogs were first isolated as natural products (antagonists of dihydrostreptomycin) from *Pseudomonas aeruginosa* (formerly called *Ps. pyocyanea*).[32] They are now produced by chemical synthesis. These compounds differ in the length of their 2-*n*-alkyl side chain (Table IV); the heptyl and the nonyl homologs are commercially available.

The following properties apply to both the heptyl and the nonyl derivative: In 1 m*M* NaOH, HQNO shows absorbance maxima at 253 n*M* ($\varepsilon_{mM^{-1}\,cm^{-1}}$ = 24) and 346 nm ($\varepsilon_{mM^{-1}\,cm^{-1}}$ = 9.45). At pH 7.4 (buffered solution), it reveals a fluorescence spectrum with an excitation maximum at 355 nm and an emission maximum at 480 nm (Table V). As in the case of antimycin, fluorescence is enhanced on binding to albumin and completely quenched in submitochondrial particles.[52]

The K_d is about 3 orders of magnitude higher than that of antimycin ($K_d = 6.4 \times 10^{-8}$ *M*); therefore the inhibitory effects are less comprehensive than those of antimycin. When determining the K_i value, a fair amount of unspecific binding has to be taken into account; e.g., HQNO is affirmed to exert a weak competitive effect at the site of inhibition by UHDBT.[65]

Binding induces only a minute red shift of the reduced heme b_h and is saturated at a titer of 1 molecule of HQNO per complex III.

Group IV

Miscellaneous Inhibitors: BAL

The electron-transferring activity of the iron–sulfur protein can be blocked by the combined action of the antitoxicant British Anti-Lewisite (2,3-dimercaptopropanol) and oxygen.[66] This irreversible inhibition of electron flow has been attributed to the destruction of the iron–sulfur cluster.[67] BAL treatment of submitochondrial particles also has a slight influence on succinate dehydrogenase activity, indicating that BAL is not

[65] S. Papa, G. Izzo, and F. Guerrieri, *FEBS Lett.* **145,** 93 (1982).

[66] E. C. Slater, *Biochem. J.* **45,** 14 (1949).

[67] E. C. Slater and S. de Vries, *Nature (London)* **288,** 717 (1980).

quite specific in its action. The nonexistence of a redox component with pool function between ubiquinol oxidation and reduction sites has been deduced from the fact that partial destruction of the iron–sulfur cluster by BAL treatment had the same effect on ubiquinol: cytochrome-*c* reductase activity as partial inhibition by HQNO[68]; this is completely in agreement with the model proposed (Fig. 1).

DTNB

5,5′-Dithiobis(2-nitrobenzoic acid), like other thiol reagents, inhibits the Q_o reaction site by binding to the iron–sulfur protein.[69] Unlike BAL, DTNB does not destroy the iron–sulfur cluster, as can be demonstrated by the effect of DTNB on the EPR signal. The EPR spectrum in the DTNB-treated preparation is not affected by the addition of the group II inhibitor HMHQQ, indicating that the inhibition by DTNB is irreversible.[69]

[68] Q. S. Zhu, J. A. Berden, S. de Vries, and E. C. Slater, *Biochim. Biophys. Acta* **680,** 69 (1982).
[69] C. A. M. Marres, S. de Vries, and E. C. Slater, *Biochim. Biophys. Acta* **681,** 323 (1982).

[25] Preparations of Electrogenic, Proton-Transporting Cytochrome Complexes of the b_6f-Type (Chloroplasts and Cyanobacteria) and bc_1-Type (*Rhodopseudomonas sphaeroides*)

By GÜNTER HAUSKA

Cytochrome *bc* complexes function as quinol-cytochrome *c*/plastocyanin oxidoreductases in many electron-transport chains. In the chloroplasts of plants and eukaryotic algae and in the prokaryotic cyanobacteria a b_6f complex[1] oxidizes plastoquinol with plastocyanin between the two photosystems. In the respiratory chain of mitochondria and many bacteria a bc_1 complex oxidizes ubiquinol with cytochrome *c*. These complexes have been isolated from a variety of sources in active form. They all

[1] Abbreviations: b_6f, cytochrome b_6f; bc_1, cytochrome bc_1; *bc*, cytochrome *bc*; FNR, ferredoxin–NADP$^+$ oxidoreductase; CF_1CF_0, chloroplast coupling factor complex; MEGA-9, *N*-methyl-*N*-nonanoylglucamide; MEGA-10, *N*-methyl-*N*-decanoylglucamide; Tris, tris(hydroxymethyl)aminomethane; Tricine, *N*-[2-hydroxy-1,1-bis(hydroxymethyl)-ethyl]glycine; SDS–PAGE, sodium dodecyl sulfate–polyacrylamide gel electrophoresis.

exhibit a branched electron-transport pathway termed "oxidant-induced reduction of cytochrome *b*" and function as electrogenic proton translocators when reincorporated into liposomes. They have the general redox center composition of 1 heme *c* in cytochrome c_1 or *f*, 2 heme *b* in cytochrome *b* or b_6, and one high-potential Fe_2S_2 cluster in the Rieske FeS-protein. The polypeptide composition differs, the mitochondrial complex being more complicated than the plastidal or the bacterial ones. A detailed review of these complexes appeared in 1983.[2] Recent progress mainly provided by gene sequencing has been summarized elsewhere.[3]

General Isolation Procedure

A procedure suitable for the isolation of *bc* complexes and other membrane protein complexes has been developed,[4-8] adapting a method for the isolation of the chloroplast coupling factor complex CF_1CF_0.[9] It comprises four basic steps:

1. Repeated washing of the membranes in high NaBr to remove extrinsic proteins[10]
2. Selective solubilization of the desired complex by an appropriate detergent mixture
3. Ammonium sulfate precipitation
4. Sucrose density gradient centrifugation

Note 1: Washing in NaBr. This step has been introduced to remove the extrinsic part CF_1 of the chloroplast coupling factor complex.[10] In chloroplasts it largely removes other extrinsic proteins from the outer surface, such as ferredoxin–$NADP^+$ oxidoreductase (FNR) and ribulose-bisphosphate (RuBP) carboxylase, and also from the inner surface, such

[2] G. Hauska, E. Hurt, N. Gabellini, and W. Lockau, *Biochim. Biophys. Acta* **726,** 97 (1983).

[3] G. Hauska, *in* "Encyclopedia of Plant Physiology, New Series" (A. Pirson and M. H. Zimmermann, eds.), "Photosynthetic Membranes" (L. A. Staehelin and C. J. Arntzen, eds.), in press, Springer-Verlag, Berlin and New York, 1985.

[4] E. Hurt and G. Hauska, *Eur. J. Biochem.* **117,** 591 (1981).

[5] E. Hurt and G. Hauska, *J. Bioenerg. Biomembr.* **14,** 405 (1982).

[6] M. Krinner, G. Hauska, E. Hurt, and W. Lockau, *Biochim. Biophys. Acta* **681,** 110 (1982).

[7] N. Gabellini, J. R. Bowyer, E. Hurt, B. A. Melandri, and G. Hauska, *Eur. J. Biochem.* **126,** 105 (1982).

[8] E. Hurt, G. Chreptun, J. Davenport, N. Gabellini, E. Herold, M. Krinner, W. Lockau, W. Nitschke, B. Paproth, W. Pinther, and U. Schöder contributed to this work, which has been supported by the Deutsche Forschungsgemeinschaft (SFB 43 C2).

[9] U. Pick and E. Racker, *J. Biol. Chem.* **254,** 2793 (1979).

[10] N. Nelson, this series, Vol. 69, p. 301.

as plastocyanin. Thus, high NaBr does not merely act by breaking ionic interactions, but also by the chaotropic effect of the polarizable bromide anions, perturbing the membrane and liberating internalized proteins. For efficient removal of extrinsic proteins the NaBr washing should be repeated and should be combined preceding and following swelling of the vesicular membrane system in low salt media.

Note 2: Selective Extraction with Detergent Mixtures. This is the critical step and should be carefully optimized for the extraction of any desired complex. A combination of 0.5% cholate and 30 m*M* octylglucoside (Sigma) in the presence of high salt (0.4 *M* ammonium sulfate) at about 1.5 mg chlorophyll/ml (about 8 mg protein/ml) is suitable for the extraction of the cytochrome b_6f complexes.[3–5] We suggest, however, that octylglucoside or the other detergent of choice be titrated in a pilot experiment for every batch of membrane material. For chloroplasts, extraction of more than 70% of the complex together with less than 5% of chlorophyll can be achieved. The optimal concentration of octylglucoside for this is between 20 and 40 m*M* for spinach, depending on source, age, and season. It is only 10 m*M* for pea.[11,12] Octylglucoside can be replaced by laurylmaltoside (Sigma) at about half the concentrations. Also the cheaper *N*-methyl-*N*-alkanoylglucamides[13] (Oxyl), known as the MEGA detergents, can replace octylglucoside. The decanoyl derivative MEGA-10 is about 1.5 times and the nonanoyl derivative MEGA-9 (see below) about 0.7 times as efficient as octylglucoside. Both are less water soluble, however. Below 8° MEGA-10 will precipitate or form gels from solutions above 10m*M*. The selective extraction in favor of the cytochrome complexes over chlorophyll proteins is dependent on the presence of cholate, which suppresses unspecific membrane solubilization, possibly by partially coating the surface. The efficiency of a given cholate/detergent combination can be increased by high salt (0.4 *M* ammonium sulfate). It is essential that ammonium sulfate is added together with detergent, not before. Other detergents, such as Triton X-100 or digitonin, also in the presence of cholate, are much less selective.

Note 3: Ammonium Sulfate Precipitation. The cytochrome complexes are precipitated between 55 and 70% saturation, but the cut for optimal yield and purity varies and should be determined in each new case. In our experience it is not essential to adjust the pH; solid ammonium sulfate or a saturated solution can be added. At 70% saturation a float instead of a precipitate might form; this can be collected by filtering through cotton

[11] A. L. Phillips and J. C. Gray, *Eur. J. Biochem.* **137,** 553 (1983).

[12] G. Hauska, unpublished.

[13] J. E. I. Hildreth, *Biochem. J.* **207,** 363 (1982).

wool. Alternatively, a precipitate is obtained by dilution with water to 65% just before sedimentation. The precipitate or float should be resuspended fast and concentrated, carrying over supernatant ammonium sulfate solution as little as possible. Ammonium sulfate increases the density and might disturb subsequent loading on sucrose density gradients. This can be helped either by dilution or by a short dialysis, or by cutting the sucrose density gradient on the top. Dilution is not desirable because centrifugation on the gradient does not reach equilibrium and therefore final purification is limited by the sample volume applied. In case of prolonged dialysis detergent is lost, which might cause precipitation. Addition of detergent to the external fluid is costly. Starting the sucrose gradient with 15 instead of 5%[4] is sufficient for most cases.

Note 4: Sucrose Density Gradient. The sample volume should not exceed 10% of the gradient volume. Depending on amount the Beckman rotors SW 60, SW 41, or SW 27 can be used. We found it convenient to use the fixed-angle rotor 60 Ti for larger amounts. The gradient can be formed by hand layering in 6 steps with partial mixing. A gradient former may be used to obtain complete linearity. The time of centrifugation should be chosen with respect to the applied *g* force and the relative sedimentation velocity of the components in the applied sample for optimal purification. It usually is between 10 and 25 hr. It should be noted that the cytochrome complexes are denser than 30% sucrose and will eventually sediment unless 50% sucrose is used at the bottom of the gradient. Soybean lecithin (Sigma), 0.1%, was found to stabilize the b_6f complex from spinach in the gradient, but was not necessary for the other preparations. Usually the detergent concentration in the gradient is kept the same as during solubilization. It can be lowered to prevent slow inactivation (Note 7), but not beyond about 10 m*M* octylglucoside or equivalent where the complexes might start to aggregate. Cholate alone, even up to 2%, is not sufficient to keep the complexes dispersed.

Note 5: Purity. Figure 1 shows the polypeptide patterns of the cytochrome complexes from the sucrose density gradient. The b_6f complexes appear quite pure; the bacterial bc_1 complex is still contaminated by the three-reaction center polypeptides. Eventually the b_6f complexes are contaminated (see Fig. 2) by FNR, photosystem II (seen by two diffuse bands at about 45 and 50 kDa), or CF_1CF_0 (seen by the α- and β-subunits at about 55 and 59 kDa). Additional purification is achieved by molecular sieve chromatography on Sepharose 4B, or BioGel P300, or Ultrogel AcA 34, or also by repeated density gradient centrifugation. Detergent should be kept as low as possible (Note 7). Chromatography on DEAE-cellulose or OH-apatite leads to inactivation and dissociation.

Note 6: Concentration and Detergent Exchange. The preparations can

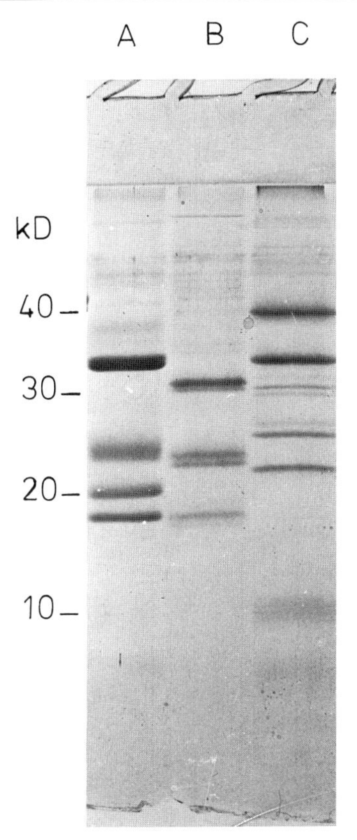

FIG. 1. Polypeptide patterns of *bc* complexes. SDS–PAGE on 15% gel after Laemmli,[18] stained with Coomassie Brilliant Blue.[34] (A) b_6f complex from spinach chloroplasts, 0.3 nmol *f*. The redox polypeptides have been identified[2]: cytochrome *f* occurring in two forms, 33 and 34 kDa; cytochrome b_6, 23 kDa; Rieske FeS-protein, 20 kDa. (B) b_6f complex from *Anabaena variabilis*, 0.1 nmol *f*. Cytochrome *f* is in two forms, 30 and 31 kDa (the 30 kDa form is not always observed[2,6]); cytochrome b_6, 23 kDa; Rieske FeS-protein, 22 kDa (this is not always resolved from cytochrome b_6[2,6]). (C) bc_1 complex from *Rhodopseudomonas sphaeroides* GA, 0.2 nmol c_1. Cytochrome *b*, 42 kDa; cytochrome c_1, 34 kDa; Rieske FeS-protein, 25 kDa; the band at 21 kDa is another form of the Rieske FeS-protein, since it cross-reacts with a specific antibody[12]; a contamination by the reaction center is seen at 27, 29, and 30 kDa.[19]

be concentrated by ultrafiltration or more conveniently by centrifugation for a few hours in a Beckman Ti 75 rotor at maximal speed, after diluting the sucrose about 3 times. Concomitant concentration of detergent during ultrafiltration should be avoided by using filters penetrable for free detergent micells, such as the Amicon filter XM100A. Repetitive application of either procedure can be used to exchange detergent. This can be obtained in one step by centrifugation into a gradient or through a cushion of sucrose solution containing the desired detergent mixture.

Note 7: Stability and Storage. An optimal detergent concentration is necessary for maximal preservation of activity and dispersion of the complexes, which is around 10 mM octylglucoside or equivalent at 10–20 μM complex. Oxidoreductase activity decreases with time. For the chloroplast b_6f complex, at 30 mM octylglucoside or MEGA-10, the half times are about 4 and 24 hr at room temperature or 4°, respectively. If frozen quickly in liquid nitrogen and stored at −13°, −80°, or in liquid nitrogen, the activity decreased to about 40, 70, and 90% within 2 weeks and then remained stable for months. Also prolonged exposure to excess detergents during purification, especially if the complexes are dilute, leads to inactivation. As a consequence, tailing to lower density or even the formation of a second, inactive, less dense band is observed on sucrose gradients. If the active band of the b_6f complex from chloroplasts from a first gradient is rerun under identical conditions on a second gradient, the inactive band is again formed, depending on the detergent concentration. At 20 and 30 mM octylglucoside or MEGA-10, 25 and 40% were found in the upper, inactive band. At 10 mM the second band was not formed; only some tailing was observed. At 30 mM, in the first gradient during our routine preparation (see below), less than 10% are usually found in the upper band. Triton X-100, originally used as detergent in the gradient,[4] inactivates more rapidly. Although the removal of the Rieske FeS-protein by chromatography in OH-apatite in the presence of high Triton X-100 has been reported,[2,14,15] the inactivation by detergents does not reflect removal, but rather dislocation of the subunits, since the polypeptide pattern (Fig. 1) is not changed in the upper band on gradients.

Note 8: Proteolytic Degradation. This has not been a problem for the isolation of the b_6f complexes, although the dual form of cytochrome f (Fig. 1, lanes A and B) might be caused by proteolysis. However, it cannot be prevented by inclusion of protease inhibitors, and the two heme polypeptides at 33 and 34 kDa are also found on SDS–PAGE if spinach chloroplasts are dissolved in boiling SDS in the presence of protease inhibitors.[12] The b_6f complexes from other plants,[12] such as pea (see Fig. 2),[11] bean, and *Oenothera,* contain only one form of cytochrome f. The use of protease inhibitors is described in detail elsewhere.[16,17]

[14] W. D. Engel, H. C. Michalsky, and G. von Jagow, *Eur. J. Biochem.* **132,** 395 (1983).

[15] E. Hurt, G. Hauska, and R. Malkin, *FEBS Lett.* **134,** 1 (1981).

[16] J. C. Gray, *in* "Methods in Chloroplast Molecular Biology" (M. Edelman, R. B. Hallick, and N.-H. Chua, eds.), p. 1093. Elsevier, Amsterdam, 1982.

[17] This series, Vols. 19, pp. 807–932, and 45, pp. 639–888.

Preparation of the b_6f Complex from Spinach Chloroplasts

Reagents (Amounts for 2 kg Leaves)

STN: 0.4 *M* sucrose, 10 m*M* Tris–HCl, pH 8.0, 10 m*M* NaCl (2 liters)
0.15 *M* NaCl (1 liter)
10 m*M* Tris–HCl, pH 8.0 (1.7 liters)
2 *M* NaBr, 0.4 *M* sucrose, 10 m*M* Tris–HCl, pH 8.0 (1.6 liters)
STKM: 0.2 *M* sucrose, 20 m*M* Tricine/NaOH, pH 8.0, 3 m*M* KCl, 3 m*M* $MgCl_2$ (1 liter)
30 m*M* Tris/succinate, pH 6.5 (250 ml)
20% (w/v) Na-cholate (recrystallized from EtOH), stock
5% Soybean lecithin (type IIS, Sigma) dispersed by sonication, stock
Solid ammonium sulfate
Solid sucrose
Solid octylglucoside (Sigma) or MEGA-9 (Oxyl)
Stock solutions and solids are added to STKM for the solubilization mixture, or to Tris/succinate for resuspension and the density gradients according to the procedure. MEGA-9 requires heating to dissolve and solutions should be cooled on ice before use.

Procedure

Fresh, young spinach leaves, 2 kg, are deveined, washed, and homogenized with a mixer for about 1 min, in 2 liters STN. The slurry is pressed through a nylon mesh plus four layers of cheesecloth. From the filtrate chloroplasts are sedimented at 16,000 *g* (av) in the Sorvall GSA rotor (10,000 rpm) for 20 min. The sediment is suspended and recentrifuged repeatedly in the following steps: (1) resuspension in 1 liter 0.15 *M* NaCl; centrifugation as above; (2) osmotic shock by resuspension in 700 ml 10 m*M* Tris–HCl, pH 8.0; after 5 min dilution 1 : 1 with 0.15 *M* NaCl; centrifugation for 30 min. (3) resuspension in 800 ml 2 *M* NaBr, 0.4 *M* sucrose, 10 m*M* Tris–HCl, pH 8.0; incubation for 30 min at 4°; dilution 1 : 1 with cold water, and centrifugation for 15 min; (4) repetition of step 3; (5) resuspension in 1 liter 10 m*M* Tris–HCl, pH 8.0; centrifugation for 20 min. The final sediment is resuspended with STKM medium to 3–4 mg chlorophyll/ml. The suspension is kept at 4° overnight. In a pilot experiment, four 1-ml portions are then diluted to 1.5 mg chlorophyll/ml with STMK containing the required additions to yield the final concentrations of 0.4 *M* ammonium sulfate (10% saturation), 0.5% cholate (w/v), and varying concentrations of octylglucoside (20–50 m*M*), or of MEGA-9 (30–70 m*M*). Warming is necessary to dissolve MEGA-9, and the solutions should be cooled to 4° before adding to chloroplasts. The pH may but need not be adjusted to 8.0. The suspensions are kept on ice for 30 min and are then centrifuged for 30 min above 40,000 *g* (Corex glass tubes in a Sorvall SS 34 rotor at 19,000 rpm). The concentration of cytochrome *f* and chlorophyll is determined in the supernatants (see Assays). The rest of the chloroplast suspension is then diluted to 1.5 mg chlorophyll/ml with the mixture above, containing octylglucoside or MEGA-9 in an amount which solubilizes more than 70% of cytochrome *f* (about 1.5 μM in the supernatant) and less than 5% of the chlorophyll (75 μg/ml). After 30 min stirring on ice the suspension is centrifuged for 1 hr above 40,000 *g* (19,000 rpm in a Sorvall SS 34 rotor). To the supernatant solid ammonium sulfate is added to 45% saturation (10% are already present!) with stirring on ice. The addition takes about 10 min and is slowed toward the end. After 5 min of further stirring the suspension is centrifuged in the Sorvall GSA rotor for 10 min at 10,000 rpm. The precipitate is discarded. From the supernatant the b_6f complex is precipitated by further addition of solid ammonium sulfate to 55% saturation and centrifugation. Supernatant ammonium sulfate solution is carefully removed from the precipitate, which is then dissolved by addition of the minimum volume (1–2 ml) of 30 m*M* Tris/succinate, pH 6.5, containing 0.5% cholate and the concentration of octylglucoside or MEGA-9 used for solubilization. The suspension, which is at least 30 μM in cytochrome *c*, is dialyzed briefly (less than 1 hr) against Tris/succinate-cholate, and is subsequently loaded on a sucrose density gradient (14–30%) containing 30 m*M* Tris/succinate, pH 6.5, 0.5% cholate, 30 m*M* octylglucose or MEGA-9, and 0.1% soybean lecithin (Sigma). Not more than 1.5 ml suspension per 24 ml gradient is applied in the tubes of a Beckman 60 Ti rotor, which is subsequently spun at 50,000 rpm overnight (12–15 hr). The brown b_6f-containing band is collected by a syringe or tubing. The final preparation, having a cytochrome *f* concentration of about 15 μM in up to 50% yield (about 0.4 μmol) with respect to the detergent extract, is quickly frozen and stored in liquid nitrogen. Oxidoreductase activity routinely is around 30 μmol plastocyanin reduced per nmol cytochrome *f* per hour, measured at pH 6.5 and with plastoquinol-1 at subsaturating concentrations (see assays). At optimal conditions and using plastoquinol-2 or -3, rates of more than 100 units have been measured. The polypeptide composition is shown in Fig. 1A.

The procedure differs somewhat from our original:[4] The washing of

[18] U. K. Laemmli, *Nature (London)* **227,** 680 (1970).

[19] G. Feher and M. Y. Okamura, *in* "The Photosynthetic Bacteria" (R. K. Clayton and W. R. Systrøm, eds.), p. 349. Plenum, New York, 1978.

the membranes has been simplified, Triton X-100 is avoided in the gradient, and MEGA-9 can be used instead of costly octylglucoside.

Preparation of the b_6f Complex from Other Sources

Other Plants

The procedure for spinach has been successfully applied to other green plants, such as Swiss chard,[20] lettuce,[20] pea,[12] barley,[12] *Oenothera*,[12] and *Chlamydomonas*.[21] It is also applicable to etioplasts from barley.[12]

An inactive b_6f complex lacking the Rieske FeS-protein, but containing FNR as an additional component, has been isolated from pea by a modification of our procedure.[11] Probably the FeS-protein is lost during electrophoresis in the presence of Triton X-100 and deoxycholate in the final step. Following our procedure, it is possible to obtain an active preparation, retaining the FeS-protein and lacking FNR, however.[12] Figure 2 shows the SDS–PAGE patterns of the fractions from the sucrose density gradient for pea, in comparison to a preparation from spinach, stained first for heme[22] and then for protein. It is clearly seen that FNR peaks above the four subunits of the b_6f complex in the gradient.

Cyanobacteria

The procedure for spinach using octylglucoside has also been adopted for *Anabaena variabilis*[6] and certainly could be modified for other cyanobacteria.

Reagents (Amounts for 10-Liter Culture)

- MTPM: 0.2 *M* mannitol, 15 m*M* Tricine/Tris, pH 8.1, 1 m*M* NaH_2PO_4, 2 m*M* $MgCl_2$ (2 liters)
- MTPM plus 2 *M* NaBr (200 ml)
- 20 m*M* Tricine/Tris, pH 8.1
- 20% (w/v) Na-cholate (recrystallized from EtOH), stock
- 20% (v/v) Triton X-100
- Saturated ammonium sulfate solution, pH 7.5 (150 ml)
- Solid ammonium sulfate
- Solid sucrose
- Solid octylglucoside (Sigma) or solid MEGA-9 (Oxyl)
- Cholate and solids are added to MTPK for the solubilization mixture,

[20] N. Nelson, unpublished.

[21] M. Sanguarmsermsi and D. S. Bendall, unpublished.

[22] P. E. Thomas, D. Ryan, and L. Wayne, *Anal. Biochem.* **75,** 168 (1976).

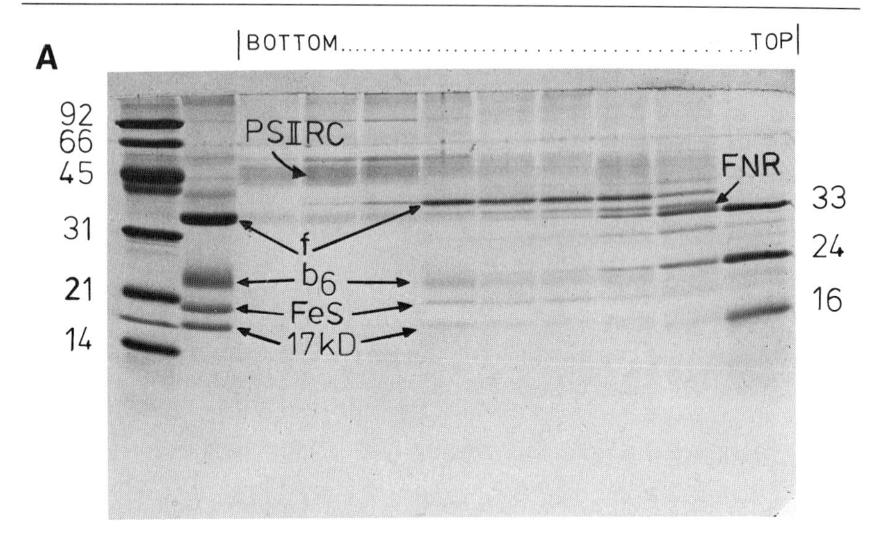

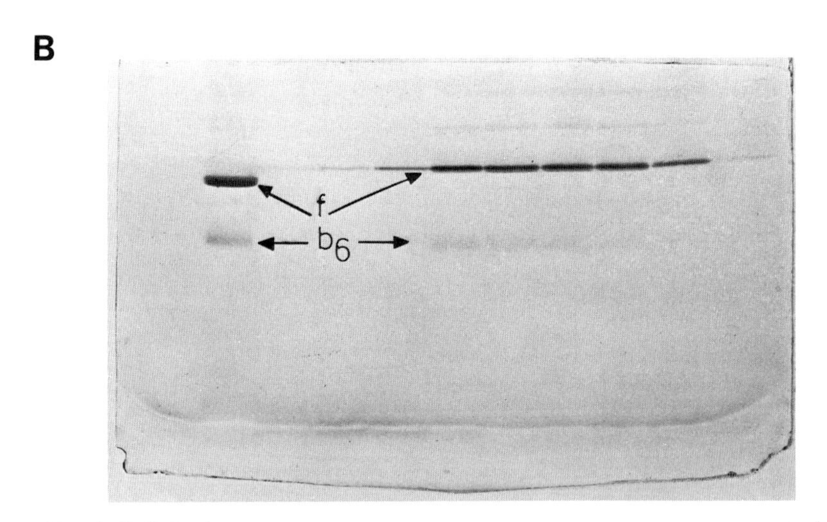

FIG. 2. SDS–PAGE of the sucrose density gradient fractions of a b_6f preparation from pea. SDS–PAGE on 15% gel after Laemmli,[18] first stained for heme[22] (B) and then with Coomassie Brilliant Blue (A).[34] On the left the gel includes a standard (14–92 kDa) and the spinach b_6f complex for comparison. The four subunits of the complex are indicated by f for cytochrome *f*, b_6 for cytochrome b_6, FeS for the Rieske FeS-protein and 17 kDa. PSIIRC is the photosystem II reaction center polypeptides, and FNR is ferredoxin–NADP$^+$ oxidoreductase (34 kDa). The top of the gradient contains three polypeptides functioning in water oxidation (33, 24, and 16 kDa).[44]

or to Tris/HCl for resuspension and the density gradients according to the procedure.

Procedure

This is a simplified version of the original.[6] MEGA-9 can be used instead of octylglucoside. *Anabaena variabilis* Kütz (ATCC 29413) is grown to the end of the logarithmic phase (3–4 days), harvested by 20 min centrifugation at 20,000 *g*, washed first with distilled water, then with MTPM-medium, suspended in this medium to 0.5 mg chlorophyll/ml, and stored frozen. All subsequent steps are performed at 4°. For the isolation of membranes, cells in the thawed suspension are broken by glass beads (0.5-mm diameter, 10 g/10 ml) using a cell homogenizer model MSK (Braun, Melsungen, FRG) at 3400 rpm for 2 min. Glass beads and debris are removed by two consecutive centrifugations for 10 min at 20,000 *g*. The membrane fraction is collected by 1 hr centrifugation at 48,000 *g* and is suspended in MTPM plus 2 *M* NaBr to a chlorophyll concentration of 1 mg/ml. After 30 min the suspension is diluted 1 : 1 with cold water and centrifuged again for 1 hr at 48,000 *g*. The sedimented membranes are washed once (centrifugation as before) and resuspended in MTPM (~20 ml) to about 3 mg chlorophyll/ml. These are subsequently diluted to 1.5 mg/ml with MTPM containing the required additions to obtain the final concentrations of 0.4 *M* ammonium sulfate, 0.5% cholate, and 30 m*M* octylglucoside, or 60 m*M* MEGA-9. After incubation for 30 min the suspension is centrifuged for 1 hr at 300,000 *g*. A solution of ammonium sulfate saturated at 4° (pH 7.5) is added to the supernatant to give 45, 55, and 70% saturation (10% is already present!). After each addition the suspension is stirred for 20 min and centrifuged for 10 min at 10,000 *g*. The material precipitating or floating (see Note 3) between 55 and 70% saturation is dissolved in a minimal volume of 20 m*M* Tricine/Tris, pH 8.1, containing 0.2% Triton X-100, and dialyzed for 1 hr against the same buffer (Triton X-100 is required for resolubilization). Insoluble material is removed by centrifugation before the fraction is loaded on linear sucrose density gradients (7–30%, w/v) containing 20 m*M* Tricine/Tris, pH 8.1, 0.5% cholate, and 10 m*M* octylglucoside, or 15 m*M* MEGA-9. The gradients are centrifuged overnight (14 hr) in a Beckman SW 60 rotor at 300,000 *g* (max). The lower brown band contains the b_6f complex. It is removed with a syringe and stored in liquid nitrogen. The polypeptide composition is shown in Fig. 1B. The final preparation should contain about 10 μM cytochrome *f* in about 20% yield with respect to the detergent extract. Oxidoreductase activity should be 20 units or more (see assays).

Preparation of the bc_1 Complex from *Rhodopseudomonas sphaeroides* GA[7]

Reagents (Amounts for 4-Liter Culture)

GG: 50 m*M* glycylglycine, pH 7.4 (1 liter)
GG plus 2 *M* NaBr (0.5 liters)
GG plus 0.5% (w/v) cholate, 30 m*M* octylglucoside (150 ml)
Saturated ammonium sulfate solution, pH 7.4 (200 ml)
GG plus 0.5% (v/v) Triton X-100, 0.25% (w/v) cholate, 30 m*M* octylglucoside, 30 mg/ml soybean lecithin (type IIS, Sigma) (5 ml); the mixture is briefly sonicated to disperse the lipid
Solid sucrose
Solid octylglucoside (Sigma)

Procedure

The bacteria are grown in four 1-liter Roux bottles and are harvested. Chromatophores are prepared as described in this series.[23] All subsequent steps are carried out at 4°. Sedimented chromatophores are suspended in 100 ml GG containing 2 *M* NaBr to 0.2 mg bacteriochlorophyll/ml. After 30 min the suspension is diluted 1 : 1 with GG and is centrifuged for 1 hr at 300,000 *g*. This step is repeated once. Then the pellet is resuspended to 0.4 mg bacteriochlorophyll/ml in GG containing 0.5% cholate and 30 m*M* octylglucoside. After 30 min the mixture is centrifuged as before. Octylglucoside is added to 60 m*M* to the supernatant, allowing better separation from the reaction center in the subsequent step. Saturated ammonium sulfate is added to give 45 and 55% saturation. After each addition the suspension is stirred for 10 min and centrifuged at 10,000 *g* for 10 min. The second precipitate is resuspended immediately in a minimal volume of GG containing 0.5% (v/v) Triton X-100, 0.25% cholate, 30 m*M* octylglucoside, and 30 mg/ml soybean lecithin. Insoluble material is removed by centrifugation and the fraction is loaded on a linear sucrose density gradient (20–50% w/v) in GG, containing 0.5% cholate and 30 m*M* octylglucoside. This is centrifuged for 21 hr in a Beckman SW 41 rotor at 40,000 rpm. The major part of the bc_1 complex bands as a brown zone at about 40% sucrose. This is collected by a syringe, frozen, and stored in liquid nitrogen. The final preparation should have about 10 μM in cytochrome c_1 and an activity of more than 10 units (see assays). The polypeptide pattern is shown in Fig. 1C. An alternative procedure yielding a similar preparation has been published.[24]

[23] A. Baccarini-Melandi and B. A. Melandri, this series, Vol. 23, p. 556.
[24] L. Yu, Q.-C. Mei and C.-A. Yu, *J. Biol. Chem.* **259,** 5752 (1984).

Assays

Chlorophyll[25,26] and bacteriochlorophyll[27] are measured in organic solvent by standard procedures. They can also be conveniently estimated by taking spectra of the aqueous extracts directly, if compared to the standard assay. Protein is measured by a Lowry procedure suitable for assay in the presence of detergents.[28]

Spectra

Cytochromes are measured by redox difference spectroscopy according to Bendall *et al.*[29] using a split-beam spectrophotometer. For routine assays the spectra are recorded between 500 and 600 nm at room temperature, assuming extinction coefficients of 20 $mM^{-1}cm^{-1}$ at the α peaks of the reduced forms. Exact values are published.[29-32] If required the sample is diluted to the maximally allowed pigment concentration (about 0.2 mg chlorophyll/ml for chloroplasts), oxidized by a grain of ferricyanide, and divided into two cuvettes. After recording the baseline, a grain of Na-ascorbate is added to the sample cuvette and cytochrome c_1 or f is recorded (high-potential cytochrome *b*-559 of chloroplasts[29] is lost in extracts). For determination of cytochrome *b*, ascorbate is added to the reference cuvette—a second baseline may be recorded—and a grain of dithionite is added to the sample cuvette. After reduction of cytochrome *b*—for cytochrome b_6 this might take several minutes—the new difference spectrum is recorded. To distinguish cytochrome b_6 from low-potential cytochrome *b*-559 in chloroplast extracts, menadiol can be used as intermediate reductant.[33] The Rieske FeS-protein can be measured by EPR.

SDS–PAGE

Polypeptide patterns on 15% gels can be obtained after Laemmli[18] and are shown in Figs. 1 and 2. The gel can be stained first for heme,[22] and after destaining for protein with Coomassie Brilliant Blue.[18] Better resolu-

[25] J. T. O. Kirk, *Planta (Berlin)* **78,** 200 (1968).
[26] K. MacKinney, *J. Biol. Chem.* **140,** 315 (1941).
[27] R. K. Clayton, *Biochim. Biophys. Acta* **75,** 312 (1973).
[28] A. Bensadoun and D. Weinstein, *Anal. Biochem.* **70,** 241 (1976).
[29] D. S. Bendall, H. E. Davenport, and R. Hill, this series, Vol. 23, p. 327.
[30] H. Almon and H. Böhme, *Biochim. Biophys. Acta* **592,** 113 (1980).
[31] A. L. Stuart and A. R. Wassermann, *Biochim. Biophys. Acta* **314,** 284 (1973).
[32] N. Gabellini and G. Hauska, *FEBS Lett.* **153,** 146 (1983).
[33] P. R. Rich, *FEBS Lett.* **96,** 252 (1978).

tion of smaller polypeptides can be obtained by gradient gels, after Cabral and Schatz,[34] or in the presence of urea.[35] Small polypeptides are preferentially stained with silver.[35,36] The migration pattern is often changed in the presence of urea, hydrophobic proteins like cytochrome *b* migrating relatively faster.[5]

Oxidoreductase Activity[4,6,7,37,38]

This is conveniently measured at room temperature and pH 6.5 following the reduction of 20 μM oxidized mammalian cytochrome *c* by addition of quinol to 50 μM, in the presence of 10–100 nM *bc* complex (Aminco dual-wavelength spectrophotometer, $\varepsilon_{549-540\ nm}$ = 20 mM^{-1}cm^{-1}). In the case of the spinach b_6f complex the presence of plastocyanin, 3 μM, is additionally required.[37] Alternatively, plastocyanin or other cytochromes *c* may be used.[4,6,7] Plastoquinol-9 and ubiquinol-10 are the physiological reductants for the b_6f and the bc_1 complexes, respectively. They can be used in the presence of Triton X-100, which should be kept to a minimum because of its inhibitory effect.[4,7] More conveniently, plasto- or ubiquinol-1 can be applied in ethanolic solution, and also duroquinol can be used. However, because of higher water solubility these quinols show a higher background reaction with cytochrome *c*, which increases with pH. Highest rates are found with plasto- and ubiquinol-2 or 3, at the pH optimum around 8.[4,7] These are about 100 μmol cytochrome *c* reduced per nmol *bc* complex (measured as cytochrome c_1 or *f*) per hour. This unit gives the TON/sec by dividing by 3.6, and since there are about 8 nmol complex per mg protein, it can be converted into IEU by multiplying by 8/60. Plastoquinone-9 and ubiquinone homologs could be obtained from Hoffmann-La Roche, Basel. Plastoquinone-1, -2, and -3 can be synthesized by a modification[12] of the procedure of Wood and Bendall.[38] Quinols are prepared from quinones after Barr and Crane[39] or Rich.[33] Activity of the bc_1 complex is sensitive to antimycin A, and that of the b_6f complex is sensitive to DMBIB. In crude extracts, especially of bacteria, counteracting cytochrome-*c* oxidases should be inhibited by 1 mM KCN.

Other assays, such as oxidant-induced reduction of cytochrome *b*,[6,7,40,41] and reconstitution into lipid vesicles, plus measurement of proton transport and membrane potential formation,[42,43] are not detailed here.

[34] F. Cabral and G. Schatz, this series, Vol. 54, p. 602.

[35] E. Harms, W. Rohde, F. Bosch, and C. Scholtissek, *Virology* **86,** 413 (1978).

[36] C. R. Merril, D. Galgman, S. A. Sedman, and M. H. Ebert, *Science* **211,** 1437 (1981).

[37] R. D. Clark and G. Hind, *J. Biol. Chem.* **258,** 10348 (1983).

[38] P. M. Wood and D. S. Bendall, *Eur. J. Biochem.* **61,** 337 (1976).

[39] R. Barr and F. L. Crane, this series, Vol. 23, p. 372.

[40] E. Hurt and G. Hauska, *Biochim. Biophys. Acta* **682,** 466 (1982).

Sources

OXYL Chemische Präparate GmbH, Peter-Henlein-Strass 11, 8903 Bobingen, West Germany; Hoffman-La Roche & Co., Dr. A. Jenni and Dr. F. Leuenberger, CH-4002 Basel.

[41] E. Hurt and G. Hauska, *Photobiochem. Photobiophys.* **4,** 9 (1982).

[42] E. Hurt, G. Hauska, and Y. Shahak, *FEBS Lett.* **149,** 211 (1982).

[43] E. Hurt, N. Gabellini, Y. Shahak, W. Lockau, and G. Hauska, *Arch. Biochem. Biophys.* **225,** 879 (1983).

[26] Reconstitution of H^+ Translocation and Photophosphorylation with Photosystem I Reaction Centers, PMS, and CF_1CF_0

By GÜNTER HAUSKA and NATHAN NELSON

Introduction

Functional integrity of isolated membrane enzyme complexes can often only be studied after reincorporation into lipid vesicles.[1] This is especially true for complexes catalyzing vectorial functions such as proton translocation. Among such complexes various components of biological electron transport chains and membrane ATPases are the most prominent examples, which together bring about ATP synthesis in respiration and photosynthesis. In plant chloroplasts photophosphorylation has been studied a great deal by investigating artificial, cyclic electron transport catalyzed by photosystem I. With an appropriate artificial redox system such as phenazine methosulfate (PMS)[2] in the presence of an inhibitor for electron transport from photosystem II, very high rates of photophosphorylation can be achieved.[3] This partial reaction catalyzed by photosys-

[1] E. Racker, B. Violand, S. O'Neal, M. Alfonzo, and J. Telford, *Arch. Biochem. Biophys.* **198,** 470 (1979).

[2] Abbreviations: PMS, *N*-methylphenazonium methosulfate; DAD, 2,3,5,6-tetramethyl-*p*-phenylenediamine; FCCP, carbonyl cyanide *p*-trifluoromethoxyphenylhydrazone; DCCD, *N,N'*-dicyclohexylcarbodiimide; PSI-RC, photosystem I reaction center; CF_1CF_0, chloroplast coupling factor complex (proton translocating ATPase); Tricine, *N*-[2-hydroxy-1,1-bis(hydroxymethyl)ethyl]glycine.

[3] G. Hauska, this series, Vol. 69, p. 648.

tem I constitutes an artificial chemiosmotic loop, because PMS translocates protons from outside to inside the chloroplast thylakoids during its redox cycle, which is driven by the light-induced charge separation across the membrane in the reaction center. The established electrochemical proton gradient provides the energy for ATP formation catalyzed by the ATPase complex. This system is an example of "artificial energy conservation," which is possible whenever artificial quinoid redox compounds can function as hydrogen carriers in electron-transporting membranes.[4] It has recently been reconstituted by us from purified components[5,6]: Photosystem I reaction centers (PSI-RC) isolated from spinach chloroplasts[7] were incorporated into soybean lipid vesicles by sonication. Light-induced quench of 9-aminoacridine in the presence of reduced PMS could be observed, which is an indication of H^+ uptake into the vesicles.[8] However, with the glass electrode under anaerobic conditions, light-induced, reversible acidification of the suspending medium was found. The system obviously contained the reaction centers in two orientations, the one causing proton liberation predominating.[5] Nevertheless, the ATPase complex CF_1CF_0, isolated from chloroplasts after Pick and Racker,[9] could be attached to the reaction center liposomes, and photophosphorylation was observed.[6] Without the proton pumping by PSI-RC plus PMS, CF_1CF_0 liposomes do not catalyze net ATP synthesis, but carry out uncoupler-sensitive ATP/P_i exchange.[9] The reaction centers oriented with the positive charge inside the vesicles, as in chloroplasts, could be functionally isolated by replacing PMS with 2,3,5,6-tetramethyl-*p*-phenylenediamine (DAD) plus internally trapped plastocyanin, or cytochrome *c*-552 from *Euglena*. This modified liposomal system then catalyzed net proton uptake measured with the glass electrode, in addition to light-induced quench of 9-aminoacridine.[10]

Preparation of PSI-RC and CF_1CF_0

Higher plant PSI-RC suitable for reconstitution can be prepared by the procedure of Bengis and Nelson,[7] which has been discussed in detail

[4] A. Trebst, *Annu. Rev. Plant Physiol.* **25,** 423 (1974).
[5] G. Orlich and G. Hauska, *Eur. J. Biochem.* **111,** 525 (1980).
[6] G. Hauska, D. Samoray, G. Orlich, and N. Nelson, *Eur. J. Biochem.* **111,** 535 (1980).
[7] C. Bengis and N. Nelson, *J. Biol. Chem.* **250,** 2783 (1975).
[8] S. Shuldiner, H. Rottenberg, and M. Avron, *Eur. J. Biochem.* **25,** 64 (1972).
[9] U. Pick and E. Racker, *J. Biol. Chem.* **254,** 2793 (1979).
[10] G. Hauska, G. Orlich, D. Samoray, E. Hurt, and P. V. Sane, *Proc. Int. Congr. Photosynth. 5th, Haldidiki, Greece, 1980* **2,** 903 (1981).

recently.[11] It comprises (1) removal of CF_1CF_0, the cytochrome b_6f complex, and other components by digitonin, (2) solubilization with Triton X-100, (3) chromatography on DEAE-cellulose, (4) ammonium sulfate fractionation, and (5) sucrose density gradient centrifugation. The preparation of Mullet *et al.*, using Triton X-100 as the only detergent, is also suitable.[12] Methods for the preparation of PSI-RC from the green alga *Chlamydomonas reinhardi*[13] and of various cyanobacteria[14–17] are available.

Functional CF_1CF_0 can be isolated after Pick and Racker,[9] which also has been scrutinized recently.[18] The procedure comprises (1) extraction of washed thylakoids with cholate/octylglucoside, (2) ammonium sulfate fractionation, and (3) sucrose density gradient centrifugation. Cholate/octylglucoside, like digitonin,[7] also removes the cytochrome b_6f complex,[19] and PSI-RC can be conveniently purified from the residue, in a combined preparation of CF_1CF_0 and PSI-RC: Chloroplasts are extracted after Pick and Racker,[9,18] with an octylglucoside concentration optimal for removal of cytochrome *f* (this is determined spectroscopically in a pilot experiment; MEGA-9 may be used instead of octylglucoside).[19] Over 80% cytochrome and less than 10% chlorophyll should be solubilized. Under these conditions practically all CF_1CF_0 is solubilized, which is purified as indicated above. PSI-RC is obtained from the residue by the following steps: (1) solubilization by homogenizing in 2% Triton X-100, 20 m*M* Tricine-NaOH, pH 7.5, to 2 mg chlorophyll per milliliter and stirring on ice for 30 min; centrifugation at 20,000 *g* for 30 min; (2) addition of Mg^{2+} to 10 m*M* of the supernatant, and centrifugation as before to sediment most of the light-harvesting chlorophyll *a*/*b* protein complex together with some PSII-RC; (3) loading on density gradients of 0.75–0.2 *M* sucrose containing 0.1% Triton X-100 and 20 m*M* Tricine-NaOH, pH 7.5; centrifugation overnight above 200,000 *g*. Three green bands form in the gradient: the top contains free chlorophyll, the middle PSII-RC plus resid-

[11] N. Nelson, *in* "Methods in Chloroplast Molecular Biology" (M. Edelman, R. B. Hallick, and N.-H. Chua, eds.), p. 907. Elsevier, Amsterdam, 1982.

[12] J. E. Mullet, J. J. Burke, and C. J. Arntzen, *Plant Physiol.* **65,** 814 (1980).

[13] R. Nechushtai and N. Nelson, *J. Biol. Chem.* **256,** 11624 (1981).

[14] P. J. Newman and L. A. Sherman, *Biochim. Biophys. Acta* **503,** 343 (1978).

[15] A. C. Stewart, *FEBS Lett.* **114,** 67 (1980).

[16] Y. Takahashi and S. Katoh, *Arch. Biochem. Biophys.* **637,** 118 (1981).

[17] R. Nechushtai, P. Muster, A. Binder, V. Liveanu, and N. Nelson, *Proc. Natl. Acad. Sci. U.S.A.* **80,** 1179 (1983).

[18] U. Pick, *in* "Methods in Chloroplast Molecular Biology" (M. Edelman, R. B. Hallick, and N.-H. Chua, eds.), p. 873. Elsevier, Amsterdam, 1982.

[19] G. Hauska, this volume [25] for the preparation of the cytochrome b_6f complex by this method.

ual chlorophyll *a*/*b* protein, and the bottom band contains PSI-RC pure enough for many purposes. It can be further purified by chromatography on DEAE-cellulose, etc.[7]

Preparation of Liposomes

PSI-RC are incorporated into vesicles from soybean lecithin (Sigma, type IIS) by sonication with a microtip for about 10 min. In a pointed glass tube cooled in ice water, as little as 0.6 ml can be processed, containing about 2 μM PSI-RC (~0.2 mg chlorophyll/ml), 25 mg/ml lipid, 20 m*M* Tricine-NaOH, pH 8.0, 100 m*M* KCl. Not more than 0.05% Triton X-100 should be present, which is carried over with PSI-RC from the sucrose gradient. This requires appropriate dilution in the sonication vial. Sucrose does not disturb. Gassing with nitrogen is not required, but foaming has to be avoided by centering the tip. The suspension should become clear. To remove buffer for the pH measurements below, the sample is applied to a Sephadex G-75 column (1.5 × 10 cm) and eluted with 100 m*M* KCl. Green fractions are collected (dilution factor ~3). For the functional isolation of right side-out PSI-RC, liposomes with internally trapped cytochrome *c*-552 from *Euglena gracilis* or plastocyanin are prepared in the same way.[10] The suspension additionally contains 0.1 m*M* of one of these proteins. After sonication external cytochrome or plastocyanin is removed by gel filtration on Sephadex G-75 as above. Plain liposomes for testing CF_1CF_0 alone by the ATP/P_i exchange reaction are prepared by sonicating 40 mg lipid in 1 ml Tricine-NaOH, pH 8.0, to clarity. CF_1CF_0 is attached to the various liposomes by simple dilution into the assay mixtures, as described below.

Assays

Light-Induced Quench of 9-Aminoacridine Fluorescence. The fluorescence of 9-aminoacridine drastically decreases with concentration. The feature can be used to indicate internal acidification in vesicular membrane systems, such as chloroplasts,[8] since 9-aminoacridine as an amine is a weak base and accumulates in acidified spaces, which causes quench of its fluorescence. The fluorescence is measured with a sensitive fluorometer fitted for illumination from the side with red light. Fluorescence is excited with 400 nm light and is measured at 455 nm. In the absence of a monochromator for emitted light, the phototube is protected from stray actinic light by an appropriate filter combination (e.g., Balzers interference filter, 447 nm, plus Schott BG 28 glass filter, 3 mm). The cuvette holder is cooled with water to 16°. The reaction mixture of 2.5 ml contains 50 m*M*

KCl, 20 mM Tricine-NaOH, pH 8.0, 1 μg valinomycin, and an aliquot of the liposome suspension corresponding to 0.2 nmol PSI-RC (~20 μg chlorophyll). In the case of the PSI-RC liposomes, 5 mM Na-ascorbate/50 μM PMS is used as the redox system. In the case of plastocyanin, or cytochrome *c*-552/PSI-RC liposomes, 0.1 mM DAD is included as the redox system. In the first case anaerobiosis is reached via autoxidation of reduced PMS, but with DAD oxygen has to be removed by flushing with nitrogen before addition of liposomes. first a baseline for fluorescence is recorded. Then 9-aminoacridine is added to 5 μM and the proper amplification of its fluorescence is chosen on the recorder. After fluorescence rose to a constant level—this may take several minutes and reflects the reduction of PMS by ascorbate—actinic light is switched on and the quench of fluorescence is recorded until a steady state is reached. Then light is switched off and the recovery of fluorescence is followed. During a second steady state of illumination, 2.5 μg of nigericin are added with mixing. This results in the immediate recovery of the fluorescence, reflecting the breakdown of the pH difference across the membrane.

Light-Induced Proton Movements Measured with a Glass Electrode. Proton translocations during redox cycles catalyzed by the PSI-RC liposomes in the light are measured as with chloroplasts.[20] The glass electrode is mounted into a 3-ml reaction cell with stirrer and cooling jacket, additionally fitted for gassing with nitrogen and an oxygen electrode. Anaerobiosis is required to avoid proton uptake during chlorophyll-sensitized photooxidation processes (f.i.: ascorbate$^-$ + O_2 + H^+ $\xrightarrow{\text{light}}$ dehydroascorbate + H_2O_2). Oxygen and pH are recorded simultaneously with a two-channel recorder fed from amplification circuits. The scale for O_2 is calibrated with O_2-saturated water ± dithionite,[21] and the scale for pH with pH standards 7.0 and 9.0. Actinic light is provided from a slide projector filtered through red glass (Schott RG610). The reaction mixture contains in 3 ml the same components as for the fluorecence measurement except that Tricine is omitted and the double amount of liposomes (0.4 nmol PSI-RC) is added. The cooling thermostat is set to 12°. After O_2 has been removed by gassing and the pH has been adjusted to about 7.0, the recorder is set to 0.1 pH full scale. As in the case of fluorescence measurement two light/dark cycles are recorded. During the second light steady state 3 μg nigericin are added, which should cause immediate reversal of the light-induced pH changes. With PMS/PSI-RC liposomes, net acidification is observed, and with DAD/cytochrome *c*-552/PSI-RC liposomes, alkalinization of the suspending medium should be observed. At the end

[20] R. A. Dilley, this series, Vol. 24, p. 68.

[21] R. W. Estabrook, this series, Vol. 10, p. 41.

of each measurement the buffer capacity of the reaction mixture is checked by addition of 2 μl 0.1 *N* HCl.

ATP/P_i Exchange of CF_1CF_0 Liposomes. The functionality of the coupling factor ATPase complex, isolated from spinach chloroplasts, can be tested by the ATP/P_i exchange reaction, which requires the attachment of the complex to lipid vesicles, is uncoupler sensitive, and indicates the microreversibility of the ATPase activity.[9,18] In each of 4 small glass centrifuge tubes 25 μl of the CF_1CF_0 preparation (~0.3 mg protein/ml) and 25 μl of preformed liposomes, prepared by sonicating 40 mg soybean phospholipids per milliliter to clarity (see above), are diluted with 100 μl water. Dithiothreitol is added to 50 m*M*, and the samples are incubated on ice for 10 min. Then they are diluted with an additional 0.35 ml of water and 0.5 ml of a solution containing 160 m*M* Tricine-NaOH, pH 8.0, 20 m*M* ATP, 20 m*M* $MgCl_2$, and 6 m*M* P_i (30 μCi ^{32}P). The second sample additionally contains 1 μM FCCP, an uncoupler, and the third sample 50 μM DCCD, an energy-transfer inhibitor. To the fourth sample 50 μl 40% trichloroacetic acid is added before incubation, giving the O time control. Then the 4 samples are incubated for 10 min in a shaking water bath thermostatted at 37°. Higher rates are obtained if $^{32}P_i$ is added after 30 min preincubation in the presence of ATP.[22,23] Subsequently, 50 μl 40% trichloroacetic acid is also added to samples 1–3, and all 4 tubes are centrifuged. From the clear supernatants 0.5 ml is transferred to larger glass tubes containing 4 ml of 5% ammonium molybdate in 1.2 *M* $HClO_4$. The resulting complex between excess P_i and molybdate is then removed by extracting twice with 7 ml of isobutanol/benzene/acetone (5/5/1), leaving any labeled ATP behind, in the aqueous phase. Its radioactivity is determined by Cerenkov counting of 1 ml of the aqueous phase in a scintillation counter.[3]

Photophosphorylation in PSI-RC/CF_1CF_0 Liposomes. In each of 4 small glass centrifuge tubes 50 μl of the CF_1CF_0 preparation (~0.3 mg protein/ml) are diluted into 1 ml of a mixture containing 50 m*M* Tricine-NaOH, pH 8.0, 50 m*M* NaCl, 5 m*M* $MgCl_2$, 3 m*M* ADP, 2 m*M* P_i (~5 μCi ^{32}P), 5 m*M* ascorbate, 50 μM PMS, 20 m*M* glucose, 2 IEU hexokinase, and 30 μl of the PSI-RC liposome suspension (see above). The second sample additionally contains 1 μM FCCP, the third sample 50 μM DCCD. The first 3 tubes are illuminated at room temperature for 10 min with light from a slide projector filtered through red glass (Schott RG610) and a layer of water. The fourth sample is kept in the dark. Then 50 μl 40%

[22] Y. Shahak and U. Pick, *Arch. Biochem. Biophys.* **223,** 393 (1983).

[23] Unpublished; Y. Shahak, H. Schöder, P. V. Sane, D. Samoray, G. Orlich, W. Lockau, and E. Hurt contributed to this work in the laboratory in Regensburg, which was funded by the Deutsche Forschungsgemeinschaft (SFB 43 C2).

trichloroacetic acid is added to all samples which are further processed as described for the ATP/P_i exchange reaction.[3]

Comments

Tests for Incorporation into Liposomes. Incorporation can be tested by gel chromatography on Sepharose 4B[5] or other matrices, or by sucrose density gradient centrifugation, proteoliposomes banding between free protein complexes and plain liposomes.[24] The amount of incorporated PSI-RC and of trapped cytochrome or plastocyanin can be measured by redox difference spectroscopy.[10]

Other Procedures of Incorporation.[1] PSI-RC could not be satisfactorily incorporated by detergent/dialysis, detergent/dilution, or other procedures for liposome formation. Only sonication as above was successful. A bath-type sonicator is not sufficient. CF_1CF_0 was originally incorporated[9,18] by the freeze/thaw/sonication method,[25] but simple detergent/dilution is at least as good.[6,22]

Lipid and Detergent. Soybean "lecithin" (Sigma, type IIS) contains galactolipids,[23] typical for chloroplast membranes, in addition to phospholipids, to which lecithin contributes 20% only.[26] Its acyl side chains are highly unsaturated. Neither egg lecithin, more saturated and lacking galactolipids, nor synthetic lecithins formed stable proteoliposomes with PSI-RC. Incorporation into vesicles from isolated chloroplast lipids was possible. Quench of 9-aminoacridine could be observed in this system, but only below 10°. Above this the vesicles were too leaky for protons. It is interesting that stable vesicles do not form from chloroplast lipids alone. Triton X-100, above 0.05% (v/v) at 25 mg/ml soybean lipid, destabilizes liposomes and finally prevents their formation. Thus concentration of the PSI-RC preparation before incorporation may be necessary, either by ultrafiltration or by sedimentation,[12] or via a small DEAE-cellulose column.[5,11] For ultrafiltration a filter which can be passed by free Triton micelles should be used (Amicon XM100). Triton also can be exchanged against other detergents, like octylglucoside, by chromatography on DEAE-cellulose.[11]

Determination of Internal Volume. This can be determined from the amount of trapped cytochrome *c*-552 (see above) or ferricyanide.[27] About 1 μl/mg lipid is found.

[24] E. Hurt, N. Gabellini, Y. Shahak, W. Lockau, and G. Hauska, *Arch. Biochem. Biophys.* **225,** 879 (1983).

[25] M. Kasahara and P. Hinkle, *J. Biol. Chem.* **252,** 7384 (1977).

[26] R. P. Casey, B. H. Ariano, and A. Azzi, *Eur. J. Biochem.* **122,** 313 (1982).

[27] A. Futami, E. Hurt, and G. Hauska, *Biochim. Biophys. Acta* **547,** 583 (1979).

PSI-RC and CF_1CF_0 from Other Sources. Very efficient proton translocation has been observed in liposomes with PMS and a five-subunit PSI-RC from the cyanobacterium *Anabaena variabilis*,[23] and presumably also other preparations[13–17] are functional. In view of the transmembrane charge separation it is interesting that proton translocation to some extent is also observed with the large subunit alone,[10] isolated from spinach PSI-RC.[7,11] CF_1CF_0 of spinach, pea, lettuce, and wheat has been tested,[18] but not of any lower organism besides a CF_1CF_0-containing preparation from the cyanobacterium *Synechococcus* 6716.[28]

Mixed Orientation of PSI-RC. The ratio of rightside-out and inside-out orientation can be estimated from the re-reduction kinetics of P700 after illumination by slowly permeating reductants, such as ferrocyanide or ascorbate.[23] From the slower part of the two-phasic reduction the proportion of the rightside-out orientation is obtained (less than 30%).

Measurement of Membrane Potential Formation in PSI-RC Liposomes. This can be conveniently measured under the conditions given for the fluorescence quench above, but by measuring absorption of cyanine dyes[29] (1–3 μM) instead of fluorescence of 9-aminoacridine. Differential absorption changes of oxonol VI, at 603–580 nm, indicate potentials inside positive, as in chloroplasts,[30] and reflect the rightside-out orientation of PSI-RC in liposomes. Absorption changes of the carbocyanines diS-C_3-(5) at 683–660 nm or of diO-C_5-(3) at 500–480 nm indicate potentials inside negative, as found for liposomes containing the cytochrome b_6f complex,[24] and reflect the inside-out orientation.

Further Activities of CF_1CF_0 Liposomes. Uncoupler-stimulated ATPase activity and acid–base phosphorylation,[9] absorption changes of oxonol VI indicating an ATP-driven membrane potential positive inside,[31] and ATP formation driven by an externally applied, electric field[32] have been observed.

Comparison to Chloroplast Activities.[5,6] The extent and rate of proton translocation in PSI-RC liposomes are similar to the ones for subchloroplast vesicles.[33] Photophosphorylation in PSI-RC/CF_1CF_0 liposomes can be as high as 15 μmol ATP per nmol P700 per hour, if 30% rightside-out

[28] H. S. van Wolraven, H. J. Lubberding, H. J. Marvin, and R. Kraayenhof, *Eur. J. Biochem.* **137,** 101 (1983).

[29] A. S. Waggoner, this series, Vol. 55, p. 689.

[30] J. J. Schuurmans, R. P. Casey, and R. Kraayenhof, *FEBS Lett.* **94,** 405 (1978).

[31] Y. Shahak, A. Admon, and M. Avron, *FEBS Lett.* **150,** 27 (1982).

[32] P. Gräber, M. Rögner, H.-E. Buchwald, D. Samoray, and G. Hauska, *FEBS Lett.* **145,** 35 (1982).

[33] R. E. McCarty, this series, Vol. 23, p. 302.

orientation of PSI-RC is assumed. This corresponds to about 20% of the rates found with PMS for CN-treated chloroplasts.[34] On the basis of CF_1CF_0 the comparison is even less favorable. This might be caused by inactivation of CF_1CF_0 during purification, but also by the fact that only the minor part of it is bound to liposomes carrying PSI-RC of proper orientation.

Steps for Extending the System. Attempts to replace the artificial redox mediator PMS or DAD by ferredoxin/plastoquinone/cytochrome b_6f complex/plastocyanin in order to reconstitute a completely native system have not been successful so far, but partial reconstitution has been achieved. In one approach the development went from the demonstration of proton translocation in PMS liposomes,[35] via PMS/PSI-RC liposomes, to DAD/plastocyanin/PSI-RC liposomes described here. Another approach went from proton translocation in plastoquinol liposomes[27] to the reconstitution of electrogenic proton transport in plastoquinol/cytochrome b_6f liposomes.[24] Additionally, fast electron transfer from the cytochrome b_6f complex to PSI-RC via plastocyanin was observed with the free complexes,[36] but not yet in lipid vesicles.[23]

[34] R. Ouitrakul and S. Izawa, *Biochim. Biophys. Acta* **305,** 105 (1973).
[35] G. Hauska and R. C. Prince, *FEBS Lett.* **41,** 35 (1974).
[36] E. Hurt and G. Hauska, *Photobiochem. Photobiophys.* **4,** 9 (1982).

[27] Construction of the Photosynthetic Reaction Center–Mitochondrial Ubiquinol–Cytochrome-c Oxidoreductase Hybrid System

By CHRISTOPHER C. MOSER, KATHLEEN M. GIANGIACOMO, KATSUMI MATSUURA, SIMON DEVRIES, and P. LESLIE DUTTON

Introduction

The photochemical reaction center (RC) from the photosynthetic bacterium *Rhodopseudomonas sphaeroides* can be combined with mitochondrial ubiquinol–cytochrome-c_1 oxidoreductase (bc_1 complex) to create an efficient light-activated electron-transfer system. The construction of this hybrid system of membrane redox proteins has been done successfully in minimal detergent solutions[1-3] as well as in phospholipid vesicles.[4] The principal virtues of this system are as follows.

1. The character of electron transfer is completely controlled. Short, intense light flashes initiate true single-turnover electron-transfer events with easily analyzed kinetics and exact electron accounting. A programmed sequence of flashes generates any number of multiple turnovers that with increasing frequency approach continuous illumination and steady-state electron-transfer kinetics.

2. There is no need to introduce a net input of reducing or oxidizing equivalents, in sharp contrast to experiments with respiratory systems. The light-initiated electron transfer is cyclic.

3. The concentration of reactants is flexible. Different stoichiometries of RC, bc_1 complex, cytochrome *c*, and quinone can be explored. Concentrations sufficient for EPR measurement can be obtained easily without the limitations encountered with native membrane suspensions.

4. Each of the electron-transfer steps can be selected for study through a combination of redox poising and the use of inhibitors.

5. Environmental effects including pH, ionic strength, viscosity, and temperature can be readily controlled over a wide range.

6. The hybrid system provides an example of a general system for light-stimulated electron transfer in bioenergetic proteins that have quinol and/or cytochrome *c* substrates, e.g., the b_6f complex,[5] cytochrome-*c* oxidase, or ubiquinol oxidases.

The Reaction Center: Light Generation of Redox Equivalents

The *R. sphaeroides* blue-green mutant (R-26) is a common source of RC. The cells are cultured anaerobically under lights, harvested by centrifugation, and broken with a French press. The pigment-containing chromatophore membranes are then collected and extracted with lauryldimethylamine oxide (LDAO) or another detergent of choice. Reaction center isolation proceeds with ion-exchange chromatography and ammonium sulfate fractionation[6,7] or with a cytochrome *c* affinity column.[8]

[1] N. K. Packham, D. M. Tiede, P. Mueller, and P. L. Dutton, *Proc. Natl. Acad. Sci. U.S.A.* **77,** 6339 (1980).

[2] K. Matsuura, N. K. Packham, P. Mueller, and P. L. Dutton, *FEBS Lett.* **131,** 17 (1981).

[3] Q. S. Zhu, H. N. Van der Wal, K. Van Grondelle, and J. A. Berden, *Biochim. Biophys. Acta* **725,** 121 (1983).

[4] P. R. Rich and P. Heathcote, *Biochim. Biophys. Acta* **725,** 332 (1983).

[5] R. C. Prince, K. Matsuura, E. Hurt, G. Hauska, and P. L. Dutton, *J. Biol. Chem.* **257,** 3379 (1982).

[6] R. K. Clayton and R. T. Wang, this series, Vol. 23, p. 696.

[7] G. Feher and M. Y. Okamura, *in* "The Photosynthetic Bacteria" (R. K. Clayton and W. R. Sistrom, eds.), Ch. 19. Plenum, New York, 1978.

Purified RC can be used directly after low-detergent low-ionic-strength dialysis. Alternatively, the detergent can be exchanged for another detergent or for lipid.[9]

The RC converts light energy into reducing equivalents at its low potential end and oxidizing equivalents at its high potential end in the following manner. Excited bacteriochlorophyll dimer reduces a quinone associated with an iron (Q_a) with a half-time of 120 psec via a bacteriopheophytin and perhaps a monomeric bacteriochlorophyll. Q_a can in turn reduce a second quinone (Q_b) with a half-time of 150 μsec. In the absence of a reductant for the oxidized bacteriochlorophyll dimer $(BChl)_2^+$, the electron will return from the quinones with potentially biphasic relaxation kinetics; if the RC has only Q_a, then the back reaction displays a 100-msec half-time; if both Q_a and Q_b are present the back reaction is ~1.5 sec. The reactions of the quinones in the RC have been reviewed.[10,11]

The reduction of Q_b generates either a semiquinone or a quinol (Q_bH_2), depending on the preexcitation oxidation state of Q_b. At redox potentials above 250 mV when the RC has been in complete darkness, the Q_b is completely oxidized and a flash generates Q_b semiquinone. A second flash can then generate fully reduced Q_bH_2. If a quinone pool is present, flash-generated quinol may leave the RC and be replaced by quinone, permitting continued oscillatory generation of quinol on subsequent flashes.[12,13] At lower potentials and under low illumination (such as that produced by a spectrophotometer measuring beam), Q_b is initially partly reduced and Q_bH_2 will be produced on the first and subsequent flashes.

The photoreduced Q_a, semiquinone, or quinol Q_b is stabilized on the reduction of $(BChl)_2^+$ at the high potential end of the RC by cytochrome *c*, which can then diffuse away from the RC. Alternatively, an artificial donor such as ferrocyanide can be added to reduce $(BChl)_2^+$.

The *bc*$_1$ Complex: Consumption of Light-Generated Ferricytochrome c and Quinol

The *bc$_1$* complex can be prepared from beef heart mitochondria by various methods.[14,15] Mitochondria can be extracted with deoxycholate

[8] G. W. Brudvig, S. T. Worlan, and K. Sauer, *Proc. Natl. Acad. Sci. U.S.A.* **80,** 683 (1983).
[9] J. M. Pachence, P. L. Dutton, and J. K. Blaise, *Biochim. Biophys. Acta* **549,** 348 (1979).
[10] C. A. Wraight, *Photochem. Photobiol.* **30,** 767 (1979).
[11] A. R. Crofts and C. A. Wraight, *Biochim. Biophys. Acta* **726,** 149 (1983).
[12] A. Vermeglio, *Biochim. Biophys. Acta* **459,** 516 (1977).
[13] C. A. Wraight, *Biochim. Biophys. Acta* **459,** 525 (1977).
[14] G. von Jagow, H. Schagger, W. D. Engel, P. Riccio, H. J. Kolb, and M. Klingenberg, this series, Vol. 53, p. 92.
[15] C. A. Yu, L. Yu, and T. E. King, *J. Biol. Chem.* **249,** 4905 (1974).

and the bc_1 complex separated from the other respiratory complexes by a succession of detergent adjustments and salt precipitations, as in the method of Hatefi and Rieske.[16,17] Alternatively, bc_1 complex can be prepared from neurospora[18] or yeast.[19]

The bc_1 complex consumes quinol and ferricytochrome *c* generated by flash-activated RC with the following reactions. At the high potential end cytochrome c_1 rereduces cytochrome *c*, and in turn, the Rieske iron-sulfur center (FeS) can rereduce cytochrome c_1. At the low potential end quinol concurrently reduces cytochrome *b* and FeS at a myxothiazol (MYX) -sensitive site called Q_z or Q_o. This quinol oxidase site contrasts with the antimycin (ANT) -sensitive site called Q_c or Q_i, which physiologically acts as a net quinol reductase. However, under some conditions the physiological reduction of quinone by the *b* cytochromes at this site can be reversed and lead to ANT-sensitive reduction of cytochrome *b* by quinol. Altogether these reactions achieve a cyclic electron flow between light-activated RC and bc_1 complex that is analogous to the cyclic electron transfer associated with proton translocation in the chromatophore membrane. The reactions in both the mitochondrial and the chromatophore bc_1 complex have been reviewed.[20-22]

Criteria for Successful Construction

Successful construction of the hybrid system requires the light-stimulated electron transfer between redox centers in the RC and the bc_1 complex through both the high-potential cytochrome *c* and the low-potential quinol ends. Physiologically significant construction exhibits the following sensitivities to inhibitors. With ANT alone, cytochrome *b* reduction is observed together with the partial rereduction of FeS, cytochrome c_1, and cytochrome *c*. With MYX alone, cytochrome *b* reduction is observed independent of FeS, cytochrome c_1 and cytochrome *c* reduction. ANT plus MYX, by acting at both the Q_c and Q_z sites, block all flash-induced cytochrome *b* reduction. *O*-Hydroxynaphthaquinones such as undecylhydroxynaphthaquinone (UHNQ) inhibit cytochrome *b* reduction and, by raising the redox midpoint of FeS, prevent electron transfer from FeS to cytochrome c_1.

[16] Y. Hatefi and J. S. Rieske, this series, Vol. 10, p. 225.
[17] Y. Hatefi and J. S. Rieske, this series, Vol. 10. p. 242.
[18] H. Weiss, B. Juchs, and B. Ziganke, this series, Vol. 53, p. 98.
[19] G. Palmer, this series, Vol. 53, p. 113.
[20] P. R. Rich, *Biochim. Biophys. Acta* **768,** 53 (1984).
[21] G. Hauska, E. Hurt, N. Gabellini, and W. Lockau, *Biochim. Biophys. Acta* **726,** 97 (1983).
[22] B. L. Trumpower and A. G. Katki, *in* "Membrane Proteins in Energy Transduction" (R. A. Capaldi, ed.), p. 89. Academic Press, New York, 1979.

General Experimental Methods

Light Initiation of Electron Transfer

Continuous illumination will result in electron transfer, but kinetic changes can be easily distinguished and redox equivalents adequately accounted for only if the illumination is brief enough to ensure a single turnover of the RC. In the hybrid system containing RC with the full complement of quinone, the flash must be significantly shorter than the cytochrome *c* reduction of $(BChl)_2^+$, which is normally on the order of 50 μsec. A xenon flash lamp (EG&G) with an infrared-pass Wratten 88A filter provides a bright brief source (8 μsec pulse width) which can be filtered from photomultipliers without blocking visible light measurements. The flash duration can be correspondingly longer if the reduction of $(BChl)_2^+$ is slowed, e.g., by substituting low concentrations of ferrocyanide for cytochrome *c* or by eliminating a $(BChl)_2^+$ reductant altogether. On the other hand, if the RC does not contain Q_b, quinol will not be generated and multiple turnovers of the RC will not be possible; hence, illumination can be of any duration.

Spectral Monitoring of Electron Transfer

Changes in redox states of the various prosthetic groups in the hybrid system can be monitored in the visual spectrum, with the exception of FeS, which can be directly monitored in its reduced form through EPR (*g* 1.90, 1.84, and 2.0).[23] EPR measurements are well characterized for cytochrome *b*-562, *b*-566, and c_1,[24,25] Q_a, Q_b,[26,27] Q_c,[28] and $(BChl)_2^+$.[7,29] The use of a double-beam spectrophotometer for visible measurements, although not necessary, considerably simplifies the interpretation of the absorbance changes.

Because several redox centers are changing simultaneously in the hybrid system, it is not possible to get true spectral isosbestics. However, by using a large excess of cytochrome *c* it is possible to rapidly rereduce $(BChl)_2^+$ and eliminate its spectral contribution. The wavelength pairs

[23] J. S. Rieske, R. E. Hansen, and W. S. Zaugg, *J. Biol. Chem.* **239,** 3017 (1964).

[24] S. de Vries, S. P. J. Albracht, and F. J. Leeuwerik, *Biochim. Biophys. Acta* **546,** 316 (1979).

[25] J. C. Salerno, *J. Biol. Chem.* **259,** 2331 (1984).

[26] C. A Wraight, *FEBS Lett.* **93,** 283 (1978).

[27] W. F. Butler, R. Calvo, D. R. Fredkin, R. A. Isaacson, M. Y. Okamura, and G. Feher, *Biophys. J.* **45,** 947 (1984).

[28] T. Ohnishi and B. L. Trumpower, *J. Biol. Chem.* **255,** 3278 (1980).

[29] J. R. Norris and J. J. Katz, *in* "The Photosynthetic Bacteria" (R. K. Clayton and W. R. Sistrom eds.), p. 397, Plenum, New York, 1978.

with the least interference for independently observing the redox changes are as follows: $(BChl)_2^+$, 602–540 nm, 0.037 μM^{-1} cm^{-1} [30]; cytochrome *c*, 550–553.5 nm, 0.011 μM^{-1} cm^{-1}; cytochrome c_1, 552.5–545.5 nm, 0.017 μM^{-1} cm^{-1}; cytochrome *b*-562, 562–572 nm, 0.03 μM^{-1} cm^{-1} [31]; quinone, 450 nm, 8.5 mM^{-1}.[32]

Redox Poising

The electron transfers that one will observe on illuminating the hybrid system can be controlled by initially setting each of the various redox centers either oxidized or reduced. This can be done by keeping the hybrid system under an oxygen-free atmosphere, such as argon gas, and adding small amounts of reducing or oxidizing agents, such as sodium dithionite or potassium ferricyanide. Because the electron transfer in the light-activated hybrid system is cyclic, the preillumination redox condition will tend to reestablish itself, and one does not need to maintain external control through a linear flow of electrons to or from oxidizing or reducing substrate.

One can establish a degree of redox control with small amounts of physiological substrates. For example, by taking advantage of residual complex II copurified with the bc_1 complex, small amounts of succinate or fumarate can be added to establish relatively reducing or oxidizing conditions.[33] However, this method tends to lead to a state of redox disequilibrium; i.e., it takes many minutes or hours for all of the redox centers to approach a common equilibrium redox potential as defined by the Nernst equation. Thus, while redox poising with small amounts of physiological substrates has the advantage that it does not interfere with light-initiated electron-transfer reactions, it has the severe disadvantage that one does not know exactly what redox state one begins from.

On the other hand, the judicious use of redox-mediating dyes speeds the redox equilibration process to the minutes time scale without interfering with biological electron transfer taking place on the milliseconds and seconds time scale.[34] Illumination and measurement can then take place on a system whose redox state has been clearly defined by a redox potential established over several minutes in the dark and measured with a

[30] P. L. Dutton, K. M. Petty, H. S. Bonner, and S. D. Morse, *Biochim. Biophys. Acta* **387,** 536 (1975).

[31] C. C. Moser and P.L. Dutton, in preparation (1986).

[32] C. A. Wraight, R. J. Cogdell, and R. K. Clayton, *Biochim. Biophys. Acta* **396,** 242 (1975).

[33] K. Matsuura, N. K. Packham, D. M. Tiede, P. Mueller, and P. L. Dutton, *in* "Function of Quinones Energy Conserving Systems" (B. L. Trumpower, ed.). Academic Press, New York, 1981.

[34] P. L. Dutton, this series, Vol. 54, p. 411.

EQUILIBRIUM pH 7 REDOX MIDPOINT POTENTIALS IN THE HYBRID SYSTEM

Em_7	Reaction center		bc_1 Complex	Mediators	Reference
440	$(BChl)_2$				*a*
430				$Fe(CN)_6$	*b*
280		Cytochrome *c*	Rieske FeS		*c,d*
240				DAD	*e*
230			Cytochrome c_1		*c*
117				Fe-EDTA	*f*
90		Q			*g*
80			Cytochrome *b*-562, Q_c		*h,i,j*
60	Q_a				*h*
5				Fe-Oxalate	*k*
−30				Pyocyanine	*e*
−60			Cytochrome *b*-566		*h*
−220				OHNQ	*e*

[a] P. L. Dutton and J. B. Jackson, *Eur. J. Biochem.* **30,** 495 (1972). [b] P. L. Dutton, this series, Vol. 54, p. 411. [c] P. L. Dutton, D. F. Wilson, and C. P. Lee, *Biochemistry* **9,** 5077 (1970). [d] J. S. Leigh and M. Erecinska, *Biochim. Biophys. Acta* **387,** 95 (1975). [e] R. C. Prince, S. J. G. Linkletter, and P. L. Dutton, *Biochim. Biophys. Acta* **635,** 132 (1981). [f] G. Schwarzenbach and J. Heller, *Helv. Chim. Acta* **34,** 576 (1951). [g] K. Takamiya and P. L. Dutton, *Biochim. Biophys. Acta* **546,** 1 (1979). [h] This work. [i] S. de Vries, J. A. Berden, and E. C. Slater, *in* "Function of Quinones in Energy Conserving Systems (B. L. Trumpower, ed.), p. 235. Academic Press, New York, 1982. [j] D. E. Robertson, R. C. Prince, J. R. Bowyer, K. Matsuura, P. L. Dutton, and T. Ohnishi, *J. Biol. Chem.* **259,** 1758 (1984). [k] L. Michaelis, and E. Friedheim, *J. Biol. Chem.* **91,** 343 (1931).

calomel–platinum electrode pair. Each of the various redox centers in the hybrid system can be set reduced or oxidized before illumination by establishing an equilibrium redox potential either below or above the equilibrium midpoint potentials listed for pH 7 in the table.

The best choice of redox mediators for the hybrid system is a combination of slowly mediating redox buffers and low concentrations of more effective redox mediators. Some of these are also listed in the table. Chelated irons, such as ferro-EDTA and ferro-oxalate, and 2-hydroxynaphthaquinone (OHNQ) are examples of slow mediators which can be used in concentrations of 10 μM or more. Diaminodurol (DAD) and pyocyanine are fast mediators that are best kept at concentrations of 1 μM or less. The phenazine dyes PMS and PES are too effective at equilibrating with the heme centers and should be avoided. Although ferricyanide can be used to raise the redox potential of the solution, at concentrations greater than 50 μM, ferrocyanide will noticeably rereduce flash-oxidized cytochrome *c* on a 100-msec time scale. On the other hand, in cytochrome *c*-free hybrid systems ferrocyanide can be used as an alternative reductant for $(BChl)_2^+$.

Detergent Solubilization

The choice and concentration of detergents in the detergent-solubilized hybrid system are critical. Although LDAO adheres tenaciously to purified RC and is difficult to remove completely by dialysis,[7] cholate-solubilized bc_1 complex easily precipitates in detergent-free buffers. Cholate (0.08%) not only stabilizes the bc_1 complex, but seems to optimize hybrid system reactions. LDAO inhibits cytochrome *b* reduction, either upon direct addition or by indirect addition with high concentrations of LDAO-purified RC. Exchanging at least part of the LDAO solubilizing the RC for cholate increases the observed rate of light-induced cytochrome *b* reduction severalfold. Such an exchange can be accomplished by loading LDAO-solubilized RC on a cellulose anion exchange column and washing extensively with cholate buffer before eluting with high salt followed by dialysis. Alternatively, LDAO can be removed directly by reducing the amine oxide with dithionite; solubilization then proceeds with another detergent or phospholipid.[9]

Observation of Specific Electron-Transfer Reactions

Cytochrome c Oxidation

Cytochrome *c* oxidation by the flash-activated RC can be observed most simply when there is no bc_1 complex to rereduce the cytochrome *c*; otherwise, pure oxidation kinetics can be seen only if cytochrome *c* oxidation is made fast relative to cytochrome *c* rereduction by cytochrome c_1. One may also want to make cytochrome *c* oxidation rapid in order to minimize the spectral contribution of $(BChl)_2^+$ when observing other hybrid system reactions.

There are several ways cytochrome *c* oxidation rates can be affected. As a result of the electrostatic nature of cytochrome *c* binding to RC, the logarithm of the half-time of cytochrome *c* oxidation rises with the square root of the ionic strength, varying from about 25 μsec at 10 mM to about 3 msec at 100 mM ionic strength. For rapid cytochrome *c* oxidation kinetics, all protein preparations should be dialyzed to remove salt before use.

Cytochrome *c* is bound more strongly to the bc_1 complex than to the RC (we find K_ds of ~0.2 and 0.5 μM respectively). Thus, unless the

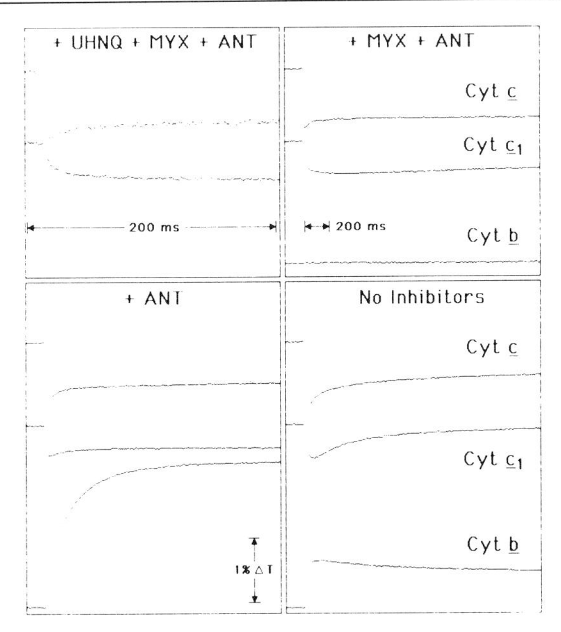

FIG. 1. Flash-induced electron-transfer kinetics in the detergent-solubilized hybrid system: effect of different inhibitor combinations and viscous medium of 50% glycerol. Cytochrome c oxidation measured at 550–553.5 nm, cytochrome c_1 oxidation at 552.5–545.5 nm, and cytochrome b reduction at 562–572 nm; 1 μM RC, 2 μM bc_1 complex, 5 μM cytochrome c, 50% glycerol, 0.04% cholate, 10 mM Tris, pH 8.0, 10 μM ferro-EDTA, OHNQ, 1 μM DAD, E_h 180 mV. When present, ANT, MYX, and UHNQ, 10 μM.

cytochrome c concentration suitably exceeds the bc_1 complex concentration, the dissociation of cytochrome c from the bc_1 complex with a half-time of about 70 msec will slow cytochrome c oxidation.[3,31]

Another way of achieving increased resolution of the electron-transfer kinetics is to slow down the interactions of flash-generated oxidized cytochrome c and quinol with the detergent-solubilized bc_1 complex by adding glycerol. Examples of cytochrome c oxidation and rereduction as well as other reactions of the hybrid system in 50% glycerol are found in Fig. 1.

Quinone Reduction

The three sorts of light-induced quinone reduction possible in the RC can be separately observed by adjusting the initial state of the RC. The reduction products are all stable for tens of seconds after a flash if a rapid electron donor to $(BChl)_2^+$ such as cytochrome c is supplied. Pure Q_a reduction occurs on light activation of RC that has only one quinone, which can be prepared, if necessary, from two quinone RC by extraction.[35]

Q_b semiquinone formation is observed upon flash activation of RC that has both oxidized Q_a and Q_b. RC deficient in quinone can be reconstituted with native ubiquinone (Q_{10}) that has been initially solubilized in detergent above the critical micelle concentration with sonication. One can assure that Q_b is initially fully oxidized by operating at potentials above 250 mV with samples dark adapted for more than 15 min. Visible observation of semiquinone at 450 nm is simplified by using a rapid nonheme donor to $(BChl)_2^+$, which has less absorbance change at this wavelength.

After $(BChl)_2^+$ has been reduced by a suitable electron donor, a second saturating flash will reduce Q_b semiquinone to quinol. In the presence of active bc_1 complex or Q pool, the Q_b quinol will be replaced by quinone and further flashes will continue a damped two-step oscillatory generation of semiquinone.[12,13] Out-of-phase oscillations will also be exhibited in reactions which are dependent on quinol generation, such as cytochrome *b* reduction. At lower redox potentials and under low levels of background illumination, a significant amount of Q_b quinol is generated on the first flash and oscillations are severely damped.

Cytochrome c_1 Oxidation and Rereduction

Cytochrome c_1 oxidation and cytochrome *c* rereduction can be most clearly observed at a redox potential of 180 mV with bc_1 complex in excess over RC and cytochrome *c* in slightly greater concentration than the sum of the bc_1 complex and RC concentrations. The electron transfer in the system can be confined to $(BChl)_2$, cytochrome *c*, and cytochrome c_1 by adding ANT and UHNQ. Using the same conditions but replacing UHNQ with MYX introduces FeS into the reaction sequence. By now omitting MYX, the electron-transfer sequence can be extended to include Q_z quinol. In each case the final oxidation states are stable for tens of seconds, permitting sampling for low-temperature spectrophotometric analysis (see Fig. 1).

Cytochrome b Reduction and Reoxidation

Cytochrome *b* can be reduced by either an ANT- or a MYX-sensitive route. With ANT as the only inhibitor, if the preillumination redox state sets the *b* cytochromes oxidized and FeS, cytochrome c_1, cytochrome *c*, and QH_2 reduced (E_h 50–150 mV), then the light-induced oxidation of cytochrome *c* by RC will initiate cytochrome c_1 and FeS oxidation, per-

[35] R. J. Cogdell, D. C. Brune, and R. K. Clayton, *FEBS Lett.* **45**, 344 (1974).

mitting the simultaneous reduction of cytochrome *b*-562 and FeS by QH_2 in the Q_z site in a process called oxidant-induced reduction. Quinol oxidation and cytochrome *b* reduction will also occur at higher redox potentials with both cytochrome c_1 and QH_2 initially oxidized, but now the reaction will depend upon the generation of QH_2 at the Q_b site in the RC and its migration to the Q_z site. Net cytochrome *b*-566 reduction will be observed if cytochrome *b*-562 is prereduced either by preflash redox equilibration below 50 mV or by previous flash reduction.[36,37] In each of these cases the *b* cytochromes go reduced with a half-time of several milliseconds and remain reduced for tens of seconds. The extent of cytochrome *b* reduction can be maximized by increasing the bc_1 complex and/or RC concentration, as long as the reaction does not become inhibited by high concentrations of LDAO, as explained above.

With MYX as the only inhibitor, cytochrome *b*-562 will go flash reduced in a separate reaction using the Q_c site that appears to be the reverse of normal electron flow in the bc_1 complex. This cytochrome *b* reduction is generally faster than the Q_z site-mediated oxidant-induced reduction and does not require oxidized cytochrome *c*, cytochrome c_1, or FeS.[38]

Cytochrome *b* reoxidation by quinone in cyclic electron transfer can be observed in the absence of inhibitors provided any mediating redox agents are at sufficiently low concentrations that they do not interfere with this slowest of electron-transfer events (half-time tens of milliseconds).

Construction of the Lipid-Solubilized Hybrid System

The bc_1 complex and the RC can be incorporated into lipid to make vesicular hybrid systems by the cholate dialysis method in which detergent-solubilized RC and bc_1 complex are mixed with a concentrated cholate/lipid solution, diluted, and dialyzed. Light-initiated transmembrane proton transport has been observed.[4]

Alternatively, a procedure used to incorporate RC into vesicles[7] can be modified to include bc_1 complex. A solution of phosphatidylcholine in ethanol is dried down under nitrogen with or without Q_{10}. Micromolar concentrations of RC are added to a molar ratio of protein to lipid of 1/1000. The solution is sonicated with a microprobe (Branson) while being

[36] S. W. Meinhardt and A. R. Crofts, *Biochim. Biophys. Acta* **723,** 219 (1983).

[37] K. Matsuura and P. L. Dutton, *in* "Chemiosmotic Proton Circuits in Biological Membranes" (V. P. Skulachev and P. C. Hinkle, eds.), p. 259. Addison-Wesley, Reading, Massachusetts, 1981.

[38] K. Matsuura, J. R. Bowyer, T. Ohnishi, and P. L. Dutton, *J. Biol. Chem.* **258,** 1571 (1983).

reduced with dithionite and held at −300 mV for several minutes to reduce the amine oxide of LDAO. The bc_1 complex is added and the solution sonicated and removed to overnight dialysis against 4 liters of 10 m*M* Tris, pH 8. The vesicles can be collected by centrifugation and resuspended to the desired concentration. All of the requirements for hybrid system construction are fulfilled, although commonly 15–20% of the RC is inserted such that the cytochrome *c* oxidation site is inaccessible to externally added cytochrome *c*. This compares with 10% cytochrome *c* inaccessible RC in the bc_1-free vesicles.[7] Lorusso *et al.* have reported that vesicles with bc_1 complex alone have 90% accessibility to cytochrome *c*.[39]

Criteria for a Quinone Pool in Lipid Systems

When Q_{10} is included in this latter vesicle preparation one observes several clear indicators of activity of a Q pool.

1. If a large excess of ferrocytochrome *c* is added to the vesicular preparation and quinol reoxidation is prevented by inhibiting the bc_1 complex with MYX and ANT and by eliminating redox mediators, then in the presence of a Q pool multiple flashes oxidize an equal amount of cytochrome *c* on each flash. If sufficient cytochrome *c* is present, flash-induced stepwise cytochrome *c* oxidation will eventually cease and so provide a measure of the size of the Q pool.
2. In the presence of ANT or MYX, the extent of single flash-induced cytochrome *b* reduction is inhibited in the presence of a Q pool. This reflects, in part, the redox buffering effect of Q pool analogous to that observed in Q-extracted chromatophores.[40]
3. In the presence of ANT, flash-induced cytochrome *b* reduction is accelerated as one lowers the potential through 100 mV in Q_{10}-containing vesicles, reflecting the preflash presence of QH_2 and providing an estimate of the equilibrium redox properties of the Q pool.

Attempts to reconstruct a Q pool in detergent-solubilized hybrid systems must face the problem of the insolubility of Q_{10} in water. Q_{10} can be solubilized by sonication in concentrated detergent solutions and added back to the hybrid system, but its interactions with RC and bc_1 seem very slow. Ubiquinone analogs with shorter isoprene tails such as Q_1 or Q_0 or other quinones such as duroquinone are much more soluble in water and can more rapidly interact with the RC and bc_1 complex in solution. However, by the very nature of their water solubility they come to resemble

[39] M. Lorusso, D. Gatti, M. Marzo, and S. Papa, *FEBS Lett.*, in press (1986).

[40] D. E. Robertson, K. M. Giangiacomo, S. de Vries, C. C. Moser, and P. L. Dutton, *FEBS Lett.* **178,** 343 (1984).

redox mediators and lose some of Q_{10}'s specific physiological properties.[20]

Concluding Remarks

The hybrid system provides a common ground for the mitochondrial and chromatophore bc_1 complexes where the remarkable similarities of the two proteins can be directly explored. In addition, RC hybrid systems provide the flexibility of flash-activated electron transfer not only for the mitochondrial bc_1 complex, but also for beef heart cytochrome oxidase through cytochrome *c* and bacterial quinol oxidase through quinol (unpublished observations). The purified hybrid system also promises application with new physical techniques, such as direct electrical measurement and voltage control of electron-transfer reactions in protein multilayers on electrodes.[41]

[41] D. M. Tiede, P. Mueller, and P. L. Dutton, *Biochim. Biophys. Acta* **681,** 191 (1982).

[28] Isolation of Ubiquinol Oxidase from *Paracoccus denitrificans*

By EDWARD A. BERRY and BERNARD L. TRUMPOWER

Introduction

Paracoccus denitrificans is a gram-negative soil bacterium which can utilize either oxygen or nitrate as a terminal electron acceptor. When grown on oxygen this bacterium elaborates a respiratory chain which is similar in function to that of eukaryotic mitochondria.[1,2] Because of this similarity, which is not common to other widely studied prokaryotes, the respiratory chain of *Paracoccus* has been of special interest to bioenergeticists seeking a bacterial model for the mitochondrial respiratory chain.

In attempting to isolate respiratory chain complexes from *Paracoccus,* we developed a method for isolation of a ubiquinol oxidase complex from membranes of this bacterium.[3] This complex consists of only 7 polypep-

[1] P. B. Scholes, and L. Smith, *Biochim. Biophys. Acta* **153,** 363 (1968).
[2] P. John and F. R. Whatley, *Adv. Bot. Res.* **4,** 51 (1977).
[3] E. A. Berry and B. L. Trumpower, *J. Biol. Chem.* **260,** 2458 (1985).

tides. It contains cytochromes *b*, c_1, *a*, a_3, and a previously unreported cytochrome *c*-552. The complex has ubiquinol–cytochrome *c* oxidoreductase, cytochrome *c* oxidase, and ubiquinol oxidase activities, with turnover numbers similar to those in the original membrane.

The ubiquinol oxidase complex of *Paracoccus* thus seems to resemble a "super complex," corresponding to the cytochromes bc_1 and oxidase complexes of mitochondria, in which the bacterial cytochrome *c*-552 fulfills the electron-transfer function associated with cytochrome *c* in mitochondria. The bacterial *c*-552 differs from mitochondrial cytochrome *c*, however, in that the former remains tightly bound to the ubiquinol oxidase complex, whereas the latter is readily dissociated from mitochondrial membranes by treatment with salt.

The isolation of this super complex, including a tightly bound *c* cytochrome, implies a degree of structural organization in the membranous respiratory chain of *Paracoccus* which either does not exist or is not apparent in mitochondria. The ubiquinol oxidase of *Paracoccus* may be an especially useful system in which to study mechanisms of electron transfer and proton translocation in the bc_1 and cytochrome *c* oxidase regions of the respiratory chain.

Materials and Analytical Methods

Serum albumin (essentially fatty acid-free), lysozyme, Brij 35, and horse heart cytochrome *c* are obtained from Sigma. Dodecyl maltoside is obtained from Calbiochem and hydroxyapatite (BioGel HT) is obtained from Bio-Rad.

Paracoccus denitrificans, ATCC 13543, is obtained as a freeze-dried culture from the American Type Culture Collection. Large-scale (550 liters) fermentations of the bacterium were carried out by the New England Enzyme Center at Tufts University Medical School, Boston, MA.

Protein is measured by the Lowry procedure and standardized against a solution of serum albumin, prepared by weight. Ubiquinol–cytochrome *c* oxidoreductase activity is measured in a reaction mixture containing 40 m*M* Na-phosphate, 20 m*M* malonate, 0.5 m*M* EDTA, 50 μM cytochrome *c*, pH 7.5, with 20 μM 2,3-dimethoxy-5-methyl-6-*n*-decyl-1,4-benzoquinol (DBH) as substrate.[4] The reduced quinone is added to the reaction mixture from a dimethyl sulfoxide rather than ethanolic solution to avoid ethanol–cytochrome *c* oxidoreductase activity found in *Paracoccus* membranes. Cytochrome *c* oxidase activity is measured in a reaction mixture containing 40 m*M* Na-phosphate, 0.5 m*M* EDTA, 20 μM cyto-

[4] B. L. Trumpower and C. A. Edwards, *J. Biol. Chem.* **254,** 8697 (1979).

chrome c, pH 7.5, to which 30 μM ascorbic acid is added to reduce the cytochrome. Rereduction of cytochrome c by excess ascorbate is slow enough to be neglected. Reduction and oxidation of cytochrome c are followed spectrophotometrically at 550 nm, using a difference extinction coefficient of 21.1.[5]

Ubiquinol oxidase activity is measured in a reaction mixture containing 40 mM K-phosphate, 0.5 mM EDTA, 10 μM DBH. Oxidation of DBH is followed at 285 versus 298 nm with a dual-wavelength spectrophotometer, using an extinction coefficient of 10.1.[3] A unit of activity in these assays is defined as 1 μmol of electrons per minute.

Isolation of Ubiquinol Oxidase

Growth of Bacteria

Paracoccus denitrificans is obtained as a freeze-dried culture and grown in peptone–yeast extract medium. Samples (at A_{750} = 1.64) are diluted with an equal volume of glycerol and stored at −20°. For preparation of membranes, cells are grown to 550 liters from these subcultures at 30° in 2.5 g/liter peptone, 2.5 g/liter yeast extract, 2.5 g/liter casamino acids, 10 g/liter glucose, 10 g/liter K_2HPO_4, adjusted to pH 6.8 with H_2SO_4. The culture is maintained aerobic and monitored by an oxygen electrode. When the absorbance at 750 nm reaches 4.5 (2% solids, 16 hr after inoculation), the medium is removed from the fermentor and cooled to 4°. The cells are harvested and stored at −20°. The yield is ~7 kg of cell paste.

Preparation of Cell Membranes

Cell membranes are prepared by lysozyme treatment and osmotic lysis. Approximately 500 g of cells are thawed in 2 liters of 500 mM sucrose, 40 mM Tris–HCl, 5 mM EDTA, pH 8.0. This suspension is put on ice and subsequent steps carried out at 4°. The suspension is homogenized in a 250-ml glass homogenizer with a loose-fitting motor-driven Teflon pestle and strained through cheesecloth. After diluting to 10 ml/g of cell paste with the same buffer, 1 g of lysozyme dissolved in 100 ml of water is added, and the suspension is stirred 15 min. Cells are then collected by centrifuging 45 min at 12,000 rpm (23,500 g_{max}) in a Sorvall GSA rotor.

The lysozyme-treated cells are resuspended in 10 mM K-phosphate, 2 mM EDTA, 0.1 mM PMSF, pH 7.5, to a volume of 10 ml/g of initial cell

[5] B. F. Van Gelder and E. C. Slater, *Biochim. Biophys. Acta* **58,** 593 (1962).

paste. The suspension is stirred for 1 hr with an overhead motor-driven stirrer to allow osmotic lysis to occur.

The viscous suspension of lysed cells is treated with a Polytron homogenizer for 2 min, after which it is not noticeably viscous, and centrifuged 60 min at 12,000 rpm (23,500 g_{max}) in the GSA rotor. The pellets consist of a chalky white lower layer and a very soft, reddish brown upper layer. As much of the supernatant is decanted as possible without losing part of the pellet, and the soft reddish layer is resuspended in the remaining supernatant by agitation on a Vortex mixer. The chalky material and a small amount of red material does not resuspend and is discarded. The resuspended pellet is centrifuged as before, and the soft red layer is resuspended in a small amount of supernatant. After adding 0.1 m*M* PMSF, the membranes are homogenized and stored at −20°.

Alkaline Wash of Cell Membranes

Membranes are thawed and diluted to 5 mg/ml protein in 10 m*M* Tris–HCl, 5 m*M* EDTA, pH 8.0. The suspension is homogenized, PMSF is added to 0.1 m*M*, and the pH is raised to 9.5 with 1 *M* Na_2CO_3. After stirring 45 min at 4° the membranes are pelleted by centrifuging 1 hr at 17,000 rpm (34,500 g_{max}) in the Sorvall SS-34 rotor. The supernatant is yellow. The pellets consist of a white button covered by a loose red layer. The red layer of membranes is resuspended in a small volume of supernatant and again adjusted to pH 9.5 and centrifuged. The pellets are then washed with 50 m*M* K-phosphate, pH 7.5, homogenized in the same buffer, and stored at −60°.

Extraction and Chromatographic Purification of Ubiquinol Oxidase

Alkaline-washed membranes are thawed and homogenized in 50 m*M* K-phosphate, pH 7.5. PMSF, in DMSO, and dodecyl maltoside, in 50 m*M* K-phosphate are added to obtain 10 g/liter of protein, 0.1 m*M* PMSF, and 1 g of dodecyl maltoside/g of protein. This mixture is stirred at 4° for 30 min and centrifuged 30 min at 28,000 rpm (78,500 g_{max}) in the Beckman Rotor 30. The chalky white pellet and a small amount of transparent, reddish material are discarded.

The dark red supernatant is made 100 m*M* in NaCl and applied to a 2.6 × 100 cm DEAE-cellulose column equilibrated with 50 m*M* K-phosphate, 100 m*M* NaCl, 2 g/liter Brij 35, pH 7.5. The sample is washed in with 25 ml of 50 m*M* K-phosphate, 100 m*M* NaCl, 1 g/liter DM, pH 7.5, followed by 200 ml of the equilibration buffer. The column is eluted with a 500-ml linear gradient of 100–400 m*M* NaCl in 50 m*M* K-phosphate, 2 g/liter Brij

TABLE I
PREPARATION OF MEMBRANES WITH HIGH UBIQUINOL–CYTOCHROME *c* OXIDOREDUCTASE ACTIVITY FROM *P. denitrificans*

	Protein content		Heme *b* content		
Fraction	(g)	(%)	(μmol/g)	(μmol)	(%)
Whole cells	33.3	100	0.26	8.76	100
Membranes	1.79	5.4	1.29	2.31	26
Alkaline-washed membranes	0.73	2.2	2.10	1.52	17

35, pH 7.5, followed by 250 ml of 400 m*M* NaCl, 50 m*M* K-phosphate, 2 g/liter Brij 35, pH 7.5.

The reddish fractions near the end of the gradient having high ubiquinol–cytochrome *c* oxidoreductase activity are combined and applied to a 2.6 × 33 cm DEAE-Sepharose 6B column equilibrated with 50 m*M* K-phosphate, 400 m*M* NaCl, 2 g/liter Brij 35, pH 7.5. The column is eluted with a 200-ml linear gradient of 400–600 m*M* NaCl in 50 m*M* K-phosphate, 2 g/liter Brij 35.

The red fractions are combined, diluted with an equal volume of distilled water, and applied to a 2.6 × 14 cm hydroxyapatite column. The column is washed with 30 m*M* K-phosphate and eluted with 200 m*M* K-phosphate, both at pH 7.5. The red fractions are combined, concentrated by pressure filtration through an Amicon PM30 membrane, and stored in aliquots at −20°.

Comments on Purification Procedure

The recoveries of protein and heme *b* during preparation of alkaline-washed membranes are shown in Table I. Recoveries of activity and specific activities at different steps in the purification of ubiquinol oxidase from alkaline-washed membranes are listed in Table II. Polypeptide profiles on SDS–gel electrophoresis of fractions from the purification are shown in Fig. 1.

The membranes prepared with lysozyme and Polytron treatment contain about 5% of the protein and 26% of the cytochrome *b* of the original cells. The yield of these might be increased by using higher centrifugal force or by minimizing formation of small membrane fragments.

The membranes were washed with alkalai to remove loosely bound membrane proteins, including proteases. This wash removes 60% of the

TABLE II
PURIFICATION OF UBIQUINOL OXIDASE FROM MEMBRANES OF *P. denitrificans*

	Protein content		Heme *b* content			Ubiquinol–cytochrome *c* oxidoreductase				Cytochrome *c* oxidase	
Purification step	(mg)	(%)	(μmol/g)	(μmol)	(%)	Units	(%)	(U/mg)	(U/nmol *b*)	Units	(U/mg)
Alkaline-washed membranes	726	100	2.10	1.52	100	8680	100	12.0	5.71	5560	7.7
DM extract	693	95	2.00	1.38	91	5449	63	7.9	3.95	6840	9.9
DEAE-cellulose fractions	107	15	4.74	0.50	33	2143	25	20.1	4.24	3252	30.4
DEAE-sepharose fractions	70	10	6.35	0.44	29	1742	20	24.8	3.91	2940	42.0
Hydroxyapatite fractions	39.8	5.5	8.55	0.34	22.5	1424	16.4	35.8	4.19	1520	38.2

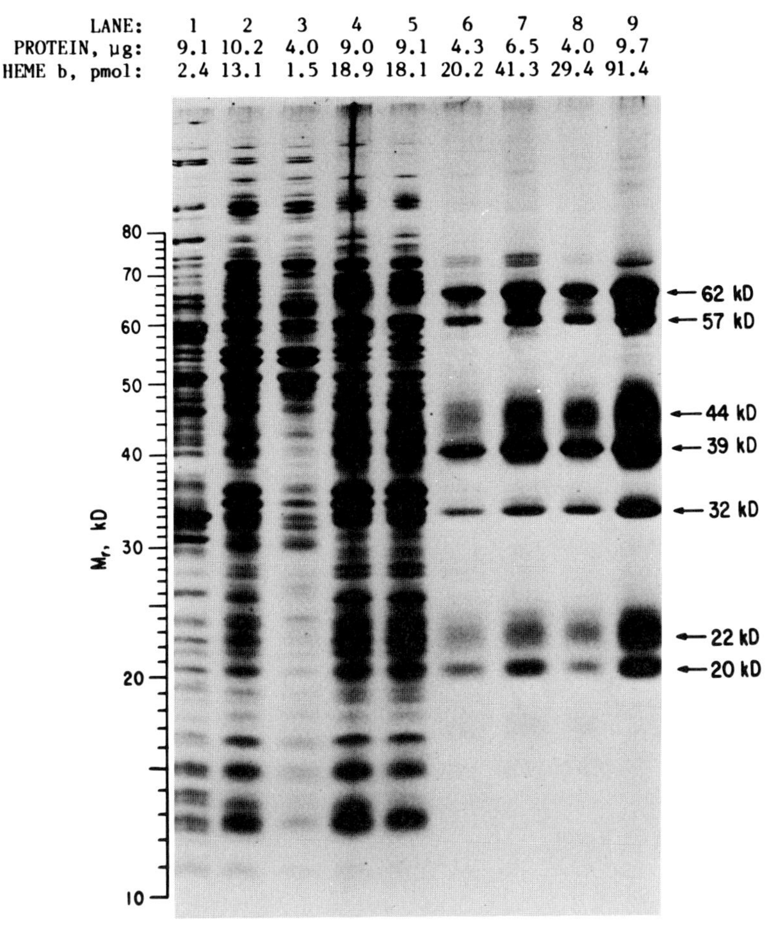

FIG. 1. SDS–polyacrylamide gel electrophoresis of samples after different steps in the purification of ubiquinol oxidase from *P. denitrificans*. Amounts of protein and heme *b* applied to each lane are indicated above the gel. The M_r scale on the left is based on mobilities of standard proteins. The M_rs of the 7 peptides of ubiquinol oxidase are indicated on the right and are the average values calculated from 5 gels of this same type. Lane 1, *Paracoccus* cells before lysis; lane 2, membranes prepared with lysozyme; lane 3, supernatant from alkaline wash of membranes; lane 4, alkaline-washed membranes; lane 5, dodecyl maltoside extract; lane 6, pooled fractions from DEAE-cellulose chromatography; lane 7, pooled fractions from DEAE-Sepharose chromatography; lane 8, ubiquinol oxidase eluted from hydroxyapatite with 30 m*M* K-phosphate; lane 9, ubiquinol oxidase eluted from hydroxyapatite with 200 m*M* K-phosphate.

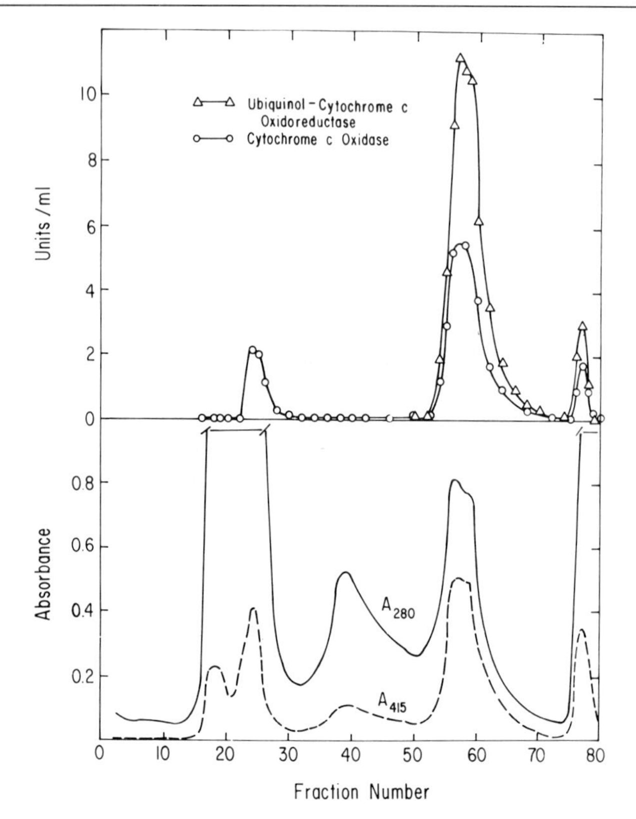

FIG. 2. DEAE-cellulose chromatography of dodecyl maltoside extract of *Paracoccus* membranes. The NaCl gradient began to elute in fraction 36. The large increase in absorbance at 280 nm is due mainly to Triton X-100.

protein, but only 35% of the cytochrome *b*, part of which may be cytochrome *o*. The proteins removed include peptides of 64, 56, 54, and 51 kDa (Fig. 1, lane 3). The 64 kDa peptide is removed by one wash at pH 9.5, but the latter three peptides are more effectively extracted by two washes. Several higher molecular weight peptides are also partially removed.

The alkaline-washed membranes are essentially completely solubilized by 1 g of dodecyl maltoside/g of protein. The elution pattern from DEAE-cellulose chromatography of this extract is shown in Fig. 2. Most of the protein, including cytochromes aa_3, a CO-reactive *b* cytochrome, and cytochrome *c*-552, elutes in the flow-through fractions. Partially purified ubiquinol oxidase, shown as coincidentally eluting peaks of ubiquinol–cytochrome *c* oxidoreductase and cytochrome *c* oxidase (Fig. 2), is recovered near the end of the gradient (fractions 55–60 in Fig. 2).

Analysis of fractions containing ubiquinol oxidase activity (lane 6 in Fig. 1) and fractions before and after this peak indicated that peptides of 62, 57, 44, 39, 32, 22, and 20 kDa are associated with the ubiquinol oxidase. Other peptides, which are present in small amounts and are not readily apparent in Fig. 2, represent contaminants. For many purposes the ubiquinol oxidase may be sufficiently pure after this step, and the higher purity that can be obtained by further treatment may not justify the decreased yield.

Ubiquinol oxidase is further purified on DEAE-Sepharose. This step removes trace amounts of a large number of polypeptides, previously present in such small amounts as to be nearly invisible on electrophoresis gels, and results in an increase of ~25% in cytochrome *c* reductase activity and heme content (Table II). Subsequent chromatography on hydroxyapatite typically resolved the ubiquinol oxidase into two populations differing only slightly in activities and heme content. When combined, the purified ubiquinol oxidase from hydroxyapatite was enriched 40–50% in heme content and electron-transfer activities relative to the material applied to this column.

Properties of Ubiquinol Oxidase

Paracoccus ubiquinol oxidase consists of 7 polypeptides of 62, 57, 44, 39, 32, 22, and 20 kDa. Eleven preparations of the complex had average contents of hemes *c*, *b*, and *a* of 6.6, 8.2, and 5.2 mol/g of protein. Since the ubiquinol oxidase can be split into a bc_1 and a c-aa_3 complex, the nonintegral ratio of heme groups suggests that the complex as isolated consists of a mixed population, including bc_1 complexes with no associated c-aa_3, and a true ubiquinol oxidase complex with hemes *c*, b, and *a* in ratios of 2 : 2 : 2 per complex. The complex contains ~40 mol of phospholipid per mol of heme *b*.

When ubiquinol oxidase is split into bc_1 and c-aa_3 complexes, the 62, 39, and 20 kDa polypeptides are associated with the bc_1 complex. The 62 kDa polypeptide has been identified as cytochrome c_1.[6] For reasons outlined in Ref. 3, the 39 kDa polypeptide is thought to be apocytochrome *b*. The 20 kDa polypeptide may be the iron-sulfur protein of the bc_1 complex, since this redox component is otherwise unaccounted for by the abovementioned polypeptides, but this identity remains to be established. In

[6] B. Ludwig, K. Suda, and N. Cerletti, *Eur. J. Biochem.* **137**, 597 (1983).

TABLE III
MIDPOINT POTENTIALS OF REDOX COMPONENTS IN *P. denitrificans* UBIQUINOL OXIDASE

Redox component	$\varepsilon_{m, 7.2}$ (mV)
Cytochrome a_3	+358
Cytochrome a	+215
Cytochrome c-552	+240
Cytochrome c_1	+188
Cytochrome b-560	+27, +122 (hp)
Cytochrome b-566	−102
Iron-sulfur protein	+250[a]

[a] Measured at pH 7.5.

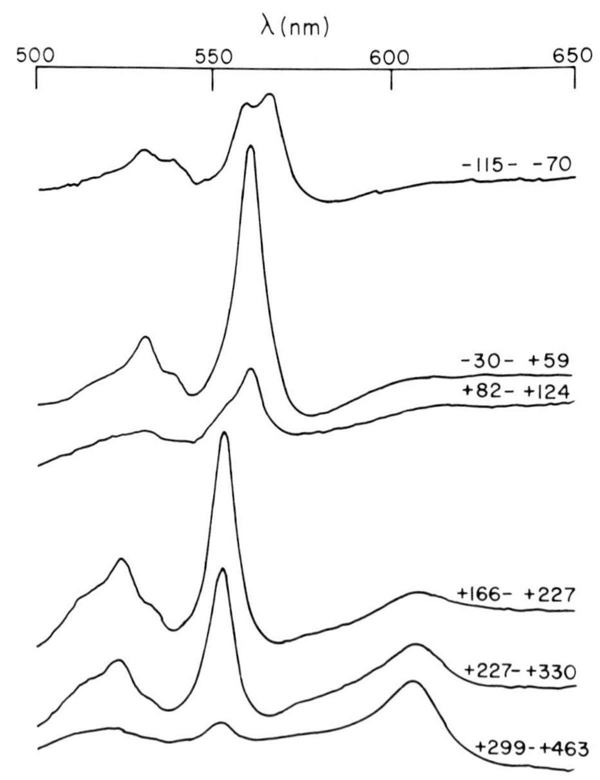

FIG. 3. Absorption difference spectra of the cytochromes of ubiquinol oxidase poised at different oxidation–reduction potentials. The numbers to the right of each spectrum indicate the oxidation–reduction potentials of the two samples used to obtain the difference spectrum shown. For example, the bottom spectrum was obtained by subtracting the spectrum of ubiquinol oxidase poised at +463 mV from that of the same sample poised at +299 mV.

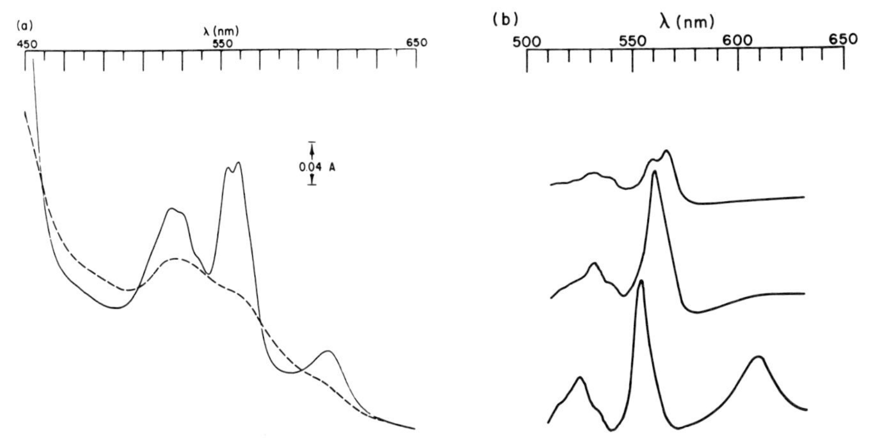

FIG. 4. Absorption spectra and difference spectra of ubiquinol oxidase from *P. denitrificans*. Spectra were obtained with samples at a protein concentration of 0.62 mg/ml in 50 m*M* K-phosphate, pH 7.5. The spectra in (a) are of the air-oxidized ubiquinol oxidase (dashed line) and the same sample after reduction with dithionite (solid line). The traces in (b) are difference spectra of the cytochromes of ubiquinol oxidase reduced sequentially with ascorbate, menaquinol, and dithionite. The bottom spectrum is a difference spectrum of the ascorbate reduced-minus-oxidized cytochromes. The middle spectrum is of the cytochromes reduced by 80 μM undecylmenaquinol minus the ascorbate-reduced cytochromes. The bottom spectrum is of the dithionite-reduced cytochromes minus the ascorbate + menaquinol-reduced cytochromes.

this regard it should be noted that the bc_1 complex, which is split from ubiquinol oxidase, is largely devoid of ubiquinol–cytochrome *c* oxidoreductase activity.

The 44, 32, and 22 kDa peptides are associated with the cytochrome c-aa_3 complex. The 44 and 32 kDa peptides correspond to those of cytochrome oxidase.[7] The 22 kDa peptide has been identified as cytochrome *c*-552,[3] the novel, membrane-bound *c* cytochrome associated with *Paracoccus* ubiquinol oxidase. The 57 kDa polypeptide, which is present in ubiquinol oxidase, is not present in either the bc_1 or c-aa_3 complex.

The redox components of ubiquinol oxidase are cytochromes a, a_3, *c*-552, c_1, *b*-560, and *b*-566. The complex also contains the Rieske iron-sulfur protein of the bc_1 complex and the antimycin sensitive, stable ubisemiquinone. The midpoint potentials of the cytochromes and iron-sulfur protein are shown in Table III. It is noteworthy that, as in mitochondria (see Ref. 8 for a review), the iron-sulfur protein is significantly more positive than cytochrome c_1. Absorption difference spectra of the cyto-

[7] B. Ludwig and G. Schatz, *Proc. Natl. Acad. Sci. U.S.A.* **77,** 196 (1980).

chromes of ubiquinol oxidase poised at different oxidation–reduction potentials are shown in Fig. 3.

The *b* cytochromes of *Paracoccus* ubiquinol oxidase consists of two spectrally distinct species, a *b*-560, which is reduced by ubiquinol or menaquinol, and a lower potential *b*-566, as shown in Fig. 4. The *b*-566 is demonstrable by a difference spectrum of the dithionite minus menaquinol-reduced cytochromes and has a split absorption band.[3] The *b*-560 is potentiometrically heterogenous and titrates with midpoint potentials of +27 and +122 mV (Table III). These spectroscopic and potentiometric properties of the *b* cytochromes and the electrophoretic properties of the apocytochrome are very similar to those of the *b* cytochromes of mitochondria (see Ref. 8 for a review).

A typical preparation of the isolated ubiquinol oxidase complex has ubiquinol–cytochrome *c* oxidoreductase, cytochrome *c* oxidase, and ubiquinol oxidase activities of 35.8, 38.2, and 12–14 units/mg, respectively. A notable feature of the ubiquinol oxidase activity is that it is manifested without addition of exogenous cytochrome *c*, presumably because the membrane-bound cytochrome *c*-552 functions to transfer electrons from the bc_1 to the aa_3 complex. The ubiquinol–cytochrome *c* oxidoreductase activity is inhibited by antimycin, 2-*n*-heptyl-4-hydroxyquinoline-*N*-oxide (HQNO), 5-*n*-undecyl-6-hydroxy-4,7-dioxobenzothiazole (UHDBT), and myxothiazol.[3] In this respect also, the bc_1 segment of the *Paracoccus* ubiquinol oxidase is like that of mitochondria.

[8] E. A. Berry and B. L. Trumpower, *in* "Coenzyme Q" (G. Lenaz, ed.), p. 365. Wiley, New York, 1985.

[29] Isolation of a Three-Subunit Cytochrome bc_1 Complex from *Paracoccus denitrificans*

By XIAOHANG YANG and BERNARD L. TRUMPOWER

Introduction

In continuing our attempts to isolate respiratory chain components from the gram-negative soil bacterium *Paracoccus denitrificans* (see Berry and Trumpower, this volume [28]), we developed a method for purification of a cytochrome bc_1 (ubiquinol–cytochrome *c* oxidoreductase) complex from membranes of this bacterium. This respiratory chain complex contains cytochromes *b* and c_1 and the Rieske-type iron-sulfur protein, has a very high turnover number in a ubiquinol–cytochrome *c* reductase assay, and this activity is fully sensitive to antimycin and myxothiazol, inhibitors of the mitochondrial complex. The *Paracoccus* bc_1 complex thus appears to have the same prosthetic groups and functional activity as the mitochondrial bc_1 complex.[1] However, in marked contrast to its mitochondrial counterpart, the *Paracoccus* bc_1 complex consists of only three polypeptide subunits.

The isolation of this three-subunit bc_1 complex, coupled with the isolation of a ubiquinol oxidase "super complex" from this same bacterium (Berry and Trumpower, this volume [28]), affords unique possibilities for elucidating the structural basis of electron transfer and energy transduction in the respiratory chain. In addition, the *Paracoccus* bc_1 complex should permit genetic approaches to understanding respiratory chain function.

Materials and Analytical Methods

Reagents and materials were obtained as described elsewhere in this volume (Berry and Trumpower [28]), except that dodecyl maltoside was obtained from Boehringer rather than from Calbiochem. We switched sources of dodecyl maltoside (Berry and Trumpower, this volume [28]) after discovering that some batches of this detergent from Calbiochem were contaminated with an impurity, possibly lauryl alcohol, to the extent that the detergent was ineffective as a dispersing agent and inhibitory to cytochrome *c* reductase activity. Preswollen, microgranular DEAE-cellulose (DE-52) was obtained from Whatman, and DEAE-Sepharose CL-6B was obtained from Sigma.

Paracoccus denitrificans, ATCC 13543, was obtained and grown aerobically as previously described (Berry and Trumpower, this volume [28]). Ubiquinol–cytochrome *c* oxidoreductase activity was measured as previously described (Berry and Trumpower, this volume [28]), except that 40 μg/ml dodecyl maltoside was added to the assay mixture before the addition of enzyme. The detergent activates the enzyme activity 10 to 15-fold. Protein was measured by the Lowry procedure. SDS–gel electrophoresis was according to Laemmli,[2] using 12.5% acrylamide.

[1] B. L. Trumpower and A. Katki, *in* "Membrane Proteins in Energy Transduction" (R. A. Capaldi, ed.), p. 89. Dekker, New York, 1979.

[2] U. K. Laemmli, *Nature (London)* **227**, 680 (1970).

Isolation of Cytochrome bc_1 Complex

Growth of Bacteria and Preparation of Cell Membranes

Bacteria are grown aerobically as previously described (Berry and Trumpower, this volume [28]). Cell membranes are prepared by lysozyme treatment and osmotic lysis (Berry and Trumpower, this volume [28]), with the modification that 0.5 m*M* diisopropylfluorophosphate and 0.5 m*M* *N*-ethylmaleimide are added to the sucrose–Tris–EDTA lysozyme treatment buffer and to the 10 m*M* phosphate lysis buffer (see Berry and Trumpower, this volume [28]).

After lysis, cell membranes are collected by centrifugation (see Berry and Trumpower, this volume [28]), suspended in 50 m*M* K-phosphate, 10 m*M* EDTA, pH 8, containing 1 m*M* diisopropylfluorophosphate to ~15 mg/ml, and stored at −80°. Note that the membranes are prepared without the alkaline washing step (see Berry and Trumpower, this volume [28]), which is unnecessary for purification of the bc_1 complex.

Extraction and Chromatographic Purification of Cytochrome bc_1 Complex

Cell membranes are thawed and homogenized at ~10 mg/ml in 50 m*M* K-phosphate, pH 8, containing 0.5 m*M* diisopropylfluorophosphate and 0.5 m*M* *N*-ethylmaleimide. Solid dodecyl maltoside is added to the suspension to obtain 1 g of detergent/g of protein. The mixture is stirred at 4° for 30 min, then centrifuged for 30 min at 50,000 rpm (206,000 g_{av}) in a Beckman 55.2 Ti rotor. The white pellets and a small amount of adherent, transparent, reddish material are discarded.

A 4-*M* solution of NaCl is added to the resulting supernate to obtain a concentration of 200 m*M* NaCl. The pH is adjusted to 8, and the mixture is stirred at 4° for 60 min.

After stirring in the presence of salt, the extract is treated with DEAE-cellulose in batch before chromatography; DEAE-cellulose resin is added directly to the stirred extract at a ratio of 0.1 g of resin/ml of extract. The suspension is stirred at 4° for an additional 5 min, and the pH is readjusted to 8 with small additions of 1 *N* HCl. The resin is separated and the nonadsorbed extract recovered by suction filtration through a sintered glass disk funnel.

After batch treatment with DEAE-cellulose, the extract is applied to a 5 × 22 cm DEAE-cellulose column, which is preequilibrated with 50 m*M* K-phosphate, 200 m*M* NaCl, pH 8, containing 0.2 g/liter dodecyl maltoside. The column is then washed with 1.2 liters of this same buffer at a flow rate of 200 ml/hr. The column is then eluted with a 1.2-liter linear

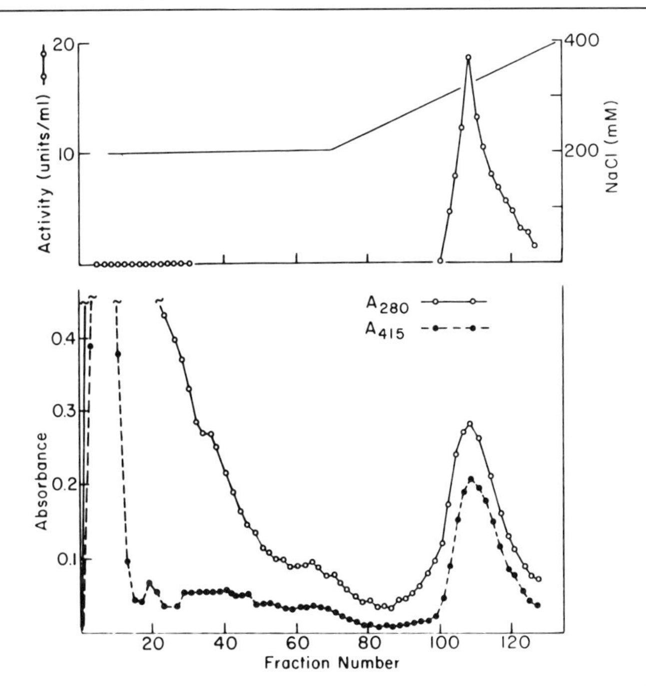

FIG. 1. DEAE-cellulose chromatography profile during purification of the *P. denitrificans* cytochrome bc_1 complex. The upper panel shows the NaCl gradient used to elute the column and the elution of ubiquinol–cytochrome *c* oxidoreductase activity. The bottom panel shows the absorbance profile at 280 nm (protein) and 415 nm (heme). Dimensions of the column and details of the elution are described in the text.

gradient of 200–400 m*M* NaCl in 50 m*M* K-phosphate, pH 8, containing 0.2 g/liter dodecyl maltoside at a flow rate of 80 ml/hr.

Fractions collected from the column are monitored for absorbance at 280 and 415 nm, for protein and heme, respectively, and for ubiquinol–cytochrome *c* oxidoreductase activity. The yellowish fractions containing activity near the end of the gradient, as shown in Fig. 1, are combined and concentrated ~5-fold with an Amicon PM10 membrane.

The concentrated sample is applied to a 2.6 × 10 cm DEAE-Sepharose column preequilibrated with 50 m*M* K-phosphate, 300 m*M* NaCl, pH 8, containing 0.2 g/liter dodecyl maltoside. The column is washed with 200 ml of the same buffer. The reddish fractions containing activity are then eluted with 50 m*M* K-phosphate, 350 m*M* NaCl, pH 8, containing 0.2 g/liter dodecyl maltoside. A profile of the DEAE-Sepharose column is shown in Fig. 2. The active fractions are combined and concentrated to ~4 mg/ml on an Amicon YM100 membrane. The concentrated sample is then diluted with an equal volume of glycerol and stored at −20°.

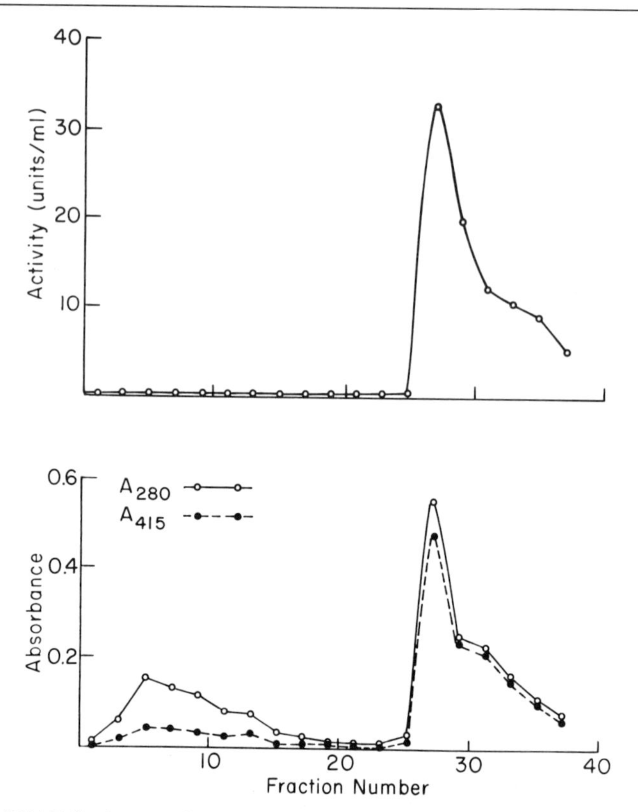

FIG. 2. DEAE-Sepharose chromatography profile during purification of the *Paracoccus* bc_1 complex. The upper panel shows the elution of activity. The bottom panel shows the absorbance profile at 280 nm (protein) and 415 nm (heme). Dimensions of the column and details of the elution are described in the text.

Comments on Purification Procedure

The method for preparation and extraction of membranes is modified from that described elsewhere (Berry and Trumpower, this volume [28]) to include the more potent protease inhibitors diisopropylfluorophosphate and *N*-ethylmaleimide. This modification is essential to prevent protease damage to the 62 kDa cytochrome c_1 peptide.

The method used for purification of the three-subunit bc_1 complex is similar in many respects to that used to purify the ubiquinol oxidase super complex (Berry and Trumpower, this volume [28]). The amounts of dodecyl maltoside used to extract the membranes are identical, and both purifications employ chromatography on DEAE-cellulose, followed by DEAE-Sepharose.

There are, however, several significant modifications which appear to be essential to isolation of the three-subunit complex, rather than the super complex. These include stirring the detergent extract with salt prior to the DEAE-cellulose column and treating the extract with DEAE-cellulose in batch form, which exposes the detergent-dispersed extract to pH 9.5–10 for a short time. In addition, the DEAE-cellulose column used to separate the three-subunit bc_1 complex is significantly shorter and larger in diameter (5 × 22 cm) than that used to purify the quinol oxidase complex (see Berry and Trumpower, this volume [28]).

The recoveries of activity and SDS electrophoresis profiles at different steps in the purification are shown in the table and Fig. 3, respectively.

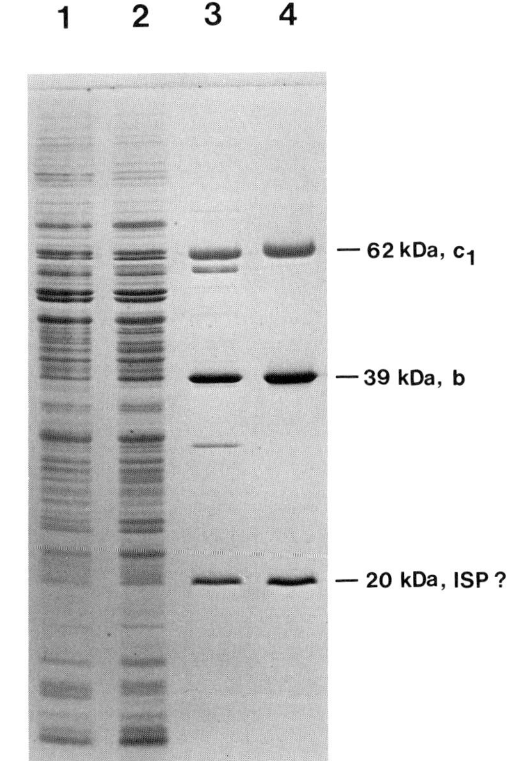

FIG. 3. SDS–gel electrophoresis showing purification of the three-subunit cytochrome bc_1 complex. Lane 1, Cell membranes; lane 2, dodecyl maltoside extract of the cell membranes; lane 3, combined fractions of partially purified bc_1 complex after DEAE-cellulose chromatography; lane 4, combined fractions of purified bc_1 complex after DEAE-Sepharose chromatography. The gel was stained with Coomassie Brilliant Blue.

PURIFICATION OF UBIQUINOL–CYTOCHROME-*c* OXIDOREDUCTASE FROM *P. denitrificans*

Purification step	Protein		Heme *b*			Activity			Turnover number
	(mg)	(%)	(μmol/g)	(μmol)	(%)	(units/mg)	Units	(%)	(sec^{-1})
Cell membranes	1210	—	—	—	—	6.3	7682	—	—
DM extract	958	100	0.49	0.47	100	10.2	9828	100	698
DEAE-cellulose	27	2.7	7.87	0.21	44.7	123	3273	33	520
DEAE-Sepharose	6	0.64	19.4	0.12	25.5	277	1687	17.1	470

The extraction recovers 80% of the membrane protein and causes a significant enhancement of the cytochrome *c* reductase activity, evident as an apparent recovery of more than 100% of the original activity. To correct for this enhancement, the activity of the DM extract is taken as 100%.

The batch treatment and chromatography on DEAE-cellulose cause a 12-fold increase in specific activity. Most of the proteins, including cytochrome *c*-552 and cytochrome oxidase, elute in the flow-through wash. This is followed by a small peak of protein and heme protein in fractions 60–70 (Fig. 2), which appears to consist of a cytochrome c-552/aa_3 subcomplex. All of the ubiquinol–cytochrome-*c* oxidoreductase activity adsorbs to the column and elutes as a single peak midway through the salt gradient. The extent of this purification can be seen by comparing lanes 2 and 3 in Fig. 3. The combined active fractions from the first column are ~50% pure (lane 3, Fig. 3), based on specific activity. It is possible that with smaller scale preparations and minor variations on this elution schedule, it may be possible to purify the three-subunit bc_1 complex with a single chromatography step.

The second chromatography column removes two relatively abundant peptide impurities and numerous others which are present in barely perceptible amounts (lanes 3 and 4, Fig. 3). The resulting complex is apparently homogenous on SDS–gel electrophoresis when the gel is stained with Coomassie Brilliant Blue (Fig. 3) or silver stain (Ref. 3, results not shown).

Properties of the *Paracoccus* Cytochrome bc_1 Complex

The purified bc_1 complex contains only three peptide subunits (Fig. 3) on SDS–gel electrophoresis. The 62 kDa peptide is cytochrome c_1,[4,5] and can be identified by staining for heme or by its fluorescence.[4] The 39 kDa peptide is cytochrome *b*, which has been identified by a single-step purification of this cytochrome from the three-subunit complex [see Payne and Trumpower, this volume [30]). The 20 kDa peptide has not been identified with certainty, but is possibly the Rieske-type iron-sulfur protein. An EPR signal characteristic of this iron-sulfur cluster is present in the isolated quinol oxidase super complex (Berry and Trumpower, this volume [28]) and is presumably retained in the active bc_1 complex.

The purified three-subunit bc_1 complex contains cytochrome c_1 and both high-potential cytochrome *b*-560 and low-potential *b*-566. Absorption spectra of the complex are shown in Fig. 4. The spectrum of the *b*

[3] R. C. Switzer, C. R. Merril, and S. Shifrin, *Anal. Biochem.* **98,** 231 (1979).

[4] B. Ludwig, K. Suda, and N. Cerletti, *Eur. J. Biochem.* **137,** 597 (1983).

[5] E. A. Berry and B. L. Trumpower, *J. Biol. Chem.* **260,** 2458 (1985).

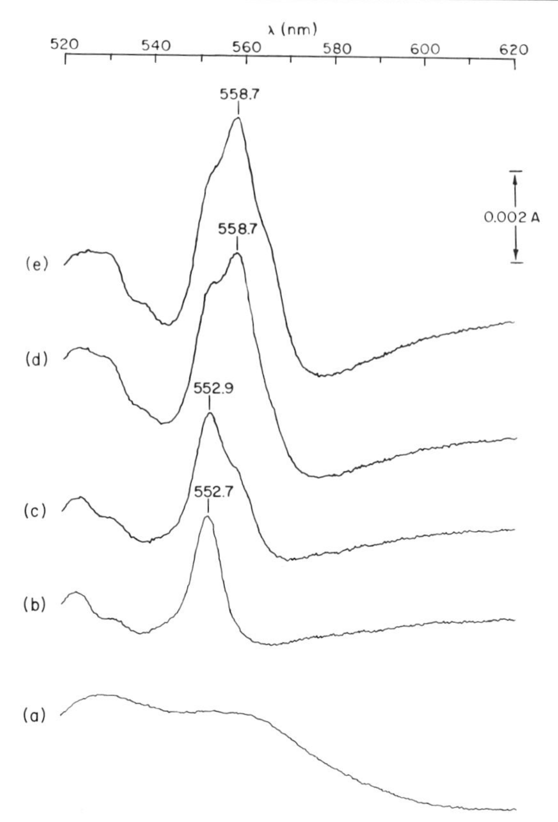

FIG. 4. Absorption difference spectra of the three-subunit cytochrome bc_1 complex. Trace (a), the oxidized complex; (b) ascorbate-reduced minus oxidized complex; (c) ubiquinol-reduced minus oxidized complex; (d) menaquinol-reduced minus oxidized complex; (e) dithionite-reduced minus oxidized complex. Note the shoulder at ~566 nm in the dithionite-reduced sample, resulting from reduction of low-potential cytochrome *b*-566.

cytochromes has a larger increment between the absorption maximum of the *b* hemes, thus the longer wavelength *b*-566 contributes a distinct shoulder at ~566 nm in the spectrum of the dithionite-reduced complex (Fig. 4).

The cytochrome *b* and c_1 contents of the purified complex are 19.4 and 13.2 nmol/mg, if one calculates these using the extinction coefficients of the mitochondrial counterparts. Assuming a bc_1 complex contains two *b* hemes and one c_1, these heme contents indicate minimum molecular weights of ~76,000–103,000 for the complex. Assuming there is one copy of each peptide in the complex, the apparent molecular weights of the peptides on SDS–gel electrophoresis indicate a minimum molecular weight of 121,000.

The three-subunit complex contains 177 nmol of phospholipid and 34 nmol of ubiquinone/mg of protein. These correspond to ~13 molecules of phospholipid and 2.6 molecules of ubiquinone/cytochrome c_1. These ratios may have to be adjusted when extinction coefficients for the bacterial cytochromes become available.

A typical preparation of the three-subunit complex has a turnover number of 470 sec^{-1} in the ubiquinol–cytochrome *c* reductase assay. Although this turnover is somewhat decreased from that of the initial extract (see the table), this is one of the highest turnover numbers reported for a bc_1 complex from any species, and is 10- to 20-fold higher than those reported for the four-subunit bc_1 complex from photosynthetic bacteria.[6] This activity is fully sensitive to antimycin, myxothiazol, UHDBT, and stigmatellin. The purified three-subunit bc_1 complex can be stored in 50% glycerol at −20° for several months without loss of activity.

[6] N. Gabellini, J. R. Bowyer, E. Hurt, B. A. Melandri, and G. Hauska, *Eur. J. Biochem.* **126,** 105 (1982).

[30] A Simple, One-Step Purification for Cytochrome *b* from the bc_1 Complexes of Bacteria

By WILLIAM E. PAYNE and BERNARD L. TRUMPOWER

Introduction

Cytochrome *b* of the ubiquinol–cytochrome *c* oxidoreductase complex (bc_1 complex) has a fundamental role in the mechanism of electron transfer through this respiratory complex.[1] However, purification of this hydrophobic integral membrane protein has proved to be extremely difficult, and therefore a detailed understanding of its chemical and physical properties has not been possible.

We describe here a simple, one-step method for purification of cytochrome *b* from the bc_1 complexes of bacteria. The entire procedure can be accomplished in 30 min and is based upon the temperature-dependent phase separation of the mild detergent, Triton X-114 (Triton X is a Röhm and Haas Company trade name). The procedure has been applied to the

[1] B. L. Trumpower, *Biochim. Biophys. Acta* **639,** 129 (1981).

bc_1 complexes of *Paracoccus denitrificans* and *Rhodopseudomonas sphaeroides*, and may be generally applicable to the simple subunit complexes of bacterial respiratory complexes.

Materials and Analytical Methods

The bc_1 complexes of *P. denitrificans* and *R. sphaeroides* were purified by a procedure developed in our laboratory (this volume [18]). Triton X-114 was obtained from Sigma and was precondensed 3 times to remove hydrophilic molecules, as described by Bordier.[2] The concentration of Triton X-114 following condensation was determined at 274 nm using an extinction coefficient of $E_{0.1\%} = 2.32$ (experimentally determined). All other reagents were of the highest grade available.

Protein was measured by the Lowry procedure as modified by Markwell *et al.*[3] and standardized against a solution of serum albumin, prepared by weight. Sodium dodecyl sulfate–polyacrylamide gel electrophoresis was performed on 12.5% slab gels according to Laemmli.[4] Samples were denatured as described elsewhere in this volume [29]. Absorption spectra were measured with an Aminco DW-2a UV/Vis spectrophotometer at 0°, and were recorded with a Nicolet 2090 digital oscilloscope.

Purification of Cytochrome *b*

Purification of Cytochrome b by Temperature-Dependent Phase Separation into Triton X-114

The procedure is based on the method of Bordier.[2] The Triton X series of nonionic detergents undergo a temperature-dependent microscopic phase separation which results in the formation of two distinct phases: an aqueous phase, which is depleted in detergent, and a detergent-enriched phase.[5] The temperature at which this phase separation occurs, termed the "cloud point," is dependent upon the number of oxyethylene groups of the detergent.[6] Thus, Triton X-100 reaches the cloud point at 64°. Triton X-114 exhibits a cloud point at 20°.[6] Upon phase separation, hydrophilic proteins partition preferentially into the aqueous phase, while hy-

[2] C. Bordier, *J. Biol. Chem.* **256,** 1604 (1981).

[3] M. K. Markwell, S. M. Haas, L. L. Bieber, and N. E. Tolbert, *Anal. Biochem.* **87,** 206 (1978).

[4] U. K. Laemmli, *Nature (London)* **227,** 680 (1970).

[5] W. N. Maclay, *J. Colloid Sci.* **11,** 272 (1956).

[6] J. Goldfarb and L. Sepulveda, *J. Colloid Interface Sci.* **31,** 454 (1969).

drophobic integral membrane proteins partition preferentially into the detergent phase.[2]

The bc_1 complex from *P. denitrificans* consists of three subunits: a 62 kDa protein which is cytochrome c_1, a 39 kDa protein which is cytochrome *b*, and a 20 kDa protein which may be the iron-sulfur protein (this volume [29]). Upon phase separation with Triton X-114, cytochrome c_1 and the iron-sulfur protein partition into the aqueous phase, and cytochrome *b* partitions into the detergent phase.

The bc_1 complex from *R. sphaeroides* consists of four subunits as isolated by our procedure: a 44 kDa protein which is cytochrome *b*, a 32 kDa protein which is cytochrome c_1, a 21 kDa protein which is the iron-sulfur protein,[7] and a 13 kDa polypeptide of unknown function. In a manner analogous to *P. denitrificans*, cytochrome c_1 and the iron-sulfur protein partition into the aqueous phase, and cytochrome *b* partitions into the detergent phase. The low-molecular-weight (13 kDa) polypeptide is found in both phases, but may be quickly and easily removed from purified cytochrome *b* by centrifugal gel filtration.

The purified bc_1 complexes from *P. denitrificans* and *R. sphaeroides* are prepared in 50 m*M* Tris–HCl, pH 7.5, at 1.0 to 1.5 mg/ml protein. NaCl (50–150 m*M*) and TX-114 (0.5–1.5%) are added, and the mixture is allowed to mix gently on a rocker table for at least 1 hr at 0°. For routine separations, a cushion of 6% (w/v) sucrose, 50 m*M* Tris–HCl, pH 7.5, 150 m*M* NaCl, and 0.06% Triton X-114 is prepared. The protein solution is visibly clear at 0° and is allowed to reach the cloud point slowly by insulation of the centrifuge tube in a foam pad.

At the cloud point the solution becomes turbid and is incubated at 30° for 3 min. The turbid sample is then overlaid on the sucrose cushion (ratio of sample to cushion 1 : 2) in a conical centrifuge tube and is immediately centrifuged for 5 min at 1200 rpm (300 *g*) at 25° in a Beckmann TJ-6 clinical centrifuge equipped with a swinging bucket rotor.

The separation is shown schematically in Fig. 1. During centrifugation, the detergent phase sediments through the sucrose cushion, while the aqueous phase remains above the cushion. The detergent phase appears as a cherry red droplet due to the presence therein of cytochrome *b*. The aqueous phase appears a faint pink due to the presence therein of cytochrome c_1. At the appropriate detergent and salt concentrations, the initial phase separation is nearly quantitative. However, an additional wash of each phase may be necessary to yield a clean preparation.

[7] N. Gabellini, U. Harnisch, J. E. G. McCarthy, G. Hauska, and W. Sebald, *EMBO J.* **4**, 549 (1985).

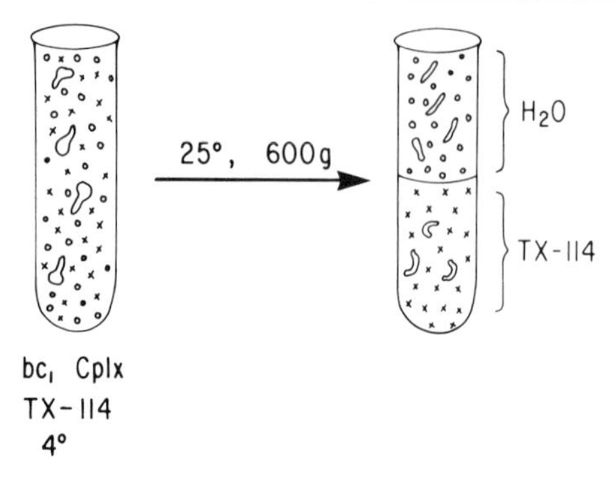

FIG. 1. Schematic representation of the purification of cytochrome *b* from bacterial bc_1 complexes by phase separation with Triton X-114. The sucrose cushion, over which the sample is laid at 4°, and which separates the aqueous and detergent phases after centrifugation at 25°, is not shown.

The separated bc_1 complex polypeptides are then analyzed by SDS–PAGE, as shown in Fig. 2, and by absorption spectroscopy as shown in Fig. 3.

Comments on the Purification Procedure

Two parameters influence the efficiency of separation in a given respiratory complex: the concentration of Triton X-114 and the concentration of salt. Some polypeptides have a tendency to partition into both the aqueous and the detergent phase. In these instances it is necessary to manipulate the detergent and salt parameters to achieve separation. (At higher salt concentrations it may be necessary to increase the specific gravity of the cushion by adding a few crystals of sucrose.) We have observed that some of the polypeptides of the more complex yeast and beef heart bc_1 complexes resist clean partitioning into a single phase. This may be an indication of stronger structural associations in the mitochondrial complexes.

Dilution of the aqueous phase is observed with each separation due to diffusion. It is thus necessary to visually remove the aqueous phase from the cushion.

Properties of Purified Cytochrome *b* from *P. denitrificans*

The biochemical and biophysical properties of the purified three-subunit bc_1 complex from *P. denitrificans* used for phase separation with Triton X-114 are described elsewhere in this volume [29]. As can be seen

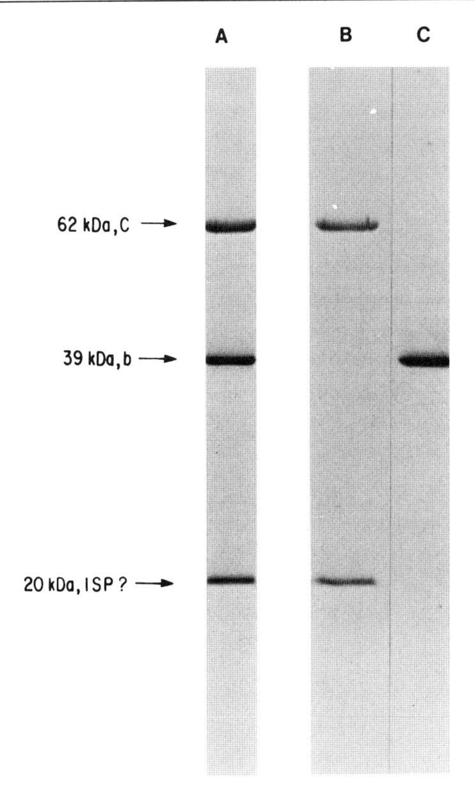

FIG. 2. SDS–polyacrylamide gel electrophoresis showing purification of cytochrome *b* by phase separation with Triton X-114. Lane A, purified bc_1 complex from *Paracoccus denitrificans* (this volume [29]); lane B, aqueous phase following phase separation containing cytochrome c_1 and the putative iron–sulfur protein; lane C, detergent phase following phase separation containing purified cytochrome *b*. Samples were denatured as described elsewhere in this volume [29].

by the SDS gels of Fig. 2, the Triton X-114 purified cytochrome *b* as well as cytochrome c_1 and the putative iron-sulfur protein migrate with molecular weights identical to those subunits of the bc_1 complex before detergent treatment.

Absorption difference spectra (dithionite-reduced minus ferricyanide-oxidized) of the purified complex and the phase-separated aqueous and detergent phases are shown in Fig. 3. Absorption spectra were recorded at 0°, since samples at room temperature are above the cloud point.

As can be seen in both Figs. 2 and 3, the separation is nearly quantitative for each subunit. The detergent phase is a bright cherry red. In addition, spectra indicate that the cytochrome is not grossly denatured,

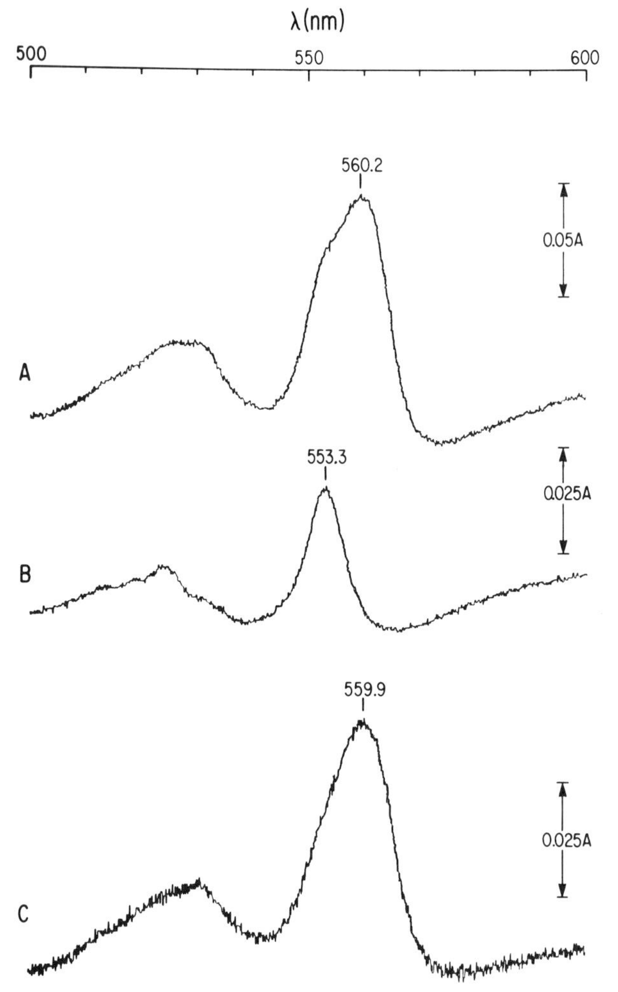

FIG. 3. Absorption difference spectra (dithionite-reduced minus ferricyanide-oxidized) of the *Paracoccus denitrificans* bc_1 complex (A) and the aqueous (B) and detergent (C) phases following phase separation with Triton X-114. Spectra were taken at 0°. The protein concentration of the purified bc_1 complex is 0.32 mg/ml.

and indeed the low-potential heme may be present in the oxidized form, since it is apparent in spectra taken immediately following reduction with dithionite, but then disappears in spectra of the same sample taken minutes later (data not shown).

Phase separation of the purified bc_1 complex from *R. sphaeroides* yields results identical to those shown in Figs. 2 and 3 for *P. denitrificans* (data not shown). It is probable that this method will allow simple purification of cytochrome *b* from a wide range of bacterial species which produce a mitochondria-like bc_1 complex.

Finally, note that treatment of bacterial bc_1 complexes with Triton X-114 is also an effective method for purification of cytochrome c_1 and the iron-sulfur protein. These polypeptides are present in an essentially detergent-free aqueous phase following treatment with Triton X-114. Hence, a simple additional gel filtration step is all that should be required for efficient purification of these subunits.

[31] Cooperative Proton-Transfer Reactions in the Respiratory Chain: Redox Bohr Effects

By S. PAPA, F. GUERRIERI, and G. IZZO

The midpoint redox potential (E_m) of various electron carriers of redox chains changes with pH, i.e., decreases as the pH of the reaction medium is raised within certain ranges.[1-4] This property, which is unexpected for pure electron carriers, indicates the occurrence of a cooperative linkage[5,6] between redox transitions of the metals and proton transfer by organic protolytic groups (in the protein or the prosthetic groups[7,8]) owing to the reversible increase in their p*K* values on reduction.

As a result of these phenomena, at pH values between the pK_a of the oxidized and the reduced form, the redox carrier functions partly as an electron carrier and partly as an effective hydrogen carrier.[6,9]

[1] P. L. Dutton and D. F. Wilson, *Biochim. Biophys. Acta* **346,** 165 (1974).
[2] S. Papa, F. Guerrieri, M. Lorusso, G. Izzo, D. Boffoli, and F. Capuano, *in* "Biochemistry of Membrane Transport" (G. Semenza and E. Carafoli, eds.), p. 503. Springer-Verlag, Berlin and New York, 1977.
[3] P. F. Urban and M. Klingenberg, *Eur. J. Biochem.* **9,** 519 (1969).
[4] V. Yu. Artzatbanov, A. A. Konstantinov, and V. P. Skulachev, *FEBS Lett.* **87,** 180 (1978).
[5] J. Wyman, *Q. Rev. Biophys.* **1,** 35 (1968).
[6] S. Papa, *Biochim. Biophys. Acta* **456,** 39 (1976).
[7] R. E. Dickerson and R. Tiukovich, *in* "The Enzymes" (P. D. Boyer, ed.), p. 397. Academic Press, New York, 1975.
[8] P. M. Callahan and G. T. Babcock, *Biochemistry* **22,** 452 (1983).
[9] S. Papa, M. Lorusso, F. Guerrieri, D. Boffoli, G. Izzo, and F. Capuano, *in* "Bioenergetics of Membranes" (L. Packer *et al.*, eds.), p. 377. Elsevier, Amsterdam, 1977.

Prototypes of protolytic linkage phenomena are the Bohr effects in hemoglobin and myoglobin[5,10,11] (alkaline and acid Bohr effects in hemoglobin,[5,10,11] oxidant Bohr effects in hemoglobin and myoglobin[5]); hence, the linkage in redox proteins is considered a redox Bohr effect.

Redox Bohr effects may represent an important operational facility of redox enzymes, provided they fall in the physiological pH range and occur at rates comparable to that of electron flow. They could play a regulatory role by adjusting the actual value of the redox potential, by controlling the kinetics of electron transfer, and/or by promoting exchange of protons between aqueous phases and protolytic redox reactions in membrane environments.[12]

Mechanisms have also been suggested in which Bohr effects could participate in vectorial proton translocation by the cytochrome system of mitochondria.[6,12–15]

Verification of these possibilities clearly depends on direct measurement and characterization of redox Bohr effects in terms of identification of the metals and protolytic organic groups involved, linkage efficiency (H^+/e^- coupling ratios), pH dependence, and kinetics.

This contribution describes direct measurement of redox Bohr effects in electron carriers of the respiratory chain.

Principle of the Measurements

Redox Bohr effects can be directly analyzed by measuring pH changes associated with oxidation or reduction of redox carriers. Artificial electron carriers such as ferricyanide–ferrocyanide can be used as oxidant or reductant.[16] We have found it convenient to measure redox Bohr effects by monitoring pH changes associated with oxidation of components of the respiratory chain by oxygen.[17,18]

Aerobic oxidation of electron carriers in soluble or membrane-bound enzymes, under conditions where vectorial processes are abolished,

[10] M. F. Perutz, *Br. Med. Bull.* **32,** 195 (1976).

[11] Y. V. Kilmartin, *Br. Med. Bull.* **32,** 209 (1976).

[12] S. Papa and M. Lorusso, *in* "Biomembranes: Dynamics and Biology" (F. C. Guerra and R. M. Burton, eds.), p. 257. Plenum, New York, 1984.

[13] S. Papa, F. Guerrieri, M. Lorusso, and S. Simone, *Biochimie* **55,** 703 (1973).

[14] M. K. F. Wikström and K. Krab, *Curr. Top. Bioenerg.* **10,** 51 (1980).

[15] G. von Jagow and W. D. Engel, *FEBS Lett.* **111,** 1 (1980).

[16] G. M. Czerlinski and K. Dar, *Biochim. Biophys. Acta* **834,** 57 (1971).

[17] S. Papa, F. Guerrieri, and G. Izzo, *FEBS Lett.* **105,** 213 (1979).

[18] F. Guerrieri, G. Izzo, I. Maida, and S. Papa, *FEBS Lett.* **125,** 261 (1980).

results in equal consumption of protons for protonation of reduced oxygen to H_2O (Eq. 1). If, however, the oxidation of the metal is accompanied by release of protons from ionizable groups in the enzyme, this acid reaction will result in a deficit of proton consumption with respect to the metals oxidized (Eq. 2).

$$n\mathrm{Fe}^{2+} + \frac{n}{2}\mathrm{O} + n\mathrm{H}^+ \rightarrow n\mathrm{Fe}^{3+} + \frac{n}{2}\mathrm{H_2O} \quad (1)$$

$$n\mathrm{Fe}^{2+} \cdot x\mathrm{H}^+ + \frac{n}{2}\mathrm{O} + (n - x)\mathrm{H}^+ \rightarrow n\mathrm{Fe}^{3+} + \frac{n}{2}\mathrm{H_2O} \quad (2)$$

The ratio x/n gives the proton/electron coupling number (s/n) which is a measure of the efficiency of the linkage phenomena.[5,17]

Determination of redox Bohr effects is therefore based on the simultaneous measurement of the oxidation–reduction of respiratory carriers and associated pH changes.

Instruments

Oxidoreduction of cytochromes and pH changes are monitored simultaneously on the same sample in a thermostatted (±0.1°) spectrophotometric cuvette under a continuous stream of nitrogen or argon. High-sensitivity instruments are required for these measurements, since the reactions analyzed are stoichiometric with respect to the catalytic centers.

Oxidoreduction is monitored with dual-wavelength spectrophotometers which allow high-sensitivity differential measurements of absorbance changes at the absorption peaks of respiratory carriers subtracted from unspecific absorption changes at isosbestic wavelengths.[19] The precision of these measurements is of ±0.001 nmol/ml even in turbid suspension with a background absorbance of 2 or 3.

The pH of the suspension is monitored with commercial combination glass reference electrodes and electrometer amplifiers fed into potentiometric pen recorders, which can allow the pH to be measured with a precision of 0.001 pH unit.[20] The potentiometric deflections are calibrated by double titration with standard HCl or KOH. With use of relatively low buffering power of the order of 0.5–2 μequiv H^+/pH, the actual proton transfer can be measured with a precision of ±0.01–0.05 ng H^+/ml.

[19] B. Chance, this series, Vol. 4, p. 273.
[20] S. Papa, F. Guerrieri, and L. Rossi Bernardi, this series, Vol. 55, p. 614.

Procedures

Measurements in Mitochondria and Submitochondrial Particles

Mitochondria[21] or submitochondrial particles are suspended in isoosmotic media supplemented with protonophoric uncouplers such as carbonyl cyanide *p*-trifluoromethoxyphenylhydrazone (FCCP) or carbonyl cyanide *m*-chlorophenylhydrazone (CCCP).

In the presence of the uncoupler, vectorial proton translocation is abolished and net pH changes resulting from scalar proton transfer reactions can be measured. The suspension is made anaerobic by oxidation of respiratory substrates, then repetitive pulses of small amounts of oxygen are applied.

Figure 1 illustrates the characteristics of proton transfer associated with redox transitions of terminal electron carriers caused by repetitive oxygen pulses of anaerobic beef heart mitochondria supplemented with antimycin A.

Aerobic oxidation of cytochromes a, a_3, c_1, and c is accompanied by reduction of b cytochromes.[22,23] Enough oxygen is administered to ensure complete oxidation of cytochromes c and cytochrome oxidase and full pH equilibration by FCCP between the mitochondrial matrix and the outer phase.

Aerobic oxidation of electron carriers results in the consumption of protons for reduction of O_2 to H_2O and their rereduction by succinate in the production of the same amount of protons. Proton uptake and release are, however, considerably lower than the sum of the electron carriers undergoing a cycle of oxidation and reduction (Table I). This difference, which is due to proton release from electron carriers associated with their aerobic oxidation and proton uptake upon their rereduction, provides an overall estimate of net Bohr effects in the cytochrome segment of the respiratory chain.

The overall number of Bohr protons per electron transferred to oxygen, which can be directly calculated in this way, amounts to 0.63 at pH 7.4 (Table I).[24–27]

[21] H. Löw and I. Vallin, *Biochim. Biophys. Acta* **69,** 361 (1963).

[22] M. Erecinska, B. Chance, D. F. Wilson, and P. L. Dutton, *Proc. Natl. Acad. Sci. U.S.A.* **69,** 50 (1972).

[23] S. Papa, M. Lorusso, G. Izzo, and F. Capuano, *Biochem. J.* **194,** 395 (1981).

[24] P. Nicholls and H. K. Kimelberg, *in* "Biochemistry and Biophysics of Mitochondrial Membranes" (G. F. Azzone *et al.*, eds.), p. 17. Academic Press, New York, 1972.

[25] P. Nicholls and B. Chance, *in* "Molecular Mechanism of Oxygen Activation" (O. Hayaishi, ed.), p. 479. Academic Press, New York, 1974.

[26] M. Erecinska, D. F. Wilson, and Y. Miyata, *Arch. Biochem. Biophys.* **177,** 133 (1976).

[27] D. F. Wilson and M. Erecinska, *Arch. Biochem. Biophys.* **167,** 111 (1975).

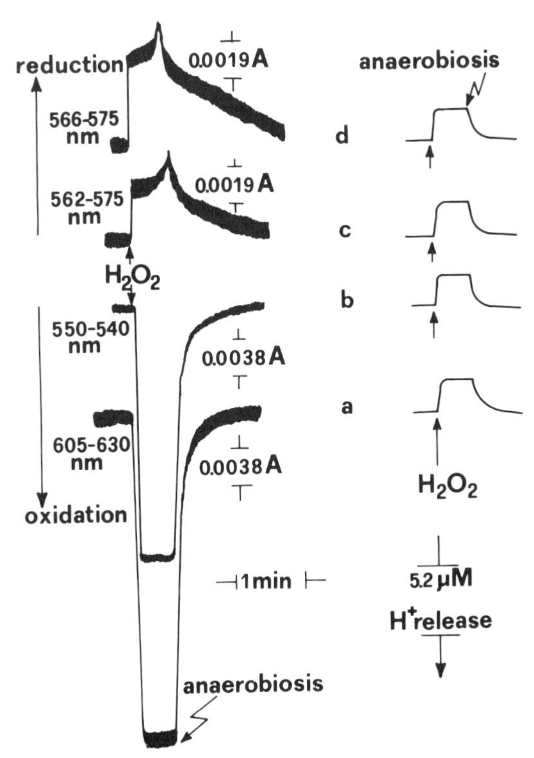

FIG. 1. Redox transitions of terminal electron carriers and associated scalar proton-transfer reactions induced by oxygen pulses of anaerobic beef heart mitochondria. Mitochondria[21] (2 mg protein/ml) were incubated in 200 m*M* sucrose, 30 m*M* KCl, 5 m*M* K-succinate, 2 μg/mg protein oligomycin, 0.5 μg/mg protein valinomycin, 0.5 μg/mg protein rotenone, 1 μg/mg protein antimycin, 3 μM FCCP, and 0.01 mg/ml catalase; final pH, 7.4; temperature, 25°. Oxygenation was brought about by repetitive additions of 5 μl 0.1% H_2O_2 to anaerobic mitochondria. The mitochondrial suspension was made anaerobic by succinate oxidation under a stream of nitrogen. Antimycin was then added and the suspension repetitively pulsed with H_2O_2. The oxygen made available was exhausted in about 30 sec by the slow antimycin-resistant respiration (from Papa *et al.*[17]). a, b, c, and d indicate repetitive H_2O_2 pulses of the same mitochondrial suspension after anaerobiosis.

It is possible to calculate the contribution to Bohr protons of the various redox transitions elicited by oxygenation. Since the reducing equivalents for the small reduction of cytochromes *b* are provided by ubiquinol,[22,28] this process should contribute stoichiometric proton release. This acidification is, however, in part neutralized (by 73% at pH 7.4) by reductive protonation of cytochromes *b*.[6,9]

[28] S. Papa, M. Lorusso, and F. Guerrieri, *Biochim. Biophys. Acta* **387,** 425 (1975).

TABLE I
ANALYTICAL EVALUATION OF PROTON-TRANSFER REACTIONS ASSOCIATED WITH REDOX TRANSITIONS OF TERMINAL ELECTRON CARRIERS CAUSED BY OXYGENATION OF ANAEROBIC BEEF HEART MITOCHONDRIA IN THE PRESENCE OF ANTIMYCIN AND FCCP[a]

Step	ng of ions or nmol / mg protein
1. H^+ uptake	1.31 ± 0.05[b]
Hemes a, a_3	1.26 ± 0.02
Copper	1.26 ± 0.02
Cytochromes c, c_1	0.69 ± 0.02
Fe-S (1.90 g)	*0.34 ± 0.01*
2. Σe^- flow	3.55 ± 0.07
3. Deficit H^+ = (step 2 − step 1)	2.24 ± 0.12
Step $3/\Sigma e^-$	0.63 ± 0.02
4. Cytochrome b reduction	0.16 ± 0.004
5. H^+ release for cytochrome b reduction by QH_2	0.045 ± 0.001
6. H^+ release for cytochrome c oxidation	0.002 ± 0.0001
7. Bohr H^+ [step 3 − (steps 5 + 6)]	2.19 ± 0.10
8. Σe^- corrected (step 2 − cytochrome c oxidized)	3.21 ± 0.07
Bohr $H^+/\Sigma e^-$ corrected	0.68 ± 0.02

[a] For experimental conditions and procedure, see legend to Fig. 1. The values are expressed in ng of ions or nmol/mg protein. The nmol of hemes a and a_3 were calculated by using a $\Delta\varepsilon(\mathrm{m}M)$ at 605–630 nm of 14,[24] cytochromes c with a $\Delta\varepsilon(\mathrm{m}M)$ at 550–540 nm of 19.1.[19] The atoms of oxidized copper are taken as equivalent to the moles of $a + a_3$ oxidized[25]; the Fe-S center, 1.90 g, as equivalent to half the amount of cytochromes c oxidized.[22] The nmol of cytochrome b reduced were calculated by assuming a contribution of b-566 to $\Delta A_{566-575\ \mathrm{nm}}$ of 60% and that of b-562 at 562–575 nm of 75%.[26,27] The $\Delta\varepsilon(\mathrm{m}M)$ of b-566 at 566–575 nm is taken as 16 and that of b-562 at 562–575 nm as 14.6.[27] The corrected Bohr $H^+/\Sigma e^-$ was obtained by subtracting from the H^+ deficit (step 3), H^+ release being associated with reduction of cytochromes b by QH_2[22,28] and oxidation of cytochrome c[16]; the nmol of oxidized cytochrome c (0.34 nmol · mg protein^{-1}) were subtracted from the Σe^- flow. The data reported are the means of three experiments ±SEM (from Papa *et al.*[17]).

[b] SEM.

Also, cytochrome c exhibits linkage between electron and proton binding; however, this phenomenon is significant only at rather alkaline pH values.[16] At pH 7.4 less than 1% of oxidoreduction of cytochrome c is proton coupled.

Correction for proton release associated with reduction of cytochromes b by ubiquinol and oxidation of cytochrome c provides an esti-

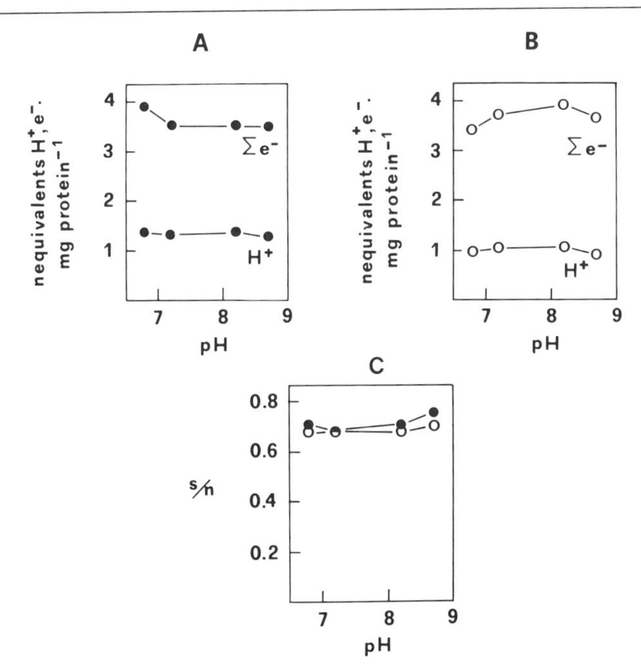

FIG. 2. pH dependence of redox transitions of electron carriers and associated scalar H^+ consumptions induced by oxygen pulses of anaerobic beef heart mitochondria (A) or submitochondrial particles (B) and pH dependence of the related H^+/e^- stoichiometry (s/n numbers) (C). Mitochondria[21] or submitochondrial particles[29] (2 mg protein/ml) were incubated as described in the legend to Fig. 1. H^+ consumption, Σe^- flow, and Bohr $H^+/\Sigma e^-$ were computed as described in the legend to Table I. Symbols: (●) beef heart mitochondria; (○), submitochondrial particles.

mate for proton transfer due to redox Bohr effects in cytochrome oxidase and the redox centers of the bc_1 complex, which are oxidized by oxygen in the presence of antimycin. The overall proton electron stoichiometry number (s/n) calculated for these redox Bohr effects amounts to 0.68 at pH 7.4.

In Fig. 2[29] the results of measurements are presented, as those illustrated in Fig. 1 and Table I, in which redox Bohr effects in the cytochrome system were determined in intact beef heart mitochondria and inside-out submitochondrial particles at various pH values. The H^+/e^- coupling numbers for redox Bohr effects measured in the two preparations coincided. This shows that the protolytic reactions analyzed in these measurements are indeed scalar and that the experimental conditions used allow full pH equilibration between the outer and the inner space separated by

[29] C. P. Lee and L. Ernster, *Eur. J. Biochem.* **3,** 391 (1968).

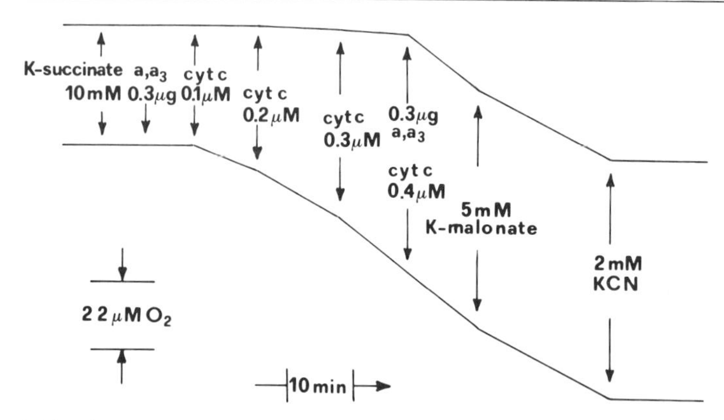

FIG. 3. Succinate-supported oxygen consumption by isolated cytochrome bc_1 complex supplemented with exogenous cytochrome *c* and purified cytochrome oxidase. Oxygen consumption was measured polarographically with a Clark electrode. bc_1 complex[31] (0.43 mg protein/ml) was incubated for 5 min in 200 m*M* sucrose, 30 m*M* KCl (pH 7.2); final volume, 1.8 ml; temperature, 25°.

the mitochondrial membrane. Any complication arising from vectorial processes is avoided. It can be noted that the overall H^+/e^- coupling number for redox Bohr effects in the cytochrome system amounts to 0.7 in the pH range 6.7–8.7. This means that at physiological pH values 70% of the oxidoreductions of electron carriers in the mitochondrial cytochrome system are cooperatively linked to H^+ transfer by ionizable groups.

Measurements in Isolated Redox Complexes

The aerobic oxidation method described for measurement of redox Bohr effects in the cytochrome system of mitochondria can also be used to measure Bohr effects in purified cytochrome-*c* reductase and cytochrome-*c* oxidase.[18]

The redox centers of the reductase are reduced by succinate, since there are traces of succinate dehydrogenase always present in the purified enzyme.[30] Aerobic oxidation of the reduced centers is effected through traces of purified cytochrome-*c* oxidase and cytochrome *c* added to the system.

Figure 3[31] shows that addition of traces of purified cytochrome-*c* oxidase to the soluble reductase, supplemented with succinate, results in a

[30] B. D. Nelson and P. Gellerfors, this series, Vol. 53, p. 80.

[31] J. S. Rieske, W. S. Zuegg, and R. E. Hansen, *J. Biol. Chem.* **239,** 3023 (1964).

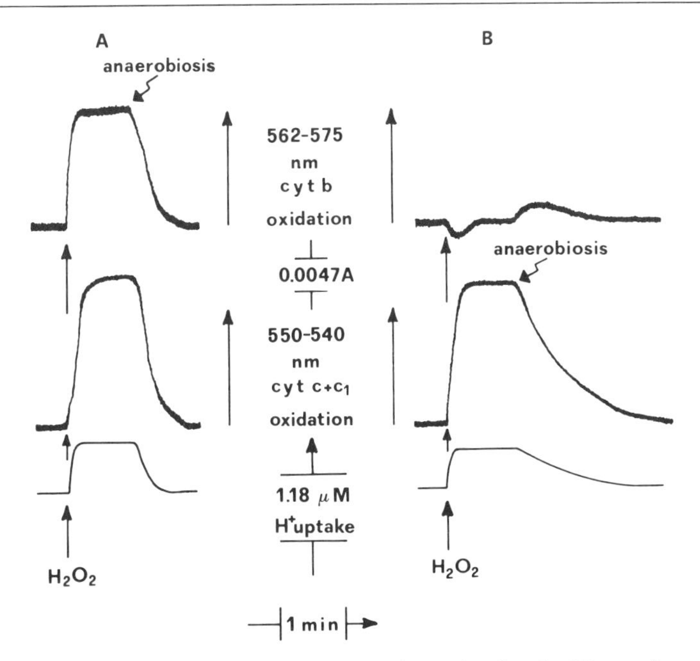

FIG. 4. Redox transitions of electron carriers and associated scalar H^+ transfer reactions induced by oxygen pulses of anaerobic isolated cytochrome bc_1 complex supplemented with exogenous cytochrome *c* and purified cytochrome-*c* oxidase. bc_1 complex (0.43 mg protein/ml) was incubated in 200 m*M* sucrose, 30 m*M* KCl, 0.3 μg purified cytochrome-*c* oxidase,[32] 0.24 μ*M* cytochrome *c* (Sigma, type IV), 9 μg purified catalase (pH 7.2); volume 2.25 ml. The incubation was carried out in a stoppered spectrophotometric cuvette for 20 min under a continuous stream of argon to lower the concentration of dissolved oxygen, then 1 m*M* K-succinate was added. Anaerobiosis was reached in about 10 min. Under these conditions cytochrome c_1 was reduced by about 80%, *b* cytochromes by about 50%. Oxygenation was brought about by repetitive addition of 1 μl 0.1% H_2O_2, in the presence of 0.5 m*M* K-malonate. (A), Control; (B), +2.8 μg antimycin A/mg protein.

cytochrome *c*-dependent respiration, which can be inhibited by KCN and malonate, a competitive inhibitor of succinate dehydrogenase.

Oxidation cycles of the cytochrome bc_1 complex supplemented with traces of added oxidase and cytochrome *c* can be produced by pulsing with oxygen the enzyme brought to anaerobic reduction by succinate respiration. Figure 4A[32] shows that net aerobic oxidation of *c* and *b* cytochromes is accompanied, as expected, by H^+ consumption for protonation of reduced oxygen to H_2O. Net rereduction of the cytochromes by succinate, which takes place upon exhaustion of oxygen, results in re-

[32] B. Errede, M. D. Kamen, and Y. Hatefi, this series, Vol. 53, p. 40.

TABLE II
ANALYSIS OF SCALAR H^+ TRANSFER REACTIONS ASSOCIATED WITH REDOX TRANSITIONS OF RESPIRATORY CARRIERS IN ISOLATED CYTOCHROME bc_1 COMPLEX[a]

Step	(A)		(B)
1. H^+ uptake	2.06		1.60
Cytochrome *b*	1.92		—
Cytochrome c_1	2.38		2.19
Fe–S (1.90 g)	2.38		2.19
Cytochrome *c*	*0.56*		*0.56*
2. Σe^- flow	7.24		4.94
3. Deficit H^+ = (step 2 − step 1)	5.18		3.34
Step 3/cytochrome *b*, Fe–S	1.20,	Step 3/Fe–S	1.52
4. H^+ release for oxidation of QH_2 to $Q^{\overline{\cdot}}$	1.64		1.51
5. Bohr H^+ (step 3 − step 4)	3.54		
Bohr H^+/cytochrome *b*, Fe–S	0.82		
6. Fraction *b* cytochrome acting as e^- carriers, 27% of value in step 2 (cf. Ref. 9)	0.52		
7. Σe^- corrected for (step 6)	5.84		
8. Bohr H^+ (Fe–S) [step 7 − (steps 1 + 4)]	2.14	Bohr H^+ (Fe–S) (step 3 − step 4)	1.83
Bohr H^+ (Fe–S)/Fe–S	0.90		0.84

[a] For experimental conditions, see legend to Fig. 4. Experiments with the addition of 1 μM FCCP or 0.2% Emasol produced the same results. The oxidation of *b* cytochromes was measured from $\Delta A_{562-575\text{ nm}}$ caused by oxygenation using a $\Delta\varepsilon(\text{m}M)$ of 20,[33] and that of cytochrome c_1 at 552–540 nm with a $\Delta\varepsilon(\text{m}M)$ of 17.5.[34] The nmol cytochrome c_1 was corrected for the contribution of cytochrome *c* oxidoreduction at 552–540 nm [$\Delta\varepsilon(\text{m}M)$ of 14]; the Fe–S protein, 1.90 g, was taken as equivalent to cytochrome c_1.[26] The values are expressed in ng of ions or nmol/mg protein. Experiment (A), control; Experiment (B), +2.8 μg antimycin A/mg protein. In this experiment the H^+/e^- ratio is computed dividing Bohr H^+ by the amount of Fe–S undergoing oxidation; there is no redox transition of *b* cytochromes.

lease of the same amount of H^+ taken up on oxygenation. It can be seen, however, that the H^+ taken up and subsequently released is much less than the sum of electron carriers undergoing oxidoreduction (Table II).[33,34] The observed deficit in the consumption and production of H^+ is a measure of H^+ release and H^+ binding, respectively, at protolytic groups in the cytochrome bc_1 complex, whose pK_a decreases upon oxidation of the redox centers and increases upon their reduction.

[33] Y. Hatefi, A. G. Haavik, and D. E. Griffiths, *J. Biol. Chem.* **237,** 1681 (1961).
[34] T. E. King, this series, Vol. 53, p. 181.

The E_m of cytochrome c_1 is pH independent.[15] Thus, the deficit of H^+ transfer has to be referred to the redox transitions of *b* cytochromes[3] and Rieske Fe-S protein[35] and exhibit an overall H^+/e^- coupling number of 1.2 (Table II, column A).

In Table II, column B a concentration of antimycin A, which prevents any net redox change of *b* cytochromes (Fig. 4B), is added. Also under these conditions H^+ transfer is less than half the sum of electron carriers undergoing oxidoreduction. The deficit of H^+ transfer gives an H^+/e^- coupling number of 1.5 when referred to redox transitions of the Rieske Fe-S protein.

An H^+/e^- coupling number of 1.5 at pH 7.2 for the Rieske Fe-S protein (Table II, column B) is much higher than that expected from the reported pH dependence of its E_m, which appears to decrease by 60 mV/pH unit increase above 7.9.[35]

In the transition of the cytochrome bc_1 complex from the reduced to the oxidized state, net H^+ release could also derive, however, from oxidation of endogenous ubiquinone, present in all the preparations of the cytochrome bc_1 complex,[30] to protein-stabilized ubisemiquinone.[36] Endogenous ubiquinone in the preparation used here amounts to 0.7 mol/mol cytochrome c_1.[30] Aerobic oxidation of ubiquinol to protein-stabilized bound ubisemiquinone,[36] for which a pK_a of 6 is reported,[37] could result in net production protons, according to

$$2QH_2 + 1/2\ O_2 \rightarrow 2(Q^{\cdot-})_n + 2(H^+)_n + 2(Q \cdot H)_{(1-n)} + H_2O \tag{3}$$

where *n* represents the percentage of the semiquinone anion at the given pH.

Assuming that endogenous ubiquinol (in the proportion of 0.7 mol/mol cytochrome c_1 oxidized) is oxidized to ubisemiquinone according to Eq. (3), one can calculate from the data in Table II the H^+ release by this reaction. The remaining H^+ production can be ascribed to deprotonation of ionizable groups in *b* cytochromes and Rieske Fe-S protein. The overall H^+/e^- coupling number for redox Bohr effects in these respiratory carriers calculates to 0.82 (Table II, column A).

When the deficit of H^+ transfer in antimycin-treated cytochrome bc_1 complex is corrected for H^+ release by Eq. (3), an H^+/e^- ratio of 0.84 for redox Bohr effect in the Rieske Fe-S protein is obtained (Table II, column B). An H^+/e^- ratio of 0.90 can be computed for Bohr effects in the Rieske Fe-S protein from the experiment in the absence of antimycin. Here the

[35] R. C. Prince and P. L. Dutton, *FEBS Lett.* **65,** 117 (1976).

[36] C. A. Yu, S. Nagooka, L. Yu, and T. E. King, *Biochem. Biophys. Res. Commun.* **82,** 1070 (1978).

[37] T. Ohnishi and B. L. Trumpower, *J. Biol. Chem.* **225,** 3278 (1980).

TABLE III
ANALYSIS OF SCALAR H^+ TRANSFER REACTIONS ASSOCIATED WITH REDOX TRANSITIONS OF ELECTRON CARRIERS IN ISOLATED CYTOCHROME-*c* OXIDASE[a]

Step	(A)	(B)
1. H^+ uptake	12.07	11.59
Hemes $a + a_3$	10.50	10.55
Copper	10.50	10.55
Cytochrome c_1	1.18	1.23
PMS	*0.49*	*0.49*
2. Σe^- flow	22.67	22.82
3. Deficit H^+ = (step 2 − step 1)	10.60	11.23
Deficit $H^+/a + a_3$	1.01	1.06
4. Bohr H^+ of bc_1 complex ($1.52 \cdot c_1$)	1.79	1.87
5. Bohr H^+ cytochrome oxidase (step 3 − step 4)	8.81	9.36
Bohr $H^+/a + a_3$	0.84	0.89

[a] Experimental conditions: (A) Purified cytochrome-*c* oxidase[32] (0.41 mg/ml) was incubated in 200 m*M* sucrose, 30 m*M* KCl, 10 μg purified catalase, and 0.2 μ*M* phenazine methosulfate (PMS) (pH 7.2); volume 2.25 ml. The incubation was carried out in a stoppered spectrophotometric cuvette for 20 min under a continuous stream of argon. Then 1 m*M* duroquinol was added. Anaerobiosis was reached in about 5 min. Under these conditions hemes *a* and a_3 were practically completely reduced. Oxygenation was brought about by addition of 1 μl of 0.1% H_2O_2. (B) Experimental conditions as reported for (A) except that instead of duroquinol, 10 m*M* ethanol was used as reductant in the presence of 30 units alcohol dehydrogenases (Boehringer), 0.5 μ*M* NAD^+, and 0.2 μ*M* phenazine methosulfate.

observed oxidation of *b* cytochromes is corrected based on the extent of effective hydrogen conduction by this carrier.[9]

Table III shows measurements of redox Bohr effects in purified cytochrome-*c* oxidase. In these experiments the oxidase is brought to anaerobiosis and reduced by (1) oxidation of duroquinol in the presence of traces of phenazine methosulfate, and (2) oxidation of ethanol in the presence of purified alcohol dehydrogenase and traces of NAD^+ and phenazine methosulfate.

Aerobic oxidation of hemes *a* and a_3 and of the other electron carriers results in a consumption of protons which is in both cases significantly lower than the sum of the electron carriers oxidized. The shortfall of H^+ uptake referred to hemes *a* and a_3 (the E_m of Cu is pH independent)[1] gives an H^+/e^- ratio of 1 with both experimental procedures. Correction for the contribution of H^+ release by unrelated protolytic reactions gave an H^+/e^- coupling number for redox Bohr effects of hemes *a* and a_3 which

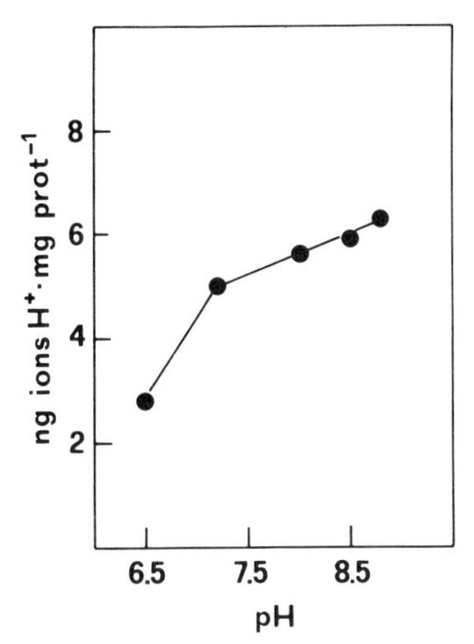

FIG. 5. pH profile of redox Bohr effects in purified bc_1 complex. bc_1 complex (0.43 mg protein/ml) was incubated in 200 m*M* sucrose, 30 m*M* KCl, 0.3 μg purified cytochrome oxidase, 0.24 μM cytochrome *c*, and 9 μM purified catalase, pH 7.2; 1 m*M* K-succinate was added. Once anaerobic, the sample was supplemented with 0.5 m*M* malonate and then pulsed repetitively with 1 μl of 0.1% H_2O_2. For other details, see legend to Fig. 4.

amounted to 0.84–0.89. It can be noted that an H^+/e^- coupling number for hemes *a* and a_3 of 0.84 is higher than expected based on the pH dependence of their midpoint redox potential measured in the mitochondrial membrane. This potential apparently amounts to a drop of 30 mV per pH unit increase[1,4] and corresponds to an H^+/e^- ratio not higher than 0.5. In light of the inherent ambiguities in redox titrations of membrane-bound multiple redox centers,[38] it must be determined whether these measurements can be applied to the isolated enzymes.

Figure 5 shows the pH profile for scalar proton release associated with aerobic oxidation of the redox centers of isolated bc_1 complexes. The number of Bohr protons increases by raising the pH from 6.5 to 9 with a pK_a of about 6.8. This corresponds to what can be predicted from the pH dependence of the potential of *b* cytochromes[3] and Fe-S protein.[35]

The method described is particularly suitable for characterization of redox Bohr effects. For example, it can be used to identify the ionizable groups involved in these effects. In this respect soluble redox complexes can be used after removal of polypeptide subunits and/or their chemical modification by amino acid reagents or digestion with specific proteolytic enzymes.

[38] D. Walz, *Biochim. Biophys. Acta* **505,** 279 (1979).

[32] Preparation of Two-Dimensional Crystals of Complex I and Image Analysis

By EGBERT J. BOEKEMA, MARIN G. VAN HEEL, and ERNST F. J. VAN BRUGGEN

NADH dehydrogenase (NADH : Q oxidoreductase, EC 1.6.99.3) is the first enzyme of the mitochondrial respiratory chain. A purified form of the enzyme that has been studied extensively is Complex I from beef heart. This was first isolated by Hatefi.[1] Despite much research in the past decade on the structure of NADH dehydrogenase, a picture of the overall structure is still lacking.[2]

A detailed analysis of the structure of integral membrane proteins with electron microscopy is a process that depends on several distinct steps. First, one needs the protein in a purified form. This is achieved mostly with the help of suitable detergents that keep the protein in a dispersed state after isolation. Second, one must prepare the material in a form suitable for electron microscopy. The subsequent recording process is a routine step, performed by a photographic emulsion. The recorded signal is difficult to interpret directly owing to its low signal-to-noise ratio. Therefore computer image analysis techniques are necessary to improve the quality of the signal. These techniques have been greatly developed over the past two decades. In principle they enable the enhancement of structural information to a resolution of better than 0.5 nm.

The bottleneck of the whole structure determination is at the level of the preparation of specimens for electron microscopy. One part of the problem is caused by the heavy metal stain used to enhance the contrast of biomacromolecules. Negative staining with heavy metal salts normally limits the resolution to 1.5–2.5 nm. This limit is set only by the way the specimen is prepared, because an electron microscope normally has a resolving power of better than 0.5 nm. If one is interested in details smaller than 1.5 nm, the only approach is to avoid conventional negative staining and to try out alternative ways of supporting and embedding the specimen.

The other part of the preparation problem is inherent to membrane proteins. These proteins are normally isolated in detergent-containing solutions and are frequently covered with remnants of strongly binding phospholipid molecules. If this material is prepared for electron micros-

[1] Y. Hatefi, this series, Vol. 53 [3].

[2] For a review of the enzyme, see C. I. Ragan, *in* "Subcellular Biochemistry" (D. B. Roodyn, ed.), Vol. 7, p. 267, Plenum, New York, 1980.

copy, one cannot recognize the shape of the molecule because the bulk of the material is in clots of protein or in a vesicle form with random orientation. Therefore one needs material in a crystalline form, preferably in two-dimensional or in thin three-dimensional crystals.

There are now several methods available for the crystallization of membrane proteins.[3,4] One approach to the problem of crystallizing membrane proteins is based on the classical methods of increasing the salt concentration.

Crystallization Procedure

Complex I is prepared from beef heart muscle according to the method of Hatefi.[1] The isolated enzyme is dissolved in 50 m*M* Tris–HCl buffer at pH 8.0 in the presence of 0.66 *M* sucrose and 0.5% (w/v) cholate at a final concentration of about 50 mg/ml. Complex I is crystallized by means of a microdialysis technique consisting of equilibrium dialysis in vertical glass capillaries of 45 mm × 1 mm inner diameter. This procedure is described by Zeppezauer *et al.*[5] The purified enzyme is diluted to 20 mg/ml in 50 m*M* sodium acetate buffer at pH 5.9. Volumes of 20 μl are introduced into the capillaries which are closed at the lower end with a dialysis membrane. The other end is sealed with a piece of plastic and the capillaries are then placed into tubes containing 10 ml of the sodium acetate buffer plus 1 *M* ammonium sulfate, but without cholate or sucrose. The tubes are stored in a cold room at 2°. After 14–20 days the capillaries are unsealed, and the enzyme solution is removed and diluted with one volume of the dialysis buffer. This stock solution contains good crystals for at least several weeks, and it can be used without further purification.

Specimens are prepared for electron microscopy with the droplet technique using aqueous 1% unbuffered uranyl acetate as a negative stain and Formvar-carbon as a supporting film. The crystalline suspension is diluted two to five times just before preparation of a specimen.

Characterization of the Crystals

Electron microscopy was carried out using a Philips EM 400 electron microscope at 80 kV with an electron optical magnification of 50,000×.[6]

[3] H. Michel, *Trends Biochem. Sci.* **8,** 56 (1983).

[4] S. Hovmöller, M. Slaughter, J. Berriman, B. Karlsson, H. Weiss, and K. Leonard, *J. Mol. Biol.* **165,** 401 (1983).

[5] M. Zeppezauer, H. Eklund, and E. S. Zeppezauer, *Arch. Biochem. Biophys.* **126,** 564 (1968).

[6] E. J. Boekema, J. F. L. Van Breemen, W. Keegstra, E. F. J. Van Bruggen, and S. P. J. Albracht, *Biochim. Biophys. Acta* **679,** 7 (1982).

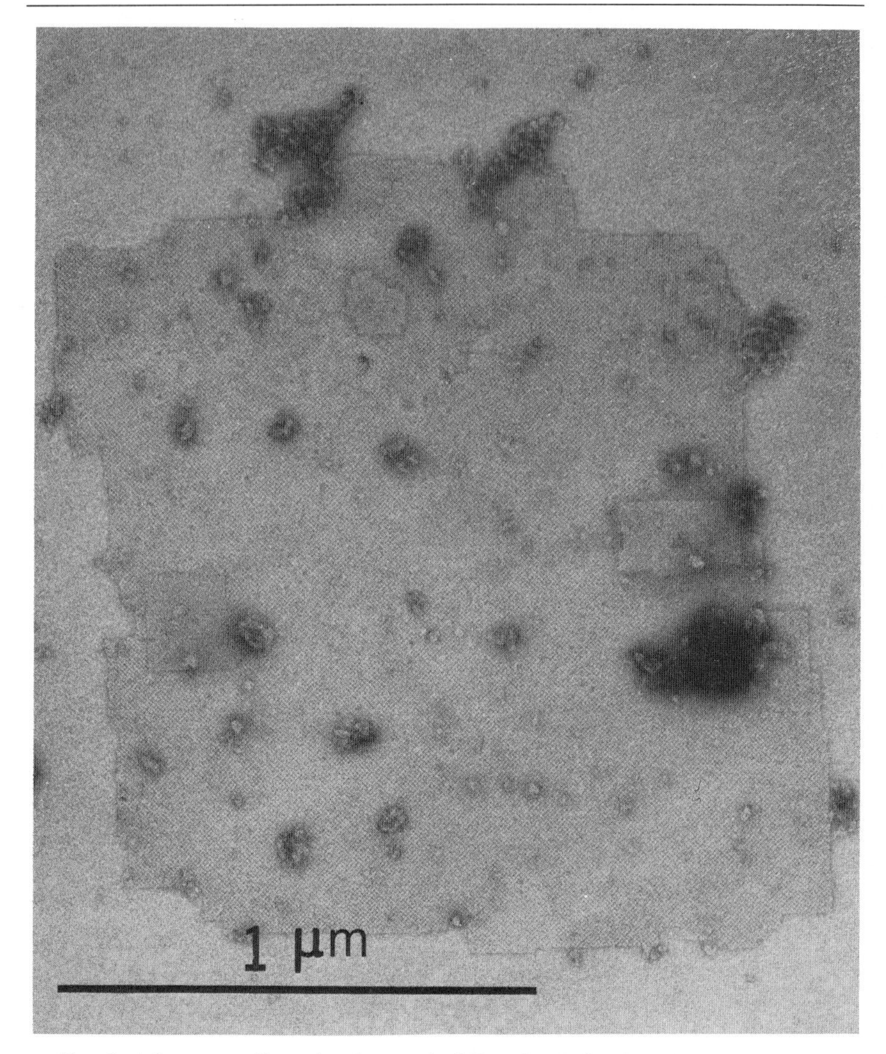

FIG. 1. A large two-dimensional crystal of Complex I, lightly stained with 1% uranyl acetate. The crystal is for the most part a monolayer. At three sites a second layer is attached. The crystal contains about 10,000 molecules.

Figure 1 shows a typical two-dimensional crystal of Complex I. The crystals are built up from single tetrameric or pseudotetrameric protein molecules and are not formed by a rearrangement of molecules in vesicles or sheets, as is the case for some other membrane proteins. The removal of cholate during dialysis is essential for the formation of the crystals. The protein concentration and pH are less critical. Good crystals are formed

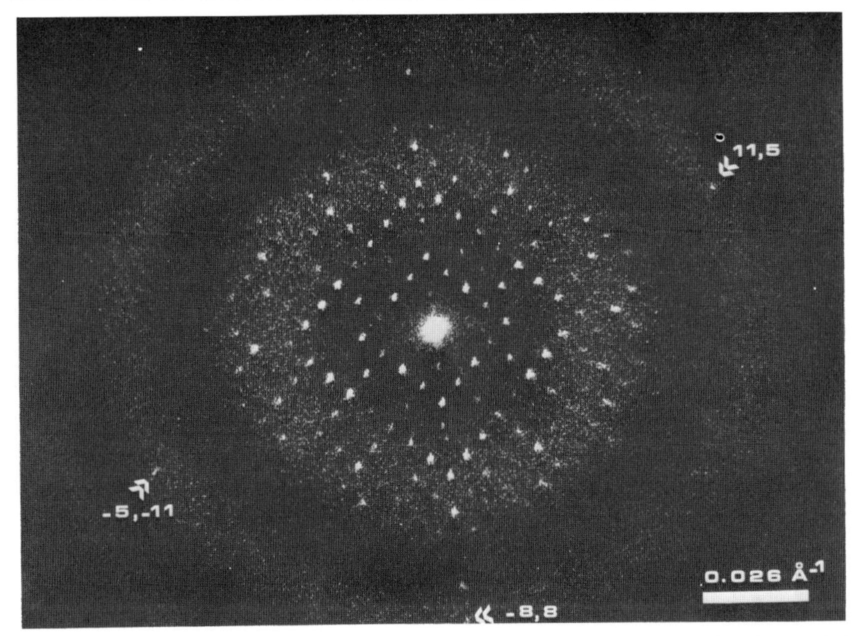

FIG. 2. Optical diffraction pattern of a double-layer part of a two-dimensional crystal of Complex I. The diffraction pattern has the same orientation as the crystal of Fig. 1.

between pH 5.3 and 7.5, though the best yield is at pH 5.9. Figure 2 shows an example of an optical diffraction pattern from a micrograph of a Complex I crystal. The optical diffraction pattern allows one to determine the reciprocal lattice constants of the crystal. The crystal has a square lattice with spacings of $a = b = 15.3$ nm and the angle between a and b is exactly 90°.

The crystals are induced in the presence of ammonium sulfate, but remnants of salt in the crystals when prepared for the electron microscope strongly diminished the resolution. This was observed from optical diffraction patterns which showed only the first four orders of reflection (3.7 nm resolution) for crystals prepared without a washing step with distilled water. The optical diffraction pattern of Fig. 2 clearly shows that medium resolution could be obtained when specimens were optimally prepared, i.e., negatively stained with 1% unbuffered uranyl acetate after one or two quick washing steps with distilled water to remove most of the ammonium sulfate. Arrows indicate two (11,5) type reflections, corresponding to 1.3 nm resolution. However, this optical diffraction pattern was made from a double-layer crystal in which the ordering was better preserved during preparation than in a single layer. Double layers are more difficult to analyze but have the same lattice parameters. In single-layer crystals only reflections up to 1.9 nm could be observed routinely in the optical diffraction pattern, and occasionally the (7,7) type reflections corresponding to 1.5 nm.

Image Processing

The ultimate aim of structural research on membrane proteins by electron microscopy is to obtain information regarding the three-dimensional shape of the molecules forming the crystal. This process can be divided into three main steps:

1. Recording of the Signal. A conventional electron microscope possesses a large depth of focus. Consequently an electron image represents the projection of the density in the specimen in the direction of the electron beam. The three-dimensional structure of an object can be calculated from a large number of different projections through the object. In the case of two-dimensional crystals the different projections are obtained by tilting the object in the microscope.
2. Filtering of the Two-Dimensional Projections. The insufficient statistical significance of an image of a single molecule must be improved by averaging over many molecules before a reliable interpretation of the studied structure is possible. To obtain noise-free images of two-dimensional crystals, Fourier peak filtering is an easy and well-known technique. The resolution of averaged projections by this method, however, is generally limited by long-range distortions of the crystal lattice and by small rotational and translational shifts of the repeating unit from its ideal position. Crystals may be perfectly ordered in aqueous solution, but attachment to the supporting film and the drying process during preparation can disturb the high order of the lattice. To correct as well as possible for these lattice imperfections, a correlation averaging technique has been developed.[7–9] It is based on the following principle: A digitized image of a large crystal is divided into small fragments. These fragments may contain one or several molecular projections, depending on the size of the motif and the amount of noise in the image. The position in which a fragment is exactly equivalent to a reference of the same size can be determined by calculating the two-dimensional cross-correlation function between the reference and each fragment. This results in small shifts of the aligned

[7] J. Frank, *Optik* **63,** 67 (1962).

[8] W. O. Saxton and W. Baumeister, *J. Microsc.* **127,** 127 (1982).

[9] M. Van Heel and J. Hollenberg, *in* "Electron Microscopy at Molecular Dimensions" (W. Baumeister and W. Vogell, eds.), p. 256. Springer-Verlag, Berlin, 1980.

fragments. After this alignment procedure the fragments are summed. If this procedure is applied to large crystals, summation over more than a thousand molecules is easily performed and the signal-to-noise ratio is considerably enhanced. Usually the first summation is used as a new reference for further alignments. Repeating the alignments is important, since the quality of the final summation is dependent on the quality of the reference.

Note: Especially in the case of very noisy images, one should use a Fourier-filtered image as a first reference instead of trying to align noisy fragments of the crystal directly on a noisy reference selected from the crystal.

3. Three-Dimensional Reconstruction. Two general algorithmic approaches for the three-dimensional reconstruction of an object from its two-dimensional projections are available. The first method is based on the central section theorem, which states that the Fourier transform of a two-dimensional projection of a three-dimensional object is identical to a "central section" through the three-dimensional Fourier transform of the object. The Fourier transform of the projections can thus be used to fill the three-dimensional Fourier space with the measured data. The three-dimensional density distribution of the object is then obtained by a three-dimensional inverse Fourier transformation. This Fourier method was originally developed by De Rosier and Klug.[10] Bacteriorhodopsin was the first enzyme in a two-dimensional crystal arrangement to be reconstructed using this method.[11]

The second family of three-dimensional reconstruction methods, on the contrary, operates in "direct" or "real space." A suitable method for electron microscopy is GSIRT (Generalized Simultaneous Iterative Reconstruction Technique).[12] It is a generalization of SIRT, which itself is an improvement of the Algebraic Reconstruction Technique (ART). The GSIRT algorithm starts the calculation with a straightforward backprojection of the measured projection data, followed by iterative corrections of the densities in the reconstruction volume, and, in contrast to SIRT, a "relaxation" or "overcorrection" is applied, a mathematical technique which is known often to lead to a considerably better convergence of iterative algorithms. The stability of the GSIRT algorithm is considerably enhanced as compared to ART by backprojecting the corrections (difference between measured and recalculated projections) relative to all projections simultaneously over the reconstruction volume.

[10] D. J. De Rosier and A. Klug, *Nature (London)* **217,** 130 (1968).

[11] This method is reviewed in L. Amos, R. Henderson, and P. N. T. Unwin, *Prog. Biophys. Mol. Biol.* **39,** 183 (1982).

[12] A. V. Lakshminarayanan and A. Lent, *J. Theor. Biol.* **76,** 267 (1979).

One of the advantages of a direct space reconstruction procedure over the Fourier method is the use of real-space data instead of data in reciprocal space as intermediate results. By displaying these real space images, one has an easy check for eventual mistakes. Nevertheless, the more flexible real-space reconstruction method and the Fourier method in principle must result in reconstructions with about the same quality if one starts with equivalent input projection data.

Experimental Procedures

Tilt series are taken with a eucentric goniometer holder for single-axis tilting. Two sets of 14 and 16 micrographs ranging in tilt angle from $-54°$ to $+57°$ show 6–8 orders of diffraction for the lower tilt images, which corresponds to a resolution of 1.9–2.5 nm. The micrographs are digitized and the images are processed using the IMAGIC software system on a VAX 780 or on a NORD 10 computer.[13]

The first step in the image-processing procedure is the enhancement of the signal-to-noise ratio of the individual micrographs by correlation averaging. The full procedure is carried out with real-space images, but as a double check for the improvement in resolution and signal-to-noise ratio, Fourier test patterns are generated during subsequent steps of the filtering procedure.[14] From the digitized image of the untilted (0°) crystal, an area of 276×276 nm (512×512 pixels), containing a total of about 250 unit cells, is Fourier transformed and filtered. The central part of 128×128 pixels is used as a reference for a first alignment of noisy fragments. The image is divided into fragments of 128×128 pixels, containing about 10 tetrameric molecules, and these fragments are aligned. From the superposition, a new reference of 64×64 pixels is selected, containing one central molecular projection and parts of six neighboring molecules. Thereafter fragments of 64×64 pixels are aligned and summed. The shift to place the untilted projection (0°) exactly in the center is calculated from a cross-correlation between the projection and its 180° rotated image. The other members of the tilt series are now averaged by alignment of their fragments relative to a reference of an averaged projection of lower tilt angle, with the untilted projection as a starting point. In this way a common origin for the filtered projections of the tilt sequence is automatically determined.

The final result of the correlation averaging procedure on one complete tilt series consisting of 16 micrographs is shown in Fig. 3. Each

[13] M. G. Van Heel and W. Keegstra, *Ultramicroscopy* **7,** 113 (1981).

[14] E. J. Boekema and E. F. J. Van Bruggen, *in* "Structure and Function of Membrane Proteins" (E. Quagliariello and F. Palmieri, eds.), p. 237. Elsevier, Amsterdam, 1983.

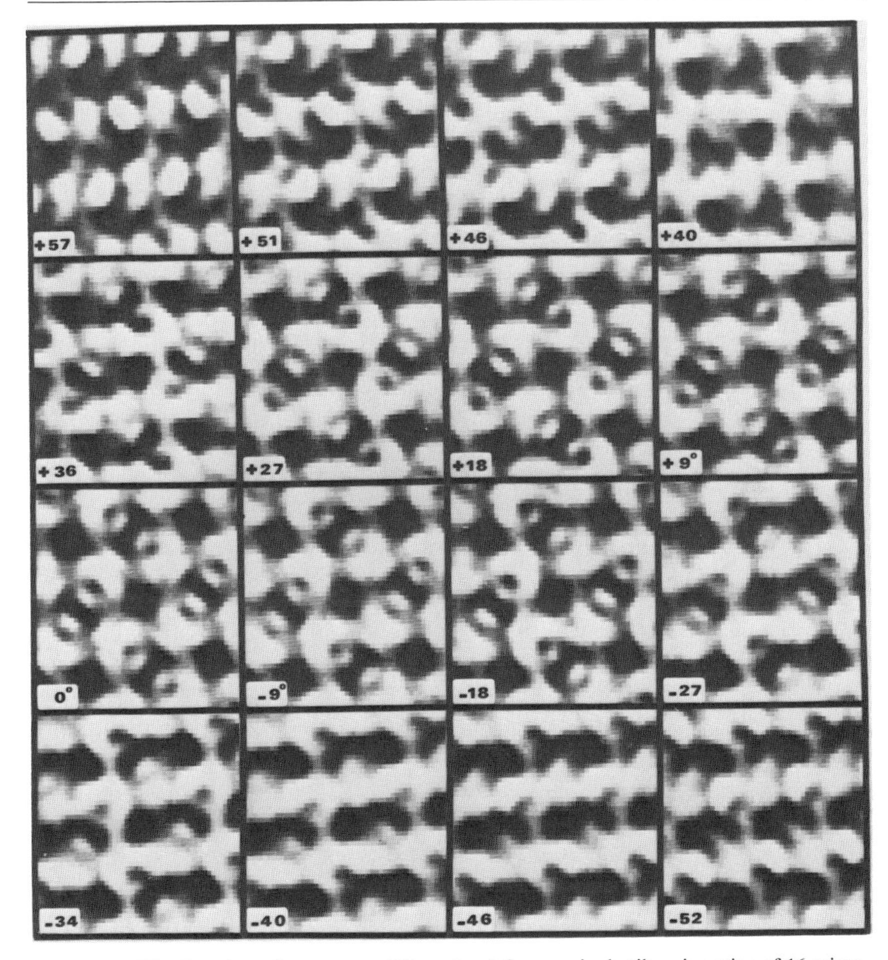

FIG. 3. Tilted projected structure of Complex I from a single tilt axis series of 16 micrographs. Each image of 26 × 26 nm (the central 48 × 48 pixels from the 64 × 64 images) is a summation of 500–2000 projected molecules, selected from the better focused part of the crystal. All images have the same orientation and have been shifted to a common origin. The tilt axis runs vertically. Protein is white; stain is dark. No symmetry has been imposed.

image results from a summation of 500–2000 aligned, noisy, projected molecules, and the 16 filtered images together contain the information from about 16,000 projections of molecules. The very large number of projections used makes the filtered images reliable. In going from a lower to a higher tilt angle the shortening in the structure can be easily followed.

Two tilt series of good quality which included 30 images of two crystals with a large monolayer region are used for calculating the three-

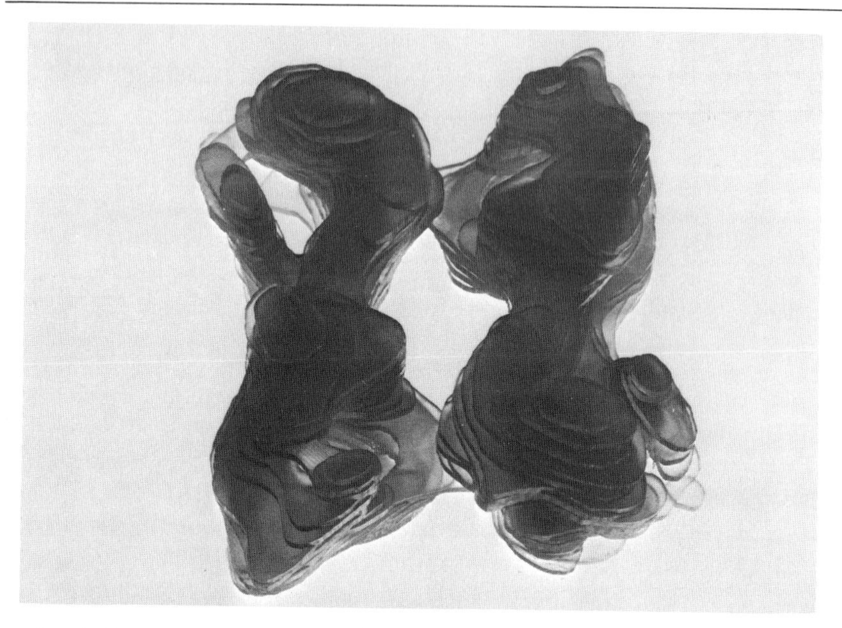

FIG. 4. View of a transparent model of the Complex I structure. It is constructed from 21 sections, each representing 0.5 nm parallel to the plane of the crystal, i.e., perpendicular to the twofold axis running through the "central pit" of the molecule. The lower side of the model is attached to the supporting film.

dimensional structure with the GSIRT algorithm. Aligned arrays of 64 × 64 pixels were used to calculate a reconstruction in a 64 × 64 × 64 format. A Perspex model built from the three-dimensional map of the combined reconstructions is shown in Fig. 4. It represents the parts of the structure that are strongly stain excluding and therefore must correspond to the backbone of the protein structure. The most remarkable feature of the reconstruction is the pore running through the center of the four monomers. Estimated from the two-dimensional projection, they have a diameter of about 1.5 nm.

The three-dimensional structure can be compared with results from biochemical and biophysical experiments.[15] A problem is that the crystal has not yet been analyzed in terms of subunit composition. In the most probable case, the whole Complex I has been crystallized. The crystal is built up from molecules which have internal twofold symmetry.[15] The

[15] E. J. Boekema, M. G. Van Heel, and E. F. J. Van Bruggen, *Biochim. Biophys. Acta* **787,** 19 (1984).

other possibility that cannot be fully excluded is crystallization of only a part of the Complex I: The square lattice of the diffraction pattern (Fig. 2) may indicate a high type of symmetry, in which case there is not enough space in the unit cell for containing the whole Complex I.

Acknowledgments

This work was supported by The Netherlands Organization for chemical Research (SON) with financial aid from The Netherlands Organization for the Advancement of Pure Research (ZWO).

[33] Purification and Reconstitution of Bovine Heart Mitochondrial Transhydrogenase

By Licia N. Y. Wu, Julia A. Alberta, and Ronald R. Fisher

Mitochondrial transhydrogenase couples the reversible transfer of a hydride ion between matrix NADH and $NADP^+$ to vectorial proton translocation from the cytosol to the matrix.[1,2] Several procedures for the purification of the enzyme to apparent homogeneity from bovine heart mitochondria have been published.[3-7] All of these preparations when reconstituted into liposomes have been shown to be capable of generating either a membrane potential or a pH gradient during transhydrogenation. Electrometric measurements with these vesicles have determined that the stoichiometry of proton translocation to hydride ion transfer ($H^+ : H^-$ ratio) is at least 1.0.[1,8] Unlike other oligomeric energy-transducing complexes of the inner membrane, i.e., NADH-CoQ reductase, ubiquinol–cytochrome-*c* reductase, cytochrome-*c* oxidase, and F_0F_1(H^+-ATPase), transhydrogenase possesses a simple subunit structure consisting of a

[1] S. R. Earle and R. R. Fisher, *in* "Pyridine Nucleotide Coenzymes" (J. Everse, B. Anderson, and K.-S. You, eds.), p. 279. Academic Press, New York, 1982.

[2] J. Rydström, *Biochim. Biophys. Acta* **463,** 155 (1977).

[3] W. M. Anderson and R. R. Fisher, *Arch. Biochem. Biophys.* **187,** 180 (1978).

[4] B. Höjeberg and J. Rydström, *Biochem. Biophys. Res. Communo* **78,** 1183 (1977).

[5] W. M. Anderson, W. T. Fowler, R. M. Pennington, and R. R. Fisher, *J. Biol. Chem* **256,** 1888 (1981).

[6] L. N. Y. Wu, R. M. Pennington, T. D. Everett, and R. R. Fisher, *J. Biol. Chem.* **257,** 4052 (1982).

[7] B. Persson, K. Enander, H.-L. Tang, and J. Rydström, *J. Biol. Chem* **259,** 8626 (1984).

[8] S. R. Earle and R. R. Fisher, *J. Biol. Chem.* **255,** 827 (1980).

dimer of apparently identical 110 kDa subunits.[9] Hence, the enzyme may generally offer certain advantages as a model system for studies on the mechanism of proton pumps. Here, we describe a very rapid, simple, and highly reproducible method for the preparation of homogeneous transhydrogenase and its reconstitution into phospholipid vesicles suitable for the electrometric determination of proton translocation.

Transhydrogenase Purification

Transhydrogenase Assay

The reduction of the 3-acetylpyridine analog of NAD^+ ($AcPyAD^+$) by NADPH is monitored continuously at 375 nm. The activity is calculated assuming a millimolar extinction coefficient of 5.1 for AcPyADH. One unit of transhydrogenase activity is defined as the amount of enzyme that produces 1 μmol of AcPyADH in 1 min at 25°. The assay mixture for solubilized transhydrogenase (3.0 ml) contains 80 m*M* potassium phosphate, pH 6.3, 170 μM $AcPyAD^+$, 150 μM NADPH, and 0.2 mg lysophosphatidylcholine (egg yolk). When membrane vesicles are assayed lysophosphatidylcholine is omitted and 0.5 μM rotenone is added to inhibit NADH oxidase.

Protein Analysis

Membrane and Triton X-100 extract protein is determined by the biuret method in the presence of 0.5% sodium cholate.[10] The amount of purified transhydrogenase is estimated by the modified Lowry method.[11]

Purification Procedure

Bovine heart mitochondria and submitochondrial particles are prepared according to Löw and Vallin.[12] The particles are suspended at 40–50 mg/ml in 10 m*M* Tris–HCl, pH 8.0, containing 0.25 *M* sucrose and stored at −70°.

Step 1: NaCl Wash. Submitochondrial particles (100 mg of protein) are thawed at room temperature and diluted to a final volume of 50 ml with 2 *M* NaCl containing 10 m*M* sodium phosphate, pH 7.5, and 5 m*M* EDTA. Protease inhibitor, phenylmethylsulfonyl fluoride (PMSF, 0.25 *M* in methanol), is added immediately to a final concentration of 0.5 m*M*. The mix-

[9] L. N. Y. Wu and R. R. Fisher, *J. Biol. Chem.* **258,** 7847 (1983).

[10] E. E. Jacobs, M. Jacobs, D. R. Sanadi, and L. B. Bradley, *J. Biol. Chem.* **223,** 147 (1956).

[11] G. L. Peterson, *Anal. Biochem.* **83,** 346 (1977).

[12] H. Löw and I. Vallin, *Biochim. Biophys. Acta* **69,** 361 (1963).

ture is incubated at 25° for 30 min and then centrifuged at 200,000 *g* for 30 min at 4°. The supernatant fraction is decanted, the pellet is rinsed with 1 ml of the above 2 *M* NaCl solution, and the rinse discarded. This step is performed to remove several peripheral proteins that otherwise contaminate the enzyme after Step 3.[6]

Step 2: Triton-X100 Extraction. The membrane pellet from the NaCl wash is suspended with homogenization to a volume of 17.5 ml at 25° in 10 m*M* sodium phosphate, pH 7.5, containing 5 m*M* EDTA. Then 40 μl of 0.25 *M* PMSF and 2.55 ml of 10% (w/v) Triton X-100 (Sigma) are added to the suspension. After incubation for 30 min at 25° with occasional shaking, the extract is centrifuged at 200,000 *g* for 30 min at 4°. The supernatant fraction containing 80–95% of the total transhydrogenase units is decanted.

Step 3: Affinity Chromatography. NAD^+ is immobilized on Sepharose 4B following the procedure of Mosbach *et al.*[13] A 3.5 × 6.5 cm column of the gel is preequilibrated at 4° with 10 m*M* sodium phosphate, pH 7.5, 1 m*M* EDTA, and 10 m*M* 2-mercaptoethanol (column buffer) containing 0.1% Lubrol-WX (Sigma). The Triton X-100 extract is divided into two equal portions which are each percolated onto an affinity column at a rate of 60 ml/hr. The column is then washed at 120 ml/hr with the following solutions: (a) 40 ml of column buffer containing 0.1% Lubrol-WX; (b) 15 ml of column buffer containing 0.4% Lubrol-WX; (c) 15 ml of column buffer containing 0.1% Lubrol-WX and 20 m*M* NaCl, and (d) 40 ml of column buffer containing 0.1% Lubrol-WX. A small amount of enzyme is not bound to the column (about 20% of the units applied) and is eluted during these washes. Wash (d) should be monitored to ensure that no transhydrogenase activity is present in the final 10 ml. To elute specifically bound transhydrogenase, 4 ml of column buffer containing 0.1% Lubrol-WX and 1 m*M* NADH is applied followed by washing at 50 ml/hr with column buffer containing 0.1% Lubrol-WX. Then 4-ml fractions are collected and those containing enzyme activity (about 20 ml) are combined. The affinity column can be used repeatedly for up to 3 months. Column regeneration is accomplished by washing with 100 ml of 2 *M* NaCl/0.05% potassium cholate followed by 500 ml of water.

A typical transhydrogenase preparation is summarized in the table. Transhydrogenase is the highest molecular weight major protein species of the submitochondrial membrane. Figure 1 illustrates that the affinity column selectively binds the enzyme (compare lanes 3 and 4). The presence of only small amounts of unbound transhydrogenase in the wash fractions (lane 4) indicates that little or no denaturation resulting in loss of

[13] K. Mosbach, H. Guilford, R. Ohlsson, and M. Scott. *Biochem. J.* **127,** 625 (1972).

TRANSHYDROGENASE PURIFICATION FROM BOVINE HEART SUBMITOCHONDRIAL PARTICLES

Fraction	Protein (mg)	Total units (μmol AcPyADH min^{-1})	Specific activity (μmol AcPyADH min^{-1} mg^{-1})	Purification factor	Yield (%)
Submitochondrial particles	100	33.0	0.33	1.0	100.0
NaCl pellet	68	36.0	0.52	1.6	109.0
Triton X-100 extract	44	34.0	0.77	2.3	103.0
Affinity column fractions	0.45	16.0	35.6	107.9	48.5

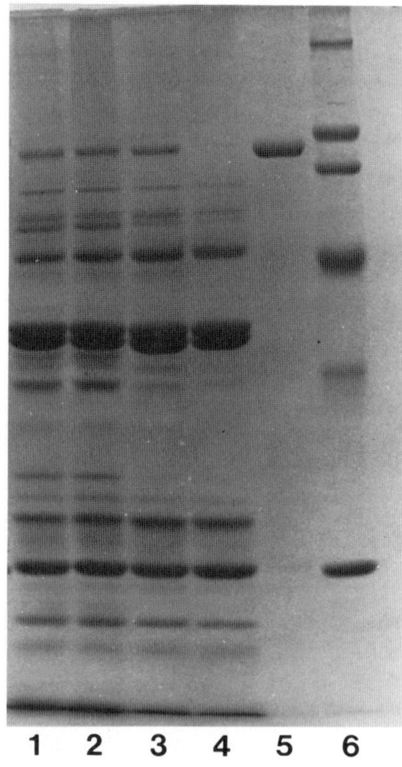

FIG. 1. SDS–polyacrylamide slab gel analysis of transhydrogenase at various steps of purification. Electrophoresis was performed on 10% (acrylamide: bisacrylamide, 30 : 8) gels. Lane 1, 68 μg of submitochondrial particles; lane 2, 67 μg of NaCl-extracted particles; lane 3, 27 μg of Triton X-100 extract; lane 4, 40 μg of protein not bound to the NAD$^+$-affinity column; lane 5, 3 μg purified transhydrogenase; land 6 contains 5 μg each of molecular weight markers: rabbit muscle myosin (205,000), *E. coli* β-galactosidase (116,000), rabbit muscle phosphorylase B (97,400), bovine albumin (66,000), egg albumin (45,000), and bovine erythrocyte carbonate dehydratase (29,000). Proteins were stained with Coomassie Brilliant Blue.

the NAD^+ binding site occurs during the preparation of the Triton X-100 extract. Enzyme eluted in the presence of NADH is apparently homogeneous (lane 5). Identical band patterns are observed when Coomassie Brilliant Blue or silver staining methods are used to visualize protein after SDS–polyacrylamide gel electrophoresis.

Comments

This preparation has several advantages when compared to others[3–5,7] with respect to percentage yield and specific activity. Yields consistently range from 35 to 50%, as opposed to 3 to 15%, and specific activities of 35–40 μmol · min^{-1} mg^{-1} are commonly observed as opposed to 5–20. It is likely that these differences arise from the instability of the enzyme. Purified transhydrogenase preparations decrease in activity by more than 50% in 2 days at 4°. Hence, it is not surprising that this rapid, one-step procedure, which can be completed in less than 4 hr rather than 2–3 days, better avoids problems related to spontaneous denaturation. In addition, other procedures[4,7] employ ion exchange and liquid-chromatographic purification steps that would not necessarily separate active from inactive enzyme.

Reconstitution of H^+-Pumping Transhydrogenase

Detergent Exchange

Insertion of transhydrogenase into liposomes is performed by a cholate dialysis procedure.[8] Prior to reconstitution the enzyme is concentrated and nondialyzable Lubrol-WX is replaced by potassium cholate. To accomplish this, transhydrogenase in the combined affinity column eluate (Step 3) is bound immediately to a minimal amount of calcium phosphate gel. Typically, 3 ml of a 5 g/50 ml gel suspension is added to 20 ml of the enzyme solution at 4°. The gel suspension is thoroughly mixed, then centrifuged at 10,000 *g* for 8 min. The pellet is washed with 10 ml of 1 m*M* sodium phosphate, pH 7.5, containing 0.05% neutralized potassium cholate (recrystallized from 70% ethanol). The bound enzyme is then eluted by suspending the gel in 2 ml of 100 m*M* sodium phosphate, pH 7.5, containing 0.05% potassium cholate followed by centrifugation. Detergent exchange results in the loss of 50–65% of the transhydrogenase units, with only a slight decrease in specific activity.

Cholate Dialysis

Egg yolk phosphatidylcholine (5 mg) (Sigma, type V-E) is suspended in a medium (0.12 ml, in a 10 × 75 mm Pyrex tube) containing 5% potassium cholate, 1 mM potassium phosphate, pH 7.2, and 100 mM potassium sulfate. The suspension is sonicated to clarity in an ice-water bath using a bath-type sonicator (Laboratory Supply Co., Hicksville, NY). Subsequently, 0.19 ml of 2 mM potassium phosphate, pH 7.2/200 mM potassium sulfate is added followed by the addition of about 15 μg of purified transhydrogenase contained in 0.19 ml of 0.05% potassium cholate/100 mM sodium phosphate, pH 7.5. After gentle agitation, the solution is dialyzed for 20 hr at 4° against 65 ml of 1 mM potassium phosphate, pH 7.2, containing 100 mM potassium sulfate and 0.1 mM EDTA, with one change of buffer after 4 hr. Dialysis tubing with a molecular weight cutoff of 12,000–14,000 is used. To remove external K^+, the proteoliposomes are further dialyzed for 6–20 hr against 90 ml of 1 mM Tricine/NaOH, pH 7.2, containing 100 mM choline-chloride, with one change after 3 hr. The vesicles should be monitored periodically after 6 hr of dialysis to attain optimal proton translocation ratios.

Determination of $H^+ : H^-$ Ratios

The uptake of protons by transhydrogenase-containing liposomes coupled to the reduction of NAD^+ by NADPH has been monitored using both permeant pH probes such as 9-aminoacridine[7,8,14,15] and by loading vesicles with impermeant probes.[14] However, these methods cannot be employed to assess the stoichiometry of H^+ translocation and have certain other limitations (see Comments). The disappearance of protons from the external medium is measured directly using a pH electrode. Proton uptake is absolutely dependent on the presence of valinomycin, which allows the efflux of K^+ from the K^+-loaded vesicles and prevents the development of an H^+-dependent membrane potential. Reaction mixtures (1.145 ml) at 25°, which are continuously stirred and contained in a 2-ml glass vessel, consist of 0.95 ml 1 mM Tricine/NaOH, pH 7.2, 100 mM choline-chloride, 0.07 ml 5.26 mM NADPH, 0.07 ml 5.82 mM $AcPyAD^+$, 0.005 ml 2×10^{-4} M valinomycin, and 0.05 ml of proteoliposomes containing about 1.5 μg of transhydrogenase. Substrates are dissolved in 1 mM Tricine/NaOH, pH 7.2, 100 mM choline-chloride, and the pH carefully adjusted to that of the medium with dilute (0.01–0.1 N) NaOH or HCl. The medium pH is determined using a combination microelectrode (Beckman 39505) and a Corning Model 12 pH meter interfaced with a Fisher Recordall 5000 recorder. Proton translocation may be initiated by the addition of NADPH, $AcPyAD^+$, or valinomycin. Vesicles should be routinely checked for the complete absence of proton uptake on addition

[14] S. R. Earle and R. R. Fisher, *Biochemistry* **19,** 561 (1980).

[15] J. Rydström, *J. Biol. Chem.* **254,** 8611 (1979).

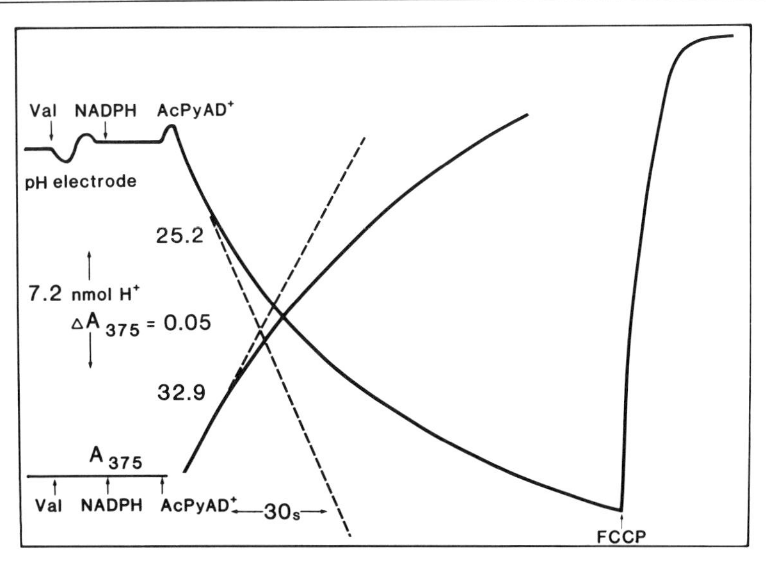

FIG. 2. Determination of $H^+ : H^-$ ratio with reconstituted transhydrogenase. Details are presented in the text. Numbers next to the traces indicate the disappearance of protons from the medium or the formation of AcPyADH in nmol min^{-1}.

of valinomycin in the absence of substrates. When properly prepared, phosphatidylcholine liposomes show no such uptake, indicating that they are inherently highly impermeable to protons. In each experiment, proton uptake is quantitated by several 10-μl additions of standardized 0.1 *N* HCl. The initial rate of proton uptake is linear for at least 10 sec. The rate of transhydrogenation between NADPH and $AcPyAD^+$ is measured under the same conditions as proton uptake. Transhydrogenation is stimulated about 5-fold on addition of valinomycin. In calculating the activity of functionally incorporated enzyme, the rate of $AcPyAD^+$ reduction in the absence of valinomycin is subtracted from that observed in the presence of the ionophore. The $H^+ : H^-$ ratio is the initial rate of H^+ uptake divided by the initial rate of transhydrogenation. A typical $H^+ : H^-$ ratio experiment is presented in Fig. 2.

Comments

$H^+ : H^-$ ratios ranging between about 0.35 and 0.9 have been routinely observed. The reason for this variability is not fully understood. However, it may be related to the observation that dialysis of the liposomes against Tricine/NaOH, choline-chloride for more than 24 hr results in preparations that show decreased ratios. Prolonged dialysis leads to complete loss of proton pumping even though valinomycin or uncoupler (FCCP) stimulation of transhydrogenation remains similar to functional liposomes. Release of activity by valinomycin or uncoupler has been assumed to indicate that the enzyme is functionally incorporated.[7,8,14,15] Clearly, this need not be true. Such stimulation may result from a preexisting membrane potential in the liposomes, which acts to maintain the enzyme in a relatively inactive conformation. Dissipation of the membrane potential could alter transhydrogenase to a more active conformation. These conformational alterations need not require that the enzyme be properly inserted for bulk proton translocation into the liposomes. Moreover, liposomes incapable of bulk proton translocation have been prepared which show apparent acidification of the intravesicular space, as indicated by quenching of 9-aminoacridine fluorescence.[14] Hence, caution should be applied to the interpretation of proton translocation assessed in this manner (cf. Ref. 16). It may be concluded that $H^+ : H^-$ ratios have thus far been measured on mixtures of liposomes containing transhydrogenase that is both functional and nonfunctional in proton pumping, and that the ratios have consequently been underestimated.

[16] C. S. Huang, S. J. Kopacz, and C.-P. Lee, *Biochim. Biophys. Acta* **722,** 107 (1983).

[34] Isolation of the Iron–Sulfur-Containing Polypeptides of NADH : Oxidoreductase Ubiquinone

By C. IAN RAGAN and YOUSSEF HATEFI

Of the 50 or so polypeptides which comprise the mammalian mitochondrial respiratory chain and the 12 iron-sulfur (FeS) clusters contained therein, 25 of the former and 8 of the latter are found in the NADH : ubiquinone oxidoreductase complex (EC 1.6.5.3) (Complex I).[1,2] The purpose of this chapter is to describe the methodology employed for the stepwise fragmentation of Complex I and isolation of subunits containing FeS clusters. The procedure for the isolation of Complex I from bovine heart mitochondria has been described elsewhere in this series.[3]

[1] Y. Hatefi, C. I. Ragan, and Y. M. Galante, *in* "The Enzymes of Biological Membranes" (A. Martonosi, ed.), 2nd Ed., p. 1. Plenum, New York, 1985.

[2] C. I. Ragan, *in* "Coenzyme Q" (G. Lenaz, ed.), p. 315. Wiley, Chichester, England, 1985.

[3] Y. Hatefi, this series, Vol. 53, p. 11.

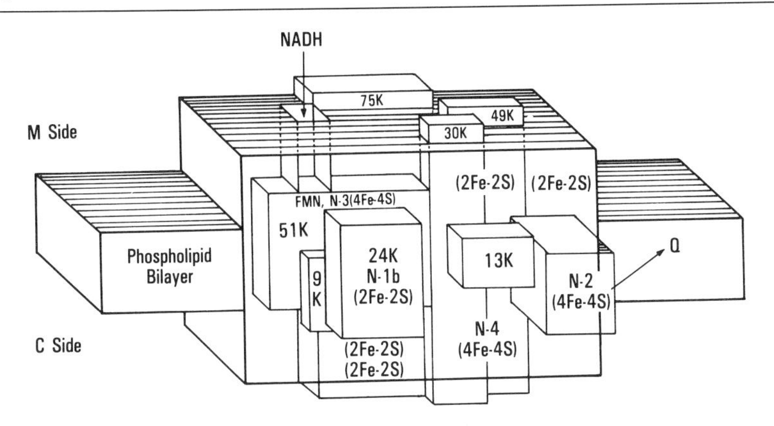

FIG. 1. Schematic representation of the arrangement of Complex I subunits. The 51, 24, and 9 kDa blocks are the polypeptide components of FP, and the 75, 49, 30, and 13 kDa blocks are the polypeptide components of IP. The large block containing the FP and IP polypeptides is the HP fraction. The relative sizes and placement of the protein blocks are not precise. For details, see Hatefi *et al.*[1]

Complex I is structurally composed of three distinct fragments, a water-soluble FeS-flavoprotein (FP), a water-soluble FeS-protein (IP), and a water-insoluble fraction containing phospholipids and hydrophobic polypeptides (HP). FP is composed of three polypeptides, contains FMN and two FeS clusters, and exhibits NADH dehydrogenase activity in the presence of quinones or ferric complexes as electron acceptors.[1-4] IP is composed of 5–6 polypeptides and contains four FeS clusters.[5] HP contains the remaining polypeptides (~16) of Complex I plus two FeS clusters.[6] The HP polypeptides form a shell around FP and IP such that FP is nearly completely and IP is largely shielded from the surroundings. Three polypeptides of IP with M_r of 75, 49, and 30 × 10^3 are transmembranous.[1,2] There is also access for substrates [NAD(H)] from the matrix side of the mitochondrial inner membrane to the largest polypeptide of FP (M_r, 51 × 10^3) which appears to carry the NAD(H) binding site.[7] These structural considerations are depicted in Fig. 1.

The resolution of Complex I into FP, IP, and HP is achieved in the presence of chaotropes, which are also used for further fragmentation of FP and IP. Chaotropes are ions of low charge density, capable of breaking

[4] Y. M. Galante and Y. Hatefi, *Arch. Biochem. Biophys.* **192,** 559 (1979).
[5] C. I. Ragan, Y. M. Galante, and Y. Hatefi, *Biochemistry* **21,** 2518 (1982).
[6] T. Ohnishi, C. I. Ragan, and Y. Hatefi, *J. Biol. Chem.* **260,** 2782 (1985).
[7] S. Chen and R. J. Guillory, *J. Biol. Chem.* **256,** 8318 (1981).

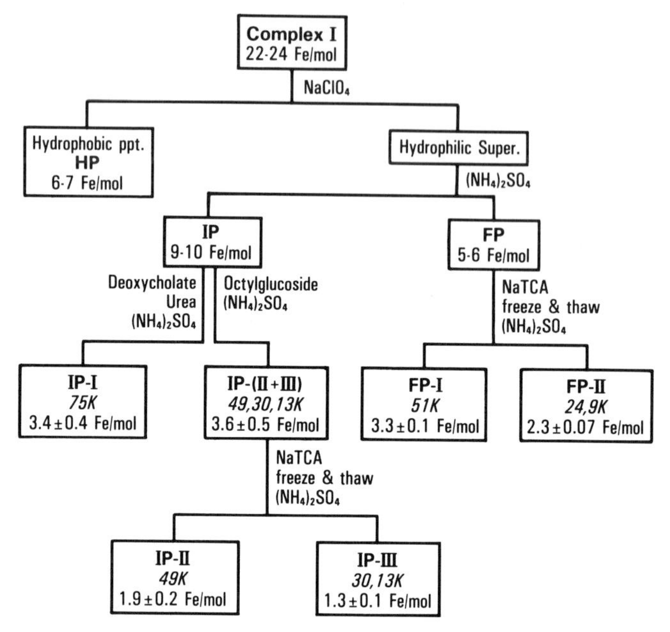

FIG. 2. Flow diagram for the resolution of Complex I, FP, and IP. The iron content of each fraction is shown. From Ohnishi *et al.*[6]

down water structure.[8,9] They destabilize membranes and increase the water solubility of hydrophobic proteins by lowering the transfer entropy of hydrophobic structures from a lipophilic surrounding into the destructured water of the aqueous phase.[8,9]

Resolution of Complex I into FP, IP, and HP

Figure 2 shows a flow diagram for the resolution of Complex I into FP, IP, and HP, and isolation of FeS-containing polypeptides from FP and IP. The procedure for the resolution of Complex I and isolation of FP and IP has been described elsewhere in this series.[10] The greenish gray residue isolated in Step 1 of the above procedure is HP. This residue should be suspended by homogenization in buffer and recentrifuged for removal of trapped FP and IP. HP is particulate and sediments easily at neutral pH, but can be rendered soluble in the presence of detergents at pH 11. Table I provides data regarding the composition of FP, IP, and HP. The catalytic

[8] Y. Hatefi and W. G. Hanstein, *Proc. Natl. Acad. Sci. U.S.A.* **62,** 1129 (1969).
[9] Y. Hatefi and W. G. Hanstein, this series, Vol. 31, p. 770.
[10] Y. M. Galante and Y. Hatefi, this series, Vol. 53, p. 15.

TABLE I
COMPOSITION OF PRINCIPAL FRACTIONS FROM CHAOTROPIC RESOLUTION OF COMPLEX I[a]

Fraction	Polypeptides	FMN[b] (nmol/mg)	Fe[b] (nmol/mg)	Fe/Complex I FMN[c] (mol/mol)
Complex I	~25	0.98	22.1	22.6
HP	~16	0.04	7.2	6.6
IP	6	0.52	48.2	9.3
FP	3	12.5[d]	77.9	6.2

[a] From Ragan *et al.*[5]
[b] Columns under FMN and Fe give measured values.
[c] The last column gives the number of Complex I iron atoms accounted for by each fragment after correction for losses and cross contamination. In all fractions, the acid-labile sulfide (S^{2-}) content was the same as the Fe content.
[d] Average values per milligram of FP are 13.5 nmol FMN, 74 nmol Fe, and 72 nmol S^{2-} (from Galante and Hatefi[4]).

properties of FP and the electron paramagnetic resonance (EPR) characteristics of the FeS clusters of FP, IP, and HP have been reported.[4,6,11]

Resolution of FP

Reagents

Tris–Cl, 50 m*M*, pH 7.8, at 4°
Dithiothreitol (DTT), 1 *M* in water; store at −20°
Saturated, neutralized ammonium sulfate at 4°
Na trichloroacetate (NaTCA), 4 *M* in water
FP, ammonium sulfate pellet; store at −50°

Procedure

FP is dissolved in Tris buffer containing 2 m*M* DTT to give a final protein concentration of 2 to 4 mg/ml. NaTCA is then added to a final concentration of 0.2 *M*, and the solution is frozen in liquid N_2 and immediately thawed at room temperature. The freezing and thawing step is repeated, and the turbid solution cooled on ice. At this stage, resolution is complete as judged by the complete absence of NADH-menadione reductase activity.[11]

The solution is then fractionated by addition of saturated, neutralized ammonium sulfate to 12, 22, and 35% saturation. After each addition of

[11] C. I. Ragan, Y. M. Galante, Y. Hatefi, and T. Ohnishi, *Biochemistry* **21,** 590 (1982).

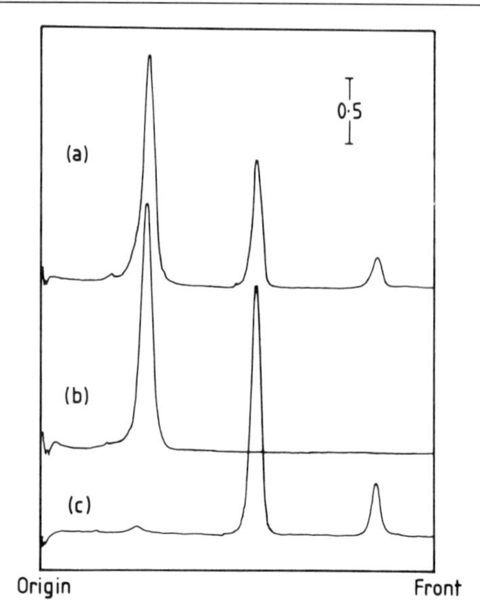

FIG. 3. Subunit composition of FP, FP-I, and FP-II analyzed by SDS–gel electrophoresis. (a), FP; (b), FP-I; (c), FP-II. From C. I. Ragan, Y. M. Galante, Y. Hatefi, and T. Ohnishi, *Biochemistry* **21,** 590 (1982).

ammonium sulfate, the solution is allowed to stand for 10 min and precipitated protein is removed by centrifugation at 90,000 *g* for 5 min. All operations are carried out at 0–4° unless otherwise indicated.

The pellet obtained at 12% saturation contains the largest subunit of FP (FP-I, M_r 51 × 10^3) in ~80% yield. The pellet obtained between 22 and 35% saturation contains the two smaller subunits (FP-II, M_r 24 × 10^3 and 9 × 10^3), again in ~80% yield.

Properties of FP-I

Solubility. FP-I is insoluble at neutral pH. For visible spectroscopy, the protein may be solubilized in Tris-DTT buffer (containing 2 m*M* DTT) adjusted to pH 11 with 1 *M* NaOH. Analyses were carried out on the material freshly homogenized in Tris-DTT buffer, and stability to storage has not been tested.

Homogeneity. FP-I is virtually free from contamination by FP-II (Fig. 3). Absolute purity depends on the homogeneity of the original FP preparation.

Iron, Acid-Labile Sulfide, and Flavin Contents. FP-I contains, on average, 3.3 mol of iron and 3.4 mol of acid-labile sulfide per 51,000 g of

TABLE II
IRON AND ACID-LABILE SULFIDE CONTENTS OF FP-I AND FP-II[a]

Sample	Fe			S^{2-}		
	nmol/mg of protein		mol/mol of protein	nmol/mg of protein		mol/mol of protein
	Range	Av.		Range	Av.	
FP-I	63–67(6)[b]	65	3.32[c]	62–71(4)	66	3.37
FP-II	67–71(4)	69	2.28	60–72(4)	68	2.24

[a] From Ragan *et al.*[11]
[b] Numbers in parentheses show the number of different preparations assayed.
[c] The molecular weights were taken as 51×10^3 for FP-I and 33×10^3 for FP-II.

protein (Table II), but is virtually free from flavin. The dithionite-reduced protein gives a broad, rather symmetric EPR signal centered around $g =$ 1.95, which has been attributed to a modified tetranuclear cluster.[6,11]

Properties of FP-II

Solubility. FP-II dissolves readily in Tris-DTT buffer at neutral pH. Again, all analyses have been carried out on freshly prepared material and storage has not been attempted.

Homogeneity. FP-II is virtually free from FP-I (Fig. 3). From stain intensity on gels, FP-II is slightly deficient in the subunit of M_r 9×10^3 compared with the original FP. Attempts to achieve complete separation of the two subunits of FP-II at higher NaTCA concentrations have not been successful.

Iron, Acid-Labile Sulfide, and Flavin Contents. FP-II contains, on average, 2.3 mol of Fe and 2.2 mol of acid-labile sulfide per 33,000 g of protein (Table II). Flavin is absent since it is quantitatively released from FP during the resolution process. The EPR spectrum of dithionite-reduced FP-II has almost the same line-shape and spin relaxation behavior as the slowly relaxing species in unresolved FP. This has been attributed to the binuclear cluster, N-1b, most probably located on the subunit of M_r 24×10^3. The higher than expected contents of iron and acid-labile sulfide may be attributed to the enrichment of the M_r 24×10^3 subunit in FP-II. This subunit has been sequenced.[13]

[12] H. Von Bahr-Lindstrom, Y. M. Galante, M. Persson, and H. Jornvall, *Eur. J. Biochem.* **134,** 145 (1983).

Resolution of IP

Reagents

Tris–Cl, 50 m*M*, pH 8.0, at 4°
Dithiothreitol (DTT), 1 *M* in water; store at −20°
Na deoxycholate, 10% w/v in water
Saturated, neutralized ammonium sulfate at 4°
Urea
Octylglucoside, 10% w/v in water
Na trichloroacetate (NaTCA), 4 *M* in water
IP, ammonium sulfate pellet; store at −50°

Procedure for Purification of IP-I

IP is dissolved in Tris buffer containing 2 m*M* DTT to give a final protein concentration of ~10 mg/ml and centrifuged at 90,000 *g* for 15 min to remove any insoluble material. The supernatant is diluted with Tris-DTT buffer to a final concentration of 2 to 3 mg/ml and Na deoxycholate is added to give a final concentration of 0.3% (w/v). The solution is then fractionated by addition of saturated, neutralized ammonium sulfate to 4% and 7.25% saturation. After addition of ammonium sulfate, the solution is allowed to stand for 10 min and precipitated protein is removed by centrifugation at 40,000 *g* for 15 min.

The protein precipitated between 4 and 7.25% saturation is dissolved in Tris-DTT buffer to give a clear solution to which solid urea is added to a final concentration of 2 *M* and a final volume one-fifth of the original volume of IP. Na deoxycholate is added to 0.3% (w/v) and ammonium sulfate fractionation is carried out as before, but at 5.25% and 8.25% saturation. The material precipitated between 5.25 and 8.25% (IP-I) is dispersed in Tris-DTT buffer to give a somewhat turbid solution which can be clarified by brief exposure to ultrasonic irradiation. All procedures are carried out at 0–4°.

Properties of IP-I

Solubility. IP-I is sparingly soluble at neutral pH. At higher pH, e.g., pH 11, it dissolves, but the iron-sulfur chromophore is rather labile. For visible spectroscopy, it is easier to use the somewhat less pure IP-I obtained from the first fractionation with ammonium sulfate, since it is readily soluble at neutral pH.

Homogeneity. IP-I consists predominantly (80%) of the subunit of M_r 75×10^3 (Fig. 4). The major impurities are subunits of the other iron-

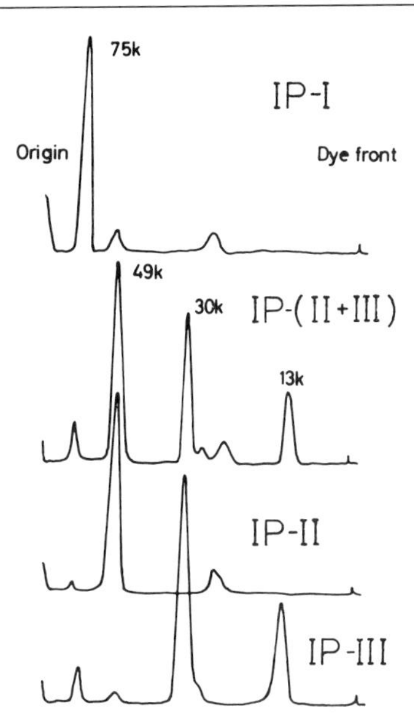

FIG. 4. Subunit composition of fractions from resolution of IP. Molecular weights in thousands are indicated. Modified from Ragan *et al.*[5]

containing proteins of IP (see below). Separation of IP-I from other subunits of IP can be achieved to varying extents by ammonium sulfate fractionation in the presence of cholate, deoxycholate, Triton X-100, or octylglucoside, but deoxycholate gives the cleanest separations.

Iron and Acid-Labile Sulfide Contents. IP-I contains, on average, 3.4 mol of iron and 3.2 mol of acid-labile sulfide per 75,000 g of protein (Table III). The dithionite-reduced protein gives an EPR spectrum similar to that of typical binuclear clusters around which protein constraint has been relaxed. The slow relaxation behavior is also characteristic of binuclear clusters. This, together with the iron content of IP-I, has given rise to the suggestion that this subunit contains two binuclear clusters.[3]

Procedure for Purification of IP-(II + III)

IP is dissolved in Tris-DTT buffer, centrifuged, and diluted to 2–3 mg of protein per milliliter as described for the purification of IP-I. Octylglucoside is added to a final concentration of 1% (w/v), followed by ammonium sulfate to 35% saturation. The precipitated material is collected by centrifugation at 40,000 *g* for 15 min and thoroughly homogenized in one-

TABLE III
IRON AND ACID-LABILE SULFIDE CONTENTS OF IP FRACTIONS[a]

	Fe		S^{2-}	
Sample	nmol/mg of protein	mol/mol of protein	nmol/mg of protein	mol/mol of protein
IP-I	45 ± 5(4)[b]	3.4[c]	43 ± 3(4)	3.2
IP-(II + III)	39 ± 5(7)	3.6	36 ± 4(4)	3.3
IP-II	38 ± 4(4)	1.9	35 ± 3(3)	1.7
IP-III	29 ± 2(5)	1.25	30 ± 3(3)	1.3

[a] From Ragan *et al.*[5]
[b] Figures are given as the mean ± standard error. Numbers in parentheses show the number of different preparations assayed.
[c] The molecular weights were taken as 75 × 10^3 (IP-I), 92 × 10^3 [IP-(II + III)], 49 × 10^3 (IP-II), and 43 × 10^3 (IP-III).

sixth the original volume of Tris-DTT buffer. After standing for 30 min the suspension is centrifuged at 40,000 *g* for 30 min and the supernatant, containing IP-(II + III), is removed. All procedures are carried out at 0–4°.

Properties of IP-(II + III)

Solubility. IP-(II + III) is completely soluble while other IP subunits are insoluble after precipitation by ammonium sulfate. This property has been exploited to provide a one-step purification which is considerably simpler than fractionation of IP with ammonium sulfate.

Homogeneity. IP-(II + III) consists mainly of the subunits of M_r 49 × 10^3, 30 × 10^3, and 13 × 10^3 present in a 1 : 1 : 1 molar ratio (Fig. 4). Small amounts of IP-I and impurities in the original IP constitute ~16% of the total protein.

Iron and Acid-Labile Sulfide Contents. IP-(II + III) contains, on average, 3.6 mol of iron and 3.3 mol of acid-labile sulfide per 92,000 g of protein (Table III). EPR spectroscopy of the dithionite-reduced protein reveals two species of iron-sulfur cluster differing in both line-shape and relaxation behavior. These have been attributed to a binuclear and a tetranuclear cluster, suggesting that IP-(II + III) becomes iron deficient during fractionation. IP-(II + III) of similar polypeptide purity can be isolated from the supernatant obtained following precipitation of IP-I. However, the specific content of iron and acid-labile sulfide is even lower in this preparation.

Procedure for Separation of IP-II from IP-III

To the solution of IP-(II + III) obtained as described above, 4 *M* NaTCA is added to a final concentration of 1 *M*. The solution is frozen in liquid N_2 and immediately thawed at room temperature. The turbid solution is cooled on ice and centrifuged at 40,000 *g* for 15 min. The precipitate is IP-II. The supernatant, containing IP-III, may be used as such or IP-III can be precipitated by addition of ammonium sulfate to 30% saturation. Unless otherwise indicated, all procedures are carried out at 0–4°.

Properties of IP-II and IP-III

Solubility. IP-II and IP-III are insoluble at neutral pH, but may be dissolved at pH 11. However, IP-III is soluble in 1 *M* NaTCA prior to ammonium sulfate precipitation.

Homogeneity. IP-II contains the subunit of M_r 49 × 10^3 (Fig. 4). The major contaminants, amounting to ~20% of the total protein, are IP-I and impurities in the original IP. IP-III contains the subunits of M_r 30 × 10^3 and 13 × 10^3 present in a 1 : 1 molar ratio (Fig. 4). Impurities, amounting to 10% of the total, are IP-I and lesser amounts of IP-II.

Iron and Acid-Labile Sulfide Contents. IP-II and IP-III contain, on average, 1.8 and 1.3 mol of iron or acid-labile sulfide per mol of protein, respectively (Table III). The sum of these values is close to that obtained from IP-(II + III), but it is not obvious which fraction contains the binuclear cluster and which contains the tetranuclear cluster. EPR spectroscopy of the dithionite-reduced proteins reveals, in both IP-II and IP-III, broad, rather symmetric signals centered at $g = 1.95$, from which it is not possible to deduce cluster structures.[6]

General Comments on the Resolution of FP and IP

The procedures described above have usually been carried out on samples containing between 4 and 10 mg of protein. The scale is necessitated by the small amounts of the starting materials generally available and has required the development of procedures requiring very simple manipulations. Improvements in the resolution of IP are clearly required, as none of the resolved proteins is completely pure. The procedures have been optimized, however, for pH, the detergents used, and the concentrations of detergent, NaTCA, urea, and ammonium sulfate. Further improvements are therefore likely to require completely new methods.

[35] Monoclonal Affinity Purification of D-Lactate Dehydrogenase from *Escherichia coli*

By EUGENIO SANTOS and H. RONALD KABACK

Active transport of many different solutes by right-side-out cytoplasmic membrane vesicles from various bacteria is driven by a proton electrochemical gradient ($\Delta\bar{\mu}_{H^+}$, interior negative and alkaline) generated by means of substrate oxidation via a membrane-bound respiratory chain.[1-3] Although vesicles have the capacity to oxidize a variety of substrates, generation of $\Delta\bar{\mu}_{H^+}$ is relatively specific for certain electron donors. Thus, D-lactate is the most effective physiological electron donor for generating $\Delta\bar{\mu}_{H^+}$ in *Escherichia coli* ML 308-225 right-side-out membrane vesicles, even though its rate of oxidation is slower than that of other electron donors, such as NADH or succinate.[4-6]

Escherichia coli membrane vesicles containing D-lactate dehydrogenase (D-LDH) catalyze the stoichiometric conversion of D-lactate to pyruvate,[7,8] and electrons derived from the reaction are transferred to oxygen through membrane-bound respiratory intermediates. Concomitant with electron flow, a transmembrane $\Delta\bar{\mu}_{H^+}$ is generated by a mechanism(s) that is (are) not completely understood.[3] In any case, D-LDH is readily solubilized from the membrane with chaotropic agents or nonionic detergents and readheres to the membrane in a functional manner upon dilution of the solubilizing agent.[9-12] D-LDH has been purified to homogeneity[13-16]; it exhibits an M_r of about 65 kDa on sodium dodecyl sulfate–

[1] H. R. Kaback, *J. Cell. Physiol.* **89,** 575 (1976).

[2] H. R. Kaback, *J. Membr. Biol.* **76,** 95 (1983).

[3] H. R. Kaback, *in* "Physiology of Membrane Disorders" (T. E. Andreoli, J. F. Hoffman, D. D. Fanestil, and S. G. Schultz, eds.), p. 387. Plenum, New York, 1985.

[4] E. M. Barnes and H. R. Kaback, *Proc. Natl. Acad. Sci. U.S.A.* **66,** 1190 (1970).

[5] S. Schuldiner and H. R. Kaback, *Biochemistry* **14,** 5451 (1975).

[6] P. Stroobant and H. R. Kaback, *Proc. Natl. Acad. Sci. U.S.A.* **72,** 3970 (1975).

[7] H. R. Kaback and L. S. Milner, *Proc. Natl. Acad. Sci. U.S.A.* **66,** 1108 (1970).

[8] E. M. Barnes and H. R. Kaback, *Proc. Natl. Acad. Sci. U.S.A.* **66,** 1190 (1970).

[9] J. P. Reeves, J. S. Hong, and H. R. Kaback, *Proc. Natl. Acad. Sci. U.S.A.* **71,** 1917 (1973).

[10] S. A. Short, H. R. Kaback, and L. Kohn, *Proc. Natl. Acad. Sci. U.S.A.* **71,** 1461 (1974).

[11] S. A. Short, H. R. Kaback, and L. D. Kohn, *J. Biol. Chem.* **250,** 4291 (1975).

[12] K. Haldar, P. J. Olsiewski, C. Walsh, G. J. Kaczorowski, A. Bhaduri, and H. R. Kaback, *Biochemistry* **21,** 4590 (1982).

[13] L. D. Kohn and H. R. Kaback, *J. Biol. Chem.* **248,** 7012 (1973).

[14] G. Kaczorowski, L. D. Kohn, and H. R. Kaback, this series, Vol. 53, p. 519.

[15] M. Futai, *Biochemistry* **12,** 2468 (1973).

[16] E. A. Pratt, L. W. M. Fung, J. A. Flowers, and C. Ho, *Biochemistry* **18,** 312 (1979).

METHODS IN ENZYMOLOGY, VOL. 126

polyacrylamide gel electrophoresis (SDS–PAGE) and is composed of a single polypeptide chain containing 1 mol of tightly bound flavin adenine dinucleotide per mol of protein. D-LDH is highly specific for D-(−)-lactic acid and is inactivated by 2-hydroxy-3-butynoic acid, which acts as a "suicide substrate."[17]

The gene encoding D-LDH has been cloned,[18] allowing amplification of the enzyme, and the amino acid sequence of the protein has been deduced from DNA sequencing.[19,20] Furthermore, studies on the *in vitro* transcription and translation of *dld* show that the enzyme is synthesized in mature form and binds to the membrane without a leader sequence.[21]

Polyclonal antibodies directed against purified D-LDH have been prepared and characterized.[22] By utilizing these antibodies in inactivation[11] and immunoadsorption experiments,[23,24] it has been demonstrated that D-LDH is associated with the cytoplasmic surface of the bacterial plasma membrane. More recently, monoclonal antibodies (Mab) against D-LDH have been described.[25] One of the Mabs (1B2a) binds exclusively to the intact D-LDH molecule and does not bind to proteolytic fragments from the enzyme.

In this chapter, a Mab immunoaffinity chromatographic procedure is described that allows purification of D-LDH to apparent homogeneity in a single step.

Experimental Procedures

D-*LDH Assay*

Oxidation of D-lactate by D-LDH is monitored by following the reduction of 2,6-dichlorophenolindophenol (DCIP) at 600 nm in the presence of phenazine methosulfate (PMS).[14] The typical assay is performed in a final volume of 1.0 ml containing 100 μM DCIP, 250 μM PMS, 20 mM lithium D-lactate, and an appropriate amount of the sample to be assayed (pH is adjusted to 7.8 with phosphate or Tris buffers). The reaction is carried out at room temperature in a 1 × 1 cm cuvette. The enzyme sample is added to the assay mixture (minus D-lactate), and a base line is

[17] C. T. Walsh, R. H. Abeles, and H. R. Kaback, *J. Biol. Chem.* **247,** 7858 (1972).
[18] I. G. Young, A. Jaworowski, and M. Poulis, *Biochemistry* **21,** 2092 (1982).
[19] H. D. Campbell, D. L. Rogers, and I. G. Young, *Eur. J. Biochem.*, in press (1985).
[20] G. S. Rule, E. A. Pratt, C. C. Q. Chin, F. Wold, and C. Ho, *J. Bacteriol.*, in press (1985).
[21] E. Santos, H. Kung, I. G. Young, and H. R. Kaback, *Biochemistry* **21,** 2085 (1982).
[22] S. A. Short, H. R. Kaback, T. Hawkins, and L. D. Kohn, *J. Biol. Chem.* **250,** 4285 (1975).
[23] P. Owen and H. R. Kaback, *Biochemistry* **18,** 1413 (1979).
[24] P. Owen and H. R. Kaback, *Biochemistry* **18,** 1422 (1979).
[25] E. Santos, S. M. Tahara, and H. R. Kaback, *Biochemistry* **24,** 3006 (1985).

recorded. The reaction is then initiated by addition of D-lactate with rapid mixing. Specific activity of D-LDH is expressed as nmol DCIP reduced/min/mg protein. The molar extinction coefficient of DCIP under these conditions is 16.5×10^3 cm^{-1} M^{-1}.

Protein Assays

For membrane vesicles or extracts thereof containing D-LDH, the method described by Lowry *et al.*[26] is used with crystalline bovine serum albumin as standard. For IgG, absorbance at 280 nm is used. A 1 mg/ml solution of IgG at neutral pH exhibits an absorbance of 1.46 at 280 nm over a light path of 1 cm. For dilute samples, such as the eluates from immunoaffinity columns, the method of Shaffner and Weissmann[27] is used.

Growth of Hybridoma Cell Line 1B2a

1B2a hybridoma cells are grown in RPMI 1640 medium (Gibco 320-1875) supplemented with 15% heat-inactivated fetal bovine serum (Gibco 200-6140), 1% glutamine (Gibco 320-5030), 1% pyruvate (Gibco 320-1360), 1% penicillin-streptomycin (Gibco 600-5140), and 50 μM 2-mercaptoethanol (Sigma).[25] Frozen stock cultures of the cells (10^7 cells in 1 ml of complete medium supplemented with 15% dimethyl sulfoxide) are thawed quickly at 42°, washed at 37° in complete medium (minus dimethyl sulfoxide), and seeded in a T25 tissue culture flask containing 4 ml of complete medium with a feeder layer of 10^4 mouse macrophages (obtained by peritoneal washing of BALB/c mice). Cultures are incubated in a tissue culture incubator set at 37° with 85–90% relative humidity and 10% CO_2. As the number of cells increases, they are transferred successively to T75 and T175 tissue culture flasks.

Ascites Fluid

BALB/c mice are primed by intraperitoneal inoculation of 0.5 ml of 2,6,10,14-tetramethylpentadecane (pristane) (Aldrich). Hybridoma cells [$1–5 \times 10^7$ in 1 ml of phosphate-buffered saline (PBS)] are injected 15–30 days later.[25] Ascites fluid is collected after 10 days with an 18-gauge needle and every third or fourth day thereafter until the animals expire. A

[26] O. H. Lowry, N. J. Rosebrough, A. L. Farr, and R. J. Randall, *J. Biol. Chem.* **193,** 265 (1951).

[27] W. Schaffner and G. Weissman, *Anal. Biochem.* **56,** 502 (1973).

total volume of 10–15 ml (20–30 mg of protein/ml) of fluid is collected from each animal. The ascites fluid is clarified by low-speed centrifugation (1000 *g* for 10 min) and stored at −20° until use.

Purification of IgG

IgG produced by 1B2a hybridoma cells is purified by either of two procedures, according to need.

Small-Scale Purification. Highly purified IgG is obtained from small volumes of ascites fluid by affinity chromatography on protein A-Sepharose 4B.[28] Essentially, 2 ml of 0.1 *M* sodium phosphate (pH 8.0) are added to 4 ml of ascites fluid, and the pH is adjusted to pH 8.1 by adding a few drops of 1 *M* Tris–HCl (pH 9.0). A column [1.5 × 15 cm (Pharmacia) or a 10-ml glass pipette] is packed to a bed volume of ~6 ml with protein A-Sepharose 4B (Pharmacia), equilibrated in 0.1 *M* sodium phosphate (pH 8.0) and washed with 30 ml of the same buffer. The sample is applied to the column, and the column is then washed with equilibration buffer until the absorbance of the eluant at 280 nm is zero (about five times the bed volume). Bound IgG is then eluted sequentially with 30 ml each of 0.1 *M* sodium citrate at pH 6.0, 5.5, 4.5, and 3.5 in order to elute IgG_1, IgG_{2a}, and IgG_{2b} isotypes, respectively. Mab 1B2a is an IgG_{2b} isotype and elutes at pH 3.5.[25] This fraction is dialyzed against 100 volumes of PBS (pH 7.2) and stored at −70° after adjusting the protein concentration to 3–5 mg/ml.

Large-Scale Purification. Clarified ascites fluid is adjusted to an absorbance of 14 at 280 nm (~10 mg of protein/ml) with PBS, and ammonium sulfate is added to 50% saturation (31.3 g/100 ml). The mixture is kept in the cold for a few hours with occasional stirring. The precipitate is collected by centrifugation and washed once with 1.75 *M* ammonium sulfate and dissolved in 10 m*M* sodium phosphate (pH 6.8) to a volume equal to that of the original ascites sample prior to precipitation. The solution is then dialyzed against 100 volumes of 10 m*M* sodium phosphate (pH 6.8) with two or three changes at 4°. The sample is applied to a column of DEAE-cellulose (DE-52; Whatman) previously equilibrated with the same buffer. The bed volume of the DEAE-cellulose column is at least as great as the volume of the original ascites sample. After sample application, the column is washed with equilibration buffer, and IgG which does not bind to the resin washes through the column and is collected. The pooled eluate is adjusted to a desired protein concentration with 10 m*M* sodium phosphate (pH 6.8) and stored at −70°.

[28] P. D. Ey, S. J. Prowse, and C. R. Jenkins, *Immunochemistry* **15,** 429 (1979).

Preparation of Immunoaffinity Resin

Affi-Gel 10 (Bio-Rad) is used as the gel matrix for immunoadsorbent chromatography.[25] Prior to use, the commercial gel slurry is washed three times with one volume of isopropanol and three times with cold water (on a sintered glass filter); fines are removed after each wash. The gel slurry is then transferred to 50-ml polypropylene Falcon tubes, and an equal volume of a solution of Mab 1B2a (20 mg/ml of protein in 0.2 *M* $NaHCO_3$/0.2 *M* NaCl, pH 8.0) is added. Coupling is achieved by gentle mixing on a rocker table at 4° overnight. The suspension is washed thoroughly with 0.1 *M* $NaHCO_3$ containing 0.1 *M* NaCl (pH 8.0) until the absorbance of the wash solution at 280 nm is zero. Unreacted sites are blocked by incubating the gel with one volume of 0.1 *M* ethanolamine–HCl (pH 8.0) for 60 min at room temperature with gentle stirring. The gel is then washed with an excess of 50 m*M* PBS containing 0.1% Triton X-100 and 0.01% sodium azide (pH 7.4). As described, the affinity gel bears about 15 mg of IgG/ml of wet gel matrix and can be stored for several months in the cold without detectable loss of affinity.

Enzyme Purification

Bacterial Growth

Escherichia coli ML 308-225 is grown aerobically at 37° in minimal medium with 1% sodium succinate (hexahydrate) as a carbon source.[13] Cells are grown until late exponential phase ($A_{560} \simeq 2.0$) and harvested by centrifugation. Alternatively, *E. coli* IY83, an overproducer strain carrying the *dld* gene on a multicopy recombinant plasmid,[18] may be used.

Preparation of Membranes

Plasma membranes contain all of the D-LDH activity in *E. coli* and are used as starting material for purification of the enzyme. Right-side-out[11,29] or inside-out[30] membrane vesicles or the membrane fraction obtained by passing cells through a French pressure cell at 20,000 psi[13] may be used.

Solubilization of D-*LDH*

Membranes are resuspended at a concentration of 8 mg of protein/ml in PBS (pH 7.4) containing 0.1% Triton X-100 and 0.5 μM phenylmeth-

[29] H. R. Kaback, this series, Vol. 22, p. 99.

[30] W. W. Reenstra, L. Patel, H. Rottenberg, and H. R. Kaback, *Biochemistry* **19,** 1 (1980).

PURIFICATION OF D-LDH BY IMMUNOAFFINITY CHROMATOGRAPHY[a]

Fraction	Protein (mg)	Specific activity (nmol DCIP reduced/min/ mg/protein)	Total activity (nmol DCIP reduced/min)	Purification factor	Yield of enzyme activity (%)
Membrane vesicles	64(100%)	107	6848	1	100
Triton X-100 extract	8.2(12.8%)	638	5232	6	76
Fractions 40–55 (Fig. 1)	0.118(0.18%)	24,600	2903	230	42

[a] Data from Santos *et al.*[25]

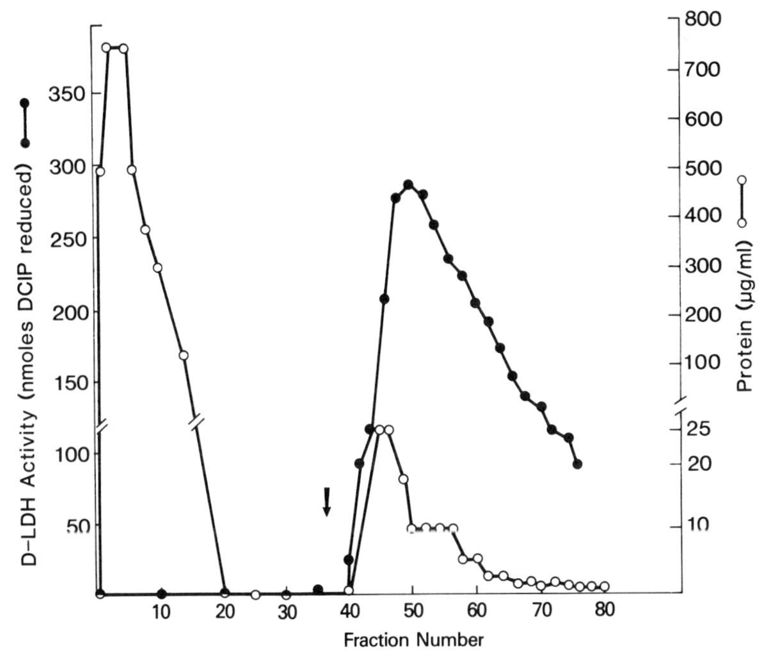

FIG. 1. Immunoaffinity chromatography of D-LDH. Right-side-out membrane vesicles from *E. coli* ML 308-225 (64 mg of protein) were resuspended in 8.0 ml of PBS/0.1% Triton X-100/0.5 μM phenylmethylsulfonyl fluoride (pH 7.4), and incubated for 1 hr at 0°. The suspension was then centrifuged in a 60Ti Beckman rotor for 60 min at 45,000 rpm. The supernatant (~8 ml) was carefully aspirated and applied to the top of a 1B2a/Affi-Gel 10 column (8.0 ml, bed volume), prepared as described. Unadsorbed material was eluted with five volumes of PBS/0.1% Triton X-100 (pH 7.4). The arrow indicates where elution with 0.1 M ethanolamine/1.1 M guanidine/0.1% Triton X-100 (pH 11.0) was begun. The fractions (1.4 ml) were immediately neutralized by adding 0.7 ml of 1.0 M Tris–HCl (pH 7.0). Protein content (○) and D-lactate : DCIP reductase activity (●) were then determined. From Santos *et al.*[25]

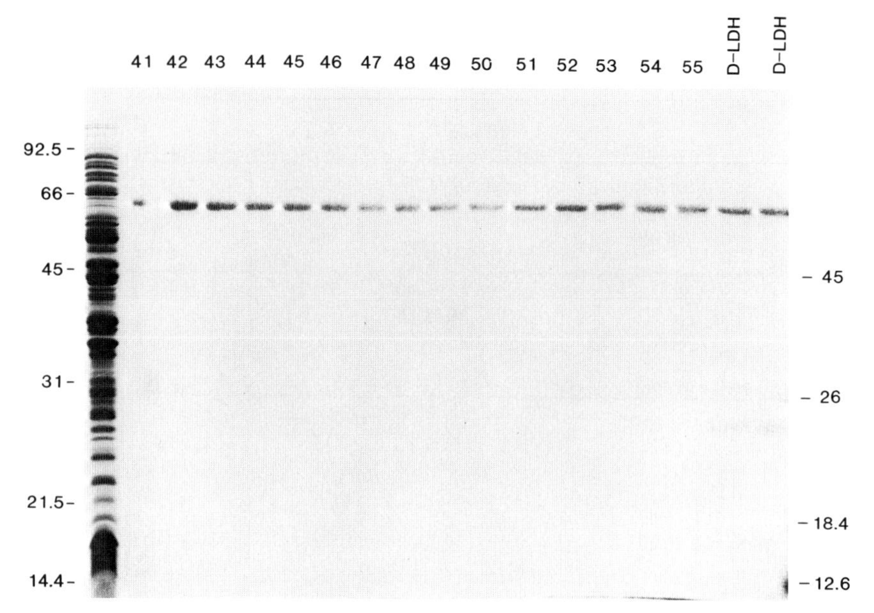

FIG. 2. SDS-PAGE of D-LDH purified by immunoaffinity chromatography. Aliquots (75 μl) of neutralized immunoaffinity chromatography fractions 41-55 (cf. Fig. 1) were subjected to SDS-PAGE with molecular weight standards (indicated on left side), 15-20 μg of protein from the original Triton X-100 extract of ML 308-225 membrane vesicles (first lane on left), and 1 μg of purified D-LDH (last two lanes on right). The gel was stained with silver. From Santos *et al.*[25]

ylsulfonyl fluoride (PMSF). The mixture is incubated at 4° for 1 hr with gentle agitation on a magnetic stirrer. Solubilized D-LDH is then separated from the debris by centrifugation at 4° in a Beckman 60Ti rotor for 60 min at 45,000 rpm. The supernatant, containing solubilized D-LDH, is used for immunoaffinity chromatography.

Immunoaffinity Chromatography

A column of appropriate dimensions is packed with Mab 1B2a immunoaffinity gel prepared as described and washed with five times the bed volume of PBS (pH 7.4) containing 0.1% Triton X-100. The Triton X-100 extract containing D-LDH is applied to the column in a volume equal to that of the column bed. After the extract has entered the column, the outlet is closed, and D-LDH (antigen) is allowed to interact with the affinity matrix for 2 hr at 4° (alternatively, the extract is incubated with the immunoaffinity resin in batch prior to pouring the column). Unadsorbed material (~95% of the total protein in the extract) is eluted from the column by washing with five volumes of PBS (pH 7.4) containing 0.1% Triton X-100. Adsorbed D-LDH is then eluted by washing with 5–10 volumes of 0.1 *M* ethanolamine containing 1.1 *M* guanidinium hydrochloride and 0.1% Triton X-100 (pH 11.0). Fractions (1.4 ml) eluted under these conditions are neutralized immediately by adding a half volume of 1.0 *M* Tris–HCl (pH 7.0) and assayed for D-lactate : DCIP reductase activity.

A typical purification of D-LDH from right-side-out membrane vesicles of *E. coli* ML 308-225 is summarized in the table. On extraction of vesicles with 0.1% Triton X-100, about 13% of the membrane protein containing about 80% of the D-lactate : DCIP reductase activity is solubilized from the membrane, and the specific activity of D-LDH increases about 6-fold in the supernatant. The extract is then applied to a 1B2a immunoaffinity column as described, and the effluent analyzed for protein and D-LDH activity (Fig. 1). During elution with PBS/0.1% Triton X-100, a major peak containing about 95% of the protein, but no D-LDH, emerges. When the column is then washed with 0.1 *M* ethanolamine/1.1 *M* guanidine/0.1% Triton X-100 at pH 11.0, ~0.2% of the original protein and over 50% of the D-LDH activity applied to the column is eluted in a single peak (fractions 40–55), and the specific activity of D-LDH in these fractions is over 200-fold higher than that of the original membrane vesicles (see the table). Furthermore, when the peak fractions from the column are subjected to SDS–PAGE, followed by silver staining, it is apparent that they contain a single major component that comigrates with authentic D-LDH (Fig. 2). The procedure described allows purification of the enzyme to apparent homogeneity in a single chromatographic step with an overall yield of over 40%.

[36] Fumarate Reductase of *Escherichia coli*

By BERNARD D. LEMIRE and JOEL H. WEINER

Introduction[1]

When the facultative anaerobe *Escherichia coli* is grown anaerobically on a glycerol-fumarate medium, a very simple electron-transport chain consisting of the anaerobic glycerol-3-phosphate dehydrogenase, a *b*-type

[1] This work was supported by grant MT 5838 from the Medical Research Council of Canada. B.D.L. was supported by a studentship from the Alberta Heritage Foundation for Medical Research.

cytochrome, and fumarate reductase is induced in the cytoplasmic membrane.[2] This chain allows the organism to grow in the absence of oxygen on a nonfermentable carbon source. Fumarate serves as the terminal electron acceptor and is reduced to succinate by fumarate reductase. We have shown that fumarate reductase is an intrinsic membrane enzyme with the catalytic site exposed to the cytoplasm.[3] The enzyme is composed of four nonidentical subunits of 69,000, 27,000, 15,000, and 13,000 Da in an equal molar ratio.[4] The four subunits comprise two major domains: the 69,000 Da covalent flavin-containing subunit[5] and the 27,000 Da nonheme iron-containing subunit[6] form the catalytic domain, while the 15,000 Da and 13,000 Da subunits form the anchor domain. A dimer composed of the two larger subunits is capable of reducing fumarate to succinate when provided with reducing equivalents from an artificial electron donor, such as benzyl viologen, and can be isolated in a highly purified, nearly detergent-free, soluble form.[7] This dimer differs from the membrane-associated tetramer in that it requires anions and dithiothreitol for optimal activity and stability. The holoenzyme form of fumarate reductase requires detergent for solubility and has two additional very hydrophobic subunits of 15,000 and 13,000 Da. These latter two subunits are most likely transmembranal and serve to anchor the catalytic dimer to the membrane.[3,4] They also stabilize the catalytic domain and eliminate the dependence on anions.

The *frd* operon located at 94 min on the circular *E. coli* chromosome has been cloned into multicopy plasmids[4,8] and bacteriophage λ.[9] The DNA sequence of the *frd* operon has been determined by Cole[6,10] and shown to code for four polypeptides of 66,000, 27,000, 15,000, and 13,000 Da, termed the *frdA*, *frdB*, *frdC*, and *frdD* gene products. The molecular weights of these gene products are in good agreement with polypeptide molecular weights determined by SDS–polyacrylamide gel electrophoresis.

Anaerobic growth of bacteria carrying a multicopy recombinant plas-

[2] B. A. Haddock and C. W. Jones, *Bacteriol. Rev.* **41,** 47 (1977).

[3] J. H. Weiner, B. D. Lemire, R. W. Jones, W. F. Anderson, and D. G. Scraba, *J. Cell Biochem.* **24,** 207 (1984).

[4] B. D. Lemire, J. J. Robinson, and J. H. Weiner, *J. Bacteriol.* **152,** 1126 (1982).

[5] J. H. Weiner and P. Dickie, *J. Biol. Chem.* **254,** 8590 (1979).

[6] S. T. Cole, T. Grundstrom, B. Jaurin, J. J. Robinson, and J. H. Weiner, *Eur. J. Biochem.* **126,** 211 (1982).

[7] J. J. Robinson and J. H. Weiner, *Can. J. Biochem.* **60,** 811 (1982).

[8] E. Lohmeier, D. S. Hagen, P. Dickie, and J. H. Weiner, *Can. J. Biochem.* **59,** 158 (1981).

[9] S. T. Cole and J. R. Guest, *Eur. J. Biochem.* **102,** 65 (1979).

[10] S. T. Cole, *Eur. J. Biochem.* **122,** 479 (1982).

mid with the *frd* operon results in a 30- to 40-fold amplification of enzyme levels.[11] This activity is localized in the cytoplasmic membrane and in novel lipid–protein tubular organelles which grow from the cytoplasmic membrane.[12] These tubules reach lengths of several microns and are composed of a helical array of fumarate reductase tetramers embedded in a lipid matrix. The catalytic dimer domain forms a 5.0 nm knob and stalk structure on the surface of the tubule.[3,11,12]

In this chapter we describe simple purification protocols for isolation of both the catalytic dimer and holoenzyme forms of fumarate reductase as well as a method for preparation of a membrane fraction enriched in inner membranes and fumarate reductase tubules.

Bacterial Strain and Growth Conditions

Escherichia coli HB101 (F^-, *hsdR, hsdM, pro, leu, gal, lac, thi, recA, rpsL*) carrying the plasmid pFRD63 (HB101/pFRD63), which encodes the entire *frd* operon, was used in all experiments.[4] This plasmid is available from the authors. Strains harboring this plasmid overproduce fumarate reductase 30- to 40-fold. Inocula for large-scale cell growth are prepared by growing the organisms aerobically in L broth[13] containing 100 μg/ml ampicillin for 12 hr on a gyratory shaker at 37°. Cells for enzyme isolation are grown anaerobically on a glycerol-fumarate medium[14] containing 100 μg/ml ampicillin to late stationary phase (O.D. 600 = 1.0). This takes about 60–72 hr for a 1% inoculum, at 37°, in a Chemap 300-liter fermentor. Cells are harvested by centrifugation in a CEPA continuous-flow centrifuge. Cells can also be grown in 20-liter plastic carboys filled to the top yielding essentially similar results. Cells can be stored frozen at −70° for at least 1 year.

The glycerol-fumarate medium contains the following per liter: KH_2PO_4, 5.44 g; K_2HPO_4, 10.49 g; $(NH_4)_2SO_4$, 2 g; $MgSO_4 \cdot 7H_2O$, 0.05 g; $MnSO_4 \cdot 4H_2O$, 5 mg; $FeSO_4 \cdot 7H_2O$, 0.925 mg; $CaCl_2$, 0.5 mg; glycerol, 0.04 *M*; sodium fumarate, 0.04 *M*.

The salts plus sodium fumarate, the metals, and glycerol are autoclaved separately and combined after cooling.

[11] B. D. Lemire, J. J. Robinson, R. D. Bradley, D. G. Scraba, and J. H. Weiner, *J. Bacteriol.* **155,** 391 (1983).

[12] J. H. Weiner, B. D. Lemire, M. L. Elmes, R. D. Bradley, and D. G. Scraba, *J. Bacteriol.* **158,** 590 (1984).

[13] J. H. Miller, *in* "Experiments in Molecular Genetics." Cold Spring Harbor Laboratory, Cold Spring Harbor, New York, 1974.

[14] M. E. Spencer and J. R. Guest, *J. Bacteriol.* **117,** 954 (1974).

Assay for Fumarate Reductase

Solutions

1. Sodium phosphate, 200 m*M* (pH 6.8); dithiothreitol, 0.5 m*M*
2. Benzyl viologen, 2.5 mg/ml, in H_2O
3. Sodium dithionite, 20 m*M*, in 200 m*M* sodium phosphate, pH 6.8. The solid dithionite is stored desiccated under vacuum until used. Solutions should be prepared fresh hourly and stored in closed containers with very little air space
4. Sodium fumarate, 500 m*M* (pH 7.0)

Fumarate reductase activity is followed by measuring the initial rate of fumarate-dependent oxidation of reduced benzyl viologen at 570 nm (molar extinction coefficient = 7800).[15] The assay is carried out in open 3-ml plastic cuvettes at 24°. Control assays performed under nitrogen atmosphere in Thunberg cuvettes give essentially identical results. The standard assay mixture contains 2.5 ml of Solution 1, 0.1 ml of Solution 2, 0.09 ml of Solution 3 and enzyme (1–50 μl). All reagents except the fumarate are added to the cuvette, which is gently inverted twice. Upon addition of 0.1 ml of Solution 4, the decrease in absorbance is followed as a function of time. In the absence of fumarate, the dithionite should be sufficient to maintain a constant absorbance for 3–4 min. Activities are expressed as micromoles of reduced benzyl viologen oxidized per minute. This assay can be reliably performed on crude cell lysates without significant interference.

Comments

Maximal reductase activities are attained at pH 6.6–6.8. Difficulties with rapid air oxidation of the reduced benzyl viologen are experienced at higher pH. This can be minimized by using an alternate donor, methyl viologen, and by performing the assays under N_2 atmosphere in degassed buffers. Fumarate reductase can also catalyze the oxidation of succinate to fumarate at about 3% the rate of the reverse reaction. The reaction can be followed spectrophotometrically by monitoring the succinate-dependent phenazine methosulfate coupled reduction of 3-(4,5-dimethylthiazolyl-2-)-2,5-diphenyltetrazolium bromide at 570 nm or as the reduction of ferricyanide at 420 nm.[16]

[15] M. E. Spencer and J. R. Guest, *J. Bacteriol.* **114,** 563 (1973).
[16] P. Dickie and J. H. Weiner, *Can. J. Biochem.* **57,** 813 (1979).

Purification of Fumarate Reductase Catalytic Dimer

The purification procedure we have developed uses the nonionic detergent Triton X-100 to solubilize the enzyme from the inner membranes and tubules. The reductase activity can also be quantitatively extracted from the membrane by a variety of detergents with hydrophobic–lipophilic balance numbers around 13.[16] Hydrophobic exchange chromatography on phenyl-Sepharose resolves the catalytic dimer from the anchor dimer. Unlike the succinate dehydrogenase of beef heart mitochondria, fumarate reductase is not sensitive to atmospheric O_2.[17] No special precautions to maintain anaerobic conditions are necessary.

Solutions

1. Buffer I: Tris–HCl, 50 m*M* (pH 7.5); NaCl, 50 m*M*; phenylmethylsufonyl fluoride (PMSF), 50 μg/ml
2. Buffer II: Tris–HCl, 50 m*M* (pH 8.0); KCl, 150 m*M*; sucrose, 250 m*M*; EDTA, 10 m*M*
3. Triton X-100, 10% (v/v)
4. Buffer IV: NH_4OAc, 17.5% (w/v); Tris-SO_4, 50 m*M* (pH 8.0); EDTA, 1 m*M*; sodium cholate, 0.5% (w/v)
5. Buffer V: NH_4OAc, 2% (w/v); Tris-SO_4, 50 m*M* (pH 8.0); EDTA, 1 m*M*; sodium cholate, 0.5% (w/v)
6. Buffer VI: Tris-SO_4, 50 m*M* (pH 8.0); Triton X-100, 2% (v/v)

Preparation of Crude Envelope Fractions

For the preparation of the catalytic dimer, a crude envelope fraction containing inner and outer membranes as well as tubules is prepared as follows. Sixty grams (wet weight) of cells are washed in 250 ml of Buffer I and collected by centrifugation at 9800 *g*. The pellet is resuspended in 250 ml of the same buffer and the cells ruptured by two passages through a French press (Aminco, Rockford, MD.) at 16,000 psi. The cell lysate is centrifuged at 11,000 *g* for 10 min in a Beckman JA20 rotor. The supernatant is centrifuged at 150,000 *g* for 60 min in a Beckman Ti50.2 rotor. The pellet obtained from the high-speed spin is washed once by resuspension with the aid of a Teflon homogenizer (Talboy's Engineering Corporation, Emerson, NJ) in 200 ml of Buffer II, centrifuged as above, and the pellet resuspended in 200 ml of 50 m*M* Tris–HCl, pH 8.0. All centrifugations are given as average *g* values.

[17] T. P. Singer, E. B. Kearney, and W. C. Kenney, *Adv. Enzymol.* **37,** 189 (1973).

Extraction

The washed and resuspended crude envelope fraction is made 0.5% (v/v) in Triton X-100 and stirred on ice for 60 min. The extract is spun at 26,000 *g* for 30 min in a JA20 rotor and the supernatant retained. In order to remove most of the Triton X-100, NH_4OAc is added slowly to the supernatant to a concentration of 17.5% (by weight) while slowly stirring at 4°. Stirring is continued for a further 45 min. The suspension is then spun at 16,000 *g* for 10 min and the pellet discarded. If the supernatant is turbid the centrifugation is repeated.

Phenyl-Sepharose Chromatography

The phenyl-Sepharose resin (Pharmacia Fine Chemicals, Dorval, Quebec) is prepared as directed by the manufacturer and degassed for 5–10 min prior to pouring. A 200-ml bed volume column (4.5 × 12 cm) is equilibrated in Buffer III without cholate by washing with 5–10 column volumes of buffer at a flow rate of 4 ml/min. Approximately 60,000 units of enzyme activity (857 mg of protein) are applied to the column. The column is then washed with four column volumes of Buffer IV or until no further protein elutes from the column. As sodium cholate has a tendency to crystallize in high salt solutions, Buffer IV is maintained at 25° until use to minimize crystallization in the buffer. Any precipitated cholate at the top of the column is removed to allow for a reasonable flow rate. This wash removes a peak of protein containing a small amount of fumarate reductase activity. At this point, a 1500-ml linear decreasing gradient from 17.5% to 2% NH_4OAc in 50 m*M* Tris-SO_4, pH 8.0, 1 m*M* EDTA, 0.5% cholate is applied at a flow rate of 1 ml/min. Fractions containing purified dimeric enzyme appear near the end of the gradient. Sufficient Buffer V is used to assure full elution of the enzyme. Pure fractions, as judged by SDS–polyacrylamide gel electrophoresis,[18] are pooled and concentrated to a volume of 50 ml using an Amicon ultrafiltration apparatus (Amicon, Lexington, MA) equipped with a PM30 membrane. To deplete the enzyme of cholate the solution is made 35% by saturation in $(NH_4)_2SO_4$ and stirred for 4° for 1 hr. The suspension is centrifuged at 48,000 *g* for 15 min in a Beckman JA20 rotor with the brake off. The detergent precipitate often floats and must be removed very carefully. The supernatant is very slowly made 60% in $(NH_4)_2SO_4$, stirred for 45 min at 0°, and centrifuged as above. The pellet can be resuspended in a buffer of choice, usually 0.2 *M* potassium phosphate, pH 6.8. A typical purification is summarized in Table I.

[18] U. K. Laemmli, *Nature (London)* **227,** 680 (1970).

TABLE I
PURIFICATION OF FUMARATE REDUCTASE CATALYTIC DIMER

Fraction	Total protein (mg)	Total activity (units)	Specific activity (units/mg)	Percentage
Crude enzyme	1440	76,000	53	100
Triton X-100 extract	857	60,000	70	79
Phenyl-Sepharose	86	43,750	509	58

Comments

The two small anchor polypeptides can be eluted from the phenyl-Sepharose column by washing with Buffer VI. However, they elute in a Triton X-100 micelle together with many contaminating proteins. The phenyl-Sepharose step described above is so powerful that the procedure can be applied to normal (nonamplified) membrane extracts with equal success. The only difference noted is that the activity elutes nearer the middle of the gradient.[16] The purified dimer shows no tendency to aggregate even in a detergent-depleted state. It binds smaller amounts of detergent than most membrane proteins, indicating that no large exposed hydrophobic domains exist.[19] It can therefore be handled as a typical soluble enzyme in the presence of anions. A wide variety of anions stimulate and stabilize the dimer. The concentration of anion necessary for maximum stimulation appears to be proportional to its charge density. The effectiveness of the anion as an activator of dimer activity increases with higher charge-to-mass ratios.[20]

Isolation of an Inner Membrane Fraction Enriched in Fumarate Reductase Tubules

Solutions

1. Buffer I: Tris–HCl, 30 m*M* (pH 8.0)
2. Buffer II: Tris–HCl, 30 m*M* (pH 8.0); sucrose, 20% (w/v); PMSF, 50 μg/ml
3. EDTA, 500 m*M* (pH 8.0)
4. $MgCl_2$, 1 *M*
5. DNase (Sigma crude), 1 mg/ml

[19] J. J. Robinson and J. H. Weiner, *Biochem. J.* **199,** 473 (1981).
[20] J. J. Robinson, Ph.D. Thesis, University of Alberta (1982).

6. RNase (Miles Laboratories), 1 mg/ml
7. EDTA, 3 m*M* (pH 7.2)
8. Sucrose, 10% (w/v); EDTA, 3 m*M* (pH 7.2)
9. Sucrose, 20% (w/v); EDTA, 3 m*M* (pH 7.2)
10. Sucrose, 44% (w/w); EDTA, 3 m*M* (pH 7.2)
11. Buffer III: Potassium phosphate, 200 m*M* (pH 6.8); PMSF, 50 μg/ml

This procedure is based on the inner membrane fractionation protocol developed by Yamato *et al.*[21] All steps are carried out at 4°. Thirty grams (wet weight) of anaerobically grown *E. coli* HB101/pFRD63 are washed in 200 ml of Buffer I and collected by centrifugation at 10,000 *g* for 10 min in a Beckman JA14 rotor. The pellet is resuspended in 200 ml of Buffer II and kept on ice for 5 min. Lysozyme (14 mg) in 2 ml of Buffer II and 4 ml of 500 m*M* EDTA are added simultaneously and the suspension is stirred for 30 min or until it becomes viscous. We have found that stationary phase anaerobically grown *E. coli* do not form spheroplasts under these conditions; however, the lysozyme treatment aids the inner membrane isolation. At this time $MgCl_2$ is added to 25 m*M* and DNase and RNase to 10 μg/ml each and stirring continued until the viscosity decreases. The suspension is then centrifuged at 10,000 *g* for 30 min in a Beckman JA20 rotor. The broken cells are suspended in 70 ml of Solution 9 with the aid of a Teflon homogenizer. The homogenate is passed twice through a French press operating at 5700 psi. The pressate is diluted with an equal volume of 3 m*M* EDTA and centrifuged at 78,000 *g* for 90 min in a Beckman Ti50.2 rotor. The supernatant is discarded and the pellet is homogenized in 100 ml of Solution 8 and centrifuged at 2000 *g* for 15 min in a Beckman JA20 rotor. The supernatant is carefully decanted and diluted with 200 ml of the same solution, then centrifuged at 78,000 *g* for 60 min in a Beckman Ti50.2 rotor. The pellet from this centrifugation is homogenized in 20 ml of 3 m*M* EDTA and centrifuged at 2000 *g* for 10 min. The supernatant is carefully decanted, and 4 ml are layered on top of 16 ml of 44% sucrose (w/w)–3 m*M* EDTA (solution 10) and centrifuged for 1 hr at 120,000 *g* in a Beckman Ti50.2 rotor. The deep brown inner membrane and tubule band, near the middle of the tube, is collected with a Pasteur pipette, diluted with 2 volumes of 3 m*M* EDTA, and spun for 2 hr at 78,000 *g* in a Beckman Ti50.2 to collect the membranes. The pellet is homogenized to a total of 7 ml with Buffer VIII and stored at −70°. This enriched fraction has about double the reductase specific activity of crude envelopes.

[21] I. Yamato, M. Futai, Y. Anraku, and Y. Nomomura, *J. Biochem.* **83,** 117 (1978).

TABLE II
PURIFICATION OF FUMARATE REDUCTASE HOLOENZYME

Fraction	Total protein (mg)	Total activity (units)	Specific activity (units/mg)	Percentage
Enriched membranes	59.9	4800	80.4	100
Triton X-100 extract	37.4	4580	122	95
Sucrose gradient	17.9	4090	230	85

Purification of the Fumarate Reductase Holoenzyme

The following two-step protocol, summarized in Table II, utilizes Triton X-100 extraction together with sucrose gradient sedimentation to keep the tetramer intact. It results in the purification of about 18 mg of nearly homogeneous fumarate reductase holoenzyme with an 85% recovery. For optimal purification we used the enriched inner membrane fraction described above.

Solutions

1. Buffer I potassium phosphate 200 m*M* (pH 6.8) PMSF 50 μg/ml
2. Triton X-100, 10% (v/v)
3. Sucrose, 20% (w/v); Triton X-100, 0.2% (v/v); potassium phosphate. 100 m*M* (pH 6.8)
4. Sucrose, 15% (w/v); Triton X-100, 0.2% (v/v); potassium phosphate, 100 m*M* (pH 6.8)
5. Sucrose, 10% (w/v); Triton X-100, 0.2% (v/v); potassium phosphate, 100 m*M* (pH 6.8)
6. Sucrose, 30% (w/v); Triton X-100, 0.2% (v/v); potassium phosphate, 100 m*M* (pH 6.8)

All steps are carried out at 4°. Sixty milligrams (protein) of enriched inner membranes and fumarate reductase tubules prepared from 4.5 g wet weight of *E. coli* HB101/pFRD63 are suspended in 2.0 ml of Buffer I. Triton X-100 is added to 0.5% final concentration and the extraction kept at 4° for 40 min with occasional swirling. The extract is centrifuged at 140,000 *g* for 1 hr in a Beckman Ti50 rotor. The supernatant (2.1 ml) is divided into two aliqouts and each is layered on top of a 33-ml sucrose step gradient prepared by layering 10-ml steps of Solutions 3, 4, and 5, respectively, above a 3-ml cushion of Solution 6. The gradients are incubated at 37° for 2 hr and then at 4° for 1 hr prior to sample application. The gradients are centrifuged for 20 hr at 64,000 *g* in a Beckman SW27 rotor equipped with large buckets and eluted with a needle from the bottom. The golden brown fractions are assayed for activity. Fractions with specific activity of 215 or greater are pooled and stored at −70° where the enzyme is stable for at least 1 year.

Comments

This procedure can be scaled up with little difficulty as long as the protein and detergent concentrations remain constant. Octylglucoside can be used in place of the Triton X-100 for both the extraction and gradient steps with only slightly lower recovery of activity. Extraction is carried out at 3% (w/v) octylglucoside and the gradients should contain at least 2% detergent to minimize aggregation.

It can be seen from comparison of Tables I and II that the purified dimer has a significantly higher specific activity than the tetramer. Both preparations appear equally pure based on SDS–polyacrylamide gel electrophoresis as determined by staining with either Coomassie Brilliant Blue or silver reagent. The discrepancy in specific activity cannot be accounted for by molecular weight differences, which amount to only 30%. The dimer, but not the tetramer, is stimulated severalfold by a wide variety of anions, including those present in the storage and assay buffers, and presumably this accounts for much of the difference.[19] It is also possible that the tetramer has some inactive enzyme in the preparation.

The membrane-bound, the holoenzyme, and the dimer forms of fumarate reductase have apparent K_m values for fumarate at 420 μM.[16] This value is unaffected by the presence of anions. However, the apparent K_m for succinate is markedly anion dependent.[19] This change in apparent K_m would shift the equilibrium in the direction of succinate formation in the presence of anions.

[37] Reconstitution of a Functional Electron-Transfer Chain from Purified Formate Dehydrogenase and Fumarate Reductase Complexes

By GOTTFRIED UNDEN and ACHIM KRÖGER

Growth of the anaerobic bacterium *Wolinella succinogenes*[1] (formerly *Vibrio succinogenes*) is catabolically sustained by reaction (1)

$$\text{Formate} + \text{fumarate} \rightarrow CO_2 + \text{succinate} \quad (1)$$

which is coupled to phosphorylation. The process exhibits all the essential properties of electron-transfer coupled phosphorylation.[2,3] The electron-transfer chain catalyzing reaction (1) consists of two membranous enzyme complexes, formate dehydrogenase complex and fumarate reductase complex, which are linked by menaquinone. In this chapter the isolation of the complexes and their incorporation into liposomes is described. The incorporation led to the restoration of the electron-transfer activity. This provided insight into the composition and organization of the electron-transfer chain.

Isolation Procedures

The isolation procedures[4,5] were derived from the technique used originally for isolating the mitochondrial adenine nucleotide carrier.[6] Similar procedures were successfully applied in the purification of other mitochondrial proteins[7] and of the hydrogenase,[8] nitrite reductase,[9] and ATP synthase[10] of *W. succinogenes*.

The procedures used for the isolation of fumarate reductase complex and formate dehydrogenase complex include solubilization of the mem-

[1] M. J. Wolin, E. A. Wolin, and N. J. Jacobs, *J. Bacteriol.* **81,** 911 (1961).
[2] C. A. Reddy and H. D. Peck, *J. Bacteriol.* **134,** 982 (1978).
[3] A. Kröger and E. Winkler, *Arch. Microbiol.* **129,** 100 (1981).
[4] A. Kröger, E. Winkler, A. Innerhofer, H. Hackenberg, and H. Schägger, *Eur. J. Biochem.* **94,** 465 (1979).
[5] G. Unden, H. Hackenberg, and A. Kröger, *Biochim. Biophys. Acta* **591,** 275 (1980).
[6] M. Klingenberg, H. Aquila, and P. Riccio, this series, Vol. 56, p. 407.
[7] G. von Jagow, H. Schägger, W. D. Engel, P. Riccio, H. J. Kolb, and M. Klingenberg, this series, Vol. 53, p. 92.
[8] G. Unden, R. Böcher, J. Knecht, and A. Kröger, *FEBS Lett.* **145,** 230 (1982).
[9] I. Schröder, A. M. Roberton, M. Bokranz, G. Unden, R. Böcher, and A. Kröger, *Arch. Microbiol.* **140,** 380 (1985).
[10] M. Bokranz, E. Mörschel, and A. Kröger, *Biochim. Biophys. Acta* **810,** 84 (1985).

TABLE I
PURIFICATION OF THE FUMARATE REDUCTASE COMPLEX

Step	Benzyl viologen radical → fumarate	
	U/mg	U
Cell homogenate	6.9	17,900
Membrane fraction	12	14,100
Triton extract	23	8,300
Chromatography on hydroxyapatite	96	6,200
Anion-exchange chromatography	180	4,100

brane with Triton X-100 and subsequent chromatography of both enzymes on hydroxyapatite and DEAE-Sepharose in the presence of Triton X-100. Formate dehydrogenase complex is further purified by sucrose density gradient centrifugation in the presence of Triton X-100.

All the buffers used are at 0° and contain 1 m*M* dithiothreitol, 0.1 m*M* phenylmethylsulfonyl fluoride, 1 m*M* malonate (with fumarate reductase complex), and 1 m*M* NaN_3 (with formate dehydrogenase complex). Malonate and azide are competitive inhibitors of the respective enzymes and stabilize the enzymatic activities. Buffer A contains 0.05% Triton X-100 and 20 m*M* Tris, pH 7.7.

Wolinella succinogenes is grown anaerobically in a medium containing formate and fumarate as substrates[11] and stored in liquid N_2.

Isolation of Fumarate Reductase Complex (Table I)

Preparation of the Triton Extract. The bacteria (30 g wet weight) are suspended and stirred for 15 min in 1 liter of a N_2-flushed buffer containing 50 m*M* Tris, 10 m*M* EDTA, and 10 m*M* formate, pH 8. Lysozyme (0.5 g/liter) and after 45 min 15 m*M* $MgCl_2$ and 1 mg/liter DNase I are added. The suspension is stirred for a further 30 min (cell homogenate) and centrifuged for 30 min at 10,000 *g*. The sediment is washed in 0.3 liter of a buffer containing 1% Tween 80 and 10 m*M* phosphate, pH 7.5, and then centrifuged for 15 min at 30,000 *g*. The sediment (membrane fraction) is extracted by homogenization and stirring for 15 min in 0.3 liter of a buffer containing 1% Triton X-100 and 10 m*M* phosphate, pH 7.5. The Triton extract is obtained after centrifugation for 20 min at 250,000 *g*.

Chromatography on Hydroxyapatite. The Triton extract (2–5 g protein/liter hydroxyapatite) is passed through a hydroxyapatite column

[11] M. Bronder, H. Mell, E. Stupperich, and A. Kröger, *Arch. Microbiol.* **131,** 216 (1982).

TABLE II
PURIFICATION OF FORMATE DEHYDROGENASE COMPLEX

Step	Formate → benzyl viologen	
	U/mg	U
Cell homogenate	0.84	2600
Membrane fraction	2.4	2300
Triton extract	4.2	1530
Chromatography on hydroxyapatite	23.3	920
Anion-exchange chromatography	180	680
Gradient centrifugation	240	390

(0.15 liter and 3 cm inner diameter) equilibrated with a buffer containing 0.05% Triton X-100, 50 m*M* imidazole, and 10 m*M* phosphate, pH 7.5. After rinsing with 0.2 liter of the same buffer, the enzyme is eluted with 0.5 liter of a linear phosphate gradient (10–300 m*M*) in the equilibration buffer at a flow rate of 50–100 ml/hr. Greater flow rates can cause tightening of the column. Fumarate reductase complex elutes at 10–70 m*M* phosphate. The fractions containing the enzyme at sufficiently high specific activity are pooled and concentrated 10-fold by ultrafiltration under N_2 through a Diaflo ultrafilter XM100 (Amicon). The concentrated sample is dialyzed for 8–16 hr against 2 × 1 liter of buffer A.

Anion-Exchange Chromatography. The dialyzed sample (10 g protein/liter anion-exchange resin) is passed through a DEAE-Sepharose CL-6B column (20 ml, inner diameter 1.6 cm) equilibrated with buffer A. After rinsing with 30 ml of buffer A, the enzyme is eluted with 80 ml of a linear NaCl gradient (0–0.15 *M*) in buffer A at a flow rate of 30 ml/hr. Fumarate reductase complex eluted at about 90 m*M* NaCl. The fractions containing the enzyme are concentrated as described above to a final concentration of 5 g protein/liter and stored in liquid N_2.

Isolation of Formate Dehydrogenase Complex (Table II)

The Triton extract is prepared and chromatography on hydroxyapatite and DEAE-Sepharose is done as described for fumarate reductase complex. Both enzymes can be purified from the same batch of bacteria, since the enzymes are separated by the chromatography on hydroxyapatite. Formate dehydrogenase complex eluted from the hydroxyapatite column at 70–200 m*M* phosphate and from the anion-exchange column at about 50 m*M* NaCl.

Gradient Centrifugation. Samples (1 ml) of the preparation obtained from anion-exchange chromatography are layered onto 39 ml of buffer A

TABLE III
PROPERTIES OF THE FUMARATE REDUCTASE COMPLEX[a]

Property	Value
M_r 79,000 subunit (μmol/g protein)[b]	6.0
M_r 31,000 subunit (μmol/g protein)[b]	4.2–6.0
M_r 25,000 subunit (μmol/g protein)[b]	9–15
Triton X-100 content (g/g protein)	0.43
Phospholipid content (g/g protein)	0.20
Molar mass of enzyme particle (g/mol)[c]	326,000
M_r of enzyme protein[d]	200,000

[a] From Unden *et al.*[5]
[b] From quantitative evaluation of photometric scans of SDS–polyacrylamide gels, as described in Unden *et al.*[5]
[c] Determined by sedimentation equilibrium.
[d] From molar mass of enzyme particle[c] after correction for bound Triton X-100 and phospholipid.

containing a linear sucrose gradient (5–20%, w/v) in a 40-ml Quickseal centrifuge tube (Beckman). Centrifugation for 3 hr is done at 206,000 *g* in a VTi50 rotor (Beckman). The contents of the tubes are fractionated by pumping 2-ml samples from the bottom of the tubes. The fractions containing formate dehydrogenase activity are pooled and stored in liquid N_2.

Properties of the Isolated Enzymes

Properties of Fumarate Reductase Complex

The fumarate reductase complex consists of three different subunits (M_r 79,000, 31,000, and 25,000) which are present at a molar ratio of ~1 : 1 : 2 (Table III) and make up more than 90% of the total protein.[5] The M_r of the enzyme protein as calculated from the molar mass of the enzyme particle and its contents of Triton and phospholipid is 200,000. This suggests that the active enzyme is composed of one of each of the two bigger subunits and two of the M_r 25,000 polypeptides. The complex can be cleaved into the soluble fumarate reductase which consists of one of each of the two bigger subunits and a lipophilic portion (M_r 25,000 in SDS–gel electrophoresis) which binds most of the Triton and phospholipid.[5,12]

[12] G. Unden and A. Kröger, *Eur. J. Biochem.* **120,** 577 (1981).

TABLE IV
COMPOSITION OF FUMARATE REDUCTASE

Component	Content[a] (μmol/g protein)	Location[b] (M_r of subunit)
Covalently bound FAD	6.0 ± 0.7	79,000
Sulfide	40.0 ± 8.0	79,000, 31,000
Nonheme iron	57.0 ± 7.0	79,000, 31,000
Cytochrome *b*	10.7 ± 1.9	25,000

[a] The values represent the average of 6–10 preparations.
[b] The location was determined after cleavage of the complex with guanidinium chloride and separation of the subunits as described in G. Unden and A. Kröger[12] and Albracht *et al.*[13]

The complex contains covalently bound FAD, sulfide, nonheme iron, and cytochrome *b* (Table IV). The FAD is bound to the M_r 79,000 peptide. The sulfide and the nonheme iron form a 4Fe-4S cluster located on the M_r 79,000 subunit, and a 2Fe-2S cluster which is situated on the M_r 31,000 peptide.[13] The M_r 25,000 subunits represent cytochrome *b*, which is present in twice the amount of FAD. Redox titration reveals that the cytochrome *b* consists of equal amounts of two different species with E_m values of -20 and -180 to -200 mV.[5] The isolated high-potential cytochrome *b* ($E_m = -20$ mV) is rapidly reduced by menaquinol analogs.[12]

Fumarate reduction by viologen radicals is catalyzed by the complex, soluble fumarate reductase (containing the M_r 79,000 and the M_r 31,000 peptides) and the isolated M_r 79,000 subunit. In contrast, fumarate reduction by menaquinol analogs is catalyzed only by the complex.[5,12] Reassociation of the soluble fumarate reductase and high-potential cytochrome *b* leads to restoration of the activity with the quinols.[12] This indicates that the active site for fumarate is localized on the M_r 79,000 subunit and that the quinols react specifically with the high-potential cytochrome *b*. Reduction by the quinols and oxidation by fumarate of the high-potential cytochrome *b* and of the iron-sulfur centers are at least as fast as the turnover of the enzyme complex.[14]

Properties of Formate Dehydrogenase Complex (Table V)

Formate dehydrogenase complex consists of three different subunits (M_r 110,000, 25,000, and 20,000) which are present at a relative molar

[13] S. P. J. Albracht, G. Unden, and A. Kröger, *Biochim. Biophys. Acta* **661**, 295 (1981).
[14] G. Unden, S. P. J. Albracht, and A. Kröger, *Biochim. Biophys. Acta* **767**, 460 (1984).

TABLE V
COMPOSITION OF THE FORMATE DEHYDROGENASE COMPLEX[a]

Component	Content (μmol/g protein)
M_r 110,000 subunit	6
M_r 25,000 subunit	3–5.6
M_r 20,000 subunit	9–15
Cytochrome *b*	2.1–4.8

[a] From Unden and Kröger.[15]

ratio of about 1 : 1 : 2. The M_r 25,000 subunit represents a low-potential cytochrome *b* (E_m = −200 to −220 mV).[15] The complex is characterized by its reactivity with the lipophilic vitamin K_1 in liposomes. A formate dehydrogenase lacking cytochrome *b* did not react with vitamin K_1, but catalyzed the reduction of 2,3-dimethyl-1,4-naphthoquinone (DMN) by formate.[15] A preparation consisting essentially of dimers of the M_r 110,000 subunit contained 9 μmol molybdenum and 160 μmol sulfide and non-heme iron per gram of protein.[4] This preparation catalyzed the reduction of various dyes by formate, but did not react with quinones.

Restoration of Electron Transfer

Preparation of Proteoliposomes

With use of the dialysis technique with octylglucoside,[16,17] the isolated enzyme complexes are incorporated into liposomes.[18,19]

Detergent Exchange. The Triton X-100 associated with the enzyme preparations is replaced by octylglucoside using sucrose density gradient centrifugation. The individual enzyme preparations (about 5 g protein/liter) are centrifuged as described for the purification of formate dehydrogenase complex, except that the centrifuge tubes contained 30 m*M* octylglucoside instead of Triton. The enzyme solutions were concentrated by ultrafiltration through a Diaflo Ultrafilter PM10 (Amicon) and stored in liquid N_2.

[15] G. Unden and A. Kröger, *Biochim. Biophys. Acta* **725,** 325 (1983).
[16] A. Helenius, E. Fries, and J. Kartenbeck, *J. Cell Biol.* **75,** 866 (1977).
[17] L. T. Mimms, G. Zampighi, Y. Nazaki, C. Tanford, and J. A. Reynolds, *Biochemistry* **20,** 833 (1981).
[18] G. Unden and A. Kröger, *Biochim. Biophys. Acta* **682,** 258 (1982).
[19] G. Unden, E. Mörschel, M. Bokranz, and A. Kröger, *Biochim. Biophys. Acta* **725,** 41 (1983).

Incorporation of the Enzyme Complexes. A phospholipid film on the surface of a test tube is prepared by blowing N_2 on the surface of a solution (1 ml) of 1 mg phosphatidylcholine (from soybean, Sigma) and 10 nmol vitamin K_1 in $CHCl_3$ and methanol (2 : 1, v/v) at room temperature, until the solvent is evaporated. The film is dissolved in 0.5 ml of a buffer (0°) containing the enzyme complexes (total 0.2 mg protein), 8 mg octylglucoside, 20 m*M* Tris, 1 m*M* NaN_3, 1 m*M* malonate, and 1 m*M* dithiothreitol, pH 7.7. The amounts of the individual enzymes are varied. The solution is dialyzed for 35–40 hr at 0° against two 1-liter volumes of a buffer (flushed with N_2) containing 10 m*M* Tris, 1 m*M* NaN_3, 1 m*M* malonate, and 0.7 m*M* dithiothreitol, pH 7.7. The preparation is immersed in liquid N_2 for 10 min and then thawed at room temperature. This is repeated three times. The preparation is stored in liquid N_2. Repeated freeze-thawing is essential for incorporation of the enzyme complexes into liposomes[19] and for restoration of electron transfer.[18]

Composition of the Liposomal Preparation

Incorporation of the enzyme complexes into the liposomes can be tested using equilibrium centrifugation on a linear sucrose gradient.[18,19] More than 80% of the protein and the enzymatic activities of the complexes were found in coincident bands with a common peak at a density of 1.07 g/ml, which corresponded to the phospholipid/protein ratio of 5 (g/g) used for preparation. SDS–gel electrophoresis indicated that the subunit composition after incorporation was identical to that of the isolated complexes.[18] This indicated that nearly complete incorporation of the intact complexes was achieved.

From the phospholipid/protein ratio, the proportions of the individual complexes, and their M_r values, the molar contents of the enzymes were calculated. As seen from Table VI, the contents based on phospholipid were of the same order of magnitude as those in the cytoplasmic membrane of *W. succinogenes*. On the basis that an average proteoliposome is made up of 10^5 molecules of phospholipid, it was estimated that each proteoliposome contains 1000 molecules of vitamin K_1 and 50 molecules of each complex.

Structural Properties of the Proteoliposomes

The proteoliposomes retained small hydrophilic molecules like glucose or taurin which were added before freeze-thawing.[19] This demonstrated the vesicular state of the preparation and the tightness of the liposomal membrane. From the amount of glucose retained, the internal volume was calculated as 2–4 ml/g phospholipid. This corresponded to an

TABLE VI
COMPOSITION OF THE PROTEOLIPOSOMES IN COMPARISON TO THAT OF THE CYTOPLASMIC MEMBRANE OF *W. succinogenes*[a]

Component	Content in: Bacterial membrane (mmol/mol phospholipid)	Content in: Proteoliposomes (mmol/mol phospholipid)	Molecules/proteoliposome
Phospholipid	≡1000	≡1000	10^5
Menaquinone, vitamin K_1	15	10	1000
Fumarate reductase	2	0.5	50
Formate dehydrogenase	0.2	0.5	50

[a] From Unden *et al.*[19] The values given for the bacterial membrane are based on a protein/phospholipid ratio of 3 g/g, those in the proteoliposomes on a ratio of 0.2 g/g with equal amounts of both enzymes.

average diameter of 60–120 nm on the basis that the proteoliposomes are unilamellar (Table VII). The unilamellar structure as well as the average diameter were confirmed by electron microscopy after negative staining of the preparation.[19] The electron micrographs also showed that proteoliposomes prepared with either fumarate reductase complex or formate dehydrogenase complex alone carried particles on their outer surface.[19] The fumarate reductase particles appeared cylindrical with 5.1 nm diameter and 7.1 nm height. The molar mass calculated from these dimensions was close to that obtained from the hydrodynamic properties of the soluble fumarate reductase (lacking cytochrome *b*). This suggested that the lipophilic cytochrome *b* is buried in the membrane where it reacts with the quinone and forms the anchor of the hydrophilic part which carries the fumarate site and protrudes into the aqueous phase. The formate dehydrogenase particles appeared as spheres (7.5 nm diameter) that were linked to the membrane by means of stalks. The molar mass of the sphere was similar to that of the dimer of the bigger subunit of the formate dehydrogenase complex.

The particle density on the surface of the proteoliposomes as calculated from the phospholipid/protein ratio was close to that seen on the outer surface in electron micrographs. This suggested that most of the enzyme molecules are inserted into the bilayer with the active sites for fumarate or formate oriented toward the outside of the liposomes. Consistently 80% of the enzyme molecules were found to be accessible from the outside to their substrates in enzymatic activity measurements.

TABLE VII
COMPARISON OF STRUCTURAL DATA OF THE PROTEOLIPOSOMES OBTAINED FROM ELECTRON MICROSCOPY AND BY OTHER METHODS[a]

Property	Electron microscopy	Other methods
Inner diameter of the proteoliposomes (nm)	100	60–120 (internal volume[b])
M_r of fumarate reductase	119,000	121,000 (hydrodynamic properties[c])
M_r of formate dehydrogenase	181,000	220,000 (hydrodynamic properties[c])
Surface area/enzyme molecule (nm^2)	250–400	350 (protein–phospholipid ratio 0.2 g/g)[d]
Orientation of the enzymes	Outside	80% outside (accessibility for dyes and substrates[e])

[a] From Unden *et al.*[19]
[b] From the contents of glucose or taurine.[19]
[c] Determined for the soluble fumarate reductase[5] and for formate dehydrogenase.[4]
[d] Calculated from the M_r of the fumarate reductase complex (200,000), of the phospholipid (1000), and of the area required by a molecule of phosphatidylcholine (0.7 nm^2).[19]
[e] From the activity of fumarate reductase and formate dehydrogenase in intact and lyzed liposomes.[19]

Enzymatic Properties of the Proteoliposomes

Recovery of Enzymatic Activity. To assess the enzymatic activity of the complexes in the proteoliposomes, their turnover numbers were determined and compared to those before (in membrane) and after isolation (in Triton) (Table VIII). The turnover numbers with benzyl viologen were regarded as representing the activities of the soluble part of the complexes. Those measured with reduced DMN ($DMNH_2$) or DMN should represent the capability of participating in the electron transfer from formate to fumarate.[12,18] Most of the activities with the viologen were found to be recovered in the proteoliposomes, while the reactivity of fumarate reductase complex with $DMNH_2$ was 52% and that of formate dehydrogenase complex with DMN was 15% of that measured in the bacterial membrane. The losses in activity were mainly caused by the purification and not by the incorporation procedure.

Restoration of Electron-Transfer Activity. The proteoliposomes catalyzed the electron transport from formate to fumarate provided that they contained vitamin K_1 (Fig. 1). Proteoliposomes containing constant amounts of each enzyme complex showed increasing electron-transfer

TABLE VIII

TURNOVER NUMBERS OF THE FUMARATE REDUCTASE AND FORMATE DEHYDROGENASE COMPLEXES IN LIPOSOMES

Enzymatic reaction	Turnover number (10^3 min^{-1})[a]		
	Membrane	Triton	Liposomes
BV radical → fumarate[b]	35	30	30
$DMNH_2$ → fumarate	23	10	12
Formate → BV	38	26	25
Formate → DMN	82	23	12

[a] The numbers refer to FAD (6 μmol/g fumarate reductase complex protein) or molybdenum (9 μmol/g formate dehydrogenase protein). The values given for the bacterial membrane are based on the purification factors obtained from Tables I and II.

[b] BV, Benzyl viologen.

activity with increasing content of the quinone. Saturation was reached with 10–20 m*M* vitamin K_1 in the phospholipid. This concentration is an order of magnitude greater than that of the enzymes and is similar to that of menaquinone in the bacterial membrane (Table VI). The apparent Michaelis constant evaluated from the saturation curve (3.5 m*M*) can

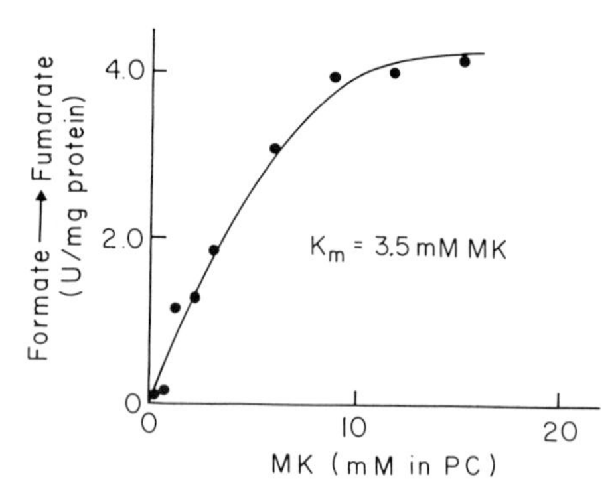

FIG. 1. Electron-transfer activity of proteoliposomes as a function of the vitamin K_1 (menaquinone) content. Proteoliposomes containing constant amounts of the formate dehydrogenase and the fumarate reductase complexes and increasing amounts of vitamin K_1 were prepared as described in the text. All the preparations catalyzed $DMNH_2$ oxidation by fumarate at a specific activity of 30 U/mg total protein. The activity of DMN reduction by formate varied between 5 and 7 U/mg total protein.

TABLE IX
RECONSTITUTION OF THE ELECTRON TRANSFER IN LIPOSOMES CONTAINING FORMATE DEHYDROGENASE AND FUMARATE REDUCTASE AT DIFFERENT RATIOS[a]

Formate → DMN	$DMNH_2$ → fumarate	Formate → fumarate
28.1	26.9	13.8
45.0	6.5	6.0
2.7	59.6	1.6

[a] Values given as U/mg protein. From Unden and Kröger.[18]

be explained on the basis of that of fumarate reductase complex for $DMNH_2$.[20]

The activity of electron transfer was related to those of the two enzymes according to Eq. (2)

$$v_{ET} = V_{FDH} V_{FR} / V_{FDH} + V_{FR} \tag{2}$$

where v_{ET} represents activity of electron transfer, V_{FDH} that of DMN reduction by formate, and V_{FR} that of $DMNH_2$ oxidation by fumarate.[18] This is shown in Table IX. The electron-transfer activity was half that of the individual enzymes, when their activity was equal. With one enzyme in large excess the electron-transfer activity was close to the activity of the limiting enzyme. This result shows that all the active enzyme complexes participate in electron transfer independently of the ratio of their molar contents.[18]

Assay of Enzymatic Activities

The activities of the reactions listed in Table X were measured photometrically using either single- or dual-wavelength recording. The measurements were made at 37° in a 0.5-cm cuvette containing the given solutions. Before addition of the enzyme, the solutions were bubbled with N_2 for 2 min and then kept under an atmosphere of N_2. Benzyl viologen radical was prepared in the anaerobic test solution by adding 10 μl of a freshly prepared $Na_2S_2O_4$ solution (10 g/liter in anaerobic H_2O at 0°).

$DMNH_2$ was prepared from DMN[21] by reduction with KBH_4. In the absence of hydrogenase activity, DMN was reduced in the test solution

[20] A. Kröger and G. Unden, *in* "Coenzyme Q" (G. Lenaz, ed.), p. 285. Wiley, Chichester, 1985.

[21] O. Kruber, *Chem. Ber.* **62,** 3046 (1929).

TABLE X

ASSAY CONDITIONS FOR ENZYMATIC ACTIVITIES OF FORMATE DEHYDROGENASE AND FUMARATE REDUCTASE

Donor[a]	Acceptor[a]	Wavelength (nm)	$\Delta\varepsilon$ (mM^{-1} cm^{-1})	Buffer	Starting substance
Formate(10)	BV(1)	546	19.5	10 m*M* Tris, pH 7.9	Enzyme or formate
Formate(10)	DMN(0.1)	270–290	15.0	50 m*M* phosphate, pH 7.4	Formate
BV radical(0.2)	Fumarate(10)	546	19.5	50 m*M* phosphate, pH 7.0	Enzyme or fumarate
$DMNH_2$(0.2)	Fumarate(1)	270–290	15.0	50 m*M* phosphate, pH 7.4	Enzyme or fumarate
Formate(10)	Fumarate(1)	260–290	0.79	50 m*M* phosphate, pH 7.4	Formate

[a] Concentrations (m*M*) are given in parentheses.

by the addition of 5 μl of a freshly prepared solution of KBH_4 (10 g/liter H_2O at 0°). In the presence of hydrogenase activity (e.g., membrane fraction of *W. succinogenes*), DMN was reduced separately in ethanolic solution (20 mM) in a sealed flask under N_2. Subsequently the solution was neutralized with HCl, degassed, and gassed with nitrogen 3 times to remove H_2.

[38] Molecular Properties, Genetics, and Biosynthesis of *Bacillus subtilis* Succinate Dehydrogenase Complex

By LARS HEDERSTEDT

Introduction

The citric acid cycle enzyme succinate dehydrogenase (SDH) (EC 1.3.99.1) is a membrane-bound iron-sulfur flavoprotein. It catalyzes the oxidation of succinate to fumarate in aerobic cells and is directly linked to the respiratory chain. Mitochondrial[1,2] and bacterial[3,4] SDH and membrane-bound fumarate reductase in anaerobic[5] and facultative[6] bacteria are very similar in composition. They consist of one larger (M_r, 60,000–79,000) and one smaller (M_r, 25,000–31,000) protein subunit and contain several prosthetic groups; an FAD, covalently bound in an 8α-[N(3)-histidyl] linkage to the large subunit, and two or three iron-sulfur centers.

SDH and also fumarate reductase can be extracted with detergent from the membrane in a complex with one or two (depending on the organism) small hydrophobic polypeptides. These small associated polypeptides are integral membrane proteins that anchor each enzyme to the membrane and are, at least in mitochondria, required for electron transfer from enzyme to quinone. In several organisms one of the small proteins has been shown to be a *b*-type cytochrome.[3,5,7]

What is known about the composition, enzymology, and membrane binding of SDH has been learned during many years of studies predomi-

[1] B. A. C. Ackrell, E. B. Kearney, and T. P. Singer, this series, Vol. 53, p. 466.
[2] Y. Hatefi, this series, Vol. 53, p. 27.
[3] L. Hederstedt and L. Rutberg, *Microbiol. Rev.* **45,** 542 (1981).
[4] C. Condon, R. Cammack, D. S. Patil, and P. Owen, *J. Biol. Chem.* **260,** 9427 (1985).
[5] G. Unden and A. Kröger, this volume [37].
[6] B. D. Lemire and J. Weiner, this volume [36].
[7] B. A. Crowe and P. Owen, *J. Bacteriol.* **153,** 1493 (1983).

nantly on bovine heart mitochondrial SDH. Information about the biosynthesis of SDH has been obtained from experiments with *Bacillus subtilis*.[3,8,9] The genetics of SDH and its binding polypeptides are mainly known from studies with *Escherichia coli,* in which the nucleotide sequence of the *sdh* genes has been determined,[10,11] and *B. subtilis,*[3,12,13] in which the *sdhA* gene has been cloned[14] and sequenced.[15]

The present understanding of the composition, structure, genetics, and biosynthesis of SDH and cytochrome *b*-558 in *B. subtilis* originates from work done largely in our laboratory.[3,8,9,12–17] Methods of central importance in that work and applicable also to other experimental systems are described in this chapter under four headings: (1) Growth of *B. subtilis* and Isolation of Membranes; (2) Immunoprecipitation and Composition of *B. subtilis* SDH Complex; (3) Genetics of *B. subtilis* SDH–protoplast fusion; and (4) Biosynthesis of SDH Complex. The technical descriptions in each section are preceded by a summary of the findings and conclusions that have been made with the aid of the methodology described.

Growth of *B. subtilis* and Isolation of Membranes

Bacillus subtilis is an obligatory aerobic, gram-positive, rod-shaped, endospore-forming bacterium which occurs naturally in soil. Properties and general experimental methods for *B. subtilis* are described in an earlier volume of this series.[18] Genetically it is the best characterized gram-positive bacterium.[19] About 500 loci have been mapped on the chromosome by transformation and transduction crosses. The well-developed genetic systems that are available in this bacterium and the presence of only one membrane in the cell are two reasons to use *B. subtilis* as a model organism in basic membrane research.

[8] L. Hederstedt, *Eur. J. Biochem.* **132,** 589 (1983).
[9] L. Hederstedt, J. J. Maguire, A. J. Waring, and T. Ohnishi, *J. Biol. Chem.* **260,** 5554 (1985).
[10] D. Wood, M. G. Darlison, R. J. Wilde, and J. R. Guest, *Biochem. J.* **222,** 519 (1984).
[11] M. G. Darlison and J. R. Guest, *Biochem. J.* **223,** 507 (1984).
[12] L. Hederstedt, K. Magnusson, and L. Rutberg, *J. Bacteriol.* **152,** 157 (1982).
[13] K. Magnusson, B. Rutberg, L. Hederstedt, and L. Rutberg, *J. Gen. Microbiol.* **129,** 917 (1983).
[14] K. Magnusson, L. Hederstedt, and L. Rutberg, *J. Bacteriol.* **162,** 1180 (1985).
[15] K. Magnusson, L. Rutberg, M. Philips, and J. R. Guest, in preparation (1985).
[16] L. Hederstedt and L. Rutberg, *J. Bacteriol.* **153,** 57 (1983).
[17] L. Hederstedt, *J. Bacteriol.* **144,** 933 (1980).
[18] R. L. Armstrong, N. Harford, R. H. Kennett, M. L. St. Pierre, and N. Sueoka, this series, Vol. 17, p. 36.
[19] P. J. Piggot and J. A. Hoch, *Microbiol. Rev.* **49,** 158 (1985).

Strains

All strains described in this chapter are derivatives of the transformable *B. subtilis* 168 strain. A collection of *Bacillus* wild-type and mutant strains is kept at the Bacillus Genetic Stock Center (The Ohio State University, Department of Microbiology, Columbus, OH) from where strains and a catalog can be ordered.

Growth Media

The membrane composition in *B. subtilis* is strongly influenced by growth conditions and growth phase. For instance, the content of SDH in cytoplasmic membranes can vary 10-fold in the same strain, depending on the medium and stage of growth. It is thus rewarding to carefully find out and control optimal growth conditions for a set of experiments.

Many different substrates can be used for cultivating *B. subtilis*. We routinely use two different liquid media, as follows. (1) Spizizen's[20] minimal glucose medium: 2 g $(NH_4)_2SO_4$, 14 g K_2HPO_4, 6 g KH_2PO_4, 1 g Na_3-citrate $\times$ 2 H_2O, 0.2 g $MgSO_4 \times 7\ H_2O$ are dissolved in 1 liter of distilled water. The solution is sterilized by autoclaving and 10 ml sterile 50% (w/v) glucose is added before use. We generally supplement the medium after sterilization with 10 ml 1 m*M* $MnCl_2$ and often 10 ml 5% (w/v) Casein hydrolyzate (casamino acids, Difco) per liter to stimulate growth. (2) NSMP[21] (nutrient sporulation medium phosphate) medium: 8 g nutrient broth (Difco) is dissolved in 950 ml distilled water and sterilized in the autoclave. The following sterile solutions are then added on the day the medium is used: 50 ml 2 *M* K-phosphate buffer, pH 6.5, 5 ml "metal mix" (140 m*M* $CaCl_2$, 10 m*M* $MnCl_2$, 200 m*M* $MgCl_2$), and 0.5 ml 2 m*M* $FeCl_3$ in 10 m*M* HCl.

Protein can effectively be uniformly radioactively labeled by adding mixtures of ^{3}H- or ^{14}C-labeled amino acids to minimal medium.[22] [^{35}S]Methionine is effectively incorporated into protein in *B. subtilis* also in the nutrient-rich NSMP medium. Flavoproteins are specifically labeled if radioactive riboflavin is added to minimal or NSMP medium.[8] Riboflavin is actively taken up by *B. subtilis* and is probably converted into flavin nucleotides at uptake. Heme can be specifically labeled in *B. subtilis* *hemA* mutants (which require 5-aminolevulinic acid for growth) by supplementing minimal medium with 5-[4-^{14}C]aminolevulinic acid.[23,24] The

[20] J. Spizizen, *Proc. Natl. Acad. Sci. U.S.A.* **44,** 1072 (1958).
[21] P. Fortnagel and E. Freese, *J. Bacteriol.* **95,** 1431 (1968).
[22] L. Hederstedt and L. Rutberg, *J. Bacteriol.* **144,** 941 (1980).
[23] E. Holmgren, L. Hederstedt, and L. Rutberg, *J. Bacteriol.* **138,** 377 (1979).
[24] L. Hederstedt, E. Holmgren, and L. Rutberg, *J. Bacteriol.* **138,** 370 (1979).

reader is asked to consult original papers for quantitative information about radioactive labeling in *B. subtilis*.

Wild-type cells and *sdh* mutants can be kept on TBAB agar slants at room temperature for shorter periods of time. For long time storage of strains freeze-drying is recommended. TBAB (tryptose blood agar base, Difco) is a complex medium which allows growth of most *B. subtilis* strains. The phenotype of *sdh* mutants is conveniently checked on PA (purification agar) slants.[25] These agar slants contain a pH indicator that changes color from purple to yellow upon acidification. Bacterial colonies accumulating and excreting acid are surrounded by a yellow halo on PA slants, whereas wild-type colonies are blue-purple. PA slants are prepared by adding 8 g nutrient broth, 20 g agar, and 15 mg of Bromocresol Purple to 1 liter of distilled water. The pH is adjusted to 6.0 and the suspension is sterilized in the autoclave for 20 min. Then 5 ml sterile metal mix (same as for NSMP) and 2 ml 50% (w/v) glucose are added before slants are cast.

Growth of Bacteria

Good aeration of cultures is critical when growing *B. subtilis*. It is obtained by using 100 ml or 500 ml medium in 1-liter Erlenmeyer or 2.8-liter Fernbach indentated flasks, respectively, and by growing the cells on a rotary shaker at 200 rpm, usually at 37°. Batch cultures can be inoculated from an overnight preculture in liquid medium or from slants. Accumulation of wild-type revertants of some *sdh* mutants in liquid precultures occurs rather frequently. To minimize this problem a single "yellow" colony from a PA plate is used to inoculate a TBAB plate for confluent growth. After incubation of the plate at 37° overnight the bacteria are suspended in a few milliliters of medium and used as inoculum. Liquid cultures of *B. subtilis sdh* mutants should always be checked for revertants on PA slants.

Bacterial growth can be monitored by measurements of light "absorption" of the culture directly, or after dilution, at 600 nm. A_{600} is proportional to cell density up to 0.4 A. Cells are best grown in NSMP and harvested after the exponential growth has declined if good yield of SDH is wanted. The specific activity of SDH in *B. subtilis* membranes increases when the growth rate decreases and is optimal 1 to 2 hr after growth has started to decline. SDH activity is low in cells grown in the presence of glucose, fumarate, or malate. Cells grown in NSMP contain 2 to 4 times more SDH than those grown in minimal glucose medium.

[25] R. A. Carls and R. S. Hanson, *J. Bacteriol.* **106,** 848 (1971).

Preparation of Membranes

The culture is harvested by centrifugation at 14,000 *g* for 20 min at 4° and the bacteria are washed once in 50 m*M* K-phosphate, pH 8.0. Membranes can be prepared directly or the bacterial pellet can be stored frozen at −20°. For the preparation of membranes 10 g (wet weight) of cells are suspended in 350 ml 50 m*M* K-phosphate, pH 8.0, prewarmed to 37° and containing 90 mg lysozyme (egg white, 3X crystallized, Sigma), 2 mg DNase (bovine pancreas, type 1, Sigma), 2 mg RNase (bovine pancreas, type 1-A, Sigma), and 3.5 ml 1 *M* $MgSO_4$ is added. The suspension is incubated at 37° for 30 min when 17.5 ml 0.3 *M* Na-EDTA, pH 7.5, is added followed a few minutes later by the addition of 10.5 ml 1 *M* $MgSO_4$. Cell debris and unbroken cells are removed by centrifugation at 5000 *g* for 30 min at 4°. All subsequent steps are done at 4° or on ice. Membranes are isolated from the opalescent supernatant by centrifugation at 20,000 rpm in a Sorvall SS-34 rotor (about 48,000 *g*) for 45 min, washed twice in 0.1 *M* K-phosphate, pH 6.6, and homogenized in 4 ml of buffer. The membrane preparation is frozen, preferably in liquid nitrogen, and stored at −80°. The yield of membrane from 10 g (wet weight) *B. subtilis* 168 grown in NSMP is about 80 mg of protein (Lowry procedure, bovine serum albumin standard), with an SDH activity of 1–2 μmol succinate oxidized per minute and per milligram of protein. SDH activity is measured at 38° in the presence of 1 m*M* KCN with phenazine methosulfate and dichlorophenolindophenol as synthetic electron acceptors, as described elsewhere.[2]

Immunoprecipitation and Composition of *B. subtilis* SDH Complex

SDH and cytochrome *b*-558 are parts of a functional unit in *B. subtilis* cytoplasmic membranes and can be solubilized as a complex with nonionic detergent. The complex isolated by immunoprecipitation with specific antibodies contains three different polypeptides in equimolar ratio (Fig. 1).[24] The composition of the complex and some properties of its components are summarized in Table I. The two larger protein subunits of the complex, Fp (M_r 65,000) and Ip (M_r 28,000), are identified as subunits of SDH proper from their resemblance to the two subunits of purified bovine heart and *Rhodospirillum rubrum* SDH.[3]

Bacillus subtilis SDH contains, as determined by EPR spectroscopy, three different iron-sulfur centers, which are analogous to the well-characterized centers S-1, S-2, and S-3 of mammalian SDH (Fig. 2).[9,26,27]

[26] J. J. Maguire and L. Hederstedt, unpublished data (1984).

[27] T. Ohnishi and T. E. King, this series, Vol. 53, p. 483.

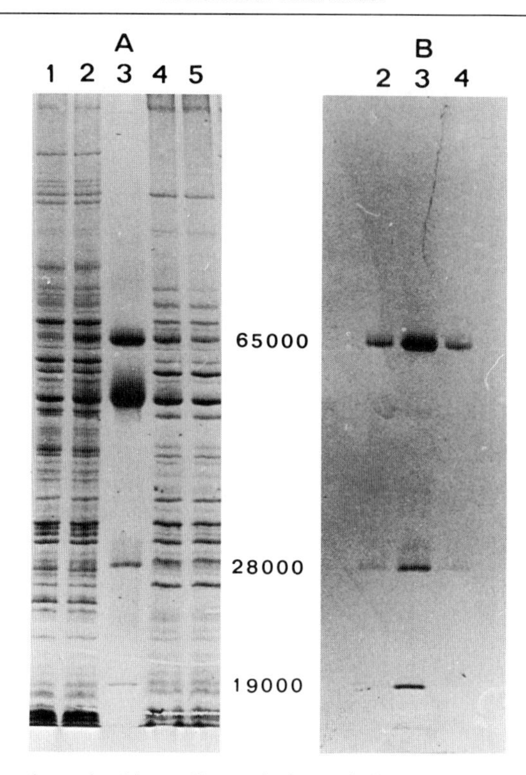

FIG. 1. SDS–polyacrylamide gradient gel of *B. subtilis* immunoprecipitated SDH complex, isolated membranes, and a Triton X-100 extract of membranes. (A) Coomassie Brilliant Blue stained gel; (B) autoradiography of gel shown in (A). The SDH complex was isolated by immunoprecipitation from membranes uniformly labeled with a mixture of ^{14}C-labeled amino acids (algal protein hydrolyzate, New England Nuclear). The slots contain (1) membranes; (2) membranes plus a small amount of radioactive SDH precipitate; (3) radioactive SDH precipitate; (4) Triton X-100 extract of membranes plus a small amount of radioactive SDH precipitate; and (5) Triton X-100 extract of membranes. The apparent molecular weights of the three subunits of the SDH complex are given at their position in the gel. From Hederstedt *et al.*[24]

Evidence for spin–spin interaction between centers S-1 and S-3 in SDH was first shown in *Micrococcus luteus*[28] and has also been found in *B. subtilis*.[26] Spin–spin interaction between centers S-1 and S-2 is seen in *B. subtilis*[9] similar to that demonstrated in SDH from many other species. The spin interactions indicate that the iron-sulfur centers are located close (less than 2 nm) to each other in the protein.

[28] B. A. Crowe, P. Owen, D. S. Patil, and R. Cammack, *Eur. J. Biochem.* **137,** 191 (1983).

TABLE I
COMPONENTS OF *B. subtilis* SDH COMPLEX

Component	Stoichiometry or concentration[a]	Properties
Polypeptides		
Fp	1 mol/mol of complex	M_r 65,000[b]
Ip	1 mol/mol of complex	M_r 28,000
Cytochrome *b*-558	1 mol/mol of complex	M_r 19,000[c]
Prosthetic groups		
Flavin	10 nmol/mg of complex protein 12 nmol/mg of Fp subunit protein	
FeS center		
S-1	1.1 mol/mol of flavin	$E_{m,7} \approx +80$ mV[d]
S-2	nd	$E_{m,7} \approx -240$ mV[d]
S-3	nd	$E_{m,7} \approx -25$ mV[d]
Protoheme	1.3–2 mol/mol of flavin	

[a] Determined on the immunoprecipitated complex or subunit; nd, not determined. Data from L. Hederstedt *et al.*,[24] T. Ohnishi, and L. Hederstedt, unpublished results (1984), and L. Hederstedt.[17]

[b] Fp contains covalently bound flavin, most probably FAD.

[c] The low-temperature EPR spectrum of oxidized cytochrome *b*-558 shows a highly anisotropic low-spin, $g_{max} \approx 3.5$, signal (L. Hederstedt and K. K. Andersson, unpublished, 1985).

[d] Midpoint redox potentials ($E_{m,7}$) were determined on membrane bound enzyme. Maguire and Hederstedt.[26]

Cytochrome *b*-558, the smallest protein subunit (M_r 19,000) of the complex, is a transmembrane protein and anchors SDH to the cytoplasmic side of the membrane.[16] The molecular weight calculated from the nucleotide sequence is 22,770.[15] The cytochrome is reduced by succinate in immunoprecipitated complex and in membranes and is probably the primary electron acceptor from SDH. Reduced cytochrome *b*-558 shows light absorption maxima at 426, 529, and 558 nm at room temperature.[17]

Solubilization of SDH Complex

Bacillus subtilis membranes, at 20 mg of protein/ml, in 24 m*M* Na-diethylbarbiturate buffer, pH 8.6, are mixed with an equal volume of 8% (v/v) Triton X-100[29] (Rhom and Haas Co., Philadelphia, PA) in the same buffer. The mixture is kept at room temperature for 10 min and Triton X-100 insoluble material is removed by centrifugation at 20,000 rpm in a Sorvall SS-34 rotor for 30 min at 4°. The straw-colored, transparent,

[29] A. Helenius, D. R. McCaslin, E. Fries, and C. Tanford, this series, Vol. 56, p. 734.

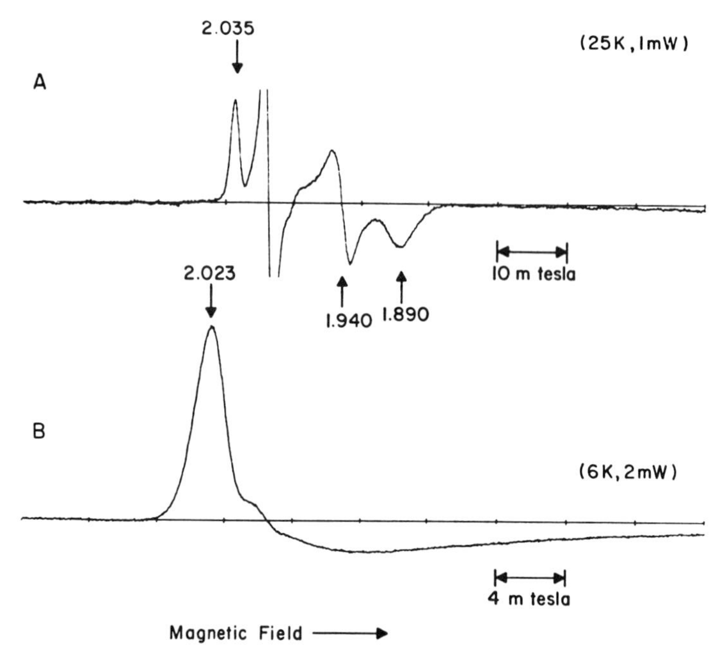

FIG. 2. EPR spectra of iron–sulfur centers S-1 and S-3 in immunoprecipitated *B. subtilis* SDH complex. (A) Center S-1 (spectrum obtained after reduction with 10 mM succinate). (B) Center S-3 [spectrum obtained after oxidation with 1 mM $K_3Fe(CN)_6$]. The arrows indicate g values. The concentration of covalently bound flavin was 10.5 μM. From Hederstedt *et al.*[9]

supernatant is transferred to a new tube and centrifuged at the same speed for 2 hr. The supernatant contains the solubilized SDH complex. (Plastic, disposable, Eppendorf 1.5-ml centrifuge tubes are useful when small volumes are to be centrifuged at up to 50,000 g in the cold. The tubes fit into Sorvall SS-34 rotor adapter part No. 00381 designed for 11 × 75 mm tubes.) About 50% of the protein and 95% of the SDH activity and cytochrome *b*-558 present in membranes are extracted by this procedure. Solubilized SDH–cytochrome *b*-558 complex in 0.05% (v/v) Triton X-100 at 4° is monodisperse and elutes from Ultrogel AcA 34 (LKB, Bromma, Sweden) columns as a protein–detergent complex with an apparent Stokes' radius of 7.0 nm.[12]

Antisera and Immunoelectrophoresis

Antimembrane antiserum is raised in rabbits by injecting the animals subcutaneously at multiple sites on the back with 1 ml of *B. subtilis* membranes at 2 mg of protein/ml in 50% (v/v) Freund's complete adjuvant. The rabbits are boostered on days 4, 7, 14, and 21 subcutaneously and then every 6 weeks intramuscularly with the same amount of antigen now in Freund's incomplete adjuvant. The rabbits can start to be bled on day 30. Contaminating antigens from the growth medium are avoided if membranes for immunization are isolated from cells grown in minimal medium rather than in a complex medium.

SDH-specific antiserum is obtained, without purification of the antigen to homogeneity, by immunizing rabbits with SDH staining immunoprecipitates cut out from wet crossed immunoelectrophoresis (CIE) agar plates containing Triton X-100 solubilized *B. subtilis* membranes run against anti-membrane antiserum.[30] Owen has written an excellent review, with technical details and advice, on the analysis of bacterial membranes by CIE.[31] Immunoelectrophoresis has also been described in a recent volume of this series.[32] CIE of Triton X-100 solubilized *B. subtilis* membranes (20–100 μg of protein) is run on 5 × 5 cm plates in 1% (w/v) agarose (Miles ME agarose, Miles Laboratories Ltd., England or SeaKem HGT, FMC Corp., Marine Colloids Division, Rockland, ME), 1% (v/v) Triton X-100, 24 m*M* Na-diethylbarbiturate buffer, pH 8.6, at 4° at 5 v/cm in the first and at 1.5 v/cm in the second dimension overnight. Triton X-100 solubilized *B. subtilis* SDH migrates toward the anode in this buffer. Purified IgG fraction should be used in immunoelectrophoresis because it results in more distinct precipitation lines, unwanted components in serum such as proteases are removed, the background after protein staining of plates is reduced, and contaminating rabbit antigens are not a problem if precipitates are to be cut out from the agarose for immunization.

After electrophoresis the CIE plate is zymogram stained for SDH activity by soaking the gel in the dark in 2,2′-di-*p*-nitrophenyl-5,5′-diphenyl-3,3′-(3,3′-dimethoxy-4,4′-diphenylene)ditetrazolium chloride (nitro blue tetrazolium, Sigma), 0.3 mg/ml, 0.1 *M* Na-succinate, 5 m*M* KCN, 50 m*M* Tris–HCl buffer, pH 7.5. Immunoprecipitates containing SDH activity are colored blue within 30 min at 37°. If the staining is faint or slow, 0.25 mg/ml of phenazine methosulfate can be included in the mixture.

The stained plate is extensively washed with sterile 0.9% NaCl and the SDH precipitate free from contaminating precipitates is cut out with a scalpel. The immunoprecipitation pattern on the CIE plate can be varied by using different types of agaroses, anti-membrane antisera from different rabbits, and solubilized membranes from cells grown in different me-

[30] B. Rutberg, L. Hederstedt, E. Holmgren, and L. Rutberg, *J. Bacteriol.* **136,** 304 (1978).

[31] P. Owen, *in* "Organization of Prokaryotic Cell Membranes" (B. K. Ghosh, ed.), Vol. 1, p. 73. CRC Press, Boca Raton, Florida, 1981.

[32] C.-B. Laurell and E. J. McKay, this series, Vol. 73, p. 339.

dia or membranes from different strains. All immunoprecipitates on a parallel plate and the plate from where the antigen was cut out are stained with Coomassie Brilliant Blue. This is a help in the cutting and acts as a control of clean excision of SDH.

The agarose pieces with SDH precipitates from several CIE plates are pooled and homogenized in a small volume of sterile 0.9% NaCl, mixed 1 : 1 (v/v) with Freund's complete adjuvant, and injected subcutaneously at multiple sites on the back of rabbits. The injections are repeated as described for immunization with membranes. SDH antigen from 100 μg or more of Triton X-100 solubilized *B. subtilis* membranes is injected per rabbit per day of injection. Crude sera can be tested for SDH antibodies in CIE against solubilized membranes followed by staining for SDH. About two-thirds of all the white New Zealand female rabbits we have immunized produce immunoprecipitating antibodies against *B. subtilis* SDH.

SDH-specific antisera should only give one symmetrical precipitate arc in CIE against Triton X-100 solubilized *B. subtilis* membranes.[30] The SDH precipitate has been shown to contain covalently bound flavin, heme, iron, and most likely contains only three bacterial polypeptides, i.e., Fp, Ip, and cytochrome *b* polypeptide.

The specificity and subspecificity of SDH antisera can be analyzed in Western blots[33] against membranes or immunoprecipitated SDH complex polypeptides separated in an SDS–polyacrylamide gel or by SDS–polyacrylamide crossed immunoelectrophoresis.[22,34] In both techniques Triton X-100 insoluble antigens can be detected and their molecular weight determined. An advantage of the latter technique is that a good estimate of the relative amount of antigen or antibody is obtained in only one gel. However, only immunoprecipitating antibodies are detected and the method requires large amounts of antibodies.

The SDH-specific antisera we have produced have all contained precipitating antibodies against the Fp subunit, often also to the Ip subunit, but never against cytochrome *b*-558.[16,22] Fp- and Ip-specific antisera can be raised by immunizing rabbits with Fp or Ip polypeptides from immunoprecipitated SDH complex separated on an SDS–polyacrylamide gel. Each polypeptide band is cut out, the gel is homogenized in a small Waring-type blender, as described,[22] and the suspension is injected into rabbits according to the procedure described above for membranes. Polypeptides to be used as antigens should not be stained before, since stains such as Coomassie Brilliant Blue are immunogenic when bound to protein.

[33] J. Renart and I. V. Sandoval, this series, Vol. 104, p. 455.

[34] C. A. Converse and D. S. Papermaster, this series, Vol. 96, p. 244.

Immunoprecipitation of SDH Complex

Bacillus subtilis SDH constitutes 2% or less of the protein in membranes and has only been isolated by immunoprecipitation of detergent-solubilized enzyme. Direct immunoprecipitation of SDH complex is accomplished by mixing Triton X-100 membrane extract in a centrifuge tube with SDH-specific, polyclonal, IgG fraction of rabbit antiserum and followed by incubation at 4° for 12–20 hr. The amount of antiserum required to precipitate SDH is determined in a preliminary experiment. The resulting red-brown immunoprecipitate is collected by centrifugation at 35,000 *g* for 15 min at 4° and washed five times in 0.1 *M* Na-phosphate buffer, pH 8.0; 1% (v/v) Triton X-100 is included in the buffer during the first two washes. The precipitate can be stored frozen at −80° in a small amount of buffer. A considerable amount of enzyme activity (up to 45 μmol of succinate oxidized per minute and per milligram of bacterial protein, at 38°) is retained in the immunoprecipitate. The activity corresponds to a turnover number of up to 5000 mol of succinate oxidized per minute per mol of covalently bound flavin. The purity and the amount of bacterial protein in the immunoprecipitate is best analyzed if radioactive uniformly labeled bacterial protein is used. The purity can be checked by SDS–electrophoresis of the radioactive precipitate and autoradiography of the resulting gel (Fig. 1). The amount of SDH is quantitated by fluorometric determination of covalently bound flavin or calculated from the amount of radioactivity in the precipitate, assuming that the specific radioactivity in SDH complex is the same as that of average cell protein and that radioactivity is only in protein in the precipitate.[24]

Genetics of *B. subtilis* SDH–Protoplast Fusion

All *B. subtilis* mutants totally lacking SDH activity carry mutations in the *sdh* locus at 255° on the circular chromosomal genetic map. This locus was previously called *citF*.[12] It contains at least three genes: *sdhA*, *sdhB*, and *sdhC*, which are part of one operon[13] and are the structural genes for cytochrome *b*-558, Fp, and Ip, respectively (Fig. 3). We have by now, by a variety of methods involving genetic, immunological, biochemical, and biophysical techniques, characterized 33 different *sdh* mutants. Based on their different phenotypes the mutants can be arranged into the 8 classes listed in Table II.

Isolation of sdh Mutants

Bacillus subtilis SDH negative mutants have been isolated as asporogenic or acid-excreting mutants from ethylmethane sulfonate (EMS)[12] or

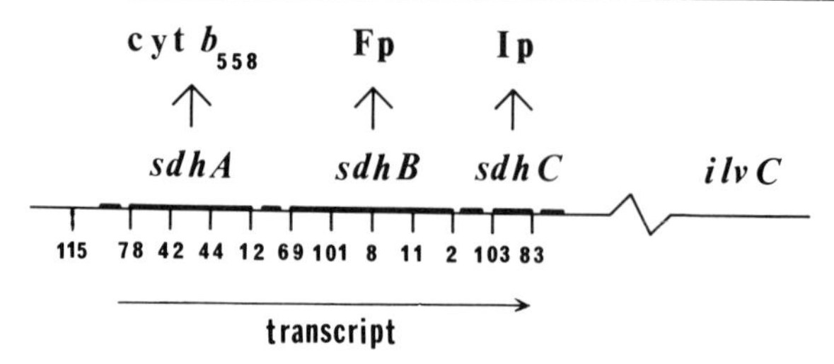

FIG. 3. Genetic organization of the *sdh* operon on the *B. subtilis* chromosome. The numbers refer to different mapped *sdh* mutations. The relative positions of the mutations are shown.

heat-treated[25] spores and from *N*-methyl-*N'*-nitro-*N'*-nitrosoguanidine-(NTG)[35,36] treated vegetative cells. Spores are practical to use, since they can easily be stored and occur separate in contrast to vegetative cells which often form chains. This latter property of spores is of importance when we want to select mutants with recessive phenotypes directly on agar slants. Furthermore, each mutant colony that grow up on an agar slant on which spores have been spread represents a unique clone.

For the isolation of *B. subtilis* mutants lacking SDH activity, spores, mutagenized with EMS according to the method of Ito and Spizizen,[37] are spread on PA agar slants at a density of about 50 viable germinating cells per plate. Acid-excreting (yellow) mutant colonies appearing overnight at 37° are restreaked on PA and finally on TBAB slants. *sdh* mutants, among the acid-excreting mutants, are screened for by a modification of the replica colony zymogram staining procedure of Edwards *et al.*[38] The colonies on TBAB are replicated to a filter paper. The bacteria on the replica are then lysed by soaking the filter in lysozyme at 4 mg/ml in 50 m*M* K-phosphate buffer, pH 7.4, at 37° for 30 min. The filter is washed with 5 × 50 ml of the phosphate buffer on a Büchner funnel, connected to a vacuum flask, and stained for SDH activity in the zymogram mixture that is also used for CIE plates and described under "Antisera and Immunoelectrophoresis." Colonies containing SDH activity are stained blue in about 5 min at 37°, whereas SDH-negative colonies are not stained. The *sdh* phenotype is confirmed by direct enzyme activity measurement on isolated membranes from cells grown in NSMP. Finally, the mutation is transformed into an isogenic background. Linkage to *ilvC* (Fig. 3) on the

[35] B. Rutberg and J. A. Hoch, *J. Bacteriol.* **104,** 826 (1979).

[36] M. Ohné, B. Rutberg, and J. A. Hoch, *J. Bacteriol.* **115,** 738 (1973).

[37] J. Ito and J. Spizizen, *Mutat. Res.* **13,** 93 (1971).

[38] D. L. Edwards, D. M. Bensole, H. J. Guzik, and B. W. Unger, *J. Bacteriol.* **137,** 900 (1979).

TABLE II
PHENOTYPES OF *B. subtilis sdh* MUTANTS

Class	Mutated subunit of SDH complex	Phenotype	Example of mutation	References
I	Fp	Membrane-bound SDH complex contains flavin and centers S-1, S-2, S-3	*sdhB101*	8,9,12
II	Fp	Same as class I, but lacks flavin	*sdhB124*	8,9,12
III	Fp	Ip in cytoplasm, cytochrome *b*-558 in membrane, cell lacks center S-1	*sdhB11*	9,12,22,36
IV	Ip	Fp in cytoplasm and contains flavin, cytochrome *b*-558 in membrane, cell contains center S-1	*sdhC83*	8,9,12,22,36
V	Ip	Same as class IV, but center S-1 is missing	*sdhC123*	8,9,12
VI	Cytochrome *b*-558	Fp and Ip in the cytoplasm, *b*-558 chromophore in the membrane, cell contains center S-1	*sdhA12*	8,9,12,22,36
VII	Cytochrome *b*-558	Same as class VI, but lacks *b*-558 chromophore	*sdhA109*	9,12
VIII	—	Lacks Fp, Ip, and cytochrome *b*-558, promoter mutation	*sdh-115*	13

chromosome can be tested at the same time.[36] Methods for genetic mapping and characterization of *sdh* mutants can be found in the references to Table II.

Complementation Analysis of sdh Mutants by Protoplast Fusions

Protoplast fusion is a very useful method for complementation studies of mutants in situations where stable diploid cells cannot easily be maintained and complementation in mixed extracts *in vitro* is not successful because labile components (like iron-sulfur centers) or unknown required factors are involved. Membrane-bound active SDH can be formed in fused protoplasts of different *B. subtilis sdh* mutants by complementation.[12,39] Three complementation groups, *sdhA, sdhB,* and *sdhC,* have been found. *sdhA* mutants are defective in binding of SDH to the membrane and accumulate nonmutated, seemingly enzymatically inactive FP and Ip subunits in the cytoplasm. The majority of *sdhB* and *sdhC* mutants

[39] L. Hederstedt and S.-Å. Franzén, *FEMS Microbiol. Lett.* **23,** 51 (1984).

contain nonmutated Ip or Fp subunits, respectively, in the cytoplasm and functional binding sites for SDH in the membrane. Membranes and cytoplasms of both cells are mixed in fused protoplasts. Thus, all three SDH subunits are present and membrane-bound SDH can be assembled in fused protoplasts of, for example, an *sdhB* and an *sdhC* mutant.

The *sdh* mutant of unknown complementation group is fused with one representative mutant each of group *sdhA, sdhB,* and *sdhC*. Mutants containing one or two functional soluble SDH subunits are expected to be complemented by mutants of two of the three groups.[12] Complementation is determined from SDH activity in membranes isolated from the fused protoplasts. The activity can be determined quantitatively in a spectrophotometer or qualitatively (which is more sensitive) by rocket immunoelectrophoresis[32] of Triton X-100 solubilized membranes run against SDH-specific antiserum and SDH zymogram staining of immunoprecipitates. Protoplasts of each participating mutant fused with itself are negative controls. The efficiency of complementation measured as SDH activity in membranes from fused *sdh* mutant protoplasts is usually a few percent and rarely exceeds 15% of the theoretical maximal activity.

Solutions for protoplast fusion are as follows:[40,41]

1. SMM buffer: 20 m*M* $MgCl_2$, 20 m*M* maleate buffer, pH 6.5, containing 0.5 *M* sucrose
2. SMM lysozyme: 4 mg lysozyme/ml in SMM
3. SMM PEG: 50% (w/v) polyethylene glycol (MW 6000, Merck) in SMM
4. SMM BSA: 1% (w/v) bovine serum albumin in SMM
5. PPI: 10% (v/v) LB in SMM BSA (LB: 1 g tryptone, 0.5 g yeast extract, 0.5 g NaCl in 100 ml distilled water, pH is adjusted to 7.0)

The two mutants to be fused together are grown separately in 600 ml of NSMP at 32° until A_{600} = 0.6 (or until the exponential growth starts to decline). Bacteria grown in NSMP with 0.5 *M* sucrose result in more stable protoplasts, but contain only about 20% the amount of SDH subunits present in cells grown in NSMP without sucrose. Each culture is harvested by centrifugation at 6000 *g* for 10 min at 15°, washed once in SMM, and suspended in 12 ml SMM BSA. SMM lysozyme (0.4 ml) is added and the bacteria are converted to protoplasts by incubation at 42° until more than 95% of the cells are spheres (30–40 min). The protoplasts, 5 ml of each mutant, are mixed and centrifuged in a Wifug clinical centrifuge at 5700 rpm for 15 min at room temperature. The protoplast pellet is

[40] P. Schaeffer, B. Cami, and R. D. Hotchkiss, *Proc. Natl. Acad. Sci. U.S.A.* **73,** 2151 (1976).

[41] C. Sanchez-Rivas and A. J. Garro, *J. Bacteriol.* **137,** 1340 (1979).

very gently suspended in 0.65 ml SMM BSA and 1 ml of the suspension is mixed with 5 ml cold SMM PEG in a beaker on ice. The protoplasts are after 1.5 min on ice diluted by the addition of 54 ml PPI (prewarmed at 37°) and incubated at 37° for 60 min with slow mixing. About 50% of the protoplasts lyse during the procedure if the *sdh* mutants have been grown in NSMP and about 25% if grown in NSMP containing 0.5 *M* sucrose. Membranes from protoplasts are isolated by centrifugation of the PPI suspension at 7000 *g* for 10 min at 15° and by homogenizing the pellet in 10 m*M* $MgSO_4$, 50 m*M* K-phosphate buffer, pH 8.0, containing 10 μg/ml of DNase and RNase. After incubation at 37° for 15 min 0.75 ml 0.4 *M* Na-EDTA, pH 7.5, is added and after another 2 min 0.6 ml 1 *M* $MgSO_4$ is added. Unlysed cells and debris are removed by centrifugation at 5000 *g* for 15 min at 4° and membranes are isolated from the supernatant by centrifugation at 20,000 rpm in a Sorvall SS-34 rotor for 30 min.

Biosynthesis of SDH Complex

We have proposed[23] and, by the use of *sdh* and heme mutants, to a large extent experimentally confirmed a model for the biosynthesis and assembly of SDH in *B. subtilis*.[3] Fp and Ip polypeptides are synthesized, probably translated from polycistronic mRNA, in the cytoplasm. The newly synthesized SDH subunits are soluble and of the same apparent molecular weight as the corresponding membrane-bound subunits.[22] Apo-cytochrome *b*-558 is proposed to be inserted across the membrane in concert with its synthesis. When heme is incorporated, resulting in the appearance of holocytochromes, SDH can be bound to the cytochrome to form active SDH complex. At heme starvation in *B. subtilis* all three polypeptides of the complex are still made, but the Fp and Ip subunits accumulate in the cytoplasm.[22]

Flavin is covalently attached, by a yet unknown mechanism, to the soluble Fp polypeptide.[22] This occurs also in the absence of Ip subunit.[8] Covalently bound flavin in the Fp subunit is required for enzymatic activity, but not for assembly of membrane-bound SDH complex containing all three iron-sulfur centers.[9] Furthermore, the apparent midpoint redox potentials of the centers and spin–spin interactions seen between the centers are the same in mutant SDH lacking flavin, as in the wild-type enzyme.[26] Center S-1 is incorporated into soluble SDH before the enzyme is bound to the membrane, but in contrast to the situation for flavin, both Fp and Ip polypeptide seem to be required for assembly of center S-1.[9]

From our present data on different *sdh* mutants it seems likely that *B. subtilis* SDH can be assembled and bound to the membrane only if iron-sulfur centers are present in the protein.

Heme-Deficient B. subtilis

The usefulness of heme mutants in studies of cytochrome-linked electron-transfer systems in bacteria and yeasts has been emphasized in earlier volumes of this series.[42–44] *Bacillus subtilis* mutants defective in different steps in heme biosynthesis have been isolated.[45] We have used a 5-aminolevulinic acid requiring (*hemA*) mutant for studies on SDH biosynthesis.[22,23] *Bacillus subtilis* is a strictly aerobic bacterium dependent on a respiratory chain with cytochromes for growth and cannot, as facultative cells, be grown in the absence of heme. To analyze effects of heme deficiency, the strategy of heme starvation has to be employed. A *B. subtilis hemA* mutant is grown in minimal glucose medium supplemented with 5-aminolevulinic acid at 2–5 μg/ml (depending on the strain). The cells, in exponential growth phase, are centrifuged at 14,000 *g* for 20 min at room temperature, suspended, and diluted in prewarmed medium now without growth factor and incubated again. Holocytochrome synthesis and membrane binding of SDH stops, but cell growth and protein synthesis can continue unaffected for up to three cell generations.[23] The cytochrome and SDH content in cell membranes increase concomitantly and reach normal levels within 15 min if 5-aminolevulinic acid is added to the starved cells. The concentration of 5-aminolevulinic acid required in the medium and the extent to which the cells can be starved varies with the strain. We have obtained best results with derivatives of *B. subtilis* strain 3G18.

Acknowledgment

Work done in the author's laboratory has been supported by grants from the Swedish Medical Research Council. I thank Lars Rutberg for criticism and Kristina Malmqvist for typing the manuscript.

[42] J. Lascelles, this series, Vol. 56, p. 172.

[43] J. R. Mattoon, J. C. Beck, E. Carvajal, and D. R. Malamud, this series, Vol. 56, p. 117.

[44] J. M. Haslam and A. M. Astin, this series, Vol. 56, p. 558.

[45] T. J. Anderson and G. Ivánovics, *J. Gen. Microbiol.* **49,** 31 (1967).

Section II

Reversible ATP Synthase (F_0F_1-ATPase)

A. Preparation and Reconstitution
Articles 39 through 61

B. Kinetics, Modification, and Other Characterization
Articles 62 through 74

[39] Preparation of a Highly Coupled H^+-Transporting ATP Synthase from Pig Heart Mitochondria

By DANIÈLE C. GAUTHERON, FRANÇOIS PENIN, GILBERT DELÉAGE, and CATHERINE GODINOT

This chapter describes a procedure to obtain a preparation of H^+-transporting ATP synthase from pig heart mitochondria as pure and as active for ATP synthesis as possible. One of the difficulties in preparing such a large membrane complex is that the number and stoichiometry of the subunits required for ATP synthesis are not well established. Besides, due to the hydrophobic or hydrophilic character of the subunits and parts of the complex, a problem often encountered is the loss of components during the purification procedure. In working out this method our purpose has been to increase the specific activity for ATP synthesis and to obtain a directly active complex without a reconstitution step. The preparation exhibits well-correlated rates of ATP hydrolysis, ATP synthesis, and proton translocation.

The methodologies described are based on our accumulated experience in this field.[1,2]

Principle of the Preparation

The basic approach of the procedure was to extract the H^+-transporting ATP synthase from the mitochondrial inner membrane after elimination of as many contaminating proteins as possible from either face of the membrane. Therefore mitoplasts are prepared from pig heart mitochondria, thoroughly washed, then sonicated to yield inverted submitochondrial particles. After the final washing of these particles the H^+-transporting ATP synthase complex is extracted with α-lysophosphatidylcholine, a mild detergent.

Purification Procedures

All steps are performed at 0–4° unless stated otherwise.

Mitochondria are prepared from homogenized fresh pig heart, spun down 20 min at 15,000 *g*, suspended in 0.25 *M* sucrose, 10 m*M* Tris–HCl,

[1] F. Penin, C. Godinot, J. Comte, and D. C. Gautheron, *Biochim. Biophys. Acta* **679,** 198 (1982).

[2] G. Deléage, F. Penin, C. Godinot, and D. C. Gautheron, *Biochim. Biophys. Acta* **725,** 464 (1983).

pH 7.4, and washed twice in the same medium before any fragmentation, according to a procedure[3] derived from that of Crane *et al.*[4]

Mitoplasts must be obtained immediately from fresh mitochondria as previously described,[5] except that mitochondria are homogenized at 10 mg protein/ml in 20 m*M* sodium phosphate instead of 10 m*M* with a loose-fitting pestle, in order to eliminate any trace of adenylate kinase. After 30 min swelling, the suspension is centrifuged at 105,000 *g* for 60 min to eliminate all intermembrane enzymes (supernatant discarded). The pellet, which contains outer and inner membranes plus matrix, is homogenized at about 20 mg/ml protein in 0.25 *M* sucrose, 10 m*M* Tris–HCl, pH 7.4, and spun down at 11,500 *g* for 10 min. The mitoplasts (inner membrane + matrix) constitute the pellet, while the supernatant containing outer membrane is discarded. Mitoplasts are washed in the same medium and pelleted at 11,500 *g*.

To prepare submitochondrial particles, mitoplasts are homogenized at 30 mg protein/ml in 0.25 *M* sucrose, 10 m*M* Tris–H_2SO_4, 1 m*M* ATP, 15 m*M* $MgSO_4$, pH 7.5. The suspension is sonicated under the following conditions: 3 min, 50 ml batch, temperature maintained between 0 and 6° in a jacketed cell connected to a cryostat, Branson sonifier B12 equipped with a 0.5-in. diameter probe, at a power of 60 W. The sonicated mitoplasts are centrifuged at 22,500 *g* for 10 min. The supernatant is centrifuged at 105,000 *g* for 75 min. The pellet of submitochondrial particles is homogenized in 0.25 *M* sucrose, 10 m*M* Tris–H_2SO_4, 10 m*M* $MgSO_4$, pH 7.5, and centrifuged again under the same conditions. Mitoplasts and submitochondrial particles can be kept at −80° and thawed later on without damage for the further operations. In this latter case the particles are washed again before use.

Extraction of the H^+-transporting ATP synthase complex is performed by treating submitochondrial particles with α-lysophosphatidylcholine as follows: The submitochondrial particles are diluted at 10 mg protein/ml in 0.25 *M* sucrose, 10 m*M* Tris–H_2SO_4, 10 m*M* $MgSO_4$, pH 7.5. A 10% solution of α-lysophosphatidylcholine (Sigma Chemical Co., type I) in the same buffer is added to a final concentration of 0.1% and 30 min later, an equal volume of 100 m*M* of 2-(*N*-morpholino)ethanesulfonic acid (Mes) adjusted to pH 6.5 with KOH is added. The mixture is gently stirred during the additions and for a further 20 min. Then it is centrifuged at 105,000 *g* for 15 min. The pellets are discarded and the upper two-thirds of the supernatants are centrifuged again at 155,000 *g* for 2 hr. The final

[3] C. Godinot, C. Vial, B. Font, and D. C. Gautheron, *Eur. J. Biochem.* **8,** 385 (1969).

[4] F. L. Crane, J. L. Glenn, and D. E. Green, *Biochim. Biophys. Acta* **22,** 475 (1956).

[5] J. Comte and D. C. Gautheron, this series, Vol. 55, p. 98.

TABLE I
PURIFICATION OF H^+-TRANSPORTING ATP SYNTHASE FROM PIG HEART MITOCHONDRIA

Step	Total protein (mg)	Ratio[a] PL/protein (mg/mg)	ATPase[b] activity	P_i–ATP[b] exchange	Purification factor	
					ATPase	exchange
Mitochondria	4032	0.34	0.24	0.07	1	1
Mitoplasts	3010	0.32	0.24	0.034	1	0.5
Submitochondrial particles	624	0.59	1.2	0.36	5	5
H^+-transporting ATP synthase	36	1.14	5.36	1.65	22	24

[a] PL, Phospholipids estimated as previously described.[16]
[b] Activity expressed in μmol/min/mg protein.

pellets containing the H^+-transporting ATP synthase complex are homogenized in a glass potter with a Teflon pestle in a minimum volume of 0.25 *M* sucrose, 10 m*M* Tris–H_2SO_4, 10 m*M* $MgSO_4$, pH 7.5. The mixture is centrifuged at 9000 *g* for 5 min to eliminate any aggregates. The final supernatant contains the H^+-transporting ATP synthase at a concentration of about 15 mg protein/ml. Aliquots stored in liquid nitrogen keep their activities for months, but repeated thawing and freezing must be avoided. Just before use, the aliquots are submitted to filtration–centrifugation through a 1-ml tuberculin syringe containing fine[6] Sephadex G-50 swollen in the sucrose buffer required for further experiments. This last step increases the rate of ATP synthesis, probably due to excess detergent elimination. All operations starting from the collection of pig hearts at the slaughterhouse can be performed the same day. The summary of a typical run of purification is given in Table I.

Assay Methods

Protein Concentration

This is estimated by the biuret method[7] in the presence of 1% deoxycholate or by the method of Lowry *et al.*[8] modified as in Ref. 9.

[6] H. S. Penefsky, this series, Vol. 56, p. 527.
[7] A. G. Gornall, C. J. Bardawill, and M. M. David, *J. Biol. Chem.* **177,** 751 (1949).
[8] O. H. Lowry, N. J. Rosebrough, A. L. Farr, and R. J. Randall, *J. Biol. Chem.* **193,** 265 (1951).
[9] A. Bensadoun and D. Weinstein, *Anal. Biochem.* **70,** 241 (1976).

Hydrolytic Activity

Principle. The rate of hydrolysis of ATP or GTP (free from any ATP) is estimated by the amount of $^{32}P_i$ liberated from [γ-^{32}P]ATP or [γ-^{32}P]GTP.

Reagents

[γ-^{32}P]ATP is prepared according to Penefsky.[10] The method was adapted to prepare [γ-^{32}P]GTP which is eluted from the Dowex AG1-X8 chloride column (1 × 3 cm) according to the following schedule: 20 ml of H_2O; 20 ml of 0.003 *N* HCl; 35 ml of 0.01 *N* HCl, 0.08 *M* NaCl; 25 ml 0.01 *N* HCl, 0.2 *M* NaCl, 0.1 *M* KCl. Labeled GTP is eluted in the latter fraction. Salts are eliminated by charcoal treatments, as in Ref. 11. The radiochemical purity of nucleotides is monitored on a radiochromatogram scanner (Packard) after thin-layer chromatography on PEI-cellulose F (Merck) using 1 *M* LiCl as a developing solvent.

Perchloric acid, 35%

Incubation mixture, stock solutions: (1) 50 m*M* Tris-acetate, 0.6 *M* sucrose, and 30 m*M* $MgSO_4$, pH 7.5; (2) potassium phosphate, 0.4 *M*, pH 7.5; (3) ATP or GTP [for GTPase activity (Boehringer, disodium salts)], 200 m*M*, pH 7.5 (Tris base); (4) ADP (Boehringer, monopotassium salt), 100 m*M*, pH 7.5 (Tris base). The incubation mixture is made extemporaneously with 10 volumes of (1), 2 volumes of (2), 1 volume of (3) and (4), and 5 volumes of distilled water containing labeled nucleotides (10^4 cpm/mol, final) and effectors, when added.

Ap_5A: P_1,P_5-diadenosine 5′-pentaphosphate (Boehringer, trilithium salt), stock solution 10 m*M*, pH 7.5 (Tris base)

Inorganic phosphate extraction medium

Stock solutions: (5) Perchloric acid, 1.25 *M* (dilute 108 ml of 70% acid to 1000 ml with distilled water); (6) isobutanol/benzene, 1 : 1 (v/v); (7) isobutanol saturated with H_2O; (8) ammonium molybdate, 4%. Final extraction medium: For each assay, 4 ml (5), 5 ml (6), 3 ml (7), and 1 ml (8) are mixed in a 25-ml stoppered test tube.

Procedure for ATPase or GTPase Activity

The H^+-transporting ATP synthase (15 mg protein/ml) after thawing and gel filtration (as described above) is diluted in 0.25 *M* sucrose, 25 m*M*

[10] H. S. Penefsky, this series, Vol. 10, p. 702.

[11] D. C. Gautheron and R. Morélis, *Bull. Soc. Chim. Biol.* **47,** 1923 (1965).

Tris–H_2SO_4, 15 mM $MgSO_4$, pH 7.5 to obtain a concentration of 1–5 mg protein/ml, and incubated for 15 min at 37°. After this incubation, which increases the specific activities, the reaction rates remain constant for 1 hr. To initiate the reaction, 50 μl of enzyme are mixed to 950 μl of incubation mixture containing routinely 0.2 mM Ap_5A to prevent any adenylate kinase activity. After 5 min at 37° a 500-μl aliquot from the above incubation mixture is added to the final extraction medium (13 ml in stoppered test tubes). After shaking the tubes vigorously for 20 sec, the upper organic phase (8 ml) containing more than 99.9% of the free inorganic phosphate is removed. Aliquots of 5 ml are directly counted for radioactivity. At the end of the counting, 5 μl of incubation medium are added to each assay in view of estimating the amount of labeled nucleotide hydrolyzed using an internal standard.

When ATPase or GTPase activities are measured in mitochondria, mitoplasts, and submitochondrial particles, after the 5-min incubation of the reaction mixture at 37°, 0.1 ml 35% perchloric acid is added at 0° to stop the reaction. The assays are centrifuged at 14,000 g for 5 min and $^{32}P_i$ liberated is measured as quickly as possible, as described above on 500 μl aliquots.

ATP Synthase Activity

Principle. The rate of ATP synthesis is estimated by the ATP–P_i exchange based on the incorporation of $^{32}P_i$ into the terminal phosphoryl group of ATP as proposed in a previous volume,[12] modified by the addition of excess ADP so as to obtain linear rates of ATP synthesis even in preparations with low rates of ATPase activity.

Reagents

The incubation mixture is that described for the hydrolytic activity except that added nucleotides are unlabeled and that $^{32}P_i$ is included (1–2.10^6 cpm/ml). $^{32}P_i$ is obtained from the Commissariat à l'Energie Atomique, C.E.A., France.

Other reagents are the same as for the measurements of hydrolytic activities.

Procedure

ATP–$^{32}P_i$ exchange activity is measured exactly under the conditions described for ATPase activity, but for the presence of $^{32}P_i$ and absence of labeled nucleotides in the initial reaction mixture. After 5 min of reaction,

[12] M. E. Pullman, this series, Vol. 10, p. 57.

inorganic phosphate is extracted from 500 μl aliquots as described above for the hydrolytic activity, but the upper organic phase is suctioned up and discarded. The lower water phase containing the nucleotides is washed successively by 10 ml of water-saturated isobutanol, then by 2 ml of diethyl ether. Aliquots of 3 ml are counted as described above and the amount of labeled nucleotide estimated with an internal standard. It was checked by chromatography on PEI-cellulose that all $^{32}P_i$ present in the water phase is incorporated into the terminal phosphoryl group of ATP.[1] Correction was made for ATP hydrolysis during the exchange, as reported in Ref. 13.

Adenylate Kinase and Nucleoside Diphosphokinase Activities

These are estimated by measuring the initial rate of ATP formation using the firefly luciferase assay as previously described.[14] The H^+-transporting ATP synthase preparation is incubated in the same reaction mixture described above for ATP hydrolysis and synthesis. To measure the adenylate kinase activity, Ap_5A was omitted and 25 μg oligomycin/ml is added to prevent ATP formation through the ATP–P_i exchange system. ADP (5 m*M*) was present and ATP absent. For the measure of nucleoside diphosphokinase activity, 10 m*M* GTP, 5 m*M* ADP, 0.2 m*M* Ap_5A, and 25 μg oligomycin/ml are added.

Proton Translocation Induced by ATP Hydrolysis

Principle. The rate of proton translocation is estimated by the quenching of fluorescence of ACMA (9-amino-6-chloro-2-methoxyacridine),[15] the membrane potential being collapsed as much as possible with the addition of potassium plus valinomycin.

Reagents

Medium buffer: 30 m*M* Tris, 10 m*M* Mes, 10 m*M* $MgSO_4$, 50 m*M* KCl, 1 m*M* dithiothreitol, 10% glycerol, pH adjusted to 7.4 with HCl
Valinomycin in ethanol, 0.15 m*M*
ACMA in ethanol, 0.25 m*M* (ACMA was kindly provided by Dr. Kraayenhof)
ATP, 10 m*M*, pH 7.4 (Tris base)
ADP, 100 m*M*, pH 7.4 (Tris base)

[13] D. L. Stigall, Y. M. Galante, and Y. Hatefi, this series, Vol. 55, p. 308.
[14] F. Penin, C. Godinot, and D. C. Gautheron, *Biochim. Biophys. Acta* **548**, 63 (1979).
[15] R. Kraayenhof and J. W. T. Fiolet, *in* "Dynamics of Energy Transducing Membranes" (L. Ernster, R. W. Estabrook, and E. C. Slater, eds.), p. 355. Elsevier, Amsterdam, 1974.

Oligomycin (Boehringer), 1 mg/ml ethanol
FCCP [carbonyl cyanide *p*-trifluoromethoxyphenylhydrazone (Sigma Chemical Co.)] in ethanol, 0.1 m*M*

Procedure

In a fluorescence cuvette are added 1.75 ml of medium buffer, 10 μl of valinomycin (final 0.75 μ*M*), 10 μl ACMA (final 1.25 μ*M*), and the final volume is adjusted to 2 ml with distilled water, taking into account protein and effectors addition. H^+-transporting ATP synthase, 110 μg protein/2 ml, is preincubated for 1 min at room temperature. The fluorescence of ACMA in the assay is taken as 100% for a full scale of the recorder. The reaction is initiated by addition of ATP (final concentration between 0.01 and 0.5 m*M*). The initial rate of fluorescence quenching of ACMA is estimated from the slope of the recorder trace in less than 30 sec and expressed by percentage of quenching per minute. The excitation and emission wavelengths are 420 and 510 nm, respectively. The measurements can be made with a Farrand manual spectrofluorimeter.

Characteristics of the Purified H^+-Transporting ATP Synthase

Structure and Composition

Freeze-fracture studies show that the H^+-transporting ATP synthase complex is integrated in lipid vesicles of 40–60 nm. The vesicular structure and the presence of lipids appear necessary for a high ATP synthesis activity. The increase in phospholipids shown in Table I[16] is mainly due to incorporation of α-lysophosphatidylcholine during extraction of the complex.

Figure 1 shows the polypeptide composition of the preparation as determined by two-dimensional polyacrylamide gel electrophoresis using sodium dodecyl sulfate (SDS) in the first dimension and tetradecyltrimethylammonium bromide (TDAB), a cationic detergent, in the second one, as described in Ref. 17. This method allows a very good separation of α from β and of small peptides. Any attempt to decrease the number of peptide subunits involved solubilization steps and resulted in a loss of ATP synthase activity.

Activities

Table I shows that both activities increase in parallel along the purification steps. Our procedure yields a complex with specific activities

[16] J. Comte, B. Maïsterrena, and D. C. Gautheron, *Biochim. Biophys. Acta* **419,** 271 (1976).
[17] F. Penin, C. Godinot, and D. C. Gautheron, *Biochim. Biophys. Acta* **775,** 239 (1984).

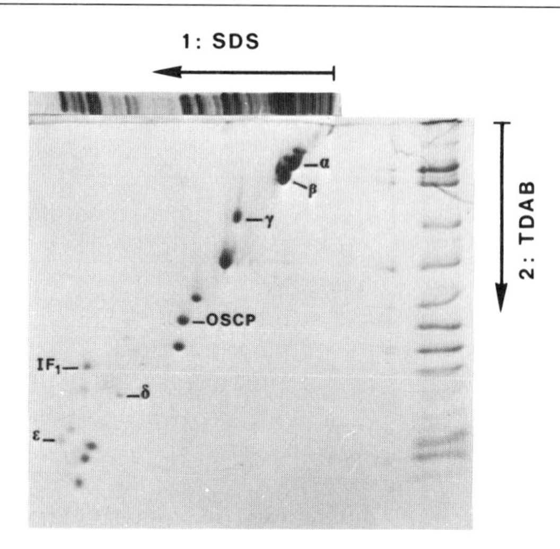

FIG. 1. Two-dimensional polyacrylamide gel electrophoresis pattern of the purified H^+-transporting ATP synthase.[17] Sodium dodecyl sulfate (SDS), an anionic detergent, is used in the first dimension in 15% polyacrylamide gel, according to U. K. Laemmli, *Nature (London)* **227,** 680 (1970).

After removal of the SDS by a mixture of isopropanol–water (1 : 3, v/v), a strip of the gel containing about 50 μg of proteins is equilibrated with 3% TDAB, 10% glycerol, 2% 2-mercaptoethanol, 50 m*M* sodium phosphate buffer, pH 4, for 2 hr. Then the electrophoresis is run in the second dimension in the presence of the cationic detergent, TDAB, 0.125% (w/v), in 85 m*M* glycine buffer adjusted to pH 3.0 with phosphoric acid, as described by A. Amory, F. Foury, and A. Goffeau, *J. Biol. Chem.* **255,** 9353 (1980). The electrophoresis is run 1 hr at 50 mA and 3 hr at 80 mA per slab (2 mm thick × 120 mm wide × 160 mm high, Bio-Rad cell, model 220). Finally, gels are stained by Coomassie Brilliant Blue R. Identification of spots is made by comigration of purified peptides labeled by ^{125}I.

about 5-fold that of the submitochondrial particles. This result had never been reported before and our preparation of H^+-transporting ATP synthase appears the most efficient for ATP synthesis. The high activity obtained (1.6–1.7 μmol ATP synthesized/mg protein/min) in addition to the close relationship between both activities indicate that most or all of the complex is purified by this method. Further support comes from the fact that a ratio close to 1 between hydrolytic and ATP synthetic activities can be obtained under the following conditions[1]:

The purified complex can hydrolyze many nucleoside triphosphates and especially GTP with a good activity (specific activity 2.4 μmol/mg protein/min), while the P_i–GTP exchange is negligible. In contrast, if GTP is hydrolyzed to supply energy in the presence of $^{32}P_i$ and ADP, a net ATP synthesis is obtained (Table II), with a ratio GTPase activity/ATP synthase activity tending to 1 as ADP concentration increases from 1 to 10

TABLE II
COMPARED ACTIVITIES OF $^{32}P_i$ INCORPORATION INTO ATP, OF ADENYLATE KINASE, AND OF NUCLEOSIDE DIPHOSPHOKINASE

Enzyme	Inhibitor present[a]	Nucleotides	Activity μmol ATP or [γ-^{32}P]ATP formed/min/mg protein
$^{32}P_i$ incorporation into ATP	0	ATP + ADP	1.7
	0	GTP + ADP	0.22
	Oligomycin	ATP + ADP	<0.008
	Ap_5A	ATP + ADP	1.65
Adenylate kinase	Oligomycin	ADP	0.020
	Oligomycin + Ap_5A	ADP	<0.006
Nucleoside diphosphokinase	Oligomycin + Ap_5A	GTP + ADP	<0.006

[a] When present, the inhibitors were used at the following concentrations: oligomycin: 25 μg/mg protein; Ap_5A: 0.2 mM. The nucleotide concentrations were 10 mM for ATP and GTP and 5 mM for ADP.

mM. It is noted that ADP competitively inhibits the hydrolytic activity while it stimulates ATP synthesis. Obviously the ratio of hydrolytic activity/ATP synthesis activity is the best criterion to characterize the integrity of H^+-transporting ATP synthase preparations.

The purified complex must be devoid of adenylate kinase and nucleoside diphosphokinase, as shown in Table II, since these enzymes can introduce errors in the estimation of ATP synthesis. The addition of Ap_5A to reaction mixtures inhibits any eventual adenylate kinase activity.

Effects of Inhibitors

The effects of oligomycin reported in Tables II and III prove that ATP synthesis as measured by ATP–P_i exchange is strictly due to the H^+-transporting ATP synthase complex. The exchange is fully inhibited by uncouplers while the ATPase activity is stimulated. Neither Ap_5A nor inhibitors of adenine nucleotide translocation (carboxyatractyloside, bongkrekic acid) affect the activities. Valinomycin plus KCl, as expected, only partially decrease the ATP synthesis without affecting ATPase activity. Nigericin alone plus KCl strongly inhibit ATP synthesis and stimulate ATPase activity. The addition of both valinomycin and nigericin plus KCl abolish ATP synthesis and strongly affect ATPase activity. All these effects are in agreement with those observed in proton translocation experiments.

TABLE III
INFLUENCE OF VARIOUS INHIBITORS ON ATP–P_i EXCHANGE AND ATPASE ACTIVITIES OF THE H^+-TRANSPORTING ATP SYNTHASE

Inhibitor[a]	Concentration	ATP–P_i exchange (% of control)	ATPase (% of control)
Oligomycin	5 μg/mg protein	< 0.5	2
FCCP	1 μ*M*	< 0.5	128
Ap_5A	200 μ*M*	97	103
Carboxyatractyloside	100 μ*M*	99	100
Bongkrekic acid	20 μ*M*	98	100
Valinomycin (+K^+)	2 μg/ml	68	95
Nigericin (+K^+)	2 μg/ml	18	156
Valinomycin + Nigericin (+K^+)	2 μg/ml, 2 μg/ml	0.6	23

[a] All inhibitors were present in the assays at the indicated concentrations. Assays with valinomycin and nigericin were made in the presence of 0.1 *M* KCl.

Proton Translocation

The profile of proton translocation routinely induced by ATP hydrolysis in the presence of valinomycin plus KCl, as measured by the quenching of fluorescence of ACMA, is given in Fig. 2. Oligomycin or FCCP

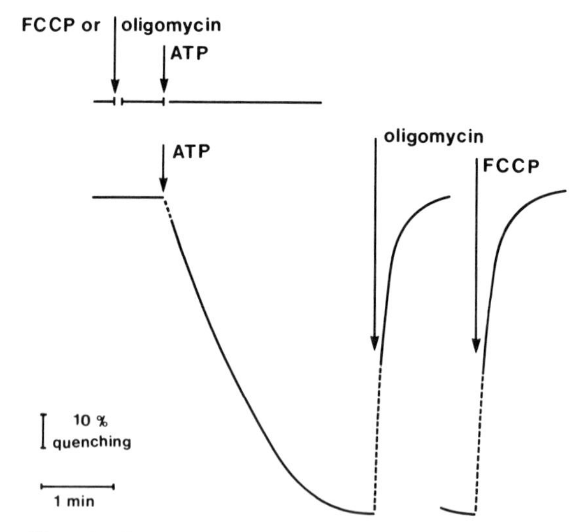

FIG. 2. Quenching of ACMA fluorescence induced by 0.04 m*M* ATP in the presence of the H^+-transporting ATP synthase (110 μg). Effects of 0.25 μ*M* FCCP and oligomycin (5 μg) per 2 ml assay.

totally prevents any ATP-dependent proton translocation. When these inhibitors are added after the maximal extent of fluorescence quenching (steady state) is reached, they fully restore the initial level of fluorescence. Oligomycin blocks any proton movement and FCCP abolishes any proton gradient.

All the results described demonstrate that the vesicular preparation of H^+-transporting ATP synthase is nonpermeant and inverted.

Correlation between Proton Movements, ATP Hydrolysis, and ATP Synthesis

The initial rate of fluorescence quenching as measured by the initial slope (Fig. 2) is dependent on ATP concentration. Double-reciprocal plots of the rates of ATP hydrolysis and initial fluorescence quenching are both linear and give the same K_m ATP of 0.1 mM. There exists a direct linear correlation between the rates of ATP hydrolysis and ATP-dependent fluorescence quenching. Besides, since the rate of ATP synthesis by the H^+-transporting ATP synthase complex depends on both ADP and P_i concentrations, the rate of ATP synthesis can be modulated as a function of P_i. Under these conditions, when the protons are supplied by ATP hydrolysis, a direct correlation can be drawn between ATP synthesis and the use of protons as measured by the changes in initial rates of fluorescence quenching of ACMA. In addition, the higher the phosphate concentration, the lower the ratio of ATP hydrolytic rate over ATP synthetic rate. This ratio is also correlated to the initial rate of fluorescence quenching of ACMA and extrapolates to 1 when the initial rate of fluorescence tends to zero. This means that a maximal efficiency of the complex occurs and suggests that all the energy from ATP hydrolysis is recovered for ATP synthesis, since no proton leak can be seen in fluorescence quenching. This could happen if the proton movements could be limited to a single molecule of H^+-transporting ATP synthase complex.

Concluding Remarks

The purification procedure presented here is a simple and quick one. It yields a very efficient enzyme with high ATP synthase activity. It is a good model to study the properties of H^+-transporting ATP synthase. It should be stressed that a ratio of hydrolytic activity/ATP synthase activity as close as possible to 1 is the best criterion to reflect both the efficiency and the integrity of the system.

[40] Resolution and Reconstitution of F_0F_1-ATPase in Beef Heart Submitochondrial Particles

By L. ERNSTER, T. HUNDAL, and G. SANDRI

Introduction

Beef heart submitochondrial particles—vesiculated fragments of the mitochondrial inner membrane, with an inside-out orientation of the membrane surface—have been widely used over the past two decades in studies of the resolution and reconstitution of the enzyme system responsible for mitochondrial respiration and oxidative phosphorylation. Introduced by Green and associates,[1] these particles were the starting material used in Racker's laboratory[2,3] for the identification and isolation of F_1-ATPase and the reconstitution of oxidative phosphorylation by means of purified F_1 and other coupling factors. Among the latter, the oligomycin sensitivity conferring protein, OSCP, purified and characterized by MacLennan and Tzagoloff,[4] and factor F_6, described by Fessenden-Raden,[5] have been extensively investigated with respect to their role in the interaction of F_1 with the membrane-bound, proton-translocating moiety (F_0) of the ATPase system, to yield an oligomycin-sensitive ATPase (often referred to as the F_0F_1-ATPase).[6–21] Recently, the primary structures of

[1] F. L. Crane, J. L. Glenn, and D. E. Green, *Biochim. Biophys. Acta* **22,** 475 (1956).
[2] M. E. Pullman, H. S. Penefsky, A. Datta, and E. Racker, *J. Biol. Chem.* **235,** 3322 (1960).
[3] H. S. Penefsky, M. E. Pullman, A. Datta, and E. Racker, *J. Biol. Chem.* **235,** 3330 (1960).
[4] D. H. MacLennan and A. Tzagoloff, *Biochemistry* **7,** 1603 (1968).
[5] J. M. Fessenden-Raden, *J. Biol. Chem.* **247,** 2351 (1972).
[6] Y. Kagawa and E. Racker, *J. Biol. Chem.* **241,** 2461 (1966).
[7] A. Tzagoloff, D. H. MacLennan, and K. H. Byington, *Biochemistry* **7,** 1596 (1968).
[8] A. F. Knowles, R. J. Guillory, and E. Racker, *J. Biol. Chem.* **246,** 2672 (1971).
[9] L. K. Russel, S. A. Kirkley, T. R. Kleyman, and S. H. P. Chan, *Biochem. Biophys. Res. Commun.* **73,** 434 (1976).
[10] A. Vàdineanu, J. A. Berden, and E. C. Slater, *Biochim. Biophys. Acta* **449,** 468 (1976).
[11] B. Norling, E. Glaser, and L. Ernster, *in* "Frontiers of Biological Energetics: From Electron to Tissues" (L. P. Dutton, J. S. Leigh, and A. Scarpa, eds.), p. 504. Academic Press, New York, 1978.
[12] T. Hundal and L. Ernster, *in* "Membrane Bioenergetics" (C. P. Lee, G. Schatz, and L. Ernster, eds.), p. 429. Addison-Wesley, Reading, Massachusetts, 1979.
[13] E. Glaser, B. Norling, and L. Ernster, *Eur. J. Biochem.* **110,** 225 (1980).
[14] T. Hundal, B. Norling, and L. Ernster, *FEBS Lett.* **162,** 5 (1983).
[15] A. M. Liang and R. J. Fisher, *J. Biol. Chem.* **258,** 4784 (1983).
[16] A. M. Liang and R. J. Fisher, *J. Biol. Chem.* **258,** 4788 (1983).
[17] A. Dupuis, M. Satre, and P. V. Vignais, *FEBS Lett.* **156,** 99 (1983).
[18] B. Norling, T. Hundal, G. Sandri, E. Glaser, and L. Ernster, *in* "H^+-ATPase (ATP

OSCP,[22,23] F_6,[24,25] and of several other subunits of the mitochondrial F_0F_1-ATPase (for review, see Ref. 26) have also been determined.

This chapter is a survey of some current methods for the resolution and reconstitution of F_0F_1-ATPase with respect to F_1,[27,28] OSCP, and F_6 in beef heart submitochondrial particles. Figure 1 illustrates in a schematic form the various steps of the resolution–reconstitution procedure described below.

Preparations

Submitochondrial Particles

The particles used in these experiments are prepared from beef heart mitochondria by sonication in a buffered medium (pH 8.6) containing 2 m*M* EDTA as described by Lee and Ernster[29]; they will be referred to in the following as E particles. These particles exhibit NADH and succinate oxidase activities and are nonphosphorylating, i.e., they give rise to little or no ATP synthesis. Addition of oligomycin induces a high degree of respiratory control, and low concentrations of oligomycin can restore a limited capacity for oxidative phosphorylation.[30] The uncoupled state of

Synthase): Structure, Function, Biogenesis" (S. Papa, K. Altendorf, L. Ernster, and L. Packer, eds.), p. 291. ICSU Press and Adriatica Editrice, Bari, 1984.

[19] T. Hundal, B. Norling, and L. Ernster, *J. Bioenerg. Biomembr.* **16,** 535 (1984).

[20] A. Dupuis, J.-P. Issartel, J. Lunardi, M. Satre, and P. V. Vignais, *Biochemistry* **24,** 728 (1985).

[21] A. Dupuis, J. Lunardi, J.-P. Issartel, and P. V. Vignais, *Biochemistry* **24,** 734 (1985).

[22] Y. A. Ovchinnikov, N. N. Modyanov, V. A. Grinkevich, N. A. Aldanova, O. E. Trubetskaya, I. V. Nazimov, T. Hundal, and L. Ernster, *FEBS Lett.* **166,** 19 (1984).

[23] V. A. Grinkevich, N. A. Aldanova, P. Kostetsky, O. E. Trubetskaya, N. N. Modyanov, T. Hundal, and L. Ernster, *in* "H^+-ATPase (ATP Synthase): Structure, Function, Biogenesis" (S. Papa, K. Altendorf, L. Ernster, and L. Packer, eds.), p. 153. ICSU Press and Adriatica Editrice, Bari, 1984.

[24] V. A. Grinkevich, N. A. Aldanova, P. V. Kostetsky, N. N. Modyanov, T. Hundal, Y. A. Ovchinnikov, and L. Ernster, *EBEC Rep.* **3,** 307 (1984).

[25] J. Fang, J. W. Jacobs, B. I. Kanner, E. Racker, and R. A. Bradshaw, *Proc. Natl. Acad. Sci. U.S.A.* **81,** 6603 (1984).

[26] J. E. Walker, V. L. J. Tybulewicz, G. Falk, N. J. Gay, and A. Hampe, *in* "H^+-ATPase (ATP Synthase): Structure, Function, Biogenesis" (S. Papa, K. Altendorf, L. Ernster, and L. Packer, eds.), p. 1. ICSU Press and Adriatica Editrice, Bari, 1984.

[27] G. Sandri, E. Suranyi, L. E. G. Eriksson, J. Westman, and L. Ernster, *Biochim. Biophys. Acta* **723,** 1 (1983).

[28] G. Sandri, L. Wojtczak, and L. Ernster, *Arch. Biochem. Biophys.* **239,** 597 (1985).

[29] C. P. Lee and L. Ernster, this series, Vol. 10 [87].

[30] C. P. Lee and L. Ernster, *Eur. J. Biochem.* **3,** 391 (1968).

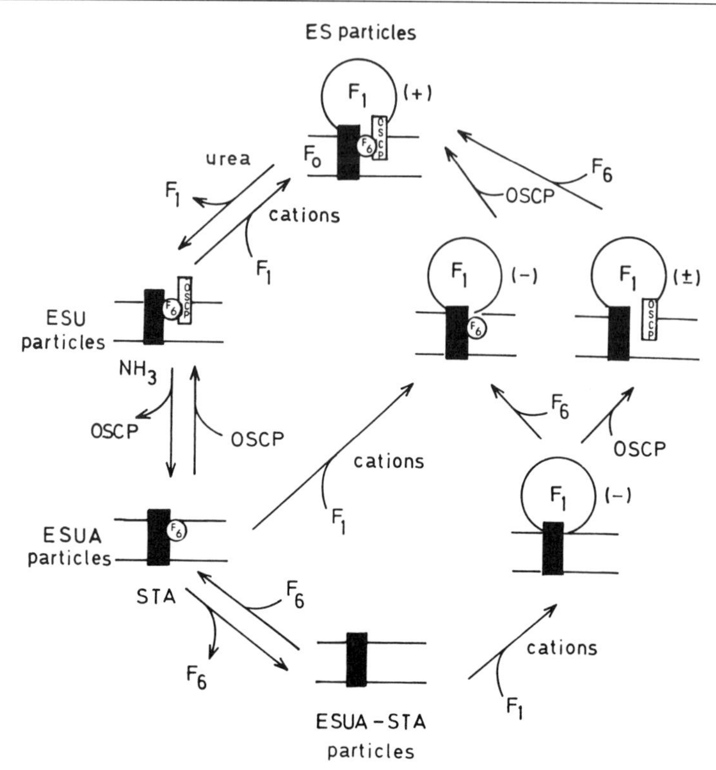

FIG. 1. Resolution and reconstitution of F_0F_1-ATPase in beef heart submitochondrial particles. The scheme illustrates different steps in the resolution and reconstitution of F_0F_1-ATPase in ES particles (E particles from which the ATPase inhibitor has been removed by treatment with Sephadex; see text). F_0 refers to the components of the F_0F_1-ATPase after removal of F_1; it consists of F_6, OSCP, and a membrane sector symbolized by the shaded bar across the membrane. The symbols (+), (±), and (−) refer to oligomycin-sensitive, partially oligomycin-sensitive, and oligomycin-insensitive ATPase activity, respectively. Cations are required for the rebinding of F_1 to F_0, their role probably being to neutralize negative charges on the membrane surface (cf. Refs. 27, 28). For further details, see text.

the particles is due to a partial (~20%) deficiency in F_1,[31] and respiratory control can likewise be induced by the addition of F_1, provided that OSCP also is added.[32] The particles as prepared exhibit relatively low ATPase activity because of the presence of the ATPase inhibitor protein of Pullman and Monroy.[33] This can be removed by various procedures as de-

[31] L. Ernster, K. Nordenbrand, O. Chude, and K. Juntti, *in* "Membrane Proteins in Transport and Phosphorylation" (G. F. Azzone, M. E. Klingenberg, E. Quagliariello, and N. Siliprandi, eds.), p. 29. North-Holland Publ., Amsterdam, 1974.

[32] K. Nordenbrand, A. Gómez-Puyou, and L. Ernster, *EBEC Rep.* **2,** 83 (1982).

[33] M. E. Pullman and G. C. Monroy, *J. Biol. Chem.* **238,** 3762 (1962).

scribed earlier in this series.[34] The ATPase activity of the particles is fully oligomycin sensitive.

F_1, OSCP, F_6

F_1 is prepared according to Horstman and Racker[35] and stored as a 50% $(NH_4)_2SO_4$ precipitate at 4°.

OSCP is purified according to Russel *et al.*[9] and F_6 according to Kanner *et al.*[36]

Resolution

Removal of ATPase Inhibitor

The particles are run through a Sephadex G-50 column to remove the naturally occurring inhibitor (ES particles, where S refers to Sephadex). By this treatment the ATPase activity of the particles is increased about 5- to 10-fold, yielding an activity ranging between 6 and 10 μmol/min/mg particle protein. The details of this procedure are described elsewhere.[34]

The Sephadex treatment is an important step in the resolution procedure, since F_1 cannot be fully removed if the inhibitor is still bound.

Removal of F_1 by Urea

The procedure is essentially according to Racker and Horstman.[37] A suspension of ES particles in 0.25 *M* sucrose (~10 mg protein/ml) is diluted with an equal volume of a solution containing 4 *M* urea, 4 m*M* EDTA, and 0.1 *M* Tris-SO_4, pH 8.0. The suspension is incubated for 30 min at 0° with stirring and then transferred to centrifuge tubes. After 40 min the suspension is centrifuged at 140,000 *g* in a Spinco 50 Ti rotor for 10 min. The pellet is washed once with 0.25 *M* sucrose and centrifuged as before. The final pellet (ESU particles) is suspended in 0.25 *M* sucrose containing 5 m*M* DTT to yield a protein concentration of ~10 mg/ml.

The above treatment removes 90–95% of the ATPase activity of the particles.

Removal of OSCP by Ammonia

The following procedure is based on that described by Tzagoloff *et al.*[7] ESU particles suspended as described above are diluted with one-half

[34] L. Ernster, C. Carlsson, T. Hundal, and K. Nordenbrand, this series, Vol. 55 [51].

[35] L. L. Horstman and E. Racker, *J. Biol. Chem.* **245,** 1336 (1970).

[36] B. I. Kanner, R. Serrano, M. A. Kandrach, and E. Racker, *Biochem. Biophys. Res. Commun.* **69,** 1050 (1976).

[37] E. Racker and L. L. Horstman, *J. Biol. Chem.* **242,** 2547 (1967).

volume of a freshly prepared solution of 1.2 *M* NH_3 in 0.25 *M* sucrose. The pH of the suspension should be 11.5–11.7. (If the pH is lower, more NH_3 is added.) The mixture is incubated with efficient stirring for 20 min at 0° and then centrifuged at 105,000 *g* for 15 min in a Spinco 50 Ti rotor. The pellet is rinsed carefully with 0.25 *M* sucrose containing 10 m*M* Tris-SO_4, pH 8.0, and then homogenized in the same solution (4–5 volumes/pellet volume). The particles are centrifuged as above and the pellet (ESUA particles) is finally suspended in 0.25 *M* sucrose to ~10 mg protein/ml. The ESUA particles are stored in small aliquots in liquid nitrogen. They are stable for 4–6 weeks. The above treatment removes about 85–90% of the OSCP.

Removal of F_6 by Silicotungstic Acid (STA)

This is done essentially according to Knowles *et al.*[8] ESUA particles are suspended in a medium containing 0.15 *M* sucrose, 20 m*M* Tris-SO_4, pH 8.0, and 1.5% STA (stock solution of 10% STA is adjusted to pH 5.5 with 1 *M* KOH) to a protein concentration of ~10 mg/ml. The suspension is incubated for 10 min at 0°. After incubation, the mixture is diluted with 9 volumes of cold 0.25 *M* sucrose and centrifuged for 30 min at 105,000 *g* in a Spinco 50 Ti rotor. The pellet is resuspended in 0.25 *M* sucrose containing 5 m*M* DTT and centrifuged as above. The particles (ESUA-STA particles) are finally suspended in the same medium to give a final protein concentration between 10 and 15 mg/ml. The suspension is stored in small aliquots in liquid nitrogen. Under these conditions the particles are stable for several weeks.

Reconstitution

In all reconstitution experiments, the F_1 used is first centrifuged, resuspended in a medium containing 0.25 *M* sucrose, 10 m*M* Tris-SO_4, pH 8.0, and 0.25 m*M* EDTA (STE buffer) to a protein concentration of 4 mg/ml, and then dialyzed against 1000 volumes of the same buffer for 3 hr at room temperature. The dialysis can be replaced by filtration through a Sephadex G-50 minicolumn equilibrated with STE buffer.

Binding of F_1 to ESU Particles

A suspension of ESU particles, prepared as described above, is diluted with 9 volumes of STE buffer containing dialyzed F_1 in an amount of 0.07 mg/mg particle protein, as well as either a monovalent cation (e.g., NH_4^+, K^+, Rb^+, Cs^+, Na^+, Li^+) in a final concentration of 50–100 m*M* or a divalent cation (Ca^{2+}, Mg^{2+}) in a final concentration of 2–3 m*M*, added as the chloride. The mixture is incubated for 15 min at room temperature and then centrifuged in a Beckman J21 centrifuge for 15 min at 15,000 *g* in a JA20 rotor. The particles obtained contain ~0.1 mg F_1/mg ESU and exhibit an oligomycin-sensitive ATPase activity of ~5–7 μmol/min/mg particle protein. For further details of this procedure, see Sandri *et al.*[27]

Binding of F_1 and OSCP to ESUA Particles

A suspension of ESUA particles, prepared as described above, is diluted with 2.5 volumes of STE buffer containing 0.075 mg F_1 and 0.015 mg OSCP/mg particle protein, as well as NH_4^+, Rb^+, or Cs^+ (added as the chloride) to give a final concentration of 75 m*M*. The mixture is incubated for 30 min at 25°. ATPase activity is measured in the absence and presence of oligomycin (5 μg/mg particle protein). Successful reconstitution results in the restoration of virtually complete oligomycin sensitivity. For further details of these procedures, see Refs. 9, 10, 13, and 28.

If the above procedure is carried out in the absence of OSCP, F_1 is still rebound to the ESUA particles, but in an oligomycin-insensitive manner. Subsequent addition of OSCP confers oligomycin sensitivity.

Binding of F_1, OSCP, and F_6 to ESUA-STA Particles

This procedure is according to Sandri *et al.*[27] A suspension of ESUA-STA particles, prepared as described above, is diluted with 9 volumes of STE buffer containing per milligram particle protein 0.07 mg F_1, 0.025 mg OSCP, and 0.05 mg F_6 as well as 75 m*M* NH_4^+, Rb^+, or Cs^+ (added as the chloride). After incubation at room temperature for 15 min, the reaction mixture is centrifuged and the reconstituted particles are resuspended in STE buffer for assay of ATPase activity.

The reconstituted system contains about 65 μg F_1/mg particle protein and exhibits an ATPase activity of 1 μmol/min/mg protein with an oligomycin sensitivity of 80–90%.

If F_6 and OSCP are omitted, cations alone can promote the binding of the same amount of F_1 to the membrane, but in this case in an oligomycin-insensitive way. Subsequent addition of F_6 causes little (less than 10%) further binding of F_1 and again no oligomycin sensitivity. The latter can be increased to a maximal extent (80–90%) by the further addition of OSCP. In the absence of F_6, addition of OSCP following the rebinding of F_1 in the presence of cations causes a ~50% oligomycin sensitivity, which can be increased to about 85% by the subsequent addition of F_6.

[41] Electron Microscopy of Single Molecules and Crystals of F_1-ATPases

By CHRISTOPHER W. AKEY, STANLEY D. DUNN, VITALY SPITSBERG, and STUART J. EDELSTEIN

Background

The proton-translocating F_1-ATPases play an important role in energy transduction in prokaryotes and eukaryotes. *In vivo,* the F_1-ATPase is coupled to the membrane-bound F_0 sector through a stalk.[1] Isolation procedures detach the F_1 headpiece from the F_1–F_0 complex, thereby forming a soluble macromolecular complex. Elucidation of the subunit structure of the F_1-ATPase should further our understanding of the mechanism of ATP synthesis.[2] Crystals suitable for structural studies by electron microscopy have been grown with F_1-ATPase isolated from PS3,[3] beef heart,[4] and rat liver.[5] A complete three-dimensional structure of F_1-ATPase at atomic resolution may eventually be obtained from the rat liver crystals, if they prove to be sufficiently well ordered.[6]

Structural studies of macromolecular complexes by electron microscopy have been greatly facilitated by employing negative staining methods[7] in conjunction with computer-based analysis.[8] Image reconstruction is a powerful tool for the elucidation of molecular structure at low resolutions (~20 Å); however, minimal dose methods of imaging are essential in preserving an interpretable specimen.[9,10] A conventional transmission electron micrograph of a thin specimen (<1000 Å) is an axial projection of the object because of the relatively large depth of focus of the microscope[11]; therefore, analysis of an electron micrograph yields a two-dimen-

[1] P. C. Hinkle and R. E. McCarty, *Sci. Am.* **238,** 104 (1978).
[2] P. Mitchell, "Chemiosmotic Coupling in Oxidative and Photosynthetic Phosphorylation." Glynn Research Laboratories, Bodmin, Cornwall, England.
[3] Y. Kagawa, N. Sone, M. Yoshida, H. Hirata, and H. Okamoto, *J. Biochem.* (*Tokyo*) **80,** 141 (1976).
[4] V. Spitsberg and R. Haworth, *Biochim. Biophys. Acta* **492,** 237 (1977).
[5] L. M. Amzel and P. L. Pedersen, this series, Vol. 55, p. 333.
[6] L. M. Amzel, M. McKinney, P. Narayanan, and P. L. Pedersen, *Proc. Natl. Acad. Sci. U.S.A.* **79,** 5852 (1982).
[7] J. L. Farrant, *Biochim. Biophys. Acta* **13,** 569 (1954).
[8] D. J. DeRosier and P. Moore, *J. Mol. Biol.* **52,** 355 (1970).
[9] R. C. Williams and H. W. Fisher, *J. Mol. Biol.* **52,** 121 (1970).
[10] T. S. Baker, *Proc. Int. Congr. Electron Microsc., 9th* **2,** 2 (1978).
[11] H. P. Erickson and A. Klug, *Philos. Trans. R. Soc. Ser. B* **261,** 105 (1971).

sional projection of the structure under study. Three-dimensional data can be obtained by collecting a tilt series and combining the resulting two-dimensional projections.[12] In this chapter, methods of sample preparation are described for studies of the projected structures of single molecules and crystals of F_1-ATPases by electron microscopy.

Single Molecules

The study of single molecules by electron microscopy is often limited by variable specimen preservation, including differential staining and flattening, coupled with random orientation of the oligomeric macromolecular complexes. The morphology of the F_1-ATPases causes these problems to be unusually severe; however, these difficulties were partially circumvented using two adaptations of existing methods of single-molecule preparation for electron microscopy which employ tannic acid. Tannic acid is effective at three levels in sample preparation: fixation, preservation, and contrast enhancement. The methods described below are generally applicable to studies of other macromolecular complexes by electron microscopy.

Adhesion Method

Negatively stained specimens of beef heart F_1-ATPase are prepared for electron microscopy by the adhesion method employing tannic acid. A 50-μl aliquot of an ammonium sulfate slurry of the F_1-ATPase (1 mg/ml) is spun in a Beckman airfuge at 5 psi for 10 min. The resulting pellet is resuspended in 10 m*M* potassium phosphate, pH 6.8, and rapidly desalted by the method of Penefsky[13] over a 1-ml Sephadex G-25 column equilibrated with the same buffer. The sample is diluted to a final protein concentration of about 0.5 mg/ml in phosphate buffer, and tannic acid is added from a 2% stock solution to bring the final concentration to 0.2%. The stock solution is prepared in 10 m*M* potassium phosphate buffer with tannic acid obtained from Mallinckrodt[14] and titrated to pH 6.8 with 0.5 *N* sodium hydroxide to prevent precipitation. A 100-μl aliquot of the mixture of beef heart F_1-ATPase and tannic acid is then desalted by the centrifuged column method[13] to separate the F_1-ATPase from unbound tannic acid. Samples are prepared for electron microscopy on carbon-coated grids using the droplet adhesion method[15]; after blotting, the grids are stained with freshly prepared 0.5% uranyl oxalate, pH 6.8.

[12] L. A. Amos, R. Henderson, and P. N. T. Unwin, *Prog. Biophys. Mol. Biol.* **39,** 183 (1982).
[13] H. S. Penefsky, *J. Biol. Chem.* **252,** 2891 (1977).
[14] C. W. Akey, M. Szalay, and S. J. Edelstein, *Ultramicroscopy* **13,** 103 (1984).
[15] A. Huxley and G. Zubay, *J. Mol. Biol.* **2,** 10 (1960).

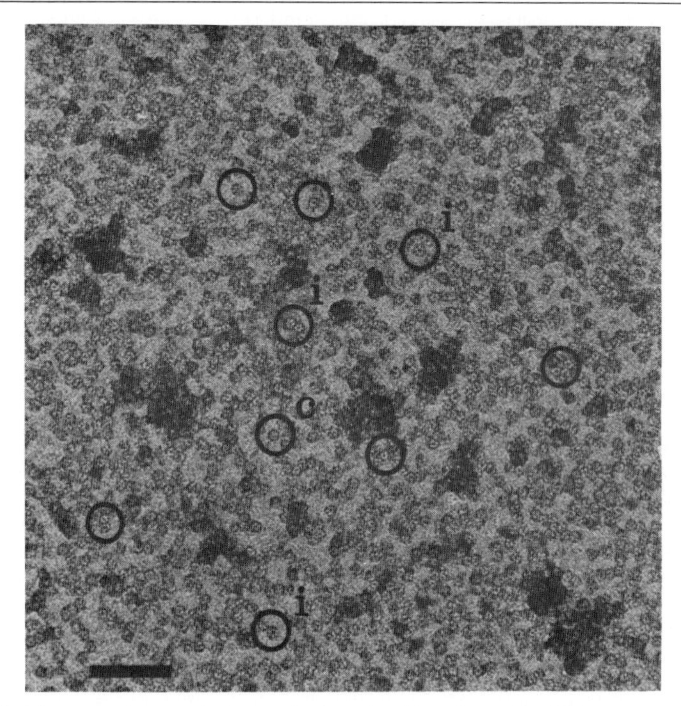

FIG. 1. Electron micrograph of negatively stained beef heart F_1-ATPase prepared by the adhesion method employing tannic acid. Incomplete (i) and complete (c) particles are circled. Scale bar is 1000 Å.

A typical field from a minimal dose micrograph is shown in Fig. 1. The beef heart F_1-ATPase molecules are randomly oriented with respect to the plane of the grid; however, some molecules demonstrate either the incomplete (hollow) or complete (solid) morphologies which have been observed in preparations of F_1-ATPases from both prokaryotic and eukaryotic sources.[16] The molecules are strongly contrasted as a result of tannic acid-mediated binding of stain to the peripheral regions and subunit interfaces of the molecules. The images can be analyzed by computer-based methods using either rotational harmonic analysis,[17] single-particle correlation,[18] or Fourier filtering.[19] Digitized and Fourier-filtered images of hollow and solid beef heart F_1-ATPase molecules are presented in Fig. 2. Interpretable detail is limited to the low-resolution pseudo-hexagonal morphology observed in both the hollow and solid particles.

[16] P. L. Pedersen, *Bioenergetics* **6,** 243 (1975).

[17] R. A. Crowther and L. A. Amos, *J. Mol. Biol.* **60,** 123 (1971).

[18] J. Frank, *Ultramicroscopy* **1,** 159 (1975).

[19] C. W. Akey, V. Spitsberg, and S. J. Edelstein, *J. Biol. Chem.* **258,** 3222 (1983).

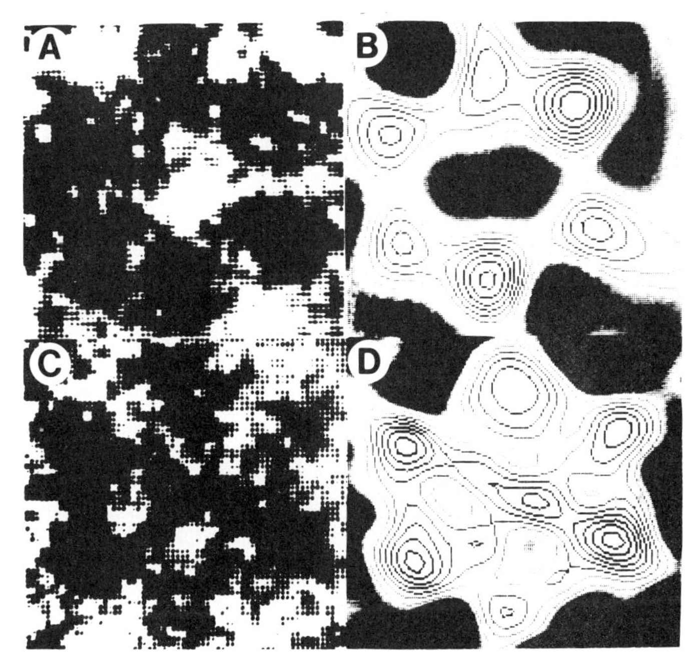

FIG. 2. Single-particle images of beef heart F_1-ATPase. (A) Digitized image of molecule with a hollow (incomplete) morphology. (B) Low-resolution Fourier-filtered reconstruction of image in (A). (C) Digitized image of molecule with a solid (complete) morphology. (D) Low-resolution Fourier-filtered reconstruction of image in (C). Significant structural detail is limited to the observation of pseudo-hexagonal symmetry in these molecules. From Akey *et al.*[19]

Mica Flotation Method

An alternative method to the adhesion procedure was developed which gives uniformly stained and close-packed arrays of F_1-ATPase molecules. This procedure represents a combination of mica flotation[20] and the tannic acid preparation technique for single molecules.[21] Three-subunit *Escherichia coli* F_1-ATPase is reconstituted from purified α, β, and γ subunits by the method of Dunn and Futai.[22] A 70-μl aliquot of the ECF_1 (0.3–0.5 mg/ml) is desalted into a buffer containing 2 m*M* PIPES, 2 m*M* ATP, pH 6.85. An aliquot of a 2% tannic acid stock solution prepared in

[20] R. W. Horne and I. P. Ronchetti, *J. Ultrastruct. Res.* **47,** 361 (1974).

[21] C. W. Akey, R. H. Crepeau, S. D. Dunn, R. E. McCarty, and S. J. Edelstein, *EMBO J.* **2,** 1409 (1983).

[22] C. W. Akey and S. J. Edelstein, *J. Mol. Biol.* **163,** 575 (1983).

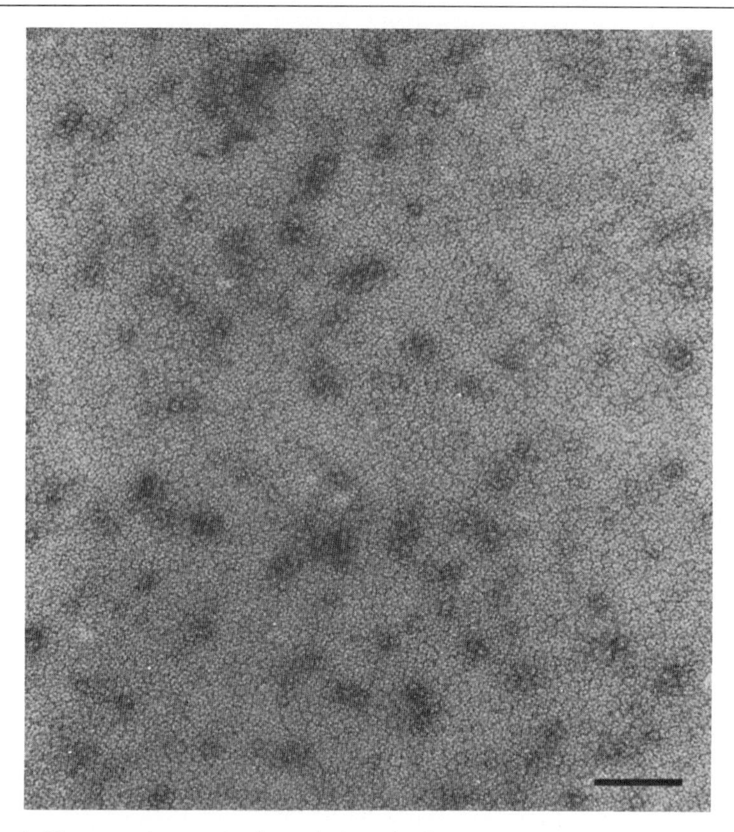

FIG. 3. Electron micrograph of negatively stained 3-subunit *E. coli* F_1-ATPase prepared by the tannic acid–mica flotation technique. The molecules form a close-packed, noncrystalline array and appear hexagonally shaped with a central area of stain accumulation. Scale bar is 1000 Å. From Akey *et al.*[21]

the same buffer (pH 6.85) is then added to a final concentration of 0.1%. The solution is layered onto the surface of freshly cleaved mica (2.5 × 5 cm) and slowly air dried at room temperature for 1 hr. Samples are carbon shadowed and small areas (0.5 × 1 cm) are used to float the carbon-stabilized sample off of the mica and onto the surface of freshly prepared 0.5% uranyl oxalate, pH 6.85. The negatively stained specimen film is then picked up onto 400-mesh copper grids from beneath the surface and the grids are carefully blotted dry. This procedure gives specimen films with large areas of close-packed F_1-ATPase molecules (see Fig. 3). The close packing of the resulting specimen is optimized by varying the amount of protein in the initial drying phase. Flotation of the sample under these conditions proves impossible if the initial protein concentration is greater than 0.5 mg/ml.

A hollow, pseudo-hexagonal morphology is clearly observed in a large percentage of the 3-subunit F_1-ATPase molecules in Fig. 3. This is the result of a preferred orientation of the ECF_1 molecules during the drying phase of specimen preparation. Tannic acid acts as a mild fixative and preserves the structure of the F_1-ATPase complex during drying in a manner analogous to glucose.[22] Calculations suggest that immediately after drying the samples are 3–10 molecular layers thick, depending on the local uniformity of drying. Subsequent flotation of the carbon-shadowed specimen onto negative stain results in a carbon-stabilized monolayer; the molecular layers not in contact with the carbon film are released into solution and can be observed visually as a diffuse orange-brown cloud of uranyl stain–tannin complex. The close-packed areas are verified to be monolayers by observing the thickness of curled edges of broken grid holes.

The specimen in Fig. 3 is an example of a two-dimensionally ordered, noncrystalline, close-packed array. Optical transforms of similar areas yield a series of concentric, hexagonally, or elliptically shaped maxima (see Fig. 4). The first-order maximum is related to the average interparticle spacing. The second maximum is composed of contributions from the second order of the first maximum and the first order of the molecular transform. The average spacing of the second maximum obtained from about 1.7×10^4 molecules (8 areas) is 1/53 Å. Model data for a hexagonal molecule were evaluated and suggested that the first-order molecular transform spacing could be corrected for the scattering geometry by the empirically derived term 1/sin (60°) to yield the average half-molecule width. The 3-subunit F_1-ATPase was therefore determined to have an average diameter of 122 (±8) Å in the pseudo-hexagonal projection, in agreement with recent results from X-ray diffraction studies of single crystals.[6]

The image in Fig. 3 can also be used for single-molecule image reconstruction employing any of the three methods described earlier. In addition, this image could be prefiltered using computer-based methods to reduce the background noise level by calculating a reconstruction from a radially filtered transform in which only the three maxima are included. A total of 47 hollow molecules have been digitized and subjected to rotational harmonic analysis. After image alignment, an average power spectrum is calculated and subsequently used to reconstruct an average pseudo-hexagonal projection of the 3-subunit ECF_1 molecule.[21] This reconstruction is presented in Fig. 5 as a combination grayscale-contour

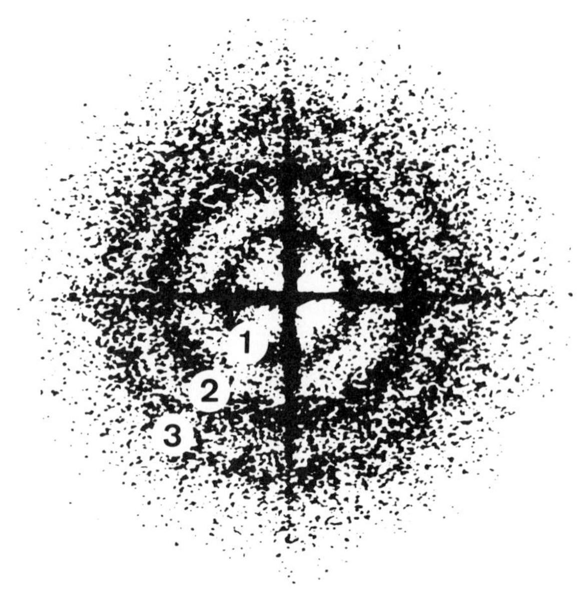

FIG. 4. Hexagonally shaped powder pattern obtained by optical diffraction from an electron micrograph of close-packed 3-subunit *E. coli* F_1-ATPase. The transformed area contained about 1500 molecules. The three maxima occur at spacings of 1/99, 1/53, and 1/37 Å. From Akey *et al.*[21]

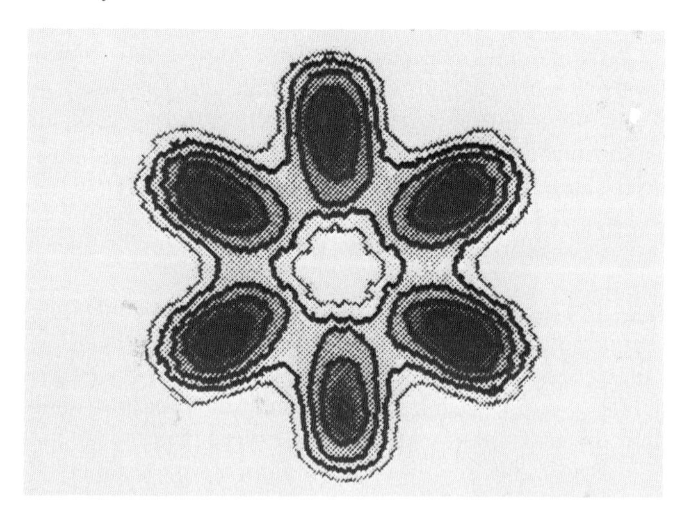

FIG. 5. Reconstruction of 3-subunit *E. coli* F_1-ATPase obtained from 47 molecules in close-packed arrays. Stain-excluding regions are indicated by heavy contours and increasingly darker grayscales. The image is not symmetry enforced, but has been reconstructed employing the 0-, 2-, 3-, and 6-fold harmonics of the average power spectrum. From Akey *et al.*[21]

map. Stain-excluding regions correspond to darker grayscales and heavy contours. When viewed in this projection the 3-subunit F_1-ATPase ($\alpha_3\beta_3\gamma$) demonstrates a strongly hexagonal substructure after averaging. The six large subunits appear to be arranged in a cyclic manner in projection, but may be staggered in an alternating fashion about the median plane of the molecule in three dimensions.[6] Differences in α and β subunits are not resolved in projection at 40 Å resolution. A comparison of reconstructions from molecules with either a hollow or solid morphology indicates that a seventh, centrally located subunit (possible γ) is variably contrasted in negatively stained preparations as a result of the presence of a central hole in the complex. The average diameter obtained from 47 molecules was 125 Å.

Crystals

Two-dimensional arrays and ultrathin three-dimensional crystals are excellent specimens for electron microscopy and computer-based image reconstruction. Unfortunately, most proteins crystallize in forms which are too thick for direct imaging in the electron microscope and hence are unsuitable for direct analysis. Recent advances in techniques for the preservation and embedding of crystalline specimens now make thin sectioning of protein crystals a viable alternative for low-resolution studies by electron microscopy. The new methods include tannic acid fixation,[19,22,23] low-temperature embedding,[24] and a combination of the two techniques.[14] The degree of structural preservation which can be achieved with these three methods varies with the crystal being studied, but resolutions of 15–30 Å are obtained under optimal conditions of fixation and thin sectioning. Large, well-formed crystals can be sectioned parallel to the principal axes and along body diagonals of the unit cell.[14,23] Smaller crystals, especially those with a platelike habit, can be pelleted before fixation to achieve partial orientation.[22]

Thin crystalline plates of beef heart F_1-ATPase are grown in 2 *M* ammonium sulfate by the method of Spitsberg,[4] using F_1-ATPase purified by the chloroform release method.[25] The crystals are processed for embedding and electron microscopy as follows. Crystals in mother liquor (80 ml) are cross-linked with 2% glutaraldehyde at 4° for 2 hr on a rotary turner. The slightly yellow suspension is diluted and pelleted in a Beckman airfuge. The resulting white pellet is then washed repeatedly with 8 rinses of 50 m*M* potassium phosphate buffer, with some repelleting as

[23] C. W. Akey, J. K. Moffat, D. C. Wharton, and S. J. Edelstein, *J. Mol. Biol.* **136,** 19 (1980).
[24] E. Carlemalm, R. M. Garavito, and W. J. Villiger, *J. Microsc.* **126,** 123 (1983).
[25] V. Spitsberg and J. E. Blair, *Biochim. Biophys. Acta* **460,** 136 (1977).

needed. The final pellet volume is ~3–5 μl and proves optimal for obtaining embedded crystals with a high degree of structural preservation. The crystals are subsequently rinsed 8 times with 4% tannic acid, 2% glutaraldehyde in 50 m*M* potassium phosphate buffer, pH 7.0, until high salt-induced precipitation of the tannic acid ceases. The crystals are stored in fixative overnight in the dark at room temperature. The pellet is rinsed with buffer and prestained with 0.5% osmium tetroxide in 50 m*M* potassium phosphate buffer, pH 6.8, for 30 min in the airfuge tube. After additional rinses in distilled water, the pellet is dislodged from the side wall of the airfuge tube and transferred to a white polyethylene cap (Pelco). The pellet is carried through a standard ethanol–water dehydration series and infiltrated with a low-viscosity embedding medium.[26] After polymerization, the blocks are sectioned with a diamond knife in a direction perpendicular to the pellet surface. The sections are picked up on 400-mesh grids and poststained with 2% (w/v) uranyl acetate in 35% (v/v) ethanol, 12.5% (v/v) methanol for 30 min.[27] The sections are rinsed with an ethanol–methanol solution, further poststained with lead citrate for 10 min,[28] and carbon shadowed for stability in the beam.

The platelike crystals of beef heart F_1-ATPase are composed of at least two related orthorhombic crystal forms. The form I crystals have been characterized as space group $P2_12_12$ with a = 164 Å, b = 324 Å, and c = 118 Å, with one molecule of F_1-ATPase per asymmetric unit.[19] The crystals tend to align with their unique 2-fold axes perpendicular to the pellet surface under these conditions of sample preparation. Therefore, the views encountered in thin sections cut perpendicular to the pellet surface tend to have the c-axis in the plane of the section. Crystal sections are screened either by using electron diffraction or by viewing the ends of the long crystal cross sections with a highly condensed beam. Adjacent areas with correct symmetry are photographed using minimal beam methods. A typical micrograph of a thin section cut parallel to the (100) plane is presented in Fig. 6. When the micrograph is viewed at a glancing angle parallel to first one and then the other edge, it becomes evident that 2-fold axes run parallel to the vertical axis (c-axis) and 2_1-screw axes run parallel to the horizontal axis (b-axis).

The symmetries present in a micrograph of a crystalline specimen can be discerned by optical diffraction. A composite optical transform of the (100) plane from two different crystals is shown in Fig. 7. The diffraction pattern has *mm* symmetry and systematic absences along the b^*-axis as

[26] A. R. Spurr, *J. Ultrastruc. Res.* **26,** 31 (1969).

[27] H. Kim, L. F. Binder, and J. L. Rosenbaum, *J. Cell Biol.* **80,** 266 (1979).

[28] E. S. Reynolds, *J. Cell Biol.* **17,** 208 (1963).

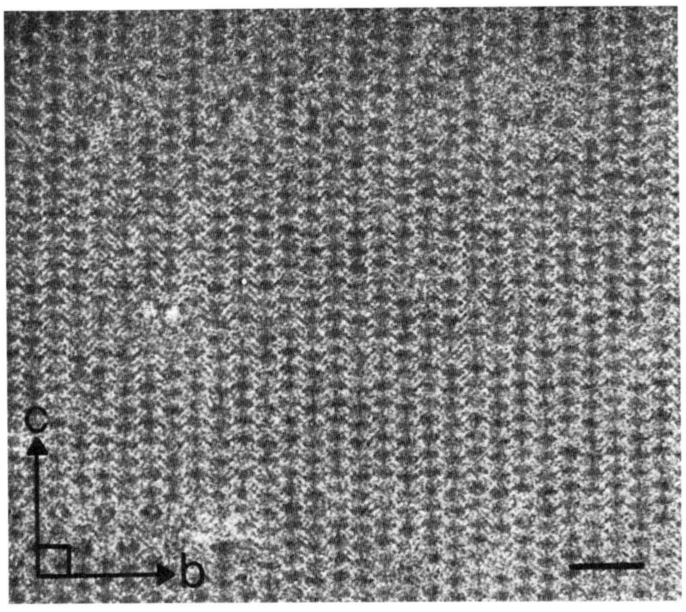

FIG. 6. Electron micrograph of a thin section cut parallel to the (100) plane of a form I crystal of beef heart F_1-ATPase. The unit cell dimensions are b = 324 Å and c = 118 Å. Scale bar is 500 Å. From Akey *et al.*[19]

expected for the *Okl* zone, with reflections extending to a resolution of 1/30 Å. The micrograph is digitized and Fourier transformed. A symmetry-enforced reconstruction is calculated as described previously[19] and is presented in Fig. 8. The preservation of the original crystal is quite good; therefore, the unenforced and symmetry-enforced reconstructions are virtually identical. Furthermore, the crystallographic symmetry is well preserved in projections of sections greater than 3–4 unit cells thick.

The asymmetric unit of the form I crystal is outlined by a parallelepiped in the lower left-hand corner of Fig. 8. In this projection, the F_1-ATPase is observed in a nearly sideways orientation. The molecular density is bipartite with a lightly stained, elongated headpiece (H) and a heavily stained tailpiece (T) aligned at right angles to the headpiece. There are three major classes of intermolecular contacts in the crystal structure. One interaction of interest is the tailpiece-to-tailpiece contact which occurs across the in-plane 2-fold axes between molecules located at the same height along the c-axis. Crystals of rat liver F_1-ATPase have recently been grown in ammonium sulfate from enzyme purified by the

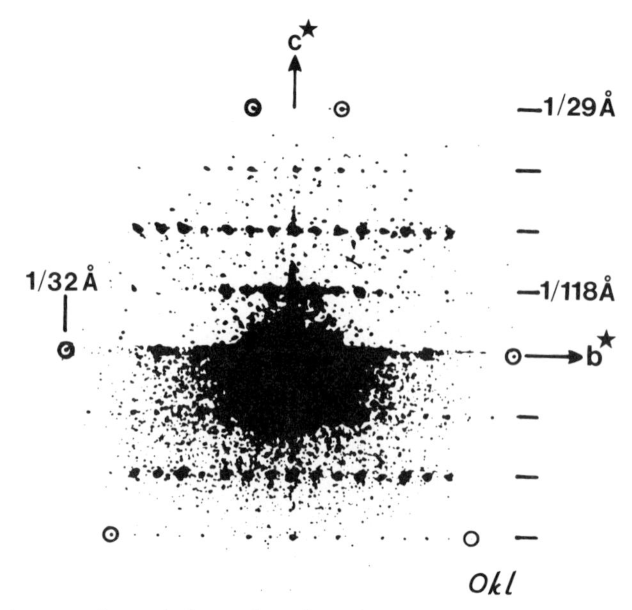

FIG. 7. A composite optical transform from the (100) plane of form I crystals of two different specimens. The overall resolution is 1/30 Å in all directions. From Akey *et al.*[19]

chloroform release method (Vitaly Spitsberg, unpublished data). The external habit of this crystal form is usually cubic, similar to the crystals currently being investigated by Amzel and co-workers using X-ray diffraction.[5,6] Reconstructions have recently been completed from sections cut parallel to a set of crystal planes which possess 2-fold axes in the plane of the projection. This projection showed a clear side view of the rat liver F_1-ATPase in the crystal. The F_1-ATPase is strongly bipartite in this view and easily demarcated into headpiece and tailpiece regions. Furthermore, one of the major intermolecular contacts in the rat liver crystal consists of a tailpiece-to-tailpiece interaction across an in-plane 2-fold axis.[29] The similarity of these two features in projections from two different crystal forms of F_1-ATPase is striking. The rat liver crystals are currently being characterized to determine their space group and similarity to the crystals being studied by X-ray diffraction in Amzel's laboratory.[5,6]

Conclusions

Information on the subunit structure of macromolecular complexes can be obtained by electron microscopy of single molecules and crystals

[29] C. W. Akey, V. Spitsberg, and S. J. Edelstein, unpublished results (1984).

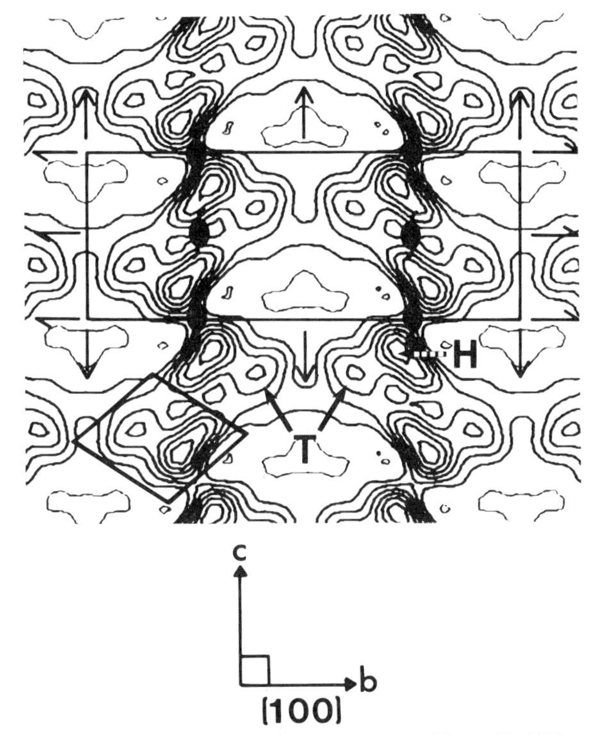

FIG. 8. Reconstruction of the (100) plane of form I beef heart F_1-ATPase crystals. The unit cell, symmetry elements, and asymmetric unit are indicated. Single molecules are oriented in the crystal to give a nearly side-on view of the F_1-ATPase in this projection. The headpiece (H) and tailpiece (T) regions of the bipartite molecule are indicated, along with the tailpiece-to-tailpiece interaction which occurs between adjacent molecules across the in-plane 2-fold axis. From Akey *et al.*[19]

employing the methods described in this chapter. When these results are combined with recent data on the stoichiometric distribution of the α subunit in *E. coli* F_1-ATPase obtained by immunoelectron microscopy,[30] an overall model for the low-resolution, large-subunit architecture of the F_1-ATPases emerges. Based on electron microscopy, the F_1-ATPases have maximal dimensions of about $120 \times 120 \times 90$ Å and possess two characteristic projections, an *en face* view down the pseudo-hexagonal axis and a bipartite side view. The large-subunit stoichiometry of the F_1-ATPases is presumably $\alpha_3\beta_3\gamma$; the α and β subunits must alternate within the structure about a centrally located γ subunit. Single-molecule studies demonstrate a variably contrasted central region in the F_1-ATPases, attributed to the presence of a central hole.

[30] H. Lunsdorf, K. Ettrig, P. Friedl, and H. U. Schairer, *J. Mol. Biol.* **173,** 131 (1984).

The data obtained by electron microscopy on single molecules and crystals of F_1-ATPases are in reasonable agreement with the morphology of the rat liver enzyme recently reported at 9 Å resolution[6] except that a centrally located, seventh subunit was not apparent in the maps obtained by X-ray diffraction and subsequent refinement. Amzel and co-workers were forced to outline other possible subunit arrangements for the F_1-ATPase based on the presence of a putative 2-fold axis in the molecule. The results obtained by electron microscopy clearly demonstrate the presence of a seventh, stalklike central subunit and the lack of a molecular 2-fold axis. The discrepancy between the results obtained by electron microscopy and X-ray diffraction may be caused by a combination of a crystal disorder[6] and the methods used to refine the phases of the structure at 9 Å resolution (see discussion of this point in Ref. 21).

Three-dimensional reconstructions have recently been reported from ultrathin sections of the M-band region of fish muscle (70 Å)[31,32] and the MYAC layer from insect flight muscle (60 Å).[33] General procedures for obtaining three-dimensional reconstructions from thin sections of protein crystals are currently being developed; therefore, low-resolution three-dimensional structures of prokaryotic and eukaryotic coupling factors may be obtained by electron microscopy of thin crystals and crystal thin sections in the near future.

Acknowledgments

The authors would like to thank M. Szalay for help with the figures, M. Rozycki for comments on the manuscript, and J. Broadhead for preparing the manuscript. This work was supported in part by National Science Foundation Grant PCM-7910462.

[31] P. K. Luther and R. A. Crowther, *Nature (London)* **307,** 566 (1984).

[32] R. A. Crowther and P. K. Luther, *Nature (London)* **307,** 569 (1984).

[33] K. A. Taylor, M. C. Reedy, L. Córdova, and M. K. Reedy, *Nature (London)* **310,** 285 (1984).

[42] Isolation and Reconstitution of Membrane-Bound Pyrophosphatase from Beef Heart Mitochondria

By I. S. KULAEV, S. E. MANSUROVA, and YU. A. SHAKHOV

Introduction

Mitochondria of different eukaryotes[1,2] as well as of *Rhodospirillum rubrum* chromatophores[3,4] are capable of carrying out inorganic pyrophosphate (PP_i) synthesis coupled to electron transport. The coupling has been shown to be performed by inorganic pyrophosphatase (PPase), confirmed by the experiments on the reconstitution of the system by using the isolated membrane enzyme.[5–9]

The present chapter deals with the method for isolation of membrane-bound PPase from beef heart mitochondria and reconstitution of the system of PP_i synthesis in submitochondrial particles (SMP).

Procedures

Beef heart mitochondria were isolated as described by Crane and coworkers[10] in 0.25 *M* sucrose containing 10 m*M* phosphate buffer, pH 7.5, and 1 m*M* EDTA. Submitochondrial particles capable of synthesizing PP_i were prepared by using gentle sonication in an ATP- and Mn^{2+}-free medium.[11] ATP and PP_i synthesis was measured after incubation of SMP (5–10 mg/ml) by vigorous stirring with a magnetic mixer for 10 min at 18–20°

[1] S. E. Mansurova, Yu. A. Shakhov, T. N. Beljakova, and I. S. Kulaev, *FEBS Lett.* **55,** 94 (1975).

[2] S. E. Mansurova, S. A. Ermakova, R. A. Zvyagilskaya, and I. S. Kulaev, *Mikrobiologiya (USSR)* **44,** 874 (1975).

[3] H. Baltscheffsky, L.-V. von Stedingk, H. W. Heldt, and M. Klingenberg, *Science* **153,** 1120 (1966).

[4] M. Baltscheffsky, H. Baltscheffsky, and L.-V. von Stedingk, *Brookhaven Symp.* **19,** 246 (1966).

[5] S. E. Mansurova, Yu. A. Shakhov, and I. S. Kulaev, *FEBS Lett.* **74,** 31 (1977).

[6] A. A. Kondrashin, V. G. Remennikov, V. B. Samuilov, and V. P. Skulachev, *Eur. J. Biochem.* **113,** 219 (1980).

[7] Yu. A. Shakhov, P. Nyrén, and M. Baltscheffsky, *FEBS Lett.* **146,** 177 (1982).

[8] P. Nyrén, K. Hagnal, and M. Baltscheffsky, *Biochim. Biophys. Acta* **766,** 630 (1984).

[9] P. Nyrén and M. Baltscheffsky, *FEBS Lett.* **155,** 125 (1983).

[10] F. L. Crane, J. L. Glenn, and D. E. Green, *Biochim. Biophys. Acta* **22,** 475 (1956).

[11] M. Hansen and A. Smith, *Biochim. Biophys. Acta* **81,** 214 (1964).

in 2 ml of medium containing 0.25 *M* sucrose, 5 m*M* Tris–HCl buffer, 37 m*M* succinate, 0.8 m*M* EDTA, 2.5 m*M* KH_2PO_4, 30 m*M* $MgCl_2$, pH 7.4. To measure the rate of ATP synthesis, the medium is supplemented with 3 m*M* ADP. The reaction was stopped by adding perchloric acid up to a final concentration of 0.5 *N*.

ATP was determined by using glucose-6-phosphate dehydrogenase and hexokinase, and PP_i was measured as described in Ref. 12, including preliminary 3-fold extraction of P_i in the form of phosphomolybdenum complex by isoamyl alcohol. The synthesis of PP_i was also confirmed by the incorporation in the above medium. High-voltage (40 V/cm) paper (Filtrak) electrophoresis (Shandon Southern, Model 24) in 0.05 *M* citrate buffer, pH 3.6, was used to separate the labeled nucleotides, P_i, and PP_i. These compounds on the electrophoregrams were located with the aid of markers. Phosphorus compounds were developed according to Hanes and Isherwood.[13] The spots were excised and the radioactivity was monitored on an automatic counter (Mark II).

ATPase and PPase activities were measured by P_i release during incubation at 30° in a medium containing 50 m*M* Tris–HCl buffer, pH 8.0 (for PPase) and 7.5 (for ATPase), 4 m*M* $MgCl_2$, 1 m*M* $Na_4P_2O_7$, or ATP. The reaction was stopped by adding perchloric acid. Orthophosphate was determined according to Berenblum and Chain.[14]

Electrophoresis of proteins was performed in a 7.5% gel at pH 7.5.[15] The samples were stained with Amido Black 10B. To identify PPase-active bands, the gel was placed in the above medium and incubated for 5 min at 37°. The orthophosphate formed was located as described in Ref. 16).

The protein content was measured by the Lowry procedure.[17]

Reconstitution of the PP_i Synthesis System by Crude PPase

While preparing PP_i-synthesizing SMP, we observed that PPase was bound to the membrane less tightly than ATPase. PPase activity in the SMP depended on the intensity and duration of sonication (Table I). Thus PPase activity in SMP and its capability to synthesize PP_i were found to be correlated.

[12] G. B. Grindey and C. A. Nichol, *Anal. Biochem.* **33,** 114 (1970).
[13] C. S. Hanes and F. A. Isherwood, *Nature* (*London*) **169,** 1107 (1949).
[14] J. Berenblum and E. Chain, *Biochem. J.* **32,** 295 (1938).
[15] B. J. Davis, *Ann N.Y. Acad. Sci.* **121,** 404 (1964).
[16] J. Sugino and Y. Migoshi, *J. Biol. Chem.* **239,** 2360 (1964).
[17] O. H. Lowry, N. J. Rosebrough, A. L. Farr, and R. J. Randall, *J. Biol. Chem.* **193,** 265 (1951).

TABLE I
EFFECT OF SONICATION ON PP_i SYNTHESIS AND PPASE ACTIVITY OF SUBMITOCHONDRIAL PARTICLES

Time (min)	ATPase activity (μmol/g protein/min)	PPase activity (μmol/g protein/min)	PP_i synthesis (μmol/g protein/min)
1.0	98	4.3	0.5
1.5	95	1.2	0.4
2.0	92	0.6	0.0

PPase can be washed from SMP with 0.25 *M* sucrose by vigorous stirring with a magnetic mixer at 0° for 20 min followed by centrifugation at 105,000 *g* for 45 min. ATPase and ATP-synthesizing activities are retained after this treatment, whereas PP_i synthesis decreases simultaneously with a decrease in PPase activity. The "washing" medium concentrated on PM10 membrane (Amicon) in an argon atmosphere was incubated with the "washed" particles in the presence of 1 m*M* succinate, and this restored PP_i synthesizing activity of SMP (Table II). It was evident from this experiment that the protein fraction washed out from SMP and containing PPase also possessed coupling activity.

A similar method for reconstitution of the ATP-synthesizing system from ATPase-containing eluate and from bacterial membrane particles was described earlier for chromatophores of *R. rubrum*.[18] In this case, the ATPase of the eluate supplemented with the washed chromatophores likewise promoted phosphorylating activity.

To verify PPase participation in PP_i synthesis, we isolated the homogeneous membrane-bound PPase of mitochondria.

Isolation and Purification of Mitochondrial PPase

Electrophoresis of protein extracts from mitochondria in a polyacrylamide gel revealed two proteins exhibiting PPase activity. One was readily solubilized by a buffer solution and the other extracted from mitochondria by the freeze-thaw technique or by treatment with detergents.

In our experiment, PPases were extracted from mitochondrial acetone powder[19] stored *in vacuo* at 4° for several weeks. The purification procedure is described in Table III. The extraction was performed by using a

[18] B. C. Jochanson, *FEBS Lett.* **20,** 338 (1972).
[19] M. L. Selvyn, *Biochem. J.* **105,** 279 (1967).

TABLE II
RECONSTITUTION OF PP_i SYNTHESIS FROM "WASHED" SMP AND CONCENTRATED ELUATE CONTAINING THE PPASE ACTIVITY[a]

Synthesized compound	Control SMP	Washed SMP	Washed SMP + eluate
PP_i (μmol/g protein/min)	1.4	0.0	0.5
ATP (μmol/g protein/min)	4.5	4.2	4.1

[a] The medium contained 1 m*M* succinate.

standard buffer containing 10 m*M* Tris–HCl, pH 7.4, and 20 m*M* 2-mercaptoethanol (1 ml/20 mg of the powder).

Proteins were fractionated by ammonium sulfate. The latter procedure should be done carefully to avoid acidification and foaming. The protein precipitate formed after 40 min at 40% saturation was removed by 20 min centrifugation at 20,000 *g*. Then ammonium sulfate was added to the

TABLE III
PURIFICATION OF PPASES FROM BEEF HEART MITOCHONDRIA

Fractions	Volume (ml)	Protein (mg)	Specific activity (μmol/mg protein/min)	Total activity (μmol/min)	Degree of purity	Yield (%)
Mitochondria	200	12,000	50	600	1.0	100
Acetone powder extract	450	900	500	450	10	75
Precipitate [40–60% $(NH_4)_2SO_4$ saturation]	2.5	225	910	205	18	34
Gel filtration, Sephadex G-150	87	61	1,100	72	22	12
Ion-exchange chromatography, DEAE-cellulose, pH 7.4	33	7.6	4,900	37	98	6.4
Ion-exchange chromatography, DEAE-cellulose, pH 6.7						
PPase I	24	1.2	10,000	12	200	2
PPase II	15	2.0	4,700	9	94	1.5
Gel filtration, Sephadex G-150: PPase II	38	0.4	12,300	5	246	0.8

supernatant to a final concentration of 70%. After 60 min the precipitate was removed under the same conditions. The pellet containing the inorganic PPases was dissolved in 5–10 ml of the standard buffer.

Gel filtration on the Sephadex G-150 (Pharmacia) was performed on columns 80 × 5 cm and 75 × 2 cm equilibrated with the standard buffer. Chromatography on the DEAE-cellulose DE-32 (Whatman) was performed in a 25 × 5 cm column. In the first case the resin was equilibrated with the standard buffer containing 1 m*M* $Na_4P_2O_7$ according to the method suggested earlier by Irie and co-workers[20] for purification of mitochondrial PPases from rat liver. The sample was loaded on the column which was then washed until the disappearance of absorption at 280 nm. Elution was performed in an NaCl gradient in the standard buffer. In the latter case the resin was equilibrated with 10 m*M* phosphate buffer, pH 6.7, in the presence of 2-mercaptoethanol, but without pyrophosphate. In this case the PPase presence in the fraction was also revealed electrophoretically.

As seen from Fig. 1, the first ion-exchange chromatography was used only for enzyme purification while the second made it possible to separate the two PPases. According to electrophoretic data, the first peak (PPase I) was a homogeneous protein fraction, and the second one (PPase II) exhibited three protein zones. An additional filtration on the column with Sephadex G-150 after concentrating the protein on membrane PM30 (Amicon) permitted obtaining homogeneous fraction of PPase II.

The fractions with PPases were collected and the protein was sedimented with 70% ammonium sulfate and kept under it in the presence of 20 m*M* 2-mercaptoethanol. Before utilization, the PPase preparations were desalted on the column with coarse Sephadex G-50 equilibrated with Tris–HCl buffer, pH 7.4.

Both enzymes were found to be Mg^{2+} dependent and extremely labile. They were stabilized by SH reagents, possessed a high specificity for PP_i, and had a similar pH optimum at 8.0,[21] but exhibited different electrophoretic mobility (Fig. 2). In addition, as shown by further experiments,[22] PPase II also contains the phospholipid lecithin.

The electrophoretic mobility of isolated PPase I coincides with the mobility of the readily extractable enzyme, while that of PPase II is simi-

[20] M. Irie, A. Jabuta, K. Kimura, Y. Shindo, and K. Tamita, *J. Biochem.* **67,** 47 (1970).

[21] N. V. Efremovich, S. E. Volk, A. A. Baykov, and Yu. A. Shakhov, *Biochimiya (USSR)* **45,** 831 (1980).

[22] Yu. A. Shakhov, V. F. Dukhovich, A. M. Velandia, V. A. Spiridonova, S. E. Mansurova, and I. S. Kulaev, *Biochimiya (USSR)* **47,** 601 (1982).

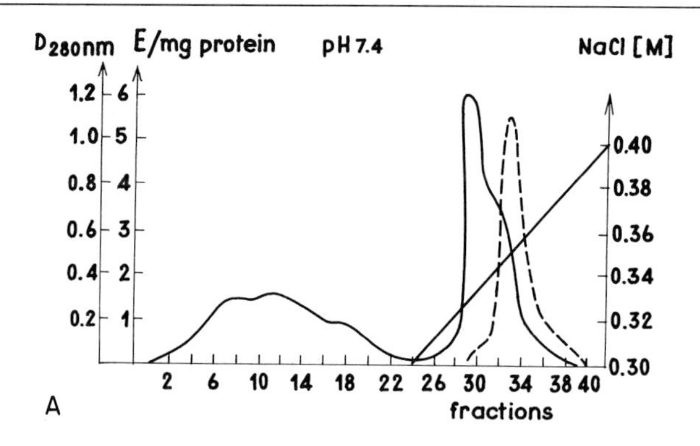

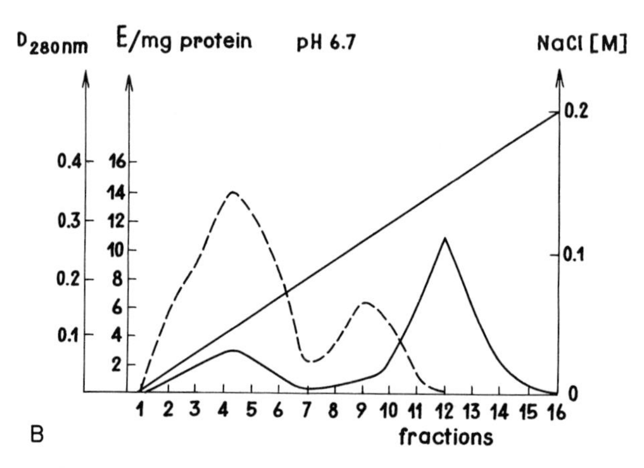

FIG. 1. Ion-exchange chromatography of the mitochondrial PPases on DEAE-cellulose. (A) pH 7.4; (B) pH 6.7. (——), Protein; (---), activity.

lar to the mobility of the enzyme extracted by the freeze-thaw technique or by detergents. We have concluded, therefore, that PPase II is the membrane enzyme involved in the coupling of respiration and phosphorylation. This is confirmed by the data presented below.

Reconstitution of Membrane-Bound PPase

The similarities in the properties of PPase I and PPase II and accumulation of PPase I during the storage of PPase II made it possible to assume that they are similar proteins. Actually, Avaeva and co-workers[23] showed recently that PPase I and PPase II possess two common catalytic sub-

[23] S. E. Volk, A. A. Baykov, E. B. Kostenko, and S. M. Avaeva, *Eur. J. Biochem.* **74,** 127 (1983).

TABLE IV
RECONSTITUTION OF PP_i SYNTHESIS IN BEEF HEART SMP BY PURIFIED PPASES

Activity	Control SMP	Washed SMP	Washed SMP + PPase I	Washed SMP + PPase II	Washed SMP + PPase I + 1 m*M* succinate	Washed SMP + PPase II + 1 m*M* succinate	Washed SMP + PPase I + 1 m*M* PP_i	Washed SMP + PPase II + 1 m*M* PP_i
Synthesis of PP_i, incorporation of [^{32}P]orthophosphate, $\times 10^{-6}$ mCi/mg protein/min	59.0	10.5	13.6	92.4	10.1	107.0	0.0	0.0
Hydrolysis of PP_i (μmol/g) protein/min)	10.7	2.5	2.7	3.8	2.8	3.9	0.0	0.0

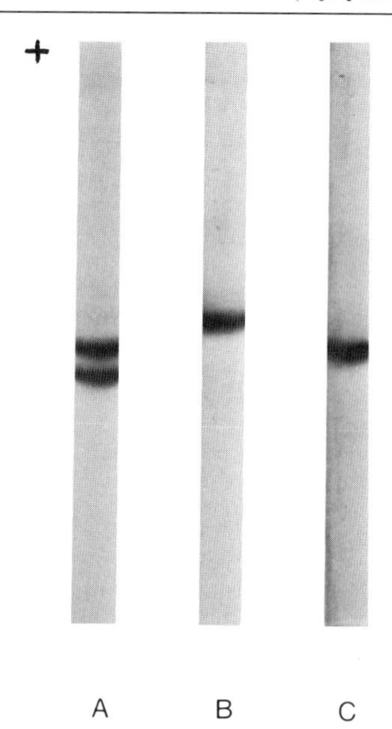

FIG. 2. Electrophoresis of mitochondrial PPases in a polyacrylamide gel (pH 7.5). A, PPase I plus PPase II; B, PPase II; and C, PPase I.

units, α and β. Unlike PPase I, PPase II proved to be a more complex protein, containing additional subunits; the functions of the latter are not clear as yet.

Phospholipid-free PPase I may interact with different phospholipids, thus acquiring an ability to incorporate into SMP membrane and to couple electron transport to PP_i synthesis, as PPase II.[24] Significantly, the activity of lipidized PPase I as well as PPase II in solution or PPase II in the membrane is controlled by the phospholipid environment.[24,25] It is quite probable that the free PPases I and II in the matrix and PPase II of the membrane are in equilibrium, depending on the functional state of the mitochondria.

[24] S. E. Mansurova, V. F. Dukhovich, V. A. Spiridonova, Yu. A. Shakhov, A. M. Velandia, N. V. Khrapova, and A. N. Tikhonov, *Biochem. Int.* **8,** 749 (1984).

[25] S. E. Mansurova, I. S. Kulaev, V. F. Dukhovich, A. P. Khohlov, and E. B. Burlakova, *Biochem. Int.* **5,** 457 (1982).

[43] Use of Monoclonal Antibodies to Purify Oligomycin Sensitivity-Conferring Protein and to Study Its Interactions with F_0 and F_1

By PHILIPPE ARCHINARD, FRANÇOIS PENIN, CATHERINE GODINOT, and DANIÈLE C. GAUTHERON

Oligomycin sensitivity-conferring protein (OSCP) is a subunit of the H^+-transporting ATP synthase complex in mitochondria which was first purified and characterized by MacLennan and Tzagoloff[1] and then by Senior.[2,3] However, the purification procedure remained delicate, tedious, and time-consuming. The major difficulty in improving the purification came from the absence of known intrinsic enzymatic activity of this peptide. In order to better know the structure and function of OSCP in the complex, anti-OSCP monoclonal antibodies were prepared.[4] These antibodies proved to be very useful tools to work out a new procedure of purification of OSCP, which is simple, rapid and efficient. This procedure is described here.

Purification

Principle

The use of monoclonal anti-OSCP antibodies permits testing the efficiency of each purification step after separation of the proteins with SDS–PAGE,[5] transfer to nitrocellulose,[6] and estimation of the amount of OSCP present in each fraction by immunodecoration.[7]

This preliminary study has shown that the procedure can directly start from mitochondria and that the preparation of submitochondrial particles and the sodium bromide extraction,[1,3] which lead to a severe loss of OSCP, must be avoided. These steps are replaced by alkaline and salt treatments of mitochondria. Then OSCP is extracted by ammonia and

[1] D. H. MacLennan and A. Tzagoloff, *Biochemistry* **7,** 1603 (1968).
[2] A. E. Senior, *Bioenergetics* **2,** 141 (1971).
[3] A. E. Senior, this series, Vol. 55 [49].
[4] P. Archinard, M. Moradi-Améli, C. Godinot, and D. C. Gautheron, *Biochem. Biophys. Res. Commun.*, **123,** 254 (1984).
[5] U. K. Laemmli, *Nature (London)* **222,** 680 (1970).
[6] H. Towbin, T. Staehlin, and J. Gordon, *Proc. Natl. Acad. Sci. U.S.A.* **76,** 4350 (1979).
[7] C. Godinot, M. Moradi-Améli, and D. C. Gautheron, this volume [73].

purified by an ion-exchange chromatography that was improved as compared to previously described methods.

Once the purification protocol is established, the use of the monoclonal antibodies is no longer necessary.

Procedure

All steps are carried out at 0–4°.

Step 1: Alkaline Extraction. Frozen pig heart mitochondria prepared as described by Smith[8] are thawed and adjusted to 15 mg protein/ml in 0.25 *M* sucrose, 1 m*M* EDTA, 1 m*M* dithiothreitol, 10 m*M* Tris–HCl, pH 7.5. The suspension is vigorously shaken with a magnetic stirrer while 6 *N* KOH is quickly added to obtain a final concentration of 50 m*M* (final pH of about 11.5). One minute later, the pH is lowered to 9.2 with 10 *N* acetic acid. The suspension is centrifuged for 10 min at 25,000 *g*. The slightly turbid supernatant contains the natural proteic inhibitor IF_1. It is saved for further purification, as described by Hortsman and Racker.[9] The pellets are homogenized with a Potter–Elvehjem in the above medium and the same treatment is repeated.

Step 2: Salt Treatment. The pellets are homogenized in 0.25 *M* sucrose, 1 m*M* EDTA, 1 m*M* dithiothreitol, 0.2 *M* KCl, 10 m*M* Tris base, pH 9.2 (KOH), at a final concentration of about 15 mg protein/ml. After centrifugation for 30 min at 100,000 *g*, the pink-yellow supernatant is discarded, the pellets homogenized again in the same buffer, and the suspension centrifuged as above.

Step 3: Ammonia Extraction. The pale yellow supernatant is discarded and the pellets are homogenized in the salt treatment buffer (about 10 mg protein/ml). Ammonia (0.9 specific gravity NH_3 solution) is added to give a final concentration of 0.4 *N*. After vigorous stirring for 10 min, the suspension is centrifuged for 1 hr at 300,000 *g*. The clear supernatant is suctioned off, adjusted to pH 8.0 with 10 *N* acetic acid, and centrifuged for 30 min at 100,000 *g* to remove any precipitate.

Step 4: CM-Trisacryl M Ion-Exchange Chromatography. The CM-Trisacryl M gel (obtained from Industrie Biologique Française or LKB) is washed with 2 volumes of 1 *M* ammonium acetate, 2 *M* KCl, pH 8.0 (KOH), and then with at least 3 volumes of equilibrating buffer prepared as follows: Ammonia (0.9 specific gravity NH_3 solution) is added to a final concentration of 0.4 *N* to a solution of 0.25 *M* sucrose, 1 m*M* EDTA, 1 m*M* dithiothreitol, 0.2 *M* KCl, 10 m*M* Tris base, pH 9.2 (KOH). The pH is

[8] A. Smith, this series, Vol. 10 [13].

[9] L. L. Horstman and E. Racker, *J. Biol. Chem.* **245,** 1336 (1970).

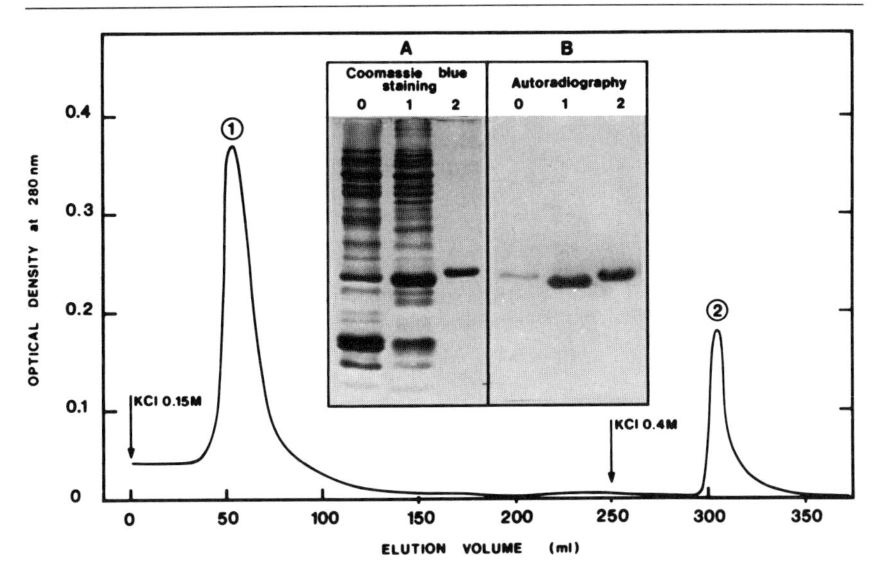

FIG. 1. Elution of OSCP from the CM-Trisacryl M column at Step 4. Insert A: SDS–PAGE electrophoresis according to Laemmli.[5] Lane 0, 30 μg protein not retained by the column; lane 1, 30 μg protein from peak 1; lane 2, 10 μg protein from peak 2 after the treatment described in the text. Insert B: Autoradiograms obtained after transfer to nitrocellulose sheets of the proteins separated in insert A followed by successive incubations with the anti-OSCP monoclonal antibody, rabbit anti-mouse immunoglobulin, and ^{125}I-labeled protein A as described in Ref. 7.

then adjusted to 8.0 with 10 *N* acetic acid, and this solution is diluted by addition of 5 volumes of cold distilled water.

The supernatant obtained at Step 3 is diluted with 5 volumes of cold distilled water and the pH eventually readjusted to 8.0. The solution is filtered through a glass fiber filter (GF/C Whatman). The equilibrated CM-Trisacryl M gel (60 ml) is poured on a sintered glass funnel and the protein solution is filtered through the gel with a high flow rate (about 1 liter/hr). Under these conditions, the amount of OSCP not retained by the column is negligible, as checked by immunodecoration, using the anti-OSCP monoclonal antibody as a probe (see Fig. 1 insert, A and B, lane 0). Then 1 volume of gel is suspended in 2 volumes of equilibrating buffer and the slurry poured into a 3.2-cm diameter column. The column is extensively washed with 3 volumes of buffer containing 0.15 *M* KCl, 0.1 m*M* EDTA, 0.1 m*M* dithiothreitol, 15 m*M* Tris-H_2SO_4, pH 7.5, at a flow rate of 2 ml/min. OSCP is then eluted with 0.4 *M* KCl, 0.1 m*M* EDTA, 0.1 m*M* dithiothreitol, 15 m*M* Tris-H_2SO_4, pH 7.5, at a flow rate of 1 ml/min. Figure 1 gives the elution profile of the chromatography. At 0.15 *M* KCl, one main peak is eluted. The electrophoresis pattern and autoradiogram show that OSCP is present in association with other peptides (peak 1). The main contaminant of about 12 kDa, difficult to eliminate by the Senior procedure,[10] comes off in this washing step. The fractions eluted at 0.4 *M* KCl corresponding to the peak 2 are pooled and concentrated by addition of solid ammonium sulfate to obtain a 70% saturation at 0°. After at least 1 hr, the suspension is centrifuged for 20 min at 100,000 *g*. The pellet is dissolved in a minimal volume of 20 m*M* Tris-H_2SO_4, pH 8.0, and centrifuged for 5 min at 10,000 *g* to eliminate any aggregated material. The supernatant contains OSCP free of contaminants as shown by the electrophoresis (Fig. 1, lane 2). It can be stored either directly in liquid nitrogen or as an ammonium sulfate suspension saturated at 70%. The yield of pure OSCP is 8–10 mg from 10 g of mitochondria, i.e., about 2-fold the yield described by Senior for beef heart OSCP[3] and about 5-fold the yield obtained by us with the Senior's method applied to pig heat OSCP. The whole procedure can be performed within 1 day, which is a great improvement.

Comments

The OSCP prepared by this procedure has a good biological activity as shown in the table.[12] It increases by about 3-fold the rate of net ATP synthesis obtained after reconstitution of urea-treated submitochondrial particles (ETP-U)[11] with purified F_1-ATPase[13] and OSCP.

In Steps 1 and 2, almost no OSCP is lost. In addition, the first supernatant obtained after alkaline treatment (Step 1) serves as a starting material for the protein inhibitor purification.[9] Besides, one ammonia extraction (Step 3) is sufficient to solubilize about 90% of the OSCP present in the membranes. An eventual second extraction introduces contaminants that tend to induce an aggregation of OSCP with other peptides and therefore is avoided. Finally, the use of CM-Trisacryl M (Step 4) permits very high flow rates and improves the separation of free OSCP from OSCP associated with other peptides. Moreover, the elution of OSCP does not require a KCl gradient.

Interactions of OSCP with F_0 and F_1

Anti-OSCP monoclonal antibodies prepared and purified as in Ref. 4 can be used to study the interaction of OSCP with F_0 and F_1 and its function in the whole complex H^+-transporting ATP synthase.

[10] A. Dupuis, G. Zaccai, and M. Satre, *Biochemistry* **22,** 5951 (1983).
[11] E. Racker and L. L. Horstman, *J. Biol. Chem.* **242,** 2547 (1967).
[12] M. E. Pullman, this series, Vol. 10 [9].
[13] F. Penin, C. Godinot, and D. C. Gautheron, *Biochim. Biophys. Acta* **679,** 198 (1982).

COMPARISON OF BIOLOGICAL ACTIVITY OF OSCP PREPARED BY THE SENIOR PROCEDURE AND BY THE PROCEDURE DESCRIBED HEREIN

Reconstitution conditions[a]	Net ATP synthesis (nmol/mg protein/min)	Activation
ETP-U	0.8	—
ETP-U + F_1	38.0	1
ETP-U + OSCP[b]	0.9	—
ETP-U + F_1 + OSCP[b]	114.0	3.0
ETP-U + F_1 + OSCP[c]	119.0	3.1

[a] Urea particles (ETP-U, 200 μg) prepared according to Racker and Horstman[11] were mixed in the presence or absence of F_1-ATPase (40 μg) prepared according to Penin *et al.*[13] and OSCP (3 μg) with a buffer containing 0.25 *M* sucrose, 10 m*M* Tris-H_2SO_4, pH 7.5, in a final volume of 95 μl. After addition of 5 μl of 100 m*M* ATP, 200 m*M* $MgSO_4$, pH 7.5, the mixture was incubated for 10 min at 30°. ATP synthesis started upon addition of 600 μl of 6 m*M* ADP, 7 m*M* $MgSO_4$, 60 m*M* glucose, 12 m*M* succinate, 23 m*M* phosphate (K), 5 μCi ^{32}Pi, 30 units hexokinase (Sigma, type F 300), 0.29 *M* sucrose, and 12 m*M* Tris-H_2SO_4, pH 7.5, and was stopped 4 min later by addition of 70 μl 35% perchloric acid. The amount of glucose 6-[^{32}P]phosphate formed corresponding to the ATP synthesized was extracted and counted as described by Pullman.[12]

[b] OSCP prepared according to Senior.[3]

[c] OSCP prepared according to the procedure described herein.

Accessibility of OSCP to the Antibodies

Various experiments were conducted to study the accessibility of the epitopes recognized by the antibodies when OSCP was reconstituted with F_1 or in submitochondrial particles (SMP). In solid-phase radioimmunoassays (SPRIA)[7] F_1 did not decrease the binding of the antibody to OSCP when purified F_1 was reassociated with purified OSCP. In addition, OSCP was accessible to the antibodies in SMP whether the accessibility was tested by direct or competitive SPRIA or by electron microscopy with the immunogold method.[14] (Thanks are due to Dr. J. Comte for performing the electron microscopy experiments.)

Effects of the Antibodies on ATP Synthesis

The antibodies had no effect on the succinate-dependent net ATP synthesis of SMP or of SMP depleted from F_1 by urea treatment and

[14] W. P. Faulk and G. N. Taylor, *Immnochemistry* **8,** 1081 (1971).

reconstituted with F_1 and OSCP except if the antibodies were preincubated with OSCP before the reconstitution. This means that the antibodies bind to OSCP integrated in the membrane without interfering with ATP synthesis or proton translocation. However, the OSCP–antibody complex cannot be used to reconstitute F_1 with depleted SMP.

Peptides Close to OSCP in the Complex

Cross-linking experiments made either by using the antibodies to identify the cross-linked products of OSCP or by direct photolabeling of the cross-linked products of OSCP revealed that major neighbors of OSCP were the α and β subunits of F_1 and at least two other peptides of about 30 and 24 kDa, not yet identified.

[44] Purification and Properties of the ATPase Inhibitor from Bovine Heart Mitochondria

By MAYNARD E. PULLMAN

Mitochondrial ATPase (F_1) inhibitor is a small heat-stable protein that was first isolated from beef heart mitochondria.[1] When complexed with either soluble or membrane-bound F_1,[1,2] the protein inhibits the ATPase activity of the enzyme, but does not interfere with F_1-dependent oxidative phosphorylation.[1] Thus, the inhibitor–F_1 complex, which exhibits no ATPase activity, retains an undiminished capacity to restore oxidative phosphorylation in F_1-deficient submitochondrial particles (SMP).[1]

The purification and properties of the inhibitor protein were described in an earlier volume of this series.[3] Since then several other purification procedures for the beef heart, rat liver, yeast, and chloroplast inhibitor have been described. Most of these procedures have been patterned after the original purification procedure which exploited the heat and acid stability of the protein. References to these procedures together with an updated description of the properties of the inhibitor protein have been presented by Ernster *et al.*[4] in Volume 55 of this series. The procedure described here represents an improvement over the original procedure[1]

[1] M. E. Pullman and G. C. Monroy, *J. Biol. Chem.* **238,** 3762 (1963).

[2] L. L. Horstman and E. Racker, *J. Biol. Chem.* **245,** 1316 (1970).

[3] G. C. Monroy and M. E. Pullman, this series, Vol. 10 [80].

[4] L. Ernster, C. Carlsson, T. H. Undal, and K. Nordenbrand, this series, Vol. 55 [51].

with respect to yield and purity. Some newer aspects of the properties of the inhibitor are also described.

Assay Method

Principle

The inhibitor content of fractions emerging from the purification procedure is determined by measuring the inhibition of the ATPase activity of either submitochondrial particles or soluble F_1. The inhibitor does not inhibit ATPase in intact mitochondria. The assay system employing SMP has several advantages and is favored by most investigators. These advantages include the fact that SMP are easier to prepare than is soluble F_1.[2] In addition the assay with SMP is less sensitive to salts[1,2] and is reported to be 4 times more sensitive[2] than that described for the soluble enzyme.[3]

The assay is carried out in two steps. The inhibitor is first incubated with submitochondrial particles, Mg^{2+}, and ATP to permit the inhibitor to bind to the particle-bound ATPase. This is followed by an assay for the remaining ATPase activity. The procedure is similar to that described previously.[2]

Reagents

0.5 *M* Tris-SO_4, pH 7.4
0.4 *M* sodium ATP, pH 7.4
0.4 *M* magnesium sulfate
0.05 *M* potassium phosphoenolpyruvate[5]
10 mg/ml pyruvate kinase[6]
0.2 *M* Bis-Tris-SO_4, pH 6.5[7]
0.025 *M* sodium ATP/0.025 *M* magnesium sulfate, pH 7.0 (ATP/Mg mixture)
0.25 *M* sucrose
1.0 *M* Na_2SO_4
10 mg/ml bovine serum albumin in 0.2 *M* Bis-Tris-SO_4, pH 6.5

[5] The monocyclohexylammonium salt of phosphoenolpyruvate was purchased from Sigma Chemical Company, St. Louis, MO and converted to the potassium salt as described in V. M. Clark and A. J. Kirby, *Biochem. Prep.* **11,** 101 (1966). The potassium salt is also commercially available, but is about twice as expensive.

[6] Pyruvate kinase is a crystalline suspension in ammonium sulfate (Boehringer-Mannheim) with a specific activity of 200.

[7] [Bis(2-hydroxyethyl)imino-tris(hydroxymethyl)methane]; purchased from Sigma Chemical Company.

Mixed Medium A

A mixture sufficient for 25 assays contains 2.5 ml of Tris-SO_4, 0.25 ml of ATP, 0.25 ml of $MgSO_4$, 2.5 ml of phosphoenolpyruvate, and 0.75 ml water. This mixture is stable at −20° for at least 6 months.

Mixed Medium B

A mixture sufficient for 25 assays contains 6.25 ml of mixed medium A, 0.08 ml of pyruvate kinase, and 12.4 ml of water. Because of the uncertain stability of pyruvate kinase in the mixture, mixed medium B is prepared fresh and discarded at the end of the day.

Inhibitor Diluent

To obtain reproducible measurements of inhibitor activity, particularly of highly purified preparations, it is essential to dilute the protein in a solution containing 20 m*M* Bris-Tris-SO_4, pH 6.5, 0.2 *M* Na_2SO_4, and 1 mg/ml of bovine serum albumin. The diluent is prepared with 10 ml of Na_2SO_4, 5 ml of the bovine serum albumin solution, and 35 ml of water.

Submitochondrial Particles

SMP are prepared by utilizing steps 1 and 2 and part of step 3 as described for the preparation of the soluble ATPase from beef heart mitochondria.[8] After incubation overnight in 0.1 *M* sucrose containing 4 m*M* ATP and 2 m*M* EDTA, pH 9.3, the particles are centrifuged at 30,000 rpm (106,000 *g*) for 90 min in the No. 30 rotor of a Spinco ultracentrifuge. The sedimented particles are resuspended in 0.1 *M* sucrose at a protein concentration of 35–40 mg/ml. The specific activity of the ATPase in these particles is between 2 and 4 and remains unchanged for at least 1 year when the preparation is stored at − 70°. To avoid unnecessary freezing and thawing, the particle suspension is stored in aliquots of 0.2 ml. Immediately before use, an aliquot is thawed and diluted with 0.25 *M* sucrose.

Procedure

SMP containing 0.2–0.3 unit of ATPase activity (0.05–0.1 mg protein) are incubated for 5 min at 30° with 5–20 μl of the inhibitor sample in the presence of 5 μl of the ATP/Mg^{2+} mixture and 25 μl of Bis-Tris-SO_4, pH 6.5. The volume is adjusted to 0.25 ml with 0.25 *M* sucrose. The final pH of the incubation mixture is between 6.3 and 6.5. The order of addition is

[8] H. S. Penefsky, this series, Vol. 55 [33].

as follows: sucrose, buffer, inhibitor, SMP, and Mg^{2+}-ATP. Because the inhibitor protein has a strong affinity for glass, the incubation is carried out in polystyrene test tubes. After 5 min of incubation at 30°, ATP hydrolysis is initiated by adding 0.75 ml of mixed medium B, prewarmed to 30°. After an additional 10 min of incubation, the reaction is terminated by the addition of 2 ml of 2.5% ammonium molybdate in 5 N H_2SO_4, and P_i is determined on the entire reaction mixture.[9] However, if the amount of protein in the inhibitor sample used exceeds 0.1 mg, the reaction is stopped with 0.1 ml 50% trichloroacetic acid. After centrifugation to remove the precipitated protein, P_i is measured on an aliquot of the supernatant. The samples are read at 660 mμ against a blank containing water in place of the assay mixture. The results are corrected for P_i in the reagents by carrying out an incubation without enzyme. The extent of ATP hydrolysis in the absence of added inhibitor is used to calculate the degree of inhibition in the inhibited samples. Titrations in which increasing amounts of inhibitor protein are added to the preincubation mixture should be carried out with each series of assays to ensure that the extent of inhibition is proportional to added inhibitor. The assay is not linear in the region above 50% inhibition. For reasons not entirely clear, a rectilinear titration curve is not always obtained even in the region below 50% inhibition. Thus, although acceptable for monitoring the progress of the purification procedure, the activity assay is less than satisfactory for more demanding quantitative measurements. A procedure using a radioimmunoassay has been developed for the inhibitor protein which is specific, sensitive, and more reliable for quantitative measurements.[10]

Estimation of Protein

The protein content of the mitochondrial suspension and of all fractions emerging from the purification procedure is measured spectrophotometrically[11] after diluting the samples in either concentrated formic acid (mitochondria) or buffer (soluble protein). The reference cell contains the corresponding solution, minus protein. The values obtained for the mitochondrial suspension agree very closely with those obtained by a biuret procedure modified for particulate protein.[12] The biuret procedure was used to measure the protein content of submitochondrial particles.

[9] K. Lohmann and L. Jendrassik, *Biochem. Z.* **178,** 419 (1926).

[10] A. E. Otoadese, H. S. Penefsky, and M. E. Pullman, *Fed. Proc., Fed. Am. Soc. Exp. Biol.* **41,** 895 (1982). Complete details of the radioimmunoassay are described in a manuscript in preparation (1986).

[11] O. Warburg and W. Christian, *Biochem. Z.* **310,** 384 (1941).

[12] E. E. Jacobs, M. Jacob, D. R. Sanadi, and L. B. Bradley, *J. Biol. Chem.* **223,** 147 (1956).

Definition of Unit

A unit of inhibitor activity is defined as the amount of protein which results in 50% inhibition of 0.2 unit of particulate ATPase under the specified assay conditions. One unit of ATPase activity is that amount of enzyme which hydrolyzes 1 μmol of ATP per minute under the specified assay conditions. Specific activity is expressed as units per milligram of protein.

Purification of ATPase Inhibitor

Beef heart mitochondria,[13] consisting of almost equal amounts of "heavy" and "light" layer mitochondria, is used as starting material. The mitochondrial preparation can be stored at $-60°$ in 0.25 sucrose/0.01 *M* Tris-SO_4, pH. 7.4 (sucrose/Tris) for at least 2 years without adversely affecting the outcome of the purification procedure.

Twenty grams of mitochondria is worked up at a time and taken through the ammonium sulfate fractionation step. The ammonium sulfate fractions can be stockpiled at $-20°$ for at least 1 year without loss of inhibitor activity.

Steps 1–4 of the purification procedure are conducted at 4° whereas steps 5–8 can be carried out at room temperature. All centrifugations are performed in a Sorval RC2-B centrifuge.

Step 1: Washed Mitochondria. Prior to use, the mitochondrial suspension (60 to 70 mg/ml) is mixed with an equal volume of sucrose/Tris and centrifuged at 10,500 rpm (18,000 × g) in the GSA rotor for 20 min. The pellet is resuspended in sucrose/Tris to give the same final volume as the initial suspension and centrifuged as above. The latter procedure is repeated once more and the resulting mitochondrial pellet is resuspended in sucrose/Tris at a final protein concentration of 40 mg/ml.

Step 2: Alkaline Extraction. To 500 ml (20 g protein) of the washed mitochondrial suspension is added an equal volume of cold glass-distilled water. While the suspension is stirred rapidly on a strong magnetic stirrer, the pH is rapidly adjusted to 11.5 with 25–30 ml of 1 *N* KOH. After 1 min the pH is readjusted to 7.5 with ~2.5 ml of 10 *N* acetic acid. The neutralized suspension is centrifuged for 10 min at 18,000 *g*, as above.

Step 3: Trichloroacetic Acid Precipitation. To the somewhat turbid and yellow supernatant (~866 ml), 96 ml of 50% tricholoroacetic acid is added. After 10 min at 4°, the precipitated protein is collected by centrifugation at 18,000 *g* for 10 min and suspended by homogenization in 150 ml distilled water. The water suspension of the precipitate ($pH < 2$) is ad-

[13] P. V. Blair, this series, Vol. 10 [12].

justed to pH 11 with 6 *N* KOH and after 1 min is returned to pH 7.4 with 10 *N* acetic acid. After centrifugation at 18,000 *g* for 10 min, the slightly turbid supernatant solution is stored overnight at −20°.

Step 4: Ammonium Sulfate Fractionation. After thawing, a large brown gelatinous precipitate, devoid of activity, is occasionally noticed and is removed by centrifugation. More often, only a slight turbidity appears and is ignored. To each 100 ml of solution, 29 g of ammonium sulfate is added, and after stirring on ice for 20 min the precipitate is removed by centrifugation and discarded. To each 100 ml of the supernatant solution is added 15.9 g of ammonium sulfate, and after 20 min of stirring on ice, the precipitate is collected by centrifugation and dissolved in 50 m*M* Tris-SO_4, pH 8.0, to give a final volume of 10–12 ml of a clear solution. The ammonium sulfate fraction may be stored at −20° until all of the mitochondria have been brought to this stage.

Step 5: Carboxymethyl (CM)-Cellulose Chromatography. Ammonium sulfate fractions obtained from 420 g of washed mitochondria are pooled and desalted on a column of Sephadex G-25 (fine), equilibrated with 10 m*M* Tris-SO_4, pH 8.5. The desalted preparation is applied at a flow rate of 2.5 ml/min to a CM-cellulose column (Whatman, CM-52; 1.6 × 6 cm), equilibrated with 10 m*M* Tris-SO_4, pH 8.5. Under these conditions, almost all of the applied protein is recovered in the effluent. Although this step does not result in a significant increase in the specific activity of the preparation, it does remove all of the cytochrome *c* present in the fraction. Cytochrome *c* can be seen to form a tight red band at the top of the column. Omission of this step results in the contamination of the final inhibitor preparation by cytochrome *c*.

Step 6: Diethylaminoethyl (DEAE)-Cellulose Chromatography. The effluent from the previous step is applied, without further treatment, to a DEAE-cellulose column (Whatman DE-52) (2.5 × 20 cm), equilibrated with 10 m*M* sodium pyrophosphate, pH 8.5. Column effluents and eluents are monitored at 280 nm with a UV detector attached to a recorder. The flow rate of the column is adjusted to 1.6 ml/min by means of a peristaltic pump. After the sample is applied, the column is washed with 10 m*M* pyrophosphate, pH 8.5, until the recorder returns to the baseline. The column is then eluted with 50 m*M* sodium pyrophosphate, pH 8.5. About 80% of the inhibitor activity applied is found in the next protein peak.

Step 7: CM-Sephadex Chromatography. The step 6 inhibitor fraction (230 mg protein) is adjusted to pH 5.2 with 3 *N* acetic acid. The slightly turbid solution is centrifuged for 10 min at 12,000 *g*. The clear supernatant solution is applied at a pumping rate of 2.0 ml/min to a CM-25 Sephadex column (2.8 × 3.5 cm; bed volume, 22 ml) equilibrated with 10 m*M* sodium acetate, pH 5.2. After the sample is applied, the column is washed with 10 m*M* sodium acetate, pH 5.2, until the recorder returns to the baseline. The sample effluent plus the wash contains 80% of the applied protein but is devoid of inhibitor activity. The column is then eluted with 2.5 bed volumes of 10 m*M* sodium acetate, 100 m*M* sodium sulfate, pH 5.2. The eluate contains less than 1% of the protein and no inhibitor activity. Virtually all of the inhibitor activity applied is recovered when the column is subsequently eluted with 7 to 8 bed volumes of 10 m*M* sodium acetate, 1 *M* sodium sulfate, pH 5.2 (150 ml). Under these conditions the inhibitor protein is only slowly desorbed from the column. The rate of desorption is accelerated somewhat by periodically stopping the column flow for 30–60 min and then resuming collection. To concentrate and remove the inhibitor from the concentrated salt solution, the active fractions are pooled and the inhibitor protein is precipitated with 5% trichloroacetic acid at room temperature. The copius white precipitate is sedimented by centrifugation and dissolved in 10 m*M* PIPES, K^+, pH 6.9. The pH of the protein solution is 2.2. With some preparations, the solution was slightly turbid at room temperature and somewhat more turbid at 4°. The insoluble protein, which exhibited no inhibitor activity, is removed by centrifugation at 4°. The crystal-clear supernatant solution is adjusted to pH 6.9 with KOH and applied to a Sephadex G-25 (fine) column, equilibrated with 10 m*M* PIPES, K^+, pH 6.9, to ensure maximal removal of salt and trichloroacetic acid. A summary of the purification procedure is given in Table I. The values shown for the protein content of

TABLE I
SUMMARY OF PURIFICATION PROCEDURE[a]

Step	Total protein (g)	Total units ($\times 10^6$)	Specific activity (units/mg)	Yield (%)
1. Washed mitochondria	420	—	—	—
2. Alkaline extract	23.1	7.6	329	100
3. Extract of trichloroacetic acid precipitate	7.7	7.4	961	97
4. Desalted ammonium sulfate precipitate	1.03	7.4	7,184	97
5. CM-52 effluent	0.99	7.4	7,475	97
6. DE-52 eluent	0.23	5.8	25,217	77
7. CM-Sephadex	0.040	5.2	130,000	68
8. Desalted trichloroacetic acid precipitate of step 7	0.036	4.9	136,111	64

[a] The protein content of all fractions was measured spectrophotometrically.[11] This method gives a low estimate for step 7 and step 8 fractions because of the unusually low content of aromatic amino acids in the inhibitor molecule (cf. text and Table II).

TABLE II
PROTEIN CONCENTRATION OF THE STEP 8 FRACTION AS DETERMINED BY SEVERAL METHODS

Method	Protein concentration (mg/ml)
Dry weight[15]	5.52
Calculated from molar extinction coefficient[14]	5.46
Spectrophotometric[11]	0.81
Lowry *et al.*[16]	6.60
Modified Lowry *et al.*[17]	6.76
Bicinchoninic acid[18]	4.40

all the fractions emerging from the purification procedure were determined spectrophotometrically.[11] This method gives a very low estimate of the purified protein (step 8) because of the unusually low content of aromatic amino acids in the molecule.[14] To correct to the value obtained from the dry weight of the protein,[15] the value obtained from the spectrophotometric procedure[11] is multiplied by 6.81. Table II[15-18] summarizes the protein concentration obtained for the step 8 fraction using several different methods. Protein concentration determined from the molar extinction coefficient of the inhibitor protein ($1.68 \times 10^3\ M^{-1}\ cm^{-1}$)[14] agrees closely with that based on its dry weight.

Comments

Two Coomassie Brilliant Blue staining bands are observed when 100 μg of protein[16] from the step 8 fraction are electrophoresed with sodium dodecyl sulfate (SDS) in polyacrylamide gels.[19] The minor band, which is usually not visible when less than 50 μg of protein are applied to disk gels, represents about 3–5% of the total protein as determined from measurements of radioactivity in slices of gels containing ^{125}I-labeled protein. Three lines of evidence suggest that the "impurity" is in fact a dimeric

[14] B. Frangione, E. Rosenwasser, H. S. Penefsky, and M. E. Pullman, *Proc. Natl. Acad. Sci. U.S.A.* **78,** 7403 (1981).

[15] M. E. Pullman and H. S. Penefsky, unpublished results.

[16] O. H. Lowry, N. J. Rosebrough, A. L. Farr, and R. J. Randall, *J. Biol. Chem.* **193,** 265 (1951).

[17] G. L. Peterson, *Anal. Biochem.* **83,** 346 (1977).

[18] Pierce Chemical Co., *Previews,* Oct. (1984).

[19] K. Weber and M. Osborn, *J. Biol. Chem.* **244,** 4406 (1969).

form of the inhibitor protein. It has a molecular weight of 18,500 (estimated from its migration in SDS–gel electrophoresis),[20] which is approximately twice that of the monomer form.[14] Moreover, eluates of the gel band inhibit the ATPase activity of SMP and show strong cross-reactivity with a specific antibody raised against the inhibitor protein.[20] Evidence for multiple molecular species of the beef heart,[21] rat liver,[22] and yeast[23] inhibitor proteins has been presented previously. In these cases, however, the multimeric forms appear to arise from the spontaneous aggregation of the protein, since interconversion of one form to another can be affected *in vitro* by salt[21] and SDS.[22]

Properties

Amino Acid Sequence

The complete amino acid sequence of the beef heart inhibitor protein has been determined.[14] The molecule contains 84 residues, accounting for a molecular weight of 9578. Lysine, arginine, glutamic acid, and aspartic acid comprise nearly 50% of the total amino acid residues, whereas apolar residues are present in low amounts. The polarity index calculated according to Capaldi and Vanderkooi[24] is 66%, making it considerably more hydrophilic than average proteins. The 40 charged amino acids are distributed along the chain in clusters which tend to occur with some regularity. A section of the chain, located at the COOH terminal end, contains several duplicated regions, the most prominent of which are pentapeptides. This section of the chain contains all of the five histidines present in the molecule. The presence of a formyl-blocking group at the NH_2 terminus has been found by some,[25] but not all investigators.[14,26]

A comparison of the bovine heart inhibitor with the corresponding protein from the yeast *Saccharomyces cerevisiae* has revealed significant sequence homology between the two proteins.[25,27,28] On the other hand,

[20] H. Reinheimer and M. E. Pullman, unpublished results.

[21] G. Klein, M. Satre, G. Zaccai, and P. V. Vignais, *Biochim. Biophys. Acta* **681,** 226 (1982).

[22] S. Chan, H. L. Tsai, C. Mohls, and D. McNeilly, *Fed. Proc., Fed. Am. Soc. Exp. Biol.* **44,** 1080 (1985).

[23] E. Ebner and L. L. Maier, *J. Biol. Chem.* **252,** 671 (1977).

[24] R. A. Capaldi and G. Vanderkooi, *Proc. Natl. Acad. Sci. U.S.A.* **69,** 930 (1972).

[25] A. C. Dianoux, A. Tsugita, and M. Przybylski, *FEBS Lett.* **174,** 151 (1984).

[26] Dr. John Walker, personal communication.

[27] H. Matsubara, K. Inoue, T. Hashimoto, Y. Yoshida, and K. Tagawa, *J. Biochem.* **94,** 315 (1983).

[28] We have used the computer program FASTP [D. L. Lipman and W. R. Pearson, *Science* **227,** 1435 (1985)] to compare the amino acid sequence of the bovine ATPase inhibitor with

the chloroplast[29] and *Escherichia coli*[30] ATPase inhibitors, which are thought to be identical with the ε subunits of the corresponding ATPases, do not appear to be related in sequence to the bovine or yeast inhibitors.

Relationship of Structure to Functional Activity

Measurements of the inhibitory activity of peptides isolated after partial digestion of the beef heart inhibitor with proteolytic enzymes indicate that the first 16 amino acid residues from the NH_2 terminus are not required for activity.[25] However, removal of 22 residues from the NH_2 terminal end or ~10 amino acids from the COOH terminal end destroys the inhibitory activity.[25,31]

Modification of two histidine residues in the inhibitor molecule by diethyl pyrocarbonate, under conditions where neither tyrosine fluorescence nor immunoreactivity is altered, results in the loss of binding to and inhibition of membrane-bound F_1.[32]

Functional Cross-Reactivity of ATPase Inhibitors

A summary of the cross-reactivity of ATPase inhibitors from various sources has been presented in a previous volume of this series.[4] Since then, some additional information in this area has become available. For example, the bovine heart inhibitor fails to inhibit the purified *E. coli* ATPase.[33] Although the ATPase inhibitor purified from the yeast *S. cerevisiae* inhibits the ATPase activity of beef heart SMP,[34] it is 20 times less effective than the homologous beef heart inhibitor.[20] On the other hand, the beef heart and *S. cerevisiae* inhibitors appear to be almost equally effective in inhibiting the ATPase activity in yeast and rat liver

---

each of the sequences in the National Biomedical Research Foundation protein sequence library. The best initial and optimized score was obtained with the ATPase inhibitor from *S. cerevisiae*.

[29] E. T. Krebbers, I. M. Larrinua, L. McIntosh, and L. Bogorad, *Nucleic Acids Res.* **10,** 4985 (1982).

[30] M. Saraste, N. J. Gay, A. Eberle, W. J. Runswick, and J. E. Walker, *Nucleic Acids Res.* **9,** 5287 (1981).

[31] A. C. Dianoux, A. Tsugita, G. Klein, and P. V. Vignais, *FEBS Lett.* **140,** 223 (1982).

[32] D. A. Haake, A. E. Otoadese, H. S. Penefsky, and M. E. Pullman, unpublished observations.

[33] A. E. Otoadese and M. E. Pullman, unpublished observations. The activity of purified *E. coli* ATPase, a gift from Dr. M. Futai, was unaffected by amounts of bovine ATPase inhibitor which were 20 times greater than that required to inhibit bovine F_1 by 50%.

[34] M. Tuena de Gomez-Puyou, U. Miller, J. Devars, A. Nava, and G. Dreyfus, *FEBS Lett.* **146,** 168 (1982).

SMP.[34] The ATPase inhibitor isolated from dog heart mitochondria is ~30% as effective as the beef heart inhibitor on beef heart SMP.[20]

Immunological Cross-Reactivity

Although comparisons of amino acid sequence indicate a genetic relationship between the mitochondrial ATPase inhibitors from *S. cerevisiae* and beef heart, their immunological characteristics differ. Antibodies against the bovine[20,34] and yeast[34] inhibitor protein do not cross-react with the heterologous yeast and beef proteins, respectively. Antibodies to the yeast and bovine inhibitors also fail to react with the inhibitor from rat liver mitochondria.[34] However, dog heart inhibitor does cross-react with the antibody against the beef heart inhibitor.[20]

[45] Purification of the Proton-Translocating ATPase from Rat Liver Mitochondria Using the Detergent 3-[(3-Cholamidopropyl)dimethylammonio]-1-propane Sulfonate

By MAUREEN W. MCENERY and PETER L. PEDERSEN

The procedure that follows for the purification of the F_0F_1-ATPase from rat liver mitochondria is a modification of the original procedure by McEnery *et al.*[1] The important feature of this protocol which distinguishes it from earlier attempts at the purification of F_0F_1-ATPase complexes[2–7] is the use of the detergent 3-[(3-cholamidopropyl)dimethylammonio]-1-propane sulfonate (CHAPS) developed by Hjelmeland[8] as an alternative to existing detergents. We have previously demonstrated that CHAPS appears to be the detergent of choice for the solubilization of the membrane-bound F_0F_1-ATPase complex.[1] In addition, by lowering the concentration of CHAPS present during velocity centrifugation in a su-

[1] M. W. McEnery, E. L. Buhle, U. Aebi, and P. L. Pedersen, *J. Biol. Chem.* **259,** 4642 (1984).

[2] J. W. Soper, G. L. Decker, and P. L. Pedersen, *J. Biol. Chem.* **254,** 11170 (1979).

[3] R. Serrano, B. I. Kanner, and E. Racker, *J. Biol. Chem.* **251,** 2453 (1976).

[4] D. L. Stiggall, Y. M. Galante, and Y. Hatefi, *J. Biol. Chem.* **252,** 956 (1978).

[5] R. Rott and N. Nelson, *J. Biol. Chem.* **256,** 9224 (1981).

[6] A. Tzagoloff and P. Meagher, *J. Biol. Chem.* **246,** 7328 (1971).

[7] D. L. Foster and R. H. Fillingame, *J. Biol. Chem.* **254,** 8230 (1979).

[8] L. M. Hjelmeland, *Proc. Natl. Acad. Sci. U.S.A.* **77,** 6368 (1980).

crose solution, the solubilized protein shifts from primarily a monodispersed state (0.6% CHAPS) to a highly aggregated state (0.2% CHAPS). It is presumed that this change in aggregation of the complex forms the basis for the purification of the F_0F_1-ATPase complex. In this chapter, a 25% sucrose solution prepared in TA buffer plus 0.2% CHAPS replaces the discontinuous sucrose solution (20, 25, and 50% sucrose solution prepared in TA buffer plus 0.2% CHAPS) used in the final step of purification of the original method. This modification further simplifies the isolation procedure and eliminates the problems which occur when trying to control for the presence of sucrose in the aliquots assayed for protein concentration via the Bradford protein assay.[9]

Solutions

Solutions Used in the Purification of the F_0F_1-ATPase

2× PA Buffer: 300 m*M* K_2HPO_4, 2 m*M* ATP, 50 m*M* EDTA, 1.0 m*M* DTT, 10% ethylene glycol, pH 7.9 (titrated with concentrated KOH). Solutions are prepared as 2× buffers and stored at −20° for several months. Diluted to 1× with distilled H_2O immediately prior to use

2× TA Buffer: 100 m*M* Tricine, 2 m*M* ATP, 50 m*M* EDTA, 1.0 m*M* DTT, 10% ethylene glycol, pH 7.9

3-[(3-cholamidopropyl)dimethylammonio]-1-propane sulfonate (CHAPS) (purchased from Calbiochem) solution: 10% CHAPS (w/v) dissolved in TA buffer prepared immediately prior to use

Sucrose solution: 50 m*M* Tricine, 1 m*M* ATP, 25 m*M* EDTA, 0.5 m*M* DTT, 5% ethylene glycol, 25% sucrose (w/v), and 0.2% CHAPS (w/v), pH 7.9; prepare 100 ml

Solutions Used in the Reconstitution of the Purified F_0F_1-ATPase

Buffer A: 100 m*M* Na-phosphate, 1 m*M* ATP, 1 m*M* DTT, 25 m*M* EDTA, pH 7.9; prepare 4 liters

Buffer B: 60 m*M* Na-phosphate, 0.5 m*M* ATP, 0.5 m*M* DTT, pH 7.4; prepare 2 liters

Phospholipid solution: Soybean asolectin (150 mg/ml) purchased from Associated Concentrates, Woodside, NY, was purified according to the procedure of Kagawa and Racker[10] and stored in $CHCl_3$ at −20° until use

[9] M. M. Bradford, *Anal. Biochem.* **72,** 248 (1976).

[10] Y. Kagawa and E. Racker, *J. Biol. Chem.* **246,** 5477 (1971).

Analytical Procedures

ATPase Assays

ATPase activity was measured spectrophotometrically at 340 nm by coupling the production of ADP to the oxidation of NADH via the pyruvate kinase and lactate dehydrogenase reactions, as first described by Pullman *et al.*[11] The reaction mixture contained in a volume of 1 ml at pH 7.5 and at room temperature, 4.0 m*M* ATP, 65 m*M* Tris-Cl, 4.8 m*M* $MgCl_2$, 2.5 m*M* KCl, 0.40 m*M* NADH, 0.60 m*M* phosphoenolpyruvate, 5 m*M* KCN, 1 unit of lactate dehydrogenase, 1 unit of pyruvate kinase, and 1–15 μg of protein. The reaction was initiated by addition of protein.

Inhibitor Assays

Methanolic stock solutions of oligomycin, venturicidin, dicyclohexylcarbodiimide (DCCD), and tricyclohexyltin were prepared and stored at 0° for several months. The inhibitor solutions (final concentration of methanol 0.5% v/v) were added to the ATPase reaction buffer immediately prior to the addition of protein.

Protein Determinations

Membrane protein was estimated by the biuret method in the presence of 0.25% sodium cholate.[12] Soluble protein was measured by the method of Bradford.[9] Bovine serum albumin was used as a standard in both cases. All samples were normalized with respect to detergent and buffer concentration.

Purification of the F_1-ATPase

The F_1-ATPase was isolated according to the procedure of Catterall and Pedersen.[13]

SDS–Gel Electrophoresis

One-dimensional gel electrophoresis was carried out under those conditions originally described by Laemmli[14] with the addition of 8 *M* urea to both the stacking and running gels.

[11] M. E. Pullman, H. S. Penefsky, A. Datta, and E. Racker, *J. Biol. Chem.* **235,** 2322 (1960).
[12] E. E. Jacobs, M. Jacobs, D. R. Sanadi, and L. B. Bradley, *J. Biol. Chem.* **223,** 147 (1956).
[13] W. A. Catterall and P. L. Pedersen, *J. Biol. Chem.* **246,** 4987 (1971).
[14] U. K. Laemmli, *Nature* (*London*) **227,** 680 (1970).

Purification of Rat Liver F_0F_1-ATPase

Day 1

Step 1. Adult, male CD albino rats (6–8) are sacrificed by decapitation. From these animals, inner membrane vesicles (IMVs) are prepared according to the procedure of Wehrle *et al.*[15] The resulting membranes (yield of ~500 mg) were resuspended at a protein concentration of 50 mg/ml in H medium (220 m*M* D-mannitol, 70 m*M* sucrose, 2 m*M* HEPES, and 0.5 mg/ml of defatted bovine serum albumin, pH 7.4) and stored in liquid nitrogen until use.

Day 2

Step 2. Approximately 250 mg of IMVs are thawed at room temperature, after which time 16 mg of IMVs are placed in each of 16 thick-walled Spinco centrifuge tubes and diluted to 2 mg/ml (total volume: 8 ml/tube) in PA buffer (the initial amount of IMV protein is 256 mg).

Step 3. The diluted IMVs are centrifuged for 45 min at 48,000 rpm at 2°. The resulting pellets are resuspended in the centrifuge tubes with a nongrooved pestle, and 7 ml of PA buffer is added. The solutions are centrifuged and the pellets resuspended as before, and again, 7 ml of PA buffer is added per tube. These solutions are again centrifuged as before (this constitutes the third spin), and these final pellets are resuspended to 15 mg/ml in TA buffer. These three-times washed membranes (3× membranes) account for usually 55–60% of the total starting protein and 75–80% of the original total ATPase activity.[1] (The 3× membranes may be stored in liquid nitrogen at this stage.)

Step 4. Prepare 10% CHAPS (w/v) solution in TA buffer. Add 24 mg of 3× membranes to each thick-walled Spinco centrifuge tubes; then add TA buffer to bring the final volume to 5.64 ml. Gently mix to distribute the 3× membranes into solution. Then, without shaking, vortexing, or mixing, add 0.36 ml of 10% CHAPS solution directly to each tube. [Note: Final volume, 6 ml; final protein concentration, 4 mg/ml; final CHAPS concentration, 0.6% (w/v).] The addition of the detergent to the diluted 3× membranes should result in the immediate clarification of the golden-yellow solution. Incubate for 15 min on ice. (At this step, there should be ~150 mg protein, or about 6 tubes of solubilized 3× membranes.)

After this time, centrifuge the samples for 1 hr at 48,000 rpm at 2° in the Sorvall T-865.1 rotor.

Step 5. Collect the soluble fraction and concentrate to half of the original volume (which corresponds to ~20 ml final volume) using the

[15] J. P. Wehrle, N. M. Cintron, and P. L. Pedersen, *J. Biol. Chem.* **253,** 8598 (1978).

PM10 filter with the Amicon Diaflo apparatus, which is kept cold via an ice bath. (Note: The solubilized F_0F_1-ATPase in the presence of 0.6% CHAPS is not stable; do not store the complex at this stage.)

Step 6. Prepare the 25% sucrose–0.2% CHAPS solution. In each of 4 Sorvall tubes, place 25 ml of sucrose solution. Layer ~5 ml of the concentrated and solubilized 3× membranes on top of each tube. Centrifuge for 9.75 hr at 25,000 at 2° in the Sorvall SS-28 rotor.

Day 3

Step 7. Remove the gradients and note the presence of a deep golden-colored band which has formed at the interface of the 25% sucrose solution and the sample buffer. Carefully remove this fraction which does not possess significant ATPase activity (specific activity for this pooled fraction is less than 1 μmol/min/mg ATP hydrolyzed).

Step 8. Collect the remaining volume of the gradients in 2-ml fractions (which corresponds to a total of 15 fractions per individual gradient) at room temperature. Store the collected fractions on ice.

Step 9. Assay each fraction for ATPase activity (add 25 μl of each fraction per assay) and determine the protein concentration via the Bradford procedure (be certain to control for the sucrose solution in the BSA standards). Finally, combine those fractions which have a specific activity of greater than or equal to 9 μmol/min/mg of ATP hydrolyzed.

Step 10. Concentrate the highly purified, pooled fractions to a final protein concentration of 150–300 μg/ml using the Amicon Diaflo apparatus equipped with a PM10 filter. Again, keep the concentrating protein on ice. Store in 1-ml aliquots in liquid nitrogen. The final preparation has an average specific activity of 10–14 μmol/min/mg ATP hydrolyzed (determined for 10 separate preparations), is stable for up to 1 year, and retains its sensitivity to oligomycin when stored under these conditions. (Note: It is possible to finish this procedure in 2 days if Steps 1–6 can be accomplished on the first day.)

Reconstitution of the F_0F_1-ATPase into Asolectin Vesicles

Insertion of F_0F_1-ATPase into Liposomes: Purified F_0F_1-ATPase was reconstituted into asolectin vesicles according to the CHAPS-dialysis procedure first reported by McEnery *et al.*[1] This procedure consists of two sequential steps: The first involves the suspension of the phospholipids in Buffer A plus 0.6% (w/v) CHAPS. A 400-μl aliquot of purified phospholipids[10] is placed in a 100-ml round-bottom flask which has been prewashed with 1 ml of anhydrous ether. The organic solvent is removed under a stream of nitrogen gas until the phospholipids form a translucent yellow film on the bottom of the flask. At this time, 4 ml of Buffer A plus

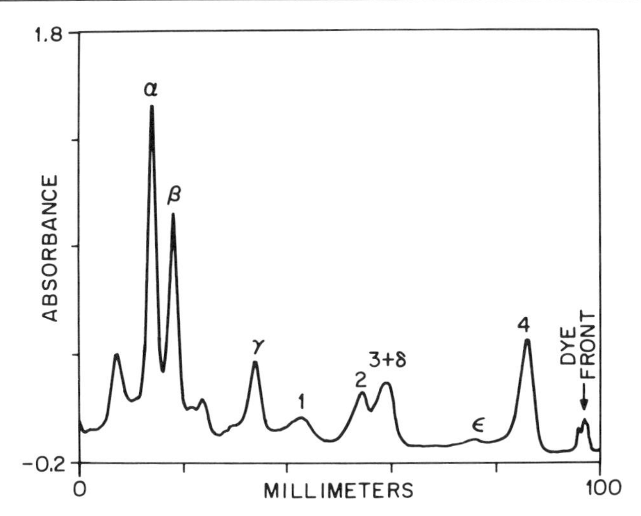

FIG. 1. SDS–gel electrophoretic profile of rat liver F_0F_1-ATPase. A 20-μg sample was acetone precipitated and redissolved in 100 μl of 3% SDS, 5% mercaptoethanol, 0.1% Bromophenol Blue, 67.5 m*M* Tris-Cl, and 10% glycerol, pH 6.7. The sample was boiled for 2 hr, cooled, and then loaded on the gel. The conditions for electrophoresis were according to the procedure of Laemmli[14]: a 4% stacking gel and a 12% running gel in the presence of 0.05% SDS. The gels were run at 10 mA/slab gel for 8 hr at room temperature. The peaks were identified according to their mobility relative to isolated F_1-ATPase. In addition to the five F_1-ATPase subunits (with the δ subunit comigrating with F_0 subunit 3), four subunits were present which correspond in their relative mobility to the F_0 subunits previously identified.[1] The apparent molecular weights of these components are as follows: subunit 1, 28,000; subunit 2, 19,000; subunit 3, 13,600; subunit 4, 6000.

0.6% CHAPS (24 mg/4 ml Buffer A) is added and evacuated under nitrogen gas for several minutes. The flask is then sealed with parafilm and sonicated in a bath sonicator (obtained from Laboratory Supply Company, Hicksville, NY) for 15 min or until dissolved. The solution is then immediately put on ice and used as soon as possible. The concentration of phospholipids is 15 mg/ml.

The second step in this reconstitution procedure is the combining of the protein and phospholipid solutions. F_0F_1-ATPase (5 ml of 100 μg/ml protein in TA buffer and 25% sucrose, 0.2% CHAPS) purified from the sucrose gradient and concentrated is placed in an 8-ml test tube. To this solution, 500 μl of the resuspended phospholipids in 0.6% CHAPS is added (ratio of 15 : 1 mg asolectin/mg protein). Finally, 26 mg of CHAPS which had been first dissolved in 1 ml of Buffer A is added to the F_0F_1-ATPase–phospholipid solution and gently mixed by inverting the tube. This important last addition of CHAPS brings the final concentration of CHAPS to 0.6% in a total volume of 6.5 ml. The protein–phospholipid suspension is then placed in prewashed dialysis tubing (Thomas Co.,

EFFECT OF VARIOUS COMPOUNDS ON F_0F_1-ATPASE[a]

Compound	Concentration	Inhibition (%)		
		Inner membrane vesicles	Purified F_0F_1-ATPase	Reconstituted F_0F_1-ATPase
Oligomycin	5 μg/ml		95(7)	98(5)
DCCD	13 μ*M*		86(4)	
	50 μ*M*	93(4)	75(6)	87(6)
	100 μ*M*		56(3)	
Venturicidin	5 μg/ml	88(3)	87(5)	72(4)
Cd^{2+}	100 μ*M*	29(2)	32(4)	36(6)
Tricyclohexyltin	13 μ*M*		83(3)	
	25 μ*M*	96(4)	94(5)	96(3)
	50 μ*M*		97(2)	
DDT	37 μ*M*	38(3)	58(3)	11(3)
Kelthane	37 μ*M*	87(3)	16(4)	69(3)

[a] Inner membrane vesicles (IMVs), purified F_0F_1-ATPase, and purified F_0F_1-ATPase which had been reconstituted into asolectin vesicles were all assayed for ATPase activity in the presence of the stated compounds. The conditions for the ATPase assay are reported in "Analytical Procedures." The amount of protein added per individual assay varied according to whether the F_0F_1-ATPase had been purified (1–5 μg protein/assay) or in membranes (3–15 μg protein/assay). The number in parentheses refers to the number of individual determinations for each average value.

16 mm/3787-F27) and dialyzed overnight in the cold against 4 liters of Buffer A. After this time, the sample is dialyzed against 2 liters of Buffer B for 2 hr. The dialyzed solution is then centrifuged in a Sorvall T-865.1 rotor for 90 min at 2°. The resulting pellet is resuspended in TA buffer at a protein concentration of 1–2 mg/ml.

Properties of the Isolated F_0F_1-ATPase

As shown in Fig. 1, the F_0F_1-ATPase isolated by this procedure is routinely comprised of nine electrophoretically distinguishable components. It is important to note that these results obtained using the Laemmli procedure[14] in the presence of 8 *M* urea are identical to our original report which contrasted the SDS–gel profiles of the purified F_0F_1-ATPase complex on several SDS–gel systems.[1] The advantage of the gel system employed in this study is the resolution obtained for the smaller subunits, most notably the 6000-molecular weight subunit which has been shown in this laboratory to be the subunit that binds [^{14}C]DCCD (data not shown). With regard to the simplified subunit composition of the F_0 portion of the complex, this preparation of the F_0F_1-ATPase is most similar to the complex isolated from *Escherichia coli*.[7]

The purified and reconstituted complex has been shown to catalyze ATP–P_i exchange and ATP-dependent proton translocation[1] and is inhibited by oligomycin, dicyclohexylcarbodiimide (DCCD), venturicidin, cadmium, tricyclohexyltin, and the pesticides 1,1,1-trichloro-2,2-bis(*p*-chlorophenyl) ethane (DDT) and 1,1,1-trichloro-2,2-bis(*p*-chlorophenyl)-ethanol (Kelthane). The results in the table indicate that the sites that mediate the inhibitory effects of these compounds, which are believed to interact with the F_0 portion of the complex, are present in the starting material (IMVs) and in the purified complex. These results suggest that this preparation of F_0F_1-ATPase isolated from rat liver mitochondria is a very highly purified, intact complex.

Acknowledgment

Work was performed when M.W.M. was a predoctoral trainee supported by NIH grant GM 31940. Work supported also by NIH grant CA 10951 to PLP.

[46] Rapid Purification of F_1-ATPase from Rat Liver Mitochondria Using a Modified Chloroform Extraction Procedure Coupled to High-Performance Liquid Chromatography

By NOREEN WILLIAMS and PETER L. PEDERSEN

The most rapid method to date for isolation of F_1-ATPase from mammalian sources is that developed by Beechey *et al.*[1] employing chloroform for solubilization of the F_1 from the membrane. The procedure described here is a modification of that method for the isolation of F_1-ATPase from rat liver. Purified inner membrane vesicles are extracted with chloroform in the presence of ATP and EDTA. The resulting F_1 has a high specific activity, is reconstitutively active, catalyzing high rates of ATP synthesis, and can be crystallized. In most cases the enzyme exhibits only the five characteristic bands: α, β, γ, δ, and ε. In those cases where an additional band is present (apparent molecular weight below the β subunit), a rapid high-performance liquid chromatography (HPLC) step may be added if desired to eliminate this band.

[1] R. B. Beechey, S. A. Hubbard, P. E. Linnett, D. A. Mitchell, and E. A. Munn, *Biochem. J.* **148,** 533 (1975).

Analytical Procedures

Reconstitution Assay

F_1 (~100 μg) is incubated with 0.25 mg urea particles in a final volume of 1 ml containing 1 mg bovine serum albumin, 4 m*M* ATP, 125 m*M* potassium phosphate, 2.5 m*M* EDTA, pH 7.5. After 2 hr at room temperature, the mixture is sedimented at 100,000 *g* for 30 min at 20°. The sediment is rinsed with H medium (described herein under "Isolation of Mitochondria") and resuspended in a final volume of 100 μl.

Urea Particles

Urea particles for the reconstitution assay (see above) are prepared by the method of Pedersen *et al.*[2] Fresh inner membrane vesicles (60 mg/ml) in H medium (see below) are diluted 1 : 1 with 6.4 *M* urea and placed on ice for 5 min. This mixture is then diluted 1 : 4 (v/v) with H medium and sedimented at 100,000 *g* for 30 min at 0–4°. The urea particles are washed once by resuspension in 8 ml H medium and centrifuged again. The final resuspension is stored in liquid nitrogen until use.

ATPase Assay

ATPase activity is determined spectrophotometrically as described by Pullman *et al.*[3] and modified by Catterall and Pedersen.[4] The reaction mixture contains 65 m*M* Tris-Cl or Tris bicarbonate, 4.8 m*M* $MgCl_2$, 0.40 m*M* NADH, 0.60 m*M* phosphoenolpyruvate, 5.0 m*M* KCN, 1 unit lactate dehydrogenase, and 1 unit pyruvate kinase.

ATP Synthesis Assay

ATP synthesis is determined by the method of Pullman and Racker.[5] The reaction contains 1 m*M* ADP, 10 m*M* potassium phosphate, 10 m*M* succinate, 50 m*M* Tris-acetate, 1 m*M* glucose, 1 m*M* $NADP^+$, 2 m*M* $MgCl_2$, 2 mg/ml bovine serum albumin, 200 m*M* sucrose, 20 units hexokinase, 7 units glucose-6-phosphate dehydrogenase, pH 7.5. The reaction is initiated by the addition of reconstituted particles.

[2] P. L. Pedersen, J. Hullihen, and J. P. Wehrle, *J. Biol. Chem.* **256,** 1362 (1981).

[3] M. E. Pullman, H. S. Penefsky, A. Datta, and E. Racker, *Science* **123,** 1105 (1956).

[4] W. A. Catterall and P. L. Pedersen, *J. Biol. Chem.* **246,** 4987 (1971).

[5] M. E. Pullman and E. Racker, *Science* **123,** 1105 (1956).

SDS–Polyacrylamide Gel Electrophoresis

Cylindrical SDS–polyacrylamide gel electrophoresis was performed as described by Weber and Osborn[6] and modified by Catterall *et al.*[7] Gels are stained and destained as described by Williams *et al.*[8]

Determination of Protein

Membrane protein is determined by the biuret method[9] in the presence of 0.33% sodium cholate. Soluble protein is measured as described by Lowry *et al.*[10] Bovine serum albumin is employed as the standard protein.

Preparative Procedures

Isolation of Mitochondria

Rat liver mitochondria are prepared by the high-yield method of Bustamante *et al.*[11] Isolation medium (H medium) contains 220 m*M* D-mannitol, 70 m*M* sucrose, 2 m*M* HEPES, and 0.5 mg/ml bovine serum albumin.

Preparation of Inverted Inner Membrane Vesicles

Inverted inner membrane vesicles are prepared by a modification of the protocol of Chan *et al.*[12] Freshly isolated mitochondria (100 mg/ml, see above) in H medium are stirred with 12 mg/ml digitonin on ice for 20 min. This incubation mixture is diluted 3 : 1 with H medium and centrifuged at 10,000 *g* (10 min, 0–4°). The sediment is washed by resuspension to half the previous volume and sedimented as above. The washed mitoplasts are resuspended (50 mg/ml) in cold, deionized H_2O and sonicated with the Bronwill Biosonik sonicator (large probe, 95% maximal intensity) for 2 min total in 15-sec intervals (0–4°). This is first sedimented at 10,000 *g* for 10 min. The resulting supernatant is then sedimented at 100,000 *g* (30 min, 0–4°) to collect the inner membrane vesicles. The vesicles are resuspended to 50 mg/ml in H medium. These inner mem-

[6] K. Weber and M. Osborn, *J. Biol. Chem.* **244,** 4406 (1969).

[7] W. A. Catterall, W. A. Coty, and P. L. Pedersen, *J. Biol. Chem.* **248,** 7427 (1973).

[8] N. Williams, J. M. Hullihen, and P. L. Pedersen, *Biochemistry* **23,** 780 (1984).

[9] E. E. Jacobs, M. Jacobs, D. R. Sanadi, and L. B. Bradley, *J. Biol. Chem.* **223,** 147 (1956).

[10] O. H. Lowry, N. J. Rosebrough, A. L. Farr, and R. J. Randall, *J. Biol. Chem.* **193,** 265 (1951).

[11] E. Bustamante, J. Soper, and P. L. Pedersen, *Anal. Biochem.* **80,** 402 (1977).

[12] T. L. Chan, J. W. Greenawalt, and P. L. Pedersen, *J. Cell Biol.* **45,** 291 (1970).

brane vesicles can be prepared in large quantities and are stable for extended periods of time when stored in liquid nitrogen until use.

Rapid Purification of Rat Liver F_1-ATPase, Preparation of F_1-ATPase Washed Membranes

Inverted inner membrane vesicles (50 mg/ml, 1.5–2.0 ml) are thawed and removed from suspension by centrifugation at 100,000 *g* (30 min, 0–4°). These vesicles are washed by resuspension in 0.25 *M* sucrose, 3 m*M* Tris, 5 m*M* EDTA, pH 7.5, to original volume (with a Potter–Elvejhem homogenizer fitted with an unserrated pestle) and sedimented as described above. Washed vesicles are resuspended as above in sucrose–Tris–EDTA to 10 mg/ml at 25°. These washed membranes show a 3-fold increase in total activity (Table I) compared to the starting inner membrane vesicles. This is probably due to release of the inhibitor protein. It should be noted that neither this preparation nor the Catterall and Pedersen[4] preparation shows significant quantities of inhibitor protein in the final F_1-ATPase preparation.

Release of the F_1-ATPase

ATP is added to the washed membranes to 100 μM final concentration. After a 5-min incubation with ATP present, chloroform is added to one-half volume (e.g., 2.5 ml chloroform to 5.0 ml resuspended vesicles). This mixture is then gently shaken by hand for 20 sec at 25° and *immediately* centrifuged at 1100 *g* in a Sorvall SS-34 rotor (10 min, 25°) to separate the aqueous and organic phases. Extended exposure to chloroform causes a significant loss of ATPase activity. The upper aqueous phase (pale brown) is carefully removed by Pasteur pipet, avoiding the membrane debris at the interface. This fraction shows only a small loss (20%) in total protein compared with the washed membranes and no change in total activity (Table I). The aqueous phase is then centrifuged at 100,000 *g* in a Spinco TY-65 rotor (45 min, 25°). This high-speed centrifugation step sediments 95% of the protein present in the chloroform-treated aqueous phase. Approximately one-third of the ATPase activity is also lost, probably due to F_1-ATPase not solubilized by the chloroform extraction. The small amount of protein (4% starting protein) remaining in this supernatant possesses a high specific activity (<40 μmol ATP hydrolyzed/min/mg). The clear supernatant is rapidly removed and concentrated to 1 ml with an Amicon Diaflo (equipped with a prewashed PM10 membrane, 30 psi). This concentrated enzyme can be used immediately or can be transferred into a stabilizing buffer for storage.

TABLE I
PURIFICATION OF RAT LIVER F_1-ATPase[a]

Fraction	Protein[b] (mg)	ATPase activity	
		Specific activity[c]	Total activity[d]
Inner membrane vesicles	80.02 (100%)	1.92	153.76
Washed membranes	77.45 (96.8%)	5.87	454.77
Chloroform-treated membranes prior to high-speed centrifugation	63.49 (79.3%)	6.60	418.93
Chloroform extract	3.25	40.85	132.73
Postlyophilization enzyme	3.1 (3.9%)	42.13	130.61

[a] From Williams *et al.*[13]
[b] Expressed as milligrams protein and in parentheses as percentage protein based on inner membrane vesicles as starting material.
[c] Micromoles of ATP hydrolyzed/min/mg.
[d] Total micromoles of ATP hydrolyzed/min.

Transfer into Stabilizing Buffer

This transfer is accomplished by elution of the enzyme from a Sephadex G-25 desalting column (2.0 × 14 cm) with 250 m*M* potassium phosphate, 5.0 m*M* EDTA, pH 7.5. The most active fractions are concentrated to a total volume of 1.0–1.5 ml, as described above. The desalting column and concentration steps require very little time (~20 min), but allow the rat liver enzyme to be lyophilized and stored at −20° for several months with no significant loss of activity. Only minimal changes in total protein, specific activity, or total activity are seen subsequent to the high-speed centrifugation step.

HPLC Purification

In preparations where the β' band (the band below the β subunit on Weber and Osborn gels) is noticeable, a one-step HPLC procedure is added to further purify the preparation if desired. Sucrose, present in the enzyme preparation after high-speed centrifugation, is a problem with the HPLC columns employed, so the desalting column is used to exchange the enzyme into a compatible buffer. HPLC is carried out using a Waters

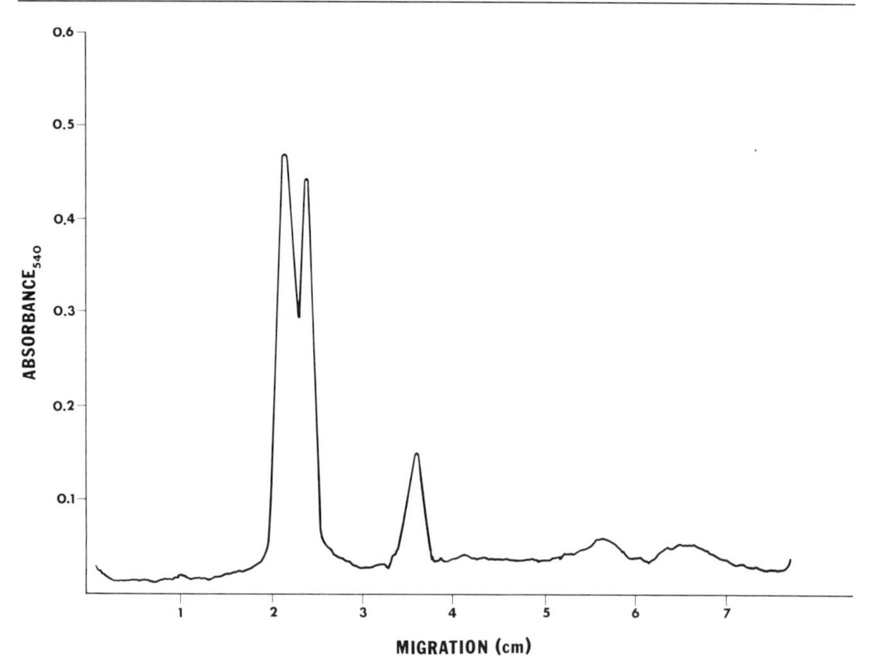

FIG. 1. SDS–polyacrylamide gel pattern of the chloroform-released F_1-ATPase. The lyophilized F_1-ATPase was resuspended in 100-μl volume and ammonium sulfate precipitated. It was then treated for SDS–gel electrophoresis. Approximately 10 μg was applied to cylindrical gels for electrophoresis. From Williams *et al.*[13]

chromatograph with a Protein Pak 300 SW column in series with an I-60 column. The elution buffer is 250 m*M* potassium phosphate, 5 m*M* EDTA, pH 7.5, with a flow rate of 0.5 ml/min. The eluant is monitored at 280 nm using the Waters Model 440 absorbance detector. The purified F_1-ATPase activity peak has an elution volume of 11.4–11.7 ml. This step yields a preparation devoid of this extra β' band.

Properties of the F_1-ATPase

Physical Properties

This preparation of rat liver mitochondrial F_1-ATPase appears to be nearly identical in structure to the previous preparation of Catterall and Pedersen.[4] It consists of five different subunits: α, β, γ, δ and ε, with variable quantities of a sixth lightly staining band occurring just below the β-subunit band on cylindrical Weber and Osborn[6] gels (Fig. 1).

TABLE II
ABILITY OF THE CHLOROFORM-RELEASED F_1-ATPase TO CATALYZE ATP HYDROLYSIS AND ATP SYNTHESIS[a,b]

Fraction	ATPase activity[c]			ATP synthesis[d]
	Without inhibitor	With oligomycin	With DCCD	
F_1[e]	20.57	—	—	—
F_1[f]	37.15	—	—	—
Urea particles	0.245	—	—	0
Urea particles + F_1[e]	2.58	0.161	0.145	170
Urea particles + F_1[f]	2.35	0.145	0.128	150

[a] From Williams *et al.*[13]
[b] F_1-ATPase was reconstituted with urea particles (0.25 mg) and assayed as described in "Analytical Procedures." The effect of oligomycin (0.1 μg) and DCCD (10 μM) was tested on the reconstituted system.
[c] Micromoles of ATP hydrolyzed/min/mg.
[d] Nanomoles of ATP formed/min/mg.
[e] F_1-ATPase prepared by the method of Catterall and Pedersen.[4]
[f] F_1-ATPase prepared by this chloroform method.

Most significantly for the purposes of our work, this preparation formed crystals[13] under the identical conditions for the Catterall and Pedersen preparation.[4] The crystals of this preparation are morphologically identical to those described earlier.[14]

Catalytic Properties

The average specific activity of six preparations after lyophilization is 42.1 μmol ATP hydrolyzed/min/mg by the Pullman *et al.* assay.[4] The use of Tris–bicarbonate rather than Tris–HCl buffer increases the specific activity by 1.9-fold. Increasing the pH of bicarbonate assay mixture to 8.2 brings the specific activity to 2.5-fold the Pullman *et al.* assay conditions, with average specific activity >100 μmol ATP hydrolyzed/min/mg. Table II also shows that this F_1-ATPase preparation is capable of binding to urea-treated membranes with good sensitivity to both DCCD (inhibited to 94% control activity) and oligomycin (inhibited to 95% control activity). This indicates that the binding of F_1 to the membranes occurs specifically

[13] N. Williams, L. M. Amzel, and P. L. Pedersen, *Anal. Biochem.* **140,** 581 (1984).
[14] L. M. Amzel and P. L. Pedersen, *J. Biol. Chem.* **253,** 2067 (1978).

to an F_0 moiety. In addition, the reconstituted system is coupled and therefore capable of ATP synthesis, indicating that the F_1 is functionally intact.

Acknowledgment

This work was supported by NIH Grant CA 10951 to P.L.P.

[47] Purification of α and β Subunits and Subunit Pairs from Rat Liver Mitochondrial F_1-ATPase

By NOREEN WILLIAMS and PETER L. PEDERSEN

The soluble F_1-ATPase is known to contain five distinct subunits: α, β, γ, δ, and ε present in 3 : 3 : 1 : 1 : 1 stoichiometry.[1] The function of each subunit has been studied extensively; however, much is yet unknown.[2] In order to more clearly define the function of the ATPase subunits we wished to separate and isolate the individual subunits to facilitate study. We describe here a procedure for separation based on cold treatment with subsequent purification.

Analytical Procedures

Assay for ATPase Activity

Soluble F_1-ATPase activity is determined by the coupled spectrophotometric assay of Pullman *et al.*[3] with modifications as described by Catterall and Pedersen.[1] A final volume of 1 ml contains 58 mM Tris, 4.7 mM KCN, 4.0 mM $MgSO_4$, 4.0 mM ATP, 0.5 mM phosphoenolpyruvate, 0.3 mM NADH, 1 unit of lactate dehydrogenase, 8 units of pyruvate kinase, at pH 7.5.

[1] W. A. Catterall and P. L. Pedersen, *J. Biol. Chem.* **246,** 4987 (1971); A. E. Senior, *Biochim. Biophys. Acta* **301,** 249 (1973); F. S. Esch and W. S. Allison, *J. Biol. Chem.* **254,** 10740 (1979).

[2] For recent reviews see A. E. Senior, *in* "Membrane Proteins in Energy Transduction" (R. A. Capaldi, ed.), p. 233. Dekker, New York, 1979; S. Dunn and L. Heppel, *Arch. Biochem. Biophys.* **210,** 421 (1981); L. M. Amzel and P. L. Pedersen, *Annu. Rev. Biochem.* **52,** 801 (1983).

[3] M. E. Pullman, H. S. Penefsky, A. Datta, and E. Racker, *J. Biol. Chem.* **235,** 3322 (1960).

Protein Determination

Soluble protein is determined by the method of Lowry[4] using bovine serum albumin as standard.

SDS–Polyacrylamide Gel Electrophoresis

Cylindrical SDS–gel electrophoresis is performed as described by Weber and Osborn[5] and modified by Catterall *et al.*[6] The staining method is described in Williams *et al.*[7]

Preparation of F_1

F_1-ATPase is prepared from rat liver mitochondria[8] essentially according to Catterall and Pedersen[1] with modifications as described in Pedersen and Hullihen.[9] F_1 is stored lyophilized in 250 m*M* potassium phosphate, 5 m*M* EDTA, pH 7.5, at −20°.

Results

Purification of F_1 Subunits or Subunit Pairs

Cold Treatment of F_1. The F_1-ATPase is resuspended to a final protein concentration of 2.0–2.5 mg/ml in 250 m*M* potassium phosphate, 5 m*M* EDTA, pH 7.5, just before cold treatment. Generally, the F_1 is then placed into the cold (0–4°) for a period of 17 hr. No precipitation of protein is observed in the cold under these conditions. If the treatment period is longer, however (several days), some cloudiness due to protein precipitation does occur.

Separation of the Cold-Treated Enzyme into Two Fractions. The F_1, once removed from the cold, is warmed immediately to 37° in a water bath. After 30 min the solution turns cloudy. Additional time does not increase the amount of protein precipitating but simply allows more time for settling of the flocculent material. After 30 min warming, the cold-treated F_1 separates into two fractions: an insoluble and a soluble fraction. The insoluble fraction (a white precipitate) is removed from suspension by centrifugation at 10,000 *g* for 20 min (25°). The supernatant that re-

[4] O. H. Lowry, N. J. Rosebrough, A. L. Farr, and R. J. Randall, *J. Biol. Chem.* **193,** 265 (1951).

[5] K. Weber and M. Osborn, *J. Biol. Chem.* **244,** 4406 (1969).

[6] W. A. Catterall, W. A. Coty, and P. L. Pedersen, *J. Biol. Chem.* **248,** 7427 (1973).

[7] N. Williams, J. M. Hullihen, and P. L. Pedersen, *Biochemistry* **23,** 780 (1984).

[8] E. Bustamante and P. L. Pedersen, *Anal. Biochem.* **80,** 401 (1977).

[9] P. L. Pedersen and J. M. Hullihen, *J. Biol. Chem.* **253,** 2176 (1978).

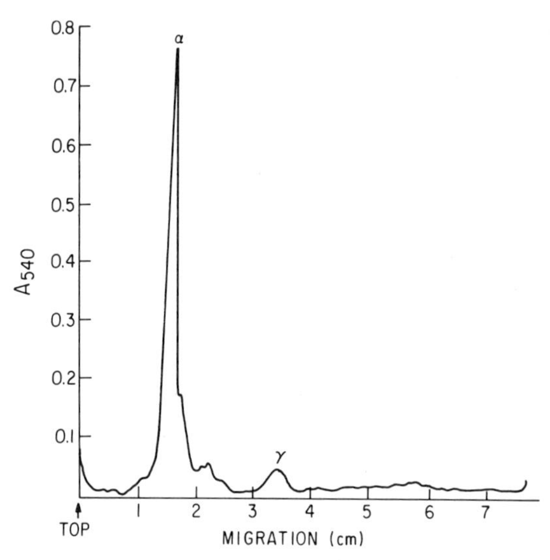

FIG. 1. SDS–polyacrylamide gel patterns of the insoluble fraction formed upon warming of cold-treated liver F_1. The soluble F_1-ATPase (250 μg/100 μl) in PE buffer was cold treated for 17 hr (4°). It was then removed from the cold and warmed to 37° in a water bath for 30 min. The insoluble fraction which formed was removed from suspension and prepared for SDS–PAGE as described in Analytical Procedures. Sample (20 μg) was applied to the cylindrical gels for electrophoresis. From Williams *et al.*[7]

mains is removed by Pasteur pipet and stored. These two fractions (supernatant and sediment) are electrophoresed in the Weber and Osborn system.[5] The sediment contains only the α and γ subunits (Fig. 1), while the supernatant contains β, δ, ε, and some γ (Fig. 2B).

Confirmation of the identity of the major subunits in supernatant and sediment is accomplished by an abbreviated cold treatment. F_1 is exposed to the cold (0–4°) for 20 min, then warmed to 37° for 30 min. The soluble fraction is electrophoresed in the above gel system to determine whether the α or β subunit is lost from the complex. After 20 min cold treatment, the α band in the soluble, supernatant fraction is considerably diminished, while the β-subunit band is unchanged (Fig. 2A).

Further Purification

Purification of Subunit α and the Subunit Pair $\alpha\gamma$. The white sediment is precipitated twice with ammonium sulfate (3.5 *M* ammonium sulfate, 10 m*M* Tris, 1 m*M* EDTA, pH 7.5) to remove the potassium phosphate buffer, and then resuspended in 5.0 m*M* Tris, 1.0 m*M* EDTA, pH 8.0. This washed sediment is then applied to a Sephadex G-75 column (1.2 × 24 cm) and eluted with 10.0 m*M* Tris, 5 m*M* $MgCl_2$, pH 8.0. The eluant is

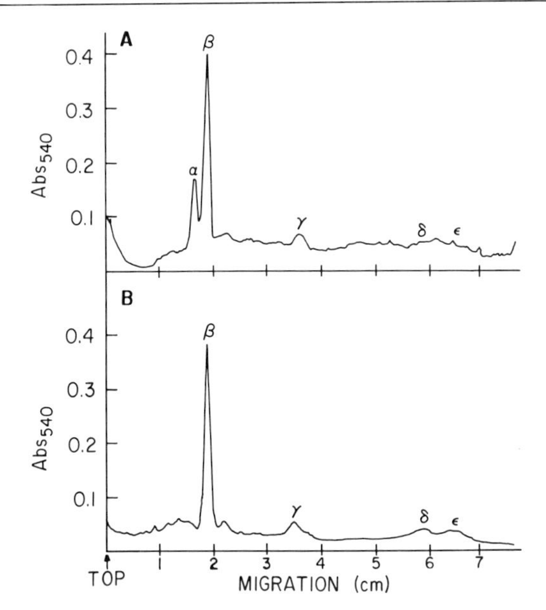

FIG. 2. SDS–PAGE pattern of the soluble fraction of cold-dissociated liver F_1. The soluble fraction remaining after removal of the sediment was prepared for SDS–PAGE as described in "Analytical Procedures." (A) The gel pattern of the soluble fraction when the enzyme was incubated for 20 min at 4°. Only about 40% of the precipitable protein was removed from the complex. (B) The electrophoresis pattern of the soluble fraction after 17 hr at 4°. The α subunit is completely absent from this fraction. Protein (10 μg) was applied to the gels in each case. From Williams *et al.*[7]

monitored at 280 nm with an ISCO UA-5 UV monitor (Lincoln, NE). The two peak fractions (Fig. 3, inset) are collected and either lyophilized or ammonium sulfate precipitated. SDS–polyacrylamide electrophoresis of the two fractions (see Fig. 3) shows that the first peak contains only the α subunit, while the second peak contains the α and γ subunits as a complex. It should be noted that a large portion of the α subunit does not elute from the column, but rather apparently aggregates at the top of the column. The $\alpha\gamma$ fraction elutes very poorly from the column unless Mg^{2+} is included in the elution buffer. The interesting fact that the $\alpha\gamma$ complex elutes from the column after α indicates a stronger adherence to the column by $\alpha\gamma$, which is weakened by Mg^{2+}. Neither portion of the sediment is soluble in high-ionic-strength buffer and is only poorly soluble (0.5 mg/ml) in the low-ionic-strength buffers employed for these experiments.

Purification of Subunit β and the Subunit Pair $\beta\gamma$

The supernatant fraction is kept in 250 m*M* potassium phosphate, 5 m*M* EDTA, pH 7.5, since it is most soluble in higher ionic strength buffer.

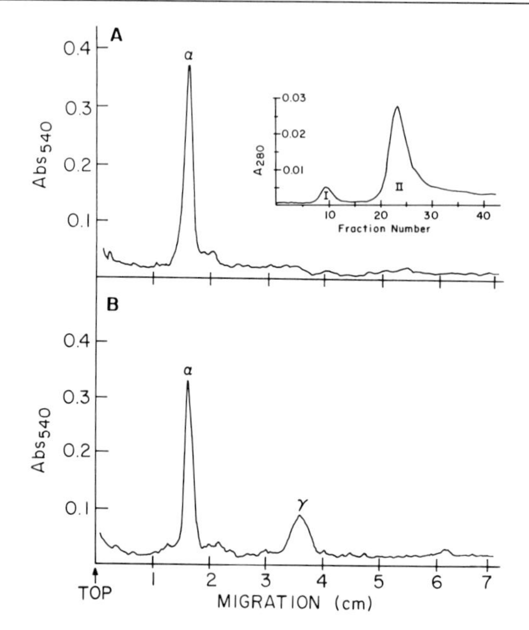

FIG. 3. Inset: Sephadex G-75 chromatography of the sediment resulting from cold treatment of liver F_1. Fraction I (the sediment) was further separated into two components on a Sephadex G-75 column (1.2 × 24 cm) eluted with 10 m*M* Tris, 5 m*M* $MgCl_2$, pH 8.0. Absorbance at 280 nm was determined with a UA-5 monitor (ISCO, Lincoln, NE), and 1-ml fractions were collected. (A,B) SDS–gel patterns of the Sephadex G-75 fractions. The peaks eluted from the Sephadex G-75 column were lyophilized and prepared for SDS–gel electrophoresis. (A) This shows that peak I contains only the α subunit. (B) The pattern of peak II containing α and γ subunits. Protein (8 and 10 μg, respectively) was employed on each gel. From Williams *et al.*[7]

This fraction is purified on a Sepharose CL-6B column (1.2 × 54 cm) employing the buffer above. Two peaks are eluted from the column (Fig. 4, inset). The first peak is composed of β and γ subunits as determined by SDS–gel electrophoresis (Fig. 4A). This $\beta\gamma$ complex is stable to repeated chromatography and ammonium sulfate precipitation. The second peak contains only the β subunit (Fig. 4B). The fate of the two smaller subunits, δ and ε, is as yet undetermined due to the difficulty in detection.

Discussion

These protocols combining cold treatment for dissociation of the rat liver mitochondrial F_1-ATPase together with chromatographic separation allow the purification of the two major subunits, α and β, and two subunit

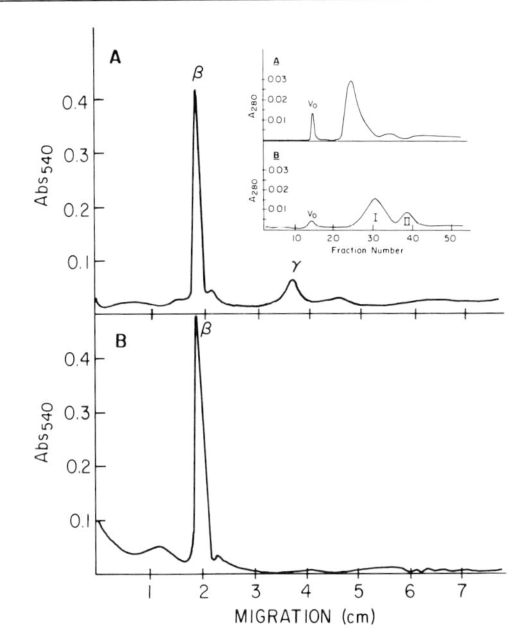

FIG. 4. Inset: Sepharose CL-6B chromatography of native liver F_1 and of the soluble fraction following cold inactivation. Chromatography was performed on a Sepharose CL-6B column (1.2 × 54 cm) with PE as the elution buffer. (*A*) The elution profile of the native enzyme. (*B*) The profile of the soluble fraction of the cold-dissociated enzyme. Peaks were pooled, lyophilized, and prepared for SDS–gels. (A,B) SDS–PAGE patterns of the Sepharose CL-6B fractions. The gel patterns of the two peaks eluted from the Sepharose CL-6B column are shown. (A) This shows peak I containing β and γ subunits. (B) This shows peak II containing only β subunit. Protein (10 μg) was employed for each gel. From Williams *et al.*[7]

complexes, $\alpha\gamma$ and $\beta\gamma$. The complexes are both quite stable to subsequent treatment and are not affected by the presence of dithiothreitol during cold treatment or purification.

Neither the individual subunits (α and β) nor the subunit complexes ($\alpha\gamma$ and $\beta\gamma$) are capable of catalysis under our conditions. However, under the appropriate conditions[7] (the addition of Mg^{2+} and ATP) the supernatant and sediment fractions reconstitute to yield 100% of the control activity, indicating that they are still competent.

Acknowledgment

This work was supported by NIH Grant CA 10951 to P.L.P.

[48] Isolation and Hydrodynamic Characterization of the Uncoupling Protein from Brown Adipose Tissue

By MARTIN KLINGENBERG and CHI-SHUI LIN

Introduction

Brown fat adipose tissue (BAT) in mammalians is now well characterized as a heat-generating tissue (for review, see Ref. 1). For this purpose BAT contains high amounts of mitochondria, which oxidize primarily fatty acids. Instead of producing ATP most of the oxidation energy is converted to heat. The unusual quality of the mitochondria from this tissue is mechanistically quite well explained by a high "H^+ conductivity" of the inner membrane.[2] H^+ ions exported by the respiratory chain are thus returned into the matrix, bypassing the ATP synthesis. This short-circuiting of the H^+ current interconverts the energy obtained from fatty acid oxidation into heat. As a result respiration of these mitochondria is largely uncoupled.

The molecular basis for this H^+ conduction seemed to be represented by a particular nucleotide binding protein, occurring in the inner membrane of brown fat mitochondria. Originally nucleotides were found to bring the normally uncoupled respiration of these mitochondria into a more inhibited coupled state.[3] On the outer surface of the inner membrane, binding sites for GDP and GTP were determined which were responsible for the control of the coupling and could be differentiated from the ADP/ATP carrier sites.[4,5] By photoaffinity labeling with azido-ATP a protein band was identified with M_r of 32,000.[6] This protein has been suggested to transport OH^- (or H^+) and also Cl^- which should be inhibited by nucleotide binding.[5] It should be noted that the striking occurrence of a prominent 32 kDa band in SDS gels of BAT mitochondria was noted early[7] and correlated to the thermogenetic activity of the tissue. Only after associating the nucleotide binding site with this protein was full attention drawn to the 32 kDa protein as the key factor in thermogenesis.

[1] D. G. Nicholls and R. M. Locke, *Physiol. Rev.* **62,** 1 (1984).

[2] D. G. Nicholls, *Eur. J. Biochem.* **77,** 349 (1977).

[3] J. Rafael, H.-J. Ludolph, and H.-J. Hohorst, *Hoppe Seyler's Z. Physiol. Chem.* **350,** 1121 (1969).

[4] J. Rafael and H. W. Heldt, *FEBS Lett.* **63,** 304 (1976).

[5] D. G. Nicholls, *Eur. J. Biochem.* **62,** 223 (1976).

[6] G. M. Heaton, R. J. Wagenvoord, and A. Kemp, *Eur. J. Biochem.* **82,** 515 (1978).

[7] D. Ricquier, G. Mory, and P. Hemon, *Can. J. Biochem.* **57,** 1262 (1979).

The similarity in molecular weight of the 32,000 protein in polyacrylamide gels with the ADP/ATP carrier of M_r 30,000 and the fact that both proteins bind ADP and ATP and appear to occur in similarly high amounts in the inner membrane suggested to us that the nucleotide binding protein from BAT mitochondria and the ADP/ATP carrier may be related. As methods developed for the isolation and purification of the ADP/ATP carrier[8-10] had been quite fruitful also for the isolation of other membrane proteins, we were encouraged to apply these to the nucleotide binding protein from BAT mitochondria.[11-13] Previously, isolation was attempted by affinity GDP-binding chromatography; however, no convincing results were obtained.[14] In our laboratory, the nucleotide binding protein was isolated from mitochondria of cold-adapted hamster and rat.[11,12] The procedure permitted solubilization and purification of this protein in a functionally intact state and at high yields. The same procedure was then applied to other rodents.[15] The nuleotide binding protein has been named by us "uncoupling protein" (UCP) in accordance with its uncoupler-type action. Also the name "thermogenin" has been proposed.[16]

Isolation and Purification

Strategy and Assay

The aim of the isolation procedure is to obtain a "functionally" intact protein at as high a yield as possible and as pure as possible. Functionally intact in the strict sense would require retention of all known or still unknown activities of the UCP in the original tissue. However, in practice this postulate can only be limited to an assay which reflects largely structural intactness, i.e., binding capacity for nucleotides. This binding is quite specific—only purine-ribose di- and triphosphates are accepted. It has a high affinity and a characteristic pH dependence. All these properties should require a highly intact protein structure. They can be conven-

[8] P. Riccio, H. Aquila, and M. Klingenberg, *FEBS Lett.* **56,** 133 (1975).

[9] M. Klingenberg, P. Riccio, and H. Aquila, *Biochim. Biophys. Acta* **503,** 193 (1978).

[10] M. Klingenberg, H. Aquila, and P. Riccio, this series, Vol. 56, p. 229.

[11] C. S. Lin and M. Klingenberg, *FEBS Lett.* **113,** 299 (1980).

[12] C. S. Lin and M. Klingenberg, *Biochemistry* **21,** 2950 (1982).

[13] M. Klingenberg, H. Hackenberg, R. Krämer, C. S. Lin, and H. Aquila, *Ann. N.Y. Acad. Sci.* **358,** 83 (1980).

[14] D. Ricquier, C. Gervais, J. C. Kader, and P. Hemon, *FEBS Lett.* **101,** 35 (1980).

[15] D. Ricquier, C. S. Lin, and M. Klingenberg, *Biochem. Biophys. Res. Commun.* **106,** 582 (1982).

[16] O. Lindberg, J. Nedergaard, and B. Cannon, *in* "Mitochondria and Microsomes" (C. P. Lee *et al.*, eds.), p. 93. Addison-Wesley, Menlo Park, California, 1981.

iently assayed in the isolated protein. On the other hand, the actual function, i.e., ion transport, can only be assayed by a reincorporation into liposomes. As we shall show,[17] this reconstitution requires for certain reasons a different isolation procedure. The present method provides a UCP preparation suitable for structural studies, for immunochemical applications, and for numerous functional studies on the nucleotide binding site.

For practical reasons GDP and GTP instead of ADP and ATP are used in the binding assays for the UCP, since they are not degraded by spurious adenylate kinase and ATPase activities present in the preparations. The retention of this GDP binding on solubilization by certain detergents was most important for the isolation of UCP.[11]

Solubilization and Selection of Detergents

For selection of detergents three parameters are important: the solubilization, the retention of nucleotide binding, and the stability of binding and homodispersity. The hierarchy of suitable detergents was found to be quite similar to that found for the ADP/ATP carrier.[9,12] Also salt addition is important for solubilization, although to a lesser degree for mitochondria from brown fat than from beef heart.

In contrast to the ADP/ATP carrier, which is advantageously purified as a tight carrier–carboxyatractylate complex, UCP is isolated in the unliganded state. The ADP/ATP carrier is identified by prior loading with the very tightly binding [^{3}H]carboxyatractylate and can be monitored with the bound [^{3}H]carboxyatratylate during the isolation steps. This cannot be applied in the case of UCP since GDP does not bind as tightly. Therefore, after prior binding of labeled GDP to the mitochondria, constant reequilibration during purification with new GDP would be required. Fortunately, the GDP-binding capability of UCP is retained in some detergents even when starting with mitochondria where UCP is unliganded.

The use of various detergents for the solubilization of UCP is illustrated in Fig. 1. The binding capacity for GDP in the solubilized supernatant is determined by equilibrium dialysis. The negative response to some detergents is for two reasons: The linear alkylpolyoxyethylene detergents, such as Brij 58 ($C_{16}E_{20}$), do not solubilize the inner mitochondrial membrane; more ionic detergents such as LAPAO and cholate solubilize but inactivate the binding. Cholate inactivates not only because it is an ionic detergent, but also due to its high critical micelle concentration (cmc). Even nonionic high cmc detergents, such as octylglucoside, rap-

[17] M. Klingenberg, M. Herlt, and E. Winkler, this series, Vol. 127 [58].

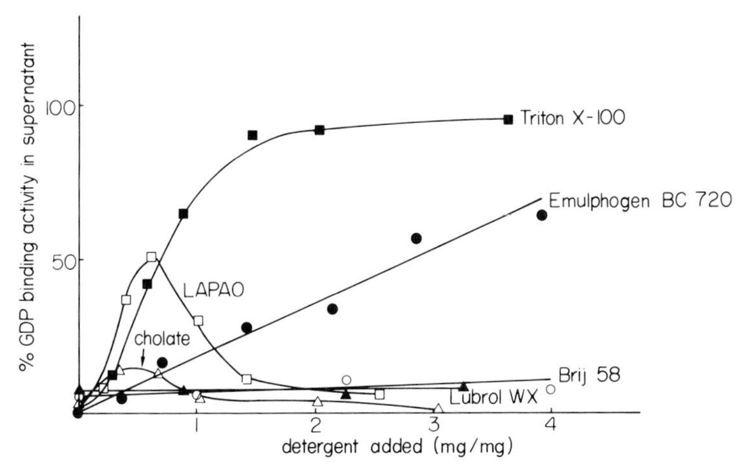

FIG. 1. The solubilization of the uncoupling protein (UCP) by different detergents (from Lin and Klingenberg[12]). Isolated mitochondria from brown adipose tissue of cold-adapted hamster. Mitochondria were incubated with 20 m*M* Na_2SO_4, 20 m*M* MOPS buffer, 1 m*M* EDTA, pH 6.7, for 30 min at 0° and with the detergent in the amounts given in the abscissa. In the 10,000 *g* supernatant the binding was measured with [^{3}H]GDP. The binding capacity in the absence of detergent is referred to as 100%. Binding was determined by equilibrium dialysis.[18]

idly inactivate the UCP. However, the similar detergent octyl-$E_{2\text{-}10}$ (octyl-POE) is less deleterious and is employed because of its high cmc in the reconstitution of the UCP.[17] Clearly Triton X-100 stands out as the most useful detergent, followed by Emulphogen BC (iso-$C_{13}E_{10}$), both for solubilization and for maintaining the nucleotide binding capacity. The standard isolation procedure is based on the use of Triton X-100.

In conclusion, the detergent requirements for UCP are similar to those for the isolation of the ADP/ATP carrier.[9] The "protected" carrier–carboxyatractylate complex, however, is more stable, since it tolerates LAPAO, octylglucoside. The unprotected ADP/ATP carrier is also rapidly degraded with these detergents.

Another critical component is the salt used with the detergent. Salt "saves" detergent and is critical for solubilizing the highly charged mitochondrial membranes with nonionic detergents. For isolating UCP less salt is applied than for the ADP/ATP carrier because anions interfere with the binding assay for nucleotides.

Extraction from Mitochondria

Critical for the extraction is the detergent/lipid ratio—not the concentration—of the nonionic low cmc detergents used. With this in mind it is best to prepare mitochondria from brown fat tissue in such a manner that they are free from excess triglycerides. The top fat layers formed during centrifugation should be carefully removed, otherwise too much detergent is sequestered by the lipids. A pretreatment with nonsolubilizing nonionic detergents such as Brij 58 ($C_{16}E_{20}$) or Lubrol WX [$C_{16}(C_{18})$-E_{17}] opens up the membranes and removes proteins which are soluble or loosely bound to the membranes. Experience shows that less Triton is then required for solubilization. For practical purposes the amount of Triton required is defined by the ratio of detergent to mitochondrial protein, although it is actually the ratio of detergent to phospholipid that matters. However, the protein content of mitochondria is generally more easily determined and well proportioned to the phospholipid content.

Precautions for Separating UCP from ADP/ATP Carrier

Since the UCP and the ADP/ATP carrier behave so similarly toward detergents and have similar structural properties, separation of UCP from ADP/ATP carrier is expected to raise problems. ADP/ATP carrier is also present in BAT mitochondria, at a molar ratio to UCP varying from 0.1 to 1 (e.g., Fig. 2). The following measures permit their separation: (1) The ADP/ATP carrier is left unprotected by not adding carboxyatractylate; (2) by addition of ATP the ADP/ATP carrier may be further labilized whereas the UCP is protected; (3) UCP is slightly more easily extracted and therefore the ratio Triton/protein = 2 is used, at which the ADP/ATP carrier is solubilized only to 70%; (4) the hydroxylapatite absorbs the ADP/ATP carrier more easily if it is "denatured." Denaturation of the unprotected ADP/ATP carrier is promoted by passing the extract through hydroxylapatite at 25° (room temperature).

Separation and Purification

Just like ADP/ATP carrier, UCP is most efficiently purified by passage through hydroxylapatite (Fig. 2). Since UCP is not adsorbed, in contrast to most other extracted proteins, it is strongly enriched in the first pass through. The yield at this stage is high, reaching more than 70% of the original UCP in the mitochondria. For many applications this preparation, which can be up to 70% pure, suffices. In the polyacrylamide gels little contamination by ADP/ATP carrier is found. Most disturbing at this stage can be the high content of Triton X-100 (Triton/protein $>$ 30 w/w) and of phospholipids. For partial removal a gel chromatography on Sephadex G-150 is recommended which gives a fair separation of the bulk Triton and

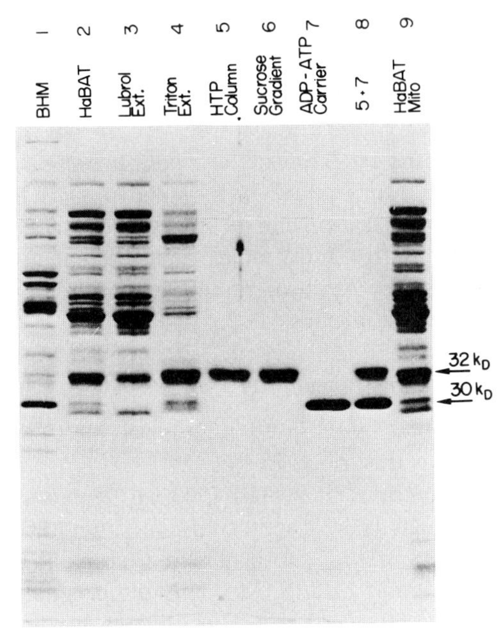

FIG. 2. Comparison of extracts from mitochondria of brown adipose tissue with those from beef heart (from Lin and Klingenberg[12]). Gradient polyacrylamide slab gel electrophoresis in sodium dodecyl sulfate. From left to right: (1) beef heart mitochondria (BHM); (2 and 9) hamster brown adipose tissue mitochondria (HaBAT); (3) Lubrol extract; (4) Triton extract; (5) hydroxyapatite column pass-through fraction (HTP column); (6) sucrose gradient fraction; (7) ADP/ATP carrier from beef heart mitochondria; and (8) hydroxylapatite column pass-through fraction plus ADP/ATP carrier.

phospholipids from the protein. A better purification and separation is obtained by sucrose density centrifugation.

Assay and Yields

The amounts of UCP protein can be estimated in SDS–polyacrylamide gels which give the sum of intact and denatured protein. Intact UCP is assayed by GDP binding. Descriptions of binding assays are given in a separate chapter.[18] Typical values for GDP binding capacity are (μmol GDP bound/g protein) in the Triton extract 3.0, in hydroxylapatite efflux 10–12, and in sucrose gradient fractions 15.

Yields obtained in the purification steps are calculated from the GDP-binding assay and refer to maximum binding in the mitochondria. These

[18] M. Klingenberg, M. Herlt, and E. Winkler, this volume [49].

are approximately after Triton X-100 extraction (solubilized) 75%, after hydroxylapatite 70%, and after sucrose gradient 30%.

Stability

GDP-binding capacity is also used as a measure of stability of UCP.[12] At 4° binding is retained for up to 7 days. At 37° it has a half-time of only about 1 hr. The UCP preparation can be stored in the frozen state in liquid nitrogen for several months. Care must be taken that the preparation is frozen and thawed rapidly. Addition of 20% glycerol can further stabilize against the freeze-thawing process.

Isolation Protocol for the Uncoupling Protein

Source. Mitochondria from brown adipose tissue (e.g., from hamster or rat) were isolated as described in Ref. 19.

Preextraction. Mitochondria protein (500 mg) stored in 10 ml isolation medium is mixed with 20 ml of preextraction medium (3% Brij 58, 30 m*M* Na_2SO_4, 15 m*M* MOPS, pH 6.7) to give a final concentration of 2% Brij 58, 20 m*M* Na_2SO_4, 10 m*M* MOPS. Mix with good agitation for 5 min at 0°. Then centrifuge at 8000 *g* for 5 min. Discard supernatant. Resuspend in 20 ml of a medium containing 20 m*M* Na_2SO_4, 10 m*M* MOPS, pH 6.1. Centrifuge at 8000 *g* for 5 min.

Extraction. Dissolve sediment in 30 ml extraction medium (3% Triton X-100, 30 m*M* Na_2SO_4, 15 m*M* MOPS, pH 6.7) by strongly shaking for 5 min. Then centrifuge at 100,000 *g* for 20 min.

Separation. The supernatant is then applied to a 70-ml hydroxylapatite column (3 × 10 cm), preequilibrated with 30 m*M* Na_2SO_4, 15 m*M* MOPS, pH 6.7, and is eluted with the same medium. The first 30 ml are pooled and concentrated by pressure dialysis to 5–8 ml. This preparation can be stored and used for many purposes.

Purification. Two procedures are described which primarily remove excess Triton and phospholipids. A fast method with limited resolution is the following. Apply the concentrated preparation to a Sephadex G-150 column (2 × 70 cm) which has been preequilibrated with a medium of 0.2% Triton, 20 m*M* Na_2SO_4, 10 m*M* MOPS, pH 6.7. The efflux is monitored for Triton X-100 at 280 nm and for protein by the Lowry method. The protein-containing fractions peak at about $1 - K_{av} = 0.2$ and the bulk of Triton at $1 - K_{av} = 0.35$. The protein fractions are pooled and pressure dialyzed. The preparation has a ratio of Triton to protein ≤ 15.

[19] B. Cannon and O. Lindberg, this series, Vol. 55, p. 65.

A better separation of Triton and phospholipids is achieved by sucrose gradient centrifugation. The concentrated hydroxylapatite eluate is applied on three 30-ml centrifuge tubes which are filled with a 5–20% linear sucrose gradient containing 0.1% Triton, 20 mM Na_2SO_4, and 10 mM MOPS, pH 6.7. Centrifugation takes place in a vertical 50 VTi rotor at 160,000 g for 14 hr. Fractions of 2.5 ml are collected and the concentration of Triton X-100 is measured at 280 nm and that of protein by the Lowry method. The protein peaks at about fraction 6 and Triton together with phospholipid at fraction 11. The fractions 4–8 are pooled and pressure dialyzed with further additions of a medium containing 0.2% Triton, 20 mM Na_2SO_4, and 10 mM MOPS in order to remove sucrose.

Hydrodynamic Characterization of the Isolated UCP[20]

In the SDS–polyacrylamide gels UCP migrates at a rather diffuse band of M_r around 32,000–33,000. The "functional" molecular weight calculated from the binding capacity of the purified protein (15 μmol of GDP/g protein) is 67,000. This would correspond to one binding site per two subunits; i.e., it would indicate a dimer structure of UCP with a "half-site" reactivity. Exactly the same situation was found for the ADP/ATP carrier, with one binding site for the inhibitors per two subunits.

All hydrodynamic studies must take into account the fact that isolated UCP is present in a mixed protein–detergent micelle with possibly some bound phospholipids. Therefore the determination of protein-bound detergent is a prerequisite for the evaluation of hydrodynamic data. This can be done either by gel filtration or by sucrose density centrifugation. In gel filtration a content of Triton X-100/protein (1.9 w/w) was found to be associated with UCP. The Stokes radius of protein-Triton micelles was determined by gel filtration to be 16 Å.

In ultracentrifugal studies the absorbance of Triton at 280 nm was utilized for the UV scans of protein sedimentation. As originally worked out for the ADP/ATP carrier,[21] at low concentration of free Triton sedimentation of the protein can be clearly differentiated due to absorbance of a high amount of protein-bound Triton. For this purpose a UCP preparation, from which the bulk of Triton has been removed by sucrose gradient centrifugation, is diluted to about 180 μg protein/ml and 0.1% Triton. Note: UCP has a stronger tendency to form higher aggregates on dilution of Triton than ADP/ATP carrier. Therefore the latter can be diluted to 0.025% Triton without aggregation which gives an A_{280} of 0.53 cm^{-1}.

[20] C. S. Lin, H. Hackenberg, and M. Klingenberg, *FEBS Lett.* **113,** 304 (1980).
[21] H. Hackenberg and M. Klingenberg, *Biochemistry* **19,** 548 (1980).

However, with 0.1% Triton, as required for UCP, an absorbance A_{288} of 1.0 cm^{-1} is obtained by scanning at 280 nm. Thus, in sedimentation velocity runs, a sedimentation coefficient $s^0_{20,w} = 4.0 \pm 0.1 \cdot 10^{-13}$ sec is obtained. The molecular weight of the protein–Triton complex is best determined in sedimentation equilibrium runs. The equilibration distribution profile of log A_{288} versus r^2 should be linear, for indicating homodispersity of the UCP–Triton micelles. The molecular weight ratio calculated from the slope is 180,000. By subtracting the Triton content, M_r of 62,000 is obtained for the UCP protein. This actually corresponds to a protein dimer, as expected on the basis of the half-site reactivity of nucleotide binding.

These hydrodynamic data, together with the high binding of detergent, are very similar to the data determined for the ADP/ATP carrier. Therefore, the UCP and Triton micelles as well as the UCP may have structures similar to those deduced for the ADP/ATP carrier. In a minimum geometric model the mixed micelle is seen as an oblate ellipsoid with a short axis corresponding to the 2-fold symmetry axis of the dimer protein.[21] The large Stokes radius of 62–65 Å reflects the large detergent envelope around the hydrophobic surface of the protein.

Acknowledgments

This work was supported by grants from the Deutsche Forschungsgemeinschaft (Kl 134/21-23). C. S. L. was a recipient of a Humboldt Foundation Fellowship.

[49] Nucleotide Binding Assay for Uncoupling Protein from Brown Fat Mitochondria

By MARTIN KLINGENBERG, MARIA HERLT, and EDITH WINKLER

Introduction

The uncoupled state of respiration is a characteristic property of brown adipose tissue mitochondria. It is the source for the heat production of brown adipose tissue. Addition of purine nucleotides to isolated brown fat mitochondria recouples and consequently slows down respiration.[1] Obviously, nucleotides are regulators of the uncoupling process and

[1] H. J. Hohorst and J. Rafael, *Hoppe Seyler's Z. Physiol. Chem.* **349**, 268 (1968).

thus of thermogenesis in these mitochondria. This inhibition was correlated to nucleotide binding to receptor sites facing the c-surface of the inner mitochondrial membrane.[2] Although this site has about equal affinity for adenine and guanine nucleotides, GDP and GTP are the preferred ligands in experimental studies in order to avoid the complications of interfering reactions of ADP and ATP with adenylate kinase and ATPase. The receptor site was identified as a protein with a molecular weight of 32,000 and clearly differentiated from the ADP/ATP carrier site.[3] It was proposed that this nucleotide binding protein is instrumental in the uncoupling process, probably channeling OH^- or H^+ into the matrix and thus short-circuiting the electrogenic H^+ generator of electron transport.[4]

This nucleotide binding protein has been isolated in the intact state in our laboratory,[5] as described elsewhere in this volume [48], and is called uncoupling protein (UCP). The isolation of UCP actually relied on the nucleotide binding and was used both in the assay for the amount and for its functional intactness of the uncoupling protein, for example, in the selection of the solubilizing detergents. Nucleotide binding was fully retained in the nonionic detergent Triton X-100, although it was highly sensitive to cholate, to octylglucoside, and to some other popular protein-solubilizing detergents. Also, after incorporation into phospholipid vesicles (proteoliposomes), nucleotide binding was retained and used as an assay for differentiating the uncoupling protein-linked proton transport activity.

General Properties of Nucleotide Binding to Uncoupling Protein

The properties of nucleotide binding to the uncoupling protein still in the original membranes are quite similar to those after isolation and reconstitution. The specificity of the binding tolerates all purine nucleotides and only the di- and triphosphate derivatives. Deoxyribose nucleotides are also accepted with slightly decreased affinity. A summary of K_d values for various nucleotides is given in the table.[6] The binding is inhibited by Mg^{2+}, i.e., only the free nucleotides are accepted. A most important feature of the nucleotide binding to the uncoupling protein is the strong pH dependence. The binding affinity increases with H^+ concentra-

[2] J. Rafael, H. W. Heldt, and H. J. Hohorst, *FEBS Lett.* **28,** 125 (1972).

[3] G. M. Heaton, R. J. Wagenvoord, R. J. Kemp, Jr., and D. G. Nicholls, *Eur. J. Biochem.* **82,** 515 (1978).

[4] D. G. Nicholls, *Biochim. Biophys. Acta* **548,** 1 (1979).

[5] C. S. Lin and M. Klingenberg, *Biochemistry* **21,** 2950 (1982).

[6] M. Klingenberg, *Biochem. Soc. Trans.* **12,** 390 (1984).

SURVEY OF DISSOCIATION CONSTANTS OF NUCLEOTIDES AND NUCLEOTIDE DERIVATIVES TO THE ISOLATED UNCOUPLING PROTEIN

Nucleotide	K_d [a]	pK_d
ATP	5×10^{-7}	6.3
ADP	2.5×10^{-6}	5.6
GTP	1.1×10^{-6}	5.95
GDP	1.3×10^{-6}	5.9
ITP	4.0×10^{-6}	5.4
AMP PNP	1.3×10^{-5}	4.9
ATP-γS	1.3×10^{-5}	4.9
DAN-ATP [b]	1.0×10^{-5}	5.0

[a] pH 6.2.
[b] DAN, 1,5-dimethylaminonaphthoyl.

tion over a broad range, as shown in Fig. 5 of Ref. 5 and Fig. 1 of Ref. 6. Anions compete with the nucleotide binding to some degree.[5] The nucleotide binding is relatively insensitive to SH reagents, but inhibited by lysine reagents. The binding is also photoinactivated but not decreased by the histidine reagent diethyl pyrocarbonate.[7]

The following applications of nucleotide binding to UCP should be mentioned: quantitative determination of UCP content in mitochondria, in solubilized extracts during purification, and in reconstituted phospholipid vesicles; affinity for nucleotides, as determined by concentration dependence, and the influence of various factors relevant to the binding activity, such as pH, anions, and amino acid reagents; kinetics of nucleotide binding and dissociation; correlation of binding with transport inhibition in mitochondria and in reconstituted vesicles; differentiation of UCP-linked H^+ transport from other modes of H^+ flux; exploration of the binding center by binding competition with nucleotide analogs, e.g., with photoaffinity and fluorescent nucleotide derivatives.

Principles of Measurement of Nucleotide Binding to UCP

Binding measurements rely on the association of labeled purine di- and trinucleotides to UCP, such as [^{3}H]- or [^{14}C]ATP, -ADP, -GTP, and -GDP, as well as [^{32}P]ATP. For determining the rates of binding and dissociation, fluorescent derivatives, such as 1,5-dimethylaminonaphthoyl-3′-*O*-ADP and -ATP are also employed.[8] In general, the binding

[7] M. Klingenberg, unpublished results.

[8] M. Klingenberg, this series, Vol. 125, p. 618.

measurements require equilibrating of the nucleotide with UCP. Depending on the applications the following four methods are employed: equilibrium dialysis, exclusion gel chromatography, rapid anion-exchange chromatography, and centrifugal sedimentation. The last method is limited to membrane-bound UCP. Since the binding has a medium affinity, the methods must maintain UCP in equilibrium with the free ligands, with the exception of rapid anion exchange, as will be elucidated below.

Procedures for Binding Measurements

Equilibrium Dialysis

In order to use only small amounts of protein, microdialysis cells are recommended. Relatively high rates of dialysis and equilibration are here obtained by a dialysis membrane which is large in relation to the dialysis volume. One commercial solution is the "Dianorm" apparatus of Dr. Weder (ETH, Zürich) with a cell chamber column of 200 μl and a membrane diameter of 15 mm. Up to 20 cells can be dialyzed in parallel. Stirring is achieved with an air bubble in the rotating cells. The cells can be immersed in a water bath for temperature control. A dialysis membrane material, acetylcellulose, is used (Visking) which is boiled for 30 min in 5% $NaHCO_3$ solution and then rinsed with H_2O. The time required for equilibration of labeled nucleotides between the two dialysis chambers by diffusion amounts to 2–3 hr. It is prolonged in the presence of Triton X-100 to 4–5 hr. Obviously the pores of the cellulose membrane are partially plugged by the Triton micelles.

The stability of the UCP limits time and temperature of dialysis. At 20° UCP loses 30% of its binding capacity within 5 hr of dialysis. The stability of the nucleotides can be a major problem. ADP and ATP are degraded easier than GDP or GTP because of the adenine specificity of adenylate kinase and ATPase. Spurious amounts of adenylate kinase and sometimes ATPase occur mainly in mitochondria and in raw extracts, but also in the purified UCP. ADP is particularly labile due to the adenylate kinase reaction. This decomposition can be partially suppressed by omitting Mg^{2+} and adding EDTA and the adenylate kinase inhibitor diadenosine pentaphosphate (AP_5A). GTP and GDP are more resistant, since they do not or only poorly react with these enzymes. Especially in binding measurements with mitochondrial membranes, GTP and GDP are to be preferred because high activities of ATPase and adenylate kinase are present here.

The amounts of protein and nucleotides employed in the binding assay depend on affinity to UCP in order to measure reasonably accurate ratios of bound to free nucleotides. The relevant mass action relation is

$$CN = C_0 - K_d(CN/N) \tag{1}$$

where, all in concentrations, C_0 is the total carrier sites, C is the free carrier, CN is the carrier–nucleotide complex, N is the free nucleotide, N_0 is the total nucleotide, and K_d is the dissociation constant. Sufficient accuracy for the determination of bound/free nucleotide (CN/N) is given only in the range $0.1 < (CN/N) < 10$. Rearranging Eq. 1 gives

$$\frac{CN}{N} = \frac{C_0 - CN}{K_d} \tag{2}$$

The concentrations C_0 and N_0 employed should be adjusted so that

$$0.1 < \frac{C_0 - CN}{K_d} = \frac{C_0 - N_0 + N}{K_d} < 10$$

From this follows that, for example, with $K_d = 10^{-6}\ M$, the amount of UCP, C_0, should be $C_0 = 0.1–1 \times 10^{-6}\ M$; with $K_d = 10^{-5}\ M$, $C_0 = 5 \times 10^{-6}\ M$; and with $K_d = 10^{-7}\ M$, $C_0 = 0.5–2 \times 10^{-7}\ M$. All values are within limits of a reasonable use of the scarce protein amounts. The nucleotide concentration N_0 should range from 0.5 to 5 μM for $K_d = 10^{-7}\ M$ and from 5 to 50 μM for $K_d = 10^{-5}\ M$.

A typical assay determining K_d and binding capacity is described as follows. Take 1 ml incubation medium containing 20 mM PIPES, pH 6.0, and add 50 μl of solubilized UCP preparation containing ~80 μg UCP protein. Pipette 200 μl each into the right chamber of four assembled microdialysis cells. Add 200 μl medium into the left chamber, together with increasing concentrations of [^{14}C]GTP at 2, 4, 8, and 16 μM. Allow to dialyze for 4 hr at 10°, then withdraw the fluid from each chamber with a micropipette and take a 100-μl aliquot for liquid scintillation counting.

The ratio of [^{14}C]GTP from the right and left chamber gives directly the ratio of bound/free nucleotides (CN/N). The bound nucleotides are calculated from the specific activity and related to the amount of protein employed. These values are evaluated in a "mass action" plot. The slope gives K_d according to Eq. (1), and the maximum binding C_0 is obtained by extrapolation to the ordinate.

For determining the pH dependency, several sets of four nucleotide concentrations are dialyzed at various pH. The protein (UCP) concentration is increased from 40 μg/ml at pH 5.0–5.8 to 80 μg/ml at pH 6.0–6.8 and 150 μg/ml at pH 7.0–7.6. The range of nucleotide concentration (N_0) increases with pH according to the K_d increase. The nucleotide concentrations are adjusted to the expected K_d as elucidated above.

Exclusion Gel Chromatography[9]

This method follows the principle of Hummel and Dreyer,[9] but has limited use because it requires relatively high amounts of protein. An example is described as follows.[5] A 1 × 30 cm Sephadex G-75 (Pharmacia) column is preequilibrated with the buffered medium (e.g., 20 mM MOPS, pH 6.7) and with 5 μM [^{3}H]GDP at 4°; then 300 μl of Triton X-100 solubilized mitochondria (2 mg/ml) are mixed with 3 nmol [^{3}H]GDP to give 10 μM [^{3}H]GDP and applied to the Sephadex column. Elution is carried out with the same buffer and 0.5 ml fractions are collected at a flow rate of 0.8 ml/min.

Anion-Exchange Chromatography

This method takes advantage of the very slow dissociation of the nucleotides from UCP. It permits a rapid removal of free nucleotides when the UCP–nucleotide complex passes through a small Dowex column before it has time to dissociate during the passage. The pass through then is taken to reflect the nucleotide bound to UCP. Another advantage is that because of its short duration it circumvents the slow degradation of nucleotides. It can also be applied to very small samples.

The following procedure is used in our laboratory. A small glass capillary column of 2.0 × 60 mm equipped with a sintered glass bottom is filled with 60 mg wet Dowex 1 X-8, 200–400 mesh Cl^- form. This column has the right size for a binding sample of 150 μl. This probe may contain 10–25 μg protein of UCP in a medium of 20 mM triethanolamine plus maleate buffer, pH 5.2–8.0, 1 mM EDTA, 20 μM diadenosine pentaphosphate, and the labeled nucleotide (e.g., GTP) in concentration from 1 to 40 μM.

The probe is applied on top of the resin with a capillary pipette (e.g., Hamilton type). Immediately it is then chased by squeezing 2 × 200 μl H_2O through the column with a plastic tip pipette. This whole procedure should take only 1 min. The eluate from the column of about 600 μl contains all the bound UCP and is entirely used for scintillation counting.

Sedimentation Methods

This elementary method relies on the collection of the binding preparation into a pellet by centrifugation. It is applied to mitochondria and derived membranes and, if possible, also to reconstituted vesicles. The following procedure is recommended. Brown fat mitochondria or particles derived therefrom are incubated at the desired pH with buffer, as

[9] J. P. Hummel and W. J. Dreyer, *Biochim. Biophys. Acta* **63**, 530 (1962).

previously mentioned, and, if necessary, for maintaining osmotic balance with 0.2 *M* sucrose. The samples are conveniently contained in 0.7-ml Eppendorf cups at 1 mg protein/0.5 ml. For binding to mitochondria the guanine nucleotides, GTP and GDP, are preferentially used at the concentration as given above. ATP and ADP would create complications due to rapid breakdown or translocation into mitochondria. The latter can be avoided by addition of 10 μM carboxyatractylate. After sedimentation by centrifugation for 8 min at 8000 *g* in the case of mitochondria and for 20 min at 50,000 *g* in the case of particles, the supernatant is decanted and the adherent fluid carefully removed by a capillary attached to a suction line. The sediment is then dissolved in 100 μl 5% SDS. Aliquots of supernatant and the sediment are assayed by scintillation counting.

Acknowledgment

This work was supported by grants from the Deutsche Forschungsgemeinschaft (Kl 134/21-23).

[50] Isolation and Characterization of an Inactivated Complex of F_1F_0-ATPase and Its Inhibitory Factors from Yeast

By KUNIO TAGAWA, TADAO HASHIMOTO, and YUKUO YOSHIDA

An intrinsic F_1-ATPase inhibitor protein found in mitochondria is considered to regulate oxidative phosphorylation. ATPase inhibitors have now been obtained in purified forms from various sources, including beef heart,[1] rat liver,[2] skeletal muscle,[3] and yeast.[4,5] The complete amino acid sequences of these proteins from yeast[6] and beef heart[7] mitochondria have been determined. No inhibitor protein is present in purified F_1-ATPase or F_1F_0-ATPase, but it is found in mitochondria and submitochondrial particles[8] in an equimolar ratio to F_1-ATPase, indicating that it

[1] M. E. Pullman and G. C. Monroy, *J. Biol. Chem.* **238,** 3762 (1963).
[2] N. M. Cintron and P. L. Pedersen, *J. Biol. Chem.* **254,** 3439 (1979).
[3] E. W. Yamada and N. J. Huzel, *Biosci. Rep.* **3,** 947 (1983).
[4] E. Ebner and K. L. Maier, *J. Biol. Chem.* **252,** 671 (1977).
[5] T. Hashimoto, Y. Negawa, and K. Tagawa, *J. Biochem.* **90,** 1151 (1981).
[6] H. Matsubara, T. Hase, T. Hashimoto, and K. Tagawa, *J. Biochem.* **90,** 1159 (1981).
[7] B. Frangione, E. Rosenwasser, H. S. Penefsky, and M. E. Pullman, *Proc. Natl. Acad. Sci. U.S.A.* **78,** 7403 (1981).
[8] T. Hashimoto, Y. Yoshida, and K. Tagawa, *J. Biochem.* **94,** 715 (1983).

is an integral subunit of the enzyme, but is readily released from the enzyme during the processes of isolation of the enzyme. A completely inactivated complex of F_1-ATPase with the inhibitor can be reconstituted by incubating the two purified proteins in the presence of ATP and Mg^{2+}. However, the inactivated complex of purified F_1-ATPase dissociates gradually when ATP and Mg^{2+} are removed from the external medium and consequently the enzyme becomes active.[5-8] In contrast to inhibition of F_1-ATPase activity, F_1F_0-ATPase in mitochondria is inhibited even in the absence of ATP and Mg^{2+}. This strongly suggests the existence in mitochondria of factors that stabilize the inactivated complex between the enzyme and the inhibitor. Recently, stabilizing factors, designated as 9K and 15K proteins, have been isolated from yeast mitochondria[8] and purified by high-performance liquid chromatography (HPLC).[9] The 9K and 15K proteins act together to stabilize the inhibited membrane-bound enzyme, and either protein alone is less effective. The factors also exert their action on isolated F_1F_0-ATPase, but not on F_1-ATPase. The 9K and 15K factors, like F_1-ATPase inhibitor, are basic, heat-stable proteins. We have also determined the complete amino acid sequences of the 9K[10] and 15K[11] proteins. Here we describe the purifications and biochemical properties of these proteins.

Preparation of ATPase Inhibitor and 9K and 15K Protein-Stabilizing Factors

Heat Extraction

Powder of pressed yeast (2 kg) is gradually added to 4 liters of boiling 1 *M* ammonium sulfate with stirring, and the suspension is allowed to stand for 10 min, keeping the temperature above 95°. It is then cooled and centrifuged to remove debris.

Adsorption and Elution on Dowex 50 Column

The heated extract is adjusted to pH 7.0 with 1 *N* ammonia and applied to a column of Dowex 50 (3 × 20 cm) equilibrated with 1 *M* ammonium sulfate at room temperature. The column is washed successively with 1 *M* ammonium sulfate, distilled water, and 7 *M* urea. The inhibitor and fac-

[9] T. Hashimoto, Y. Yoshida, and K. Tagawa, *J. Biochem.* **95,** 131 (1984).

[10] H. Matsubara, K. Inoue, T. Hashimoto, Y. Yoshida, and K. Tagawa, *J. Biochem.* **94,** 315 (1983).

[11] Y. Yoshida, S. Wakabayashi, H. Matsubara, T. Hashimoto, and K. Tagawa, *FEBS Lett.* **170,** 135 (1984).

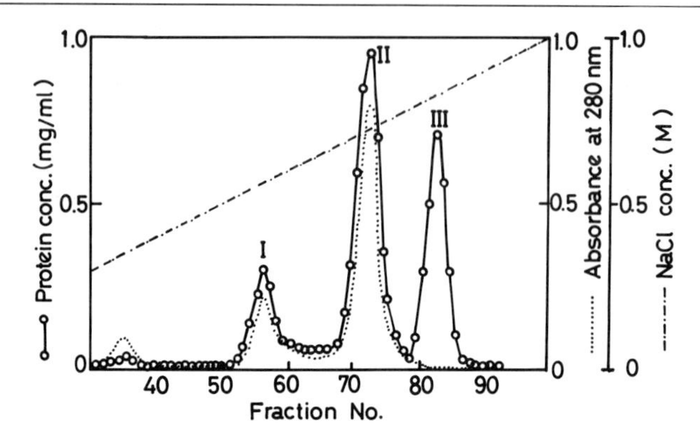

FIG. 1. Separation of ATPase inhibitor from stabilizing factors on a CM-cellulose column. About 80 mg of protein containing ATPase inhibitor and stabilizing factor are loaded on a CM-cellulose column (4 × 20 cm). The starting buffer (500 ml) is 50 m*M* sodium acetate, pH 5.0, and the elution buffer (500 ml) contains 1.0 *M* NaCl in addition (Ebner and Maier[4]). Fractions of 10 ml are collected and the protein concentration of each fraction is measured.

tors are eluted with 7 *M* urea containing 0.1 *M* Tris-SO_4, pH 7.3. Fractions of 15 ml are collected and assayed for the inhibitor, which is eluted together with the 9K and 15K protein-stabilizing factors, after about 2.5 column volumes.

Separation of ATPase Inhibitor from Stabilizing Factors by CM-Cellulose Column Chromatography

The eluate (200 ml) containing ATPase inhibitor and stabilizing factors is chilled and poured into 4 volumes of ethanol previously cooled to −10°. The precipitate is collected by centrifugation (5000 rpm for 10 min), dissolved in about 10 ml of 0.05 *M* sodium acetate, pH 5.0, and chromatographed on a CM-cellulose column (4 × 20 cm) equilibrated with the 0.05 *M* acetate buffer. Other conditions of chromatography are essentially as described by Ebner and Maier,[4] applying a linear gradient of 0–1.0 *M* sodium chloride in the buffer. Figure 1 shows a typical chromatogram. Peak I does not contain either ATPase inhibitor or the stabilizing factors. Peak II contains strong stabilizing activity for the F_1F_0-ATPase–inhibitor complex contaminated with small amounts of ATPase inhibitor and impurities. Peak III contains pure ATPase inhibitor. Its purity is easily judged from its absorption at 280 nm because the inhibitor from yeast mitochondria does not contain either tryptophan or tyrosine. The yield of purified inhibitor is about 20 mg from 2 kg of pressed yeast. The purified ATPase

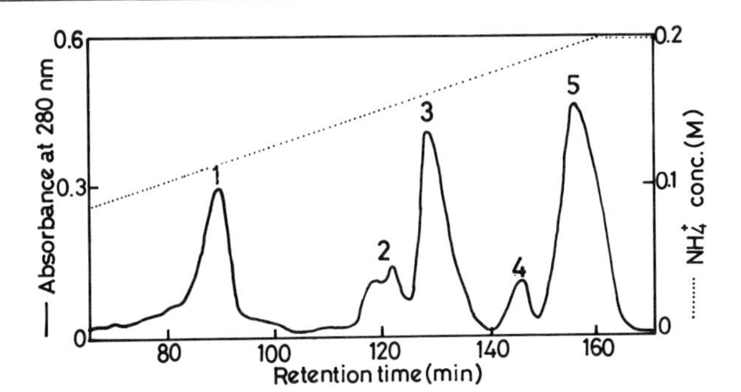

FIG. 2. Purification of the 9K and 15K proteins by IEX-535 chromatography. About 40 mg of protein recovered in peak II from the CM-cellulose column is loaded on an IEX-column (6 × 150 mm) equilibrated with 25 mM ammonium phosphate, pH 5.7. Linear gradient chromatography is carried out at a flow rate of 1 ml/min at room temperature, raising the buffer concentration to 200 mM in 160 min. Fractions of eluate are monitored for absorbance at 280 nm.

inhibitor obtained has a specific activity of 40,000 units, when one unit of ATPase inhibitor activity is defined as the amount of inhibitor required for 50% inhibition of 0.2 unit of ATPase.[1]

Purification of 9K and 15K Proteins

The 9K and 15K protein-stabilizing factors in peak II are precipitated with trichloroacetic acid at a final concentration of 10%. The precipitate containing about 40 mg of protein is dissolved in 5 ml of 25 mM ammonium phosphate, pH 5.7, and charged on a cation exchanger, IEX-535 column (6 × 150 mm, Toyo Soda Co. Ltd., Tokyo), equilibrated with the same buffer. Linear gradient chromatography is carried out at a flow rate of 1 ml/min at room temperature, raising the buffer concentration to 200 mM in 160 min. Figure 2 shows a typical chromatogram with five major peaks. Peaks 2, 3, and 4 each contained more than one of the three proteins and are not clearly separated from each other, but the 9K and 15K proteins are recovered in purified forms in peaks 1 and 5, respectively. The yield of each of these proteins is about 10 mg from 2 kg of pressed yeast. The 9K and 15K proteins are each precipitated with trichloroacetic acid, dialyzed against 10 mM ammonium carbonate, pH 7.0, and lyophilized. The 9K and 15K proteins and the ATPase inhibitor are very stable at room temperature and the lyophilized preparation can be stored for more than 2 years without loss of activity.

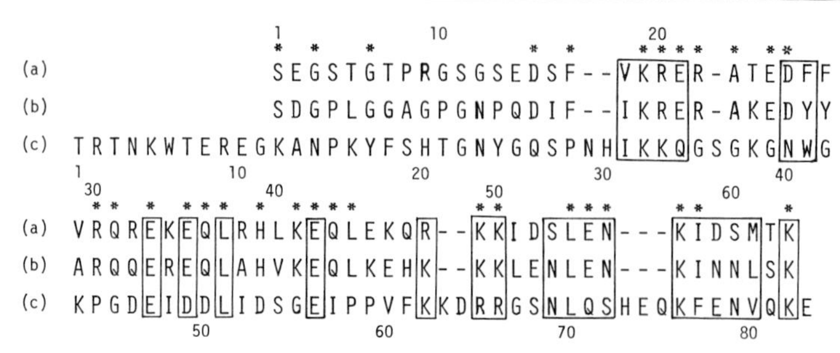

FIG. 3. Comparison of amino acid sequences of yeast ATPase inhibitor and 9K and 15K proteins. (a) ATPase inhibitor (Matsubara *et al.*).[6] (b) 9K protein (Matsubara *et al.*[10]). (c) 15K protein (Yoshida *et al.*[11]). Identical and similar amino acid residues in the three proteins are enclosed in solid boxes. Asterisks indicate identical amino acid residues in the ATPase inhibitor and 9K protein. Dashes indicate gaps, making alignments highly homologous.

Assay of Stabilizing Activity

Either submitochondrial particles or purified F_1F_0-ATPase can be used as enzyme for assay of stabilizing activity on the inactivated enzyme complex. When submitochondrial particles are used, however, F_1F_0-ATPase must be fully activated by preincubation with 0.5 *M* Na_2SO_4 containing 50 m*M* Tris-SO_4, pH 7.5. ATPase (about 30 units) is incubated with purified ATPase inhibitor (800 units) and the test sample in 0.1 ml of medium containing 25 m*M* Tris-SO_4, pH 7.2, 1 m*M* ATP, and 1 m*M* $MgSO_4$. After complete inactivation of the enzyme, which is accomplished within 30 min, external ATP and Mg^{2+} are removed by centrifugation on a column packed with about 1 ml of Sephadex G-50 by the method of Penefsky.[12] A sample of 30 μl of the effluent (about 0.1 ml) is diluted with 25 m*M* Tris-SO_4, pH 7.3, to a final volume of 0.2 ml and incubated at 25°.[13] When samples contain neither of the stabilizing factors, the enzyme is fully activated after 3–4 hr, but the inactivated enzyme is not reactivated within 4 hr in the presence of sufficient amounts of both the 9K and 15K proteins.

Properties of the ATPase Inhibitor, and 9K and 15K Protein-Stabilizing Factors

Figure 3 shows the complete amino acid sequences of the ATPase inhibitor and the 9K and 15K proteins from yeast. These proteins are all heat-stable, basic proteins. The inhibitor and 9K proteins both have 63

[12] H. S. Penefsky, *J. Biol. Chem.* **252,** 2891 (1977).

[13] The sample material may be added in the second incubation after column centrifugation instead of in the first incubation.

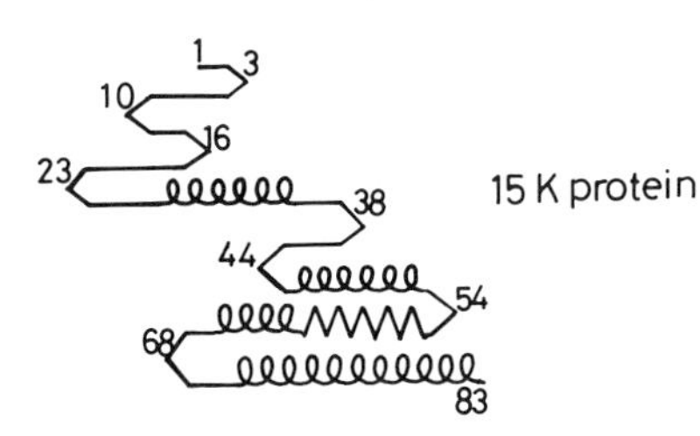

FIG. 4. Secondary structures of ATPase inhibitor and 9K and 15K proteins. The secondary structures of ATPase inhibitor and stabilizing factors were determined by the method of Chou and Fasman (Chou and Fasman).[14] α-Helix (ℓℓℓ); β sheet (~~); β turn (⊐>). Numbers are those of amino acid residues from the amino terminal.

amino acid residues. In their sequences 31 residues are identical and 16 differ conservatively. The 15K protein has 83 residues and shows partial homology with the other two proteins, especially in the vicinity of the carboxy terminal. The inhibitor protein from yeast is highly homologous with the bovine inhibitor and also shows significant homology with the ε subunit of *Escherichia coli*.[10] Therefore, it seems highly likely that all these proteins, including the 9K and 15K proteins, were derived from a common ancestral gene and then diverged with different functions. The secondary structures of these three proteins predicted by the method of Chou and Fasman[14] are very characteristic (Fig. 4). The α-helical contents of the ATPase inhibitor and 9K protein are 80% and 57%, respectively. In the helical regions of the two proteins, basic, acidic, and hydrophobic clusters are found to be localized on the surface of the three-dimensional molecule, suggesting a suitable structure to form a stable complex with other subunits of F_1F_0-ATPase (Fig. 5). The α-helical content of the 15K protein is much lower (about 32%) than those of the other two proteins, but clusters of basic, acidic, and hydrophobic residues are also found in its helical regions.

ATPase in submitochondrial particles is completely inhibited by addition of the ATPase inhibitor in the presence of ATP and Mg^{2+}, but is

[14] P. Y. Chou and G. D. Fasman, *Adv. Enzymol.* **47,** 45 (1978).

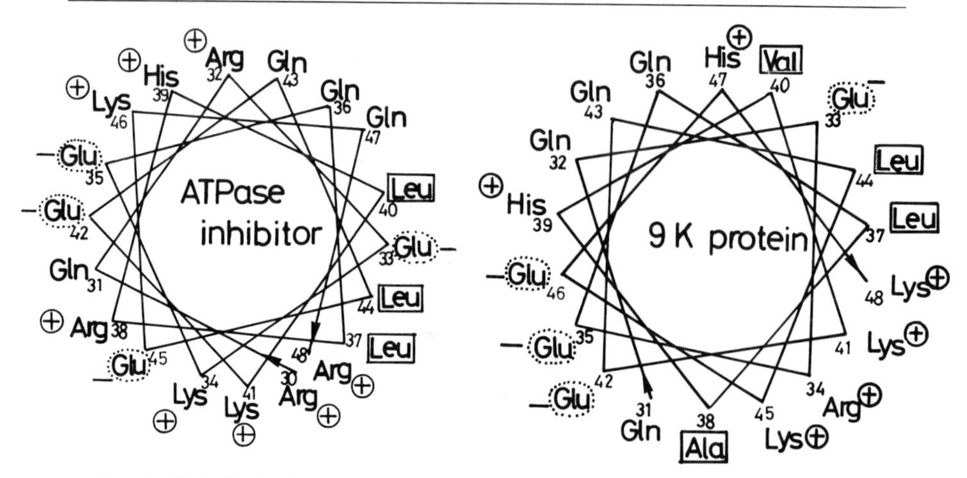

FIG. 5. Helical wheel representations of ATPase inhibitor and 9K protein. Amino acid residues 30–48 of ATPase inhibitor and 31–48 of 9K protein are projected as the helical wheel according to Schiffer and Edmundson [M. Schiffer and A. B. Edmundson, *Biophys. J.* **7,** 121 (1967)].

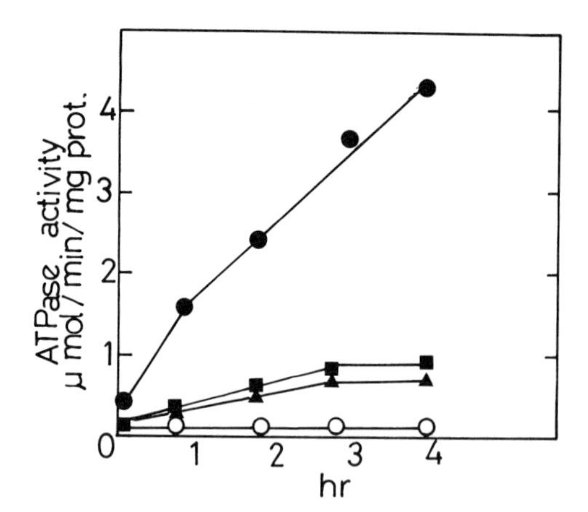

FIG. 6. Stabilization of inactivated F_1F_0-ATPase bound to submitochondrial particles. F_1F_0-ATPase bound to submitochondrial particles (100 units) is incubated with ATPase inhibitor (4000 units) in 0.5 ml of medium containing 25 m*M* Tris-SO_4, pH 7.2, 1 m*M* ATP, and 1 m*M* $MgSO_4$. After complete inactivation of the enzyme, external ATP and Mg^{2+} are removed by the centrifuge column method (Penefsky[12]). The inactivated ATPase (about 10 units) is incubated at 25° with 20 μg of 9K (■), 20 μg of 15K (▲), 10 μg each of 9K and 15K (○), or without the factors (●) in 0.2 ml of medium containing 25 m*M* Tris-SO_4, pH 7.3. Samples of 10 μl are taken at the indicated times for measurement of ATPase activity.

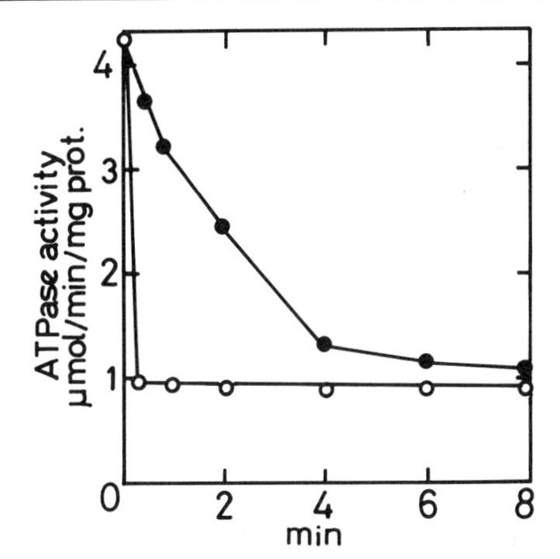

FIG. 7. Rapid inactivation of F_1F_0-ATPase in the presence of 9K and 15K proteins. F_1F_0-ATPase (10 units) bound to submitochondrial particles activated by Na_2SO_4 treatment was incubated at 0° in 0.2 ml of 25 m*M* Tris-SO_4, pH 7.2. Reactions were initiated by the simultaneous additions of 0.5 μmol of ATP-Mg^{2+} and 3 μg (120 units) of ATPase inhibitor in the presence (○) and absence (●) of 10 μg each of the 9K and 15K proteins. Samples of 10 μl were taken at the indicated times for measurement of ATPase activity.

gradually reactivated on removal of external ATP and Mg^{2+}. Figure 6 shows that the reactivation is completely suppressed by the additions of both the 9K and 15K proteins. Addition of either protein alone causes incomplete stabilization of the inactivated enzyme. The two stabilizing proteins, in addition to the stabilizing activity, exert their effect by facilitating binding of the inhibitor protein to F_1F_0-ATPase. When a limited amount of the inhibitor protein is added alone to submitochondrial particles in the presence of ATP and Mg^{2+}, maximum inhibition is observed only after several minutes. This slow binding of the inhibitor to the enzyme raises doubt about its regulatory action *in situ*.[15] However, when the 9K and 15K proteins are added with the inhibitor, formation of the inactivated complex is complete within a few seconds (Fig. 7). Furthermore, the binding is accomplished much faster when the two factors are preincubated with the enzyme before the addition of the inhibitor.

Characteristics of the Inactivated Complex of F_1F_0-ATPase with Inhibitor and the 9K and 15K Proteins

The stable inactivated complex of F_1F_0-ATPase with the purified inhibitor and the 9K and 15K stabilizing factors probably represents the

[15] P. L. Pedersen, K. Schwerzmann, and N. Cintron, *Curr. Top. Bioenerg.* **11**, 149 (1981).

actual form of the enzyme in the mitochondrial membrane *in situ*. In mitochondria *in vivo,* the enzyme catalyzes only synthesis of ATP, not ATP hydrolysis, although the hydrolysis can be brought about by coupling with other energy-requiring reactions in isolated mitochondria under nonphysiological conditions where respiratory activity is blocked. Furthermore, all three proteins are found to be present in mitochondria in equimolar ratios to the F_1 enzyme. Therefore, it may be concluded that in mitochondria F_1F_0-ATPase forms a regulatory complex composed of ATPase inhibitor and 9K and 15K proteins, in addition to the F_1 and F_0 complex. The proposed structure is valid only for the mitochondrial enzyme, but not for bacterial F_1F_0-ATPase, since no proteins corresponding to the three protein factors are found in bacteria.[16] Another characteristic feature of the inactivated complex of ATPase is its mode of nucleotide binding as substrate and product. Formation of the complex takes place in the presence of ATP and Mg^{2+}, but does not require the hydrolysis of ATP.[5] This suggests that the inhibitor does not bind to the enzyme alone, but to the enzyme–substrate complex. Actually, the inactivated complex was shown to bind 1 mol each of ATP and ADP,[5,17] indicating that two of three substrate-binding sites of the enzyme molecule are occupied by substrate and product, respectively, and the remaining site is vacant. This nucleotide binding pattern coincides with that proposed by Gresser *et al.*,[18] assuming cooperative interactions between the three catalytic sites, on F_1-ATPase. Thus, it is most probable that F_1F_0-ATPase is fixed in an active state in the inactivated complex.

[16] M. Futai and H. Kanazawa, *Microbiol. Rev.* **47,** 285 (1983).
[17] G. Klein, J. Lunardi, and P. V. Vignais, *Biochim. Biophys. Acta* **636,** 185 (1981).
[18] M. J. Gresser, J. A. Myers, and P. D. Boyer, *J. Biol. Chem.* **257,** 12030 (1982).

[51] Isolation and Reconstitution of CF_0-F_1 from Chloroplasts

By URI PICK

Proton-translocating ATPases (CF_0-F_1)[1] are the universal ATP-synthesizing machinery of energy-transducing membranes which utilize the protonmotive force created by electron-transport reactions for ATP syn-

[1] Abbreviations: Chl, chlorophyll; CF_0-F_1, chloroplast ATP synthase complex; crude CF_0-F_1, CF_0-F_1 partially purified by precipitation between $37\frac{1}{2}\%$ and 45% ammonium sulfate;

thesis. Proton–ATPase complexes were isolated and purified from a variety of energy-transducing membranes including mitochondria, chloroplasts, algae, yeast, and bacteria. All proton ATPase resemble each other in their basic structure, in their sensitivity to the inhibitor dichlorohexylcarbodiimide, and probably also in their catalytic mechanisms. The aim of this work is to provide a detailed description of the isolation, purification, and reconstitution of CF_0-F_1 from chloroplasts of higher plants. Comprehensive reviews about the structure[2] and properties[3] of CF_0-F_1 were published before. A special section will be devoted to specific lipid requirements for reconstitution of chloroplast CF_0-F_1.

Purification of CF_0-F_1

Preparation of Broken Chloroplasts

The following procedure adapted from Strotmann *et al.*[4] is recommended for preparation of thylakoids from higher plants and algae, since it removes most ribulose-bisphosphate carboxylase, which is the major contamination in CF_0-F_1 preparations.

Washed leaves (200 g) are ground with 150 ml ice-cold solution consisting of 0.2 *M* sucrose, 0.1 *M* NaCl, 50 m*M* Na-Tricine (pH 8), 10 m*M* Na ascorbate, and 10^{-4} *M* *p*-chloromercuribenzene sulfonate (a protease inhibitor). The slurry is filtered through three layers of cheesecloth and one layer of Miracloth paper. Thylakoid membranes are collected at 6000 g (10 min, 2°), the supernatant is discarded, and the thylakoids are resuspended in 10 m*M* Na pyrophosphate, pH 8 (original volume), and collected at 9000 g (10 min, 2°). The pyrophosphate wash is repeated 5 times, followed by a final wash in 0.2 *M* sucrose, 50 m*M* Na-Tricine (pH 8), 3 m*M* $MgCl_2$, 3 m*M* KCl (8000 *g*, 10 min, 2°). The chloroplasts are resuspended in a minimal volume of the sucrose/Mg buffer containing also 50 m*M* DTT, chlorophyll (Chl) concentration is determined, and the final membrane concentration is adjusted to 4 mg Chl/ml. The chloroplast-DTT suspension is left on ice for 15–30 min before starting the solubilization.

Na-cholate–SG, CF_0-F_1 purified on an Na-cholate sucrose gradient; Triton X-100–SG, CF_0-F_1 purified on a Triton X-100 sucrose gradient; BR, bacteriorhodopsin; DTT, dithiothreitol; DCCD, dicyclohexylcarbodiimide; PC, phosphatidylcholine; PE, phosphatidylethanolamine; PS, phosphatidylserine; PG, phosphatidylglycerol; MG, monogalactosyldiacylglycerol; DG, digalactosyldiacylglycerol; SQ, sulfoquinovosyldiacylglycerol.

[2] B. A. Baird and G. G. Hammes, *J. Biol. Chem.* **251,** 6953 (1976).

[3] N. Nelson *Curr. Top. Bioenerg.* **11,** 1 (1981).

[4] H. Strotmann, H. Hesse, and K. Edelman, *Biochim. Biophys. Acta* **314,** 202 (1973).

Solubilization

The chloroplast suspension is placed on ice with gentle mixing and the following solutions are added in the order listed below.

Solution	ml/100 ml	Final concentration
Chloroplast-DTT suspension	50.0	2 mg Chl/ml
Sucrose buffer		
0.8 *M* sucrose		0.2 *M*
12 m*M* $MgCl_2$	12.5	3 m*M*
12 m*M* KCl		3 m*M*
Saturated ammonium sulfate	10.0	10%
H_2O	16.8	
DTT (0.5 *M*)	5.0	50 m*M*
20% Na-cholate	1.7	0.33%
Octylglucoside (0.5 *M*)	4.0	20 m*M*

The presence of ammonium sulfate during solubilization is essential for higher plants,[5,6] but should be avoided in most algal species.[7] After 15 min the thylakoid membranes are removed by centrifugation at 230,000 *g* for 45 min at 2°. The volume of the straw-colored supernatant, which contains the solubilized CF_0-F_1 and most of the lipids, is determined. Saturated ammonium sulfate solution is added to a final concentration of 37.5%. After 5 min at 4° the precipitate is removed by centrifugation (12,000 *g* for 10 min at 2°), and the supernatant is adjusted to 45% ammonium sulfate. After 15 min at 4° the precipitate which contains CF_0-F_1 is collected by centrifugation (12,000 *g* for 10 min at 2°), carefully dried with Whatman filter paper to remove traces of detergents, dissolved in a small volume (about 3 ml) of 0.2 *M* sucrose, 20 m*M* Na-Tricine (pH 8), 3 m*M* $MgCl_2$ (final protein concentration of 20–25 mg/ml), and stored in liquid nitrogen at −196°. This preparation (crude CF_0-F_1) contains 30–60% CF_0-F_1 ATPase protein, the cytochrome b_6-f complex, traces of chlorophyll-binding proteins, and ribulose bisphosphate carboxylase. It also contains about 10% lipids (Table II). It can be stored for at least 12 months without significant loss of activity. This procedure is applicable for preparation of CF_0-F_1 from higher plants[5,6] and from algae.[7]

[5] U. Pick and E. Racker, *J. Biol. Chem.* **254,** 2793 (1979).

[6] U. Pick, *in* "Methods in Chloroplast Molecular Biology" (M. Edelman, R. B. Hallick, and N. H. Chua, eds.), p. 873. Elsevier, Amsterdam, 1982.

[7] M. Finel, U. Pick, S. Selman-Reimer, and B. R. Reimer, *Plant Physiol* **74,** 766 (1984).

TABLE I
PURIFICATION OF CF_0-F_1 FROM SPINACH CHLOROPLASTS

Preparation	Protein (mg)	$^{32}P_i$–ATP exchange (nmol/mg protein/min)	Recovery (%)
Chloroplast thylakoid	1330	50[a]	(100)
Crude CF_0-F_1	130	150	29.3
Na-cholate–SG CF_0-F_1	35	420	22.0
Triton X-100–SG CF_0-F_1	40	260	15.6

[a] Light + DTT-triggered activity in thylakoid.

Purification on a Sucrose Gradient

Samples of the crude CF_0-F_1 (3 mg protein) are diluted with 0.4 ml containing 20 m*M* Tris-succinate (pH 6.5), 0.5 m*M* Na-EDTA, 0.1 m*M* ATP, and 0.2% Triton X-100. The mixture is layered onto 4 ml of a sucrose gradient in Spinco SW-60 T rotor tubes. Two possible gradients can be used; a linear 7–30% sucrose gradient with 0.2% Triton X-100 and 0.1% sonicated soybean lipids or a linear 7–40% sucrose gradient with 0.4% Na-cholate and 0.1% sonicated soybean lipids and ingredients as in the dilution medium (above). The gradients are centrifuged 5 hr at 150,000 *g* at 2°. The peak CF_0-F_1 protein appears in the Triton X-100 gradients as a sharp band (1–2 mg/ml) around 40% down the tube and in the Na-cholate gradients as a diffuse band (0.3–0.5 mg/ml) around 60–80% down the tube. The enzyme can be stored in liquid nitrogen without further treatment or after reconstitution with phospholipids (see below) and is stable for several months. A summary of the purification steps of spinach CF_0-F_1 is demonstrated in Table I. The purified ATPase complex from higher

TABLE II
LIPID CONTENT AND COMPOSITION OF SPINACH CF_0-F_1 PREPARATIONS[16]

Preparation	Polar lipid content (%)	Lipid/protein (w/w)	Polar lipid composition (mol%)[a]			
			MG	DG	PG	SQ
Chloroplast thylakoids	(100)	0.5	45	25	13	8
Crude CF_0-F_1	1.6	0.08	40.5	13	17	28
Triton X-100–SG CF_0-F_1	0.1	0.012	0	0	0	100

[a] MG, Monogalactosyldiacylglycerol; DG, digalactosyldiacylglycerol; PG, phosphatidylglycerol; and SQ, sulfoquinovosyldiacylglycerol

plants consists of eight different polypeptides: five CF_1 polypeptides, α, β, γ, δ, and ε with MW 58,000, 56,000, 37,000, 17,500, and 13,500, respectively, and three CF_0 polypeptides I, II, and III (the DCCD binding protein) with MW 15,500, 12,500, and 7,500, respectively.[5,6] Purified CF_0-F_1 also contains a small amount of tightly bound sulfolipids (about 5 mol/mol enzyme) which comigrate with the enzyme on the sucrose gradient (Table II).

This procedure with minor modifications is applicable for purification of the proton–ATPase complexes from chloroplasts of most higher plants[6] of the algae *Dunaliella*[7] and *Chlamydomonas* and from mitochondria of yeast and rat liver.[3]

Reconstitution Procedures

The most stringent criterion for the functional integrity of CF_0-F_1 is the capacity to catalyze ATP synthesis coupled to proton translocation which requires the reconstitution of the enzyme into lipid vesicles.

CF_0-F_1 proteoliposomes catalyze several partial reactions which are coupled to proton translocation and are sensitive to uncouplers and to DCCD: $^{32}P_i$–ATP exchange,[5] ATP synthesis driven by an artificial pH gradient[5] or by light in bacteriorhodopsin (BR)-proteoliposomes,[8] and ATP-induced proton uptake.[9] $^{32}P_i$-ATP exchange is recommended as a standard assay because the reconstitution is simple and the assay is more sensitive than other assays.

Since all three preparations of CF_0-F_1 are useful for different purposes and since there are significant differences in the effectiveness of reconstitution of these preparations by different techniques, we describe below three recipes which in the author's hands give the best results with the different preparations of CF_0-F_1. A fourth procedure describes the coreconstitution of BR and CF_0-F_1 into the same proteoliposomes for measurement of light-driven ATP formation.

Reconstitution of Crude CF_0-F_1 Preparation

This partially purified preparation is useful for studies of the catalytic properties of the enzyme. The highest rates of $^{32}P_i$–ATP exchange are obtained by reconstitution by the cholate-dilution procedure.[10]

[8] G. D. Winget, N. Kanner, and E. Racker, *Biochim. Biophys. Acta* **460,** 490 (1977).

[9] A. Admon, U. Pick, and M. Avron, *J. Membr. Biol.* **86,** 45 (1985).

[10] E. Racker, T. E. Chien, and A. Kandrach, *FEBS Lett.* **57,** 14 (1975).

Acetone-washed soybean phospholipids[11] are sonicated to clarity under nitrogen in a bath-type sonicator in 1 ml containing 1.4% Na-cholate, 0.2 *M* sucrose, and 20 m*M* Na-Tricine (pH 8). To 850 μl sonicated phospholipids are added 3 μl of 1 *M* $MgCl_2$ and 15 μl freshly thawed enzyme (about 3 mg protein). The mixture is incubated 30 min at 4° and the assay is initiated by dilution of 20 μl into 1 ml $^{32}P_i$–ATP exchange medium. Typical rates of $^{32}P_i$–ATP exchange obtained by this reconstitution procedure are 100–200 nmol $^{32}P_i$ incorporated/mg protein/min (Table I).

Reconstitution of Na-Cholate–SG Purified CF_0-F_1

This preparation, which has the highest specific activity in $^{32}P_i$–ATP exchange, can be reconstituted immediately after the sucrose gradient purification step by exclusion of the cholate on a Sephadex column.

To 2 ml purified CF_0-F_1 recovered from the Na-cholate sucrose gradient (about 0.8 mg protein) are added 0.2 ml of 1 *M* Na-Tricine (pH 8) and 60 μl of 100 m*M* $MgCl_2$; the mixture is incubated 30 min at 4° and applied to a 25 × 1 cm Sephadex G-50 (course) column preequilibrated with 0.2 *M* sucrose, 20 m*M* Na-Tricine (pH 8), and 3 m*M* $MgCl_2$ at 23°. CF_0-F_1 proteoliposomes are eluted from the column at a flow rate of 3 ml/min as a slightly turbid suspension. Before the assay of $^{32}P_i$–ATP exchange the enzyme is incubated for 30 min with 50 m*M* DTT at 4°.[5] This step is essential for reactivation of the enzyme since during the sucrose gradient purification the enzyme reverts to the latent ATPase state due to oxidation of thiol groups in the γ subunit. Samples of 100 μl containing 10–20 μg protein are taken to measure $^{32}P_i$–ATP exchange.[5,6] Typical rates are 215–400 nmol $^{32}P_i$ incorporated/mg protein/min.

Reconstitution of Triton X-100–SG Purified CF_0-F_1

The method of choice for reconstitution of this preparation is the freezing-thawing technique.[12]

Crude soybean phospholipids (40 mg/ml) are sonicated to clarity in 30 m*M* Na-Tricine (pH 8), 0.5 m*M* Na-EDTA. Sonicated phospholipids (0.2 ml) are mixed with 10 μl purified enzyme (containing 10–30 μg protein) at 4°, and the mixture is rapidly frozen in liquid nitrogen and left to thaw at room temperature. When completely thawed, the turbid suspension is transferred to 4°, incubated with 50 m*M* DTT, and assayed as described

[11] Y. Kagawa and E. Racker, *J. Biol. Chem.* **246,** 5477 (1971).
[12] M. Kasahara and P. C. Hinkle, *J. Biol. Chem.* **252,** 7384 (1977).

above. Typical rates of $^{32}P_i$–ATP exchange catalyzed by these proteoliposomes are 150–300 nmol $^{32}P_i$ incorporated/mg protein/min.

Co-reconstitution of CF_0-F_1 and Bacteriorhodopsin

This two-step procedure is useful for reconstitution of either crude or purified CF_0-F_1 with BR for measurements of light-driven ATP formation.

Soybean phospholipids (10 mg) are sonicated to clarity in 0.5 ml containing 30 m*M* Na-HEPES (pH 7.5) and 50 m*M* KCl. The sonicated phospholipids are mixed with 0.5 ml containing 2 mg BR in the same buffer; the mixture is rapidly frozen in liquid nitrogen, thawed at room temperature, and sonicated for 3 min. The process of freezing–thawing–sonication is repeated once again. To 1 ml BR proteoliposomes are added 3 μl $MgCl_2$/1 *M*, 50 μl Na-cholate/20%, and 0.5 mg CF_0-F_1, and the mixture is incubated for 30 min at 4°. Samples of 100 μl are transferred through Sephadex G-50 columns contained in 1-ml tuberculin syringes and pre-equilibrated with 30 m*M* Na-HEPES (pH 7.5), 0.2 *M* sucrose at 4° by the centrifugation procedure.[13] The resulting turbid suspension eluted from the columns contains the BR-CF_0-F_1 proteoliposomes. Typical rates of $^{32}P_i$ incorporation are 30–60 nmol/mg CF_0-F_1/min.

Assays

$^{32}P_i$–ATP Exchange[5,6]

The incorporation of $^{32}P_i$ into ATP is measured by incubation of CF_0-F_1 proteoliposomes (containing 10–50 μg protein) in 1 ml $^{32}P_i$–ATP exchange medium for 30 min at 37°. The medium contains 80 m*M* Na-Tricine (pH 8), 6 m*M* $MgCl_2$, 6 m*M* ATP, 5 m*M* DTT, and 3 m*M* Na-$^{32}P_i$ (phosphate, 1 μCi/μmol). The reaction is terminated by addition of 5% trichloroacetic acid, and the amount of $^{32}P_i$ incorporated into ATP is determined by extraction of the free $^{32}P_i$ with isobutanol–benzene.[14]

The incorporation of $^{32}P_i$ into ATP at 37° is linear with time for at least 30 min except for an initial lag of about 2 min. K_m (ATP) is 0.9 m*M* (soybean phospholipids) or 0.45 m*M* (chloroplast glycolipids)[15] and K_m (P_i) is 1.2 m*M* (soybean phospholipids).

Light-Dependent ATP Synthesis in CF_0-F_1–BR Proteoliposomes

ATP synthesis is measured by the incorporation of $^{32}P_i$ into ADP in the presence of hexokinase in the light. Samples of proteoliposomes contain-

[13] H. Penefsky, *J. Biol. Chem.* **252,** 2891 (1977).
[14] M. Avron, *Anal. Biochem.* **2,** 535 (1961).
[15] U. Pick, K. Gounaris, A. Admon, and J. Barber, *Biochim. Biophys. Acta* **765,** 12 (1984).

TABLE III
LIPID REQUIREMENTS FOR RECONSTITUTION OF CF_0-F_1[a]

	Rate (nmol/mg protein/min)	
Lipid composition[b]	$^{32}P_i$–ATP exchange	Mg-ATPase
Soybean phospholipids	115	200
PC	20	50
PE	15	—
PS	10	35
PC : PE (1 : 1)	45	—
PC : PE : PS (4 : 4 : 2)	62	—
Chloroplast lipids	80	540
Chloroplast lipids : PS (4 : 1)	125	460
MG	15	370
DG	5	120
SQ	3	30
MG : DG (2 : 1)	38	440
MG : DG : SQ (6 : 3 : 1)	95	390

[a] CF_0-F_1 (crude preparation) reconstituted by the cholate dilution reconstitution procedure.
[b] The ratios of lipid mixtures are weight ratios.

ing 50 μg CF_0-F_1 and 200 μg BR are incubated for 30 min at 30° in a thermostated waterbath illuminated by white light (10^6 ergs/cm^2/sec). The phosphorylation medium (1 ml) contains 40 mM Na-Tricine (pH 8), 30 mM KCl, 1 mM $MgCl_2$, 0.5 mM ADP, 1 mM $^{32}P_i$ (phosphate, 1 μCi/μmol), 20 mM glucose, and hexokinase (10 U/ml). The reaction is terminated by addition of trichloroacetic acid (5%) and incorporation of $^{32}P_i$ into glucose 6-phosphate is determined.[15]

Specific Lipid Requirements

Selection of lipids for the reconstitution of CF_0-F_1 is not a simple task, since the catalytic properties of the enzyme are influenced by the type of lipids used for reconstitution.

The reconstitution of $^{32}P_i$–ATP may be influenced by the permeability of the lipid vesicles to protons, by the efficiency of incorporation of the enzyme, and by specific interactions with lipids in the membrane.

Table III demonstrates a few examples of the effect of different phospholipids and glycolipids on Mg-ATPase and on $^{32}P_i$–ATP exchange activities of CF_0-F_1.

[16] U. Pick, K. Gounaris, M. Weiss, and J. Barber, *Biochim. Biophys. Acta* **808**, 415 (1985).

Chloroplast glycolipids activate higher rates of ATP hydrolysis and lower rates of $^{32}P_i$–ATP exchange than soybean phospholipids. The differences are due to a specific activation of CF_0-F_1 by MG, the major chloroplast glycolipid, on one hand, and to the higher permeability of chloroplast glycolipids to protons, on the other hand.[15] Chloroplast glycolipids supplemented with 10–20% PS have a reduced permeability to protons and catalyze higher rates of $^{32}P_i$–ATP exchange. Individual phospholipids or glycolipids are inefficient in reconstitution of $^{32}P_i$–ATP exchange. Optimal reconstitutions are obtained by mixtures of PC : PE : PS (4 : 4 : 2) or of MG : DG : SQ (6 : 3 : 1).

Acknowledgment

This work was supported by a grant from the United States–Israel Binational Science Foundation, Jerusalem, Israel.

[52] Extraction, Purification, and Reconstruction of the Chloroplast *N,N'*-Dicyclohexylcarbodiimide-Binding Proteolipid

By ANGELO AZZI, KRISTINE SIGRIST-NELSON, and NATHAN NELSON

Introduction

The ATPase complex of chloroplasts, mitochondria, and bacterial membranes is composed of two distinct parts: F_1, a peripheral membrane protein complex, provided with the ATPase activity, and F_0, an intrinsic hydrophobic protein complex necessary for the ATP synthase activity. The peripheral part of the ATPase can be removed from the membrane by mild treatment while the intrinsic sector can be isolated by procedures using detergents or nonaqueous solvents.[1] The identification of F_0 has been made originally through the use of *N,N'*-dicyclohexylcarbodiimide (DCCD), a reagent which was shown to inhibit ATPase activity of the F_0-F_1 complex, but not that of the F_1 alone.[2] DCCD was also shown to reduce the proton permeability of the membrane from which F_1 had been

[1] N. Nelson, *Curr. Top. Bioenerg.* **11,** 1 (1981).

[2] C. T. Holloway, A. M. Robertson, I. G. Knight, and R. B. Beechey, *Biochem. J.* **100,** (1966).

removed, suggesting that the action of the inhibitor was to block a site(s) of the F_0 that participates in proton translocation.[3]

It has been shown that DCCD binds only to one of the F_0 subunits, which is called DCCD-binding proteolipid, or simply the proteolipid.[4] An extraction, purification, and reconstitution procedure for the chloroplast proteolipid will be described below. The isolation is initiated by a one-step single-phase 1-butanol extraction.[5]

Other lipophilic components of the chloroplast membrane, such as chlorophyll, are simultaneously extracted with the proteolipid. This preparation, when used to form liposomes, however, is suitable to demonstrate proteolipid-mediated proton permeability.[6] Spectroscopic measurements involving fluorescent probes, optical absorption, and spin labels are, in contrast, difficult if not impossible with such a preparation. Separation of the proteolipid from coextracted agents while retaining functional activity may be achieved by a relatively easy chromatographic procedure.

Preparation and Labeling of the Chloroplast Membranes

Preparation of Chloroplasts

Lettuce (*Lactuca scarida* var. *longifolia*) chloroplast membranes are isolated according to a published method.[7] The chloroplast membranes are suspended in 10 m*M* Tricine (pH 8), at chlorophyll concentration of about 3 mg/ml.

Labeling of Membranes with DCCD

The proteolipid of chloroplast membranes can be most conveniently followed by the presence of covalently bound *N,N'*-dicyclohexyl[^{14}C]carbodiimide.

For this purpose, isolated chloroplast membranes are incubated with *N,N'*-dicyclohexyl[^{14}C]carbodiimide (45 mCi/mmol, 30 nmol/mg of protein) for 2 hr at 25°. They are recovered by centrifugation and washed twice with 10 m*M* Tricine (*N*-[2-hydroxy-1,1-bis(hydroxymethyl) ethyl]glycine), pH 8.

[3] L. Patel, S. Schuldiner, and H. R. Kaback, *Proc. Natl. Acad. Sci. U.S.A.* **72,** 3387 (1975).

[4] K. Cattell, C. Lindop, I. Knight, and R. Beechey, *Biochem. J.* **125,** 169 (1971).

[5] H. Sigrist, K. Sigrist-Nelson, and C. Gitler, *Biochem. Biophys. Res. Commun.* **74,** 178 (1977).

[6] N. Nelson, E. Eytan, B. Notsani, H. Sigrist, K. Sigrist-Nelson, and C. Gitler, *Proc. Natl. Acad. Sci. U.S.A.* **74,** 2375 (1977).

[7] A. Kamienietzky and N. Nelson, *Plant Physiol.* **55,** 282 (1975).

Purification of the Proteolipid

1-Butanol Extraction of the Proteolipid[5,6]

Two milliliters of a chloroplast (12 mg protein) suspension in water is injected, using a tuberculin syringe, into 100 ml of 1-butanol at 0–4°, under vigorous stirring. The precipitate is removed by centrifugation (two times, at 10,000 *g* for 10 min) and filtered through a Sartorius filter (11305, 0.6 μM) to ensure complete removal of precipitated protein. The butanol, containing the proteolipid and lipid, is concentrated in a rotary evaporator with the condenser cooled to −20° under high vacuum.

Diethyl Ether Precipitation of the Proteolipid

Diethyl ether (precooled at −20°) is added to the 1-butanol supernatant (5 volumes per volume of 1-butanol). The precipitated proteolipid is recovered by centrifugation (10,000 *g* for 10 min at 4°). The precipitate is washed with water, 1-butanol, and cold diethyl ether in three subsequent washes.

Chromatographic Purification of the 1-Butanol-Extracted Proteolipid[8]

DEAE-cellulose (Cellex D, Bio-Rad, Bromley, Kent, UK) is treated essentially as described.[9] The resin is washed with 1 *M* HCl, water, and with 0.1 *M* KOH, again followed by water. The procedure is repeated three times. The resin is then further washed with 3 volumes of acetic acid and 3 volumes of methanol, after which it is dried. After washing, DEAE-cellulose is converted into the acetate form by suspension in acetic acid and the column packed. The column is then equilibrated with 1-butanol and is ready for use.

The concentrated butanol supernatant of *N,N'*-dicyclohexyl[^{14}C]carbodiimide-labeled membranes is applied to the DEAE-cellulose column (column volume 19 ml) and eluted with 1-butanol. The first eluted peak (Fig. 1) is broad and has an intense green color. At least 2 column volumes of 1-butanol should be applied to the column until no additional radioactivity or increase in absorbance is noted. The column is then eluted with 1-butanol/formic acid (50 : 1, v/v). A second peak (Fig. 1) of radioactivity and a corresponding small increase in material absorbing at 645 nm is eluted. Further washing of the column with 1-butanol/formic acid does not elute additional material.

[8] K. Sigrist-Nelson and A. Azzi, *Biochem. J.* **177,** 687 (1979).

[9] G. Rouser, G. Kritcchenvsky, A. Yamamoto, G. Simon, C. Galli, and A. Baumann, this series, Vol. 14, p. 272.

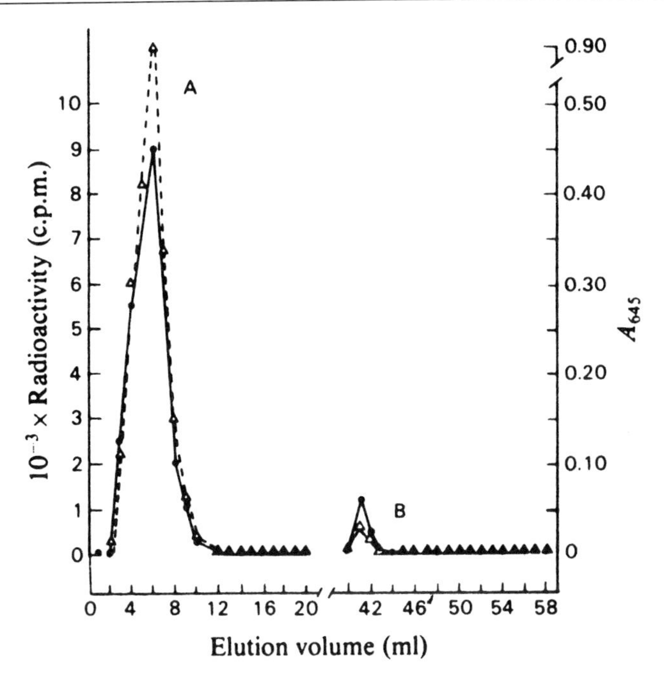

FIG. 1. DEAE-cellulose chromatography of the butanol-extracted dicyclohexylcarbodiimide-binding proteolipid. Chloroplast membranes (12 mg) were labeled with dicyclohexyl-[^{14}C]carbodiimide and the proteolipid was extracted with 1-butanol, as described in the text. Butanol supernatant was applied to the column (column volume 19 ml) and eluted with 1-butanol. Samples of column fractions were taken for A_{645} (△) and radioactivity (●) measurements. (A) The material eluted with 1-butanol. After at least 2 column volumes the eluent was changed. (B) The peak eluted with 1-butanol/formic acid (50 : 1, v/v). From Sigrist-Nelson and Azzi[8] with permission.

Characterization of the Extracted Proteolipid

It is possible to detect the ^{14}C-labeled proteolipid by electrophoresis in slab gels followed by fluorography; however, the proteolipid has the tendency to diffuse out from these gels and therefore, sodium dodecyl sulfate (SDS)–polyacrylamide gel electrophoresis is better carried out, as described by Weber and Osborn.[10] Two methods are used to ascertain the specificity of the *N,N'*-dicyclohexyl[^{14}C]carbodiimide labeling. In one case, when *N,N'*-dicyclohexyl[^{14}C]carbodiimide-labeled samples are electrophoresed, identical samples are applied to separate gels. One gel is stained; the duplicate is not stained, but is immediately cut into 1-mm slices. The individual gel slices are extracted with 10 m*M* Triton X-100

[10] K. Weber and M. Osborn, *J. Biol. Chem.* **244,** 4406 (1969).

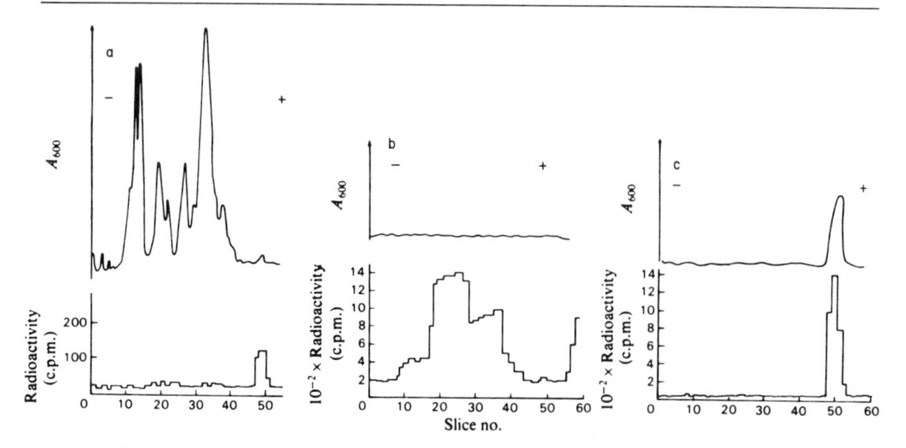

FIG. 2. Isolation of the dicyclohexyl[^{14}]carbodiimide-binding protein from chloroplast membranes: Distribution of protein radioactivity in sodium dodecyl sulfate–polyacrylamide gels. (a) Butanol precipitate (60 μg of protein); (b) ether supernatant (equivalent of 10 μg of protein); (c) ether precipitate (10 μg of protein). The cathode (–) and anode (+) are indicated. From Sigrist-Nelson and Azzi[8] with permission.

overnight. Scintillation fluid is added and the slices are counted for radioactivity. The extraction of radioactivity from Coomassie Brilliant Blue-stained gels is also carried out. After staining and destaining procedures the gels are cut into 1-mm slices and the following mixture is added: Lipoluma/Lumasolve/water (50 : 5 : 1, by volume). The gel slices are extracted either overnight at room temperature (20°) or for 5 hr at 40–50°. The slices are then counted for radioactivity directly.

1-Butanol-Extracted and Ether-Precipitated Proteolipid

Dodecyl sulfate–polyacrylamide gels of the labeled chloroplast membranes and of the butanol extract after diethyl ether precipitation are shown in Fig. 2a,b. Distribution of radioactivity is indicated underneath each trace.

The small peak in the chloroplast membrane gel pattern with associated radioactivity is extracted and precipitated by addition of ether and corresponds directly with the Coomassie Brilliant Blue staining proteolipid band whose apparent molecular weight is 8000 (Fig. 2c). Purity and homogeneity of the *N,N'*-dicyclohexylcarbodiimide-binding proteolipid have been demonstrated by high-voltage electrophoresis and N-terminal amino acid determination.[6,8,11]

[11] K. Sigrist-Nelson and A. Azzi, *J. Biol. Chem.* **255,** 10638 (1980).

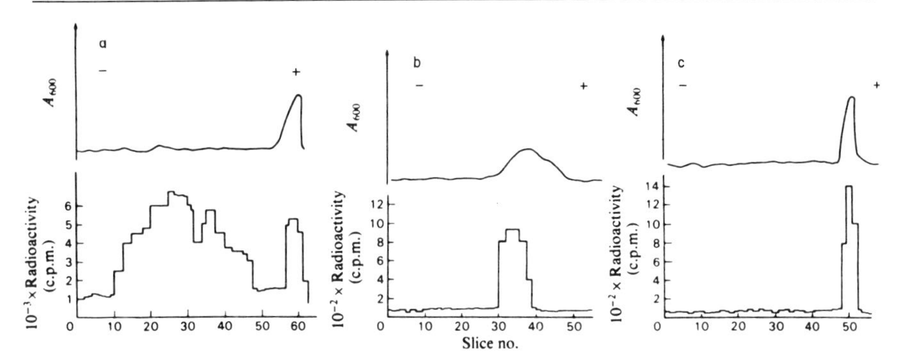

FIG. 3. Electrophoretic analysis of fractions eluted from the DEAE-cellulose column. Chloroplast membranes were incubated with dicyclohexyl[^{14}C]carbodiimide, extracted with 1-butanol, and applied to the ion-exchange column, as indicated in Fig. 1. Radioactivity was extracted immediately after electrophoresis; duplicate gels were stained. Distribution of radioactivity is shown underneath each densitometric trace. The cathode (−) and anode (+) are indicated. (a) The peak obtained (A, Fig. 1) with 1-butanol as eluent; (b) the second peak (B, Fig. 1) obtained with butanol/formic acid (50 : 1, v/v) (15 μg of protein); (c) the ether-precipitated fraction of peak B (Fig. 1) (10 μg of protein). See the text for details. From Sigrist-Nelson and Azzi[8] with permission.

DEAE-Cellulose Chromatography Purified Proteolipid[8]

SDS–polyacrylamide gels of the fractions eluted from the DEAE-cellulose column (see Fig. 1) indicate that fraction A contains a green chlorophyll band (the absorbance visible near the anode), but no Coomassie Brilliant Blue staining band (Fig. 3a). The radioactivity present, in the form of three large peaks when the gels are sliced and extracted directly after electrophoresis, disappears when the gels are first stained and destained before extraction of radioactivity, indicating that the radioactivity is not covalently associated with protein. Fraction B contains a single peak of radioactivity whose mobility corresponds to a diffuse Coomassie Brilliant Blue staining band (Fig. 3b). The apparent molecular weight of the band ranges from 12,000 to 24,000. If the two peaks of radioactivity, A and B, are separately combined with 5 volumes of diethyl ether, neither radioactivity nor protein is precipitated from fraction A, while from fraction B, a precipitate containing approximately 1–2% of the total membrane protein and corresponding radioactivity is obtained (Fig. 3c).

The fraction eluted by butanol/formic acid is primarily composed of the dicyclohexylcarbodiimide-binding proteolipid, but contains also small amounts of lipid (possibly tightly associated with the protein: monogalactosyldiacylglycerol, two glycolipids, and two chlorophylls).

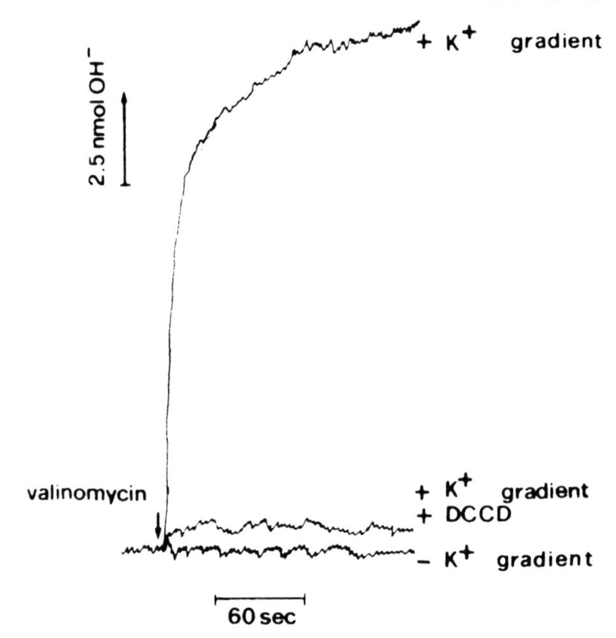

FIG. 4. Potentiometric measurement of proton uptake in proteolipid vesicles. K^+-loaded proteolipid vesicles (0.05 ml containing 10 nmol of proteolipid/5 μmol of lipid) were added to a water-jacketed cuvette containing 3 ml of 500 m*M* sucrose, 2.5 m*M* $MgSO_4$, 0.2 m*M* Tricine, pH 7.5. An aliquot of the same vesicle population was pretreated with DCCD as described in the text. Additionally, proton conduction was measured with vesicles (0.05 ml, same protein/liquid ratio as above) which were added to 3 ml of 296 m*M* KCl, 2.5 m*M* $MgSO_4$, 0.2 m*M* Tricine, pH 7.5. Proton uptake was started by addition of valinomycin. From Sigrist-Nelson and Azzi[11] with permission.

The apparent molecular weight of the proteolipid as inferred from polyacrylamide gel electrophoresis is ~8,000.

H^+ Conduction of the Reconstituted Proteolipid

Reconstitution of the Proteolipid[11]

Proton conduction mediated by the proteolipid reconstituted in phospholipid vesicles can be shown. An H^+ conduction has been demonstrated to be associated also with the proteolipid from yeast mitochondria, reconstituted in lipid planar membranes.

Partially purified soybean phospholipids, 10 mg, are weighed out and added to a glass vessel containing 5 mg of isolated glycolipid and 1 ml of a solution containing cholate (1.6%), deoxycholate (0.8%), EDTA (0.2

mM), Tricine (30 mM, pH 8), and dithiothreitol (7.5 mM). Precautions should be taken that cholate and deoxycholate are recrystallized before usage and that dithiothreitol is freshly weighed out for each preparation. The mixture is sonicated in a bath-type sonicator for 2 min at 22° and then can be added to the vessel containing 300 μg of chromatographically isolated proteolipid protein. The lipid, detergent, and proteolipid are sonicated together again for 20 sec under nitrogen in the mentioned sonicator. The mixture is transferred to dialysis tubing and dialyzed against 1 liter of a solution containing 10 mM Tricine, 2.5 mM $MgSO_4$, 0.3 mM dithiothreitol for 2 hr at room temperature. After changing the dialysis medium, the dialysis is continued overnight at 4°. The lipid vesicles (0.5 ml) can be loaded with KCl by incubating them in 3 ml of 300 mM KCl, 0.1 ml of 100 mM EDTA, pH 7.0, and 0.02 ml of 500 mM dithiothreitol, pH 7.0. After centrifugation (103,000 g for 30 min), the above solution is added again and the vesicles are incubated for 30 min at 30° and then cooled on ice for 15 min. $MgSO_4$ is added to a final concentration of 2.5 mM and pelleted at 103,000 g for 30 min. A suspension of the vesicles in 0.2 ml of 500 mM sucrose, 2.5 mM $MgSO_4$, is made.

Proton Conduction Measurements

The most direct way to measure the proton permeability of the reconstituted vesicles is by following the changes of pH in the suspension medium using a pH electrode after addition of valinomycin. The effect of the ionophore in initiating a rapid, uncompensated K^+ efflux from the vesicles is that of creating a transmembrane diffusion potential, negative inside. If the membrane is rendered permeable to protons due to the presence of the proteolipid, a charge compensation by this species will occur, which, flowing inside the vesicles, will produce an increase of the pH of the incubation medium. Such an effect should be sensitive to DCCD. Such a result is shown in Fig. 4.

Conclusions

The techniques described in this chapter are suitable for isolating and reconstituting in an active form the proteolipid subunit of chloroplast ATPase. The question remains open as to the possibility that the isolated proteolipid is only partly active after reconstitution and that other subunits are needed to confer to it full catalytic competence.

[53] Selective Extraction and Reconstitution of F_1 Subunits from *Rhodospirillum rubrum* Chromatophores

By ZIPPORA GROMET-ELHANAN and DANIEL KHANANSHVILI

The F_1-ATPase isolated from membranes of mitochondria, bacteria, and chloroplasts is very similar, containing five nonidentical polypeptide subunits: α, β, γ, δ, and ε.[1] Elucidation of its mechanism of action requires a precise determination of the function of each of its individual subunits. This has become possible with the development of methods for the isolation of single subunits, in a native active state, that can be reconstituted into an active complex. Two main methods have been developed so far. The first involves dissociation of the F_1 complex into its individual subunits.[2–4] After their separation and purification a soluble ATPase activity has been reconstituted by reassociation of mixtures of such isolated subunits.[3,4] This method has been developed in various bacterial systems, but has also been applied to mitochondrial F_1.[5,6] The second method involves a sequential removal of single subunits from the membrane-bound F_1, leaving a fully reconstitutable membrane lacking only the removed subunits.[7–10] Reconstitution of the extracted purified subunits into the depleted membranes leads to restoration of their ATP synthesis as well as hydrolysis activities.[8–11] This method enables also a determination of energy-linked reactions, such as light-induced proton uptake, that remain active in the depleted membranes.[7,8,10,11] It has been developed in chromatophores of the photosynthetic bacterium *Rhodospirillum rubrum* and has also been tested in chloroplasts[12] and submitochondrial particles.[13]

[1] L. M. Amzel and P. L. Pedersen, *Annu. Rev. Biochem.* **52,** 801 (1983).
[2] G. Vogel and R. Steinhart, *Biochemistry* **15,** 208 (1976).
[3] M. Yoshida, N. Sone, H. Hirata, and Y. Kagawa, *J. Biol. Chem.* **252,** 3480 (1977).
[4] M. Futai, *Biochem. Biophys. Res. Commun.* **79,** 1231 (1977).
[5] G. J. Verschoor, P. R. Van Der Sluis, and E. C. Slater, *Biochim. Biophys. Acta* **462,** 438 (1977).
[6] N. Williams, J. M. Hullihen, and P. L. Pedersen, *Biochemistry* **23,** 780 (1984).
[7] Z. Gromet-Elhanan, *J. Biol. Chem.* **249,** 2522 (1974).
[8] A. Binder and Z. Gromet-Elhanan, *Proc. Int. Congr. Photosynth. 3rd,* **II,** 1163 (1974).
[9] S. Philosoph, A. Binder, and Z. Gromet-Elhanan, *J. Biol. Chem.* **252,** 8747 (1977).
[10] S. Philosoph, D. Khananshvili, and Z. Gromet-Elhanan, *Biochem. Biophys. Res. Commun.* **101,** 384 (1981).
[11] D. Khananshvili and Z. Gromet-Elhanan, *J. Biol. Chem.* **257,** 11377 (1982).
[12] N. Nelson and R. Broza, *Eur. J. Biochem.* **69,** 203 (1976).
[13] I. A. Kozlov, Y. M. Milgrom, and I. S. Tsyborski, *Biochem. J.* **192,** 483 (1980).

Using *R. rubrum* chromatophores, two F_1 subunits, β and γ, have been extracted in two consecutive steps, leaving all other F_1 subunits attached to the chromatophore membrane. The procedures developed for the extractions, for purification of the isolated subunits, and for their reconstitution into the depleted chromatophores are described here.

Assay of the Reconstitutive Activity of the Isolated β and γ Subunits

The isolated β and γ subunits have no activity by themselves,[9,11] although their extraction leads to complete loss of photophosphorylation and ATPase activities of the depleted chromatophores.[7–11] The subunits are therefore identified during the extraction and purification procedures by their reconstitutive activity. This is defined as their capacity to rebind to depleted chromatophores and restore their lost activities. The experimental system for measuring the reconstitutive activity involves two steps: (a) reconstitution of the isolated subunits into the depleted chromatophores (β-less or β,γ-less), and (b) assay of the reconstituted chromatophores for restored activities.

Reagents

Reaction mixture for reconstitution of β subunits into β-less chromatophores: 50 m*M* Tricine-NaOH (pH 8.0), 25 m*M* $MgCl_2$, 4 m*M* ATP, β-less chromatophores containing 10 μg of bacteriochlorophyll (Bchl) and between 30 and 300 μg of β subunit at various stages of purification, in a total volume of 0.5 ml. Control chromatophores are incubated in the same reaction mixture, except that no β is added.

Reaction mixture for reconstitution of β and γ subunits into β,γ-less chromatophores: 10 m*M* Mes-NaOH (pH 6.5), 25 m*M* $MgCl_2$, 4 m*M* ATP, β,γ-less chromatophores containing 10 μg of Bchl, 100 μg of purified β subunit, and between 30 and 100 μg of γ subunit at various stages of purification, in a total volume of 0.7 ml.

Reaction mixture for photophosphorylation: 80 m*M* Tricine-NaOH (pH 8.0), 5 m*M* $MgCl_2$, 4 m*M* sodium phosphate containing $2–5 \times 10^6$ cpm of ^{32}P, 2 m*M* ADP, 15 m*M* glucose, 10 units of hexokinase, 66 μM *N*-methylphenazonium methosulfate, and chromatophores containing 10 μg of Bchl, in a total volume of 3.0 ml.

Reaction mixture for ATP hydrolysis: 50 m*M* Tricine-NaOH (pH 8.0), 4 m*M* ATP containing $2–5 \times 10^5$ cpm of [γ-^{32}P]ATP, either 2 m*M* $MgCl_2$ or 4 m*M* $CaCl_2$, and chromatophores containing 2–10 μg of Bchl in a total volume of 1 ml.

Perchloric acid, 50%

Reconstitution. The reaction is started by addition of chromatophores and incubated for 60 min at 35°. Reconstitution of β into β-less chromatophores requires the presence of at least 2 mM ATP[8,9] and is optimal at a pH range of 6.5–8.0.[14] It follows a simple saturation curve, so that β-less chromatophores containing 10 μg of Bchl are fully reconstituted with 100 μg of purified β.[11] Reconstitution of β,γ-less chromatophores, which is optimal at pH 6.5,[14] requires the presence of both β and γ subunits.[10] Their order of addition is not important, but their relative concentrations are crucial for obtaining maximal restoration of ATP-linked activities.[11,14] Therefore, a saturating concentration of 100 μg of β subunit is added together with the stated concentrations of γ subunit. At least 20 mM $MgCl_2$ is required for optimal reconstitution of both β-less and β,γ-less chromatophores.[7,14] This concentration has been found to inhibit ATP synthesis and hydrolysis in *R. rubrum* chromatophores.[7,15] The reconstitution step is therefore carried out in a small volume that is diluted by the assay mixture to reach the stated optimal concentrations of all components.

When the reconstitution is carried out at pH 6.5, concentrated Tricine-NaOH (pH 8.0) is added to a final concentration of 80 mM and the suspension is equilibrated for 15 min at 4° before being assayed. The reconstituted chromatophores can also be centrifuged at 4° for 30 min in the microfuge (Eppendorf 5414). The pellet is resuspended in 50–100 μl of Tricine-NaOH (pH 8.0) and assayed. The recovery of Bchl is usually 90–95%, and control rates of photophosphorylation or ATPase activity are not decreased after this centrifugation step. This treatment is useful for removal of all excess unbound β or γ subunits or of $MgCl_2$ when assaying Ca^{2+}-ATPase activity.

Assay of Restored ATP Synthesis and Hydrolysis. Both reactions are carried out at 35°. Photophosphorylation is started by illumination after a 5-min equilibration allowing the conversion of ATP into ADP by the hexokinase–glucose trap. It is stopped after 3 min by turning off the lights and adding 0.2 ml of 50% perchloric acid. After centrifugation at 2000 g for 20 min, 1.0 ml of a deproteinized supernatant is measured for [γ-^{32}P]ATP formation according to Avron.[16] ATPase is started by addition of [γ-^{32}P]ATP and stopped after 10 min by 0.1 ml of 50% perchloric acid. After centrifugation as described above the released $^{32}P_i$ is measured as outlined by Shahak.[17]

[14] Z. Gromet-Elhanan, S. Philosoph, and D. Khananshvili, *in* "Energy Coupling in Photosynthesis" (B. Selman and S. Reimer-Selman, eds.), p. 323. Elsevier, Amsterdam, 1981.

[15] R. Oren and Z. Gromet-Elhanan, *Biochim. Biophys. Acta* **548,** 106 (1979).

[16] M. Avron, *Biochim. Biophys. Acta* **40,** 257 (1960).

[17] Y. Shahak, *Plant Physiol.* **70,** 87 (1982).

TABLE I
PURIFICATION PROCEDURE OF THE F_1 β SUBUNIT

Purification steps	Total protein (mg)[a]	Specific reconstitutive activity[b]	Total activity: Reconstitutive units	Total activity: Yield (%)	Number of bands[c]
1. LiCl extract	620	12	7440	100	9
2. 50% $(NH_4)_2SO_4$	410	16	6560	88	7
3. DEAE-Sephadex	105	39	4095	55	3
4. DEAE-cellulose	15	56	840	11	1

[a] Total protein extracted from two batches of chromatophores, each of them containing 400 mg of Bchl.

[b] In units/mg of protein. One unit is the amount of protein that restores 30% photophosphorylation in β-less chromatophores containing 10 μg of Bchl under the described reconstitution and assay conditions.

[c] Number of bands revealed on SDS–polyacrylamide gel electrophoresis.

Extraction and Purification of the β Subunit

The β subunit is extracted by LiCl from chromatophores obtained from *R. rubrum* cells ruptured in the Yeda press in TS buffer: 50 m*M* Tricine-NaOH (pH 8.0) and 0.25 *M* sucrose containing also 5 m*M* $MgCl_2$, ribonuclease, and deoxyribonuclease at a final concentration of 10 μg/ml.[7,11] Before extraction, these chromatophores are washed three times in TS buffer containing 0.5 m*M* EDTA. This treatment reduces their protein/Bchl ratio to about 12 : 1 without decreasing their photophosphorylation or ATPase activities and simplifies the extraction and purification procedures.

Step 1: LiCl Extraction. All extractions and purification procedures are carried out at 4°. Washed chromatophores (about 400 mg of Bchl) are suspended in TS buffer containing 8 m*M* ATP and 8 m*M* $MgCl_2$ at 0.6–0.8 mg Bchl/ml, and stirred gently for 2 hr. An equal volume of 4 *M* LiCl in TS buffer is slowly added to the chromatophore suspension to give a final concentration of 0.3–0.4 mg Bchl/ml, 2 *M* LiCl, 50 m*M* Tricine-NaOH (pH 8.0), 0.25 *M* sucrose, 4 m*M* ATP, and 4 m*M* $MgCl_2$. After 30 min incubation the suspension is centrifuged at 180,000 *g* for 6 hr. The pelleted β-less chromatophores are washed twice in TS buffer and stored in 50% glycerol at 1.5–2.0 mg Bchl/ml in liquid air. The supernatant that contains the extracted β subunit is subjected to step 2; 1 ml of it is dialyzed against 200 ml of buffer A: 100 m*M* Tricine-NaOH (pH 8.0), 4 m*M* ATP, 4 m*M* $MgCl_2$, and 10% glycerol for 4–6 hr and stored in liquid air for checking the specific reconstitutive activity of the LiCl extract (Table I).

Step 2: Ammonium Sulfate Precipitation. Solid ammonium sulfate is added to the LiCl extract to reach 60% saturation. The suspension is allowed to stand overnight and centrifuged for 30 min at 50,000 *g*. This step is essential for complete removal of LiCl. The precipitate is dissolved in buffer A at a concentration of 3–5 mg of protein/ml and centrifuged for 1 hr at 50,000 *g* to remove all aggregated proteins. The clear supernatant is brought to 50% ammonium sulfate saturation, incubated for 1 hr, and centrifuged. The precipitate is dissolved in buffer A at a concentration of 10–15 mg of protein/ml, dialyzed for 4–6 hr against 100 volumes of buffer A, and centrifuged for 1 hr at 50,000 *g;* then 100–200 μl of the clear supernatant is checked for the specific reconstitutive activity (Table I) and the rest is chromatographed on DEAE-Sephadex A-50. It can be stored in liquid air before being chromatographed. Steps 1 and 2 are performed with two batches of chromatophores and the supernatants from step 2 are combined for step 3.

Step 3: DEAE-Sephadex Chromatography. DEAE-Sephadex A-50 must be equilibrated with ATP before application of the solution containing the β subunit, otherwise the β subunit loses its reconstitutive activity during the chromatography. DEAE-Sephadex A-50 is therefore allowed to swell by suspending it at 1 g/100 ml of 100 m*M* Tricine-NaOH (pH 8.0). After swelling is complete, 0.5 *M* MgATP is added with stirring until the volume of the settled slurry is reduced to about 15 ml/1 g of material. The slurry is washed with buffer A until full equilibration of their ATP concentration is obtained. The combined supernatant from step 2 is applied on a column (4.5 × 16 cm) of equilibrated DEAE-Sephadex A-50. The column is washed with 500 ml of buffer A containing 0.16 *M* sodium chloride, followed by 500 ml of buffer A containing 0.30 *M* sodium chloride. Fractions (10 ml each) are collected at a flow rate of about 100 ml/hr. Two main peaks are eluted, but only the second peak contains the β subunit reconstitutive activity; 300–500 μl of each fraction are used for assay of the specific reconstitutive activity. The rest is frozen in liquid air, since β activity is lost when stored at 4° in buffer A for more than 24 hr. Fractions containing β activity are pooled and precipitated at 50% ammonium sulfate saturation. The precipitate is dissolved in buffer B containing 50 m*M* Tricine-NaOH (pH 8.0), 4 m*M* ATP, 4 m*M* $MgCl_2$, and 10% glycerol at a concentration of 10–15 mg of protein/ml, and dialyzed against 200 volumes of buffer B for 4–6 hr to remove all traces of ammonium sulfate. All insoluble materials are removed by centrifugation for 2 hr at 350,000 *g* in a Spinco SW 65 rotor. The active supernatant can be stored overnight at 4° or for longer periods in liquid air.

Step 4: DEAE-Cellulose (DE-23) Chromatography. This anion exchanger must also be equilibrated with ATP, so it is washed with buffer B until full equilibration of the ATP concentration is obtained. The supernatant from step 3 is applied on a column (1.5 × 60 cm) of DE-23. The column is washed with 100 ml of buffer B containing 0.16 *M* sodium chloride, and a linear gradient of 0.16–0.4 *M* sodium chloride in 500 ml of buffer B is applied. The flow rate is adjusted to about 40 ml/hr and fractions of 3 ml are collected. β activity is eluted between 0.18 and 0.24 *M* sodium chloride.[11] Active fractions are pooled and concentrated by ultrafiltration to 2–3 mg of protein/ml. The solution is dialyzed exhaustively against buffer B to remove all traces of sodium chloride and stored in liquid air. The purification procedure is summarized in Table I.

Extraction and Purification of the γ Subunit

The γ subunit is extracted by LiBr from washed β-less chromatophores. Before extraction, the glycerol present in their storage buffer is removed by diluting them 5-fold in TS buffer: 50 m*M* Tricine-NaOH (pH 8.0) and 0.25 *M* sucrose, and centrifuging for 3 hr at 180,000 *g*.

Step 1: LiBr Extraction. All extraction and purification procedures are carried out at 4°. The pellet of β-less chromatophores (about 400 mg of Bchl) is suspended in TS buffer containing 8 m*M* ATP and 8 m*M* $MgCl_2$ at 0.5–0.6 mg of Bchl/ml and stirred gently for 1 hr. An equal volume of 4 *M* LiBr in TS buffer is slowly added to give a final concentration of 0.25–0.3 mg of Bchl/ml, 2 *M* LiBr, 50 m*M* Tricine-NaOH (pH 8.0), 0.25 *M* sucrose, 4 m*M* ATP, and 4 m*M* $MgCl_2$. The concentrations of LiBr and chromatophores and the pH of the suspension during extraction are crucial for achieving maximal extraction of the γ subunit.[11] With <2 *M* LiBr or >0.35 mg of Bchl/ml, not all of the γ subunit is extracted, and with >2 *M* LiBr or <0.2 mg of Bchl/ml, additional components are extracted. Also, although pH 6.5 is optimal for reconstitution of γ into β,γ-less chromatophores, the extraction and purification are optimal at pH 8.0. The suspension is incubated for 30 min and centrifuged at 180,000 *g* for at least 8 hr. The pelleted β,γ-less chromatophores are washed, resuspended, and stored as described for the β-less chromatophores. The supernatant that contains the extracted γ subunit is subjected to step 2; 1 ml of it is dialyzed against 200 volumes of buffer C: 10 m*M* Mes-NaOH (pH 6.5), 150 m*M* NaCl, 2 m*M* $MgCl_2$, and 5% glycerol for 4–6 hr and stored in liquid air for checking the specific reconstitutive activity of the LiBr extract (Table II).

Step 2: Ammonium Sulfate Precipitation. The protein in the LiBr extract is freed of all LiBr by precipitation at 60% ammonium sulfate saturation as described above for the β subunit. The precipitate is dissolved in buffer D: 50 m*M* Tricine-NaOH (pH 8.0), 1.0 *M* KCl, 2 m*M* $MgCl_2$, and 10% glycerol at a concentration of 1 mg of protein/ml, and

TABLE II
PURIFICATION PROCEDURE OF THE F_1 γ SUBUNIT

Purification steps	Total protein (mg)	Specific reconstitutive activity[a]	Total activity: Reconstitutive units	Total activity: Yield (%)	Number of bands[b]
1. LiBr extent	120	21	2520	100	5
2. 45% $(NH_4)_2SO_4$	89	28	2492	98	3
3. Hydroxyapatite	7	69	483	19	1

[a] In units/mg of protein. One unit is the amount of protein which restores 20% photophosphorylation in β,γ-less chromatophores containing 10 μg of Bchl under the described reconstitution and assay conditions.
[b] Number of bands revealed on SDS–polyacrylamide gel electrophoresis.

centrifuged at 270,000 *g* for 2 hr to remove all insoluble material. The clear supernatant is brought to 45% ammonium sulfate saturation. The precipitate is dissolved in buffer D at 1 mg of protein/ml, dialyzed against 100 volumes of buffer D for 6 hr with two changes of external buffer, and centrifuged for 2 hr at 270,000 *g*. The supernatant can be stored in liquid air before being chromatographed. This supernatant (1 ml) is dialyzed against buffer C and checked for the specific reconstitutive activity (Table II).

Step 3: Hydroxyapatite Chromatography. For optimal chromatographic resolution the hydroxyapatite (BioGel HT, shipped in 0.01 *M* sodium phosphate buffer) is equilibrated with 50 m*M* Tricine-NaOH (pH 8.0) and heated for 15 min at 80°. After cooling the fines are decanted and a column (3.0 × 19 cm) is prepared and washed with three volumes of 50 m*M* Tricine-NaOH (pH 8.0) at a flow rate of 150 ml/hr (with a peristaltic pump). The supernatant from step 2 is applied on the column. It is washed with 100 ml of buffer D and a linear gradient of 0.0–0.3 *M* potassium phosphate in 400 ml of buffer E is applied. The buffer contains 50 m*M* Tricine-NaOH (pH 8.0), 1.0 *M* KCl, 1.0 *M* urea, and 10% glycerol. The urea should be recrystallized from a 1 : 1 ethanol–acetone solution before being used. The flow rate is adjusted to 100 ml/hr and fractions of 4 ml are collected. The presence of protein is quickly assayed at 280 nm, and all protein-containing fractions are immediately dialyzed against 100 volumes of buffer C for 6 hr with three changes of external buffer. The chromatography and dialysis should be carried out as rapidly as possible, since in the presence of the high concentrations of salt and urea of buffer E, γ activity is lost within 8–10 hr at 4°. Active fractions are pooled and concentrated by vacuum dialysis against buffer C to no more than 0.4 mg of protein/ml, since at higher concentrations γ tends to aggregate. The

TABLE III
ACTIVITIES OF COUPLED, DEPLETED, AND RECONSTITUTED *R. rubrum* CHROMATOPHORES

Type of depleted chromatophores	Subunits added during reconstitution	Assayed activities			
		$\Delta\bar{\mu}_{H^+}$ (mV)[a]	ATP synthesis[b]	ATP hydrolysis[b] with	
				Mg^{2+}	Ca^{2+}
Coupled	None	240	810	234	132
β-less	None	216	4	3	1
β,γ-less	None	180	3	1	1
β-less	β	212	747	200	105
β-less	γ	210	4	2	1
β,γ-less	β	182	7	4	5
β,γ-less	γ	184	4	5	2
β,γ-less	$\beta + \gamma$	183	542	172	95

[a] $\Delta\bar{\mu}_{H^+}$ was measured by following changes in the fluorescence of 9-aminoacridine and anilinonaphthalenesulfonic acid.[11]

[b] μmol/hr/mg bacteriochlorophyll.

solution is stored in liquid air after addition of glycerol to a final concentration of 10% to avoid aggregation during freezing and thawing. The purification procedure is summarized in Table II.

Properties of the Depleted Chromatophores and Isolated β and γ Subunits

β-less Chromatophores. Rhodospirillium rubrum chromatophores obtained by extraction of β with LiCl lose practically all their photophosphorylation and ATPase activities (Table III). They retain, however, >85% of the capacity to take up protons during light-induced electron transport,[8,10,14] resulting in formation of an electrochemical proton gradient ($\Delta\bar{\mu}_{H^+}$). These chromatophores are defined as β-less because results from several types of experiments indicate as follows that they have lost only their β subunit, and all of it, while all other F_1 subunits remained attached to the membrane:

1. Reconstitution of purified β into these β-less chromatophores restores about 90% of all their ATP synthesis and hydrolysis activities (Table III).

2. A four-subunit F_1 complex composed of only α, γ, δ, and ε has been solubilized from these β-less chromatophores.[9] It has no ATPase activity

by itself, but its reconstitution with purified β results in restoration of a soluble Ca^{2+}-ATPase.[18]

3. An antibody prepared against the purified β subunit agglutinates control-coupled *R. rubrum* chromatophores, but not β-less ones.[19] It also forms precipitin lines with the β subunit and with isolated *R. rubrum* RrF_0F_1 and RrF_1, but not with the four-subunit β-less RrF_1.

β Subunit. The purified β subunit has a molecular weight of 50,000[20] and appears as a single band on sodium dodecyl sulfate–polyacrylamide gel electrophoresis even when the gel is loaded with 200 μg of protein. Different preparations of purified β restore between 70 and 95% of both photophosphorylation and ATPase activities of β-less chromatophores.[9,11,14] The degree of reconstitution of a fixed concentration of β-less chromatophores increases with increasing amounts of purified β subunit, so that about 100 μg of it is able to saturate β-less chromatophores containing 10 μg of Bchl.[11] With crude β preparation, especially from steps 1 and 2, this linear correlation holds up to only about 40% reconstitution.[9] Therefore, the reconstitutive unit is defined as the amount of β that reconstitutes 30% photophosphorylation (Table I). The reconstitutive activity of the isolated β subunit is dependent on the presence of ATP and $MgCl_2$ during all stages of extraction, purification, storage, and reconstitution.[8,9,11] Inactivation of purified β by removal of MgATP can be partially prevented by 10–20% glycerol.[21] Purified β subunit can be stored in liquid air for 3 months in 20% glycerol without MgATP and for at least 1 year in buffer A.

Several lines of evidence indicate that this β subunit is the catalytic subunit of the RrF_0F_1 ATP synthase complex:

1. Its removal eliminates all ATP-linked activities and its reconstitution restores them (Table III).

2. An antibody prepared against it inhibits ATP synthesis and hydrolysis activities of coupled and reconstituted *R. rubrum* chromatophores as well as RrF_0F_1 and RrF_1 ATPase activities.

3. Direct binding studies with labeled substrates have demonstrated two binding sites for ATP[22] or ADP[23] and one binding site for P_i[24] on the purified β subunit.

[18] S. Philosoph and Z. Gromet-Elhanan, *Isr. J. Med. Sci.* **15,** 109 (1979).
[19] S. Philosoph and Z. Gromet-Elhanan, *Eur. J. Biochem.* **119,** 107 (1981).
[20] C. Bengis-Garber and Z. Gromet-Elhanan, *Biochemistry* **18,** 3577 (1979).
[21] D. Khananshvili and Z. Gromet-Elhanan, *J. Biol. Chem.* **258,** 3714 (1983).
[22] Z. Gromet-Elhanan and D. Khananshvili, *Biochemistry* **23,** 1022 (1984).
[23] D. Khananshvili and Z. Gromet-Elhanan, *FEBS Lett.* **178,** 10 (1984).
[24] D. Khananshvili and Z. Gromet-Elhanan, *Biochemistry* **24,** 2482 (1985).

4. Studies with various chemical modifiers revealed that histidine and carboxyl residues are essential for the binding of P_i and of one of the two ATP molecules that bind to isolated β.[25] The same residues are also involved in the capacity of purified β to restore the catalytic activity of β-less chromatophores.

β,γ-less Chromatophores. Further extraction of β-less chromatophores by LiBr removes one additional subunit, the γ subunit.[10,11,14] The resulting β,γ-less chromatophores have, of course, no ATP-linked activities, but retain 70% of their capacity to take up protons in the light and form an $\Delta\tilde{\mu}_{H^+}$ (Table III). They are defined as β,γ-less chromatophores because their ATP-linked activities can be restored to about 70% by reconstitution with both purified β and γ subunits, but not with either one of them (Table III). Moreover, optimal reconstitution of β,γ-less chromatophores is obtained when β and γ are added in amounts that are close to their molar ratio in F_1.[14] β,γ-less as well as β-less and coupled chromatophores can be stored in 50% glycerol in liquid air for at least 2 years with no detectable loss of their light-induced proton uptake or their capacity to be reconstituted by the missing subunits.

γ Subunit. The purified γ subunit appears as a single band on SDS–polyacrylamide gel electrophoresis (Table II) with a molecular weight of 35,000.[11] The γ subunit, unlike the β subunit, remains active in the absence of ATP. It tends to aggregate at concentrations above 0.4 mg of protein/ml in low salt concentrations (buffer C), but stays soluble at up to 1 mg of protein/ml in high salt (buffer D). It can be kept for a few days at 4° in 2 *M* LiBr or 45–60% $(NH_4)_2SO_4$, and for at least 6 months in liquid air in buffer C.

γ as well as β can rebind separately to β,γ-less chromatophores,[11,14] but this separate binding does not restore any ATP-linked activity (Table III). ATP synthesis and hydrolysis can be restored to a very similar extent by a separate but consecutive rebinding of both subunits. However, reconstitution of β and γ together leads to a 50% increase in the extent of restoration.[11] These results indicate that γ is required together with β for the operation of ATP synthesis and hydrolysis. But, whereas β has been shown to be absolutely necessary for the catalysis itself, it cannot be decided as yet whether γ is needed for the catalysis or for the correct rebinding of β.

The fact that β,γ-less chromatophores retain their capacity to generate an $\Delta\tilde{\mu}_{H^+}$ during light-induced electron transport (Table III) suggests that the β and γ subunits cannot be an integral part of an H^+ gate in the *R*.

[25] D. Khananshvili and Z. Gromet-Elhanan, *Proc. Natl. Acad. Sci. U.S.A.* **82,** 1886 (1985).

rubrum chromatophore membrane. This function could be fulfilled by a combination of α, δ, and ε.

Acknowledgment

Development of the methods described in this article was supported by grants from the United States–Israel Binational Science Foundation (BSF), Jerusalem, Israel.

[54] Preparation and Reconstitution of the Proton-Pumping Membrane-Bound Inorganic Pyrophosphatase from *Rhodospirillum rubrum*

By MARGARETA BALTSCHEFFSKY and PÅL NYRÉN

The membrane-bound inorganic pyrophosphatase (PPase) in chromatophores from the photosynthetic bacterium *Rhodospirillum rubrum* is an alternative coupling factor to the corresponding ATPase. It catalyzes the synthesis of inorganic pyrophosphate (PP_i) coupled to light-induced electron transport.[1] The enzyme also hydrolyzes PP_i and can mediate the energy released in this hydrolysis to various energy-requiring reactions, such as ATP synthesis,[2,3] cytochrome redox changes,[4] carotenoid band shift,[5] succinate-linked pyridine nucleotide reduction,[6] and transhydrogenation.[7] In analogy with the F_0F_1-ATPase complex also, the PPase has been shown to act as a proton pump, both when bound to the chromatophore membrane[8] and when purified and reconstituted into liposomes.[9] A partly purified preparation of the enzyme has also been shown

[1] H. Baltscheffsky, L.-V. von Stedingk, H.-W. Heldt, and M. Klingenberg, *Science* **153,** 1120 (1966).

[2] D. L. Keister and N. J. Minton, *Arch. Biochem. Biophys.* **147,** 330 (1971).

[3] M. Baltscheffsky and H. Baltscheffsky, *in* "Oxidation Reduction Enzymes" (Å. Åkesson and A. Ehrenberg, eds.), p. 257. Pergamon, Oxford, 1972.

[4] M. Baltscheffsky, *Biochem. Biophys. Res. Commun.* **28,** 270 (1967).

[5] M. Baltscheffsky, *Arch. Biochem. Biophys.* **130,** 646 (1969).

[6] D. L. Keister and N. J. Yike, *Arch. Biochem. Biophys.* **121,** 415 (1967).

[7] D. L. Keister and N. J. Yike, *Biochemistry* **6,** 3847 (1967).

[8] J. Moyle, R. Mitchell, and P. Mitchell, *FEBS Lett.* **23,** 233 (1972).

[9] Y. A. Shakov, P. Nyren, and M. Baltscheffsky, *FEBS Lett.* **146,** 177 (1982).

to act as a PP_i-dependent electric generator[10] when incorporated into a planar phospholipid membrane.

The PPase is an integral membrane protein, which probably spans the chromatophore membrane. Earlier attempts to solubilize and purify this enzyme have only been partially successful.[11] Recently we have been able to obtain an apparently pure enzyme in reasonable yield,[12] giving us the possibility to eventually perform a closer characterization of this protein, a characterization which, due to the analogy between the PPase and the ATPase and the apparent comparative simplicity of the former, also may yield closer insight into essential aspects of the functioning of coupling factor proteins in general.

Method of Solubilization and Purification

Preparation of Chromatophores

Rhodospirillum rubrum, strain S-1, is grown anaerobically in the light in batch culture for 40 hr at 30° on the synthetic medium described by Bose *et al.*[13] The bacteria are harvested by centrifugation, resuspended in 0.02 *M* glycylglycine buffer, pH 7.4, and washed once. The bacterial pellet is resuspended in ice-cold 0.2 *M* glycylglycine, pH 7.4, in a concentration of about 20 g of bacteria (wet weight) per 100 ml suspension and then disrupted in a Ribi cell fractionator at 20,000 psi. The cell homogenate is supplied with DNase and RNase and then centrifuged 60 min at 10,000 *g* in a Servall refrigerated centrifuge. The supernatant from this centrifugation is then recentrifuged for 90 min at 100,000 *g* in a Spinco preparative ultracentrifuge in order to sediment all membranous material, the chromatophores. The chromatophores are resuspended and washed twice with 0.2 *M* glycylglycine, pH 7.4, with centrifugation for 60 min at 100,000 *g* in each wash. The washed chromatophores are finally suspended in a minimal volume of 0.2 *M* glycylglycine and kept on ice until used. After breaking the cells, all operations are carried out as close to 0° as possible.

Solubilization of the Inorganic Pyrophosphatase

Chromatophores (3 ml) (60–70 mg protein/ml, about 1.5 m*M* bacteriochlorophyll) are mixed with 21 ml of Tris–HCl buffer, pH 8.4, containing

[10] A. A. Kondrashin, V. G. Remennikov, V. D. Samuilov, and V. P. Skulachev, *Eur. J. Biochem.* **113,** 219 (1980).

[11] P. V. Rao and D. L. Keister, *Biochem. Biophys. Commun.* **84,** 465 (1978).

[12] P. Nyren, K. Hajnal, and M. Baltscheffsky, *Biochim. Biophys. Acta* **766,** 630 (1984).

[13] S. K. Bose, H. Gest, and J. G. Ormerod, *J. Biol. Chem.* **236,** 13 (1961).

2.5% (v/v) Triton X-100, 0.75 M $MgCl_2$, 25% ethylene glycol, and 0.2 mM dithiothreitol (DTE). The suspension is kept on ice and gently stirred for 20 min, after which it is centrifuged for 60 min at 215,000 g. The supernatant, containing the solubilized PPase is decanted and kept on ice until further use, or alternatively kept frozen at −70°.

Chromatography on Hydroxylapatite

The solubilized enzyme (6 ml) is desalted on a Sephadex G-25 column, 3.2 × 8.5 cm, equilibrated with 50 mM Tris–HCl, pH 8.4, 25% (v/v) ethylene glycol, 0.1% Triton X-100, 1 mM $MgCl_2$, and 0.2 mM DTE. This is done in order to reduce the ionic strength of the solution and enable the binding of the enzyme to the hydroxylapatite. BioGel HTP (8 g) is rehydrated in 50 mM Tris–HCl buffer, pH 8.4, containing 0.1% (v/v) Triton X-100, 25% (v/v) ethylene glycol, 1 mM $MgCl_2$, and 0.2 mM DTE. The rehydrated gel is packed into a 3.2 × 10 cm column and equilibrated with the same buffer. The sample is applied on the column with a flow rate of 1–2 ml/min. After application of the sample, the flow is stopped for 2 min, then the column is washed first with 50 ml of the equilibration buffer with 10 mM $MgCl_2$, then with 30 ml of the same buffer with 130 mM $MgCl_2$. The enzyme is then eluted with 0.2 M $MgCl_2$ in the above-described buffer and fractions of 6 ml are collected. Those fractions showing the highest PPase activity (in a typical experiment, fractions 5–8) are pooled and concentrated to a final volume of 1 ml by ultrafiltration through an Amicon YM30 (50 psi) or XM100A (25 psi) membrane. This preparation is about 80–90% pure as judged by gel electrophoresis and may be used directly or stored frozen at −70°. The specific activity of the preparation, before concentration, expressed as micromoles of PP_i hydrolyzed in 1 min by 1 mg protein, is typically 24. Some of the activity is lost during concentration, with higher losses with the XM100A than with the YM30 filter. Since the XM100A filter allows the passing through of Triton X-100 micelles, the final product has a lower content of detergent and is better suited for incorporation into liposomes than the one concentrated with the YM30 filter, which, on the other hand, has higher activity and thus may be better suited for other purposes.

Incorporation of the PPase in Liposomes

The enzyme preparation after concentration is desalted on a Sephadex G-25 (coarse) column, 0.5 × 25 cm. The column is equilibrated with 50 mM Tris–HCl buffer, pH 8.0, 25% (v/v) ethylene glycol, 5 mM $MgCl_2$, 0.2

mM DTE, and 0.04% Triton X-100, and the enzyme is eluted with the same buffer.

The freeze-thaw technique of Kasahara and Hinkle[14] for preparing liposomes has been used with good results. Soybean phospholipids (40 mg) are supplemented with 1 ml of medium containing 10 mM Tris–HCl, pH 7.5, 0.5 mM DTE, 0.5 mM EDTA, and 0.05% Na-cholate. The suspension is flushed with nitrogen and sonicated in a bath-type sonicator, model G112SP1T (Laboratory Supplies, Hicksville, NY). The PPase preparation, 0.2 ml, containing 50–100 μg protein is then added to 0.3–0.4 ml of liposomes, sonicated for 10 sec and rapidly frozen in a dry-ice–ethanol bath. After thawing at room temperature, the preparation is stored on ice.

If liposomes with both the F_0F_1 complex and the PPase are desired, the procedure is the same as above, with the exception that together with the PPase preparation, 0.1 ml containing 10–20 μg protein, is added 0.1 ml of a purified ATPase preparation (50 μg protein), prepared according to Pick and Racker[15] with some minor modifications.[16]

In the liposomes the PPase has regained certain characteristics, typical for proton pumping activity, such as stimulation of hydrolysis by uncouplers or ionophores.[9] An account of the effects of various compounds on the PPase activity before and after incorporation is given in Table I.

Determination of Pyrophosphatase Activity

The PPase activity is assayed in a reaction mixture containing 0.75 mM $MgCl_2$, 0.5 mM $Na_4P_2O_7$, 1 ml 0.1 M Tris–HCl, pH 7.5, 0.2 mg ml^{-1} asolectin or 25 μg ml^{-1} cardiolipin, when required, and H_2O, in a total volume of 2 ml. Asolectin is prepared by sonication in 0.1 M Tris–HCl, pH 7.5, until a clear suspension is obtained. The assay mixture is incubated at 30° and the reaction is terminated after the desired time by addition of 1 ml 10% trichloroacetic acid. In the blanks the trichloroacetic acid is added before the enzyme sample. If, as sometimes is the case with crude preparations, a protein precipitate forms, the mixture is chilled and centrifuged, and the supernatant fluid is collected. Then 0.4 ml of the assay mixture or the supernatant is analyzed for P_i according to Rathbun *et al.*[17] This method is reliable, provided that the remaining concentration of PP_i is lower than 0.35 mM. At higher concentrations the sensitivity is

[14] M. Kasahara and P. Hinkle, *J. Biol. Chem.* **252,** 7384 (1977).

[15] U. Pick and E. Racker, *J. Biol. Chem.* **254,** 2793 (1979).

[16] P. Nyren and M. Baltscheffsky, *FEBS Lett.* **155,** 125 (1983).

[17] W. B. Rathbun and W. M. Betlach, *Anal. Biochem.* **28,** 436 (1969).

TABLE I

EFFECT OF VARIOUS COMPOUNDS ON THE PPASE ACTIVITY BEFORE AND AFTER ITS INCORPORATION INTO LIPOSOMES[a]

Activity	Control	Addition							
		FCCP (1.5 μ*M*)	Valinomycin (10 μg/ml)	Nigericin (10 μg/ml)	Valinomycin + nigericin	Oligomycin (10 μg/mg protein)	DCCD (100 μ*M*)	NaF (10 m*M*)	IDP (1 m*M*)
Incorporated PPase									
nmoles P_i min^{-1}	37	50	46	54	55	36	36	6	21
Percentage	100	135	124	146	149	97	97	16	57
Solubilized PPase									
nmoles P_i min^{-1}	40	39	40	40	39	38	39	4	18
Percentage	100	98	100	100	98	95	98	10	45

[a] PPase activity was measured in the same mixture with phospholipids added after (solubilized PPase) and before (incorporated PPase) freezing-thawing. The assay medium contained 5×10^{-2} *M* KCl, and 50 μl of PPase preparation was added. From Shakov *et al.*[9]

gradually lost, and misleading results may be obtained if special attention is not given to the remaining PP_i concentration.

Polyacrylamide Gel Electrophoresis

Native gel electrophoresis may be performed with desalted enzyme on 5% gels prepared according to Davis,[18] with the exception that 20% ethylene glycol (v/v), 0.1 m*M* DTE, and 0.1% Triton X-100 were included in the gel. The electrolyte buffer contains 0.1% Triton X-100. The PPase activity may be localized by incubating the gel in a solution identical to that used for assaying the enzyme activity for 20 min at 30°, after which it is rinsed with distilled water immediately immersed in the triethylamine-molybdate reagent described by Sugino and Miyoshi,[19] which specifically precipitates P_i. A sharp, discrete zone corresponding to the enzyme activity in the gel appears within a few minutes.

Inhibitors of the Pyrophosphatase Activity

Table II shows the action of a number of compounds on the PPase activity, both when the enzyme is membrane bound, either to the chromatophores or incorporated into liposomes, and in the solubilized state. Classical uncouplers and ionophores, as has been mentioned, only are effective with the membrane-associated enzyme, whereas inhibitors, which can be assumed to act at or close to the catalytic site, are effective both with the membrane-bound and the solubilized enzyme. Another observation is that the catalytic inhibitors usually are much more efficient with the solubilized enzyme, which may indicate that the catalytic site in the original chromatophore membrane resides somewhat buried below the membrane surface.

In order to elucidate the subunit composition of the enzyme, SDS–polyacrylamide gel electrophoresis, essentially according to Weber and Osborn,[20] has been used. To dissociate the enzyme into subunits, 1 part of purified enzyme solution is incubated overnight at 37° with 1 part of 0.25 *M* Tris–HCl, pH 6.9, 20% glycerol, 2% SDS, and 2% mercaptoethanol. About 20 μg of protein is then run on 11.25% gels which include 0.1% SDS. The bands are stained with Coomassie Brilliant Blue. This procedure yields 6–7 discrete bands with apparent molecular weights of, respectively, 64,000, 52,000, 41,000, 31,000 (25,000), 20,000, and 15,000. The 25K band is usually very weak or absent and may not be part of the

[18] B. J. Davis, *Ann. N.Y. Acad. Sci.* **121,** 404 (1964).

[19] Y. Sugino and Y. Miyoshi, *J. Biol. Chem.* **239,** 2360 (1964).

[20] K. Weber and M. Osborn, *J. Biol. Chem.* **244,** 4406 (1969).

TABLE II
EFFECT OF SOME SUBSTANCES ON MEMBRANE-BOUND AND PURIFIED PPASES[a]

Additions	Concentration	Activity of PPase (% of control) dependent on: Membrane-bound	Purified
FCCP	1.5 μM	150	100
DCCD (0°)	100 μM	30	97
NaF	5 mM	83	32
	10 mM	64	11
	20 mM	45	4
MDP	0.1 mM	84	64
	0.2 mM	74	36
IDP	0.1 mM	48	24
	0.2 mM	34	12
Nbf-Cl (0°)	0.25 mM	33	23
	0.50 mM	18	7
Nbf-Cl (30°)	0.25 mM	82	12
	0.50 mM	61	1
MalNET (0°)	1 mM	30	20
	2 mM	13	13
MalNET (30°)	1 mM	99	84
	2 mM	85	76
Dio-9 (0°)	15 μg/ml	80	40
	30 μg/ml	60	30

[a] Particles corresponding to 120 μg protein and purified enzyme corresponding to 15 μg protein were assayed as described in the text except for the MDP and IDP treatment for which 12.5 m*M* $MgCl_2$ was used instead of 0.75 m*M*. The DCCD, MalNET, and Nbf-Cl treatments were performed by incubation of particles and enzyme for 10 min at 0° or 30°. Dio-9 was incubated with particles and enzyme for 20 min at 0°. The PPase reaction was initiated by adding PP_i. From Nyren *et al.*[12]

enzyme. Also, it cannot at present be excluded that some of these bands are dimers of others, especially since the protein is rather difficult to dissociate.

Substrate Specificity

The hydrolyzing activity of the PPase is very specific for PP_i. The activity with a number of other compounds containing the pyrophosphate moiety, ATP, ADP, AMP, IDP, MDP, tetrapolyphosphate, and 3-glycerophosphate is zero. With tripolyphosphate there is a low activity. In the case of tri- and tetrapolyphosphate there is an apparent initial activity which seems to be due to PP_i contamination in the reagents. Time curves of the PPase activity with these substrates show that the activity ceases after 10–30 min, whereas the activity with PP_i is more or less linear for 80 min. With tripolyphosphate the low remaining activity is 6% of that with PP_i. Imidodiphosphate and methylene diphosphonate act as competitive inhibitors of PP_i hydrolysis.

[55] Purification of F_1F_0 H^+-ATPase from *Escherichia coli*

By ROBERT H. FILLINGAME and DAVID L. FOSTER

This chapter[1] summarizes a method for purification of the F_1F_0 H^+-ATPase of *Escherichia coli* to homogeneity. The procedure given is a slight modification of that originally reported by Foster and Fillingame.[2] Also described are modifications of the basic method for purification of F_1F_0 from membranes in which the complex is overproduced by 6- to 8-fold.[3] Both procedures generate very pure F_1F_0 preparations composed of eight types of subunits, all of which are now known to be coded for by genes of the *unc* operon and hence are authentic components of the complex. Successful application of the method described may depend upon use of an equivalent strain, growth conditions, and possibly the method of membrane preparation.

Materials and General Procedures

Strains of E. coli. The monoploid strain used is strain AN180 (*argE3, thi, rpsL*).[4] Strain KY7485 is lysogenic for a thermally inducible λ-*unc*$^+$-transducing phage, λasn5 (λ *cI857S7* [*bglR*$^-$C$^+$, *glm*S$^+$, *unc*$^+$, *asnA*$^+$]); it

[1] The work described was supported by Public Health Service Grant GM-23105.

[2] D. L. Foster and R. H. Fillingame, *J. Biol. Chem.* **254,** 8230 (1979).

[3] D. L. Foster, M. E. Mosher, M. Futai, and R. H. Fillingame, *J. Biol. Chem.* **255,** 12037 (1980).

[4] J. D. Butlin, G. B. Cox, and F. Gibson, *Biochem. J.* **124,** 75 (1971).

is also lysogenic for λ*cI857S7* and carries the *asnA31*, *asnB32*, *thi*, and *rif* chromosomal markers.[5]

Growth of Cells. The strains above are routinely grown in 10-liter batches in a 14-liter New Brunswick fermentor with aeration at 6.0 liters/min and stirring at 400 rpm. The minimal medium contains 0.1 *M* K^+-phosphate (pH 7.5), 46.5 m*M* NH_4Cl, 0.4 m*M* Na_2SO_4, 0.8 m*M* $MgCl_2$, 3.2 μ*M* $FeSO_4$, and the supplements described below. Additional minerals (1/20 volume 0.93 *M* NH_4Cl, 8 m*M* Na_2SO_4, 16 m*M* $MgCl_2$) were added to the fermentor when growing cells reach an optical density (*A* at 550 nm) of 2–3 units. These additional salts, which cannot be added initially due to formation of an Mg-phosphate precipitate, are required for growth to maximal density.

Strain AN180 is grown at 37° in the above medium supplemented with 1.4% glucose, 5 m*M* L-arginine, and 5 mg/liters thiamine. Cells reach a maximal optical density (A_{550}) of >9, equivalent to 2.4×10^{10} cells/ml. We routinely harvest cells shortly before stationary phase, as indicated by decreased O_2 consumption monitored with an O_2 electrode.

Strain KY7485 is grown in the above medium supplemented with 1.4% glucose and 5 μg/ml thiamine at 32° with the supplementary minerals added prior to the heat induction. At an optical density of 2–3 units, the λ-transducing phage is induced by raising the temperature to 42°, which requires 17 min with the fermentor heater. After 30 min at 42°, the temperature is reduced to 37° over a period of 3 min. Aeration and stirring are continued for 3 hr and the cells harvested. The above minimal medium can be supplemented with 1.0 g/liter Bacto yeast extract and 0.5 g/liter Bacto tryptone to hasten growth, and good overproduction can still be achieved. However, we have observed greatest overproduction when cells are grown and induced in minimal medium.

Cells are routinely collected by centrifugation or with a Millipore Pellicon cell-harvesting filtration system and washed once with 50 m*M* tris-hydroxymethylaminomethane (Tris)–HCl (pH 7.5), 5 m*M* $MgCl_2$, 1 m*M* dithiothreitol, 10% (v/v) glycerol, and stored as a paste at −80°.

Other Materials. All chemicals used are of reagent grade. The source of those listed below may be important: sodium cholate (Calbiochem-Behring, La Jolla, CA) or cholic acid (Sigma, St. Louis, MO) twice recrystallized from 70% ethanol and neutralized with NaOH; sodium deoxycholate (Calbiochem-Behring, La Jolla, CA); phenylmethylsulfonyl fluoride (PMSF) (Sigma, St. Louis, MO) is added to buffers as a solid

[5] T. Miki, S. Hiraga, T. Nagata, and T. Yura, *Proc. Natl. Acad. Sci. U.S.A.* **75**, 5099 (1978).

immediately before use or from a fresh 100 m*M* stock in ethanol; *p*-aminobenzamidine · 2HCl (Aldrich, Milwaukee, WI). The 100% saturated ammonium sulfate stock refers to a saturated solution at 4°.

Escherichia coli phospholipids were prepared by extraction of a paste of *E. coli* cells with chloroform–methanol (1 : 2) according to the procedure of Bligh and Dyer.[6] The crude preparation dissolved in chloroform–methanol (2 : 1) was precipitated with 10 volumes of acetone, collected by centrifugation, and dryed *in vacuo*. The dry phospholipid was dispersed in 50 m*M* Tris–HCl (pH 7.5), 1 m*M* dithiothreitol at 4 mg/ml by sonication for 8 min at 2° with a Heat Systems-Ultrasonics Model W200R sonicator with microprobe, and stored at −20°.

Buffers. The following buffers are used routinely. All pH adjustments were made at room temperature (20–25°). With buffers containing glycerol, the pH was adjusted prior to the addition of glycerol. The pH of all buffers containing *p*-aminobenzamidine was adjusted after its addition.

Buffer A: 50 m*M* Tris–HCl (pH 7.5), 5 m*M* $MgSO_4$, 1 m*M* dithiothreitol, 10% (v/v) glycerol, 6 m*M* *p*-aminobenzamidine · 2HCl

Buffer B: 100 m*M* Tris–HCl (pH 7.5), 5 m*M* $MgSO_4$, 2 m*M* dithiothreitol, 10% (v/v) methanol, and 12 m*M* *p*-aminobenzamidine · 2HCl

Buffer C: 50 m*M* Tris–HCl (pH 7.5), 1 m*M* $MgCl_2$, 1 m*M* dithiothreitol, 10% (v/v) glycerol, 6 m*M* *p*-aminobenzamidine · 2HCl

Buffer D: 50 m*M* Tris–HCl (pH 7.5), 5 m*M* $MgCl_2$, 1 m*M* dithiothreitol, 10% (v/v) methanol, and 6 m*M* *p*-aminobenzamidine · 2HCl. Methanol rather than glycerol is included in this buffer so that the density of the sample applied to sucrose gradients does not exceed the density of the top of the gradient.

Buffer E: 50 m*M* Tris–HCl (pH 7.5), 1 m*M* dithiothreitol

Preparation of Membranes. Frozen cells are resuspended in buffer A containing 1 m*M* PMSF (using 5 ml/g cell paste) and disrupted by a single pass through a French press at 18,000 psi. Unbroken cells and large debris are removed by two sequential centrifugations at 8000 rpm in a Sorvall SS-34 rotor (7700 g_{max}). The supernatant is typically centrifuged at 40,000 rpm in a Beckman 45 Ti rotor (186,000 g_{max}) for 75–90 min to collect the membrane. The membrane is washed once by resuspension in buffer A using a Potter–Elvehjem homogenizer and centrifuged. The washed membrane pellet is resuspended in buffer A at ~30 mg/ml and stored at −80°. The ATPase activity of the membrane is stable for months to years.

[6] E. G. Bligh and W. J. Dyer, *Can. J. Biochem. Physiology* **37,** 911 (1959).

General Assays

ATPase Assay. ATPase activity is quantitatively determined by measuring the formation of $[^{32}P]P_i$ from 0.4 m*M* $[\gamma\text{-}^{32}P]$ATP at 30° and pH 7.8. The assay is described elsewhere in this series.[7] One unit of ATPase activity equals 1 μmol of ATP hydrolyzed per minute.

Sucrose gradient fractions containing ATPase activity can easily be located and peak fractions discriminated using a colorimetric method.[8] The assay mixture of 0.3 ml, 50 m*M* Tris–HCl (pH 8), 2.5 m*M* $MgCl_2$, 5 m*M* Na_2ATP, and varying amounts of gradient fractions (1–20 μl) are incubated in the wells of white ceramic plates for 5–10 min at 37°. Five drops of 0.18 *M* $FeSO_4$ and 13 m*M* $(NH_4)_6Mo_7O_{24}$ in 1 *N* H_2SO_4 are added and the intensity of the resulting blue color is estimated by eye.

Assay of $[^{32}P]P_i$–ATP Exchange. $[^{32}P]P_i$–ATP exchange activity is measured by a modification of the procedure used by Schairer *et al.*[9] Membrane (0.1–0.5 mg protein) or reconstituted F_1F_0-ATPase (5–20 μg protein) is added to 1 ml of 50 m*M* 2-(*N*-morpholino)ethanesulfonic acid/NaOH (pH 6.5), 10 m*M* $MgCl_2$, 2 m*M* Na_2ATP, 0.5 m*M* KH_2PO_4 ($[^{32}P]$phosphate at 10 Ci/mol), 250 m*M* sucrose, and 1 mg/ml of bovine serum albumin at 30°. The reaction is terminated after 10–20 min by the addition of 0.5 ml of ice-cold 1.5 *M* perchloric acid. The precipitated protein is removed by centrifugation at 2000 rpm for 5 min in a clinical centrifuge at 4°. Aliquots of 0.5 ml are transferred to 18 × 150 mm test tubes, and nonesterified phosphate extracted by the procedure to Avron.[10] To each tube 1.2 ml of acetone, 2 ml of water saturated with isobutanol/benzene (1 : 1), 0.8 ml of 5% (w/v) ammonium molybdate in 4 *N* H_2SO_4, and 7 ml of isobutanol/benzene (1 : 1) are added. Each tube is mixed for 20 sec on a Vortex mixer. The upper phase is removed and discarded. To the remaining lower phase 20 μl of 10 m*M* KH_2PO_4 and 7 ml of isobutanol/benzene (1 : 1) are added. After mixing and a 15-min incubation at room temperature, the radioactivity present in a 0.5-ml aliquot of the bottom phase is determined by liquid scintillation counting. Radioactivity in the lower phase, calculated after subtraction of a blank, is assumed to represent $[^{32}P]P_i$ esterified in $[\gamma\text{-}^{32}P]$ATP.

Protein Assays. The method used for quantitating protein is a modification of the Lowry assay.[11] Samples are dissolved in 0.2 ml 0.5 *N* NaOH

[7] R. H. Fillingame, this series, Vol. 56, p. 163.

[8] T. Tsuchiya and B. P. Rosen, *J. Biol. Chem.* **250,** 8409 (1975).

[9] H. U. Schairer, P. Friedl, B. I. Schmid, and G. Vogel, *Eur. J. Biochem.* **66,** 257 (1976).

[10] M. Avron, *Biochim. Biophys. Acta* **40,** 257 (1960).

[11] O. H. Lowry, N. J. Rosebrough, A. L. Farr, and R. J. Randall, *J. Biol. Chem.* **193,** 265 (1951).

containing 2.5% sodium dodecyl sulfate, diluted with 1.0 ml of copper reagent (1 part 0.5% $CuSO_4 \cdot 5H_2O$ in 1% Na_3 citrate mixed with 50 parts 2% Na_2CO_3), and after 10 min mixed with 0.1 ml of a 1 : 1 dilution of commercial Folin–Ciocalteau reagent. Bovine serum albumin is used as a standard. Due to the interference by sucrose in the above assay, the Coomassie Brilliant Blue dye binding assay of Bradford[12] is conveniently used for determination of relative protein values in fractions of sucrose gradients.

Sodium Dodecyl Sulfate (SDS)–Polyacrylamide Gel Electrophoresis. The subunits of the *E. coli* F_1F_0-ATPase are well resolved by the discontinuous buffer system described by Laemmli.[13] We most consistently obtain high resolution and uniformly sharp bands using 0.6 × 10 cm disc gels composed of 13% acrylamide, 0.41% bisacrylamide with 0.4% SDS in the gel forming solution. The Tris–glycine running buffer is made 0.1% in SDS. We generally run these gels at 1.5 mA per tube with the gel tubes being cooled by buffer in the surrounding reservoir. For reasons that are still not obvious, we have never consistently obtained as high a resolution and band sharpness using slab gels. Smaller subunits are generally more diffuse. The best resolving slab gel system we have used utilizes 0.75 mm thick slabs of 14 cm width and 13–25 cm length composed of 15% acrylamide, 0.25% bisacrylamide with 0.4% SDS. These are run at 10 mA per slab at room temperature with cooling via circulation of tap water through a heat sink.

Samples for electrophoresis are routinely prepared by a 1 : 1 dilution of sucrose gradient fractions of F_1F_0 with 125 m*M* Tris–HCl, pH 6.8, 4% SDS, 5% 2-mercaptoethanol, and 20% (v/v) glycerol, and immediate heating in a boiling water bath for 3 min. Samples are frozen at $-20°$ until use and then generally heated at 50° for 15–30 min before application to the gels. We have seen no difference using other heating regimens.[14]

We have found that staining of gels with Coomassie Brilliant Blue R-250 to be very reliable and relatively sensitive. The fixation method used increases the intensity of staining. Gels are first fixed with 15% (w/v) trichloroacetic acid, and the trichloroacetic acid is removed by two prolonged rinses of 3–8 hr in methanol : glacial acetic acid : H_2O (5 : 1 : 5). The gels are stained in 0.25% Coomassie Brilliant Blue R-250 (Bio-Rad, Richmond, CA) dissolved in 9% glacial acetic acid/45% methanol for 6–12 hr. Destaining is done initially with 7.5% glacial acetic acid/25% methanol, followed by 7.5% glacial acetic acid/10% methanol. Destained disc gels

[12] M. M. Bradford, *Anal. Biochem.* **72,** 248 (1976).

[13] U. K. Laemmli, *Nature (London)* **227,** 680 (1970).

[14] D. L. Foster and R. H. Fillingame, *J. Biol. Chem.* **257,** 2009 (1982).

are stored in 7.5% glacial acetic acid. Slab gels are dryed after equilibration in 7.5% glacial acetic acid/2% glycerol. For slab gels we also routinely use the sensitive silver staining procedure of Giulian *et al.*[15] after prefixation of the gel with trichloroacetic acid and acetic acid/methanol as described above. In our hands, there is considerably more variation in the relative staining intensity of the eight subunits of F_1F_0 by silver staining in comparison to the Coomassie Brilliant Blue staining method.

Purification of F_1F_0 from Monoploid Cells

Solubilization with Deoxycholate. Membranes suspended at 20 mg protein/ml in buffer A are stirred at 0° and 100 m*M* PMSF and solid KCl added to 1 m*M* and 1 *M* final concentrations, respectively. Na-deoxycholate and Na-cholate are then added from 10% (w/v) stock solutions, each to a final concentration of 0.5%. The suspension is stirred for 10 min at 0° and then centrifuged at 40,000 rpm (193,000 g_{max}) for 75 min in a Beckman 50.2 Ti rotor at 4°. The clear, amber supernatant is immediately diluted by addition of one volume of buffer A and dialyzed against 50–100 volumes of buffer A for 17 hr, usually with one change of buffer at 6 hr. The resultant turbid suspension is centrifuged as described above and the material in the pellet homogenized in a volume of buffer B equal to one-fifth that used in the extraction of the membrane.

More than 80% of the ATPase activity of monoploid membranes is solubilized by deoxycholate in the range of 0.35–0.50% (Fig. 1). However, the concentration required for optimal solubilization varies somewhat from one membrane preparation to another. Since the membrane preparations used in these studies contains both outer and inner membrane, it is possible that the amount of detergent necessary for optimal extraction may be related to the proportion of inner membrane present. The use of 0.5% Na-deoxycholate in combination with 0.5% Na-cholate led to uniformly high extraction yields. ATPase activity is somewhat unstable in the presence of 0.5% Na-deoxycholate, but more stable when 0.5% Na-cholate is simultaneously present. The solubilized ATPase is diluted prior to dialysis to reduce inactivation by the detergents.

Ammonium Sulfate Fractionation. The suspended extract in buffer B is adjusted to ~11 mg protein/ml and 10% (w/v) Na-deoxycholate is added to 0.6%. After 2 min at 0° the clarified solution is brought to 25% saturation with $(NH_4)_2SO_4$ by the dropwise addition of 1 volume of 50% saturated $(NH_4)_2SO_4$ solution while constantly stirring. After 30 min at 0° the suspension is centrifuged at 15,500 rpm (30,000 g_{max}) for 30 min in a

[15] G. G. Giulian, R. L. Moss, and M. Greaser, *Anal. Biochem.* **129,** 277 (1983).

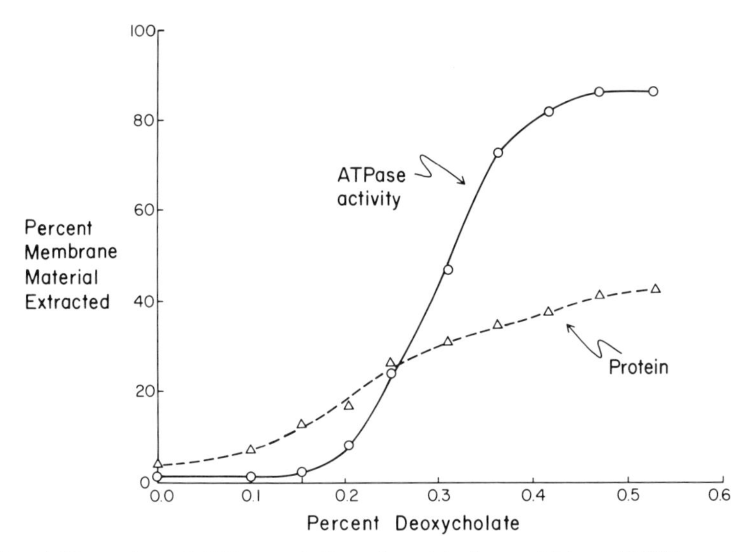

FIG. 1. Extraction of ATPase activity and protein from wild-type AN180 membranes. Membrane at 20 mg protein/ml in buffer A was made 1 *M* in KCl and extracted with the indicated concentration of Na-deoxycholate. The percent ATPase activity (○) and percent protein (△) solubilized are indicated.

Sorvall SS-34 rotor at 4°. The volume of the resulting supernatant solution is measured, and while stirring as before, 0.154 volumes of 100% $(NH_4)_2SO_4$ is added dropwise to bring the final $(NH_4)_2SO_4$ concentration to 35%. After 30 min at 0° the precipitate is removed by centrifugation as described above. The precipitated material is resuspended by homogenization in a volume of buffer C equal to 5–10% the volume used in extraction of the membrane. Sufficient detergent is present in the precipitate so that the homogenate is nearly clear. Although this material was usually applied immediately to sucrose gradients, it is relatively stable during storage at −20°, the ATPase activity decaying by 50% after 40 days.

Occasionally some of the precipitate formed with 35% $(NH_4)_2SO_4$ floats rather than sediments during centrifugation. This problem was largely eliminated by the use of methanol rather than glycerol in the buffers employed in this step. If it does occur, try to decrease the $(NH_4)_2SO_4$ concentration by 2–3% by adding buffer A and then centrifuge again.

Sucrose Gradient Centrifugation. Linear 10–40% (w/v) sucrose gradients are prepared in 50 m*M* Tris–HCl (pH 7.5), 0.5 m*M* $MgSO_4$, 1 m*M* DTT, 6 m*M* *p*-aminobenzamidine, 0.28% (w/v) deoxycholate, and 1 mg/ml of *E. coli* phospholipid. Inclusion of phospholipid in the gradients is

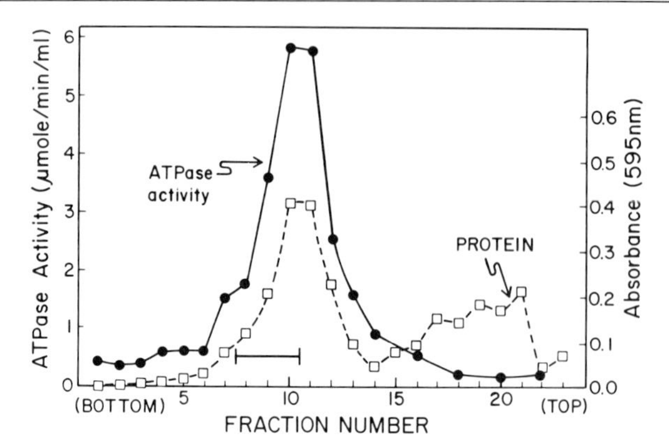

FIG. 2. Sucrose density gradient centrifugation of $(NH_4)_2SO_4$ fractionated ATPase complex. The fractions pooled (8–10 as indicated by the bar) are composed of pure F_1F_0 complex. The ATPase activity (●) and relative protein content as estimated by the Bradford assay (□) are indicated for each fraction. Adapted from Foster and Fillingame.[2]

essential for retention of the reconstituted activities described in the following section.[2] The 10% or 40% gradient solutions lacking phospholipid are added to *E. coli* lipid (dried from chloroform–methanol) at 1 ml/mg lipid, and the lipid dispersed with a bath sonicator at room temperature. The gradient solutions are cooled to 4° and the linear 5-ml gradients generated at 4° in cellulose nitrate tubes shortly before application of the samples. The resuspended 25–35% $(NH_4)_2SO_4$ cut is clarified by centrifugation for 10 min at 50,000 rpm in a Beckman 50 Ti rotor, and 0.2 to 0.3-ml aliquots applied to the 5-ml 10–40% sucrose gradients. The gradients are centrifuged at 50,000 rpm for 9 hr at 4° in a Beckman SW 50.1 or SW55Ti rotor. Following centrifugation the gradients are fractionated into 23 fractions of 0.22 ml following puncture of the tube bottom. The qualitative ATPase "spot" assay is usually used to quickly judge the location of the peak of ATPase activity. The fraction exhibiting maximum ATPase activity and the next 2 or 3 faster sedimenting fractions are combined (see Fig. 2). The fractions immediately preceding the fraction with peak ATPase activity are usually slightly contaminated with polypeptides, showing apparent molecular weights of 76,000, 26,000, and 15,000 on SDS–polyacrylamide gel electrophoresis. These contaminating proteins appear to be associated with an orange/yellow-colored band visible in the gradients about 5 mm above a slightly hazy, colorless band that coincides with the position of peak ATPase activity. If absolute purity is required, these slower sedimenting fractions should be discarded.

TABLE I
PURIFICATION OF F_1F_0-ATPASE FROM WILD-TYPE *E. coli* MEMBRANE

Step	Protein (mg)	ATPase (units)	Yield (%)	Specific activity (units/mg)
Membrane	400.0	188	100	0.47
Centrifuged dialyzed extract	164.0	156	83	0.95
25–35% $(NH_4)_2SO_4$ cut	20.2	53	28	2.60
Sucrose gradient pool	1.3	17	9	12.90

Yield and Purity. The purification procedure is summarized in Table I. The yield of purified product is typically low, <10% of the ATPase activity of the beginning membrane preparation. We should note that the purification scheme was developed using membranes from cells grown on succinate rather than glucose.[2] The yield from these membranes is higher, but the product is invariably contaminated with minor amounts of polypeptides that are not components of F_1 or F_0.[2] The yield shown in Table I is reduced by half due to omission of the slow-sedimenting fractions containing minor contaminants (Fig. 2). The purity of the product from glucose-grown cells is demonstrated in Fig. 3. Contaminating polypeptides (the tiny blips on the scan shown) represent <2% of the Coomassie Brilliant Blue staining material.

Purification of F_1F_0 from Induced λ-unc$^+$ Cells

Heat induction of λ*asn5*, the λ-*unc*$^+$ lysogen of strain KY7485, results in a 4- to 8-fold increase in the ATPase activity of membranes relative to monoploid cells.[3] The highest level of ATPase is observed with cells grown on minimal medium. The high concentration of ATPase in the membrane required that the purification procedure be modified. Purification from properly induced cells is considerably easier than from monoploid cells, the yield greater (see Table II), and the F_1F_0 preparation nearly as pure.

The membrane is extracted as described above and the solubilized extract dialyzed. The particulate pellet from the dialyzed extract is resuspended by homogenization at 20 mg/ml in buffer D and resolubilized by addition of solid KCl to 0.25 *M*, Na-deoxycholate to 1.25%, and Na-cholate to 0.5%. After stirring at 0° for 5 min, the resolubilized extract is centrifuged for 20 min at 50,000 rpm in a Beckman 50 Ti rotor (227,000

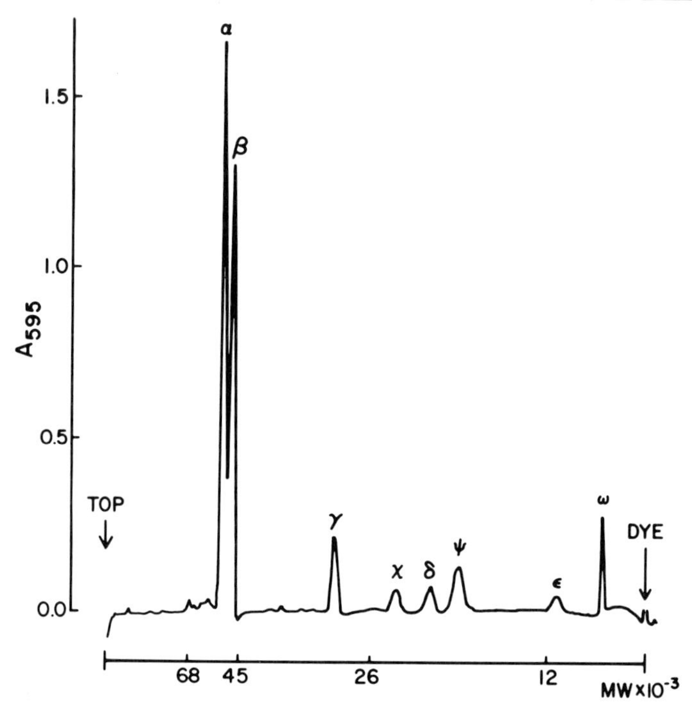

FIG. 3. Electrophoresis of purified F_1F_0 complex purified from wild-type membranes through acrylamide gels containing SDS. The sample shown (5 μg protein) is from the pooled fractions indicated in Fig. 2. The gel was fixed and stained with Coomassie Brilliant Blue as described in the text, and after destaining scanned at 595 nm with a Gilford spectrophotometer equipped with a linear transporter. "Dye" indicates the position of the Bromphenol Blue tracking dye on the gel. The subunits of F_1 are indicated by α, β, γ, δ, and ε, and subunits of F_0 by χ, ψ, and ω. The apparent M_r of subunits, based upon the migration of standards, are indicated on the horizontal axis. Adapted from Foster and Fillingame.[2]

TABLE II
PURIFICATION OF F_1F_0-ATPASE FROM INDUCED λ-*unc*$^+$ *E. coli* MEMBRANES

Step	Protein (mg)	ATPase (units)	Yield (%)	Specific activity (units/mg)
Membrane	100	300	100	3.0
Centrifuged dialyzed extract	28	144	48	5.1
Resolubilized dialyzed extract	—	114	38	—
Sucrose gradient pool	4	54	18	13.5

g_{max}). Aliquots of 0.2–0.3 ml from the resultant supernatant are applied to the 5-ml 10–40% sucrose gradients, which are centrifuged and fractionated as described above.

Reconstitution of Function

To 1.0 ml of the pooled sucrose gradient fractions (~250 μg protein) are added 0.164 ml buffer E and 1.25 ml of a sonicated suspension of 4 mg/ml *E. coli* phospholipid in buffer E. This mixture is adjusted to 0.38% (w/v) Na-deoxycholate and 1 mM $MgSO_4$ by addition of 66 μl 10% Na-deoxycholate and 20 μl 100 mM $MgSO_4$, and the clarified suspension dialyzed against 100 volumes buffer A at 4° for 48 hr, with changes of buffer at 6, 12, and 30 hr. If the protein concentration of the sucrose gradient fractions is much greater than 250 μg/ml, the material should be diluted to maintain a lipid/protein ratio >15.

Dicyclohexylcarbodiimide(DCCD)-Sensitive ATPase. Reconstituted F_1F_0 in ATPase assay buffer is treated with 20 μM DCCD at 30° for 20 min prior to addition of [γ-^{32}P]ATP. DCCD should reduce ATPase activity by 70–85%. Under these assay conditions the activity of F_1 is only marginally affected by DCCD (0–10% inhibition).

$^{32}P_i$–ATP Exchange. The reconstituted F_1F_0 gives an exchange activity of 50 nmol/min/mg protein,[2] which is 10–12 times the specific activity of the native monoploid membrane.

ATP-Dependent H^+ Pumping. Reconstituted F_1F_0 liposomes (0.1 ml) are added to 1.0 ml 10 mM HEPES/KOH, pH 7.5, containing 5 mM $MgCl_2$ and 0.3 M KCl. Quinacrine (Sigma) is added to 1.5 μg/ml and the fluorescence measured in an Aminco Bowman spectrophotofluorometer with excitation and emission wavelengths of 450 nm and 505 nm, respectively. Addition of ATP to 0.8 mM results in a quenching of fluorescence (30–50% of initial fluorescence) that is thought to result from accumulation of the dye inside the liposomes in response to the ΔpH generated by the H^+-pumping ATPase. Quenching is reversed by the protonophore, carbonyl cyanide M-chlorophenylhydrazone (CCCP), or by DCCD.[2]

ATP Synthesis. We have reconstituted F_1F_0 in liposomes with bacteriorhodpsin and demonstrated light-dependent ATP synthesis.[16] Since we believe that the methods used are less than optimal, only a brief description will be given here. Purple membrane is prepared from *Halobacterium halobium* strain S9 by the method of Oesterhelt and Stoeckenius.[17] Liposomes containing bacteriorhodopsin are prepared by suspending 80 mg

[16] D. L. Foster, Ph.D thesis, University of Wisconsin, Madison (1980).

[17] D. Oesterhelt and W. Stoeckenius, this series, Vol. 31, p. 667.

acetone-precipitated soybean phospholipids (asolectin) and 4 mg purple membrane protein in 3 ml 10 m*M* Tricine/NaOH (pH 8.0), 25 m*M* K_2SO_4, 0.2 m*M* Na_2EDTA, 5% (v/v) methanol. After 60 min at room temperature, the suspension is sonicated in a water-jacketed vessel at 18° for five 3-min periods separated by 2-min intervals using a microprobe and Heat Systems Ultrasonic W200R sonicator. Purified F_1F_0 (1.3 ml), taken directly from the sucrose gradient, is added to 1.5 ml of the bacteriorhodopsin liposomes and mixed with 0.31 ml 10% Na-cholate. The mixture is dialyzed versus 150 volumes of 10 m*M* Tricine/NaOH (pH 8.0), 25 m*M* K_2SO_4, 1 m*M* dithiothreitol, 1 m*M* $MgCl_2$, 5% (v/v) methanol at 4° for 48 hr with three changes of external buffer at 6, 12, and 30 hr. The resultant liposomes (0.8 ml) are mixed with 4 ml 20 m*M* Na-phosphate (pH 8.0), 30 m*M* D-glucose, 100 m*M* sucrose, 100 m*M* Na_2SO_4, 0.1 m*M* Na_2EDTA, 1 m*M* $MgSO_4$ containing 100 units hexokinase (Sigma), and 65 μCi Na[^{32}P]-phosphate. Tubes are incubated at 37° and illuminated with a 500-W slide projector lamp at 15 cm. ATP synthesis is initiated by addition of 0.2 ml 30 m*M* ADP. Aliquots of 0.5 ml are withdrawn at 20- to 30-min intervals and the amount of esterified [$^{32]P}$phosphate determined as in the [^{32}P]P_i–ATP exchange assay.

Under these conditions the reconstituted F_1F_0 synthesizes 390 nmol ATP/mg F_1F_0 protein in the initial 30 min, which corresponds to roughly 200 mol ATP/mol enzyme and turnover number of 6 ATP/min/F_1F_0. The low efficiency may be due to poor incorporation of F_1F_0 into the bacteriorhodopsin liposomes by cholate dialysis, a problem previously reported by Ryrie *et al.* for yeast mitochondrial F_1F_0.[18]

Preparation and Reconstitution of F_0. An F_0 preparation that is still partially contaminated by F_1 (~20%) can be prepared from the purified F_1F_0 complex as described by Negrin *et al.*[19] When reconstituted into liposomes by sonication, this F_0 preparation exhibits high rates of H^+ translocation (10–12 μmol/min/mg protein) driven by a valinomycin-induced K^+ diffusion potential. Altendorf and co-workers have subsequently described a better procedure for preparing F_0, which is summarized in this volume.[20]

Reproducibility and Comparison to Other Methods

In our laboratory, the purification method described consistently yields product of the purity indicated, and the procedure has been repro-

[18] I. J. Ryrie, C. Critchley, and J. E. Tillberg, *Arch. Biochem. Biophys.* **198,** 182 (1979).
[19] R. S. Negrin, D. L. Foster, and R. H. Fillingame, *J. Biol. Chem.* **255,** 5643 (1980).
[20] E. Schneider and K. Altendorf, this volume [57].

duced by at least one other laboratory.[21] Other workers have had difficulty with the method (H. Kanazawa, Okayama University, personal communication). Any of the following factors may be crucial: (1) The strain of *E. coli* used. The efficiency of extraction of the ATPase with detergents does vary with the strain used.[22] (2) The growth conditions are of critical importance. Pure F_1F_0 is obtained from monoploid cells only after growth on glucose minimal medium. Pure F_1F_0 is only obtained from the induced λ-*unc*$^+$ strain when membrane ATPase exceeds the monoploid level by >4-fold. (3) We suspect that the $(NH_4)_2SO_4$ fractionation may be critically affected by variations in protein concentration and/or ATPase content. (4) The preparation of lipid used in the gradient may be crucial in resolving F_1F_0 from contaminants. We have substituted soybean phospholipid for *E. coli* phospholipid and in many experiments achieved equivalent resolution. However, for reasons that are not clear, successful resolution of the complex depended upon the preparation of soybean phospholipid used.

The major advantage of the method described is the consistent purity of product. The ratio of subunits in the final product is reasonably constant from preparation to preparation.[14] The obvious drawbacks of the methods described are the relatively low yields and inherent difficulty in significantly increasing the scale of the procedure.

Coincident with publication of our original purification procedure, Friedl *et al.*[23] reported an alternative method that generated a product with major amounts of several contaminating polypeptides. With minor revisions this method now yields a product of nearly equivalent purity,[24] although reports from other laboratories still indicate significant amounts of these impurities.[25,26] The most troublesome contaminant is a polypeptide of M_r 26,000 on SDS–gel electrophoresis that we had difficulty in eliminating when we began with cells grown on succinate.[2] The polypeptide is a major impurity of the slower sedimenting sucrose density gradient fractions that we normally discard, and can easily be mistaken for the chi (χ) subunit. The Friedl *et al.* method has the advantage of higher yields and is easier to scale up. A large-scale version of this alternative method is described elsewhere in this volume.[27]

[21] H. R. Lotscher, C. deJong, and R. A. Capaldi, *Biochemistry* **23,** 4128 (1984).
[22] J. H. Verheijen, Ph.D. thesis, University of Amsterdam, (1980).
[23] P. Friedl, C. Friedl, and H. U. Schairer, *Eur. J. Biochem.* **100,** 175 (1979).
[24] P. Friedl and H. U. Schairer, *FEBS Lett.* **128,** 261 (1981).
[25] E. Schneider and K. Altendorf, *Eur. J. Biochem.* **126,** 449 (1982).
[26] J. P. Aris and R. D. Simoni, *J. Biol. Chem.* **258,** 14599 (1983).
[27] P. Friedl and H. U. Schairer, this volume [58].

[56] Use of λ-*unc* Transducing Phages in Genetic Analysis of H^+-ATPase Mutants of *Escherichia coli*[1]

By ROBERT H. FILLINGAME and MARY E. MOSHER

The structural genes coding the subunits of the F_1F_0 H^+-ATPase of *Escherichia coli* are organized in an operon designated *unc*.[2,3] Mutants of the H^+-ATPase have been reported by a number of laboratories. An understanding of the lesions in these mutants should provide insight into the mechanism of how the enzyme works. An essential prerequisite to understanding these mutants is definition of the gene altered by mutation. The altered gene can in many cases be defined by genetic complementation. Diploid *unc*/*unc* strains, with the two copies of the *unc* operon originating from independent mutants, are generated in a recombination-negative (*recA*) background. If the two mutations lie in different genes, some wild-type enzyme should be produced from the products of the "complementing" trans wild-type genes of the diploid. As described previously in this series,[4] Gibson and co-workers developed the first genetic complementation system for *unc* by use of an F′ episome carrying *unc* for generation of *unc*/*unc* diploid cells.[5]

We have developed a system for genetic complementation using λ-*unc* transducing phages.[6] The λ-*unc* transducing phage is lysogenized in a cell carrying a different *unc* mutation to generate the heterodiploid cell. Using the strains described, a λ-transducing phage carrying any *unc* mutation can be generated. The series of λ-*unc* transducing phages described here can be used to genetically define the mutationally altered *unc* gene in other *E. coli* host strains. The system has the advantage that the λ-*unc*/*unc* homodiploid strain can be used to amplify the mutant *unc* DNA by heat induction of λ-*unc* prophage replication. A convenient source of mutant *unc* DNA is thus provided which can be used for cloning and DNA sequencing of the altered gene. Amplification of the *unc* DNA also results

[1] The work described was supported by United States Public Health Service Grant GM-23105.

[2] Eight structural genes have been identified and are designated *B, E, F, H, A, G, D,* and *C* in order of transcription. The genes code, respectively, the χ, ω, ψ subunits of F_0, and δ, α, γ, β, and ε subunits of F_1 (see Fillingame and Foster, this volume [55] for composition).

[3] A ninth gene, designated *uncI,* of unknown function lies between the promoter and the *B* gene [N. J. Gay, *J. Bacteriol.* **158,** 820 (1984)].

[4] G. B. Cox and J. A. Downie, this series, Vol. 56, p. 106.

[5] F. Gibson, G. B. Cox, J. A. Downie, and J. Radik, *Biochem. J.* **164,** 193 (1977).

[6] M. E. Mosher, L. K. Peters, and R. H. Fillingame, *J. Bacteriol.* **156,** 1078 (1983).

TABLE I
E. coli STRAINS USED IN GENERAL CONSTRUCTIONS

Strain	Genotype or description	Reference
MM284	*ilv* :: *Tn10, bglB* :: λ*cI857S7*, *bglR*, Δ[*gal-bio*]	6
ER	*asnA31, asnB32, thi-1*	9
ER(λ)	λ*cI857S7* lysogen of strain ER	—
AN346	*ilvC7, entA403, pyrE41, argH1, rpsL109, supE44*	10
MH2869	*recA56, srl-1300* :: *Tn10, ilv-318, thr-300, thi-1, spc-300*, [P1 *clr100 cml*]	6

in overproduction of the mutant H^+-ATPase complex and can aid in its biochemical characterization. The rationale and protocols described here closely follow the procedures used by Miki *et al.*[7] for generation of λ-*asn5*, a λ*unc*$^+$ transducing phage.

Materials and General Procedures

Media. The M63 minimal medium and LB-rich medium that we routinely use are described by Miller.[8] We routinely add thiamine–HCl (0.2 μg/ml) to all minimal media. The M63 minimal buffer and salts are supplemented with 0.2 m*M* uracil, 20 μ*M* 2,3-dihydroxybenzoic acid, 0.2 m*M* L-arginine, 0.67 m*M* L-asparagine, 0.4 m*M* thymine, and 0.4 μ*M* biotin as required. D-Glucose (0.2%) is routinely used as a carbon source. Alternate carbon sources used are 0.6% sodium succinate hexahydrate for distinguishing *unc*$^+$ from *unc*$^-$, 0.5% arbutin as indicator of *bglR*$^-$,*C*$^+$, and 0.5% salicin as indicator of *bglR*$^-$,*B*$^+$. Minimal soft agar contains 0.1% NaCl, 10 m*M* $MgSO_4$, and 0.8% agar.

Strains of E. coli. The basic strains needed for the constructions outlined are listed in Table I.[6,9,10] New strains are generated by generalized transduction with P1 *clr100 cml* according to the protocol of Miller.[8] When the recipient strain is a λ-*cI857S7* lysogen, the transduction is done at 32° and chloramphenicol-resistant transductants (P1 lysogens) are discarded. Strains are made lysogenic for λ*cI857S7* by infection on LB plates at 32°. Cells from the turbid area at the site of infection are tested for λ

[7] T. Miki, S. Hiraga, T. Nagata, and T. Yura, *Proc. Natl. Acad. Sci. U.S.A.* **75,** 5099 (1978).

[8] J. H. Miller, "Experiments in Molecular Genetics" Cold Spring Harbor Laboratory, Cold Spring Harbor, New York, (1972).

[9] J. Felton, S. Michaelis, and A. Wright, *J. Bacteriol.* **142,** 221 (1980).

[10] F. Gibson, G. B. Cox, J. A. Downie, and J. Radik, *Biochem. J.* **162,** 665 (1977).

prophage by cross-streaking with Charon 1 and $\lambda v_1v_2v_3$ lysates.[6] Charon 1-immune (repressor plus), $\lambda v_1v_2v_3$-sensitive (receptor plus) λ lysogens are purified.

Construction of λ-*uncX* Transducing Phage

λ-Transducing phage carrying any *unc* mutation, designated *uncX* here, can be generated by the following protocol. We begin with strain MM284 in which λ*cI*857*S*7 is inserted in the *bglB* gene, the bglB gene product being required for growth on salicin. This λ strain is nonlytic due to the *S*7 mutation and heat inducible for replication due to the *cI*857 mutation. The MM284 strain retains a $bglC^+$ gene, the product of *bglC* being required for growth on arbutin, and is $bglR^-$, a mutation that results in expression of *bglB* and *bglC*. The *bgl* locus lies near and to the left of *unc* at 84 min on the *E. coli* chromosome map. To the right of *unc* are located the *asnA* (asparagine) and *ilv* (isoleucine/valine) loci, which cotransduce with *unc* with a frequency of ~50%. Strain MM284 is ilv^- due to *Tn*10 insertion in the *ilv* operon. The *uncX* gene is transferred into MM284 by cotransduction with ilv^+ from the donor. The λ-*uncX* transducing phage is generated by a rare abnormal excision of λ from the chromosome, where the $bglR^-,C^+$ to *asnA* region of the chromosome is recombined into the λ DNA and packaged into the phage particle. An essential portion of the λ genome is lost during the recombination, giving rise to the transducing phage, and a normal "helper" phage is required for subsequent replication or lysogeny of the transducing phage DNA.

Step 1. The *uncX* mutation from an ilv^+ donor strain is transduced into strain MM284 (*ilv* :: *Tn*10, *bglB* :: λ-*cI*857*S*7) by selecting for ilv^+ recombinants on minimal medium plates supplemented with D-glucose and biotin at 32°. The ilv^+ recombinants are screened for cotransduction of *uncX* by plating on minimal medium supplemented with succinate and biotin. The strain should retain an arbutin(+)/salicin(−) phenotype for growth that is indicative of the $bglR^-$, $bglC^+$, *bglB* :: λ*cI*857*S*7 genotype.

Step 2. Heat treatment of the *uncX* transductant of strain MM284 promotes excision of λ from the chromosome and the infrequent generation of transducing phage. A 250-ml culture of the *uncX* transductant is grown to ~0.5 O.D. (550 nm) on LB medium, and while the cells are still growing exponentially, the culture is shifted to 42° for 25 min. Shaking is continued at 37° for 3 hr. The cells are centrifuged and resuspended in 5 ml 20 m*M* trishydroxymethylaminomethane (Tris–HCl), pH 7.4, 10 m*M* $MgSO_4$, 100 m*M* NaCl, 0.1% gelatin (Knox), and lysed by mixing with a few drops (~0.1–0.2 ml) chloroform. After removal of cell debris by centrifugation, the supernatant is heated briefly at 37° to remove residual chloroform and stored at 4°. This cell lysate is referred to subsequently as the "low-frequency transducing lysate."

The rare λ-*uncX* transducing phage is selected from the low-frequency transducing lysate by transduction of the *asnA*$^+$ locus carried on the transducing phage into an *asnA*$^-$ recipient.[11] In the presence of helper λ phage, the transducing phage will lysogenize at the λ-*att* site on the chromosome in tandem with the λ helper.

Step 3. A recipient strain carrying *uncX* at the normal chromosomal locus must be constructed for the λ transduction described above. This is because recombination can occur between the transducing phage DNA and homologous regions of the chromosome, even in a *recA* background. This is indicated by experiments where mixtures of λ-*uncX* and λ-*unc*$^+$ transducing phage were generated when the λ-*uncX* was lysogenized in a *unc*$^+$, *recA* background.[6]

The appropriate *uncX* recipient is constructed in two steps. A spontaneously generated *bglR*$^-$ derivative of an *uncX* strain is selected based upon growth of such derivatives on minimal medium containing arbutin as the carbon source. The *bglR*$^-$ and *uncX* markers are transduced from this donor to strain ER, selecting initially for growth of the ER transductant on minimal medium supplemented with asparagine and arbutin. Single colonies are picked and streaked on arbutin plus asparagine minimal plates, and single colonies are then screened for *uncX* (lack of growth on succinate) and *asnA* (asparagine auxotrophy). The *bglR*$^-$, *uncX*, *asnA* recombinants are found at a frequency of ~2%.

Step 4. The *bglR*$^-$, *uncX*, *asnA*, *asnB* recipient is transduced to Asn$^+$ with the low-frequency λ-transducing lysate (from Step 2). A single colony of the recipient strain is grown to late exponential phase in 5 ml LB medium, centrifuged, and resuspended in 0.4 ml 10 m*M* $MgSO_4$.[12] After a further incubation with shaking at 37° for 45 min, 12.5 μl of the recipient cells are mixed with 0.5 ml of the low-frequency transducing lysate and 10^{10} PFU of λ*cI*857*S*7. After 20 min at 32°, 0.1-ml aliquots are plated on minimal medium supplemented with glucose and incubated 3 days at 32°.

The Asn$^+$ colonies generated above are initially screened for temperature-sensitive growth (32° v 42°) to ensure that they are λ*cI*857*S*7 lysogens. They are then screened for capacity to produce transducing phage complementing strain ER (λ) (the λ*cI*857*S*7 lysogen of strain ER). The

[11] The recipient is also *asnB*$^-$, since mutations in both genes are required for asparagine auxotrophy.

[12] A fresh culture of the recipient, which has not reached stationary phase, is used here to avoid Asn$^+$ revertants. The recipient culture is usually plated on glucose minimal medium as a control to ensure that most of the Asn$^+$ colonies from the λ transduction arise due to transduction of *asnA*$^+$ on the transducing phage.

Asn$^+$ colonies are spread in patches on minimal glucose plates and after 2 days at 32° replica plated with velvet onto LB plates. The LB plates are incubated at 42° for 4 hr to induce phage production and the patches of cells lysed by exposure to chloroform vapor for 5 min. The lysed patches are replica plated onto minimal glucose plates on which the ER (λ) recipient cells are spread after resuspension in 10 m*M* $MgSO_4$. As discussed below, several Asn$^+$ transductants which test positively for λ-*asnA*$^+$ transducing phage should be saved, since the frequency of transducing phage production varies widely from strain to strain.

High-frequency transducing lysates of the λ-(*asnA*$^+$,*uncX*)/*uncX* diploid, selected in Step 4 above, are prepared by the same method as given in Step 2 and should generate a lysate with $>10^9$ PFU of helper phage per milliliter.

Genetic Complementation

Construction of Recipients. The recipients used for complementation tests with λ-*unc* are made *recA* so that only intergenic complementation is detected. The recipients are also made lysogenic for λ*cI*857*S*7 to prevent extensive killing by the helper phage in the λ-*unc* high-frequency lysate. We routinely test for complementation in the AN346 background (Table I).[13] The mutation that is to be tested in the recipient (referred to here as *uncY*) is transduced into AN346 by selection of *ilv*$^+$ recombinants with subsequent scoring for lack of growth on succinate minimal medium. The *recA* mutation is transferred by P1 transduction from donor strain MH2869 (*recA, srl* :: *Tn*10) with selection of tetracycline-resistant recombinants. Approximately 50% of the tetracycline-resistant transductants are *recA* as determined by sensitivity to UV light (exposure to 15-W General Electric G15T8 germicidal lamp at 60 cm for 1 min). The strains are then lysogenized with λ*cI*857*S*7 as described in the general methods section.

Spot Test for Complementation. Recipient *uncY* cells are grown overnight in LB medium, and 0.1 ml of the saturated culture mixed with 2.5 ml of minimal soft agar and poured onto a succinate minimal medium plate. A wire loop (3–5 μl) of high-frequency transducing *uncX* lysate is spotted on the surface and growth scored after 4 days at 32°. Heavy growth at the spot of mixing is indicative of genetic complementation.

Verification of Complementation. Some combinations of λ-*uncX*/*uncY* show no growth or very slow growth in the complementation spot

[13] This strain reportedly carries the suppressor mutation, *supE44* (*E. coli* Genetic Stock Center, Yale University). Several mutations that we have tested (e.g., *uncB206, uncH5*) appear to be suppressed when transduced into this background.

test described above. We have used other tests to confirm that such a result should usually be scored as negative. In the case of weak growth, cells at the site of the spot can be streaked for single colonies on arbutin minimal medium at 32°. The λ-*uncX* transductants of the *uncY* recipient (in the AN346 background) should show growth due to λ transduction of *bglR*$^-$, *bglC*$^+$. Alternatively, or in the case of no growth in the spot complementation test, arbutin-positive transductants can be selected directly. The *uncY* recipient strain is plated in minimal soft agar on an arbutin minimal medium plate, and 5–10 μl of the λ-*uncX* high-frequency transducing lysate spotted on the plate. Cells growing at the spot of infection are streaked onto another minimal arbutin plate, and single colonies are tested for production of *asnA*$^+$ transducing phage, as described above.

The degree of complementation in such λ-*uncX*/*uncY* diploids is tested in two ways: (1) the time required for formation of single colonies on succinate minimal medium at 32°, and (2) the growth yield on limiting glucose (0.05%) liquid minimal medium. Diploid strains showing strong complementation form single colonies of >0.2 mm on succinate medium within 2 days and show an increase in growth yield from 50 to >80%, the percentages being relative to the isogenic *unc*$^+$ strain.

Lysates Giving Weak or Negative Complementation. The high-frequency transducing lysates formed by different λ-(*asnA*$^+$, *uncX*)/*uncX* isolates can vary considerably in titer of transducing phage. Such strains also differ in the extent of overproduction of F_1F_0 on heat induction. Examples of the variability are indicated in Table II. For a given strain, the relative proportion of transducing phage in different lysates may vary by 5- to 10-fold (not shown). The frequency of transducing phage production by different λ-Asn$^+$ transductants carrying the same *unc* allele may vary by >100-fold. It is important to pick a transductant which gives a high titer of helper phage (>10^9 PFU/ml), and also a relatively high frequency of Asn$^+$ transducing phage production.

Interpretation of Genetic Complementation Analysis

Nomenclature. According to established and widely used convention,[14] each mutant allele of a given phenotype (e.g., *unc* for H^+-ATPase) is assigned a new number, e.g., *unc-401*. This assigned allele number is used to unambiguously identify the mutant allele in any strain to which it is subsequently transferred. Capital letters following the three-letter mnemonic abbreviation for phenotype are used to designate different genetic complementation groups (genes), e.g., *uncA401* once the *unc-401* muta-

[14] B. J. Bachmann, *Microbiol. Rev.* **47,** 180 (1983).

TABLE II
COMPARISON OF Asn$^+$ TRANSDUCING FREQUENCY WITH HIGH-FREQUENCY TRANSDUCING LYSATES FROM DIFFERENT λ-*unc*/*unc* DIPLOID STRAINS

Strain	*unc* genotype	High-frequency transducing lysate[a] (PFU/ml) × 10^9	(Asn$^+$/PFU) × 10^{-3}
MM549	λ-*uncA401*/*uncA401*	3.3	0.6[b]
MM917	λ-*uncB108*/*uncB108*	10	4–28[c]
MM1092	λ-*uncE114*/*uncE114*	2.1	0.003
MM1095	λ-*uncE114*/*uncE114*	3.3	0.3
MM1097	λ-*uncE114*/*uncE114*	1.3	6–12[c]
MM1098	λ-*uncE114*/*uncE114*	2.3	0.2
MM1028	λ-*uncF469*/*uncF469*	6.0	0.7–2.5[c]
MM1197	λ-*uncF469*/*uncF469*	4.4	7.6
MM1198	λ-*uncF469*/*uncF469*	4.7	15
MM1200	λ-*uncF469*/*uncF469*	3.0	1.2
MM1203	λ-*uncF469*/*uncF469*	1.5	273

[a] The high-energy frequency-transducing lysate was titered for helper phage (expressed as plaque-forming units, PFU) according to Miller with strain Y_{mel}.[8] The efficiency of Asn$^+$ transducing phage formation was estimated by mixing serial dilutions of the transducing lysate with 5 × 10^8 cells of strain ER (λ) and 1.5 × 10^9 λ*cI857S7* helper phage. After incubation at 32° for 20 min in 10 m*M* $MgSO_4$, the cells were plated on glucose minimal medium and Asn$^+$ transductants scored. The frequency of Asn$^+$ transduction is normalized to the PFU in each lysate.

[b] Lysates with this frequency of transducing phage are minimally satisfactory for the complementation spot test.

[c] Range of two or three determinations with the same lysate.

tion has been assigned to the *A* complementation group. The occasional and misleading practice of referring to *unc* mutants as "*uncA*" or "*uncB*," depending upon whether F_1 or F_0 is defective, should be abandoned.

Mutations Showing Simple Complementation Pattern. A number of mutant *unc* alleles show a simple complementation pattern indicative of a single defective gene. Representative λ-*unc* diploid strains and *unc* recipients are given in Table III.[15–20] For example, the λ-*uncB402* transducing

[15] R. H. Fillingame, M. E. Mosher, R. S. Negrin, and L. K. Peters, *J. Biol. Chem.* **258,** 604 (1983).

[16] M. E. Mosher, L. K. White, J. Hermolin, and R. H. Fillingame, *J. Biol. Chem.* **260,** 4807 (1985).

[17] L. K. White and R. H. Fillingame, unpublished.

[18] B. P. Rosen, *J. Bacteriol.* **116,** 1124 (1973).

TABLE III
λ-*unc* DIPLOID STRAINS AND *unc* RECIPIENT STRAINS SUITABLE FOR GENETIC COMPLEMENTATION ANALYSIS

unc Allele of diploid or recipient	λ-*unc*/*unc* Diploid strain (Ref.)[a,b]	*unc* Recipient (Ref.)[c]
unc+	MM598(15)[a]	MM180(6)
uncA401	MM549(6)[a]	MM495(6)
uncB108	MM917(6)[b]	MM444(6)
uncB402	MM502(15)[a]	MM449(6)
uncC424	—	MM990(6)[d]
uncD11	MM946(6)[b]	MM841(6)
uncD409	—	MM991(6)
uncE106	MM913(6)[b]	MM586(6)
uncE114	MM1097(16)[b]	MM994(6)
uncE123	LW67(17)[b]	MM1045(6)
uncE429	—	MM992(6)
uncF120	LW132(17)[b]	MM1063(6)
uncF469	MM1203(6)[b]	MM993(6)
uncG70	—	MM455[e]
uncH239	—	RH339(λ)(19)[f]

[a] Diploid strain derived from strain AN346 by cotransduction of *unc* with *ilv*$^+$, and λ-*unc* transduction.

[b] Diploid strain derived from strain ER as described in text.

[c] With the exception of RH339(λ), all recipient strains are *ilv*$^+$, *unc* cotransductants of strain AN346 made *recA56*, *srl* :: *Tn10* and lysogenic for λ*cI857S7*.

[d] This recipient strain grows slowly on succinate minimal medium.

[e] The *unc* mutation from strain NR-70 (Ref. 18) was transduced into strain AN346. The mutation shows a B$^+$E$^+$F$^+$A$^+$G$^-$D$^+$C$^+$ complementation pattern with the F′-*unc* system used by Gibson and co-workers (F. Gibson and L. Langman, personal communication).

[f] Strain RH339 (*recA56*, *srl* :: *Tn10*, *bglR*, *thi-1*, *rel-1*, *Hfr*P01) was made lysogenic for λ*cI857S7*.

phage complements each recipient strain listed, with the exception of the *uncB402* and *uncB108* recipients. In the reciprocal cross, the *uncB402* recipient is complemented by each of the λ-*unc* transducing phages listed, except λ-*uncB402* and λ-*uncB108*.

[19] R. Humbert, W. S. A. Brusilow, R. P. Gunsalus, D. J. Klionsky, and R. D. Simoni, *J. Bacteriol.* **153,** 416 (1983).

[20] D. A. Jans, L. Hatch, A. F. Fimmel, F. Gibson, and G. B. Cox, *J. Bacteriol.* **160,** 764 (1984).

The complementation groups established by λ-*unc* transductions faithfully match those established by Gibson and co-workers using F′-*unc* episomes as complementation vectors.[6] The *uncA401, uncB402, uncC424, uncD409, uncE429,* and *uncF469* alleles listed in Table III were isolated and genetically categorized by Gibson and co-workers and are routinely used in their genetic analyses.[20]

Mutations Showing Dominance. Several of the *unc* recipients that we have tested do not show complementation in the spot test with any of the λ-*unc* transducing phage. Examples are the *uncE105* and *uncE107* recipients.[6] In these two cases, this is due to dominance by the mutant *uncE* gene of the recipient.[6,21] Both mutants generate a normal F_1-ATPase, and should be complemented by λ-*uncA* and λ-*uncB* transducing phage. Heterodiploid strains of *uncE107* have been examined more extensively for indications of complementation.[6] The λ-unc^+/*uncE*107 heterodiploid shows weak growth in the spot test for complementation, which would normally be scored as negative. However, this heterodiploid forms single colonies on succinate within 2 days at 32° and shows a growth yield on limiting glucose equal to wild type. These properties of the λ-unc^+/*uncE*107 heterodiploid are exceptional in that all other heterodiploids showing weak growth in the complementation spot test consistently show reduced growth yields and do not form or require prolonged incubations for single-colony formation on succinate minimal medium. Both the λ-*uncA401*/*uncE107* and λ-*uncE107*/*uncA401* heterodiploids show weak or negligible growth in the complementation spot test and growth yields no greater than either mutation in monoploid. However, both heterodiploids give small single colonies on succinate minimal medium after a prolonged (9 day) incubation at 32°. Neither of the monoploid mutant strains gives single colonies, even after this extended incubation. The latter result suggests a legitimate but very weak complementation between the *uncE107* and *uncA401* alleles. However, the extent of complementation is so low that an increase in functional F_1F_0 complexes cannot be detected biochemically.[6]

The dominance of the *uncE105* and *uncE107* mutations can be rationalized by the multiple copies of the uncE gene product, ω, in F_0. The current best estimate of stoichiometry is 10 ± 1 ω per F_0.[22] The probability of forming a purely wild-type F_0 in an $uncE^+/uncE^-$ heterodiploid is then $(1/2)^{10} = 1/64$. However, this simple rationale does not account for the differences in extent of complementation observed on comparing λ-unc^+/*uncE107* and λ-*uncA401*/*uncE107* heterodiploids.

[21] P. Friedl, C. Friedl, and H. U. Schairer, *FEBS Lett.* **119,** 254 (1980).

[22] D. L. Foster and R. H. Fillingame, *J. Biol. Chem.* **257,** 2009 (1982).

As discussed below, the *uncE105* and *uncE107* mutations are complemented well by multicopy plasmids carrying the *unc*$^+$ gene, as judged by colony formation on succinate and growth yields on limiting glucose.[23] Although the complementation remains only partial, as analyzed biochemically by function of F_1F_0,[23] it can be rationalized by the increase in ratio of wild-type to mutant genes.

In summary, the complementation system described does in some cases fail to identify the altered gene and may give misleading patterns of complementation. In many mutants showing strong genetic complementation, the mutant gene products are not incorporated into F_0 or F_1, and thus do not compete for assembly with the wild-type gene product in the diploid. Of the mutations listed in Table III, this is true for the *uncB108, uncB402, uncE106, uncE429, uncF469* mutations of F_0,[6,15,20,24] and the *uncD409* of F_1.[25] However, all mutations showing a simple genetic complementation pattern are not defective in assembly. For example, an F_1 of normal size and composition is formed in *uncA401* and *uncD11* mutants,[26,27] and a normal F_0 is assembled in the case of the *uncE114* mutant.[16]

Comparison to Complementation by Other Methods. As indicated above, the complementation results obtained with *unc* heterodiploids formed with transducing phages are generally consistent with those obtained with F′-*unc* heterodiploids. More recently, genetic analyses have been carried out with multicopy plasmids carrying cloned segments of the *unc*$^+$ operon.[19,28] Positive complementation in a *recA* background requires that the cloned gene be transcribed from a promoter endogenous to the plasmid DNA. The efficiency of transcription from different promoters varies and may thus affect the results obtained.

We have limited information comparing complementation of the *uncB* and *uncE* genes with the two types of systems. An *uncB*$^+$ plasmid, pBP120 (containing the wild-type *uncB* gene in a segment of DNA from the *Hin*dIII site prior to the *uncB* gene to the *Ava*I site in the *uncE* gene, cloned behind the Tet^R promoter in plasmid pBR322), complements both *uncB108* and *uncB402* as judged by growth on succinate minimal medium.

[23] R. H. Fillingame, L. K. Peters, L. K. White, M. E. Mosher, and C. R. Paule, *J. Bacteriol.* **158,** 1078 (1984).

[24] J. A. Downie, G. B. Cox, L. Langman, G. Ash, M. Becker, and F. Gibson, *J. Bacteriol.* **145,** 200 (1981).

[25] D. R. H. Fayle, J. A. Downie, G. B. Cox, F. Gibson, and J. Radik, *Biochem. J.* **172,** 523 (1978).

[26] H. Kanazawa, S. Saito, and M. Futai, *J. Biochem. (Tokyo)* **84,** 1513 (1978).

[27] H. Kanazawa, Y. Horiuchi, M. Takagi, Y. Ishino, and M. Futai, *J. Biochem. (Tokyo)* **88,** 695 (1980).

[28] T. Noumi and H. Kanazawa, *Biochem. Biophys. Res. Commun.* **111,** 143 (1983).

Plasmid pCP35 containing the *uncE*$^+$ gene (DNA from the *Bam*HI site in *uncB* to the *Hpa*I site between the *uncE* and *uncF* genes) cloned into the Tet^R gene of pBR322 complements all *uncE* mutants tested.[16,23] The *uncE*$^+$ gene in pCP35 is presumed to be transcribed from the Tet^R promoter. There is an unusual difference in degree of complementation by pCP35 versus various λ-*unc* transducing phages, which merits consideration. The *uncE106* mutation is complemented strongly by λ-*unc* phage, but extremely poorly by plasmid pCP35, whereas the *uncE105* and *uncE107* mutations are complemented extremely weakly by λ-*unc* phage, but very well by plasmid pCP35.[23] A reasonable explanation for this difference is not apparent at this time. The example does emphasize the limitations of any single method in defining an altered *unc* gene and emphasizes that negative or weakly positive complementation results should be interpreted with caution.

We have cloned the *uncE*$^+$ gene behind the *lac* promoter in plasmid pUC8. In a *lacI*^{Q} host, expression of the gene is dependent upon the concentration of inducer of the *lac* promoter, isopropylthio-β-D-galactoside (IPTG). A positive complementation response (growth on succinate) by transformants of different *uncE* host strains depends upon the level of expression of the *uncE*$^+$ gene on the plasmid, as regulated by IPTG concentration.[29] Clearly then, at least for mutant *uncE* alleles, a positive or negative complementation response will depend upon the level of expression of the wild-type *uncE* gene.

Conclusions

The λ-*unc* genetic complementation system described here is useful in defining the gene of the *unc* operon altered by mutation. The degree of intergenic complementation by heterodiploid *unc*$^+$ alleles in a trans configuration will vary with the type of mutation and degree of expression of the *unc*$^+$ allele. Hence, negative complementation tests should be interpreted with caution. A positive genetic complementation response can be used to identify the gene altered by mutation as well as genes that are normal. The normalcy indicated by growth in the complementation test does not rule out the possibility of secondary mutations, which may be insufficient by themselves in generating a succinate-minus phenotype.

[29] D. Fraga and R. Fillingame, unpublished.

[57] Proton-Conducting Portion (F_0) from *Escherichia coli* ATP Synthase: Preparation, Dissociation into Subunits, and Reconstitution of an Active Complex

By ERWIN SCHNEIDER and KARLHEINZ ALTENDORF

The F_0 part of the ATP synthase complex (F_1F_0) of *Escherichia coli* is composed of three different polypeptides (a, b, and c or χ, ψ, and ω)[1-3] with the proposed stoichiometry of $1:2:10 \pm 1$.[4] The complex is thought to serve as a proton channel. Whether the F_0 complex undergoes conformational changes during proton translocation remains to be established.

Starting from purified F_1F_0 complex the F_0 part can be isolated in high yield. Furthermore, the F_0 complex can be dissociated into the individual subunits, which can be recombined to form an active F_0 complex.

Assays

Principles. After reconstitution into liposomes a functional F_0 complex can be tested either by measuring passive proton translocation or by adding back F_1. Passive proton translocation through F_0 is measured using K^+-loaded liposomes. Addition of valinomycin elicits K^+ efflux, thereby generating a K^+ diffusion potential across the membrane, internally negative. This provides a driving force for H^+ uptake, which in its turn is monitored by a pH electrode.[5]

ATPase activity of reconstituted F_1F_0 complexes is measured by determining the liberated orthophosphate according to the method of Fiske and Subbarow[6] in a continuous flow apparatus.[7] The proper binding between F_0, incorporated into liposomes, and F_1 is examined by measuring ATP-dependent quenching of 9-amino-6-chloro-2-methoxyacridine (ACMA) fluorescence.[8,9]

[1] D. L. Foster and R. H. Fillingame, *J. Biol. Chem.* **254,** 8230 (1979).
[2] P. Friedl and H. U. Schairer, *FEBS Lett.* **128,** 261 (1981).
[3] E. Schneider and K. Altendorf, *Eur. J. Biochem.* **126,** 149 (1982).
[4] D. L. Foster and R. H. Fillingame, *J. Biol. Chem.* **257,** 2009 (1982).
[5] H. Hirata, K. Altendorf, and F. M. Harold, *J. Biol. Chem.* **249,** 2939 (1974).
[6] C. H. Fiske and Y. Subbarow, *J. Biol. Chem.* **66,** 375 (1925).
[7] A. Arnold, H. U. Wolf, B. P. Ackermann, and H. Bader, *Anal. Biochem.* **71,** 209 (1976).
[8] G. Vogel, this series, Vol. 55F [86].
[9] R. Kraayenhof and J. W. T. Fiolet, *in* "Dynamics of Energy-Transducing Membranes" (L. Ernster, R. W. Estabrook, and E. C. Slater, eds.), p. 355. Elsevier, Amsterdam, 1974.

Proton translocation through F_0, ATPase activity of reconstituted F_1F_0 complexes, and ACMA fluorescence quenching are specifically inhibited by *N,N'*-dicyclohexylcarbodiimide (DCCD), which covalently binds to an aspartic acid residue in subunit c.[10] Therefore, inhibition by DCCD of the above-listed activities is routinely assayed.

Protein is determined by the method of Lowry *et al.*[11] with the modification by Dulley and Grieve.[12]

Passive Proton Translocation

Materials

Buffer A: 0.2 m*M* *N*-[2-hydroxy-1,1-bis(hydroxymethyl)ethyl]glycine (Tricine)–NaOH
5 m*M* $MgSO_4$
0.2 *M* Na_2SO_4
Final pH 7.0

Stock solutions: 10 μg/ml valinomycin (in ethanol)
200 μ*M* 4,5,6,7-tetrachloro-2-trifluoromethylbenzimidazole (TTFB) (in ethanol)
40 m*M* DCCD (in ethanol)
1 m*M* HCl

Procedure

Passive proton translocation is monitored continuously with a Radiometer electrode (GK2321C) connected to a pH meter (Radiometer model PHM84) and a recorder. The medium in a glass vessel is stirred and kept at 25°.

K^+-loaded proteoliposomes (50 μl) are suspended in a final volume of 2 ml in buffer A. The initial pH is around 7. After an equilibration period between 10 and 20 min proton uptake is started by the addition of valinomycin (2 μl). K^+ loading of control liposomes lacking protein is checked by adding an uncoupler (1 μl of TTFB). Inhibition of H^+ uptake by DCCD (2 μl) is monitored after preincubation for 20 min at 25° in buffer A.

For calibration, 5 μl of 1 m*M* HCl are used.

[10] W. Sebald and J. Hoppe, *Curr. Top. Bioenerg.* **12,** 1 (1981).

[11] O. H. Lowry, N. J. Rosebrough, A. L. Farr, and R. J. Randall, *J. Biol. Chem.* **193,** 265 (1951).

[12] J. R. Dulley and P. A. Grieve, *Anal. Biochem.* **64,** 136 (1975).

ATPase Activity

Materials

Buffer B: 50 m*M* Tris–HCl, pH 8.0
Stock solutions: 100 m*M* ATP
100 m*M* $MgCl_2$
40 m*M* DCCD (in ethanol)

Procedure

ATPase activity of reconstituted F_1 proteoliposomes (5–20 μl) is measured in 10 ml of buffer B, in the presence of 1 m*M* ATP and 1 m*M* $MgCl_2$ at 37°. One unit of ATPase activity is defined as the amount of enzyme that catalyzes the turnover of 1 μmol of ATP per minute. For inhibition with DCCD (80 μM) identical aliquots are incubated in 1 ml buffer B for 30 min at 37° and subsequently assayed as mentioned above.

ATP-Dependent Quenching of ACMA Fluorescence

Materials

Buffer C: 20 m*M* Tricine–NaOH, pH 8.0
300 m*M* KCl
10 m*M* $MgCl_2$
Stock solutions: 1 m*M* valinomycin (in ethanol)
100 m*M* ATP
200 μM ACMA
200 μM TTFB (in ethanol)
80 m*M* DCCD (in ethanol)

Procedure

Reconstituted F_1 proteoliposomes (50–100 μl) are sonicated in a bath-type sonicator (Branson B220) for 30 sec. Subsequently they are diluted with buffer C to a final volume of 2 ml and incubated with valinomycin (3 μl) for 5 min at 37°. The reaction mixture is then transferred to a cuvette and the fluorescence intensity (measured in a spectrofluorimeter, Aminco SPF500) at an excitation wavelength of 410 nm and an emission wavelength of 490 nm is set to zero. After the addition of ACMA (20 μl), the final fluorescence intensity is chosen as 100% (full scale of the recorder). The quenching of fluorescence is monitored after the addition of ATP (40 μl) at 37°. The reaction is uncoupled by the addition of TTFB (10 μl) or inhibited by incubation with DCCD (2 μl) prior to the valinomycin treatment for 20 min at 37°.

Preparation of F_0 by Hydrophobic Interaction Chromatography

Principle. Purified F_1F_0 is bound via the F_0 portion to a column of agarose-coupled poly(L-lysine)-deoxycholic acid.[13] The enzyme complex is then dissociated by urea[14] and F_1 is eluted. Finally, F_0 is eluted by washing with a detergent-containing buffer.[3]

Materials

Bacteria: *E. coli* K12 (λ) is grown at 37° with vigorous aeration in a 100-liter fermenter in a minimal salt medium[15] supplemented with 0.4% glucose to late logarithmic phase and harvested by centrifugation.

Enzyme: F_1F_0 is prepared according to the procedure published by Friedl and Schairer (this volume [58][14,16]). The active ATPase fractions from DEAE Sepharose CL-6B chromatography (the final step in the purification of F_1F_0 according to Friedl *et al.*[16]) are collected and centrifuged for 15 hr at 220,000 *g*. The enzyme pellet is removed from the tube by using a spatula, thoroughly resuspended in buffer (see Step 2) by homogenization to a final concentration of 20–50 mg/ml, and stored in liquid nitrogen.

Immobilized poly(L-lysine)-deoxycholic acid (Pierce)

Buffers and solutions:

D 10 m*M* Tris–HCl, pH 8.0
150 m*M* NaCl
0.2 m*M* phenylmethylsulfonyl fluoride (PMSF)

E 10 m*M* Tris–HCl, pH 8.0
150 m*M* NaCl
5 m*M* EDTA
5 m*M* dithiothreitol (DTT)
0.1% (w/v) deoxycholate
7 *M* urea (2× crystallized from ethanol)
0.2 m*M* PMSF
20% (v/v) glycerol

[13] P. Cresswell, *J. Biol. Chem.* **254,** 414 (1979).
[14] E. Schneider and K. Altendorf, *FEBS Lett.* **116,** 173 (1980).
[15] B. D. Davis and G. S. Mingioli, *J. Bacteriol.* **60,** 17 (1950).
[16] P. Friedl, C. Friedl, and H. U. Schairer, *Eur. J. Biochem.* **100,** 175 (1979).

F 10 m*M* Tris–HCl, pH 8.0
150 m*M* NaCl
0.2 m*M* PMSF
10% (v/v) glycerol
2% (w/v) cholate

G 10 m*M* Tris–HCl, pH 8.0
150 m*M* NaCl
2% (w/v) deoxycholate

0.3 *M* NaCl

Saturated ammonium sulfate solution at 25° adjusted to pH 8.0 by NH_4OH

Procedure

Step 1. Packing of the Column. A slurry of immobilized poly(L-lysine)-deoxycholic acid in buffer D is poured into a column (1 × 20 cm) and allowed to pack under gravity. The upper flow adapter is then adjusted (about 1 mm above the gel surface) and the buffer is suctioned through the column by a peristaltic pump (flow rate 1.5 ml/hr). The material is washed with at least 1 bed volume of buffer D.

Step 2. Removal of F_1 and Elution of F_0. (These steps are carried out at 4°.) Purified ATP synthase (14.1 mg) in 0.5 ml 50 m*M* Tris–HCl, pH 8.0, 10 m*M* taurodeoxycholate, 1 m*M* $MgCl_2$, 0.2 m*M* dithiothreitol, 0.2 m*M* EGTA, 0.1 m*M* phenylmethylsulfonyl fluoride, 6 m*M* *p*-aminobenzamidine, 20% (v/v) methanol, and 50 μg/ml soybean phospholipids is diluted and adjusted to 150 m*M* NaCl by the addition of an equal volume of 300 m*M* NaCl. (Note: If too much detergent is present the F_1F_0 complex does not bind to the column.) This particulate suspension is layered onto the column and washed with the following buffers: (1) two bed volumes of buffer D (removal of excess of protein); (2) one bed volume of buffer E (elution of F_1 and minute amounts of whole complex); (3) two bed volumes of buffer D; (4) one bed volume of buffer F (elution of F_0).

Step 3. Ammonium Sulfate Precipitation. Protein-containing fractions eluted by buffer F are pooled and F_0 is precipitated by adding ammonium sulfate to a final concentration of 40% saturation (0.67 ml/ml of a saturated solution) at room temperature. After incubation for 30 min on ice, F_0 is collected by centrifugation for 20 min at 40,000 *g*. Finally, the precipitate is suspended in ~200 μl of buffer F and stored in liquid nitrogen. A summary of the purification steps is shown in the table.

Step 4. Regeneration of the Column. The column is regenerated at room temperature by washing with several bed volumes of buffer G and subsequent reequilibrated with buffer D.

PURIFICATION OF F_0[a]

Step	Protein (mg)	Yield (%)
F_1F_0 (applied to the column)	14.1	
F_1F_0 (bound to the column)	11.0	100
Buffer E eluate	8.6	78
Buffer F eluate	1.5	14

[a] From Schneider and Altendorf.[3]

Properties

On sodium dodecyl sulfate (SDS)–polyacrylamide gel electrophoresis the F_0 preparation reveals three different polypeptides (Fig. 1) with the apparent molecular weights of 24,000 (a), 19,000 (b), and 8,000 (c), respectively.[3] Based on DNA sequencing data the exact molecular weights have been determined as 30,276 for (a), 17,265 for (b), and 8,288 for (c).[17–19]

Dissociation of the F_0 Complex into Subunits[28]

Materials: Sephadex G-150 superfine (Pharmacia)

Buffers

H 10 m*M* Tris–HCl, pH 8.0
1 m*M* EDTA
1 m*M* DTT
3 *M* trichloroacetic acid (adjusted to neutral pH by NaOH prior to the addition of the remaining components)
1% (w/v) deoxycholate
5% (w/v) Zwittergent 3-14 (Calbiochem)

J 10 m*M* Tris–HCl, pH 8.0
1 m*M* EDTA
1 m*M* DTT
2% (w/v) cholate
1 *M* trichloroacetic acid (adjusted as above)

[17] N. J. Gay and J. E. Walker, *Nucleic Acids Res.* **9,** 3919 (1981).
[18] H. Kanazawa, K. Mabuchi, T. Kayano, T. Noumi, T. Sekiya, and M. Futai, *Biochem. Biophys. Res. Commun.* **103,** 613 (1981).
[19] J. Nielsen, F. G. Hansen, J. Hoppe, P. Friedl, and K. von Meyenburg, *Mol. Gen. Genet.* **184,** 33 (1981).

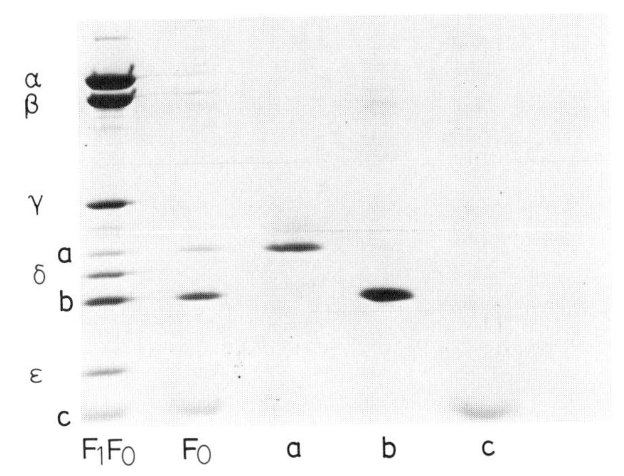

FIG. 1. Subunit composition of *E. coli* ATP synthase (F_1F_0). F_1F_0 (40 μg), purified F_0 (10 μg), and isolated subunits a, b, and c (5 μg of each) were run on a gradient slab gel (7.5–17.5% acrylamide) in the presence of SDS, according to Douglas *et al.*[20] The proteins were visualized with Coomassie Brilliant Blue. Protein bands designated α–ε are subunits of F_1.

K 10 m*M* Tris–HCl, pH 8.0
150 m*M* NaCl
10% (v/v) glycerol
0.2 m*M* PMSF
1% (w/v) cholate

Procedure

Step 1. Dissociation. F_0 (5 mg) is stirred in buffer H at a protein concentration of 1 mg/ml for 20 hr at room temperature.

Step 2. Separation of Subunits. (This step is carried out at 4°.) Subsequently the mixture is layered onto a column (2.6 × 95 cm) containing Sephadex G-150 superfine, equilibrated with buffer J. The dissociated polypeptides are eluted with buffer J which is suctioned through the column by a peristaltic pump at a flow rate of 3–5 ml/hr.

Undissociated protein as well as minor contaminants, mainly minute amounts of α and β (subunits of F_1), still present in the F_0 preparation[3] are running with the void volume. The F_0 subunits are eluted in the order of b, a, and c. Additional absorption peaks (at 280 nm) do not contain any protein and are probably due to detergent micelles.

Step 3. Concentration and Rechromatography. Fractions containing isolated subunits (pure as judged by SDS–gel electrophoresis) are pooled and dialyzed against 50 volumes of buffer K for 20 hr at 4°. Finally, the fractions are concentrated by ultrafiltration (Amicon membrane YM10 in the case of subunits a and b and YM5 in the case of subunit c) to final volumes of ~0.5 ml and stored in liquid nitrogen. From 4 mg F_0, about 230 μg of a, 360 μg of b, and 350 μg of c can be obtained. Routinely, isolated subunits are checked for homogeneity by rechromatography on Sephadex G-150 superfine (1 × 50 cm), equilibrated with buffer K.

Properties

As shown in Fig. 1,[20] the isolated subunits run on SDS gels with the same mobility as in the purified F_0 complex.

Preparation of Proteoliposomes

Procedure I: Cholate Dialysis[21,22] (for F_0 and Isolated Subunits)

Materials

Reagents: L-α-Phosphatidylethanolamine (PE, Sigma P5399)
L-α-Phosphatidylcholine (PC, Sigma P2772)
L-α-Phosphatidylserine (PS, Sigma P6641)
Stock solution: 20% (w/v) cholate
Buffers: K 10 m*M* Tris–HCl, pH 8.0
150 m*M* NaCl
10% (v/v) glycerol
0.2 m*M* PMSF
1% (w/v) cholate
L 10 m*M* Tricine–NaOH, pH 8.0
2.5 m*M* $MgSO_4$
0.2 m*M* EDTA
0.2 m*M* DTT

A phospholipid mixture (8 mg) (PE/PC/PS as 5 : 3 : 2, weight ratio) in chloroform–methanol is dried under a stream of nitrogen and subsequently connected to a vacuum pump for 5 min. Appropriate amounts of protein (F_0: 100 μg; a: 20 μg; b: 23 μg; c: 56 μg or mixtures of subunits; based on stoichiometry and exact molecular weights; see above) in buffer K are added and the mixture is adjusted to 4% (w/v) cholate. The final volume is 250 μl. The sample is thoroughly vortexed, sonicated to clarity for 5–10 min in a bath-type sonicator (Branson B220), and subsequently dialyzed for 20 hr at 4° against 250 ml of buffer L.

[20] M. Douglas, D. Finkelstein, and R. A. Butow, this series, Vol. 56G [6].
[21] Y. Kagawa and N. Sone, this series, Vol. 55F [44].
[22] E. Schneider and K. Altendorf, *Proc. Natl. Acad. Sci. U.S.A.* **81,** 7279 (1984).

Procedure II: Freeze/Thaw[3,23] *(for Intact F_0 Only)*

Materials

Reagent: L-α-Phosphatidylcholine from soybeans (Sigma P 5638)
Stock solution: 100 m*M* $MgCl_2$
Buffer: M 20 m*M* 4-morpholinoethanesulfonic acid (Mes)–NaOH, pH 6.3
0.25 m*M* EDTA

Soybean phosphatidylcholine is acetone washed as described recently.[24] A stock solution (40 mg/ml) is prepared in buffer M by sonication for 10 min in an ice bath under a stream of nitrogen (Branson B14, microtip, pulses of 0.5 sec, output 7). Subsequently, 100 μl of the suspension are mixed with F_0 (40 μg), 1.5 μl $MgCl_2$, and buffer M to give a final volume of 150 μl. The mixture is vortexed, frozen in liquid nitrogen for at least 5 min, and thawed at room temperature (20 min).

Reconstitution of DCCD-Sensitive ATPase Activity and ATP-Dependent Quenching of ACMA Fluorescence[3,25]

Materials: Proteoliposomes, prepared according to procedures I or II

F_1 ATPase (60–80 U/mg) is prepared from *E. coli* strain ML 308-225, as previously described,[26] with the following modification: The gel filtration step is carried out applying Sepharose CL-6b[7] instead of BioGel A 0.5 m.[26]

Buffer: N 50 m*M* Tricine–NaOH, pH 8.0
2 m*M* $MgSO_4$
2 m*M* DTT

Procedure

Proteoliposomes (50 μl) (procedure I; 20 μg of protein) are sonicated in a bath-type sonicator for 30 sec, subsequently suspended in 1 ml of buffer N, and incubated with F_1 (60 μg) for 20 min at 37°. After centrifugation for 15 min at 200,000 *g*, the pellet is washed twice in buffer N (1 ml). Finally, the liposomes are resuspended in 100 μl of buffer N and assayed. F_0-containing liposomes (50 μl; 13 μg of protein), prepared according to

[23] M. Kasahara and P. C. Hinkle, *J. Biol. Chem.* **252,** 7384 (1977).
[24] N. Sone, M. Yoshida, H. Hirata, and Y. Kagawa, *J. Biochem.* (*Tokyo*) **81,** 519 (1977).
[25] K. Steffens, E. Schneider, B. Herkenhoff, R. Schmid, and K. Altendorf, *Eur. J. Biochem.* **138,** 617 (1984).
[26] M. Futai, P. C. Sternweis, and L. A. Heppel, *Proc. Natl. Acad. Sci. U.S.A.* **71,** 2725 (1974).

procedure II, are suspended in 2 ml of buffer N and incubated with F_1 (40 μg) for 20 min at room temperature. Subsequently, the reconstituted F_0F_1 liposomes are collected by centrifugation for 30 min at 300,000 *g*, washed once, and finally resuspended in 100 μl of buffer N.

Reconstitution of Passive H^+-Translocating Activity

Loading of Proteoliposomes with Potassium[27]

Materials: Proteoliposomes, prepared according to procedure I
Buffer: O 20 m*M* sodium phosphate, pH 7.2
0.4 *M* KCl

Procedure

Proteoliposomes (150 μl) are diluted in an Eppendorf tube by one volume of buffer O, thoroughly vortexed and sonicated for 10 sec in an ice-water bath [Branson sonifier B14, special microtip (part no. 101-148-063), 20 pulses/0.5 sec] at an output of 3. Subsequently the sample is frozen in liquid nitrogen for 10 min and thawed at room temperature (20 min). After being sonicated as described above for a second time, the sample is stored on ice until use.

Properties

Purified F_0 incorporated into liposomes either by the dialysis procedure (I) or by the rapid freeze/thaw procedure (II) can bind F_1-ATPase as demonstrated by DCCD-sensitive ATPase activity and ATP-dependent quenching of ACMA fluorescence.[3,25]

Passive proton translocation via F_0 can be shown by using K^+-loaded proteoliposomes prepared by procedure I. The initial rate of H^+ uptake is about 6–9 μmol H^+/min and milligram of protein. By replacing the phospholipid mixture by acetone-washed soybean phospholipids, the rate is reduced by one-half.[3]

Reconstitution of a functional F_0 complex is only achieved by incorporation of all three kinds of subunits into phospholipid vesicles.

[27] N. Sone, T. Hamamoto, and Y. Kagawa, *J. Biol. Chem.* **256,** 2873 (1981).
[28] E. Schneider and K. Altendorf, *EMBO J.* **4,** 515 (1985).

[58] Preparation and Reconstitution of F_1F_0 and F_0 from *Escherichia coli*

By PETER FRIEDL and HANS ULRICH SCHAIRER

The ATP synthase, F_1F_0, is extracted from everted membrane vesicles of *Escherichia coli* by detergent and purified by ion-exchange chromatography and precipitation with polyethylene glycol (PEG).[1] The proton conductor, F_0, is isolated from F_1F_0 by dissociation of F_1 with KSCN and ultracentrifugation of the insoluble F_0.[2] The enzymatic activities of F_1F_0 being assayed are ATPase, ATP–$^{32}P_i$ exchange, and ATP-dependent quenching of acridine dye fluorescence. F_0 is tested for proton conduction by direct pH electrode measurements and by quenching of acridine dye fluorescence. These activities are inhibited by dicyclohexylcarbodiimide (DCCD). The easiest assay is the assay of ATPase activity. It is not disturbed by detergents as long as some residual phospholipid is present. The assay is performed as described by Vogel and Steinhart.[3] One unit of ATPase activity is defined as the amount of enzyme hydrolyzing 1 μmol of ATP per minute at 28°. The other activities are dependent on a vesicular structure, so proteoliposomes have to be reconstituted from the enzyme preparation and phospholipids.

Reconstitution of Proteoliposomes

Method A[1,4]

Reagents

Asolectin (phosphatidylcholine type II-S from Sigma)
Buffer pH 6.3 (20 m*M* Mes, 0.25 m*M* EDTA)
$MgCl_2$, 100 m*M*

Procedure. Buffer (1 ml) was added to 40 mg of soybean lipids in a conical glass tube. The tube was placed in an ice bath and the mixture sonicated with the Microtip of a Labsonic 1510 (Braun, Melsungen) at 50 W output under a gentle stream of nitrogen for 20–30 min. Then 100 μl of the phospholipid suspension are immediately added to 10–30 μl of

[1] P. Friedl, C. Friedl, and H. U. Schairer, *Eur. J. Biochem.* **100,** 175 (1979).
[2] P. Friedl and H. U. Schairer, *FEBS Lett.* **128,** 261 (1981).
[3] G. Vogel and R. Steinhart, *Biochemistry* **15,** 208 (1976).
[4] P. Friedl, B. Schmid, and H. U. Schairer, *Eur. J. Biochem.* **73,** 461 (1977).

enzyme preparation; 1 μl of 100 m*M* $MgCl_2$ is added and the mixture incubated for 10 min at room temperature.

Method B[2,5]

Reagents

Asolectin
Buffer pH 8 (10 m*M* Tricine, 0.2 m*M* EDTA, 0.8% deoxycholate, 1.6% cholate)
Buffer pH 7.8 (10 m*M* Tricine, 2.5 m*M* $MgSO_4$, 50 μ*M* $CaCl_2$)

Procedure. Buffer pH 8 (1 ml) is added to 25 mg asolectin in a conical glass tube. The mixture is sonicated with the Microtip of a Labsonic 1510 at 50 W output for 3–5 min at room temperature. Then 50 μl of the enzyme preparation are added to 200 μl of the phospholipid suspension and the sample dialyzed for 12–18 hr against a 1000-fold volume of buffer pH 7.8 at 12°.

Method A is very fast and reliable; it results in a yellowish silky solution. Method B is a modification of the Kagawa dialysis method. It needs more time and yields somewhat higher enzymatic activities; the resulting vesicle preparation is white and milky.

Assay of ATP–$^{32}P_i$ Exchange[1,4]

Reagents

Buffer pH 7.3 [20 m*M* MOPS, 10 m*M* $MgCl_2$, 300 m*M* KCl, 0.5 m*M* potassium [^{32}P]phosphate (15 Ci/mol)]
ATP, pH 7.0, 100 m*M*
$HClO_4$, 1.5 *M*
Buffer pH 2 (250 m*M* potassium phosphate, 50 m*M* potassium pyrophosphate)
Charcoal filters, 2.5 cm diameter (Binzer, Hatzfeld)
Buffer pH 7 (500 m*M* potassium phosphate, 100 m*M* potassium pyrophosphate)

Procedure. The charcoal filters are pretreated by soaking in buffer pH 7 for 30 min. The buffer is discarded and wet filters dried and stored at 60° until use. Membranes or proteoliposomes (maximal volume 20 μl) are added to 200 μl buffer pH 7.3 in an Eppendorf reaction tube. After a 10-min incubation at 25°, the reaction is started by addition of 5 μl ATP stock

[5] H. Okamoto, N. Sone, H. Hirata, M. Yoshida, and Y. Kagawa, *J. Biol. Chem.* **252,** 6125 (1977).

solution. After 0, 5, 10, and 15 min, 50-μl aliquots of the test sample are pipetted into 5 μl $HClO_4$. The samples are stored on ice for at least 5 min and then centrifuged for 4 min at 11,000 *g*. Then 10 μl of the supernatant are spotted on a charcoal filter. After washing the filter four times with 5 ml buffer pH 2, the radioactivity of the wet filters is determined in a liquid scintillation counter measuring the Cerenkov radiation.

Assay of ATP-Dependent Quenching of Acridine Dye Fluorescence[1]

Reagents

Valinomycin, 1 m*M* methanolic solution
Buffer pH 7.3 (20 m*M* MOPS, 10 m*M* $MgCl_2$, 300 m*M* KCl)
9-Amino-6-chloro-2-methoxyacridine (ACMA), 250 μM ATP, pH 7.0, 200 m*M*

Procedure. Membranes or proteoliposomes (maximum volume 50 μl) are added to 1 ml test buffer; after addition of 2 μl valinomycin, the sample is incubated for 5 min at room temperature. The fluorescence (excitation 410 nm, emission 490 nm) of the sample is calibrated with the zero line on the recorder. After addition of 10 μl ACMA solution, the fluorescence is adjusted to a full-scale deflection on the recorder. When the inner volume of the vesicles is not in pH equilibrium with the outer cuvette volume, there will be a slow increase or decrease of the fluorescence signal. The test must not be started unless the fluorescence signal is constant. Then 10 μl of the ATP solution are added to the sample with rapid mixing. The paper of the recorder runs with a speed of 12 cm/min. The initial rate of fluorescence quenching is extrapolated to 1 min and expressed as units of fluorescence quenching activity. One unit of fluorescence quenching activity (U_{fl}) is defined as the amount enzymatic activity that quenches the fluorescence of ACMA with an initial rate of 100% (1 full scale) in 1 min.

It should be stressed that this test is a quantitative one for proton translocation; for F_1F_0 the fluorescence quenching activity is fully equivalent to the ATP–$^{32}P_i$ exchange and for F_0 fully equivalent to the pH conduction measured by a pH electrode.[2,5-7] ACMA may be substituted by other acridine dyes such as atebrine or 9-aminoacridine, although then the test is less sensitive.

[6] P. Friedl, C. Friedl, and H. U. Schairer, *FEBS Lett.* **119,** 254 (1980).

[7] P. Friedl, G. Bienhaus, J. Hoppe, and H. U. Schairer, *Proc. Natl. Acad. Sci. U.S.A.* **78,** 6643 (1981).

Inhibition by DCCD[4]

Reagents

Buffer pH 8.0 (50 m*M* Tris, 10 m*M* $MgCl_2$)
DCCD, 5 m*M* methanolic solution

Procedure. Membranes are diluted to 1 mg protein per milliliter buffer pH 8; the enzyme preparations are adjusted to 1 mg phospholipid per milliliter by addition of sonicated phospholipid (method A). Then 10 μl DCCD are added and the samples incubated for 45 min at 25°. The enzyme preparations can then be reconstituted to proteoliposomes for assay of ATP–$^{32}P_i$ exchange and quenching of ACMA fluorescence.

Preparation of K^+-Loaded Vesicles[6]

Reagents

Buffer A, pH 7.5 (250 m*M* K_2SO_4, 0.1 m*M* EDTA)
Buffer B, pH 7.5 (250 m*M* Na_2SO_4, 5 m*M* $MgSO_4$)
$MgSO_4$, 1 *M*

Procedure. Membranes (2–5 mg protein) or reconstituted proteoliposomes are sedimented by ultracentrifugation (200,000 *g*, 30 min, 4°) and resuspended in 1 ml buffer A. The suspension is sonicated for 45 sec with the Microtip of a Labsonic 1510 (30 W output) at room temperature and then incubated at 37° for 20 min. After addition of 5 μl $MgSO_4$, the vesicles are sedimented, very gently resuspended in 250 μl ice-cold buffer B, and immediately assayed for H^+ conduction.

Assay of H^+ Conduction[6]

Method A

Reagents

Buffer pH 7.5 (5 m*M* MOPS, 250 m*M* Na_2SO_4, 5 m*M* $MgSO_4$)
ACMA, 250 μM
Valinomycin, 100 μM methanolic solution

Procedure. The setup conditions are identical to those for the measurement of ATP-dependent quenching of ACMA fluorescence. Differences are as follows: 1–10 μl of loaded vesicles are added to 1 ml buffer pH 7.5, and the test is started by the addition of 2 μl valinomycin solution.

Method B

Procedure. Loaded vesicles (50 μl) are added to 3 ml of 250 m*M* Na_2SO_4, 5 m*M* $MgSO_4$. The suspension is gently stirred with a magnetic stirrer and maintained at 25°. The pH of the suspension is measured continuously with an Ingold complex electrode LOT 421 connected to a pHM 64 research pH meter (Radiometer Copenhagen) and a Servogor S recorder. As soon as a stable or steadily drifting baseline is established, the test is started by addition of 2 μl valinomycin solution. At the end of the test the assay is calibrated by addition of 5–10 μl 1 m*M* HCl.

Purification of F_1F_0[1,2]

Growth of Cells

Escherichia coli K12 wild type (in our laboratory strain A 1)[1] or the diploid strain CM 845[8] (a wild type carrying a λ-d*unc* prophage) with twice the amount of ATP synthase in its membranes is used. The cells are grown in Vogel–Bonner[9] mineral salt medium with 1% succinate as the sole carbon source. The cells are fermented in a 50- or 200-liter fermenter at 37° (strain A 1) or 32° (strain CM 845), with maximum aeration at a constant pH of 7. The cells are harvested in the late logarithmic growth phase; the yield of a 50-liter/200-liter fermentation is about 500 g/2.5 kg (wet weight) of cells.

Medium-Scale Preparation

Reagents

- Buffer A, pH 7.0 [50 m*M* MOPS, 175 m*M* KCl, 10 m*M* $MgCl_2$, 0.2 m*M* EGTA, 0.2 m*M* dithiothreitol (DTT), 0.1 m*M*, phenylmethylsulfonyl fluoride (PMSF), 6 m*M* *p*-aminobenzamidine (PAB)]
- Buffer B, pH 8 (5 m*M* Tris, 5 m*M* $MgCl_2$, 0.2 m*M* EGTA, 0.2 m*M* DTT, 0.1 m*M* PMSF, 6 m*M* PAB, 10% glycerol)
- Buffer C, pH 8 (50 m*M* Tris, 1 m*M* $MgCl_2$, 0.2 m*M* EGTA, 0.2 m*M* DTT, 0.1 m*M* PMSF, 6 m*M* PAB)
- Buffer D, pH 8, same as buffer C including 100 m*M* KCl, 20% methanol, 25 m*M* Aminoxid WS 35, 50 μg/ml azolectin lipids
- Sodium cholate, 200 m*M*

[8] F. G. Hansen, J. Nielsen, E. Riise, and K. von Meyenburg, *Mol. Gen. Genet.* **183,** 463 (1981).

[9] H. J. Vogel and D. M. Bonner, *J. Biol. Chem.* **218,** 97 (1956).

The pH of the buffers has to be adjusted at 0°.

KCl, 4 *M*

$MgCl_2$, 1 *M*

Polyethylene glycol 6000 (Serva), 50% (w/v)

Aminoxid WS 35: Acyl(C_8–C_{18})amidopropyldimethylamine oxide; the detergent used has an acyl composition of 7% C_8, 6% C_{10}, 50% C_{12}, 18% C_{14}, 9% C_{16}, 10% C_{18}. The concentration of the stock solution is roughly 1 *M*. It is a commercial detergent delivered by Theo Goldschmidt AG., Postfach 17, 4300 Essen 1, FRG. The company also synthesizes a charge with defined acyl chain length and delivers small samples of defined detergent or commercial batches without charging. The authors can assist those having difficulty obtaining the detergent.

The size-limiting step in the procedure is the available rotor volume for ultracentrifugation. About 7–7.5 g (wet weight) of tightly packed cell sediment results later on in a volume of roughly 10 ml for ultracentrifugation. We are routinely using two 45 Ti rotors (Beckman) with polycarbonate tubes of a total volume of ~840 ml, which allows the processing of maximally 600 g of cells.

Preparation of Membranes

The sedimented cells are suspended in buffer A at an end volume of ~2.5 ml/g wet weight of cells; larger amounts are suspended using an Ultrathurax. Bovine pancreas DNase (40 units) (Boehringer, Mannheim) is added per 100 ml. Cells are broken by passage through a Ribi press at a pressure of 18,000 lb × in^{-2} at 10–15°; an appropriately cooled French press may also be used. Volumes larger than 600 ml are passaged twice through a Gaulin–Manton press at a pressure of 700 kg × cm^{-2} in the cold room (4°); after the first passage the suspension has a temperature of ~20°, and before the next passage the suspension is incubated for 30 min at 4° with stirring. Cell debris is removed by centrifugation at 20,000 *g* for 30 min. The supernatant is centrifuged for 2 hr at 235,000 *g*.

Washing of Membranes

The membrane pellet is suspended in the original volume in buffer B, incubated for 20 min at 0°, and sedimented again by centrifugation at 235,000 *g* for 1.5 hr. The supernatant is discarded and the pellet may be stored in the centrifuge bottle overnight on ice. The last washing step is repeated and the final sediment is resuspended in buffer C (including 10% glycerol) at a protein concentration of 10 mg/ml. Then 1/20 volume of cholate solution is added and the suspension is stirred at 0° for 20 min before centrifugation at 235,000 *g* for 1.5 hr.

Extraction of F_1F_0

The sediment is resuspended in 90% of the original volume in buffer C, and 1% of Aminoxid WS 35 solution is added. The suspension is stirred at 0° for 20 min before centrifugation at 235,000 *g* for 1 hr. The supernatant contains about 50% of the original ATPase activity. Cold methanol is added to a final concentration of 20%, and Aminoxid WS 35 is adjusted to 25 m*M* and KCl to 100 m*M*. The pH of the ice-cold solution is readjusted to pH 8.

DEAE-Sepharose CL-6B Chromatography

The extract is adsorbed on a DEAE-Sepharose CL-6B column previously equilibrated with buffer D. For the extract from 600 g wet weight of cells a 16.5 × 5 cm column (314 ml) is used. The column is washed with 3 volumes of buffer D (including 125 m*M* KCl instead of 100 m*M* KCl). The enzyme is eluted in a 125–300 m*M* KCl gradient in buffer D (a 6-fold column volume). For a 16.5 × 5 cm column the flow rate is 250 ml/hr, the fraction size 20 ml. Usually fractions with an ATPase activity >3 U/ml are pooled, and much of the yellow material eluting shortly before the ATPase activity is avoided.

Concentration of the Enzyme Preparation

A PEG precipitation is tried in a pilot assay: A small sample is taken from the pool, adjusted to 6 m*M* $MgCl_2$ and 12.5% (w/v) PEG 6000, and stirred at 0° for 5 min before centrifugation at 25,000 *g* for 15 min. The sediment is resuspended in buffer D at a protein concentration of about 10 mg/ml. If more than 50% of the original ATPase activity is recovered, the rest of the enzyme solution is also precipitated by PEG. Otherwise the enzyme solution is concentrated by a second chromatography on DEAE-Sepharose CL-6B (0.25× the volume of the first column).

Comments

The purification procedure described here is modified compared to the original procedure: More protein is loaded on the first column. This normally results in an enzyme pool with a protein concentration high enough for a direct precipitation with PEG and with a recovery of more than 80% of the original ATPase activity. The concentrated enzyme solution is

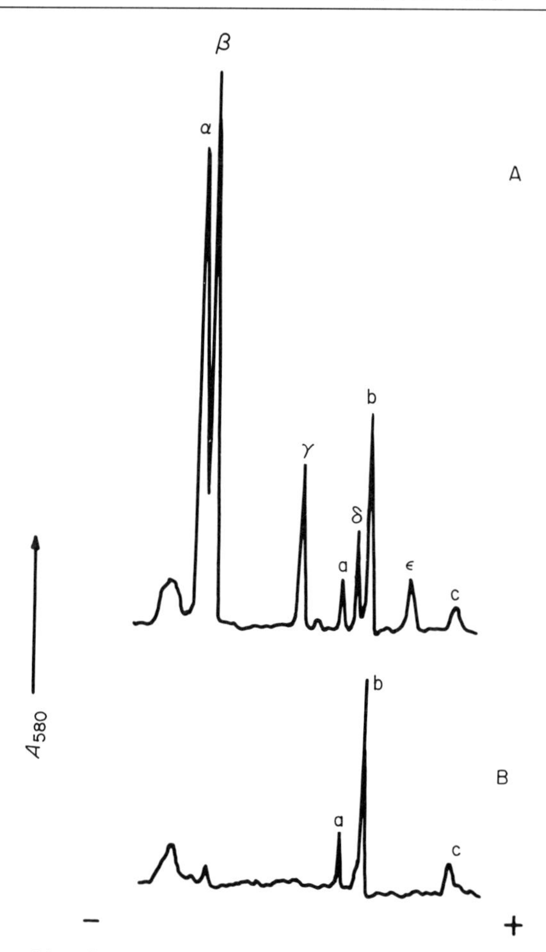

FIG. 1. Polypeptides of (A) the F_1F_0 and (B) the F_0 preparation. The enzyme preparations (15 μg F_1F_0, 4 μg F_0) were subjected to SDS–gel electrophoresis. Protein was stained with Coomassie Brilliant Blue R-250 and the absorbance at 580 nm was recorded. Greek letters denote F_1 subunits; latin letters denote F_0 subunits. Approximate M_r of F_1 subunits: α, 56,000; β, 52,000; γ, 32,000; δ, 21,000; ε, 14,000; approximate M_r of F_0 subunits: a, 24,000; b, 19,000; c, 8,500. The approximate M_r differs from the exact M_r calculated from the primary structures.

aliquoted in 2 or 4 ml polypropylene vials with screw caps and frozen in liquid nitrogen. The vials are stored without loss of enzymatic activities in the gas phase over liquid nitrogen for several years. Repeated thawing and freezing should be avoided.

Properties of F_1F_0

The purification typically has a yield of 30–40% of the ATPase activity of the original membranes. The specific activities of the purified enzyme are ATPase, 22 U/mg; ATP-dependent quenching of ACMA fluorescence, 2000–3000 U_{fl}/mg; ATP–$^{32}P_i$ exchange, 40–60 nmol/min/mg. Figure 1 shows the typical subunit pattern of the enzyme after SDS–polyacrylamide gel electrophoresis.

Large-Scale Preparation

Reagents needed in addition to those for the small and medium-scale procedure: Buffer E pH 8 (5 m*M* Tris–Cl, 6 m*M* PAB, 0.2 m*M* PMSF, 0.2 m*M* DTT, 1 m*M* EDTA).

The procedure demands two refrigerated centrifuges, each with a rotor volume of 2.5 liters, e.g., Sorvall centrifuges with GS3 rotors and 2 ultracentrifuges, each with a rotor volume of 420 ml.

Cells (2.5 kg wet weight) are suspended in 6 liters of buffer A (plus DNase), broken with a Gaulin–Manton press, and the cell debris is removed (as described for the medium-scale procedure). The cell free supernatant, about 4 liters, is adjusted to 10% (w/v) PEG 6000 and stirred for 10 min at 0° before centrifugation at 9000 rpm (Sorvall, GS3 rotor) for 30 min. The sediment is suspended in 4 liters buffer E and stirred for 10 min before the addition of 100 m*M* KCl and 10% PEG 6000. After another 10 min incubation, the membranes are sedimented by centrifugation at 9000 rpm for 30 min. The sediment is resuspended in 4 liters buffer E, stirred for 10 min at 0°, adjusted to 100 m*M* KCl, 5% PEG 6000, stirred for another 10 min, and centrifuged at 9000 rpm for 1.5 hr. The slightly turbid supernatant is discarded and the sediment resuspended in 3.6 liters of buffer E. Cholate (2 m*M*) is added and the suspension is stirred for 30 min at 0° before centrifugation at 9000 rpm for 15–18 hr. The clear yellow supernatant is discarded, and the sediment is resuspended in 1.4 liters buffer C; then 12 m*M* Aminoxid WS 35 is added to one-half of the membrane suspension. The suspension is stirred for 20 min at 0° before centrifugation at 235,000 *g* for 1 hr. Meanwhile, the rest of the membranes are extracted with Aminoxid WS 35. The supernatant is adjusted to 20% methanol, 100 m*M* KCl, 25 m*M* Aminoxid WS 35, as described above. The extract is stirred at 0° for 1 hr and then centrifuged again for 30 min at 235,000 *g*.

About 1.2 g of enzyme is purified from the supernatant by ion-exchange chromatography and PEG 6000 precipitation, as described above. The size of the DEAE-Sepharose CL-6B column is 24 × 8 cm (1.2 liters of gel), the flow rate 500 ml/hr, the fraction size 20 ml, the volume of buffer D (+125 m*M* KCl) 3 liters, gradient volume 5 liters.

Preparation of F_0[2]

Reagents

Buffer F, pH 7 (20 m*M* MOPS, 10 m*M* $MgCl_2$, 100 m*M* KCl, 12 m*M* taurodeoxycholate, 20% (w/v) glycerol)
Buffer G, pH 7.8 (50 m*M* Tris, 0.2 m*M* $MgCl_2$)
Buffer H, pH 7.8 (50 m*M* Tris, 1 m*M* $MgCl_2$, 1 m*M* DTT)
KSCN, 4 *M*
EDTA, 100 m*M*
DTT, 100 m*M*
PEG 400 (Serva)

F_1F_0 (10 mg/ml) is diluted to 2 mg/ml in buffer F, precipitated by addition of an equal volume PEG 400, incubated at 0° for 5 min, and centrifuged at 40,000 *g* for 10 min. The sediment is resuspended in the original volume buffer F, and the precipitation is repeated. The final pellet is resuspended in twice the original volume buffer G and centrifuged at 220,000 *g* for 10 min. The supernatant is decanted and adjusted to 1.2 m*M* EDTA, 1 m*M* DTT, 1 *M* KSCN, and incubated at 0° for 20 min before centrifugation at 220,000 *g* for 45 min. The sediment is resuspended in buffer H at roughly 2 mg/ml.

The preparation looks turbid and silky and contains F_0 probably in a particulate form. The specific activities of H^+ conduction/translocation after reconstitution of proteoliposomes are pH electrode, 870 nmol H^+/min/mg; quenching of ACMA fluorescence, 2000–2600 U_{fl}/mg; ATP-dependent quenching of ACMA fluorescence after addition of F_1, 2000–2300 U_{fl}/mg. The typical subunit pattern of F_0 after SDS–polyacrylamide gel electrophoresis is shown in Fig. 1.

[59] Use of Isolated Subunits of F_1 from *Escherichia coli* for Genetic and Biochemical Studies

By MASAMITSU FUTAI and HIROSHI KANAZAWA

In analyzing the structure and mechanism of action of multisubunit enzymes such as F_0F_1, resolution and reconstitution of the subunits seems to be one of the most fruitful approaches. Isolation of the five subunits of F_1 and their reconstitution have been achieved only with F_1

from two bacterial species, the thermophilic bacterium PS3 and *Escherichia coli*. The F_1 from the thermophilic bacterium PS3 was dissociated with 8 *M* guanidine–HCl, and the pure subunits were separated by column chromatography on CM- and DEAE-cellulose in the presence of 8 *M* urea.[1] However, procedures using urea or guanidine–HCl were unsuccessful for *E. coli* F_1 (EF_1), mainly because the enzyme was denatured irreversibly by these reagents. The δ and ε subunits of EF_1 could be obtained by treatment with 50% pyridine,[2] but the other subunits were irreversibly precipitated by this treatment. Taking advantage of the cold sensitivity of F_1, a property which has been recognized since the first purification of beef heart F_1,[3] EF_1 was dissociated and its α, β, and γ subunits were obtained in reconstitutively active forms.[4,5] Nucleotide binding sites were found in the isolated α[5] and β[6] subunits. The $\alpha\beta\gamma$ complex has ATPase activity,[4,5] and the original F_1 was reconstituted by adding δ and ε to this $\alpha\beta\gamma$ complex.[5] The isolated subunits have also been useful in identifying defective subunits in mutants of F_1.[6–11] The procedure used for EF_1 is much milder than that developed for the thermophilic enzyme. Thus, the isolated *E. coli* subunits, especially subunits α and β, do not seem to be unfolded, and the reconstitution process may not involve renaturation. Isolated subunits from mutants and the wild-type strain may give further information on the structure and mechanism of action of the enzyme. We may be able to form hybrid enzymes from mutant and wild-type subunits and analyze their mechanisms of action, just as hybrid enzymes could be formed from subunits of F_1 of the thermophilic bacterium, *E. coli*, and *Salmonella typhimurium*.[12–14] Here we

[1] M. Yoshida, H. Okamoto, N. Sone, H. Hirata, and Y. Kagawa, *Proc. Natl. Acad. Sci. U.S.A.* **74,** 936 (1977).

[2] J. B. Smith and P. C. Sternweis, *Biochemistry* **16,** 306 (1977).

[3] M. E. Pullman, H. S. Penefsky, A. Data, and E. Racker, *J. Biol. Chem.* **235,** 3322 (1960).

[4] M. Futai, *Biochem. Biophys. Res. Commun.* **79,** 1231 (1977).

[5] S. D. Dunn and M. Futai, *J. Biol. Chem.* **255,** 113 (1980).

[6] M. Hirano, K. Takeda, H. Kanazawa, and M. Futai, *Biochemistry* **23,** 1652 (1984).

[7] H. Kanazawa, S. Saito, and M. Futai, *J. Biochem. (Tokyo)* **84,** 1513 (1978).

[8] S. D. Dunn, *Biochem. Biophys. Res. Commun.* **82,** 596 (1978).

[9] H. Kanazawa, Y. Horiuchi, M. Takagi, Y. Ishino, and M. Futai, *J. Biochem. (Tokyo)* **88,** 695 (1980).

[10] H. Kanazawa, T. Noumi, M. Futai, and T. Nitta, *Arch. Biochem. Biophys.* **223,** 521 (1983).

[11] H. Kanazawa, T. Noumi, N. Oka, and M. Futai, *Arch. Biochem. Biophys.* **227,** 596 (1983).

[12] M. Futai, H. Kanazawa, K. Takeda, and Y. Kagawa, *Biochem. Biophys. Res. Commun.* **96,** 227 (1980).

[13] K. Takeda, M. Hirano, H. Kanazawa, N. Nukiwa, K. Kagawa, and M. Futai, *J. Biochem.* **91,** 695 (1982).

[14] S.-Y. Hsu, M. Senda, H. Kanazawa, T. Tsuchiya, and M. Futai, *Biochemistry* **23,** 988 (1984).

discuss the preparation of subunits of EF_1 for genetic and biochemical studies.

Preparation of Subunits of EF_1

EF_1 in an EDTA extract of membranes (defined below) can be purified by chromatography on DEAE-Sepharose CL-6B and gel filtration through BioGel A-0.5 m, essentially according to a previous procedure,[15] and used for isolation of subunits. This procedure gives F_1 with 4 subunits (without δ) from *E. coli* K12 strains, and with 5 subunits from strain ML308-225. Recently, we used a diploid strain KY7485 (*asn31, thi, rif*/λ*asn*-5 (cI857-S7, *bglR, asn, unc*)/λ*c*I857S7)[16] for preparation of wild-type F_1. For this procedure cells are grown aerobically at 30° in a synthetic medium (50 liters)[17] supplemented with 1.0% glucose and 2 μg/ml of thiamine.[18] In the middle of the logarithmic phase, the defective transducing phage λ*asn*-5 (carrying genes for F_0F_1) is induced by shifting the temperature of the culture to 42°, and incubation is continued for 30 min. Then the temperature is decreased to 37° and incubation is continued for 2 hr. The cells are then collected and used for preparation of F_1. A slightly modified procedure[9] can be used for purification of mutant F_1.[7–11] The procedure described in detail below is carried out at 4° unless otherwise specified.

Preparation of α, β, and γ Subunits

Dissociation of EF_1.[4,5] Purified EF_1 from *E. coli* K12 is precipitated by the addition of solid ammonium sulfate to 70% saturation, redissolved in dissociation buffer [50 m*M* succinate-Tris (pH 6.0), 1.0 *M* NaCl, 250 m*M* $NaNO_3$, 0.1 m*M* dithiothreitol, and 1.0 m*M* EDTA] at a concentration of 3–5 mg/ml and dialyzed overnight against 100 volumes of the same buffer with one change of the buffer. NaCl and $NaNO_3$ can be replaced by KCl and KNO_3, respectively, with essentially the same results. Extensive dialysis is required to remove the ATP used during purification of F_1. Insoluble matter, if any, appearing during dialysis is removed by centrifugation at 15,000 *g* for 10 min, and the clear solution is rapidly frozen in a dry ice–ethanol bath and stored at −80° for at least 2 days before isolation of subunits.

[15] M. Futai, P. C. Sternweis, and L. A. Heppel, *Proc. Natl. Acad. Sci. U.S.A.* **71,** 2725 (1974).

[16] H. Kanazawa, T. Miki, F. Tamura, T. Yura, and M. Futai, *Proc. Natl. Acad. Sci. U.S.A.* **76,** 1126 (1979).

[17] S. Tanaka, S. A. Lerner, and E. C. C. Lin, *J. Bacteriol.* **93,** 642 (1967).

[18] M. Senda, H. Kanazawa, T. Tsuchiya, and M. Futai, *Arch. Biochem. Biophys.* **220,** 398 (1983).

Hydroxyapatite and DEAE-Sepharose Chromatography.[5] The solution of dissociated F_1 (40–60 mg protein in dissociation buffer) is diluted with an equal volume of buffer A [50 m*M* Tris–HCl (pH 8.0), 0.1 m*M* EDTA, and 0.1 m*M* dithiothreitol] and applied to a column (1.4×25 cm) of hydroxyapatite (BioGel HTP, Bio-Rad). Fractions of 4 ml are collected. The column is washed with 50 ml of buffer A to elute the salt in the dissociation buffer. Complete washing of thc column can be monitored by following the absorbancy of the elute at 280 nm due to nitrate. Then a protein peak containing the α and β subunits is eluted with 300 ml of 30 m*M* sodium phosphate (pH 8.0) in buffer A. Fractions with absorbancy at 280 nm of more than 0.05 are eluted within 50 ml with this buffer, and the recovery of the protein in these fractions is about 70%. However, the entire 300 ml of 30 m*M* sodium phosphate in buffer A is necessary for removal of all traces of α and β subunits; this extensive washing is essential for obtaining pure γ subunit. The column is then washed with 50 ml of buffer A. A linear gradient of 0–150 m*M* sodium phosphate (pH 8.0) in 400 ml of buffer A containing 3 *M* urea is then applied. Three protein peaks are obtained: the third peak, eluted with about 70 m*M* phosphate, contains the γ subunit. Fractions with absorbancy at 280 nm of more than 0.025 are pooled and immediately dialyzed extensively against buffer A containing 10% glycerol and 200 m*M* NaCl. The γ subunit eluted constitutes about 5% of the protein applied. Sometimes the three peaks are not completely separated. Thus, we usually subject portions of each fraction of the third peak to polyacrylamide gel electrophoresis (in the presence of sodium dodecyl sulfate) and pool fractions containing the pure γ subunit for use in experiments. The first and second peaks from the column are the ε subunit contaminated with the α and β subunits and residual α and β, respectively. Pure ε can be obtained from the first peak by gel filtration through Sephadex G75 (Pharmacia).

The fraction of α plus β obtained from the hydroxyapatite column is applied to a column (2.0×20 cm) of DEAE Sepharose CL-6B (Pharmacia). The column is developed with a linear gradient of 0–400 m*M* NaCl in 500 ml of buffer A. The α subunit (about 20% of the dissociated F_1) is eluted as a sharp first peak with 0.15 *M* NaCl and the β subunit (about 25% of the dissociated F_1) as a broader second peak with about 0.25 *M* NaCl. These subunits are dialyzed against buffer A, concentrated with an Amicon UM10 (Amicon Corp.) membrane, and stored at $-80°$. This procedure is also applicable to F_1 from *S. typhimurium* with essentially the same results (14). The hydoxyapatite column can be used only once. Bragg *et al.*[19] scaled down the procedure, applying 12 mg of dissociated F_1

[19] P. D. Bragg, H. Stan-Lotter, and C. Hou, *Arch. Biochem. Biophys.* **213,** 669 (1982).

to a hydroxyapatite column of 1 × 1.8 cm. They also reported a modified procedure for isolation of a defective α subunit from an *uncA* mutant.

Hydrophobic Column Chromatography. The three subunits can be purified by hydrophobic column chromatography of the dissociated F_1[4]: The γ subunit can be obtained by applying the dissociated F_1 to a butyl-agarose (Miles-Yeda Ltd.) column and the α and β subunits can be isolated by phenyl-Sepharose (Pharmacia) column chromatography. However, chromatographic conditions for these hydrophobic resins have not yet been well standardized, and this method gives lower recoveries of the subunits than the method using hydroxyapatite and DEAE-Sepharose.

Preparation of δ and ε Subunits

Pyridine Treatment and Column Chromatography. The following procedure was developed by Smith and Sternweis.[2] About 20 mg of F_1 from *E. coli* strain ML308-225 is brought to 65% saturation of ammonium sulfate by adding solid salt. The precipitate is dissolved in 5 ml of 10 m*M* glycine-NaOH and 1 m*M* EDTA, pH 9.0, and 5 ml of pyridine is added dropwise with stirring at 23°. The mixture is stirred for 10 min and then diluted with 15 ml of water. Then 26 μl of saturated ammonium sulfate is added and the suspension is incubated at 4° for 4–5 hr. The following steps are carried out at 4°. The mixture is centrifuged at 20,000 *g* for 15 min. The supernatant is dialyzed against 5 liters of 5 m*M* Tris–HCl, 1 m*M* EDTA, and 0.1 m*M* dithiothreitol (pH 8.0) overnight to remove pyridine, and concentrated with an Amicon UM10 membrane to about 2 ml. The concentrated fraction is applied to a Sephadex G-75 column (1.5 × 85 cm) which is equilibrated before use with 50 m*M* Tris–HCl, 1 m*M* EDTA, and 0.1 m*M* dithiothreitol (pH 8.0). The position of elution of the δ subunit can be determined by measuring ATP-driven transhydrogenase activity on incubation of a portion of each fraction (50 μl) with the 4 subunits of F_1 lacking δ (1.0 unit) and membrane vesicles depleted of F_1 (0.16 mg protein). Results by this assay show that the δ subunit is eluted in two peaks, the second peak containing homogeneous δ subunit. The position of the ε subunit can be located by assaying the ATPase inhibitor activity of each fraction with 0.02 unit of purified EF_1. Details of these assays were reported by Smith and Sternweis.[2]

Reconstitution of EF_1 and Properties of Isolated Subunits

Reconstitution of F_1 and Identification of Defective Subunits in Mutants

Reconstitution of F_1 and Hybrid F_1 Complexes. The ATPase complex ($\alpha\beta\gamma$ complex) can be reconstituted by dialyzing a mixture of the α, β, and

γ subunits against a buffer containing ATP and Mg^{2+} ions. Reconstitution is maximal when a mixture of the α, β, and γ subunits (total protein concentration, 0.1–0.3 mg/ml; molar subunit ratio, 3 : 3 : 1) in 10 m*M* succinate-Tris, pH 6.0, 10% glycerol, and 0.1 m*M* dithiothreitol is dialyzed at 23° for 8 hr against reconstitution buffer [50 m*M* succinate-Tris (pH 6.0), containing 10% glycerol, 1 m*M* dithiothreitol, 0.5 m*M* EDTA, 5 m*M* ATP, and 5 m*M* $MgCl_2$]. Active EF_1 can be reconstituted by mixing the $\alpha\beta\gamma$ complex (16–100 μg) with the δ (2 μg) and ε (1.3 μg) subunits. Essentially the same procedure can be used for the formation of hybrid ATPase from subunits of *E. coli* and the thermophilic bacterium[12,13] or *S. typhimurium*.[14]

Identification of Defective Subunits in Purified Mutant F_1. Defective subunits from mutant F_1 can be identified by dialyzing mixtures of dissociated mutant F_1 and one of the isolated subunits against reconstitution buffer. The defective α, β, and γ subunits in strains AN120 (*uncA401*),[7,8] KF11 (*uncD11*),[9] and KF43[11] and KF12,[10] respectively, were identified. A typical protocol used for KF11 was as follows: Purified defective F_1 from this strain was dissociated as described above and 30 μg portions were mixed with various amounts of isolated subunits (α, β, or γ) from the wild type at molar ratios (mol subunit/mol mutant F_1) of 1 and 10 in 300 μl of 10 m*M* succinate-Tris (pH 6.0) containing 0.2 *M* KCl, 10% glycerol, and 0.1 m*M* dithiothreitol. The mixtures were dialyzed against reconstitution buffer at 25° for 8 hr. Reconstitution of ATPase activity was observed only when dissociated EF_1 from KF11 was mixed with the isolated β subunit.

Identification of Defective Subunits in Crude Extracts from F_1 Mutants. Assay with an EDTA extract of membranes is also a convenient method for preliminary characterization of an F_1 mutant. Strains AN120,[7,20] DL54,[21] and NR70[20] were characterized in this way. For instance, membranes from 1 g (wet weight) of AN120 cells were suspended in 10 ml of 0.5 m*M* EDTA containing 1 m*M* Tris–HCl (pH 8.0), 10% glycerol, and 0.1 m*M* dithiothreitol. After incubation at 23° for 1 hr, the suspension was centrifuged.[7,20] The supernatant fraction (EDTA extract) was concentrated, dialyzed against 50 m*M* succinate-Tris (pH 6.0) containing 1.0 *M* KCl, 0.10 *M* KNO_3, and 0.1 m*M* dithiothreitol at 4° and stored at −80°. The EDTA extract containing dissociated subunits (0.56 mg protein) was thawed and mixed with isolated α (13 μg), β (15 μg), or γ (2 μg) subunit in 200 μl of 10 m*M* succinate-Tris (pH 6.0) containing 10% glycerol and 0.1 m*M* dithiothreitol. ATPase activity was reconstituted

[20] M. Futai and H. Kanazawa, *in* "Cation Flux Across Biomembranes" (Y. Mukohata and L. Packer, eds.), p. 291. Academic Press, New York, 1979.

[21] H. Kanazawa and M. Futai, *FEBS Lett.* **109**, 104 (1980).

only by dialyzing the mixture of the EDTA extract with the wild-type α subunit. However, assays using a crude extract should be interpreted carefully. As shown previously,[7] the isolated γ subunit (wild-type) stimulated reconstitution of ATPase from α and the extract, but had no effect on reconstitution from the α subunit and purified AN120 F_1. This may be due to loss of a substantial amount of the active γ subunit in the mutant extract. A trace of protease, if present, may act on the subunit, because treatment to dissociate or reconstitute F_1 involves fairly lengthy incubation of the extract. Similar experiments were carried out with strain NR70,[19] confirming the absence of inactive F_1 attached to the membrane.[7,22]

Properties of Isolated Subunits

The properties of isolated subunits have been reviewed in detail[23–25] and so are mentioned here only briefly. The isolated α subunit binds ATP and ADP (1 mol of nucleotide per mol α) with K_d values of 0.1 μM and 0.9 μM, respectively, as determined by equilibrium dialysis[5] or enhancement of fluorescence of the ATP analog.[26] The β subunit has a low affinity binding site for ATP and ADP with a K_d value of 10^{-4}–$10^{-5}M$.[6] This low affinity site can be detected by measuring quenching of the fluorescence of 8-anilinonaphthalene 1-sulfonate upon addition of nucleotide.[6] The β subunit binds aurovertin D (1 mol of aurovertin/mol of β) with a K_d value of 3 μM, as measured by fluorescence enhancement of aurovertin with the subunit.[5]

[22] B. P. Rosen, *J. Bacteriol.* **116,** 1124 (1973).
[23] M. Futai and H. Kanazawa, *Curr. Top. Bioenerg.* **10,** 181 (1980).
[24] M. Futai and H. Kanazawa, *Microbiol. Rev.* **47,** 285 (1983).
[25] S. D. Dunn and L. A. Heppel, *Arch. Biochem. Biophys.* **210,** 421 (1981).
[26] I. Matsuoka, K. Takeda, M. Futai, and Y. Tonomura, *J. Biochem.* (*Tokyo*) **92,** 1383 (1982).

[60] Analysis of *Escherichia coli* Mutants of the H^+-Transporting ATPase: Determination of Altered Site of the Structural Genes

By HIROSHI KANAZAWA, TAKATO NOUMI, and MASAMITSU FUTAI

We describe four procedures for location of a mutation, mainly taking mutations in the β subunit of F_1-ATPase as examples: (1) rapid mapping of a mutation site within a defined domain of a subunit by a genetic recombination test, (2) cloning of a mutation site, (3) detection of anomalous migration of DNA fragments carrying a mutation, and (4) determination of an altered nucleotide sequence.

The proton-translocating ATPase of *Escherichia coli* (EF_1F_0) has a key role in the energy-transducing system and is present in cytoplasmic membranes.[1] The enzyme has eight subunits (α, β, γ, δ, ε, a, b, and c). All the structural genes form a cluster as an operon (named *unc, pap,* or *atp*) at about 83 min on the *E. coli* linkage map,[2] and their complete nucleotide sequences have been determined.[3-12] Sequence analysis also showed the presence of another gene in the operon coding for a protein of 14,000 daltons,[1] although its function is not yet clear. Genetic and biochemical analyses of mutations in EF_1F_0 may give clues to the correlate between the function and structure of the subunits. Information on the complete

[1] M. Futai and H. Kanazawa, *Microbiol. Rev.* **47,** 285 (1983).
[2] F. Gibson, J. A. Downie, G. B. Cox, and J. Radik, *J. Bacteriol.* **134,** 728 (1978).
[3] H. Kanazawa, K. Mabuchi, T. Kayano, F. Tamura, and M. Futai, *Biochem. Biophys. Res. Commun.* **100,** 219 (1981).
[4] K. Mabuchi, H. Kanazawa, T. Kayano, and M. Futai, *Biochem. Biophys. Res. Commun.* **102,** 172 (1981).
[5] H. Kanazawa, T. Kayano, K. Mabuchi, and M. Futai, *Biochem. Biophys. Res. Commun.* **103,** 604 (1981).
[6] H. Kanazawa, K. Mabuchi, T. Kayano, T. Noumi, T. Sekiya, and M. Futai, *Biochem. Biophys. Res. Commun.* **103,** 613 (1981).
[7] H. Kanazawa, T. Kayano, T. Kiyasu, and M. Futai, *Biochem. Biophys. Res. Commun.* **105,** 1257 (1982).
[8] H. Kanazawa, K. Mabuchi, and M. Futai, *Biochem. Biophys. Res. Commun.* **107,** 568 (1982).
[9] N. J. Gay and J. E. Walker, *Nucleic Acids Res.* **9,** 2187 (1981).
[10] N. J. Gay and J. E. Walker, *Nucleic Acids Res.* **9,** 3919 (1981).
[11] M. Saraste, N. J. Gay, A. Eberle, M. J. Runswick, and J. E. Walker, *Nucleic Acids Res.* **9,** 5287 (1981).
[12] J. Nielsen, F. G. Hansen, J. Hoppe, P. Friedl, and K. von Meyenberg, *Mol. Gen. Genet.* **184,** 33 (1981).

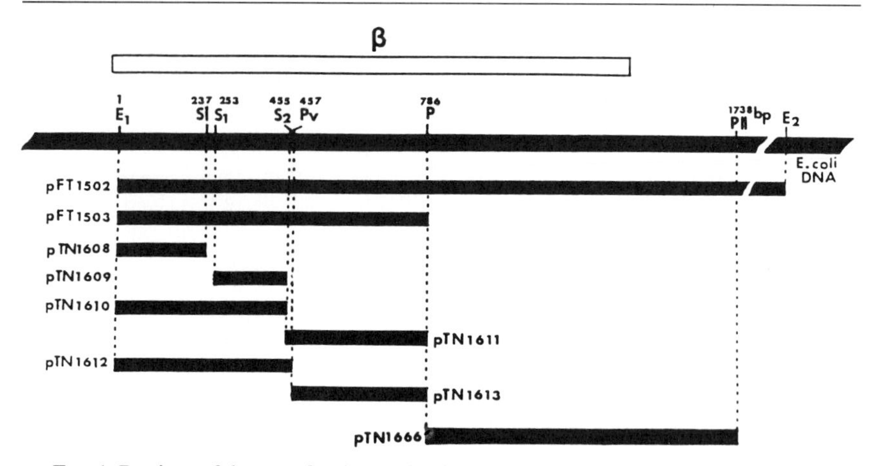

FIG. 1. Portions of the gene for the β subunit carried by hybrid plasmids. At the top of the figure, the reading frame of the gene for β subunit is shown as an open box. Cleavage sites by various restriction endonucleases with the number of the nucleotide residue taking that by *Eco*RI (E_1) as 1 are indicated on *E. coli* DNA, shown by a solid bar. E_1 and E_2, *Eco*RI; P, *Pst*I; Sl, *Sal*I; S_1 and S_2, *Sau*3A; Pv, *Pvu*I; PII, *Pvu*II. The genomic DNAs carried by the indicated plasmids are shown by solid bars. From Noumi *et al.*[17]

nucleotide sequence and a detailed physical map of the genes facilitated mapping of the mutations and enabled us to pinpoint mutated bases.

Many mutations of an EF_1F_0 have been assigned to a certain gene of a subunit by genetic complementation or recombination studies using F′ plasmids, hybrid plasmids, or λ-transducing phages carrying various portions of the gene cluster.[1] We located several mutations within a defined portion of the cistron for the α, β, γ, or b subunits by genetic recombination tests with plasmids carrying various portions of each gene.[13-16] The portion of the gene in which the mutation was located was cloned on a hybrid plasmid. In most cases studied, small DNA fragments around the mutation site estimated by the genetic procedure were subjected to polyacrylamide gel electrophoresis under conditions separating double strands of the fragments into single strands. Only the single strands containing a mutation (a single base substitution) showed anomalous migration com-

[13] H. Kanazawa, T. Noumi, M. Futai, and T. Nitta, *Arch. Biochem. Biophys.* **223,** 521 (1983).

[14] H. Kanazawa, T. Noumi, N. Oka, and M. Futai, *Arch. Biochem. Biophys.* **227,** 596 (1983).

[15] H. Kanazawa, T. Noumi, I. Matsuoka, T. Hirata, and M. Futai, *Arch. Biochem. Biophys.* **228,** 258 (1984).

[16] T. Noumi and H. Kanazawa, *Biochem. Biophys. Res. Commun.* **111,** 143 (1983).

pared with those from the wild type.[17] These results not only supported genetic results, but also eliminated tedious determination of the complete nucleotide sequence of a gene to find the single base change; sequencing of only the portion of the gene carrying a mutation was sufficient for determination of the mutated base, even in the larger subunits α and β.

In another volume of this series (Volume 97), general methods for construction of recombinant plasmids and sequencing of the *unc* region are described.[18-20] For details of digestion of DNA fragments with restriction endonucleases, agarose, or polyacrylamide gel electrophoresis, extraction of DNA fragments from the gel matrix, ligation of DNA fragments, and other methods related to the recombinant DNA technique, which are not described in detail here, the reader should refer to Volume 65 and Volume 68 of this series.

Mapping of a Mutation Site within a Defined Portion of a Structural Gene by Genetic Recombination

Construction of Hybrid Plasmids Carrying Various Portions of the β-Subunit Gene

A restriction map of the β subunit gene (*uncD* or *papB*) and plasmids carrying various portions of the constructed gene are shown in Fig. 1.[17] Plasmids pFT1502, pFT1503, and pTN1666 were constructed by inserting genomic DNA fragments prepared from a transducing phage λ*asn*-5 DNA[21] into a vector plasmid pMCR561[22] in the cases of pFT1502 and pFT1503 or into pBR322 in the case of pTN1666. Other plasmids were constructed by inserting an appropriate DNA fragment prepared from pFT1503 into pBR322, or pMCR561 in the case of pTN1611.

One liter of a lysogen of transducing phage λ*asn*-5 (KY7485) in L-broth medium[21] was cultured at 30° to the mid-log phase of growth, and the phage with a heat-labile repressor was induced by incubation of the cells at 42° for 30 min.[21] After incubation for 3 hr with shaking at 37°, the cells were harvested and a crude lysate of the phage was prepared. Phages

[17] T. Noumi, M. E. Mosher, S. Natori, M. Futai, and H. Kanazawa, *J. Biol. Chem.* **259,** 10071 (1985).

[18] D. A. Jans and F. Gibson, this series, Vol. 97, p. 176.

[19] W. S. Bruislow, R. P. Gunsalus, and R. D. Simoni, this series, Vol. 97, p. 188.

[20] J. E. Walker and N. J. Gay, this series, Vol. 97, p. 195.

[21] H. Kanazawa, T. Miki, F. Tamura, T. Yura, and M. Futai, *Proc. Natl. Acad. Sci. U.S.A.* **76,** 1126 (1979).

[22] F. Tamura, H. Kanazawa, T. Tsuchiya, and M. Futai, *FEBS Lett.* **127,** 48 (1981).

were purified by CsCl density gradient centrifugation, and the phage DNAs were recovered in the aqueous phase on shaking with phenol. Plasmid DNAs were prepared from cells incubating in a 2-liter culture and purified by CsCl density gradient centrifugation.[23] Candidate clones harboring hybrid plasmids were first checked for resistance to antibiotics on L-broth medium agar with an appropriate antibiotic (20–50 μg/ml). Then a crude lysate of plasmids was prepared by the procedure of Birnboim and Doly (rapid procedure).[24] For identification of a plasmid DNA, restriction patterns should be analyzed by polyacrylamide gel electrophoresis after digestion with appropriate restriction endonucleases.

Genetic Recombination Test

A set of EF_1F_0 mutants (KF series) was obtained by localized mutagenesis using P1 phage.[25] None of the mutants could use succinate as a sole carbon source (succinate-minus phenotype), whereas the wild-type cells could utilize this carbon source. The mutants could use glucose as the sole carbon source.

Competent cells of the mutants were prepared as follows: Mutant cells were cultured in 20 ml of L-broth at 37° with 0.5 *M* sucrose and harvested in the mid-log phase (OD_{600} = 0.4–0.5) of growth. In the following procedures, the cells were kept in an ice bath. The cells were washed twice with 5 ml of 10 m*M* NaCl in 0.5 *M* sucrose and then with 50 m*M* $CaCl_2$ in 0.5 *M* sucrose. Finally, they were suspended in 2 ml of 50 m*M* $CaCl_2$. At this stage they could be kept frozen with glycerol (17%) at −80° until use. An aliquot of plasmid lysate prepared as described above or purified plasmid DNA (about 2 μg) was mixed with 1 ml of competent cells, and the mixture was allowed to stand for 60 min in an ice bath. Then, it was incubated at 42° for 2 min and mixed with 5 ml of L-broth. Cells were cultured for 2 hr at 37° to allow replication of plasmids, harvested, and resuspended in 1 ml of saline. An aliquot of the suspension was spread on L-broth with 20 μg/ml of an appropriate antibiotic. Transformants that were resistant to the antibiotic were streaked on minimal medium (Tanaka medium)[21] agar supplemented with 0.4% succinate or 0.2% glucose and incubated for 24–48 hr at 37°. Since the frequency of appearance of recombinants among the antibiotic-resistant clones was low, scarcely any clones of the succinate-plus phenotype appeared on the succinate plates. This rare appearance of clones is a typical feature of recombinants. For

[23] J. A. Mayers, D. Sanchez, L. P. Elwell, and S. Falkow, *J. Bacteriol.* **127,** 1529 (1976).

[24] M. C. Birnboim and J. Doly, *Nucleic Acids Res.* **7,** 1513 (1979).

[25] H. Kanazawa, F. Tamura, K. Mabuchi, T. Miki, and M. Futai, *Proc. Natl. Acad. Sci. U.S.A.* **77,** 7005 (1980).

judging the phenotype, incubations of plates should not exceed 48 hr, because some mutations are leaky and increase the background on longer incubation periods.

Applications and Examples of Results

Mutants defective in the β subunit, KF-11, -39, and -43,[14] gave recombinants of succinate-plus with plasmid pKF1503 (Fig. 1), suggesting that the mutations in these strains exist between amino acid residues 17 and 279. Mutants defective in the β subunit, KF-20, -26, -27, -29, -30, -32, -33, -37, -38, and -40, gave succinate-plus recombinants with plasmids pFT1502 and pTN1666, but not with pFT1503, suggesting that these mutations exist between residues 280 and 459.[14] Mutations in KF-11 and KF-43 were analyzed further by hybrid plasmids carrying a narrower portion of the amino terminal half of the cistron. Both mutations gave recombinants with pTN1611 and pTN1613, suggesting that the mutations are located between residues 169 and 279. As described below, a substitution of thymine for cytosine at nucleotide residue 524 was found in KF-11 by determination of the nucleotide sequence,[17] supporting the above results.

None of the *recA*$^-$ derivatives of the mutants described here gave transformants or recombinants of the succinate-plus phenotype. Plasmids carrying entirely different portions of the gene cluster, such as the α subunit gene, consistently gave no succinate-plus transformants or recombinants. These control experiments were essential for the genetic analysis described here.

Mapping of mutation sites by essentially the same approach as that described above has been successful for several mutations of α, γ, or *b* subunit genes.

Cloning of a DNA Fragment Containing an Altered Base

To date we have cloned three mutated genes from KF-11, AN120, and KF-3, respectively, by independent approaches.

KF-11

The mutated gene of KF-11 (*uncD11*) was cloned on transducing phage λ*uncD11* by Mosher *et al.*[26] Their procedure is described elsewhere in this volume. We subcloned the DNA fragment between the *Eco*RI (E_1) and *Pst*I (P) site (Fig. 2). This hybrid plasmid pKF11 gave recombinants with KF-43, but did not with KF-11.

[26] M. E. Mosher, L. K. Peters, and R. H. Fillingame, *J. Bacteriol.* **156,** 1078 (1983).

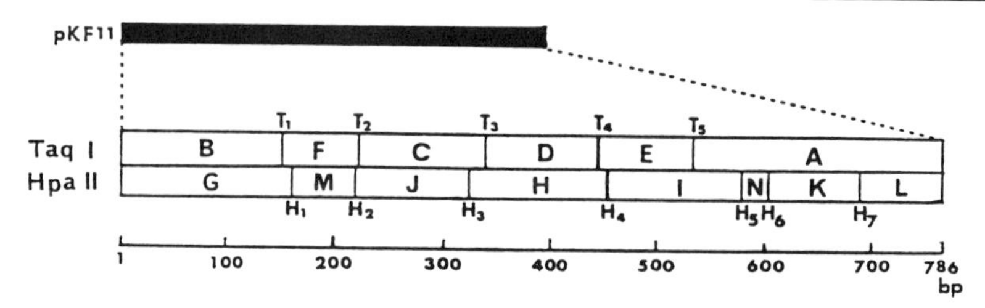

FIG. 2. Restriction map of the portion of the β subunit gene carried by plasmid pKF11. The genomic DNA fragment cloned on pKF11 is shown by a solid bar. Cleavage sites in the fragment are shown by T (*Taq*I site) and H (*Hpa*II site). DNA fragments cleaved by *Taq*I or *Hap*II are shown by open boxes. bp, Base pairs. From Noumi *et al.*[17]

AN120

A mutant AN120 was originally isolated by Butlin *et al.*[27] and was shown to be defective in the α subunit as described in the following article of this volume. The mutation site was located between amino acid residues 370 and 387 by a genetic recombination test with hybrid plasmids carrying various portions of the α subunit gene.[15]

Total DNA of AN120 was prepared by the procedure of Saito and Miura,[28] and an aliquot of this DNA (150 μg) was digested with *Eco*RI and *Bgl*II and subjected to electrophoresis on polyacrylamide gel (4.8% acrylamide, 0.17% bisacrylamide). The fraction of DNA fragments corresponding in size to the fragment between the *Bgl*II site at the amino acid residue 353 of the α subunit gene and the *Eco*RI site at the amino acid residue 17 of the β subunit gene was cut out and eluted from the gel matrix. The DNA fragments (2 μg), obtained by digestion with *Eco*RI plus *Bgl*II, were ligated with pBR322 by T4-DNA ligase. The religated hybrid plasmids were introduced into KF-12 (*recA*$^+$),[13] which is defective in the γ subunit, and the resulting ampicillin-resistant transformants were tested for tetracycline resistance. The tetracycline-sensitive clones thus obtained were then tested for the succinate-plus phenotype. Plasmid DNAs from the succinate-plus clones were prepared and analyzed by polyacrylamide gel electrophoresis after digestion with appropriate restriction endonucleases.

The results are described here briefly. In all, 657 ampicillin-resistant transformants were obtained and 23 of these had the tetracycline-sensitive phenotype. Finally, two clones with the succinate-plus phenotype

[27] J. D. Butlin, G. B. Cox, and F. Gibson, *Biochem. J.* **124,** 75 (1971).

[28] H. Sato and K. Miura, *Biochim. Biophys. Acta* **72,** 619 (1963).

were obtained. One of these clones was shown to harbor a plasmid carrying the mutation site by a genetic recombination test and analysis of restriction sites of the plasmid DNA. Determination of DNA sequence showed that this plasmid (pAN120) carried a mutated base.[29]

KF-3

The mutation site in KF-3 was mapped in the gene for the *a* or *c* subunit by a genetic recombination test. For cloning of the mutated gene, the mutated site was transferred to a hybrid plasmid carrying the wild-type allele by *in vivo* recombination and then the plasmid carrying the mutation was selected as described below.[30] The experimental process involved four steps:

1. Plasmid pMCR533[25] carrying the genes for the *a*, *c*, *b*, δ, and α subunits was introduced into competent cells of KF-3. Ampicillin-resistant transformants were isolated and from them succinate-plus recombinants were selected. All recombinants should have the wild-type allele on the chromosome, whereas the hybrid plasmid in these cells should have the mutant allele after recombination.
2. Plasmid lysate was prepared from a mixture of the succinate-plus recombinants by the rapid procedure. Some of the plasmids in this lysate should carry the mutant allele.
3. An aliquot of the lysate was again introduced into competent KF-3 cells and ampicillin-resistant transformants were selected. Then clones of the succinate-minus phenotype, which was expected to harbor a hybrid plasmid carrying the mutation, were selected on minimal medium (Tanaka medium) agar as described above.
4. Plasmid DNAs from each clone obtained were prepared by the rapid procedure and their genetic properties and the restriction map of their DNA were analyzed. A desired plasmid carrying the mutation should not give a recombinant with KF-3 (*recA*$^+$), but should give one with a mutant of subunit *c*, *b*, δ, or α. The plasmid should have the various restriction sites observed in pMCR533.

The results after step 3 were as follows: Of 86 ampicillin-resistant transformants, 14 had the succinate-minus phenotype. Plasmid DNAs from all the succinate-minus clones gave succinate-plus recombinants with a mutant KF-5 defective in the gene for the δ subunit.[16] One of these plasmids with appropriate restriction sites was named pKF3.[30]

[29] T. Noumi, M. Futai, and H. Kanazawa, *J. Biol. Chem.* **259,** 10076 (1985).
[30] H. Kanazawa, N. Oka, T. Noumi, and M. Futai, submitted (1985).

Detection of Anomalous Migration of Single Strands of a DNA Fragment Carrying an Altered Base

For plasmids pKF11, pAN120, and pKF3, a DNA fragment (100–200 base pairs) carrying the mutation showed anomalous migration on polyacrylamide gel electrophoresis. As an example, the detection of anomalous migration is described for pKF11.[17]

The DNAs of plasmid pKF11 (Fig. 2) and PFT1503 (Fig. 1) (50 μg each) were digested with a mixture of *Eco*RI and *Pst*I, and DNA fragments of 786 base pairs of the *Eco*RI(E_1)-*Pst*I(P) region (Fig. 1) were separated by polyacrylamide gel electrophoresis (4.8% acrylamide, 0.17% bisacrylamide). On further digestion of the fragments with *Taq*I and *Hpa*II, six and eight DNA fragments, respectively, were obtained (Fig. 2). The 5′ ends of these DNA fragments were labeled with [γ-^{32}P]ATP with T4 polynucleotide kinase and solubilized in 40 μl of dye solution consisting of 30% dimethyl sulfoxide, 1 m*M* EDTA, 0.05% xylene cyanole (XC), and 0.05% Bromphenol Blue (BPB). This solution was heated at 90° for 5 min, rapidly cooled in an ice bath, and subjected to polyacrylamide gel electrophoresis (9.7% acrylamide, 0.3% bisacrylamide, $20 \times 40 \times 0.2$ cm gel) at a constant voltage at 300 V. During electrophoresis double-stranded DNA fragments (about 100 base pairs) were separated into single strands when the tracking dye BPB reached to the bottom of the gel. The gel was then wrapped in Saran Wrap and subjected to autoradiography with X-ray film (Fuji RX 100).

As shown in Fig. 3, the mobility of fragment E (Fig. 2) differed from that of the wild type (Fig. 3). In this experiment, the marker dye (BPB) migrated to the bottom of the gel and the two single strands of DNA fragments of more than 150 base pairs (A, B, C, and D in Fig. 3) were not well separated. For better separation of these fragments (200–300 base pairs), longer electrophoresis should be performed under the same conditions or the concentrations of polyacrylamide and bisacrylamide should be decreased.

Determination of an Altered Nucleotide Sequence

Plasmid DNA carrying a mutation was prepared from a 1-liter culture of cells harboring this plasmid. The cells were cultured in L-broth supplemented with 0.2% glucose, and the plasmids were amplified with addition of chloramphenicol (180 μg/ml). Plasmid DNAs were purified by CsCl density gradient centrifugation. Approximately 1 mg of plasmid DNA was obtained.

DNA fragments carrying a mutation (100–200 base pairs) identified as described above were prepared by gel electrophoresis after digestion with

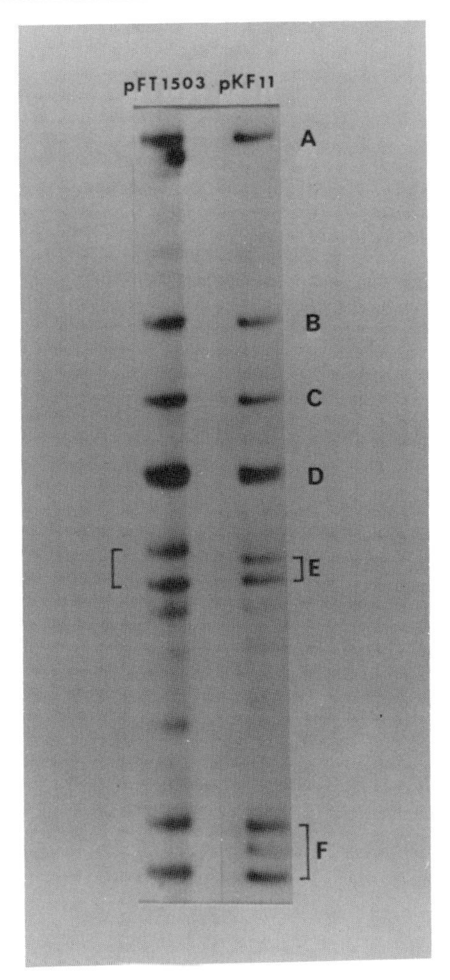

FIG. 3. Strand separation of fragments derived from the *Eco*RI(E_1)-*Pst*I(P) region of pFT1503 and pKF11 by digestion with *Taq*I. The *Eco*RI(E_1)-*Pst*I(P) fragment (Fig. 1) was prepared from pFT1503 and pKF11. Subsequent digestion of this fragment with *Taq*I generated 6 fragments (Fig. 2). Labeling of these fragments and gel electrophoresis are described in the text. From Noumi *et al.*[17]

appropriate restriction endonucleases. In the case of pKF11, fragment E (Fig. 2) was recovered and ~1 μg of the fragment was used for phosphorylation at the 5′ end. Its nucleotide sequence determined by the procedure of Maxam and Gilbert showed that cytosine at nucleotide residue 524 was replaced by thymine. It should be noted that both strands of a DNA fragment should be sequenced to obtain a final conclusion on the location of an abnormal residue.

[61] Proton Permeability of Membrane Sector (F_0) of H^+-Transporting ATP Synthase (F_0F_1) from a Thermophilic Bacterium

By NOBUHITO SONE and YASUO KAGAWA

The membrane moiety of H^+-transporting ATP synthase (F_0F_1), which catalyzes ATP synthesis utilizing $\Delta\bar{\mu}_{H^+}$, is the H^+ pathway across the membrane.[1,2] This F_0 moiety can be easily prepared by treating F_0F_1 with urea or a chaotropic anion, and the resulting F_0, when inlayed in phospholipid vesicles, shows appreciable H^+ conductivity[2-4]; i.e., protons are taken up when valinomycin is added to K^+-loaded F_0 vesicles. The velocity of H^+ uptake is almost proportional to the amount of F_0 added. Stable preparations of TF_0F_1 (H^+-transporting ATP synthase from the thermophilic bacterium PS3)[5] together with a simple procedure[6] to prepare the H^+-conducting TF_0 moiety result in an interesting system[7] to study H^+ translocation through membrane proteins, which may exist in respiration-driven and light-driven H^+ pumping enzymes generating $\Delta\bar{\mu}_{H^+}$.

Preparation of TF_0

Preparation of F_0 moiety (TF_0) from H^+-ATPase from the thermophilic bacterium PS3 (TF_0F_1) is carried out by dissociating TF_1 moiety with urea. The improved procedure described here gives a more active TF_0 preparation than before.[2]

Procedure[6]

A solution (20 ml) of purified TF_0F_1[8] in 7 *M* urea/0.2 *M* sucrose/5 m*M* EDTA/5 m*M* dithiothreitol/50 m*M* Tris–H_2SO_4, pH 8.0, is stirred for 2.5–

[1] V. Shchipakin, E. Chuchlova, and Y. Evtodienko, *Biochem. Biophys. Res. Commun.* **69,** 123 (1976).
[2] H. Okamoto, N. Sone, H. Hirata, M. Yoshida, and Y. Kagawa, *J. Biol. Chem.* **252,** 6125 (1977).
[3] R. Negrin, D. L. Oster, and R. Fillingame, *J. Biol. Chem.* **255,** 5643 (1981).
[4] P. Friedl and H. U. Schairer, *FEBS Lett.* **128,** 261 (1981).
[5] N. Sone, M. Yoshida, H. Hirata, and Y. Kagawa, *J. Biol. Chem.* **250,** 7910 (1975).
[6] N. Sone, M. Yoshida, H. Hirata, and Y. Kagawa, *Proc. Natl. Acad. Sci. U.S.A.* **75,** 4219 (1978).
[7] N. Sone, T. Hamamoto, and Y. Kagawa, *J. Biol. Chem.* **256,** 2873 (1981).
[8] Y. Kagawa and N. Sone, this series, Vol. 55 [44].

3 hr at 5°, diluted with distilled water (6 ml) to reduce the concentration of urea, and centrifuged at 200,000 *g* for 40 min. This process is repeated once more, and the resultant precipitate (TF_0) is suspended in 10 m*M* *N*-tris(hydroxymethyl)methylglycine (Tricine)–NaOH (pH 8.0), containing 2 m*M* EDTA, and stored at −80° until use. About one-fifth of TF_0F_1 was recovered as TF_0 on a protein basis.

Preparation of TF_0 Vesicles

A dialysis method with cholate plus deoxycholate gives the best results, as in the case of TF_0F_1 reconstitution.[5]

Procedure

Partially purified soybean phospholipids[9] (500 mg) are suspended in 10 ml of solution containing 200 mg of sodium cholate, 100 mg of sodium deoxycholate, 100 μmol of Tricine–NaOH, pH 8.0, 50 μmol of dithiothreitol, and 2 μmol of EDTA, and subjected to sonic oscillation in an ice bath for 2–5 min until the suspension becomes almost clear. If the resulting solution is slightly turbid, it is centrifuged at 104,000 *g* for 10 min. A small amount of TF_0F_1 (10–150 μg protein) and 0.4 ml of phospholipid solution in a final volume of 0.5 ml are briefly sonicated (5–10 sec) and are then placed in cellophane tubing of 0.25 in. diameter and 0.002 in. thickness and dialyzed against 250 ml of 10 m*M* Tricine–NaOH buffer, pH 8.0, containing 2.5 m*M* $MgSO_4$, 0.25 m*M* dithiothreitol, and 0.2 m*M* EDTA, with stirring for 20 hr at 30°.

Assay of Proton Conductivity

Proton uptake by the vesicles in response to an artificial membrane potential imposed by K^+ diffusion mediated by valinomycin (inside negative) is followed with a pH meter. Vesicles are loaded with KCl by a freeze-thaw sonication procedure[10] instead of the heat treatment method used previously.[2]

Procedure

Equal volumes (usually 0.5 ml) of TF_0 vesicles and 0.4 *M* KCl in 20 m*M* sodium phosphate buffer, pH 7.2, are mixed in a small test tube, and the mixture on ice water is sonicated with a Branson sonifier (model 200)

[9] N. Sone, this volume [14].

[10] M. Kasahara and P. C. Hinkle, *J. Biol. Chem.* **252,** 7384 (1977).

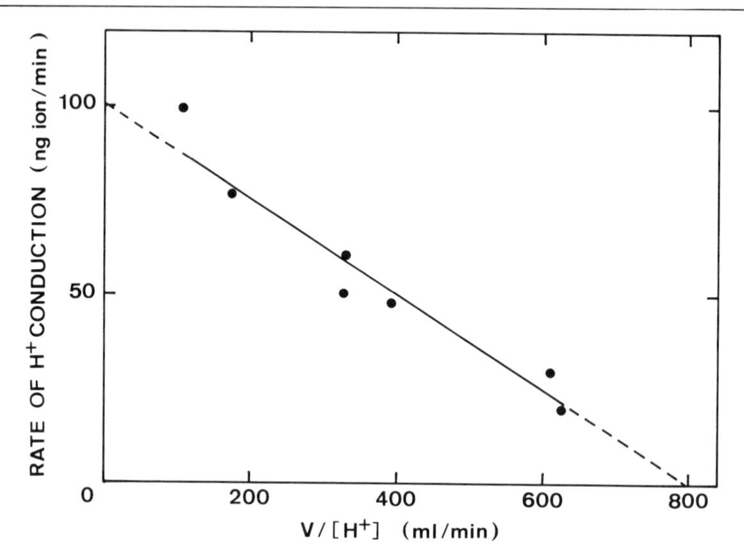

FIG. 1. Plot of H^+ permeation rate (V) vs V/[s].

equipped with a microprobe for 10 sec (20 pulses of 0.5 sec) at an output of 2. The resulting mixture is then frozen at −80° for more than 10 min, thawed at room temperature, and sonicated again as before. The resulting vesicles are referred to as K^+-loaded vesicles.

The H^+ conductivity of vesicles is measured at 25° by adding valinomycin (20 ng) after 4–5 min of preincubation of K^+-loaded vesicles (25 μl) in 2 ml of the reaction mixture composed of 0.2 M choline–HCl, 2 mM

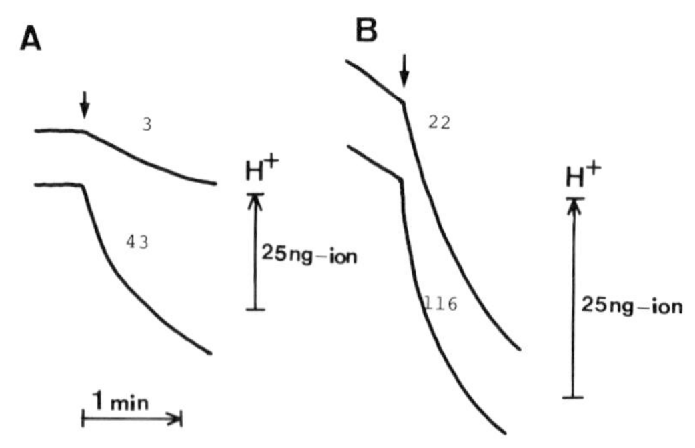

FIG. 2. Measurement of proton permeation. The upper traces show the results with simple liposomes and the lower traces with TF_0 vesicles. The number shows the initial rate of H^+ permeation (ng ion/min). The fraction due to drift of pH meter and CO_2 extrusion is subtracted. (A) pH 7.2; (B) pH 6.3.

$MgSO_4$, and 1 m*M* Tricine–NaOH at pH 7.2. The initial velocity (5 sec) of H^+ uptake is used to calculate H^+ conductivity of vesicles.

H^+ Permeation through TF_0 Vesicles

Proportionality and Effect of TF_1 Addition

The initial velocity of H^+ uptake is proportional to the amount of TF_0, up to 150 μg in 0.5 ml of reconstituted vesicles and blocked by the addition of TF_1.[2,7] Titration by TF_1 shows that 5 times as much protein of TF_1 is necessary to block H^+ permeation through TF_0.[7]

Effect of $\Delta\bar{\mu}_{H^+}$

The rate of H^+ uptake is proportional to the K^+ diffusion potential imposed. Dependence of the H^+ permeation rate through TF_0 on H^+ concentration (pH) fits a Michaelis–Menten equation. Figure 1 shows an Eadie–Hofstee plot of a typical result. V_{max} values at 25° vary from preparation to preparation (20–100 μg ion/min × mg TF_0). A K_m value around 0.1 μg ion/liter is observed. Figure 2 shows the measurement of H^+ permeability of TF_0 vesicles and simple liposomes containing no TF_0 at pH 7.2 and pH 6.2. Nonspecific H^+ permeation of simple liposomes seems to be proportional to H^+ concentration of the medium at this pH range. TF_0 preparations lose their H^+ permeability gradually, even when reconstituted into liposomes.

Effect of Chemical Modification

H^+ permeability of TF_0 is also blocked by the addition of DCCD (*N,N'*-dicyclohexylcarbodimide), which covalently binds to DCCD-binding proteolipid (the very hydrophobic smallest subunit of F_0).[11] Amino acid modifiers, such as tetranitromethane for a tyrosyl residue and phenylglyoxal for an arginine residue, block H^+ permeability almost completely.[12] By contrast, acetic anhydride, which derivatizes amino groups and blocks TF_1-binding activity of TF_0, does not inhibit H^+ permeability of TF_0.[12]

[11] N. Sone, M. Yoshida, H. Hirata, and Y. Kagawa, *J. Biochem.* **85,** 503 (1979).

[12] N. Sone, K. Ikeba, and Y. Kagawa, *FEBS Lett.* **97,** 61 (1979).

[62] Rate Constants and Equilibrium Constants for the Elementary Steps of ATP Hydrolysis by Beef Heart Mitochondrial ATPase

By HARVEY S. PENEFSKY

Introduction

The elucidation of the reaction mechanism for ATP hydrolysis by the soluble ATPase from beef heart mitochondria is based upon direct analysis of the amounts of substrate and products bound in catalytic sites during the reaction.[1,2] The success of the experimental approach further depends upon the strong catalytic site cooperativity exhibited by the enzyme[1-4]; i.e., the rate of hydrolysis of substrate, whether trinitrophenyl-ATP (TNP-ATP)[5] or ATP, is very low when the concentration of substrate added to reaction mixtures is sufficient to fill only one of the two readily available catalytic sites on the molecule.[1,3] Addition of substrate sufficient to occupy a second catalytic site results (in the case of ATP) in an acceleration of the net rate of hydrolysis of at least 10^6-fold.[2] It is under the conditions of single-site catalysis (sufficient substrate to fill only one or less than one catalytic site) that it is possible to quantitate the elementary steps of the reaction mechanism.[1,2] Single-site catalysis can be carried out with almost a 1 to 1 mole ratio of enzyme to TNP-ATP,[3] but when ATP is substrate, a mole ratio of enzyme to substrate of 3 or more is required to ensure binding of ATP predominantly in a single catalytic site.[1]

Three major experimental approaches are utilized. These include the cold chase or isotope trap,[1,3,6] chemical quench methods,[1,2] and the centrifuge column.[7]

[1] C. Grubmeyer, R. L. Cross, and H. S. Penefsky, *J. Biol. Chem.* **257,** 12092 (1982).

[2] R. L. Cross, C. Grubmeyer, and H. S. Penefsky, *J. Biol. Chem.* **257,** 12101 (1982).

[3] C. Grubmeyer and H. S. Penefsky, *J. Biol. Chem.* **256,** 3718 (1981).

[4] C. Grubmeyer and H. S. Penefsky, *J. Biol. Chem.* **256,** 3728 (1981).

[5] Abbreviations used are: TNP-ATP, 2′,3′-*O*-(2,4,6-trinitrophenyl ATP; CDTA, trans-1,2-diaminocyclohexane-*N,N,N′,N′*-tetraacetic acid; F_1, soluble mitochondrial ATPase; Mg buffer, a buffer containing 0.25 *M* sucrose, 10 m*M* $MgSO_4$, 1 m*M* P_i, 40 m*M* 4-morpholinoethane sulfonic acid, and 40 m*M* Tris. The pH was adjusted to 7.5 with KOH. CDTA buffer is identical with Mg buffer except that Mg^{2+} is replaced by 2.5 m*M* CDTA.

[6] For a review, see I. A. Rose, this series, Vol. 64, p. 47.

[7] H. S. Penefsky, this series, Vol. LVI, p. 527.

General Considerations

Because the enzyme is used in substrate quantities, only enzyme preparations of high purity and specific activity should be used in these measurements. Thus, for beef heart F_1, a specific activity of 110–120 units/mg, measured with a regenerating system,[8] appears to be adequate. Preparations of significantly lower specific activity must contain catalytic sites that are masked or silent for other reasons. If inactive catalytic sites nevertheless bind substrate or products, it may be difficult to interpret experimental results.

It is also important that chemical quench experiments be carried out at constant flow velocity; i.e., the aging times used to provide data for a single rate constant should be varied by changing the length of the aging hose while the flow velocity remains constant. Ideally, rate constants should be calculated from data obtained at each of at least two flow velocities. If two such series of measurements at different flow velocities produce the same rate constant, one can be reasonably confident that artifacts due to inadequate mixing are not present in the experiment and that, in fact, the observed rate constant is independent of flow velocity. It is not sufficient simply to show that the chemical quench instrument and mixer in use reproduces previously published values of, e.g., the rate of alkaline hydrolysis of 2,4-dinitrophenyl acetate.[9]

Manual mixing of substrate amounts of F_1 with solutions containing radioactive ATP must be done rapidly and efficiently in order to avoid the possibility of local high concentrations of ATP relative to F_1. If the latter conditions prevail, even momentarily, a significant population of enzyme molecules in the reaction mixture will contain more than one molecule of substrate and 2-site catalysis will ensue. Manual mixing is facilitated if enzyme solutions are injected from a gas-tight Hamilton syringe with a bent tip on the needle into a rapidly stirred reaction mixture. The bent tip considerably enhances mixing.[10]

A quick and simple means of determining whether single-site catalysis can be studied in any given form of the ATPase is to carry out a cold chase experiment. The experiment can be done by hand using manually added reagents and magnetic stirring for mixing. The following description is for beef heart F_1. The homogeneous enzyme is prepared as described.[8] The enzyme is desalted on a Sephadex centrifuge column[7] equilibrated with CDTA buffer. The reaction mixture contains 0.25 ml of Mg buffer (identi-

[8] H. S. Penefsky, this series, Vol. LV, p. 304.

[9] R. E. Hansen, *in* "Chemical Pathways," Vol. 1, p. 1. Update Instrument, Inc., Madison, Wisconsin, 1984.

[10] W. J. Ray and J. W. Long, *Biochemistry* **15,** 3990 (1976).

cal to CDTA buffer except that 10 mM $MgSO_4$ replaces CDTA) and 0.1 nmol [γ-^{32}P]ATP, specific activity at least 10^6 counts/min/nmol. The reaction is started by adding 50 μl (0.3 nmol) of a solution of F_1 in CDTA buffer. The reaction mixture is stirred vigorously with a small magnetic stirring bar, and the enzyme is added from a 50-μl Hamilton syringe with a bent tip. Room temperature is used (26°). A time curve is carried out with incubations incremented by 10 sec up to a total of 60 sec. These are "acid quench" experiments. A second series of incubations is carried out in which, at the end of each incubation, 1 mM ATP is added and the reaction allowed to proceed for an additional 5 sec (cold chase). Reactions are stopped by adding 0.15 ml of 60% perchloric acid. Unhydrolyzed [γ-^{32}P]ATP and the $^{32}P_i$ present in the deproteinized reaction mixture are separated by a modified version[3] of the P_i precipitation method of Sugino and Mioshi[11] or by extraction with isobutanol-benzene.[12] Radioactivity is measured in a scintillation counter using a Triton–toluene-based counting cocktail[1] or, alternatively, Liquiscint (National Diagnostics) may be used. The difference between the $^{32}P_i$ formed in the cold chase and in the acid quench is a measure of the amount of [γ-^{32}P]ATP bound in catalytic sites. If a molar excess of enzyme over radioactive substrate is used in the experiment (e.g., 3 : 1) and if a substantial fraction of the added [γ-^{32}P]ATP is bound in catalytic sites (e.g., 60–70%), the enzyme preparation is likely to be suitable for studying single-site catalysis.

Measurement of Single-Site Catalysis

The mechanism of ATP hydrolysis catalyzed by beef heart F_1 is presented as a 4-step reaction[1]:

$$F_1 + ATP \underset{1}{\rightleftharpoons} F_1 \cdot ATP \underset{2}{\rightleftharpoons} F_1 \cdot ADP \cdot P_i \overset{P_i}{\underset{3}{\rightleftharpoons}} F_1 \cdot ADP \overset{ADP}{\underset{4}{\rightleftharpoons}} F_1 \tag{1}$$

In step 1, ATP binds in catalytic sites on the enzyme. Step 2 is the catalytic step. Hydrolysis takes place, but products remain bound in the catalytic site. Steps 3 and 4 are product release steps. The ordered release of P_i and ADP shown is based only on the observation that the rate of P_i release is about 10 times faster than ADP release.[1]

Step 1. The equilibrium constant for step 1, which is an affinity constant, is calculated from the ratio of the individual rate constants, $K_1 = k_1/k_{-1}$. The rate at which ATP binds in catalytic sites, k_1, is obtained from

[11] Y. Sugino and Y. Mioshi, *J. Biol. Chem.* **239,** 2360 (1964).

[12] O. Lindberg and L. Ernster, in "Methods of Biochemical Analysis" (D. Glick, ed.), Vol. 3, p. 1. Wiley (Interscience), New York, 1956.

experiments that make use of both chemical quench and isotope trap techniques. The chemical quench apparatus was purchased from Update Instrument, Inc., Madison, WI and utilizes a 4-grid Wiskind mixer provided by the same company. Both the push rate, and thus the flow velocity, as well as the distance moved by the syringe plungers are incrementally adjustable via front panel controls. Measurement of k_1 utilizes two syringes and one mixer. The length of the tubing on the outflow side of the mixer determines the age of the mixed reactants. The nozzle on the outflow end of the aging hose contains 2 holes, each 0.2 mm in diameter. The resulting fine jets provide adequate mixing when directed into a receiving vessel containing acid quench or cold chase solutions as described below. Syringe 1 contains 0.25–1.0 μM F_1 in Mg buffer. Syringe 2 contains 0.09–0.35 μM [γ-^{32}P]ATP (specific activity at least 10^6 counts/min/nmol) dissolved in the same buffer. Radioactive ATP is purchased commercially or prepared by enzymatic exchange.[13] The β-phosphoryl position must be free of label. Equal volumes from the two syringes (64 μl or more) are mixed and the mixed reactants allowed to age for varying periods of time (25–800 msec). Incubation times longer than 200 msec are obtained in the "push-push" mode.[1] In the push-push mode, reactants enter the aging hose during the first push, age there for a predetermined delay period, and are expelled into a receiving vessel by a second push. The receiving vessel contains 0.8 ml of Mg buffer and 0.2 ml of 50 mM ATP (cold chase). The enzyme is allowed to turn over in the cold chase for ~5 sec (about 3000 turnovers) before the addition of 0.1 ml of 60% perchloric acid. $^{32}P_i$ formed in the reaction can arise only from that [γ-^{32}P]ATP bound in catalytic sites at the time of mixing with the cold chase. The very large dilution of radioactive by nonradioactive ATP precludes dissociation, rebinding, and subsequent hydrolysis of the labeled form. Since the aging times utilized permit only fractional occupancy of catalytic sites by radioactive ATP and occupancy increases with aging time, the experiment measures the rate at which [γ-^{32}P]ATP binds in catalytic sites. The data from an experiment with 1 μM F_1 and 0.3 μM [γ-^{32}P]ATP are shown in Fig. 1A. $^{32}P_i$ formed is expressed as the percentage of the total [γ-^{32}P]ATP added. The bimolecular rate constant (k_1) is best determined from several experiments, with different concentrations of F_1 and ATP, by plotting the cold chase $^{32}P_i$ data, as shown in Fig. 1B. Lines are fitted to the points by linear least squares analysis. $F_{1(0)}$ and $ATP_{(0)}$ are the original concentrations of enzyme and substrate. k_1 = slope/($F_{1(0)}$ − $ATP_{(0)}$). The total hydrolysis of ATP occurring in the cold chase should be determined in order to show that significant hydrolysis of unbound [γ-^{32}P]ATP does not

[13] I. M. Glynn and J. B. Chappell, *Biochem. J.* **90,** 147 (1964).

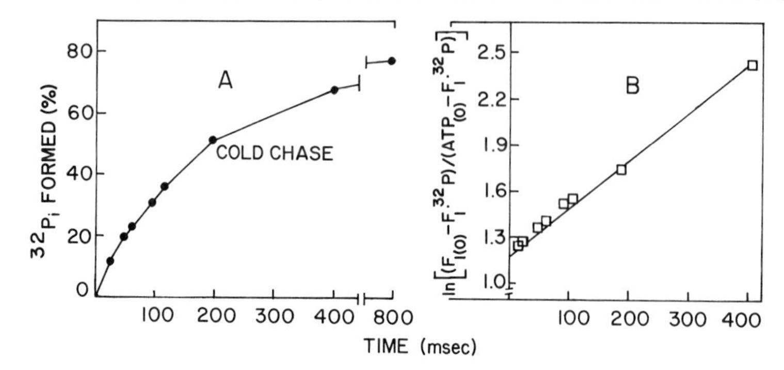

FIG. 1. Rate of binding of ATP in high-affinity catalytic sites (k_1). (A) The time course of binding. (B) Graphical determination of k_1. Details are given in the text.

take place. A control experiment is carried out in which the substrate syringe contains nonradioactive ATP and the cold chase in the receiving vessel contains a known amount of 10 m*M* [γ-^{32}P]ATP. The correction obtained is subtracted from the apparent amount of [γ-^{32}P]ATP bound in catalytic sites.

The rate at which ATP dissociates from catalytic sites on F_1 (k_{-1}) is difficult to measure because it is very slow and because the total amount dissociated is very small. The procedure includes incubation of a molar excess of enzyme with [γ-^{32}P]ATP for formation of the enzyme–substrate complex, removal of unbound radioactive ATP on a centrifuge column, and incubation of the enzyme–substrate complex in the effluent with hexokinase, glucose, and Mg^{2+}. The hexokinase added is sufficient to convert any [γ-^{32}P]ATP that dissociates into glucose-6-^{32}P. The latter is readily distinguished from undissociated [γ-^{32}P]ATP and from any $^{32}P_i$ in the reaction mixture.

F_1 (1.6 μM) is incubated with 0.6 μM [γ-^{32}P]ATP in Mg buffer for 45 sec. The volume is 150 μl. Any unbound ^{32}P in the samples is removed on a centrifuge column equilibrated with the same buffer. The effluents are collected in a single, 15-ml conical centrifuge tube containing 0.5 ml Mg buffer, 50 μmol of glucose, and 200 units of hexokinase. A second control incubation is identical with the first except that hexokinase is omitted from the receiving tube. The enzyme–substrate complex is incubated with the hexokinase trap for 40 min, after which 1 ml of a solution containing 2 *N* HCl and 0.25 m*M* P_i is added and the tubes heated in a boiling water bath for 7 min. The tubes are cooled in ice and denatured protein is removed. Esterified ^{32}P in the deproteinized supernatant is separated from $^{32}P_i$ by precipitation of P_i[11] or by extraction with isobutanol–benzene.[12] Two additional controls are required to establish that the hexokinase

TABLE I
RATE OF DISSOCIATION OF [γ-^{32}P]ATP FROM HIGH-AFFINITY CATALYTIC SITES ON $F_1(k_{-1})$[a]

Additions	[γ-^{32}P]ATP added (cpm)	^{32}P bound (cpm)	Glucose-6-^{32}P (cpm)	Glucose-6-^{32}P (%)
Hexokinase	247,000	125,000	448	0.36
None	200,000	102,000	27	0.02

[a] In the absence of hexokinase, line 2, ~2% of the total radioactivity is found in the supernatant after precipitation of $^{32}P_i$. Other details are given in the text.

functions properly. Add 0.5 ml of 0.6 μM [γ-^{32}P]ATP in Mg buffer to 0.5 ml of a solution containing 1.6 μM F_1, 200 units of hexokinase, and 50 mM glucose. The hexokinase is sufficiently competitive with F_1 to convert 75% of the radioactive ATP to glucose-6-^{32}P. In a second control experiment, carried out without F_1, 98% of the added [γ-^{32}P]ATP is converted to glucose-6-^{32}P. The results of a dissociation experiment are shown in Table I. After 40 min of incubation of the enzyme–substrate complex, about 90% of the radioactivity is dissociated from the enzyme. The actual amount of [γ-^{32}P]ATP dissociated, corrected for incomplete conversion by hexokinase is 0.49%. Since radioactivity can dissociate from F_1 either as radioactive ATP or as $^{32}P_i$ and since neither can rebind under the conditions of the experiment, the rate of [γ-^{32}P]ATP dissociation can be calculated from the known rate of $^{32}P_i$ dissociation, 0.9×10^3 M^{-1} sec^{-1}, as discussed below, and the partition of $^{32}P_i$ and [γ-^{32}P]ATP:

$$\frac{\%\ [\gamma\text{-}^{32}\text{P}]\text{ATP dissociated}}{\%\ ^{32}\text{P}_i\text{ dissociated}} = \frac{[\gamma\text{-}^{32}\text{P}]\text{ATP ``off'' rate}}{^{32}\text{P}_i\text{ ``off'' rate}}$$

The "off" rate for [γ-^{32}P]ATP is calculated to be 4.5×10^{-6} sec^{-1}. k_{-1} is calculated from this rate by adjusting for K_2, i.e., by multiplying by 1.5. The value of k_{-1} is 6×10^{-6} sec^{-1}. The affinity for ATP in catalytic sites, K_1, is equal to k_1/k_{-1} or 10^{12} M^{-1}.[1]

Step 2. The equilibrium distribution of substrate and products in the catalytic site (K_2) is determined by directly measuring the amounts of $^{32}P_i$ and total radioactivity bound in catalytic sites under the conditions of single-site catalysis. F_1 (0.325 μM) is incubated with 0.3 μM [γ-^{32}P]ATP in Mg buffer for 1 min in order to form the enzyme–substrate complex. The volume is 0.5 ml. The complex is then freed of unbound radioactive ligand by passage of 100 μl samples of the reaction mixture through individual centrifuge columns (equilibrated with Mg buffer) into a dry receiv-

TABLE II
EQUILIBRIUM DISTRIBUTION OF SUBSTRATE AND PRODUCTS IN HIGH-AFFINITY CATALYTIC SITES[a]

Time after mixing F_1 + [γ-^{32}P]ATP	Fraction of ^{32}P bound at 1 min	Bound $^{32}P_i$/Total bound ^{32}P
1	1.00	0.30
4	0.62	0.31
7	0.42	0.31
10	0.32	0.32

[a] Details are described in the text.

ing tube. Thereafter, the enzyme–substrate complex is aged for 4, 7, or 10 min, as shown in Table II, and samples of the complex are freed of dissociated ligand by passage through a second centrifuge column, also equilibrated with Mg buffer. The column effluents are collected in a tube containing 2 ml of 0.5 *M* perchloric acid and 1 mm NaP_i. The shortest incubation time shown (1 min in Table II) is obtained by centrifuging a sample of the original incubation mixture immediately after it is formed into a tube containing perchloric acid-P_i. Mixing of column effluents in the perchloric acid quench is facilitated by mounting a 23-g syringe needle with a bent tip on the column and positioning the bent tip near the bottom of a round bottom receiving tube. $^{32}P_i$ in the perchloric acid quench is determined by precipitation of P_i[11] or by extraction with isobutanol–benzene[12] and the ratio of bound $^{32}P_i$ to total bound radioactivity in the quench solution is calculated. The ratio, total radioactivity bound per mole of F_1 (^{32}P) at 1 min (0.55), is set equal to a fraction of 1.00. As shown in Table II, the ratio of $^{32}P_i$ bound/total bound ^{32}P remains essentially constant at 0.3 in spite of the fact that the total amount of radioactivity bound to enzyme decreases with increasing incubation time. The system thus behaves as though it were at equilibrium. The equilibrium constant (K_2) is 0.5., i.e., under the conditions of the experiment twice as much ATP as products is bound in catalytic sites.

The values for the individual rate constants governing the catalytic step, k_2 and k_{-2}, are determined from computer modeling of the rate of approach to an equilibrium distribution of substrate and products in the catalytic site. The experimental data are obtained from the first 200 msec of a chemical quench experiment of the same kind used to measure the rate of binding of [γ-^{32}P]ATP in catalytic sites (Fig. 1). $^{32}P_i$ formed in the acid quench is taken as a measure of the amount of $F_1 \cdot ADP \cdot {}^{32}P_i$ bound in catalytic sites. This is appropriate because the rate of dissociation of $^{32}P_i$

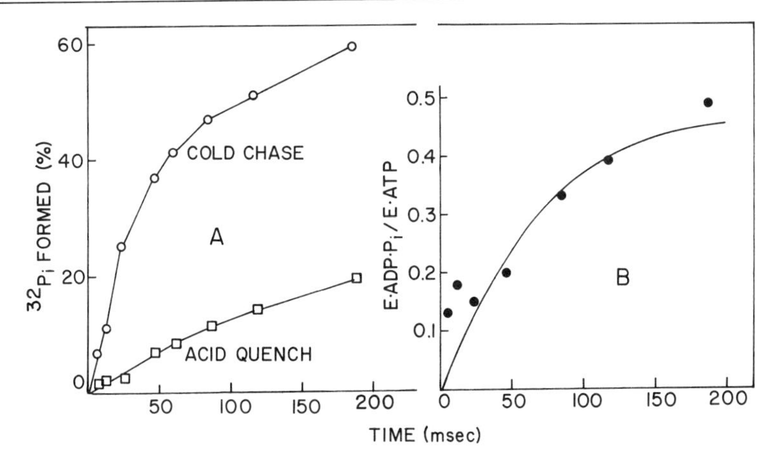

FIG. 2. Estimation of k_2 and k_{-2} from computer modeling of the rate of approach to an equilibrium distribution of substrate and product in high-affinity catalytic sites. (A) [γ-^{32}P]ATP bound in high-affinity catalytic sites is calculated as $^{32}P_i$ formed in the cold chase minus $^{32}P_i$ formed in the acid quench. (B) The ratio of E · ADP · P_i to E · ATP present in catalytic sites and calculated as described in the text is plotted versus time (●). The solid line is obtained from numerical integration of the rate equation described in the text and assuming that k_2 = 12 sec^{-1} and k_{-2} = 24 sec^{-1}.

from catalytic sites is very slow and the rate of dissociation of ADP is even slower.[1] These measurements are discussed below. Thus, over a sufficiently small time period, as used in the experiment, products are unlikely to dissociate and any $^{32}P_i$ found in the deproteinized reaction mixture is a reasonable measure of bound ADP $^{32}P_i$. The amount of [γ-^{32}P]ATP bound in catalytic sites is calculated as the difference between the cold chase and the acid quench (Fig. 2A). Control experiments to determine corrections for any hydrolysis of free radioactive ATP in the cold chase also should be carried out. The increase in the ratio $^{32}P_i/F_1$ · [γ-^{32}P]ATP with increasing times of incubation is shown in Fig. 2B (solid circles). The smooth curve is determined by numerical integration of the rate equation based on Eq. (1) using concentrations for F_1 of 2.3 μM and for [γ-^{32}P]ATP of 1.15 μM and a value for k_1, calculated from the cold chase values in Fig. 2A, of 4.6×10^6 M^{-1} sec^{-1}. Since the dissociation of [γ-^{32}P]ATP is negligible and, as explained, $^{32}P_i$ dissociation also is negligible, k_{-1} and k_{-3} are set equal to zero. The best fit to the data points is obtained when k_2 and k_{-2} are set equal to 12 sec^{-1} and 24 sec^{-1}, respectively; i.e., the model is restricted to a value for K_2 of 0.5.

Steps 3 and 4: Product Release and Rebinding. The rate of release of products ($^{32}P_i$ and [^{3}H]ADP) from high-affinity catalytic sites is measured by first forming the enzyme–substrate complex with F_1 and labeled ATP and removing unbound ligand on a centrifuge column. The complex collected in the eluate is aged for measured periods of time before dissociated products are removed on a second centrifuge column. The amount of enzyme-bound radioactivity remaining in the second eluate is determined.

F_1 (0.31 μM) is incubated with 0.25 μM [γ-^{32}P]ATP in 1 ml of Mg buffer for 1 min. Unbound ligand is removed by centrifuging 100-μl aliquots through centrifuge columns equilibrated with Mg buffer. The eluates are collected and combined. F_1 in the combined eluates contains 0.41 mol of ^{32}P/mol of protein (zero time). This value is set equal to 1.0. The enzyme–substrate complex is aged for as much as 60 min and, at 5 min intervals, 100 μl of combined eluate are withdrawn and passed through a second centrifuge column equilibrated with Mg buffer. Enzyme-bound $^{32}P_i$ is determined by measuring both the amount of radioactivity and protein in the column eluates. Results are expressed as the fraction of the zero time value. A semilog plot of the fraction of bound ^{32}P remaining versus time gives a rate of ^{32}P dissociation of 0.9×10^{-3} sec^{-1}. This value, which is taken to be the rate of $^{32}P_i$ dissociation, would be overestimated if appreciable [γ-^{32}P]ATP dissociated from catalytic sites during the measurement. However, the measurement of k_{-1} indicates that dissociation of radioactive ATP is negligible. Nevertheless, the procedures described for measurement of [γ-^{32}P]ATP dissociation should be applied to determine if significant release of nucleotide occurs during the measured release of $^{32}P_i$. The presence of 1 mM P_i in the reaction mixture effectively precludes rebinding of dissociated $^{32}P_i$ to the enzyme. The rate constant for $^{32}P_i$ dissociation (k_3) is obtained by multiplying the rate by 3, since only 30% of the radioactivity bound to F_1 is present as $^{32}P_i$ (K_2 = 0.5). Thus, k_3 in this measurement is 2.7×10^{-3} sec^{-1}.

Attempts to measure k_{-3}, the rate of binding of $^{32}P_i$ in catalytic sites under the conditions of single site catalysis, have not been successful.[1] Although k_{-2} is fast, 24 sec^{-1} (Fig. 2), the actual binding of $^{32}P_i$ in a catalytic site would appear to be unfavorable. The experiments of Sakamoto,[14] who measured the formation of [γ-^{32}P]ATP on F_1 from ADP and P_i, are relevant to this question. It appears that k_{-3} is considerably less than 5 M^{-1} sec^{-1}.[1]

The rate of release of product ADP from catalytic sites (k_4) is determined under conditions similar to those for measurement of product $^{32}P_i$ release except that the radioactive ligand is [^{3}H]ATP. F_1 (1 nmol) is rapidly mixed, using a Hamilton syringe with a bent tip, with 0.2 nmol of [^{3}H]ATP in a final volume of 1.5 ml of Mg buffer containing 1 mg/ml of

[14] J. Sakamoto, *J. Biochem.* **96,** 475 (1984).

bovine serum albumin. After 1 min of incubation, unbound ligand is removed from 135-μl aliquots of the reaction mixture by passage through centrifuge columns equilibrated with Mg buffer containing 1 mg/ml bovine serum albumin. The albumin serves to reduce loss of F_1 via nonspecific binding to the Sephadex of the column.[15] The mole ratio of tritium-labeled nucleotide to F_1 in the column effluent is 0.14. This value is taken as 100% binding. The column effluents are combined and aged either at the F_1 concentration of the mixed effluents or the mixture is diluted so as to provide several concentrations of the enzyme–substrate complex for the aging step. In this experiment, the effluents are diluted at 3.5 min (zero time) to 0.274 μM, 0.053 μM, and 0.014 μM F_1 with Mg buffer containing 1 mg/ml albumin. The mole ratio of ^3H bound to F_1 was determined after incubation periods up to 50 min by passing 100-μl aliquots of the aged reaction mixtures through Sephadex centrifuge columns equilibrated with Mg buffer–albumin. The log of the fraction of nucleotide that remained bound to protein was plotted versus time. The rate of [^{3}H]ADP dissociation obtained from the linear portion of the curves is multiplied by 3 (only one-third of the bound label is in the form of dissociable products) to give a rate constant of 3.6×10^{-4} sec^{-1}. The failure of the linear portion of the dissociation rate curves to extrapolate to 100% bound nucleotide at zero time[1] may be due to losses of F_1 on passage through the second centrifuge column. Under the conditions of this experiment, rebinding of dissociated [^{3}H]ADP is insignificant. This point should be checked in control experiments in which 0.1 μM nonradioactive ADP is added to the aging reaction mixtures. The rate of release of labeled ADP should not be affected under these conditions.[1]

The rate of binding of [^{3}H]ADP by $F_1(k_{-4})$ is now discussed. Two precautions are important. First, the preparation of F_1 should be washed once so as to reduce any possibility that nucleotides might be present in catalytic sites. A sample of the ammonium sulfate suspension of the enzyme[8] sufficient for the experiment is centrifuged and the sediment resuspended in 70% ammonium sulfate, pH 8. The enzyme is sedimented again, dissolved in Mg buffer, and passed through a centrifuge column equilibrated with the same buffer. A second precaution is to remove any adventitious ATP from [^{3}H]ADP stock solutions. Two milliliters of a 70-μM solution of [^{3}H]ADP, 285 counts/min/pmol, are added to the sediment obtained after centrifuging 0.4 ml of washed agarose-bound hexokinase (Sigma). The sediment contains 25 units of hexokinase. After addition of 2 μl of 1 M glucose, the reaction is continued for 3 min and the mixture filtered on a Millipore filter (0.45 μm) to remove the agarose-

[15] R. L. Cross and C. M. Nalin, *J. Biol. Chem.* **257**, 2874 (1982).

bound hexokinase. The procedure reduces an initial 1.4% contamination by ATP ~60-fold.

The reaction mixture contains 0.25 μM F_1, 0.25 M sucrose, 40 mM 4-morpholinoethanesulfonic acid, 40 mM Tris, pH 7.5, 2.5 mM $MgSO_4$, and 1 mg/ml bovine serum albumin. The volume is 1.15 ml. After 20 sec at room temperature, 100 μl of 57 μM [^{3}H]ADP, freed of ATP, is added with rapid mixing on a vortex mixer. The final concentration of MgADP is 3.9 μM, calculated as described[1] from the known concentrations of ADP, Mg^{2+}, and other ligands in the reaction mixture. At 15-sec intervals, over a period of 150 sec, unbound ligand is removed by passing aliquots of the reaction mixture through centrifuge columns equilibrated with the same buffer containing albumin. Both protein and radioactivity in the column effluents are determined and the ratio of [^{3}H]ADP/F_1 (mol/mol) is calculated and plotted versus time to show the time course of binding. A second-order plot of the data is constructed by plotting ln[$ADP_{(0)} - F_1 \cdot ADP/F_{1(0)} - F_1 \cdot ADP$] versus the time. $ADP_{(0)}$ is the initial concentration of Mg[^{3}H]ADP^-, $F_{1(0)}$ is the initial concentration of F_1, and $F_1 \cdot ADP$ is the experimentally determined concentration of the $F_1 \cdot ADP$ complex. The bimolecular rate constant calculated from the slope, k_{-4} = slope/$ADP_{(0)} - F_{1(0)}$, is 1.3×10^3 M^{-1} sec^{-1}. Inclusion of 1 mM P_i in the incubation mixtures increases the rate of ADP binding.[1]

[63] Refinements in Oxygen-18 Methodology for the Study of Phosphorylation Mechanisms

By KERSTIN E. STEMPEL and PAUL D. BOYER

Introduction[1]

The analysis of volatile derivatives of phosphate for ^{18}O content by gas chromatography/mass spectrometry has proved to be a valuable tool for probing the details of many enzyme reaction mechanisms, especially the ATPases. The distribution of ^{18}O in the product phosphate or ATP formed in a reaction gives far more information about the mechanisms involved than measurement of the average ^{18}O enrichment of the P_i or medium

[1] Preparation of this chapter and research in this laboratory were supported by U.S. Public Health Service Grant GM11094, National Science Foundation Grant PCM 81-00817, and Department of Energy Contract DE-AT03-76ER70102.

water after various types of oxygen exchanges.[2] Similar techniques can also be applied to the measurement of oxygen and deuterium exchanges in carboxyl groups with enzymes such as citrate synthase.[3]

Several changes have been made in our procedures for the analysis of volatile phosphate derivatives since a previous publication from this laboratory on this topic.[2] This chapter will described these refinements, including the synthesis of *N*-ethyl-*N*-nitroso-*p*-toluenesulfonamide (the starting material for the preparation of diazoethane, which is not commercially available); a convenient method for the preparation of diazoethane from *N*-ethyl-*N*-nitroso-*p*-toluenesulfonamide; the derivatization of H_3PO_4 to triethyl phosphate using diazoethane; and the gas chromatography of triethyl phosphate and its analysis for ^{18}O content by specific ion monitoring on a rapid-scanning, quadrupole mass spectrometer (Hewlett Packard 5995 GC/MS). Some changes in our previously published methods[2] for the preparation and purification of [^{18}O]P_i and the purification of H_3PO_4 will also be described, as will an improved method of preparation and purification of [^{18}O]ATP, a method for the rapid approximation of the ^{18}O content of small amounts of highly enriched $H^{18}OH$, and a convenient method for the preparation of [^{18}O]phosphoenolpyruvate.

Derivatization of H_3PO_4 for Analysis by Gas Chromatography/Mass Spectrometry

Triethyl phosphate serves as a convenient derivative of phosphate to use for gas chromatography/mass spectrometry (GC/MS) analysis. It is preferable to the trimethylsilyl (TMS) derivative described by us in our previous publication of this methodology[2] because no complex spillover corrections need to be applied for the silicone isotopes. Triethyl phosphate chromatographs more rapidly at lower temperatures than the TMS derivative and does not contaminate the mass spectrometer source as rapidly. It is preferable to the trimethyl derivative[4,5] because its fragmentation pattern under electron impact ionization yields two major peaks, either one of which can be maximized to be the base peak, depending on the tuning of the instrument, and they both contain all the oxygens of the phosphate and are therefore suitable for analysis of the ^{18}O distribution.[6]

[2] D. D. Hackney, K. E. Stempel, and P. D. Boyer, this series, Vol. 64, p. 60.
[3] J. A. Myers and P. D. Boyer, *Biochemistry* **23,** 1264 (1984).
[4] T. R. Sharp and S. J. Benkovic, *Biochemistry* **18,** 2910 (1979).
[5] C. F. Middlefort and I. A. Rose, *J. Biol. Chem.* **251,** 5881 (1976).
[6] The use of this derivative was suggested to us by Dr. George Popjak of the Brain Research Institute at UCLA to whom we are extremely grateful for the use of his facilities, in which much of our preliminary work on this problem was performed.

Under the same conditions trimethyl phosphate presents a pattern where none of the major peaks contains all the oxygens of the phosphate moiety, and one must therefore use the molecular ion which produces a much smaller signal. With chemical ionization, sensitivity may not be a problem.

Triethyl phosphate can conveniently be prepared in small quantities by the action of diazoethane ($C_2H_4N_2$) on phosphoric acid. Although $C_2H_4N_2$ can be prepared by essentially the same methods as diazomethane (CH_2N_2), it is less easy to prepare. The commercially available 1-ethyl-3-nitro-1-nitrosoguanidine, which can be used in the diazomethane generator, available from Aldrich, gives poor yields compared with the methyl compound, probably because of the lower volatility of $C_2H_4N_2$. The system, because of its small scale, is only suitable for the preparation of <1 mmol of the compound. Also, 1-ethyl-3-nitro-1-nitrosoguanidine is a potent mutagen and a flammable solid, altogether a very nasty item to handle on a routine basis.

An alternative starting material for the preparation of larger amounts of CH_2N_2 is *N*-methyl-*N*-nitroso-*p*-toluenesulfonamide, available from Aldrich as Diazald. The equivalent ethyl compound to make $C_2H_4N_2$ is not available commercially, but we have found its synthesis in the laboratory to be a simple one-day procedure that yields a year's supply of material. *N*-Ethyl-*N*-nitroso-*p*-toluenesulfonamide is an irritant, but is less toxic, nonflammable, and not explosive, and is therefore less hazardous to use than 1-ethyl-3-nitro-1-nitrosoguanidine.

Other workers have used the propyl derivative of toluenesulfonamide as the starting material to make tripropyl phosphate, which has a fragmentation pattern similar to that of triethyl phosphate. However, the synthesis and purification of the starting material are more problematical, and it is no longer used as a routine procedure.[7]

Preparation of N-Ethyl-N-nitroso-p-toluenesulfonamide

Principle. This synthesis is based on the procedure for the synthesis of *N*-methyl-*N*-nitroso-*p*-toluenesulfonamide.[8] Aqueous ethylamine is reacted with *p*-toluenesulfonyl chloride to produce *N*-ethyl-*p*-toluenesulfonamide. The ethylamine hydrochloride that forms is neutralized with NaOH to reform ethylamine and thus maintains the reaction. A precipitate of NaCl results. The reaction is then acidified with glacial acetic acid and the nitroso compound is formed by the addition of $NaNO_2$.

[7] D. D. Hackney, personal communication.
[8] *Org. Synth.* **34,** 96 (1954).

The reaction scheme is as follows:

$$CH_3C_6H_4SO_2Cl + 2C_2H_5NH_2 \rightarrow CH_3C_6H_4SO_2NHC_2H_5 + C_2H_5NH_2 \cdot HCl$$

$$(C_2H_5NH_2 \cdot HCl + NaOH \rightarrow C_2H_5NH_2 + NaCl + H_2O)$$

$$CH_3C_6H_4SO_2NHC_2H_5 + HNO_2 \rightarrow CH_3C_6H_4SO_2NNOC_2H_5 + H_2O$$

Materials

p-Toluenesulfonyl chloride (98% pure, Aldrich Chemical Co.)
Ethylamine, 33% aqueous solution prepared by diluting 70% stock ethylamine (99% pure, Sigma Chemical Co.)
NaOH, 50% solution, by weight
$NaNO_2$, 7.2 *M*
Glacial acetic acid

Procedure. The entire synthesis should be done in a well-ventilated fume hood. The *p*-toluenesulfonyl chloride is weighed out in three aliquots of 47.5 g, 22.5 g, and 10 g each. The first aliquot of 47.5 g is added slowly with stirring to 52.5 ml of 33% ethylamine in a 3-neck, 500-ml round-bottom flask on a heating mantle. The reaction is exothermic and the mixture is allowed to heat up to 70–80° in order to keep the *N*-ethyl-*p*-toluenesulfonamide in a molten condition (MP 64°), but do not allow it to boil. A reflux condenser in the center neck can help to minimize the loss of ethylamine (BP 16.6°). After all the sulfonyl chloride has been added, the mixture is stirred vigorously and tested with pH paper to determine if it becomes acidic. As soon as it does, 12.5 ml of 50% NaOH is added slowly with stirring, from a constant addition funnel through the third neck of the flask. (If the solution does not become acidic after completion of each addition of sulfonyl chloride, do not wait longer than 5 min before adding the NaOH.) Immediately, the 22.5-g aliquot of sulfonyl chloride is added slowly with stirring, as before. The pH is checked and 6.25 ml NaOH is added. This sequence is repeated with the last 10-g aliquot of sulfonyl chloride and 6.25 ml more of NaOH. The pH is checked once more. If it is not basic, indicating excessive loss of ethylamine, sufficient ethylamine is added to make it basic. The walls of the flask are rinsed with 5 ml of water and the reaction is completed by heating to 80–90° for 15 min with vigorous stirring. Heating is slow up to 70–75° and then increases rapidly.

While hot, the reaction mixture is poured into 375 ml of glacial acetic acid in a 2-neck 1000-ml round-bottom flask. The smaller flask is rinsed with 65 ml of glacial acetic acid, which is added to the larger flask. The mixture is cooled to 0–5°. This amount of acetic acid is required to keep the *N*-ethyl-*p*-toluenesulfonamide in solution at this low temperature and to acidify the $NaNO_2$.

With constant stirring, 63 ml of 7.2 *M* $NaNO_2$ is added dropwise over a period of 45 min. While keeping the temperature below 10°, the reaction mix is stirred for an additional 15 min after addition of the $NaNO_2$ is complete. The nitroso compound, *N*-ethyl-*N*-nitroso-*p*-toluenesulfonamide (ENTS), separates as a light-yellow crystalline solid. Highly toxic NO_2 gas may be evolved during this step.

The mixture is diluted with 250 ml of cold water and filtered on a Buchner funnel. The ENTS is washed with 200 ml cold water on the funnel. The crystals are transferred to a beaker with 200 ml of cold water, stirred well to wash, and refiltered. The washing procedure is repeated until the filtrate is no longer acidic.

The ENTS is dried in an evacuated desiccator over sulfuric acid in the cold. The yield from this procedure is 65–70 g or about 70%. The melting point of the product is diffuse and starts at about 32–33°. It should be stored in the dark at 4° because the crystals can melt and may undergo some decomposition at room temperature. We have successfully used preparations that are well over 2 years old.

ENTS may show the low and diffuse melting point because it is still impure, the principal impurity most likely being *N*-ethyl-*p*-tolulene-sulfonamide. This can be removed by following the water washes with a wash of cold 1 *N* NaOH (the nitroso compound is not attacked by cold aqueous alkali), followed by water to neutrality again. Recrystallization can be accomplished by dissolving the ENTS in boiling ether (1 ml/g), adding an equal volume of low-boiling petroleum ether, and refrigerating overnight. Purification would probably improve the long-term storage of ENTS, but we have not found it to be necessary, as the presence of *N*-ethyl-*p*-toluenesulfonamide does not interfere with the generation of diazoethane.

The Generation of Diazoethane from N-Ethyl-N-nitroso-p-toluenesulfonamide

Principle. Diazoethane is generated by the methanolic base-catalyzed decomposition of *N*-ethyl-*N*-nitroso-*p*-toluenesulfonamide in ether solution with 2-(2-ethoxyethoxyethanol) present as an additional catalyst.

The reaction scheme is as follows:

$CH_3C_6H_4$-SO$_2$-N(NO)(C_2H_5) → $CH_3C_6H_4$-SO$_2$-N(C_2H_5)(ON)

$CH_3C_6H_4$-SO$_2$-O-N=N-CH(H)-CH_3 (:Base) $\xrightarrow{H^+}$ $CH_3C_6H_4$-SO$_2$-OH + C_2H_4=$^{+}$N=$^{-}$N

Materials

- *N*-Ethyl-*N*-nitroso-*p*-toluenesulfonamide
- Saturated KOH in 50% methanol
- Glass-distilled diethyl ether (Burdick and Jackson) stored over 4 Å molecular sieve
- 2-(2-ethoxyethoxy)ethanol, Chem Services 0-173; this compound is peroxidizable and it is therefore desirable to purchase it in small quantities even though this is slightly more expensive.

Procedure. The following cautions should be noted: Diazoethane is highly toxic and may be carcinogenic; wear gloves when carrying out this procedure and work in a well-ventilated fume hood. Diazoethane is also potentially explosive. Scratches and rough edges on glass are known to promote explosion; therefore, one should be careful to use only new, unscratched glassware, Clear-Seal joints (Wheaton), and all rough edges should be fire-polished. Work behind a safety shield.

The apparatus shown in Fig. 1 was constructed in our departmental glass-blowing facility and works very well for the production of several millimoles of $C_2H_4N_2$. A stream of nitrogen is saturated with ether as it bubbles through the first tube. The stream then carries the $C_2H_4N_2$ generated in the second tube over to the third tube where it is trapped in ice-cold ether. Any untrapped $C_2H_4N_2$ is vented to the back of the hood. A satisfactory substitute for the apparatus shown can be constructed from two standard test tubes, one 40-ml screw-capped glass tube (Corning 8122) with a Teflon-lined cap for the final receiving tube, 2-mm i.d. glass tubing, and three number 1 rubber stoppers. The condensing coil between tubes 2 and 3 helps to prevent entrainment of the reaction components over into the final product where their presence can interfere with the analysis, especially with very small samples.

ENTS, 1.4 g, is weighed into tube 2 and dissolved in 2.5 ml of ether and 2.5 ml of ethoxyethanol with gentle vortexing. About 5 ml of ether is added to tube 1 and 4 ml to tube 3. The apparatus is secured at the Clear-Seal joints with rubber bands and a gentle stream of nitrogen, about 2

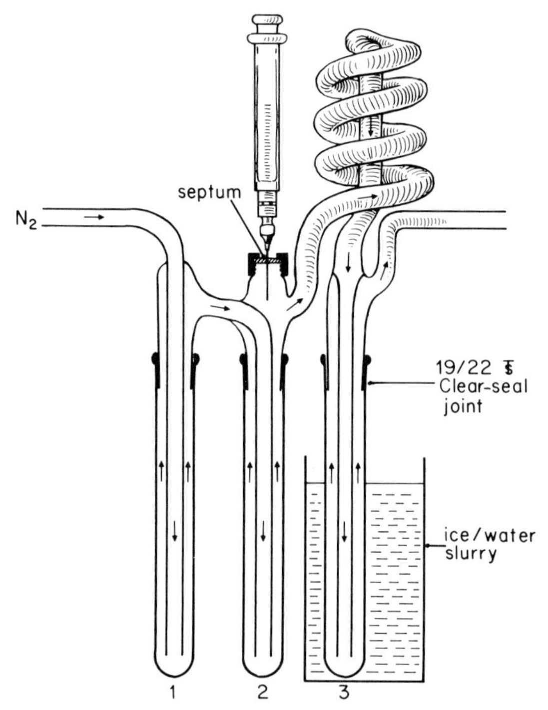

FIG. 1. Apparatus for the preparation of diazoethane. The Teflon-lined silicone septum is inserted in a 13-425 screw-threaded septum cap taken from the vial that is fused into the top of tube number 2. The glass tubing is 6 mm o.d. and 3 mm i.d. The tubes are 15 cm long, and the overall height of the apparatus is 28 cm.

bubbles/sec, is used to flush the system. The collection tube is placed in a 250-ml beaker with an ice/water slurry. If the room temperature is below 20°, it may also be necessary to immerse the reaction tube in warm water in order to promote a rapid reaction. The reaction is initiated by adding 3 ml of methanolic KOH from a 3-ml disposable plastic syringe equipped with a 22-gauge stainless-steel needle through the septum. The bubbling rate of the nitrogen should be increased at this point in order to promote mixing of the ether with the methanol or else the reaction is slow and the yield poor. The reaction will turn a dark yellow-brown and will bubble vigorously as the initial evolution of gas takes place. At this point the nitrogen flow is reduced so as to avoid carryover of the reaction components. The bubbling of the reaction itself now will help to provide sufficient mixing. Three-milliliter aliquots of water and methanol can now be added alternately through the syringe needle. This seems to drive the reaction to completion and prevents the needle from becoming clogged. The nitrogen is bubbled slowly through the system until the reaction no longer forms bubbles when the stream is stopped, but no longer than 15 min. The final solution of $C_2H_4N_2$ should be a deep yellow color. With the N_2 flow still on, the third tube is removed and capped or the solution is poured into an acid-washed, unscratched, screw-capped glass tube capped tightly with a Teflon-lined cap. Protect the solution from light, since strong light promotes explosion, and store it in an explosion-proof freezer. It will keep for up to 2 weeks, depending on the initial strength. If it is significantly yellow, it will still work.

Preparation of Triethyl Phosphate from Diazoethane and Phosphoric Acid

Principle. The phosphate must be fully protonated for the reaction with diazoethane to take place. The reaction is as follows:

$$3C_2H_4N_2 + H_3PO_4 \rightarrow (C_2H_5)_3PO_4 + 3N_2$$

Procedure. The samples are prepared in conical glass Reactivials with Minnert valve tops (Pierce Chemical Co.). The 5-ml size is convenient for collecting the relatively large elution volume from the column purification of the phosphate,[2] which can then be lyophilized to dryness and redissolved at the bottom of the cone in the much smaller volume required for analysis. For very small samples, less than 40 or 50 nmol, the samples can be transferred to smaller Reactivials after the initial lyophilization and relyophilized before derivatization. The derivatization reaction must be carried out in a well-ventilated fume hood. The diazoethane solution in ether is added in 25- or 50-μl aliquots from a microliter syringe until the solution remains yellow for 5 min after the addition. The vials should be capped between additions because the solution is volatile. The evolution of small bubbles of nitrogen can be observed as the reaction takes place. After the reaction is complete, the excess diazoethane in ether is evaporated in with a gentle stream of nitrogen until the observable liquid just disappears, then glass-distilled dichloromethane (stored over 4 Å molecular sieve) is added rapidly to solubilize the triethyl phosphate at a final concentration of about 1 to 2 nmol/μl. The triethyl phosphate is less volatile than the diazoethane in ether solution, but excessive drying may result in some loss of sample. Although septum tops of Teflon-lined silicone will also work for these vials, we have found that the Minnert valves, although expensive, are much more satisfactory for preservation of samples, especially after the septum has been pierced.

Gas Chromatography of Triethyl Phosphate

Triethyl phosphate is a highly polar compound and does not chromatograph easily. Part of the problem seems to be retention of $(C_2H_5O)_3PO$ by the support material leading to excessive tailing. In the past, we have had success with 2 mm × 6 ft glass packed columns of 3% OV-275 on 40–60 Chromosorb T, but more recently purchased columns no longer seem to give symmetrical peaks, probably due to minor changes by the manufacturer of the support or liquid phase. We have therefore adapted the GC to accept a Supelco 30 m × 0.75 mm i.d. wide-bore glass capillary column coated with 0.2 μm of SP-2330, a polar cyanosilicone phase. This column gives nearly symmetrical peaks of triethyl phosphate. We routinely inject 1 μl of sample at a concentration of 1 nmol/μl in dichloromethane (range, 0.1–10 nmol/μl). Triethyl phosphate has a retention time of about 6 min and a peak width of about 30 sec under the following conditions: injection port at 250°, oven temperature at 144° for 7.5 min followed by a 15°/min ramp to 180° and hold for 15 min, He flow through column of 5 ml/min with make-up gas flow of 20 ml/min added in order to operate the jet separator interface to the mass spectrometer. The solvent peak elutes by 4.5 min. The temperature ramping is necessary to clean the column between samples. Although the turn-around time is longer than with the packed columns, much better separation of minor contaminating peaks is achieved and so more accurate integrations of minor ^{18}O-containing peaks can be obtained. We have not tried a fused silica wide-bore capillary column coated with a polar phase but this could reduce tailing of the phosphate even more.

We have some limited experience with capillary column separation of triethyl phosphate and have obtained adequate separations on a 0.25-mm i.d., 30 m fused silica column with a 0.25-μm film of DB-5 run at 140° isothermal with a split injector (40 : 1 split ratio) and the injection port at 250°. The triethyl phosphate elutes at about 3.5 min. This tends to confirm that at least part of the problem with the packed columns is the support and not the liquid phase, since DB-5 is a nonpolar column, but the triethyl phosphate peak still shows more tailing than is desirable and a more polar phase would probably produce better results. However, the extremely narrow peaks achieved on capillary columns are not really ideal for quantitative analysis of individual masses since insufficient scans can be collected over the elution time of the sample to give an accurate integration of mass peak size.

Mass Spectrometry of Triethyl Phosphate

A rapid-scanning quadrupole-type mass spectrometer is advantageous for accurate quantitation of any given mass present in a sample. Single scans taken at the GC elution peak show too much variability from injection to injection. Even multiple scans taken across the peak are frequently too variable to give consistent results even after multiple injections of the same sample. The Hewlett Packard 5995 GC/MS has proved to be a satisfactory instrument for our purposes. In the specific ion monitoring (SIM) mode it can monitor up to 20 individual ions at any one time, though only 6 can be integrated at once. Normally we monitor 5 or 6 at any one time with a 100-msec dwell time on each ion. This means a total of 500–600 msec per scan and, with a peak width of about 30 sec at 144°, 50–60 scans can be accumulated for each mass in any one run.

Figure 2 shows typical mass spectrometry scans of triethyl phosphate of three different ^{18}O enrichments produced by electron impact ionization. Depending on the tuning of the instrument, different regions of the scan can be enhanced, e.g., the m/e 99 region, representing the fully protonated phosphate moeity, H_4PO_4, rather than the m/e 155 region, the fully protonated diethyl phosphate fragment, $(C_2H_5O)_2H_3PO_2$, or the molecular ion at m/e 182. We have chosen to use the 155 region over the other two for several reasons. It is about 10 times larger than the molecular ion and, although about the same size or slightly smaller than the base peak at 99, it is in a cleaner region of the spectrum and there are fewer minor fragments to interfere with ^{18}O analysis. However, this is not true for ^{17}O analysis because of the presence of significant 154 and 156 peaks. For ^{17}O the 99 family is better. Since the HP 5995 is not a high-resolution instrument, one cannnot isolate the phosphate masses from contaminating fragments of close to the same mass by specifying masses to the fourth decimal place.

The table presents the raw data from SIM scans of natural abundance, 44% and 99% $[^{18}O]P_i$ at the three possible analysis areas. It is obvious from the integrated area per microliter that the sensitivity is much reduced with the molecular ion. The 99–107 region consistently shows a peak at 107 (PO_4), which would be expected if there were four ^{18}O present in the phosphate molecule; this is impossible in a natural abundance sample which contains only 0.204 atom% ^{18}O per oxygen. Similarly, a highly enriched sample which has been shown by analysis at the 155 region to have <1% unenriched ions (PO_0) shows the presence of 2% mass 99, presumably due to a kick up in mass of a fragment of lower mass that contains one or more oxygens. This also occurs with the 155 region, but there is only one significant contaminant due to the presence of a fragment of m/e 153, which is about 5% of the sum of the 155 and 157 masses in a natural abundance sample. This fragment contains all four oxygens and probably represents the diethyl phosphate fragment minus 2 protons. In enriched samples this fragment becomes kicked up to mass

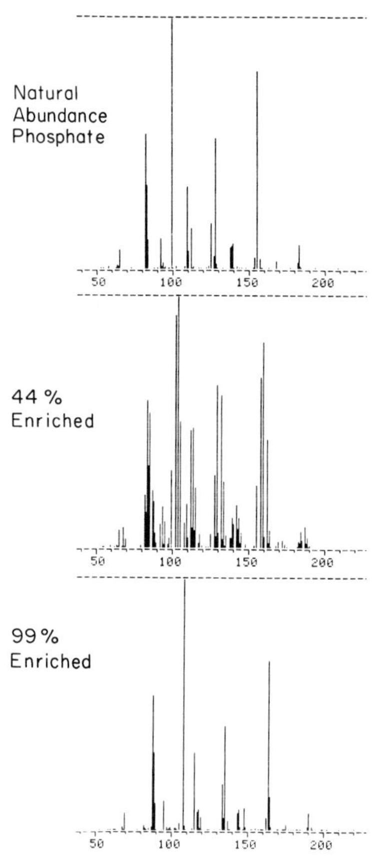

FIG. 2. Mass spectral scans of triethyl phosphate of natural abundance (0.20%), and of 44 and 99% average enrichment with ^{18}O. The mass spectrometer tuning is optimized for the region between m/e 150 and 169.

161 (PO_3) and therefore the overall enrichment and the PO_4 to PO_3 ratio appear to be too low. This can be corrected for by running a standard of natural abundance P_i on the day of analysis to determine the size of the 153 peak, which varies slightly according to the tuning of the MS and the cleanness of the source. Over the past 3 years it has averaged 5.05 ± 0.35% (N = 184). This number can then be used to apply a spillover correction to the raw data obtained from the SIM scans. It is assumed that the 163 peak is correct, but that the 161 peak is too big by 5% of the 163,

SIM ANALYSIS AT THREE MASS RANGES OF TRIETHYL PHOSPHATE STANDARDS OF DIFFERENCE ^{18}O ENRICHMENT

Sample	Mass range	Distribution of ^{18}O species[a] (% of total)					Integrated area/μl $\times 10^{-5}$
		PO_0	PO_1	PO_2	PO_3	PO_4	
Natural	99–107	98.83	0.69	0	0	0.48	28
abundance	155–163	99.18	0.82	0	0	0	22
(0.2%)	182–190	99.08	0.92	0	0	0	4
44% enriched	99–107	10.62	31.07	36.12	18.43	3.76	267
	155–163	11.16	31.63	35.84	17.92	3.45	415
	182–190	10.63	31.31	35.88	18.53	3.64	30
99% enriched	99–107	1.94	0.75	0.36	2.87	94.08	22
	155–163	0.40	0.12	0.46	7.01	92.01	77
	182–190	1.34[b]	0.78[b]	0.90[b]	3.83	93.15	8

[a] For convenience the different ^{18}O species are designated PO_0, PO_1, . . . , PO_4, rather than $P^{16}O_4$, $P^{16}O_3{}^{18}O_1$, . . . , $P^{18}O_4$.

[b] These numbers are roughly two times too large due to the presence of unresolved leading shoulders on these three peaks.

so 5% of 163 is subtracted from 161 to give a corrected area. Then 5% of this corrected area is subtracted from the 159 peak, and so on for each mass. The final values are renormalized to 100%, and this is taken as the correct distribution. The 155–163 and 99–107 distributions for the 99% sample agree quite well if this correction is made, especially if the excess PO_0 and PO_1 are also eliminated from the 99–107 series. There are still small deviations from the theoretical values for a binomial distribution of a sample of known average enrichment, but they are small, $\pm 0.3\%$.

Another disadvantage of the molecular ion, aside from the lack of sensitivity, is the presence of small amounts of contaminants that are not separable under our standard GC conditions, and therefore appear as shoulders on the 182, 184, and 186 peaks in highly enriched samples. They are too small to show up in moderately enriched samples, or they are a very small percentage of the total integrated area, but they make the molecular ion region totally unsuitable for analysis of highly enriched P_i or ATP.

It will be noted from the table that the distribution of ^{18}O in the natural abundance sample is not exactly as expected from the known average enrichment of phosphate present in the environment. The observed value of 0.82 for the 157 peak is closest to the expected value of 0.204 atom%

per oxygen × 4 = 0.816% of the phosphate being $P^{18}O_1$ rather than $P^{18}O_0$, but this is not necessarily always the case. The average values for the natural abundance distribution of the same standard solution injected over a period of 3 years (N = 184) are 0.891 ± 0.069% 157 ($p^{16}O_3{}^{18}O_1$) and 99.109 ± 0.070% 155 ($P^{16}O_4$). Thus, the 157 peak tends to be too large. This may be due to the presence of a small contaminant fragment with mass 157 which cannot be separated from the phosphate fragment without high resolution. It may also be a bias of the instrument, since a recent change in the electron multiplier has resulted in a consistently lower value for 157 than our previous cumulative average. However, it is a consistent, reproducible, and measurable bias which can be allowed for in calculations where natural abundance contamination must be subtracted from the overall distribution.

Repeated injections of the same sample give reproducible distribution patterns even on different days. However, we have noted that equivalent samples of the same 99% [^{18}O]P_i standard prepared on the same day, using the same reagents, but lyophilized in different vials, show larger than expected variation, and the differences in distribution appear to be due to exchange taking place either during lyophilization or derivatization. Lyophilization of unenriched P_i in enriched water showed no detectable incorporation of ^{18}O into the phosphate; therefore the exchange is probably not occurring with medium water. Silanization of the Reactivials with dimethyldichlorosilane greatly reduced the variability in the analyses, and so it seems likely that the exchange is taking place with active OH groups on the surface of the glass vials. Whether this occurs during lyophilization or derivatization is not known. We routinely wash the vials by soaking them overnight in a 1 : 1 mixture of concentrated H_2SO_4 and HNO_3, and this may contribute to the activation of the surface even while it removes all traces of phosphate. In any case, for samples where small changes in distribution are critical to interpretation of the data, it is recommended that the vials be silanized or siliconized with Aqua Sil (Pierce Chemical Co.) before use. The latter reagent is easier to handle, being in aqueous rather than in toluene solution.

Improvements in and Modifications of Previously Published ^{18}O Methodology

The following section will briefly outline some major and minor (but important) changes in the procedures that were previously published from this laboratory.[2]

Modifications in the Preparation of [^{18}O]KH_2PO_4 from PCl_5 and $H^{18}OH$

The most important criterion for the preparation of P_i with maximum ^{18}O enrichment by this method[2] is that the PCl_5 should be pure and *dry*. We have found it difficult to obtain this chemical in small quantities and, even more important, packaged so as to minimize reaction with atmospheric moisture. Alfa Products is the only company we have found that sometimes supplies Ultrapure phosphorus(V) chloride in 25-g lots and ships it in a sealed can under dry argon. We open the can and bottle in a dry anaerobic chamber and weigh it in 2-g aliquots into glass vials with Teflon-lined screw caps which are stored in a desiccator over P_2O_5 in the same chamber. When it is needed for a preparation, a Schlenk tube (Kontes Scientific Glassware) is passed into the chamber through the air lock and one vial of PCl_5 is emptied into it. The Schlenk tube can then be transferred into a second, less stringent chamber (dry but not necessarily anaerobic) or a glove bag for the actual reaction with $H^{18}OH$.

If the water is added through a constant addition funnel, which is capped with a drying tube packed with KOH pellets before starting the addition, the HCl gas which is evolved during the reaction is trapped and the reaction is easier to handle.

The $H^{18}OH$ is more easily recovered if the vacuum transfer is performed before addition of the imidazole buffer. The receiving flask is primed with an excess of sodium glycinate which has been ground and dried over P_2O_5. If the transfer is done with liquid nitrogen, a good deal of the HCl is also transferred and neutralized by the glycinate when the flasks are thawed. The $H^{18}OH$ can then be purified by successive vacuum transfers. This prevents contamination of the recovered water by imidazole, since sodium glycinate does not sublime.

Separation of Salt-Free [^{18}O]KH_2PO_4

Principle. The ^{18}O-P_i is isolated from the acid AG1-X4 column eluent by conversion to $MgNH_4PO_4$, which is then removed by filtration and reconverted to H_3PO_4 by treatment with a cation exchange resin.

Materials

3 *N* NH_4Cl in 15 *N* NH_4OH
5 *N* NH_4OH
Magnesium mix: 0.54 *M* $MgCl_26H_2O$, 3.76 *M* NH_4Cl, 3 *M* NH_4OH
A50W-X8 cation exchange resin (100-200, H^+), Bio-Rad, freshly recycled with 2 *N* KOH and 2 *N* HCl

Acid-washed charcoal
95% redistilled ethanol

Procedure. The P_i-containing fractions from the 30 m*M* HCl elution of the 2 × 19 cm AG1-X4 ion-exchange column[2] are pooled and lyophilized overnight to reduce the volume. Ice-cold 15 *N* NH_4OH, 3 *N* NH_4Cl is added to the measured volume such that the final concentration of NH_4OH is 3 *N* and that of NH_4Cl is 1 *N*. Then an amount of cold magnesium mix equal to 20% of this combined volume is added, and the mixture is chilled thoroughly to promote precipitation of the phosphate. The precipitate is collected on a fine sintered glass funnel and washed with cold 5 *N* NH_4OH.

The $MgNH_4PO_4$ is resolubilized by adding 50 ml settled volume of A50W-X8, which has been freshly recycled just before use, to the sintered glass funnel. Then 50-ml aliquots of water are added to the funnel and the precipitate is stirred gently to solubilize the P_i, which is allowed to drip through into an acid-washed suction flask. The washing is continued until all visible precipitate is dissolved; the remaining solution is removed by aspiration. To ensure complete conversion to H_3PO_4, the filtrate can be passed rapidly through a 1 × 5 cm column of freshly prepared A50W-X8, H^+ form.

To remove the UV-absorbing contaminants that are apparently introduced with the resin, the solution is now treated with about 1 g of acid-washed charcoal and filtered on a 0.45-μm membrane filter.

The $[^{18}O]H_3PO_4$ is titrated to pH 4.5 with 5 *N* KOH, shell frozen, and lyophilized to dryness. It should be light and fluffy at this stage and is essentially free of salts. If the preparation is sticky after lyophilization, it can be further purified by recrystallization from 50% ethanol prepared from redistilled 95% ethanol.

Purification of H_3PO_4 on AG1-X4 Columns

The separation of P_i from the molybdate on AG1-X4 columns after extraction from reaction mixtures as the phosphomolybdate complex, as described in our earlier publication,[2] can be facilitated by the addition of small amounts of 20 m*M* $MgCl_2$ to the column following the application of sample and the subsequent water wash. Two milliliters is usually sufficient to move most of the P_i away from the molybdate so they do not recombine on acidification; however, the formation of magnesium phosphate can also greatly hinder the recovery of small amounts of phosphate, so the $MgCl_2$ wash is followed by a wash with 2 ml of 20 m*M* KCl, which removes the Mg^{2+}. After another brief water wash, the P_i is eluted with 10 and 30 mM HCl treatment, as described previously. Using this procedure, samples as small as 10–20 nmol have been recovered for GC/MS analysis. Complete removal of the Mg^{2+} is important because as well as interfering with the actual recovery of P_i from the column, it can also interfere with the derivatization of the H_3PO_4 by diazoethane.

Preparation of [γ-^{18}O]ATP by the Use of the Carbamate Kinase, ADP, and [^{18}O]Carbamoyl Phosphate

Principle. This procedure is based on the method described by Mokrasch *et al.*[9] for the synthesis of [^{32}P]adenine nucleotides which uses carbamate kinase to form ATP from carbamoyl phosphate and ADP. This reaction takes place without significant exchange of oxygens with medium water. The introduction of contaminating unenriched phosphate into the reaction can be more easily controlled than in the previously described method[2] based on Penefsky *et al.*,[10] which tends to introduce considerable unlabeled ATP into the final preparation. The ^{18}O-labeled carbamoyl phosphate is formed by the nonenzymatic reaction of [^{18}O]P_i with KCNO at acid pH. A disadvantage is that slight oxygen exchange may occur at this stage of the preparation.

Materials

KCNO, 2 M
[^{18}O]P_i
HCl, 1 M
Tris, 1 M
Mg acetate, 120 mM
ADP, 120 mM
Carbamate kinase, 1000 U/ml (Sigma Chemical Co.)
AG MP-1 ion-exchange resin (200–400 mesh, Cl^-), Bio-Rad
HCl, 50 mM
HCl, 150 mM
KOH, 2 M in Tris, 1 M
Sephadex G-10-120, swollen in water
$AgNO_3$, 0.5 N
All glassware used in this procedure for making up solutions and running the reaction should be acid-washed or disposable plastic should be used to minimize P_i contamination. The HCl solutions

[9] L. C. Mokrasch, J. Carvaca, and S. Grisolia, *Biochim. Biophys. Acta* **37,** 442 (1960).
[10] H. S. Penefsky, M. E. Pullman, A. Datta, and R. Racker, *J. Biol. Chem.* **235,** 3330 (1960).

should be prepared from redistilled, phosphate-free starting material.

Procedure. Two milliliters of 2 *M* KCNO are placed in a convenient container with a small stirring bar and a pH electrode. About 800 μmol of [^{18}O]P_i is added and the pH adjusted to between 5.5 and 6.0 by the addition of 50-μl aliquots of HCl. The vigorous evolution of CO_2 indicates that the reaction is taking place. The pH is maintained at <6 by addition of HCl for about 10 min. A further 1 mmol of solid KCNO is added at this point, and the further evolution of bubbles indicates more reaction. The pH is maintained at <6 for a total of 20 min and the reaction is then stopped by adjusting the pH to 8.3 with 1 *M* Tris. Then 1.4 ml of 120 m*M* magnesium acetate is added followed by 3 ml of 120 m*M* ADP. The reaction tube is placed in a 37° water bath and about 50 U of carbamate kinase is added. The reaction mixture is incubated at 37° for 30 min, and the reaction is stopped by vortexing with 1 ml of chloroform and chilling on ice.

Purification of [γ-^{18}O]ATP. Purification of the [γ-^{18}O]ATP is based on the method described by Axelson *et al.*[11] for high-performance liquid chromatography of nucleotides in a volatile solvent. Their method, which uses trifluoracetic acid for the elution, is based on an earlier description by Hsu and Chen[12,13] of the use of AG MP-1 resin to separate nucleotides. Because the scale of this preparation is much larger than that described for either of the other systems, we have used a batch elution rather than a gradient technique.

The cold reaction mix is centrifuged to break the emulsion and pellet the precipitated magnesium phosphate. The top layer is removed and the bottom layer is washed with 3 ml of water; the wash is then combined with the original supernatant. This mixture is applied, in the cold, to a 2 × 20 cm column of AG MP-1 ion-exchange resin, a macroporous Dowex-1 resin. This resin is much more effective than Dowex 1, AG1-X4, for the separation of ADP and ATP. The phosphate and AMP do not stick to the resin at pH 8 and the ionic strength of the reaction mix as it is applied to the column, and they are eluted in the first fractions of the water wash following sample application. The ADP and ATP elute at a much lower acidity than on AG1-X4, a resin previously used. The column is washed with water until the eluent is around pH 5 and then 50 m*M* HCl is applied. The ADP begins to elute as soon as the eluent becomes significantly acidic (<pH 3), and most of it is eluted by the time 100 ml of 50 m*M* HCl have

[11] J. T. Axelson, J. W. Bodley, and T. F. Walseth, *Anal. Biochem.* **116,** 357 (1981).

[12] D. Hsu and S. S. Chen, *J. Chromatogr.* **192,** 192 (1980).

[13] S. S. Chen and D. Hsu, *J. Chromatogr.* **198,** 500 (1980).

passed through the column. About 50–100 ml of 150 mM HCl must now be passed through the column before the ATP begins to elute. The ATP comes off in the next about 150 ml. Hydrolysis can be prevented by neutralizing each fraction with a solution of 2 M KOH in 1 M Tris as it comes off the column. The ATP-containing fractions are pooled, lyophilized to concentrate the ATP, and then desalted by passage over a 1 × 100 cm column of Sephadex G-10-120 which has been equilibrated with water. About 80% of the ATP elutes from the column before the salt begins to elute with the ATP, as determined by testing for Cl^- with $AgNO_3$. (A 10-μl sample of each fraction is added to a 25-μl drop of 0.5 N $AgNO_3$ on a glass slide.) The remaining portion of the ATP fraction can be pooled, lyophilized, and reapplied to the column to improve the recovery of labeled ATP. The separation of this smaller amount of nucleotide from the salt on the Sephadex G-10 column is almost complete.

Comments. The ^{18}O content of the final ATP is reduced by any nonisotopic P_i that becomes incorporated into the γ-phosphoryl group. One source of such P_i is the phosphate that is commonly present in commercial preparations of ADP. If the ADP is added to the nonenzymatic reaction before the pH is adjusted, this unenriched P_i may be incorporated into the carbamoyl phosphate and thus into the ATP. Adjusting the pH to 8 first, adding the magnesium acetate, which precipitates the P_i as magnesium phosphate, thus taking it out of the reaction, and then adding the ADP minimizes the chance of this incorporation taking place.

The precipitated magnesium phosphate, still highly enriched in ^{18}O, can be recovered by acidification to dissolve it and reprecipitation as $MgNH_4PO_4$, as described for the final purification of the [^{18}O]P_i. If such recovery is planned, it is beneficial to purify the commercial ADP used in the preparation by passage over an AG MP-1 column first to remove any contaminating phosphate which will otherwise contaminate the recovered [^{18}O]P_i even if it does not find its way into the ATP.

It is preferable to use HCl for the column elution and neutralize the ATP fractions as they elute to prevent hydrolysis rather than eluting with trifluoroacetic acid and lyophilizing it away before neutralization, as described by Axelson *et al.*[11] Significant hydrolysis of the ATP occurs under these conditions, leading to an ADP content which is unacceptable for certain ^{18}O exchange experiments. Also, Sephadex G-10 does not seem to separate trifluoracetate from ATP completely, and it is difficult to tell whether separation is complete, so we reverted to the HCl elution described in the contributions by Hsu and Chen.[12,13]

Assay of [γ-^{18}O]ATP. Before the [γ-^{18}O]ATP can be used for exchange experiments, the following parameters need to be determined: the final concentration of ATP, the ^{18}O content of the γ-phosphoryl group, the amount of hydrolysis that occurred during isolation as determined by contaminating levels of ADP and enriched phosphate, and the amount of unenriched phosphate contamination.

The absorbance at A_{259} gives the total nucleotide present and the ADP can be determined by a coupling assay with the following final concentrations in a final volume of 1 ml: Tricine/NaOH buffer, 40 m*M*, pH 8; Mg^{2+}, 5 m*M*; K^+-phosphoenolpyruvate, 3 m*M*; NADH, 200 μM; lactate dehydrogenase, 100 μg/ml; and pyruvate kinase, 300 μg/ml. Aliquots of the ATP solution are added and the decrease in A_{340} gives a measure of the contaminating ADP.

The ^{18}O analysis is done by cleaving the ATP with glycerol kinase and D-glyceraldehyde in the presence of Mg^{2+}, a reaction that occurs without exchange.[14] The cleaved P_i is isolated as the molybdate complex and purified on AG1-X4 columns, as described earlier[2] and in this chapter. The levels of enriched and unenriched contaminant P_i can be determined at the same time by adding known amounts of highly enriched and unenriched P_i to cleaved and uncleaved samples of the ATP. The dilution of the $P^{18}O_4$ peak of the added enriched P_i gives a measure of the amount of total or contaminating phosphate present, since there should be no $P^{18}O_4$ in the phosphoryl group of ATP. Strictly speaking, it should not be necessary to add unenriched P_i to uncleaved ATP, but if the amount of spontaneous hydrolysis and therefore the amount of P_i is small, it is easier to detect the presence of enriched masses of P_i on a background of natural abundance than it is to isolate and accurately determine the distribution of a small amount of hydrolyzed P_i by itself.

The following final concentrations in a 1-ml volume are suitable for the cleavage of ATP by glycerokinase: Tris/HCl buffer, 50 m*M*, pH 8.1; Mg^{2+}, 2.4 m*M*; D-glyceraldehyde, 1 m*M*; [^{18}O]ATP, 0.1 m*M*; and glycerokinase, 20 μg/ml (prepared at a concentration of 1 mg/ml in a buffer containing 100 m*M* Tes, 10 m*M* EDTA, and 1 m*M* 2-mercaptoethanol, pH 7.0 and passed over a centrifuge/Sephadex G-50 column[15] equilibrated with the same buffer to remove P_i contamination. Incubation at room temperature for 30 min gives about a 50% yield. The reaction is quenched by chilling on ice and adding 1 ml of 2 *N* HCl. To minimize acid hydrolysis, the reaction mix is kept cold until all the P_i has been extracted into isobutanol/benzene (1 : 1). The controls are acid quenched and chilled before addition of enzyme. P_i introduced in the reaction components is determined by controls that contain everything except ATP. After

[14] M. R. Webb, *Biochemistry* **19,** 4744 (1980).

[15] H. S. Penefsky, *Adv. Enzymol.* **49,** 233 (1979).

quenching, highly enriched P_i standard, 50–100 nmol, is added to 2 of 4 hydrolyzed ATP samples, 2 of 4 unhydrolyzed ATP samples, and all controls without ATP. Unenriched P_i standard is added to the other two unhydrolyzed ATP samples.

A Method for the Rapid Approximation of the ^{18}O Content of Small Amounts of Water

Principle. The same principle that is used for the synthesis of $[^{18}O]P_i$ can be used to give a rapid analysis for the ^{18}O content of highly enriched water.

Procedure. In a dry box or anaerobic chamber, a couple of crystals of PCl_5 are placed in the bottom of a Reactivial. Then 20 μl of the water in question is pipetted onto the side of the vial which is capped and then tapped to bring the drop of water down onto the PCl_5 where it rapidly reacts to form H_3PO_4 and HCl. The vial is frozen on dry ice, lyophilized, and derivatized as described earlier in this chapter. This procedure can only give a minimum value because it is totally dependent on the dryness of the PCl_5. Therefore, it is necessary to analyze several replicates of each sample. This is feasible since the procedure requires only small amounts of water.

Preparation of $[^{18}O]$Phosphoenolpyruvate

For experiments at low levels of substrate ATP it is convenient to use a regenerating system of phosphoenolpyruvate and pyruvate kinase to maintain the level of substrate constant. If the ATP is enriched, the enriched γ-phosphoryl group is replaced with unenriched phosphate as the regeneration proceeds, greatly complicating interpretation of the results of exchange experiments. If, however, the label is in the phosphoenolpyruvate and the reaction is initiated by the addition of a small amount of ADP, then all the ATP generated will have the same enrichment as the phosphoenolpyruvate and will not change during the experiment except by enzymatic exchange.

Principle. It was found by O'Neal *et al.*[16] that when phophoenolpyruvate is heated in 1 *N* HCl, the carboxyl and phosphate oxygens exchange with water oxygens much more rapidly than the enolic oxygen and that this exchange rate is 16-fold greater than the hydrolysis rate at 98°. This provides a convenient method for the synthesis of $[^{18}O]$phosphoenolpyruvate.

[16] C. C. O'Neal, G. S. Bild, and L. T. Smith, *Biochemistry* **22**, 611 (1983).

Materials

Phosphoenolpyruvate, K-salt, dried over P_2O_5
Cylinder of HCl gas or a high-vacuum line, dry PCl_5, and unenriched water
$H^{18}OH$, 97–99% enriched
$^{32}P_i$
Tris, 1 *M*
AG1-X4 ion-exchange resin (Bio-Rad)
HCl, 30 m*M* and 100 m*M* prepared from glass-redistilled HCl
Na glycinate, dried over P_2O_5
KOH, 1 *N*, P_i free

Procedure. The principal technical difficulty with this procedure is making 1 *N* HCl that is highly enriched with ^{18}O. It can be done in two ways. A stream of HCl gas can be played over the surface of a small volume of highly enriched water in a fume hood. Because the gas is hygroscopic and the system is necessarily exposed to the atmosphere, the final HCl becomes diluted with atmospheric water. Because it is impossible to control the acidity of this solution, it is necessary to titrate and dilute the enriched acid. It is also very unpleasant to handle cylinders of HCl gas. If a vacuum line is available, there is an alternative method of generating a known amount of HCl gas. A small known volume of water is vacuum transferred to a large flask containing a large excess of PCl_5. The known amount of HCl gas thus generated can then be vacuum transferred to a known volume of a solution of phosphoenolpyruvate in highly enriched water to produce the appropriate final concentration of 1 *N* HCl without any dilution of the enrichment of the water. The following amounts of reagents are convenient for a typical preparation: 2.2 mmol phosphoenolpyruvate, 4.5 ml 98% $H^{18}OH$, 2 g PCl_5, and 96 μl water.

After transfer of the HCl generated from the PCl_5 to the solution of phosphoenolpyruvate in $H^{18}OH$, the flask, closed with a stopcock, can be removed from the vacuum line and, because it is under vacuum, can be heated in a boiling water bath for the required 10 min without making allowance for the escape of steam, which would risk exposure to unenriched water. The enriched water can be recovered from the reaction by vacuum transfer onto excess Na glycinate, as described in the section on the preparation of $[^{18}O]P_i$. Note that because PCl_5 sublimes, it is necessary to keep it frozen in dry ice while the HCl gas is transferred to the phosphoenolpyruvate with liquid N_2 or else the PCl_5 will transfer also.

After recovery of the enriched water, the phosphoenolpyruvate can be handled normally. It is neutralized with 1 *M* Tris to pH 8, a small spike of $^{32}P_i$ is added, and the solution is diluted to 130 ml and applied to a 1 × 14 cm AG1-X4 column at a flow rate of 2 ml/min. The phosphate is eluted with 30 m*M* HCl. Its removal can be monitored by the $^{32}P_i$ or by a spot test with Sumner reagent (25 μl of sample is mixed with 25 μl of a 2× concentrated solution of Sumner phosphate reagent on a white ceramic test plate. A blue color indicates the presence of phosphate). After removal of the P_i, the column is eluted with 400 ml of 100 m*M* HCl which is collected in a round-bottom flask, shell frozen, and lyophilized. Complcte lyophilization is difficult unless the solution is degassed by thawing under vacuum after the initial freeze. The lyophilized phosphoenolpyruvate is dissolved in cold water and neutralized to pH 7–8 with 1 *N* KOH. Since complete removal of the HCl is possible by lyophilization, the amount of KCl in the preparation should be minimal. Phosphoenolpyruvate is more resistant to hydrolysis during lyophilization than ATP, so it is not necessary to neutralize first. The typical yield from this procedure is about 50% or about 1 mmol. The solution is stored in small aliquots as a 10 m*M* solution at −80°.

Assay of [^{18}O]Phosphoenolpyruvate. There are four parameters to be determined about the enriched phosphoenolpyruvate before it can be used for exchange experiments: the actual concentration of phosphoenolpyruvate in the final preparation, the distribution of the ^{18}O in the phosphoryl group, the level of contamination with unenriched phosphate, and the level of contamination with enriched phosphate due to hydrolysis during the purification steps.

The concentration of phosphoenolpyruvate can most easily be determined by the same coupling assay used to determine the level of contaminating ADP in [^{18}O]ATP, described earlier in this chapter except that ADP at a final concentration of 1.3 m*M* is substituted for the phosphoenolpyruvate, aliquots of the enriched preparation are added in limiting amounts, and the decrease in A_{340} is recorded.

The ^{18}O content of the phosphoryl group is determined by cleavage with mercury. $HgCl_2$, 0.12 ml of 0.29 *M* brought to pH 1.9 with HCl, is added to 100 nmol of [^{18}O]phosphoenolpyruvate in a final volume of 0.5 ml. Appropriate controls with added enriched and unenriched standards as described for the analysis of [^{18}O]ATP are set up, and the phosphate is extracted and analyzed by gas chromatography/mass spectrometry, as described earlier.

[64] ATP Formation in Mitochondria, Submitochondrial Particles, and F_0F_1 Liposomes Driven by Electric Pulses

By YASUO KAGAWA and TOSHIRO HAMAMOTO

Assay Methods

Principle

Pioneering work by Witt established ATP formation in subchloroplast particles.[1,2] The advantage of using electric pulses over artificially imposed ion gradients is that time resolution is excellent and the energy component is simple; i.e., only the $\Delta\psi$ component of $\Delta\bar{\mu}_{H^+}$. In order to confirm that the observed ATP synthesis takes place on the proton-translocating ATPase (F_0F_1) without participation of other components, F_0F_1 liposomes were reconstituted and their electric pulse-driven ATP synthesis was measured.[3-5] The theoretical background of this method was described previously.[1,6,7] Although the pulse-driven ATP synthesis could be observed in F_0F_1 microliposomes by applying high-voltage pulses which decay exponentially,[4,5] rectangular pulses are preferable for quantitative estimation of the applied energy.[8] This method is also applicable to intact mitochondria, submitochondrial particles, and F_0F_1 liposomes which have large diameters.[8] The background on the F_0F_1 liposomes[9] and proton-motive ATP synthesis has been described elsewhere.[10]

Reagents

Volumes are per one assay, in the central compartment.

Vesicle suspension: 200 μl containing mitochondria (6–9 mg) or F_0F_1

[1] H. T. Witt, E. Schlödder, and P. Gräber, *FEBS Lett.* **69,** 272 (1976).
[2] H. T. Witt, *Biochim. Biophys. Acta* **505,** 355 (1979).
[3] M. Rögner, K. Ohno, T. Hamamoto, N. Sone, and Y. Kagawa, *Biochem. Biophys. Res. Commun.* **91,** 362 (1979).
[4] P. Gräber, M. Rögner, H.-E. Buchwald, D. Samoray, and G. Hauska, *FEBS Lett.* **145,** 35 (1982).
[5] B. E. Knox and T. Y. Tsong, *J. Biol. Chem.* **259,** 4757 (1984).
[6] P. Graber, *Curr. Top. Membr. Transp.* **16,** 215 (1982).
[7] Y. Kagawa, *Curr. Top. Membr. Transp.* **16,** 195 (1982).
[8] T. Hamamoto, K. Ohno, and Y. Kagawa, *J. Biochem.* **91,** 1759 (1982).
[9] Y. Kagawa, C. Ide, T. Hamamoto, M. Rögner, and N. Sone, *in* "Membrane Reconstitution" (G. Poste and G. L. Nicolson, eds.), p. 137. Elsevier, Amsterdam, 1982.
[10] Y. Kagawa, *in* "New Comprehensive Biochemistry, Bioenergetics" (L. Ernster, ed.), p. 149. Elsevier, Amsterdam, 1984.

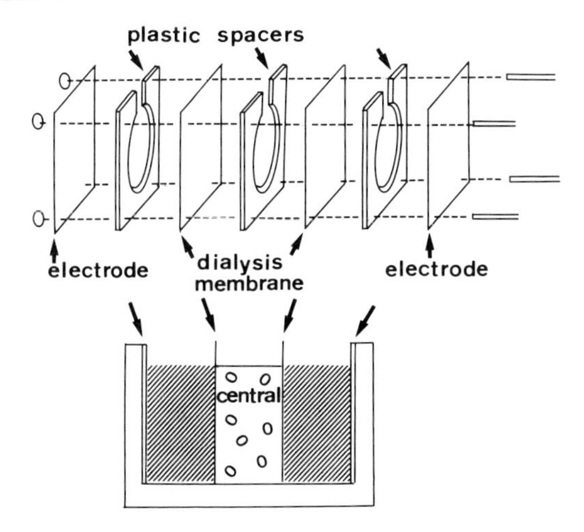

FIG. 1. Electrical apparatus used for F_0F_1 macroliposomes.

liposomes (1.5 mg F_0F_1),[3,9] 0.25 *M* sucrose, and 2 m*M* Tricine-NaOH, pH 7.5

Hexokinase solution: 500 μl containing 18 units hexokinase (Type F-300, baker's yeast, crystallized, Sigma) previously dialyzed against 5 m*M* Tricine-NaOH, pH 7.5, and 36 m*M* glucose

Rotenone: 5 μl of 0.4 mg/ml rotenone, in ethanol

Antimycin A: 5 μl of 0.5 mg/ml antimycin A, in ethanol

Aurovertin: 10 μl of 1 mg/ml aurovertin, in ethanol

ADP-P_i solution: 200 μl containing 4.5 m*M* K_2HPO_4 (about 1 μCi ^{32}P), 11.4 m*M* ADP, 2.8 m*M* $MgSO_4$, and 5 m*M* Tricine-NaOH, pH 7.5

Ammonium molybdate solution: 2 ml, 2% ammonium molybdate in 1.6 *N* $HClO_4$

Triethanolamine

Isobutanol–benzene mixture (1 : 1, v/v), saturated with H_2O

50% ethanol

Electrical Apparatus[11]

The electrical apparatus shown in Fig. 1[8] is used for F_0F_1 macroliposomes.[3] However, for F_0F_1 microliposomes,[12] the simple cell described by Kagawa[7] or Knox and Tsong[5] is used to apply high-voltage pulses. In

[11] Y. Kagawa and N. Sone, this series, Vol. 55(F), p. 364.

[12] Y. Kagawa and E. Racker, *J. Biol. Chem.* **246,** 5477 (1971).

Fig. 1, two stainless-steel plate electrodes (5 × 6 cm) are placed parallel and 4.4 mm apart in a polyacrylic cell.[8] Direct contact of the electrodes with the vesicles or mitochondria is avoided by dividing the space between the electrodes into three compartments with two sheets of dialysis membrane (seamless cellulose tubing 36/32, Visking Co.); the mitochondrial suspension is placed in the central compartment. An electric condenser (5 mF) charged by a high-voltage power supply (Mitsumi Scientific Industries Co., Tokyo) is connected to the electrodes *via* a switching circuit which is switched on and off by a low-voltage function generator (FG163S, NF Circuit Co., Yokohama). The pulse duration is calibrated with a crystal time-base generator (T 107-001, Thinky, Tokyo). The electric potential in the solution is measured with a copper single electrode connected to a transient recorder (DL905, Datalab, UK). Other components of the electrical apparatus are as described in the previous reports.[3-8]

Preparations

The preparations of F_0F_1 macroliposomes[9] and F_0F_1 microliposomes[12] have been described previously. In fact, for high-voltage experiments,[5] the classical F_0F_1 microliposomes prepared by cholate dialysis[13] are satisfactory.

For the preparation of the mitochondria, fresh rat liver is homogenized in 0.25 *M* sucrose, 2 m*M* Tricine-NaOH, pH 7.5, 0.2 m*M* EDTA, and the mitochondria are isolated by the method of Hogeboom.[13] The mitochondria are washed twice with the same medium, suspended in 0.25 *M* sucrose and 2 m*M* Tricine-NaOH, pH 7.5, and stored at 4° at a protein concentration of 40 mg/ml for 1–4 hr before use. Their phosphorylation activity and respiratory control ratio (4.5 with succinate) are well maintained during the procedures.

Submitochondrial particles called ETPH(Mg) are prepared from rat liver mitochondria, as described elsewhere,[14] and washed and suspended in the same medium as used for the mitochondria.

Analytical Procedures

F_0F_1 liposomes and mitochondria are preincubated with respiratory inhibitors and subjected to electric pulse irradiation. The central compartment of Fig. 1 is filled with the indicated volumes of the vesicle suspension, hexokinase solution, and if necessary, inhibitors, and incubated at 25° for 5 min. The outer compartments are filled with solutions without

[13] G. H. Hogeboom, this series, Vol. 1, p. 16.
[14] R. E. Beyer, this series, Vol. 10, p. 186.

vesicles. Then the cell is cooled to 4° for 10 min, and the ADP-P_i solution (prechilled to 4°) is added. External electric field pulses of 30 msec, 760 V/cm, double reciprocal rectangular shape and 30-sec intervals, are applied to the central compartment of the cell (Fig. 1), unless otherwise specified.

ATP synthesis is followed by measuring the esterification of P_i. The reaction mixture (0.5 ml) in the central compartment is mixed with 2 ml of ammonium molybdate solution. Addition of triethanolamine (0.05 ml) results in the precipitation of a complex of P_i-molybdate and vesicles, which are removed by centrifugation at 1000 *g* for 10 min. The supernatant is extracted twice with 2 ml of isobutanol–benzene mixture to remove the remaining $^{32}P_i$, and the radioactivity of glucose [^{32}P]phosphate in the water phase is measured with a Packard liquid scintillation counter (model 3330, the dial is set at ^{3}H channel for the Cerenkov rays emitted). The amount of P_i esterified per milligram protein is calculated as the difference between the value obtained with and without electric pulses. The electric field strength was found to be 680 V/cm throughout the cell, even near the dialysis membrane, and the pulse shape was very nearly rectangular. The electric charge passing through the reaction medium with a 760 V/cm, 20-msec rectangular phase was 5×10^{-2} coulomb. The temperature rise of the medium caused by this current (760 V/cm) is only about 2°, as reported previously, and the temperature returns to the original level within the 30-sec pulse interval.

[65] Synthesis of Enzyme-Bound Adenosine Triphosphate by Chloroplast Coupling Factor CF_1

By RICHARD I. FELDMAN

Isolated chloroplast CF_1-ATPase and the thylakoid ATP synthase complex, in the absence of protonmotive force, can synthesize enzyme-bound ATP from medium P_i.[1,2] The ATP product does not dissociate from the enzyme, demonstrating that the formation of the γ-phosphoryl bond of ATP during net photophosphorylation occurs spontaneously, while protonmotive force promotes the tight binding of substrates and the release of the bound product.[3] The synthesis of enzyme-bound ATP appears

[1] R. I. Feldman and D. S. Sigman, *J. Biol. Chem.* **257,** 1676 (1982).
[2] R. I. Feldman and D. S. Sigman, *J. Biol. Chem.* **257,** 12178 (1983).
[3] P. D. Boyer, *in* "Membrane Bioenergetics" (C. P. Lee, G. Schatz, and L. Ernster, eds.), p. 461. Addison-Wesley, Reading, Massachusetts, 1979.

to be a partial reaction of photophosphorylation and will be a valuable tool for dissecting the individual steps of this process. Since the characteristics of ATP formation by isolated CF_1 and the membrane-bound $F_0 \cdot CF_1$ ATP synthase are very similar,[2] information derived from more easily studied coupling factors can be applied to the mechanism of the intact enzyme complex.

ADP remaining tightly bound to the CF_1-ATPase during its isolation and storage, most likely at a catalytic site, is the substrate for the synthesis of enzyme-bound ATP. This point is discussed in depth elsewhere.[1,2,4]

The synthesis of enzyme-bound [^{32}P]ATP by CF_1 from $^{32}P_i$ is measured by extracting the bound [^{32}P]ATP with acid and isolating the product from $^{32}P_i$. The equilibrium constant of bound ATP formation from bound ADP and P_i is about 0.5.[1,2] Since relatively low levels of [^{32}P]ATP are produced, contaminating activities that can form medium [^{32}P]ATP from $^{32}P_i$ may interfere with the assay. In particular, polynucleotide phosphorylase (PNPase) is present in significant quantities in CF_1 preparations and catalyzes a $P_i \rightleftharpoons$ ATP exchange reaction and a more active $P_i \rightleftharpoons$ ADP exchange reaction.[5] Adenylate kinase activity was also detected in CF_1 preparations. The properties of these contaminants relevant to the measurement of ATP by CF_1 have been described in detail.[5]

The formation of medium [^{32}P]ATP from $^{32}P_i$ by activities contaminating CF_1 preparations is eliminated by freeing the reaction mixture of medium nucleotides that can undergo exchange reactions. Added ADP is not required for the synthesis of enzyme-bound ATP by CF_1. Alternatively, the buildup of medium [^{32}P]ATP in the $^{32}P_i$ reaction mixtures is prevented by including hexokinase and glucose (30 mm), which rapidly converts medium [γ-^{32}P]ATP to ADP and [^{32}P]Glu-6-P. Enzyme-bound [^{32}P]ATP synthesized by CF_1 is not accessible to the action of hexokinase.

Procedure

Although CF_1 can be isolated by a variety of methods, I prefer the EDTA extraction procedure of Binder *et al.*[6] The enzyme was stored as a precipitate in 50% saturated $(NH_4)_2SO_4$, 2.0 m*M* EDTA, 1 m*M* ATP, 50 m*M* Tricine-NaOH, pH 7.6, 1.0 m*M* dithiothreitol at 4°.

To measure ATP synthesis by CF_1, the enzyme is dissolved at about 10 mg/ml in 25 m*M* Tricine, 30 m*M* K^+-acetate, 10 m*M* $MgCl_2$, pH 6.0, at room temperature, and is passed through two consecutive Sephadex G-50

[4] R. I. Feldman and P. D. Boyer, *J. Biol. Chem.* **260,** 13088 (1985).
[5] R. I. Feldman and D. S. Sigman, *Eur. J. Biochem.* **143,** 583 (1984).
[6] A. Binder, A. Jagendorf, and E. Ngo, *J. Biol. Chem.* **253,** 3094 (1978).

centrifuge columns,[7] equilibrated in the same buffer. This is done to remove residual $(NH_4)_2SO_4$ as well as medium and loosely bound nucleotides that can undergo exchange reactions with medium $^{32}P_i$. About 40% of the protein is usually lost in this procedure.

The ATP synthesis reaction is carried out by mixing 200 μl of CF_1 (about 100–200 μg) with 200 μl reaction mixture containing 25 m*M* Tricine, 30 m*M* KAc, 10 m*M* $MgAc_2$, 20 m*M* $^{32}P_i$ (0.5–1.0×10^7 cpm/μmol P_i) at pH 6.0 and 22°. After 30 min, the reaction is stopped by adding 400 μl 1 *M* perchloric acid. In addition, 1.0 μmol ATP and ADP are added as nucleotide carriers and an internal standard for losses of ATP during its isolation. Several blanks, with buffer substituted for CF_1, are included to determine the ^{32}P background. The formation of bound [^{32}P]ATP reaches equilibrium within about 5–10 min, although some CF_1 preparations with less heat-activated Ca^{2+}-ATPase activity synthesized ATP more slowly. The ATP synthesis reaction can be prolonged for at least 1 hr with no decrease in the yield of [^{32}P]ATP. After acid quenching the reaction, the extracted ATP is relatively stable for several hours at 25°. However, for storage up to several days, the samples can be neutralized with 2 *M* Tris base and kept at 4°, unless hexokinase was included in the reaction mixture. Hexokinase can regain activity when the perchloric acid is neutralized, resulting in the loss of the extracted [^{32}P]ATP.

To quantitate the formation of [^{32}P]ATP by CF_1, the extracted product is isolated from $^{32}P_i$ (and [^{32}P]Glu-6P_i) using a combination of charcoal and Dowex chromatography, as described by Vinkler *et al.*[8] Charcoal columns elute very slowly unless suction is applied. This can be accomplished by placing a small column (1.5 × 10 cm) through a rubber stopper which is fit into an opening of chamber that can be evacuated. Some vacuum manifolds designed for collecting material onto Millipore-type filter disks are easily modified to hold a large number of charcoal columns by removing the filter disk support and adding a rack to hold collection vials under each column. Scintillation vials (20 ml) are convenient disposable receptacles. Celite 503 (2 ml 100 mg/ml) (from J. T. Baker Chemical Co.) is added to each column and allowed to settle, followed by 0.7 ml 100 mg/ml washed activated charcoal. In all steps, the columns are allowed to run dry under continuous suction. The columns are washed with 2 ml of solution A (0.3 *M* perchloric acid, 0.1 *M* H_3PO_4, and 0.025 *M* PP_i). Before adding the acid-quenched ^{32}P reaction mixture, it is diluted with 2–5 ml solution A and is spun in a clinical centrifuge to remove denatured protein. After [^{32}P]ATP has been absorbed to the column, it is washed with 20

[7] H. S. Penefsky, this series, Vol. 56, p. 527.

[8] C. Vinkler, G. Rosen, and P. D. Boyer, *J. Biol. Chem.* **253,** 2507 (1978).

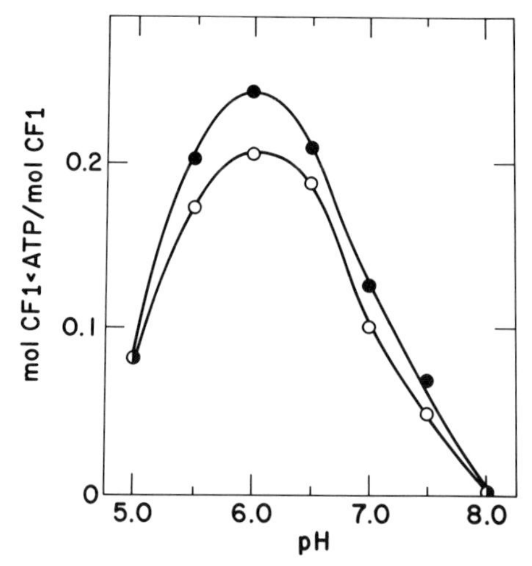

FIG. 1. The pH dependence of CF_1 < ATP formation. The final 100-μl reaction mixtures contained 1.0 mg/ml latent CF_1, 25 m*M* Tricine-KOH, 30 m*M* KAc, 10 m*M* $MgAc_2$, 100 μM ADP, 30 m*M* glucose, 3.5 units of hexokinase, 80 m*M* [^{32}P]P_i (1.0×10^{10} cpm/mmol) (○) or 160 m*M* [^{32}P]P_i (5.0×10^9 cpm/mmol) (●), at the pH indicated. After 40 min of incubation, the reaction was quenched with 1.0 ml of 1.0 *M* PCA, and the [^{32}P]ATP produced was quantitated as described under "Procedure." Temperature, 25°.

ml of solution A, and the [^{32}P]ATP is eluted into clean receptacles with 5 ml 1 *M* NH_4OH/ethanol (60 : 40 v/v).

The charcoal column eluate is applied to a Dowex AG-1-X4 anion exchange column (0.7×2 cm). The column is washed successively with 10 ml H_2O and 20 ml 60 m*M* HCl, and the ATP is eluted into a clean vial with 10 ml of 1 *M* HCl. The entire ATP fraction is counted in the tritium channel of a liquid scintillation counter. The recovery of ATP in the isolation is calculated from the A_{257} of the ATP fraction. The recovery averages about 60%. Background of about 40 cpm can be obtained from reaction mixtures containing about 5×10^7 cpm $^{32}P_i$. Some batches of $^{32}P_i$ gave high background and could be cleaned up on a small Dowex AG1-X4 column.[1]

It is a good idea to confirm that the counts per minute measured in the ATP fraction are from [γ-^{32}P]ATP. One criterion of [^{32}P]ATP formation is the ability of the extracted product to serve as a substrate for hexokinase, which is very specific for ATP. The kinetics of hexokinase action on the [^{32}P]ATP product should be identical to that of an [^{3}H]ATP internal standard.[1] The extracted [^{32}P]ATP product can be most easily subjected to hexokinase treatment before its isolation from $^{32}P_i$. To accomplish this, the perchloric acid-quenched reaction mixture is neutralized with KOH

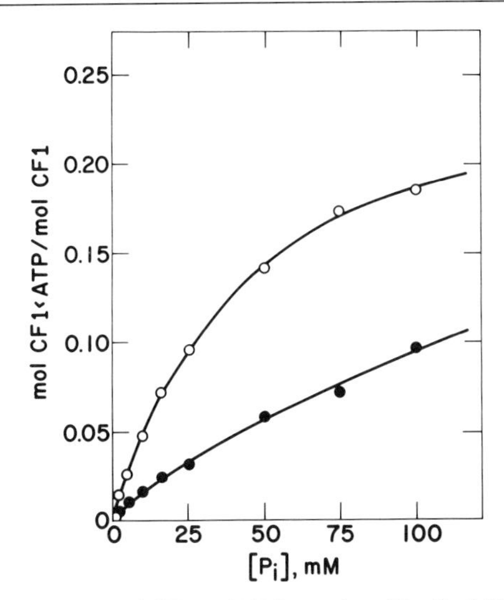

FIG. 2. The $[P_i]$ dependence of CF_1 < ATP formation. The final 100-μl reaction mixture contained 2.8 mg/ml latent CF_1, 25 m*M* Tricine-KOH, pH 6.0 (○) or 7.0 (●), $[^{32}P]P_i$ (9.4×10^7 cpm) as indicated, 10 m*M* $MgAc_2$, 30 m*M* KAc. The reaction was quenched by the addition of 100 μl of 1.0 *M* PCA after 24 min of incubation. The $[^{32}P]$ATP produced was quantitated as described under "Procedure."

and the solid K^+-perchlorate formed is removed by centrifugation. The solution is then diluted with 5 volumes of 25 m*M* Tricine-KOH, pH 7.0, 30 m*M* glucose, 10 m*M* $MgAc_2$, 200 μM $[^3H]$ATP, and divided into equal portions: One sample serves as a control, while the other samples are incubated with about 0.5 units hexokinase for various times at 22°. The hexokinase treatment is stopped by adding concentrated perchloric acid to give 0.5 *M;* then 1 μmol carrier ATP and ADP is also added and the ATP fraction is isolated as described above. The loss of ^{32}P counts recovered in the ATP fraction should parallel the loss of ^3H counts from the $[^3H]$ATP standard.

Characteristics of ATP Synthesis by CF_1

ATP synthesis is maximal at pH 6 (Fig. 1). Raising the pH increases the dissociation constant of P_i, but does not alter the equilibrium between bound ADP and P_i and bound ATP.[1,2] The P_i concentration dependence is shown in Fig. 2. ATP synthesis was routinely measured at 10 m*M* $^{32}P_i$ where the lower amount of product formed at subsaturating P_i is offset by the higher specific activity of $^{32}P_i$ achievable. The kinetics of $[^{32}P]$ATP formation are shown in Fig. 3. The binding of P_i appears to be the rate-

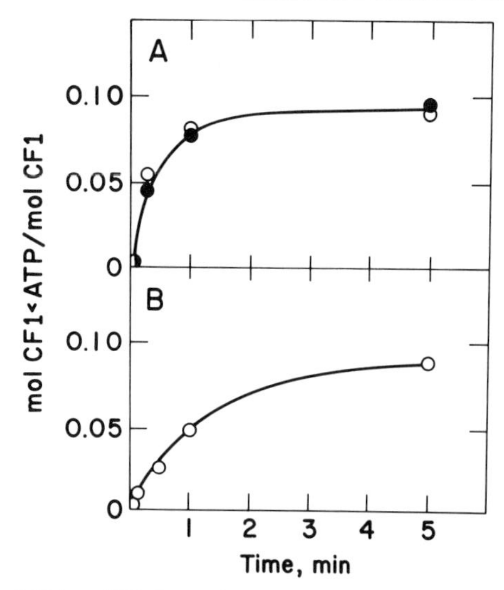

FIG. 3. Kinetics of $CF_1 < ATP$ formation at pH 6.0. (A) The reaction mixture contained 2.4 mg/ml latent CF_1, 2.0 m*M* ADP, 20 m*M* [^{32}P]P_i (6.5 × 10^9 cpm/mmol), 30 m*M* KAc, 10 m*M* $MgAc_2$, 25 m*M* Tricine-KOH, pH 6.0, 25 m*M* glucose at 25° (○), 125 units of hexokinase (●). At the indicated times, 300-μl aliquots were quenched with 300 μl of 1.0 *M* PCA and the [^{32}P]ATP formed was quantitated as described under "Procedure." (B) CF_1 was preincubated in a reaction mixture containing 3.6 mg/ml latent CF_1, 2.0 m*M* ADP, 20 m*M* P_i, 30 m*M* KAc, 25 m*M* Tricine-KOH, pH 6.0, 25 m*M* glucose, 14 units of hexokinase at 25°. After 10 min, 1.2 μl of carrier-free [^{32}P]P_i was added to yield a [^{32}P]P_i specific activity of 6.4 × 10^9 cpm/mmol. At the indicated times after the addition of [^{32}P]P_i, 110-μl aliquots were quenched with 110 μl of 1.0 *M* PCA and the [^{32}P]ATP formed was quantitated as described under "Procedure."

limiting step.[4] The latent Ca-ATPase activity of CF_1 can be activated by heat treatment in the presence of dithiothreitol,[9] without appreciable effect on the synthesis of enzyme-bound ATP. Enzyme-bound ATP, in equilibrium with medium P_i, is rapidly chased if enzyme turnover is induced by addition of medium ATP. Several inhibitors of Ca-ATPase activity of heat-activated CF_1 such as 7-chloro-4-nitro-Z,1,3-benzoxydiazole (NBD-Cl), efrapeptin, or NH_4Cl, do not inhibit the synthesis of enzyme-bound ATP, and thus probably inhibit a different step that is only needed for complete enzyme turnover.

The synthesis of enzyme-bound ATP by beef heart mitochondrial F_1 (MF_1) in the presence of DMSO has also been demonstrated.[10] Including

[9] S. Lien and E. Racker, this series, Vol. 23, p. 547.

[10] J. Sakamoto and Y. Tonomura, *J. Biochem.* **93,** 1601 (1983).

40% (w/v) DMSO reduces the P_i concentration giving half-maximal ATP formation from greater than 400 m*M*, to 0.7 m*M*, and appears to increase the amount of enzyme-bound ATP formed from bound ADP and P_i about 3-fold.[11] Added ADP was required (K_d = 2 μM) for the synthesis of bound ATP if the enzyme used was stored in 50% glycerol, which can induce the release of tightly bound nucleotide. However, MF_1, isolated by the method of Penefsky[12] and freed of medium and loosely bound nucleotide with Sephadex G-50 centrifuge columns, retained the bound ATP substrate (R. Kandpol, personal communication). The conditions used for the synthesis of enzyme-bound ATP by MF_1 were 10% (w/v) glycerol, 35% (w/v) DMSO, 5 m*M* $MgCl_2$, 0.8 m*M* EDTA, 0.1 *M* 2-(*N*-morpholino)ethanesulfonic acid (MES)-Tris, 0.2 m*M* $^{32}P_i$, 2 m*M* ADP, 0.8 mg/ml MF_1 at pH 6.7 and 30° in a volume of 100 μl. The reaction was stopped after 60 min.[10]

Acknowledgment

This work was supported by United States Public Health Service Grant 21199.

[11] J. Sakamoto, *J. Biochem.* **96,** 483 (1984).
[12] H. S. Penefsky, this series, Vol. 55, p. 304.

[66] Divalent Azido-ATP Analog for Photoaffinity Cross-Linking of F_1 Subunits

By HANS-JOCHEN SCHÄFER

Photoaffinity labeling has been applied successfully to study the interactions of specific receptor sites of proteins with their biological ligands such as substrates, effectors, or coenzymes.[1-4] Photoaffinity labels possess some advantages over affinity labels with conventional groups. They do not react in the dark. The reactivity of a photoaffinity label remains masked until irradiation. This fact enables one to control easily the time

[1] Abbreviations most commonly used: F_1 or F_1-ATPase, coupling factor 1, adenosinetriphosphatase; 3'-arylazidoadenosine nucleotides, 3'-*O*-{3-[*N*-(4-azido-2-nitrophenyl)-amino]propionyl}adenosine nucleotides; DiN_3ATP or 3'-arylazido-β-alanine-8-azido-ATP, 3'-*O*-{3-[*N*-(4-azido-2-nitrophenyl)amino]propionyl}-8-azido-ATP.
[2] J. R. Knowles, *Acc. Chem. Res.* **5,** 155 (1972).
[3] H. Bayley and J. R. Knowles, this series, Vol. 46, p. 69.
[4] H. Bayley, "Photogenerated Reagents in Biochemistry and Molecular Biology." Elsevier, Amsterdam, 1983.

and the rate of labeling by the illumination procedure. Furthermore, kinetic investigations analogous to studies with natural ligand can be performed in the dark. Besides carbenes, free radicals, or triplet states, the nitrenes are highly reactive photogenerated reagents suitable for photoaffinity labeling. Nitrenes are formed on irradiation of azido derivatives. Aromatic azido compounds are used because of their relative stability in the dark. Upon activation the photogenerated highly reactive nitrenes react with amino acid residues of the specific binding site or of its vicinity. This results in an easy identification of the binding site.

Bifunctional reagents are useful to cross-link proteins.[5] The cross-linking may occur inter- or intramolecularly. It can be applied to stabilize the tertiary structure, to determine the distances between reactive groups in proteins, or to study protein–protein interactions. A very important application of bifunctional reagents, however, is to investigate the spatial arrangement of compounds in biological systems such as multisubunit enzymes,[6,7] ribosomes,[8–10] or membranes.[11–13] Cross-linking experiments on F_1-ATPases [EC 3.6.1.3] of different origins have resulted in spatial models for this enzyme complex formed by five different subunits, α–ε.[14,15]

Cross-linking can also be performed by reagents with one photosensitive and one conventional functional group, or with two photosensitive groups (photo-cross-linking).[4] The introduction of two highly reactive functional groups into a biological ligand creates a tool to study the vicinity of the corresponding specific binding site by cross-linking (affinity cross-linking). A biological ligand bearing two photosensitive groups represents a photoaffinity label capable of specifically cross-linking proteins upon irradiation (photoaffinity cross-linking).

Photoaffinity cross-linking should occur when the ligand's binding site at one enzyme subunit is located in a proper distance to a second subunit, especially upon its localization at the interface of two subunits. Yet,

[5] G. E. Means and R. E. Feeney, "Chemical Modification of Proteins." Holden-Day, San Francisco, 1971.
[6] J. R. Coggins, E. A. Hopper, and R. N. Perham, *Biochemistry* **15,** 2527 (1976).
[7] J. R. Coggins, J. Lumsden, and A. D. B. Malcolm, *Biochemistry* **16,** 1111 (1977).
[8] A. J. Peretz, H. Towbin, and D. Elson, *Eur. J. Biochem.* **63,** 83 (1976).
[9] T. A. Bickle, J. W. B. Hershey, and R. R. Traut, *Proc. Natl. Acad. Sci. U.S.A.* **69,** 1327 (1972).
[10] H.-G. Wittmann, *Eur. J. Biochem.* **61,** 1 (1976).
[11] K. Wang and F. M. Richards, *J. Biol. Chem.* **249,** 8005 (1974).
[12] T. J. Steck, *J. Mol. Biol.* **66,** 295 (1972).
[13] H. M. Tinberg and L. Packer, this series, Vol. 56, p. 622.
[14] P. D. Bragg and C. Hou, *Eur. J. Biochem.* **106,** 495 (1980).
[15] R. D. Todd and M. G. Douglas, *J. Biol. Chem.* **256,** 6984 (1981).

photoaffinity cross-linking experiments often show a very low yield of cross-link formation compared with a high inactivation of the enzyme. This seems to be contradictory at first. However, various explanations are possible:

1. The major part of the cross-linking analog is bound covalently only once to the catalytic site of the enzyme, and merely a minor part of the label is located in a position which allows cross-linking of two subunits upon irradiation. The bifunctional label reacts mainly (with one photoactivated azido group) at the catalytic site inactivating the enzyme, whereas the other functional azido group reacts with water, forming a hydroxylamino derivative. This is due to the high reactivity and the short lifetime of the formed nitrene. It is likely that the nonoptimal size of the label is responsible for the mainly one-sided (single) labeling. However, it is also conceivable that the distance between the catalytic site on one subunit and the second subunit changes during the enzymatic process, so that the distance is proper for cross-linking only at one moment. It might be possible to increase the cross-link formation by changing the distance between the two azido groups.
2. Both azido groups react with the very same subunit. Such a double-labeled subunit cannot be discriminated from unlabeled or single-labeled subunits on SDS electrophoresis gels.
3. A great part of the formed cross-link is split off again if easily cleavable bifunctional reagents are used.

The ATP molecule contains two distinct structural groups where photosensitive azido groups have been introduced successfully. Modifications at the adenine base resulted in 8-azido-ATP[16] and 2-azido-ATP.[17] Esterification of the 3′-hydroxyl group of the ribose sugar with different arylazido acids yielded 3′-arylazidoadenosine nucleotides.[18,19] All three photolabile ATP analogs have been successfully applied in the investigation of F_1-ATPases from mitochondria,[20-22] bacteria,[23,24] or chloroplasts.[25]

[16] B. E. Haley and J. F. Hoffman, *Proc. Natl. Acad. Sci. U.S.A.* **71**, 3367 (1974).
[17] N. J. Cusack and M. Planker, *Br. J. Pharmacol.* **67**, 153 (1979).
[18] S. J. Jeng and R. J. Guillory, *J. Supramol. Struct.* **3**, 448 (1975).
[19] R. J. Guillory and S. J. Jeng, this series, Vol. 46, p. 259.
[20] R. J. Guillory, *Curr. Top. Bioenerg.* **9**, 267 (1979).
[21] R. J. Wagenvoord, I. van der Kraan, and A. Kemp, *Biochim. Biophys. Acta* **460**, 17 (1977).
[22] J. Lunardi, M. Satre, and P. V. Vignais, *Biochemistry* **20**, 473 (1981).
[23] P. Scheurich, H.-J. Schäfer, and K. Dose, *Eur. J. Biochem.* **88**, 253 (1978).
[24] J. H. Verheijen, P. W. Postma, and K. van Dam, *Biochim. Biophys. Acta* **502**, 345 (1978).
[25] J. J. Czarnecki, M. S. Abbott, and B. R. Selman, *Proc. Natl. Acad. Sci. U.S.A.* **79**, 7744 (1982).

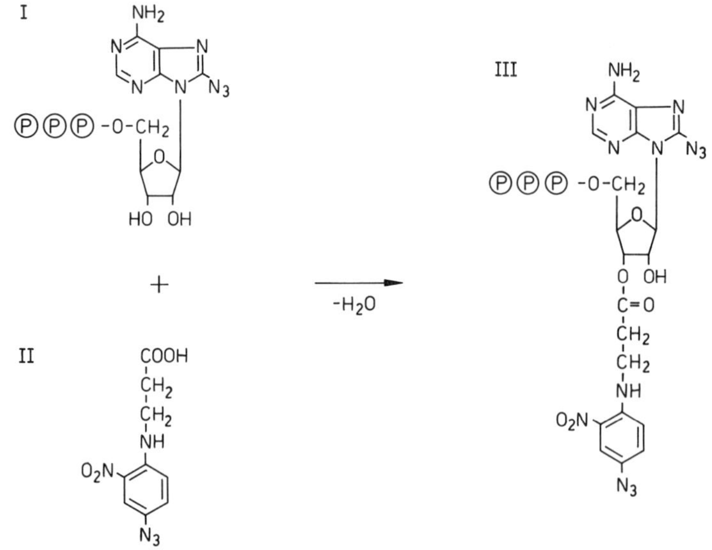

FIG. 1. Synthesis of DiN_3ATP.

These photoaffinity labeling experiments resulted in the specific labeling of α and/or β subunits, indicating that the catalytic nucleotide binding site could be situated at the interface between α and β subunits. In this case, a bifunctional ATP analog should cross-link α and β. We have synthesized 3′-arylazido-β-alanine-8-azido-ATP (DiN_3ATP) (**III**)[26] for photoaffinity cross-linking by esterification of 8-azido-ATP (**I**) with *N*-4-azido-2-nitrophenyl-β-alanine (**II**) (Fig. 1). The synthesis was performed in analogy to the preparation of other 3′-arylazidoadenosine nucleotides as described by Guillory and Jeng.[18,19] The use of DiN_3ATP for photoaffinity cross-linking is illustrated with F_1-ATPase from *Micrococcus luteus* prepared according to Risi *et al.*[27]

Methods

Synthesis of 3′-O-{3-[N-(4-azido-2-nitrophenyl)amino]propionyl}-8-azidoadenosine 5′-triphosphate (3′-arylazido-β-alanine-8-azido-ATP, DiN_3ATP)

8-Azido-ATP (**I**) (triethylammonium salt) is synthesized directly from ATP via 8-bromo-ATP, as described earlier,[28] according to the prepara-

[26] H.-J. Schäfer, P. Scheurich, G. Rathgeber, K. Dose, A. Mayer, and M. Klingenberg, *Biochem. Biophys. Res. Commun.* **95,** 562 (1980).

[27] S. Risi, M. Höckel, F. W. Hulla, and K. Dose, *Eur. J. Biochem.* **81,** 103 (1977).

[28] H.-J. Schäfer, P. Scheurich, and K. Dose, *Liebigs Ann. Chem.* 1749 (1978).

tion of 8-azido-AMP described by Haley and Hoffman.[16] The synthesis of *N*-4-azido-2-nitrophenyl-β-alanine (**II**) is performed according to Guillory and Jeng.[18,19] 4-Fluoro-3-nitroaniline is first diazotized and subsequently treated with sodium azide to yield 4-fluoro-3-nitrophenyl azide. Then the azide is condensed with β-alanine, resulting in the formation of **II**.

DiN_3ATP (**III**) is prepared by esterification of 8-azido-ATP (**I**) with *N*-4-azido-2-nitrophenyl-β-alanine (**II**). For this purpose, 63 mg of **II** (0.25 mmol) and 48.5 mg carbodiimidazole (0.3 mmol) are dissolved in 0.2 ml of dried dimethylformamide. The solution is stirred for 15 min at room temperature in the dark. Then a solution of 42.5 mg 8-azido-ATP triethylammonium salt (**I**) (0.05 mmol) in 1.0 ml of water is added. The reaction mixture is stirred for a further 7 hr. The solvent is evaporated in the vacuum. The residue is repeatedly washed with dry acetone to remove the major part of unreacted **II** and the excess carbodiimidazole, centrifuged, and vacuum dried. The solid residue is redissolved in 0.2 ml of water. Undissolved material is removed by centrifugation. The clear solution is subjected to descending paper chromatography on a sheet of paper (18 × 40 cm, type 2040 B, Schleicher and Schüll) with *n*-butanol : water : acetic acid (5 : 3 : 2 v/v). The chromatography reveals two major orange bands ($R_f = 0.4$ and $R_f = 0.9$) and a colorless UV absorbing band ($R_f = 0.1$). The orange band ($R_f = 0.4$) contains DiN_3ATP (**III**). It is eluted from the paper with water and lyophilized. The residue is stored at $-20°$. The yield is 3.5–5.0% (spectroscopically). The colorless band ($R_f = 0.1$) and the orange band ($R_f = 0.9$) are identified as the starting materials **I** and **II**, respectively.

Stability of DiN_3ATP

DiN_3ATP is relatively unstable, especially in alkaline solution. Incubation at alkaline pH resulted in the cleavage of the ester linkage to form the starting materials. In addition to the hydrolysis of **III**, the formation of an insoluble (in water or acetone) red material was sometimes observed. This occurs upon longer storage of DiN_3ATP as solid or in aqueous solution possibly due to the polymerization of the bifunctional compound. For these reasons, it is recommended to use freshly synthesized DiN_3ATP for all experiments.

Spectral Characteristics and Photochemical Reactivity of DiN_3ATP

The infrared spectrum of DiN_3ATP shows two characteristic bands for N_3 stretching vibrations at 2130 cm^{-1} and 2170 cm^{-1}, indicating the presence of two azido groups. The position of the 2170 cm^{-1} band is identical to the position of the N_3 stretching vibration band of 8-azido-ATP.[28] The absorption spectrum shows two maxima at 263 nm and 475 nm (Fig. 2,

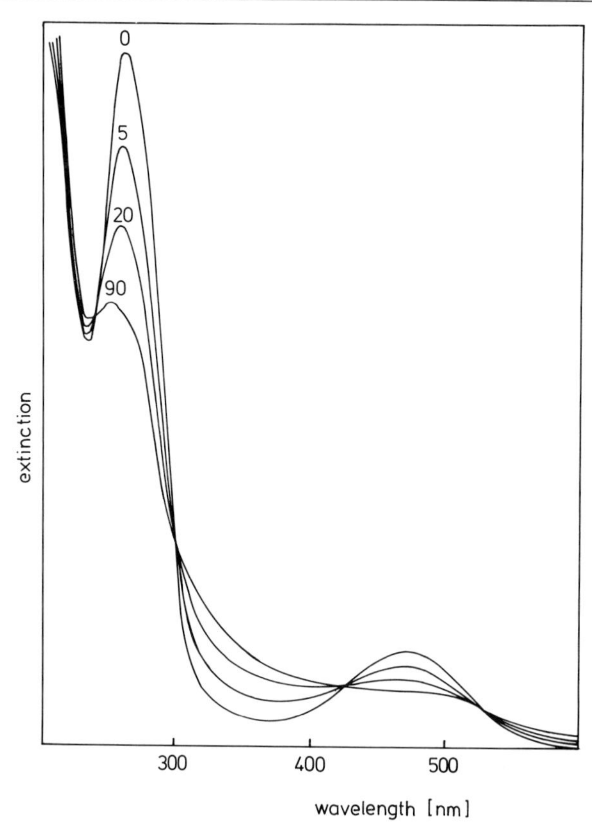

FIG. 2. The absorption spectra of DiN_3ATP in aqueous solution (pH 6.1) as a function of irradiation time. The sample was irradiated at 20°. Irradiation times: 0 min, 5 min, 20 min, 90 min.

spectrum 0). It is quite similar to the absorption spectrum of 3′-arylazido-β-alanine ATP.[18,19]

The photochemical reactivity of **III** was demonstrated by two independent experiments. First, DiN_3ATP bound irreversibly to cellulose on thin-layer plates when it was irradiated with UV light prior to the development of the chromatogram. A red-colored spot remained at the origin of the chromatogram after its development, in contrast to the nonirradiated control. Second, the photolysis of DiN_3ATP in aqueous solution resulted in a quick decomposition of the compound, which was followed spectroscopically (Fig. 2). The sample was placed in a quartz cuvette (1 cm). It was irradiated laterally with a mineralight ultraviolet lamp UVSL 25 at long wavelength. The distance between the sample and the ultraviolet lamp was 4 cm. Spectra were taken before the irradiation (spectrum 0) and after irradiation at times of 5, 20, and 90 min. The absorption maximum at 475 nm decreased and the maximum at 263 nm decreased and shifted to 250 nm during irradiation.

Biochemical Interactions

Specific Binding of DiN_3ATP to F_1-ATPase

The specific interaction of the modified substrate analog with the investigated protein is a necessary precondition for an efficient affinity label.[2-4] In the dark a useful photoaffinity label should be a substrate, or at least a competitive inhibitor. These criteria are fulfilled by DiN_3ATP. F_1-ATPase from *M. luteus* hydrolyzed DiN_3ATP in the dark to the corresponding diphosphate in the presence of Mg^{2+} ions. The specific activity was found to be 0.65 μmol/min/mg protein. This is very poor activity compared with the hydrolysis of the natural substrate ATP (CaATP: 30–40 μmol/min/mg protein, MgATP: 7 μmol/min/mg protein). On the other hand, the apparent K_m value for the hydrolysis of Mg-DiN_3ATP was very low in comparison with ATP or 8-azido-ATP. This is analogous to the behavior of other 2′- and/or 3′-*O*-substituted adenosine nucleotides.[29-31] The specific interaction of DiN_3ATP with the bacterial F_1-ATPase has also been demonstrated by competition experiments. The hydrolysis of CaATP was inhibited competitively by addition of even small amounts of DiN_3ATP as indicated by a Lineweaver–Burk plot (K_i = 0.025 m*M*). Both experiments demonstrated the specific interaction of DiN_3ATP with a catalytic site of F_1-ATPase from *M. luteus*, indicating that DiN_3ATP is a suitable photoaffinity label for this enzyme.

Light-Induced Inactivation of F_1-ATPase by DiN_3ATP

Photoaffinity labeling (photoaffinity cross-linking) of F_1-ATPase with DiN_3ATP is performed in an apparatus described earlier.[32] In this apparatus three samples can be irradiated simultaneously and independently under equal conditions. It allows an irradiation at constant temperatures. The irradiation is performed with a mineralight ultraviolet lamp UVSL 25 at long wavelengths. This long-wave UV light allows the photoactivation of both azido groups of DiN_3ATP without any remarkable photodamage

[29] G. Schäfer, G. Onur, and M. Schlegel, *J. Bioenerg. Biomembr.* **12,** 213 (1980).
[30] C. Grubmeyer and H. S. Penefsky, *J. Biol. Chem.* **256,** 3718 (1981).
[31] C. Grubmeyer and H. S. Penefsky, *J. Biol. Chem.* **256,** 3728 (1981).
[32] H.-J. Schäfer, P. Scheurich, G. Rathgeber, and K. Dose, *Anal. Biochem.* **104,** 106 (1980).

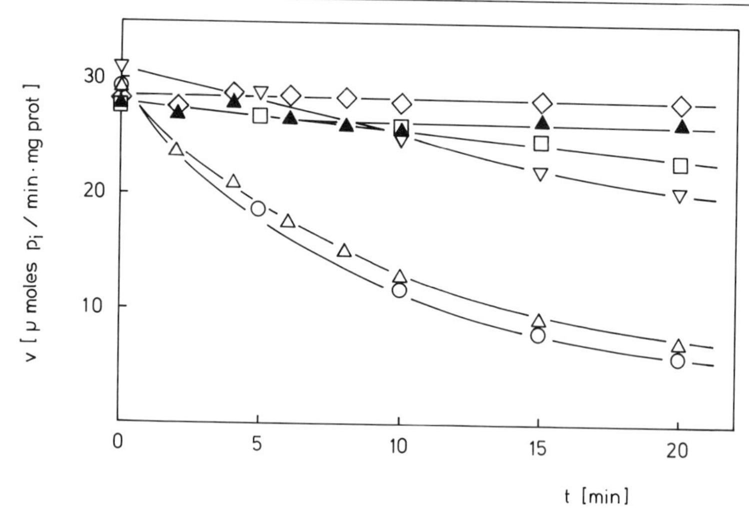

FIG. 3. Light-induced inhibition of F_1-ATPase. Irradiation in the presence of 0.5 m*M* Mg-DiN$_3$ATP (△), light control in the absence of DiN$_3$ATP (◇), dark control in the presence of 0.5 m*M* Mg-DiN$_3$ATP (▲), irradiation in the presence of 0.5 m*M* Mg-DiN$_3$ATP and 1 m*M* MgAMP (○), 1 m*M* MgADP (▽), or 1 m*M* MgATP (□). The enzymatic activity was determined in a 10-ml test solution containing 2 μg F_1, 100 m*M* Tris–HCl (pH 8.0), 5 m*M* Ca^{2+}, and 1 m*M* ATP. Prior to the addition of Ca^{2+} and ATP, the test solution was incubated at 37° for 1 hr.

to the protein. The ultraviolet lamp is placed above the samples in a distance of 4 cm. The energy flux at the position of the sample is 4 W/m^2. Generally 250 μg of F_1-ATPase are diluted in 250–500 μl Tris–HCl buffer (100 m*M*, pH 8.0). Then Mg-DiN$_3$ATP (0.5 m*M*) is added. The samples are stirred vigorously on a magnetic stirrer. They are kept at 37° during irradiation. After distinct irradiation times, small portions (1–2 μg protein) are taken for the ATPase assay.[33]

The irradiation of F_1-ATPase in the presence of DiN$_3$ATP results in an inhibition of ATPase activity (Fig. 3). This inhibition is not observed in control experiments. Neither the irradiation of the protein in the absence of DiN$_3$ATP (light control) nor the incubation of the enzyme in the presence of label (dark control) yields inactivation. The observed inhibition of the enzyme by photoaffinity labeling with DiN$_3$ATP is analogous to a noncompetitive inhibition as shown by Lineweaver–Burk plots of the native and of a partially inactivated enzyme. This indicates a covalent binding of the label to the protein.

[33] A. Arnold, H. U. Wolf, B. Ackermann, and H. Bader, *Anal. Biochem.* **71,** 209 (1976).

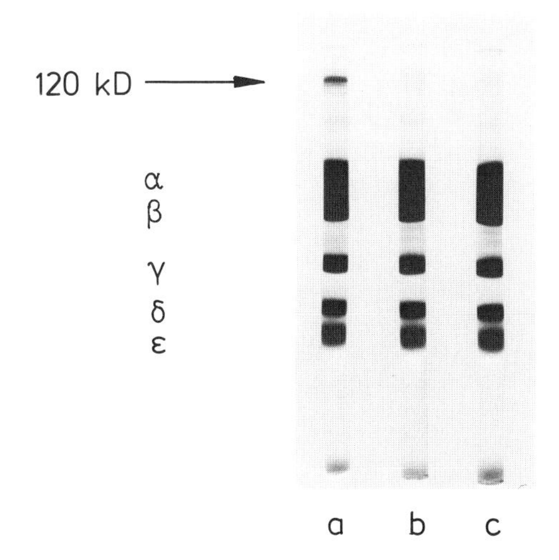

FIG. 4. Photoaffinity cross-linking of F_1-ATPase. SDS electrophoresis gels of labeled (cross-linked) F_1: (a) F_1 labeled by 0.5 m*M* Mg-DIN$_3$ATP; (b) F_1 labeled by 0.5 m*M* Mg-8-azido-ATP; (c) native F_1 (control).

Nonreactive specific ligands such as ATP or ADP (not AMP) compete with DiN$_3$ATP for the nucleotide binding sites of the enzyme without inactivating it. Therefore the light-induced inactivation of F_1-ATPase is prevented by the addition of ATP or ADP before the irradiation of the sample (enzyme, label, Mg^{2+}). AMP does not influence inactivation.

Photoaffinity Cross-Linking of F_1-ATPase

F_1-ATPase (250 μg) is photoinactivated in the presence of Mg-DiN$_3$ATP (0.5 m*M*), as described previously. After irradiation SDS (1%) and 2-mercaptoethanol (1%) are added. Then the sample is applied directly to the SDS–gel electrophoresis on 7.5% (w/v) gels according to Weber *et al.*[34] The current is 5 mA per gel. After electrophoresis the gels are stained with 0.25% (w/v) Coomassie Brilliant Blue R-250 (Serva) in acetic acid/methanol/water. Destaining is performed by diffusion against an aqueous solution of acetic acid (7.5%) and methanol (5%).

The SDS electrophoresis gel of the labeled enzyme shows a new protein band in the higher molecular weight region ($m > 100,000$) in addition to the subunits α–ε (Fig. 4, gel a). This cross-linked protein can be de-

[34] K. Weber, J. R. Pringle, and M. Osborn, this series, Vol. 26, p. 3.

tected very easily, since almost no other proteins migrate above the α subunit (m = 65,000). The cross-link was not observed in control experiments. The electrophoretic separation of neither the native enzyme (gel c) nor of a sample labeled and inactivated by the monovalent 8-azido ATP (gel b) results in formation of the cross-linked protein.

Cross-linking may occur inter- or intramolecularly. Striking evidence for an intramolecular cross-link formation is obtained by photoaffinity labeling at three different protein concentrations (200 μg/200 μl; 200 μg/1000 μl; 200 μg/2000 μl). The highest amount of cross-linked protein is observed in the most diluted solution.

In agreement with the inactivation experiments discussed previously (Fig. 3), the addition of ATP or ADP prior to irradiation in the presence of the label reduces the formation of the cross-link drastically. Addition of AMP does not affect cross-linking.

Structure of the Cross-Link

The first evidence for composition of the cross-linked product is obtained from its approximate molecular mass estimated from SDS–gel electrophoresis of the labeled protein. The plot of the logarithms of the known molecular masses of the subunits α–ε versus their migration indicates a molecular mass of about 120,000. This suggests the participation of two of the larger subunits, α and/or β. α-α, α-β, or β-β present possible structures of the cross-linked product. Certainty about the subunit composition of cross-linked product can be obtained if a cleavable bifunctional reagent is used. The product may be cleaved hydrolytically by incubation in alkaline solution due to the lability of the ester linkage of DiN_3ATP. Figure 5 illustrates the cleavage of the product. A labeled sample (500 μg protein) is subjected to SDS–gel electrophoresis, as described previously. After staining and destaining of the gel, the slice containing the cross-linked protein is isolated and incubated in 300 μl of weak alkaline solution [$NaHCO_3$ (0.05 M), SDS (1%)] at 37° for 24 hr. Due to the acidic pH of the stained gel slice, the incubation medium has to be adjusted repeatedly to weak alkaline pH by addition of NaOH. After incubation the supernatant is applied to a further SDS–gel electrophoresis.

Hydrolytic cleavage and subsequent SDS–gel electrophoresis of the cross-linked product demonstrate that the cross-linked protein is almost entirely split into two proteins with nearly identical concentration. These two proteins correspond in their electrophoretic mobility exactly with the α and β subunits of the native enzyme. This result demonstrates the subunit composition α-β for the cross-linked protein.

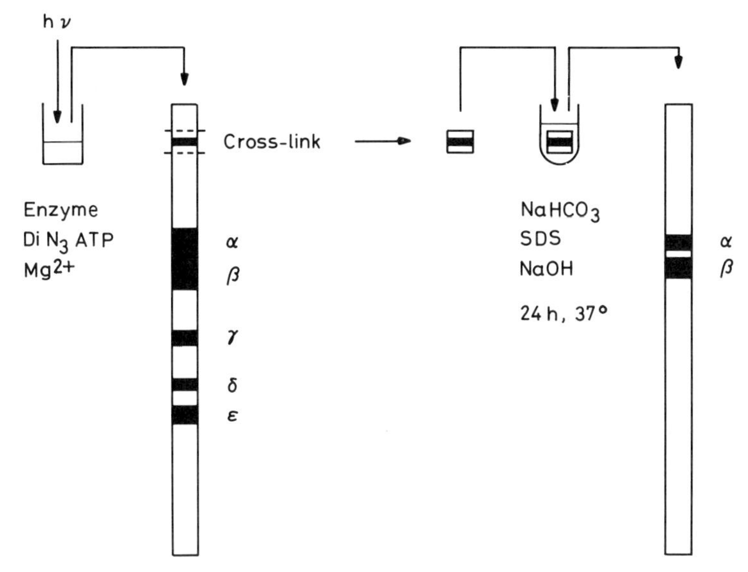

FIG. 5. Hydrolytic cleavage of the cross-linked product.

In principle, there are two possibilities for the orientation of the bifunctional reagent between the α and β subunits. DiN_3ATP may label β with its 8-azidoadenine moiety and α with its 3′-arylazido group, or vice versa. To obtain information about this orientation, we prelabeled F_1-ATPase at the β subunit with monovalent 8-azido-ATP before photoaffinity cross-linking with divalent DiN_3ATP. For this purpose, 250 μg F_1-

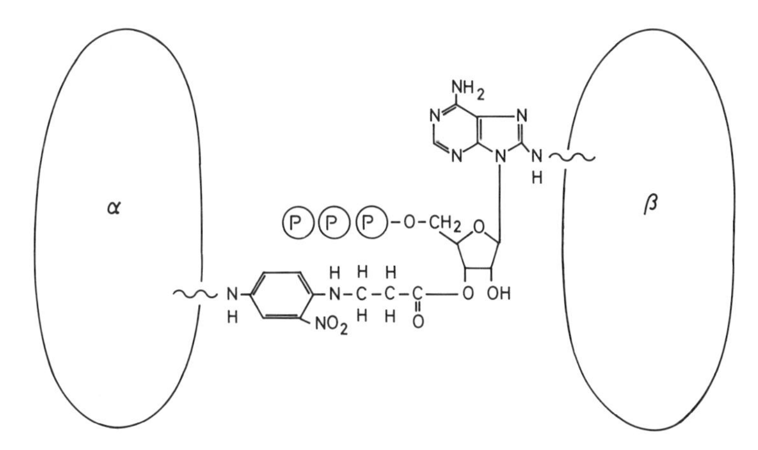

FIG. 6. Structure of the α-β cross-link.

ATPase are irradiated in the presence of monovalent Mg-8-azido-ATP (1 m*M*) for 30 min (prelabeled sample). This procedure results in a Mg^{2+}-dependent specific labeling of β.[23] Simultaneously, a control is irradiated under identical conditions in the absence of the monovalent label. Then Mg-DiN_3ATP (0.5 m*M*) is added to both samples. The samples are irradiated for another 15 min to perform cross-linking. The labeled proteins are separated electrophoretically, as described previously. The formation of the cross-link is decreased remarkably in the prelabeled sample, indicating that DiN_3ATP labeled the same nucleotide binding site as 8-azido-ATP at the β subunit. Further evidence for such an orientation is supplied by the Mg^{2+} dependence of the cross-linking and its decrease upon addition of ATP or ADP, whereas AMP does not show any remarkable effect. All these results are analogous to the specific binding of 8-azido ATP to the nucleotide binding site at the β subunit.[23] Figure 6 shows the most probable structure of the α-β cross-link.[35]

Acknowledgments

The excellent technical assistance of Mrs. G. Rathgeber is gratefully acknowledged. The author thanks Mrs. M. Mittelmann-Sicurella for editing the manuscript and Dr. K. Dose for helpful discussion. This work was supported by the Deutsche Forschungsgemeinschaft.

[35] H.-J. Schäfer and K. Dose, *J. Biol. Chem.* **259,** 15301 (1984).

[67] Synthesis and Use of an Azido-Labeled Form of the ATPase Inhibitor Peptide of Rat Liver Mitochondria

By KLAUS SCHWERZMANN and PETER L. PEDERSEN

The binding and dissociation of the mitochondrial ATPase inhibitor peptide[1] by the ATPase complex is governed by the energy state of the inner mitochondrial membrane.[2] In order to be able to study the interaction of the inhibitor peptide with the ATPase in a quantitative fashion, we describe here the synthesis of a radioactive reagent that binds covalently to the inhibitor peptide without affecting its activity. In addition, the same reagent can be used to introduce a covalent cross-link between the inhibitor peptide and its binding site(s) on the ATPase complex.

[1] M. E. Pullman and G. C. Monroy, *J. Biol. Chem.* **238,** 3762 (1963).
[2] P. L. Pedersen, K. Schwerzmann, and N. Cintrón, *Curr. Top. Bioenerg.* **11,** 149 (1981).

Analytical Procedures

Measurement of ATPase and Inhibitor Activity. The assay for the inhibitor activity is essentially as described by Horstman and Racker.[3] Mitochondrial inner membrane vesicles are incubated at room temperature with varying amounts of inhibitor peptide. The incubation mixture contains 50 μg of inner membrane vesicles, 5 mM imidazole (pH 6.7), 0.5 mM $MgCl_2$, 5 mM Na_2ATP, and 250 mM sucrose up to a total volume of 0.25 ml. After 10 min, an aliquot is transferred into a cuvette containing 1 ml of ATPase assay medium whereby the hydrolysis is enzymatically coupled to the oxidation of NADH [4 mM ATP, 0.5 mM phosphoenolpyruvate, 0.3 mM NADH, 4 mM $MgCl_2$, 5 mM KCN, 58 mM Tris–HCl (pH 7.5), 1 unit of lactate dehydrogenase, and 8 units of pyruvate kinase].

Definition of Unit Inhibitor Activity. One unit of inhibitory activity is defined as that amount of protein required to inhibit 50% of 0.2 units of ATPase activity under the conditions above.

Protein Measurement. Protein is measured with the method of Lowry *et al.*[4] after precipitation with trichloroacetic acid (TCA). Fatty acid free bovine serum albumin (BSA) was used as standard.

Preparative Procedures

Inverted inner membrane vesicles are obtained from water-lysed mitoplasts of rat liver mitochondria.[5]

F_1-ATPase and inhibitor peptide are both isolated from frozen rat liver mitochondria. For the isolation of F_1-ATPase, the method of Catterall and Pedersen[6] is applied. The inhibitor peptide is isolated according to Cintrón and Pedersen[7] with the modifications introduced later.[8]

Synthesis of *N*-Hydroxysuccinimidyl-4-[7-^{14}C]azidobenzoate

A scheme of the synthesis pathway is shown in Fig. 1. In a first step, the azido derivative of the aminobenzoic acid is synthesized according to Hixson and Hixson.[9] Then 3 mg of 4-aminobenzoic acid and 0.25 mCi of the 7-^{14}C-labeled compound (Research Products International, 54 mCi/

[3] L. L. Horstman and E. Racker, *J. Biol. Chem.* **245,** 1336 (1970).

[4] O. H. Lowry, N. J. Rosebrough, A. L. Farr, and R. J. Randall, *J. Biol. Chem.* **193,** 265 (1951).

[5] C. R. Hackenbrock and K. M. Hammon, *J. Biol. Chem.* **250,** 9185 (1975).

[6] W. A. Catterall and P. L. Pedersen, *J. Biol. Chem.* **246,** 4987 (1971).

[7] N. M. Cintrón and P. L. Pedersen, this series, Vol. 55, p. 408.

[8] K. Schwerzmann and P. L. Pedersen, *J. Biol. Chem.* **257,** 9555 (1982).

[9] S. H. Hixson and S. S. Hixson, *Biochemistry* **14,** 4251 (1975).

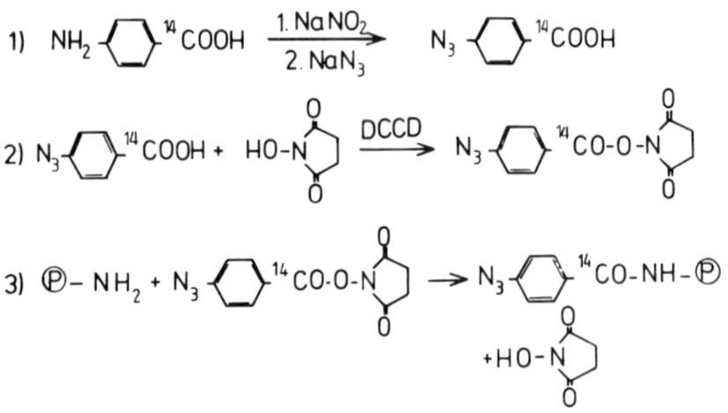

FIG. 1. Synthesis of *N*-hydroxysuccinimidyl-4-[7-^{14}C]azidobenzoate and its reaction with proteins. For details see text.

mmol, 0.63 mg) are suspended in 1 ml of chilled 11% (w/v) aqueous sulfuric acid, and 9.5 mg of $NaNO_2$ dissolved in 0.2 ml of water is added slowly under continuous stirring (30 min).

To avoid activation of the azido group by light, the following steps are performed under dim light. The reaction mixture is overlayered with 1.5 ml of diethyl ether. The mixture is stirred vigorously and 14 mg of NaN_3 dissolved in 0.1 ml water is added at once. After 30 min, the organic phase is withdrawn with a Pasteur pipet. The aqueous phase is extracted again with 1 ml of ether. The combined ether phases are first washed with 1 ml of saturated NaCl in water and then dried by filtration over 2 g of solid $MgSO_4$. The ether is evaporated under a stream of nitrogen gas.

For the synthesis of the succinimidyl ester the method of Lewis *et al.*[10] is adapted. 4-[7-^{14}C]Azidobenzoic acid (3.8 mg) is dissolved in 0.5 ml of dioxane and reacted with 2.7 mg of *N*-hydroxysuccinimide in the presence of 4.8 mg of dicyclohexylcarbodiimide (DCCD). The reaction is allowed to go to completion (10–12 hr). Then precipitates are removed from the reaction mixture by filtration and the dioxane evaporated under a stream of nitrogen. The product is recrystallized twice from cold dioxane with ether.

The course of the reaction can be followed by thin-layer chromatography on silica gel in a system butanone : acetone : water (65 : 25 : 15). The end product has an R_f of ~0.8, whereas the starting materials have R_f values of 0.5 or less. The overall molar yield of the final product is more than 50%, and ~30% in terms of radioactivity (this depends largely on the purity of the radioactive starting material). The product is stable over several months when kept dry in the dark under nitrogen.

[10] R. V. Lewis, M. F. Roberts, E. A. Dennis, and W. S. Allison, *Biochemistry* **16,** 5650 (1977).

BINDING OF HYDROXYSUCCINYL-4-[^{14}C]AZIDOBENZOATE TO PURIFIED INHIBITOR PEPTIDE[a]

	Reaction mixture	Labeled material
Protein (mg)	0.28	0.23
Specific activity (units/mg protein)	2500	2200
[^{14}C]HS-AB (mol/mol)	30	2.6
Total radioactivity (cpm)	1.94×10^6	1.04×10^5

[a] The labeling of the inhibitor peptide was performed as indicated in the text. The specific radioactivity of [^{14}C]HS-AB used in this experiment was 2210 cpm/nmol, and the concentration of the stock solution was 9.4 m*M*.

Characterization of the Product. The UV spectrum shows an absorption maximum at 289 nm (in 25 m*M* potassium phosphate, pH 7.5). Exposure to UV light causes a rapid decrease of absorption, whereas high pH (above 9) causes a shift of the absorption maximum to 275 nm, indicating hydrolysis of the ester. Addition of lysine at neutral pH has the same effect.

The imido ester reacts primarily with free amino groups in proteins, but also with histidine and free sulfhydryl groups.[11] The azido residue forms a reactive nitrene under irradiation with UV light.[11]

Labeling of the Inhibitor Peptide with [^{14}C]HS-AB

Procedure. To avoid activation of the azido group, all the following operations should be performed under dim light. Lyophilized inhibitor peptide (0.25–0.30 mg) is dissolved in a small amount of buffer (0.075 ml of 0.1 *M* NaCl in 0.1 *M* NaP_i, pH 7.6). Three 10-μl aliquots of a solution of 9.8 m*M* [^{14}C]HS-AB in dioxane are added in 30-min intervals. After 90 min, the inhibitor peptide is sedimented with an equal volume (105 μl) of saturated ammonium sulfate and the protein collected by centrifugation. The sediment is dissolved in 0.1 ml of buffer and again precipitated with ammonium sulfate as before. Alternatively, the protein can be precipitated with 5% TCA after a short incubation with an excess of lysine to remove unreacted [^{14}C]HS-AB. The labeled protein is then dissolved in the buffer of choice and can be stored frozen in the dark over a period of weeks.

Characterization of [^{14}C]HS-AB-Labeled Inhibitor Peptide. The results of a labeling experiment are shown in the table. To get a high

[11] T. H. Ji, *Biochim. Biophys. Acta* **559,** 39 (1979).

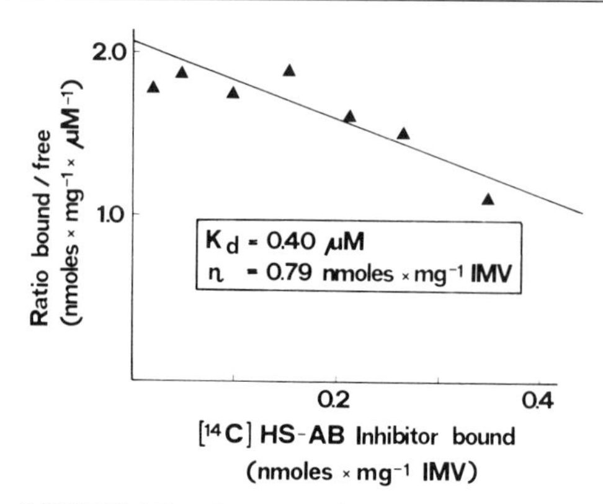

FIG. 2. Binding of [^{14}C]HS-AB to inner membrane vesicles from rat liver mitochondria. Inner membrane vesicles (1.28 mg protein) were incubated with various amounts of radiolabeled inhibitor peptide (0.46 mg protein/ml, 14,600 cpm/nmol) in 2 ml of buffer containing 0.19 *M* sucrose, 0.5 m*M* MgATP, and 5 m*M* HEPES-KOH (pH 6.7). After 30 min incubation, the suspension was centrifuged (100,000 *g*, 30 min) and the radioactivity determined in the supernatant and the pellet. The results are presented in the form of a Scatchard plot.[12] K_d is the dissociation constant of the binding site, and n the number of binding sites per 1 mg of protein inner membrane vesicles.

degree of labeling, [^{14}C]HS-AB has to be present in a large molar excess. Up to 5 mol of label can be incorporated per mol of inhibitor peptide (M_r 12,500)[7] without altering its specific activity.

SDS–gel electrophoresis of the labeled inhibitor peptide reveals one major band containing most of the radioactivity corresponding to the native inhibitor peptide (not shown). Peaks of higher molecular weight are occasionally found corresponding in their molecular weights to dimeric and trimeric forms of the inhibitor peptide.

Use of [^{14}C]HS-AB-Labeled Inhibitor Peptide

Binding of Inhibitor Peptide by Inverted Inner Membrane Vesicles. In the dark, [^{14}C]HS-AB can be used as a simple radiolabel for the inhibitor peptide in order to study its binding to membrane-bound mitochondrial ATPase. The results of such a binding study are presented as a Scatchard plot[12] (Fig. 2). The linear character of the plot indicates the presence of only one class of binding sites with a K_d of 0.4 μM and a total of 0.79 nmol of binding sites per milligram of inner membrane vesicle protein. Assuming that the F_1-ATPase constitutes 15–20% of the protein in the inner

[12] G. Scatchard, *Ann. N.Y. Acad. Sci.* **51,** 660 (1949).

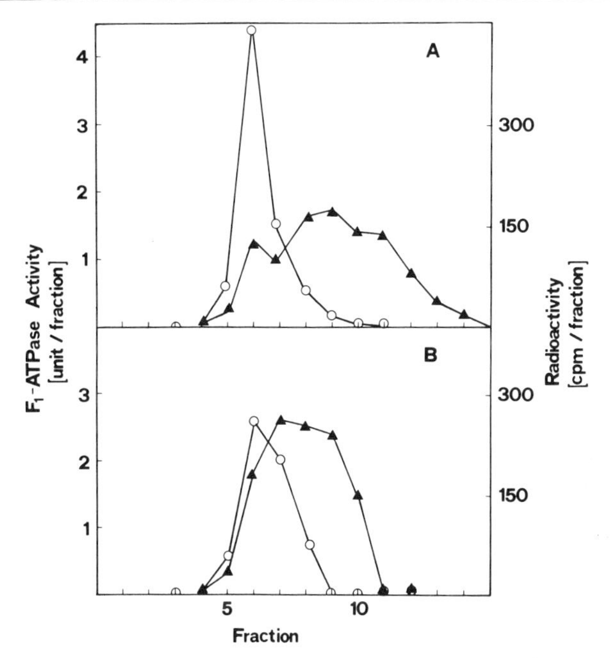

FIG. 3. Purified F_1 (0.25 mg protein) and [^{14}C]HS-AB-labeled inhibitor peptide (0.08 mg protein, 4900 cpm/nmol) are incubated in 0.6 ml of buffer as used for measuring inhibitor activity. After 15 min incubation in the dark, an aliquot of 0.3 ml is transferred on a Sephadex G-100 column and eluted with 20 m*M* KP_i (pH 7.0). Fractions are collected and analyzed for ATPase activity and radioactivity (A). (B) The other aliquot of 0.3 ml is subjected to irradiation with UV light (mineral light, 259 nm, 4 × 30 sec) prior to chromatography. ○, ATPase activity; ▲, radioactivity.

membrane of rat liver mitochondria, one would obtain a ratio of 2–3 inhibitor binding sites per mol of F_1, a value that is in line with the number of active sites, i.e., β subunits per F_1.[13]

Covalent Cross-Linking of Inhibitor Peptide to Isolated F_1-ATPase by [^{14}C]HS-AB. The azido group of the radiolabel can be activated by low doses of UV light to form a reactive nitrene which has a broad reactivity.[11] Preliminary results also confirm that [^{14}C]HS-AB synthesized as described here and inserted into the purified inhibitor peptide can cross-link the inhibitor peptide with F_1-ATPase. This is demonstrated in Fig. 3. When [^{14}C]HS-AB-labeled inhibitor peptide is bound to purified F_1 in the dark and then passed over a Sephadex G-100 column, most of the radioactivity elutes with the free inhibitor, indicating that the complex of inhibi-

[13] L. M. Amzel, *J. Bioenerg. Biomembr.* **13,** 109 (1981).

tor peptide and F_1 readily dissociates as expected.[8] If prior to the column chromatography photolysis is induced by UV irradiation, a considerable portion of the radioactivity elutes together with the ATPase activity, which is partially inhibited. This is a clear indication for the cross-link induced by photolysis of the azido group.

Final Notes

A great many photosensitive heterobifunctional reagents for cross-linking proteins have been described in the literature.[11,14] Few of them, however, are available in radioactive form.[15] [^{14}C]HS-AB is such a reagent that can be easily synthesized from radioactive precursors in any biochemistry laboratory. Although it may not always be necessary to have the radioactive marker in the cross-linker itself, such a label has a distinct advantage. It eliminates the need to introduce in a second step the radiolabel, e.g., by radioiodination, into the protein, which may reduce its functional activities or even render it functionally inactive, as in the case of this ATPase inhibitor peptide from rat liver mitochondria (unpublished observation). Therefore, a cross-linking radioactive marker as described here may be very useful in the study of protein–protein interactions where the protein of interest is sensitive to chemical reactions or cannot be radioactively labeled otherwise.

Acknowledgments

This work was performed when K.S. was a recipient of a Fogarty International Fellowship (1-F05 TWO 2755) in P.L.P.'s laboratory. This work was also supported by NIH grant CA 10951 to P.L.P.

[14] K. Peters and F. M. Richards, *Annu. Rev. Biochem.* **46,** 523 (1977).

[15] G. Klein, M. Satre, A.-C. Dianoux, and P. V. Vignais, *Biochemistry* **20,** 1339 (1981).

[68] Benzophenone-ATP: A Photoaffinity Label for the Active Site of ATPases

By NOREEN WILLIAMS, SHARON H. ACKERMAN, and PETER S. COLEMAN

Introduction[1]

Photoaffinity labels have enjoyed an increasing popularity in enzymology since their introduction by Westheimer and colleagues.[2,3] When successfully designed and applied, these molecular probes seek out selective domains on macromolecular targets for which they possess a substrate-like affinity, and upon irradiation with actinic light, form covalent cross-linkages at these specific binding sites. Such an approach can be extremely valuable for enzyme structure/mechanism studies. The site-specific affinity of photoreactive substrate analogs leads to covalent occupancy of the catalytic site of enzymes, which causes irreversible enzyme inhibition and provides a unique opportunity for examining the primary sequence at the active site.

Due to the very large number of enzymes requiring adenine nucleotides (especially ATP) as primary or cosubstrate, there has been much interest in designing ATP derivatives that are photochemically active. Most of the available photoaffinity probes contain the azide group ($—N_3$) as the photoactive center.[4–7] While there is no doubt that azido derivatives of ATP (and of other substrates or ligands) have proved very useful, the photochemical properties of the intermediate nitrene species display complicated reaction pathways that are not always easily predictable or understood mechanistically, and thus the azido compounds may offer the experimentalist some undesired difficulties.[8–10]

[1] The original work on BzATP and rat liver F_1 ATPase was funded, in part, by Biomedical Research Support Group Grant RR07062.

[2] F. H. Westheimer, *Ann. N.Y. Acad. Sci.* **346,** 134 (1980).

[3] V. Chowdry and F. H. Westheimer, *Annu. Rev. Biochem.* **48,** 293 (1979).

[4] B. E. Haley and J. Hoffman, *Proc. Natl. Acad. Sci. U.S.A.* **71,** 3367 (1974).

[5] H. Bayley and J. R. Knowles, this series, Vol. 46, p. 69.

[6] R. J. Guillory and S. J. Jeng, this series, Vol. 46, p. 259.

[7] A recent collection of symposium papers that deals with photoaffinity labeling with either carbene or nitrene precursor analogs is given in *Fed. Proc., Fed. Am. Soc. Exp. Biol.* **42,** 2825 (1983).

[8] H. Bayley and J. R. Knowles, *Biochemistry* **17,** 2414 (1978).

[9] J. V. Staros, *Trends Biochem. Sci.* **5,** 320 (1980).

[10] P. V. Vignais, A.-C. Dianoux, G. Klein, G. Lauquin, J. Lunardi, R. Pougeois, and M. Satre, *Prog. Clin. Biol. Res.* **102B,** 439 (1982).

In this chapter we present, together with methods, arguments in support of the use of benzophenone as a preferred photoactive moiety in the design of photoaffinity substrate analogs, referring specifically to 3′-*O*-(4-benzoyl)benzoyl-ATP (so-called benzophenone-ATP or BzATP). Here we shall limit discussion to our experience with BzATP and the F_1-ATPase of rat liver mitochondria,[11,12] which has given rise to more recent work by other laboratories where successful photolabeling has been achieved with BzATP (or BzADP) and different ATPase enzymes.[13–16,16a] One of the most attractive aspects of BzATP as a photoaffinity probe, in addition to the unique photochemical properties of benzophenone described below, is the relative ease with which it is synthesized. These factors should help to encourage the use of BzATP (and other "Bz-ligand" derivatives) as being particularly well suited for studies of ATP-utilizing (and other) enzymes.

Readers should note, however, that the benzophenone functional group is attached to the adenine nucleotide moiety via a 3′-ester linkage (see Fig. 1). Despite evidence that this ester bond seems to remain stable under mild acid/base conditions[11,12] (cf. Characterization of BzATP, pH Stability), recent experience indicates that the bond is labile under the more extreme conditions required by some laboratory protocols, such as those involving protein sequencing. Consequently, the most effective and unambiguous use of benzophenone-derivatized photoaffinity probes probably necessitates methods for the prior synthesis of [4-^{14}C]carboxybenzophenone,[16,17] or derivitization of the nucleotide with [3-^{3}H]-4-benzoylbenzoic acid, which has recently become available commercially (Cat. #TNC-455, Rotem Industries Ltd., P. O. Box 9046, Beer Sheva, Israel).

Some General Considerations

The following discussion is presented in order to clarify the underlying logic pertaining to the use of benzophenone-derivatized ligands for receptor domains on diverse biological macromolecules in an aqueous reaction medium. Formalized treatments of the photochemistry underlying these concepts are available in several excellent monographs.[18–20]

[11] N. Williams and P. S. Coleman, *J. Biol. Chem.* **257,** 2834 (1982).
[12] N. Williams, Ph.D. thesis, New York University, 1981.
[13] D. Bar-Zvi, M. Tiefert, and N. Shavit, *FEBS Lett.* **160,** 233 (1983).
[14] N. G. Kambouris and G. G. Hammes, *Proc. Natl. Acad. Sci. U.S.A.* **82,** 1950 (1985).
[15] M. B. Cable and F. N. Briggs, *J. Biol. Chem.* **259,** 3612 (1984).
[16] R. Mahmood and R. G. Yount, *J. Biol. Chem.* **259,** 12956 (1984).
[16a] M. F. Manolson, P. A. Rea, and R. J. Poole, *J. Biol. Chem.* **260,** 12273 (1985).
[17] D. Licht and P. Coleman, unpublished results (1985); also, K. Nakamaye and R. G. Yount, *J. Label. Compnd. Radiopharm.* **22,** 607 (1985).
[18] N. J. Turro, "Modern Molecular Photochemistry." Benjamin/Cummings, Reading, Massachusetts, 1978.
[19] D. O. Cowan and R. L. Drisko, "Elements of Organic Photochemistry." Plenum, New York, 1976.

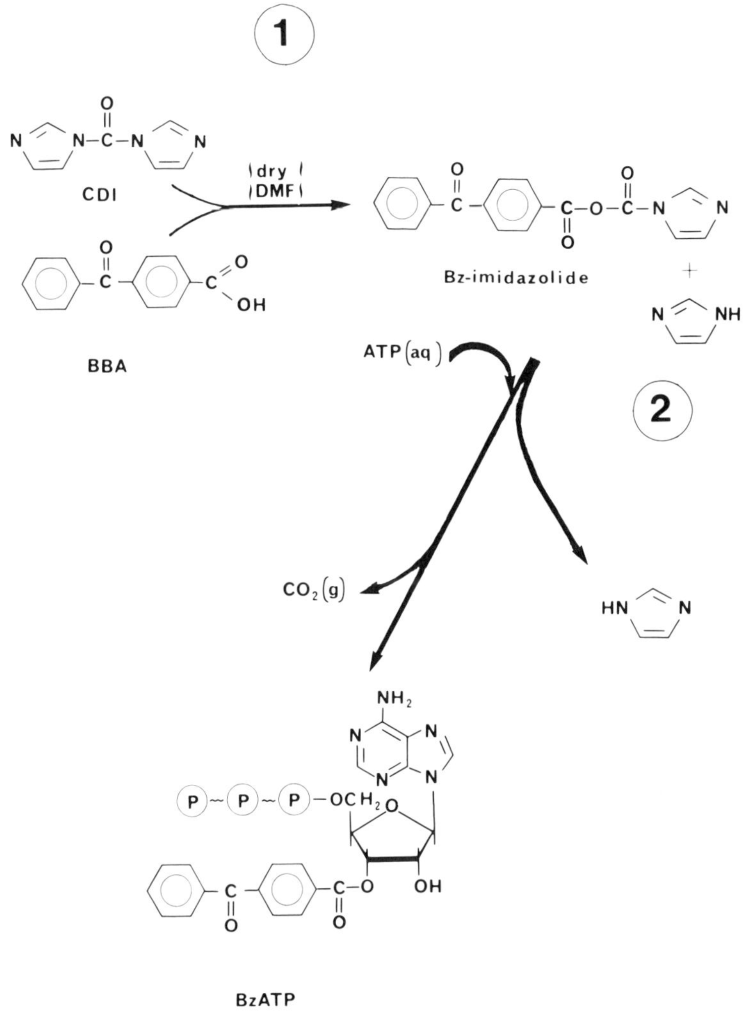

FIG. 1. Reaction route for BzATP synthesis.

There appear to be at least three physicochemical considerations that, when satisfied, will determine whether photoaffinity labeling with the benzophenone-ATP (or any other photoaffinity) probe occurs at a ligand binding site on the target enzyme. Such specific site labeling is, of course, the desired outcome with any affinity label, in contrast to a less site-specific and more random covalent cross-linking. These considerations are as follows: (1) the "residence time" (i.e., the lifetime of the probe : enzyme complex) of the BzATP triplet intermediate at the specific binding site(s) on the enzyme *prior* to covalent cross-linking; (2) the magnitude of

[20] J. G. Calvert and J. N. Pitts, "Photochemistry." Wiley, New York, 1966.

the second-order rate constant of hydrogen atom abstraction from the binding site domain by the benzophenone triplet intermediate, compared with the first-order rate constant (i.e., the lifetime^{-1}) for the decay of the photoinduced triplet; and (3) the use of experimental conditions that guard against the possibility that any as yet unbound BzATP triplets, diffusing away from the ATP binding site, would lead to fortuitous labeling at other domains on the enzyme not directly involved with catalysis.

For multimolecular reactions whose mechanisms consist of many intermediate steps, one can argue that the lifetime of the reaction complex (such as an ES transition state complex) is the principal consideration. Clearly, the so-called residence time of the BzATP triplet at the catalytic site of an ATPase would be dominated by the affinity of the probe for that site. First, consider, as an example at one extreme, the dissociation of nonspecific complex AB (AB $\rightarrow$ A + B). If there is no structurally based affinity of A for B, the AB association can display half-times estimated to be as short as 10^{-13} sec.[21,22] At the other extreme, the mean lifetimes of ES complexes have been found to be quite durable, between 10^{-7} and 10^{-4} sec, indicative of "substrate anchoring" due to a specific affinity of the enzyme for the ligand.[22–27] Therefore, if other physicochemical circumstances required for photochemical cross-linking are optimized (see below), specific site labeling may be achieved, or at least may be heavily favored, by stipulating that the residence time of the BzATP at the ATP binding site of the enzyme manifests a duration in the same general range of otherwise productive E : ATP complexes.

A second major concern is the rate constant for the decay of the benzophenone triplet state in degassed H_2O, which was found to be about 10^4 sec^{-1} via flash photolysis spectroscopy.[28] For photoaffinity labeling, one must ask whether such a triplet lifetime is "long" or "short" relative to the rate of its productive reaction via hydrogen abstraction from the target enzyme to be labeled. By means of a flash photolysis experiment in a degassed aqueous system, this laboratory[29] measured a rate constant of

[21] M. Frost and R. Pearson, "Kinetics and Mechanism," Chap. 11. Wiley, New York, 1961.

[22] J. Reuben, *Proc. Natl. Acad. Sci. U.S.A.* **68,** 63 (1971).

[23] J. T. Gerig, *J. Am. Chem. Soc.* **90,** 2681 (1968).

[24] J. T. Gerig and J. D. Reinheimer, *J. Am. Chem. Soc.* **92,** 3147 (1970).

[25] B. D. Sykes, P. D. Schmidt, and G. R. Stark, *J. Biol. Chem.* **245,** 1180 (1970).

[26] A. G. Marshall, *Biochemistry* **7,** 2450 (1968).

[27] A. S. Mildvan and M. C. Scrutton, *Biochemistry* **6,** 2978 (1967).

[28] M. B. Ledger and G. Porter, *J. Chem. Soc., Faraday Trans. I* **68,** 539 (1972).

[29] V. A. Kuzmin and P. S. Coleman, unpublished results (1981). BzATP solutions (pH 7.0) in deionized water were thoroughly degassed on a high-vacuum line by means of the freeze-pump-thaw method. The flash absorption spectroscopy apparatus was kindly provided by the Department of Chemistry, New York University. J. Navarro assisted with the experiments.

$6.4 \pm 0.8 \times 10^8\ M^{-1}\ sec^{-1}$ for the dissipation of triplet BzATP, presumably by hydrogen abstraction to yield the ketyl radical, the hydrogen being provided by either another ground or another triplet state BzATP in the environment. The value we obtained is not far from that for the diffusion rate constant of small solute species in water ($\sim 10^9\ M^{-1}\ sec^{-1}$)[18] and agrees nicely with rate constants that have been found for hydrogen abstraction by triplet benzophenone from some solvent donors.[30] The important fact is that a comparison of these rate constants indicates that in water, to which the benzophenone triplet is nearly inert,[28] even random collision of solute molecules by diffusion yields a rate of hydrogen abstraction that is more than four orders of magnitude larger than the lifetime of the benzophenone triplet. Since the probability for hydrogen abstraction would be enhanced to an even greater extent by a stabilized spatial proximity between the triplet benzophenone and its target (as found in E : BzATP complexes), the phenomenon of diffusion is eliminated altogether from the argument. Therefore, with residence times for BzATP and an ATPase on a time scale similar to those estimated for many long-lived ES complexes, we can expect site-specific covalent coupling with the BzATP triplet to be a highly probable and efficient reaction.

Regarding conditions that would help preclude fortuitous, random-site labeling, we comment upon the low energy of the benzophenone triplet state (69 kcal/mol),[18] which, although ensuring its relative inertness toward reaction with water,[28,31,32] still leaves it vulnerable to collisional triplet–triplet quenching by the molecular oxygen (a ground state triplet)

$$:\dot{\underset{\cdot\cdot}{O}}—\dot{\underset{\cdot\cdot}{O}}:$$

dissolved in our aqueous enzyme assay medium. In air-equilibrated water at room temperature ($[O_2] \simeq 0.1\ mM$), the rate constant for the oxygen quenching of triplet benzophenone (and by analogy, of any triplet state BzATP "free" in air-equilibrated aqueous solution) is probably close to $4 \times 10^8\ M^{-1}\ sec^{-1}$.[28] However, it seems unlikely that both molecular oxygen and BzATP would occupy, simultaneously, exactly the same location at the specific nucleotide binding site on the ATPase enzyme. Consequently, only that fraction of the BzATP triplet concentration

[30] An interesting example is the rate constant ($2.5 \times 10^8\ M^{-1}\ sec^{-1}$) for the photoreduction of the benzophenone triplet by a primary amine, such as 2-butylamine [S. G. Cohen, A. Parola, and G. Parsons, Jr., *Chem. Rev.* **73,** 141 (1973)]. 2-Butylamine may be taken as a conceivable structural analog for the 6-NH_2 group "vicinity" on the purine ring of ATP, a source of donatable hydrogen atoms for a colliding triplet state benzophenone. For isopropanol as hydrogen-donating solvent, see C. Walling and M. Gibian, *J. Am. Chem. Soc.* **86,** 3902 (1964).

[31] A. Beckett and G. Porter, *Trans. Faraday Soc.* **59,** 2038 (1963).

[32] V. A. Kuzmin and A. K. Chibisov, *Teor. Eksp. Khim.* **7,** 403 (1971).

which is *not* anchored at the ATP binding site on the enzyme, but is freely diffusing in the aqueous medium, would be susceptible to rapid annihilation to the ground state upon collision with dissolved O_2. Consonant with this reasoning, we believe that with an air-equilibrated enzyme assay medium, nonspecific photo-cross-linking at random surface domains of the ATPase enzyme would probably not occur because of the ample presence of molecular oxygen as triplet quencher. On the basis of this argument and our empirical results, we have found it unnecessary to add radical scavenger molecules, as are often employed to eliminate random-site photo-cross-linking.[33]

It is important to note that the benzophenone triplet probably does not undergo intramolecular structural rearrangement.[34] In contrast, arylazido probes readily rearrange from the excited singlet state and react covalently as electrophilic agents only at nucleophilic functional groups on the target enzyme.[35] Furthermore, by insertion into water, the singlet nitrene introduces a "new" (chemically inert) species into the system that competes for the substrate-specified target site. When such undesirable side reactions occur with photolabile probe molecules, the yield of covalent incorporation at the desired target site is lowered considerably.[36] Although substantially less information is available regarding the photochemical mechanisms that predominate with benzophenone-linked affinity probes, they appear to be relatively free of such complications.

Finally, we wish to emphasize that introduction of a new photoaffinity probe into the arsenal of useful molecular tools employed by the biochemist must await its successful application by other laboratories. In this regard, the apparent usefulness of benzophenone-derivatized photoaffinity probes (particularly Bz-adenine nucleotides for studies on the general category of ATPase enzymes) is supported empirically by the following information. Different laboratories have recently demonstrated that while BzATP may (or may not) act as a substrate, it is unquestionably a good nucleotide-site ligand for at least five distinct ATP-hydrolyzing enzymes: the rat liver F_1[11]; the chloroplast F_1[13,14]; the Ca^{2+},Mg^{2+}-ATPase of the sarcoplasmic reticulum[15]; the myosin SF_1-ATPase[16]; and the tonoplast ATPase of beet root.[16a] Recent preliminary evidence also seems to indicate that BzATP may be a substrate for certain phosphotransferases, such as creatine kinase.[37,38]

[33] A. E. Ruoho, H. Kiefer, P. E. Roeder, and S. J. Singer, *Proc. Natl. Acad. Sci. U.S.A.* **70**, 2567 (1973).

[34] The possibility of forming peroxydiradicals in aerated aqueous solution cannot be excluded, however.

[35] B. DeGraff, D. Gillespie, and R. Sundberg, *J. Am. Chem. Soc.* **96**, 7491 (1974).

[36] P. E. Nielsen and O. Buchard, *Photochem. Photobiol.* **35**, 317 (1982).

[37] A. Vinitzky and C. Grubmeyer, *Ann. N.Y. Acad. Sci.* **435**, 222 (1984).

[38] It would also appear that the enzyme pyruvate kinase can utilize BzADP as substrate.

Preparation of 3′-*O*-(4-Benzoyl)benzoyl-ATP (BzATP)[39]

Synthesis

N,N′-Dimethylformamide (DMF; Gold Label, Aldrich) is distilled and then stored over $MgSO_4$, or better, over a molecular sieve (type 4A, 8–12 mesh, Aldrich) in an evacuated desiccator until used. The carboxyl group activator, 1,1′-carbonyldiimidazole (CDI), the purity of which must be accounted for in this synthesis (10.46 g, 64.5 mmol, Aldrich), and 4-benzoylbenzoic acid (BBA; 4.75 g, 21 mmol, Aldrich) are allowed to react in 25 ml of the anhydrous DMF in a light-shielded, 500-ml round-bottom flask at room temperature, with vigorous stirring. The ensuing reaction thickens to an opaque white mass within 15 min. A 125-ml solution of 0.03 *M* ATP (disodium salt in deionized water, Sigma, Grade I) is then added slowly with good stirring.[40] The mixture immediately evolves CO_2 (Fig. 1), but retains its opaque white appearance for the first hour of reaction. Clearing of the reaction occurs with overnight stirring to yield a yellowish, straw-colored solution. After the reaction has finished, the volume (150 ml) is reduced to about 15 ml by rotary evaporation under vacuum with very mild heating. The crude product is precipitated in the flask with about 200 ml of acetone. The off-white precipitate is repeatedly washed with acetone on Whatman #1 filter paper by Büchner funnel vacuum filtration to remove the majority of unreacted CDI and BBA. The powdered crude product on the filter is dried by vacuum desiccation and stored desiccated at −20°.

Alternatively, the 15-ml viscous solution is precipitated in the reaction flask with acetone, as described above. Then the flask is cooled on ice, the hygroscopic (somewhat sticky) fine precipitate is allowed to settle, and the acetone is decanted. Fresh acetone (100 ml) is added and the slurry is transferred to 30-ml Corex tubes (Corning) and repeatedly washed by centrifugation (10,000 *g*, 5 min, 0°). The acetone supernatant, which con-

This enzyme was found capable of catalyzing the regeneration of the BzATP employed in the F_1-ATPase assay (cf. Ref. 47), but it was necessary to add it to the assay cocktail in excess (9–10 additional units) to preclude the coupled enzyme assay itself from becoming rate limiting for the evaluation of F_1-ATPase activity (see Ref. 11).

[39] The abbreviations used are as follows: DMF, *N,N′*-dimethylformamide; BBA, 4-benzoylbenzoic acid; Bz-, any 4-carboxybenzophenone-derivatized reagent; BzATP, 3′-*O*-(4-benzoyl)benzoyl-adenosine 5′-triphosphate; CDI, 1,1′-carbonyldiimidazole; SMP, sonicated submitochondrial particles.

[40] The rationale for using a 1 : 5 DMF : water volume ratio as reaction solvent in this system has been discussed by B. P. Gottikh, A. A. Krayevsky, N. B. Tarussova, P. P. Purygin, and T. L. Tsilevich, *Tetrahedron* **26,** 4419 (1970), and reiterated by R. J. Guillory and S. J. Jeng, this series, Vol. 46, p. 259, as the means for promoting preferential derivatization of the ATP at the ribose hydroxyl functions.

tains diminishing amounts of starting materials (Fig. 1, step 1), is discarded. The resulting crude product, which has lost much of its stickiness, is dried in the presence of P_2O_5 under vacuum desiccation and stored at $-60°$ until used.

One of the benefits of the synthesis of BzATP given above is that it is accomplished via one simple reaction setup. There are no stable intermediates to be isolated and characterized en route. Radioactive BzATP (^3H and ^{32}P) is synthesized according to the same protocol from commercially obtained ATP labeled with either ^3H or ^{32}P.[11]

Purification of BzATP

Originally,[11] our purification of the crude product was performed on a light-shielded column of Sephadex LH-20 (Pharmacia Fine Chemicals) with a bed volume of 1800 ml. In this method, crude product (150–200 mg) is dissolved in ~5 ml elution buffer (0.1 *M* ammonium formate, pH 7.4), the flow rate adjusted to 2 ml/min, and the elution profile monitored at 260 nm. The fourth of five peaks (elution volume = 1620 ml) consists of the purified BzATP as determined by TLC analysis (see below).

An equally satisfactory, but significantly more rapid procedure employs a reversed-phase "flash" column chromatography method.[41,42] Two approaches have proved successful. With the first, the stationary phase sorbent consists of octadecyl (C_{18}) chains bonded to silica gel (40 μm particle size, 60 A pore size, J.T. Baker). A heavy-walled glass column (1.9 cm i.d. × 35.6 cm high) is packed with the dry bead sorbent to a height of 15 cm (42 ml bed volume), and the purified product is resolved by elution with 35% methanol : 65% 0.75 *M* ammonium formate (by volume), pH 8.5, passed through the column under 20 psi N_2 pressure at a flow rate of 20 ml/min. Crude product (30–50 mg) is mixed with a minimal volume of water, loaded on the flash chromatography column, and 5-ml fractions are collected by hand (due to the rapidity of the flow rate). The absorbance of the eluate at 260 nm is measured (Fig. 2). The peak fractions containing the purified BzATP, eluting after 250–325 ml (13–17 min), are pooled and lyophilized. Several lyophilizations are performed after redissolving the product in deionized water.

In the second method, the stationary phase sorbent is the silica-bonded octyl (C_8) moiety. Elution is performed discontinuously. First, after crude product (~50 mg) is loaded onto the column, a solvent containing 30% methanol : 70% 0.175 *M* ammonium formate (by volume), pH 7.6, is passed through, under 20 psi N_2 pressure, and both ATP and BBA

[41] W. C. Still, M. Kahn, and A. Mitra, *J. Org. Chem.* **43,** 2923 (1978).
[42] L. J. Crane, M. Zief, and J. Horvath, *Am. Lab.* **13,** 128 (1981).

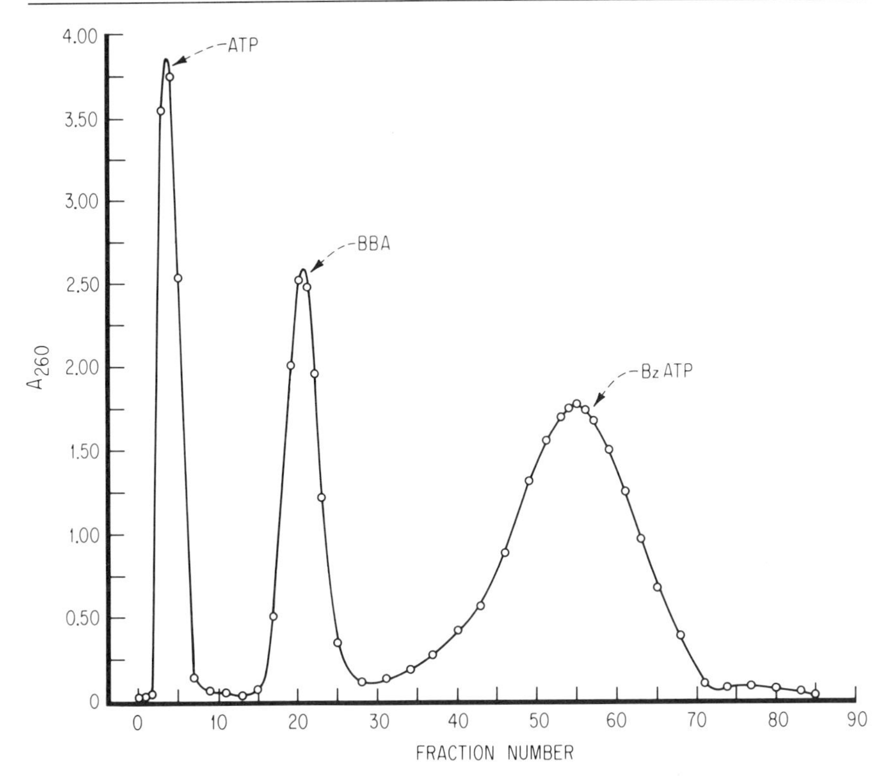

FIG. 2. "Flash" chromatography column elution profile (5-ml fractions) of a typical purification; 31 mg of crude, acetone-precipitated product was loaded onto the column. The actual elution volume for the BzATP peak is sensitive to the column pressure and may vary somewhat. We find that the ratio of the peak elution volumes corresponding to BzATP versus BBA is always between 2.6 and 3.2, despite column pressure fluctuations. Peak identification was made by TLC against standards. See text for further details.

are eluted, in that order, with about 400 ml solvent. After the BBA peak has been fully eluted (absorbance at 260–261 nm approaches the base line), the flow is temporarily interrupted. The original low salt–methanol elution solvent is quickly substituted for one containing no salt, which consists of 40% methanol : 60% water (by volume), with pH held to about 7.5. The N_2 pressure is reapplied and the BzATP is eluted in the ensuing fractions totaling 75–100 ml, which are pooled, lyophilized twice, and stored at −70°. This method results in the recovery of product essentially as the free acid.

With either the LH-20 or the flash column procedure, the yield is usually about 20% relative to the starting amount of ATP. Upon lyophilization, the purified BzATP is dried over P_2O_5 in a vacuum desiccator and stored at −70°.

Characterization of BzATP

Thin-Layer Chromatography

Two TLC procedures are used. In the original method, microcrystalline TLC plates with fluorescent indicator (Avicel, Analtech) are developed with 1-butanol/acetic acid/H_2O (5 : 1 : 3 by volume). The R_f values obtained are ATP, 0.12; BzATP, 0.63; benzoylbenzoic acid, 0.81. In the second procedure, reversed-phase TLC on 1 × 3 in. MKC_{18} F plates (Whatman, 200 μm thickness) are developed with a buffer containing 55% methanol : 45% 0.75 *M* ammonium formate (by volume), pH 8.5. Analysis of the crude product with this reversed-phase system reveals three spots: BzATP (R_f = 0.16); benzoylbenzoic acid (R_f = 0.30); and unreacted ATP (R_f = 0.62). Occasionally, a fourth spot, R_f = 0.58, indicative of a small amount of ATP dephosphorylation to ADP, may be observed. If, prior to developing the TLC plate (with either procedure), the BzATP at the spotting origin is wetted with water and exposed for a few minutes to long-wavelength irradiation from a UVSL-25 mineralight (Ultraviolet Products), a substantial portion of the spotted material remains immobilized at the origin subsequent to development, with no evidence of breakdown products observed. This provides preliminary assurance that the BzATP product is covalently photoreactive. We generally run TLC analyses in pairs: one plate having been exposed to actinic UV light as indicated, and the other run normally for verification of R_f values.

pH Stability of BzATP

Purified BzATP (0.01 *M*) was incubated for 30 min (25°) over a broad pH range, 4–10, in 10 m*M* Tris–maleate. Subsequent TLC analysis, with ATP and 4-benzoylbenzoic acid as standards, showed no evidence of ester hydrolysis.

Elemental Analysis

Commercial elemental analysis of the purified BzATP (as the ammonium salt) confirmed a unit stoichiometric addition of the benzophenone moiety to each mole of ATP in the final product. The results were as follows: Calculated: C 36.60; H 4.36; N 12.45; P 12.97; Experimentally determined: C 36.30; H 5.10; N 12.40; P 12.63.

Proton NMR

Proton NMR spectra[12] of purified BzATP in D_2O indicate two regions of interest, each of which corresponds to the NMR profile of pure ATP and benzoylbenzoic acid, respectively. One region is identifiable with exchangeable protons on the ribosyl substituent of ATP between about 4 and 6 ppm. A second (downfield) region between 7.1 and 8.5 ppm corresponds to exchangeable protons affiliated with aromatic (benzophenone and purine) substituents. All resonance peaks can be accounted for by comparison with ATP and benzoylbenzoic acid, except for a chemical shift from about 4.3 ppm to nearly 6.25 ppm, observed with BzATP but not with ATP, which is attributable to a substitution at the 3′ position of the ribose moiety.[6]

UV Absorbance

The λ_{max} for BzATP in phosphate buffer, pH 7.0, is 261.5 nm. As indicated earlier, we have also found it possible to purify the BzATP product from the flash chromatography column under low- to no-salt conditions by employing a discontinuous elution solvent protocol. The BzATP thus obtained, upon lyophilization, may be taken to be fully protonated and assumed to contain 2 mol of bound H_2O (MW = 751). In 10 m*M* KH_2PO_4 buffer, pH 7.0, the ε_M at 261 nm for the purified, salt-free BzATP was determined to be $3.192 \pm 0.137 \times 10^4\ M^{-1} \times cm^{-1}$, the value we use in our enzyme studies. It should be noted, however, that purification of BzATP in high salt (0.75 *M* ammonium formate/methanol, see Fig. 2) seems to help preserve the stability of the lyophilized product upon storage.

General Remarks

It is difficult to state unequivocally that BzATP exists mainly as the 3′-*O*-substituted derivative, but the evidence at hand appears to indicate that this is the case. Although the 2′-hydroxyl is probably preferred as the kineticially favored esterification site, the 3′-substituted isomer is thermodynamically more stable, and it is reasonable to conclude that acyl migration occurs during the synthetic reaction.[43,44] A small amount of 2′-isomer contamination would probably not be detectable by our analytical measurements, and indeed, for experiments usually performed with BzATP, the adenine ring system and phosphate groups are the more significant

[43] P. G. Zamecnik, *Biochem. J.* **85,** 257 (1962).

[44] C. S. McLaughlin and V. M. Ingram, *Biochemistry* **4,** 1442, 1448 (1965).

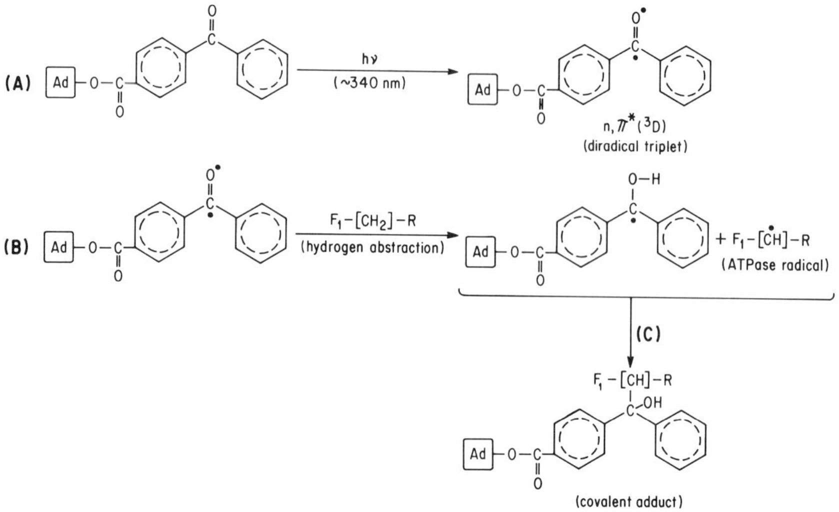

FIG. 3. Probable reaction pathway for covalent photoaffinity labeling of F_1-ATPase by BzATP.

elements correlating with affinity of the probe/substrate for the ATP-utilizing enzyme.

Photolabeling ATPase Enzymes with BzATP

Mechanism

The $n \rightarrow \pi^*$ absorption band for BzATP has a wavelength maximum at about 350 nm in water, with a very weak extinction coefficient ($\leq 160\ M^{-1}\ cm^{-1}$) characteristic of this type of electronic transition.[18] Upon irradiation with wavelengths >340 nm, the excited singlet state of benzophenone rapidly ($\simeq 10^{-11}$ sec) undergoes intersystem crossing to the metastable triplet state with nearly 100% efficiency. This triplet state posseses diradicaloid characteristics and may undergo productive photolabeling by means of the postulated two-step sequence given in Fig. 3: (1) hydrogen atom abstraction from the target molecule to yield a ketyl radical of the benzophenone plus a target molecule free-radical species; (2) radical-radical coupling of the ketyl and target species intermediates to generate a covalent C—C bond between the components. The requirement for irradiation of benzophenone derivatives at about 350 nm (≤82 kcal/mol) nearly eliminates the potential for photodestruction of proteins, which may occur at shorter wavelengths.

Methods

For our early work,[11,12] we used the hand-held, low-intensity UVSL-25 mineralight (Ultraviolet Products), employing the long-wavelength setting. A Pyrex filter placed over the illuminating surface ensures transmission of only wavelengths above 300 nm. The low-intensity radiation emission provided by this lamp proved sufficient to yield good photoincorporation of the BzATP probe, obviating the requirement for a more elaborate apparatus. However, an alternative and more efficient irradiation setup is provided by a high-pressure HBO 200 W mercury arc lamp (Osram), with Universal lamp housing and a model MTr 14 power supply (Wild-Heerbrugg). This apparatus possesses a very high-intensity light source, especially around 334 and 366 nm, and is equipped with an iris/diaphragm that can be closed down to irradiate a relatively small area (about 1 cm in diameter at a distance of 30 cm), allowing for shorter sample irradiation times (rarely exceeding 1 min). It should be emphasized, however, that under appropriate conditions, either of these light sources yields good photoincorporation of BzATP.

Our photolysis incubations with purified rat liver F_1-ATPase and sonicated submitochondrial particle (SMP) preparations were performed in 1-cm, 3-ml quartz fluorescence cuvettes, positioned 5 cm from the UVSL-25 mineralight. The effector molecule (BzATP, ATP, or BBA) was added to the cuvette 30 sec prior to irradiation. Photolysis time was 10 min or less. With SMP (1.25 mg/ml), incubations were performed at 10° during photolysis, and after irradiation the particles were sedimented at 140,000 *g* (45 min). The pellet was rinsed twice with cold deionized water to remove unbound ligand, then resuspended in cold deionized water to 20 mg/ml and assayed for ATPase activity. With the soluble F_1-ATPase, the temperature was maintained at 25° during irradiation, after which the reaction was dialyzed against 250 m*M* potassium phosphate, 5 m*M* EDTA, pH 7.5, to remove unbound ligand. The dialysate was concentrated by application of dry Sephadex G-25 to the outside of the dialysis tubing before being assayed for ATPase activity.

SMP preparations and SMP-ATPase assays were accomplished via the methods of Kaplan and Coleman.[45] Rat liver mitochondrial F_1 was prepared essentially according to Catterall and Pedersen.[46] The isolation medium prior to sonication contained 4 m*M* ATP. Following the concentration of the F_1 enzyme after purification, it was lyophilized in 250 m*M* potassium phosphate, 5 m*M* EDTA, pH 7.5, and stored at −60°. F_1-

[45] R. S. Kaplan and P. S. Coleman, *Biochim. Biophys. Acta* **501,** 269 (1978).

[46] W. A. Catterall and P. L. Pedersen, *J. Biol. Chem.* **246,** 4987 (1971).

TABLE I
KINETIC CONSTANTS FOR THE ATPase OF SMP AND F_1 IN THE ABSENCE OF ILLUMINATION[a]

System	Substrate	K_m (mM)	V_{max} (μmol/min/mg)	K_I^{ADP} (mM)
SMP	ATP	0.16	3.20	0.61
	BzATP	0.13	0.36	0.07
F_1	ATP	0.83	20.71	0.36
	BzATP	0.94	2.51	0.06

[a] With SMP, medium contained 10 mM Tris–maleate (pH 7.2), 1.0 mM $MgCl_2$, ATP or BzATP (0.05–1.0 mM), ADP (0.05–1.0 mM). When ATP was substrate, incubations contained 0.50 mg SMP protein/ml. Reactions were run for 10 sec at 28° and terminated with 6% (v/v) perchloric acid. With isolated F_1, the assay was performed essentially as described according to Ref. 48 with ATP or BzATP concentrations = 0.3–2.0 mM. When ATP was substrate, incubations contained 1.5 μg F_1 protein/ml. When BzATP was substrate, incubations contained 3.0 μg F_1 protein/ml plus an additional 9 units of pyruvate kinase. Assays for competition studies were performed essentially as described in Ref. 47 in a total volume of 1.0 ml that contained 50 mM Tris–HCl (pH 7.5), varying concentrations of both ADP and either ATP or BzATP, a $MgCl_2$ concentration equal to total adenine nucleotide concentration, and 9 μg purified F_1-ATPase. Reactions were run at room temperature for 2 min.

ATPase activity was measured with an ATP regenerating assay system according to established procedures.[46,47]

Results with Rat Liver Mitochondrial F_1-ATPase[11]

BzATP as Substrate

Studies with both SMP-ATPase and the soluble F_1 established BzATP as a functional substrate for the ATPase enzyme complex in the absence of actinic illumination (Table I). The apparent K_m values for BzATP relative to ATP are identical with either the membrane-bound or soluble enzyme. Such results may imply a similar binding affinity for both substrates at the catalytic site(s). The V_{max} with BzATP as substrate for both enzyme preparations, although substantially decreased compared to control ATPase activity, is still 11–12% that with ATP. The rate of hydrolysis of both ATP and BzATP (to ADP and BzADP, respectively) is effectively inhibited by ADP with both the SMP and soluble ATPase; however, different K_I values for ADP were obtained, depending on the use of ATP or

[47] M. E. Pullman, H. S. Penefsky, A. Datta, and E. Racker, *J. Biol. Chem.* **235,** 3322 (1960).

TABLE II

CONTROLS FOR BzATP PHOTOLABELING WITH SMP-ATPase AND F_1-ATPase[a]

Conditions		% ATPase activity remaining	
Illumination	System variable	SMP-ATPase	F_1-ATPase
1. No	4 μmol BzATP/mg	99.8	98.1
2. Yes	4 μmol ATP/mg	95.7	95.6
3. Yes	2.5% (v/v) ethanol	97.0	93.2
4. Yes	4 μmol BBA/mg in 2.5% (v/v) ethanol	94.2	91.0
5. Yes	4 μmol BzATP/mg plus 4 μmol ATP/mg	—	91.0
6. Yes	4 μmol BzATP/mg	23.0	34.1

[a] Experiments were performed according to the methods given in the text and Ref. 11. Photolysis time was always 10 min (UVSL 25 lamp).

BzATP as substrate. The latter is an interesting result that merits further investigation, for it may comment, indirectly, upon mechanistic restrictions that apply to ATP hydrolysis by this ATPase.

Photoinactivation of ATPase by BzATP

Photoinactivation experiments require that stringent controls be performed to demonstrate unequivocally that a loss in enzyme activity occurs as a direct result of specific photoincorporation of the analog at the

TABLE III

INHIBITION OF MITOCHONDRIAL ATPase BY PHOTOAFFINITY LABELING WITH BzATP[a]

	SMP-ATPase		F_1-ATPase	
Photolysis conditions	% Inhibition	K_m^{ATP} (mM)	% Inhibition	K_m^{ATP} (mM)
1. ATP plus Mg^{2+}	0	0.16	0	0.83
2. BzATP minus Mg^{2+}	51	0.13	—	—
3. BzATP plus Mg^{2+}	77	0.12	70	0.84

[a] For SMP experiments (1.25 mg protein), the BzATP concentration was 4.0 μmol/mg SMP protein. Photolysis was performed for 10 min at 10° in 1.0 ml. ATP concentration range for SMP-ATPase assay subsequent to photolysis was 0.3–1.0 mM. For soluble F_1 experiments (0.15 mg protein), the BzATP concentration was again 4.0 μmol/mg F_1 protein. Photolysis conditions were identical to those with SMP, except the temperature was 25°. The ATP concentration range for the F_1-ATPase assay subsequent to photolysis was 0.3–2.0 mM. See text and Williams and Coleman[11] for further details.

ATPase catalytic site. All other irrelevant processes, such as nonspecific labeling, interference by any unanticipated photogenerated intermediates different from BzATP, and photodestruction of the protein, must be ruled out as the mechanism behind the observed effect. The results of these control experiments are shown in Table II. Both the SMP and soluble F_1-ATPase, irradiated in the absence of effector, indicate little loss in activity. No evidence was obtained for the ability of BzATP to form an inhibitory complex in the dark with either membrane-bound or soluble enzyme, inasmuch as no loss of ATPase activity accompanied such incubations. When either SMP or soluble enzyme was irradiated with only the photolabile moiety of the BzATP, i.e., 4-benzoylbenzoic acid (BBA), virtually no loss in subsequently assayed ATPase activity was observed. This is an important finding, for it demonstrates that the triplet benzophenone itself is not site-specifically directed with respect to the enzyme. Therefore, only a photochemical reaction at the ATP binding site of the enzyme (and not elsewhere) is expected to be inhibitory. A substrate protection experiment conclusively supports this interpretation (Table II); the presence of equimolar ATP and BzATP together during irradiation of the enzyme leads to a $>90\%$ retention of ATPase activity.

Table III summarizes the following results. Irradiation of SMP with BzATP, without added Mg^{2+}, yielded a 51% decrease in the subsequently assayed V_{max} of the ATPase, and a 77% decrease in the V_{max} when Mg^{2+} was present during photolysis. Soluble F_1, irradiated with BzATP, followed by removal of unbound ligand from the system (via dialysis) and kinetics assays over a range of ATP concentrations, showed an unaltered K_m despite a V_{max} dramatically reduced by 70%. Collectively, these data support the catalytic site selectivity of BzATP on the rat liver mitochondrial ATPase complex, whether the enzyme is membrane bound or soluble.

[69] Use of ADP Analogs for Functional and Structural Analysis of F_1-ATPase

By GÜNTER SCHÄFER, UWE LÜCKEN, and MATHIAS LÜBBEN

General Considerations

The use of substrate analogs which can bind without undergoing catalytic conversion or can be used as covalent markers or reporter molecules at the catalytic domains of proteins is a classical approach to enzyme structure and function. This approach has been applied to mitochondrial F_1-ATPase in this chapter to elucidate nucleotide interactions and kinetic properties of the enzyme, although its molecular mechanism is not yet fully understood. The structure of several ATP synthases has been determined based on primary amino acid sequences as reported in a series of reviews.[1-3] A continuing interest in molecular probes with well-defined chemical, physical, and biochemical properties exists. The analogs described have been used with membrane-bound F_1 of submitochondrial particles (SMP), thylakoid membranes, and isolated F_1 from beef heart mitochondria.

The basic requirements for an analog are not only that its structure closely resembles that of the natural substrate, but also that binding occurs to the proper catalytic site with comparable affinity. In this respect modifications of the purine moiety are a critical manipulation because specificity of enzyme–substrate recognition largely depends on this part of the molecule. In the case of adenine nucleotides, for example, introduction of a C_8 substituent leads to an analog unable to assume the favorable *anti*-conformation of the purine ring relative to the ribose moiety, a situation which drastically lowers the affinity of these derivatives to F_1-ATPase compared to the natural substrates. Surprisingly, at the ribose moiety of adenine nucleotides, even large substituents are tolerated by F_1-type ATPases without loss of affinity. In several cases even much higher affinities than with ADP or ATP could be found.[4,5]

A comprehensive description of chemical modifications has been given by Scheit,[6] including substitutions at the ribose ring. While these usually require multistep synthetic procedures in order to avoid a configurational transition, simple derivatizations are achieved by esterification of the 3′,2′-hydroxyls in the sugar. The first analog of this type, of interest for the study of F_1-ATPases, was described by Guillory and Jeng[7] in a photoaffinity labeling approach. The kinetic properties of these 3′(2′) esters, however, were only insufficiently determined, and it was studies from our laboratory[4,8] which for the first time reported on a systematic investigation of 3′ esters as inhibitors of F_1-type ATPases. The general structure of these analogs is given in Fig. 1.

[1] M. Futai and H. Kanazawa, *Microbiol. Rev.* **47,** 285 (1983).
[2] A. E. Senior and J. Wise, *J. Membr. Biol.* **73,** 105 (1983).
[3] A. E. Senior, *Biochim. Biophys. Acta* **726,** 81 (1983).
[4] G. Schäfer and G. Onur, *Eur. J. Biochem.* **97,** 415 (1979).
[5] Ch. Grubmeyer and H. Penefsky, *J. Biol. Chem.* **256,** 3118 (1981).
[6] K. H. Scheit, "Nucleotide Analogs." Wiley, New York, 1980.
[7] S. J. Jeng and R. J. Guillory, *J. Supramol. Struct.* **3,** 448 (1975).
[8] G. Onur, G. Schäfer, and H. Strotmann, *Z. Naturforsch.* **38c,** 49 (1983).

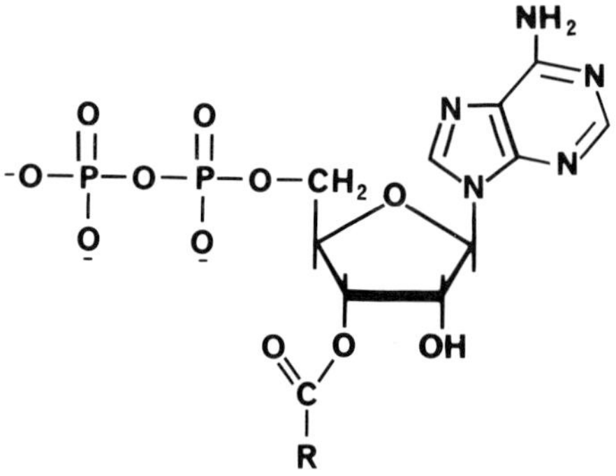

FIG. 1. General structure of the analogs.

A wide range of compounds can be used in kinetic studies of oxidative phosphorylation or photophosphorylation. Table I gives an overview of 3′ esters having different capabilities as energy-transfer inhibitors. The half-maximal inhibitory concentrations, c_i [50], in the presence of 100 μM ADP are shown. The ATP or AMP esters which are prepared for comparison with ADP esters are about 10 times less effective (ATP derivatives) or ineffective (AMP derivatives) as energy-transfer inhibitors.

All analogs of this type behave as extremely strong inhibitors of oxidative phosphorylation and photophosphorylation. Their inhibitory effect on isolated F_1-ATPases is generally about 10–100 times less. Therefore, they are considered to be conformation-specific probes, acting differently on the enzyme in its nonenergized form (isolated F_1), and its energized, ATP-synthesizing form in the membrane-bound state.[8a,9]

Another important property of the analogs is the fact that they bind strongly to F_1-ATPases, but are not, or only negligibly hydrolyzed (in the case of the ATP derivatives) or phosphorylated (ADP derivatives). The usefulness of the 3′ esters in kinetic studies relies on this fact which allows reversible binding to be investigated in the absence of catalysis. In addition, variations in length and structure of the 3′-carboxyl side chain offer the possibility of mapping nucleotide-binding regions in proteins.

In general, these analogs provide high flexibility for the design of probes by proper choice of the 3′-acyl residue. Aliphatic, aromatic, or araliphatic acyls can be introduced, which can be used as carriers for covalently reacting groups like azido functions, fluorescent groups like dimethylaminonaphthoyl, or spin-labeled residues. Their synthesis and applications are described in the article.

[8a] G. Schäfer, G. Onur, and M. Schlegel, *J. Bioenerg. Biomembr.* **12,** 213 (1980).
[9] G. Schäfer, *FEBS Lett.* **139,** 271 (1982).

TABLE I
STRUCTURE OF 3'-*O*-ACYL DERIVATIVES OF ADP AND INHIBITORY ACTIVITY IN OXIDATIVE PHOSPHORYLATION AND PHOTOPHOSPHORYLATION[a]

Compound no.	Side chain of 3'-acyl: R—	c_i [50]; half-maximal inhibitory concentration [μM] Photophosphorylation	Oxidative phosphorylation
1	$-CH_3$	85	55
2	$-C_5H_{11}$	20	1.7
3	$-C_6H_{13}$	—	2.7
4	$-C_7H_{15}$	41	1.7
5	$-CH_2-C(CH_3)_3$	22	1.5
6	$-C_6H_5$ (phenyl)	26	6.0
7	$-CH_2-C_6H_5$	9	3.6
8	$-(CH_2)_3-C_6H_5$	12	1.3
9	$-(CH_2)_2-C_6H_3(NO_2)-N_3$	20	1.3
10	$-(CH_2)_3-C_6H_3(NO_2)-N_3$	43	—
11	$-(CH_2)_3-NH-C_6H_4-NO_2$	2.5	0.76
12	$-(CH_2)_3-NH-C_6H_4(NO_2)$	4.5	0.55
13	$-(CH_3)_3-NH-C_6H_3(NO_2)-N_3$	6	2.0

(*continued*)

TABLE I (*continued*)

Compound no.	Side chain of 3′-acyl: R—	c_i [50]; half-maximal inhibitory concentration [μM]	
		Photophosphorylation	Oxidative phosphorylation
14	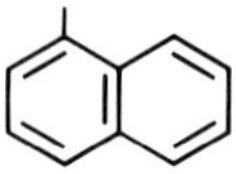	0.3	0.35
15	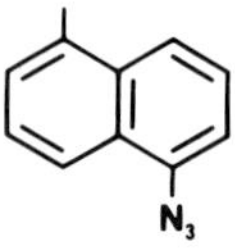 N_3	—	0.4
16	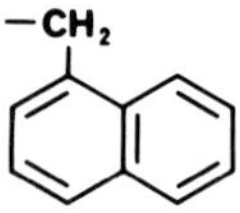 $-CH_2$	—	0.86
17	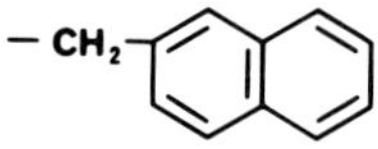 $-CH_2$	—	0.9
18	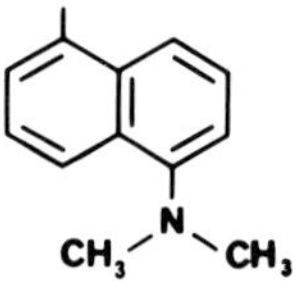 CH_3 N CH_3	0.4	0.6
19	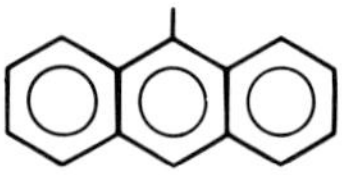	—	5.9
20	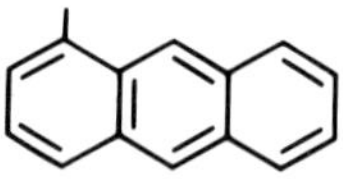	0.6	0.56
21	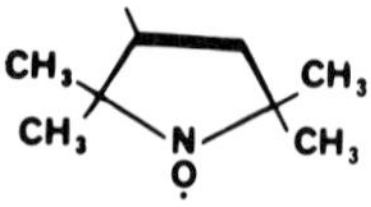 CH_3 CH_3 CH_3 CH_3 N O	—	1.6

[a] Biochemical data are compiled from Schäfer and Onur,[4] Onur *et al.*,[8] and Schäfer *et al.*[8a] Some newly measured values were determined according to the methods described in the respective references.

Materials and General Methodology

All chemicals were obtained from commercial sources in the highest degree of purity. Thin-layer plates used were either 0.2 mm silica gel (Merck 60 F254; No. 5554) or 0.1 mm microcrystalline cellulose on glass plates (Merck 5716), from which the bands could be easily scraped off and extracted. Isotopically labeled compounds (phenyl[^{3}H]propionic acid, and ^{14}C-labeled nucleotides) were purchased from Amersham–Buchler (Braunschweig).

Submitochondrial particles (SMP) were prepared from beef heart mitochondria by sonication according to standard procedures,[4] except that sonication was carried out in 1-sec pulses for a total of 10–20 sec, applied to a total sample volume of 10 ml (protein 30 mg/ml), using a Branson sonifier at maximal power output. Beef heart F_1 was isolated from SMP as described by Penin *et al.*[10] for the pig heart enzyme, with slight modifications as noted by Tiedge *et al.*[11] The resulting enzyme is devoid of endogenous nucleotides and stored in 50% (v/v) glycerol/2 m*M* EDTA, 25 m*M* Tris, pH 8. ATPase activity was determined as described by Vogel and Steinhardt[12] using a coupled enzymatic assay to monitor NADH absorbancy. Oxidative phosphorylation with SMP was measured by ^{32}P incorporation into ATP as described in detail elsewhere.[3]

Chloroplasts were isolated from spinach leaves according to Strotmann *et al.*[13] and photophosphorylation was measured by standard methods.[14]

AC64 membranes (Schleicher & Schüll) were used to measure AD(T)P binding to F_1 by pressure filtration. For elution centrifugation (centrifuged column method) determinations Sephadex G-50 fine was used as gel matrix. For aliphatic compounds a buffer containing 20 m*M* MOPS, 5 m*M* $MgCl_2$, 125 m*M* KCl, pH 7, is recommended. With aromatic compounds Tris, pH 8, replaces MOPS. It has to be stressed that only with AD(T)P is pressure filtration without problems. Due to unspecific binding none of the analogs to be described can be filtered through cellulose nitrate membranes. N-ADP and ANA-ADP also stick to cellulose acetate membranes, but less firmly. In the concentration range studied nonspecific binding to cellulose acetate is 12% with DMAN derivatives, 6% with ANP derivatives, and 4% with ADP. In centrifuged column experiments BSA as a tracking protein has to be omitted because it also binds nonspecifically the 3′-ester derivatives, especially those with aromatic residues.

[10] F. Penin, C. Godinot, and D. Gautheron, *Biochim. Biophys. Acta* **548,** 63 (1979).
[11] H. Tiedge, U. Lücken, J. Weber, and G. Schäfer, *Eur. J. Biochem.* **127,** 291 (1982).
[12] G. Vogel and R. Steinhart, *Biochemistry* **15,** 208 (1976).
[13] H. Strotmann and S. Bickel-Sandkötter, *Biochim. Biophys. Acta* **460,** 126 (1977).
[14] U. Franek and H. Strotmann, *FEBS Lett.* **126,** 5 (1981).

NMR spectra were recorded using PF-NMR on a Varian XL-200. Chemical shifts are given in ppm and coupling constants J in Hz. Solvent systems empolyed for carboxylic acids were DMSO-d_6 (totally deuterated dimethyl sulfoxide), 40°; for 3′(2′) esters of nucleotides DMSO-d_6 with 1–4% D_2O (v/v). The notation describing the properties of resonances attributed to distinct protons is given in parentheses using s for singlet; d, doublet; dd, double doublet; t, triplet; m, multiplet. Generally the higher value of the coupling constants corresponds to the next neighbor (aromatic *o*-protons). In some cases where no attribution of protons is given, the number of protons located at the indicated chemical shift is given as the first figure in parentheses.

Synthetic Procedures

The general instructions given for the synthesis of the probes mentioned above apply as well to preparation of ADP, ATP, and AMP analogs. Although the necessary organic precursor acids can usually be obtained in high yield, the overall chemical yield of 3′ esters is often rather low, rarely reaching 20%.

The low yields result from the fact that depending on the conditions, several products are formed in the esterification step, ranging from the monoesters (2′- and 3′-) to the diester and the α-phosphate acylated nucleotide. These products can be separated by thin-layer chromatography. It is advisable that the identity of the isolated products is carefully determined by NMR and UV spectroscopy as well as phosphate analysis prior to use. For radiolabeled analogs synthesis has to be conducted in microscale preparations, which are given in addition to the usual procedures. Radioactive labeling is normally performed using ^{14}C- or ^{3}H-labeled AD(T)P. Introduction of radioactivity into the 3′-acyl residue in some cases is difficult because the starting material is not commercially available in labeled form. Only the synthesis of some newly introduced precursor acids is described in more detail. Esterification procedures are given at the end of this chapter.

Synthesis of Precursor Acids

The precursor acids of compounds **1–8** (Table I), **14**, **16**, and **19** are available commercially. Most of them can also be obtained in isotopically labeled form. Compounds **11–13** are included in Table I for comparison of their biological activity. Their synthesis has been described in full by

Guillory and Jeng[15] and has been verified in our laboratory and by Vignais[16a,b,c] *et al.*

Other precursors have been synthesized as described below.

N-(2-Nitrophenyl)4-aminobutyric Acid. 4-Aminobutyric acid [0.631 g (6 mmol)] and 1.5 g Na_2CO_3 (12 mmol) are dissolved in 12 ml H_2O and mixed with a solution of 0.720 g (5 mmol) 2-fluoronitrobenzene. The mixture is stirred under reflux at 75° for 12 hr in thc dark. Then the volume is reduced to one-third under rotation and 15 ml H_2O are added; after filtration the solution is extracted twice with 50 ml diethyl ether. The aqueous phase is acidified with HCl to pH 2 and extracted 3 times with 100 ml ethyl ether. The ether phase is washed twice with saturated NaCl solution and dried with $MgSO_4$. After evaporation of the ethyl ether, the residue is recrystallized from toluene. Then 0.65 g of product (60%) is obtained as yellow-orange crystals; mp = 115–118°; λ_{max} 233, 278, and 426 nm; ε_{426} = 7124. Purity is checked on *TLC* plates (Merck silica gel F254) with dichloromethane/methanol 9 : 1 (v/v).

N-(4-Nitrophenyl)4-aminobutyric Acid. This compound is synthesized exactly as described for the 2-nitrophenyl derivative, using 4-fluoronitrobenzene as the starting material. This procedure is used to synthesize the respective derivatives from 3-aminopropionic acid.

Anthracene-1-carboxylic Acid. This compound is not available commercially. Its esterification with AD(T)P was performed in order to synthesize a fluorescent analog. Preparation of the acid is possible from 1,2,3-benzenetricarboxylic acid according to the classical publications by Gräbe and Leonhardt[17] and Gräbe and Blumenfeld.[18]

1,2,3-Benzenetricarboxylic acid (10 g) (47.8 mmol) and 10.85 g (52.6 mmol) of DCCI are mixed with 280 ml of dry acetone and stirred under moisture-free conditions for 6 hr. The solution is separated from the precipitate, which is washed with acetone several times. The combined filtrates are evaporated to dryness. The white residue is dried at 110–120°. The white product is pure 3-carboxyphthalic anhydride, mp 187° (lit[17,18] 196°). It is chromatographically pure (silica gel plates, Merck F254; toluene/acetic acid 4 : 1 v/v) with a yield of 84–95%.

The anhydride can also be formed by heating 1,2,3-benzenetricarboxylic acid to 200°. Most of the anhydride sublimates, however.

[15] J. R. Guillory and S. J. Jeng, this series, Vol. 46, p. 259.

[16a] J. Lunardi, G. J. M. Lauquin, and P. V. Vignais, *FEBS Lett.* **80,** 317 (1977).

[16b] J. Lunardi, M. Satre, and P. V. Vignais, *Biochemistry* **20,** 473 (1981).

[16c] J. Lunardi and P. V. Vignais, *Biochim. Biophys. Acta* **682,** 124 (1982).

[17] C. Gräbe and M. Leonhardt, *Ann. Chem.* **280,** 221 (1896).

[18] C. Gräbe and S. Blumenfeld, *Ber. Dtsch. Chem. Ges.* **30,** 1115 (1897).

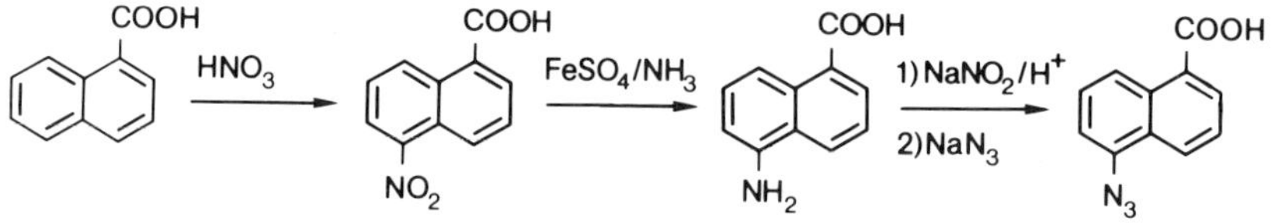

COOH $\xrightarrow{HNO_3}$ COOH, NO_2 $\xrightarrow{FeSO_4/NH_3}$ COOH, NH_2 $\xrightarrow[2)\,NaN_3]{1)\,NaNO_2/H^+}$ COOH, N_3

SCHEME 1. Synthesis of 5-azido-1-naphthoic acid.

Condensation with benzene is performed by heating 8.6 g (44.76 mmol) of 3-carboxyphthalic anhydride with 172 g dry benzene for 30 min (reflux condensor fitted with $CaCl_2$ tube). After cooling to room temperature, 9.46 g $AlCl_3$ (70.95 mmol) are quickly added and the mixture is refluxed again for exactly 25 min. An orange-brown mixture with a light-colored precipitate forms. After cooling to room temperature, the whole mixture is poured into 400 ml ice water and 5 ml concentrated HCl are added.

The mixture is placed in the refrigerator overnight; a precipitate forms which is filtered off and the aqueous phase is concentrated to one-fourth of the original volume and kept at 4°. A white precipitate with mp 248–250° forms and is separated. The solution is adjusted to pH 3 using Na_2CO_3 and allowed to stand overnight; the precipitate remaining after this treatment is also collected. The combined precipitates are recrystallized from water. The raw yield is about 51%; after recrystallization 3.05 g (25%) of pure white product is collected [mp 229°; pure in TLC on silica gel plates (Merck F254) with toluene/acetic acid (4 : 1 v/v)].

Cyclization of the benzoyl derivative to the anthraquinone is performed using 3 g (11 mmol) of the derivative to which 30 g concentrated sulfuric acid is added, resulting in formation of a yellow solution. The mixture is kept at 140–150° for 10 min and after cooling poured into 200 ml ice water. The bright yellow precipitate is recrystallized twice from ethanol [mp = 294–295° (lit 294°); purity is checked on TLC plates (Merck silica gel 254 F254) with toluene/acetic acid 4 : 1 v/v].

Reduction of the anthraquinone is achieved with $ZnCl_2$ in ammonium hydroxide. To 1 g of the preceding intermediate product (11.8 mmol) 2.8 g Zn powder and 28 ml 2 *N* NH_3 are added and the mixture is refluxed for 30–40 min. After cooling and filtration (filter washed several times with diluted NH_3), the solution is acidified with HCl. The free acid precipitates as yellow crystals; after standing for 2 hr at 0–4°, the product is collected by filtration and recrystallized from ethanol. The yellow crystals show a strong blue fluorescence in UV light. [Yield 53%; mp 233° after first recrystallization (lit 245°); λ_{max} excitation 385 nm, emission 484 nm.]

5-Azido-1-naphthoic Acid (ANA). The synthesis of 5-azido-1-naphthoic acid from 1-naphthoic acid is given in Scheme 1. The 5-amino acid is synthesized via 5-nitro-1-naphthoic acid and is then converted to the 5-azido compound. This acid is the precursor for a photoaffinity analog of AD(T)P to naphthoyl-AD(T)P, which is an especially interesting ligand for the high-affinity sites of F_1-ATPase from mitochondria and chloroplasts.

Using the method of Bell and Morgan,[19] 17.2 g (0.1 mol) 1-naphthoic acid are heated under reflux with 110 ml concentrated HNO_3 and 50 ml H_2O for 3 hr. Then the mixture is poured into 1 liter of ice water and filtered. The residue is extracted with 200 ml saturated hot Na_2CO_3 solution and separated from undissolved material. The solution is acidified at room temperature to pH 4 with concentrated HCl, producing a yellow precipitate of isomeric nitro compounds. The 5-nitro derivative is obtained by fractionated crystallization. The dried precipitate is dissolved in ethanol, boiled for a short time, and then kept at 60° for 3–4 hr. 5-Nitro-1-naphtoic acid slowly crystallizes. Yield after recrystallization 25%; R_f = 0.45–0.5 in toluene/acetic acid 4 : 1 (v/v) on silica gel cards; UV (dioxane) λ_{max} 259 nm, ε = 26200; 326 nm, ε = 5420; NMR: H_8 9.17 (dd, J = 8; 1), H_4 8.43 (dd, J = 8; 1), H_6 8.28 (dd, J = 7; 1), H_2 8.24 (dd, J = 7; 1), H_3 7.81 (t, J = 7; 8), H_7 7.78 (t, J = 7; 8).

5-Nitro-1-naphthoic acid [3.7 g (17 mmol)] are dissolved in 30 ml 2 *N* NH_3 and added in small aliquots to 100 ml of hot 0.9 *M* $FeSO_4$ solution. After addition of 70 ml 25% NH_3, the mixture is boiled for another 10 min under reflux and filtered. The filtrate is acidified with concentrated HCl to pH 4, and the precipitating product is collected on a filter by suction. Recrystallization from a large volume of water yields 40% of 5-amino-1-naphthoic acid. [R_f = 0.31 on silica gel cards with toluene/acetic acid 4 : 1 (v/v); UV (methanol, 0.5% 2 *N* NH_4OH) λ_{max} 218 nm, ε = 24790, 244 nm, ε = 12100; 330 nm, ε = 3700; NMR: H_8 8.37 (dd, J = 8; 1), H_4 and H_2 7.96 (2, dd, J = 8;1 / 7;1), H_3 7.39 (dd, J = 8; 7), H_7 7.28 (dd, J = 7; 8), H_6 6.72 (dd, J = 7; 1).]

5-Amino-1-naphthoic acid [0.75 g (4 mmol)] are diazotized by suspension in 10 ml 4.1 *M* HCl and dropwise addition of 1.6 ml 2.5 *M* $NaNO_2$ in an ice bath. After dilution with 100 ml H_2O, the solution is filtered and further steps carried out in the dark. Then 1 ml 4 *M* NaN_3 is added dropwise (slowly) and the mixture is allowed to rest on ice for 30 min. The precipitate is collected by suction and washed with ice-cold water. The yield is 80% [R_f = 0.54 on silica gel with toluene/acetic acid 4 : 1 (v/v), or R_f = 0.33 with dichloromethane/methanol 5 : 1 (v/v); UV: (methanol, 0.2% 2 *N* NH_4OH) λ_{max} 229 nm, ε = 44000; 306 nm, ε = 94000; λ_{max} 304 nm in 0.1 *N* NaOH, ε = 10,000; λ_{max} = 313 nm in methanol, ε = 8420; IR:

[19] F. Bell and W. D. H. Morgan, *J. Chem. Soc.* 1716 (1954).

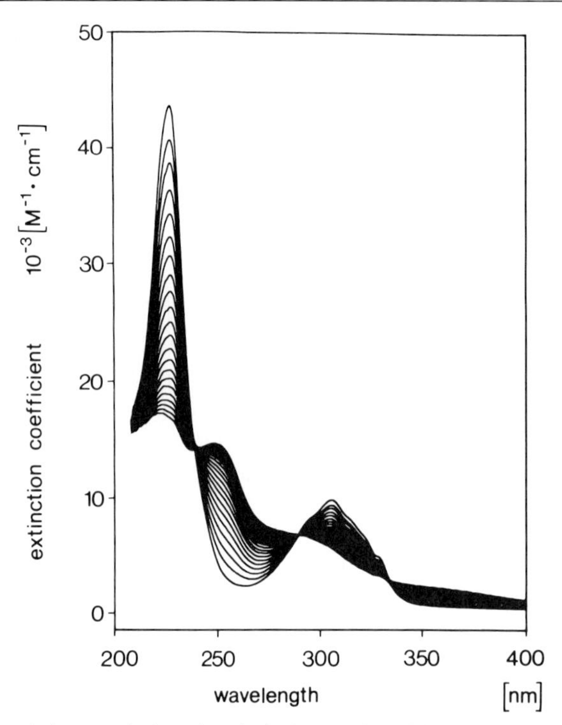

FIG. 2. Spectral changes during photolytic decay of 5-azido-1-naphthoic azid. A 27-μM solution in MeOH with 0.005% NH_3 was irradiated directly in the stoppered quartz cell between recordings of spectra, using a 4W HANAU UV handlamp from a distance of 2.5–3 cm. Spectra were recorded in a Hewlett-Packard 8450 diode-array spectrophotometer after each of the 3-sec irradiation periods. Recording of a spectrum required only 1 sec.

2110 cm^{-1} (azido group), 1690 cm^{-1} (carboxyl); NMR in DMSO-d_6: H_8 8.66 (1, dd, J = 1.0, 8.0), H_2 8.26 (1, dd, J = 1.0, 8.0), H_4 8.19 (1, dd, J = 1.0, 7.0), H_3 7.67 (1, dd, J = 7.0, 8.0), H_7 7.62 (1, dd, J = 7.0, 8.0), H_6 7.51 (1, dd, J = 1.0, 7.0)].

Spectral changes during photolytic decay of the azido derivative are given in Fig. 2.

5-Dimethylamino-1-naphthoic Acid (DMAN). 5-Amino-1-naphthoic acid is used as starting material. Methylation can be achieved with methyl iodide in a sealed tube at increased pressure (70°) as described previously.[8] This procedure generates substantial amounts of side products, such as iodinated aromatic, monomethyl derivatives, and methyl esters. For microscale preparation with [^{14}C]methyl iodide, 40 μmol of acid can be reacted in a closed apparatus and the reaction products separated on preparative TLC plates [2 mm thickness of SiO_2 layer (Merck 5745); toluene/acetic acid 4 : 1].

In a more elegant way the derivative can be obtained as follows: 38 mg (200 μmol) of 5-amino-1-naphthoic acid and 20.6 mg $NaSO_3$ are dissolved in 817 μl of saturated $KHCO_3$ in a Wheaton microflask. Then 400 μl methanol and 450 μl methyl iodide are added and the mixture is kept at 60° for 90 min under reflux (condenser flushed with cooling fluid of −4° from a kryostate). After cooling, the mixture is adjusted to pH 4.3 with 1 *N* HCl (beige/pink color and turbidity occurs). Extraction with ethyl ether continues until no more fluorescent material is extracted into the ether (UV handlamp) and the combined extracts are dried with Na_2SO_4. The ether is removed and the residue recrystallized from petroleum ether (bp 40–60°). The product forms yellow needles [mp 134–135°; R_f = 0.41–0.45 with toluene/acetic acid 4 : 1 (v/v) on silica gel cards].

UV (methanol): λ_{max} 218 nm, ε = 34,000; 247 nm, ε = 10,200; 319 nm, ε = 3850; MS: *m/e* = 215; IR (KBr), 3200–2500 (broad), 1695, 1580, 1470, 1285; NMR (DMSO-d_6): H_8 8.45 (dd, *J* = 8; 1), H_2 8.41 (dd, *J* = 7; 1), H_4 8.02 (dd, *J* = 8; 1), H_3 7.57 (t, *J* = 7, 5; 8, 5), H_7 7.53 (t, *J* = 8, 5; 7, 5), H_6 7.19 (dd, *J* = 7; 1).

Methyl ester forms as a side product [R_f = 0.52 in the same solvent, λ_{max} 252, 336 nm (methanol shift from 247 to 252 induced by esterification)]. It is soluble in cold petroleum ether. [MS: *m/e* = 229.]

An alternate synthesis of 5-dimethylamino-1-naphthoic acid has been reported by Mayer *et al.*[20] following a procedure described by Friedlander.[21] No spectroscopic data were given however.

3-(4-Azido-2-nitrophenyl)propionic Acid (ANP). The synthesis of ANP is given in Scheme 2 and includes the partial reduction of dinitrophenyl compounds described by Gabriel and Zimmermann[22] in 1880. This procedure applies to 4-phenylbutyric acid, or phenylacetic acid as well.

3-Phenylpropionic acid [15 g (100 mmol)] is added in small aliquots to 70 ml of nitrating acid (30 ml concentrated HNO_3 + 40 ml concentrated H_2SO_4) kept at 60°. The temperature is raised to 80° and the mixture is stirred under reflux for 1 hr. The yellow solution is poured into 2 liters of ice water and the yellow precipitate removed. The aqueous solution is extracted with 4 or 5 liters of diethyl ether (in fractions of 500 ml) and the combined extracts are dried over Na_2SO_4. The ether is evaporated *in vacuo* and the resulting bright yellow residue is dried under vacuum. The

[20] I. Mayer, A. S. Dahms, W. Riezler, and M. Klingenberg, *Biochemistry* **23,** 2436 (1984).
[21] W. Friedländer, *Chem. Ber.* **21,** 3122 (1878).
[22] S. Gabriel and J. Zimmermann, *Ber. Dtsch. Chem. Ges.* **13,** 1680 (1880).

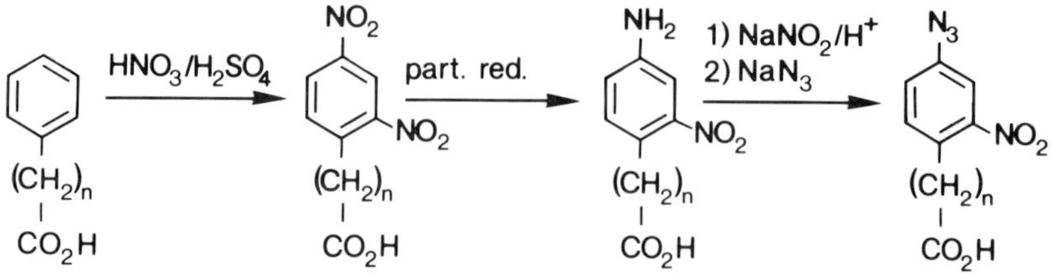

$$\text{C}_6\text{H}_5(\text{CH}_2)_n\text{CO}_2\text{H} \xrightarrow{\text{HNO}_3/\text{H}_2\text{SO}_4} (\text{NO}_2)_2\text{C}_6\text{H}_3(\text{CH}_2)_n\text{CO}_2\text{H} \xrightarrow{\text{part. red.}} \text{NH}_2(\text{NO}_2)\text{C}_6\text{H}_3(\text{CH}_2)_n\text{CO}_2\text{H} \xrightarrow[\text{2) NaN}_3]{\text{1) NaNO}_2/\text{H}^+} \text{N}_3(\text{NO}_2)\text{C}_6\text{H}_3(\text{CH}_2)_n\text{CO}_2\text{H}$$

SCHEME 2. Synthesis of 3-(4-azido-3-nitrophenyl)propionic acid.

product is recrystallized from 1.5–1.6 liters of water, yielding 18.5 g (78%) of purified 3-(2,4-dinitrophenyl)propionic acid [mp 124° (lit. 126°)].

$(NH_4)_2S$ solution (28 ml 40%) (165 mmol) is diluted with 100 ml water; 12 g (50 mmol) of the dinitro derivative are added slowly and carefully (vigorous exothermic reaction). The resulting dark red solution is heated under reflux on an oilbath at 100–120° for 1 hr. Sulfur precipitates and is filtered off after cooling; the filter is washed with 10% NH_3 solution. The filtrate is adjusted to pH 3–4 with diluted H_2SO_4 (the pH has to be carefully controlled; at lower pH the ammonium salt of the product is formed which cannot be extracted with ether); a bright yellow precipitate is removed. The aqueous solution is extracted in portions with 4–5 liters of diethyl ether; the combined extracts are dried over Na_2SO_4 and the ether is evaporated. The orange residue is dried under vacuum and recrystallized twice. The raw yield is 88%. After recrystallization up to 60% pure product can be obtained (mp 134–137°).

Conversion to the azido derivative is performed according to Fleet *et al.*[23] with large-scale preparations and the products (propionyl or butyryl derivatives) recrystallized from water with yields of 60–70%.

4-Amino-2-nitropropionic acid (1.7 g, 8.1 mmol) was dissolved in 40 ml concentrated HCl while heating. Then the solution was cooled in ice/NaCl to −10° and 552.7 mg (8.8 mmol) of $NaNO_2$ dissolved in 8 ml water were added dropwise while stirring. Then 520 mg NaN_3 (8.8 mmol) in 8 ml water were added slowly and stirred for about 30 min until nitrogen liberation ceased (0°). The bright yellow precipitate is separated by suction and washed with ice-cold water. The product is recrystallized from about 400 ml water and dried. Yield of pure product 60%; mp 127–128° with decomposition. The azido compound should be synthesized in the dark or in red light! 3-(4-Azido-2-nitrophenyl)propionic acid [mp 127–128°; uv λ_{max} (methanol) 245 nm, $\varepsilon = 16,400$; IR (KBr) 3300–2600 (broad), 2130, 1705,

[23] G. W. Fleet, J. R. Knowles, and R. Porter, *Biochem. J.* **128,** 499 (1972).

1535, 1435, 1495, 1330; MS m/e = 236; NMR (DMSO-d_6) 7.63 (1, d, J = 2.5), 7.55 (1, d, J = 8.0), 7.41 (1, dd, J = 2.5, 8.0), 3.02 (2, t, J = 7.5), 2.58 (2, t, J = 7.5)].

4-(4-Azido-2-nitrophenyl)butyric acid [mp 106–107° (decomposition), λ_{max} (methanol) 245 nm, ε = 16500; IR (KBr) 3200–2700 (broad), 2150, 1705, 1535, 1495, 1435, 1330; MS, m/e = 250; NMR (DMSO-d_6), 7.64 (1, d, J = 1.8), 7.54 (1, d, J = 8.5), 7.44 (1, dd, J = 1.8, 8.5), 2.78 (2, t, J = 8.0), 2.25 (2, t, J = 8.0), 1.79 (2, quintet, J = 8.0).

Small-scale synthesis with isotopically labeled phenylpropionic acid was performed as follows. ^{3}H-Labeled acid was obtained from Amersham, prepared by catalytic hydration of cinamic acid. Then 30.1 mg (0.2 mmol) of phenylpropionic acid is dissolved in 0.7 ml nitration mix by careful and slow addition of the latter to the acid (mix consists of 0.3 ml concentrated HNO_3, 0.4 ml concentrated H_2SO_4) and is kept at 60° for 1 hr and then poured into 30 ml ice water. The aqueous solution is extracted with ethyl acetate; after evaporation a yellowish brown residue is isolated. *TLC* control shows only one spot. The raw product (38 mg; 79%) is directly used for reduction of the 4-nitro group.

The dinitro compound [48 mg (0.2 mmol)] is dissolved in 5 ml 12.5% NH_4OH; 225 μl of 20% $(NH_4)_2S$ solution (0.66 mmol) are added and the mixture heated on an oil bath (100–120°) for 90 min under reflux. After cooling sulfur is removed by filtration and the filter is washed with a few milliliters NH_4OH. The pH is adjusted to 3–4 and the mixture extracted with ethyl acetate. After evaporation 17 mg (40%) of light brown raw product are retained. For purification the product is dissolved in 0.5 ml ethyl acetate, transferred on preparative thin-layer plates 2 mm thickness of SiO_2 layer (Merck 60 F254; No. 5717)], and developed with toluene/acetic acid (95 : 5 v/v). The lowest band is eluted with methanol. Since considerable amounts of silicic acid are also dissolved, after drying the product has to be reextracted with ethyl ether.

Conversion into the 4-azido derivative is performed in the dark directly using the product from the preceding step. Then 30 mg 4-amino derivative dissolved in 5 ml 4 N HCl is treated in the cold with 1 ml 1 M $NaNO_3$ (dropwise), and subsequently with 1 ml 1 M NaN_3. After completion of the reaction, the aqueous solution is extracted with ethyl acetate; after evaporation of the solvent the residue (25 mg; 80%) is used for esterification without further purification. TLC (silica gel, toluene/acetic acid 95% : 5%) shows only one spot.

3-Carboxy-2,2,5,5-tetramethylpyrrolidine-1-oxyl. The precursor acid for compound 21 is prepared from 3-carbamoyl-2,2,5,5-tetramethylpyrrolidine-1-oxyl by hydrolysis of the amide. If this is not commercially available, it can be synthesized according to Rozantsev[24] from 2,2,5,5-tetramethylpyrrolidine-3-carboxamide by oxidation with H_2O_2 under mild conditions in 80–90% yield.

The 3-carboxypyrrolidine derivative is synthesized as follows: 0.93 g of the carbamoyl compound and 1.58 g of $Ba(OH)_2$ octahydrate are boiled under reflux in 3 ml water for 20 hr in a microscale apparatus (Wheaton). After this time liberation of ammonia is complete. The reaction mixture is diluted with 25 ml water, heated to 60°, and saturated with carbon dioxide to precipitate $BaCO_3$. The filtrate is again flushed with CO_2 until no more $BaCO_3$ precipitates. The supernatant is acidified with HCl to pH 2 and extracted with ether. The ether extract is dried with anhydrous sodium sulfate, and after removal from the desiccant is concentrated slowly under a stream of nitrogen until yellow crystals of product are formed. Recrystallization from chloroform/methanol (1 : 1 v/v) yields 0.8 g of product (86% of theoretical yield) [mp 193° under decomposition; MW 186.23; λ_{max} 230 nm; ε (230) = 2900 M^{-1} cm^{-1}; ε(259) = 1300 M^{-1} cm^{-1}].

Esterification of Nucleotides

The usual procedure for large-scale preparations has been described in detail by Guillory and Jeng[15] and has been successfully confirmed in several laboratories.[8,16a,b] It is based on carboxyl activation by carbodiimidazole[25,26] and allows the formation of 3′ and 2′ esters of nucleotides without interference of the 6-amino group in the heterocycles when suitable mixtures of water with weakly solvating aprotic media are selected. In general, 1 : 1.5 molar amounts of carboxylic acid and carbodiimidazole were stirred with dry DMF for 30 min at room temperature under moisture-free conditions. The reaction is controlled by TLC [dichloromethane/ethyl acetate, 1 : 4 (v/v)]. This solution was added to aqueous solutions of the nucleotides and stirred for 4 hr at room temperature. Then the solvent was removed under reduced pressure and the residue treated with small amounts of acetic acid/acetone (30 : 70) and sedimented. The sediment was dissolved in water and the products separated by preparative thin-layer chromatography on cellulose, as described by Onur *et al.*[4,8]

The relative yields of monoesters, diesters, or α-phosphate-acylated derivatives largely depend on the ratio of nucleotide/imidazolide and of H_2O/DMF. The reaction is controlled by TLC on microcrystalline cellulose-coated glass plates; *n*-butanol/isopropanol/H_2O 1 : 1 : 1 or 1 : 2 : 1, re-

[24] E. G. Rozantsev, "Free Nitroxyl Radicals." Plenum, London, 1970.

[25] H. A. Staab and A. Mannschreck, *Chem. Ber.* **95,** 1284 (1962).

[26] B. P. Gottikh, A. A. Krayevsky, N. B. Tarussova, P. D. Purygin, and T. L. Tsilevich, *Tetrahedron Lett.* **26,** 4419 (1970).

spectively. Lowering the relative proportion of nucleotides supports formation of diester; increasing favors formation of α-phosphate acylation.

Although the 2′ ester is considered the kinetically controlled product, the final result in aqueous solution is a mixture of 70% 3′- and 30% 2′-isomers. This follows from the well-known acyl migration in *cis*-1,2-diols[27,28] that depends on the nature of the substituent. The rate of isomerization is temperature- and pH-dependent. For naphthoyl derivatives the reaction half-time is 50 min at 40° at pH 9. The final state of equilibration has been investigated by Onur[8] in NMR studies using pure adenosine reference compounds.

The stability of 3′ (2′) esters depends on the acyl residue: aromatic $\gg$ aliphatic $>$ aliphatic amino acids. In any case, freshly prepared substances should be used for kinetic and photolabeling studies.

The synthesis of nucleotide analogs specifically used for kinetic and mechanistic studies of F_1 in our laboratory is given as a *microprocedure* for radiolabeled derivatives. The procedures given for ADP analogs also apply for AMP or ATP derivatives.

Activation of carboxylic acids is carried out on a slightly larger scale. For all reactions conical 1-ml Wheaton microvials with a magnetic microstirring bar are used. The vials are closed by septum caps on top of a microstopcock. Carboxylic acid and (62.5 μmol) CDI (100 μmol) are mixed in a total volume of 50 μl absolute DMF and stirred at room temperature for 1 hr under moisture-free conditions. The progress of the reaction is controlled by TLC on silica gel with dichloromethane: ethyl acetate 1 : 4 (v/v); R_f of imidazolides is 0.65–0.69. The reaction vessel is kept closed and aliquots of activated acid are removed as required through the septum under moisture-free conditions with a Hamilton syringe. The mixture is stable at −20° for several days.

For esterification the ^{14}C-labeled nucleotide solution is neutralized to pH 7 and placed in a microvial, as described, and the activated acid is added using a Hamilton syringe. The concentration of nucleotide never should be lower than 0.5 *M*. Average reaction time is 3 hr. The reaction process is controlled by TLC. Purification of reaction products is carried out directly from the reaction mix on 0.1 mm cellulose-coated glass plates by TLC. The bands are eluted with water and the solutions filtered through Amicon PM10 membranes for removal of microcrystalline cellulose and lyophilized.

3′-O-[Naphthoyl(1)]ADP: N-ADP. [^{14}C]ADP (5 μmol) are added to 40 μl H_2O (neutralized); then a single addition is made of 10 μmol activated

[27] C. B. Reese and D. R. Trentham, *Tetrahedron Lett.* **29,** 2467 (1965).

[28] C. S. McLaughlin and V. M. Ingram, *Biochemistry* **4,** 1442 (1965).

1-naphthoic acid in 8 μl DMF. This is stirred at room temperature for 3 hr. Separation of the products on cellulose takes place with *n*-butanol : isopropanol : water 1 : 2 : 1 (v/v); R_f of products are N-ADP = 0.44, N_2-ADP (and traces of N-AMP) = 0.55; α-phosphate-acylated N-ADP = 0.2; yield: 26–30%. UV : λ_{max} 259 nm, ε = 16,900; fluorescence $\lambda_{excitation}$ 298 nm; $\lambda_{emission}$ = 392 nm/in aqueous media); NMR: aromatic protons as in the respective carboxylic acids; ribose protons (60°, DMSO-d_6 D_2O) H_1, 6.09 (d, J = 7.5), H_2, 5.1 (t, J = 7.5; 5.5), H_3, 5.73 (dd, J = 5.5; 1.8), H_4, 4.5 (dd, J = 1.8), $2H_5$, 4.2 (m). The respective resonances of the 2′ ester are 6.32, 5.81, 4.85, 4.2, 4.08.

3′-O-[5-Azidonaphthoyl(1)][^{14}C]ADP: "ANA-ADP." Synthesis has to be carried out in the dark or in red light. It is synthesized exactly as N-ADP; however, 5 μmol [^{14}C]ADP and 7.5 μmol activated acid are used. Reaction time is 4 hr [R_f = 0.25 (1 : 2 : 1) and 0.54 (1 : 1 : 1) UV (phosphate buffer pH 7): 248 nm, ε = 28000; 325 nm, ε = 7400; ε was determined from specific radioactivity with great accuracy; IR: 2110 (azido band); yield 26%]. NMR (DMSO-d_6D_2O, 40°): H_1, 6.06 (d, J = 7.5), H_2, 5.19 (t, J = 7.5; 5.5), H_3, 5.73 (dd, J = 5.5; 1.8), H_4, 4.47 (dd, J = 1.8), $2H_5$, 4.1 (m). The respective resonances of the 2′ ester are 6.32, 5.89, 4.78, 4.22, 4.1.

3′-O-[5-Dimethylaminonaphthoyl(1)][^{14}C]ADP: "DMAN-ADP." This compound is synthesized exactly as N-ADP. R_f on 0.1 mm cellulose plates (as an example all reaction products are listed): free acid 1.0, imidazolium salt 0.85, $(DMAN)_2$-ADP 0.63, DMAN-ADP 0.36, α-phosphate acylated DMAN-ADP 0.25, AMP 0.18, ADP 0.12. UV (pH 7 in phosphate buffer) 209 nm, ε = 62,700; 252 nm, ε = 30,000; 324 nm, ε = 6,000. The molar extinction coefficients are strongly pH-dependent due to protonization of the amino nitrogen; at 252 nm typical values are 28,500 (pH 6), 30,600 (pH 7), 32,000 (pH 9), 29,000 (0.1 *N* NaOH, hydrolysis occurs). Yield of synthesis ~30%. NMR (DMSO-d_6 D_2O, 40°): H_1, 6.05 (d, J = 7.5), H_2, 4.99 (t, J = 7.5; 5.5), H_3, 5.65 (dd, J = 5.5; 1.8), H_4, 4.5 (dd, J = 1.8), $2H_5$, 4.15 (m). The respective resonances of the 2′ ester are 6.35, 5.86, 4.68, 4.25, 4.15.

3′-O-[3-(4-Azido-2-nitrophenyl)propionyl][^{14}C]ADP: "ANP-ADP." Reaction to be performed in the dark or at dim red light. Neutralized [^{14}C]ADP (1 μmol) (130 Ci/mol) in a mixture of 10 μl H_2O and 3 μl DMF are placed in a Wheaton vial. Then 4 additions are made of 1 μmol activated acid in 1 μl, each after a period of 30 min. After 3 hr the reaction mix is separated on TLC cellulose, as above, with *n*-butanol : isopropanol : water 1 : 1 : 1 (v/v). R_f ANP-ADP = 0./6; $(ANP)_2$-ADP = 0.7; UV (aqueous buffer pH 7): 252 nm, ε = 27,200; yield 6–15%.

The radioactive concentration in the solution is of critical importance for the stability of the product. Thus, an equally concentrated solution of

[U-^{14}C]-labeled product (540 Ci/mol), which may reach a radioactive concentration of 6000 μCi/ml, was much less stable than that of [8-^{14}C]-labeled product (50 Ci/mol).

Biochemical Applications and Interactions

While all analogs of Table I have been identified as inhibitors of oxidative phosphorylation or photophosphorylation, detailed kinetic and thermodynamic parameters have been determined only with selected compounds, useful as fluorescence probes or as photoaffinity labels. One of the analogs, DMAN-ADP, has been identified also as a conformation-specific fluorescent ligand to the mitochondrial adenine nucleotide carrier.[8a,29] Examples of their application will be summarized, demonstrating how these analogs can be applied as tools for identification of different types of nucleotide-binding sites on F_1, for determination of their number and location, and for their possible function. Although the basic design of these compounds is identical, their interactions with F_1 may differ significantly. A lengthy description of individual experiments has been given in a series of original communications.[4,8,8a,11,30,31]

Kinetics of ATP Synthesis and Hydrolysis

The inhibitory capability of 3′ esters is modulated significantly by the substituent, as seen from Table I. Structural comparison indicates that aromatic or araliphatic compounds are stronger inhibitors than those with short-chain aliphatic acyl residues. The mobility of the acyl residue relative to the ribose moiety is essential, as concluded from the attenuating effect of at least one methylene group between carboxyl and the remainder of the substituent. For the same reasons among analogs bearing condensed aromatic ring systems, those with a 1-carboxyl group show stronger effects than others (compare compounds **19** and **20**; 2-naphthoyl also inhibits less than 1-naphthoyl-ADP, whereas naphthyl-1-acetyl and naphthyl-2-acetyl produce the same inhibitory effect).[8a]

In any case, compounds derived from 1-naphthoic acid are the strongest inhibitors found so far and have been investigated in great detail because they also exhibit fluorescent properties.

In general, the relative activity of 3′ esters always follows the order ADP analog > ATP analog >> AMP analog (practically inactive). A comparison is given in Table II, demonstrating in addition that the action

[29] G. Schäfer and G. Onur, *FEBS Lett.* **117,** 269 (1980).

[30] G. Schäfer and J. Weber, *J. Bioenerg. Biomembr.* **14,** 479 (1982).

[31] M. Lübben, U. Lücken, J. Weber, and G. Schäfer, *Eur. J. Biochem.* **143,** 483 (1984).

TABLE II
COMPARISON OF 3′*O*-ACYLATED AMP, ADP, AND ATP DERIVATIVES AS INHIBITORS OF ATP SYNTHESIS AND HYDROLYSIS BY MEMBRANE-BOUND (SMP) AND ISOLATED F_1

Nucleotide analog	Oxidative phosphorylation K_i (μM)	ATPase (SMP) K_i (μM)	F_1 (soluble) K_i (μM)
DMAN-AMP	Inactive	Inactive	Inactive
-ADP	0.08	9.8	9.5
-ATP	0.60	12	19.5
N-AMP	Inactive	Inactive	Inactive
-ADP	0.02[a]	9–11	4.6
-ATP	0.35[a]	8	8–13

[a] The respective values in photophosphorylation are 0.3–0.8 (depending on light intensity),[8] and 17–18 μM, respectively. The inhibitory constants were calculated by evaluations of Lineweaver–Burk plots or Dixon plots on the basis of competitive type of inhibition.

on oxidative phosphorylation in all cases is much stronger than on ATP hydrolysis by membrane-bound or soluble F_1. The same holds true for CF_1.[8,32]

This leads to the interpretation that these probes interact differently with F_1 in the energized, membrane-bound state than with nonenergized solubilized F_1.[9] Because this effect is exactly the reverse of that found with the analog TNP-AD(T)P,[9,30] both groups of 3′ analogs may be considered to represent conformation-specific probes, differentiating between the proposed two states of the enzyme.

The type of inhibition of all compounds investigated is competitive versus ATP in ATP hydrolysis and ADP in ATP synthesis. It appears to be noncompetitive versus inorganic phosphate in ATP synthesis. In double-reciprocal plots of oxidative phosphorylation or photophosphorylation versus ADP concentration, a slight deviation from linearity could be observed in the presence of N-ADP or DMAN-ADP, indicating positive cooperativity.[8] Apparent Hill coefficients of 1.4–1.5 have been reported,[30] suggesting that at least two adenine nucleotide sites are involved in the process of ATP formation.

In addition, with N-ADP and DMAN-ADP a lag phase in oxidative phosphorylation has been observed when the inhibitory analogs had been added prior to the phosphate acceptor. This effect also points to participa-

[32] G. Schäfer, G. Onur, and H. Strotmann, *in* "Energy Conservation in Biological Membranes" (G. Schäfer and M. Klingenberg, eds.), p. 220. Springer-Verlag, Berlin and New York, 1978.

tion of more than one catalytic site and to a slow replacement from the binding locus by the natural substrate. It is released specifically by ATP rather than by ADP.[33]

If the ADP analogs are phosphorylated at all [only extremely slowly (400 times less than ADP) by illuminated thylakoid membranes],[8] no phosphorylation by submitochondrial particles could be detected. This finding, however, supports the view that the 3′ analogs are ligands of the catalytic sites, which are considered to be the so-called high-affinity sites on F_1-type enzymes.[29]

Fluorescent 3′ Esters of Adenine Nucleotides

N-AD(T)P, the DMAN derivatives, and anthranoyl derivatives (compounds **14**, **18**, **19**, **20**) have been investigated as fluorescent probes. Unexpectedly they do not respond by a fluorescence increase on binding to F_1-ATPase, although it has been supposed that hydrophobic interactions of the aromatic substituent largely contribute to their high affinity. Thus, they are inadequate as hydrophobicity probes of F_1 binding sites. Moreover, their fluorescence properties are essentially different.

N-ADP shows a typical fluorescence spectrum (excitation 298 nm, emission 392 nm); binding to F_1 occurs without a change in quantum yield or any shift of the emission band. The lifetime of the excited state is 6.7 nsec[11] in bound and free state. However, fluorescence polarization can be used as a sensitive measure of immobilization in the bound state. A sensitive spectrofluorometer in T configuration can be used and mobility is determined as fluorescence anisotropy,[34] with excitation at 298 nm and emission measured at 392 nm. The anisotropy of free N-ADP in aqueous solution is 0.009, whereas that of F_1-bound N-ADP is 0.23. Binding to F_1 can be determined by plotting fluorescence anisotropy versus the concentration of F_1 titrated into the cuvette as reported by Tiedge *et al.*,[11] yielding the number of binding sites and the thermodynamic K_d for N-ADP (see below). The technique also allows determination of K_d values for N-ADP binding to F_1 residing in the membrane of submitochondrial particles, as shown by Schäfer and Weber,[30] yielding K_d = 17–29 nM, which is in agreement with the values for isolated F_1. Stoichiometries cannot be determined in the latter case because the number of F_1 molecules exposed on the membrane cannot be calculated exactly.

Interference to binding of the nucleotide carrier, which has 50 times lower affinity than the analog, is negligible.

[33] G. Schäfer, *FEBS Lett.* **131,** 45 (1981).

[34] D. M. Jameson, G. Weber, R. D. Spencer, and G. Mitchell, *Rev. Sci. Instrum.* **49,** 510 (1978).

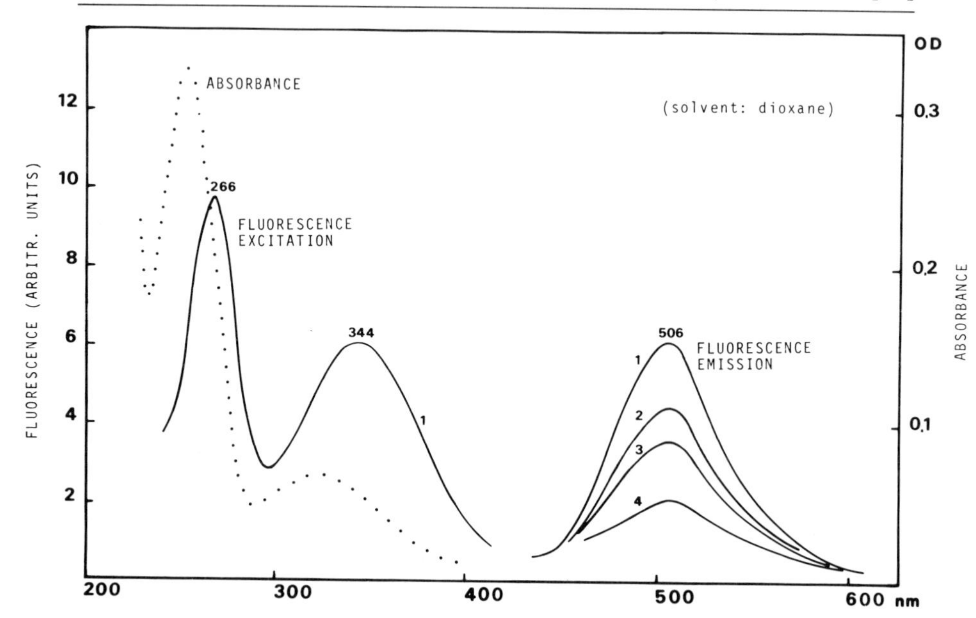

FIG. 3. Fluorescence spectra of DMAN-ADP in media having different dielectric constants. Spectra were taken in a Schoeffel RRS-1000 ratio spectrofluorometer at room temperature. The spectra of a 10.9-μM solution were recorded in (1) dioxane, (2) dioxane + 2.2% H_2O (v/v), (3) dioxane + 3.6% H_2O, (4) dioxane + 5.4% H_2O. The dotted line gives an absorption spectrum of a 10.9-μM solution in 0.1 *M* potassium phosphate buffer, pH 7.2.

Direct monitoring of fluorescence polarization can be used to record the kinetics of displacement of bound N-ADP from its sites on isolated or membrane-bound F_1. Anisotropy is recorded directly as the ratio of fluorescence intensities parallel and perpendicular to the polarization of the excitation beam.

An excess of ADP (60 μM) added to either F_1 or submitochondrial particles loaded with N-ADP results in a biphasic process. From F_1, loaded with 2 mol N-ADP/mole F_1, 1 mol of the analog is displaced rapidly within mixing time, while 1 mol is exchanged very slowly with $t_{1/2}$ of about 1 min.[11,30]

DMAN-ADP shows almost no fluorescence in aqueous media. Fluorescence yield is about 0.018. In ethanol it is 0.096 and it increases to 0.28 in dioxane. The fluorescence spectrum in dioxane is depicted in Fig. 3. Emission intensity strongly depends on the polarity of the medium without a significant shift of the emission maximum at 506 nm.

Although binding to F_1 is not accompanied by a fluorescence change, submitochondrial particles generate a dramatic increase in fluorescence quantum yield to 0.45 when added to DMAN-ADP. This is due to binding to the adenine nucleotide carrier as described below.

The fluorescence of DMAN-AD(T)P is easily quenched by a number of phenyl derivatives (such as Phe, Tyr, Tyr-OCH_3). Lack of fluorescence changes on F_1 binding to the analog may result either from the fact that the fluorophore is exposed to a site having high dielectric constant or to the aqueous surface (though being strongly immobilized). Alternatively, it may directly interfere with one of the tyrosine residues on F_1, presumably present in the catalytic site.[35,36] This is supported by the finding that intrinsic tyrosine fluorescence of F_1 is quenched much more efficiently by DMAN-ADP than by ADP, which causes a conformational change upon binding to the enzyme, accompanied by a weak quench of intrinsic tyrosine fluorescence.[11]

The quenching constants [according to (8a)] in increasing order are water 7.01 10^{-1}; Tyr-O-CH_3 6.4 10^{-3}; Phe 8.01 10^{-3}; Tyr 2.73 10^{-4} *M*.

Binding of Analogs to Isolated F_1

In our laboratory binding studies have been conducted with nucleotide-depleted beef heart F_1 in order to provide a reproducible material for determination of stoichiometry and thermodynamic binding parameters to high-affinity sites. As reported elsewhere[11] the binding data are in agreement with a model assuming three *a priori* identical high-affinity sites exhibiting strong negative cooperativity of nucleotide binding.[36a] Such negative cooperativity has also been reported by Cross *et al.*,[37] which results in positive cooperativity of catalysis.[4,38,39]

To prevent confusion of high-affinity sites with those described as high-affinity, nonexchangeable, noncatalytic sites,[37,40] we emphasize our definition for reversible high-affinity binding. It includes three reversible sites with K_d 50 nM–3 μM, of which two sites assume K_d values >50 nM to 3 μM only after occupation of the first site. The term low affinity is suggested for those with K_d values $\gg$10 μM, also found on nucleotide-depleted F_1.[41]

Binding can be studied by four different methods as follows:

[35] F. S. Esch and W. S. Allison, *J. Biol. Chem.* **253,** 6100 (1978).
[36] L. C. Cantley and G. G. Hammes, *Biochemistry* **14,** 2976 (1975).
[36a] U. Lücken, F. Peters, and G. Schäfer, *ICSU Short Rep.* **3,** 338 (1985).
[37] R. L. Cross and C. M. Nalin, *J. Biol. Chem.* **257,** 2874 (1982).
[38] Ch. Grubmeyer and H. S. Penefsky, *J. Biol. Chem.* **256,** 3728 (1981).
[39] R. L. Cross, Ch. Grubmeyer, and H. S. Penefsky, *J. Biol. Chem.* **257,** 12101 (1982).
[40] P. V. Vignais and M. Satre, *Mol. Cell. Biochem.* **60,** 33 (1984).
[41] J. Weber, U. Lücken, and G. Schäfer, *Eur. J. Biochem.* **148,** 41 (1985).

1. By use of isotopically labeled 3′ analogs applying the method of centrifuged column filtration as described by Penefsky.[42] This method, however, is not a true equilibrium method and yields reliable values of n and K_d only for sites of very high affinity.

2. By use of pressure filtration according to Paulus.[43] However, one has to take into account that certain analogs may bind unspecifically to the filter membrane, which has to be selected in control experiments accordingly. For example, DMAN-ADP strongly binds to nitrocellulose membranes.

3. By intrinsic fluorescence of isolated F_1. With beef heart F_1 tyrosine fluorescence can be used, since no tryptophan is present.[44] With F_1 from the thermophilic bacterium PS3 either tryptophan or tyrosine fluorescence can be used.[45]

4. By fluorescence anisotropy of analogs such as N-AD(T)P or anthranoyl-ADP.

It is essential to note that divergent results may be obtained by optical methods and isotope methods, as has been shown with beef heart F_1. The reason is that intrinsic fluorescence of F_1 responds to occupation of only one or two of the three high-affinity sites, whereas isotope binding allows monitoring of binding to sites which are "optically silent." Though this might appear as a disadvantage, it is not the case. Binding to the first site, as in the case of ADP, can be monitored with great accuracy with respect to K_d, although a total of three high-affinity sites exists.

The intrinsic K_d values for ADP and for several ADP analogs to isolated beef heart F_1 are given in Table III, as well as the total number of binding sites emerging from the titrations. It should be stressed that according to Klotz *et al.*[46] only intrinsic binding or dissociation constants are physically meaningful for the description of multisite ligand interactions. They may differ significantly from the apparent K_d values. The constants given in Table III were derived from nonlinear fitting procedures,[47,47a] but not from Scatchard Plots, owing to the large inaccuracy of the latter.

While with ADP only one site can be titrated by monitoring intrinsic fluorescence of F_1, three sites are found with isotope studies. The number of sites for N-ADP is two, determined with intrinsic fluorescence or via

[42] H. S. Penefsky, *J. Biol. Chem.* **252,** 2891 (1977).
[43] H. Paulus, *Anal. Biochem.* **32,** 91 (1969).
[44] H. S. Penefsky and R. C. Warner, *J. Biol. Chem.* **240,** 4694 (1965).
[45] M. Rögner, Ph.D. Thesis, Techn. Universität Berlin (1984).
[46] I. M. Klotz and D. L. Hunston, *Arch. Biochem. Biophys.* **193,** 314 (1979).
[47] F. Peters and A. Pingoud, *Int. J. Biomed. Comput.* **10,** 401 (1979).
[47a] F. Peters and U. Lücken, this volume [71].

TABLE III
BINDING OF ADP AND ANALOGS TO HIGH AFFINITY SITES OF F_1

Nucleotide	Number of sites			Site dissociation constants (in nM)			Model[d]	Stoichiometric dissociation constants calculated (in nM)		
	N[a]	i[b]	f[c]	k_{d_1}	k_{d_2}	k_{d_3}		K^{d}_{1}[e]	K_{d_2}	K_{d_3}
ADP	3	3	1	150	2600	950	5	50	2500	2850
				(3448)	(3300)	(66)	(3)	(63)	(1730)	(6810)
N-ADP	2	2	2	20	20	—	4′	10	40	—
DMAN-ADP	2	2	2	620	620	—	4′	310	1240	—
				(312)	(2350)	—	(3′)	(270)	(2650)	—
ANA-ADP	3	3	1	450	420	180	5	150	420	540
				(320)	(314)	(317)		(105)	(316)	(950)
ANP-ADP	0	—	—	Only three low affinity sites $K_d > 10$ mM						

[a] N denotes the total number of high affinity sites/mol F_1.

[b] i denotes the number of sites detectable by isotope binding studies.

[c] f denotes the number of sites detectable by changes of endogenous protein fluorescence or fluorescence anisotropy of the analogs.

[d] Three binding models were used which imply the following relations between constants (see Ref. 47a). Model 5: Three a priori identical sites with dependency; $K_{d_1} = k_{d_1}/3$, $K_{d_2} = k_{d_2}$, $K_{d_3} = 3k_{d_3}$. Model 4′: Two a priori equal and independent sites; $K_{d_1} = k_{d_1}/2$, $K_{d_2} = 2k_{d_2}$. Model 3: Three permanently different and independent sites; $K_{d_1} = (k_{d_1}k_{d_2}k_{d_3})/(k_{d_2}k_{d_3} + k_{d_1}k_{d_2} + k_{d_1}k_{d_3})$, $K_{d_2} = (k_{d_1}k_{d_2} + k_{d_1}k_{d_3} + k_{d_2}k_{d_3})/k_{d_1} + k_{d_2} + k_{d_3})$, $K_{d_3} = k_{d_1} + k_{d_2} + k_{d_3}$. Model 3 is given for comparison with model 5 only. In both cases it can be excluded (in parentheses) on the basis of error factors indicating that the best fit is obtained with model 5 which includes site cooperativity. Model 3′ corresponds to model 3 for only two sites.

[e] K_d normally describes the apparent dissociation constant read from a titration experiment. In the table, K_d denotes the macroscopic dissociation constant, derived from the intrinsic site-dissociation constants k_d which, however, depend on the model to which the best fit of titration data is obtained. Their relation is given below and is theoretically based on the treatment of multisite systems according to Klotz *et al.*[46,47a]

fluorescence anisotropy. Only two high-affinity sites were found for DMAN-ADP, monitored by intrinsic F_1 fluorescence. Like ADP, ANA-ADP shows one high-affinity site when determined with intrinsic F_1 fluorescence, but three sites when measured by isotope techniques (centrifuged column). Thus, it behaves as a true analog for the high-affinity sites occupied by the natural ligand AD(T)P. The affinity of ANP-ADP is weak. As shown below, it is a ligand to the low-affinity sites, capable of photolabeling.

Table III includes the dark-binding constants for the photoaffinity analogs ANA-ADP and ANP-ADP. ANA-ADP is a photoaffinity analog to the high-affinity ligand N-ADP. Determination of dark-binding behavior is essential prior to application of photolabeling agents. Besides our own studies with 3′ esters, only one report by Lunardi *et al.*[16a,b,c] gives such data, though the photolabeling approach has been widely used by others.

Photoaffinity Labeling by ANA-ADP and ANP-ADP

Both probes were designed for photoaffinity labeling and allow tagging specifically either the *high*-affinity sites (ANA-ADP), or the *low*-affinity sites (ANP-ADP).

Photoinactivation of F_1 is performed in glycerol-containing buffer for stabilization of the enzyme [100 m*M* Tris-sulfate, 2.5 m*M* EDTA, 50% (v/v) glycerol; pH 8.0]. The result of photolabeling is essentially independent of the presence of 10 m*M* $MgCl_2$. ADP in high concentrations (1–4 m*M*) protects the enzyme against photoinactivation and incorporation of the label. A suitable procedure for photolabeling is to incubate 1 ml of enzyme (1–2 μM) with the photolabile reactant in small glass dishes (15 mm diameter) placed on a cooled support; illumination is carried out with a 4 VA UV handlamp (366 nm) from a distance of 2–4 cm for a variable number of time periods. The process of photoinactivation is monitored by withdrawal of samples and ATPase assay during the dark periods. The total illumination time for complete photoinactivation depends on the concentration of the label and may vary when other illumination devices are used.

Best results are obtained and nonspecific labeling is avoided when the photoreactive analog is added stepwise during the labeling experiment, maintaining the concentration of freshly added, unphotolyzed label approximately constant. In a typical example, photoinactivation was started with 50 μM ANA-ADP, and increased 5 times by 100 μM after each dark period, instead of applying 500–600 μM from the beginning.

Separation of the photolabeled enzyme for determination of incorporation of label (^{14}C or ^3H) and exact residual enzymatic activity is recommended by centrifuged column filtration followed by diafiltration with glycerol buffer (PM10 membranes, Amicon) in order to remove any noncovalently bound label or photolytic products. An alternate procedure, used with ANP-ADP, is repetitive precipitation by addition of 3 volumes of a solution containing 50 m*M* Tris-Cl, pH 8.0, 50 m*M* KCl, 50 m*M* $MgCl_2$, and 15% (w/w) polyethylene glycol. ATPase activity is usually assayed in a coupled enzymatic test, as described by Vogel and Steinhart.[12] Protein determination should be performed by the method of Bradford[48] calibrated with F_1 or with BSA (factor $f = 0.8$).

[48] M. M. Bradford, *Anal. Biochem.* **72,** 248 (1976).

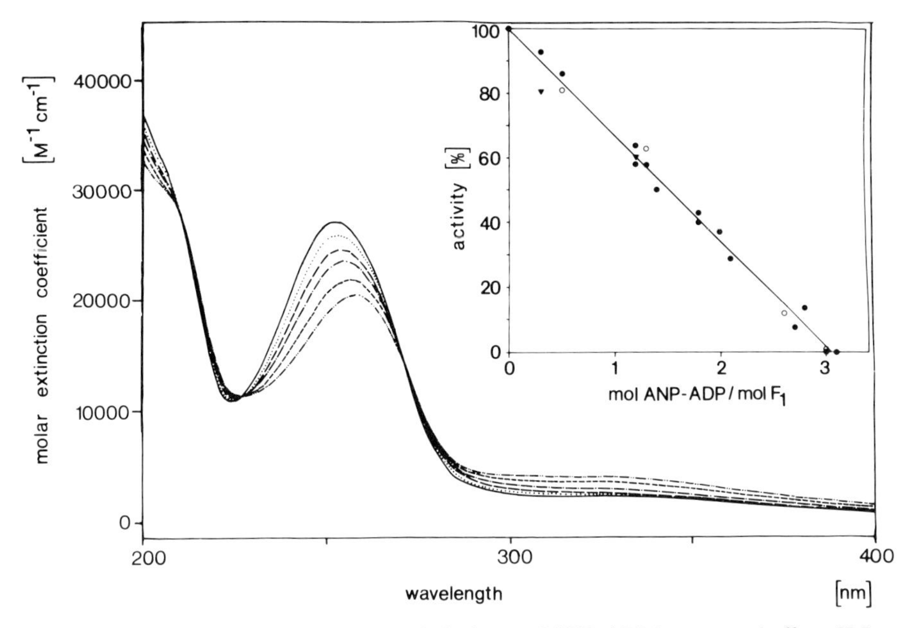

FIG. 4. Spectral changes during photolytic decay of ANP-ADP in aqueous buffer, pH 7. A 25-μM solution of the analog was illuminated with a UV handlamp (HANAU, 4W) between single spectral measurements; spectra 1–6 were taken after 0, 0.5, 1, 1.5, 3, and 5 min of illumination. It should be noted that the half-time of photolytic decomposition is much longer for the 3′ ester than for the respective carboxylic acid dissolved in methanol. Inset: ATPase activity of beef heart F_1 as function of covalent incorporation of the photolabel ANP-ADP, according to Weber *et al.*[41]

Figure 4 shows spectra taken during the photolytic decay of ANP-ADP in aqueous solution.

Irradiation with a UV handlamp is sufficient for photoactivation of the azido group. With a filter cutting off light below 300 nm, a xenon lamp may be used for photoactivation, resulting in shorter illumination periods. The inset of the figure shows the relative activity of mitochondrial F_1-ATPase as a function of the incorporated amount of label, indicating that complete inactivation is achieved at a labeling stoichiometry of 3 mol ANP-ADP/mol F_1. This is the maximum amount of label which can be incorporated covalently. With ANA-ADP complete inactivation is achieved already after covalent incorporation of only 2 mol per mol F_1.

This is the maximum obtainable degree of labeling with this analog, in agreement with its reversible binding behavior (Table III). However, one high-affinity site remains unlabeled, retaining the capability of reversible nucleotide binding.[31]

TABLE IV
REVERSIBLE NUCLEOTIDE BINDING TO F_1 MODIFIED COVALENTLY BY PHOTOLABELING WITH ANA-ADP OR ANP-ADP

Nucleotide bound	F_1[a]	ANA-F_1[b]			ANP-F_1[c]		
	n	n	K_d[nM]	Δn[d]	n	K_d[nM]	Δn[d]
ADP	3	1	1,700	2	3	>10,000	0
N-ADP	2	0	—	2	2	100	0
ANA-ADP	3	1	nd	2	—	nd	—

[a] Denotes native F_1, depleted of endogenous nucleotides as described; n, the number of binding sites only refers to high-affinity sites; together with the low-affinity sites labeled by ANP-ADP, the total number of sites on F_1 is 6! All data are compiled from those of Table III and from Refs. 31 and 41.

[b] Denotes F_1 covalently labeled with 2 mol ANA-ADP/mol F_1, catalytically inactive.

[c] Denotes F_1 covalently labeled with 3 mol ANP-ADP/mol F_1, and catalytically inactive.

[d] Denotes the difference of high-affinity sites between modified and native F_1.

The properties of F_1 covalently modified by ANA-ADP or ANP-ADP are summarized in Table IV. Distribution of radioactivity (^{14}C) in covalently modified F_1 shows that only the larger subunits α and β become labeled. The location of radioactivity on subunits is resolved best with urea gels, as described by Knowles and Penefsky.[49] The labeling pattern after full enzyme inactivation under the above conditions is 45% α/55% β with ANA-ADP and 30% α/70% β with ANP-ADP.

Reactivation of Photolabeled Enzyme

F_1 inactivated completely by ANP-ADP could be reactivated to almost original activity by hydrolysis of the 3′ ester bond at the covalently bound analog.[41] Reactivation is performed by incubation of the photolabeled enzyme at pH 9.0 in Tris buffer containing 50% (v/v) glycerol at room temperature. If photoinactivation was carried out by ANP-[^{14}C]ADP, the progress of hydrolysis can be followed, measuring the amount of ^{14}C activity remaining on the enzyme. With beef heart F_1 ~60% of label was hydrolyzed within 24 hr. After 15 days, ~10% of label remained unhydrolyzed.

[49] A. F. Knowles and H. S. Penefsky, *J. Biol. Chem.* **247,** 6617 (1972).

This experiment is of importance and may be advantageous also to other applications of the 3′ esters of nucleotides. On the one hand, it demonstrates that the natural ligand ADP, kept in place by the covalently linking group, is the inhibitor proper, rather than the auxiliary covalent anchor. Thus, the covalent insertion of this group into the polypeptide is harmless to the enzyme. On the other hand, it allows modification and blocking of specific low-affinity sites so that other sites can be studied independently.

DMAN Analogs and the Adenine Nucleotide Carrier

DMAN-AD(T)P[49a] binding to F_1 is not accompanied by a change in fluorescence. However, when added to submitochondrial particles (SMP), a dramatic increase of quantum yield from 0.018 to about 0.45 occurs. This 25-fold increase of fluorescence is due to binding of the analog to the adenine nucleotide carrier, as shown for the first time by Schäfer and Onur.[29] Since the analog is practically nonfluorescent in water, the signal can be used to titrate carrier binding, using a simple filter fluorometer with excitation at 366 (Hg) nm and a bandfilter 450–3000 nm for emission monitoring. Although DMAN-ADP inhibits mitochondrial adenine nucleotide transport competitively, binding to mitochondria does not cause fluorescence enhancement. Fluorescence increase is specific for SMP (inverted) membranes. The fluorescence of DMAN-ADP with SMP is quenched immediately by bongkrekic acid, displacing the bound analog. Atractylate has only a weak and slowly appearing effect.

Thus, the analog was identified as truly a conformation-specific probe, binding to the adenine nucleotide carrier with concomitant fluorescence enhancement only in the m-state, as exposed to the inner surface of the mitochondrial membrane. The typical fluorescence enhancement is retained also with SMP devoid of F_1 by urea treatment.[8a]

This property is unique and has been widely used in later studies by Klingenberg *et al.*[20,50] on the kinetics of carrier transitions between different conformational states.

Specificity of carrier binding with SMP is almost inverse to that of F_1-ATPase, with DMAN-AMP exhibiting the highest affinity. The apparent K_d values measured for the three DMAN analogs for BKA-sensitive binding to SMP membranes are DMAN-AMP 2.5 μM, DMAN-ADP 6–10 μM, and DMAN-ATP ~40 μM when measured by fluorescence titration.

[49a] Diverging abbreviations have been used in the preceding literature for the same type of analogs, such as F-AD(T)P because of the fluorescence, DAN-ADP, or DMAN-ADP. It is suggested to use only DMAN-AD(T)P in agreement with the initials of the correct chemical term: 3′-*O*-dimethylaminonaphthoyl-AD(T)P.

[50] M. Klingenberg, I. Mayer, and A. S. Dahms, *Biochemistry* **23,** 2442 (1984).

ADP competes for binding to the carrier on the surface of SMP with a K_i(ADP) = 2.7 m*M*.

Because of its particular properties, it is likely that DMAN-AD(M)P is an analog to BKA rather than to AD(T)P with respect to the conformation of the carrier stabilized after binding of these ligands.

Spin-Labeled ADP Analogs

The spin-labeled ADP analog (compound **21** from Table I) as well as similar compounds have been preliminarily investigated. It is a ligand to F_1 as seen from its strong inhibitory action on oxidative phosphorylation. As usually observed with 3′ esters, the ATP analog is a weaker inhibitor; half-maximal inhibitory concentrations are 12.7 μM for the ATP and 1.6 μM for the ADP derivative, respectively (with 100 μM ADP present in the assay of oxidative phosphorylation with beef heart SMP).

Binding to SMP membranes is reflected by a dramatic broadening of its ESR absorption bands, indicating immobilization of the probe similar to that of N-ADP seen by fluorescence anisotropy. Its affinity to F_1 certainly is much higher than to the adenine nucleotide carrier. Therefore, the immobilized state is not released by addition of bongkrekate to SMP incubated in the presence of nonsaturating concentrations of the probe.

The spin-labeled ADP analog can be applied to differentiate between nucleotide binding at the surface of SMP membranes either to F_1 or to the adenine nucleotide carrier by combined use together with the fluorescent DMAN-AMP. The latter does not bind to F_1, but has a high affinity to the carrier, whereas the spin-labeled 3′ ester binds firmly to F_1 only.

Comments

The kinetic mechanism of F_1-catalyzed reactions has been investigated by means of analogs in many ways, either using isolated F_1 and studying ATP hydrolysis, or using SMP or chloroplasts and studying ATP synthesis. A comprehensive review by Vignais and Satre[40] is referred to for details. At present nothing is known on a molecular level about the linkage of proton flux to formation or hydrolysis of ATP. Undoubtedly, however, hydrolysis and synthesis use the same catalytic sites and occur via the same principal mechanism. Questions remaining are the function of the inhibitor peptide[40,50–53] and the problem of cooperativity between the multiple nucleotide binding sites.

[51] M. E. Pullman and G. C. Monroy, *J. Biol. Chem.* **238,** 3762 (1963).

[52] P. L. Pedersen, K. Schwerzmann, and N. Cintron, *Curr. Top. Bioenerg.* **11,** 149 (1981).

[53] G. Klein, J. Lunardi, and P. V. Vignais, *Biochim. Biophys. Acta* **636,** 185 (1981).

The analogs described here were designed mainly to study this latter problem. They may be used for additional investigations, only performed in preliminary studies so far. For example, differentiation of affinities of ADP versus ATP analogs has not been fully investigated, and the interaction with reconstituted systems studied under controlled conditions of membrane energization is still lacking. Actually, the problem as to whether there exist permanently different catalytic and control sites, as well as the question of whether the catalytic mechanism involves three sites in a sequential reaction[54,55] have not been solved conclusively. These considerations also include the question of whether the F_1 molecule is a permanently symmetrical structure as revealed by electron microscopy[56,57] or if it exhibits intrinsic assymmetry as suggested by X-ray studies[58] at a 9 Å resolution.

From the use of the above analogs in kinetic and photolabeling studies, a number of conclusions may be drawn. F_1-ATPase has a total of six nucleotide binding loci; three sites can be covalently labeled by ANP-ADP, whereas three other sites are still available for reversible nucleotide binding. These reversible sites are high-affinity sites as revealed by their specificity to bind N-ADP, a typical high-affinity ligand.[11] The identity of these latter sites is further documented by the fact that they can be covalently labeled by ANA-ADP and that the labeling stoichiometry exactly reflects that of reversible binding of the parent compound, the 3′ analog N-ADP. Thus, 2 mol/mol F_1 are sufficient for complete enzyme inactivation, whereas a stoichiometry of 3 mol/mol F_1 is required when the low-affinity sites are labeled. For the low-affinity sites, a regulatory role may be assumed; this is reasonable also from the viewpoint that such a role for the high-affinity sites is questionable, since they would be steadily occupied under *in vivo* nucleotide concentrations. In conclusion, the two photoreactive analogs, ANP-DP and ANA-ADP, allow differentiation between the two types of sites: *high*-affinity and *low*-affinity sites on F_1.

The general model for nucleotide binding derived from our kinetic studies with ADP and ADP analogs has been described[11,59] and applies to the high-affinity sites, which are assumed to be the catalytic ones.[5,37,38] It suggests the presence of three *a priori* equivalent high-affinity sites, exhibiting pronounced negative cooperativity of binding. It is in line with the

[54] R. L. Cross, *Annu. Rev. Biochem.* **50,** 681 (1981).

[55] M. Gresser, J. A. Myers, and P. D. Boyer, *J. Biol. Chem.* **257,** 12030 (1982).

[56] H. Tiedge, G. Schäfer, and F. Mayer, *Eur. J. Biochem.* **132,** 37 (1983).

[57] H. Lünsdorf, K. Ehrig, P. Friedel, and H. U. Schairer, *J. Mol. Biol.* **173,** 131 (1984).

[58] L. M. Amzel and P. L. Pedersen, *Annu. Rev. Biochem.* **52,** 801 (1983).

[59] G. Schäfer, *in* "H^+-ATPase: Structure, Function and Biogenesis" (S. Papa, K. Altendorf, L. Ernster, and L. Padier, eds.). ICSU Press and Adriatica Editrice, Bari, 1984.

observation of positive cooperativity of catalysis as demonstrated by the use of other analogs[38] or of ^{18}O exchange as a probe of cooperativity in the catalytic mechanism.[60,61] Whether two or three catalytic sites are competent for completion of a catalytic cycle, a final comment would only be speculative. The result of photoinactivation with the high-affinity ligand ANA-ADP suggests that two interacting sites are a minimum requirement, because enzyme activity is practically zero when two high-affinity sites are blocked covalently, although one reversible high-affinity site persists.[31] Therefore, it has been suggested that a stochastic usage of the three available sites may occur during *in vivo* ATP synthesis rather than a sequential mechanism involving all three high-affinity sites in a cyclic manner.[31,59] Final proof of this issue, however, requires further knowledge of molecular structure and kinetics of this complicated enzyme.

[60] R. L. Hutton and P. D. Boyer, *J. Biol. Chem.* **254,** 9990 (1979).
[61] W. E. Kohlbrenner and P. D. Boyer, *J. Biol. Chem.* **258,** 10881 (1983).

[70] Modifiers of F_1-ATPases and Associated Peptides

By MICHEL SATRE, JOËL LUNARDI, ANNE-CHRISTINE DIANOUX, ALAIN DUPUIS, JEAN PAUL ISSARTEL, GÉRARD KLEIN, RICHARD POUGEOIS, and PIERRE V. VIGNAIS

Chemical modifiers of proteins can be conveniently divided into two classes, the group-directed reagents and the active site-directed reagents. The latter group consists of affinity and photoaffinity labels.[1-3] The former one comprises chemical reagents capable of binding to side-chain residues in the investigated protein. In this case, the reagent on binding can inactivate the protein and thereby provide information on amino acid residues playing a strategic role in the functioning of this protein; alternatively, it may leave the protein fully active, and then it plays the role of a reporter applicable for topographical and functional studies.[4] In this chapter on chemical modifications of F_1-ATPases, we shall address essentially group-directed reagents.

[1] F. Wold, this series, Vol. 46, p. 3.
[2] R. J. Guillory and S. J. Jeng, this series, Vol. 46, p. 259.
[3] H. Bayley, *in* "Photogenerated Reagents in Biochemistry and Molecular Biology: Laboratory Techniques in Biochemistry and Molecular Biology" (T. S. Work and R. H. Burdon, eds.). Elsevier, Amsterdam, 1983.
[4] R. L. Lundblad and C. M. Noyes, *in* "Chemical Reagents for Protein Modification," Vols. I and II. CRC Press, Boca Raton, Florida, 1984.

The mitochondrial, bacterial, and chloroplast ATPases to be discussed are H^+-dependent ATPases, called also ATP synthases, since they catalyze both ATP hydrolysis and synthesis. They are made of two sectors, a catalytic hydrophilic sector called F_1 and a hydrophobic membrane-embedded sector called F_0 which plays the role of a proton channel.[5] Besides these two major sectors, the mitochondrial ATPase is characterized by the presence of a well-defined inhibitory peptide (IF_1)[6] and a protein conferring the oligomycin sensitivity (OSCP).[7] IF_1 binds only to F_1. OSCP binds to both F_1 and F_0. F_1 can be detached from F_0 and when isolated, F_1 exhibits only hydrolytic activity. Most of the studies on chemical modifications have been made with isolated F_1.

It is noteworthy that most of the group-directed reagents are nonselective. Fortunately in many enzymes, and ATPases are no exception, certain residues located at strategic sites exhibit an enhanced reactivity toward chemical reagents. The observed reactivity is largely different from that of the other residues of the same class within the protein or from the reactivity estimated on the basis of model reactions; this is probably due to a special microenvironment. In this chapter, we discuss applications of inactivating and noninactivating group-directed reagents to F_1, IF_1, and OSCP. For inactivating modifiers of F_1, we have chosen to focus on carboxyl and tyrosyl group reagents. For noninactivating modifiers playing the role of reporters, we have addressed IF_1 and OSCP, which both bind to F_1.

Inactivating Modifications by Chemical Reagents

Inactivation rates with several of the chemical reagents considered later, at least those which are stable under the incubation conditions, followed pseudo-first-order kinetics. They are characterized by their half-times of inactivation ($t_{1/2}$) and the apparent inactivation constants ($k'_{\text{inact}} = \ln 2/t_{1/2}$). At a given concentration of the reagent, about 75% and 88% inactivation values are obtained after incubation for a length of time equal to two $t_{1/2}$ and three $t_{1/2}$, respectively. Other conditions being fixed, $t_{1/2}$ is divided by two when the concentration of the reagent is doubled. These simple rules must be kept in mind to fix or modify inactivation conditions derived from published protocols. Excessive amounts of chemical reagents should always be avoided.

The medium to be used for the modification reactions should not react with the chemical reagent, but instead should favor as much as possible

[5] P. V. Vignais and M. Satre, *Mol. Cell. Biochem.* **60,** 33 (1984).
[6] M. E. Pullman and G. C. Monroy, *J. Biol. Chem.* **238,** 3762 (1963).
[7] A. E. Senior, *J. Bioenerg.* **2,** 141 (1971).

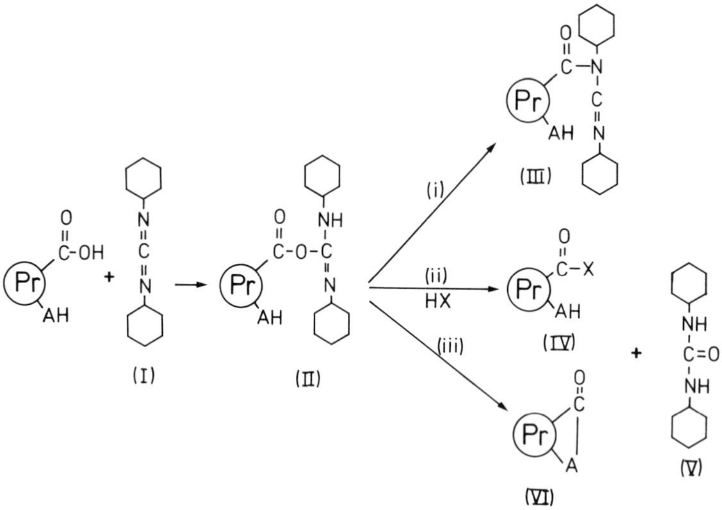

FIG. 1. Modification of carboxyl groups by DCCD. Details are given in the text.

the selectivity of the reaction, for example, by ensuring a suitable pH. At the same time, the intrinsic properties of the enzyme should be maintained, such as its stability. This is followed easily by measuring ATPase activity in the reaction medium in the absence of modifiers. F_1-ATPase is equilibrated before modification with the required medium by dialysis or gel filtration. The chromatography–centrifugation procedure described by Penefsky[8] is convenient. Tuberculin syringes (1 ml) filled with Sephadex G-50 (fine) are used with a volume of enzyme solution of 0.1 ml. Larger volumes can be processed in larger syringes.

In the absence of protecting agents such as glycerol or methanol, the cold sensitivity of the isolated F_1 should not be forgotten.

Modification of Carboxylic Groups in F_1-ATPases

Carbodiimides

Mechanism of Reaction. The mechanism of interaction of a typical carbodiimide reagent, dicyclohexylcarbodiimide (**I**), with a carboxyl group in a protein is shown in Fig. 1.[9] An *O*-acylisourea (**II**) is first formed. It is unstable and it can further rearrange into a stable *N*-acylurea (**III**). The extent of this reaction pathway, (*i*), can be accurately measured by means of a radioactive carbodiimide. Alternatively (pathway *ii*), *O*-acylisourea can react with an externally added nucleophile (HX) to give

[8] H. S. Penefsky, this series, Vol. 56, p. 527.

[9] A. Williams and I. T. Ibrahim, *Chem. Rev.* 589 (1981).

the derivative (**IV**) and dicyclohexylurea (DCU) (**V**). When the nucleophile is a water molecule, the native carboxyl group of the protein is restored. Several amines such as lysine, proflavin, or glycine ethyl ester have been used as nucleophiles to ascertain that the carbodiimide was bound to a carboxyl group in the protein. The reaction can also occur with a nucleophile residue in the protein (pathway *iii*), for example, the ε-amino group of a lysine residue, provided that this lysine is close to the carbodiimide-activated carboxyl; in this case (**VI**), a stable "zero-length" cross-link is formed.[10-14] There is no absolute selectivity of carbodiimides for carboxyl groups, and other amino acid residues, such as serine, tyrosine, or thiols, may react with carbodiimide in native proteins.[9]

Modification Procedures. Three carbodiimides have been used as modifiers of F_1-ATPases; among them, two are hydrophilic: 1-ethyl-3-(3-dimethylaminopropyl)carbodiimide (EDAC) and 1-cyclohexyl-3-(2-morpholinoethyl)carbodiimide (CMCD); and one is hydrophobic: dicyclohexylcarbodiimide (DCCD). Radioactive [^{14}C]DCCD is available commercially (CEA, Saclay, France) and the synthesis of [^{14}C]EDAC and [^{3}H]CMCD have been described.[11,13] A spin-labeled carbodiimide, 1-cyclohexyl-3-(2,2,6,6-tetramethylpiperidyl-1-oxyl)carbodiimide (NCCD), has also been synthesized and used for the measurement of distance between sites in the ATPase complex.[15]

In modification experiments, 0.1–0.5 *M* stock solutions of CMCD, EDAC, and DCCD in methanol are used; they can be stored at $-20°$ in tightly closed vials. The melting point of DCCD is 34–35°, and commercial DCCD appears as a solid mass at room temperature. Sampling can be done easily after melting DCCD by heating at 50°. This allows also the removal of any insoluble DCU (melting point: 232°).[16] [^{14}C]DCCD is stored as a 10-m*M* stock solution in methanol at $-20°$. Its radiochemical purity is checked by thin-layer chromatography on aluminum oxide with benzene as developing solvent, followed by autoradiography.[17] When [^{14}C]DCCD solutions are diluted with unlabeled DCCD, the above-described purification step to remove DCU is critical. If care is taken to avoid contact with atmospheric moisture during the handling procedure,

[10] E. A. Imedidze, I. A. Kozlov, V. A. Metelskaya, and Y. M. Milgrom, *Biokhimiya* **43,** 1404 (1978).

[11] I. A. Kozlov and V. P. Skulachev, *Biochim. Biophys. Acta* **463,** 29 (1977).

[12] R. Pougeois, M. Satre, and P. V. Vignais, *Biochemistry* **18,** 1408 (1979).

[13] R. L. Timkovich, *Anal. Biochem.* **79,** 135 (1977).

[14] R. Timkovich, *Biochem. Biophys. Res. Commun.* **74,** 1463 (1977).

[15] A. Azzi, M. A. Bragadin, A. M. Tamburro, and M. Santato, *J. Biol. Chem.* **248,** 5520 (1973).

[16] L. F. Fieser and M. Fieser, *in* "Reagents for Organic Synthesis," Vol. 1, p. 231. Wiley, New York, 1967.

[17] R. H. Fillingame, *J. Bacteriol.* **124,** 870 (1975).

DCCD solutions should remain stable for more than 6 months. Dilutions of the reagents are prepared in methanol as required and used immediately; they are not stored.

Experimental conditions published for the modification of various F_1-ATPases by DCCD, EDAC, or CMCD are given in Table I.[10,12,18–23] In all cases, the inactivation rates are reported to follow pseudo-first-order kinetics; inactivation is always much more effective at acidic pH values. In our hands, the following standard medium has been found suitable for studies on F_1 inactivation by carbodiimides: 50 m*M* MOPS-NaOH, 2 m*M* ATP, and 1 m*M* EDTA, pH 6.5. ATP potentiates the inactivation of F_1 by DCCD and it also stabilizes the enzyme when long incubation periods are required. Due to its limited solubility in water, the upper limit of concentration for DCCD is about 0.2 m*M*.

Inactivation assays with carbodiimides can be terminated by eliminating the free reagent by gel exclusion chromatography, using, for example, the elution centrifugation method of Penefsky.[8] Dilution of the inactivation mixture in buffers with basic pHs (pH 8.5) is also effective in slowing down the rate of inactivation to a negligible value.

Gel exclusion chromatography as a means to separate the free reagent from the modified F_1 is the method of choice to follow incorporation of either radioactive carbodiimides (pathway *i*) or a radioactive nucleophile (pathway *ii*) (Fig. 1). The number of carbodiimide molecules incorporated in F_1-ATPases from different species for an extrapolated inactivation of 100% are given in Table II[10–12,18,20–22,24–28]. Labeling is always associated with one or two β subunits of F_1, except in the case of F_1-ATPase from *Mycobacterium phlei* where the α and γ subunits are labeled. Labeling of one or two β subunits out of the three present in F_1 for full inactivation to occur is typical of a partial-site reactivity.[29]

[18] V. Shoshan and B. R. Selman, *J. Biol. Chem.* **255,** 384 (1980).

[19] A. N. Malyan, S. D. Zakharov, and I. I. Proskuryakov, *Biochem. Physiol. Pflanzen* **176,** 828 (1981).

[20] M. Satre, J. Lunardi, R. Pougeois, and P. V. Vignais, *Biochemistry* **18,** 3134 (1979).

[20a] H. R. Loetscher, C. de Jong, and R. A. Capaldi, *Biochemistry* **23,** 4134 (1984).

[21] N. Agarwal and V. K. Kalra, *Biochim. Biophys. Acta* **764,** 105 (1984).

[22] D. Khananshvili and Z. Gromet-Elhanan, *J. Biol. Chem.* **258,** 3720 (1983).

[23] M. Yoshida and W. S. Allison, *J. Biol. Chem.* **258,** 14407 (1983).

[24] H. R. Loetscher and R. A. Capaldi, *Biochem. Biophys. Res. Commun.* **121,** 331 (1984).

[25] M. Yoshida, W. S. Allison, F. S. Esch, and M. Futai, *J. Biol. Chem.* **257,** 10033 (1982).

[26] J. L. Arana and R. H. Vallejos, *FEBS Lett.* **123,** 103 (1981).

[27] E. Ceccarelli and R. H. Vallejos, *Arch. Biochem. Biophys.* **224,** 382 (1983).

[28] M. Yoshida, J. W. Poser, W. S. Allison, and F. S. Esch, *J. Biol. Chem.* **256,** 148 (1981).

[29] P. V. Vignais, A. Dupuis, J. P. Issartel, G. Klein, J. Lunardi, M. Satre, and J. J. Curgy, *in* "H^+-ATPsynthase: Structure, Function. The F_0–F_1 Complex of Coupling Membranes" (S. Papa, K. Altendorf, L. Ernster, and L. Packer, eds.), pp. 205–217. Adriatica Editrice, Bari, 1985.

TABLE I
INACTIVATION OF F_1-ATPases BY CARBODIIMIDES

Source of F_1-ATPase	Protein (mg/ml)	Incubation medium	Carbodiimide	Temperature (°C)	$t_{1/2}$ (min)	Reference
Beef heart mitochondria	0.5	50 m*M* MOPS-NaOH, 4 m*M* ATP, 2 m*M* EDTA, pH 6.5	200 μ*M* DCCD	24	12	12
	Not given	500 m*M* MOPS-NaOH, pH 6.8	1000 μ*M* CMCD	Not given	9	10
Spinach chloroplasts	0.4	100 m*M* MOPS-NaOH, 1 m*M* EDTA, pH 7.0	33 μ*M* DCCD	37	13	18
	0.12	40 m*M* Mes, pH 6.1	2000 μ*M* CMCD	Room temperature	26	19
Escherichia coli	0.5	25 m*M* MOPS-NaOH, pH 6.5	10 μ*M* DCCD	37	15	20
	0.5	25 m*M* MOPS-NaOH, pH 6.5	500 μ*M* CMCD	37	9	20
	0.5	25 m*M* MOPS-NaOH, pH 6.5	250 μ*M* EDAC	37	30	20
	0.8	20 m*M* MOPS, pH 6.5	4 m*M* EDAC	Room temperature	7	20a
Mycobacterium phlei	1.0	50 m*M* Mes-NaOH, pH 6.0	200 μ*M* DCCD	37	24	21
Rhodospirillum rubrum	1.0	50 m*M* Mes-NaOH, 2 m*M* EDTA, 50 m*M* NaCl, 20% (v/v) glycerol, pH 6.0	40 μ*M* DCCD	30	30	22
Thermophilic bacterium PS3	0.1	50 m*M* triethanolamine sulfate, 1 m*M* EDTA, 2 m*M* ADP, pH 7.3	150 μ*M* DCCD	20	28	23

TABLE II
INCORPORATION OF CARBODIIMIDES DURING INACTIVATION OF F_1-ATPASES (VALUES EXTRAPOLATED TO 100% INACTIVATION)

Source of enzyme	Reagent incorporated (mol/mol F_1)	Remarks	Reference
Beef heart mitochondria	2 DCCD	Only 1 DCCD incorporated in the presence of 20 m*M* glycine ethyl ester	12
Beef heart mitochondria	1 CMCD	In the presence of 10 m*M* lysine or 1 m*M* proflavine: 1 lysine or 1 proflavine incorporated per F_1	10, 11
Spinach chloroplasts	2 DCCD	Only 1 DCCD incorporated per F_1 in the presence of 50 m*M* $CaCl_2$	18, 26
Escherichia coli	1 DCCD	No DCCD incorporated in the presence of 20 m*M* glycine ethyl ester	20, 24, 25
Mycobacterium phlei	22 DCCD	Biphasic DCCD incorporation curve 4 DCCD binding sites per F_1 correlated with the initial inactivation curve	21
Rhodospirillum rubrum	1 DCCD		22
	or 2 DCCD	Only 1 DCCD incorporated per F_1 in the presence of 30 m*M* $MgCl_2$; no inactivation of the enzyme	27
Thermophilic bacterium PS3	2 DCCD		28

Application. Under controlled conditions leading to about 80% inactivation, and avoiding excess of chemical reagent, DCCD is the compound of choice to label specifically the β subunit of F_1-ATPases. The position of the DCCD-modified glutamic acid residue in the amino acid sequence of the β subunit has been determined in F_1 of beef heart mitochondria and *Escherichia coli;* it corresponds to Glu 199 and 192, respectively.[25,30] A similar sequence has been found in the same region of the β subunit of chloroplast F_1-ATPase. The DCCD binding region of the β subunit in F_1-ATPases has no counterpart in the α subunit despite sequence analogies between the α and β subunits.[31,32]

[30] F. S. Esch, P. Böhlen, A. S. Otsutka, M. Yoshida, and W. S. Allison, *J. Biol. Chem.* **256,** 9084 (1981).

[31] H. Deno and M. Sugiura, *FEBS Lett.* **172,** 209 (1984).

[32] J. E. Walker, A. Eberle, N. J. Gay, M. J. Runswick, and M. Saraste, *Biochem. Soc. Trans.* **10,** 203 (1982).

The use of DCCD as a chemical reagent to block the proton conduction in the F_0 channel of the ATPase complex at the level of the so-called DCCD-binding protein is well documented,[17,33] but the efficient and highly selective reaction of DCCD with the F_1 sector at slightly acidic pH (pH 6.5) is often neglected. It is noteworthy that a basic pH (pH 8.0) is required for DCCD to react selectively with the F_0 sector when DCCD is added to membranes or solubilized preparations of F_0F_1-ATPases.[34,35] A suitable incubation medium for selective labeling of F_0 is 50 m*M* Tris-Cl and 10 m*M* $MgCl_2$, pH 8.0. Besides the alkaline pH, which is unfavorable for F_1 modification, the presence of magnesium protects F_1 against DCCD inactivation.

N-(Ethoxycarbonyl)-2-ethoxy-1,2-dihydroquinoline (EEDQ) and N-(isobutoxycarbonyl)-2-isobutoxy-1,2-dihydroquinoline (IIDQ)

Mechanism of Reaction. The activation reaction of a carboxyl group by EEDQ or IIDQ (**I**) is shown in Fig. 2.[36] The postulated intermediate (**II**) gives the mixed carbonic anhydride (**III**). Compound **III** is of limited stability in water solutions, but it is likely to be stabilized in an environment excluding water such as an hydrophobic pocket of an enzyme. Similarly to the carbodiimide-activated carboxyl groups, the mixed anhydride can further react with a nucleophilic residue (HX) to give rise to an adduct (**V**) or to a "zero-length" cross-linking product (**VI**).

Application. EEDQ and IIDQ are prepared as 0.1 *M* stock solutions in methanol and stored at −20°. Handling care is as described for carbodiimides. If the solutions turn yellow, they must be discarded. Radioactive EEDQ labeled by ^{14}C in the ethoxycarbonyl moiety has been used for direct measurements of incorporation of the ethoxycarbonyl moiety into chymotrypsin and was shown to be linearly correlated with the degree of inactivation.[37] Similar experiments with F_1-ATPase have not yet been reported. An indirect method to assay the EEDQ-binding stoichiometry consists of displacing the bound ethoxycarbonyl group by a radiolabeled nucleophile. For example, *E. coli* F_1-ATPase incubated simultaneously in the presence of EEDQ and the radioactive nucleophile [^{14}C]glycine ethyl ester binds the radioactive glycine derivative covalently at the level of the β subunit. Complete inactivation corresponds to the covalent incorporation of one glycine ester moiety per F_1-ATPase.[38]

[33] R. B. Beechey, A. M. Robertson, T. Holloway, and I. G. Knight, *Biochemistry* **6**, 3867 (1967).

[34] P. Friedl, B. I. Schmid, and H. U. Schairer, *Eur. J. Biochem.* **73**, 461 (1977).

[35] R. Pougeois, M. Satre, and P. V. Vignais, *FEBS Lett.* **117**, 344 (1980).

[36] B. Belleau and G. Malek, *J. Am. Chem. Soc.* **90**, 1651 (1968).

[37] W. T. Robinson and B. Belleau, *J. Am. Chem. Soc.* **94**, 4376 (1972).

[38] M. Satre, A. Dupuis, M. Bof, and P. V. Vignais, *Biochem. Biophys. Res. Commun.* **114**, 684 (1983).

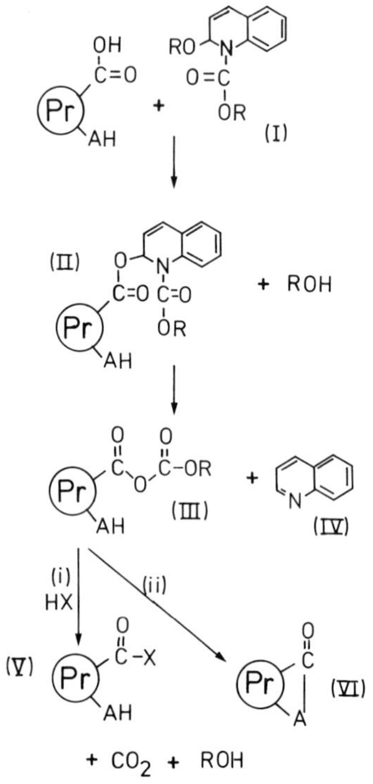

FIG. 2. Modification of carboxyl groups by EEDQ (R = CH_3—CH_2—) or IIDQ (R = $(CH_3)_2$—CH—CH_2). Details are given in the text.

Experimental conditions published for inactivation of various F_1-ATPases are reported in Table III.[20–22,39,40] In all cases, inactivation is faster at acidic pH values. At a given pH, it is slowed down by divalent cations. To stop inactivation by EEDQ or IIDQ, the unreacted modifier is eliminated by gel exclusion chromatography (see previous section on carbodiimides). Incubation can also be terminated by dilution with an alkaline buffer to bring the pH to 8.5–9.0. Alternatively, the addition of 1 m*M* hydroxylamine or thiol compounds such as 3 m*M* 2-mercaptoethanol or 1 m*M* dithiothreitol stops very efficiently the inactivation process. No reactivation of the inactivated enzyme molecules is observed.[39] EEDQ and DCCD react with different carboxyl groups both on beef heart mitochondrial F_1 and *E. coli* F_1.[12,20,41] EEDQ-activated carboxyl might be directly involved in the binding of Mg^{2+} at catalytic sites.[41]

[39] R. Pougeois, M. Satre, and P. V. Vignais, *Biochemistry* **17,** 3018 (1978).

[40] Y. S. Chung and M. R. J. Salton, *FEMS Lett.* **8,** 183 (1980).

[41] R. Pougeois, *FEBS Lett.* **154,** 47 (1983).

TABLE III
MODIFICATION OF F_1-ATPases BY EEDQ AND IIDQ

Source of F_1-ATPase	Protein (mg/ml)	Incubation medium	EEDQ or IIDQ	Temperature (°C)	$t_{1/2}$ (min)	Reference
Beef heart mitochondria	0.5	50 m*M* MOPS-NaOH, pH 7.0	400 μM EEDQ	24	15	39
Escherichia coli	0.5	25 m*M* MOPS-NaOH, pH 6.5	20 μMEEDQ	37	16	20
	0.5	25 m*M* MOPS-NaOH, pH 6.5	20 μM IIDQ	37	7	20
Micrococcus lysodeikticus	0.1	100 m*M* Tris-Cl, 8 m*M* $CaCl_2$, pH 7.5	800 μM EEDQ	37	30	40
Mycobacterium phlei	1.0	50 m*M* Mes-NaOH, pH 6.0	50 μM EEDQ	37	30	21
Rhodospirillum rubrum	1.0	50 m*M* Mes-NaOH, 2 m*M* EDTA, 50 m*M* NaCl, 20% (v/v) glycerol, pH 6.0	40 μM EEDQ	30	30	22

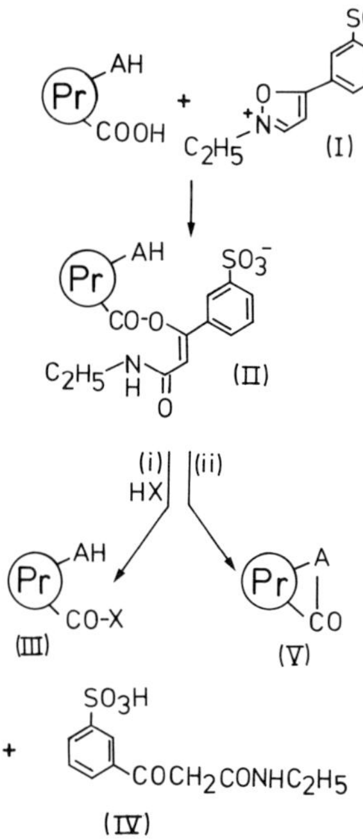

FIG. 3. Modification of carboxyl groups by Woodward's reagent K.

EEDQ and IIDQ also react with the F_0 sector of chloroplast ATPase; however, the rate of the reaction is 20 times lower than with DCCD. Prior incubation of thylakoid membranes with EEDQ reduces subsequent incorporation of [^{14}C]DCCD.[42]

N-Ethyl-5-phenylisoxazolium 3'-sulfonate (Woodward's Reagent K)

Mechanism of Reaction. As shown in Fig. 3, Woodward's reagent K (**I**) reacts with a carboxyl group to form an activated ester (**II**) which on reaction with a nucleophilic residue (HX) gives a covalent derivative and a water-soluble by-product (**IV**).[43] A zero-length product (**V**) can be formed.

Application. Experimental conditions are given in Table IV.[26,27,44] Inactivation occurs efficiently only at basic pH; it is negligible at pH 6.0–6.5.[20,22] Inactivation rates with the Woodward's reagent K follow pseudo-

[42] Y. K. Ho and J. H. Wang, *Biochemistry* **19**, 2650 (1980).

[43] L. F. Fieser and M. Fieser, *in* "Reagents for Organic Synthesis," Vol. 1, p. 384. Wiley, New York, 1967.

[44] J. L. Arana, M. Yoshida, Y. Kagawa, and R. H. Vallejos, *Biochim. Biophys. Acta* **593,** 11 (1980).

TABLE IV
INACTIVATION OF F_1-ATPases BY WOODWARD'S REAGENT K

Source of enzyme	Protein (mg/ml)	Incubation medium	Woodward's K (m*M*)	Temperature (°C)	$t_{1/2}$ (sec)	Reference
Spinach chloroplast	0.5	40 m*M* Tricine-NaOH, pH 7.9	1	25	30	26
Rhodospirillum rubrum	Not given	250 m*M* Sucrose, 5 m*M* $MgCl_2$ 20 m*M* Tricine-NaOH, pH 8.0	0.625	25	41	27
Thermophilic bacterium PS3	Not given	100 m*M* Tricine-NaOH, pH 8.0	1	25	60	44

first-order kinetics only for a limited period of time because the reagent itself is unstable in the presence of water.

"Zero-Length" Cross-Linking with Carboxyl Group Reagents

All the carboxyl group reagents are potential zero-length cross-linkers provided that reactive residues, mostly ε-amino groups of lysine, are located in close proximity of the activated carboxyl groups. Cross-linking of F_1-ATPases is easily followed by sodium dodecyl sulfate–polyacrylamide gel electrophoresis. Intermolecular cross-linking by random collisions[14] between independent F_1 molecules is not predominant at the protein concentrations used for chemical modification (about 1 mg/ml). Intramolecular cross-linking can establish bridges in a single subunit (hairpin formation) or between two adjacent subunits. The latter type of cross-linking involves sites of contact between subunits which are thought to be of critical importance in the catalytic mechanism of oligomeric enzymes.

Cross-linking requires higher concentrations of reagents than inactivation. With DCCD or IIDQ, two highly hydrophobic reagents, almost no cross-linking is observed at concentrations which fully inactivate F_1.[20] EEDQ and EDAC have been used to cross-link beef heart mitochondrial F_1-ATPase to IF_1 and OSCP (see "Noninactivating Modifications").

Modification of Tyrosine Residues in F_1-ATPases

4-Chloro-7-nitrobenzofurazan (Nbf-Cl), Called Also 7-Chloro-4-nitrobenzofurazan or 7-Chloro-4-nitrobenzo-2-oxa-1,3-diazole, (NBD-Cl)

Mechanism of Reaction. Nbf-Cl has been used as a labeling reagent in the studies of a number of proteins. Reaction with thiol, amino, and tyrosyl groups have been observed.[45] The nitrobenzofurazan moiety becomes covalently attached to a protein nucleophilic atom which replaces the chlorine at position C-4. This is illustrated in Fig. 4 by the binding of Nbf-Cl to a tyrosyl residue of a protein. Nbf bound to a tyrosyl group in a protein can be transferred at alkaline pH to a suitably aligned amino group in the same protein.[46]

Application. Nbf-Cl is prepared as a 50-m*M* stock solution in methanol and stored at −20°. Because of the propensity of Nbf to photolytic decomposition, Nbf-Cl solutions must be protected from strong light irradiation. Radioactive [^{14}C]Nbf-Cl is available commercially (CEA, Saclay, France).

[45] D. J. Birkett, N. C. Price, G. K. Radda, and A. G. Salmon, *FEBS Lett.* **6,** 346 (1970).
[46] S. J. Ferguson, W. J. Lloyd, and G. K. Radda, *Eur. J. Biochem.* **54,** 127 (1975).

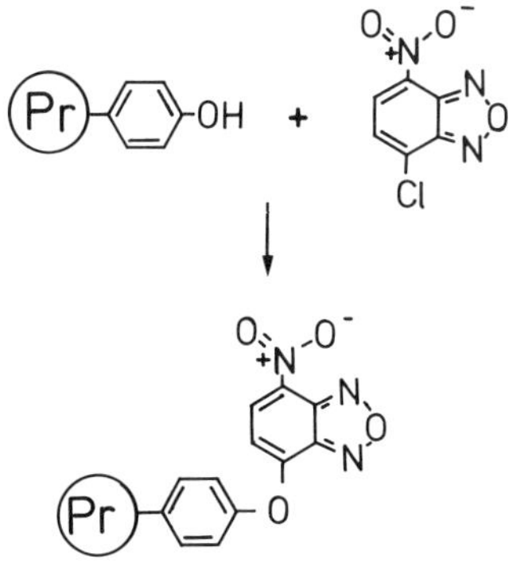

FIG. 4. Modification of tyrosine residues by Nbf-Cl.

Experimental conditions for the modification of various F_1-ATPases by Nbf-Cl are given in Table V.[21,47–50] The inactivation rates follow pseudo-first-order kinetics. Binding of Nbf-Cl to a tyrosine residue is conveniently followed by the increase in absorbance at 385 nm. The stoichiometry of incorporation is calculated using an absorption coefficient of 11,600 M^{-1} cm^{-1}.[51] Alternatively, incorporation of radioactive Nbf-Cl can be directly measured.

Incorporation of 1 mol Nbf per mol F_1 is sufficient for full inactivation to occur as reported for F_1 of beef heart,[47] yeast,[48] *E. coli*,[49] and *Rhodospirillum rubrum*.[50] This is consistent with a third of the site reactivity.

At neutral pH, the tyrosyl-bound Nbf (O-Nbf) is efficiently displaced by thiols. This is the basis for a method of measurement of the content of F_1 in submitochondrial particles. The strategy is to inhibit a defined proportion of the ATPase molecules in submitochondrial particles by reaction with Nbf-Cl and then to add cysteine to remove the Nbf group from the enzyme and form a thiol-Nbf derivative (S-Nbf). After centrifugation of the particles, the S-Nbf group in the supernatant can be measured spectrophotometrically at 425 nm.[52]

The O-Nbf derivative can be stabilized at acidic or neutral pH by reduction.[53,54] There are conflicting results on the position of the Nbf

[47] S. J. Ferguson, W. J. Lloyd, M. H. Lyons, and G. K. Radda, *Eur. J. Biochem.* **54,** 117 (1975).

[48] R. Gregory, D. Recktenwald, and B. Hess, *Biochim. Biophys. Acta* **635,** 284 (1981).

[49] J. Lunardi, M. Satre, M. Bof, and P. V. Vignais, *Biochemistry* **18,** 5310 (1979).

[50] D. Khananshvili and Z. Gromet-Elhanan, *J. Biol. Chem.* **258,** 3714 (1983).

[51] A. A. Aboderin, E. Boedefeld, and P. C. Luisi, *Biochim. Biophys. Acta* **238,** 20 (1973).

[52] S. J. Ferguson, W. J. Lloyd, and G. K. Radda, *Biochem. J.* **159,** 347 (1976).

[53] J. W. Ho and J. H. Wang, *Biochem. Biophys. Res. Commun.* **116,** 599 (1983).

[54] W. W. Andrews, F. C. Hill, and W. S. Allison, *J. Biol. Chem.* **259,** 8219, 1984.

TABLE V
INACTIVATION OF F_1-ATPases WITH Nbf-Cl

Source of F_1-ATPase	Protein (mg/ml)	Incubation medium	Nbf-Cl (μM)	Temperature (°C)	$t_{1/2}$ (min)	Reference
Beef heart mitochondria	3.3	250 m*M* sucrose, 50 m*M* triethanolamine-KOH, 4 m*M* ATP, 4 m*M* EDTA, pH 7.5	100	30	11	47
Yeast mitochondria	5.4	200 m*M* sucrose, 50 m*M* triethanolamine, 4 m*M* EDTA, pH 7.5	300	Room temperature	6	48
Escherichia coli	2.0	250 m*M* sucrose, 30 m*M* triethanolamine-KOH, 2 m*M* EDTA, pH 7.5	50	30	20	49
Mycobacterium phlei	1.0	50 m*M* Tris-acetate, 150 m*M* KCl, 4 m*M* $MgCl_2$, pH 8.0	100	37	36	21
Rhodospirillum rubrum	1.0	50 m*M* Tricine-NaOH, 2 m*M* EDTA, 50 m*M* NaCl, 20% (v/v) glycerol, pH 7.6	50	30	30	50

tyrosine in the amino acid sequence of the β subunit of beef heart mitochondrial F_1; the modified tyrosine was reported to be Tyr-197[53] and Tyr-311.[54]

At alkaline pH, the reversibly modified form of F_1, O-Nbf-F_1, is transferred intramolecularly to a lysine residue in the β subunit to form a stable derivative, N-Nbf-F_1, with specific fluorescent properties.[46,47]

The transfer to a lysine residue of a β subunit is more difficult to achieve in *E. coli* F_1 than in beef heart F_1.[49] With F_1 from some *uncA* mutants, Nbf-Cl reacts equally well with the α and β subunits, in contrast to the wild-type F_1 where β is preferentially labeled.[55]

Tetranitromethane (TNM)

Mechanism of Reaction. Nitration of tyrosine residues in proteins with TNM gives a 3-nitrotyrosine derivative. The reaction is very slow at acidic pH and rapid at basic pH. Nitration proceeds through a free-radical mechanism, and side reactions such as cross-linking can occur. Amino acid residues like cysteine, tryptophan, histidine, or methionine may react as well with TNM.[56]

Application. Diluted TNM solutions are prepared in anhydrous ethanol and used immediately. The following experimental conditions have been used for nitration of F_1.[57] Beef heart mitochondrial F_1 (1 mg/ml) in 50 mM Tris-sulfate, 1 mM EDTA, 60 mM K_2SO_4, pH 8.0, is incubated at 30° with 80 μg/ml TNM. The molar ratio, TNM/F_1, is 150. The kinetics of reaction are complex; they are further complicated by the rapid breakdown of TNM. The reaction is terminated by gel filtration. After 15 min of reaction at pH 8.0, about 90% of the ATPase activity has disappeared. There is a downward shift in the pH optimum of ATPase activity of the nitrated protein; in fact, when assayed at pH 6.0, the modified enzyme hydrolyzes ATP faster than the native enzyme.[58,59] To determine the extent of nitration, the enzyme is treated by 0.1 M Tris-sulfate, 6 M urea, pH 8.0, and the nitrotyrosine content is calculated using an absorption coefficient of 4100 M^{-1} cm^{-1} at 428 nm. Under the above conditions, there are about six nitrated tyrosine residues per F_1; the unique tyrosine residue which is modified by Nbf-Cl is not nitrated.[52] Using a higher molar excess of TNM (molar ratio TNM/F_1 = 250), 9–10 tyrosine residues are nitrated and inactivation of ATPase activity becomes higher than 95%.

[55] A. E. Senior, L. R. Latchney, A. M. Ferguson, and J. G. Wise, *Arch. Biochem. Biophys.* **228,** 49 (1984).

[56] M. Sokolovsky, J. F. Riordan, and B. L. Vallee, *Biochemistry* **5,** 3582 (1966).

[57] A. E. Senior, *Biochemistry* **12,** 3622 (1973).

[58] S. Risi, C. Schroeder, H. J. Kaiser, J. Carreira, and K. Dose, *Hoppe-Seyler's Z. Physiol. Chem.* **359,** 37 (1978).

[59] L. J. Dorgan and S. M. Schuster, *J. Biol. Chem.* **256,** 3910 (1981).

1-Fluoro-2,4-dinitrobenzene (FDNB)

Mechanism of Reaction. FDNB is a classical reagent of amino groups,[60] but it also reacts with tyrosine, histidine, and cysteine residues. It is the reaction of FDNB with a tyrosine residue which is of interest in the case of F_1 inactivation.

Application. FDNB is prepared as a 0.1-*M* stock solution in ethanol. Experimental conditions for the modification of beef heart mitochondrial and *E. coli* F_1 have been described.[61–63] The reagent is employed at pH 8.0. Inactivation follows pseudo-first-order kinetics. It is stopped by elimination of the free reagent by gel exclusion chromatography, as previously mentioned. It is possible to reactivate FDNB-treated F_1 by incubation with high concentrations (90 m*M*) of dithiothreitol. The same tyrosine residue is modified by Nbf-Cl and FDNB,[62,63] but it is another tyrosine residue that binds covalently the nucleotide analog 5′-*p*-fluorosulfonylbenzoyladenosine.[64–66]

Modification of Arginine and Lysine Residues

The α,α'-dicarbonyl compounds are selective reagents of arginine residues which are often involved in binding of nucleotides at the binding sites of phosphotransferases.[67] Arginine reagents such as butanedione or phenylglyoxal inactivate F_1-ATPases. Radioactive substituted glyoxals allow a direct analysis of the correlation between the loss of ATPase activity and the reagent incorporation. Labeling is extensive, but it is possible to detect 1 or 2 arginine residues per mol of F_1 reacting more rapidly than the others or which could be protected against modification either by nucleotides or by efrapeptin, a polypeptide inhibitor of F_1-ATPases.[44,68–71]

Evidence for the presence of essential lysine residues in F_1-ATPases has been obtained with reagents such as pyridoxal phosphate or 2,4,6-

[60] K. Narita, H. Matsuo, and T. Nakajima, *in* "Protein Sequence Determination" (S. B. Needleman, ed.), p. 32. Springer-Verlag, Berlin and New York, 1975.

[61] L. P. Ting and J. H. Wang, *Biochemistry* **19,** 5665 (1980).

[62] L. P. Ting and J. H. Wang, *Biochemistry* **21,** 269 (1982).

[63] W. W. Andrews and W. S. Allison, *Biochem. Biophys. Res. Commun.* **99,** 813 (1981).

[64] F. S. Esch and W. S. Allison, *J. Biol. Chem.* **253,** 6100 (1978).

[65] F. S. Esch and W. S. Allison, *J. Biol. Chem.* **254,** 10740 (1979).

[66] E. De Benedetti and A. Jagendorf, *Biochem. Biophys. Res. Commun.* **86,** 440 (1979).

[67] J. F. Riordan, K. D. McElvany, and C. L. Borders, *Science* **195,** 884 (1977).

[68] F. Marcus, S. M. Schuster, and H. A. Lardy, *J. Biol. Chem.* **251,** 1775 (1976).

[69] W. E. Kohlbrenner and R. L. Cross, *J. Biol. Chem.* **253,** 7609 (1978).

[70] A. M. Viale, C. S. Andreo, and R. H. Vallejos, *Biochim. Biophys. Acta* **682,** 135 (1982).

[71] T. Takabe, E. De Benedetti, and A. Jagendorf, *Biochim. Biophys. Acta* **682,** 11 (1982).

trinitrobenzene sulfonate. The selectivity of the modifiers appears to be poor, since complete inactivation usually requires the binding of 6–10 mol reagent per mol F_1.[72–76] In this context, the dial nucleotides (oAMP, oADP, and oATP) have been found to behave as nonselective lysine reagents.[77,78]

Noninactivating Modifications

Noninactivating modifications by radiolabeled reagents have been applied to OSCP and IF_1, two peptides belonging to the mitochondrial ATPase complex. Through the use of these radiolabeled and still fully active peptides, it has been possible to determine their binding stoichiometry with respect to F_1 and the nature of the interacting subunits in F_1.

Radiolabeled Oligomycin-Sensitivity Conferring Protein (OSCP)

Principle. OSCP is an elongated peptide (axial ratio >3) of M_r 20,967 which links the F_1 and F_0 sectors of the mitochondrial ATPase complex.[7,79,80] It makes the ATPase activity sensitive to oligomycin, an antibiotic which binds to the F_0 sector of the ATPase complex. Because OSCP is devoid of biological activity per se, exploration of its interaction with F_1 and F_0 requires an appropriate labeling of this molecule. Advantage has been taken of the fact that OSCP possesses a single cysteine residue[80] and that this cysteine residue can be alkylated with *N*-[^{14}C]ethylmaleimide (NEM) without any alteration in the functioning of OSCP.[81]

Radiolabeling of OSCP. OSCP (4 mg) in 0.05 *M* Tris-sulfate, pH 7.4, prepared as described by Senior,[7] is dialyzed against 0.1 *M* Na-phosphate, pH 7.0, for 5 hr, and then incubated at 0° for 1 hr at 2.5 mg/ml with 8 m*M* [^{14}C]NEM (specific radioactivity, 10^8 dpm/μmol). Alkylation is stopped with 10 m*M* dithiothreitol. [^{14}C]NEM-OSCP is then precipitated at 50% saturation ammonium sulfate at 0° for 1 hr. The precipitate is

[72] C. Godinot, F. Penin, and D. C. Gautheron, *Arch. Biochem. Biophys.* **192,** 225 (1979).

[73] Y. Sugiyama and Y. Mukohata, *FEBS Lett.* **98,** 276 (1979).

[74] H. Peters, S. Risi, and K. Dose, *Biochem. Biophys. Res. Commun.* **97,** 1215 (1980).

[75] P. G. Koga and R. L. Cross, *Biochim. Biophys. Acta* **679,** 269 (1982).

[76] L. P. Ting and J. H. Wang, *Biochem. Biophys. Res. Commun.* **101,** 934 (1981).

[77] D. Fernandes de Melo, M. Satre, and P. V. Vignais, *FEBS Lett.* **169,** 101 (1984).

[78] P. D. Bragg, H. Stan-Lotter, and C. Hou, *Biochem. Biophys. Res. Commun.* **207,** 290 (1981).

[79] A. Dupuis, G. Zaccaï, and M. Satre, *Biochemistry* **22,** 5951 (1983).

[80] Y. A. Ovchinnikov, N. N. Modyanov, V. A. Grinkevich, N. A. Aldanova, O. E. Trubetskaya, I. V. Nazimov, T. Hundal, and L. Ernster, *FEBS Lett.* **166,** 19 (1984).

[81] A. Dupuis, J. P. Issartel, J. Lunardi, M. Satre, and P. V. Vignais, *Biochemistry* **24,** 728 (1985).

recovered by centrifugation for 4 min at 10,000 *g* and solubilized in 0.05 *M* Tris-sulfate, pH 7.5. Any insoluble material is removed by centrifugation. Finally, the [^{14}C]NEM-OSCP preparation is either dialyzed against 0.05 *M* Tris-sulfate, pH 7.5, or chromatographed on a column (25 × 1 cm) filled with Sephadex G-50 (fine) and equilibrated in 0.05 *M* Tris-sulfate, pH 7.5. Under these conditions, 1 mol [^{14}C]NEM binds covalently to 1 mol OSCP. [^{14}C]NEM-OSCP mimics unlabeled OSCP, so that both labeled and unlabeled OSCP have exactly the same efficiency to confer to ATPase activity the sensitivity to oligomycin in a reconstituted F_1F_0 complex.[81,82]

Application. [^{14}C]NEM-OSCP has been used to determine the binding stoichiometry of OSCP with respect to F_1. Binding measurements by equilibrium dialysis have shown three OSCP binding sites on beef heart F_1. These sites differ in affinity; one of these sites has a much higher affinity ($K_d = 0.08\ \mu M$) than the other two ($K_d \simeq 6$–$8\ \mu M$).[81] By means of cross-linking with the zero-length cross-linkers, EDAC and EEDQ (see above), OSCP has been found to interact with both the α subunit and the β subunit of mitochondrial F_1.[83]

Radiolabeled IF_1

Principle. The mitochondrial IF_1[6,84,85] is a heat- and acid-stable protein of $M_r \simeq 9600$[86]; the N-terminal amino acid of beef heart IF_1 is blocked by a formyl residue.[87] IF_1 binds to F_1 and may play a regulatory function in oxidative phosphorylation.[88–90] To elucidate the mechanism of this function in terms of molecular interactions, a radiolabeled, fully active IF_1 is required. In the following, two procedures of radiolabeling for IF_1 are described. One is based on the use of phenyl [^{14}C]isothiocyanate (Fig. 5a), the other on that of [^{14}C]methylazidobenzoimidate (Fig. 5b). IF_1 is purified by the method of Horstman and Racker,[91] as modified by Klein *et al.*[92]

Radiolabeling of IF_1 with Phenyl [^{14}C]isothiocyanate ([^{14}C]PITC). [^{14}C]PITC (100 μCi, 10 m*M* final concentration) in dimethyl sulfoxide is

[82] A. Dupuis, M. Satre, and P. V. Vignais, *FEBS Lett.* **156,** 99 (1983).
[83] A. Dupuis, J. Lunardi, J. P. Issartel, and P. V. Vignais, *Biochemistry* **24,** 734 (1985).
[84] L. Ernster, C. Carlsson, T. Hundal, and B. Nordenbrand, this series, Vol. 55, p. 399.
[85] P. L. Pedersen, K. Schwerzmann, and N. Cintron, *Curr. Top. Bioenerg.* **11,** 149 (1981).
[86] B. Frangione, E. Rossenwasser, H. S. Penefsky, and M. E. Pullman, *Proc. Natl. Acad. Sci. U.S.A.* **78,** 7403 (1981).
[87] A. C. Dianoux, A. Tsugita, and M. Przybylski, *FEBS Lett.* **174,** 151 (1984).
[88] R. J. Van de Stadt and K. Van Dam, *Biochim. Biophys. Acta* **347,** 240 (1974).
[89] D. A. Harris, V. Von Tscharner, and G. K. Radda, *Biochim. Biophys. Acta* **548,** 72 (1979).
[90] G. Klein and P. V. Vignais, *J. Bioenerg. Biomembr.* **15,** 347 (1983).
[91] L. L. Horstman and E. Racker, *J. Biol. Chem.* **245,** 1336 (1970).
[92] G. Klein, M. Satre, G. Zaccaï, and P. V. Vignais, *Biochim. Biophys. Acta* **681,** 226 (1982).

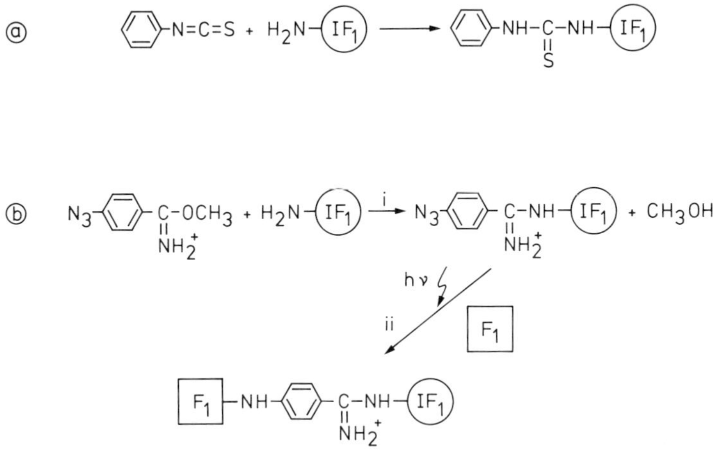

FIG. 5. (a) Modification of IF_1 by PITC. (b) Modification of IF_1 by MABI. IF_1 is labeled by MABI in the dark. Photoirradiation of MABI-IF_1 in the presence of F_1 results in covalent labeling of F_1 by MABI-IF_1.

added to 1 mg of IF_1 in 100 m*M* triethanolamine buffer, pH 8.0, final volume 1 ml. The final concentration of dimethyl sulfoxide is 10% (w/v). After a 2-hr incubation under continuous stirring at room temperature, the reaction is stopped by addition of an excess of Tris to trap unreacted PITC (0.1 ml of 2 *M* Tris–HCl, pH 9.0). Excess reagent is then eliminated by dialysis against 1 liter of 40 m*M* sodium acetate and 10 m*M* Tris-sulfate, pH 7.4, for 3 hr with two changes followed by chromatography on a Sephadex G-25 (medium) column (20 × 1.2 cm) equilibrated with the same buffer. The [^{14}C]PITC-IF_1 is eluted in the void volume (10 ml). The fractions containing the ATPase inhibitor activity are concentrated by precipitation with 10% trichloroacetic acid (w/v). The protein pellet is suspended in 20 m*M* Tris–HCl and the pH adjusted to 7.4 with 1 *N* KOH.

The [^{14}C]PITC-IF_1 prepared according to this procedure[93] contains about 1 mol [^{14}C]PITC per mol IF_1. There is a total of 10 lysine residues per mol IF_1 from beef heart.[86] The N-terminal amino acid residue is blocked by a formyl residue.[87] As PITC reacts with amino groups in their unprotonated form, the extent of labeling can be controlled by pH. In fact, the number of mol [^{14}C]PITC bound per mol IF_1 can be substantially increased by raising the pH of the reaction mixture to 9.

Incorporation of less than 6 mol [^{14}C]PITC per mol IF_1 does not modify the inhibitory efficiency of IF_1 on the ATPase activity of F_1. How-

[93] G. Klein, M. Satre, A. C. Dianoux, and P. V. Vignais, *Biochemistry* **19,** 2919 (1980).

ever, raising the number of bound PITC molecules above six results in an abrupt decrease in the inhibitory efficiency of IF_1.

Radiolabeling of IF_1 with [^{14}C]Methylazidobenzimidate (MABI). [^{14}C]MABI (5×10^7 dpm/μmol) is synthesized by CEA, Saclay, France. Labeling of IF_1 by [^{14}C]MABI is carried out in the dark. [^{14}C]MABI is added to 1 mg IF_1 in 100 m*M* triethanolamine buffer, pH 9.0, to a final concentration of 4.7 m*M* (final volume 1 ml). The mixture is incubated for 1 hr at 37° under constant stirring; the nonreacted [^{14}C]MABI is inactivated by addition of Tris-sulfate (10 μl of 0.5 *M* Tris-sulfate, pH 8.0). [^{14}C]MABI-IF_1 is separated from unreacted [^{14}C]MABI on a Sephadex G-25 (medium) column (20×1.5 cm) equilibrated in 40 m*M* sodium acetate, 10 m*M* Tris-sulfate, pH 7.4, and eluted in the void volume. When [^{14}C]MABI-IF_1 is irradiated under UV light, the peak at 273 nm characteristic of the MABI moiety is gradually reduced in accordance with the photoactivation of the azido group. [^{14}C]MABI-IF_1 prepared under the above conditions[94] is fully inhibitory with respect to F_1-ATPase. It contains about 1 mol [^{14}C]MABI per mol F_1. Up to 5 mol MABI can be incorporated into 1 mol IF_1 without loss of efficiency.

Application. Fully active IF_1 partially labeled with [^{14}C]PITC can be used to determine the binding stoichiometry of IF_1 with respect to F_1. The ATPase activity of F_1 is measured as a function of the bound [^{14}C]PITC-IF_1. By extrapolation of the titration data, the ATPase activity appears to be totally inhibited when 1 mol IF_1 is bound to 1 mol F_1.[93] In photolabeling experiments, [^{14}C]MABI-IF_1 has been used to investigate the localization of the IF_1 binding sites on F_1. Photoactivation of [^{14}C]MABI-IF_1 is obtained with a Mineralight UVS 11 lamp placed at a distance of 5 cm from the sample. In routine experiments, [^{14}C]MABI-IF_1 (7 μg) and F_1 (75 μg) are preincubated at 25° in the dark for 15 min before UV irradiation in 0.25 ml of a medium containing 0.25 *M* sucrose, 1 m*M* $MgCl_2$, 0.5 m*M* ATP, and 10 m*M* MOPS, final pH 6.5, for 5 min. Upon photoactivation, [^{14}C]MABI-IF_1 binds selectively to the β subunit of soluble and membrane-bound F_1. Cross-linking of mitochondrial F_1 with [^{14}C]PITC-IF_1 by means of EDAC or EEDQ shows also that IF_1 binds to the β subunit of mitochondrial F_1.[93,94]

[94] G. Klein, M. Satre, A. C. Dianoux, and P. V. Vignais, *Biochemistry* **20,** 1339 (1981).

[71] A Nonlinear Approach for the Analysis of Different Models of Protein–Ligand Interaction: Nucleotide Binding to F_1-ATPase

By FRENS PETERS and UWE LÜCKEN

Theoretical Basis and Fitting Procedure

The fractional binding ν of a univalent ligand L to a multivalent protein P may be described by Adair's equation:

$$\nu = \frac{[L]_{bound}}{[P]_{total}} = \frac{K_1[L] + 2K_1K_2[L]^2 + \cdots + n[L]^n \prod_{i=1}^{n} K_i}{1 + K_1[L] + K_1K_2[L]^2 + \cdots + [L]^n \prod_{i=1}^{n} K_i} \qquad (1)$$

where [L] is the free ligand concentration and K_i the stoichiometric equilibrium constant of binding of the ith ligand to a protein with $(i - 1)$ ligands bound.

In the case of equivalent independent binding sites, this relation leads to

$$\nu/[L] = nk - \nu k \qquad (2)$$

(k is the affinity of an isolated binding site), which is often used for graphical determination of the affinity and stoichiometry of the binding sites (Scatchard plot). The pitfalls of this procedure have been frequently described[1–5] with the general conclusion that it is an insufficient procedure when applied to more complex models of binding. Nevertheless, it is frequently used for the analysis of complex equilibria producing very often erroneous results.

The stoichiometric binding constants used in Adair's equation can be expressed by the site binding constants[6] which define binding of a ligand to the protein at the ith site ($_i$P):

$$k_i = \frac{[_iPL]}{[_iP][L]} \qquad (3)$$

[1] D. A. Deranleau, *J. Am. Chem. Soc.* **91,** 4050 (1969).

[2] F. Peters and A. Pingoud, *Int. J. Biomed. Comput.* **10,** 401 (1979).

[3] D. Rodbard and H. A. Feldman, this series, Vol. 36, p. 3.

[4] J. G. Norby, P. Ottolenghi, and J. Jensen, *Anal. Biochem.* **102,** 318 (1980).

[5] F. Peters and V. A. Pingoud, *Biochim. Biophys. Acta* **714,** 442 (1982).

[6] I. M. Klotz and D. L. Hunston, *Arch. Biochem. Biophys.* **193,** 314 (1979).

They may be influenced by the state of saturation of the other sites. Thus, in general the number of site binding constants exceeds the number of stoichiometric constants.

This is exemplified for a bivalent system in which the affinity of each site may depend on whether the other site is occupied; because of microscopic reversibility, the number of independent constants is reduced to three: k_1, k_2, k_{12}, where k_{12} is the affinity of site 2 after saturation of site 1:

$$K_1 = \frac{[\cdot \mathrm{PL}] + [\mathrm{LP}\cdot]}{[\mathrm{P}][\mathrm{L}]} = k_1 + k_2 \tag{4}$$

$$K_2 = \frac{[\mathrm{LPL}]}{([\cdot \mathrm{PL}] + [\mathrm{LP}\cdot])[\mathrm{L}]} = \frac{k_1 k_{12}}{k_1 + k_2} \tag{5}$$

($\cdot$ means a not occupied binding site). For a distinct binding model, the number of constants may be reduced: For two independent sites ($k_{12} = k_2$) two constants remain, for identical independent sites ($k_1 = k_2 = k_{12}$) only one constant remains. Relationships for the dependence of $K_i = f(k_j, k_k, k_l, \ldots)$ have been derived by Klotz and Hunston[6] for two to four binding sites. Based on their conclusions, we have developed a nonlinear procedure for the determination of the stoichiometric as well as the site binding constants using Adair's equation (Eq. 1). For a set of starting values of K_i, the program computes the degree of saturation ν_i^{theor} for any given concentration and compares it to the measured ν_i. The minimum of the (weighted) sum of squares is found by variation of the K_i (see Fig. 1, dashed path). The stoichiometric constants describe a macroscopic behavior and yield an optimal fit of binding parameters to measured titration data, but in general they have no direct physical meaning; beyond that it is generally not possible to derive from them the site binding constants k_i. In order to obtain information about the microscopic binding behavior, we have varied and determined the site binding constants themselves from which the corresponding K_i values are calculated in any step of the iteration using a special binding model (see Fig. 1, solid path).

The implementation of special binding models into the program is possible without difficulties. Up to now the following models have been implemented (see Appendix for the relationships between K_i and k_i and further explanations):

Two sites (3 k_i's in the general case)

1. Independent, equivalent sites (1 k_i)
2. Independent, nonequivalent (2 k_i's)
3. Dependent, nonequivalent (3 k_i's)
4. A priori equivalent, cooperative (2 k_i's)

Three sites (7 k_i's in the general case)

1. Independent, equivalent sites (1 k_i)

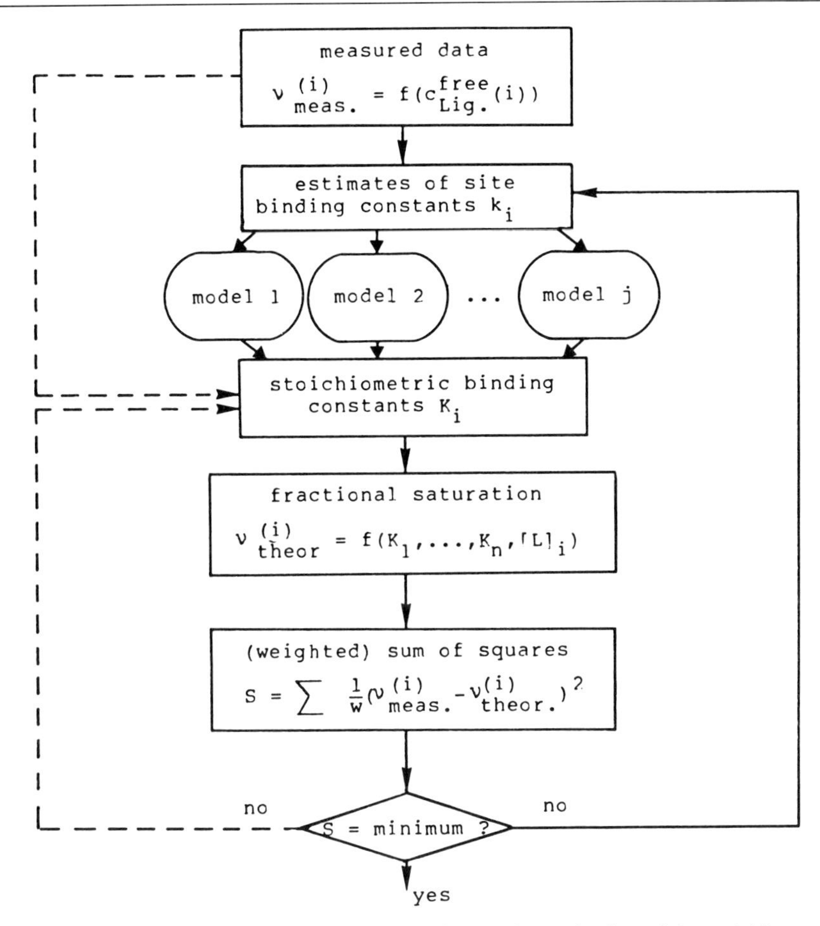

FIG. 1. Flow chart of the fitting procedure for the determination of the stoichiometric (dashed path) and the site binding constants (solid path).

2. Two classes of independent sites (2 k_i's)
3. Three nonequivalent, independent sites (3 k_i's)
4. A priori equivalent, equivalent cooperative (2 k_i's)
5. A priori equivalent, nonequivalent cooperative (3 k_i's)

Four sites (15 k_i's)

1. Independent, nonequivalent (4 k_i's)
2. In pairs equivalent, independent (2 k_i's)
3. A priori equivalent, cooperative on subunit (2 k_i's)
4. A priori two equivalent sites, cooperative on subunit (2 k_i's)
5. A priori equivalent, stepwise cooperative (4 k_i's)

Discrimination between Binding Models

Many aspects must be considered before one can restrict the analysis to a special model. A statistical test such as the F test including the definition of acceptable confidence limits is not sufficient. Knowledge of the structure and biochemical behavior must be taken into account to exclude all models that are not realistic. A comparison of binding behavior between natural ligands and ligand analogs is helpful for the decision. A graphical representation gives the best impression as to whether deviations between experimental data and the fitted curves are significant with respect to actual experimental errors. The (weighted) sum of squares and the error factors of binding constants as well as correlation coefficients are also helpful. The site binding constants of an adequate model should give similar stoichiometric constants as the directly fitted values of K_i. Normally, it is not possible to make a decision only using thermodynamic equilibrium data; kinetic experiments and competition studies must support the favored model.

Nucleotide Binding of F_1-ATPase

The approach described above has been applied to nucleotide-depleted mitochondrial F_1-ATPase, which normally exhibits three reversible high-affinity sites[7,8] (K_d = 0.05–3 μM). The low-affinity nucleotide binding sites[9] and the endogenous bound nucleotides[10] could be neglected under experimental conditions. ADP and the photoreactive ADP analog ANA-ADP[11] were used as ligands.[12] It has been shown by competition experiments that these ligands bind to the same sites. Titration data (see Fig. 2) were analyzed on the basis of the trivalent models discussed above. Because dissociation constants are usually reported for F_1, we adapted this expression using dissociation constants rather than binding constants.

The analysis of stoichiometric dissociation constants generally yields the best fit. The binding behavior of neither of the ligands can be sufficiently interpreted by a binding model with independent binding sites [model 1: dashed lines; models 2 and 3: dashed (ANA-ADP) and dashed/dotted lines (ADP)]. The site dissociation constants of models 2 and 3 (see

[7] Ch. Grubmeyer and H. S. Penefsky, *J. Biol. Chem.* **256,** 3728 (1981).

[8] H. Tiedge, U. Lücken, J. Weber, and G. Schäfer, *Eur. J. Biochem.* **127,** 291 (1982).

[9] J. Weber, U. Lücken, and G. Schäfer, *Eur. J. Biochem.* **148,** 41 (1985).

[10] L. Cross and C. M. Nalin, *J. Biol. Chem.* **257,** 2874 (1982).

[11] ANA-ADP = 3′-*O*-[5-azidonaphthoyl]ADP.

[12] M. Lübben, U. Lücken, J. Weber and G. Schäfer, *Eur. J. Biochem.* **143,** 483 (1984).

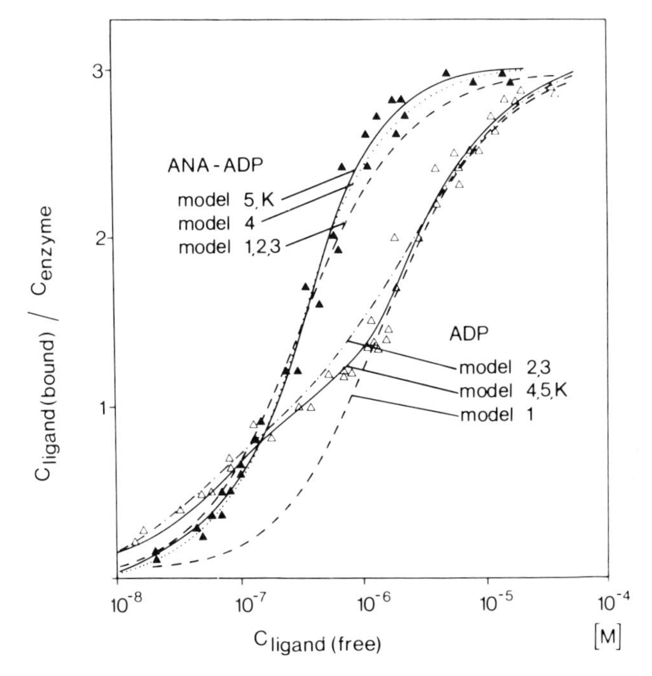

FIG. 2. Titration data obtained for [^{14}C]ADP (△) by an ultrafiltration method [H. Tiedge, U. Lücken, J. Weber, and G. Schäfer, *Eur. J. Biochem.* **127,** 291 (1982)]. At saturation concentration (20–50 μM), the centrifuged column technique [H. S. Penefsky, *J. Biol. Chem.* **252,** 2891 (1977)] was used with slight modifications. F_1 (0.4–1 μM) was incubated for 5 min in 240 μl buffer containing 10 mM Tris–HCl, 5 mM $MgCl_2$, 125 mM KCl, desired ^{14}C-labeled ligand concentration at pH 8, and room temperature. Two 100-μl samples were transferred on a probe applicator of prepared centrifuged columns and the procedure was carried out as described. Protein and radioactivity of the eluent were measured, and ν was calculated assuming a molecular weight of 360,000. Under the chosen conditions sites with fast dissociation rates in relation to the centrifugation elution time (about 10 sec) could not be detected.

Data for ANA-[^{14}C]ADP (▲) were obtained only from the centrifuged column technique with 0.1–0.4 μM F_1 executed in the dark. Single points are shown. The dashed ADP curve represents a simulation for equivalent independent sites.

the table) for ADP yield similar stoichiometric constants. Accordingly, the titration curves have the same shape. The flattened binding curve is optimally fitted by model 5 (solid line) and exhibits a significant negative cooperativity represented by k_{d1} = 150 nM, k_{d2} = 2500 nM, and k_{d3} = 950 nM.

Binding of ANA-ADP shows a steeper slope than the curves fitted by models 1–3 (dashed line); the site dissociation constants for these models turned out to be equal (see the table). The curve based on model 5 (solid

SITE DISSOCIATION CONSTANTS (k_d) AND STOICHIOMETRIC DISSOCIATION CONSTANTS (K_d)[a]

Model	k_{d1}	k_{d2} (nM)	k_{d3}	K_{d1}	K_{d2} (nM)	K_{d3}
ADP						
Model 2	68(1.2)	2900(1.08)		65	1500	5900
Model 3	3448(>10)	3300(>10)	66(>10)	63	1730	6810
Model 5	150(1.1)	2500(1.05)	950(1.07)	50	2500	2850
ANA-ADP						
Model 1	330(>10)	—	—	110	330	1000
Model 3	330(>10)	330(>10)	330(>10)	110	330	1000
Model 5	450(1.08)	420(1.07)	180(1.1)	150	420	540

[a] Obtained from computerized fit of the data from Fig. 2 for different binding models. Qualitative error factors in parentheses.

line) agrees well with the optimal fit of the stoichiometric dissociation constants, whereas model 4 (dotted line) deviates only slightly and cannot be excluded. The k_d values and the steeper slope are indicative of a modest positive cooperative interaction.

Binding of ADP and ANA-ADP to F_1-ATPase is interpreted by a model with a priori equivalent and dependent sites. Dissociation kinetics strongly support an asymmetry of the occupied binding sites. Because of the symmetrical composition of subunits ($\alpha_3\beta_3\gamma\delta\varepsilon$), the asymmetry of binding seems to be induced by the ligands.

Acknowledgments

The authors would like to thank Dr. G. Maass (Hannover) and Dr. G. Schäfer (Lübeck) for their kind assistance, and Mrs. D. Mutschall and Mr. K. Stieglitz for their technical and programming assistance.

Appendix: Relationships between Stoichiometric and Site Binding Constants for Special Binding Models with Two to Four Binding Sites

A. *Two Sites*

$K_1 = k_1 + k_2$, $K_2 = k_1 k_{12}/(k_1 + k_2)$

1. Independent, equivalent sites (1 site binding constant)
   $k = k_1 = k_2 = k_{12}$
   $K_1 = 2k$, $K_2 = k/2$

2. Independent, nonequivalent (2 constants)
   $k_2 = k_{12}$
   $K_1 = k_1 + k_2,\ K_2 = k_1k_2/(k_1 + k_2)$
3. Dependent, nonequivalent (3 constants)
   $K_1 = k_1 + k_2,\ K_2 = k_1k_{12}/(k_1 + k_2)$
4. A priori equivalent, cooperative (2 constants)
   $k_1 = k_2$
   $K_1 = 2k_1,\ K_2 = k_{12}/2$

B. Three Sites

$K_1 = k_1 + k_2 + k_3$
$K_2 = (k_1k_{12} + k_1k_{13} + k_2k_{23})/K_1$
$K_3 = (k_1k_{12}k_{123})/K_1K_2$

1. Independent, equivalent sites (1 constant)
   $k = k_1 = k_2 = k_3 = k_{12} = k_{13} = k_{23} = k_{123}$
   $K_1 = 3k,\ K_2 = k,\ K_3 = k/3$
2. Two classes of independent sites (2 constants)
   Class 1: site 1; class 2: sites 2 and 3
   $k_2 = k_3 = k_{12} = k_{13} = k_{23} = k_{123}$
   $K_1 = k_1 + 2k_2$
   $K_2 = (2k_1k_2 + k_2^2)/K_1$
   $K_3 = k_1k_2^2/K_1K_2$
3. Three nonequivalent sites (3 constants)
   $k_2 = k_{12},\ k_3 = k_{13} = k_{23} = k_{123}$
   $K_1 = k_1 + k_2 + k_3$
   $K_2 = (k_1k_2 + k_1k_3 + k_2k_3)/K_1$
   $K_3 = (k_1k_2k_3)/K_1K_2$
4. A priori equivalent, equivalent cooperative (2 constants)
   $k_{II} = k_1 = k_2 = k_3$
   $k_{II} = k_{12} = k_{13} = k_{23} = k_{123}$
   $K_1 = 3k_I$
   $K_2 = 3k_Ik_{II}/K_1$
   $K_3 = k_Ik_{II}^2/K_1K_2$
5. A priori equivalent, stepwise cooperative (3 constants)
   $k_I = k_1 = k_2 = k_3$
   $k_{II} = k_{12} = k_{13} = k_{23}$
   $k_{III} = k_{123}$
   $K_1 = 3k_I,\ K_2 = k_{II},\ K_3 = k_{III}/3$

C. Four Sites

$K_1 = k_1 + k_2 + k_3 + k_4$
$K_2 = (k_1k_{12} + k_1k_{13} + k_1k_{14} + k_2k_{23} + k_2k_{24} + k_3k_{34})/K_1$

$K_3 = (k_1k_{12}k_{123} + k_1k_{12}k_{124} + k_1k_{13}k_{134} + k_2k_{23}k_{234})/K_1K_2$
$K_4 = (k_1k_{12}k_{123}k_{1234})/K_1K_2K_3$

1. Independent, nonequivalent (4 constants)
   $k_2 = k_{12},\ k_3 = k_{13} = k_{23} = k_{123}$
   $k_4 = k_{14} = k_{24} = k_{34} = k_{124} = k_{134} = k_{234}$
   $K_1 = k_1 + k_2 + k_3 + k_4$
   $K_2 = (k_1k_2 + k_1k_3 + k_1k_4 + k_2k_3 + k_2k_4 + k_3k_4)/K_1$
   $K_3 = (k_1k_2k_3 + k_1k_2k_4 + k_1k_3k_4 + k_2k_3k_4)/K_1K_2$
   $K_4 = k_1k_2k_3k_4/K_1K_2K_3$
2. In pairs equivalent, independent (2 constants)
   $k_1 = k_2 = k_{12}$
   $k_3 = k_4 = k_{13} = k_{14} = k_{23} = k_{24} = k_{34} =$
   $k_{123} = k_{124} = k_{134} = k_{234} = k_{1234}$
   $K_1 = 2k_1 + 2k_3$
   $K_2 = (k_1^2 + 4k_1k_3 + k_3^2)/K_1$
   $K_3 = 2(k_1^2k_3 + k_1k_3^2)/K_1K_2$
   $K_4 = (k_1^2k_3^2)/K_1K_2K_3$
3. A priori equivalent, cooperative on subunit (2 constants) (subunit 1: sites 1 and 3; subunit 2: sites 2 and 4)
   $k_{\mathrm{I}} = k_1 = k_2 = k_3 = k_4 = k_{12} = k_{14} = k_{23} = k_{34} = k_{134}$
   $k_{\mathrm{II}} = k_{13} = k_{24} = k_{123} = k_{124} = k_{234} = k_{1234}$
   $K_1 = 4k_{\mathrm{I}},\ K_2 = (4k_{\mathrm{I}}^2 + 2k_{\mathrm{I}}k_{\mathrm{II}})/K_1$
   $K_3 = 4k_{\mathrm{I}}^2k_{\mathrm{II}}/K_1K_2,\ K_4 = k_{\mathrm{I}}^2k_{\mathrm{II}}^2/K_1K_2K_3$
4. A priori two equivalent sites, cooperative on subunit (2 constants)
   $k_{\mathrm{I}} = k_1 = k_2 = k_{12}$
   $k_{\mathrm{II}} = k_{13} = k_{24} = k_{123} = k_{124} = k_{1234}$
   $k_3 = k_4 = k_{14} = k_{23} = k_{34} = k_{134} = k_{234} = 0$
   $K_1 = 2k_{\mathrm{I}}$
   $K_2 = (k_{\mathrm{I}}^2 + 2k_{\mathrm{I}}k_{\mathrm{II}})/K_1 = (k_{\mathrm{I}} + 2k_{\mathrm{II}})/2$
   $K_3 = 2k_{\mathrm{I}}^2k_{\mathrm{II}}/K_1K_2 = 2k_{\mathrm{I}}k_{\mathrm{II}}/(k_{\mathrm{I}} + 2k_{\mathrm{II}})$
   $K_4 = k_{\mathrm{I}}^2k_{\mathrm{II}}^2/K_1K_2K_3 = k_{\mathrm{II}}/2$
5. A priori equivalent, stepwise cooperative (4 constants)
   $k_{\mathrm{I}} = k_1 = k_2 = k_3 = k_4$
   $k_{\mathrm{II}} = k_{12} = k_{13} = k_{14} = k_{23} = k_{24} = k_{34}$
   $k_{\mathrm{III}} = k_{123} = k_{124} = k_{134} = k_{234}$
   $k_{\mathrm{IV}} = k_{1234}$
   $K_1 = 4k_{\mathrm{I}},\ K_2 = 3k_{\mathrm{II}}/2$
   $K_3 = 2k_{\mathrm{III}}/3,\ K_4 = k_{\mathrm{IV}}/4$

[72] Identification of Essential Residues in the F_1-ATPases by Chemical Modification

By WILLIAM S. ALLISON, DAVID A. BULLOUGH, and WILLIAM W. ANDREWS

The isolated F_1-ATPases have been subjected to a variety of chemical modifications that have been intended to shed some light on the molecular mechanism of ATP synthesis catalyzed by the intact ATP synthase complexes, which, owing to their greater complexity, are much less amenable to such studies. This approach assumes that the same functional groups participate in ATP synthesis catalyzed by the intact, membrane-bound ATP synthase complex as participate in the hydrolytic reaction catalyzed by the isolated F_1-ATPase.

All the essential residues that have been identified in the primary sequences of various F_1-ATPases to date reside in the β subunit. With the numbers designating the sequence of the β subunit of MF_1,[1] these residues are as follows: Tyr-368, which is labeled during the inactivation of the MF_1- and YF_1-ATPases with 5′-*p*-fluorosulfonylbenzoyladenosine (FSBA)[2,3]; Tyr-311, which is labeled during the inactivation of the MF_1-ATPase with 7-chloro-4-nitro[^{14}C]benzofurazan ([^{14}C]Nbf-Cl)[4,5]; Lys-162, which is labeled with the [^{14}C]Nbf group by migration from the [^{14}C]Nbf-*O*-Tyr residue in the MF_1- and TF_1-ATPases under slightly alkaline conditions[6–9]; Glu-188, which is labeled during the inactivation of the TF_1-ATPase with dicyclohexyl[^{14}C]carbodiimide ([^{14}C]DCCD)[10]; Glu-199, which is labeled during the inactivation of the MF_1- and EF_1-ATPases with [^{14}C]DCCD[11,12]; and β-Arg-295, which is labeled during the inactiva-

[1] M. J. Runswick and J. E. Walker, *J. Biol. Chem.* **258,** 3081 (1983).
[2] F. S. Esch and W. S. Allison, *J. Biol. Chem.* **253,** 6100 (1978).
[3] K. G. Bital, *Biochem. Biophys. Res. Commun.* **109,** 30 (1982).
[4] S. J. Ferguson, W. J. Lloyd, M. H. Lyons, and G. K. Radda, *Eur. J. Biochem.* **54,** 117 (1975).
[5] W. W. Andrews, F. C. Hill, and W. S. Allison, *J. Biol. Chem.* **259,** 8219 (1984).
[6] S. J. Ferguson, W. J. Lloyd, and G. K. Radda, *Eur. J. Biochem.* **54,** 127 (1975).
[7] W. W. Andrews, F. C. Hill, and W. S. Allison, *J. Biol. Chem.* **259,** 14378 (1984).
[8] W. W. Andrews, M. Yoshida, F. C. Hill, and W. S. Allison, *Biochem. Biophys. Res. Commun.* **123,** 1040 (1984).
[9] R. Sutton and S. J. Ferguson, *FEBS Lett.* **179,** 283 (1985).
[10] M. Yoshida, J. W. Poser, W. S. Allison, and F. S. Esch, *J. Biol. Chem.* **256,** 148 (1981).
[11] F. S. Esch, P. Böhlen, A. S. Otsutka, M. Yoshida, and W. S. Allison, *J. Biol. Chem.* **256,** 9084 (1981).
[12] M. Yoshida, W. S. Allison, F. S. Esch, and M. Futai, *J. Biol. Chem.* **257,** 10033 (1982).

tion of the CF_1-ATPase with [^{14}C]phenylglyoxal.[13] Inactivation of MF_1 with 1-(*N*-ethoxycarbonyl)-2-ethoxyl-1,2-dihydroquinoline (EEDQ) in the presence of [^{3}H]aniline leads to the formation of the [^{3}H]anilide of β-Glu-199, the same residue that is modified when the enzyme is inactivated with [^{14}C]DCCD.[14]

All the reagents that have been used to modify the F_1-ATPases in the studies cited above are known to inactivate a variety of enzymes, among which are other membrane-bound ATPases. Therefore, the experimental approaches and some specific procedures that were used to identify essential residues in the F_1-ATPases might be useful for other enzymologists, and thus, are presented in detail here.

General Considerations

Specific Radioactivity of Labels. HPLC techniques are now available for the rapid purification of peptides in nanomole quantities. Methods are also available for the automatic sequence analysis of peptides at the level of 1 nmol or less. In order to use these techniques for the identification of labeled residues in the ATPases, the specific radioactivity of the labeling reagents should be 10 cpm/pmol or higher to allow sensitive detection of radioactive peptides during isolation procedures.

Subunit Isolation. Since five different subunits comprise the F_1-ATPases, it is sometimes advantageous to isolate labeled subunits before initiating fragmentation procedures. Until recently the subunits were most reliably isolated on the preparative level by a combination of cation- and anion-exchange chromatography in the presence of 8 *M* urea.[15-17] Since the anion-exchange steps are carried out under alkaline conditions, carbamoylation of primary amino groups of proteins is always a problem with this method for separating subunits. The fact that labeled derivatives containing ester or other alkali-sensitive linkages are subject to hydrolysis under alkaline conditions is another hazard of the ion-exchange method of subunit isolation. A useful alternative to the older ion-exchange procedures is to carry out subunit separations by reversed-phase HPLC on short-chain hydrocarbons attached to silica. Figure 1 illustrates the small-scale separation of the subunits of MF_1 on a 4.5 × 250 mm Vydac C-4 column. To achieve this separation, a 10-μl sample of a 10 mg/ml solution of MF_1 in 6 *M* guanidine–HCl, pH 7.9, containing 50 m*M* dithiothreitol

[13] A. M. Viale and R. H. Vallejos, *J. Biol. Chem.* **260,** 4958 (1985).
[14] P. K. Laikind, F. C. Hill, and W. S. Allison, *Arch. Biochem. Biophys.* **240,** 904 (1985).
[15] A. F. Knowles and H. S. Penefsky, *J. Biol. Chem.* **247,** 6617 (1972).
[16] N. Nelson, D. W. Deters, H. Nelson, and E. Racker, *J. Biol. Chem.* **248,** 2049 (1973).
[17] M. Yoshida, N. Sone, H. Hirata, and Y. Kagawa, *J. Biol. Chem.* **252,** 3480 (1977).

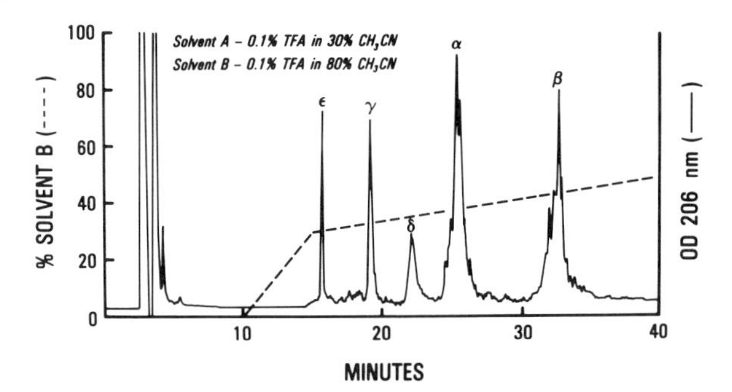

FIG. 1. Separation of the subunits of MF_1 by reversed-phase HPLC on a C_4 column. The procedures used to prepare the enzyme for chromatography and to develop the column are described in detail in the text.

was injected onto the column which was equilibrated with 30% acetonitrile in 0.1% trifluoroacetic acid 20 min after depolymerization was initiated. The column was eluted at 1.0 ml/min with increasing concentration of acetonitrile in 0.1% trifluoroacetic acid, as indicated on the chromatogram shown. Buffer A contained 0.1% trifluoroacetic acid in 30% acetonitrile and Buffer B contained 0.1% trifluoroacetic acid in 80% acetonitrile. Yoshida has described different conditions for the resolution of the subunits of MF_1 and for the separation of the subunits of TF_1 and CF_1 also using a Vydac C-4 reversed-phase HPLC column.[18]

Identification of Amino Acid Residues in Enzymes That React with 5′-Fluorosulfonylbenzoyladenosine (5′-FSBA)

Synthesis and Storage of Radioactive 5′-FSBA. The synthesis and properties of 5′-FSBA have been described in detail by Colman *et al.*[19] The syntheses of 5′-FSBA labeled in the adenosine moiety with ^{3}H[19] or in the benzoyl moiety with ^{14}C,[20] each with low specific radioactivity, have also been described. The cost of *p*-amino[7-^{14}C]benzoic acid, the starting material for the synthesis of 5′-[^{14}C]FSBA labeled in the benzoyl moiety, prohibits the synthesis of this material with sufficiently high specific radioactivity to be useful for the isolation of radioactive peptides from small amounts of labeled protein using HPLC methods. However, 5′-[^{3}H]FSBA can be synthesized from [2,8-^{3}H]adenosine with sufficiently high specific

[18] M. Yoshida, *in* "H^+-ATPase (ATP Synthase), Structure, Function, Biogenesis of the F_0F_1 Complex of Coupling Membranes" (S. Papa, K. Altendorf, L. Ernster, and L. Packer, eds.), p. 147. ICSU Press, Adriatica Editrice, Bari, Italy, 1984.

[19] R. F. Colman, P. K. Pal, and J. L. Wyatt, this series, Vol. 46, p. 240.

[20] A. DiPietro, C. Godinot, and D. C. Gautheron, *Biochemistry* **20,** 6312 (1981).

radioactivity to be used in the small-scale isolation of labeled peptides by HPLC. A procedure for this synthesis is outlined below.

Synthesis of 5'-[^{3}H]FSBA. Transfer of [2,8-^{3}H]adenosine (10 mCi with a specific radioactivity of about 30 Ci/mmol from its shipping vial is accomplished by first dissolving the solid in 200 μl of warm hexamethylphosphoramide. As pointed out by Colman *et al.*,[19] ^{3}H exchanges from the 8-position of the purine ring under these conditions. Thus, appropriate precaution should be taken. The resulting solution is then transferred, using a 1-ml glass tuberculin syringe equipped with a stainless-steel needle, to a 5-ml Pierce Reactivial containing 0.25 mmol (67 mg) of nonradioactive adenosine. The nonradioactive adenosine had been dried in a vacuum oven at 90° for 2 hr and then cooled to room temperature in a vacuum desiccator. The vial from which the [2,8-^{3}H]adenosine was transferred was washed three times with 100-μl portions of warm hexamethylphosphoramide. The washes were added to the Reactivial containing the [2,8-^{3}H]adenosine which was then capped and swirled in an oil bath at 50° until the adenosine was dissolved. After the vial was cooled to room temperature, 0.35 mmol (77 mg) of *p*-fluorosulfonylbenzoyl chloride was added. As pointed out by DiPietro *et al.*,[20] higher yields are obtained if the commercial *p*-fluorosulfonylbenzoyl chloride is treated with thionyl chloride as described for the synthesis of *p*-fluorosulfonylbenzoyl chloride from *p*-fluorosulfonylbenzoic acid.[21] The Reactivial was then capped tightly, swirled to dissolve the acyl chloride, and then placed at 23° for 18 hr.

The reaction mixture was then extracted with three 0.5-ml portions of petroleum ether (35–60° fraction), which had been dried over molecular sieves. After each addition of petroleum ether, the capped Reactivial was shaken vigorously and then was subjected to low-speed centrifugation in a tabletop clinical centrifugation to facilitate phase separation. The petroleum ether layers were removed with a 3-ml glass syringe equipped with a stainless-steel needle. After the third extraction, the product was precipitated by the slow addition of ethyl acetate : diethyl ether (1 : 1) which had been dried over molecular sieves. The product was then collected as a firm pellet by centrifugation and the mother liquor was removed with a syringe. The product was then dissolved in 0.5 ml of dimethylformamide which had been dried over molecular sieves. The final product was then crystallized by the slow addition of dry ethyl acetate : diethyl ether (1 : 1). The crystals were collected by centrifugation and the mother liquor was removed with a 3-ml glass syringe equipped with a stainless-steel needle. The crystals were then dissolved in 2.0 ml of redistilled dimethyl sulfoxide which had been dried over molecular sieves. After diluting samples in

[21] F. S. Esch and W. S. Allison, *Anal. Biochem.* **84,** 642 (1978).

absolute ethanol, the concentration of this solution was determined to be 80 m*M* from its absorbance at 259 nm using the molar extinction coefficient of 1.58×10^4.[19] The specific radioactivity of the final product was 20 cpm/pmol when subjected to liquid scintillation counting in cocktail 3a70B purchased from Research Products International. The final product was free of adenosine as assessed by gel permeation chromatography on a 7.5×600 mm Toya Soda G-2000SW column which was equilibrated and eluted with 0.2 *M* $NaPO_4$, pH 6.0, at 1.0 ml/min. The column effluent was monitored at 254 nm. Under these conditions the retention time of adenosine was 27.5 min and that of 5′-FSBA was 32.6 min.

The 80 m*M* solution of 5′-[^{3}H]FSBA in a tightly capped 5-ml Reacti-vial was stored in a capped bottle containing silica gel desiccant, which in turn was stored in a refrigerator at 5°. The desiccated bottle was warmed to room temperature before it was opened to remove samples of the 5′-[^{3}H]FSBA stock solution. When stored under these conditions, the 5′-[^{3}H]FSBA showed no appreciable decomposition over a period of 2 years. The decomposition products of 5′-FSBA were monitored by HPLC. The free sulfonic acid, SBA^-, which is formed on hydrolysis of the sulfonyl fluoride moiety, has a retention time of 10.5 min when subjected to isocratic elution on a 7.5×300 mm Toya Soda IEX-540K anion-exchange column with 0.20 *M* $NaPO_4$, pH 6.0, at a flow rate of 0.8 ml/min. Under these conditions, 5′-FSBA and adenosine are eluted from the anion-exchange column at 5.3 min. Adenosine is resolved from 5′-FSBA by high-performance gel permeation chromatography, as described above.

Stability of the Sulfonyl Fluoride and Stability of the Incorporated Label. The decomposition of the sulfonyl fluoride moiety of FSBA is conveniently monitored with the use of a fluoride-specific electrode. However, owing to the influence of organic solutes on fluoride-specific electrodes, in our hands this method has not produced accurate results. We have monitored the formation of [^{3}H]SBA^- during the hydrolysis of [^{3}H]$FSBA^-$ in 50 m*M* triethanolamine-H_2SO_4 buffers at pH 7.0, 7.5, and 8.0 by ion-exchange HPLC as described above. The pseudo-first-order rate constants obtained at 23° at these pH values were, respectively, 0.8×10^{-3} min^{-1}, 4.3×10^{-3} min^{-1}, and 9.2×10^{-3} min^{-1}.

Since adenosine is esterified to the benzoyl moiety of the reagent, the inactivation of enzymes with 5′-[^{14}C]FSBA leads to the incorporation of radioactivity which is not removed from labeled protein or peptides by hydrolysis during separation procedures. The [^{3}H]adenosine moiety is, of course, susceptible to removal from labeled protein or peptides by hydrolysis during workup to identify modified residues after inactivation of enzymes with 5′-[^{3}H]FSBA. This hydrolysis can be minimized if the pH is maintained between 3 and 7 during manipulation of the labeled protein or labeled peptides derived from it. The [^{3}H]adenosine moiety is lost from labeled proteins or peptides by hydrolysis during the conditions of cyanogen bromide cleavage which requires long incubation at low pH.

Conditions for Labeling the F_1-ATPases with Radioactive 5′-FSBA. Optimal rates of inactivation of MF_1 by 5′-FSBA are obtained after removing nucleotides and, presumably, Mg^{2+} from catalytic sites by gel-filtering the enzyme on centrifuge columns[22] equilibrated with the inactivation buffer containing 1 m*M* CDTA. When samples of MF_1, which were purified and stored as described by Knowles and Penefsky,[15] were prepared for inactivation by gel filtration on centrifuge columns in 50 m*M* triethanolamine sulfate, pH 7.5, containing (1) no additions, (2) 1 m*M* EDTA, and (3) 1 m*M* CDTA, the nucleotide content of a particular preparation of the enzyme was found in moles per mole of enzyme to be (1) 2.18 ADP and 2.12 ATP, (2) 1.55 ADP and 2.12 ATP, and (3) 0.78 ADP and 1.55 ATP. Although the nucleotide content of other MF_1 preparations was somewhat different, the pseudo-first-order rate constants obtained for the initial rates of inactivation of these enzyme preparations, each at 1.0 mg/ml by 0.8 m*M* 5′-FSBA at 23°, were about (1) 0.031 min^{-1}, (2) 0.038 min^{-1}, and (3) 0.058 min^{-1} at 23°. Although the inactivation of MF_1 proceeded considerably faster at pH 8.0 than it did at pH 7.0 or 7.5, modification with 5′-[^{3}H]FSBA was much more selective at the lower pH values.[2]

The inactivation of MF_1 by 5′-FSBA exhibits biphasic kinetics.[2,23] When the CDTA-treated enzyme was inactivated with 0.8 m*M* 5′-FSBA at pH 7.5, the resulting semilogarithmic plot of the fractional activity vs time showed a fast linear phase which transformed to a slower linear phase at about 50% inactivation. Biphasic kinetics were also observed in the presence of 20% glycerol where inactivation was slow, and in phosphate buffer where inactivation rates were faster than those observed in Tris and triethanolamine buffers at pH 7.5.[2,23] It has been suggested that the modification of regulatory sites is responsible for the fast phase of inactivation by 5′-FSBA and that the modification of catalytic sites occurs during the slow phase.[23] However, by following the appearance of labeled residues during the inactivation of MF_1 by 5′-[^{3}H]FSBA, in recent experiments it is clear that the rate of modification of β-Tyr-368 correlates with the rate of loss of ATPase activity during both kinetic phases (D.A. Bullough and W. S. Allison, unpublished experiments). Therefore, inactivation of MF by 5′-FSBA might occur by modification of β-Tyr-368 in both catalytic sites and regulatory sites.

[22] H. S. Penefsky, this series, Vol. 56, p. 527.

[23] A. DiPietro, C. Godinot, and D. C. Gautheron, *Biochemistry* **18,** 1738 (1979).

Isolation of Labeled Peptides and Identification of Labeled Amino Acid Residues after Inactivating F_1-ATPases with Radioactive 5′-FSBA. A rapid procedure for the purification of peptides labeled with 5′-[^{14}C]FSBA has been developed which is based on the cleavage of the ester linkage of the reagent covalently bound to the side chains of tyrosine and lysine residues in peptides. The three-step procedure described below was developed for the purification of the tryptic peptide derived from the β subunit of MF_1 in which β-Tyr-368 was labeled with 5′-[^{14}C]FSBA.[24] In the first step a tryptic digest of the labeled β subunit was subjected to anion-exchange chromatography on a column of DEAE-Sephadex A-25 which was eluted with a pH gradient. In the second step the material in the major, radioactive peak which was eluted from the ion-exchange column was lyophilized and then treated with 0.1 *M* NaOH for 4 hr at room temperature to hydrolyze the ester bond of the covalently incorporated reagent. The NaOH was removed from the mixture of tryptic peptides by gel filtration on a column of Sephadex G-50 which was equilibrated and eluted with 0.05 *M* NH_4OH. In the third step, the material in the major radioactive peak which eluted from the gel permeation column was subjected to anion-exchange chromatography on a column of DEAE Sephadex A-25 using a pH gradient under the same conditions used in the first step. Since cleavage of the ester bond of the bound reagent in the second step released adenosine and specifically introduced an additional negative charge onto the ^{14}C-labeled peptide, the radioactive peptide was resolved in the third step from the nonradioactive peptides that contaminated it.

The above procedure has also been successfully applied to a tryptic digest of the catalytic subunit of the porcine skeletal muscle cAMP-dependent protein kinase which had been inactivated with 5′-[^{14}C]FSBA.[25] The peptide purified from the labeled protein kinase contained the [^{14}C]carboxybenzenesulfonyl lysine ([^{14}C]CBS-Lys).

The inactivation of different enzymes with 5′-FSBA has been found to be caused by the modification of the side chains of tyrosine,[2,3] lysine,[25,26] and cysteine[27] residues. Myosin ATPase is apparently inactivated by the reaction of 5′-FSBA with an essential cysteine side chain to form a thiosulfonate derivative. Then SBA^- is displaced from the thiosulfonate by another thiol group during the formation of an intramolecular disulfide bond.[27]

[24] F. S. Esch and W. S. Allison, *Anal. Biochem.* **95,** 39 (1979).

[25] M. J. Zoller and S. S. Taylor, *J. Biol. Chem.* **254,** 8363 (1979).

[26] J. A. Schmidt and R. F. Colman, *J. Biol. Chem.* **259,** 14515 (1984).

[27] C. T. Togashi and E. Reisler, *J. Biol. Chem.* **257,** 10112 (1982).

When subjected to acid hydrolysis with 6 *M* HCl at 105°, proteins or peptides sulfonylated at tyrosine and lysine residues with 5′-FSBA liberate the corresponding 4-carboxybenzenesulfonyl derivatives of these amino acids, CBS-Tyr and CBS-Lys. Procedures have been described for the synthesis of authentic CBS-Tyr and CBS-Lys from either the *N-tert*-butyloxycarbonyl[2] or the *N*-acetylamino acids[19] for use as standards for amino acid analysis. The synthesis of the phenylthiohydantoin derivative of CBS-Tyr has also been described for use as a standard for the Edman degradation of peptides derived from enzymes inactivated with radioactive 5′-FSBA.[2]

Identification of Residues Modified by Nbf-Cl in the F_1-ATPases

In 1975 it was reported by Ferguson *et al.* that the inactivation of MF_1 by Nbf-Cl was accompanied by the modification of a single tyrosine residue per mol of enzyme.[4] The stoichiometry of inactivation was based on spectrophotometric determinations. The development of absorption at 385 nm when Nbf-Cl was added to the enzyme paralleled inactivation of the enzyme. The O-Nbf derivative of *N*-acetyltyrosine ethyl ester also absorbs at 385 nm and its molar extinction coefficient at that wavelength was determined to be 11,600 M^{-1} cm^{-1}.[4] With this extinction coefficient it was determined that the modification of a single tyrosine residue per mol of MF_1 was sufficient to inactivate the enzyme completely. The inactive Nbf-*O*-tyrosine derivative of the enzyme was found to be very sensitive to thiolysis by dithiothreitol. Treatment of Nbf-*O*-tyrosine derivative with dithiothreitrol reactivated the enzyme and led to the concomitant loss of the absorption peak at 385 nm.[4] In another report it was also shown by Ferguson *et al.* that under slightly alkaline conditions an O → N migration takes place in which the Nbf group is transferred from the initially reacting tyrosine to a lysine residue that resides in the β subunit.[6] The O → N migration was also monitored spectrophotometrically. The migration was accompanied by the loss of an absorption peak at 385 nm and the appearance of an absorption peak at 475 nm, the wavelength at which the N^{ε}-Nbf derivative of L-lysine has an absorption maximum with a molar extinction coefficient of 26,000 M^{-1} cm^{-1}. The Nbf-*N*-lysine derivative of the enzyme was also found to be inactive.[6]

It was evident from the early work cited above that the Nbf-*O*-tyrosine derivative of the enzyme was not sufficiently stable to be identified by conventional methods of protein chemistry. It was found that the Nbf-*O*-tyrosine adduct could be stabilized by reducing the nitro group of the covalently bound reagent with dithionite.[5] Treatment of the [^{14}C]Nbf-*O*-tyrosine derivative of MF_1 with sodium dithionite was found to be accompanied by partial reactivation of the enzyme. However, when dithiothreitol was added to the labeled enzyme after treatment with dithionite, further reactivation was not observed. On addition of dithionite to the [^{14}C]Nbf-*O*-tyrosine derivative of the enzyme, an amount of ^{14}C was displaced which was equivalent to the degree of reactivation obtained in this step. This indicated that dithionite or its aqueous decomposition products partially displaced the [^{14}C]Nbf group under the conditions of reduction. The amount of reactivation and concomitant loss of bound ^{14}C decreased with decreasing pH of the buffer in which the reduction was performed. Addition of dithionite to a final concentration of 10 m*M* to the Nbf-*O*-tyrosine derivative of MF_1 as a function of pH gave the following results: pH 6.0 (MES), 9.4% reactivation; pH 7.0 (MOPS), 30% reactivation; and pH 8.0 (HEPES), 58% reactivation. That part of the incorporated label, which, after dithionite addition, was resistant to removal and reactivation by dithiothreitol is assumed to have undergone reduction with the conversion of the 4-nitrobenzofurazan derivative having been converted to the 4-aminobenzofurazan derivative (Abf derivative).

Preparation and Identification of the [^{14}C]Abf-O-tyrosine Derivative of MF_1. For preparation of MF_1 for large-scale inactivations, the enzyme was removed by centrifugation from its storage suspension in 0.55 saturated ammonium sulfate, pH 7.0, containing 4 m*M* ATP and 1 m*M* EDTA. The enzyme was dissolved at a concentration of about 12–14 mg/ml in 50 m*M* MOPS–HCl, pH 7.4, which contained 2 m*M* EDTA. The dissolved enzyme was then dialyzed against the same buffer at 23° for 4 hr, at which time the protein concentration decreased to about 10 mg/ml. To this solution was added 4 mol of [^{14}C]Nbf-Cl per mol of MF_1, and inactivation was allowed to proceed until less than 10% of the original activity remained. At this point essentially all of the enzyme activity could be recovered by including 10 m*M* dithiothreitol in the assay reaction mixture. At the end of the inactivation, the enzyme was precipitated by the slow addition of saturated ammonium sulfate, adjusted to pH 7.0 with NaOH, until 55% saturation was attained. The resulting protein precipitate was removed by centrifugation and was then dissolved in sufficient 100 m*M* MES–NaOH, pH 6.0, containing 2 m*M* EDTA to give a final protein concentration of 20–25 mg/ml. Then, for each 1.0 ml of the resulting protein solution, 0.01 ml of a freshly prepared solution of 1 *M* sodium dithionite in 0.1 *M* NaOH was added slowly with stirring to reduce the nitro groups of the [^{14}C]Nbf incorporated into the enzyme.

At the end of each step, unbound radioactive reagents were removed from samples by gel filtration on 1-ml centrifuge columns of Sephadex G-50, as described by Penefsky.[22] The gel-filtrated samples were subjected to protein determinations and liquid scintillation counting. These analyses showed that the inactivation of the enzyme by [^{14}C]Nbf-Cl was accompanied by the incorporation of 1.5 mol of [^{14}C]Nbf per mol of MF_1. Treatment of a sample of the inactivated enzyme with dithiothreitol removed about 1.0 mol of [^{14}C]Nbf per mol of enzyme, indicating that of the 1.5 mol of [^{14}C]Nbf incorporated on inactivation, about 0.5 mol was present on lysine residues and about 1.0 mol was present on tyrosine residues. About 10% reactivation occurred on the addition of dithionite, which was accompanied by an equivalent loss of [^{14}C]Nbf from tyrosine residues. The radioactivity that remained bound to the enzyme after the addition of dithionite could not be removed by treating the labeled protein with dithiothreitol. Therefore, it is assumed that the reduction step displaced about 0.1 mol of [^{14}C]Nbf from tyrosine residues and converted by reduction about 0.9 mol of [^{14}C]Nbf-*O*-Tyr to [^{14}C]Abf-*O*-Tyr. Isolation of the subunits from the reduced, labeled enzyme by the method of Knowles and Penefsky[15] revealed that the majority of the [^{14}C]Abf was in the β subunit (about 0.35 mol/mol). The α subunit contained about 0.20 mol of [^{14}C]Abf per mol, while the minor subunits contained an insignificant amount of the radiolabel. When tryptic digests of the purified, labeled α subunit were subjected to reversed-phase HPLC, several radioactive peptides were resolved, none of which accounted for a large percentage of the radioactivity recovered from the column, which was about 50% of that applied. Therefore, it appears that the label incorporated into the α subunit represents the random reaction of [^{14}C]Nbf-C1 with surface lysine and/or tyrosine resides.

The tryptic peptide containing the [^{14}C]Abf derivative of β-Tyr-311 proved very difficult to purify by HPLC. This is the longest tryptic peptide in the β subunit of MF_1 and contains 36 amino acid residues. This peptide resisted purification by reversed-phase HPLC under a variety of acidic conditions and by anion-exchange chromatography. It was finally purified by first fractionating a cyanogen bromide digest of the labeled β subunit on a column of Sephadex G-75 which was equilibrated and eluted with 0.5% NH_4HCO_3. The cyanogen bromide fragment in which β-Tyr-311 resides contains 65 amino acid residues. The majority of the radioactivity (83%) eluted from the Sephadex G-75 column appeared in a single peak near the void volume. The radioactive material in this peak was lyophilized and then digested with trypsin. The tryptic digest was fractionated on a column of Sephadex G-75. The majority of the radioactivity (60%) eluted from this column in a single peak which appeared near the void volume. The radioactive material in this peak was subjected to final purification on a C_{18} reversed-phase HPLC column which was equilibrated with 0.5% NH_4CO_3. The column was subjected to gradient elution with increasing concentrations of acetonitrile. About 14% of the radioactivity which was applied to this column eluted at 22% acetonitrile in a sharp peak, while another 12% of the radioactivity applied smeared over a larger volume at a higher acetonitrile concentration. Automatic Edman degradation of the radioactive material in the peak eluting at 22% acetonitrile revealed that it contained the presumed [^{14}C]Abf derivative of β-Tyr-311. The yield of radioactivity released with the anilinothiazoline during the tenth step of this degradation, which corresponds to the position of β-Tyr-311, was approximately equivalent to the yields of PTH amino acids obtained at other steps of the degradation.[5] These results are inconsistent with those of Ho and Wang, who have reported that the [^{14}C]Abf derivative of β-Tyr-197, presumably formed on reduction of MF_1 with granular zinc in the presence of methyl viologen and EDTA, is unstable to the Edman degradation.[28] It is interesting that β-Tyr-311 along with β-Lys-301 and β-Ile-304 have been identified to be labeled during the photoinactivation of MF_1 with 8-azido[^{3}H]ATP.[29]

Identification of the Lysine Residue to Which the Nbf Group Migrates in the F_1-ATPases. Conditions similar to those described by Ferguson *et al.*[6] were used to promote the O → N migration after the inactivation of MF_1 with [^{14}C]Nbf-Cl. After 20 mg of MF_1 was removed from its storage suspension in 0.55 saturated ammonium sulfate, pH 7.0, containing 4 m*M* ATP and 1 m*M* EDTA by centrifugation, the protein pellet was dissolved in 1.5 ml of 50 m*M* triethanolamine-H_2SO_4, pH 7.4, containing 1 m*M* CDTA and 200 m*M* sucrose. The protein solution was dialyzed against 1 liter of the same buffer for 4 hr at 23°. All subsequent steps were carried out in the dark or, when not possible, in subdued light. The dialyzed enzyme was treated with 4 mol of [^{14}C]Nbf-Cl per mol of MF_1, and inactivation was allowed to proceed at 23° until less than 10% of the original activity remained. The inactivated enzyme was then subjected to gel filtration to remove excess [^{14}C]Nbf-Cl on a 1.5 × 20 cm column of Sephadex G-50, which was equilibrated and eluted with the buffer used for the inactivation, to which 1 m*M* ATP was added. To initiate the migration, sufficient 2 *M* Tris base in 30 μl portions was added to the pooled protein fractions from the column to raise the pH to 9.0.

To monitor the O → N migration, samples of the enzyme incubated at pH 9.0 were removed with time and were assayed in the presence and absence of 10 m*M* dithiothreitol, as illustrated in Fig. 2. During the incubation, a control, which consisted of a sample of the inactivated enzyme

[28] J. W. Ho and J. H. Wang, *Biochem. Biophys. Res. Commun.* **116,** 509 (1983).

[29] M. Hollemans, M. J. Runswick, I. M. Fearnley, and J. E. Walker, *J. Biol. Chem.* **258,** 9307 (1983).

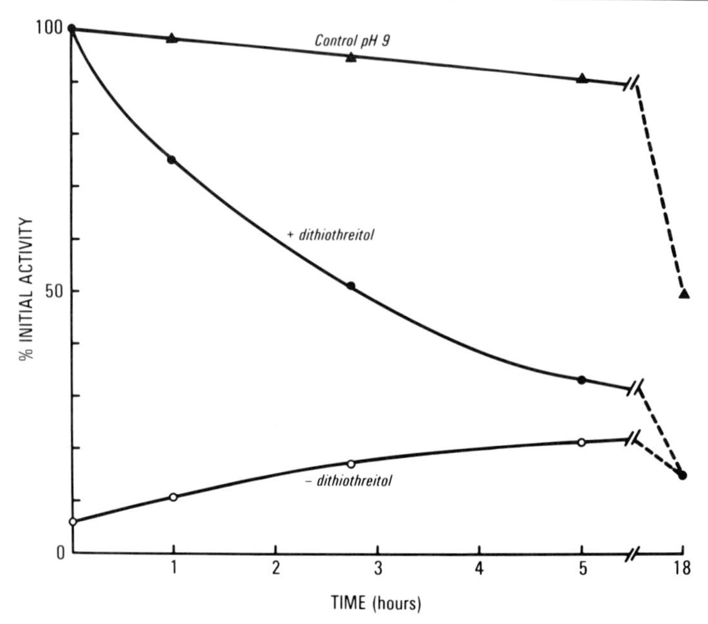

FIG. 2. The kinetics of the O → N migration of the Nbf group promoted by incubation at pH 9.0. MF_1 (20 mg) was inactivated with [^{14}C]Nbf-Cl, separated from excess reagent, and then adjusted to pH 9.0 as described in detail in the text. Immediately after adjusting the pH to 9.0, 10% of the enzyme solution was removed to be used as a control and was reactivated by the addition of dithiothreitol to a final concentration of 10 m*M*. The reactivated control and the untreated enzyme solution were then left at 23° in the dark. At the times indicated, samples were withdrawn from both solutions which were assayed in the presence and absence of 10 m*M* dithiothreitol. (▲), Control assayed in the absence of dithiothreitol; (●), modified enzyme assayed in the presence of dithiothreitol; (○), modified enzyme assayed in the absence of dithiothreitol.

that was reactivated by the addition of dithiothreitol to a final concentration of 20 m*M* immediately after the pH was adjusted, lost very little activity after 5 hr. However, as shown in Fig. 2, the control lost about 50% of its activity in the next 13 hr. Figure 2 also shows that incubation of the inactive enzyme at pH 9.0 led to about 20% reactivation in the absence of dithiothreitol after 5 hr, indicating that significant hydrolysis of the Nbf-*O*-Tyr derivative of the enzyme had occurred. The activity of samples assayed in the presence of dithiothreitol decreased with time, as shown in Fig. 2. This reactivation was used as an indicator of the O → N migration. The amount of radioactivity, which could be removed when the enzyme was treated with dithiothreitol and then gel-filtered on centrifuge columns, declined in parallel with the loss of reactivation promoted by dithiothreitol.

After 18 hr when the activity of the enzyme became equal in the presence and absence of dithiothreitol, the migration was terminated by precipitating the protein by the addition of saturated ammonium sulfate, pH 7.5, to a final concentration of 55% saturation. The precipitated, labeled enzyme was denatured and then digested with trypsin. The tryptic digest of the labeled enzyme was then subjected to HPLC on a C_{18} column which was equilibrated with 0.1% HCl and eluted with a gradient of increasing acetonitrile concentration in 0.1% HCl. Of the total radioactivity applied to this column, 31% was recovered in a sharp peak that eluted at 72% acetonitrile, which accounted for 53% of the total radioactivity recovered. The remainder of the radioactivity recovered was associated with several peaks that eluted at 20–50% acetonitrile. The material in the sharp peak of radioactivity was shown to contain a single peptide by automatic sequence analysis.

The results of the automatic Edman degradation of the labeled tryptic peptide clearly showed that the $[^{14}C]$Nbf group migrated to β-Lys-162. The lysine residue in TF_1, which is homologous with β-Lys-162 in MF_1, was shown to be labeled when the TF_1-ATPase was inactivated with $[^{14}C]$Nbf-Cl and then incubated at pH 9.0 to promote the O → N migration, under similar conditions described above for MF_1. When a tryptic digest of the $[^{14}C]$Nbf derivative of the isolated β subunit of TF_1 was applied to a C_{18} reversed-phase HPLC column, which was equilibrated with 0.1% HCl and then eluted with a gradient of increasing acetonitrile concentration in 0.1% HCl, a single peptide was eluted at 68% acetonitrile. The peptide contained the majority of the radioactivity recovered from the column. Automatic amino acid sequence analysis showed that this peptide contained 38 residues, the first 15 of which were identical to the sequence around Lys-162. This homologous sequence is as follows: -I-G-L-F-G-G-A-G-V-G-K*-T-V-L-I-; where K* represents the $[^{14}C]$Nbf-*N*-Lys derivative of the β subunits of MF_1 and TF_1 resulting from the O → N migration of the $[^{14}C]$Nbf group at alkaline pH.[7,8]

It has been reported by Guillory that illumination of the inactive Nbf-*O*-Tyr derivative of MF_1 with direct sunlight destroyed its capacity to be reactivated with dithiothreitol.[30] Complete irreversible inactivation, which was accomplished in a matter of minutes, was accompanied by the generation of a difference spectrum between a dark control and the illuminated sample, which showed a maximum around 475 nm, the wavelength at which N-Nbf derivatives have absorption maxima.[6] From these observations it was concluded that bright sunlight is a more effective means to promote the O → N migration of the Nbf group than is incubation of the

[30] R. J. Guillory, *Curr. Top. Bioenerg.* **9,** 268 (1979).

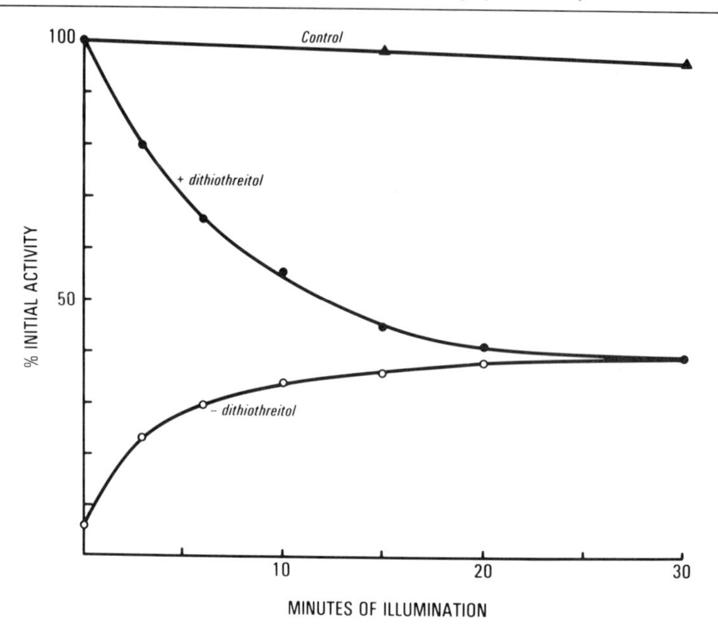

FIG. 3. The kinetics of the O → N migration promoted by illumination at 350 nm. A solution containing 1 mg of MF_1 in 100 μl of 50 m*M* triethanolamine-H_2SO_4, pH 7.4, containing 2 m*M* CDTA, which had been subjected to gel filtration on a 1-ml centrifuge column (Penefsky[22]), was inactivated with a 4-fold molar excess of [^{14}C]Nbf-Cl. After greater than 90% inactivation had occurred, excess reagent was removed by gel filtration on a centrifuge column containing Sephadex G-50 equilibrated with the same buffer at pH 7.5. The effluent was diluted to 1 ml with the same buffer at pH 7.5. To half of the diluted protein solution, which was to be used as a control, dithiothreitol was added to a final concentration of 10 m*M*. The reactivated control and the untreated enzyme solution were then illuminated at 350 nm in a Rayonet Photochemical Reactor. At the times indicated, 2-μl samples were withdrawn and assayed for ATPase activity in the presence and absence of 10 m*M* dithiothreitol. (▲), Control assayed in the absence of dithiothreitol; (●), modified enzyme assayed in the presence of dithiothreitol; (○), modified enzyme assayed in the absence of dithiothreitol.

Nbf-*O*-Tyr derivative of the enzyme at pH 9.0, which requires several hours to complete.[6,7] During studies to identify essential lysine residues labeled by [^{14}C]Nbf, observations were made that show that the O → N migration promoted by alkali, on the one hand, and by illumination, on the other, generate different products. This is an important consideration for those who wish to prepare F_1-ATPases with β-Lys-162, or its equivalent, modified with Nbf for functional studies.

The irreversible inactivation of MF_1 containing the [^{14}C]Nbf derivative of β-Tyr-311 at pH 7.4 promoted by illumination at 350 nm is illustrated in Fig. 3. The control, which had been inactivated with Nbf-Cl and then reactivated by dithiothreitol before illumination, lost no activity over the course of the experiment when it was assayed in the presence and absence of dithiothreitol. When the enzyme modified at β-Tyr-311 with [^{14}C]Nbf was illuminated, about 30% reactivation occurred in 30 min in the absence of added thiols. When the modified enzyme was assayed in the presence of dithiothreitol during the course of the inactivation, the degree of reactivation observed decreased with time. After 30 min illumination, the same amount of activity was observed when the modified enzyme was assayed in the presence and absence of dithiothreitol. Tryptic digests of samples of MF_1 irreversibly inactivated by illumination at 350 nm for 30 min, on the one hand, and by incubation at pH 9.0 for 18 hr, on the other, were subjected to reversed-phase HPLC under the conditions described above for the isolation of the tryptic peptide containing the [^{14}C]Nbf derivative of β-Lys-162. Although migration promoted by alkali or by illumination was about equally efficient, as shown in Figs. 2 and 3, the profiles of radioactivity observed in chromatograms, obtained after subjecting tryptic digests of the two preparations to HPLC, were very different. Comparison of the radioactive profiles obtained after chromatographing equal amounts of the two samples showed differences. The chromatograms of both samples showed sharp peaks of radioactivity eluting at 72% acetonitrile, the concentration at which the tryptic peptide containing the [^{14}C]Nbf derivative of β-Lys-162 elutes. However, the peak eluting at 72% acetonitrile when the illuminated sample was chromatographed contained only 30% as much ^{14}C as eluted at this position when the sample treated at pH 9.0 was chromatographed. Furthermore, there was a considerable increase in the amount of radioactivity that eluted between 20 and 50% acetonitrile in the chromatogram of the illuminated sample. These results suggest that illumination induces reactions in addition to the O → N migration. The photochemical decomposition of benzofurazan is known to generate intermediates and products capable of reacting covalently with amino acid side chains in proteins.[31] Thus, the chromatogram of the tryptic digest of the illuminated sample described above probably reflects extensive photochemical decomposition of the [^{14}C]Nbf group. Therefore, O → N migration promoted by illumination is not recommended for structure–function studies.

Identification of Essential Carboxyl Groups in the F_1-ATPases

Chemical modification studies have indicated that more than one carboxyl group is essential for the hydrolytic reaction catalyzed by the F_1-

[31] W. Heinzelmann and P. Gilgen, *Helv. Chim. Acta* **59,** 2727 (1976).

ATPases. The F_1-ATPases from a variety of sources have been inactivated with DCCD,[10–12,32–35] EEDQ,[14,36–38] and Woodward's reagent K,[39–41] all of which are known to react with carboxyl groups to form activated intermediates that are reactive with nucleophiles. The inactivation of TF_1 by DCCD is accelerated 7-fold upon binding ADP to a single catalytic site.[10,42] Again, with reference to the numerical sequence of the β subunit of MF_1, the selective modification of β-Glu-188 has been shown to occur when TF_1 is inactivated with [^{14}C]DCCD in the presence or absence of ADP.[10] ADP has a slight stimulatory effect on the inactivation of MF_1 with DCCD.[11] Inactivation of MF_1 with [^{14}C]DCCD in the presence or absence of ADP led to the modification of β-Glu-199. In contrast to what is observed with TF_1 and MF_1, ADP has a slight protective effect on the inactivation of the *Escherichia coli* F_1-ATPase by DCCD.[33] However, inactivation of EF_1 with [^{14}C]DCCD leads to the selective modification of β-Glu-192, the glutamic acid residue which is homologous to β-Glu-199 in MF_1.[12] It has yet to be determined whether both of these glutamic acid residues are essential for the activities of all F_1-ATPases. It is interesting to note that Sakamoto and Tonomura have shown that the inactivation of the ATPase activity of MF_1 does not severely alter its capacity to synthesize enzyme-bound ATP at a single catalytic site in the presence of dimethyl sulfoxide at slightly acidic pH.[43] On the other hand, inactivation of the ATPase activity of TF_1 with DCCD, which is accompanied by the modification of the equivalent of β-Glu-188 rather than β-Glu-199, also abolishes the capacity of the enzyme to synthesize enzyme-bound ATP at a single catalytic site[42] under the conditions described by Yoshida.[44] These results suggest that the DCCD-reactive glutamic acid residue of TF_1 has a direct catalytic role in ATP synthesis, while the DCCD-reactive glutamic acid residue of MF_1 does not.

[32] R. Pougeois, M. Satre, and P. V. Vignais, *Biochemistry* **18,** 1408 (1979).
[33] M. Satre, J. Lunardi, R. Pougeois, and P. V. Vignais, *Biochemistry* **18,** 3134 (1979).
[34] V. Shoshan and B. R. Selman, *J. Biol. Chem.* **255,** 384 (1980).
[35] D. Khananshvili and Z. Gromet-Elhanan, *J. Biol. Chem.* **258,** 3720 (1983).
[36] R. Pougeois, M. Satre, and P. V. Vignais, *Biochemistry* **17,** 3018 (1978).
[37] M. Satre, A. Dupuis, M. Bof, and P. V. Vignais, *Biochem. Biophys. Res. Commun.* **114,** 684 (1983).
[38] R. Pougeois, *FEBS Lett.* **154,** 47 (1983).
[39] J. L. Arana and R. H. Vallejos, *FEBS Lett.* **113,** 319 (1980).
[40] J. Arana, M. Yoshida, Y. Kagawa, and R. H. Vallejos, *Biochim. Biophys. Acta* **593,** 11 (1980).
[41] E. Ceccarelli and R. H. Vallejos, *Arch. Biochem. Biophys.* **224,** 382 (1983).
[42] M. Yoshida and W. S. Allison, *J. Biol. Chem.* **258,** 14407 (1983).
[43] J. Sakamoto and Y. Tonomura, *J. Biochem. (Tokyo)* **93,** 1601 (1983).
[44] M. Yoshida, *Biochem. Biophys. Res. Commun.* **114,** 907 (1983).

Comparison of the characteristics of the inactivation of the F_1-ATPases by DCCD with the characteristics of their inactivation by Woodward's reagent K suggests that the two reagents are specific for different essential carboxyl groups. Vallejos and his colleagues have shown that the inactivation of CF_1,[39] TF_1,[40] and the F_1-ATPase from *Rhodospirillum rubrum* chromatophores[41] by Woodward's reagent K is inhibited by ADP or ATP, but not by Mg^{2+} or Ca^{2+}. On the other hand, Mg^{2+} protects each of these enzymes against inactivation by DCCD, suggesting that the DCCD-reactive carboxyl group participates in Mg^{2+} binding and the carboxyl group that reacts with Woodward's reagent K does not. However, it should not be concluded that this is indeed the case until it can be shown directly that the two reagents do indeed react with different carboxyl groups. For example, based on indirect evidence, Pougeois has concluded that the modification of different carboxyl groups occurs when DCCD, on the one hand, and EEDQ, on the other, inactivate MF_1.[38] However, when the inactivation of MF_1 by EEDQ was carried out in the presence of [^{3}H]aniline, analysis of the labeled enzyme for covalently bound ^{3}H showed that the formation of the [^{3}H]anilide of Glu-199 was responsible for the inactivation. Thus, the same carboxyl group is modified when DCCD and EEDQ inactivate MF_1.[11,14]

Conditions for Labeling the F_1-ATPases with [^{14}C]DCCD. After breaking the seal of the vial in which [^{14}C]DCCD (Research Products International) is shipped, the pentane in which it is dissolved is allowed to evaporate in a fume hood at room temperature. Immediately after the pentane is evaporated, the [^{14}C]DCCD is diluted with nonradioactive reagent in absolute ethanol to the desired specific radioactivity. DCCD reacts slowly with water. Therefore, stock solutions are stored at $-20°$ in tightly sealed vials in sealed jars containing desiccant. To remove samples from the stock solutions, the storage jars are warmed to room temperature before opening to avoid condensation of water vapor.

The rate of inactivation of the F_1-ATPases with DCCD increases as the pH is decreased from 8 to 6.[10,32,33] It is not clear what is responsible for the increased rate as the pH is lowered. The accumulated evidence suggests that the initial product of the reaction of DCCD with a carboxyl group is an *O*-acylurea formed by the attack of the unprotonated carboxyl group on protonated DCCD.[45] Therefore, the pK_a between 6 and 7 associated with the inactivation of the F_1-ATPases by DCCD is not likely to be that of the carboxyl group undergoing modification. It is possible that the pH dependencies for the inactivations of the F_1-ATPases by DCCD reflect the presence of a common acidic group that protonates the bound carbo-

[45] K. Kurzer and K. Dowaghi, *Chem. Rev.* **67,** 107 (1967).

diimide, thus converting it to a form capable of reacting with a neighboring carboxylate. With TF_1 being the notable exception, most F_1-ATPases are not stable for prolonged times below pH 6.5. Therefore, the F_1-ATPases are usually inactivated with DCCD between pH 6.5 and 7.0.

The F_1-ATPases show different sensitivities to DCCD. To achieve approximately the same rates of inactivation of MF_1, TF_1, and EF_1, the respective concentrations of DCCD required were 200 μM, 100 μM, and 10 μM. The inactivations of the three F_1-ATPases by DCCD are also affected differently by chlorpromazine.[46,47] Whereas chlorpromazine protected MF_1 and EF_1 against inactivation by DCCD, it had no effect on the rate of inactivation of TF_1 by DCCD at 37° and slightly stimulated the rate of inactivation of TF_1 at 23°. Whether these observations are related to the fact that different essential glutamic acid residues are modified by [^{14}C]DCCD in the subunits of TF_1,[10] on the one hand, and MF_1[11] and EF_1,[12] on the other, has yet to be elucidated.

Stoichiometry of Labeling with [^{14}C]DCCD and Identification of Labeled Residues. The stoichiometry of ^{14}C incorporation into essential carboxyl groups during the inactivation of the F_1-ATPases with [^{14}C]DCCD is somewhat controversial. The large-scale inactivation of MF_1 at 14 mg/ml by [^{14}C]DCCD at pH 7.0 in the presence of 2 mM ADP was accompanied by the incorporation of 1.9 g atoms of ^{14}C when 70% inactivation was attained.[11] On isolation of the subunits, it was shown that greater than 85% of the incorporated radioactivity was bound to the β subunit which contained 0.56 g atom of ^{14}C per mol. Analysis of a cyanogen bromide digest of the labeled β subunit revealed that at least 60% of the radioactivity incorporated into it resided on β-Glu-199. The remainder of the incorporated ^{14}C was associated with several cyanogen bromide fragments. From these results it is estimated that about 1.0 mol of β-Glu-199 per mol of MF_1 was modified when the enzyme was inactivated by 70% with [^{14}C]DCCD. Pougeois *et al.*[32] have shown that inactivation of MF_1 at 0.7 mg/ml with [^{14}C]DCCD at pH 7.0 in the presence of 4 mM ATP led to the incorporation of 1.6 g atoms of ^{14}C when the ATPase was inactivated by 85%. When the inactivated enzyme was subjected to polyacrylamide gel electrophoresis in the presence of sodium dodecyl sulfate, the majority of the radioactivity was found to be associated with the β-subunit.[32] Pougeois *et al.* also reported that while glycine ethyl ester did not affect the rate of inactivation of MF_1 by [^{14}C]DCCD, it decreased the amount of

[46] P. K. Laikind, T. M. Goldenberg, and W. S. Allison, *Biochem. Biophys. Res. Commun.* **109,** 423 (1982).

[47] D. A. Bullough, M. Kwan, P. K. Laikind, M. Yoshida, and W. S. Allison, *Arch. Biochem. Biophys.* **236,** 567 (1985).

radioactivity incorporated into the enzyme by about 50%. From these results it was tentatively concluded that two sites are modified by [^{14}C]DCCD, a hydrophilic site, which is accessible to glycine ethyl ester, and a hydrophobic site, which is not.[32,48] Another interpretation of these results is that glycine ethyl ester binds to a saturable site where it attacks the *O*-acylisourea derivative of β-Glu-199 at about the same rate as the rearrangement of the *O*-acylisourea to the stable *N*-acylurea proceeds. Bearing on this argument is the observation that during inactivation of the enzyme with EEDQ in the presence of [^{3}H]aniline, the incorporation of ^{3}H into MF_1 exhibits saturation behavior with respect to the concentration of [^{3}H]aniline.[14] More recently, Wong *et al.*[49] have reported that the inactivation of MF_1 at 1 mg/ml with [^{14}C]DCCD in the presence of 4 m*M* ATP at pH 6.8 led to the incorporation of 1 g atom of ^{14}C per mol of enzyme on inactivation of the ATPase by about 95%. The relationship between ^{14}C bound and inactivation observed was reported to be linear up to about 90–95% inactivation. A much larger amount of ^{14}C was incorporated during the loss of the last 5–10% of activity. About 2 g atoms of ^{14}C were incorporated covalently on 99% inactivation. By subjecting the inactivated enzyme to polyacrylamide gel electrophoresis in the presence of sodium dodecyl sulfate after 1.0 and 1.65 g atoms of ^{14}C were incorporated per mol of MF_1, it was shown that only the β subunit was modified.[49]

Some variation in the amount of ^{14}C incorporated per mol of enzyme has been observed when different laboratories have inactivated the EF_1-ATPase with [^{14}C]DCCD. When EF_1 at 22 mg/ml was inactivated by 83% at pH 6.5 with [^{14}C]DCCD, 0.9 g atom of ^{14}C was bound after excess reagent was removed by precipitating the inactivated enzyme with ammonium sulfate, which was then followed by gel filtration on Sephadex G-50. However, since about half of the ^{14}C was removed when the gel-filtered enzyme was denatured with sodium dodecyl sulfate, urea, or guanidine–HCl, only about 0.4 g atom of ^{14}C became covalently bound per mol of enzyme during the inactivation. Of the radioactivity covalently incorporated into the enzyme, 0.13 g atom was bound per mol of the isolated β subunit and 0.01 g atom was bound per mol of the isolated γ subunit. The majority of the radioactivity in the β subunit was shown to be associated with β-Glu-193, which is homologous with β-Glu-199 of MF_1.[12] Satre *et al.*[33] have reported that the inactivation of EF_1 at 0.7 mg/ml by about 90% with [^{14}C]DCCD at pH 6.3 led to the binding of about 1 g atom of ^{14}C per mol when it was inactivated by about 90%. Excess radioactive reagent was removed by centrifugation elution on columns of Sephadex G-50.

[48] P. V. Vignais and M. Satre, *Mol. Cell. Biochem.* **60,** 33 (1984).

[49] S.-Y. Wong, A. Matsuno-Yagi, and Y. Hatefi, *Biochemistry* **23,** 5004 (1984).

Although it was not reported whether any of the bound radioactivity was lost on denaturation, it was shown that nearly all of the covalently bound radioactivity was associated with the β subunit by subjecting the labeled enzyme to polyacrylamide gel electrophoresis in the presence of sodium dodecyl sulfate. Lotscher and Capaldi[50] have also examined the inactivation of EF_1 by [^{14}C]DCCD. When the enzyme at 0.7 mg/ml was inactivated by about 90%, they observed that about 1 g atom of ^{14}C was bound per mol when excess radioactive reagent was removed by centrifugation elution on columns of Sephadex G-50.[22] They also found that about 25% of the radioactivity that remained with the enzyme after the gel-filtration step was removed from the protein when it was denatured with sodium dodecyl sulfate.

The inactivation of the TF_1-ATPase by about 90% with [^{14}C]DCCD can be consistently correlated with the covalent modification of about 1.5 mol/mol of enzyme of the glutamic acid residue, which is homologous with β-Glu-188 of MF_1. The modification of TF_1 by [^{14}C]DCCD is rather selective. A small amount of nonspecific labeling of the β subunit accompanies the modification of β-Glu-188 in about half of the β subunits. When TF_1 was inactivated by 90% with DCCD, dissociated with 8 *M* urea, and then diluted, no activity was recovered, while a control, when subjected to the same dissociation–reassociation procedure, regained 80% of its activity on dilution.[10] These results suggest that the inactivation of TF_1 with DCCD is caused by modifications in addition to the conversion of β-Glu-188 to an *N*-acyldicyclohexyl[^{14}C]urea derivative, the product of which is isolated and identified after the inactivation.

Some observations made during a study of the inactivation of MF_1 with EEDQ in the presence of [^{3}H]aniline[14] are pertinent to the controversial results and apparent dilemmas that have arisen from the investigations summarized above on the inactivation of the F_1-ATPases by [^{14}C]DCCD. When MF_1 at 1.7 mg/ml was inactivated with 0.9 m*M* EEDQ in the presence of 1.7 m*M* [^{3}H]aniline at pH 7.0, the [^{3}H]anilides of three glutamic acid residues were formed in sufficient amounts to be identified. The [^{3}H]anilides and the moles of each formed per mole of MF_1 are as follows: β-Glu-199, 1.5; β-Glu-341, 0.3; and α-Glu-402*, 0.7, where position 402* refers to the sequence of the α subunit of EF_1.[51,52] The amounts of the [^{3}H]anilides of β-Glu-341 and α-Glu-402* formed when MF_1 was inactivated by 90% by EEDQ in the presence of [^{3}H]aniline varied as the protein concentration of the inactivation mixtures varied, while the

[50] H. R. Lötscher and R. A. Capaldi, *Biochem. Biophys. Res. Commun.* **121,** 331 (1984).

[51] N. Gay and J. E. Walker, *Nucleic Acids Res.* **11,** 2185 (1981).

[52] H. Kanazawa, K. Matuchi, K. Toshiaki, F. Tamura, and M. Futai, *Biochem. Biophys. Res. Commun.* **100,** 219 (1981).

amount of the [^{3}H]anilide of β-Glu-199 did not. From these results and the fact that the modification of β-Glu-199 appears also to be responsible for the inactivation of MF_1 by DCCD, it has been concluded that the modification of β-Glu-199 is responsible for the inactivation of MF_1 by EEDQ.

The activated carboxyl groups formed on reaction of MF_1 with EEDQ have a transient lifetime.[14] When MF_1 was inactivated by 90% with EEDQ, gel-filtered to remove excess reagent, and then treated immediately with [^{3}H]aniline, only about half as much ^{3}H was incorporated per mol of enzyme as was incorporated when [^{3}H]aniline was present when the inactivation was initiated. Furthermore, when [^{3}H]aniline was added 3 hr after the inactivated enzyme was gel-filtered, ^{3}H was not incorporated into the enzyme. Since reactivation was not observed to occur in the interval between the end of treatment with EEDQ and the addition of [^{3}H]aniline or during the subsequent incubations with [^{3}H]aniline, hydrolysis of the mixed carbonic anyhydride of β-Glu-199 was not responsible for these results. Therefore, it appears that in the absence of [^{3}H]aniline or another suitable nucleophile, the mixed carbonic anhydride formed by reaction of EEDQ with β-Glu-199 reacts with a neighboring nucleophilic side chain to form an intramolecular cross-link. From these results it can be argued that the *O*-acylisourea derivative formed on the reaction of β-Glu-199 with DCCD in MF_1 is also capable of reacting with the same nucleophile to form an intramolecular cross-link. Therefore, until methods are developed to determine the efficiency of the O $\rightarrow$ N rearrangement during the modification of an essential carboxyl group with [^{14}C]DCCD, the correlation of the incorporation of ^{14}C with the degree of inactivation of an F_1-ATPase must be interpreted cautiously.

[73] Monoclonal Antibodies to F_1-ATPase Subunits as Probes of Structure, Conformation, and Functions of Isolated or Membrane-Bound F_1

By CATHERINE GODINOT, MAHNAZ MORADI-AMELI, and DANIÈLE C. GAUTHERON

Antibodies able to recognize sequences of only a few amino acids can serve as tools to define structure–function relationships in enzymes. Conventional antisera are complex mixtures of antibodies of different classes with different affinities for various antigenic determinants. Therefore they cannot provide precise information on the domains or subdomains of proteins. Furthermore, because of the complexity of the immune response, no conventional antiserum is exactly like another even if both come from the same species. In contrast, monoclonal antibodies are attractive reagents since once the cell line is established, it provides, in principle, a permanent source of a well-defined antibody. In addition, since monoclonal antibodies recognize a single antigenic determinant, they are suitable as specific probes of protein assembly and permit detailed studies of the topology of defined regions. This chapter describes the preparation of monoclonal antibodies against pig heart mitochondrial F_1-ATPase subunits and their use to determine the stoichiometry, the conformations, and the functions of the isolated or membrane-integrated enzyme.[1,2]

Preparation and Control of Monoclonal Antibodies (McAb) to Mitochondrial F_1-ATPase Subunits

Immunization of Mice

Animals. BALB/c mice from IFFA-CREDO, L'Arbresle, France

Reagents

Potassium phosphate buffer, 0.1 *M*, pH 7.5
Mitochondrial F_1-ATPase prepared as an ammonium sulfate suspension, as described by Penin *et al.*[3]
Glutaraldehyde, 25% aqueous solution, electron microscopy grade (Merck)

Procedure. The ammonium sulfate suspension of pig heart mitochondrial F_1-ATPase was spun down at 9000 *g* for 5 min. The pellet was dissolved in phosphate buffer (2 mg protein/ml). After centrifugation at 9000 *g* for 5 min to eliminate any insoluble material, glutaraldehyde was added to the supernatant at a final concentration of 0.05%. Then 30 min later, the cross-linked F_1-ATPase was dialyzed for 2 hr against 500 volumes of phosphate buffer with two changes of buffer and emulsified with an equal volume of complete Freund's adjuvant. The enzyme (0.2 mg protein[4]) was then injected intraperitoneally to the mice. This injection was repeated biweekly (2 to 5 times) in the absence of Freund's adjuvant until the titer of antibodies was high enough to be detected at a dilution of

[1] M. Moradi-Améli and C. Godinot, *Proc. Natl. Acad. Sci. U.S.A.* **80,** 6167 (1983).
[2] P. Archinard, M. Moradi-Améli, C. Godinot, and D. C. Gautheron, *Biochem. Biophys. Res. Commun.* **123,** 254 (1984).
[3] F. Penin, C. Godinot, and D. C. Gautheron, *Biochim. Biophys. Acta* **548,** 63 (1979).
[4] O. H. Lowry, N. J. Rosebrough, A. L. Farr, and R. J. Randall, *J. Biol. Chem.* **193,** 265 (1951).

at least 1/1000 by solid-phase radioimmunoassay (SPRIA, see below). To evaluate this titer, the blood was collected from the retroorbital sinus with a Pasteur pipette and the serum exuded from the blood clot was tested. The mouse exhibiting the highest titer of antibody was selected for hybridoma production and received a last injection 3 days prior to fusion.

Preparation of Monoclonal Antibodies (McAb)

McAb against mitochondrial F_1-ATPase were obtained as described in Ref. 1 with the method of Galfré and Milstein.[5] After fusion of the spleen cells of immune mice with hypoxanthine phosphoribosyltransferase-deficient NS_1 myeloma cells (obtained from Dr. J. Huppert, INSERM, U51, Lyon, France), the hybrid cells were selected in hypoxanthine, azaserine medium and then screened (SPRIA or ELISA) for the production of antibodies. The antibody-producing cells were cloned by limiting dilution fractionation.[5]

The cloned hybridoma cells were grown to a large scale either by culture in 800-ml tissue culture bottles or by injection into BALB/c mice to produce ascitic tumors.[5] Culture supernatants or ascitic fluid were mixed with an equal volume of saturated ammonium sulfate to precipitate immunoglobulins. After centrifugation for 30 min at 8000 *g*, the pellet of antibodies was dissolved and dialyzed in 0.14 *M* potassium phosphate buffer at 0–4° and then purified on protein A–Sepharose CL4B (Pharmacia) following the procedure of Ey *et al.*[6]: The column was equilibrated with 0.14 *M* sodium phosphate buffer, pH 8.0, and the McAb was eluted in 0.14 *M* sodium phosphate buffer, pH 6.0. The purified McAb could be concentrated with an immersible CX membrane (Millipore). Unless stated otherwise, the final concentration of McAb was calculated from the absorbance at 280 nm ($\varepsilon_{1\ cm}$ = 14 for 1% solution, w/v).

Screening Methods

Reagents

Mitochondrial F_1-ATPase,[3] 0.1 mg protein/ml of 0.1 *M* potassium phosphate buffer, pH 7.5 (cf. "Immunization of Mice")

Washing buffer: 0.1 *M* glycine adjusted to pH 7.4 with NaOH, 1% bovine serum albumin (fraction V, 96–99%, Sigma Chemical Co.)

Rabbit anti-mouse immunoglobulin (Sera-Lab, Crawley Down, Sussex, England) diluted 1 : 50 in 0.1 *M* phosphate buffer, pH 7.5, 0.9% NaCl (PBS)

[5] G. Galfré and C. Milstein, this series, Vol. 73, p. 3.

[6] P. L. Ey, S. J. Prowse, and C. R. Jenkin, *Immunochemistry* **15,** 429 (1978).

Sheep anti-mouse immunoglobulin conjugated to peroxidase (Biosys, Compiègne, France) diluted 1 : 1000 in PBS

Iodinated proteins: Protein A (Industrie Biologique Française or LKB) or F_1-ATPase, 0.1 mg in 0.1 ml of PBS, was mixed with 0.3 mCi of $Na^{125}I$ (I-125-S4, Commissariat à l'Energie Atomique, France) in a stoppered glass tube (6 × 50 mm) coated with 10 μg of IODO-GEN[7] (Pierce Chemical Co.). After 5 min, the labeled protein A was separated from free ^{125}I by filtration–centrifugation,[8] using a 1-ml tuberculin syringe containing Sephadex G-50 fine (Pharmacia) equilibrated with PBS or with 50 m*M* Tris–H_2SO_4, pH 7.5

2,2'-Azino-di(3-ethylbenzthiazoline sulfonate) (ABTS), 0.25 m*M* in 0.1 *M* phosphate buffer mixed with 0.18 m*M* H_2O_2 just before use

Sodium dodecyl sulfate (SDS), 2.5%

Films for autoradiography, Kodak, direct exposure

Microtiter plates, 96 wells

Procedure. The amount of free McAb present in hybridoma cultures or in assays can be estimated either by solid-phase radioimmunoassay (SPRIA) or by enzyme-linked immunosorbent assay (ELISA).

1. Dispense 50 μl of F_1-ATPase in each well of a microtitration plate (flat-bottom flexible vinyl for SPRIA, polystyrene for ELISA) and let stand for 16 hr at room temperature; empty the wells.

2. Fill the plate with the washing buffer; wait at least 10 min to saturate with BSA the remaining binding sites. Empty the wells.

3. Repeat step 2 three times. These plates can be stored at 0–4° for 2–3 weeks.

4. Introduce 50 μl of hybridoma culture supernatant or antibody-containing assays adequately diluted in each well except in control assays, which are filled up with culture medium or buffer. Let stand overnight.

5. Repeat step 2.

6. Dispense 50 μl of diluted rabbit anti-mouse immunoglobulin (SPRIA) or anti-mouse immunoglobulin conjugated to peroxidase (ELISA). Incubate 1–2 hr at 37°; empty the wells; repeat step 2.

7. SPRIA: Dispense 50 μl of ^{125}I-labeled protein A (2 × 10^5 cpm). Incubate 1 hr at 37°; repeat step 2. Autoradiograph or cut out each well and count for radioactivity.

8. ELISA: Dispense 50 μl of ABTS–H_2O_2 solution. The positive wells become green after about 30 min. The reaction is stopped by addition of 10 μl SDS. The absorbancy can be quantitated at 420 nm.

[7] P. J. Fracker and J. C. Speck, Jr., *Biochem. Biophys. Res. Commun.* **80,** 849 (1978).

[8] H. S. Penefsky, *J. Biol. Chem.* **252,** 2891 (1977).

Determination of Specificity by Immunodecoration

Principle. The specificity of the McAb was determined after separation of F_1-ATPase subunits by electrophoresis in the presence of a cationic detergent, TDAB, transfer to nitrocellulose sheets, incubation of the sheets in the presence of McAb, and immunodecoration in the presence of ^{125}I–protein A.

Reagents

Electrophoresis gel slab: 14% acrylamide, 0.1% TDAB (tetradecyltrimethyl ammonium bromide) according to Refs. 9 and 10, with a 15-cm-wide well

Electrophoresis buffer: TDAB 0.125% (w/v) in 85 m*M* glycine buffer adjusted to pH 3.0 with phosphoric acid, as in Ref. 10

F_1-ATPase: The samples were prepared by mixing 20 μl of 50 m*M* sodium phosphate buffer (pH 4.0) containing 50 μg F_1-ATPase,[3] 40 mg urea, 5 μl 2-mercaptoethanol, 60 μl of 10% TDAB (w/v), and 2 μl of 0.1% Bromphenol Blue; the mixture was heated for 5 min at 90° in a water bath

Transfer buffer: 20 m*M* Tris, 150 m*M* glycine, pH 8.3, 20% methanol

Incubation medium: 5% bovine serum albumin in 50 m*M* Tris–HCl, 150 m*M* NaCl, pH 7.4

McAb: Ammonium sulfate-purified pellets dissolved in the above buffer containing only 3% bovine serum albumin

Iodinated protein A: See above, "Screening Methods"

Rabbit anti-mouse immunoglobulins: See "Screening Methods," 1/100 dilution

Films for autoradiography: See "Screening Methods"

Nitrocellulose sheets, 0.45-μm pore, Schleicher and Schuell

Procedure. The subunits of F_1-ATPase (150 μg) were separated by electrophoresis in TDAB at 25 mA/slab for 1 hr and 50 mA/slab for 4 hr.[9] This procedure permits a very good separation of the subunits, especially α and β, which is a prerequisite to ascertain the specificity of McAb. Then the subunits were electrophoretically transferred for 105 min at 190 mA from the gels onto nitrocellulose sheets (10 × 15 cm) by the method of Towbin *et al.*[11] After transfer, the sheets were maintained in the incubation medium for 1 hr at 40° and overnight at 4°. Then they were cut in bands (1 × 10 cm) and gently shaken for 16 hr at room temperature with 2.5 ml of McAb. After washing in the same buffer without serum albumin,

[9] F. Penin, C. Godinot, and D. C. Gautheron, *Biochim. Biophys. Acta* **775,** 239 (1984).

[10] A. Amory, F. Foury, and A. Goffeau, *J. Biol. Chem.* **255,** 9353 (1980).

[11] H. Towbin, T. Staehlin, and J. Gordon, *Proc. Natl. Acad. Sci. U.S.A.* **76,** 4350 (1979).

the sheets were incubated with rabbit anti-mouse immunoglobulin for 1 hr. After two additional washes, the sheets were finally incubated at room temperature for 1 hr with ^{125}I–protein A (5×10^5 cpm/ml) in the incubation buffer, including 3% bovine serum albumin. After 6 washings the sheets were dried and autoradiographed.

Results

Among 300 hybrid cell cultures screened by using SPRIA, 14 gave a positive result. The five lines giving the highest positive response ($5G_{11}$, $14D_5$, $19D_3$, $20D_6$, and $7B_3$) were cloned, found to be stable, and further tested. By immunodecoration it was shown that $20D_6$ and $7B_3$ specifically recognized the F_1–ATPase α subunit, while $5G_{11}$, $14D_5$, and $19D_3$ recognized the β subunit.

Use of McAb to Study the Structure, Conformation, and Functions of F_1-ATPase

Stoichiometry of Subunits in Active F_1-ATPase

Principle. The method consists of measuring the maximal number of mol of McAb bound per mol of F_1-ATPase, assuming that each subunit can bind only 1 mol of McAb when the related epitope is accessible. Known amounts of F_1-ATPase are adsorbed at the bottom of microtiter plate wells and the amount of McAb bound is determined by SPRIA.

Methods

1. To determine the amounts of F_1-ATPase adsorbed in the wells, 50 μl of various concentrations of ^{125}I-labeled F_1-ATPase are added to the wells and let to stand for 16 hr. Then the wells are emptied, washed 4 times, dried, and cut out for counting the γ radiations. The treatment by iodine + IODO-GEN increases the binding of F_1-ATPase to the wells by 13%. This must be taken into account for the calculation.

2. Titration of McAb: Because protein A–Sepharose purification of immunoglobulin species from hybridoma culture supernatants is expected to result in some contamination of the final fraction with immunoglobulins derived from the fetal calf serum, the concentration of purified McAb was estimated by comparison with a calibration curve made with pure mouse immunoglobulins. First, the concentration of pure mouse immunoglobulins was determined by the Lowry method[4] using crystalline bovine serum albumin (Sigma Chemical Co., Fraction V) as a standard. Different concentrations of purified McAb anti-F_1-ATPase subunits and of pure mouse

CHARACTERIZATION OF FOUR MONOCLONAL ANTIBODIES

Clones	Subunit recognition[a]	Scatchard plot	Mol IgG per mol F_1	K_d (M)	F_1 recognition form[b]	
					Native	Dissociated
$19D_3$	β	Linear	2.5	4.7×10^{-8}	++++	+
$5G_{11}$	β	Linear	1.4	3.3×10^{-9}	++++	+
$14D_5$	β	Linear	1.3	6.5×10^{-10}	++++	++++
$20D_6$	α	Linear	2.2	1.1×10^{-9}	++++	++++

[a] As determined by immunodecoration technique, i.e., after removal of dissociating agents.

[b] F_1-ATPase dissolved in 0.1 *M* sodium phosphate buffer, pH 7.5 (native), or in 8 *M* urea, 1% SDS, 0.1 *M* phosphate buffer, pH 7.5 (dissociated), was adsorbed on nitrocellulose disks and dried by evaporation. The disks were then immunodecorated as the nitrocellulose sheets used in electrotransfer (see text). The number of + indicates the intensity of the autoradiograms.

immunoglobulins (Sera-Lab, Crawley Down, Sussex, England) were coated in separate wells on flexible polyvinyl microtest plates and let to stand for 16 hr. After 4 washings, rabbit anti-mouse immunoglobulins and ^{125}I–protein A were successively added and the radioactivity of the wells measured in a γ counter, as described in "Screening Methods."

3. Calibration curve of McAb bound to F_1-ATPase versus ^{125}I–protein A: The wells were coated with a large excess of F_1-ATPase (2.5–25 μg) to permit the complete binding of McAb[12] added in various amounts to the wells and let to stand at room temperature for 16 hr. Then the SPRIA method was applied. Finally, the counts were related to the amounts of McAb bound to F_1-ATPase and the calibration curve drawn.

4. Scatchard plots: Low concentrations of F_1-ATPase (250–500 ng/50 μl) were used to coat the wells so as to obtain a binding of 100–200 ng protein per well (cf. method 1). Various dilutions of McAb were added and the SPRIA method was used. The radioactivity due to the binding of labeled protein A was counted and used to calculate the amount of McAb bound to F_1-ATPase, in each case using the calibration curve (cf. method 3). Assuming molecular weights of 150,000 and 380,000,[13] respectively, for McAb and F_1-ATPase, the Scatchard plots[14] for the binding of McAb were drawn.

5. Results: The table summarizes the characteristics of four McAb. While the affinity was reported to be similar for antigens, whether immobilized or in solution, the capacity of immobilized antigens to bind McAb

[12] S. J. Kennel, *J. Immunol. Methods* **55,** 1 (1982).

[13] A. Di Pietro, C. Godinot, M. L. Bouillant, and D. C. Gautheron, *Biochimie* **57,** 959 (1975).

[14] D. Scatchard, *Ann. N.Y. Acad. Sci.* **51,** 660 (1949).

could be decreased.[12] Therefore the number of moles of McAb bound per mol of F_1-ATPase may be underestimated. Since 2.5 mol of the anti-β-$19D_3$ and 2.2 mol of the anti-α-$20D_6$ at least are bound per mol of F_1-ATPase, the stoichiometry of these subunits must be $\alpha_3\beta_3$ in the enzyme.

Under the same conditions, F_1-ATPase binds only 1.4 and 1.3 of the anti-β-$5G_{11}$ and $14D_5$, respectively. This might reflect some asymmetry in the spatial arrangement of the β subunits in the active F_1-ATPase. The anti-β-$19D_3$ and $5G_{11}$ appear essentially conformational, since they hardly bind to F_1 treated by urea and sodium dodecyl sulfate. In contrast, the same treatment does not modify the recognition of F_1 by $14D_5$ and $20D_6$.

Recognition of Conformations Related to the Activities of F_1ATPase

The McAb raised against F_1-ATPase subunits were tested on the various functions of the enzyme, isolated, or integrated in the membrane (submitochondrial particles): ATP hydrolysis, regulation of ATP hydrolysis, ATP synthesis. ATP hydrolysis and synthesis are measured as described in Ref. 15. The regulation of ATP hydrolysis concerns the ADP- and P_i-induced inhibition of isolated F_1-ATPase after preincubation with the effectors.[16]

The anti-β-$19D_3$ inhibits the ATPase activity of the isolated enzyme, but only in the presence of inorganic phosphate at pH 6.6, i.e., in the presence of monovalent P_i. It has been shown that monovalent P_i induces a significant conformational change.[17] In contrast, $19D_3$ does not affect either the ATPase activity or the ATP synthesis of F_1 integrated in submitochondrial particles.

Preincubation of F_1-ATPase with ADP and Mg^{2+} induces a hysteretic inhibition which slowly develops during MgATP hydrolysis.[16] During the setting up of hysteretic inhibition, two slow conformational changes can be detected. The presence of phosphate is compulsory to obtain the second conformational change that leads to the inhibited form of F_1-ATPase.[17] The anti-β-$5G_{11}$ prevents the hysteretic inhibition, while it does not affect the ATPase activity of F_1-ATPase isolated or in submitochondrial particles. In contrast, $5G_{11}$ strongly inhibits ATP synthesis (net ATP synthesis and ATP–P_i exchange) in these particles. Therefore these McAb are useful tools to detect the various conformational changes of the

[15] D. C. Gautheron, F. Penin, G. Deléage, and C. Godinot, this volume [39].

[16] A. Di Pietro, F. Penin, C. Godinot, and D. C. Gautheron, *Biochemistry* **19,** 5671 (1980).

[17] D. C. Gautheron, A. Di Pietro, F. Penin, M. Moradi-Améli, G. Fellous, and C. Godinot, *in* "H^+-ATP synthase: Structure, Function, Regulation" (S. Papa *et al.*, eds.), p. 219. ICSU Press, Adriatica Editrice, Bari, 1984.

enzyme during its catalytic process or regulation; more precisely it allows distinguishing a conformation competent for ATP synthesis which is different from that of ATP hydrolysis. It is interesting to note (see the table) that F_1-ATPase binds only 1.4 mol of $5G_{11}$ per mol, indicating that $5G_{11}$ binds to a maximum of two β subunits. This could be related to an asymmetric functioning of F_1 in the complex, as revealed by the structure.

Epitopes Well Conserved in the Phylogenic Scale

These anti-β-McAb from pig heart $19D_3$ and $5G_{11}$ that recognize specific conformations of F_1-ATPase cross-reacted with all F_1-ATPase tested: *Escherichia coli, Saccharomyces cerevisiae, Aspergillus niger,* spinach chloroplasts, rat liver, and beef heart.[2] This indicates that the epitopes reacting with these McAb are well conserved in the phylogenic scale and are involved in parts essential for the activities of F_1. Therefore these antibodies will be useful tools to identify these essential epitopes.

The anti-β-$14D_5$ and the anti-α-$20D_6$ only recognized the F_1 subunits of mammalian species.

Determination of Epitopes Recognized by McAb

To determine the various epitopes once the subunit recognized by the McAb is identified, one needs to know the amino acid sequence of the subunit and to apply various fragmentation procedures (chemical, enzymatic) to the subunits to localize the epitopes. The sequence of all beef heart subunits and methods of fragmentation have been published by Walker *et al.*[18,19] After applying these fragmentation methods to the pig heart subunits, the peptides were separated by gel electrophoresis and tested with the anti-α and anti-β-McAb. The identification of the peptides giving a positive response is under investigation.

[18] J. E. Walker, M. J. Runswick, and M. Saraste, *FEBS Lett.* **146,** 393 (1982).

[19] J. E. Walker, I. M. Fearnley, N. J. Gay, B. W. Gibson, F. D. Northrop, S. J. Powell, M. J. Runswick, M. Saraste, and V. L. Tybulewicz, *J. Mol. Biol.* **184,** 677 (1985).

[74] Determination of the Stoichiometry and Arrangement of α and β Subunits in F_1-ATPase Using Monoclonal Antibodies in Immunoelectron Microscopy

By KARIN EHRIG, HEINRICH LÜNSDORF, PETER FRIEDL, and HANS ULRICH SCHAIRER

The stoichiometry and arrangement of the five subunits in the F_1-ATPase so far have not been unequivocally elucidated. The methods used for the determination of the stoichiometry, i.e., staining intensities of bands after sodium dodecyl sulfate–polyacrylamide gel electrophoresis,[1-3] metabolic labeling with radioactive precursors,[3-9] cross-linking,[10-12] chemical labeling of sulfhydryl groups,[13,14] titration of the subunits using radioactively labeled antibodies and evaluation of the binding with a Scatchard plot[15] and reconstitution experiments,[16-19] are based on the molecular weight of the enzyme, statistical behaviour of the enzyme–reagent interactions and a high homogeneity of the enzyme preparation.

It should be possible to determine the subunit stoichiometry unequivocally, if one directly counts the different subunits of a statistical significant number of individual enzyme complexes. This is the basic principle of the approach described here for the determination of the subunit stoi-

[1] W. A. Catterall, W. A. Coty, and P. L. Pedersen, *J. Biol. Chem.* **248,** 7427 (1973).

[2] A. Senior, *in* "Membrane Proteins in Energy Transduction" (R. A. Capaldi, ed.). Dekker, New York, 1978.

[3] D. L. Foster, PhD thesis, University of Wisconsin (1980).

[4] P. D. Bragg and C. Hou, *Arch. Biochem. Biophys.* **167,** 311 (1975).

[5] D. L. Foster and R. M. Fillingame, *J. Biol. Chem.* **257,** 2009 (1982).

[6] Y. Kagawa, N. Sone, M. Yoshida, H. Hirata, and H. Okamoto, *J. Biochem.* **80,** 141 (1976).

[7] M. Hubermann and M. R. J. Salton, *Biochim. Biophys. Acta* **547,** 230 (1979).

[8] R. D. Todd, T. A. Griesenbach, and M. G. Douglas, *J. Biol. Chem.* **255,** 5461 (1980).

[9] E. Stutterheim, M. A. C. Menneke, and J. A. Berden, *Biochim. Biophys. Acta* **64,** 271 (1981).

[10] P. D. Bragg and C. Hou, *Eur. J. Biochem.* **106,** 495 (1980).

[11] R. Enns and R. S. Criddle, *Arch. Biochem. Biophys.* **183,** 742 (1977).

[12] B. A. Baird and G. G. Hammes, *J. Biol. Chem.* **252,** 4743 (1977).

[13] A. G. Senior and J. C. Brocks, *FEBS Lett.* **17,** 327 (1971).

[14] M. Yoshida, N. Sone, H. Hirata, Y. Kagawa, and M. Ui, *J. Biol. Chem.* **254,** 9525 (1979).

[15] M. Moradi-Améli and C. Godinot, *Proc. Natl. Acad. Sci. U.S.A.* **80,** 6167 (1983).

[16] G. Vogel and R. Steinhart, *Biochemistry* **15,** 208 (1976).

[17] P. C. Sternweis and J. B. Smith, *Biochemistry* **16,** 4020 (1977).

[18] P. C. Sternweis, *J. Biol. Chem.* **253,** 3123 (1978).

[19] S. D. Dunn and M. Futai, *J. Biol. Chem.* **255,** 113 (1980).

chiometry: Monoclonal antibodies are used to label a specific subunit only once and the number of antibodies bound per molecule F_1-ATPase is determined by electron microscopy.[20,21] The DNA sequence data of the *Escherichia coli* ATP synthase operon exclude repetitive sequences within one subunit or sufficiently homologous sequences in different subunits.[22–24] Thus it is conclusive that one bound antibody corresponds to one subunit.

Procedures

Immunization of Mice and Isolation of Hybridoma

Female BALB/c mice were immunized with pure F_1[16] according to the following scheme: Day 0: an emulsion of 50 μg F_1 in 150 μl H_2O and 100 μl Freund's adjuvant complete was injected intraperitoneally; day 28 and 56: an emulsion of 50 μg F_1 in 150 μl and 100 μl Freund's adjuvant incomplete was injected intraperitoneally; 4 days before sacrificing the mouse: 100 μg F_1 in 150 μl PBS (8 g NaCl, 0.2 g KCl, 1.44 g $Na_2HPO_4 \cdot 2H_2O$, and 0.2 g KH_2PO_4/liter buffer) was injected intravenously (tail).

The spleen cells of the immunized mouse were fused with the myeloma cell line NS1[25] by the procedure of Gefter *et al.*[26] The production of antibodies was screened with an ELISA,[27] and monoclonal hybridoma were obtained after cloning under limiting dilution conditions.

Immunological Methods

ELISA.[27] F_1 was fixed on 96-well immunoplates (NUNC) by incubating 1 μg F_1 in 100 μl PBS per well for at least 2 hr at room temperature. Free protein binding sites of the well were blocked by buffer A (2% BSA, 0.05% Tween 20 in PBS). F_1-specific antibodies were bound to the fixed F_1 by incubating the F_1-coated wells with mouse serum (dilution 1:20–1:1280 in buffer A, 100 μl/well) or with culture supernatants of hybridoma (100 μl/well) for at least 2 hr at room temperature. The bound

[20] H. Lünsdorf, K. Ehrig, P. Friedl, and H. U. Schairer, *J. Mol. Biol.* **173,** 131 (1984).

[21] H. Tiedge, H. Lünsdorf, G. Schäfer, and H. U. Schairer, *Eur. Bioenerg. Conf., 3rd Short Rep.* **3A,** 31 (1984).

[22] J. N. Gay and J. E. Walker, *Nucleic Acids Res.* **9,** 3919 (1981).

[23] H. Kanazawa, K. Mabuchi, T. Kagano, T. Noumi, T. Sekiya, and M. Futai, *Biochem. Biophys. Res. Commun.* **103,** 613 (1981).

[24] J. Nielsen, F. G. Hansen, J. Hoppe, P. Friedl, and K. von Meyenburg, *Mol. Gen. Genet.* **184,** 33 (1981).

[25] G. Köhler and C. Milstein, *Eur. J. Immunol.* **6,** 511 (1976).

[26] M. Gefter, D. H. Margulies, and M. D. Scharff, *Somat. Cell Genet.* **3,** 231 (1977).

[27] E. Engvall, this series, Vol. 70, p. 419.

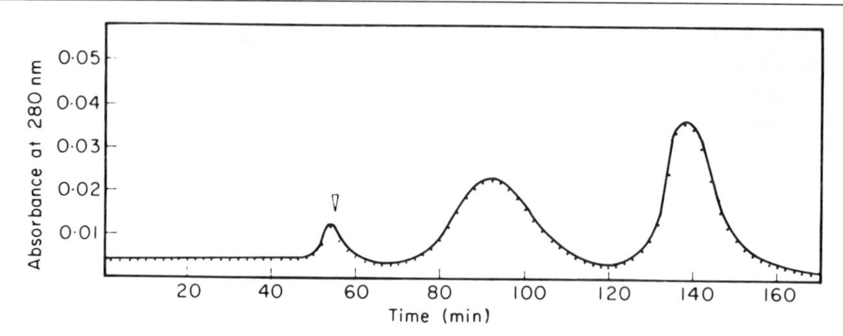

FIG. 1. Gel exclusion chromatography of immune complexes. The arrowhead indicates the void volume fractions containing the immune complexes.

murine antibodies were visualized after incubation with alkaline phosphatase-labeled goat anti-mouse Ig (source: Medac, Sigma).

Subunit Specificity, Ig Class, and Purification of the Antibodies. The specificity of the antibodies for a certain subunit was determined using the immune blot technique.[28] The subunits of the F_1-ATPase were separated by SDS–polyacrylamide gel electrophoresis (12.5% polyacrylamide, Laemmli system[29]) and subsequently transferred onto nitrocellulose sheets by electrophoresis. As many hybridoma supernatants had to be tested, the whole width of the gel was used for the separation. After the blot, the nitrocellulose sheet was cut vertically into strips which were 2 mm wide. The protein bands on one strip were stained with Amido Black as a control. The free protein binding sites of the other strips were blocked by incubating in buffer A. These strips were incubated for 16 hr with 1-ml hybridoma supernatants which contained antibodies against F_1. In order to detect the bound antibodies, the strips were incubated subsequently for 2 hr with fluorescein-labeled anti-mouse Ig (source: Medac, Nordic, Dianova, Sigma). After washing with PBS, the bound fluorescein-labeled antibodies were visualized under UV light (340 nm).

The Ig class of the antibodies was determined by the Ouchterlony test,[30] with Ig class-specific antisera (source: Nordic) after a 10-fold concentration of the hybridoma supernatants with Ig class-specific antisera (source: Nordic). IgG was purified by affinity chromatography on protein

[28] H. Towbin, T. Staehelin, and J. Gorden, *Proc. Natl. Acad. Sci. U.S.A.* **76,** 4350 (1976).
[29] U. K. Laemmli, *Nature (London)* **227,** 680 (1970).
[30] Ö. Ouchterlony, *Acta Pathol. Microbiol. Scand.* **26,** 507 (1949).

FIG. 2. Electron micrographs of immune complexes of F_1 with anti-α (A) or anti-β monoclonal antibodies (B). Distinct immune complexes are encoded and marked by numbers 1–3, indicating complexes with 1, 2, or 3 antibodies. Individual monoclonal IgG antibodies are marked by arrowheads.

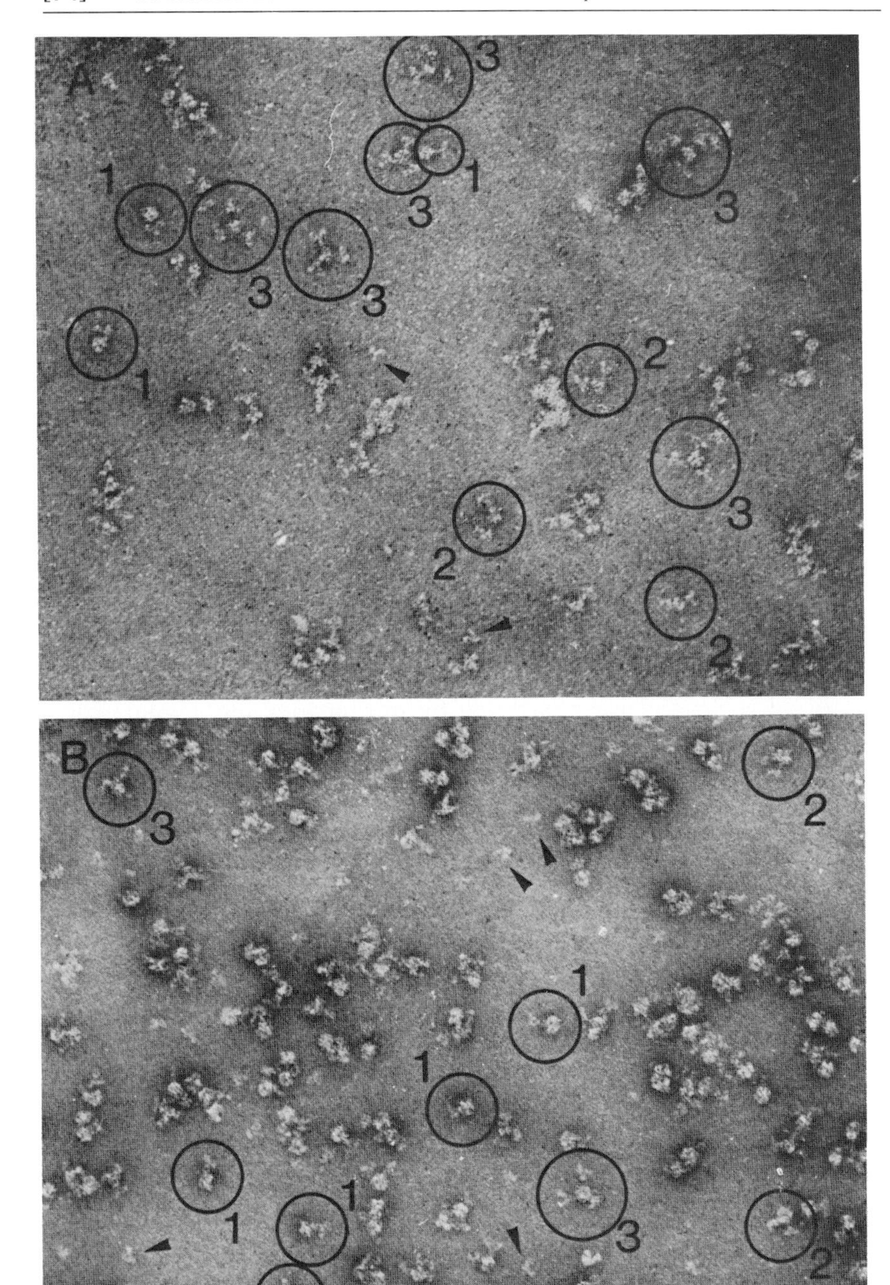
A
B
1
2
3
100 nm

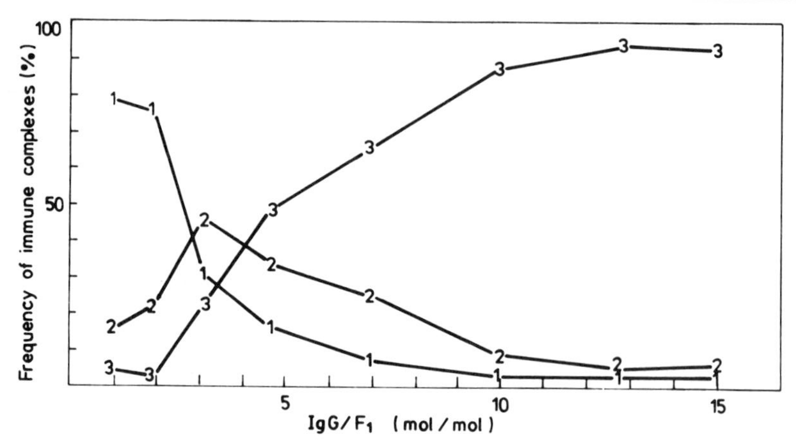

FIG. 3. Distribution curves of immune complexes containing one (-1-1-), two (-2-2-), or three (-3-3-) antibodies per molecule F_1. About 200 immune complexes for the different molar ratios were counted.

A–Sepharose (Pharmacia) from hybridoma supernatants containing 2.5% (v/v) fetal calf serum in Dulbecco's modified Eagle's medium.[31] About 5–20 mg IgG was obtained from 1 liter of supernatant.

Preparation of Immune Complexes. Binding of monoclonal antibodies to the native F_1-ATPase was tested by precipitating the immune complexes with a second antibody. F_1 was incubated for 3 hr with a 10-fold molar excess of monoclonal antibodies in buffer B [2% (v/v) *n*-butanol, 200 m*M* Tris–HCl, 500 m*M* glycine pH 7.6]. The immune complexes were precipitated with a rabbit serum raised against murine IgG. The amount of the rabbit serum needed to precipitate the murine IgG was determined before. The precipitates were analyzed by SDS–polyacrylamide gel electrophoresis for F_1 subunits.

For electron microscopy the immune complexes were obtained after incubation of 100 μg F_1-ATPase with different molar ratios of monoclonal antibodies for 14 hr at 4° in 0.5 ml buffer B. The immune complexes were separated from the bulk proteins by chromatography in 100 m*M* Tris–HCl, pH 7.5, 5% (v/v) methanol on a BioGel A 1.5 m (50 × 1 cm, 0.25 ml/min, 0.5 ml/fraction). Figure 1 shows the elution profile. The immune complexes were always found in the void volume.

Electron Microscopy

For electron microscopy, the void volume fractions were used directly for negative stain preparations[32] with 4% (w/v) uranyl acetate at pH 4.5.

[31] D. Delacroix and J. P. Vaerman, *Mol. Immunol.* **16,** 837 (1979).

[32] R. C. Valentine, B. M. Shapiro, and E. R. Stadtman, *Biochemistry* **7,** 2143 (1968).

The electron micrographs presented in Fig. 2 show a typical survey of immune complexes of F_1 with monoclonal antibodies against subunit α (Fig. 2A) or subunit β (Fig. 2B). Figure 3 shows the distribution curves of immune complexes using different molar ratios of F_1 and anti-α antibodies.

Conclusion

By titrating the binding sites of F_1 for monoclonal antibodies against α or β, it has been shown unequivocally that the complex contains three α and three β subunits. In immune complexes containing 2 or 3 antibodies against α or against β, the angle between the binding sites of the antibodies was always about 120°. This clearly indicates that the three α and three β subunits of the native F_1 are arranged symmetrically in a hexagon.

Author Index

Numbers in parentheses are footnote reference numbers and indicate that an author's work is referred to although the name is not cited in the text.

A

Abbott, M. S., 651
Abeles, R. H., 371
Aboderin, A. A., 725
Ackermann, B. P., 569, 577(7), 656
Ackers, G. K., 243
Ackrell, B. A. C., 399
Adman, E. T., 30
Admon, A., 292, 516, 518, 519(15), 520(15)
Aebi, U., 470, 473(1), 474(1), 475(1), 476(1), 477(1)
Agarwal, N., 716, 717(21), 718(21), 720(21), 721(21), 726(21)
Akerlund, H.-E., 280(44), 284(44), 285
Akey, C. W., 435, 436, 437, 438, 439(21, 22), 440, 441(14, 19, 22), 442(19), 443, 444, 445, 446(21)
Al-Ayash, A. I., 45, 46(5), 59(5), 60(5)
Albracht, P. J., 259
Albracht, S. P. J., 297, 345, 391
Aldanova, N. A., 429, 729
Alfonzo, M., 78, 285, 291(1)
Alizai, N., 45, 46(5), 59(5), 60(5)
Alkema, J. Y. E., 166
Allison, W. S., 484, 485, 662, 703, 716, 717(23), 718(25, 28), 725, 727(54), 728, 741, 742, 744, 746(2), 747, 748(2, 5), 751(5), 753(7, 8), 754(7), 756(10, 11, 12, 14), 757(10, 11, 14), 758(10, 11), 759(12), 760(10, 14), 761(14)
Almon, H., 283
Altendorf, K., 556, 557, 569, 572(3), 574(28), 576, 577(3), 578(3, 25)
Amory, A., 765
Amos, L. A., 349, 435, 436, 439(6)
Amzel, L. M., 3(47), 6, 8(47), 439, 441(6), 444(5, 6), 446(6), 481(13), 482(13), 483, 484, 528, 665, 711
Anderson, S., 45, 226(21), 235
Anderson, T. J., 414
Anderson, W. F., 378, 379(3)
Anderson, W. M., 353, 355(5)
Andersson, B., 280(44), 284(44), 285
Ando, K., 254, 267(30, 31), 268(30, 31)
Andreo, C. S., 728
Andrews, E. C., 22, 33
Andrews, W. W., 725, 727(54), 728, 741, 748(5), 751(5), 753(7, 8), 754(7)
Anfinsen, C. B., 254, 263(26), 265(26)
Angiolillo, P., 29
Anholt, R., 128, 130(17)
Anke, T., 254, 258(14, 15, 16, 19, 20), 260(16), 261(20, 36), 269(15, 36)
Anraku, Y., 4, 95, 96(7), 97(5), 99(4, 5), 101(4, 5), 105(4, 5), 106(4, 5), 111(4, 5, 7), 112, 112(4, 5), 113, 114, 119(6), 384
Aparicio, P. J., 211
Aquila, H., 226(16), 234, 387, 491, 492(9), 493(9)
Arad, T., 192
Arana, J. L., 716, 718(26), 722(26), 723(26, 44), 728(44), 756, 757(39, 40)
Archinard, P., 455, 458(4), 762, 769(2)
Ariano, B. H., 20, 171, 291
Arima, K., 254, 267(30), 268(30)
Aris, J. P., 557
Armstrong, R. L., 400
Arnold, A., 569, 577(7), 656
Arntzen, C. J., 287, 291(12)
Artzatbanov, V. Yu., 331, 343(4)
Asai, M., 2
Asanger, M., 78
Aschroff, J. R., 108
Ash, G., 567
Astin, A. M., 414
Avaeva, S. M., 452

Avron, M., 286, 288(8), 292, 516, 518, 530, 548
Axelson, J. T., 634, 635
Axen, R., 202
Azzi, A., 13, 14(8), 15, 18(13), 20, 21(15), 22, 26, 33, 45, 47, 48(17), 52(17), 53(17), 59(17), 60(17), 64(17), 64, 67, 70(8), 71, 73, 74, 75, 76(4), 80, 81, 82(24), 139, 140(6), 142(6), 144(6), 145(6) 158, 159, 160(6), 163, 164, 165(6), 167(6), 169(6), 171, 174, 291, 522, 523, 524, 525, 526, 715, 716(15)

B

Babcock, G. T., 29, 331
Baccarini-Melandi, A., 282
Bachmann, B. J., 563
Bader, H., 569, 577(7), 656
Baird, B. A., 513, 770
Baker, T. S., 434
Ball, E. G., 254, 263(26), 265(26)
Baltscheffsky, H., 447, 538
Baltscheffsky, M., 447, 538, 539, 541(9), 542(9), 544(12)
Barber, J., 518, 519(15), 520(15)
Barchi, R. L., 9
Bardawill, C. J., 419
Barnabeu, C., 43
Barnes, E. M., 370
Barr, R., 284
Bar-Zvi, D., 668, 672(13)
Battie, C. A., 173, 178(7)
Baum, H., 217, 268
Baumeister, W., 348
Baxter, J., 46, 48(13), 141, 161, 181
Baykov, A. A., 451, 452
Bayley, H., 649, 650(4), 655(3, 4), 667, 712
Bearden, A. J., 211
Beattie, D. S., 173, 178(7, 8), 179, 180(20, 22)
Beck, J. C., 414
Becker, M., 567
Becker, W. F., 254, 259(36), 261(36), 269(35, 36)
Beckett, A., 671
Beechey, R. B., 477, 520, 521, 719
Beinert, H., 33, 204
Beljakova, T. N., 447
Bell, F., 691
Belleau, B., 719
Bendall, D. S., 174, 253(10), 254, 257(10), 279, 283, 284
Bengis, C., 286, 287(6), 288(7), 292(7)
Bengis-Garber, C., 536
Benkovic, S. J., 619
Bennett, J. C., 43
Bensadoun, A., 283, 419
Bensole, D. M., 410
Berden, J. A., 253, 254(43, 44–46, 52), 255, 259, 264(43), 265(43), 266(5, 43, 44), 268(46), 269(43, 44, 45, 46, 52), 270(52), 271, 294, 299, 301(3), 428, 433(10), 770
Berenblum, J., 448
Bergsma, J., 160, 166
Berriman, J., 192, 194(7), 345
Berry, E. A., 253(9), 254, 305, 307(3), 316(3), 323
Bershak, P. A., 211
Besl, H., 254, 258(19)
Betlach, W. M., 541
Beyer, R. E., 642
Bhaduri, A., 370
Bhown, A. S., 43
Bickel-Sandkötter, S., 687
Bickle, T. A., 650
Bieber, L. L., 182, 326
Bienhaus, G., 581
Biewald, R., 27, 36, 37(36), 44(36)
Bild, G. S., 637
Bill, K., 33, 45, 67, 70(8), 71, 73, 74, 75, 76(4), 139, 158, 163, 174
Binder, A., 287, 292(17), 528, 529(8, 9), 530(8, 9), 535(8, 9), 536(8, 9), 644
Binder, L. F., 442
Birchmeier, W., 30, 74
Birkett, D. J., 724
Birnboim, M. C., 598
Bisschop, A., 160, 167(10)
Bisson, R., 26, 28(15, 21), 30, 40, 64, 81
Bital, K. G., 741
Blair, J. E., 441
Blair, P. V., 464
Blaise, J. K., 295, 300(9)
Blake, R. II, 93
Bligh, E. G., 547
Blumberg, W. E., 29
Blumenfeld, S., 689

Bodley, J. W., 634, 635(11)
Böcher, R., 387
Boedefeld, E., 725
Boekema, E. J., 345, 350, 352
Bof, M., 719, 725, 726(49), 727(49), 756
Boffoli, D., 331, 335(9), 342(9)
Bogorad, L., 2, 3(16), 8(16), 469
Boheim, G., 128, 129(20)
Böhlen, P., 718, 741, 756(11), 757(11), 758(11)
Böhme, H., 283
Bokranz, M., 387, 392, 393(19), 394(19), 395(19)
Bolli, R., 45, 47, 48(17), 52(17), 53(17), 59(17), 60(17), 64(17)
Bonaventura, C., 45, 46(4), 59(4), 60(4)
Bonaventura, J., 45, 46(4), 59(4), 60(4)
Bonner, D. M., 583
Bonner, H. S., 298
Bonner, W. D., 211
Bonner, W. M., 41(47), 43
Borchart, U., 226(16, 17, 20), 234(20), 235(17), 236(17)
Borders, C. L., 728
Bordier, C., 326, 327(2)
Bosch, C. G., 178, 180(20)
Bosch, F., 284
Bose, S. K., 539
Bosshard, H. R., 40, 64
Bottomley, W., 2, 3(15), 8(15)
Boublik, M., 7
Bouillant, M. L., 767
Boveris, A., 208
Bowyer, J. R., 174, 191, 211, 222, 223(35), 254(40, 42), 255, 264(40), 265(40, 42), 269(40, 42), 272, 282(7), 284(7), 299, 303, 325
Boyer, P. D., 512, 619, 625(2), 630(2), 632(2), 633(2), 636(2), 643, 644, 645, 648(4), 711, 712
Bradford, M. M., 471, 472, 549, 706
Bradley, L. B., 354, 463, 472, 479
Bradley, R. D., 379
Bradshaw, R. A., 429
Bragadin, M. A., 715, 716(15)
Bragg, P. D., 87, 102, 106, 108, 591, 650, 729, 770
Brandl, C. J., 2, 7(12)
Brandon, J. R., 254(51), 255, 269(51)
Brauer, D., 45
Brautigan, D. L., 30, 37, 38, 38(39), 39(39, 41), 40(39), 64, 168
Breitenberger, C., 201
Brierley, G., 4
Briggs, F. N., 668
Briggs, M. M., 27, 30(25)
Brisson, A., 31
Brocklehurst, J. R., 254(51), 255, 269(51)
Brocklehurst, K., 67
Brocks, J. C., 770
Brocks, S. P. G., 169
Broger, C., 33, 67, 70(8), 73, 76(4), 81, 139, 158, 163, 174
Bronder, M., 388
Broza, R., 26, 528
Brudvig, G. W., 29, 294(8), 295
Brune, D. C., 302
Brunner, J., 78
Brunoir, M., 45, 46(4), 59(4), 60(4)
Brusilow, W. S. A., 564(19), 565, 567(19), 597
Buchard, O., 672
Büchel, D. E., 2, 8(7)
Buchwald, H.-E., 292, 640, 642(4)
Büge, U., 32(40), 33(40), 37, 45(12)
Buhle, E. L., 470, 473(1), 474(1), 475(1), 476(1), 477(1)
Bukow, R. A., 231
Bullough, D. A., 758
Burke, J. J., 287, 291(12)
Burke, J. M., 201
Burlakova, E. B., 454
Burnett, W., 144
Buse, G., 24, 26, 27, 28(15), 29, 30, 32, 36, 37(36), 44(36), 45(11), 64, 81, 153, 158(9)
Bustamante, E., 479, 485
Butler, W. F., 297
Butlin, J. D., 545, 600
Butow, R. A., 576
Byington, K. H., 428, 431(7)

C

Cable, M. B., 81, 668
Cabral, F., 231, 275(34), 280(34), 284
Callahan, P. M., 331
Calvert, J. G., 668
Calvo, R., 297
Cami, B., 412

Cammack, R., 399, 404
Campbell, H. D., 371
Cannon, B., 491, 496
Cantley, L. C., 703
Capaldi, R. A., 22, 23(7), 24(5, 7, 8), 26(12), 27, 28(14, 15, 17, 20, 21, 28), 29(2, 17), 30(2, 25), 31(12, 13, 17), 32, 33, 40, 45, 46, 47(6), 51(11), 52(6, 11), 54(11), 57(6, 11), 60(6), 64, 75, 468, 557, 716, 717(20a), 718(24), 760
Capkova, J., 254, 258(18)
Capuano, F., 331, 334, 335(9), 342(9)
Carafoli, E., 13, 18(7), 20, 70, 80, 81, 153, 154(3), 158(3, 10)
Carey, M. C., 59
Carlemalm, E., 441
Carls, R. A., 402, 410(25)
Carlsson, C., 431, 460, 469(4), 730
Carlsson, J., 67
Carrasco, N., 122, 123, 130(8), 137(8)
Carreira, J., 727
Carroll, R. C., 26, 33
Cartledge, T. G., 211
Carvaca, J., 633
Carvajal, E., 414
Casey, R. P., 13, 14(6, 8), 15, 16(16), 18(13, 16), 20(6), 21(15), 45, 70, 71, 80, 81, 139, 140(6), 142(6), 144(6), 145(6), 158, 159, 171, 291, 292
Caslin, P. R. M., 193
Castor, L. N., 94
Cattell, K., 521
Catterall, W. A., 472, 478, 479, 480, 482, 483, 484, 485, 661, 679, 680(46), 770
Ceccarelli, E., 716, 718(27), 722(27), 723(27), 756, 757(41)
Cerletti, N., 313, 323
Chain, E., 448
Chan, S., 468
Chan, S. H. P., 26, 46, 47(10), 54(10), 59(10), 60(10), 428, 431(9), 433(9)
Chan, S. I., 29
Chan, T. L., 479
Chance, B., 29, 94, 208, 333, 334, 335(22), 336(19, 22)
Chang, J. Y., 45
Changeux, J. P., 1
Chappell, J. B., 15, 16(16), 18(16), 21(15), 81, 611
Cheetham, P. S. J., 78, 79(9)
Chen, P. S., 48, 53
Chen, S., 361
Chen, S. S., 634, 635
Cheneval, D., 80
Cherry, R. J., 81
Chiang, Y. L., 243, 246, 248(15)
Chibisov, A. K., 671
Chien, T. E., 516
Chin, C. C. Q., 371
Ching, Y., 29
Choc, M. G., 211
Choo, W. M., 254, 263(24), 266(24)
Chou, P. Y., 509
Chowdry, V., 667
Chreptun, G., 272
Christian, W., 463, 466(11), 467(11)
Chu, L. V., 123
Chuchlova, E., 604
Chude, O., 430
Chung, Y. S., 720, 721(40)
Cintron, N. M., 473, 504, 511, 660, 661, 664(7), 710, 730
Citterich, M. H., 201
Clark, R. D., 284
Clark, V. M., 461
Clayton, R. K., 283, 294, 298, 302
Clejan, L., 178, 180(20)
Cogdell, R. J., 298, 302
Coggins, J. R., 650
Cohen, G. N., 89
Cohen, S. G., 671
Cohen-Bazire, G., 182
Cohn, E. J., 53
Cohn, M., 7
Coin, J. T., 16, 20(17)
Cole, L., 147
Cole, S. T., 378
Coleman, P. S., 211, 668, 670, 672(11), 673(11), 674(11), 679(11), 680(11), 681(11)
Colman, P. M., 30
Colman, R. F., 743, 744, 745(19), 747, 748(19)
Colonna, R., 64
Comte, J., 417, 418, 419(16), 422(1), 423, 424(1)
Conde, F. P., 43
Condon, C., 399
Converse, C. A., 408
Cooper, O., 254, 263(26), 265(26)

Corbley, M. J., 73
Córdova, L., 446
Cornforth, J. W., 254, 267(32), 269(32), 270(32)
Coronado, R., 128, 129(18), 130(18)
Coty, W. A., 479, 485, 770
Coulson, A. R., 45, 226(21), 235
Cowan, D. O., 668
Cox, G. B., 545, 558, 559, 564(20), 565, 566(20), 567(20), 595, 600
Crane, F. L., 22, 23(9), 33, 284, 418, 428, 447
Crane, L. J., 674
Crane, R. K, 1, 7(3)
Crepeau, R. H., 437, 438(21), 439(21), 440(2), 446(21)
Cresswell, P., 572
Criddle, R. S., 770
Crimi, M., 235
Critchley, C., 556
Crofts, A. R., 210, 211, 264, 295, 303
Crook, E. M., 67
Cross, R. L., 3(48), 6, 7(48), 608, 610(1), 611(1), 613(1), 615(1), 616(1), 617(1), 618(1), 703, 711(37), 728, 729, 736
Crowe, B. A., 399, 404
Crowther, R. A., 436, 446
Cuatrecasas, P., 65
Culik, K., 254, 258(18)
Cullis, R. P., 81, 87
Cunningham, C., 217
Cunningham, W., 22, 33
Curgy, J. J., 716
Cusack, N. J., 651
Czarnecki, J. J., 651
Czerlinski, G. M., 332, 336(16)

D

Dahms, A. S., 693, 709(20), 710
Dalley, J. R., 178
Danielli, J. F., 4
Dar, K., 332, 336(16)
Darley-Usmar, V. M., 22, 26, 28(15), 29(2), 30(2), 32, 40, 45, 46(4, 5), 47(6), 52(6), 57(6), 59(4, 5), 60(4, 5, 6)
Darlison, M. G., 400
Darszon, A., 128
Datta, A., 428, 472, 478, 484, 633, 673(47), 680
Davenport, J., 272, 283
David, M. M., 419
Davis, B. D., 572
Davis, B. J., 448, 543
Davson, H. J., 4
Day, E. P., 211(23), 212
Deatherage, J. F., 24, 26, 31(12, 13)
DeBenedetti, E., 728
de Bruijn, M. H. L., 45, 226(21), 235
Decker, G. L., 470
de Cuyper, M., 18
Degli Eposti, M., 235
DeGraff, B., 672
De Hollander, J. A., 105
Deisenhofer, J., 9
deJong, C., 26, 28(15), 30, 40, 557, 716, 717(20a)
Dekker, H. L., 167, 169(29)
Delacroix, D., 774
de la Rosa, F. F., 173, 174(6), 182, 211
Deléage, G., 417, 768
Dennis, E. A., 166, 662
Deno, H., 718
Deranleau, D. A., 733
DeRosier, D. J., 349, 434
Deters, D. W., 742
Deul, D. H., 266
Devars, J., 469, 470(34)
De Vries, S., 54, 197, 200(13), 201(13), 203, 211, 253, 254(43), 255, 259, 264(43), 265(43), 266(5, 43), 269(43), 270, 271, 297, 299, 304
de Vrij, W., 70, 159, 160(6), 164, 165(6), 167(6), 169(6)
Dewey, R. S., 254, 267(29), 269(48)
Dianoux, A. C., 468, 469(25), 666, 667, 730, 731(87), 732(93)
Dickerson, R. E., 331, 332(7)
Dickie, J. P., 254(48), 255, 267(29), 269(48)
Dickie, P., 378, 380, 381(16), 383(16), 386(16)
Dilley, R. A., 289
Di Pietro, A., 743, 744, 746, 767, 768
Dixon, M., 109
Docktor, M. E., 30
Doly, J., 598
Doolittle, R. F., 235, 237
Dorgan, L. J., 727

Dose, K., 651, 652, 653(28), 655, 660(23), 727, 729
Douglas, M., 231, 576
Douglas, M. G., 2, 3(18), 8(18), 770
Dowaghi, K., 757
Downer, N. W., 26, 27, 28(20, 28), 33, 45
Downie, J. A., 558, 559, 567, 595
Drachev, L. A., 123
Dreyer, W. J., 503
Dreyfus, G., 469, 470(34)
Drisko, R. L., 668
Dujon, B., 201
Dukhovich, V. F., 451, 454
Dulley, J. R., 570
Dunham, W. R., 211(23), 212
Dunn, S. D., 437, 438(21), 439(21), 440(21), 446(21), 484, 589, 590(5, 8), 591(5), 593(8), 594(5), 770
Dunshee, B. R., 254, 267(28)
Dupuis, A., 428(20, 21), 429, 458, 716, 717, 729, 730(8), 756
Dutton, P. L., 105, 138, 211, 254(42), 255, 257, 264(54), 265(42), 269(42), 294, 295, 298, 299, 300(9), 301(31), 303, 304, 305, 331, 334, 335(22), 336(22), 341, 343(1, 35)
Dyer, W. J., 547

E

Earle, S. R., 353, 357(8), 358(8), 360(8, 14)
Ebashi, S., 4
Eberle, A., 2, 3(14), 7(14), 8(14), 469, 595, 718
Ebert, M. H., 284
Ebner, E., 468, 504, 506
Edelman, K., 513
Edelstein, S. J., 56, 57(29), 435, 436, 437(19), 438(21), 439(21, 22), 440(21), 441(14, 19, 22), 442(19), 443(19), 444(19), 445(19), 446(21)
Edgell, M. H., 33
Edmundson, A. B., 510
Edsall, J. T., 53
Edwards, C. A., 211, 212(5), 213(4), 216(5, 25), 217(5, 25, 37, 38), 218(5, 25), 222(24), 223(35), 254(40, 41), 255, 264(40), 265(40), 269(40, 41), 306
Edwards, D. L., 410
Efremovich, N. V., 451
Ehrig, K., 711, 771
Eklund, H., 345
Elferink, M. G. L., 166
Elmes, M. L., 379
Elson, D., 650
Elwell, L. P., 598
Enander, K., 353, 357(7), 358(7), 360(7)
Endo, H., 254, 267(31), 268(31)
Engel, W. D., 174, 211, 219(6), 225, 236, 254, 259(34), 266(34), 269(34), 276, 295, 332, 341(15), 387
Engvall, E., 771
Enns, R., 770
Eperon, I. C., 45, 226(21), 235
Erdweg, M., 27, 36, 37(36), 44(36)
Erecinska, M., 208, 299, 334, 335(22), 336(22, 26, 27), 340(26)
Erickson, H. P., 434
Eriksson, L. E. G., 429, 430(27), 433(27)
Ermakova, S. A., 447
Ernbach, S., 202
Ernster, L., 254(50), 255, 269(50), 270(50), 337, 428(19), 429, 430(27, 28), 431, 433(13, 27, 28), 460, 469(4), 610, 612(12), 614(12), 729, 730
Errede, B., 339, 342(32)
Esch, F. S., 484, 485, 703, 716, 718(25, 28), 728, 741, 744, 746(2), 747, 748(2), 756(10, 11, 12), 757(10, 11), 758(10, 11), 759(12), 760(10)
Estabroock, R. W., 83, 289
Ettrig, K., 445
Evans, M. C. W., 211
Everett, T. D., 353
Evtodienko, Y., 604
Ey, P. D., 373
Ey, P. L., 763
Eytan, E., 521, 522, 524(6)
Eytan, G. D., 26

F

Falk, G., 429
Falkow, S., 598
Fang, J., 429
Farr, A. L., 37, 38(38), 44(38), 51, 55(22), 89, 372, 419, 448, 467, 479, 485, 548, 570, 660, 661, 762, 766(4)

Farrant, J. L., 434
Fasman, G. D., 509
Faulk, W. P., 459
Faure, M., 231
Fayle, D. R. H., 567
Fearnley, I. M., 3(61), 8, 751, 769
Fee, J. A., 13, 159, 167, 211(23), 212
Feeney, R. E., 650
Feher, G., 144, 275(19), 278, 294, 297(7), 300(7), 303(7), 304(7)
Feldman, H. A., 733
Feldman, R. I., 643, 644(1, 2), 648(1), 647(1, 2), 648(4)
Fellous, G., 768
Felton, J., 559
Ferguson, A. M., 727
Ferguson, S. J., 6, 7(55), 724, 725, 726(47), 727(46, 47), 741, 748(4, 6), 751, 753(6), 754(6)
Ferguson-Miller, S., 30, 33, 37, 38, 38(39), 39(39, 41), 40(39), 46, 48(13), 56, 64, 66, 75, 141, 161, 166, 168, 181
Fernandes de Melo, D., 729
Fessenden-Raden, J. M., 428
Fieser, L. F., 715, 716(16), 722
Fieser, M., 715, 716(16), 722
Fillingame, R. H., 470, 477(7), 545, 548, 549, 552, 553(2, 3), 554, 555(2), 556, 557(2, 14), 558, 559(6), 560(6), 561(6), 564, 566(6), 567(6, 15, 16), 568(16, 23), 569, 599, 604, 715, 716(17), 719(17), 770
Fimmel, A. F., 564(20), 565, 566(20), 567(20)
Findling, K., 159, 211(23), 212
Finel, M., 514
Finkelstein, D., 231, 576
Fiolet, J. W. T., 569
Fiolet, W. T., 422
Fisher, H. W., 434
Fisher, R. J., 428
Fisher, R. R., 353, 354, 355(5), 357(8), 358(8), 360(8, 14)
Fiske, C. H., 569
Fleet, G. W., 694
Fleischer, S., 4, 5
Flowers, J. A., 370
Folkers, K., 212, 254(43), 255, 263(23, 24, 27), 264(43), 265(43), 266(24, 43), 269(43)
Font, B., 418
Fortnagel, P., 401
Foster, D. L., 7, 114, 470, 477(7), 545, 549, 552, 553(2, 3), 554, 555(2), 556, 557(2, 14), 566, 569, 604, 770
Foury, F., 765
Fowler, L. R., 33
Fowler, W. T., 353, 355(5)
Fracker, P. J., 764
Fraga, D., 568
Franek, U., 687
Frangione, B., 467, 468(14), 504, 505(7), 730, 731(86)
Frank, J., 348, 436
Franzén, S.-Å, 411
Fredkin, D. R., 297
Freeman, H. C., 30
Freese, E., 160, 401
Friedheim, E., 299
Friedl, C., 557, 566, 572, 579, 580(1), 581(1), 582(6), 583(1)
Friedl, P., 445, 548, 557, 566, 569, 572, 574, 579, 580(1, 2, 4), 581(1), 582(4, 6), 583(1, 2), 595, 604, 711, 719, 771
Friedländer, W., 693
Friedmann, M. D., 254, 263(23)
Fries, E., 193, 392, 405
Frost, M., 670
Fujita, Y., 8, 9(64)
Fukuda, K., 8, 9(64)
Fukumori, Y., 152, 153
Fuller, S. D., 22, 23(7), 24(7, 8), 30
Fung, L. W. M., 370
Furth, A. J., 78
Futai, M., 2, 3(13), 7(13), 8(13), 370, 384, 512, 528, 545, 553(3), 567, 574, 577, 589, 590(4, 5, 7, 9–11), 591(5), 592(4), 593(7, 9, 10, 11, 12, 13, 14), 594(5, 6, 7), 595, 596(1, 17), 597(14), 598, 599(14, 17), 600(13, 15, 17), 601(25), 602(17), 603(17), 683, 716, 718(25), 741, 756(12), 759(12), 760, 770, 771
Futami, A., 291, 293(27)

G

Gabellini, N., 174, 191, 224, 253, 272, 275(2), 276(2), 282(7), 283, 284(7, 43), 285, 291, 292(24), 293(24), 296, 325, 327
Gabriel, S., 693

Galante, Y. M., 360, 361(1), 362, 363(5), 364, 365(11), 367, 368(5), 422, 470
Galfré, G., 763
Galgman, D., 284
Galli, C., 522
Garavito, R. M., 441
Garland, P. B., 155
Garro, A. J., 412
Gatti, D., 304
Gautheron, D. C., 417, 418, 419(16), 420, 422(1), 423, 424(1, 17), 455, 457(7), 458(4), 569(7, 13), 687, 729, 743, 744(20), 746, 762, 765, 767, 768, 769(2)
Gay, N. J., 2, 3(14), 7(14), 8(14), 429, 469, 558, 574, 595, 597, 718, 760, 769, 771
Gazzotti, P., 79, 87(12)
Gefter, M., 771
Gellerfors, P., 173, 217, 338, 341(30)
Gennis, R. B., 13, 14(8), 15, 70, 71, 88, 89(2), 91(2), 93, 94(10), 106, 113, 114(2), 115(2), 120, 122(2), 139, 140, 142(6), 144, 145(6), 159
Georgevich, G., 26, 28(14, 17), 29(17), 31(17), 45, 47, 52(6), 57(6), 60(6), 75
Gerig, J. T., 670
Gerth, K., 254, 258(11, 12), 259(11)
Gervais, C., 491
Gest, H., 539
Giangiacomo, K. M., 304
Gibson, B. W., 769
Gibson, F., 2, 558, 559, 564(20), 565, 566(20), 567(20), 595, 600
Gibson, Q. H., 158
Gilgen, P., 755
Gillespie, D., 672
Gitler, C., 521, 522, 524(6)
Giulian, G. G., 550
Glanville, R., 32, 45(11)
Glaser, E., 428, 433(13)
Glenn, J. L., 418, 428, 447
Glynn, I. M., 611
Godde, B., 254(41), 255, 269(41)
Godinot, C., 417, 418, 422(1), 423, 424(1, 17), 455, 457(7), 455(4), 459(7, 13), 687, 729, 743, 744(20), 746, 762, 763(1), 765, 767, 768, 769(2), 770
Goffeau, A., 765
Goldenberg, T. M., 758
Goldfarb, J., 326
Gomez-Puyou, A., 430
Gorden, J., 772
Gordon, J., 455, 765
Gornall, A. G., 419
Gottikh, B. P., 673, 696
Gounaris, K., 518, 519(15), 520(15)
Gräbe, C., 689
Gräber, P., 292, 640, 642(4, 6)
Grabo, M., 158
Graf, P., 254(41), 255, 257, 260(53), 264(53), 269(41)
Gray, J. C., 273, 276(11), 279(11)
Greaser, M., 550
Green, D. E., 418, 428, 447
Green, G. N., 115
Green, N. M., 2, 7(12)
Greenawalt, J. W., 479
Greengard, P., 6
Gregor, I., 158
Gregory, R., 725, 726(48)
Gresser, M., 512, 711
Gribble, G. W., 260
Griesenbach, T. A., 770
Grieve, P. A., 178, 570
Griffiths, D. E., 33, 173, 211, 340
Grindey, G. B., 448
Grinkevich, V. A., 429, 729
Grisolia, S., 633
Gromet-Elhanan, Z., 528, 529(7–11), 530(7–11), 531(7, 11), 533(11), 535(8, 9, 10, 11, 14), 536(8, 9, 11, 14), 537(10, 11, 14), 716, 717(22), 718(22), 720(22), 721(22), 722(22), 725, 726(50), 756
Gronenborn, B., 2, 8(7)
Gronow, M., 9
Groot, G. S. P., 173, 174(5)
Grubmeyer, C., 608, 610(1), 611(1), 613(1), 615(1), 616(1), 617(1), 618(1), 655, 672, 683, 703, 711(5, 38), 712(38), 736
Grundstrom, T., 378
Guerrieri, F., 270, 331, 332, 333(17), 335(9, 17), 336(17, 28), 338(18), 342(9)
Guest, J. R., 378, 379, 380, 400, 405(15)
Guilford, H., 355
Guillory, R. J., 361, 428, 432(8), 651, 652, 653, 654(18, 19), 667, 673, 677(6), 683, 689, 696, 712, 753
Gunsalus, R. P., 564(19), 565, 567(19), 597
Guss, J. M., 30
Gutweniger, H., 26, 64, 81
Guy, H. R., 8
Guzik, H. J., 410

H

Haake, D. A., 469
Haas, S. M., 182, 326
Haavik, A. G., 173, 211
Hackenberg, H., 387, 390(5), 391(5), 392(4), 395(4), 491, 497, 498(21)
Hackenbrock, C. R., 661
Hackney, D. D., 619, 620, 625(2), 630(2), 632(2), 633(2), 636(2)
Haddock, B. A., 94, 108, 378
Hagen, D. S., 378
Haggerty, J. G., 254, 264(38), 265(38), 269(38)
Hagihara, B., 109
Hajnal, K., 539, 544(12)
Haldar, K., 370
Haley, B. E., 651, 653, 667
Hamamoto, T., 5, 6(40), 122, 123, 130(7, 8), 137(8), 578, 604, 607(7), 640, 641(3, 8), 642(3, 8)
Hamill, O. P., 125, 130(13), 135(13)
Hammes, G. G., 513, 668, 672(14), 703, 770
Hammon, K. M., 661
Hampe, A., 429
Hanes, C. S., 448
Hanke, W., 128, 129(20)
Hansen, F. G., 574, 583, 595, 771
Hansen, M., 447
Hansen, R. E., 211, 297, 338, 609
Hanson, R. S., 402, 410(25)
Hanstein, W. G., 362
Harms, E., 284
Harnisch, U., 200, 204, 205(7), 226(22), 235, 327
Harold, F. M., 569
Harris, D. A., 730
Hartford, N., 400
Hartmann, R., 27, 32, 33(4), 36(4), 40(4), 41(4), 43(4), 44(4), 45(11), 46(8), 70, 72, 75
Hartzell, C. R., 33
Hase, T., 504, 505(6), 508(6)
Hashimoto, T., 468, 504, 505(5, 6, 8), 508(6, 10, 11), 509(10), 512(5)
Haslam, J. M., 414
Hatch, L., 564(20), 565, 566(20), 567(20)
Hatefi, Y., 33, 173, 211, 220, 296, 339, 340, 342(32), 344, 345, 360, 361(3), 362(6), 363(5, 6), 364, 365(6, 11), 367, 368(5), 369(6), 399, 403(2), 422, 470, 759
Hauser, H., 78
Hauska, G., 174, 191, 211, 224, 253, 272, 273(3, 4, 5), 274(4), 275(2, 6, 12), 276(2, 4, 12), 279(6, 12), 281(6), 282(7), 283, 284(4, 5, 6, 7, 41–43), 285, 286, 287, 288(10), 290(3), 291(3, 6, 10), 292(5, 6, 10, 24), 293(24, 27), 294, 296, 325, 327, 640, 642(4)
Hawkins, T., 371
Haworth, R., 434, 441(4)
Hayashi, H., 22
Hayashida, H., 2, 7(10), 8(9)
Hearshen, D. O., 211(23), 212
Heathcote, P., 294, 303(4)
Heaton, G. M., 490, 499
Hecht, J., 254, 258(20), 261(20)
Heckmann, J. E., 201
Hederstedt, L., 399, 400(3), 401, 403(3, 9, 24), 404(9, 26), 405(16), 406(12), 407, 408(16, 22, 30), 409(12, 13, 24), 411(8, 9, 12, 13, 22), 412(12), 413(3, 8, 9), 413(22, 23, 26), 414(22, 23)
Heinzelmann, W., 755
Heldt, H. W., 447, 490, 499, 538
Helenius, A., 193, 392, 405
Heller, J., 299
Hellingwerf, K. J., 166, 171, 172(35)
Hemon, P., 490, 491
Hempfling, W. P., 94
Henderson, L. E., 43
Henderson, P. J. F., 5
Henderson, R., 22, 23(7), 24(5, 7, 8), 26(12), 31(12, 13), 192, 349, 435
Heppel, L. A., 484, 577, 590, 594
Herkenhoff, B., 577, 578(25)
Herlt, M., 492, 493(18), 495
Hermolin, J., 564, 567(16), 568(16)
Herold, E., 272
Hershey, J. W. B., 650
Hess, B., 725, 726(48)
Hesse, H., 513
Hildreth, J. E. I., 273
Hill, F. C., 725, 727(54), 741, 742, 748(5), 751(5), 753(7, 8), 754(7), 756(14), 757(14), 760(14), 761(14)
Hill, R., 283
Hille, R., 211(23), 212
Hind, G., 284
Hinkle, P. C., 5, 16, 20(17), 106, 109(15), 123, 145, 151(4), 153, 291, 434, 517, 541, 577, 605
Hiraga, S., 546, 559

Hirano, M., 589, 594(6)
Hirata, H., 5, 6(39), 14, 151, 434, 528, 569, 577, 580, 581(5), 589, 596, 600(15), 604, 605(2, 5), 607(2), 742, 770
Hirose, T., 2, 7(10), 8(9), 9(64)
Hixson, S. H., 661
Hixson, S. S., 661
Hjelmeland, L. M., 470
Hjerten, S., 225
Ho, C., 370, 371
Ho, J. W., 725, 727(53), 751
Ho, Y. K., 722
Hoch, J. A., 400, 410, 411(36)
Höckel, M., 652
Hodgkin, A. L., 1
Hoffman, F., 110
Hoffman, J. F., 651, 653, 667
Höfle, G., 254, 258(13), 262(21, 22, 37), 269(37)
Hogeboom, G. H., 642
Hohorst, H.-J., 490, 498, 499
Höjeberg, B., 353, 357(4)
Hollemans, M., 3(61), 8, 751
Hollenberg, J., 348
Holloway, C. T., 520
Holloway, P. S., 193, 194(10), 206
Holloway, P. W., 79
Holloway, T., 719
Holmgren, E., 401, 403(24), 404(24), 407, 408(30), 409(24), 413(23), 414(23)
Holsappel, M., 172
Hong, J. S., 370
Hongoh, R., 21
Hon-nami, K., 152, 153
Hope, M. J., 81, 87
Hoppe, J., 570, 574, 581, 595, 771
Hopper, E. A., 650
Horiuchi, Y., 567, 589, 590(9), 593(9)
Horne, R. W., 437
Horstman, L. L., 431, 456, 458(9), 459, 460, 461(2), 661, 730
Horvath, J., 674
Hotchkiss, R. D., 412
Hou, C., 591, 650, 729, 770
Hovmöller, S., 192, 194(7), 197(4, 6), 201(6), 203(2), 206, 210(8), 345
Hsu, D., 634, 635
Hsu, S.-Y., 589, 593(14)
Huang, C. S., 360
Hubbard, S. A., 477
Huber, R., 9
Hubermann, M., 770
Huganir, R. L., 6
Hugentobler, G., 79, 87(12)
Hulla, F. W., 652
Hullihen, J., 478, 479, 485, 486(7), 487(7), 488(7), 489(7), 528
Humbert, R., 564(19), 565, 567(19)
Hummel, J. P., 503
Hundal, T., 428(19), 429, 431, 729, 730
Hunston, D. L., 704, 705(46), 733
Hunter, F., 43
Hurt, E., 174, 191, 211, 224, 253, 272, 273(4, 5), 274(4), 275(2, 6), 276(2, 4), 279(6), 281(6), 282(7), 284(4, 5, 6, 7, 41–43), 285, 286, 288(10), 290, 291(10), 292(10, 23, 24), 293(23, 24, 27), 294, 296, 325
Hutchinson, C. A., 33
Hutton, R. L., 712
Huxley, A., 435
Huxley, A. F., 1
Huzel, N. J., 504

I

Ibrahim, I. T., 714, 715(9)
Ide, C., 640
Ikeda, T., 2, 8(9), 607
Ikegami, A., 81
Imedidze, E. A., 715, 716(10), 717(10), 718(10)
Imoto, K., 8, 9(64)
Inayama, S., 2, 7(10), 8(9), 9(64)
Ingledew, W. J., 106
Ingram, V. M., 677, 697
Innerhofer, A., 387, 392(4), 395(4)
Inoue, K., 468, 505, 508(10), 509(10)
Irie, M., 451
Irschik, H., 254, 258(11), 259(11)
Isaacson, R. A., 297
Isherwood, F. A., 448
Ishino, Y., 567, 589, 590(9), 593(9)
Issartel, J.-P., 428(20, 21), 429, 716, 729, 730(81)
Ito, J., 410
Ivánovics, G., 414
Izawa, S., 293
Izzo, G., 270, 331, 332, 333(17), 334, 335(9, 17), 336(17), 338(18), 342(9)

J

Jabuta, A., 451
Jackson, J. B., 299
Jacob, F., 1
Jacobs, B., 30
Jacobs, E. E., 22, 23(9), 33, 185, 354, 463, 472, 479
Jacobs, J. W., 429
Jacobs, M., 354, 463, 472, 479
Jacobs, N. J., 387
Jagendorf, A., 644, 728
James, A. T., 254, 267(32), 269(32), 270(32)
Jameson, D. M., 701
Jans, D. A., 564(20), 565, 566(20), 567(20), 597
Jansen, R., 172
Jansson, C., 280(44), 284(44), 285
Jarausch, J., 26, 27, 32, 34(4), 36(4), 40(4), 41(4), 42(13), 43(4), 44(4), 45(12), 46(8), 70, 72, 75
Jasaitis, A. A., 123
Jaurin, B., 378
Jaworowski, A., 371
Jendrassik, L., 463
Jeng, S. J., 651, 652, 653, 654(18, 19), 667, 673, 677(6), 683, 689, 696, 712
Jenkins, C. R., 373, 763
Jensen, J., 733
Jensen, L. H., 30
Ji, T. H., 663, 666(11)
Jochanson, B. C., 449
Johansson, T., 173
John, P., 138, 153, 305
Johnson, D., 34
Jones, C. W., 94, 378
Jones, G. D., 45, 46(5), 59(5), 60(5)
Jones, K. R., 210
Jones, M. N., 6
Jones, O. T. G., 138
Jones, R. W., 378, 379(3)
Joniau, M., 18
Jornvall, H., 365
Juchs, B., 65, 66(5), 296
Juntti, K., 430

K

Kaback, H. R., 1, 7, 113, 114(2, 5), 115(2, 5, 8), 118(8), 119(5), 121(5), 122(2), 123, 124, 125(11, 12), 128(11, 12), 130(8), 137(8), 160, 166, 173, 370, 371(11, 14), 372(25), 373(25), 374(11, 13, 25), 375(25), 376(25), 521
Kaczorowski, G. J., 370, 371(14)
Kadenbach, B., 26, 27, 32, 33, 34(4), 36(4, 17), 37, 37(14), 38(8, 9), 40(4, 15), 41(4), 42(13, 13a), 43(4), 44(4), 45(11, 12), 46, 63(7), 70, 72, 75, 153, 231
Kader, J. C., 491
Kagano, T., 771
Kagawa, Y., 2, 3(19, 33, 46), 4, 5(33), 6(39, 40), 8(19, 46), 9(46), 13, 14, 78, 80, 145, 146, 151(1), 179, 428, 434, 471, 517, 528, 576, 577, 578, 580, 581(5), 589, 593(12, 13), 604, 605(2, 5), 607(2, 7), 640, 641(3, 8, 9), 642(3, 7, 8, 9, 12), 722, 723(44), 728(44), 742, 756, 757(40), 770
Kahn, M., 674
Kaiser, H. J., 727
Kakiuchi, S., 4
Kalra, V. K., 716, 717(21), 718(21), 720(21), 721(21), 726(21)
Kambouris, N. G., 668, 672(14)
Kamen, M. D., 339, 342(32)
Kamienietzky, A., 521
Kaminski, L. S., 246, 248(15)
Kamo, N., 21
Kanaoka, Y., 2
Kanazawa, H., 2, 3(13), 7(13), 8(13), 512, 567, 574, 589, 590(7, 9–11), 593(7, 9, 10, 11, 12, 13, 14), 594(6, 7), 595, 596(1, 17), 597(14), 598, 599(14, 17), 600(13, 15, 17), 601(16, 25), 602(17), 603(17), 683, 760, 771
Kandrach, A., 516
Kandrach, M. A., 431
Kangawa, K., 2
Kanner, B. I., 429, 431, 470
Kanner, N., 516
Känzig, W., 59
Kaplan, R. S., 679
Karlsson, B., 192, 194(7), 197(6), 201(6), 203(2), 345
Kartenbeck, J., 392
Kasahara, M., 5, 95, 106, 109(15), 110, 113, 151, 291, 517, 541, 577, 605
Katan, M. B., 173, 174(5)
Katki, A., 212, 216(25), 217(25, 27), 218(25, 27), 220, 296, 317

Katoh, S., 287, 292(16)
Katz, J. J., 297
Kaulen, A. D., 123
Kawai, K., 200, 224, 226(3), 235(3)
Kawakami, K., 2, 7(10)
Kawamura, M., 2, 7(10)
Kawato, S., 81
Kayano, T., 2, 8(9), 574, 595
Kayne, F. K., 7
Kearney, E. B., 381, 399
Keegstra, W., 345, 350
Keilin, D., 159
Keister, D. L., 538, 539
Keitt, G. W., 254, 267(28)
Kemmer, T., 254, 262(21)
Kemp, A., 490, 651
Kemp, R. J., Jr., 499
Kempner, E. S., 56
Kennel, S. J., 767, 768(12)
Kennett, R. H., 400
Kenney, W. C., 381
Kent, T. A., 211(23), 212
Khananshvili, D., 528, 529(10, 11), 530(10, 11), 530(11), 533(11), 535(10, 11, 14), 536(11, 14), 537(10, 11, 14), 716, 717(22), 718(22), 720(22), 721(22), 722(22), 725, 726(50), 756
Khohlov, A. P., 454
Khomutov, G. B., 266
Khrapova, N. V., 454
Kiefer, H., 672
Kierstan, M. P. J., 67
Kilmartin, Y. V., 332
Kim, C. H., 200, 224, 226(3, 18), 234(18), 235(3), 238, 240(1, 2), 241, 243(2), 244(1), 245(8), 246(1), 247(1), 248(1)
Kim, H., 442
Kim, I. C., 178
Kimelberg, H. K., 334, 336(24)
Kimura, K., 451
King, T. E., 33, 48, 73, 173, 200, 217, 224, 226(3, 18, 19), 234(18, 19), 235(3), 238, 239, 240(1, 2, 5), 241(7), 243(2), 244(1), 245(8), 246(1), 247(1), 248(1, 15), 249, 295, 340, 341, 403
Kinoshita, K., Jr., 81
Kirby, A. J., 461
Kirk, J. T. O., 283
Kirkley, S. A., 428, 431(9), 433(9)
Kirkpatrick, F. H., 33
Kita, K., 95, 96(7), 97(5), 99(4, 5), 101(4, 5), 105(4, 5), 106(4, 5), 111(4, 5, 7), 112, 112(4, 5, 7), 113, 114, 119(6)
Kitagawa, T., 151
Kiyasu, T., 595
Klein, G., 468, 469, 512, 666, 667, 710, 716, 730, 731, 732(93)
Kleyman, T. R., 428, 431(9), 433(9)
Klingenberg, M., 35, 174, 295, 331, 343(3), 387, 447, 491, 492, 492(9, 11, 12), 493(9, 18), 495, 497, 498(21), 499, 500, 500(5, 6, 7), 503(5), 538, 652, 693, 709(20), 710
Klionsky, D. J., 564(19), 565, 567(19)
Klotz, I. M., 704, 705, 733
Klouwen, H., 4
Klug, A., 349, 434
Knaff, D. B., 211
Knecht, J., 387
Knight, I. G., 520, 521, 719
Knowles, A. F., 428, 432, 742, 750
Knowles, J. R., 649, 655(2, 3), 667, 694, 708
Knox, B. E., 640, 641, 642(5)
Kobatake, Y., 21
Koga, P. G., 729
Kohlbrenner, W. E., 712, 728
Kohler, C. E., 30, 74
Köhler, G., 771
Kohn, L. D., 370, 371(11, 14), 374(11, 13)
Koland, J. G., 88, 93, 94(10), 106, 113
Kolb, H. J., 174, 192, 201, 202(1), 225, 295, 387
Kolb, J., 33, 66
Kondrashin, A. A., 123, 447, 539
König, B. W., 167, 169(29), 242
Konigsberg, W., 43
Konings, W. N., 70, 159, 160(6), 164, 165(6), 166, 167(6, 10), 169(6), 171, 172(35)
Konishi, K., 95, 97(5), 99(4, 5), 101(4, 5), 105(4, 5), 106(4, 5), 111(4, 5), 112(4, 5), 113, 114, 119(6)
Konstantinov, A., 266, 331, 343(4)
Kopacz, S. J., 360
Koppenol, W. H., 3, 7(23), 39
Korczak, B., 2, 7(12)
Kostenko, E. B., 452
Kostetsky, P., 429
Kozlov, I. A., 528, 715, 716(10, 11), 717(10), 718(10, 11)
Kraayenhof, R., 292, 422, 569
Krab, K., 7, 8(58), 13, 16(4, 5), 22, 32, 46,

54(9), 72, 81, 123, 136(2), 137(2), 153, 154, 154(5a), 332
Krämer, R., 491
Kranz, R. G., 91, 120
Krayevsky, A. A., 673, 696
Krebbers, E. T., 2, 3(16), 8(16), 469
Krinner, M., 174, 272, 275(6), 279(6), 281(6), 284(6)
Kritcchenvsky, G., 522
Kröger, A., 387, 388, 390(5), 391(5), 392(4), 393(18, 19), 394(19), 395(4, 12, 18, 19), 397, 399
Kruber, O., 397
Ksenzenko, M., 266
Kuhn-Nentwig, L., 32, 37, 42(13a), 45(12)
Kulaev, I. S., 447, 451
Kunapuli, S. P., 2, 3(18), 8(18)
Kundig, W., 4, 5
Kung, H., 371
Kuno, M., 8, 9(64)
Kunze, B., 254, 262(21, 22, 37), 269(37)
Kuramitsu, S., 226(19), 234(19), 235
Kurasaki, M., 8, 9(64)
Kurzer, K., 757
Kuzmin, V. A., 670, 671
Kwan, M., 758
Kyte, J., 235, 237

L

Labarca, P., 130
Laemmli, U. K., 90, 143, 165, 182, 231, 275, 278, 280, 317, 326, 382, 424, 455, 457(5), 472, 475, 476, 549, 772
Laikind, P. K., 742, 756(14), 757(14), 758, 760(14), 761(14)
Lakshminarayanan, A. V., 349
Lalla-Maharajh, W., 45, 46(4), 59(4), 60(4)
Langen, J. J., 182
Langman, L., 567
Lardy, H., 34
Lardy, H. A., 728
Larrinua, I. M., 2, 3(16), 8(16), 469
Larsson, C., 280(44), 284(44), 285
Lascelles, J., 414
Laskey, R. A., 41(47), 43
Latchney, L. R., 727
Latorre, R., 128, 129(18), 130(18)
Lauquin, G. J. M., 667, 689, 696(16a), 706(16a)
Laurell, C.-B., 407, 412
Lawford, H. G., 155
Leben, C., 254, 267(28)
Ledger, M. B., 670, 671(28)
Lee, C. P., 254(51), 255, 269(51), 299, 337, 360, 429
Lee, J. C., 53
Leeuweric, F. J., 105
Leeuwerik, F. J., 297
Leigh, J. S., 22, 24(5), 211, 299
LeMaire, M., 47
Lemire, B. D., 378, 379(3, 4), 399
Lenaz, G., 235
Lennon, V. A., 5
Lent, A., 349
Leonard, K., 192, 194(7), 197(4, 6), 200(12, 13), 201(6, 13), 203(2), 206, 210(8), 211, 345
Leonhardt, M., 689
Lerner, S. A., 590
Leung, K. H., 212
Levin, O., 225
Levin, W., 143
Levis, R. A., 135
Lewis, R. V., 662
Li, Y., 197, 200(12, 13), 201(13), 203, 211
Liang, A. M., 428
Licht, D., 668
Lichtenberg, D., 166
Lien, S., 648
Light, P. A., 155
Lin, C. S., 491, 492(11, 12), 493, 495, 497, 499, 500(5), 503(5)
Lin, E. C., 590
Lindberg, O., 491, 496, 610, 612(12), 614(12)
Lindop, C., 521
Lindsay, J. G., 211
Lindstrom, J., 123, 128(6), 129(6), 130(6, 17)
Lingrel, J. B., 2, 7(11)
Link, Th. A., 210, 226(16, 20), 234(20), 235, 237, 253(8), 254
Linke, P., 192, 193(8), 236
Linkletter, S. J. G., 299
Linnane, A. W., 254, 263(24), 266(24)
Linnett, P. E., 477
Lipman, D. L., 468
Lipmann, F., 4
Lipton, S. H., 217, 268

Liveanu, V., 287, 292(17)
Ljungberg, U., 280(44), 284(44), 285
Ljungdahl, P. O., 257, 260(53), 264(53)
Lloyd, D., 211
Lloyd, W. J., 724, 725, 726(47), 727(46, 47), 741, 748(4, 6), 751(6), 753(6), 754(6)
Lochrie, M. A., 27(48), 31, 32
Lockau, W., 174, 224, 253, 275(2, 6), 276(2), 279(6), 281(6), 284(6, 43), 285, 290, 291(23), 292(23, 24), 293(23, 24), 296
Locke, R. M., 490
Lohmann, K., 463
Lohmeier, E., 378
Lolkema, J. S., 171, 172(35)
Long, J. W., 609
Loomans, M. E., 254(48), 255, 267(29), 269(48)
Lord, A. V., 211
Lorence, R. M., 159, 211
Lorusso, M., 304, 331, 332, 334, 335(9), 336(28), 342(9)
Lötscher, H. R., 557, 716, 717(20a), 718(24), 760
Love, B., 46, 47(10), 54(10), 59(10), 60(10)
Löw, H., 334, 337(21), 354
Lowe, A. G., 6
Lowry, O. H., 37, 38(38), 44(38), 51, 55, 89, 372, 419, 448, 467, 479, 485, 548, 570, 661, 762, 766(4)
Lübben, M., 699, 707(31), 708(31), 712(31), 736
Lubberding, H. J., 292
Lücken, U., 687, 699(11), 701(11), 702(11), 703(11), 704, 705(47a), 707(31, 41), 708(31, 41), 711(11), 712(31), 736, 737
Ludolph, H.-J., 490
Ludwig, B., 13, 14(8), 18(7), 26, 28(20), 29(29), 32, 33, 45, 70, 71, 72, 139, 140(6), 142(6), 144(6), 145(6), 153, 154(3), 156, 158(3, 9, 10, 12), 159, 313, 315, 323
Luisi, P. C., 725
Lumsden, J., 650
Lunardi, J., 428(20, 21), 429, 512, 651, 667, 689, 696(16a, 16b), 706, 710, 716, 717(20), 718(20), 720(20), 721(20), 722(20), 724(20), 725(20), 726(49), 727(49), 729, 730(81), 756, 757(33), 759(33)
Lundblad, R. L., 712
Lunden, M., 217
Lünsdorf, H., 445, 711, 771
Lustig, A., 158
Luther, P. K., 446
Lyons, M. H., 725, 726(47), 727(47), 741, 748(4)

M

Mabuchi, K., 574, 595, 598, 601(25), 771
McCarthy, J. E. G., 327
McCarty, R. E., 292, 434, 437, 438(21), 439(21), 440(21), 446(21)
McCaslin, D. R., 405
McConkey, E. H., 62
McElvany, K. D., 728
McEnery, M. W., 470, 473(1), 474, 475(1), 476(1), 477(1)
McEven, C. R., 61, 63
Machleidt, I., 225
Machleidt, W., 225, 226(17, 20), 234(20), 235(17), 236(17)
Macino, G., 201
McIntosh, L., 2, 3(16), 8(16), 469
McKay, E. J., 407, 412
MacKinney, K., 283
McKinney, M., 434, 439(6), 441(6), 444(6), 446(6)
McLaughlin, C. S., 677, 697
Maclay, W. N., 326
MacLennan, D. H., 2, 7(12), 211, 428, 431(7), 455
McNeilly, D., 468
Madden, T. O., 81, 87
Magnusson, K., 400, 405(15), 406(12), 409(12, 13), 411(12, 13), 412(12)
Maguire, J. J., 400, 403(9), 404(9, 26), 406(9), 411(9), 413(9, 26)
Mahmood, R., 668, 672(16)
Maida, I., 332, 338(18)
Maier, K. L., 504, 506
Maier, L. L., 468
Maisterrena, B., 419(16), 423
Makino, S., 53, 58
Malamud, D. R., 414
Malatesta, F., 22, 26, 28(15, 17), 29(2, 17), 30(2), 31(17), 32, 45, 46, 47(6), 52(6), 57(6), 60(6), 75
Malcolm, A. D. B., 650

Malek, G., 719
Malkin, R., 211, 276
Mallon, D. E., 182
Malyan, A. N., 716, 717(19)
Mannschreck, A., 696, 709(25)
Manolson, M. F., 668
Mansurova, S. E., 447, 451, 454
March, S., 65
Marcus, F., 728
Margoliash, E., 3, 7(23), 30, 37, 38, 38(39), 39(39, 41), 40(39), 64, 168, 249
Margulies, D. H., 771
Marjorie, J., 147
Markwell, M. K., 182, 326
Marres, C. A. M., 232, 259, 271
Marshall, A. G., 670
Martin, C. T., 29
Marty, A., 125, 130(13), 135(13)
Marvin, H. J., 292
Marzo, M., 304
Marzuki, S., 254, 263(24), 266(24)
Massey, V., 14
Matsubara, H., 200, 224, 226(3, 18, 19), 234(18, 19), 235(3), 241, 245(8), 468, 504, 505(6), 508(11), 509(10)
Matsuno-Yagi, A., 759
Matsuo, H., 2, 728
Matsuoka, I., 594, 596, 600(15)
Matsushita, K., 113, 114(2, 5), 115(2, 5), 119(5), 121(5), 122(2), 123, 124, 125(11, 12), 128(11, 12), 130(8), 137(8)
Matsuura, I., 254(42), 255, 265(42), 267(31), 269(42)
Matsuura, K., 294, 298, 299, 303
Mattenberger, M., 79, 87(12)
Mattoon, J. R., 414
Matuchi, K., 760
Mayer, A., 652
Mayer, I., 693, 709(20), 710, 711
Mayers, J. A., 598
Mazer, N. A., 59
Means, G. E., 650
Mei, Q.-C., 282
Meinecke, L., 27, 36, 37(36), 44(36), 153
Meinhardt, S. W., 210, 303
Meister, N., 85
Melandri, B. A., 174, 191, 272, 282(7), 284(7), 325
Mell, H., 388
Mennecke, M. A. C., 770
Merle, P., 27, 32, 33, 34(4), 36(4, 17), 37(14), 40(4), 41(4), 43(4), 44(4), 45, 46(8), 63(7), 70, 72, 72, 153, 231
Merlie, J. P., 9
Merril, C. R., 284, 323
Metelskaya, V. A., 712, 716(10), 717(10), 718(10)
Methfessel, C., 128, 129(20)
Michaelis, L., 299
Michaelis, S., 559
Michalski, C., 211, 219(6), 236
Michalsky, H. C., 276
Michel, H., 9, 191, 345
Michels, P. A. M., 166
Middlefort, C. F., 619
Migoshi, Y., 448
Miki, K., 159
Miki, T., 546, 559, 590, 597, 598, 601(25)
Mildran, A. S., 670
Miles, G. L., 159
Milgrom, Y. M., 528, 715, 716(10), 717(10), 718(10)
Mille, M., 7
Miller, J. H., 379, 559, 564(8)
Miller, M. J., 88, 89(2), 91(2), 93, 94(10), 106, 113
Miller, U., 469, 470(34)
Millett, F., 30, 40
Milligan, R. A., 31
Milner, L. S., 370
Milstein, C., 763, 771
Mimms, L. T., 78, 392
Minamio, N., 2
Mingioli, G. S., 572
Minton, N. J., 538
Mioshi, Y., 610, 612(11), 614(11)
Mishina, M., 8, 9(64)
Mitchell, D. A., 477
Mitchell, G., 701
Mitchell, P., 3, 4, 6, 9, 13, 18, 20(19), 123, 136(1), 137(1), 210, 253, 434, 538
Mitchell, R., 538
Mitra, A., 674
Miura, K., 600
Miyata, T., 2, 7(10), 8(9)
Miyata, Y., 334, 336(26), 340(26)
Miyoshi, Y., 543
Mocek, U., 254, 258(19)
Modyanov, N. N., 429, 729
Moffat, J. K., 441

Mohls, C., 468
Mokrasch, L. C., 633
Mole, J. E., 43
Moller, J. V., 47
Monod, J., 1
Monroy, G. C., 430, 460, 461(3), 504, 507(1), 660, 710, 713, 730(6)
Montal, M., 122, 123, 128(6), 129(6, 21), 130(6, 7, 8, 17), 137(8)
Montecucco, C., 26, 40, 64, 81
Moor, H., 85
Moore, P., 434
Moradi-Améli, M., 455, 457(7), 458(4), 459(7), 762, 763(1), 768, 769(2)
Moreland, R. N., 30
Morélis, R., 420
Morelli, G., 201
Morgan, W. D. H., 691
Morimoto, Y., 8, 9(64)
Morowitz, H. J., 7
Mörschel, E., 387, 392, 393(19), 394(19), 395(19)
Morse, S. D., 298
Morton, R. A., 264
Mory, G., 490
Mosbach, K., 355
Moser, C. C., 298, 301(31), 304
Moser, K. K., 254(49), 255, 269(49), 270(49)
Mosher, M. E., 545, 553(3), 558, 559(6), 560(6), 561(6), 564, 566(6), 567(6, 15, 16), 568(16, 23), 596(17), 597, 599(17), 600(17), 602(17), 603(17)
Moss, R. L., 550
Moyle, J., 6, 18, 20(19), 538
Mueller, P., 128, 129(21), 294, 298, 305
Muijsers, A. O., 18
Mukohata, Y., 729
Müller, M., 80, 85
Müller-Hill, B., 2, 8(7)
Mullet, J. E., 287, 291(12)
Munck, E., 211(23), 212
Munkres, K. D., 231, 239
Munn, E. A., 477
Murakami, H., 112
Murata, M., 30
Muratsugu, M., 21
Musilek, V., 254, 258(17), 260(17)
Muster, P., 287, 292(17)
Myers, J. A., 512, 619, 711

N

Nagano, K., 2, 7(10)
Nagata, T., 546, 559
Nageswara, Rao, B. D., 7
Nagle, J. F., 7
Nagooka, S., 341
Nakajima, T., 728
Nakamaye, K., 668
Nakayama, H., 2, 8(9)
Nałęcz, K. A., 45, 47, 48(17), 52(17), 53(17), 59(17), 60(17), 64(17)
Nalin, C. M., 617, 703, 711(37), 736
Narayanan, P., 437, 439(6), 441(6), 444(6), 446(6)
Narita, K., 728
Narlock, R., 56
Natori, S., 596(17), 597, 599(17), 600(17), 602(17), 603(17)
Nava, A., 469, 470(34)
Nawata, Y., 254, 267(31), 268(31)
Nazaki, Y., 392
Nazimov, I. V., 429, 729
Nechushtai, R., 287, 292(13, 17)
Nedergaard, J., 491
Needleman, S. B., 728
Negawa, Y., 504, 505(5), 512(5)
Negrin, R. S., 556, 564, 567(15), 604
Neher, E., 1, 125, 130(13, 14), 135(13, 14)
Nelson, B. D., 173, 217, 254(50), 255, 269(50), 270(50), 338, 341(30)
Nelson, H., 742
Nelson, N., 128, 130(17), 174, 272, 279, 286, 287(7), 288(7), 291(6, 11), 292(6, 7, 11, 13, 17), 470, 513, 516(3), 520, 521, 522, 524(6), 528, 742
Nerud, F., 254, 258(17), 260(17)
Neumann, J., 174
Newbold, J. E., 33
Newman, M. J., 114
Newman, P. J., 287, 292(14)
Ngo, E., 644
Nichol, C. A., 448
Nicholls, D. G., 490, 499
Nicholls, P., 146, 150(6), 151(6), 152, 169, 334, 336(24)
Nicolson, G. L., 4
Niederman, R. A., 182
Nielsen, J., 574, 583, 595, 771
Nielsen, P. E., 672

Nieva-Gomez, D., 15
Nikaido, H., 5
Nishibayashi-Yamashita, H., 217
Nitschke, W., 273
Nitta, T., 589, 590(10), 589(10), 596, 600(13)
Noda, M., 2, 7(10), 8(9)
Noguchi, S., 2, 7(10)
Nojima, H., 2, 7(10)
Noll, H., 62
Nomomura, Y., 384
Norby, J. G., 733
Nordenbrand, B., 730
Nordenbrand, K., 430, 431, 460, 469(4)
Norling, B., 428(19), 429, 433(13)
Norris, J. R., 297
Norris, V. A., 30
Northrop, F. D., 769
Notsani, B., 521, 522, 524(6)
Noumi, T., 567, 574, 589, 590(10, 11), 593(10, 11), 595, 596, 597(14), 599(14, 17), 600(13, 15), 601(16), 602(17), 603, 771
Noyes, C. M., 712
Nozaki, Y., 47, 58, 59(18), 78
Nukiwa, N., 589
Numa, S., 2, 7(10), 8(9), 9(64)
Nyrén, P., 447, 538, 539, 541(9), 542(9), 544

O

Oberwinkler, F., 254, 258(14, 15), 269(15)
O'Brien, T. A., 15, 93
Oda, T., 22
Oesterhelt, D., 555
Ohlsson, R., 355
Ohné, M., 410, 411(36)
Ohnishi, T., 211, 212, 216(25), 217(25), 218(25), 222(24), 224, 235, 254(40, 42), 255, 257, 260(53), 263, 264(40, 53), 265(40, 42), 266, 268, 269(40, 42), 297, 299, 303, 341, 361, 362, 363(6), 364, 365(6, 11), 369(6), 400, 403(9), 404(9), 406(9), 411(9), 413(9)
Ohno, K., 5, 6(40), 640, 641(3, 8), 642(3, 8)
Ohta, S., 2, 3(19), 8(19)
Ohta, T., 2, 7(10)
Ohyama, T., 145
Oka, N., 589, 590(11), 593(11), 596, 597(14), 599(14), 601
Okamoto, H., 434, 580, 581(5), 589, 604, 605(2), 607(2), 770
Okamura, M. Y., 144, 275(19), 278, 294, 297(7), 300(7), 303(7), 304(7)
Okumura, M., 66
Okunuki, K., 159
Okytomi, T., 254, 267(31), 268(31)
Olsiewski, P. J., 370
Oltmann, L. F., 105
O'Neal, C. C., 637
O'Neal, S., 78, 285, 291(1)
Onur, G., 655, 683, 684, 686(8a), 687(4), 692(8), 696(4, 8), 697, 699(4, 8, 8a), 700(8), 701(8, 29), 703(4), 709(8a)
Oppliger, W., 153, 158(9)
Oren, R., 530
Orgura, T., 151
Orii, Y., 81, 145, 151(1)
Orlich, G., 286, 288(10), 290, 291(6, 10, 23), 292(5, 6, 10, 23), 293(23)
Ormerod, J. G., 539
Oroszlan, S., 43
Osborn, M., 239, 240(3), 241(3), 467, 479, 484, 486, 523, 543, 657
O'Shea, P. S., 15, 21(15), 45, 80, 81, 82(24)
Oshino, R., 208
Otoadese, A. E., 463, 469
Otsutka, A. S., 718, 741, 756(11), 757(11), 758(11)
Ottolenghi, P., 733
Ouchterlony, Ö., 772
Ouitrakul, R., 293
Ovchinnikov, Y. A., 429, 729
Overath, P., 1, 7(4), 8(4)
Owen, P., 166, 371, 399, 404, 407
Oya, H., 112
Ozawa, T., 66

P

Pachence, J. M., 295, 300(9)
Packer, L., 650
Packham, N. K., 294, 298
Pal, P. K., 743, 744(19), 745(19), 748(19)
Palmer, G., 29, 173, 174(6), 182, 211, 296
Palmieri, F., 21, 153, 243, 350
Papa, S., 243, 270, 304, 331, 332(6), 333(17), 334, 335(6, 9), 336(28), 338(18), 342(9)
Papahadjopoulos, D., 78

Papermaster, D. S., 408
Paproth, B., 272
Parikh, I., 65
Parola, A., 671
Parsons, G., Jr., 671
Parsons, P., 66
Patel, L., 113, 114(2, 5), 115(2, 5), 119(5), 121(5), 122(2), 124, 125(11, 12), 128(11, 12), 374, 521
Patil, D. S., 399, 404
Patterson, T. E., 27(48), 31, 32
Paule, C. R., 567, 568(23)
Paulson, L., 30, 40
Paulus, H., 704
Pearson, R., 670
Pearson, W. R., 468
Peck, H. D., 387
Pedersen, P. L., 3(47), 6, 8(47), 434, 436, 439(6), 441(6), 444(5, 6), 446(6), 470, 472, 473(1), 474(1), 475(1), 476(1), 477(1), 478, 479, 480, 481(13), 482, 482(13), 483, 484, 485, 486(7), 487(7), 488(7), 489(7), 504, 511, 528, 660, 661, 664(7), 666(8), 679, 680(46), 710, 711, 730, 770
Peisack, J., 29
Penefsky, H. S., 110, 146, 419, 420, 428, 435, 462, 463, 467, 468(14), 469, 472, 478, 484, 504, 505(7), 508, 510, 589, 608, 609(7), 610(1), 611(1), 613(1), 615(1), 616(1), 617(1, 8), 618(1), 633, 636, 645, 649, 655, 673(47), 680, 683, 703, 704, 708, 711(5, 38), 712(38), 714, 716, 730, 731(86), 736, 737, 742, 746, 749, 750, 754, 760(22), 764
Penin, F., 417, 422(1), 423, 424(1, 17), 458, 459, 687, 729, 762, 765, 768
Pennington, R. M., 353, 355(5)
Penttilä, T., 16, 20(18), 21(18), 33, 46, 47(12), 51(12), 52(12), 57(12), 60(12), 75, 165
Peretz, A. J., 650
Perham, R. N., 650
Perkins, S. J., 197, 206, 210(9)
Persson, B., 353, 357(7), 358(7), 360(7)
Persson, M., 365
Perutz, M. F., 332
Peters, F., 703, 704, 705(47a), 733
Peters, H., 729
Peters, K., 666
Peters, L. K., 558, 559(6), 560(6), 561(6), 564, 566(6), 567(6, 15), 568(23), 599
Peterson, G. L., 354, 467
Peterson, J. A., 94
Petrone, G., 80, 81, 82(24)
Petty, K. M., 298
Philips, M., 400, 405(15)
Phillips, A. L., 273, 276(11), 279(11)
Philosoph, S., 528, 529(9, 10), 530(9, 10), 535(9, 10, 14), 536(9, 14), 537(10, 14)
Pick, U., 272, 286, 287, 290(9, 18), 291(9, 18, 22), 292(9, 18), 514, 516(5, 6), 517(5, 6), 518(5, 6), 519(15), 520(15), 541
Piggot, P. J., 400
Pingoud, A., 704, 733
Pinther, W., 272
Pitts, J. N., 668
Planker, M., 651
Pool, L., 173, 174(5)
Poole, R. J., 668
Poole, R. K., 32, 153, 159
Poolman, B., 164
Porath, J., 202
Porter, G., 670, 671(28)
Porter, R., 694
Porter, T. H., 254(43), 255, 263(23, 24, 27), 264(43), 265(43), 266(24, 43), 269(43)
Poser, J. W., 716, 718(28), 741, 756(10), 757(10), 758(10), 760(10)
Posner, H. P., 211
Postma, P. W., 651
Potter, S. S., 33
Pougeois, R., 667, 715, 716(12), 717(12, 20), 718(12, 20), 719, 720(12, 20), 721(20, 39), 722(20), 724(20), 725(20), 756, 757(32, 33, 38), 758, 759(32, 33)
Poulis, M., 371
Powell, G. L., 81
Powell, S. J., 769
Power, S., 173, 174(6), 182, 211
Power, S. D., 27(48), 31, 32
Powers, L., 29
Poyton, R. O., 27(48), 31, 32
Pratt, E. A., 370, 371
Preisig, R., 59
Prezbindowski, K. S., 22, 23(9), 33
Price, N. C., 724
Prince, R. C., 211, 264, 293, 294, 299, 341, 343(35)
Pringle, J. R., 657

Prochaska, L., 26, 28(15, 21)
Proskuryakov, I. I., 716, 717(19)
Prowse, S. J., 373, 763
Przybylski, M., 468, 469(25), 730, 731(87)
Pudek, M., 106, 108
Pullman, M. E., 421, 428, 430, 458, 459, 460, 461(3), 463, 467, 468(14), 469(20), 470(20), 472, 478, 484, 504, 505(7), 507(1), 589, 633, 660, 673(47), 680, 710, 713, 730(6), 730(86)
Purygin, P. P., 673, 696
Püttner, I., 154, 158(10)

Q

Quagliariello, E., 21, 153, 243, 350

R

Racker, E., 3(33), 4, 5(33), 13, 26, 33, 78, 80, 91, 179, 212, 217, 272, 285, 286, 287, 290(9), 291(1, 9), 292(9), 428, 429, 431, 432(8), 456, 458(9), 459, 460, 461(2), 470, 471, 472, 478, 484, 514, 516(5), 517(5), 518(5), 541, 633, 641, 642(12), 648, 661, 673(47), 680, 730, 742
Radda, G. K., 724, 725, 726(47), 727(46, 47), 730, 741, 748(4, 6), 751(6), 753(6), 754(6)
Radik, J., 558, 559, 567, 595
Rae, J. L., 135
Rafael, J., 490, 498, 499
Raftery, M. A., 2
Ragan, C. I., 344, 360, 361(1, 2), 362(6), 363(6), 364, 365(6), 367, 368, 369(6)
Rajbhandary, V. L., 201
Ramos, S., 114
Ramshaw, J. A. M., 30
Randall, R. J., 37, 38(38), 44(38), 51, 55(22), 89, 372, 419, 448, 467, 479, 485, 548, 570, 661, 762, 766(4)
Rao, P. V., 539
Rascati, R. J., 66
Rathbun, W. B., 541
Rathgeber, G., 652, 655
Ray, W. J., 609
Raymond, K., 147
Rea, P. A., 668
Recktenwald, D., 725, 726(48)
Reddy, C. A., 387
Reed, D. W., 211
Reedy, M. C., 446
Reedy, M. K., 446
Reenstra, W. W., 374
Reese, C. B., 697
Reeves, J. P., 370
Reeves, S. G., 211
Regenass, M., 158
Reichardt, J., 158
Reichenbach, H., 209, 254, 258(11), 259(11), 262(21, 22, 37), 269(33, 37)
Reid, G. A., 106
Reifenstahl, G., 254, 258(12)
Reimer, B. R., 514
Reinheimer, H., 468, 469(20), 470(20)
Reinheimer, J. D., 670
Reisler, E., 747
Remennikov, V. G., 447, 539
Renart, J., 408
Reuben, J., 670
Revzin, A., 56, 166
Reynolds, E. S., 442
Reynolds, J. A., 47, 53, 54, 56(27), 57(30a), 58, 59(18), 78, 79, 392
Riccio, P., 174, 295, 387, 491, 492(9), 493(9)
Rice, C. W., 94
Rich, P. R., 207, 210(11), 253(10), 254, 257(10), 264(7), 283, 284, 294, 296, 303(4), 305(20)
Richards, F. M., 650, 666
Richardson, S. H., 33
Rickenburg, H. W., 89
Ricquier, D., 490, 491
Rieder, R., 40, 64
Rieske, J. S., 106, 114, 173, 201, 211, 217, 268, 296, 297, 338
Riezler, W., 693, 709(20)
Riise, E., 583
Riordan, J. F., 727, 728
Risi, S., 652, 727, 729
Rivas, E., 47
Roberts, H., 254, 263(24), 266(24)
Roberts, M. F., 662
Robertson, A. M., 387, 520, 719
Robertson, D. E., 299, 304
Robinson, J. J., 378, 379(4), 383, 386(19)
Robinson, N. C., 27, 28(28), 46, 51(11), 52(11), 54(11), 57(11)
Robinson. W. T., 719

Robson, R. J., 166
Rodbard, D., 733
Roeder, P. E., 672
Rogers, D. L., 371
Rögner, M., 5, 6(40), 292, 640, 641(3), 642(3, 4), 704
Rohde, W., 284
Ronchetti, I. P., 437
Rose, I. A., 608, 619
Rosebrough, N. J., 37, 38(38), 44(38), 51, 55(22), 89, 372, 419, 448, 467, 479, 485, 548, 570, 661, 762, 766(4)
Roseman, S., 4, 5
Rosen, B. P., 548, 564, 594
Rosen, G., 645
Rosenbaum, J. L., 442
Rosenbusch, J. P., 158
Rosenwasser, E., 467, 468(14), 504, 505(7)
Rosevaer, P., 46, 48(13), 141, 161, 181
Rossenwasser, E., 730, 730(86)
Rossi Bernardi, L., 333
Rott, R., 470
Rottenberg, H., 123, 286, 288(8), 374
Rouser, G., 522
Rozantsev, E. G., 696
Rule, G. S., 371
Runswick, M. J., 2, 3(14, 61), 7(14), 8(14), 469, 595, 718, 741, 751, 769
Ruoho, A. E., 672
Russel, L. K., 428, 431, 433(9)
Russell, P., 93
Rutberg, B., 400, 408(30), 409(13), 410, 411(13, 36)
Rutberg, L., 399, 400(3), 401, 403(3, 24), 404(24), 405(15, 16), 406(12), 407, 408(16, 22, 30), 409(12, 13, 24), 411(12, 13, 22), 412(12), 413(3, 22, 23), 414(22, 23)
Ruuge, E., 266
Ryan, D., 143, 279, 280(22), 283(22)
Rydström, J., 353, 357(4, 7), 358(7), 360(7, 15)
Ryrie, I. J., 556

S

Saari, H. T., 13
Saeki, T., 254, 267(30, 31), 268(30, 31)
St. Pierre, M. L., 400
Saito, A., 5
Saito, S., 567, 589, 590(7), 593(7), 594(7)
Sakamoto, J., 616, 648, 649(10), 756
Sakmann, B., 125, 130(13, 14), 135(13, 14)
Saleem, I., 29
Salerno, J. C., 297
Salmon, A. G., 724
Salton, M. R. J., 720, 721(40), 770
Saltzgaber-Muller, J., 2, 3(18), 8(18)
Samoray, D., 286, 288(10), 290, 291(6, 10, 23), 292(6, 10, 23), 293(23), 640, 642(4)
Samuilov, V. B., 447, 539
Sanadi, D. R., 185, 354, 463, 472, 479
Sanato, M., 715, 716(15)
Sanchez, D., 598
Sanchez-Rivas, C., 412
Sandoval, I. V., 408
Sandri, G., 428, 429, 430(27, 28), 433(28)
Sane, P. V., 286, 288(10), 290, 291(10, 23), 292(10, 23), 293(23)
Sanger, F., 45, 226(21), 235
Sanguarmsermsi, M., 279
Santos, E., 371, 372(25), 373(25), 374(25), 375, 376
Saraste, M., 2, 3(14), 7(14), 8(14, 58), 22, 32, 33, 46, 47(12), 51(12), 52(12), 54(9), 57(12), 60(12), 72, 123, 136(2), 137(2), 153, 154(5a), 165, 469, 595, 718, 769
Sasaki, H., 254, 267(31), 268(31)
Sato, H., 600
Satre, M., 428(20), 429, 458, 468, 651, 666, 667, 689, 696(16b), 703, 706(16b), 710, 713, 715, 716(12), 717(12, 20), 718(12, 20), 719, 720(12, 20), 721(20, 39), 722(20), 724(20), 725(20), 726(49), 727(49), 729, 730(81), 731, 732(93), 756, 757(32, 33), 758(32), 759(32)
Sauer, K., 294(8), 295
Saus, J., 235
Saxton, W. O., 348
Scatchard, D., 767
Scatchard, G., 664
Schachman, H. K., 56, 57(29)
Schaeffer, P., 412
Schäfer, G., 683, 684, 686(8), 687(4), 692(8), 696(4, 8), 697(8), 699(4, 8, 8a, 11), 700(8, 9, 30), 701(8, 11, 29), 702(11, 30), 703(4, 11), 707(31, 41), 708(31, 41), 709(8a), 711(11), 712(31, 59), 736, 737, 771

Schäfer, H.-J., 651, 652, 653(28), 655, 660(23)
Schaffner, W., 372
Schägger, H., 174, 210, 219, 225, 226(16, 17, 20), 234(20), 235(17), 236(17), 295, 387, 392(4), 395(4)
Schairer, H. U., 445, 548, 557, 566, 569, 572, 579, 580(1, 2, 4), 581(1, 2), 582(4, 6), 583(1, 2), 604, 711, 719, 771
Scharff, M. D., 771
Schatz, G., 26, 30, 74, 156, 158(12), 159, 231, 275(34), 280(34), 284, 315
Schechter, N. M., 47, 59(18)
Scheit, K. H., 683
Scherka, R., 43
Scheurich, P., 651, 652, 653(28), 655, 660(23)
Schiffer, M., 510
Schilder, L. T. M., 242
Schindler, H., 128, 129(15, 16, 19)
Schirmer, R. H., 8
Schlegel, M., 655, 684, 686(8a), 699(8, 8a), 709(8a)
Schlödder, E., 640
Schmid, B., 579, 580(4), 582(4)
Schmid, B. I., 548, 719
Schmid, R., 577, 578(25)
Schmidt, J. A., 747
Schmidt, P. D., 670
Schneider, E., 556, 557, 569, 572(3), 574(28), 576, 577(3), 578(3, 25)
Schöder, H., 290, 291(23), 292(23), 293(23)
Schöder, U., 272
Scholes, C. P., 29
Scholes, P. B., 305
Scholtissek, C., 284
Schramm, G., 254, 258(14, 15, 16, 20), 260(16), 261(20), 269(15)
Schrock, H. L., 93
Schröder, I., 387
Schroeder, C., 727
Schuerholz, T., 128, 129(19)
Schuldiner, S., 114, 173, 370, 521
Schuly, G. E., 8
Schürtenberger, P., 59
Schuster, S. M., 727, 728
Schuurmans, J. J., 292
Schwalge, B., 254, 258(16), 260(16)
Schwartz, A., 2, 7(11)
Schwarzenbach, G., 299
Schwendener, R. A., 78
Schwerzmann, K., 511, 660, 661, 666(8), 710, 730
Scott, M., 355
Scraba, D. G., 378, 379(3)
Scrutton, M. C., 670
Sebald, W., 200, 204, 205(7), 226(22), 235, 327, 570
Sedman, S. A., 284
Sedmera, P., 254, 258(17), 260(17)
Seelig, A., 158
Seelig, J., 158
Seijen, H. G., 166
Seki, S., 22
Sekiya, T., 574, 595, 771
Sekuzu, I., 159
Selman, B. R., 651, 716, 718(18), 756
Selman-Reimer, S., 514
Selvyn, M. L., 449
Semenov, A. Y., 123
Senda, M., 589, 590, 593(14)
Senior, A. E., 22, 455, 459, 484, 683, 687(3), 713, 727, 729(7), 770
Sepulveda, L., 326
Serrano, R., 431, 470
Sevarino, K. A., 32
Severina, I. I., 123
Shahak, Y., 284(42, 43), 285, 290, 291(22, 23), 292(23, 24), 293(23, 24), 530
Shakhov, Yu. A., 447, 451, 454, 538, 541(9), 542
Shapiro, B. M., 774
Sharp, T. R., 619
Sharpe, A., 45, 46(5), 59(5), 60(5)
Shavit, N., 668, 672(13)
Shchipakin, V., 604
Sheldrick, W. S., 254, 258(13)
Sherman, L. A., 287, 292(14)
Shifrin, S., 323
Shimizu, S., 2, 8(9)
Shimonishi, Y., 226(19), 234(19), 235
Shindo, Y., 451
Shinozaki, K., 2, 3(17), 8(17)
Shipp, W. S., 89
Short, S. A., 370, 371(11), 374(11)
Shoshan, V., 716, 718(18), 756
Shuldiner, S., 286, 288(8)
Shull, G. E., 2, 7(11)
Sidhu, A., 173, 178(8)
Siedow, J. N., 173, 174, 182, 211

Sigel, E., 20, 81
Sigman, D. S., 643, 644(1, 2), 646(1), 647(1, 2)
Sigrist, H., 521, 522, 524(6)
Sigrist-Nelson, K., 521, 522, 523, 524(6), 525, 526
Sigworth, F. J., 125, 130(13), 135(13)
Siliprandi, N., 21, 243
Silman, H. I., 217
Simon, G., 522
Simone, S., 332
Simoni, R. D., 557, 564(19), 565, 567(19), 597
Sims, P. J., 110
Singer, S. J., 4, 672
Singer, T. P., 381, 399
Sistrom, W. R., 138, 182
Skabral, P., 78
Skou, J. C., 4
Skulachev, V. P., 123, 331, 343(4), 447, 539, 715, 716(11), 718(11)
Slater, E. C., 5, 21, 232, 243, 253, 254(43, 45, 46, 47), 255, 259, 264(43), 265(43), 266(5, 43), 268(46), 269(43, 45, 46, 47), 270, 271, 299, 307, 428, 433(10), 528
Slaughter, M., 192, 194(7), 345
Slawtterback, D. B., 4
Smith, A., 447, 456
Smith, A. L., 34, 69, 182, 225
Smith, J. B., 589, 592(2), 770
Smith, L., 305
Smith, L. T., 637
Smith, S. C., 254, 263(24), 266(24)
Smyth, D. G., 43
Snozzi, M., 210
Sokolovsky, M., 727
Solioz, M., 13, 18(7), 70, 153, 154(3), 158(3, 10)
Sone, N., 5, 6(39, 40), 14, 124, 128(9, 10), 145, 146, 147(3), 150(5, 6), 151(1, 3, 4, 5, 6), 152(5), 153, 159, 434, 528, 576, 577, 578, 580, 581(5), 589, 604, 605(2, 5), 607(2, 7), 640, 641(3), 642(3), 742, 770
Soper, J. W., 470, 479
Sorgato, M. C., 6, 7(55)
Speck, J. C., Jr., 764
Spencer, M. E., 379, 380
Spencer, R. D., 701
Spiridonova, V. A., 451, 454
Spitsberg, V., 434, 436, 437(19), 441(19), 442(19), 443(19), 444(19), 445(19)
Spizizen, J., 401, 410
Spurr, A. R., 442
Staab, H. A., 696, 709(25)
Stadtman, E. R., 774
Staehlin, T., 455, 765, 772
Stanier, R. Y., 182
Stan-Lotter, H., 591, 729
Stark, G. R., 43, 670
Staros, J. V., 667
Steck, T. J., 650
Steele, J. C. H., 54, 56
Steffan, B., 254, 258(16), 260(16)
Steffens, G. C. M., 24, 26, 27, 28(15), 30, 36, 37(36), 44(36), 64, 81, 153, 158(9)
Steffens, K., 577, 578(25)
Steglich, W., 254, 258(14, 15, 16, 19, 20), 260(16), 261(20, 36), 269(15, 36)
Steiner, L. A., 144
Steinhart, R., 528, 579, 687, 706(12), 770
Stempel, K. E., 619, 625(2), 630(2), 632(2), 633(2), 636(2)
Sternweis, P. C., 577, 589, 590, 592(5), 770, 771(16)
Stevens, C. F., 1
Stevens, T. H., 29
Stewart, A. C., 287, 292(15)
Stigall, D. L., 422, 470
Still, W. C., 674
Stoeckenius, W., 555
Stone, C. D., 268
Stotter, P. L., 254, 263(23)
Stotz, E., 46, 47(10), 54(10), 59(10), 60(10)
Stouthamer, A. H., 105, 154
Strijker, R., 166
Stroh, A., 32, 40(15)
Strong, F. M., 254(48), 255, 267(28, 29), 269(48)
Stroobant, P., 370
Strotmann, H., 513, 683, 686(8), 687, 692(8), 696(8), 697(8), 699(8), 700(8), 701(8)
Stuart, A. L., 283
Stupperich, E., 388
Stutterheim, E., 770
Suarez, M., 56, 166
Suarez-Isla, B. A., 123, 128(6), 129(6), 130(6)
Subbarow, Y., 569

Suda, K., 313, 323
Sueoka, N., 400
Sugino, J., 448
Sugino, Y., 543, 610, 612(11), 614(11)
Sugiura, M., 2, 3(17), 8(17), 718
Sugiyama, Y., 729
Sun, F. F., 22, 23, 33
Sundberg, R., 672
Suranyi, E., 429, 430(27), 433(27)
Sutton, R., 741
Suzuki, S., 254, 267(30), 268(30)
Swaisgood, M., 166
Swank, R. T., 231, 239
Sweet, W. J., 94
Switzer, R. C., 323
Sykes, B. D., 670
Szalay, M., 435, 441(14)
Szoka, F., 78

T

Tagawa, K., 151, 468, 504, 505(5, 6, 8), 508(6, 10, 11), 509(10), 512(5)
Tahara, S. M., 371, 372(25), 373(25), 374(25), 375(25), 376(25)
Takabe, T., 2, 728
Takagi, M., 567, 589, 590(9), 593(9)
Takahashi, H., 8, 9(64)
Takahashi, T., 2, 7(10), 8(9), 9(64)
Takahashi, Y., 287, 292(16)
Takai, T., 2, 8(9)
Takamiya, K., 105, 299
Takao, T., 226(19), 234(19), 235
Takashima, H., 2
Takeda, H., 226(18), 234(18), 235, 241, 245(8)
Takeda, K., 589, 593(12, 13), 594(6)
Takemori, S., 249
Tamburro, A. M., 715, 716(15)
Tamita, K., 451
Tamura, F., 590, 595, 597, 598, 601(25), 760
Tamura, G., 254, 267(30, 31), 268(30, 31)
Tanabe, T., 2, 8(9)
Tanaka, K., 8, 9(64)
Tanaka, S., 590
Tanford, C., 3(49), 6, 8(49), 47, 53, 54, 56(27), 57(30a), 58, 78, 79, 193, 243, 392, 405
Tang, H.-L., 353, 357(7), 358(7), 360(7)
Tarr, G. E., 211(23), 212
Tarussova, N. B., 673, 696
Taylor, G. N., 459
Taylor, K. A., 446
Taylor, S. S., 747
Telford, J., 78, 285, 291(1)
Tervoort, M. J., 242
Thelen, M., 81, 82(24)
Thierbach, G., 209, 254, 262(37), 269(33, 37)
Thomas, P. E., 143, 279, 280(22), 283(22)
Thompson, D. A., 33, 56, 66, 75, 166
Thompson, D. H., 160
Thorn, M. B., 266
Tiede, D. M., 294, 298, 305
Tiedge, H., 687, 699(11), 701(11), 702(11), 703(11), 711(11), 736, 737, 771
Tiefert, M., 668, 672(13)
Tikhonov, A. N., 266, 454
Tillberg, J. E., 556
Timasheff, S. N., 53
Timkovich, R. L., 715, 716(13, 14), 724(14)
Tinberg, H. M., 650
Ting, L. P., 728, 729
Tiselius, A., 225
Tiukovich, R., 331, 332(7)
Tobert, N. E., 182
Tobimatsu, T., 8, 9(64)
Todd, R. D., 650, 770
Togashi, C. T., 747
Tolbert, N. E., 326
Tonk, D. W., 6
Tonomura, Y., 594, 648, 649(10), 756
Toribara, T. Y., 48, 53(21)
Toshiaki, K., 760
Towbin, H., 455, 650, 765, 772
Toyosato, M., 2
Tracy, R. P., 26
Trapp, M., 43
Traut, R. R., 650
Trebst, A., 286
Trentham, D. R., 697
Trowitzsch, W., 254, 258(11, 12, 13), 259(11)
Trubetskaya, O. E., 429, 729
Trumpower, B. L., 211, 212(5), 213(4), 216(5, 25), 217(5, 25, 27, 38), 218(5, 25, 27), 220, 222(24), 223(35), 253(9), 254(40, 41), 255, 257, 260(53), 264(38, 40, 53), 265(38, 40), 268, 269(38, 39, 40,

41), 296, 297, 305, 306, 307(3), 316(3), 317, 323, 325, 341
Tsai, H. L., 468
Tsilevich, T. L., 673, 696
Tsong, T. Y., 640, 641, 642(5)
Tsuchiya, T., 548, 589, 590, 593(14), 597
Tsugita, A., 468, 469(25), 730, 731(87)
Tsyborski, I. S., 528
Tuenade Gomez-Puyou, M., 469, 470(34)
Turro, N. J., 668, 671(18), 678(18)
Tybulewicz, V. L. J., 429, 769
Tzagoloff, A., 33, 428, 431, 455

U

Undal, T. H., 460, 469(4)
Unden, G., 387, 390, 391(5), 392, 393(18, 19), 394, 395(12, 18), 397, 399
Unger, B. W., 410
Ungibauer, M., 32, 45(12)
Unwin, N., 192
Unwin, P. N. T., 31, 349, 435
Urban, P. F., 331, 343(3)

V

Vàdineanu, A., 428, 433(10)
Vaerman, J. P., 774
Valentine, R. C., 774
Vallee, B. L., 727
Vallejos, R. H., 716, 718(26, 27), 722(26, 27), 723(26, 27, 44), 728(44), 742, 756, 757(39, 40, 41)
Vallin, I., 334, 337(21), 354
VanAken, T., 46, 48(13), 141, 161, 181
van Ark, G., 254(52), 255, 269(52), 270(52)
Van Breemen, J. F. L., 345
Van Bruggen, E. F. J., 345, 350, 352
van Buuren, K. J. H., 18
van Dam, K., 651, 730
Vanderkooi, G., 22, 468
vander Kraan, I., 651
Van Der Sluis, P. R., 528
van der Wal, H. N., 254(44), 255, 266(44), 269(44), 294, 301(3)
Van de Stadt, R. J., 730
van Gelder, B. F., 18, 33, 167, 168, 169(29), 242, 307
van Grondelle, R., 254(44), 255, 266(44), 269(44), 294, 301(3)
Van Heel, M., 348, 350, 352
Vanneste, W. H., 182
van Rijn, J. M. L. L., 168
van Steelant, J., 29
Van Tamelen, E. E., 254(48), 255, 267(29), 269(48)
van Verseveld, H. W., 154
van Wielink, J. E., 105
van Wolraven, H. S., 292
Vazquez, D., 43
Veenhuis, M., 160, 167(10)
Veerman, E. C. I., 167, 169(29)
Velandia, A. M., 451, 454
Veno, M., 79
Verheijen, J. H., 557, 651
Vermeglio, A., 295, 302(12)
Vermeulen, C. A., 160, 167(10)
Verschoor, G. J., 528
Veukatuppa, M. P., 30
Vial, C., 418
Viale, A. M., 728, 742
Vickery, L. E., 29
Vignais, P. V., 428(20, 21), 429, 468, 469, 512, 651, 666, 667, 689, 696(16a, 16b), 703, 706(16a, b, c), 710, 713, 715, 716(12), 717(12, 20), 718(12, 20), 719, 720(12, 20), 721(20, 39), 722(20), 724(20), 725(20), 726(49), 727(49), 729, 730(81), 731, 732(93), 756, 757(32, 33), 758(32), 759(32, 33)
Viitanen, P. V., 114, 115(8), 118(8)
Villalobo, A., 179, 180(22)
Villiger, W. J., 441
Vinitzky, A., 672
Vinkler, C., 645
Violand, B., 78, 285, 291(1)
Vogel, G., 528, 548, 569, 579, 583, 706, 770, 771(16)
Volk, S. E., 451, 452
Von Bahr-Lindstrom, H., 365
Vondracek, M., 254, 258(17, 18), 260(17)
von Jagow, G., 35, 174, 210, 211, 219(6), 224, 225, 226(3, 16, 17, 20), 234(20), 235(17), 236(17), 237, 255(8), 254, 257, 259(34, 36), 260(53), 261(36), 262, 263, 264(53), 266(34), 268, 269(34, 35, 36), 276, 295, 332, 341(15), 387
von Meyenburg, K., 574, 583, 595, 771

von Stedingk, L.-V., 447, 538
Von Tscharner, V., 730

W

Wagenvoord, R. J., 490, 499, 651
Waggoner, A. S., 110, 114, 292
Wakabayashi, S., 200, 224, 226(18, 19), 234(18, 19), 235(3), 241, 242, 245(8), 505, 508(11)
Walasek, O. F., 249
Walker, J., 468
Walker, J. E., 2, 3(14, 61), 7(14), 8(14), 429, 469, 574, 595, 597, 718, 741, 751, 760, 769, 771
Walseth, T. F., 634, 635(11)
Walsh, C. T., 370, 371
Walter, P., 254(49, 50), 255, 269(49, 50), 270(49, 50)
Walz, D., 343
Wan, J. P., 254, 263(27)
Wan, K., 123, 128(6), 129(6), 130(6)
Wan, W. P., 212
Wang, C. H., 110
Wang, H., 29
Wang, J. H., 722, 725, 727(53), 728, 729, 751
Wang, K., 650
Wang, R. T., 294
Wang, T.-Y., 226(19), 234(19), 235
Warburg, O., 463, 466(11), 467(1)
Waring, A. J., 400, 403(9), 404(9), 406(9), 411(9), 413(9)
Warner, H., 48, 53(21)
Warner, R. C., 704
Wassermann, A. R., 283
Wayne, L., 279, 280(22), 283(22)
Webb, E. C., 109
Webb, M. R., 636
Webb, W. W., 6
Weber, G., 701
Weber, J., 687, 699(11), 700(30), 701(11), 702(11, 30), 703(11), 707(31), 708(31, 41), 711(11), 712(31), 736, 737
Weber, K., 239, 240(3), 241(3), 467, 479, 484, 486, 523, 543, 657
Weder, H. G., 78
Wehrle, J. P., 473, 478
Weigand, K., 59
Weiner, J. H., 378, 379(3, 4), 380, 381(16), 383(16), 386(16, 19), 399
Weinstein, D., 283, 419
Weiss, H., 33, 65, 66(5), 174, 192, 193(8), 194(7), 197(4, 6), 200(12, 13), 201(6, 13), 202(1), 203(2), 204, 205(7), 206, 210(8, 9), 211, 226(22), 235, 236, 296, 345
Weiss, M., 519
Weiss, R. A., 173, 178(7)
Weissman, G., 372
Wendel, W. B., 254, 263(25), 265(25)
Westheimer, F. H., 667
Westman, J., 429, 430(27), 433(27)
Whale, F. R., 138
Wharton, D. C., 33, 441
Whatley, F. R., 138, 153, 305
White, L. K., 564, 567(16), 568(16, 23)
Whitfeld, P. R., 2, 3(15), 8(15)
Wikström, M., 7, 8(58), 13, 16(4, 5), 21, 22, 32, 33, 46, 47(12), 51(12), 52(12), 54(9), 57(12), 60(12), 72, 81, 123, 136(2), 137(2), 153, 154(5a), 165, 332
Wilde, R. J., 400
Wilkins, J., 147
Williams, A., 714, 715(9)
Williams, N., 479, 481, 482, 483, 485, 486, 487, 488, 489, 528, 668, 672(11), 673(11), 674(11), 677(12), 679(11, 12), 680(11), 681(11)
Williams, R.-C., 434
Williams, R. H., 212
Williams, R. J. P., 6
Wilms, J., 167, 168, 169(29), 242
Wilmsen, V., 128, 129(20)
Wilschut, J., 172
Wilson, D. F., 138, 257, 264(54), 299, 331, 334, 335(22), 336(22, 26, 27), 340(26), 343(1)
Wilson, M. T., 29, 45, 46(4, 5), 59(4, 5), 60(4, 5)
Wilson, T. H., 114
Winget, G. D., 516
Wingfield, P., 192
Winkler, E., 387, 392(4), 395(4), 492, 493(18), 495
Wise, J., 683, 727
Witt, H. T., 6, 640
Wittmann, H.-G., 650
Wittmann-Liebold, B., 45
Wojtczak, L., 429, 430(28), 433(28)

Wold, F., 371, 712
Wolf, H. U., 569, 577(7), 656
Wolin, E. A., 387
Wolin, M. J., 387
Wong, C. T., 5
Wong, S.-Y., 759
Wood, D., 400
Wood, P. M., 174, 284
Worlan, S. T., 294(8), 295
Wraight, C. A., 295, 297, 298, 302(13)
Wray, V., 254, 258(12)
Wright, A., 559
Wright, J. K., 1, 7(4), 8(4)
Wu, L. N. Y., 353, 354
Wyatt, J. L., 743, 744(19), 745(19), 748(19)
Wyman, J., 331, 332(5), 333(5)

Y

Yagi, K., 66
Yamada, E. W., 504
Yamamoto, A., 522
Yamanaka, T., 105, 152, 153
Yamato, I., 95, 96(7), 384
Yanagita, Y., 124, 128(10), 145, 147(3), 150(5), 151(3, 5), 152(5), 153, 159
Yike, N. J., 538
Yonentani, T., 33
Yoshida, M., 5, 6(39), 146, 151, 434, 528, 577, 580, 581(5), 589, 604, 605(2, 5), 607(2), 716, 717(23), 718(25, 28), 722, 723(44), 728(44), 741, 742, 753(8), 756(10, 11, 12), 757(10, 11, 40), 758(10, 11), 759(12), 760(10), 770
Yoshida, S., 81
Yoshida, T., 13, 14, 159, 167, 211(23), 212
Yoshida, Y., 468, 504, 505(8), 508(10), 509(10)
Yoshida-Momoi, M., 5
Young, I. G., 45, 226(21), 235, 371
Yount, R. G., 668, 672(16)
Yu, C., 48, 73
Yu, C.-A., 173, 191, 217, 243, 249, 282, 295, 341
Yu, L., 48, 73, 173, 191, 217, 243, 249, 282, 295, 341
Yura, T., 546, 559, 590, 597

Z

Zaccai, G., 458, 468, 729, 730
Zakharov, S. D., 716, 717(19)
Zalman, L. S., 5
Zamecnik, P. G., 677
Zampighi, G., 78, 392
Zanotti, A., 64
Zaugg, W. S., 211, 297
Zeppezauer, E. S., 345
Zeppezauer, M., 345
Zhang, Y.-Z., 26, 28(14)
Zhang, Z. P., 226(19), 234(19), 235
Zhu, Q. S., 253, 254(43, 44), 255, 264(43), 265(43), 266(5, 43, 44), 269(43, 44), 271, 294, 301(3)
Zief, M., 674
Ziganke, B., 65, 66(5), 296
Zimmermann, J., 693
Zoller, M. J., 747
Zorzin, C., 254, 262(22)
Zubay, G., 435
Zuegg, W. S., 338
Zumbuehl, O., 78
Zurawski, G., 2, 3(15), 8(15)
Zuurendonk, P. F., 18
Zvyagilskaya, R. A., 447

Subject Index

A

Acetylcholine receptor, 8
Acid–base cluster, of transmembrane structure, 8
Active-site directed reagents, 712
Active transport, 2
  secondary, 1
Adair's equation, 733
Adenine nucleotide
  analogs, 683–684
    binding to isolated F_1, 703–706
      study methods, 703–704
    biochemical applications and interactions, 699–712
    esterification, 688
    3′-esters as inhibitors of F_1-ATPases, 683–684
    flexibility for design of probes by choice of 3′-acyl residue, 684
    high-affinity binding, 704–705
    low affinity binding, 703
    microprocedure for, 697
    precursor acids, 688–696
    radiolabeled, 688, 697
    reversible high-affinity binding, 703
    synthesis, 688–699
      general methodology, 687–688
      materials, 687
    ^{14}C-labeled solution, esterification, 697
  fluorescent 3′ esters, 701
Adenine nucleotide carrier, binding of DMAN analogs to, 709–710
Adenosine diphosphate
  3′-*O*-acyl derivatives
    inhibitory activity in oxidative phosphorylation and photophosphorylation, 685–686
    structure, 684–686
  analogs
    requirements for, 683
    spin-labeled, 710
    synthesis, 688
    for structural analysis of F_1-ATPase, 682–712
  binding
    to F_1-ATPase, 736–738
      site dissociation constants, 736–738
      stoichiometric dissociation constants, 736–738
    to high affinity sites of F_1, 704–705
Adenosine monophosphate, analogs, synthesis, 688
Adenosine triphosphate
  analogs, synthesis, 688
  azido derivatives, 667
  enzyme-bound
    substrate, 644
    synthesis
      by beef heart mitochondrial F_1, 648–649
      by chloroplast coupling factor CF_1, 643–649
    measurement, 644
  formation, 640–643
  photolabile analogs, 651
  photosensitive azido groups introduced into, 651
  [γ-^{18}O]
    assay, 635–637
    preparation by use of carbamate kinase, ADP, and [^{18}O]carbamoyl phosphate, 633–637
    purification, 634–635
Adenylate kinase, in CF_1 preparations, 644
Affinity cross-linking, 650
Affinity label, 712
Algebraic Reconstruction Technique, 349
Alkylhydroxybenzothiazole, inhibition of cytochrome bc_1 complex, 264–265
Alkylhydroxynaphthoquinone, inhibition of cytochrome bc_1 complex, 265

Amberlite XAD-2, 79
p-Aminobenzamidine · 2HCl, source, 547
Aminoxid WS 35, 584
ANA-ADP, 711, 712
binding
to F_1-ATPase, 736–738
site dissociation constants, 736–738
stoichiometric dissociation constants, 736–738
to high-affinity sites of F_1, 705–706
photoaffinity labeling with, 706–708
synthesis, 698
Anabaena variabilis, cytochrome b_6f complex, 275, 279–281
ANP, synthesis, 693–695
ANP-ADP, 711
binding to high affinity sites of F, 705–706
photoaffinity labeling with, 706–708
synthesis, 698–699
Anthracene-1-carboxylic acid, synthesis, 689–690
Antimycin, 256
inhibition of cytochrome bc_1 complex, 267–268
molecular formula, 267
physicochemical data, 269
structural formula, 267
Antimycin A, 284
Antiport, definition, 4
ART. *See* Algebraic Reconstruction Technique
3′-Arylazido-β-alanine-8-azido-ATP
light-induced inactivation of F_1-ATPase, 655–657
photochemical reactivity, 654–655
specific binding to F_1-ATPase, 655
spectral characteristics, 653–654
stability, 653
synthesis, 652–653
ATPase. *See also* F_0-ATPase; F_1-ATPase; F_0F_1-ATPase
activity assay, 661
beef heart mitochondrial
acid quench experiments, 610
ATP hydrolysis, 608–618
cold chase experiments with, 609–610
single-site catalysis, 608, 609
measurement, 610–618
Ca^{2+}, 4, 5, 7
H^+-translocating, 3
lipid-dependent proton-translocating (F_0F_1), 5
Mg^{2+}, 4
mitochondrial, stabilizing factors
assay of stabilizing activity, 508
preparation, 505–507
properties, 508–511
secondary structure, 509
myosin, inactivation with 5′-FSBA, 747
Na^+,K^+, 4, 7
oligomycin-sensitive. *See* F_1F_0-ATPase
photoaffinity labeling with benzophenone-ATP, 678–680
mechanism, 678
methods, 679–680
proton-translocating, reconstitution into liposomes, 5
residues with enhanced reactivity toward chemical reagents, 713
ATPase inhibitor, 504
beef heart mitochondria, 460–470
amino acid sequence, 468–469
assay, 461–463
definition of unit, 464
dimeric form, 467–468
estimation of protein, 463
properties, 460, 468–470
protein concentration, 467
purification, 460, 464–468
relationship of structure and functional activity, 469
chloroplast, amino acid sequence, 469
E. coli, amino acid sequence, 469
functional cross-reactivity, 469–470
immunological cross-reactivity, 470
multiple molecular species, 468
preparation, 505–507
properties, 508–511
removal from submitochondrial particles, 431
secondary structure, 509
sources, 504
yeast, amino acid sequence, 468–469
ATPase inhibitory peptide, 713
activity assay, 661

azido-labeled form, 660–666
definition of unit, 661
function, 710
[^{14}C]HS-AB-labeled
binding by inverted inner membrane vesicles, 664–665
characterization, 663–664
use, 664–666
isolation, 661
labeling with [^{14}C]HS-AB, 663–664
protein measurement, 661
radiolabeled, 730–732
radiolabeling
application, 732
with phenyl [^{14}C]isothiocyanate, 730–732
with [^{14}C]methylazidobenzimidate, 732
principle, 730
ATP synthase. *See also* F_1F_0-ATPase
highly coupled mitochondrial H^+, 417–427
from pig heart mitochondria, 417–427
activities, 423–425
adenylate kinase activity estimation, 422
assay, 419–423
ATP synthase activity assay, 421–422
composition, 423, 424
correlation between proton movements, ATP hydrolysis, and ATP synthesis, 427
effects of inhibitors, 425–426
hydrolytic activity assay, 420–421
nucleoside diphosphokinase activity estimation, 422
preparation, principle, 417
protein concentration determination, 419
proton translocation, 426–427
induced by ATP hydrolysis, estimation, 422–423
purification, procedures, 417–419
purified, characteristics, 423–427
structure, 423
Azido-ATP analog, for photoaffinity cross-linking of F_1 subunits, 649–660
5-Azido-1-naphthoic acid
photolytic decay, spectral changes during, 692
synthesis, 690–692
3′-*O*-[5-Azidonaphthoyl(1)][^{14}C]ADP. *See* ANA-ADP
3′-*O*-{3-[*N*-(4-Azido-2-nitrophenyl)amino]-propionyl}-8-azidoadenosine 5′-triphosphate. *See* 3′-Arylazido-β-alanine-8-azido-ATP
3-(4-Azido-2-nitrophenyl) propionic acid. *See* ANP
3′-*O*-[3-(4-Azido-2-nitrophenyl)propionyl]-[^{14}C]ADP. *See* ANP-ADP

B

Bacillus stearothermophilus, 145
Bacillus subtilis
168 strain, 401
cytochrome-*c* oxidase, 159–173
purification, 70
growth, 160, 400–402
growth media, 401–402
heme-deficient, 414
membranes, preparation for succinate dehydrogenase isolation, 403
membrane vesicles, 160
isolation, 160–161
radioactive labeling of proteins, 401
respiratory chain, 159
sdhA gene, 400
sdh mutants, 409
classes, 409, 411
complementation analysis by protoplast fusions, 411–413
isolation, 409–411
sdh operon, genetic organization, 409–410
succinate dehydrogenase–protoplast fusion, genetics, 409–413
succinate dehydrogenase activity, and growth conditions, 402
succinate dehydrogenase complex, 399–414
Bacteriorhodopsin, 7
co-reconstitution with F_1F_0-ATPase, 518

F_1F_0-ATPase reconstituted with, ATP synthesis, 555–556
Benzophenone-ATP, 667–682
characterization, 676–677
dissipation of triplet by hydrogen abstraction, 670–671
elemental analysis, 676
3′-ester linkage, lability, 668
nucleotide-site ligand for ATP-hydrolyzing enzymes, 672
photoreduction of triplet by 2-butylamine, 671
pH stability, 676
preparation, 673–676
proton NMR, 677
purification, 674–676
residence time at catalytic site of ATPase, 670–671
specific site labeling, physicochemical considerations, 669–670
substrate for phosphotransferases, 672
synthesis, 668, 673–674
reaction route, 669
thin-layer chromatography, 676
UV absorbance, 677
Benzophenone-linked affinity probes, photochemical mechanisms, 672
3′-*O*-(4-Benzoyl)benzoyl-ATP. *See* Benzophenone-ATP
Bio-Beads SM-2, 79
Biomembrane, structure, 4
British Anti-Lewisite, and oxygen, inhibition of cytochrome bc_1 complex, 270–271
Brown adipose tissue
ADP/ATP carrier, 490–491, 498
mitochondria
H^+ conduction, 490
thermogenesis, 490, 498–499
uncoupling protein. *See* Uncoupling protein
Butanedione, modification of F_1-ATPase, 728

C

Carbodiimide, modification of carboxylic groups in F_1-ATPases, 714–719
application, 718–719
mechanism of reaction, 714–715
procedures, 715–718
3-Carboxy-2,2,5,5-tetramethylpyrrolidine-1-oxyl, synthesis, 695–696
Carrier, definition, 1
CF_1CF_0. *See* Chloroplast coupling factor CF_1CF_0
Channel, definition, 1
CHAPS. *See* 3-[(3-Cholamidopropyl)dimethylammonio]-1-propane sulfonate
Chlamydomonas reinhardi, photosystem I reaction center, 287
4-Chloro-7-nitrobenzofuran
modification of residues of F_1-ATPase, 724–727, 748–755
radiolabeled, inactivation of F_1-ATPase, 741
7-Chloro-4-nitrobenzofuran. *See* 4-Chloro-7-nitrobenzofuran
7-Chloro-4-nitrobenzo-2-oxa-1,3-diazole. *See* 4-Chloro-7-nitrobenzofuran
Chloroplast
broken, preparation, 513
cytochrome b_6f complex, 271–285
extrinsic proteins, removal, by NaBr washing, 272–273
H^+ translocation, reconstitution with photosystem I reaction centers, PMS, and CF_1CF_0, 285–293
lettuce, preparation, 521
membrane, labeling with DCCD, 521
photophosphorylation, reconstitution with photosystem I reaction centers, PMS, and CF_1CF_0, 285–293
PSI-RC/CF_1CF_0 liposomes
determination of internal volume, 291
photophosphorylation in, 290–291, 292–293
spinach, F_1-ATPase
inactivation by carbodiimides, 717
incorporation of carbodiimides during inactivation, 718
modification by Woodward's reagent K, 723
spinach leaf, isolation, 687
Chloroplast ATP synthase complex. *See* F_1F_0-ATPase, chloroplast
Chloroplast coupling factor CF_1
ATP formation

characteristics, 647–649
kinetics, 647–648
measurement, 644–645
P_i concentration dependence, 647
pH dependence, 646–647
[^{32}P]ATP formation, measurement, 645–647
isolation, 644
Chloroplast coupling factor complex CF_1CF_0, 272
attachment to liposomes, 288
incorporation into liposomes
alternative procedures, 291
lipid and detergent, 291
tests for, 291
preparation, 286–288
in reconstitution of chloroplast photophosphorylation, 286–293
removal of extrinsic CF_1 part, 272–273
sources, 292
Chloroplast coupling factor CF_1CF_0 liposome
activities, 292
ATP/P_i exchange, 290
comparison to chloroplast activities, 292–293
Chloroplast vesicles, H^+ translocation
assay
by light-induced proton movements measured with glass electrode, 289–290
by light-induced quench of 9-aminoacridine fluorescence, 288–289
3-[(3-Cholamidopropyl)dimethylammonio]-1-propane sulfonate, purification of rat liver F_1F_0-ATPase with, 470–477
Cholic acid, source, 546
Chromone inhibitor, 256, 261–263
physicochemical data, 269
CMCD, modification of carboxylic groups in F_1-ATPases, 715–719
Complex I, 344–353
composition, 363
crystallization, 352–353
procedure, 345
crystals
characterization, 345–348
electron microscopy, 345–348
electron microscopy, image processing, 350–353
FP (water-soluble FeS-flavoprotein), 361
composition, 363
FeS-containing polypeptides, 362
iron content, 362
resolution, 363–365, 369
FP-I
acid-labile sulfide content, 364–365
flavin content, 364–365
homogeneity, 364
iron content, 364–365
properties, 364–365
solubility, 364
FP-II
acid-labile sulfide content, 365
flavin content, 365
homogeneity, 365
iron content, 365
properties, 365
solubility, 365
HP (water-insoluble fraction containing phospholipids and hydrophobic polypeptides), 361
composition, 363
IP (water-soluble FeS-protein), 361
composition, 363
FeS-containing polypeptides, 362
iron content, 362
resolution, 366–367
IP-I
acid-labile sulfide content, 367, 368
homogeneity, 366–367
iron content, 367, 368
properties, 366–367
purification, 366
solubility, 366
IP-(II + III)
acid-labile sulfide content, 368
homogeneity, 368
iron content, 368
properties, 368
purification, 367–368
solubility, 368
IP-II and IP-III
acid-labile sulfide content, 369
homogeneity, 369
iron content, 368, 369
properties, 369
separation, 369
solubility, 369

resolution into FP, IP, and HP, 361–369
structure, transparent model, 352
subunits, arrangement, 361
tilted projected structure, 350–351
Complex III, 211. *See also* Cytochrome bc_1 complex
beef heart
preparation, 173–180
purification, 173
isolation, 173, 174
large-scale preparation, 174–178
N. crassa, 173–174
purification, 174
rat liver, 173
reaction catalyzed, 173
resolution of iron–sulfur protein and cytochrome bc_1 subcomplex from, 218–221
restoration of cytochrome-*c* reductase activity to, by reversibly dissociated iron–sulfur protein, 223–224
sources, 173
ubiquinol–cytochrome-*c* oxidoreductase activity, 173
yeast, 173
activity, 178
aluminum sulfate fractionation, 176–177
binding of DCCD to cytochrome *b* in, 180
bioenergetic properties, 179–180
choice of strain for preparation of, 175
enzymatic assays, 178
H^+ ejection and electron flow, effect of DCCD on, 180
polypeptide composition, 178
potassium cholate solubilization, 176–177
properties, 178–179
protein content, 179
reconstituted liposomes, preparation for H^+ ejection and respiratory control measurement, 179–180
reconstitution into liposomes, 179
subunits, p*I*, 178
succinate dehydrogenase assay, 178
ubiquinol–cytochrome-*c* oxidoreductase activity, assay, 178
yield, 178
Creatine kinase, benzophenone-ATP as substrate, 672
Crystal, two-dimensional, electron microscopy, 348–349
Cyanobacteria
cytochrome b_6f complex, 271–285
photosystem I reaction center, 287
1-Cyclohexyl-3-(2-morpholinoethyl)carbodiimide. *See* CMCD
Cytochrome, oxidoreduction, monitoring, 333
Cytochrome *a*, aerobic oxidation, 334
Cytochrome a_1, 87–88, 105
prosthetic group, 88
Cytochrome a_3, aerobic oxidation, 334
Cytochrome *b*, 224, 255
amino acid sequence, 235
double kill of, 266
isolation, 225
from cytochrome bc_1 complex, 230
oxidant-induced reduction, 257, 266, 272, 284
P. denitrificans
absorption difference spectra, 329–330
properties, 328–331
reduction, 334–335
reduction and reoxidation, observation in hybrid system, 303–304
single-step purification from bacterial cytochrome bc_1 complex, 323, 325–331
analytical methods, 326
materials, 326
by temperature-dependent phase separation into Triton X-114, 325–330
yield, from cytochrome bc_1 complex, 229
Cytochrome b_6, distinguishing from cytochrome *b*-559, 283
Cytochrome *b*-556, 88
Cytochrome *b*-558, 87–88, 94, 119
B. subtilis, 400, 403
properties, 405
prosthetic group, 88
Cytochrome *b*-558–*d* complex, *E. coli*
chemical composition, 100, 101
oxidase activity, effect of oxygen concentration, 109
oxidation–reduction potential, 105–106
oxygen affinity, 109

polypeptides, 101
purification, 97–100
reconstituted liposomes, formation of membrane potential by, 109–111
reconstitution into liposomes, 109–110
spectral properties, 100, 103–105
ubiquinol oxidase activity, 106–108
Cytochrome *b*-559, chloroplast, 283
Cytochrome *b*-560, 316, 323
Cytochrome *b*-562, 255
Cytochrome *b*-562–*o* complex, *E. coli*
chemical composition, 99–100
oxidase activity, effect of oxygen concentration on, 109
oxidation–reduction potential, 105
oxygen affinity, 109
polypeptide composition, 99–100
purification, 95–98
reconstituted liposomes, formation of membrane potential by, 109–111
reconstitution into liposomes, 109–110
spectral properties, 100–103
ubiquinol oxidase activity, 106–108
Cytochrome *b*-563, 119
Cytochrome *b*-566, 255, 316, 323
Cytochrome bc_1
iron–sulfur protein. *See also* Iron–sulfur protein
resolution and reconstitution, 211–224
mitochondrial, 316
of *Paracoccus* ubiquinol oxidase, 316
Cytochrome *bc* complex
functions, 271
isolation, 272–276
ammonium sulfate precipitation, 273–274
concentration and detergent exchange, 274–276
selective extraction with detergent mixtures, 273
sucrose density gradient, 274
preparations, 271–285
purity, 274
stability, 276
storage, 276
Cytochrome bc_1 complex. *See also* Complex III
6.4 kDa protein, 227–229
isolation, 236
7.2 kDa protein, 229
8 kDa DCCD-binding protein, 235
8 kDa protein, 229, 230
folding pattern, 236
11 kDa protein, 229, 230
isolation, 236
13.4 kDa protein, 229, 230
activity, 191
analytical methods, 181–183
antimycin-sensitive site, 296
assays, 283
beef heart, 181, 186–187, 189
subunits, 224–237
cleavage into three fractions, 227–229
consumption of light-generated ferricytochrome *c* and quinol, 295–296
core protein, isolation, 230, 231, 236
cytochrome-*c* oxidase activity, 182
from diverse species, purification, 181–191
inhibitors, 253–271
groups, 256–257
points of action, 256
isolation, 183–185
mitochondrial
inhibitors
groups, 254–255
physicochemical data, 254–255
structural data, 254–255
model, 255
subunits, 255
mitochondrial vs. chromatophore, 304–305
myxothiazol-sensitive site, 296
optical spectra, 184, 186
P. denitrificans, 181, 188
absorption difference spectra, 323–324, 329–330
activity, 317
components, 317
inhibitors, 317, 325
isolation
analytical methods, 317
materials, 317
method, 318–323
properties, 323–325
storage, 325
subunits, 321, 323
turnover numbers, 325
preparation in Triton X-100, 225–227
purification, 181
materials, 181–182

purity, 274
Q_o center, 235, 255, 296
Q_i center, 255, 296
quinone-binding protein, 235
R. sphaeroides, 181, 186–191, 275
activity, 191
phase separation, 330–331
preparation, 282
red-shift spectra, 257, 259
scalar H^+ transfer reactions associated with redox transitions of respiratory carriers in, analysis, 340
SDS-PAGE, 231–234, 283–284
electrophoresis conditions, 233–234
gel preparation, 233
sample preparation, 233
staining, 234
stock solutions, 233
single-step purification of cytochrome *b* from, 323, 325–331
sources, 181
spectra, 283
storage, 227
subunit composition, bacterial vs. mitochondrial, 184–185, 188
subunits
amino acid sequences, 234–235, 237
buffers for, 227–228
functional roles, 237
isolation, 225–226
membrane anchoring domains, 237
preparation, 227–231
yield, 229
supplemented with exogenous cytochrome *c* and purified cytochrome-*c* oxidase
redox Bohr effects, 339
succinate-supported oxygen consumption, 338
three-subunit, from *P. denitrificans,* 316–325
transition from reduced to oxidized state, oxidation of endogenous ubiquinone to protein-stabilized ubisemiquinone, 341
ubiquinol–cytochrome-*c* oxidoreductase, isolation, 183
ubiquinol–cytochrome-*c* oxidoreductase activity, 182
ubiquinone reduction, inhibitors, 266–270
yeast, 181, 185–187
Cytochrome bc_1 subcomplex
cleavage, 229
from cytochrome bc_1 complex, 227–229
isolation, 203–204
membrane crystals, 193, 197–198
Neurospora, 202
resolution from complex III, 218–221
SDS–gel electrophoresis, 204
structural properties, 205–206
three-dimensional structure, 197, 199
yield, 229
Cytochrome b_6f complex, 174, 224, 271–285
A. variabilis, 275, 279–281
assays, 283
barley, 279
Chlamydomonas, 279
from cyanobacteria, preparation, 279–281
function, 271
lacking iron–sulfur protein, from pea, 279
lettuce, 279
Oenothera, 279
oxidoreductase activity, 284
pea, 279, 280
proteolytic degradation, 276
purity, 274–275
SDS–PAGE, 283–284
spectra, 283
from spinach, 274, 275
preparation, 277–279
stability, 276
storage, 276
Swiss chard, 279
Cytochrome *c*
aerobic oxidation, 334
affinity column, regeneration, 164
horse heart, linked to CNBr-activated Sepharose 4B, 65–67
linked to affinity gels, 64–65
oxidation, observation in hybrid system, 300–303
oxidoreduction, 336
proton-consuming oxidation, by cytochrome oxidase vesicles, 18–19

yeast
linked to Affi-Gel 102, 68, 71–72
linked to thiol-Sepharose 4B, 67–71, 139
preparation, 161, 164
Cytochrome c_1, 224, 229–230, 235, 255
aerobic oxidation, 334
amino acid composition, 244–245
amino acid sequence, 235
DEAE cellulose elution profile, 250–251
isolation, 225
oxidation and rereduction, observation in hybrid system, 303
preparation, 249–251
purification, from bacterial cytochrome bc_1 complex with Triton X-114, 331
purification profile, 250, 252
purity, 250, 252
two-band, 238
preparation, 240
yield, 250, 252
from cytochrome bc_1 complex, 229
Cytochrome c_1 complex, isolation of hinge protein from, 241
Cytochrome c_1-cytochrome c complex
dissociation, 248
effect of hinge protein on, 245–247
formation
effect of hinge protein on CD spectra, 246–249
molecular ellipticity, as function of hinge protein, 247–249
Cytochrome c-552, 306, 315
Cytochrome-c oxidase, 353
aa_3-type, 159
activity, dependence on assay conditions, 38–40
affinity chromatography purification, 64–72
aggregation state, effect of detergents, 46–47
B. subtilis, 159–173
accessibility, compared with accessibility of succinate dehydrogenase, 166–167
activity, role of detergents and phospholipids, 170
affinity chromatography purification, 70
crossed immunoelectrophoresis, 166
electrical potential measurement, 171–173
electron donors, 169
enzymatic properties, 167–170
gel filtration, 166
inhibitors, 169
ionic strength effects on activity, 167–169
orientation, 170–171
effect of phospholipid composition, 171
reconstruction, 170–173
effect of phospholipid composition, 171
solubilization, 161–162, 164
spectral properties, 167–168
structural properties, 165–167
bacterial, 32, 72
affinity chromatography purification, 72–73
beef heart, 167
affinity chromatography purification, 69–71
and beef liver, kinetic differences, 37
depletion of subunit III, 74–77
N-terminal amino acid sequences, 44
polypeptides, 27
purification, 66
splitting into subcomplexes, 73–74
structure, 22–31
two-dimensional crystals, 22
beef liver, N-terminal amino acid sequences, 44
Bohr effects, measurement, 338
bovine
binding of detergent, 52
interconversion from monomeric to dimeric form, 47
lipid content, 52
catalytic subunits, 32
centrifugation techniques, 53–64
crystals, from purified enzyme and lipid, 24
cytochrome c binding sites, 167–169
density measurements, 54–56

dimer, amount of bound detergent, determination, 51–52
domains, 24–26
gel filtration study, 47–53
isolation, 34–35
from mitochondria, 33–34
isozymes, 32–45
immunological characterization, 41–43
kinetic properties, 37–39
naturally occurring, 46
polypeptide composition, 36–37
properties, 36–39
spectral properties, 37
mammalian, tissue-specific isozymes, 32
detection, 32
mitochondrially encoded subunits, 32–33
molecular weight, 46, 64
of different species, 59–60
monomer, amount of bound detergent, determination, 52–53
monomer–dimer association, 45–64
monomer–dimer equilibrium, 50
monomers and dimers, separation, 49–51
N. crassa, purification, 66
N-terminal amino acid sequences, 44–46
nuclear-encoded subunits, 32, 38
regulatory function, 32
P. denitrificans, 145, 153–159
activity measurement, 158
affinity chromatography, 70, 158
amino acid sequence analysis, 158
EPR studies, 158
gel electrophoresis, 158
immune precipitation, 158
kinetic data, 158
metal quantitations, 158
protein estimation, 158
proton pumping measurements, 158
purification, 156–158
reconstitution, 158
sedimentation analysis, 158
spectral heme determination, 158
subunits, 153–154
partial specific volume
calculation, 54–55
measurement, from equilibrium centrifugation in H_2O/D_2O mixtures, 56
phospholipid content, calculation, 53
pig
comparison of kinetic properties in different tissues, 39
properties, 38
pig heart, N-terminal amino acid sequences, 44
pig liver, N-terminal amino acid sequences, 44
polypeptides
folding, 28
SDS–gel electrophoresis, 45–46
from PS3, 123–124
membrane potential generation, 131–132
preparation, 124
reconstitution, 124
subunits, 124
purification
by affinity chromatography, 161–164
by ammonium-sulfate precipitation, 164–166
R. sphaeroides, 71
aa_3-type, purification, 138–145
activity analysis, 141–142
affinity chromatography purification, 70
characterization, 141
extracted
characterization, 142–143
oxidase activity, 142
immunological analysis, 144–145
membrane-bound, characterization, 142
preparation, 139–141
purified, activity, 143
reduced-minus-oxidized spectra, 140–142
spectral analysis, 140–142
structural analysis, 143–144
yield, 141
rat liver
depletion of subunit III, 75
polypeptide pattern, 36–37
purification, 35, 66
reaction catalyzed, 22, 64

reconstitution into phospholipid vesicles, 78–87
cholate dialysis technique, 78, 80, 81–83
detergent dialysis technique, 81–83
factors affecting, 85–87
hydrophobic adsorption technique, 79–81
preparation of lipids, 79–80
procedures, 78
redox Bohr effects in
measurement, 342
pH profile of, 343
resolution, 72–77
sedimentation coefficient, of different species, 59–60
sedimentation equilibrium centrifugation, 57
sedimentation velocity measurement, 59–60
structure, from study of two-dimensional crystals, 24–25
subunits, 32, 45–46, 72
isolation, 43–44
labeling by EDC and [^{14}C]glycine ethyl ester, 40–41
reactivity of carboxylic groups in absence and presence of cytochrome *c*, 39–41
separation by SDS–gel electrophoresis, 34
study, 72–73
sucrose gradient centrifugation, 62–64
T. thermophilus, 167
topography of polypeptide components, biochemical approaches, 25–29
two-dimensional crystals, preparation
with deoxycholate as detergent, 23
by Triton detergent treatment of mitochondria, 22–23
yeast, 32, 145, 154
polypeptides, 27
Cytochrome-*c* oxidase complex
high-affinity site for cytochrome *c*, 30–31
prosthetic groups, locus of, 29–30
Cytochrome-*c* oxidase vesicles
characterization, 82–85
cytochrome-*c* oxidase orientation in, 87
measurement, 84–85
electron microscopy, 85–87
polarographic assay, 83–84
proton pump, potentiometric measurement, 84
respiratory control ratio, 82, 85
spectroscopic assay, 82–83
Cytochrome-*c* reductase, 70
Bohr effects, measurement, 338
Cytochrome *d*, 113
E. coli, 94
spectrophotometric assay for, 88
Cytochrome *d* terminal oxidase, *E. coli,* 87–94
characterization, 90–92
prosthetic group, 88
protein assay, 89
purification, 89–91
organism strain and growth conditions for, 89
reconstitution into proteoliposomes, 91–93
subunits, molecular weights, 90–91
ubiquinol-1 oxidase activity, 88, 91
Cytochrome *f*, 275
proteolytic degradation, 276
Cytochrome *o*, 88, 113. *See also* Cytochrome *o*-type oxidase
E. coli, 87, 94
proteoliposomes containing, characterization, 121–122
Cytochrome *o*-type oxidase. *See also* Cytochrome *o*
E. coli, 113–122, 124
assay methods, 114
characteristics, 119–121
DEAE-Sepharose column chromatography, 116–117
inhibitors, 120
kinetic properties, 120–121
membrane potential generation, 132
molecular properties, 119–120
pH optimum, 120
preparation, 124–125
purification, 113, 115–118
in situ, 116
reaction catalyzed, 113
reconstituted into proteoliposomes, lactose transport measurement, 114–115

reconstitution into proteoliposomes, 113, 118, 124–125
SDS–PAGE, 120
Sephacryl S-200 column chromatography, 117–118
solubilization, 116
subunits, 119, 124
Cytochrome oxidase
aa_3-type, 153
bacterial, 159
subunits, 153
electrogenic activity, 123
assay, 131–137
control experiments, 135
effect of glass pipet composition on, 135
effect of lipid/protein ratio on amplitude of signal, 133
effect of membrane resistance on amplitude of signal, 133–134
inhibitors, 133
specificity, 132–133
substrates, 132–133
electrical measurements, 123
H^+ pumping activity, measurement
as extravesicular pH changes, 15–20
experimental sample, 15
pH measuring system, 15–16
mammalian, CO-difference spectrum, 150
membrane formation, assay, 130
mitochondrial, 159
reconstituted into vesicles, 13
subunits, 153–154
from PS3, 145–153
absorption spectrum, 146
assay, 146
catalytic activity, 150–151
cytochrome oxidase activity, 146
H^+ pump activity, 146
measurement, 152–153
properties, 149–151
prosthetic groups, 149–150
purification, 146–149
reconstitution into vesicles capable of H^+ pumping, 151–153
spectra, 149–150
stability, 151
subunits, 145, 150–151
reconstituted in lipid bilayers
advantages, 137
effect of glass pipet composition on activity of, 135
efficiency, 136
limitations, 137
membrane potentials generated by
amplitude, 137
predicted vs. measured, 136–137
membrane stability, 130
reliability, 136
control experiments for assessing, 135
reconstitution
in planar lipid bilayers, 123–138
apparatus, 125–127
materials, 124
into lipid bilayers at tip of patch pipets, 128–130
into phospholipid vesicles, 14–15
detergent dialysis method, 14
Cytochrome oxidase vesicles
assay of functional activity, 128
assay of membrane formation, 130
ferrocytochrome *c*-induced H^+ extrusion from, 16–18
H^+ pumping activity, 13–21
charge-translocation stoichiometry, determination, 20–21
determined relative to O_2 consumption, 20
O_2 pulse measurements, 18–20
reductant pulse measurements, 16–18
transformation into monolayers at air–water interface, 129–130
Cytochrome reductase. *See also* Ubiquinol–cytochrome-*c* reductase
cleavage, 203–205
cleavage products, structural properties, 205–206
core complex, 203, 204
isolation, 204–205
structural properties, 205–206
core proteins, 199–200
cytochrome *b*, 201
cytochrome c_1, 200–201
duroquinol : cytochrome-*c* reductase activity, 208–209
electron transfer activities, 207–208

iron–sulfur subunit, 200–201, 203, 204
isolation, 204–205
structural properties, 205–206
isolation, 202–203
membrane crystals
electron micrographs, 194–198
factors influencing formation of, 193–194
multilayer membrane crystals, 193
electron microscopy, 194, 196–197
Neurospora, ubiquinol reductase site, 210
orientation, 200–201
quinol : ferricytochrome-*c* reductase activity, 207–209
quinol : quinone oxidoreductase activity, 208–210
reconstitution, from cleavage parts, 206–207
SDS–gel electrophoresis, 204
single-layer membrane crystals, 192–193
electron microscopy, 194–195
structural properties, 205–206
subunit III, 201
subunits
I and II, 199–200
I, II, and V. *See also* Ubiquinol–cytochrome-*c* reductase, subunits
IV and V, 200–201
splitting off of, 193
topography, 197–201
three-dimensional structure analysis, 197
and ubiquinone-10, incorporation into phospholipid membranes, 206

D

DCCD, 520–521
F_0 liposomes inhibited by, 8
modification of F_1-ATPases, 714–719, 756–761
radiolabeled, inactivation of F_1-ATPase, 741
DCCD-binding protein, 719
chloroplast, 520–527
apparent molecular weight, 526
1-butanol-extracted
characterization, 524
chromatographic purification, 522–523
1-butanol extraction, 522
characterization, 523–524
DEAE-cellulose chromatography purified, characterization, 525–526
diethyl ether precipitation, 522
ether-precipitated, characterization, 524
purification, 522–523
reconstituted
H^+ condition of, 526–527
proton conduction measurements, 527
reconstitution, 526–527
5-*n*-Decyl-2,3-dimethoxy-6-hydroxybenzoquinone, inhibition of cytochrome bc_1 complex, 266
Dial nucleotides, modification of F_1-ATPase, 729
Diaminodurol, redox mediator of photosynthetic reaction center–mitochondrial ubiquinol–cytochrome bc_1 complex hybrid, 299
Diazoethane
generation from *N*-ethyl-*N*-nitroso-*p*-toluenesulfonamide, 622–625
preparation, 619, 620
apparatus for, 623–624
Dicyclohexylcarbodiimide. *See* DCCD
N,N'-Dicyclohexylcarbodiimide. *See* DCCD
Diffusion constant, 58
5-Dimethylamino-1-naphthoic acid. *See also* DMAN-nucleotides
synthesis, 692–693
3'-*O*-[5-Dimethylaminonaphthoyl(1)] [^{14}C]ADP. *See* DMAN-ADP
DiN_3ATP. *See* 3'-Arylazido-β-alanine-8-azido-ATP
5,5'-Dithiobis-(2-nitrobenzoic acid), inhibition of cytochrome bc_1 complex, 271
DMAN-ADP, 699
binding, to high affinity sites of F, 705–706
as fluorescent probe, 702–703

inhibition of F_1-ATPase ATP synthesis and hydrolysis, 700
synthesis, 698
DMAN-AMP, inhibition of F_1-ATPase ATP synthesis and hydrolysis, 700
DMAN analog, and adenine nucleotide carrier, 709–710
DMAN-ATP, inhibition of F_1-ATPase ATP synthesis and hydrolysis, 700
DMBIB, 284
DTNB. *See* 5,5′-Dithiobis(2-nitrobenzoic acid)

E

EDAC, modification of carboxylic groups in F_1-ATPases, 715–719
EEDQ, modification of F_1-ATPase, 719–722, 742, 756, 760–761
Efrapeptin, modification of F_1-ATPase, 728
Electron microscopy. *See also* F_1-ATPase, electron microscopy; Membrane protein, electron microscopy
image reconstruction, 434–435
Enzyme II, 5
Escherichia coli
atp operon. *See Escherichia coli, unc* operon
cytochrome *d*-deficient mutant, 115
cytochrome *o*-type oxidase, 113–122
electron-transport chain allowing growth in absence of oxygen on nonfermentable carbon source, 377–378
F_1-ATPase
inactivation by carbodiimides, 717
inactivation with 4-chloro-7-nitrobenzofuran, 725–726
incorporation of carbodiimides during inactivation, 718
modification by EEDQ and IIDQ, 721
F_1F_0-ATPase, 545–557
F_1F_0-ATPase mutants, 595–603
genetic recombination, results, 599
frd operon, 378, 379
fumarate reductase, 377–386
genetic complementation study, comparison of methods, 567–568
genetic complementation with λ-*unc* transducing phage, 562–563
construction of recipients, 562
interpretation of analysis with, 563–564
nomenclature, 563–564
spot test for, 562
verification, 562–563
weak or negative, lysates giving, 563–564
genetic recombination test, 598–599
growth
for cytochrome *b*-562–*o* complex purification, 95
for F_1F_0-ATPase purification, 546, 583
for isolation of cytochrome *o*-type oxidase, 115
for purification of cytochrome *b*-558–*d* complex, 97
growth conditions, and purification of F_1F_0-ATPase, 557
H^+-ATPase mutants, 558–568
showing dominance, 566–567
showing simple complementation pattern, 564–566
HB101, growth, 379
inner membrane fraction enriched in fumarate reductase tubules, 379
isolation, 383–385
inner membrane vesicles, preparation, for cytochrome *b*-562–*o* complex purification, 95–96
IY83, 374
KF-11, cloning of mutated genes from, 599
KF-3, cloning of mutated genes from, 601–602
KF-series mutants, 598–599
λ-*unc* strains, suitable for genetic complementation analysis, 565
λ-*unc*/*unc* diploid strains, variability of complementation, 563, 564
membrane potential, determination, 114
membrane preparation
for F_1F_0-ATPase purification, 547, 584
for isolation of cytochrome *o*-type oxidase, 115–116

for lactate dehydrogenase isolation, 374
membranes, washing, for F_1F_0-ATPase purification, 584
membrane vesicles, conversion of D-lactate to pyruvate, 370
ML 308-225, growth, 374
oxidative energy conservation, 112–113
pap operon. *See Escherichia coli, unc* operon
pH gradient, determination, 114
respiratory chain, 113
arrangement of cytochromes in, 111–112
sdh genes, 400
strain AN120, cloning of mutated genes from, 600–601
strain AN180, growth, 546
strain KY7485, growth, 546
strains
for F_1F_0-ATPase purification, 545–546, 557
for genetic study of H^+-ATPase mutants, 559–560
β subunit gene
construction of hybrid plasmids carrying, 597–598
restriction map of, 596–597
terminal oxidase complexes, reconstituted into liposomes
assay method, 110
calibration of membrane potential, 110–112
effects of protonophore uncouplers and oxidase inhibitors, 111
formation of membrane potential by, 109–111
procedure, 109–110
terminal oxidases, 87, 94–113
transmembrane electrochemical gradient of hydrogen ion, 113
unc mutant, 2
unc operon, 545, 558, 595
multicopy plasmids carrying, genetic analyses using, 567–568
unc recipient strains, suitable for genetic complementation analysis, 565
N-(Ethoxycarbonyl)-2-ethoxy-1,2-dihydroquinoline. *See* EEDQ
1-Ethyl-3-(3-dimethylaminopropyl)carbodiimide. *See* EDAC
N-Ethyl-*N*-nitroso-*p*-toluenesulfonamide, 620
preparation, 620–622
synthesis, 619
N-Ethyl-5-phenylisoxazolium 3′-sulfonate. *See* Woodward's reagent K

F

F_0-ATPase, 5, 7, 520, 713
Facilitated diffusion, 2
Factor F_6, 428–429
preparation, 431
removal from submitochondrial particles, by silicotungstic acid, 432
F_1-ATPase, 428, 520, 713
[^{14}C]Abf-*O*-tyrosine derivative, preparation and identification, 749–751
ADP analogs used for structural analysis of, 682–712
arginine and lysine residues, modification, 728–729
arginine residues, modification, 741
ATPase activity assay, 484
ATP synthesis and hydrolysis, kinetics, inhibitory capability of nucleotide analogs, 699–701
beef heart
amount of [γ-^{32}P]ATP bound, calculation, 615
ATP hydrolysis, mechanism, 610
high-affinity catalytic sites
binding of ATP in, 611–612
equilibrium distribution of substrate and products in, 613–614
release of products from, 615–617
rate of ATP dissociation from catalytic sites on, 612–613
rate of binding of [^{3}H]ADP, 617–618
beef heart mitochondria
inactivation
by carbodiimides, 717
with 4-chloro-7-nitrobenzofuran, 725–726
incorporation of carbodiimides during inactivation, 718

modification by EEDQ and IIDQ, 721
synthesis of enzyme-bound ATP, 648–649
carboxyl group reagents, zero-length cross-linking with, 715, 719, 722, 724
carboxyl groups
identification, 755–761
modification, 713, 714–724
by Woodward's reagent K, 722–724
catalytic sites, 712
cooperativity of catalysis, 711–712
α–β cross-link, structure, 659–660
cross-linked product
hydrolytic cleavage, 658–659
structure, 658–660
cross-linking experiments, 650
crystals, 441–444
forms, 442
orientation, 443
preparation for electron microscopy, 441–442
reconstructions, 444–446
symmetry, 443
X-ray diffraction studies, 444, 446
dissociation into subunits, and reconstitution by reassociations of isolated subunits, 528
E. coli
hybrid complexes, reconstitution, 592–593
inactivation
by carbodiimides, 717
with 4-chloro-7-nitrobenzofuran, 725–726
incorporation of carbodiimides during inactivation, 718
modification by EEDQ and IIDQ, 721
mutant
crude, identification of defective subunits in, 593–594
purified, identification of defective subunits in, 593
reconstitution, 592–593
subunits, 588–594
α, β, and γ, 589, 590–592
β, mutations, 595
δ and ε, preparation, 592
isolated, properties, 594
preparation, 590–592
reconstitution, 589
three subunit
reconstruction, 437–440
structure, 440, 441
electron microscopy, 434–446
staining and preparation for, mica flotation method, 437–441
essential residues
identification, 741–761
specific radioactivity of labels, 742
Glu residues
labeling, 741
modification, 756, 758–761
high- and low-affinity binding sites, 711
inactivating modifications by chemical reagents, 713–729
isolation, 661
labeling with radioactive DCCD
conditions, 757–758
identification of labeled residues, 758–761
stoichiometry, 758
labeling with radioactive 5′-FSBA
conditions, 746–747
isolation of labeled peptides and identification of labeled amino acid residues after, 747–748
light-induced inhibition, with DiN_3ATP, 655–657
lysine residues
modification, 741
to which Nbf group migrates, identification, 751–755
M. lysodeikticus, modification by EEDQ and IIDQ, 721
M. phlei
inactivation
by carbodiimides, 716–717
with 4-chloro-7-nitrobenzofuran, 725–726
incorporation of carbodiimides during inactivation, 718
modification by EEDQ and IIDQ, 721

mitochondrial
[14C]Abf-*O*-tyrosine derivative, preparation and identification, 749–751
modification by Woodward's reagent K, 757
modified covalently by photolabeling with ANA-ADP or ANP-ADP, reversible nucleotide binding to, 708
modifiers, 712–732
morphology, 446
negatively stained specimens, preparation for electron microscopy, adhesion method, 435–436
nonactivating modifications, 729–732
nucleotide binding
high-affinity, 705–708
loci, 711
low-affinity sites, 706–708, 736–738
nonlinear approach for analysis, 736–738
reversible high-affinity sites, 711, 736–738
sites, cooperativity, 710–711
18O exchange as probe of cooperativity, 712
photoaffinity cross-linking, 649–660
photoaffinity labeling with benzophenone-ATP, reaction pathway, 678
photolabeled, reactivation, 708–709
preparation, 431
protein determination, 485
PS3
inactivation by carbodiimides, 717
incorporation of carbodiimides during inactivation, 718
modification by Woodward's reagent K, 723, 757
subunits, 589
R. rubrum
inactivation
by carbodiimides, 717
with 4-chloro-7-nitrobenzofuran, 725–726
incorporation of carbodiimides during inactivation, 718
modification
by EEDQ and IIDQ, 721
by Woodward's reagent K, 723, 757
rat liver
benzophenone-ATP as substrate, 680–681
catalytic properties, 483–484
crystals, 483
HPLC purification, 481–482
photoinactivation by benzophenone-ATP, 681–682
physical properties, 482–483
preparation, 485
properties, 482–484
purification, 472
rapid purification, 480–481
structure, 482
subunits, 482
transfer into stabilizing buffer, 481
rat liver mitochondrial, 477–484
ATPase assay, 478
ATP synthesis assay, 478
chloroform solubilization, 477
preparative procedures, 479
purification, determination of protein, 479
rapid isolation, 477
reconstitution assay, 478
SDS–PAGE, 479
release, 481
removal from submitochondrial particles, by urea, 431
residues modified by Nbf-Cl, identification, 748–755
SDS–PAGE, 485
single molecules, electron microscopy, 435
spinach chloroplast
inactivation by carbodiimides, 717
incorporation of carbodiimides during inactivation, 718
modification by Woodward's reagent K, 723
subunit arrangements, 446
subunit pairs
$\alpha\gamma$, 489
purification, 486–487
$\beta\gamma$, 489
purification, 487–488

purification, 485–486
subunits, 484, 528
α, 489
purification, 486–487
α and β, 441, 445, 460
photoaffinity labeling, 652
arrangement, 770–775
β, 489, 741
extraction, 531
properties, 536–537
purification, 487–488, 531–533
β and γ
isolated, assay of reconstitutive activity, 529–530
reconstituted, assay of ATP synthesis and hydrolysis, 530
reconstitution, 530
conformations related to enzyme activities, 768–769
γ, 445
extraction, 533
purification, 533–535
isolation, 742–743
monoclonal antibodies, 761–769
binding to native enzyme, 774
characterization, 767
determination of specificity by immunodecoration, 765–766
electron microscopic study of immune complexes with, 773–775
ELISA, 771–772
epitopes reacting with, conservation in phylogenetic scale, 769
epitopes recognized by, determination, 769
Ig class, 772–774
preparation, 762–763, 771
purification, 772–774
screening methods, 763–764, 771–774
specificity, 772
in study of structure, conformation, and functions of enzyme, 766–769
purification, 485–486
reconstitution into depleted membranes, 528
sequential removal, 528
stoichiometry, 766–768, 770–775
subunit structure, 434
tyrosine residues, modification, 724–728, 741, 748
washed membranes, preparation, 480
yeast mitochondria, inactivation with 4-chloro-7-nitrobenzofuran, 725–726
F_1-ATPase inhibitor, 504–505
Ferredoxin–$NADP^+$ oxidoreductase, 272
Ferricyanide, redox mediator of photosynthetic reaction center–mitochondrial ubiquinol–cytochrome bc_1 complex hybrid, 299–300
Ferro-EDTA, redox mediator of photosynthetic reaction center–mitochondrial ubiquinol–cytochrome bc_1 complex hybrid, 299
Ferroxalate, redox mediator of photosynthetic reaction center–mitochondrial ubiquinol–cytochrome bc_1 complex hybrid, 299
F_1F_0-ATPase (H^+), 3, 5, 353
activity, 477
ATPase assays, 472
from beef heart mitochondria, 428–433
components, 430
reconstitution, 430
resolution, 430
bound to submitochondrial particles, inactivated, stabilization, 509–511
chloroplast, 512–520
co-reconstitution with bacteriorhodopsin, 518
crude preparation, reconstitution, 516–517
liposomes, activity, 516
liposomes with bacteriorhodopsin, light-dependent ATP synthesis, 518–519
Na-cholate–SG purified, reconstitution, 517
$^{32}P_i$–ATP exchange assay, 518
purification, 513–516
reconstitution
lipid requirements, 519–520
into proteoliposomes, 516–518
solubilization, 514
Triton X-100–purified, reconstitution, 517–518

E. coli, 477, 545–557
activity, inhibition by DCCD, 570
assays, 579
of $^{32}P_i$–ATP exchange, 548
ATPase activity, 548, 579
ATP-dependent H^+ pumping, 555
ATP-dependent quenching of acridine dye fluorescence
assay, 579, 581
unit, 581
DCCD-sensitive, 555
F_0, 569–578
assays, 569–570
ATPase activity assay, 571
dissociation into subunits, 574–576
incorporated into liposomes, binding with F_1, 569
isolation, 579
passive H^+-translocating activity, reconstitution, 578
passive proton translocation, 569–570
polypeptides, 569, 574–575, 586
preparation, 588
preparation and reconstitution, 556
preparation by hydrophobic interaction chromatography, 572–574
preparation of proteoliposomes, 576–577
protein determination, 570
proton translocation assay, 579
reconstituted, properties, 578
reconstituted into liposomes, 569
SDS–PAGE, 574–575
subunits, 586
F_1, subunits, 586
H^+ conduction assay, 582–583
inhibition by DCCD, 582
K^+-loaded vesicles, preparation, 582
large-scale preparation, 587–588
medium-scale preparation, 583–587
$^{32}P_i$–ATP exchange, 555
assay, 579, 580–581
polypeptides, 586
preparation, 579–588
properties, 587
protein assays, 548–549
purification, 583–588
buffers, 547
comparison of methods, 556–557
from induced λ-*unc*$^+$ cells, 553–555
from monoploid cells, 550–554
purity, 553
reproducibility, 556
yield, 553
purity, 557
reconstituted
ACMA fluorescence quench assay, 569, 570
ATPase activity assay, 569
ATP-dependent quenching of ACMA fluorescence, 571–572, 577–578
with bacteriorhodopsin, ATP synthesis, 555–556
reconstitution into proteoliposomes, 579–580
reconstitution of DCCD-sensitive ATPase activity, 577–578
reconstitution of function, 555–556
SDS–PAGE, 549–550
subunits, 595
analysis of mutations, 597
mutations, 596
inactivated, 504–512
inactivated complex with inhibitor and stabilizing factors, characteristics, 511–512
inactivation, in presence of stabilizing factors, 511
inhibition, 477
inhibitor assays, 472
mitochondrial
inhibition, 505
stabilizing factors, 505
PS3, F_0
preparation, 604–605
proton permeability, 604–607
vesicles
addition of DCCD, effect on proton permeability, 607
addition of F_1, effect on proton permeability, 607
H^+ permeability through, 607
preparation, 605
proton conductivity assay, 605–606

purification with CHAPS
protein determinations, 472
solutions used, 471
rat liver
effect of Cd^{2+} on, 476–477
effect of DCCD on, 476–477
effect of DDT on, 476–477
effect of kelthane on, 476–477
effect of oligomycin on, 476–477
effect of tricyclohexyltin on, 476–477
effect of venturicidin on, 476–477
properties, 476–477
purification, procedure, 472–474
SDS–gel electrophoretic profile, 475–476
rat liver mitochondria, purification using CHAPS, 470–477
reconstitution, 432–433
into asolectin vesicles, 474–476
spinach chloroplast
lipid content and composition, 515–516
purification, 515
F_1F_0-ATPase liposomes
ATP synthesis, analytical procedures, 642–643
electric pulse-driven ATP synthesis, 640
F_0F_1 macroliposomes
electrical apparatus used for, 641–642
preparation, 642
F_0F_1 microliposomes, preparation, 642
1-Fluoro-2,4-dinitrobenzene. *See also* FDNB
modification of tyrosine residues of F_1-ATPase, 728
5′-*p*-Fluorosulfonylbenzoyladenosine, 728
[^{3}H], synthesis, 744–745
identification of amino acid residues in enzymes that react with, 743–748
inactivation of F_1-ATPase, 741
incorporated label, stability, 745–746
radioactive
labeling F_1-ATPases with
conditions, 746
isolation of labeled peptides and identification of labeled amino acid residues after, 747–748
storage, 745
synthesis, 743–744
sulfonyl fluoride moiety, stability, 745
Formate dehydrogenase complex
composition, 392
electron-transfer activity, 387
enzymatic activity, assay, 397–399
incorporation into liposomes, 392–393
isolation, 389–390
procedures, 387–388
liposomal preparation
composition, 393–394
enzymatic properties, 395
recovery of enzymatic activity, 395
restoration of electron-transfer activity, 395–397
structural properties, 393–395
properties, 391–392
subunits, 391–392
Fourier transformation, three-dimensional inverse, 349–350
Fumarate reductase, 399
E. coli, 377–386
assay, 380
catalytic dimer, 378–379
purification, 381–383
holoenzyme, purification, 385–386
orientation, 378
pH maxima, 380
subunits, 378
Fumarate reductase complex
active site for fumarate, 391
composition, 391
electron-transfer activity, 387
enzymatic activity, assay, 397–399
incorporation into liposomes, 392–393
isolation, 388–389
procedures, 387–388
liposomal preparation
composition, 393–394
enzymatic properties, 395
recovery of enzymatic activity, 395
restoration of electron-transfer activity, 395–397
structural properties, 393–395
molecular weight, 390
properties, 390–391
subunits, 390
Funiculosin, 256
inhibition of cytochrome bc_1 complex, 268–270

molecular formula, 267
physicochemical data, 269
structural formula, 267

G

Galactosc-binding protein, 4
Galactose/H^+ symporter, 5
β-Galactosidase, 1
Generalized Simultaneous Iterative Reconstruction Technique, 349
Glucose carrier, 5
Group-directed reagents, 712, 713
GSIRT. *See* Generalized Simultaneous Iterative Reconstruction Technique

H

Hemoglobin
Bohr effects in, 332
proton-induced affinity change, 8
Heptadecylmercaptohydroxyquinoline quinone, 271
inhibition of cytochrome bc_1 complex, 265–266
molecular formula, 263
physicochemical data, 269
structural formula, 263
Heptylhydroxyquinoline-*N*-oxide, 271
inhibition of cytochrome bc_1 complex, 270
molecular formula, 267
physicochemical data, 269
structural formula, 267
Hinge protein, 229–230, 235–236, 238–253
amino acid composition, 241, 244–245
contamination with putative one-band cytochrome c_1, 242
determination, 249–250
calculation, 252–253
method, 251–252
principle, 249
reagents, 249–251
isoelectric point, 245
molecular weight, 243–244
preparation
method, 238–241
principle, 238
primary structure, 245
properties, 242–245
requirement for interaction of cytochrome c_1 with cytochrome c, 245–248
self-associated polymers, 243
spectral properties, 241
sulfhydryl and disulfide groups, 245
HMHQQ. *See* Heptadecylmercaptohydroxyquinoline quinone
HQNQ. *See* Heptylhydroxyquinoline-*N*-oxide
Hybridoma cell line 1B2a
growth, 372
IgG produced by, purification, 373
2-Hydroxynaphthaquinone, redox mediator of photosynthetic reaction center–mitochondrial ubiquinol–cytochrome bc_1 complex hybrid, 299
Hydroxyquinoline quinone, inhibition of cytochrome bc_1 complex, 265–266
Hydroxyquinone
analogs, 256
inhibition of reoxidation of iron-sulfur protein and of reduction of cytochrome b_1, 263–266
N-Hydroxysuccinimidyl-4-azidobenzoate
binding to purified inhibitor peptide, 663
characterization, 663
labeling inhibitor peptide with, 663–664
N-Hydroxysuccinimidyl-4-[7-^{14}C]azidobenzoate
covalent cross-linking of inhibitor peptide to isolated F_1-ATPase by, 665–666
reaction with proteins, 662
synthesis, 661–662

I

IF_1. *See* ATPase inhibitory peptide
IIDQ, modification of carboxyl groups of F_1-ATPase, 719–722
Iron–sulfur protein, 224–225, 255, 257, 272, 296
activity
assay, 212–214
succinate–cytochrome-*c* reductase assay, 212–213

ubiquinol–cytochrome-*c* reductase
assay, 212–213
cleavage from cytochrome bc_1 complex, 227–229
distribution, 211
isolation, 236
N. crassa, amino acid sequence, 235
properties, 211
purification, from bacterial cytochrome bc_1 complex with Triton X-114, 331
reconstitution to succinate–cytochrome-*c* reductase, 220–223
reconstitutively active, purification, 211–212, 214–218
resolution, 212
from complex III, 218–221
yield, from cytochrome bc_1 complex, 229
Iron-sulfur subunit, of *Neurospora* cytochrome reductase, 200
locus, 200–201
sequence, 200
N-(Isobutoxycarbonyl)-2-isobutoxy-1,2-dihydroquinoline. *See* IIDQ

K

KH_2PO_4, [^{18}O]
preparation from PCl_5 and $H^{18}OH$, 631
salt-free, separation of, 631–632

L

lac permease
proteoliposomes containing cytochrome *o* with, characterization, 122
reconstitution into proteoliposomes, simultaneously with cytochrome *o*-type oxidase, 113–114, 118–119
lac promoter, inducer, 568
D-Lactate dehydrogenase, *E. coli,* 370–377
amino acid sequence, 371
assay, 371–372
gene encoding, 371
immunoaffinity chromatographic purification, 371–377
ascites fluid obtained for, 372–373
preparation of immunoaffinity resin, 374
procedure, 375–377
inactivator, 371
molecular weight, 370
monoclonal antibodies, 371
polyclonal antibodies, 371
polypeptides, 371
protein assay, 372
purification, 370–371
solubilization, 374–376
substrate specificity, 371
Lactose permease, 1
Lipid bilayer, formation, from monolayers at tip of patch pipet, 128–129

M

MEGA detergent, 273
Membrane potential, of reconstituted lipid bilayers, recording system, 125–127
Membrane protein
crystallization, 345
electron microscopy, 344
filtering two-dimensional projections, 348–349
image processing, 348–350
preparation of specimens, 344–345
recording signal, 348
three-dimensional reconstruction, 349–350
three-dimensional crystals, 191
two-dimensional crystals, 192
β-Methoxyacrylate, 256
Micrococcus luteus, succinate dehydrogenase complex, 404
Micrococcus lysodeikticus, F_1-ATPase, modification by EEDQ and IIDQ, 721
Mitochondria
beef heart, preparation, 182
beef heart, F_1-ATPase
inactivation
by carbodiimides, 717
with 4-chloro-7-nitrobenzofuran, 725–726
incorporation of carbodiimides during inactivation, 718
modification by EEDQ and IIDQ, 721
electric pulse-driven ATP synthesis, 640

outer membrane, nonspecific channel, 5
preparation, 642
rat liver
inverted inner membrane vesicles, preparation, 479–480
preparation for F_1-ATPase isolation, 479
respiratory chain, 344
bacterial model, 305
yeast, F_1-ATPase, inactivation with 4-chloro-7-nitrobenzofuran, 725–726
Molecular ellipticity, calculation, for hinge protein determination, 252
Molecular weight, 58
Mucidin. *See* Strobilurin A
Mycobacterium phlei, F_1-ATPase
inactivation
by carbodiimides, 716–717
with 4-chloro-7-nitrobenzofuran, 725–726
incorporation of carbodiimides during inactivation, 718
modification by EEDQ and IIDQ, 721
Myoglobin, Bohr effects in, 332
Myxothiazol, 267
mitochondrial cytochrome bc_1 complex inhibitor, 259–260
molecular formula, 258
physicochemical data, 268
structural formula, 258

N

NADH : Q oxidoreductase. *See* Complex I
NADH : ubiquinone oxidoreductase. *See* Complex I
NADH-CoQ reductase, 353
NADH dehydrogenase. *See* Complex I
NADH dehydrogenase ubiquinone. *See also* Complex I
iron–sulfur-containing polypeptides, 360–369
N-ADP
binding to high affinity sites of F, 705–706
as fluorescent probe, 701–702
inhibition of F_1-ATPase ATP synthesis and hydrolysis, 700
N-AMP, inhibition of F_1-ATPase ATP synthesis and hydrolysis, 700
Nanometer force, 9
3′-*O*-[Naphthoyl(1)]ADP. *See also* N-ADP
synthesis, 697–698
N-ATP, inhibition of F_1-ATPase ATP synthesis and hydrolysis, 700
E-β-Nethoxyacrylate inhibitor, inhibitor of UQH_2-oxidation, 257–261
Neurospora
mitochondrial membranes, preparation, 202
ubiquinol–cytochrome-*c* reductase, 191–201
reconstitution, 201–210
Neurospora crassa
complex III, 173–174
cytochrome-*c* oxidase, purification, 66
Nitrene formation, 650, 667
N-(2-Nitrophenyl)4-aminobutyric acid, synthesis, 689
N-(4-Nitrophenyl)4-aminobutyric acid, synthesis, 689
Nonylhydroxyquinoline-*N*-oxide
inhibition of cytochrome bc_1 complex, 270
molecular formula, 267
physicochemical data, 269
structural formula, 267
NQNO. *See* Nonylhydroxyquinoline-*N*-oxide
Nucleotide, esterification, 696–699

O

Oligomycin sensitivity conferring protein, 428–429, 713
accessibility to antibodies, 459
antibodies, effect on ATP synthesis, 459–460
biological activity, with different preparative procedures, 459
interactions with $F_1$and F_0, 458–460
monoclonal antibody probes, 455–460
peptides close to, 460
preparation, 431
purification, 455
using monoclonal antibodies, 455–458
radiolabeled, 729–730
removal from submitochondrial particles, by ammonia, 431–432

Oudemansiella mucida, 260
Oudemansin, 259
Oudemansin A
  mitochondrial cytochrome bc_1 complex inhibitor, 260–261
  ORTEP stereo plot, 261
  physicochemical data, 269
  structural formula, 258
Oudemansin B
  mitochondrial cytochrome bc_1 complex inhibitor, 260–261
  molecular formula, 258
  physicochemical data, 269
  structural formula, 258
Oxidative phosphorylation
  chemical hypothesis, 5
  inhibitory activity of ADP analogs in, 685–686
Oxygen-18
  exchange, as probe of cooperativity in F_0 catalytic mechanism, 712
  methodology for study of phosphorylation, 618–639

P

Paracoccus denitrificans
  cell breakage, 155
  cell membranes
    alkaline wash, for ubiquinol oxidase complex isolation, 308–312
    with high ubiquinol–cytochrome-*c* oxidoreductase activity, preparation, 309
    preparation
      for isolation of cytochrome bc_1, 318
      for ubiquinol oxidase complex isolation, 307–308
    crude membranes, isolation, 155–156
  cytochrome bc_1 complex, 181, 188, 316–325
    cytochrome *b* from, 326
  cytochrome-*c* oxidase, 145, 153–159
    purification, 70
  growth, 155
    for isolation of cytochrome bc_1 complex, 318
    for ubiquinol oxidase complex isolation, 307
  respiratory chain, 138, 305–306, 316–317
  succinate growth medium, 155
Partial specific volume, estimation, 53–56
Passive transport, 2
Patch pipet
  bilayer formation from monolayers at tip of, 128–129
  composition, effect on activity of reconstituted cytochrome oxidase, 135
  fabrication, 125–127
Penicillium funiculosum Thom., 268
Pentadecylhydroxydioxobenzothiazole
  inhibition of cytochrome bc_1 complex, 264–265
  molecular formula, 263
  physicochemical data, 269
  structural formula, 263
PHDBT. *See* Pentadecylhydroxydioxobenzothiazole
Phenazine methosulfate, in reconstitution of chloroplast H^+ translocation and photophosphorylation, 285–293
Phenylglyoxal
  F_0 liposomes inhibited by, 8
  modification of F_1-ATPase, 728
  radiolabeled, inactivation of F_1-ATPase, 742
Phenylmethylsulfonyl fluoride, source, 546
Phosphoenolpyruvate, [^{18}O]
  assay, 639
  preparation, 637–639
Phosphoenolpyruvate-glucose-phosphotransferase system, 4
Phospholipid, in electron transport, 4
Phosphoric acid
  derivitization, for analysis by gas chromatography/mass spectrometry, 619–625
  purification, on AG1-X4 columns, 632–633
Phosphorus(V) chloride, Ultrapure, source, 631
Phosphorylation, oxygen-18 methodology for study of, 618–639
Photoaffinity cross-linking, 650–651

Photoaffinity labeling, 649–650, 666, 667, 712
fortuitous, random site, conditions precluding, 671–672
specific site labeling, physicochemical considerations, 669–670
Photo-cross-linking, 650
Photophosphorylation, inhibitory activity of ADP analogs in, 685–686
Photosynthetic reaction centers, 70–71
R. sphaeroides
isolation, 294
light generation of redox equivalents, 295
Photosynthetic reaction center–mitochondrial ubiquinol–cytochrome bc_1 complex hybrid system
advantages, 294
construction, 293–305
criteria for, 296
detergent solubilization, 399
electron transfer, specific reactions, 300–303
equilibrium pH 7 redox midpoint potentials, 299
light initiation of electron transfer, 297
lipid solubilized, construction, 303–304
redox poising, 298–300
spectral monitoring of electron transfer, 297–298
Photosystem I, artificial, cyclic electron transport catalyzed by, 285–286
Photosystem I reaction center
A. variabilis, 292
incorporation into liposomes
alternative procedures, 291
lipid and detergent, 291
tests for, 291
incorporation into soybean lipid vesicles, 288
large subunit, from spinach, proton translocation, 292
in liposomes
comparison to chloroplast activities, 292–293
measurement of membrane potential formation in, 292
mixed orientation, 292
preparation, 286–288
from spinach, incorporated into soybean lipid vesicles, 286
Plasmid
DNA, determination of altered nucleotide sequence, 602–603
pAN120
cloning of DNA fragment containing altered base, 600–601
detection of anomalous migration of DNA fragment carrying altered base, 602
pFRD63, 379
pKF11
cloning of DNA fragment containing altered base, 599–600
detection of anomalous migration of DNA fragment carrying altered base, 602
pKF3
cloning of DNA fragment containing altered base, 601
detection of anomalous migration of DNA fragment carrying altered base, 602
pUC8, cloning of $uncE^+$ gene behind *lac* promoter in, 568
Plastocyanin, 273
Plastoquinone, synthesis, 284
PMS. *See* Phenazine methosulfate
Polynucleotide phosphorylase, in CF_1 preparations, 644
Porin, mitochondrial, 8
Protein kinase, cAMP dependent, inactivated with 5′-[^{14}C]FSBA, tryptic digest of catalytic subunit, 747
Protein–ligand interaction
binding models, 734–735
discrimination between, 736
fitting procedure, 733–735
relationships between stoichiometric and site binding constants, 738–740
site binding constants, 733–735
stoichiometry of binding sites, 733–735
theoretical basis, 733
Protein–lipid–detergent, equilibrium equation, 56
Proteoliposome, active, reconstitution, 5–6
Proteus vulgaris, terminal oxidases, 94

Proton–ATPase complex
Chlamydomonas, purification, 516
chloroplast, purification, 516
Dunaliella, purification, 516
rat liver mitochondrial, purification, 516
yeast mitochondrial, purification, 516
Protonmotive force, 6–7
chemiosmotic theory, 2–4, 13
definition, 6
Proton transfer, associated with redox transitions of terminal electron carriers caused by repetitive oxygen pulses of anaerobic beef heart mitochondria, 334–336
PS3
culture, 146
cytochrome-*c* oxidase, 123–124
cytochrome oxidase, 145–153
F_1-ATPase
inactivation by carbodiimides, 717
incorporation of carbodiimides during inactivation, 718
modification by Woodward's reagent K, 723
subunits, isolation, 589
F_1F_0-ATPase, F_0, proton permeability, 604–607
Pseudomonas aeruginosa, 270
Pseudomonas putida, terminal oxidases, 94
Pump, definition, 1
Pyocyanine, redox mediator of photosynthetic reaction center–mitochondrial ubiquinol–cytochrome bc_1 complex hybrid, 299
Pyridoxal phosphate, modification of F_1-ATPase, 728
Pyrophosphatase
beef heart mitochondrial, 447–454
isolation, procedures, 447–448
crude, reconstitution of inorganic phosphate synthesis by, 448–450
membrane-bound, reconstitution, 452–454
mitochondrial
isolation, 449–450
properties, 451
purification, 449–452
R. rubrum, 538–545
activity, 538
before and after incorporation into liposomes, effects of various compounds, 541, 542
determination, 541–543
chromatography on hydroxylapatite, 540
effect of DCCD, 544
effect of Dio-9, 544
effect of FCCP, 544
effect of IDP, 544
effect of MalNET, 544
effect of MDP, 544
effect of NaF, 544
effect of Nbf-Cl, 544
incorporation into liposomes, 540–541
inhibitors, 543–544
polyacrylamide gel electrophoresis, 543
SDS–PAGE, 543–544
solubilization, 539–540
substrate specificity, 544–545
Pyruvate : oxygen oxidoreductase chain, reconstitution, 93–94
Pyruvate kinase, benzophenone-ATP as substrate, 672–673
Pyruvate oxidase, *E. coli,* 93

Q

Quinone
pool, in detergent-solubilized hybrid systems, 304–305
reduction, observation in hybrid system, 301–302
Quinone analogs, 256

R

Redox Bohr effects, 331–343
measurement, 332
instrumentation, 333
in isolated redox complexes, 338–343
in mitochondria, 334–338
principle, 332–333
s/n numbers, 337–338
in submitochondrial particles, 334–338

role, in redox enzyme function, 332
Redox potential, midpoint, change with pH, 331
Respiratory control ratio, definition, 15
Reversible ATP synthase. *See* F_1F_0-ATPase
Rhodopseudomonas sphaeroides
cell paste, 138–139
cytochrome bc_1 complex, 181, 186–191, 271–285
cytochrome *b* from, 326
cytochrome-*c* oxidase, 71
purification, 70
cytoplasmic membranes, preparation, 139
growth, 138–139, 182
photosynthetic apparatus, 138
preparation, for isolation of cytochrome bc_1 complex, 182–183
reaction center, combined with mitochondrial ubiquinol–cytochrome bc_1 complex, 293–305
respiratory chain, 138
Rhodospirillum rubrum
chromatophores
β-less, properties, 535–536
β, γ-less, properties, 537–538
coupled, depleted, and reconstituted, activities, 535
F_1 subunits, 528–538
H^+ gate, 537–538
preparation, 539
F_1-ATPase
inactivation
by carbodiimides, 717
with 4-chloro-7-nitrobenzofuran, 725–726
incorporation of carbodiimides during inactivation, 718
modification
by EEDQ and IIDQ, 721
by Woodward's reagent K, 723
mitochondria, inorganic phosphate synthesis, 447
pyrophosphatase, 538–545
Ribulosebisphosphate carboxylase, 272
Rieske iron–sulfur protein. *See* Iron–sulfur protein
Rossmann fold, 8

S

Secondary transport, 4
Sedimentation coefficient, 58
for sucrose concentrations, 61–62
Sedimentation equilibrium measurements, 56–57
principle, 56
Sedimentation rate, calculation, 61–62
Sedimentation velocity measurements, 57–58
calculations, 58
principle, 57–58
Simultaneous Iterative Reconstruction Technique, 349
Single-channel recording, 1
SIRT. *See* Simultaneous Iterative Reconstruction Technique
Sodium cholate, source, 546
Sodium deoxycholate, source, 546
Stigmatella aurantiaca, 261
Stigmatellin, 256
Stigmatellin A
mitochondrial cytochrome bc_1 complex inhibitor, 261–263
molecular formula, 262
physicochemical data, 269
structure, 261–262
Stigmatellin B
mitochondrial cytochrome bc_1 complex inhibitor, 261–263
molecular formula, 262
structure, 261–262
Strobilurin, 259
Strobilurin A
mitochondrial cytochrome bc_1 complex inhibitor, 260
molecular formula, 258
physicochemical data, 269
structural formula, 258
Strobilurin B
mitochondrial cytochrome bc_1 complex inhibitor, 260
molecular formula, 258
physicochemical data, 269
structural formula, 258
Strobilurin C
mitochondrial cytochrome bc_1 complex inhibitor, 260
molecular formula, 258

structural formula, 258
Strobiluris tenacellus, 260
Submitochondrial particle
ATPase, photolabeling with benzophenone-ATP, 679–680, 681–682
beef heart, 428
F_1F_0-ATPase, 428–433
preparation, 447, 687
electric pulse-driven ATP synthesis, 640
F_1-ATPase, ATP synthesis and hydrolysis, comparison of nucleotide analogs as inhibitors of, 700
preparation, 429–431, 642
spin-labeled ADP analog binding, 710
Succinate–cytochrome-*c* reductase complex
beef heart, preparation, 214, 216–218
iron–sulfur protein–depleted, purification, 216–218
reconstitution of iron-sulfur protein to, 220–223
Succinate dehydrogenase, accessibility, compared with accessibility of cytochrome-*c* oxidase, 166–167
Succinate dehydrogenase complex
B. subtilis, 399–414
activity, 399
biosynthesis, 413–414
composition, 399, 403–405
immunoprecipitation, 403–405, 413
antisera for, 406–408
immunoelectrophoresis, 406–408
purity, 409
solubilization, 405–406
subunits, 399
E. coli, 400
Sucrose gradient centrifugation
calculations, 61–62
principle, 61
Symport, definition, 4

T

2,3,5,6-Tetramethyl-*p*-phenylenediamine, in reconstitution of chloroplast H^+ translocation and photophosphorylation, 286–293
Tetranitromethane. *See also* TNM-nucleotide
modification of tyrosine residues of F_1-ATPase, 727
Thermogenin. *See* Uncoupling protein
Thermus thermophilus, cytochrome-*c* oxidase, 167
TNP-AD(T)P, 700
Transducing phage
λ-*unc*
on genetic analysis of *E. coli* H^+-ATPase mutants, 558–568
genetic complementation using, 558
λ-*uncX,* construction, 560–562
Transhydrogenase
activity, 353
assay, 354
mitochondrial, beef heart, 353–360
protein analysis, 354
purification, 353, 354
procedure, 354–357
yield, 357
reconstituted
$H^+:H^-$ ratios, determination, 358–360
quench of 9-aminoacridine fluorescence, 360
reconstitution, 353, 357–360
subunit structure, 353–354
Translocator
conformational change, during solute transport, 8
crystallographic analysis, 9
genetic concept, 1–2
indirect action of driving ions, 7
ion flux and conformation change, 7–8
lipid-dependent, biochemical isolation, 4–5
molecular mechanisms, 8–9
physiological concept, 1
reconstitution into proteoliposome, 5
structure, 2, 7
three-dimensional crystals, 9
unifying principles of, 7–8
Triethyl phosphate
analysis for ^{18}O content, 619
gas chromatography, 619, 626
mass spectrometry, 626–630
at different ^{18}O enrichments, 627–629

preparation, from diazoethane and phosphoric acid, 625
2,4,6-Trinitrobenzene sulfonate, modification of F_1-ATPase, 728–729

U

Ubiquinol-1 oxidase activity, of cytochrome *d* terminal oxidase complex, 88, 91
Ubiquinol–cytochrome-*c* oxidoreductase complex. *See also* Cytochrome bc_1 complex
beef heart, 189
R. sphaeroides, 190
yeast, 187
Ubiquinol–cytochrome-*c* reductase, 353. *See also* Cytochrome reductase
and iron–sulfur protein, 222
Neurospora, 191–201
core proteins, 201–202
molecular weight, 201
reconstitution, 201–210
subunits, 201–202
Q-cycle mechanism, 210
Ubiquinol oxidase activity, of *E. coli* terminal oxidase complexes, 106–108
assay, 106
electron donors and kinetics, 106–107
inhibitors, 107–108
Ubiquinol oxidase complex
activities, 306, 316
cytochromes, 313–316
inhibitors, 316
isolation, 305–306
analytical methods, 306–307
materials, 306
method, 307–313
P. denitrificans, 305–316
absorption spectra and difference spectra, 315–316
redox components, redox potentials, 314–316
polypeptides, 305–306, 313–315
properties, 313–314
UHDBT. *See* Undecylhydroxydioxobenzothiazole
UHNQ. *See* Undecylhydroxynaphthoquinone
Uncoupling protein, 490–498
assay, 491–492, 495
extraction from mitochondria, 493–494
isolated, hydrodynamic characterization, 497–498
isolation, 491–497
selection of detergent, 492–493
nucleotide binding assay, 498–504
nucleotide binding to
general properties, 499–500
measurement
anion-exchange chromatography, 503
equilibrium dialysis, 501–502
exclusion gel chromatography, 503
principles, 500–501
procedures, 501–504
sedimentation methods, 503–504
pH dependency, determination, 502
purification, 494–495
separation from ADP/ATP carrier, 494–495
precautions, 494
solubilization, 492
stability, 496
yield, 495–496
Undecylhydroxydioxobenzothiazole
inhibition of cytochrome bc_1 complex, 264–265
molecular formula, 263
physicochemical data, 269
structural formula, 263
Undecylhydroxynaphthaquinone, inhibition of cytochrome *b* reduction, 296
Undecylhydroxynaphthoquinone
inhibition of cytochrome bc_1 complex, 265
molecular formula, 263
physicochemical data, 269
structural formula, 263
Uniporter, definition, 4

W

Water, ^{18}O content, rapid approximation of, 637
Wollinella succinogenes
cytoplasmic membrane, composition, 393–394

growth, 387, 388
Woodward's reagent K, modification of F_1-ATPase, 722–724, 756, 757

X

Xerula longipes, 260
Xerula melanotricha, 260

Y

Yeast
cell breakage, 175–176
mitochondria, preparation, 176
strain, for complex III preparation, 175
submitochondrial particles, preparation, 176, 182
ubiquinol–cytochrome-*c* oxidoreductase, 187